ION IMPLANTATION TECHNOLOGY 2008

ION IMPLANTATION TECHNOLOGY 2008

17th International Conference on Ion Implantation Technology

IIT 2008

Monterey, California *8 – 13 June 2008*

EDITORS

Edmund G. Seebauer
Yevgeniy V. Kondratenko
University of Illinois, Urbana-Champaign
Urbana, Illinois

Susan B. Felch
Spansion, Inc.
Sunnyvale, California

Amitabh Jain
Texas Instruments, Inc.
Dallas, Texas

SPONSORING ORGANIZATIONS

Plansee Metal GmbH
Evans Analytical Group
Varian Semiconductor Equipment
ATMI Inc.
Axcelis Technologies, Inc.
KLA-Tencor
Matheson Tri-Gas
Nissin Ion Equipment Co., Ltd.
SEN Corporation, an SHI-Axcelis Company
Advanced Ion Beam Technology - AIBT
Cameca
Core Systems
High Voltage Engineering
Innovion
Mattson
Praxair Electronics
American Vacuum Society

AMERICAN INSTITUTE OF PHYSICS

Melville, New York, 2008

AIP CONFERENCE PROCEEDINGS ■ VOLUME 1066

Editors:

Edmund G. Seebauer
114 Roger Adams Laboratory
600 South Mathews Street
Urbana, IL 61801
USA

E-mail: eseebaue@illinois.edu

Amitabh Jain
Texas Instruments, Inc.
13121 TI Boulevard
MS 365
Dallas, TX 75243
USA

E-mail: a-jain1@ti.com

Susan B. Felch
Spansion, Inc.
915 DeGuigne Drive
MS 49, P.O. Box 3453
Sunnyvale, CA 94088-3453
USA

E-mail: susan.felch@spansion.com

Yevgeniy V. Kondratenko
103 Roger Adams Laboratory
600 South Mathews Street
Urbana, IL 61801
USA

E-mail: kondrate@illinois.edu

L.C. Catalog Card No. 2008937900
ISBN 978-0-7354-0597-4
ISSN 0094-243X

Printed in the United States of America

CONTENTS

PROCESS CONTROL AND YIELD

MATERIALS SCIENCE

IMPLANT TECHNOLOGY

MOLECULAR AND CLUSTER ION BEAMS

PLASMA IMMERSION

NON-DOPING AND NOVEL APPLICATIONS

17th International Conference on Ion Implantation Technology (IIT 2008)

June 8-13, 2008
Monterey, California

IIT 2008 was held on June 8-13, 2008, at the Monterey Conference Center and the Portola Hotel & Spa, adjacent to Monterey Bay, Fisherman's Wharf, and historic downtown Monterey. Nearly 170 abstracts were submitted and 265 attendees from 20 countries enjoyed an exciting week of technical presentations and social activities. Traditionally IIT is held in 3 major centers: USA, Europe and Asia. Attendance figures from this year's conference consisted of USA (59%), Europe (14%) and Asia (27%). This conference series is the premier world meeting for the presentation of the latest advances in all aspects of ion implantation, from the fundamentals of ion-solid interactions to manufacturing applications of the latest machines and systems. The General Chair was Susan Felch (Spansion).

The Conference was preceded by a School featuring Ion Implantation Science and Technology. The IIT School took place on June 5-7, 2008, in the Portola Hotel & Spa. 37 attendees learned the basics of ion implantation, from Source Physics & Maintenance to the Metrology of Ion Implantation, from a team of 14 instructors. The School was coordinated by the organizational efforts of Jim Ziegler and Michael Current.

The Technical Program took place in the Monterey Conference Center and consisted of one plenary talk given by Dave Armour of the University of Salford, the Chuck McKenna Award (for an outstanding young researcher in the field of ion implantation) lecture delivered by Richard Lindsay of Infineon, 9 invited talks, 57 oral contributed talks, and 110 posters presented in 3 sessions. These presentations covered a wide range of topics including Doping Processes, Implant Technology, Materials Science, Process Control and Yield, and Novel Applications. Ultra-Shallow Junctions was a major feature covering recent advances, including Cluster Beams and Plasma Immersion. The Best Student Paper Award was shared by Sai Hooi Yeong of the National University of Singapore (and working closely with Chartered Semiconductor), and Nicholas Rudawski of the University of Florida. The Technical Program Chair was Amitabh Jain (Texas Instruments), and the Publications Chair was Ed Seebauer (University of Illinois).

A strong Social Program enabled participants to sample the history and culture of the Monterey Bay Area. The Program began with a Welcome Reception on Sunday evening, continued with an excursion on Wednesday afternoon, and culminated with a banquet amidst the serene aquatic atmosphere of the famous Monterey Bay Aquarium on Thursday night. The excursion choices ranged from wine tasting to sightseeing along the Seventeen Mile Drive or golf at the Pacific Grove Golf Links. In addition, partners of conference attendees enjoyed a memorable program, including opportunities to explore Monterey and the surrounding regions. The Partners Program was organized by Carol Current and Ann Ziegler.

The high level of sponsorships received from the worldwide semiconductor industry was essential to the success of IIT 2008. The Conference had one Gold Sponsor (Plansee), two Silver Sponsors (Evans Analytical Group and Varian Semiconductor Equipment Associates), 6 Bronze Sponsors, and 7 Other Sponsors. The strong participation in the Trade Exhibition was another key contribution to the Conference. 17 Exhibits featured ion implantation equipment and services in the Sierra Ballroom of the Monterey Conference Center throughout the week. Sponsors and exhibitors were recruited by Aaron Vanderpool (Intel) and Mitch Taylor (Applied Materials).

In such a rapidly advancing field, speedy publication of the conference proceedings is essential. We are therefore indebted to the primary reviewing team at the University of Illinois (Yevgeniy Kondratenko, Katherine Thurman and Brandon Reizman). We would like to thank the AVS (Della Miller and Heather Korff) for their tremendous organizational, administrative, and on-site support. We also wish to acknowledge our gratitude for the help and support provided by Mary Waters of AIP. Finally we would like to thank most of all the conference participants, organizers, invited speakers and authors, for making IIT 2008 such a rewarding and memorable event.

We now look forward to IIT 2010, the 18th conference in this series, to be held in Japan during the summer of 2010. This conference will be co-chaired by Jiro Matsuo (Kyoto University), Masataka Kase (Fujitsu), Masao Naito (Nissin), and Michiro Sugitani (SEN).

Susan Felch
Amitabh Jain
Ed Seebauer
Yevgeniy Kondratenko

General Conference Chair
Susan Felch (Spansion, USA)

Technical Program Chair
Amitabh Jain (Texas Instruments, USA)

Publications Chair
Ed Seebauer (University of Illinois, USA)

Sponsorship Chairs
Aaron Vanderpool (Intel, USA)
Mitchell Taylor (Applied Materials, USA)

Conference Office
AVS, 110 Yellowstone Dr., Suite 120
Chico CA 95973
Phone: 530-896-0477, Fax: 530-896-0487,
E-mail: della@avs.org,
Web: www.iit2008.com

Technical Program Committee
John Chen (AIBT, USA)
Paul Chu (City University, Hong Kong)
Michael Current (Current Scientific, USA)
Dietmar Henke (Qimonda, Germany)
Dale Jacobson (SemEquip, USA)
Masataka Kase (Fujitsu, Japan)
Karen Kirkby (University of Surrey, UK)
Takashi Kuroi (Renesas, Japan)
Larry Larson (SEMATECH, USA)
Tony Renau (Varian, USA)
Lenny Rubin (Axcelis, USA)
Seiichi Shishiguchi (NEC, Japan)
Mitchell Taylor (Applied Materials, USA)

Local Organizing Committee
Michael Current (Current Scientific)
Emi Ishida (Spansion)
Che-Hoo Ng
Ron Eddy (Innovion)
Walt Wriggins (Core Systems)
Kavita Murthi (Matheson Tri-Gas)
John Schuur (Innovion)
Charles Evans (Consultant)
Gary Mount (Evans Analytical Group)
Leonard Weitman (Applied Materials)
Walt Johnson (KLA-Tencor)
Paul Timans (Mattson)
Ray Pong (Spansion)
Mike Vella (Consultant)

International Committee
International Committee Chair: Kevin Jones (University of Florida, USA)
Robert Brown (ATMI, USA)
Michael Current (Current Scientific, USA)
Jozsef Gyulai (KFK Institute, Hungary)
Yoshitomo Hidaka (Sumitomo Eaton Nova, Japan)
Dirk Mous (High Voltage Engineering, The Netherlands)
Larry Larson (SEMATECH, USA)
Koji Matsuda (Nissin, Japan)
Adrian Murrell (Applied Materials, UK)
Tony Renau (Varian, USA)
Wesley Weisenberger (Consultant, USA)
Peter Rose (Epion, USA)
Geoffrey Ryding (Orion, USA)
Heiner Ryssel (Fraunhofer Institute, Germany)
T.C. Smith (Consultant, USA)
Mikio Takai (Osaka University, Japan)
Anatoli Vyatkin (Russian Ministry of Science, Russia)
James Stanislaus Williams (Australian National University, Australia)
Andrew Wittkower (SOITEC, USA)
Isao Yamada (Hyogo University, Japan)
James Ziegler (U.S. Naval Academy, USA)
Russell Gwilliam (University of Surrey, UK)
Mitchell Taylor (Applied Materials, USA)
Masataka Kase (Fujitsu, Japan)
Paul Chu (City University, Hong Kong)
John Chen (AIBT, Taiwan)
Lenny Rubin (Axcellis, USA)
Dave Chivers (Consultant, France)
Nathan Cheung (U.C. Berkeley, USA)

CONFERENCE SPONSORS

Gold Sponsor

PLANSEE		www.plansee.com

Silver Sponsors

Evans Analytical Group		www.eaglabs.com
Varian		www.vsea.com

Bronze Sponsors

ATMI		www.atmi.com
Axcelis		www.axcelis.com
KLA-Tencor		www.klatencor.com
Matheson Tri-Gas		www.mathesontrigas.com
Nissin Ion Equipment Co., Ltd.		www.nissin-ion.co.jp
SEN Corp. an SHI-Axcelis Company		www.senova.co.jp.

Other Sponsors

Advanced Ion Beam Technology, Inc.		www.aibt-inc.com
Cameca		www.cameca.fr
Core Systems		www.coresystems.com
High Voltage Engineering		www.highvolteng.com
Innovion		www.innovion.cc
Mattson		www.mattson.com
Praxair		www.praxair.com

DOPING PROCESSES

The Development and Evolution of Ion Implanters in the Semiconductor Industry

Dave G Armour

Joule Physics Laboratory, Institute of Materials Research, University of Salford, Salford M5 4 WT, UK

Abstract. By the end of the 1960's, the development of ion beam systems for isotope separation and materials research had reached the stage at which knowledge bases in the areas of ion beam formation and transport and the physics of atomic collisions in solids made it practical to consider the use of ion implantation as a means of modifying the near surface properties of solid materials. The beam currents and energies available made the technique particularly compatible with the doping requirements of the silicon devices being produced at that time. However, incorporation of the technique into a high volume manufacturing environment required the immediate development of new target handling facilities and improvements in machine reliability. While the manner in which ion implanters have evolved over the past forty years has continued to be dictated by the changing demands of the silicon processing industry, the dramatic reduction in transistor size and the increase in integrated circuit complexity have had significant implications for the qualities of the ion beams themselves, particularly in high current, ultra-low energy applications. Since the first commercial implanters were introduced, highly developed medium current, high current and high energy machines have evolved. In the medium current and high energy sectors, well understood ion optical principles have enabled ingenious and highly effective beam formation and transport systems to be designed. As these machines evolved, extensive studies of the implanted material using ion beam based techniques such as Rutherford backscattering and channelling provided a growing understanding of the fundamental radiation damage and annealing processes that are inevitably associated with the implantation process. For high current machines, particularly those operating in the so-called eV implantation range, beam formation and transport processes become considerably more complex and established ion optical design principles must be combined with detailed considerations of the roles of emittance, space charge and beam plasma characteristics if beams compatible with production worthy ultra-shallow junction formation are to be obtained. For these shallow, high dose implants, the proximity of the surface has a significant effect on the radiation damage build-up and annealing processes. To develop an understanding of the physics of some aspects of these processes, high depth resolution analytical techniques such as medium energy ion scattering have been applied. In the present review, the process based evolution of ion implantation equipment is discussed alongside a consideration of the contribution made by the growth in understanding of the physics of both the beams themselves and the ion collection and radiation damage processes occurring in the implanted silicon.

Keywords: Plasmas, ion beam transport, ion implanters.
PACS: 41.75.Ak, 41.85.Ja, 61.72.uf, 68.55.Ln.

INTRODUCTION

The ability of energetic ions to change the properties of solid materials was recognised many years before the physics of atomic collisions in solids and the possible applications of ion implantation became the subjects of really intensive study in the early 1960's. The development of nuclear reactors led to major interest in radiation damage in reactor materials and to the use of energetic ions to simulate the effects of fast fission products created by neutron irradiation. The development of machines capable of producing the ions required for both isotope separation and radiation damage studies laid down all the basic ground rules for the design of machines suitable for ion implantation into semiconductors.

At the time the first dedicated ion implanters came onto the market, ion beam systems had been used to carry out comprehensive studies of ion ranges [1], channelling [2,3] and damage formation [4]. These experimental studies were underpinned by a growing understanding of the atomic collision physics associated with the slowing down of fast particles in solids [5] and atomic displacement processes [6]. Although any future relevance to the use of ion implantation into silicon was not widely appreciated at the time, some of these studies covered incident ion energies as low as 100 eV. As a consequence, the

CP1066, *Ion Implantation Technology*, edited by E. G. Seebauer, S. B. Felch, A. Jain, and Y. V. Kondratenko

basic information required to assess the viability of the so-called eV implantation technique that would be used over thirty years later was obtained.

These experimental achievements owed much to the availability of a range of high quality ion beam machines. The exacting demands of isotope separators in terms of the quantity of beam and the quality of the focus at the receiver had provided a wealth of knowledge concerning the design requirements of ion sources, extraction lenses and mass analysis systems. A range of different ion optical elements such as acceleration/deceleration lenses and electrostatic and magnetic scanning systems had also been developed on research machines. By the mid-sixties, driven by different demands in terms of the quantities of material required and the enrichment factors, isotope separators were specifically designed either with circular extraction systems to deliver low currents (below about 100μA) or with slit extraction geometries to deliver medium currents up to several mA. Together with highly specialized high energy accelerators, these general classes of machine are mirrored in the range of commercial ion implanters currently available.

The technological changes responsible for the transformation of the early research accelerators into production worthy industrial tools in a market place that has changed continuously, and with increasing rapidity in some areas, for over forty years, have already been extensively reviewed [7-8]. In the present review, this process based evolution of ion implantation equipment, with particular reference to high current, low energy machines, is discussed in the context of the contribution made by the growth in understanding of the physics of both the beams themselves and the dopant atom accommodation and radiation damage processes occurring in the implanted silicon.

ATOMIC COLLISION PHYSICS CONCEPTS

The basic ability to introduce atoms of any species and defects into well defined regions of a solid target without the immediate need to consider normally accepted equilibrium criteria was seen as an exciting area of materials research that could have fascinating applications in a wide range of materials technologies. When, in the decade following the far-reaching patent taken out by Shockley in 1954 [9], the potential of ion implantation as a device doping technology started to be realized, this was widely, and for the most part wrongly, seen as just the first of a what would be a wide range of industrially successful applications of the technique. Consequently, major research effort went in to answering some of the basic questions associated with the impingement of energetic ions onto solid materials. For example: What energy do they require to penetrate? How far do they travel before coming to rest? How many atoms are displaced per incident ion and what is the spatial distribution of this damage within the solid? How many atoms are ejected (sputtered) from the surface per incident ion? How important is the angle of incidence? How critically do these processes depend on energy? The theoretical and experimental research that answered these questions not only confirmed the viability of ion implantation for the production of high quality doped layers but also provided the basic specifications of the ion beams in terms of energy range, energy and mass purity, angular spread and current density. In fact, the early dedicated ion implanters evolved almost directly from the machines developed to carry out this research

The question of the threshold energy for penetration was considered theoretically by calculating the cross-sections for scattering through angles greater than 90^0 for ions incident normally onto the surface of interest and experimentally by measuring the trapping of rare gas ions during very low energy bombardment using thermal desorption mass spectrometry [10]. Rare gas ions were chosen because they do not stick to the surface at room temperature. Theoretical studies were complicated by the lack of a precise knowledge of the form of the interatomic potentials for most sets of collision partners and the need to consider multiple, rather than simple binary interactions at the lowest energies. However, it was clearly established that for all the ion species relevant to the doping of silicon, the threshold energy for penetration at normal or near-normal incidence would be below about 200 eV. It was also established that heavy ions can produce damage without penetrating the lattice and light ions, such as helium, can penetrate the lattice without causing damage.

The early studies of range and damage profiles provided the greatest insight into the beam requirements. The work of Lindhard, Scharff and Schiótt [5] used the concepts of electronic and nuclear stopping to develop a comprehensive theory for the ranges of ions in solids and provided the means to obtain a clear insight into the nature of the energy loss processes responsible for the slowing down of the main implant species in device applications–boron, phosphorus and arsenic. These early calculations indicated that for boron energies above about 17 keV, the dominant loss process is electronic, while for arsenic, nuclear collisions dominate up to energies in excess of about 200 keV. This result has significant implications for the nature and distribution of the damage in many practical implants since electronic stopping does not result in the production of structural damage and, as would be established later, the significant differences in energy deposition density played a crucial role in determining the nature of the crystalline to amorphous transition during light and heavy ion implantation. This work inspired an enormous growth in interest in the measurement of implant and damage distributions at a time when techniques such as Rutherford backscattering and channelling were in their infancy and SIMS depth profiling had not been developed. Consequently, early profiling measurements involved implantation with a radioactive isotope of the species of interest followed by the successive removal of layers of accurately controlled depth by anodic stripping, vibratory polishing or sputter erosion. The significance of this early work from the point of view of ion implantation was that it demonstrated that LSS theory could predict ranges to within a factor of two and that the effects of channelling were quite dramatic when the incident beam direction was within a few degrees of an open crystallographic direction. It was thought at first that this effect could be exploited to achieve deep implants without the need for very high energy beams. Profile shape and damage considerations quickly dispelled this idea and, in practical implants, channelling is normally avoided by bombarding in a direction that is outside the critical angle for axial channelling and does not coincide with any of the planar channels. The 'reasonable' agreement between experimental data for implantation at an angle about 7^0 to open channel directions and LSS based range calculations seen in many of these studies became the basis for the use of this off-set angle in commercial ion implanters.

Although the dose range considered in these studies more than covered that associated with the different parametric and conductive implants being used at the time, Moore's law had already been published and there were early indications that machines with higher current and lower energy capabilities than those being used for this implantation related research would eventually be required. This interest in extending the operating range of machines, together with the need to improve levels of control over beam uniformity and angular spread as wafer sizes increased led to an increase in the background level of interest in the more fundamental aspects of beam formation and transport.

BEAM CONSIDERATIONS IN MACHINE DEVELOPMENT

The experimental studies described above used purpose built machines designed for isotope separation, materials modification or ion beam analysis. In many cases, the development of appropriate ion sources and beam transport systems was an essential first step in this field of research since commercial systems with the range of capabilities required were not available. Although machines designed to deliver ion beams of specified mass and energy to a target appear, at first sight, to be similar, the breadth of research interests and applications led to the study, development and use of a multitude of different types of ion source and transport system. Details of many of these have been discussed in reviews by McKenna [7], Freeman [8] and in the book by Ryssel and Ruge [11]. In the vast majority of systems built before the mid-1970's, the concentration on the energy range between about 20 keV and 200 keV and the use of currents up to about 1 mA (in the implanter industry – medium current machines), enabled the same basic layout, comprising an ion source with extraction and pre-acceleration optics, a mass analyser and a target stage, to be quite readily adapted to any particular application. In general, this involved the selection of a source capable of supplying adequate fluxes of the desired ion species and the use of well established ion optical design rules to specify the nature of the transport system required to deliver the beam to the target. Because this procedure worked very well for machines operating in the energy/beam current regime of interest, major consideration could be given to the industrial aspects of the design and comparatively little attention needed to be paid to the precise nature of the beams themselves and the fundamental processes affecting their transport. Consequently, in the sector of the ion implanter market that established the compatibility of the technique with the requirements of a high volume manufacturing industry, it turned out to be comparatively straightforward to tailor the beam transport systems to meet the wafer size and throughput demands by adding, where appropriate, post-analysis ion optical systems for further acceleration, deceleration and scanning or deflection of the beam.

The later developments in these machines benefited greatly from the availability of an increasing array of ion optical simulation codes [12]. Some of these codes can include the effects of finite beam emittance and space charge and their usefulness increased as beam currents in this medium current sector crept up to accommodate the increase in wafer size. Although these codes enable the effects of space charge to be assessed, there are currently no codes available that can simulate the space charge neutralization processes in a realistic way. The growing interest in high current machines in the mid-1970's brought implanter beams completely out of their comfort zone and immediately highlighted the fundamental importance of space charge and emittance in beam transport. The subsequent need for high currents at ultra-low energies in so–called eV implantation further increased the influence of these fundamental issues on machine design and led to a considerable increase in the use of beam and beam plasma diagnostics during the machine development programs. However, unlike similar and more comprehensive beams studies that had long been established in the beam heating programs associated with fusion research, the implanter related work was still being carried out in an industrial context in which the beam dump often represented the main, if not the only, beam diagnostic. Process and cost issues continued to dictate key aspects of the overall machine configuration.

While consideration of space charge and emittance issues was comparatively new to the commercial implanter industry, it was a fact of life in the high current isotope separator community. It had already been established that instabilities in the source plasma led to beams with excessively high emittance and that effective space charge neutralization could not be achieved consistently in such beams. Although not an inherent feature of the fundamental beam formation and transport processes, ion optical issues associated with the passage of these high current beams through regions of rapidly spatially varying electric or magnetic fields were also known to be a problem.

These issues were recognised and addressed in the far-sighted design of the first Lintott implanters that evolved from the isotope separator program at Harwell. The University of Salford took delivery of the first of these machines to be sold in 1969. The machine delivered mA beams of a wide range of species and was based on a Freeman source that had been specifically designed to overcome some of the instability and emittance issues that had been observed in machines using the Calutron and Bernas sources. By dealing with these issues and producing a highly stable, uniform beam, the ability of the beam to space charge neutralize was enhanced and mA beams could be transported routinely through an approximately 2 m long beam line. However, in common with all other beam systems available at that time, the transport efficiency of the system fell rapidly at low energies.

The fundamental difference between the Freeman source and the other hot cathode plasma sources that had underpinned the development of high current isotope separators can be seen in Figure 1. Not only does the Freeman design ensure that the plasma is uniform over the entire length of the extraction slit (this was one of the principal design aims) but the formation of a defined electron beam within the plasma region is also eliminated. The major 'hash' problems experienced in the Calutron ion sources were believed to be associated with the presence of a magnetically confined ionizing electron beam in which the electron velocities were significantly greater than the mean thermal velocity of the plasma electrons. The hot cathode, low pressure arcs in Calutron and Bernas type sources provide a particularly favourable environment for the excitation of plasma oscillations, especially if a reflecting anti-cathode is used. In isotope separator applications requiring very high resolution, the Bernas source was always operated with the anti-cathode connected to the anode, even though this reduced the available beam current. In high current B^+ ion implantation applications, this mode of operation has never been an option since both electron reflection and a magnetic confining field are essential if adequate cracking of the parent BF_3 molecule in the source is to be obtained.

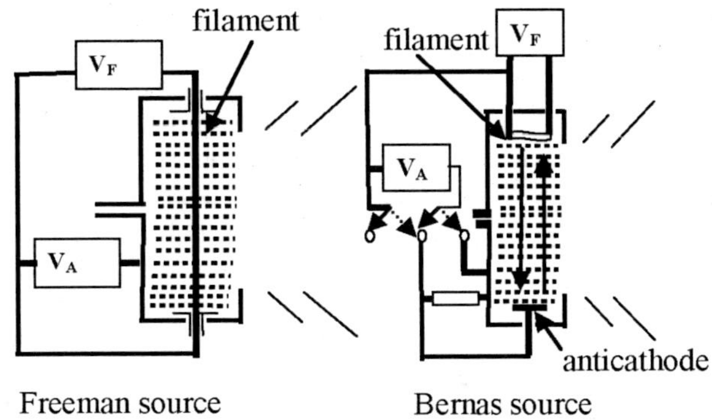

FIGURE 1. High Current, Hot Cathode Ion Sources.

Oscillations in the source plasma cause the supply of ions to the plasma sheath in the extraction opening to be modulated at the frequencies involved. This modulation degrades the extracted ion beam in two ways: firstly, it introduces density variations into the beam resulting in a reduced ability to reach a stable space charge neutralized condition and secondly, it leads to time varying changes in the shape of the plasma boundary from which the ions emerge at initially very low energies. This has the effect of increasing the spread in the transverse energies of the ions and manifests itself as an increase in emittance. Essentially the apparent ion temperature in the source rises to values well above the 0.3 eV expected for plasmas of the type produced in these sources. The emittances of 10 keV Ar+ beams extracted from Freeman sources operated in the medium current mode are typically between 5 and 8 π mm. mrad. while, for 40 keV B+ beams, those for Bernas type sources operated with the anti-cathode connected to the anode (no electron reflection) are consistently found to be in the range from 8 to 20 π mm. mrad. In the normally used electron reflection mode in which the anti-cathode is connected to the cathode, emittance values rise to between 15 and 40 π mm. mrad. The use of an indirectly heated cathode in this mode increases the emittance further into the range between 20 and 60 π mm. mrad., again for 40 keV B+ beams. For all these sources, the emittances of Ar+ and B+ beams are consistently observed to be higher than those for equivalent P+ and As+ beams. The measurements refer to the horizontal transverse plane of beams extracted from identical slit apertures. If it is assumed that the emittances are entirely due to the effective ion temperature in the source, these data indicate that the transverse energies correspond to ion temperatures in the range from about 2 eV up the 60 eV. These are totally un-physical values and must reflect quite dramatic plasma boundary effects. It is interesting to note that emittances measured in the vertical transverse plane (the longer slit dimension) yield effective ion temperatures of the order of 0.5 eV.

From a practical point of view, these emittance measurements are only of use if they enable the machine designer to exploit the information they contain. Although the practical relationship between emittance and beam transportability is complicated by the fact that emittance, ion optical aberrations and space charge neutralization efficiency are interdependent, it is possible to derive conceptually useful design information on the basis of a simple rule of thumb that relates the distance over which a beam with a given rms emittance can be transported for specified beam dimensions at the source and target positions. This rule is expressed as:

$$\varepsilon_m = 1000\frac{r_s r_t}{z}$$

where the emittance is in mm.mrad and r_s and r_t are beam half widths in mm at the source and target and z is the beamline length in mm. In conjunction with the relationship between the rms emittance and the effective ion temperature, this rule makes it possible to construct an emittance design chart from which the maximum distance over which a beam with a given emittance at a specified energy can be transported effectively for specified source and target dimensions can be deduced [13]. Such a chart is shown in Figure 2 for the energy region below 10 keV. It is clear that for beam energies below 1 keV, the transport efficiency starts to fall if the effective ion temperature in the source is greater than about 1 eV.

This type of chart provides a possible explanation for at least part of the low energy current drop-off seen in the early Lintott machines. Despite the higher quality of the beams extracted from the Freeman source, the specified ion currents obtainable at low energies on this machine and on the later series 2 and 3 machines were surprisingly low. The manufacturer's data is shown in Fig 3, where the performance is compared with that of the most recent Applied Materials' machine. Reference to the emittance design

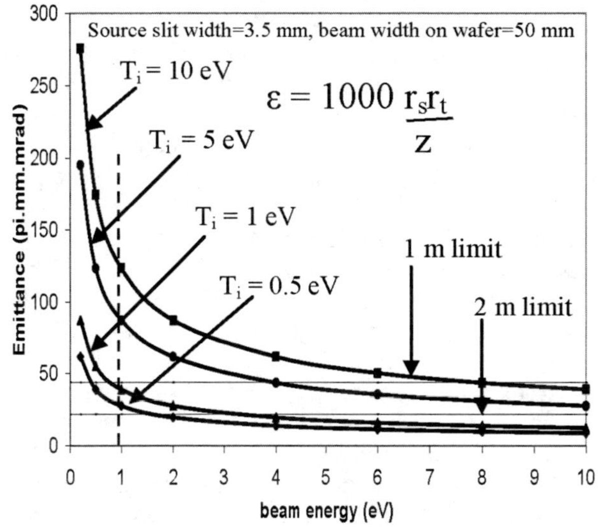

FIGURE 2. Emittance Design Chart.

chart shows that for effective ion temperatures between about 1 and 5 eV that are characteristic of the Freeman source running in medium current mode, a drop-off in transmission would be expected to start at energies above 10 keV (close to 20keV) for an ion temperature of 5 eV and at about 4 keV for an ion temperature of 1 eV. Reducing the length of the beamline to 1 m would prevent the beam loss down to energies of 4 keV and less than 1 keV for the same two temperatures. This observation gives a clear indication that if the emittance itself cannot be reduced without unacceptable loss of current, the solution to the emittance problem is to reduce the length of the beamline. This parameter is the main difference between the early Lintott machines and all the modern high current machines.

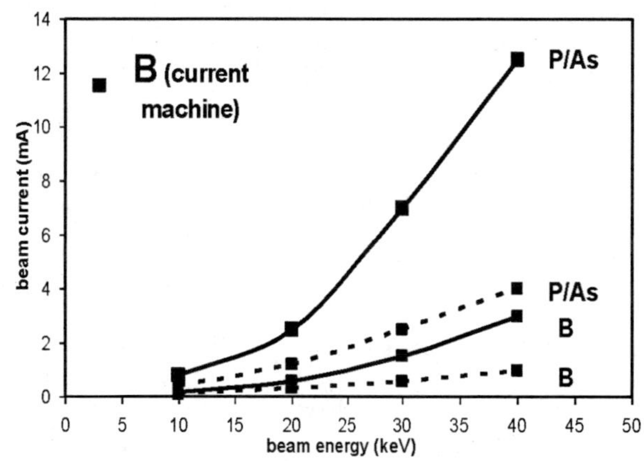

FIGURE 3. Low energy currents in Lintott Machines.

While this drop off in performance was not of immediate concern to the industry in the early to mid 1970's, if Moore's law was obeyed, the number of devices on a processor chip would increase from about four thousand in 1970 to over a billion in 2006. The reductions in device critical dimensions associated with this progression would, according to the SIA Roadmap, ultimately require junction depths of less than 50 nm. For boron implants, this implies beam energies down to 200 eV. Even though shortening and simplifying beamlines on the basis of emittance considerations made a major contribution to the reduction in beam loss at low energies, on their own, they did not produce the improvements required to enable the current generation of high current, ultra-low energy machines to produce such shallow junctions at a commercially viable rate. Detailed refinements to the ion optical performance of extraction, acceleration and deceleration lenses and major improvements in the capabilities of plasma flood systems for space charge neutralization have also been incorporated into the latest generation of machines.

The potentially crucial role of space charge in high current ion implanters stems from the nature of the source plasma used in this type of machine. The hot cathode discharges from which the ion beams are extracted are characterized by electron densities between about 5E17 and 5E18 m^{-3} and electron temperatures between about 5 and 10 eV. If the Bohm criterion is obeyed, the ion flux density at a normal plasma boundary in such plasmas significantly exceeds the maximum current density that can be extracted from the source under space charge limited conditions (Child-Langmuir) for the species and energies of interest in ion implantation into silicon. For example, for ions of mass 11 amu extracted from a plasma characterized by an electron density of 1E18 m^{-3} and an electron temperature of 5 eV, the plasma could supply a current of about 185 mA through an extraction aperture of area 1.75 cm^2. For an acceleration energy of 20 keV, the space charge limited current is approximately 80 mA.

In practice, due mainly to the geometry of the plasma boundary, the currents extracted from high current ion sources are not always at the space charge limit calculated from the simple form of Child's Law. However, in the absence of any neutralization of the beam space charge, the current densities are always high enough for the space charge related electric fields to induce significant expansion of the beam envelope. The extent of this expansion can be calculated by solving the envelope equations for a uniform, un-neutralised beam of zero emittance [14]. The result is the well known universal beam divergence curve on the basis of which the spreading of beams of practical interest, such as B^+ beams, can be calculated. For energies in the region of 1 keV, even a 100 μA beam of this light ion will expand by a factor of three in a distance of about 25 cm. It is clear that to obtain milli-ampere currents at energies in the region of 1 keV, it is essential that the beam is able to neutralize its inherent space charge by acquiring negatively charged particles.

For high energy beams, above about 20 keV, effective neutralization of the beam space charge appears to occur naturally by the production of electrons and slow ions in the beam environment via ionizing collisions between the fast beam ions and the background gas molecules. The apparent ability of the beam to retain electrons very efficiently enables the process to be effective down to pressures as low as about 1E-6 mbar in some cases. However, for high current, low energy (below about 10 keV) beams being transported in implanter beamlines in which the pressure is of the order of 1E-5 mbar, the electron generation (including secondary electron emission resulting from beam strike on surfaces) and trapping processes appear to be inadequate. Under these circumstances, enhancement of the natural supply of electrons by direct electron injection (not usually effective in implanter beams) or plasma bridging is required. In practice, the build-up of the concentration of electrons in the beam to the point where the beam space charge is almost fully neutralized is always accompanied by an associated increase in the density of slow positive ions. In the absence of a bridging plasma, these ions result from both ionizing and charge exchange collisions between the beam ions and the background gas. Despite the fact that they would be expected to drift out of the beam quite rapidly, beam plasma is formed in which, at high beam line pressures (1E-4 mbar), the slow ion density is expected to be similar to that of the beam ions. At lower pressures, 1E-5 mbar, this figure drops to less than 10% [14]. The presence of stable beam plasma is essential if the space charge neutralization process is to be effective, and it appears that the plasma potential of this plasma dictates the minimum potential that can be achieved in the beam environment.

From the point of view of optimizing beam transport, the beamline pressure and the parameters of any auxiliary source of electrons and slow ions need to be set to values that produce the most appropriate conditions in the beam plasma. Unfortunately, realistic analysis of both the natural and plasma assisted neutralization processes is complicated. The natural process is totally dependent on the parameters of the beam and calculation of the parameters for the steady state beam plasma requires consideration of the rates of generation and loss of the plasma constituents and of the nature of the spatial distribution of potential as functions of gas pressure, ionization and charge exchange collision cross-sections, the beam parameters and the electron temperature. Since the electron temperature depends critically on the energy balance, the energy transfer from the beam ions (the only source of energy) to the plasma electrons via Coulomb drag and the energy loss from the plasma due mainly to the ejection of slow positive ions must also be considered. The development of a theoretical model to describe the operation of the overall interaction of the beam with a bridging plasma is also very complicated. Both plasma-plasma and beam-plasma interactions are involved. It is known that the interaction of the beam with the plasma is not simply governed by the interaction of the beam ions with the particles of the plasma. Local fluctuations in the space charge lead to the interaction of the beam ions with many plasma particles at the same time. This type of collective phenomenon is typical of many plasma situations. In the context of an ion implanter operating with a plasma flood system, (an arrangement that has contributed significantly to improvements in beam transport at energies below about 3 keV in addition to achieving its primary purpose of eliminating wafer charging damage), plasma from the external plasma flood discharge chamber placed close to the beam streams into the region occupied by the beam. Conceptually, if the electrons in this plasma stream are to be accelerated into the beam and effectively trapped, the beam potential cannot be less than that in the plasma with which it interacts. This potential, the plasma potential of the flood plasma, is always positive with respect to the surrounding surfaces so that it can retain its electrons.

Use of this plasma technique has enabled very high levels of space charge neutralization to be obtained and, in general, its effectiveness is not strongly dependent on the detailed parameters of the beam. However, the precise way in which the beam, the beam plasma and the neutralizing plasma interact is not fully understood and the conceptual picture described above does not always seem to describe the beam-plasma interaction accurately. Unfortunately, the interaction region is difficult to study. Diagnostic tools such as Langmuir probes do not provide reliable data for beam plasmas as the application of even small positive potentials leads, for a wide range of beams, to an immediate rise in the beam potential as the space charge neutralization is perturbed. The distortion of the positive part of the probe characteristic is such that neither the plasma potential nor the electron saturation current (and hence the electron density) can be determined reliably. The potentials in the beam, the flood plasma and the coupled system can, however, be measured reliably by monitoring the energy distributions of the slow ions that drift out in a direction perpendicular to the beam direction. Using a conventional, magnetically confined, hot cathode, DC discharge created in xenon as the flood plasma, these potential measurements demonstrate how complex the interaction can be. This type of plasma flood system can be operated with the discharge chamber (the anode) at either the same potential as the beamline (usually ground) or at the value of the arc voltage (typically 20 to 40 V) positive with respect to ground. When the discharge chamber (located outside the beam) is at a positive potential, without a discharge, the beam potential rises in a similar way to that observed when a probe in the beam is made positive. The peak energy of the slow ions drifting out of a typical 10 keV, 10 mA Ar^+ beam, increases from about 10 eV up to 22 eV for a voltage of 20 V. When the discharge is run, this energy increases further and a second, higher energy peak appears at the same energy as the Xe^+ peak measured for the plasma flood system acting alone (typically about 35 V). It appears that when the bridging plasma is sufficiently dense, its plasma potential and that of the beam plasma become equal at a value that is significantly higher than the original beam plasma potential. In this mode of operation, no increase in beam current is observed. If the discharge chamber is grounded, operation of the discharge leads to a significant drop in the peak energy of the slow ions in the beam from the undisturbed value of about 10 eV down to about 4 eV. For low arc currents, e.g. 1 A, there is very little tendency for the beam plasma potential to be pulled up to the plasma potential of the bridging plasma. In fact, counter-intuitively, the potential of the beam plasma remains well below both the original beam potential and the plasma potential of the bridging plasma even at arc currents as high as 4 A. The establishment of this condition appears to reflect a significant improvement in the level of space charge neutralization and operation of the neutralizing discharge in this mode always leads to an increase in the beam current of between 50 and 100% depending on the arc current in the flood plasma. This increase in beam current is more marked for low energy beams.

Potential measurements of this type have shown that for many of the beams run in high current ion implanters, the beam potential is highly sensitive to the presence of positive voltages on any flight tube component in its vicinity. This probably reflects the fact that many of the routine beams are not fully space charge neutralized. The ability of a bridging plasma to reduce the beam potential to a value that is significantly below both its initial value and the plasma potential of the bridging plasma itself indicates that very high levels of space charge neutralization can be achieved under appropriate operating conditions. This is reflected in the sometimes quite spectacular increases in beam current, up to 7 mA for 1 keV B+, that can be obtained.

SUMMARY

Over the past forty years, ion implanters have developed to meet the increasing demands of the semiconductor device industry. Their capabilities in terms of energy purity, angular control and definition and in the high current area, beam transport efficiency at low energies have improved dramatically during this time. Many of these improvements have been achieved on the basis of intuitive, incremental developments driven by a combination of performance and productivity demands. However, the growing understanding of the ion-solid interaction processes and the physics of the ion beams themselves combined with the more widespread use of powerful ion optical design tools have made a very significant contribution. A clearer understanding of the roles of emittance and space charge neutralization was of fundamental importance in the conceptual design of the first ultra-low energy, high current machines. It was this concept that underpinned the evolution of the current generation of eV implanters.

ACKNOWLEDGEMENTS

It is a pleasure to thank Harry Freeman, Freddie Plumb, Jiri Fiala, Jaap van den Berg and Mitch Taylor for providing information used in this paper and for their help and support with many different aspects of the work. I would also like to thank George Carter for introducing me to the area of atomic collisions in solids and the ion beam technology associated with it.

REFERENCES

1. G. Dearnaley, J. H. Freeman, G. A. Gard and M. A. Wilkins, *Canadian Journal of Physics*, **46**, 587-595 (1968).
2. W. L. Brown, "Analysis of Crystalline Defects by Ion Beam Techniques," in *Radiation Damage and Defects in Semiconductors*-1972, edited by J. E. Whitehouse Institute of Physics Conference Series No. 16, 1972, pp.416-428.
3. F. H. Eisen, *Canadian Journal of Physics*, **46**, 561-574 (1968).
4. F. H. Eisen, B. Welch, J. E. Westmoreland and J. W. Mayer, " Lattice Disorder in Silicon by Boron Ion Implantation," in *Atomic Collision Phenomena in Solids-1969*, edited by D. W. Palmer *et al.*, North Holland Publishing Company, London, 1970, pp. 111-127.
5. J. Lindhard, M.Scharff and H.E.SchiØtt, Kgl. Danske Videnskab Selskab, *Mat.-Fys. Medd*, **33**, 14, (1964).
6. G. H. Kinchin and R.S. Pease, *Rep. Prog. Phys.*, **18**, 1-11 (1955).
7. C. M. McKenna, "A Personal Historical Perspective of Ion Implantation Equipment for Semiconductor Applications," in *Ion Implantation Technology-2000*, edited by H. Ryssel *et al.*, IEEE 2000, pp. 1-19.
8. J. H. Freeman, Radiation Effects, **100**, 161-248 (1968).
9. W. Shockley, US Patent No. 2787564 (1964)10. E. V. Kornelsen, Canadian Journal of Physics, **42**, 364-370 (1964).
11. H. Ryssel and I. Ruge, *"Ion Implantation"*, New York: John Wiley and Sons,(1986).
12. P. Spadke, "Computer Modeling" in *The Physics and Technology of Ion Sources*, edited by I. G. Brown, New York: John Wiley and Sons, 1989, pp.107-135.
13. S. Povall, D. Burgin, C. E. A. Cooke, R. D. Goldberg, A. J. T. Holmes, D. Arnold, J. A. van den Berg and D. G. Armour, "Emittance Considerations in Ion Source Design and Operation," in *Ion Implantation Technology-2000*, edited by H. Ryssel *et al.*, IEEE 2000, pp. 530-533.
14. A. J. Holmes, *Radiation Effects*, **44**, 47-58 (1979).

Ultra-Shallow Junction Formation – Physics and Advanced Technology

B. Colombeau[1], S. H. Yeong[2], D. X. M. Tan[2], A. J. Smith[3], R. M. Gwilliam[3], C. M. Ng[2], K. R. C. Mok[2], F. Benistant[2], and L. Chan[2]

[1]*Varian Semiconductor Equipment Associates, 35 Dory Road, Gloucester, MA 01930, USA.*
benjamin.colombeau@vsea.com
[2]*Chartered Semiconductor Manufacturing Ltd., 60 Woodlands street 2, Singapore*
[3]*Ion Beam Centre – Advanced Technology Institute, University of Surrey, Guildford, GU2 7XH, UK*

Abstract. In this paper, the advanced technology in the ultra-shallow junction formation and its physical understanding for sub-nano CMOS devices are presented. After quickly presenting the device issues for the next generation of CMOS devices, we shall focus on the formation of highly activated Ultra-Shallow Source Drain extension. In fact, the formation of ultra-shallow junctions (USJ) for future integrated circuit technologies requires achieving high activation levels and abrupt profiles. To achieve the challenging targets set out in the semiconductor roadmap, it is crucial to reach a much better understanding of the basic physical processes taking place during USJ processing. Subsequently, we review current understanding of dopant-defect interactions during thermal processing of device structures – interactions which are at the heart of the dopant diffusion and activation anomalies seen in USJ leading to device performance degradation. Based on the physical understanding, we shall review and discuss some promising methods, such as co-implants (C, F, N), cluster/molecular implants and defect engineering as well as alternative thermal processes, for further downscaling of source-drain resistance and junction depth. Emphasize on the impact on both PMOS and NMOS device performance will be presented. Finally, we will demonstrate that TCAD (Technology Computer Aided Design) can be successfully used for optimizing the new junction architecture to meet the nano-CMOS device performance requirements.

Keywords: Ion Implantation, Dopant diffusion, Extended defects, Ultra-Shallow Junctions, device optimization, TCAD.

INTRODUCTION

The semiconductor industry is at the heart of the present society and is undergoing regular and fast changes. It fulfils the needs of today life by providing for example wireless mobile technology and ever-faster processors for computers.

As CMOS evolves down into the 32 nm technology, the scaling requirements have become increasingly more stringent and difficult to achieve. In fact, when the transistor dimensions decrease, there is a clear need of a continuous improvement of the Short Channel Effect (SCE) control as well as enhanced transistor performance. In terms of requirements for further downscaling of CMOS devices can be listed below with the possible solutions:

- Increased gate capacitance (Thinner oxide, high-K metal gate).
- Increased body dopant concentration (Halo implant and channel dopant engineering).
- Reduced Gate Depletion (Heavily doped polysilicon, and metal gate).
- Enhanced carrier mobility (strained silicon and novel channel materials).
- Reduced junction depth (formation of Ultra-Shallow Junctions (USJ).

The formation of USJ in the source/drain extension poses one of the main challenges. It is well known that dopants exhibit anomalous transient enhanced diffusion (TED) and the clustering mechanism via interstitial or vacancy [1-4]. These two anomalous behaviors have been well recognized as the underlying causes of the increase in final junction depth (Xj) and sheet resistance (Rs) during the post-implantation annealing process. The conventional challenges in USJ formation are how to achieve high dopant activation, a shallow penetration of the dopant profile, and high junction steepness. However for the next generation of CMOS devices, new challenges need to be considered such as elimination of residual defects, strong control of lateral diffusion, junction stability after post-annealing processes as well as junction formation in new materials (such as SiGe/SiGeC/SiC) used during the device fabrication.

CP1066, *Ion Implantation Technology,* edited by E. G. Seebauer, S. B. Felch, A. Jain, and Y. V. Kondratenko

To achieve the challenging targets set out in the semiconductor roadmap, it is crucial to reach a much better understanding of the basic physical processes taking place during USJ processing. This paper shall review current understanding of dopant-defect interactions during thermal processing of device structures – interactions which are at the heart of the dopant diffusion and activation anomalies seen in USJ leading to degradation in device performance. Based on the physical understanding, we shall review and discuss some promising methods, such as co-implants (C, F, N), cluster/molecular implants and defect engineering as well as alternative thermal processes, for further downscaling of source-drain resistance and junction depth. The impact of the some of the USJ approaches on both PMOS and NMOS device performance will be emphasized.

Finally, we shall briefly highlight that accurate TCAD (Technology Computer Aided Design) can be a powerful tool for optimizing new junction architecture to meet the nano-CMOS device performance requirements.

Point defect, Extended Defects and Dopant Interactions

Ion implantation leads to the formation of a band of interstitial defects at or beyond the implant projected range. These defects are self-interstitial clusters, which on thermal annealing emit free interstitials and dissolve and/or ripen into larger, more stable defects. For sub-amorphizing implants the defect band is quite broad, while for amorphizing implants it is much narrower and situated at the "end of range" of the amorphizing implant, just beyond the depth of the original amorphous-crystalline (a/c) interface, as can be seen in figure 1. As figure 2 illustrates, an amorphizing Ge implant causes the region below the a/c interface to be heavily supersaturated with Si recoil interstitials. During the initial phase of annealing, both defect nucleation and solid phase epitaxial regrowth occur. The former being the faster phenomenon, and the self-diffusivity of Si being smaller in amorphous Si than in the crystal, the defects stay located below the original interface [5]. The total number of Si atoms bound to the defects is approximately equal to the number of excess interstitial atoms left behind the a/c interface after the PAI implant [6]. This model has been called the "excess interstitial" model [7]. During thermal annealing, these agglomerate into extended self-interstitial defects that emit free interstitials back into the system during thermal annealing [8].

A lot of studies have been done to obtain a better understanding of dopant diffusion and clustering mechanisms in silicon. For the case of p-type dopants, the main dopant of interest is Boron (B). For this dopant, diffusion and clustering mechanisms have been extensively studied in literature and a good understanding has been achieved [3]. B diffuses by the interstitialcy mechanism and clusters via the formation of inactive, immobile boron-interstitial clusters, i.e. B_nI_m complexes with "n" B atoms and "m" Si interstitials. The free interstitials emitting during the extended defect evolution drive transient enhanced diffusion (TED) and deactivation of the B profile.

It is clear that by avoiding dopant-interstitial pair formation, this turns off diffusion and clustering reaction chains leaving the B atoms to remain active with minimum diffusion. Such ideas have been the driving force for further improvement of USJ formation.

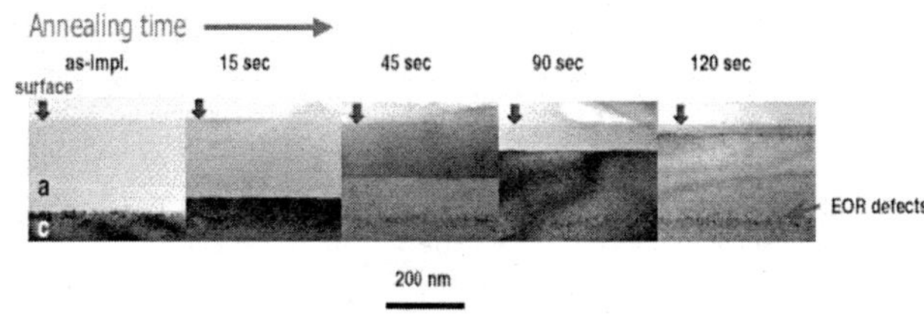

FIGURE 1. SPE regrowth at 600°C of Si substrate preamorphized with Ge^+ ions implanted at 150 keV to a dose of $2x10^{15}$ ions/cm^2. The thickness of the amorphous layer is originally 183nm [3].

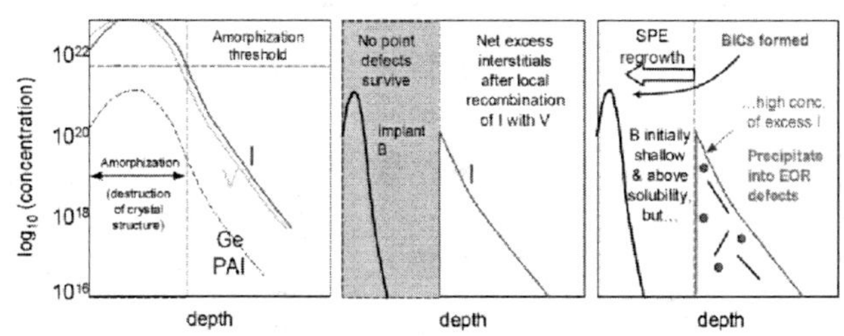

FIGURE 2. Schematic illustration of physics underlying for simulation of defect evolution and diffusion in the crystalline phase [3].

USJ TECHNIQUES

In this section, we shall review some of the "common" techniques proposed to form Ultra-Shallow Junctions.

Pre-amorphization Implant (PAI) and Solid Phase Epitaxy Regrowth (SPER)

This technique consists in pre-amorphizing the Si substrate with heavy ions (such as Si or Ge), after which dopant ions are implanted at very low energy to a very high dose. It has been reported that preamorphization prior to B implant not only avoids B channeling but overcomes the traditional limitation of solid solubility by placing B atoms directly onto substitutional sites. This increased substitutionality is a result of solid-phase epitaxial regrowth (SPER) of the preamorphized region containing the B implant [9-11].

A significant drawback of PAI, however, is that it leaves a damage band, rich in self-interstitials, I, just beyond the amorphous/crystalline interface (see figure 1 for experimental evidence of the SPER process). As previously mentioned, these defects agglomerate into extended self-interstitial defects that emit free interstitials back into the system during thermal annealing [8]. The free interstitials drive transient enhanced diffusion (TED) and deactivation of the B implant. In addition, the remaining EOR defects generate high junction leakage, degrading the device performance [12].

Recently, Pawlak *et al.* [13, 14] have extensively investigated the dopant activation behavior. Isochronal annealing and characterization by sheet resistance were used to extract detailed information on the thermal stability of the boron activation. In their study, B was implanted at 1.5keV to a dose of $1.5x10^{15}$ cm^{-2} into wafers initially implanted with Ge to a dose of 1×10^{15} cm^{-2} at energies of 8, 12, and 25 keV (producing amorphous layer depths of 15, 22 and 42 nm, respectively). Figure 3 shows detailed activation studies at temperatures between T = 650 °C and T = 950 °C for 1 min on SPER junctions with three initial a-Si depths.

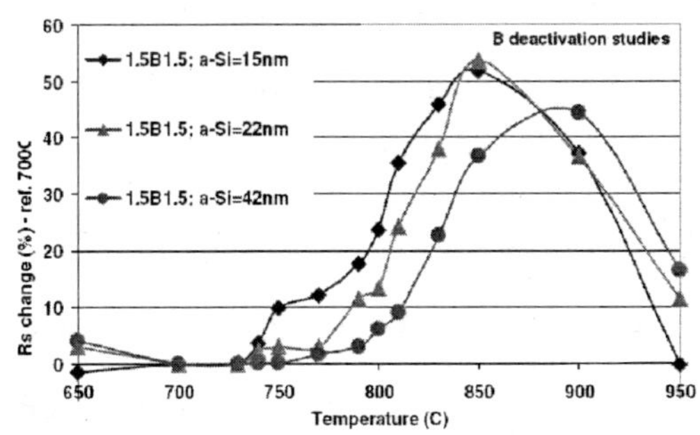

FIGURE 3. Rs change versus temperature for SPER junctions activated at various temperatures for t = 1 min. The a-Si depths were 15, 22 and 42 nm for samples doped with 1.5 keV boron to a dose of $1.5x10^{15}$ cm^{-2} [13, 14].

At temperatures 650 °C < T < 750 °C (see [13] for details) each individual type of junction is completely regrown and all measurements in this temperature range showed a similar, high level of maximum boron activation (minimum Rs value). In this temperature regime the differences in Rs between junctions with three different a-Si depths are due to the activation of the as-implanted B profile, restricted only to the original a-Si layer. When the annealing temperature is increased, electrical deactivation of the junction occurs, as seen as an increase in Rs. Dopant reactivation is finally observed for temperatures above 850°C. This behavior is typical of a 'reverse annealing' effect. As previously mentioned, it has been clearly demonstrated that the main cause for such behaviors is that Si interstitial atoms released from the EOR defect region interact with boron substitutional atoms, leading to the formation of inactive boron interstitial clusters (BICs), and/or to out-diffusion of boron interstitials to the surface.

Co-implantation

As previously mentioned, one of the main issues for the formation of advanced junction is the anomalous diffusion and activation of dopants during annealing. One very popular technique to avoid these phenomena is to co-implant with impurities [15-18]. Specific co-implantations of carbon (C) and fluorine (F) have been engineered to reduce boron diffusion on the one hand and enhance its activation on the other. Figure 4 presents the impact of C and F on B profiles [19]. Clearly, significant improvements in reduced diffusion under a standard spike anneal. The underlying idea for the reduction of B diffusion is that the incoming interstitials are trapped by the impurity band therefore reducing the amount of dopant-interstitial pairs. In the case of C, it has been evidenced that high C concentration mainly formed Carbon-Interstitial clusters [20]. For Fluorine, most recently it has been shown that during the SPER, the Fluorine atoms agglomerate with Vacancy to form Fluorine-Vacancy clusters [21, 22]. Nitrogen is also thought to follow a similar clustering but is less established. In any cases, the impurity clusters formed interact with the emitting interstitial from the extended defects decreasing dopant diffusion. In terms of activation, a lot of studies have been performed to demonstrate its improvement. Recently, Yeong *et al.* have extensively studied the activation behavior of co-implanted USJ with C and F [23]. From this study a clear picture on the optimized conditions for USJ formation with minimum deactivation. In addition, Phosphorous-doped N-type USJ created with the same approach has also been fabricated and similar effects were seen. One main benefit of this approach is to extend the use of standard spike annealing.

In the device aspect, co-implantation is not only considered for the lateral diffusion control of the PMOS Source Drain Extension, but in addition it has been demonstrated to be very efficient in improving the NMOS device performance. For the NMOS, impurities are implanted in order to generate super steep p-type halo as well as to decrease the doping concentration in the channel. Recent works revealed that improvement of Halo can be achieved with either

C or N co-implantation [24, 25]. The combination of these is clearly beneficial for the technology curve.

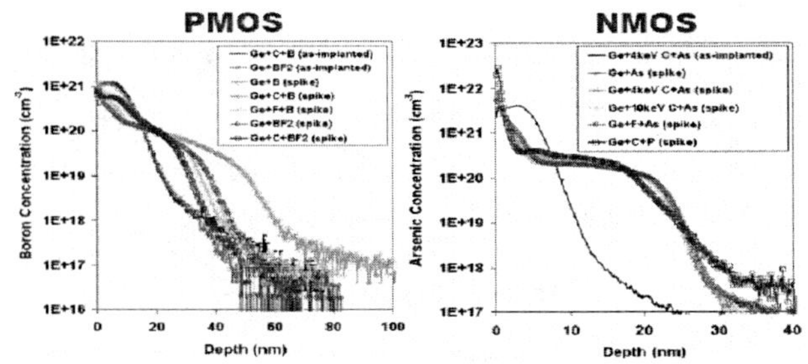

FIGURE 4. SIMS profiles showing impact of different co-implant conditions on PMOS and NMOS junction formation.

Advanced Annealing Techniques

Trade-off between the dopant activation and diffusion limits the processing window of the annealing thermal budget. Conventional ion implantation coupled with spike rapid thermal annealing (RTA) may not be an ideal option for the implementation in the forthcoming generations of devices. Advanced annealing techniques, such as laser and flash lamp annealing (FLA), offer extremely high temperature and short duration thermal process. Comparing to the spike annealing with a soak time in 1s range, flash lamp and laser annealing reduces the timing at the peak temperature down to "milli" and "micro" second ranges, respectively. Therefore, the overall thermal budget of the advanced annealing techniques is a few orders lower and subsequently lesser B TED. One reason is that the interactions of dopant-defect has been greatly reduced due to its ultra-short thermal process; while the evolution of the extended defects in the EOR region for the Ge-PAI case has changed and resulting in smaller silicon supersaturation of silicon interstitials towards the surface doped layer.

Physical junction improvement in terms of junction depth, abruptness and dopant activation has been extensively demonstrated by using flash lamp and laser annealing [26, 27]. However, those benefits cannot be converted into the devices performance directly. Recent works showed that applying FLA [28] and LSA [29] instead of spike in the source drain extension region poses certain challenges. In the case of FLA, it has been shown that there is a high pattern dependent effect [30]. There are other studies also revealed that the improvement by FLA in devices are induced from the higher poly-activation, lower contact resistance as well as better stress effect rather than the gain in the junction of source drain extension region. [29, 30]. One of the main concerns for laser annealing is on its power control, which generally leads to large device variation within wafers. Innovative thermal and phase-shift layer have been proposed to make laser annealing technique to be integrated in the industry production, but careful optimization of the SDE is inevitably necessary to achieve superior device performance when laser annealing is being applied. Recently, T. Yamamoto *et al.* reported performance gain in both PMOS and NMOS devices using a combination of spike and LSA schemes [31]. The reason of incorporating of spike is thought to enlarge the thermal process windows as well as to remove the extended defects at the EOR region caused by PAI. The latter is supported by the finding from Cristiano, where it was claimed that thermal budget of the two advanced annealing techniques are insufficient to remove the EOR defects completely. [27] One main issue of the EOR defects is it causes high junction current leakage at device level. Another concern of the residual defects is its impact on the junction stability, which is closely related to the dopant deactivation when the activated junction is subjected to post-thermal treatment, such as spacer formation and silicidation thermal processes.

Nevertheless, reduction of thermal budget is inevitably required in the downscaling of devices, thus advanced annealing is a highly potential option but optimization for low leakage, highly stable as well as low variation junctions in source drain extension region is mandatory.

Plasma and Molecular Doping

One of the conventional ways to shallower junction depth is by reducing the implantation energy. However, this approach has reached its bottleneck where with a further energy reduction it will lead to a strong degradation of the throughput and the uniformity of fabrication process. Although deceleration mode implantation has been successfully implemented for the case of B, however, the drawback of implant energy contamination will become even serious when the energy is further reduces. Therefore, a couple of new innovation doping techniques have been researched actively in recent years.

Plasma doping (PLAD) or plasma immersion ion implantation has been studied extensively since the late 1990s for the application in USJ formation [32]. The use of plasma in this doping technique enables an ultra-low energy implantation at a high beam current, leading to high doping dose rate and throughput. Junctions with high doping concentration and ultra-shallow profiles, in the region of sub-10 nm have been demonstrated [33]. By using a BF3 gas precursor, PLAD doping combining the spike can be used to fabricate advanced MOS devices at the 65 nm technology node [34]. Lallement extends the application of PLAD association with advanced annealing, and it has been shown that extended defects induced by plasma doping can be removed by the thermal budget from both flash and laser annealing [35]. The result was further verified by electrical junction study, in which same order of magnitude for junction current leakage was obtained as compared to the standard reference process.

Another doping approach which has attracted a great interest is the molecular implantation. Keeping in mind that larger cluster ion can provide high mass ratio in relaxing the ion beam energy, thus a larger clusters ion could benefit continue scaling of equivalent single ion B implant energy. In early, $B_{10}H_{14}+$ has been developed and used previously for shallow junctions. Unfortunately, the issues of stability, tuning procedure, and the dissociative nature of the ion, prevent this ion source to be considered seriously in manufacturing [36, 37].

There has been renewed interest in using the $B_{18}H_{22}+$ clusters ions recently after it was proven that the throughput of the beam has been greatly improved (scaled with equivalent beam current) [38]. As-implanted profiles with junction 10 nm are easily being fabricated without applying the acceleration/deceleration mode when performing the cluster implantation. The diffusion study showed that there is no particular impact of the co-implanted species in the post-annealed dopant profiles and its electrical characteristics. In addition, fabricated devices with $B_{18}H_{22}+$ implanted junctions revealed similar performance as compare to those obtained via B+ implants [39]. In another avenue, cluster Carbon ($C_{16}H_{10}+$) has also been proposed to suppress the B TED for USJ formation. However, no extensive study has been shown in the crystalline damages/defects generated by cluster ions as well as its effect on junction current leakage. It is suspected that the resulting defects could be significantly different from the monomer dopant implants and this will impose some detrimental effects to the junction electrical properties as well as the contact formation process windows at device level. Last but not least, considerations of ion beam tuning time, ion source lifetime, beam cleaning procedure, etc. should be carefully quantified.

Carborane ($C_2B_{10}H_{12}$, CBH) ion is also one cluster source which has been considered widely in the industry. The feasibility of implanting CBH as a molecular species for the formation of ultra-shallow junctions in pFETs has been assessed using 65 nm high performance devices [40]. The devices fabricated using CBH ion plantation exhibit excellent performance characteristics and are well matched to the POR. This approach is well suited to meet the advanced USJ requirements for 32nm technology and beyond. Additionally, the thermal stability of CBH and its ease of use have been proven to be an attracted molecular implant solution.

NOVEL USJ TECHNIQUES

Vacancy Engineering

The main principle of vacancy engineering is to generate an excess of vacancies in the vicinity of the doping region via high energy co-implantation process, typically preceding the desired dopant (B and P) implant. This technique differs from other competing approaches in that it eliminates the supersaturation of interstitials induced from dopant implantation via an interstitial-vacancy annihilation mechanism.

Raineri *et al.* [41] studied the effect of high energy co-implantation in relation to dopant enhanced diffusion. They applied a 1 MeV silicon co-implant followed by a boron implant in bulk substrate leading to reduction in TED. Roth *et al.* [42] and Venezia *et al.* [43] used Silicon On Insulator (SOI) substrates to investigate a similar effect induced by high energy silicon co-implantation. They studied the origin of diffusion reduction by physically de-coupling the excess vacancies (near to surface) and interstitials rich (away from surface) regions (resulting from high energy co-implant) through what was thought as a diffusion barrier. Among their studies, the buried silicon oxide (BOX) layer in SOI structure was used to restrict the back diffusion of silicon interstitials to the top layer, leaving the top layer rich in vacancies, while majority of silicon interstitials are beneath the BOX layer. Much better boron TED retardation was achieve in this case as compared to the observation by Raineri *et al.* in bulk Si.

It was not until recently that vacancy engineering has been applied in the USJ. Some works [44, 45, 46] showed that not only improvement was observed in boron TED, improvement in Rs was noticed at low temperature as well. However, the resulting Rs was not able to rival with the current standard requirement. To take the full advantage of this technique in SOI substrate, Smith *et al.* [47, 48] has proposed to use vacancy generating implant to create rich vacancy concentration in the active layer (the top silicon layer on BOX). Figure 6 shows the distribution of the point defects, interstitials and vacancies, in a SOI wafer structure obtained from simulation. It was indicated that the optimized high energy implant has to avoid the full amorphization to the top silicon layer of SOI such that the highly damage crystalline layer, rich in vacancies, remains as "seed" layer for SPE re-growth. The rich vacancy surface region is designed to annihilate the interstitial supersaturation induced by the low energy boron implantation. Excellent USJ characteristics were achieved such as a "diffusionless" dopant profile and the associated dopant activation was greatly enhanced, and beyond the conventional B solubility level.

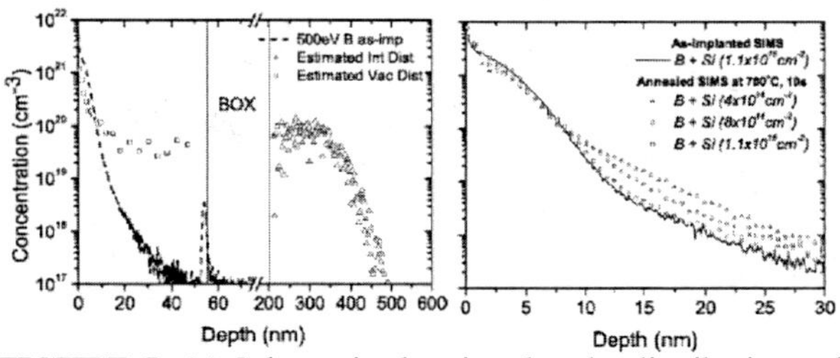

FIGURE 5. (a) Schematic showing that the distributions of excess vacancies and interstitials resulting from the high 160 keV silicon co-implant coupled with the 500 eV B implant simulated by Monte Carlo simulations. High net vacancy concentration of vacancy can be obtained in the case of SOI structure. (b) B SIMS profiles (annealed at 700°C, 60 s) showing the optimum Si co-implant condition (160keV and $1.1x10^{15}cm^{-2}$), "diffusionless" B profile can be obtained [47].

Thus, optimized vacancy engineering is highly attractive since standard tools are sufficient to fabricate the advanced USJ and no additional process step is involved. However, the reliability of the gate electrode and dielectric are affected since they are implanted with the high energy ions. So far this technique has not been comprehensively demonstrated at the device level and it is also limited to SOI substrate.

Surface Engineering

Recently, Seebauer and co-workers have shown by experiments and simulations that silicon surfaces can be chemically manipulated in a controllable way to offer a particularly efficient pathway for the annihilation of interstitials, leading to reduced diffusion of implanted species and increased activation [48-50]. The manipulation seeks to create dangling bonds to which interstitials can add with little activation barrier. Creation of an atomically clean surface leads to large number of dangling bonds, which act as a large sink that removes Si interstitials selectively over dopant interstitials. Such removal favorably impacts dopant TED and activation and should also promote the dissolution of extended EOR defects. The former work pertained to crystalline silicon without PAI, however, the extension of these concepts to PAI-induced amorphous surfaces has remained unclear. Also, former work did not examine the most technologically problematic case of the dopant boron and did not address EOR defect evolution. Very recently, Yeong *et al.* demonstrated experimentally that surface effects can also be exploited for B-implanted silicon with Ge-PAI yielding improvements in TED and boron activation as well a substantial reduction of residual EOR damage [51]. The chemical state of surface has a significant impact not only on dopant diffusion and activation but also on the removal of extended defects induced by Ge-PAI. Dangling bonds at the surface open a major alternative pathway for annihilating the excess interstitials emitted from the EOR defect band. The de-/reactivation of the dopant is also strongly influenced by the surface conditions.

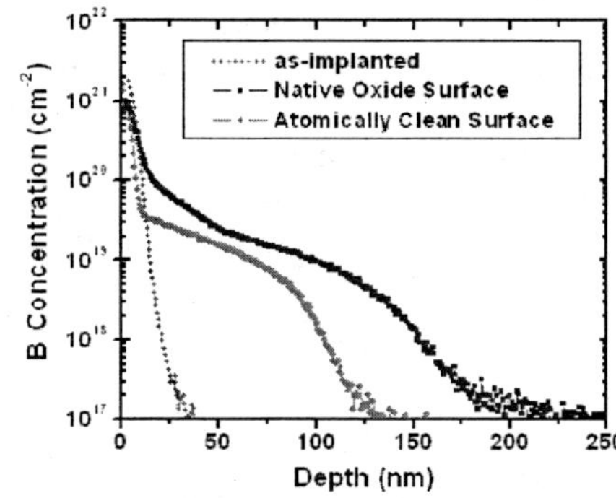

FIGURE 6. SIMS profiles of 500 eV B implant with prior 15keV Ge pre-amorphizing implant. annealed at 800°C for 60 minutes. The atomically clean surface leads to reduced TED.

The surface effects should operate regardless of the specific annealing technique and time scale employed i.e., seconds to milliseconds. The presence of substrate strain for improved carrier mobility in device should not interfere with the benefits, and the surface effect should operate in tandem with other methods of defect engineering such as dopant cluster implantation or co-implantation of impurities e.g., carbon, fluorine, and nitrogen.

Laser Vacancy Engineering

Laser annealing has been considered to be a promising candidate to produce a highly activated, ultra-shallow, and box-like junction profile due to its near-zero thermal budget. High fluence excimer laser irradiation melts the surface of the substrate and the melted surface layer undergoes a liquid phase epitaxy regrowth [52]. Recrystallization of the melted layer creates a point defect deformation beyond the liquid/solid interface. The laser-induced point defects are confirmed by double-crystal x-ray diffraction measurements to be vacancy type. It has been recently proposed that excess concentration of vacancies generated by laser irradiation is able to trap and reduce the silicon interstitial point defects that drive the boron TED.

Very recently, Tan *et al.* have demonstrated that the impact of pre-irradiated laser-generated vacancies in boron-doped, pre-amorphized silicon substrate not only on dopant diffusion and activation but also on residual defects [53]. The authors have tailored the ion amorphization depth of a particular ion implantation energy with the maximum melt depth of a corresponding pre-implant laser irradiation, it can suppress TED and BICs effect during the formation of the p+/n junction. Figure 7 shows the schematics illustrating interaction between laser-induced vacancies with the free interstitials created by PAI from the EOR defects. It has been proposed that this phenomenon is due to the reduction of the residual EOR defects via the recombination of the interstitials caused by the PAI and the vacancies created by the laser pre-irradiation.

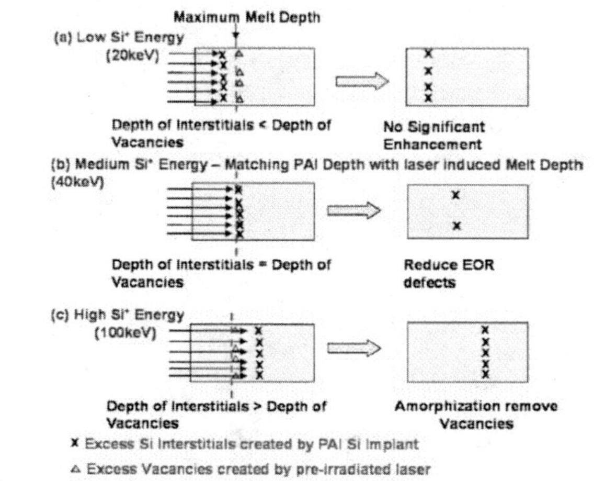

FIGURE 7. Schematics showing the mechanism associated with the interaction of laser-induced vacancies with the free interstitials created by PAI from the EOR defects.

Optimization of USJ formation using predictive TCAD simulation

In addition of demonstrating successful USJ fabrication, the process integration and device optimization are clearly needed. To reduce cost, TCAD is of great interest as the impact of the integration of such new process alternatives could be evaluated at device level and optimized. Atomistic kinetic Monte Carlo (kMC) has been regarded as a promising simulation technique for predictive TCAD simulations as it can handle the many different, complex and interacting mechanisms that occur during crucial process steps such as ion implantation and diffusion [54]. Accurate simulations of dopant SIMS profiles have been achieved for some time, although this has been mainly demonstrated for Boron and Arsenic [55-57]. However, simulations of 1D dopant profiles do not allow for the direct transfer to device simulator for the simulation of device characteristics to be performed. Moreover, accurate simulation of current and future nano-CMOS devices requires predictive 2D/3D doping simulations, including cases of novel co-implants and/or annealing scheme for device optimization. Very recently, it has been demonstrated and validated the methodology enabling the combination of advanced kMC based process modeling and device simulation [58, 59]. The methodology is explained of Figure 8.

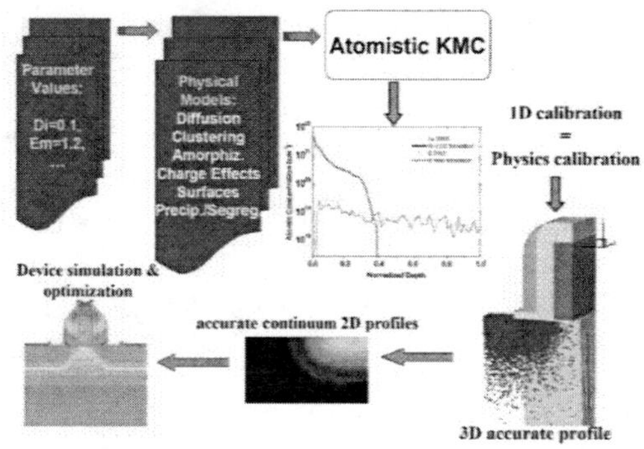

FIGURE 8. Schematic representation of the methodology used to generate accurate 2D/3D profiles for device simulation using kMC simulator.

Figure 9 also presents accurate simulation of several co-implant cases. The general advantage of this methodology is that different device characteristics, like overlap capacitance Vt_{lin}, Vt_{sat}, Id_{sat}, I_{off}, as well as roll-off behavior, can be directly predicted and optimized. Accurate simulated dopant profiles in the source/drain region can also be used for the prediction and optimization of sheet resistance and junction capacitance as shown in Figure 10. Indication of junction leakage can also be obtained from the eventual presence of defects and their relative positions with respect to the depletion region. In addition, the impact of novel processes for junction optimization can be directly evaluated at the device level [60].

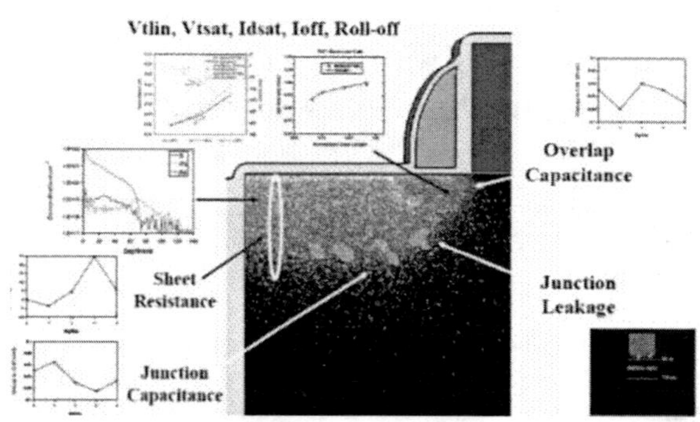

FIGURE 9. Schematic showing the various device characteristics can be simulated and predicted to design and optimize the advanced ultra-shallow using TCAD simulation.

CONCLUSIONS

In this paper, we have presented the latest achievements on the formation of highly activated Ultra-Shallow Source/Drain extension. It has been shown that a better understanding of the basic physical processes is crucial for USJ formation. This paper reviews the current understanding of dopant-defect interactions in USJ, which has significant impact on device performance in terms of dopant diffusion and activation. In addition, various conventional and innovative methods, such as impurity co-implantation, cluster/molecular implant, defect engineering as well as alternative thermal processes, have been illustrated for further downscaling of junction depth and resistance. Finally, accurate TCAD has been demonstrated and revealed that it can be a powerful tool for optimizing and designing new junction architecture to meet the future nano-CMOS device performance requirements.

REFERENCES

1. P. Stolk *et al.*, *J. Appl. Phys.* **81**, 6031 (1997).
2. N. Cowern *et al. J. Appl. Phys.* **68**, 6191 (1990).
3. B. Colombeau, *et al.*, Mat. Res. Soc. Symp. Proc., 810, C3.6.3 (2004)
4. B. J. Pawlak *et al.*, *Appl. Phys. Lett.*, **89**, 062110 (2006).
5. E. Lampin *et al.*, *J. Appl. Phys.*, **85**, 8137 (1999).
6. L. Laânab *et al.*, Mat. Res. Soc. Symp. Proc., 279, 381 (1993).
7. A. Claverie *et al.*, *Solid State Phenomena*, 47-48, 195 (1996).
8. A. Claverie *et al.*, Mat. Sci. Semicond. Proc., 3, 269 (2000).
9. E. Landi *et al.*, *Appl. Phys. A* :Mater. Sci. Proc., **A47**, 359 (1998).
10. R. Lindsay *et al.*, Mat. Res. Soc. Symp. Proc., 717, C2.1 (2002).
11. R. Lindsay *et al.*, 7th International Workshop on: Fabrication, Characterization, and Modeling of Ultra-Shallow Doping Profiles in Semiconductors, Santa Cruz, 65 (2003).
12. P. J. Timans *et al.*, in *Solid State Tech.*, May (2004).
13. B.J. Pawlak *et al.*, 7th International Workshop on: Fabrication, Characterization, and Modeling of Ultra-Shallow Doping Profiles in Semiconductors, Santa Cruz, 227 (2003).
14. B.J. Pawlak *et al.*, *Appl. Phys. Lett.*, **84**, 2055 (2004).
15. T. Murakami *et al.*, *Nucl. Instr. Meth. B*, **127**, 257 (1997).
16. N. Cowern *et al.*, *Appl. Phys. Lett.*, **86**, 101905 (2005).
17. A. Vanderpool *et al.*, *Nucl. Instr. Meth. B*, **237**, 142 (2005).
18. H. Graoui *et al.*, *Mater. Sci. Eng. B*, **124**, 188 (2005).
19. Benjamin *et al.*, 7th Int. Workshop on Junction Tech., IEEE, 13 (2007).
20. S. Mirabella *et al.*, *Phys. Rev. B*, **65**, 045209 (2002).
21. N. E. B. Cowern *et al.*, *Appl. Phys. Lett.*, **86**, 101905 (2005).
22. G. Impellizzeri *et al.*, *Appl. Phys. Lett.*, **84**, 1862 (2004).
23. S. H. Yeong *et al.*, *J. Electrochem. Soc*, **155**, H69 (2008).
24. C. F. Tan *et al.*, Int. Symp. on VLSI - Technology, Systems, and Applications, IEEE (2008).
25. N. Auriac *et al.*, 7th Int. Workshop on Junction Tech., IEEE, 13 (2007).
26. W. S. Yoo *et al.*, *Nucl. Instr. Meth. B*, **237**, 12 (2005).
27. W. Lerch *et al.*, *Mater. Sci. Eng. B*, **124–125**, 24 (2005).
28. Oishi *et al.*, VLSI Tech. Symp., (2004).
29. A. Shima *et al.*, VLSI Tech. Symp., (2004).
30. Adachi *et al.*, VLSI Tech. Symp., (2005).
31. T. Yamamoto *et al.*, VLSI Tech. Symp., (2007).
32. S. B. Felch *et al.*, *Mat. Chem. and Phys.*, **54**(1-3) (1998).
33. D. Lenoble *et al.*, 1998 Int. Conf. Ion Implantation Technology Proc., 1222 (1999).
34. F. Lallement *et al.*, VLSI Tech. Symp., (2004).
35. F. Lallement, Ph.D thesis, INSA Toulouse (2005).
36. K. Goto *et al.*, IEDM Tech. Dig., 435 (1996).
37. T. F. Chong *et al.*, *Appl. Phys. Lett.*, **76**, 3197 (2000).
38. D. Jacobson, Int. Workshop on Junction Tech. Proc. (2005).
39. M. Ishibashi *et al.*, Junction Technology, Extended Abstracts of the Fifth International Workshop, Volume 7-8 (2005).
40. C. F. Tan, Mat. Res. Soc. Symp., (2008).
41. V. Raineri *et al.*, *Appl. Phys. Lett.*, **58**, 4, 922 (1991).
42. E. Roth *et al.*, *J. Elec. Mat.*, **26**, 11, 1349 (1997).
43. V. Venezia *et al.*, *Appl. Phys. Lett.* **74**, 9, 1299 (1999).
44. Lin Shao *et al.*, *Appl. Phys. Lett.* **84**, 3325 (2004).
45. K. Kyllesbech Larsen *et al.*, *Nucl. Inst. Meth. B.*, **112**, 139 (1996).
46. S. Saito *et al.*, *Appl. Phys. Lett.*, **197** (1993).
47. A. J. Smith *et al.*, *Appl. Phys. Lett.*, **88**, 082112 (2006).
48. N. E. B. Cowern *et al.*, IEDM Tech. Dig., 968 (2005).
49. E. G. Seebauer *et al.*, *Phys. Rev. Lett.*, **97**, 055053 (2006).
50. Ramakrishnan Vaidyanathan *et al.*, *Appl. Phys. Lett.*, **89**, 152114 (2006).
51. S.H. Yeong *et al.*, *Appl. Phys. Lett.*, **92**, 102112 (2007).
52. Y. F. Chong *et al.*, *Appl. Phys. Lett.*, **77**, 19 (2000).
53. D. X. M. Tan *et al.*, *Appl. Phys. Lett.*, **92**, 203107 (2008).
54. M. Jaraiz, in Predictive Simulation of Semiconductor Processing, Springer-Verlag, Berlin, pp. 73, 2004.
55. M. Aboy *et al.*, *Appl. Phys. Lett.*, **88**, 191917,(2006).
56. T. Noda *et al.*, IEEE Tech Digest, s14 p1 (2006).
57. R. Pinacho *et al.*, *Appl. Phys. Lett.*, **86**, 252108 (2005).
58. B. Colombeau *et al.*, Int. Symp. on VLSI - Technology, Systems, and Applications, IEEE, (2006).
59. B. Colombeau *et al.*, IEDM Tech. Dig, s14, (2006).
60. K. R. C. Mok B. Colombeau, F. Benistant, R. S. Teo, S. H. Yeong, B. Yang, M. Jaraiz and S. –F. S. Chui, IEEE Trans. Elec. Devices, **54**, 9, 2155 (2007).

The Reduction of TED in Ion Implanted Silicon

Amitabh Jain

External Development and Manufacturing, Texas Instruments, 13121 TI Boulevard, MS 365, Dallas, TX 75243, USA

Abstract. The leading challenge in the continued scaling of junctions made by ion implantation and annealing is the control of the undesired transient enhanced diffusion (TED) effect. Spike annealing has been used as a means to reduce this effect and has proven successful in previous nodes. The peak temperature in this process is typically 1050°C and the time spent within 50°C of the peak is of the order of 1.5 seconds. As technology advances along the future scaling roadmap, further reduction or elimination of the enhanced diffusion effect is necessary. We have shown that raising the peak temperature to 1175°C or more and reduction of the anneal time at peak temperature to less than a millisecond is effective in eliminating enhanced diffusion. We show that it is possible to employ a sequence of millisecond anneal followed by spike anneal to obtain profiles that do not exhibit gradient degradation at the junction and have junction depth and sheet resistance appropriate to the needs of future technology nodes. We have implemented millisecond annealing using a carbon dioxide laser to support high-volume manufacturing of 65 nm microprocessors and system-on-chip products. We further show how the use of molecular ion implantation to produce amorphousness followed by laser annealing to produce solid phase epitaxial regrowth results in junctions that meet the shallow depth and abruptness requirements of the 32 nm node.

Keywords: doping, ion implantation, transient enhanced diffusion, activation, laser annealing
PACS: 61.72.uf; 85.40.Ry

INTRODUCTION

Scaling of ULSI devices requires the reduction of MOSFET source/drain extension junction depth and this poses a significant challenge. Ultra-high temperature (UHT) annealing using lasers or flash lamps is an advanced technique which allows dopant activation while limiting diffusion. Future device technologies can benefit from this technique. Of particular concern in existing technologies is the problem of end-of-range (EOR) disorder when ion implantation is used. This leads to various undesirable effects including limits on junction scaling and lack of profile abruptness.

Figure 1 shows the effect of EOR disorder on the profile evolution of implanted boron in silicon. At intermediate temperatures, there is the phenomenon of transient enhanced diffusion (TED) which results in a long tail. At higher temperatures, the source of interstitials causing this enhancement is depleted and the subsequent profile shape is governed by concentration-enhanced diffusion as shown. The TED tail generally sets a lower limit to the achievable junction scaling. Concentration-enhanced diffusion recovers the profile abruptness and this has not been a significant problem in recent technologies. Future scaling, however, requires junction depth values well below 30 nm and this can only be achieved if TED is reduced significantly.

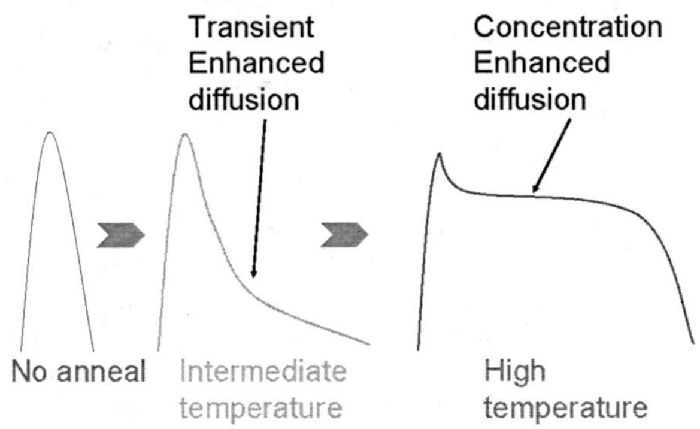

FIGURE 1. Evolution of an ion implanted boron profile in silicon during annealing.

We have developed a convenient formalism for estimating the amount of diffusion in the presence of TED when ramping up temperature at a known ramp rate [1]. The diffusion length after ramping up from say, 600°C to temperature T_f (if TED is over at an

CP1066, *Ion Implantation Technology,* edited by E. G. Seebauer, S. B. Felch, A. Jain, and Y. V. Kondratenko

intermediate temperature y, before it reaches the temperature T_f) is given by

$$(D_{ob}/ramprate)\left[\int_{873}^{y(ramprate)} \exp\{-(E_{mb}+E_{is})/kT\}\,dT + \int_{y(ramprate)}^{T_f} \exp\{-E_{mb}/kT\}\,dT\right],$$

where

D_{ob} = Pre-exponential constant for boron diffusion

E_{mb} = Migration energy for boron diffusion

E_{is} = Change in the energy of a loop when an interstitial is emitted. This is negative because the energy of the loop decreases.

The temperature y is obtained by solving the following,

$$\tau(y)\left[1-(1/ramprate)\int_{873}^{y}(1/\tau(T))\,dT\right] = 0$$

$\tau(T)$ is the TED duration at temperature T, and a simple expression given by Stolk *et al.* [2] may be used in deriving trends. The method is easily extended to include the effects of varying cool-down rate and varying spike anneal sharpness and can be used to predict what might be expected from various advanced anneal processes. Solution of these equations leads to the expectation that ramping up the temperature to a sufficiently high temperature in a sufficiently short time will reduce TED. This study explores the use of laser annealing to realize this theoretical prediction in practice. The use of co-implantation to further reduce TED is also presented.

EXPERIMENTAL CONDITIONS

Silicon wafers of diameter 200 mm and n-type conductivity were used in this study. Ion implantation was carried out on standard production equipment. Germanium ion implantation for pre-amorphization was performed at 20 keV to a dose of 3.5E14/cm^2. BF_2 ion implantation was performed at 2 keV to a dose of 1.4E15/cm^2. The sheet resistance, Rs, was measured on a KLA-Tencor 4 point probe. UHT anneals were performed using a carbon dioxide laser.

Pieces of each wafer were profiled by SIMS, which was carried out on a Cameca 6F with 500 eV O_2^+ ions and oxygen leak for boron.

RESULTS AND DISCUSSION

Figure 2 shows the results of experiments in which laser annealing was employed in an effort to reduce TED. The boron depth profile before annealing is shown for reference. After a 950°C spike anneal, a long TED tail develops for the reasons given above. The plot shows profiles that are obtained after a 1270°C or 1300°C laser anneal. In these cases, one can see the TED tail is absent. In order to verify that the source of interstitials and therefore the propensity of TED has been reduced, these anneals were followed by a 950°C spike anneal as a test. It is seen that this spike does not produce the tail that it did when a prior laser anneal was not used. This verifies the effectiveness of the laser anneal in reducing TED.

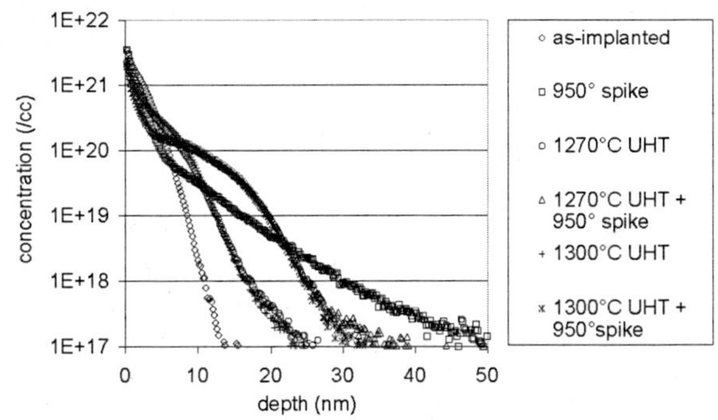

FIGURE 2. The diffusion behavior of ion implanted boron in silicon as a function of various types of anneals.

Although the technique just described is successful in reducing TED and thereby reducing junction depth and enhancing profile steepness, further improvements in junction abruptness are desirable for future generations of ULSI technology. Increased abruptness depends on reducing TED to negligible levels and additional mechanisms are required to reduce the silicon interstitial population. The use of carbon ion implantation has been employed as a means of improving profile abruptness. The effectiveness of carbon depends on a significant fraction of it being in solid solution in the silicon. This may be achieved by solid phase epitaxial regrowth of an amorphous layer containing carbon. Since carbon is a light species, this requires the use of a separate amorphizing implant. We have shown by Transmission Electron Microscopy that the use of molecular carbon ion implantation enables amorphization and therefore a high degree of carbon substitutionality without the need for a separate amorphizing implant [3]. Elimination of the separate implant that is just for the sake of amorphization should aid in the reduction of TED by removing one source of interstitials.

Figure 3 shows high-resolution XRD data to compare atomic and molecular carbon ion implantation after an 1175°C laser anneal. The peak for the molecular case shows a clear shift and therefore a higher level of carbon substitutionality in comparison to the non-molecular case. No preamorphizing implant was used in these experiments.

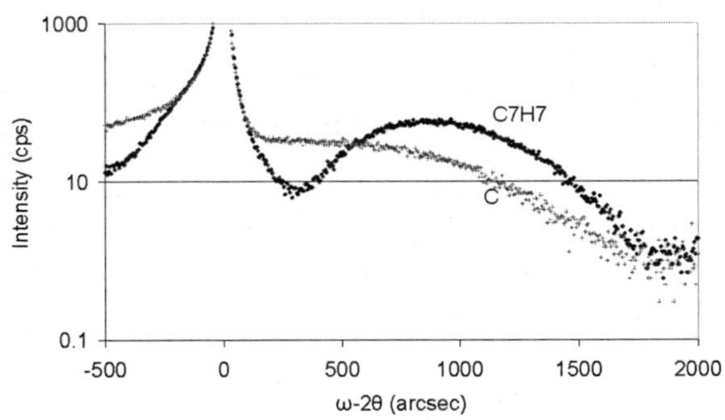

FIGURE 3. X-ray diffractograms showing the improvement in carbon substitutionality when using molecular rather than atomic ion implantation.

Figure 4 shows the SIMS profiles of boron implants for various anneals and for atomic and molecular carbon ion implantation. The final junction as formed after a spike anneal subsequent to a laser anneal has over 1nm/decade improvement in the abruptness when using the molecular species. This is attributed to the well-known ability of substitutional carbon to reduce the silicon interstitial population.

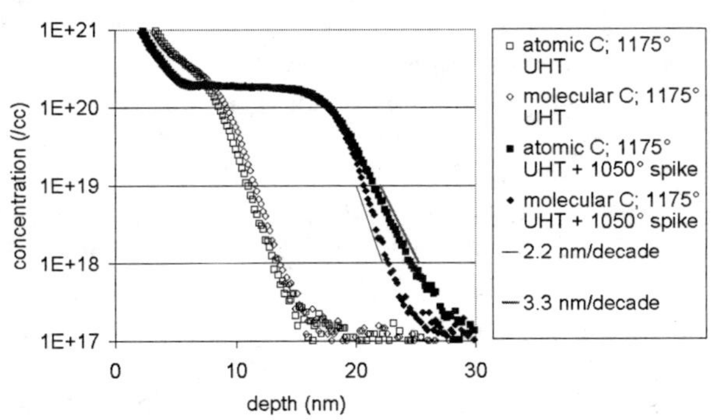

FIGURE 4. The effect of using molecular carbon rather than atomic carbon, on the diffusion of implanted boron.

It is important to address the concern of slip when incorporating laser anneal into a process flow. We have found that it is possible to take advantage of the viscoelastic nature of silicon and reduce dwell time as one goes higher in peak temperature in a process in order to reduce slip. Figure 5 shows the results of tests to determine the slip threshold. As the dwell time is reduced, one can use higher peak temperature values in the laser annealing process while avoiding slip.

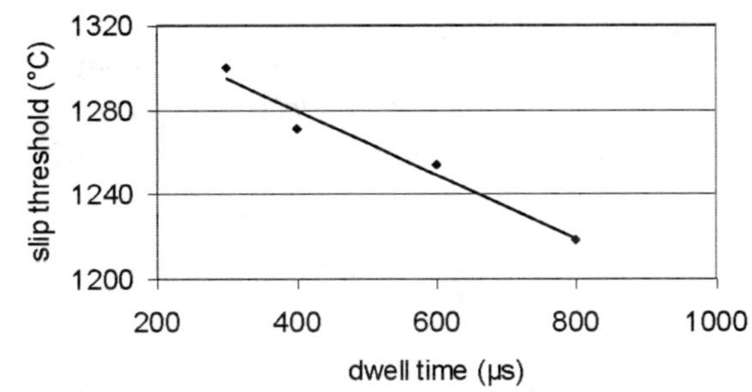

FIGURE 5. Slip threshold for laser annealed silicon.

We have previously shown that the activation of boron at high anneal temperatures is not compromised by a reduction in dwell time [4].

CONCLUSIONS

This study has demonstrated the use of ultra-high temperature annealing and molecular carbon ion implantation in reducing transient enhanced diffusion to negligible levels. This allows scaling of junction depth and achievement of highly abrupt profiles for future device generations.

REFERENCES

1. Amitabh Jain, *Materials Science Forum* **573-574** 305-318 (2008)
2. P.A.Stolk *et al.*, *J. Appl. Phys.* **81** 6031- 6050 (1997)
3. Andrei Li-Fatou, Amitabh Jain, Wade Krull, Michael Ameen, Mark Harris and Dale Jacobson, *ECS Trans.* **11** 125 (2007)
4. Amitabh Jain, *Doping Engineering for Device Fabrication, Mat. Res. Soc. Symp. Proc.* **912** Materials Research Society, C04-04, 143 (2006)

Junction Leakage Analysis of Vacancy Engineered Ultra-Shallow p-type Layers

A. J. Smith[1], L. D. Antwis[1], S. H. Yeong[2], A. P. Knights[3], B. Colombeau[4], B. J. Sealy[1], R. M. Gwilliam[1]

[1] *Ion Beam Centre – Advanced Technology Institute, University of Surrey, Guildford, GU2 7XH, UK*
[2] *Dept. of Chemical and Biomolecular Eng., National University of Singapore, Singapore*
[3] *McMaster University, Dept. of Engineering Physics, Hamilton, Ontario, Canada*
[4] *Chartered Semiconductor Manufacturing Ltd, 60 Woodlands, Industrial Park D, Street 2, Singapore 738406*

Abstract. This study examines the effect of a silicon Vacancy Engineering Implant (VEI) on a boron doped ultra-shallow junction, in terms of sheet resistance, junction depth and for the first time junction leakage - via conventional diode measurements and newer non-contact techniques. It is demonstrated that by using a VEI combined with a 10s, 800°C anneal, it is possible to electrically activate the dopant to a level 5x higher than without a VEI, whilst sustaining minimal boron diffusion. Furthermore, this thermal budget is significant enough to reduce the implantation damage that resides within the depletion region to a level that produces a junction leakage value of ~35 $\mu A/cm^2$, which is competitive with alternative techniques such as pre-amorphization and solid phase epitaxial regrowth.

Keywords: Vacancy Engineering, Leakage Measurements, Boron
PACS: 61.72.jd, 61.72.jj, 61.72.Cc, 61.72.uf, 85.40.Ry, 85.30.Kk

INTRODUCTION

As CMOS transistors have evolved over the last 40 years, many monumental technological road blocks have been overcome. Combined with the aggressive down scaling of the physical dimensions of such devices, the improvement factors that have been achieved are unrivaled by any other industry. However, maintaining such a trend has become a challenge within itself. Today, as the devices evolve into the 45nm node and beyond, there has become a distinct separation in device architecture depending on whether the final device is intended for a high performance or a low power application [1]. Such a deviation has been initiated by the significant increase in off state current flow for high performance devices which would be unpractical for electronic products that rely on a limited battery supply.

A significant portion of the total device leakage can be attributed to the techniques used to form the crucial ultra-shallow source/drain extensions. Typically, these regions have been characterized by their metallurgical junction depth and sheet resistance. As the technology nodes advance, the ability to produce ultra-thin, highly conducting regions becomes increasingly difficult, pushing not only the fundamental limits of the starting substrate but also the ability of the equipment used to fabricate such layers.

The dominant dopant for the PMOS transistor is boron. However, the anomalous diffusion and clustering mechanisms of boron hinder the achievable PMOS junction specification. The traditional implantation and anneal has been surpassed by junction engineering techniques such as pre-amorphization implant (PAI) and solid phase epitaxial regrowth (SPER), 'cocktail' implants with fluorine and/or carbon or vacancy engineering implants (VEI), combined with advanced activation schemes such as flash or laser assisted anneals. However, residual defects from these processes can often lead to an undesired increase in junction leakage.

Vacancy engineering has been proven previously to be a viable alternative to the more popular PAI and SPER when used in conjunction with, and optimized for, silicon on insulator (SOI) substrates, creating near diffusionless, highly stable boron doped layers with dopant electrical activation well-above its equilibrium solubility [2]. However, to date, the ability to create high specification p-type Ultra-Shallow Junctions (USJs) within bulk silicon using this technique has not been proven.

CP1066, *Ion Implantation Technology,* edited by E. G. Seebauer, S. B. Felch, A. Jain, and Y. V. Kondratenko

This study examines the effect of a vacancy engineering implant on a subsequent boron implant in bulk silicon in terms of not only Rs and junction depth but also for the first time junction leakage, measured using conventional electrical characterization and non-contact methods.

EXPERIMENTAL DETAILS

A 1-10 ohm-cm, n-type (100) bulk silicon wafer was implanted with a silicon VEI at an energy of 160keV to a dose of $1\text{x}10^{15}$ /cm^2. A subsequent ultra-low energy 500eV boron implant was performed to a dose of $1\text{x}10^{15}$/cm^2. Both implants were performed at the same implant geometry–7° tilt and 22° rotation. For comparison (and to act as a reference), a second wafer of the same specification only received the boron implant.

16 cm^2 samples from each wafer were annealed using a 10s isochronal scheme from 700°C to 900°C using a Process Products Corporation rapid thermal annealer (RTA), with the ramp rate kept constant at 50°C/s.

To examine the sheet electrical properties, four-point-probe (4pp, set at a weighting of 45g with a tip radius of 500µm) and Hall measurements using the Van der Pauw (VdP) geometry were performed to examine the sheet resistance (Rs), Hall mobility and active carrier density (Ns). The Hall scattering factor was assumed to be unity for all samples.

Conventional junction leakage measurements were performed by examining the reverse leakage current of multiple diodes with an area of 0.01 cm^2 at a bias potential of -0.2V. Non-contact electrical measurements were performed independently at Frontier Semiconductor Manufacturing (FSM) [3] and SemiLab [4] on the same samples.

The dopant chemical profiles were analyzed ex-situ by secondary ion mass spectrometry (SIMS) using a Cameca IMS 6f instrument. A primary beam of O2+ ions with a net energy of 0.5 keV at 56° incidence was scanned over an area of 250 µm × 250 µm and the crater depth measurement was done using a Tencor Alpha-Step 500 profilometer.

RESULTS AND DISCUSSIONS

To begin this study the initial sheet electrical characteristics as a function of anneal temperature are discussed and presented within figure 1, illustrating the Rs (A), Ns (B) and Hall mobility (C).

In terms of Rs, the reference boron only curve (open squares) indicates that at 700°C the Rs is relatively high at ~8500 ohms/sq. As the anneal temperature is increased the Rs deceases monotonically to ~950 ohm/sq at 900°C, which corresponds to over an order of magnitude increase in Ns. In comparison, the boron layer combined with a VEI (open circles) demonstrates a negligible difference in Rs at 900°C when compared to the reference sample. However, as the anneal temperature is decreased to 800°C, one can see a distinct reduction of ~3x when compared to the reference curve, which equates to over a 5x increase in the electrically active boron fraction (Ns). This trend has been shown previously in SOI substrates to be a characteristic of vacancy engineered junctions, where the improvement factors dominate at low annealing temperatures [5]. As the anneal temperature is reduced to 700°C it is possible to see that the Rs is reduced even further to ~430 ohms/sq, which is ~20x lower than the reference Rs. However, this value must be treated with caution as the corresponding Hall coefficient (not shown) was determined to have a negative sign, indicating that the majority carriers are electrons and not holes, as would be expected for a p-type implanted layer.

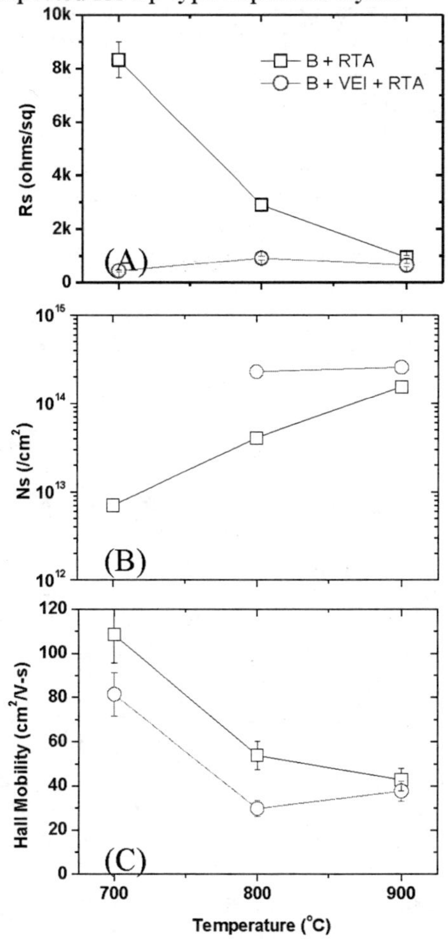

FIGURE 1. VdP and Hall measurements presenting the sheet electrical properties of a boron doped layer with (open circles) and without (open squares) a VEI in terms of Rs (A), Ns (B) and Hall mobility (C).

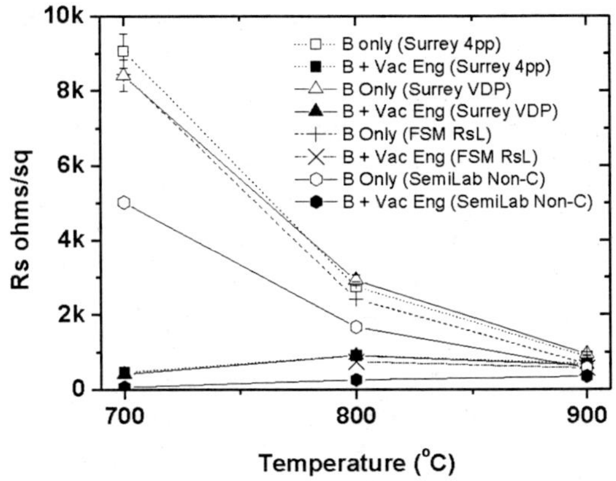

FIGURE 2. Summary of the Rs measurements performed on the boron doped layer, with (solid symbols) and without (open symbols) a VEI as a function of temperature and measurement technique.

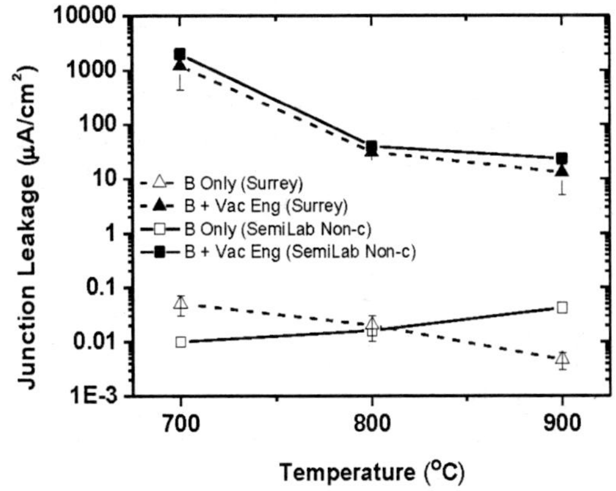

FIGURE 3. Junction leakage measurements of the boron doped layers with (solid symbols) and without (open symbols) the VEI, measured using conventional diode structures (triangles) and Semi-Labs non-contact (circles) technique.

This ambiguity in measured carrier type could be due to two effects. Firstly, the probes used to make the electrical measurements have punched through the implanted layer and therefore are making direct contact to the n-type substrate and the p-type implanted layer [6]. Secondly, in bulk silicon the electrical measurements rely on the formation of a p/n junction to isolate the implanted layer from the substrate. However, in the case of engineered junctions it is possible that remaining defects within the depletion region provide a low resistive path to the substrate. In both cases (sheet measurements using 4pp and Hall techniques) it is possible for the measured value to be a combination of parallel resistances and compensating carrier types, where the substrate component may dominate.

To remove concern associated with probe penetration, 16 cm^2 samples, with and without the vacancy engineering implant, were first sent to FSM and then the same samples were sent to SemiLab to measure the Rs and junction leakage as a function of anneal temperature, using their own non-contact techniques. Figure 2 illustrates a summary of all the Rs results gathered, with (solid symbols) and without a VEI (open symbols) from FSM (crosses), Semi-Lab (hexagons) and University of Surrey using conventional electrical characterization techniques. Comparing the VdP (triangles) and 4pp (squares) measurements performed at Surrey, one can clearly see that the trends are near identical and the measured values are in very close agreement, and certainly within the experimental uncertainties of the techniques used. If one also compares the FSM non-contact measurements with Surrey's more conventional methods it is possible to see that in the boron only case the trend is near identical and the absolute values are within 20% of each other. This is also apparent when comparing the samples that received the VEI. However, at 700°C FSM were not able to complete a measurement, which is discussed in more detail further on in this paper. Semi-Lab's non-contact measurements demonstrate a similar trend compared to Surrey and FSM, with and without a co-implant, however, the absolute values demonstrate a significant variation compared to all the other techniques. It is interesting to note that in the boron only case Semi-Lab's values are consistently 40% lower. Such a consistent variation is thought to be related to a calibration issue as these initial measurements were performed using a preliminary fitting algorithm.

One of the advantages of non-contact measurements is the ability to extract a junction leakage parameter with minimal sample preparation. Therefore, to complete the electrical analysis the following section discusses the junction leakage parameters extracted from non-contact measurements compared to the more conventional electrical measurements. Table 1 presents all the junction leakage measurements performed by FSM, whereas, figure 3 illustrates all the values determined by SemiLab (squares) and Surrey (triangles), with (solid symbols) and without a VEI (open symbols). FSM calculated that all except one sample had a junction leakage of <0.1 μA/cm^2, which corresponds to the lower limit of their technique. The only sample to have a value of >10000 μA/cm^2, which is the upper limit of their technique, had the VEI combined with the low 700°C thermal process.

TABLE 1. Junction leakage values determined by FSM

Sample	Temperature (°C)	Jo (μA/cm^2)
B + RTA	700	<0.1
B + RTA	800	<0.1
B + RTA	900	<0.1
B + VEI + RTA	700	>10000
B + VEI + RTA	800	<0.1
B + VEI + RTA	900	<0.1

SemiLab and Surrey's measurements demonstrate good agreement with FSM when comparing the boron layers without the VEI, all indicating junction leakage values below the FSM 0.1 μA/cm^2 limit. However, when comparing the boron layers with a VEI both SemiLab and Surrey illustrate a trend that is not apparent in FSM's measurements, indicating that at 700°C the leakage is high, but as the anneal temperature is increased the leakage values reduce to around 30 μA/cm^2 and 20 μA/cm^2 at 800°C and 900°C, respectively.

It must be emphasized that the main focus of this study is to examine the effect of a VEI on the electrical properties of boron doped layers and not promote one non-contact measurement over another. It should be noted that these non-contact measurements were performed under non-ideal, nor identical conditions and with equipment under development. It is possible to conclude that the difficulty in measuring these engineered junctions following a low temperature (≤700°C) anneal is related to large junction leakage and not excessive probe penetration which is demonstrated by all the techniques described here. Therefore neglecting this value and examining the 800°C and 900°C anneal conditions, it is possible to compare the current work with a recent study performed by Timans *et al.* [7]. These authors demonstrated that when a 500eV 10^{15} B/cm^2 doping implant was used in conjunction with a germanium PAI and a SPER process at 650°C for 5s (also without a Halo implant), the junction leakage measured by FSM, was determined to be approximately 600 μA/cm^2. If a 1050°C spike anneal was used this was reduced to below the detection limit of the technique but resulted in excessive diffusion. However, if a flash assisted RTP was implemented it was possible to reduce the leakage current and minimize the diffusion simultaneously. Whereas, in the present study, when a VEI is employed it is possible to achieve the same level, if not better, of junction leakage without the requirement of an advanced anneal. Moreover, figure 4 illustrates the SIMS analysis of the boron layers with a VEI as a function of anneal temperature, clearly demonstrating that at 800°C the shallow junction is retained and what diffusion is observed only corresponds to ~5nm, when taking a reference point at 3×10^{18} B/cm^3.

SUMMARY

In summary, these initial results illustrate that even without optimizing the VEI for bulk silicon it is possible to produce boron doped ultra-shallow junctions with a Rs of ~900 ohms/sq with a reasonably low level leakage compared to a standard PAI and SPER process. To further this study it would be necessary to optimize the VEI process with respect to a halo implant, which is known to increase junction leakage.

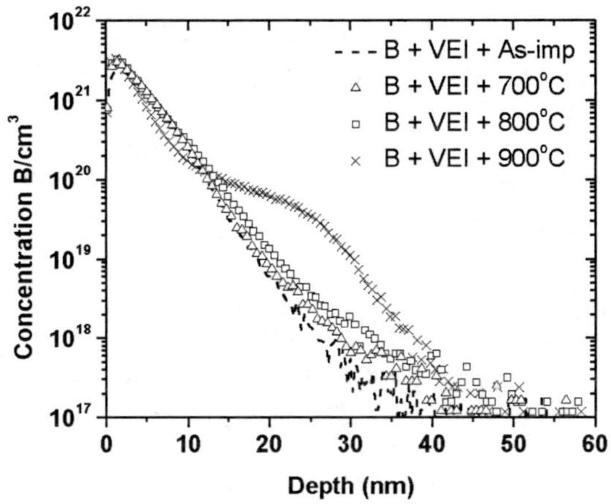

FIGURE 4. SIMS analysis of the boron layers with a VEI before (dotted line) and after a 700°C (triangles), 800°C (squares) and 900°C (crosses) 10s anneal.

ACKNOWLEDGMENTS

The authors would like to thank Eric Don and Aron Pap from Semilab Co Ltd, and Michael Current from FSM, for performing the non-contact measurements and providing many stimulating discussions.

REFERENCES

1. ITRS 2007 edition – Process Integration, Devices, and Structures, http://www.itrs.net/
2. A. J. Smith, N. E. B. Cowern, R. Gwilliam *et al.* App. Phys. Lett. 88, 082112 (2006)
3. V. N. Faifer, M. I. Curret and D. K. Schroder, App. Phy. Lett. 89, 151123 (2006)
4. US Patent 5,442,297 Verkuil, Roger.L. Granted 15th August 1995.
5. A.J. Smith, B. Colombeau, N. Bennett *et al.* in *Semiconductor Defect Engineering—Materials, Synthetic Structures and Devices*, edited by S. Ashok, J. Chevallier, B.L. Sopori, M. Tabe, and P. Kiesel (Mater. Res. Soc. Symp. Proc. 864, San Franscisco, CA, 2005), E7.1
6. T. Clarysse, D. Vanhaeren and W. Vandervorst, J. Vac. Sci. Technol. B 20, 1071 (2002)
7. P. J. Timans, Y. Hu, J. Gelpey, S. Mcoy *et al.* in *Doping Engineering for Front-End Processing*, edited by B.J. Pawlak, M. Law, K. Suguro, M.L. Pelaz (Mater. Res. Soc. Symp. Proc. 1070, San Francisco, CA, 2008

Ion Beam Control and Transistor Characteristics for 45nm Source-Drain Extension Formation

K. Okabe, M. Nishikawa, K. Ikeda[1], H. Kurata[1], T. Mori, S. Satoh[1], and M. Kase

Device Development Division Fujitsu Microelectronics Ltd., [1]Fujitsu Laboratories Ltd.,:1500, Mizono, Tado, Kuwana, Mie 511-0192, Japan,

Abstract. The beam trajectory and divergence angle is systematically examined with ion source parts degradation and focusing condition. We clarified that the variation and drive current of NMOSFET transistor characteristics are related the beam angle conditions, using our advanced logic transistor of 45 nm low stand-by power and high performance device technology. We can propose the practical condition to become insensitive to ion beam angle degradation and some gain of drive current of MOSFET is achieved.

Keywords: Ion beam, angle, Trajectory, divergence, 45nm, lens
PACS: 61.72.Uj

INTRODUCTION

The ion beam angle effect is quite important for stable manufacturing of advanced logic and memory devices. In the past conferences, the trajectory and divergence of the ion beam were highlighted. They discussed the influences to MOSFET characteristics such as the asymmetry of drive current and gate overlap variation [1-6]. We examined the beam trajectory and divergence in detail using single high current implanter and its influences on the characteristics of MOSFET using our advanced 45 nm logic transistor including the gate length of less than 35 nm.

MAIN CAUSE OF TRAJECTORY AND DIVERGENCE

We find that the change of ion source electrode shape and change of focus lens has strong impact on such beam properties. Such as the arc chamber, suppression and ground electrodes, the ion source part shape is varied depending on the running timing. The condition of lens part is changed depending on each beam setup. As a result, trajectory and divergence of ion beam is usually changed.

Figure 1 shows the change of ion source electrode shape after the running time of 120 and 200 hours. The opening slit on the electrode is becoming wider as the source running timing is progressed due to sputtering effect. After 200 hours of operation, we find the position five is particularly becoming wider. This situation is dependent on the type of implanter, ion source itself and the ion source operation condition. This degradation of source parts is influenced by the density of extracted beam and some of focusing lens parameters enhance this phenomenon strongly.

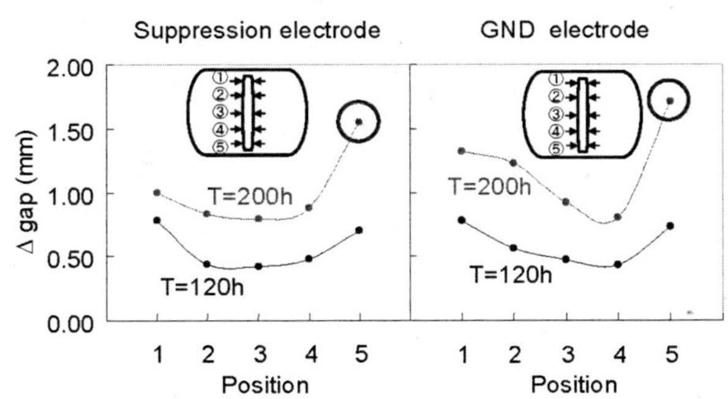

FIGURE 1. The change of ion source electrode shape after the running time of 120 and 200 hours.

INFLUENCE OF TRAJECTORY AND DIVERGENCE

The important properties of ion beam angle are its direction and spreading. To identify these, we examine the trajectory and divergency of beam using 2 dimentional profiler of the ion beam. The beam size and center of gravity are measured at the point of wafer plane and the point behind a distance *L* from wafer plane. The trajectory and divergence angle are

CP1066, *Ion Implantation Technology,* edited by E. G. Seebauer, S. B. Felch, A. Jain, and Y. V. Kondratenko

calculated from equation (1) and (2) respectivily using the beam center of gravity (*CofG*) and the difference of beam width (*W*) at the wafer plane (front) and the point behind a distance *L* (back). The beam size is defined at the point of the beam intensity of 98 %.

$$Trajectory(\text{deg}) = \tan^{-1}\left\{\frac{CofG(\text{back}) - CofG(\text{front})}{L}\right\} \quad (1)$$

$$Divergence(\text{deg}) = \tan^{-1}\left\{\frac{W(\text{back}) - W(\text{front})}{2L}\right\} \quad (2)$$

Figure 2 shows the beam profiles and angle data of ultra low energy As ion beam as function on the ion source running time. A deceleration mode is used and focus condition is fixed. For 120 hours of operation, the left figure shows the practical beam divergence and same center of gravity between the front and back position, and it has a good beam trajectory. The right figure shows the different profile of front and back, so that the ion beam has shifted beam axis and its divergence becomes wider after the running time of 200 hours. Even if the ion source setting is the same, the parts degradation causes this beam angle degradation.

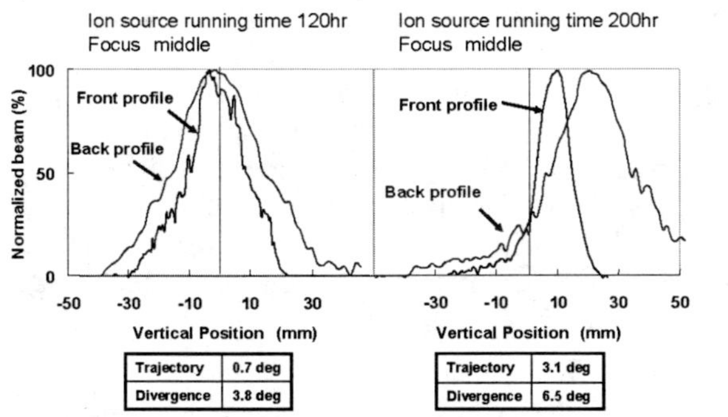

FIGURE 2. Beam profiles and angle data of ultra low energy Arsenic ion beam as function on the ion source running time.

Figures 3 shows the forward and reverse transistor Ion-Ioff characteristics on the low stand-by power type NMOS FET of 45 nm device generation, which is adopted the ultra low energy As implant for the extension formation after the ion source running time is 200 hours. To clarify the beam angle influence, the tilt angle of 0° and 1 direction implant are used. The difference between forward and reverse drive current is quite large, so it is 9% of Ion. The advanced logic transistor would be designed to be a small overlap between the gate and extension, so the angle variation makes a strong impact to the MOSFET characteristics.

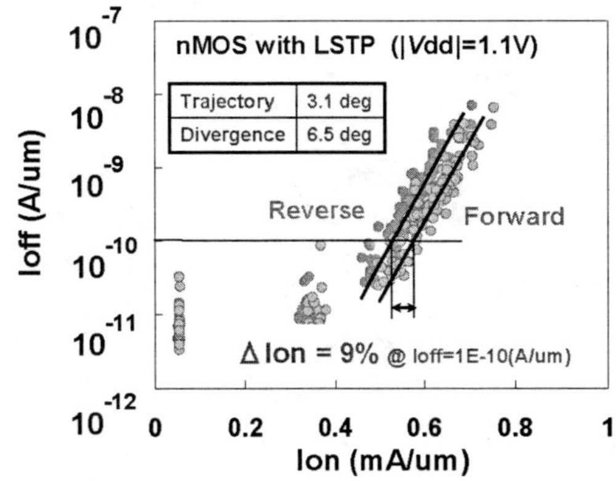

FIGURE 3. Forward and Reverse transistor *I*on-*I*off characteristics.

To clarify the influence of divergence angle, we make the beam which has a large divergence angle by tuning the focus lens condition. The beam profile results are shown in Figure 4. A new ion source is used. We success to make that the trajectory angle is 0 ° and the divergence angle is 4.1 and 6.9°.

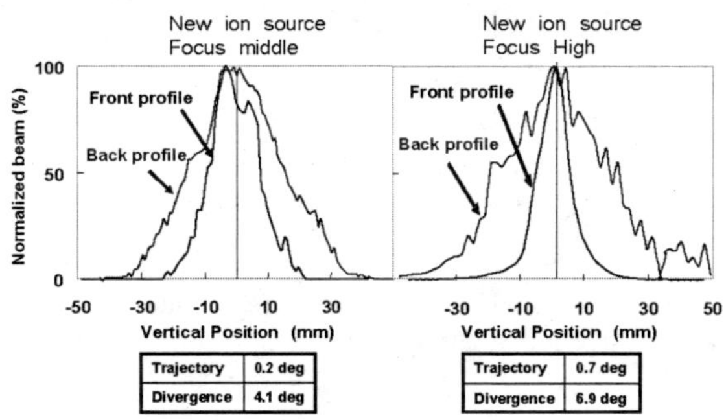

FIGURE 4. Change of divergence angle by tuning the focus lens condition.

Figure 5 shows the difference between forward and reverse drive current as function on the implant tilt angle. To clarify the beam divergence influence, the 1 direction implants are used, again. The result shows that the asymmetry of drive current is twice higher in case of 4.1° divergence. The low divergence beam makes shape lateral profile, so it makes large dependence on tilt angle. In the case of broaden beam, the dose loss may be occurred. It makes transistor performance lower. As a result, to minimize the variation of transistor characteristics, Tilt angle should be controlled preciously in mechanical and beam trajectory and divergence should be stable and repeatable.

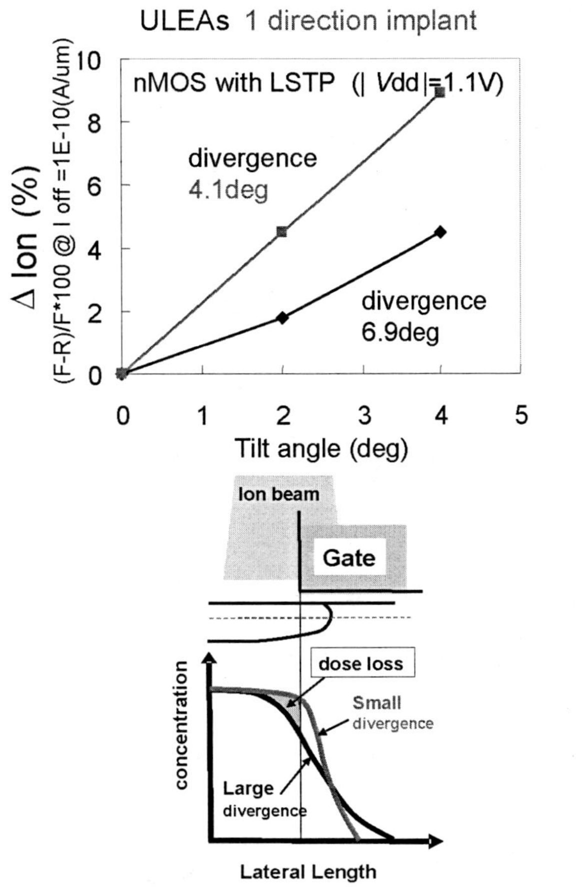

FIGURE 5. Difference between forward and reverse drive current as function on the implant tilt angle and schematic view of gate overlap of lateral profile

Figures 6 shows the *L*min-*I*on characteristics of transistor when the quad implant is adopted. *L*min is the minimum gate length of roll-off characteristics at the threshold voltage of 0 V, and *I*on is the transistor drive current. The transistor is high performance type NMOS FET of 45 nm device generation. The quad implant is used to form the extension implant. The tilted angle of wafer is 0° and 5°. The divergence angle is 4 and 7°. The trajectory angle is 0.3°. In the case of 0° tilt implant, the *I*on and *L*min is strongly depended on the beam divergence, however in the 5° tilted implant case, the dependence to the beam divergence is much small. This means that there is good balance of overlap of gate and extension, and parasitic resistance, and we can achieve small variation of transistor performance with small degradation of drive current. This result of tilted angle and implanted dose might be depended on the process flow such as side wall design, pocket hallow implant and thermal budget.

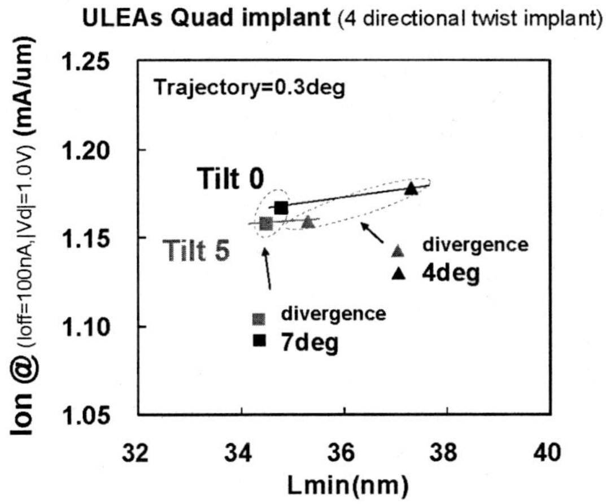

FIGURE 6. *L*min-*I*on characteristics of high performance type NMOSFET when the quad implant.

Figure 7 shows the case of PMOSFET. The result shows less dependence on the beam angle of tilt and divergence. The B profile of PMOSFET extension is different from the As profile of NMOSFET like as lower abruptness or smoother shape. Based on the mechanism described on Figure 7, we adopt the fine tuning to the implant parameters, as a results we can get 8% gain on *I*on using tilt implant at 10°. The reason why of gaining the drive current, is the B part of implanted B area does not overlap with the gate electrode. And the B part dose are more than the A part dose. Therefore the parasitic resistance is lower and the Vth roll-off does not changed. By the tilted angled and quad implant a advantage on higher *I*on and low variation are achieved.

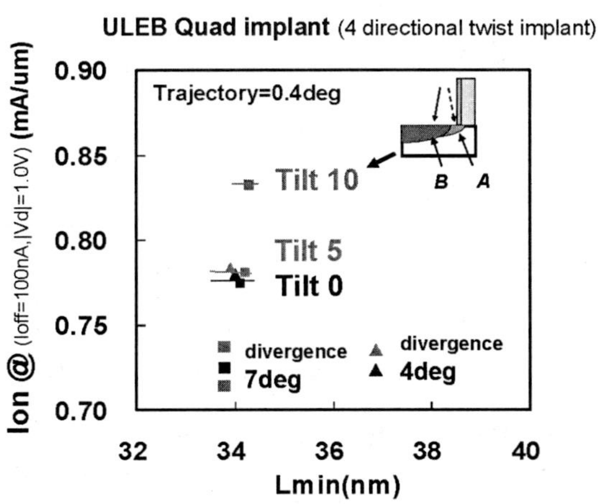

FIGURE 7. *L*min-*I*on characteristics of high performance type PMOSFET when the quad implant.

WHAT IS THE BEST TUNING OF IMPLANTER TO MINIMIZE THE VARIATION?

The thermal budget of next generation transistor fabrication flow is minimized due to the requirement of shallow junction and extension layer is formed using diffusion less annealing technique. By such the extension formation, a variation of as-implanted profile should influence the transistor characteristics. The beam tuning of implanter is directly reflected the dopant profile, so it becomes increasingly important to stabilize the transistor performance at the mass production. Figure 8 shows that the beam size and its trajectory angle depends on the focus voltage after ion source running time of 200 hours. We can see to minimize the beam size on the focus voltage of just focus value, but the trajectory angle is maximized. This is caused from bad condition of ion source. When the implantation is performed on such condition with the focus point, large trajectory angle could be degraded the transistor asymmetry. Even if quad implant is used, the variation is not suppressed enough in this case. To decrease the trajectory angle, we can choose either lower or higher voltage. Higher voltage has a possibility of glitch issue, so we choose the lower voltage area. We must decrease a beam current about 10%. Figure 9 is the divergence angle as function on the focus voltage. In the low focus voltage area the divergence angle is small. The divergence will be decided with balance of space charge effect of beam density and convergence of focus lens. As a result, we use low focus voltage and quad implant, so we can achieve stable device fabrication process condition.

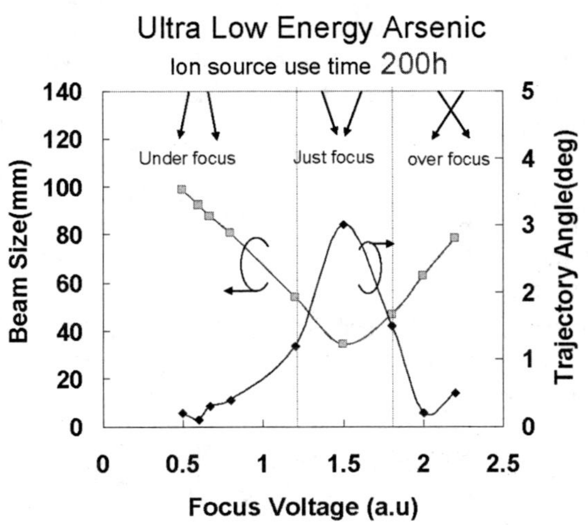

FIGURE 8. Lmin-Ion characteristics of high performance type PMOSFET with the quad implant

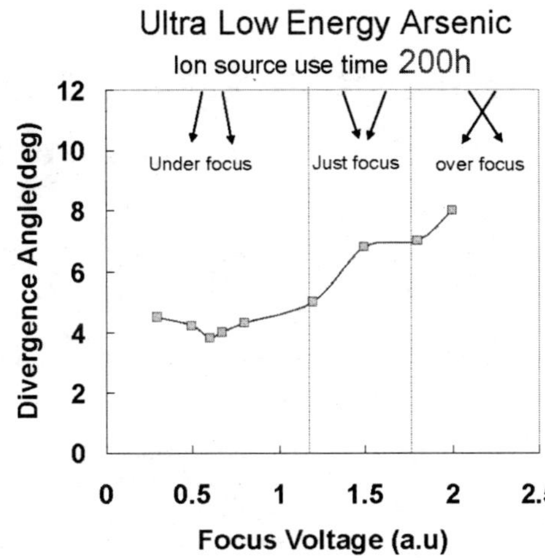

FIGURE 9. Lmin-Ion characteristics of high performance type PMOSFET with the quad implant.

SUMMARY

The beam trajectory and divergence angle is systematically examined with ion source parts degradation and focusing condition. We clarified that the variation and drive current of NMOSFET transistor characteristics are related the beam angle conditions, using our advanced logic transistor of 45 nm low stand-by power and high performance device technology. We can propose the practical condition to become insensitive to ion beam angle degradation and some gain of drive current of MOSFET is achieved. Finally, it is strongly required that the implanter can manage and stabilize automatically the beam angle parameters such as trajectory and divergence.

REFERENCES

1. Nam-Sung Kim, *et al.*, *Conf. Proc. on Ion Implantation Technology 2000*, IEEE, p.50-53.
2. Ukyo Jeong, *et al.*, *Conf. Proc. on Ion Implantation Technology 2000*, IEEE, p.64-68.
3. Susan Felch, *et al.*, *Proc. on West Coast Junction Technology Group*, April 2004, AVS, http://www.avsusergroups.org/wcjug/jtg_2007_7proceedings.htm
4. Yuri Erokhin, *et al.*, *Conf. Proc. on Ion Implantation Technology 2006*, AIP, p.520-523.
5. Vincent Kaeppelin, *et al.*, *Conf. Proc. on Ion Implantation Technology 2006*, AIP, p.353-357.
6. Hans-Joachim L. Gossmann, *et al.*, *Journal of Vacuum Science & Technology B*, **24**, 1705-1710 (2006).

Dose Rate Effects: the Impact of Beam Dynamics on Materials Issues and Device Performance

M. S. Ameen, M. A. Harris, C. Huynh, and R. N. Reece

Axcelis Technologies, 108 Cherry Hill Drive, Beverly, MA 01950

Abstract. Different implant architectures will exhibit disparities in scan rates, beam density, and temperature. Unless well understood and compensated for, these disparities can affect process results – especially when new architectures are incorporated into existing processes. In this article we will systematically examine the interaction between ion beam parameters related to beam scanning architectures and process results dependent on dose rate and thermal effects. The parameters of scan rate, duty cycle, beam density and beam current will be evaluated for effect on wafer temperature, thickness of amorphous layer, damage, and dopant profiles.

Keywords: Ion Implantation, Dose Rate, Damage.
PACS: 61.72.U.

INTRODUCTION

A switch from multi-wafer to single-wafer high-current implanters occurred around the 65 nm process node. Historically, the productivity gains associated with multi-wafer operation allowed this platform to dominate >100 nm node manufacturing. As of this writing, the majority of 300 mm high-current implants are performed on single-wafer platforms.

Based on many factors, including productivity, control, and cleanliness, two dominant standard implanter tools have become available in the market today, with broad beam non-scanned systems occupying niche applications for non-critical implants. The two architectures available for high-current single-wafer include a ribbon beam [1] based platform, "point to parallel to parallel," and the "rotary dual axis mechanical scan" implanters [2].

There are advantages and disadvantages associated with each of these beamline architectures. Both implanters cover the same application space and can be used interchangeably to build advanced devices (45 nm at this time). However, there is dissimilarity in the way that the ion beam is formed and transported to the wafer, as well as differences in the dynamics of the wafer motion under the beam. Therefore, there are quantifiable differences in the beam density reaching the wafer. This effect, coupled with dynamic scan conditions, results in ion-induced damage and thermal profiles that can lead to measurable discrepancy in the process results. This is especially true for high-dose amorphizing implants, where thickness of the resulting amorphous layer may vary. The process implications of this are discussed below.

BEAM AND SCANNING DYNAMICS

The scanning dynamics of single-wafer implanters are complex and difficult to describe without a real-time representation. For architectures being considered in this work, the wafer is translated mechanically through the beam. The ribbon beam uses a 1D motion of the wafer, and the spot beam a 2D mechanical translation. Both types of systems have a slow scan direction of the wafer; the 2D systems also have a fast scan motion orthogonal to the slow scan mechanical translation. Both types of systems have a slow scan direction of the wafer; the 2D systems also have a fast scan motion orthogonal to the slow scan direction. The slow scan speed for the ribbon beam and spot beam systems is very nearly the same, on the order of ~5 cm/s. For spot beam systems, the fast scan speed is dramatically different between the multi-wafer and single-wafer systems, the multi-wafer scan speed being a factor of ~20X faster than the single-wafer systems (in which the wafer sweeps through the beam at several Hz). For convenience, we define the

CP1066, *Ion Implantation Technology,* edited by E. G. Seebauer, S. B. Felch, A. Jain, and Y. V. Kondratenko

TABLE 1. Conventions and calculations for describing the fast and slow scan dynamics of implanters. Values other than beam size have been normalized to the peak value in the row.

PARAMETER	SYMBOL	2D mechanical	2D mechanical	1D mechanical
Implanter Type		13w batch	single wafer	single wafer
Ion Beam		spot	spot	ribbon
Rectangular Beam X-Width (FWHM) (mm)	$w = (W/3)(2\ln 2)^{1/2}$	~25	~25	~320
Rectangular Beam Y-Width (FWHM) (mm)	$h = (H/3)(2\ln 2)^{1/2}$	~20	~25	~25
Rectangular Beam Density (uA/cm^2)	$j = I/(wh)$	1	1	~.06
Ion Flux (ions/cm^2/sec)	$J = j/qe$	1	1	0.1
Fast Scan Average Ion Flux (ions/cm^2/sec)	$J_{avf} = J\xi$	0.1	1	1
Fast Scan Duty Cycle	$\xi = T_{on f}/T_f$	0.007	0.06	1

duty cycle of a ribbon beam as 1, the percent of the total implant time that a portion of the wafer is getting implanted. It is of interest to compare this percentage to that of a 2D mechanically scanned system, which rapidly sweeps the wafer through a higher-density stationary beam. In this convention, we will be comparing the slow scan direction of the ribbon beam to the fast scan of the 2D mechanically scanned architecture.

It is important to understand that for multiple scans of a wafer (all wafers pass under the beam at least twice); each point on the wafer will have a different beam-on and beam-off exposure due to the nature of the scanning system. Near the bottom of the wafer the time period for the beam to return to the same point is shorter than for a point in the center of the wafer. These times are typically on the order of seconds for the slow scan direction, and this time scale is much larger than ion-cascade and defect-relaxation mechanisms in the silicon [3]. Cooling is very rapid during the beam-off period, and most bulk thermal heating issues are associated with the turn-around point in the slow scan direction.

The key differences between the spot and ribbon beam single-wafer implants can be seen in the Table 1. The beam density is a factor of ~15 to 20 times higher for the spot beam, leading to an ion flux more than an order larger for the spot beam systems. On the other hand, the fast-scan duty cycle is significantly smaller for the spot beam systems. Stationary spot beam implants will have the characteristic that a higher flux of ions ("peak density") will be implanted into the silicon onto a rapidly moving wafer, followed by a beam off time of ~150 ms. A ribbon beam system implants a lower flux over a much longer period of time, with less relaxation period during the time of implant. While the "effective" dose rate, or ions delivered per complete pass of the wafer, of both systems is similar, the mechanics of delivery of the beam to the wafer are quite different. This leads to different ion-cascade and damage accumulation properties, and different thermal signatures for each toolset. Figure 1 illustrates the differences in ion flux for a single-wafer ribbon beam and spot beam system. This representation is for a single sweep of a point on the wafer through the entire beam; most high-current implants consist of 2 to 4 passes of the beam over the wafer.

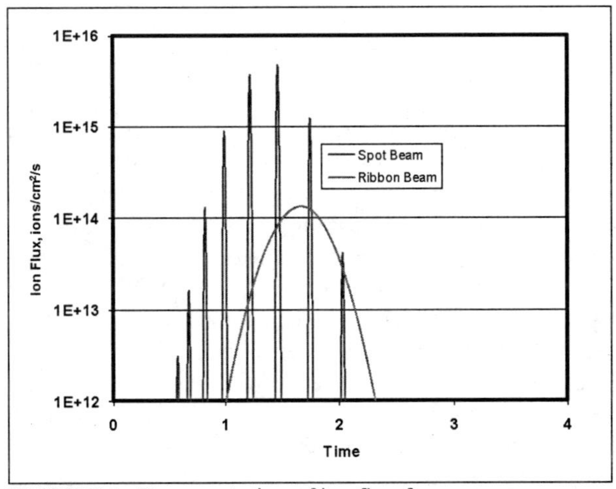

FIGURE 1. Representation of ion flux for one pass on a single-wafer ribbon and spot beam implanter, for a 1E15 ions/cm^2, 10.5 mA implant.

The temperature of the wafer is governed by the rate of power input and cooling rate as determined by the chucking mechanism [4]. Single-wafer implanters use an electrostatic clamping mechanism and are typically able to maintain the wafer temperature well below 100°C for 1,500 W implants. (High-current multi-wafer systems are mechanically clamped and in contact with a pliable elastomer, giving thermal contact to the wafer. Peak temperatures are close for the two platforms.) The time-dependent profiles that a point on the wafer experiences depend on the scanning mechanisms, as illustrated in modeling results shown in Figure 2. In general, a ribbon beam will generate a higher peak temperature during the scan due to the longer dwell time on the wafer with no cool-down period during the cycle of beam-on time.

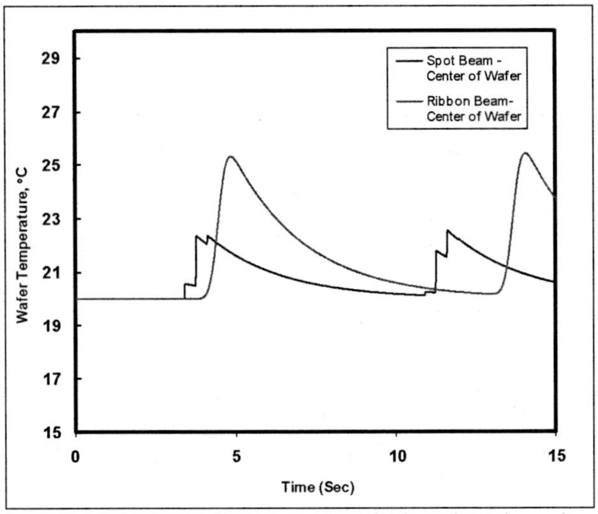

FIGURE 2. Modeling results of the time-dependant temperature profile at the center of the wafer for a spot vs. ribbon beam single-wafer implanter.

PROCESS EFFECTS

The combination of a unique damage accumulation rate and thermal profile during the implant is responsible for the differences in process results that have been reported when matching tools with different scanning and ion beam properties [5-11]. It is well known that the amorphous layer thickness is dependent on the temperature of the implant. This effect, which has been explained as dynamic annealing, leads to a balance of damage generation and lattice self-repair during the implant. Models that account for the exponential temperature dependence of point defect diffusion and annihilation have successfully predicted the dependence of the amorphous layer thickness on dose rate and thermal variation [12, 13].

Figure 3 illustrates the effect of varying beam current and duty cycle on the amorphous layer thickness of a high-dose BF_2 implant. A 5% change in amorphous layer thickness can translate into a variation in device properties [7, 9]. The thickness of the amorphous layer intersects the damage profile of the implant; a thinner amorphous layer will have a higher concentration of interstitial vacancy pairs in the end-of-range (EOR) region beyond the amorphous-crystalline (a/c) interface. Upon annealing, the activation and diffusion depth, as well as pinning of diffusing species, will depend on the accumulated damage concentration. (See Lopez [3] for a detailed discussion of dose rate and the effect on amorphous layer thickness and residual damage.)

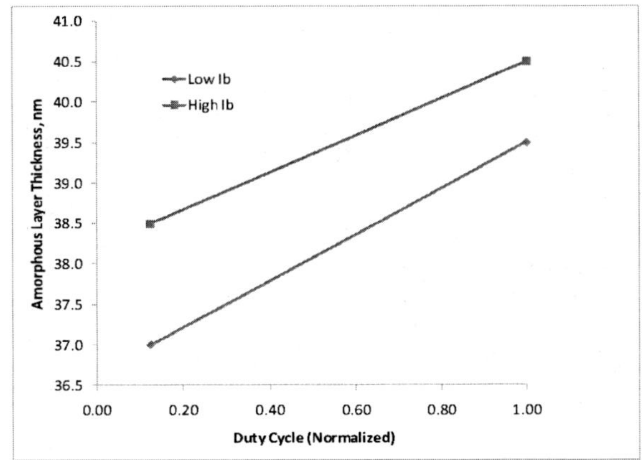

FIGURE 3. Amorphous layer formation as measured by TEM. The implant was a 20 keV BF_2, 2E15 ions/cm^2 on a single-wafer spot beam system. High I_b is 10X low I_b in this case.

The consequences of different amorphous layer thickness on a boron (from BF_2) SIMS profile are shown by example in Figur 4. The second peak of boron in the profile that accumulates in the 300 to 400 Å region depends on the thickness of the amorphous layer and concentration of damage in the EOR region. This leads to differences in activation profiles, sheet resistance Rs, and occasionally in transistor characteristics.

There are a number of important observations to be made. First, the multi-wafer systems generate a thinner amorphous layer than single-wafer systems for a given species and beam current. Since the peak temperature of the wafers on a spinning disk system are slightly lower (which would favor a thicker amorphous layer), and the flux density of the beam is the same (see Table 1), it becomes clear that a major driver for the differences in performance is the fast-scan average ion flux, or ions delivered per pass of the beam over the wafer, which is an order of magnitude larger for the single-wafer than the multi-wafer system. Second, we note that this ion flux in a ribbon beam system and a single-wafer spot beam system are similar. Thus, process differences between the single-wafer spot and single-wafer ribbon beam will be much smaller than when compared to the multi-wafer systems. When matching a single-wafer platform to a multi-wafer platform, somewhat extreme changes to implant conditions have been used to match device performance, such as beam de-focusing [10]. Integrating different single-wafer tool platforms will require less drastic methods to match device performance.

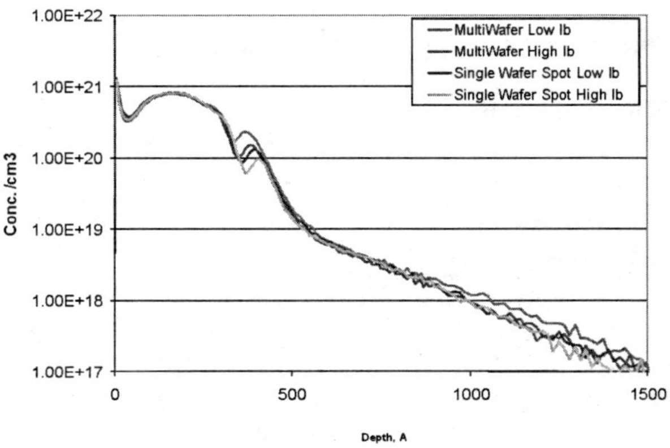

FIGURE 4. Post-anneal comparison of Boron accumulation in the EOR damaged region of a BF_2 high-dose implant for a stationary (spot) beam on multi-wafer and single-wafer platforms.

The effect of dose rate can be more directly observed using ion channeling. In an amorphizing implant, ions will experience channeling up to a point where a critical level of damage is reached in the Si-crystal, after which channeling stops. The final depth of the implant is determined by the damage accumulation rate, which is governed by the interaction of the ion flux with the wafer temperature during implant. Figure 5 illustrates two implants of similar beam current and species (30 keV 8E13 ions/cm^2 As) into c-Si and pre-amorphized silicon and compared between multi- and single-wafer platform. The implant single-wafer platform has a junction depth approximately 10 nm shallower at a concentration of 1E18 ions/cm^3 due to the faster a-layer formation. Associated with this shallower junction depth will be a thicker amorphous layer for the implants into c-Si.

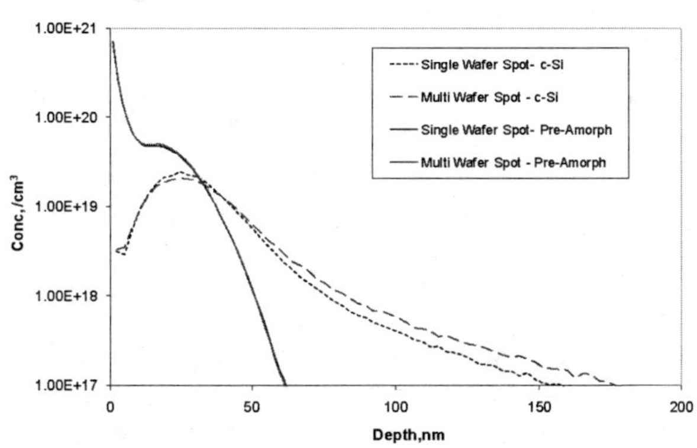

FIGURE 5. Comparison of as-implanted SIMS profiles for 30 keV As into c-Si and pre-amorphized Si.

CONCLUSIONS

We have shown that differences in the properties of the ion beam combined with scanning architecture will impact the process performance of high-current implants, where the diffusion and activation of dopants is directly dependent on the as-implanted conditions. The challenges associated with matching multi-wafer spot beams to single-wafer ribbon or spot beam are more difficult because of significant differences in damage accumulation and final amorphous layer thickness. While differences exist between a ribbon and spot beam single-wafer system, matching implant performance is more manageable due to the relatively close processing temperatures and ion fluxes.

ACKNOWLEDGEMENTS

The authors wish to acknowledge contributions and discussions with many members of the technical community at Axcelis, especially to P. Splinter, B. Vanderberg, I. Berry, R. Rathmell, S. Kondratenko, L. Rubin, and E. Collart.

REFERENCES

1 C. Campbell, *et al.*, Proceedings of the 14th International Conference on IIT 2002, p. 193.
2 P. Splinter, unpublished.
3 P. Lopez, L.s Pelaza, L. Marques *et al.*, Materials Science and Engineering: B EMRS 2005, Symposium D, "Materials Science and Device Issues for Future Technologies," **124-125**, 379-382 (2005).
4 M. Mack and M. Ameen, "Wafer Cooling and Charging in Ion Implantation," Ion Implantation Science and Technology (Ed. by Ziegler, J.F.), pp. 549-552, (2000).
5 R. Doherty, B. McComb, R. Ting *et al.*, Ion Implantation Technology: 16th International Conference on Ion Implantation Technology , 4 (2006).
6 N. Cagnat, L. Frioulaud, C. Laviron *et al.*, *ibid*, 11-16.
7 D. Franke, M. Schmeide, H. M. Brussow *et al.*, Proc. IIT 2004, Part II., 102-106 (2004).
8 M. Sano, *et.al*, Nucl. Inst. and Meth. B 237, 77-82 (2005).
9 K. Okabe *et al.*, Ion Implantation Technology: 16th International Conference on Ion Implantation Technology 645-648 (2006).
10 V. Kaeppelin, *et al.*, ibid, 353-356 (2006).
11 J. Williams, R. Goldberg, R. Elliman, *Nucl. Inst. and Meth. B* **106**, 6 (1995).
12 L. Pelaz, L. Marqués, J. Barbolla., *J. Appl. Phys.*, **96**(11) 2004, p. 5947.
13 K. Mok, M. Jaraiz, I. Martin-Bragado *et al.*, *J. Appl. Phys.* **98**(4) 2005.

Defect Engineering for Ultra-Shallow Junctions Using Surfaces

E. G. Seebauer[1]*, C. T. M. Kwok[1], R. Vaidyanathan[1], Y. V. Kondratenko[1], S. H.Yeong[2], M. P. Srinivasan[2], Benjamin Colombeau[3] and Lap Chan[3]

[1]*Department of Chemical & Biomolecular Engineering, University of Illinois, Urbana, IL 61802 USA*
* *eseebaue@illinois.edu*
[2]*Department of Chemical & Biomolecular Engineering, National University of Singapore, Singapore*
[3]*Chartered Semiconductor Manufacturing Ltd, Singapore*

Abstract. Formation of extremely shallow pn junctions with very low electrical resistance is a major stumbling block to the continued down scaling of microelectronic devices. Recent work in our laboratory has shown that the behavior of defects within silicon (and therefore dopants) can be changed significantly by controlling the chemical state of the surface. Certain chemical treatments of the surface induce it to act as an active "sink" for point defects that removes diffusing Si interstitials selectively over impurity interstitials, leading to less dopant diffusion and better electrical activation. The present work demonstrates such effects experimentally for dopants such as boron and arsenic in both crystalline and Ge pre-amorphized silicon wafers. Surface-based defect engineering was studied by annealing implanted Si according to various protocols with varying degrees of surface activity toward point defects. After implantation, specimens were heated to stimulate diffusive spreading of the implanted profile, with SIMS employed after annealing to monitor the spreading. SIMS measurements showed that the active surface caused less diffusion for boron and arsenic in crystalline silicon (c-Si). However, the boron dose loss in the case of the active surface was significantly higher than for the native oxide surface. This caused higher sheet resistance for the clean surface. Yet the percentage of activated boron was also higher for the clean surface. Thus, the atomically clean surface reduced transient enhanced diffusion and improved the percentage of dopant activation despite higher sheet resistance. Similar trends were observed in Ge preamorphized silicon. Additionally, the active surface dramatically reduced the number of end-of-range defects observed by electron microscopy.

Keywords: clean surface, dopant behavior during annealing, Boron diffusion in silicon, Arsenic diffusion in silicon, diffusion modeling, Ge pre-amorphized silicon.
PACS: 61.72.Cc, 68.47.Fg, 85.40.Ry

INTRODUCTION

Forming increasingly shallow *pn* junctions in silicon-based microelectronic logic devices is critical to accomplish as device dimensions continue to diminish. The amount of dopant in these regions that is electrically active will simultaneously need to increase. Current technology for junction formation relies mainly on ion implantation followed by rapid thermal annealing to introduce dopants into the substrate. Post-implant annealing technologies have struggled in their ability to simultaneously increase dopant activation and decrease post-implant diffusion, especially for the key dopant boron. In addition, end-of-range defects left over from pre-amorphizing implants need to be reduced.

Recent work in our laboratory [1-2] has shown that the behavior of point defects within silicon can be changed significantly by controlling the chemical state at the surface. Experiments with Si isotopes combined with modeling of boron diffusion/activation have suggested that certain chemical treatments of the surface induce it to act as a large controllable "sink" for defects that removes Si interstitials selectively over impurity interstitials, leading to less diffusion and better dopant activation.

There are two separate mechanisms for such effects. The first mechanism involves the reflection of charged interstitials from the surface due to electrically active surface defects that set up a repulsive electric field. The driving force for reflection is surface band bending that resulting from dangling bonds at a free Si surface (no oxide) or a damaged Si-oxide interface [3-5]. The opposing field can transform the interface

CP1066, *Ion Implantation Technology,* edited by E. G. Seebauer, S. B. Felch, A. Jain, and Y. V. Kondratenko

from a significant sink into a good reflector, which changes the average concentration of interstitials in the underlying bulk, which in turn influences the degree of dopant activation and transient-enhanced diffusion (TED).

The second mechanism, which is the primary focus of the present paper, involves insertion of interstitials into dangling bonds at the surface. Surfaces differ markedly in their ability to annihilate defects. For example, adsorption exerts significant effects. This effect was suggested by quantum calculations [6,7] by Kirichenko *et al.* An atomically clean surface can annihilate interstitial atoms by simple addition of the interstitials to dangling bonds. However, if the same surface becomes saturated with a strongly bonded adsorbate, annihilation requires the insertion of interstitials into existing bonds. Such insertion should have a higher activation barrier and a correspondingly reduced probability of occurrence.

Behavior of dopants such as boron during annealing is determined primarily by the interplay between lone interstitials and interstitial clusters that render the lone interstitials immobile. During implantation, numerous lone interstitials are created that diffuse quickly and accrete into clusters. Subsequent annealing dissociates these clusters. More heating dissociates more clusters. However, the Si and dopant interstitials are released together. The Si interstitials tend to keep the dopant atoms from entering and remaining in the desired substitutional atomic sites, and also promote the unwanted diffusion of the dopant atoms. Annealing protocols that avoid these problems also leave most of the dopant locked within clusters, rendering it useless.

A large additional "sink" that removes Si interstitials selectively over dopant interstitials would solve this problem. The loss rate of bulk interstitials through incorporation into the surface ranges from very high at an atomically clean surface with many dangling bonds to nearly zero at a surface whose dangling bonds are already saturated with a tightly bonded adsorbate. When the surface loss probability is adjusted to be large, Si interstitials diffuse toward the surface about 100 times faster than the dopant because dopant interstitials are impeded by exchange with the bulk lattice atoms in a way that Si interstitials are not. This preferential loss of Si interstitials keeps electrically active dopant fixed in the lattice by inhibiting the "kick-out" reaction that makes such dopant atoms mobile and inactive.

The present work demonstrates such effects experimentally for arsenic implanted into crystalline silicon and boron implanted into Ge pre-amorphized silicon.

EXPERIMENTAL PROCEDURE

To examine the effects of bond insertion on the surface loss probability, we developed an experimental technique based upon ion implantation. One set of experiments was performed on <100>-oriented, Czochralski (CZ) grown, n-type silicon wafers with a resistivity of 6~9 ohm·cm. Ge was implanted at 15 keV to a dose of 3×10^{14} atoms/cm^2, resulting in the formation of a continuous amorphous layer at a depth approximately 22 nm (measured from Rutherford backscattering (RBS) channeling measurements and transmission electron microscopy (TEM)). The wafers were then implanted with boron at 500 eV to a dose of 1×10^{15} atoms/cm^2. All implants were performed in single-quad mode at 0° tilt and 0° twist, and with native oxide thickness of 11±2 Å. After implantation, some samples were pre-treated with 49% aqueous HF to remove the native oxide and thus create the atomically clean surfaces.

Annealing was carried out in an ultrahigh vacuum environment using Ta clips for resistive heating. The pressure in the chamber was maintained around 10^{-8} ~ 10^{-9} torr during annealing to prevent formation of native oxide and contamination of the surface. In contrast, many annealing studies reported in the literature have been performed in an inert N_2 ambient at atmospheric pressure. Such environments was observed typically contain low, ill-defined levels of reactive gases such as oxygen or moisture [8]. This is especially important since any form of adsorption may deactivate the dangling bonds at the atomically clean surface. The annealing conditions in this study were in the range of 750°C to 900°C for 60 minutes.

Boron profiles were analyzed ex-situ with CAMECA secondary ion mass spectroscopy (SIMS) with oxygen source. Sheet resistance (R_s) was measured by standard four point probe; while the active carrier concentration (N_s) was obtained from Hall measurement by assuming a unity Hall scattering factor. Cross-sectional TEM (XTEM) was performed to analyze the extent of amorphization and the induced end-of-range (EOR) defects.

Another set of experiments used P-type prime Si (100) wafers (5 Ohm·cm) implanted with arsenic at 2 keV and dose of 2x10^{15} ions/cm^{-2} using Applied Materials Quantum X high current ion implanter. There was no surface treatment of the wafers prior to the implantation, and the native oxide thickness was 11±3 Å. Specimens of approximate dimensions 1.7 cm × 0.7 cm were cut from the wafers and mounted in an ultrahigh vacuum chamber using Ta clips for resistive heating. Temperature was monitored with a chromel-alumel thermocouple junction pressed into a small pit drilled into the silicon. To induce diffusion, all specimens were annealed at 750°C for 1 hr.

In arsenic diffusion experiments, specimens were exposed only to atmospheric oxidation, and were not subjected to HF treatment to remove oxide before annealing. In one case, the oxidation was simply the native oxide that built up at room temperature over a period of several weeks after implantation. In another case, native oxide was removed by degreasing and exposure to dilute HF. Thermal oxide was then grown at 630°C for 20 min in moist air. Ellipsometric measurements determined the resulting SiO_2 thickness to be 6 nm.

Arsenic diffusion profiles were measured *ex situ* with secondary ion mass spectroscopy (SIMS) using a Cameca 4f, with a 700 eV Cs beam incident at 50°. Sheet resistance (R_s) was measured with a four-point probe, accurate to within about 10%.

RESULTS AND DISCUSSION

Figure 1 shows data which illustrates the difference in arsenic diffusion based on the type of oxide (native vs grown). The native oxide promotes significantly more profile spreading than the grown oxide.

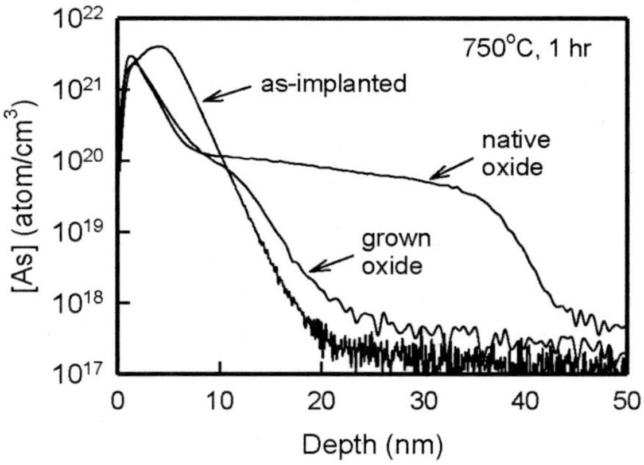

FIGURE 1 SIMS profile of implanted arsenic diffusion in silicon for various types of surface oxide. The grown oxide was 6 nm thick.

The grown oxide has higher R_s than the native oxide, meaning that both the diffusional spreading and dopant activation decreased for the grown oxide. This dataset clearly shows that the surface chemical bonding state exerts large effects on dopant diffusion in the bulk.

Figure 2 shows the boron SIMS profiles annealed at a higher temperature of 800°C. Boron TED is observed in both samples and is due to the emitted interstitials from EOR bands diffusing back towards the surface during annealing. The emitted Si interstitials cause boron diffusion via the well known kick-out and boron-interstitial pairing mechanism [9,10]. A 30% reduction in junction depth occurs for the sample with the atomically clean surfaces compared to the one with the native oxide at a dopant concentration of 1×10^{18} atoms/cm^2. Similarly, its sheet resistance R_s is lower to the same extent when compared with that of the native oxide sample. Hall measurement results also show an increase in the active carrier concentration (N_s) from 2.85×10^{14} cm^2 to 4.16×10^{14} cm^2 for the active surface case with an approximately constant mobility value. The reduction of R_s can be explained by enhancement in the dopant activation level.

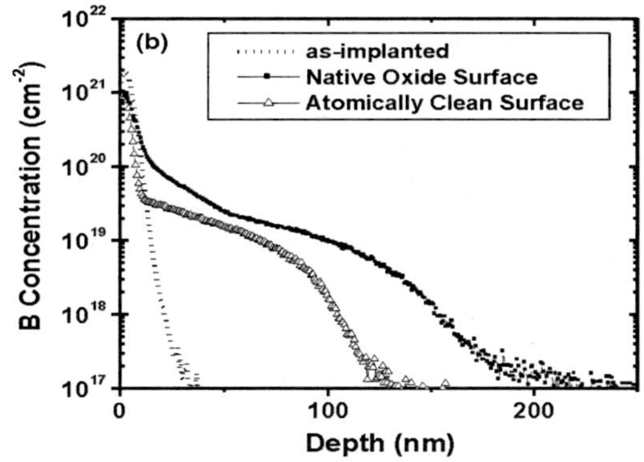

FIGURE 2. SIMS profiles of 500eV B implant with prior 15keV Ge pre-amorphizing implant. Annealing was performed at 800°C for 60 minutes. The atomically clean surface leads to reduced TED in both cases.

Isochronal annealing was performed in vacuum as a function of temperature in the range of 700°C to 950°C. Figure 3 shows that for native oxide-covered Si that has undergone pre-amorphization implantation with Ge, R_s increases from the initial temperature of 700°C and peaks at 850°C, followed by a reduction at higher temperature. This curve reveals the de/reactivation of boron during the thermal cycle involving the evolution and ripening of EOR defects. In the early stage after epitaxial regrowth of the Si, free interstitials are released from the EOR band. These will diffuse towards the B-rich region, interact with boron to form the boron-interstitial clusters (BICs) and deactivate B in the process. As temperature is further increased, the dissolution of BICs will be observed which leads to dopant reactivation and hence reduction in R_s.

The atomically clean surface yields a totally different de/reactivation curve. It starts out with a higher R_s value at 700°C that could be due to a higher dopant loss. Unlike the native oxide surface, R_s continuously decreases up to 800°C and then peaks at a lower temperature of 825°C. The activated dangling bonds at the clean surface may contribute to a higher "sink" efficiency in the removal of free silicon interstitials. The lower deactivation level, as seen from overall lower R_s values for the clean surface indicates the possibility of higher EOR defect dissolution rates, resulting in smaller sizes or lesser amounts of extended defects remaining after thermal annealing.

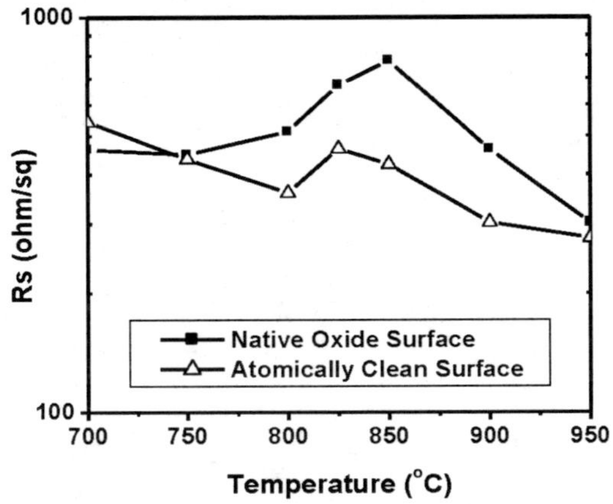

FIGURE 3 Sheet resistance (Rs) for boron at 1×10^{15} atom/cm^2 as a function of isochronal annealing temperature. The active surface leads to lower Rs than the oxided surface in almost all cases.

The higher rate of interstitial annihilation at the active surface promotes higher EOR defect dissolution rates, resulting in smaller numbers and sizes of these defects after epitaxial regrowth. XTEM imaging revealed that the EOR defects show up clearly for the native oxide-covered case. For the active, atomically clean surface, no EOR defects were evident.

In conclusion, the chemical state of surface is seen to have a significant impact not only on arsenic and boron diffusion and activation, but also on the removal of extended defects induced by Ge-PAI. The dangling bonds at the surface play an important role in annihilating the excess interstitials emitted from the EOR defect band. This results in the change of dopant-defect interactions and defect evolution behaviors. The de/re-activation of the dopant was also shown to be strongly influenced by the surface conditions. For the first time, we have shown that such a surface chemistry approach can be applied in improving Ge-PAI B doped layer, which is advantageous in the formation of p-type ultra-shallow junctions. The surface effect is believed to be applicable regardless of the annealing techniques such as spike, flash or sub-melt laser annealing provided the surface state can be maintained throughout the annealing process. The approach offers promise for simultaneous reduction in TED and increase in dopant activation, as well as reducing problems with EOR defects. It should also work in the presence of strain (for SiGe) as well as co-implanted species such as C and F. Moreover, the effect of surface sink should be felt strongly in 3-dimensions (*i.e.*, under the gate), though only 1-dimension profiles have been studied in this work.

ACKNOWLEDGEMENTS

This work was supported by National University of Singapore (NUS), University of Illinois (UIUC) and Chartered Semiconductor Manufacturing (CSM) research collaboration program. Characterization supports from Center for Microanalysis of Materials, University of Illinois, and Institute of Material Research and Engineering (IMRE), Singapore are also gratefully acknowledged. R. V. acknowledges support from an Applied Materials Fellowship. Partial support for this work came from the Petroleum Research Fund of the American Chemical Society.

REFERENCES

1. E. G. Seebauer, K. Dev, M. Y. L. Jung, R. Vaidyanathan, C. T. M. Kwok, J. W. Ager, E. E. Haller, and R. D. Braatz, *Phys. Rev. Lett.*, **97**, 055053 (2006).
2. X. Zhang, M. Yu, C. T. M. Kwok, R. Vaidyanathan, R. D. Braatz, and E. G. Seebauer, *Phys. Rev. B* **74**, 235301 (2006).
3. K. Dev, M. Y. L. Jung, R. Gunawan, R. D. Braatz and E. G. Seebauer, *Phys. Rev. B*, **68**, 195311 (2003).
4. K. Dev and E. G. Seebauer, *Surface Sci.*, **550**, 185 (2004).
5. M. Y. L. Jung, R. Gunawan, R. D. Braatz and E. G. Seebauer, *J. Appl. Phys.*, **95**, 1134 (2004).
6. T. A. Kirichenko, S. Banerjee, and G. S. Hwang, *Phys. Rev. B* **70**, 045321 (2004).
7. T. A. Kirichenko, S. Banerjee, and G. S. Hwang, *Phys. Status Solidi B* **241**, 2303 (2004).
8. H. -J.Gossman, C. S. Rafferty, F. C. Unterwald, T. Boone, T. K. Mogi, M. O. Thompsom, and H. S. Luftman, *Appl. Phys. Lett.* **67**, 1558 (1995).
9. P. Stolk, H. -J. Gossmann, D. Eaglesham, D. Jacobson, C. Rafferty, G. Gilmer, M. Jaraiz, J. Poate, H. Luftman, and T. Haynes, *J. Appl. Phys.* **81**, 6031 (1997).
10. N. Cowern, K. Janssen, and H. Jos, *J. Appl. Phys.* **68**, 6191 (1990).

The Formation of Ultra-Shallow Phosphorous Doped Layers Using Vacancy Engineering

A. J. Smith[1], S. H. Yeong[2], B. Colombeau[3], B. J. Sealy[1], R. M. Gwilliam[1]

[1] *Ion Beam Centre – Advanced Technology Institute, University of Surrey, Guildford, GU2 7XH, UK*
[2] *Dept. of Chemical and Biomolecular Eng., National University of Singapore, 9 Engineering Drive 1, 1175776, Singapore*
[3] *Chartered Semiconductor Manufacturing Ltd, 60 Woodlands, Industrial Park D, Street 2, Singapore 738406*

Abstract. Vacancy Engineering has previously been shown to be highly efficient in improving the junction properties of a p-type boron implant. This study examines the effect of a Vacancy Engineering Implant (VEI) prior to a low-energy n-type phosphorous implant. These initial results indicate that an excess of vacancies not only reduces the observed enhanced diffusion but also reduces the inherent electrical deactivation/reactivation process.

Keywords: Vacancy Engineering, Phosphorous, SOI, USJ
PACS: 61.82.Fk, 74.62.Dh, 61.72.-y, *61.72.j,* 61.72.J-, *61.72.j,* 61.72.U-

INTRODUCTION

Source/drain extensions are crucial components within a MOSFETs device architecture. Forming these highly conducting, ultra-shallow layers is known to be a challenge due to anomalous diffusion [1] and clustering mechanisms [2], which hinder the final junction depth and sheet resistance, respectively. Therefore, innovative technologies have to be implemented to overcome these detrimental atomic interactions. Today, the most popular methods involve one of, or a combination of, Pre-Amophisation Implants (PAI) and Solid Phase Epitaxial Re-growth (SPER), fluorine and carbon co-implantation or an advanced annealing scheme. However, our previous studies have shown that using a Vacancy Engineering Implant (VEI) prior to a doping implant is a viable alternative to the preferred techniques, producing highly electrically active, stable p-type ultra-shallow junctions [3]. However, there is a distinct deficit within the scientific literature on the effect of a VEI on n-type dopants.

This study examines the effect of a VEI on a subsequent n-type phosphorous doping implant. Phosphorous is an obvious selection as it is well known to be an interstitial mediated diffuser [4] similar to boron. Arsenic is currently the preferred n-type silicon dopant, however with the improvements made in p-type junction technology, n-type junctions also need to be optimized, making phosphorous a potential alternative due to its high solubility level as indicated by recent studies [5,6].

An advanced CMOS device will undergo a plethora of fabrication steps during its creation. The complementary configuration of P and N MOS devices can often require different process technologies for each component. An example of this is illustrated within the latest 45 nm technology node where a different high k, metal gate combination for each device type is required [7]. Therefore, creating a P and N MOS compatible process can lead to significant economic gain.

EXPERIMENTAL DETAILS

A 300 keV silicon VEI to a dose of $1.5x10^{15}/cm^2$ was implanted into a Silicon On Insulator (SOI) substrate with a 110/200 nm (silicon over layer/buried oxide) structure. This positions the projected range of the implant at around 450 nm and therefore leaving the excess silicon interstitial distribution beneath the Buried OXide (BOX) A subsequent phosphorous doping implant was performed at an energy of 2 keV to a dose of $10^{15}/cm^2$. All implants were performed with a 7° tilt and a 22° rotation. A reference wafer of the same specification only received a phosphorous implant to act as a comparison. Figure 1 illustrates the experimental design with respect to the SOI structure. The net excess defect distributions were estimated

CP1066, *Ion Implantation Technology,* edited by E. G. Seebauer, S. B. Felch, A. Jain, and Y. V. Kondratenko

from Monte Carlo simulations using the following simple expression:

$$C_{D(x)} = [Si_{I(x)} - Si_{V(x)}] + Si_{D(x)} \quad (1)$$

Where C_D is the defect distribution, Si_D is the additional silicon atoms from the implant, Si_V and Si_I are the vacancy and interstitial distributions, respectively.

An actual Secondary Ion Mass Spectroscopy (SIMS) profile of an as-implanted phosphorous implant used within this study is represented via the open triangles.

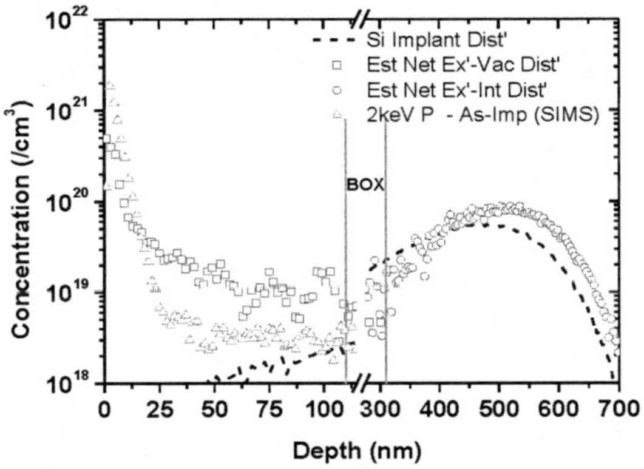

FIGURE 1. A schematic representation of the experimental design, illustrating the estimated net excess vacancies (open squares) and interstitials (open circles) after a 300keV $1.5x10^{15}$ Si/cm^2 VEI (dotted line) in a SOI substrate. The phosphorous implant determined via SIMS is indicated by the open triangles.

Post implant thermal processing was kept to a 10 s isochronal scheme from 600°C to 1000°C within a nitrogen ambient, using a Uniaxis Addax XM80 rapid thermal annealer (RTA). Sheet resistance (Rs) measurements were determined using the Van der Pauw method. Hall measurements were used to measure the carrier density (Ns) and Hall mobility where the Hall scattering factor has been assumed to be one. SIMS analysis was performed by time-of-flight secondary ion mass spectrometry (TOFSIMS) using the dual beams of 0.5 keV Cs+ and 5 keV Ga+ to study the phosphorous diffusion.

Cross-sectional transmission electron microscopy analyses were done using a JEM 2010 TEM system to examine the as-implanted damage.

RESULTS AND DISCUSSIONS

To begin this study it is important to examine the starting condition of the SOI substrate after implantation. Figure 2 presents an XTEM image of the top silicon over layer, after the VEI and subsequent phosphorous implant. It is possible to see that the majority of the silicon top layer remains crystalline. However, two distinct amorphous bands are present, one ~5 nm wide extending from the surface, due to the phosphorous implant self-amorphisation. Whereas the second 5 nm band from the Si/BOX interface is a result of the damage created by the VEI, which is known to regrow via a reverse SPER leaving no TEM detectable defect band in the over layer [8]. Therefore, it is possible to conclude that the top silicon layer is vacancy rich as the VEI has not completely amorphised up to the surface. This was also confirmed by Positron Annihilation Spectroscopy (PAS), which determined the vacancy concentration to be ~$2x10^{19}$ /cm^3, over the 110 nm top silicon layer, after a 300°C 10 s anneal to promote local interstitial - vacancy recombination [9].

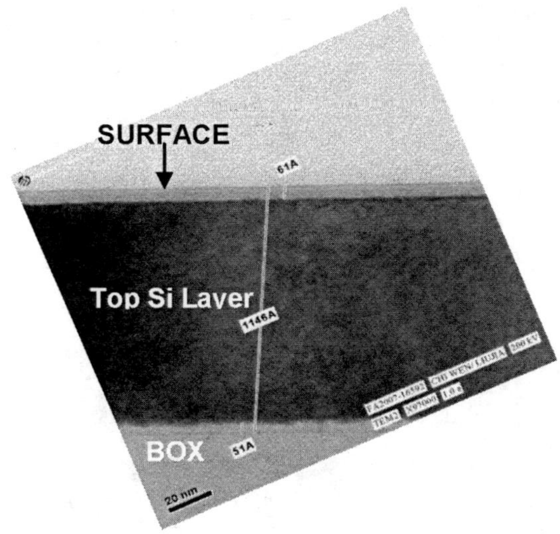

FIGURE 2. XTEM image of the top silicon layer of the SOI structure.

Phosphorous is well known to be susceptible to TED, even more so than boron [10], and it is for this reason that arsenic has always been the preferred n-type dopant. Therefore, to begin the analysis on the phosphorous layer in terms of diffusion, figure 3(a) presents the SIMS profiles of the phosphorous distribution as a function of temperature without a VEI. At low temperatures, 600°C and 700°C, a negligible amount of diffusion is observed. When the temperature is increased to 800°C a significant amount of movement in the tail region occurs, resulting in a 'kink' at ~10^{19} P/cm^3, which separates an immobile peak with a fast diffusing tail. Above 800°C the tail region exhibits an exponential trend characteristic of enhanced diffusion concomitant with the kink increasing to $7x10^{19}$ P/cm^3 and 10^{20} P/cm3 at 850°C and 900°C, respectively. However, when a VEI implant is used, as shown in figure 3(b), there are distinct differences in the phosphorous diffusion. It is possible to see that at 700°C there is a slight movement in the tail of around 5 nm, taken at a concentration of 10^{19} P/cm^3, indicating that a higher fraction of the phosphorous profile is mobile at lower temperatures, and hence less clustering has occurred. This is also indicated by the increased position of the 'kink' which now corresponds to ~5x10^{19} P/cm3. However, a more pronounced effect is observed at 800°C and 850°C. Taking a reference point at 6x10^{18} P/cm3, at 800°C, the tail of the phosphorous distribution corresponds to a depth of ~40 nm, which is a 20 nm reduction in depth compared to the phosphorous profile without a VEI. Interestingly, at 850°C, there is also a 20 nm reduction in depth when a VEI is used. Therefore, it is possible to conclude that the VEI has reduced the amount of enhanced diffusion. However, to gain a more complete understanding to the effect of a VEI on a phosphorous implant it is necessary to examine the

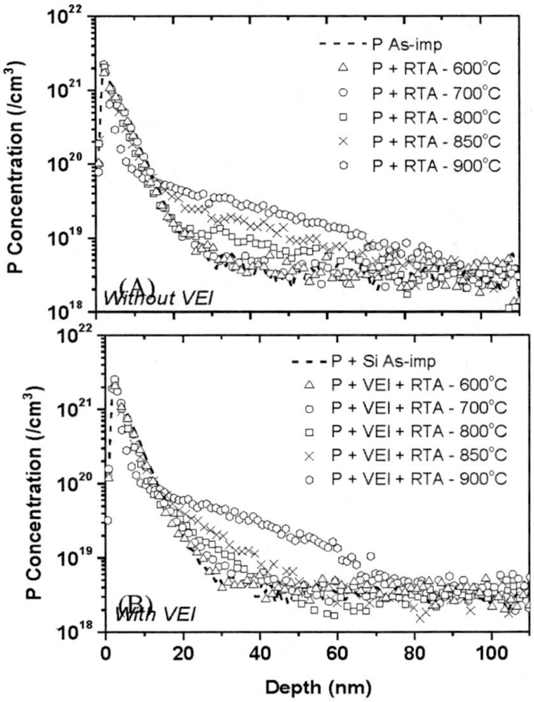

FIGURE 3. SIMS analysis of the phosphorous doping implant without (A) and with the VEI (B) as a function of anneal temperature.

electrical properties of such layers. Hence, the following section is focused on the sheet electrical measurements, as represented in Figure 4, in terms of Rs (A), carrier density (B) and Hall mobility (C) as a function of temperature. The phosphorous only curves (open circles) illustrate that in terms of Rs over the 600°C to 700°C range, the Rs is relatively constant and what subtle change that is apparent in the Ns is compensated by a reduction in mobility. After 700°C, the phosphorous layer undergoes a de-activation process where the Ns is reduced by roughly four times, resulting in a significant increase in Rs. A further increase in temperature to 800°C initiates the Rs recovery as the Ns fraction of the phosphorous implant increases. It is important to note that it is at this temperature where a significant amount of phosphorous tail diffusion is observed (figure 3(a)) combined with an increase in the concentration where a 'kink' occurs.

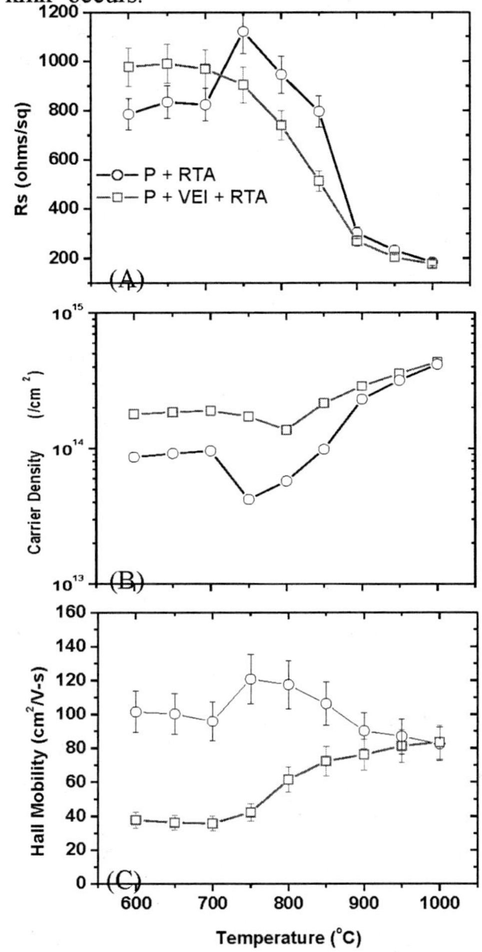

FIGURE 4. Sheet electrical measurements presenting the Rs (A), Ns (B) and Hall mobility (C), with (open squares) and without (open circles) the VEI as a function of anneal temperature.

When a VEI implant is used (open squares) the Rs at low temperatures (600°C to 700°C) is higher than the reference curve. However, the corresponding Ns clearly indicates that the Ns is far greater, which therefore, demonstrates that the difference in Rs is as a result of the large difference in mobility. Even though the mobility will be lower due to the increase in ionized impurity scattering as the Ns is higher, for this temperature range it is thought that a significant contribution to the total scattering mechanisms is due to remaining implant damage. Conversely, after 700°C the Rs starts to decline. As the Ns is also reducing it is possible to deduce that this reduction is as a result of the increase of mobility due to damage removal. From this point, the Rs monotonically decreases showing no "deactivation" in Rs as seen in the reference curve.

This effect is more pronounced in the Ns curves when a VEI is used, the reduction seen in electrical activation at 800°C, is only 25% compared to the reference curve at 750°C for which there is a 55% reduction.

Eventually, at high temperatures, >900°C, the effect of the VEI diminishes, indicated by all the VEI curves merging with the phosphorous reference curves. In a similar manner to previous studies on boron ultra-shallow junctions, this is determined to be as a result of cluster dissolution and diffusion kinetics starting to dominate at high thermal budgets [11].

Providing a full understanding of this study is certainly a non-trivial task, as the underlying mechanisms in which phosphorous clusters have been under debate for a number of years. The following section presents a discussion with some tentative conclusions.

It is assumed that as the phosphorous is self-amorophising, the first 5 nm of the distribution is contained within an amorphous region where no excess defect distribution would survive. Therefore at an anneal temperature of 600°C this surface region regrows via SPER, leaving an excess silicon interstitial band around the peak of the phosphorous implant. As this is in such a close proximity to the surface this band may not evolve into a full EOR defect band. However, as it is known that phosphorous diffusion is mediated by silicon interstitials, the SIMS analysis tends to suggest there is a high enough supersaturation of interstitials to cause significant enhanced diffusion, which is greatly reduced when a VEI is used. Therefore it is possible to conclude that the VEI has been efficient in mopping up excess interstitials via a vacancy – interstitial annihilation mechanism, within the tail region. Either during post-implant annealing or even dynamically during the implantation process itself [12].

The reference phosphorous electrical data clearly indicates an apparent deactivation/reactivation process, similar to boron which has been pre-amorphised [13]. However, recent studies indicate that phosphorous clusters preferentially with vacancies [14]. Therefore, it is intuitive that the addition of a VEI would result in a higher fraction of the phosphorous implant being held in phosphorous-vacancy complexes, and furthermore, reduce its electrically active fraction. However, the results in figure 4(B) clearly indicate that a VEI increases the Ns at low temperatures and actually reduces the amount of deactivation. A possible reasoning for this phenomenon is due to a significant portion of the high concentration part of the phosphorous profile being activated during the SPER and that the excess vacancies in the tail region help reduce phosphorous-interstitial clusters that may occur in the lower concentration of the phosphorous distribution. To provide a greater understanding, further investigation would be required in terms of not only well designed experiments but also simulations.

SUMMARY

In Summary, a silicon VEI has been shown to reduce the enhanced diffusion and assist in reducing the deactivation/reactivation process of a subsequent low energy phosphorous implant. These initial results show promise in that it is possible to use a VEI on not only p-type junctions but also n-type junctions. It is thought that by optimizing the VEI and the phosphorous doping implant, much larger improvement factors will be achieved.

REFERENCES

1. W. Hofker, H. Werner, D. Oosthoek and H. de-Grefte, *App. Phys.* **2**, 265 (1973).
2. N. Cowern, K. Janssen and H. Jos, *J. Appl. Phys.* **68**, 6191 (1990).
3. A. J. Smith, N. E. B. Cowern, R. Gwilliam *et al. App. Phys. Lett.* **88**, 082112 (2006).
4. A. Ural, P. B. Griffin and J. D. Plummer, *J. App. Phys.* **85**, 6440 (1999).
5. A. Vanderpool, A. Budrevich, M. Taylor, in: *Proceedings of the 16th International Conference on Ion Implantation and Technology*, Marseille, France 2006. AIP Conf. Proc. 866, 41 (2006).
6. B. J. Pawlak, R. Duffy, T. Janssens *et al. App. Phys. Lett.* **89**, 062102 (2006).
7. Intel Unveils 45nm Chip With High-K, Metal Gates *Solid State Technology*, **50** March (2007).
8. A. J. Smith, B. Colombeau, R. Gwilliam, N. E. B. Cowern *et al. Mat. Sci. Eng. B.* **124-125**, 210 (2005).
9. A. J. Smith, A. P. Knights, P. Coleman *et al.* To be Published.
10. J. S. Christensen, H. H. Radamson, A. Yu. Kuznetsov and B. G. Svensson, *App. Phys. Lett.* **82**, 2254 (2003).
11. A. J. Smith, N. E. B. Cowern, B. Colombeau *et al.* in: *Proceedings of the 16th International Conference on Ion Implantation and Technology*, Marseille, France 2006. AIP Conf. Proc. 866, 84 (2006).
12. R. Gwilliam, N. E. B. Cowern, B. Colombeau *et al. Nucl. Inst. Meth. Phys. Res. B* **261**, 600 (2007).
13. B. J. Pawlak, R. Surdeanu, B. Colombeau *et al. App. Phys. Lett.* **84**, 2055 (2004).
14. Y. Takamura, P. B. Griffin, J. D. Plummer, *J. App. Phys.* **92** 235 (2002).

Non-Fluorine Plasma Strip of HDI Resist for 45 nm

A. Zakharov[1], V. Grimm[1], C. Krueger[1*], K. Han[2], I. Berry[2], R. Sonnemans[2], and S. Luo[2]

[1] AMD Fab36 LLC & Co. KG, D-01109 Dresden, Germany
[2] Axcelis Technologies, Inc., Beverly, MA 01915, USA
**christian.krueger@amd.com*

Abstract. High dose implant resist cleaning for ultra-shallow junction at 45 nm and beyond requires complete photo-resist and residual removal with near-zero substrate loss and minimized dopant loss. Conventional fluorine-containing plasma ashing results in significant dopant bleaching and a new partial plasma strip process was developed in this work to reduce dopant loss.

INTRODUCTION

Ion implantation causes the formation of a highly cross-linked carbon crust layer on the top surface of the photo resist (PR) mask. Its thickness and composition depends on the type of PR, implant species, energy, dose, temperature during implantation and other factors. The crust is often difficult to remove and its relative thickness tends to increase with the continuous shrink of technology nodes as implant doses increase. The conventional process for stripping PR after high dose implantation (HDI) is using a fluorine-containing plasma ash to completely remove resist with crust on top and then using wet clean to remove any residue left after plasma strip. The drawbacks of conventional fluorine-based plasma ashing are increased substrate loss and dopant bleaching, while wet cleaning by itself is typically not effective for removing implanted PR with doses higher than $1x10^{15}$ ions/cm^2. This paper introduces a new partial plasma strip (PPS) process which consists of a partial non-fluorine plasma ash to break through the crust, followed by a conventional wet clean to remove the resist underneath. This process combines both the crust-breaking ability of the plasma process and the selectivity of the wet clean, resulting in reduced dopant bleaching.

HISTORY AND MOTIVATION

Dose precision requirements of <1% are common in implant technologies. To maintain this dose control, the retention of the implanted dopants during strip and clean must be considered and controlled. For ultra-shallow junction (USJ) implant conditions, where the silicon is highly damaged (very chemically reactive) and the dopants lie close enough to the surface to be within the interaction zone of the strip and clean processes, significant dopant loss can result.

Figure 1 shows the various mechanisms for substrate and dopant loss caused by the interactions of plasma and subsequent wet clean with the exposed substrate surface.

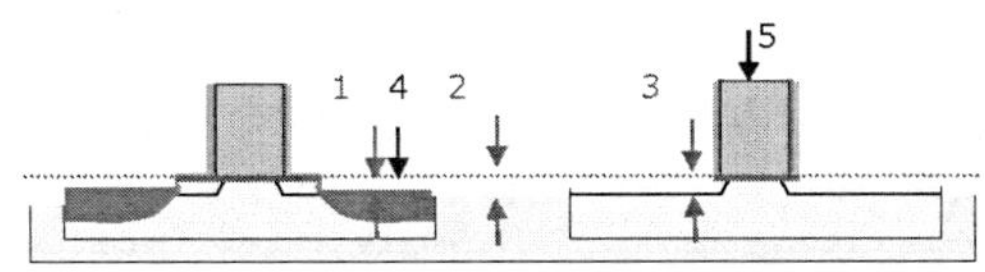

FIGURE 1. Main surface interactions for dopant loss: 1) Embedded SiGe attack, 2) STI oxide loss, 3) Active silicon loss, 4) Surface dopant bleaching, 5) Poly dopant bleaching.

To maintain dose control, loss of dopant during the strip and clean processes must be considered. For a 1 keV B implant for example, about 10% of the implanted boron lies within 20 Å of the surface, increasing to about 35% at 300 eV implant energy [1]. Therefore a loss of only a few mono-layers from the surface can remove more than 1% of the dopant. Further compounding the problem, pre-amorphization and co-implants can enhance surface chemical reactivity, which enhance surface loss and oxidation in both plasma strip and wet chemical clean processes.

Figure 2 is the result of a 45 nm node device test to determine the dopant loss effects from a fluorine (CF_4) containing plasma strip process which was found to remove < 8 Å of the junction. In this test a device lot split was subjected to 1, 2, and 3 additional plasma strip processes. The increased sheet resistance and lowering of Miller capacitance indicates significant loss of active dopant within the extension implant region [2]. While it maybe possible to tune the implant conditions to compensate for the dopant loss, provided the loss in the strip and clean is consistent and no re-works are used, it is doubtful that adequate dose control can be maintained. Testing determined that the loss of dopant was a consequence of the fluorine used to remove the implant crust. Removal of the fluorine nearly

CP1066, *Ion Implantation Technology*, edited by E. G. Seebauer, S. B. Felch, A. Jain, and Y. V. Kondratenko

eliminates the dopant loss but controlling post strip and clean defects/residues becomes challenging.

The objective of this work is to develop a resist strip and clean processes that reduces or eliminates the dopant loss while maintaining a low-defectivity, high yield and manufacturable process. Dopant loss testing revealed that minimal dopant loss could be achieved through the use of a fluorine-free plasma strip process to break-up the implanted crust material followed by a dilute APM + SPM clean chemistry.

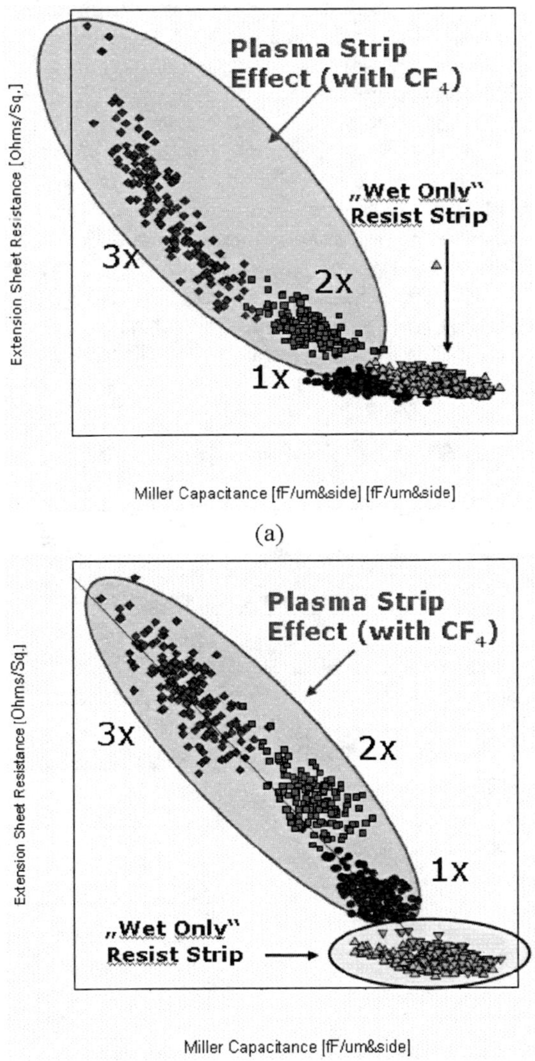

FIGURE 2. Impact of fluorine containing plasma strip process on surface and near surface dopant concentration: (a) NMOS, (b) PMOS.

PARTIAL PLASMA STRIP EXPERIMENTS

A DUV photoresist was printed on 300 mm wafers with a CVD coated SiO_2 surface. The mask was implanted with As at 30 keV and a dose of $3x10^{15}$ ions/cm^2. The mask contained structures with different feature width ranging from 440 nm to > 2 µm. The SEM cross-section in Figure 3 shows the profile of a mask feature. The crust on top of the PR lines can be clearly recognized. It comprises about 1/3 of the PR thickness. Note that the crust on the sidewall is thinner than on top.

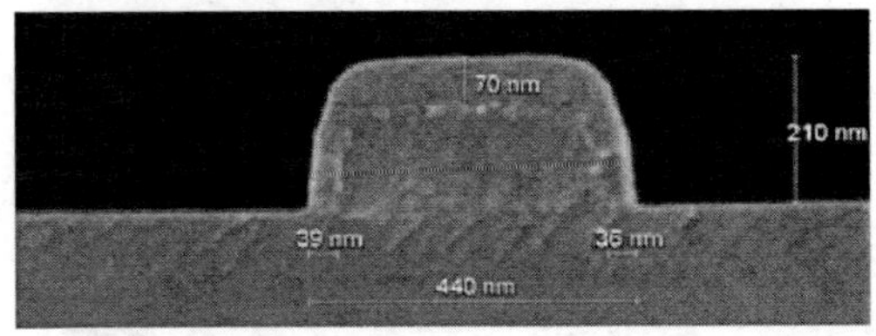

FIGURE 3. Cross-section SEM of an as-implanted *PR*.

The wafers were subjected to a plasma process in O_2 and N_2/H_2 gas mixture at a wafer temperature of <200°C for different durations. Wafers were then cleaved and the cross section was examined under SEM. Figure 4 shows a resist after a partial plasma strip process of 50 sec. The implanted crust is removed from the side wall which allows the following wet clean process to undercut the underlying bulk resist and lift off the remaining crust.

FIGURE 4. Cross section of a partially ashed photo resist.

The wet clean which followed the plasma strip was a standard SPM (Piranha, 120°C) + SC-1 (60°C) conducted in a batch tool. The wet clean was identical in all experiments.

RESULTS

The partial plasma strip process can have three different outcomes, depending on the plasma process duration. Figures 5 and 6 show SEM inspection before and after wet clean.

i. Plasma process is too short (e.g., 30 sec): Sidewall crust breakthrough has not yet occurred and the wet clean is not able to remove the encapsulated PR (Figure 5b).

ii. Plasma process reaches crust sidewall breakthrough (e.g., 50 sec): The wet clean removes PR by undercutting the bulk resist (Figure 6b). The remaining crust is also removed in the wet clean process.

iii. Plasma process too long (e.g., 90 sec): The narrow crust-features start to collapse due to bulk resist being completely consumed by the plasma ash. Crust that falls down on the hot substrate surface "bakes on" during the plasma process and the remaining residues are difficult to be removed during the subsequent wet clean.

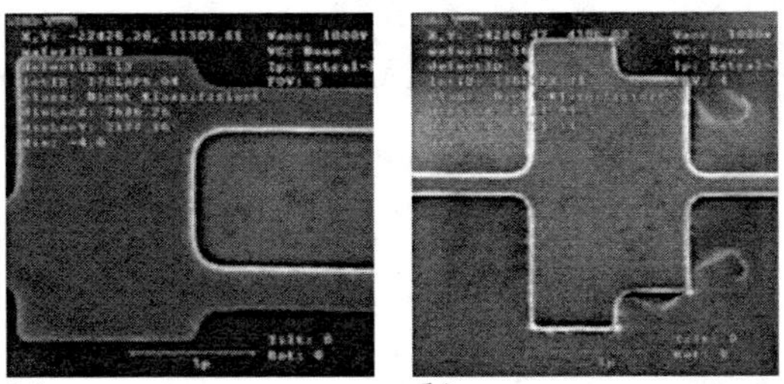

(a) (b)

FIGURE 5. Top down SEM after 30sec PPS (a) and after 30 sec PPS + wet clean (b).

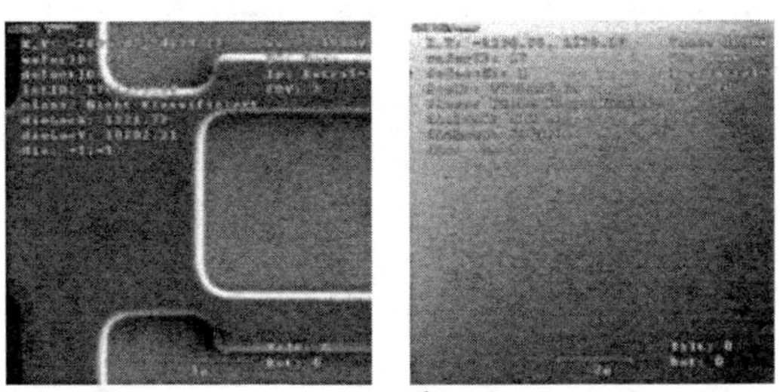

(a) (b)

FIGURE 6. Top down SEM after 50sec PPS (a) and after 50 sec PPS + wet clean (b).

Not only is the duration of the partial plasma strip process shorter than the conventional fluorine-based plasma ash, but more importantly the fluorine-free plasma process showed no impact on device performance, in contrast to the CF_4-containing strip process, which resulted in substantial dopant bleaching.

SUMMARY

Increased dopant bleaching was observed for conventional HDI resist cleaning processes in which fluorine-containing plasmas are used. Removal of the fluorine (CF4), shortening the duration of a plasma ash process and restricting the plasma process to the crust breakthrough achieved complete resist removal after wet clean and resulted in no observable dopant loss. The low defectivity can be explained by the opening of the sidewalls by the plasma processes allowing the wet clean access to remove the now exposed photoresist underneath the crust, and eventually lifting it off of the surface. This process sequence prevents the crust from collapsing and "burning" onto the substrate surface which makes the crust residues difficult to remove by the wet clean process.

ACKNOWLEDGEMENTS

All SEM Pictures are made at the AMD Material Analysis Lab. Fab36 in Dresden, Germany

REFERENCES

1. I.L. Berry, C. Waldfried, K. Han, S. Luo, R. Sonnemans, M. Ameen, Proc. IJWT Workshop, Shanghai, China, May 2008.
2. F. Wirbeleit, et.al. MRS Spring Meeting, San Francisco, California 2008.

DOPING PROCESSES (ANNEALING)

An Extensive Study on the Boron Junctions Formed by Optimized Pre-Spike / Multiple-Pulse Flash Lamp Annealing Schemes: Junction Formation, Stability and Leakage

S. H. Yeong[1,2], D. X. M. Tan[3], B. Colombeau[1], C. H. Poon[1], K. R. C. Mok[1], A. See[1], F. Benistant[1], K. L. Pey[3], C. M. Ng[1], L. Chan[1], and M. P. Srinivasan[2]

[1] *Chartered Semiconductor Manufacturing Ltd, 60 Woodlands Industrial Park D, Street 2, Singapore 738406.*
[2] *Department of Chemical and Biomolecular Engineering, National University of Singapore, 9 Engineering Drive 1, Singapore 117576.*
[3] *Microelectronics Center, School of Electrical and Electronic Engineering, Nanyang Technological University, Nanyang Avenue, Singapore 63979.*

Abstract. In this work, the electrical activation of Boron in Germanium pre-amorphized silicon substrate upon flash lamp annealing (FLA) is investigated. We demonstrate that FLA helps in the reduction of the EOR defects, resulting in minimal transient enhanced diffusion and dopant deactivation effect. It has also been observed that the junction stability improves with the increasing number of flash pulses, which is clearly reflected by the dopant deactivation level upon post-thermal treatment. In another FLA scheme, the spike rapid thermal annealing (RTA) performed prior to the flash further enhances the junction stability. However, this pre-spike RTA step induces extensive dopant diffusion and an overall degradation in sheet resistance. The above observations are concluded to be due to the different extent of silicon interstitial supersaturation that can be explained by the interactions between the extended defects and dopants. Lastly, leakage current for the junctions formed under different FLA schemes are compared. Typical single pulse FLA junction shows high leakage current and it can be reduced through the additional pulses of FLA or effectively suppressed by the pre-spike RTA flash scheme. In addition, it is also found that the junction leakage can be correlated to the FLA residual defects.

Keywords: Flash Lamp Annealing, Boron, Activation, Deactivation, Residual defects
PACS: 61.72.jd, 61.72.jj, 61.72.Cc, 61.72.uf, 85.40.Ry, 85.30.Kk

INTRODUCTION

Recent advancement in ultra-shallow junction formation has provided various possibilities in the annealing approaches. As the CMOS devices scale down to the sub-nm technology, conventional implantation process coupled with rapid thermal annealing (RTA) does not cater to the stringent requirements in the advanced ultra-shallow junction formation. Pre-amorphizing implant (PAI) was later developed to be integrated into the formation of ultra-shallow junction to inhibit the ion channelling effect but however, it creates a rich end-of-range (EOR) defect band beyond the amorphous/crystalline (a/c) interface [1]. These EOR defects which mainly comprise of free interstitials migrate towards the substrate surface during the thermal annealing cycle and cause dopant deactivation and transient enhanced diffusion (TED) [2-7].

Flash Lamp Annealing (FLA) has proven to be one of the most potential techniques for the integration in state-of-the-art CMOS process due to its reduced thermal budget. FLA offers the ability to form highly activated shallow junctions with a much shorter soak annealing duration at higher temperature [8]. Although ultra-shallow junctions with minimum TED can be optimized using FLA [9], the dopant activation level and density of residual defects after FLA are the main concerns. In the device level, these residual defects contribute to high junction leakage [10-13]. The junction stability and leakage current are intimately related to the presence of these residual defects, which eventually affect the overall performance of the devices. In this paper, we demonstrate the stability of the B junctions in terms of dopant activation and diffusion formed by FLA in Ge-PAI silicon substrates. The effects of multiple-pulse FLA and pre-spike RTA are also shown. In addition, the junction leakage study associated with the various FLA schemes is studied with the diodes.

CP1066, *Ion Implantation Technology,* edited by E. G. Seebauer, S. B. Felch, A. Jain, and Y. V. Kondratenko

EXPERIMENTAL DETAILS

The p+/n B junctions were formed by performing Ge-PAI on the Czochralski grown (100) n-type silicon wafers at 15 keV with a dose of 5 × 1014 cm-2 followed by 1 keV B implant with a dose of 2 × 1015 cm-2. The front side of the wafer surface was mounted facing the xenon flash lamps with the backside in contact with the hot plate. The implanted wafers were raised to an intermediate temperature of 500°C by the hot plate before the xenon flash lamps were triggered for irradiation. There is no re-crystallization of the Ge-PAI induced amorphous layer at the intermediate temperature of 500°C. (verified by TEM). FLA was carried out in N2 ambient for one, three, and six pulses with the intensity of 26 J/cm2 and a pulse duration of 0.8 ms. The peak wafer temperature was estimated to be around 1120oC to 1150oC. In addition, some of the wafers were subjected to a 950oC spike RTA to anneal out the residual defects before the FLA. Post-RTA isochronal annealing of temperature ranging from 600oC to 1050oC for 60s, was conducted on some samples to study the junction stability through the dopant de/re-activation behavior. The RTP annealing temperature was monitored and measured using thermocouple. The dopant diffusion profiles were characterized by secondary ion mass spectrometry (SIMS) using the Cameca IMS 6f instrument. A primary beam of O_2^+ ions with net energy of 1 keV at 56° incidence was scanned over an area of 250 x 250 μm^2. High resolution transmission electron microscopy (TEM) was performed to analyze the degree of amorphization, as well as to investigate the remaining residual defects. The electrical characterizations were performed using the 4-point probe and Hall measurements were used for the investigation of junction activation and stability. P+/n diodes with Ni silicide contacts were also fabricated for junction leakage measurement.

RESULTS AND DISCUSSION

Optimized FLA conditions can be developed to achieve the desired ultra-shallow junction characteristics. Junction formation with the implementation of FLA on the Ge-PAI and non-PAI B doped silicon substrates (solid symbols and lines) was carried out. In figure 1(a), Rs values of those samples underwent Ge-PAI and non-PAI with different number of flash pulses are compared. It is noticed that generally with the integration of Ge-PAI, the junction appears to be more activated than the non-PAI case. The Ge-PAI has resulted in an amorphous region of 26 nm being formed in the top surface layer, as seen form the XTEM (not shown). During the fast ramp FLA thermal cycle, the amorphous layer undergoes a solid phase epitaxial regrowth (SPER), achieving higher dopant solid solubility and better activation than the non-PAI crystalline counterparts.

When the multiple-pulse FLA scheme is performed, the Rs trend for the non-PAI and Ge-PAI cases exhibits the opposite phenomenon. In the case of non-PAI, a continuous Rs reduction with the increase of flash pulses has been observed and attributed to the enhanced thermal budget. However, it is interesting to see that the Rs characteristic of Ge-PAI samples is indeed increases with the pulses of FLA being applied. This observation indirectly implies that the different number of flash pulses can affect the interactions between the EOR defects and the dopants, lowering the B activation (higher Rs).

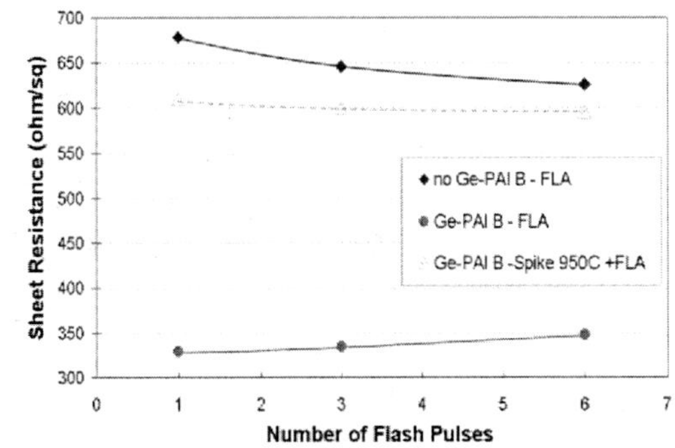

FIGURE 1. Sheet resistance as a function of the number of flash pulses for (a) non-Ge-PAI and Ge-PAI boron-doped samples, which annealed with (b) FLA and (c) pre-spike RTA 950°C followed by FLA.

In addition, figure 1 also shows the Rs trend of the PAI samples when an addition of 950°C pre-spike RTA treatment was included before the FLA. The main purpose of this scheme is to remove or reduce the EOR defects by pre-spike RTA. It resembles that of the non-PAI case since the amorphous layer created by Ge-PAI has been recrystallized by the prior RTA step before the subsequent flash annealing was carried out. This explains the continuous Rs reduction characteristic with respect to the flash pulses on this pre-spike FLA scheme. It is also noticed that the overall Rs level is higher than that of the direct FLA on the PAI samples without pre-spike RTA. The reason is due to the fact that the amorphous layer induced by the Ge-PAI has been recrystallized during the pre-spike RTA, but not during the FLA. Therefore, larger Rs is achieved due to the lower dopant activation level comparing to the direct FLA. The observation is consistent with some of the published findings, claiming that reduced BI clustering and higher dopant activation are achieved with the faster SPER process (higher temperature ramping rate) [16-17]. Nevertheless, the application of FLA on Ge-PAI B doped samples, either on the multiple or pre-spike flash annealing schemes, is not simply dependent on the thermal budget being used but on the dopant-defect interactions during FLA. It is essential to further understand the underlying involved physical mechanisms.

Figure 2 shows the Rs values of the samples performed with the various flash pulses of one, three and six as well as the pre-spike flash scheme. These samples were subsequently treated with post-RTA annealing from 600°C to 1050°C for 60 s. Clearly from figure 2 for the case of one pulse FLA, Rs at the post-RTA annealing temperature of 800°C has the largest value. This suggests that at this temperature, the dopants are noted to have the maximum degree of de-activation. This de-activation phenomenon can be explained by the formation of the boron-interstitials cluster (BICs) created by the upward migration of the free silicon interstitials from the EOR defects during the post-annealing cycle. With a more detail analysis on the Rs trend, it is noticed that the Rs values typically register an initial de-activation up to the peak temperature of 800°C then followed by a latter re-activation behavior with the Rs decreases. The enhancement in dopant activation at higher temperatures is attributed to the dissolution of the immobile BICs and B diffusion at the tail.

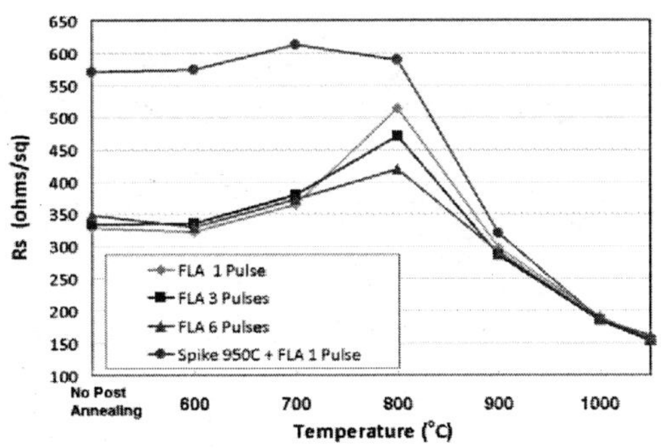

FIGURE 2. Sheet resistance value (Rs) as a function of 60s isochronal annealing temperature.

With the increment of the flash pulses, similar Rs characteristic is also been acquired. However, it is interesting to note that when the samples are subjected to more number of flash pulses, the peak Rs value at 800°C drops accordingly. With comparison made to the de-activation variation curve of the pre-spike RTA flash scheme, it is clearly seen that larger Rs is obtained during the initial low temperature post-RTA (600°C). However, the relative de-activation trend (normalized to the sample without post-annealing) is observed to be among the least compared to other FLA schemes. The reason pertaining to this behavior is due to the removal of the EOR defects by the preceding 950°C spike annealing. Hence with the annihilation of most of the EOR defects, the supersaturation of silicon interstitials towards the surface from the EOR defect region has been greatly reduced. This in turn, lessens the dopant deactivation due to the BI clustering effect. Therefore, it can be deduced that with the integration of pre-spike RTA on top of the flash annealing regime could allow for the effective dissolution of the EOR defects.

Figure 3 illustrates the junction depth obtained for the various samples in correspondence to the post-RTA treatment temperatures. Apparently with the low post-RTA temperature of 700°C, the junction depths of the samples previously irradiated with the various flash pulses demonstrate no significant dopant diffusion. It has however been observed that the sample performed with the pre-spike RTA has attained the highest junction depth value. This is due to the reason that the high temperature 950°C pre-spike RTA has resulted in the substantial diffusion of dopants deep into the substrate. Interestingly, junctions with the various FLA schemes and treated with post-RTA 800°C exhibit similar junction depths. Nevertheless, the eventual dopant diffusion trend can be realized by making close observations to the higher post-RTA temperature at 900°C. The sample subjected to the pre-spike RTA is shown to have the shallowest junction depth. This is because the pre-spike annealing has caused the removal of the existing EOR defects. Therefore, in the subsequent annealing cycle, the supersaturation of free interstitials towards the surface has been greatly reduced. In addition, junction depths with the multiple-pulse FLA scheme decrease with the increasing number of flash pulses, which could be also possibly attributed to the lowering level of interstitial supersaturation due to the smaller density of extended defects at the EOR region.

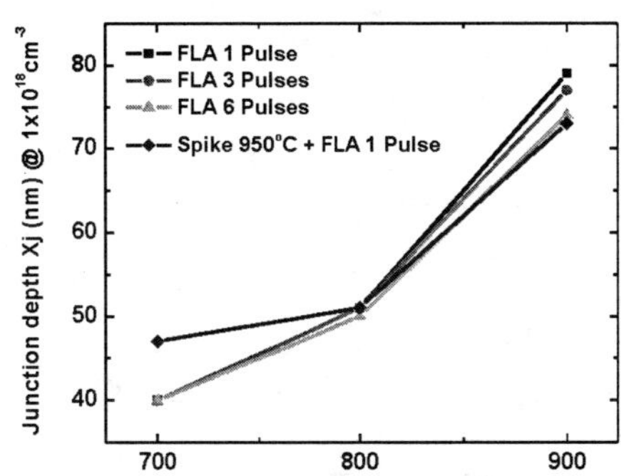

FIGURE 3. Junction depths of the Ge-PIA B doped samples subjected to the different FLA schemes plotted against the range of post-RTA temperatures.

On this note, with the results obtained from both figures 2 and 3, it can be evidenced that the dopant activation and diffusion behavior is closely related to the interactions between dopants and EOR defects. Optimized multiple-pulse FLA is able to achieve higher junction activation and shallower junction depth profiles by reducing the existing EOR defects. Although the pre-spike RTA treatment is able to acquire the shallowest junction depth as compared with the other annealing schemes, the Rs value of the junction is however noted to be the highest. Therefore, it has no absolute advantage over the direct FLA in terms of dopant activation even though the incorporation of pre-spike RTA is beneficial for dissolving the residual defects.

With the understanding of the effects of FLA on the dopant de/re-activation being established, the junction leakage characteristics of the p+/n diodes fabricated by the identical FLA schemes are being investigated. Figure 4 illustrates the I-V characteristics of the p+/n diodes. The sample subjected to the pre-spike RTA has proven to display the best junction current leakage properties, referring to the reversed biased at -2V. This observation can be explained by the fact that the pre-spike RTA annealing treatment has removed majority of the EOR defects induced by the PAI. With the low concentration of residual defects present within the silicon substrate, there are literally lesser alternative pathways of leakage for the current to flow. In addition, the degradation of the junction leakage current becomes more severe as the number of flash pulses reduces. It is therefore understandable that every addition of flash pulses irradiated onto the samples aids in the removal of the EOR defects and leads better lattice repairing.

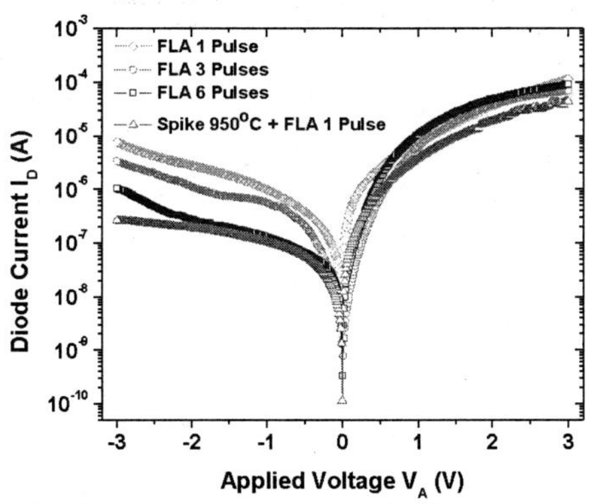

FIGURE 4. I-V characteristic of p+/n diodes (B junctions in n-type silicon) subjected to the FLA with (a) different number for flash pulses and (b) prior spike RTA schemes.

CONCLUSION

In conclusion, this study provides a detailed insight on the various FLA schemes for the dopant diffusion and activation. The physical mechanisms involved during FLA has been discussed and demonstrated through the junction stability and leakage measurements.

ACKNOWLEDGEMENTS

This research work was supported by the National University of Singapore and Special Project Group at Chartered Semiconductor Manufacturing.

REFERENCES

1. R. A. Camillo-Castillo, M. E. Law, K. S. Jones, L. Radic, R. Lindsay, and S. McCoy, *Appl. Phys. Lett.*, **88**, 232104 (2006).
2. C. Bonafos, D. Mathiot, and A. Claverie, *J. Appl. Phys.*, **83**, 3008 (1998).
3. A. Claverie, B. Colombeau, G. Ben Assayag, C. Bonafos, F. Cristiano, M. Omri, and B. de Mauduit, *Mat. Sci. Semicond. Proc.*, **3**, 269 (2000).
4. P. A. Stolk, J. H. -J Gossmann, D. J. Eaglesham, D. C. Jacobson, C. S. Rafferty, G. H. Gilmer, M. Jariz, J. M. Poate, H. S. Luftman and T. E. Haynes, *J. Appl. Phys.*, **81**, 6031 (1997).
5. M. Diebel, S. Chakravarthi, S. T. Dunham, C. F. Machala, S. Ekbote, and A. Jain, *Mater. Res. Soc. Symp. Proc.*, **765**, D6.15.1 (2003).
6. B. J. Pawlak, R. Surdeanu, B. Colombeau, A. J. Smith, N. E. B. Cowern, R. Lindsay, W. Vandervorst, B. Brijs, O. Richard, and F. Cristiano, *Appl. Phys. Lett.*, **84**, 2055 (2004).
7. G. Impellizzeri, J. H. R. dos Santos, S. Mirabella, F. Priolo, E. Napolitani, and A. Carnera, *Appl. Phys. Lett.*, **84**, 1862 (2004).
8. J. A. Sharp, N. E. B. Cowern, R. P. Webb, K. J. Kirkby, D. Giubertoni, S. Gennaro, M. Bersani, M. A. Foad, F. Cristiano, and P. F. Fazzini, *Appl. Phys. Lett.*, **89**, 192105 (2006).
9. K. A. Gable, L. S. Robertson, A. Jain, and K. S. Jones, *J. Appl. Phys.*, **97**, 044501 (2005).
10. W. Lerch, S. Paul, J. Niess, S. McCoy, T. Selinger, J. Gelpey, F. Cristiano, F. Severac, M. Gavelle, S. Boninelli, P. Pichler, and D. Bolze, *Mater. Sci. Eng.*, **B** 124–125, 24 (2005).
11. K. S. Jones, S. P. Crane, C. E. Rose, T. Malmborg, D. Downey, and Arevalo, E. Arevalo, S. McCoy and J. Gelpey, in the 14th International Conference on Ion Implantation Technology Proceedings, p.76 (2003).
12. S. H. Jain, P. B. Griffin, J. D. Plummer, S. McCoy, J. Gelpey, Selinger, T. Selinger, and D. F. Downey, *IEEE Transactions on Electron Devices*, **52**, 7, 1610 (2005).
13. B. Bayha, S. Paul, W. Lerch, D. F. Downey, E. A. Arevalo, X. Hebras, and N. Cherkashin, in the14th International Conference on Ion Implantation Technology Proceedings, p.618 (2003).
14. S. P. McCoy, E. A. Arevalo, J. C. Gelpey, and D. F. Downey, in the12th IEEE International Conference on Advanced Thermal Processing of Semiconductors, p. 99 (2004).
15. S. H. Jain, P. B. Griffin, J. D. Plummer, S. McCoy, J. Gelpey, T. Selinger, and D. F. Downey, *J. Appl. Phys.*, **96**, 7357 (2004).
16. H. S. Chao, P. B. Griffin and J. D. Plummer, *Mat. Res. Soc. Symp. Proc.*, **469**, p. 347-352 (1997).
17. R. Lindsay, B. Pawlak, J. Kittl, K. Henson, *et al.*, *Mater. Res. Soc. Symp. Proc.*, **765**, 261 (2003).

A Comparative Study of Interaction of End of Range (EOR) Defect Band with Upper Buried Oxide (BOX) Interface for B and BF_2 Implants in SOI and Bulk Silicon with Pre-Amorphizing Implant.

M. Kah[1], A. J. Smith[1], J. J. Hamilton[1], S. H. Yeong[2], B. Columbeau[2], R. Gwilliam[1], R. P. Webb[1], and K. J. Kirkby[1]

[1]*Surrey Ion Beam Centre, Advanced Technology Institute, Faculty of Engineerig and Physical Sciences, University of Surrey, Guildford, Surrey, GU2 7XH, UK*
[2]*Charted Semiconductor Manufacturing Ltd., 60 Woodlands, Industrial Park D, Street 2, Singapore, 738406*

Abstract. Highly active, ultra-shallow and abrupt dopant profiles are required for future generations of CMOS devices. A possible way to achieve this is to use pre-amorphization implantation (PAI) and solid phase epitaxial re-growth. B and BF_2 implants were studied in bulk silicon and silicon-on-insulator with Ge PAI. Results show that buried oxide (BOX) in SOI can be used a sink for silicon interstitials contributing to boron Transient Enhanced Diffusion and Boron-Interstitial Clustering. The BF_2 implants show high mobility values and no deactivation.

Keywords: B, BF_2, pre-amorphization, silicon-on-insulator.
PACS: 64.70.kg, 81.70.Jb, 61.72.U 61.72.uf

INTRODUCTION

Highly active, abrupt and ultra-shallow dopant profiles are required by the ITRS [1] to continue the downscaling of CMOS devices. Low Rs values, shallow junction depths and junction stability are the prime concerns. One way of achieving this is to use pre-amorphization implantation (PAI) and solid phase epitaxial re-growth (SPER) which in conjunction with Silicon-on-Insulator (SOI) have already provided encouraging results [6]. PAI and SPER do however, result in excess silicon self interstitials which combine into an end-of-range (EOR) defect band which forms just below the original amorphous/crystalline interface [3]. These EOR defects evolve during subsequent annealing via an Ostwald ripening process releasing silicon interstitials which diffuse towards the surface and in the case of SOI the upper BOX interface. Consequently, boron transient enhanced diffusion (TED) [4] occurs, which degrades both the junction depth and abruptness of the profile. Moreover, when the migrating silicon self interstitials impinge on the implanted boron profile they can form boron-interstitial clusters (BIC's) and so effectively electrically deactivate the boron [5] B surface segregation also reduces the amount of electrically active dopant.

The near surface region in bulk silicon is the dominant sink for the interstitials released from the EOR defects band. The buried oxide (BOX) in SOI material provides an additional sink. Hamilton *et al.* [6] showed that by positioning the EOR defect band closer to the BOX, by increasing the PAI energy, results in a reduced flux of interstitials towards the surface which produces lower deactivation of the dopant during subsequent annealing [6]. It was proposed that the interface between the silicon over layer and the BOX could be acting as a viable sink for interstitials. Moreover, the overlap of the interstitial profile with the BOX results in a significant portion becoming effectively trapped within the BOX and thus reducing the amount of interstitials available to facilitate TED and BIC formation.

An alternative technique which has also been used to reduce detrimental dopant-defect interactions is to introduce a co-implant of a species such as carbon and/or fluorine. However, there has been some controversy in the literature about the effects of fluorine in pre-amorphized silicon. Some studies have shown that it increases boron TED [7,8] while others have shown the converse [15]. More recently it has been shown that by carefully tailoring the co-implant conditions B TED can be drastically reduced [9, 15].

Boron difluoride has been widely used for shallow pn junction manufacturing [10], where reduction in B TED [11] and increased electrical activation [12, 13] have also been reported. Results for dopant deactivation and reactivation for Ge pre-amorphized B

CP1066, *Ion Implantation Technology,* edited by E. G. Seebauer, S. B. Felch, A. Jain, and Y. V. Kondratenko

and BF_2 implants into bulk silicon and SOI are presented in this paper.

EXPERIMENTAL DETAILS

The experiments were performed on n-type <100> Cz bulk silicon wafers with a resistivity of 10 to 20 Ω.cm, and on SOI wafers with a 145 nm-thick buried oxide and a 55 nm-thick p-type Si over-layer. A Ge PAI was performed using an Axcelis GSD 200 implanter, while the B and BF_2 implants were completed on Applied Materials Quantum Leap implanter. In all cases the B dose was kept to 2×10^{15} B cm^{-2} with implant energy of 500 eV and 2.2 keV for the B and BF_2, respectively. Such energies were chosen so that the peak of the B profile occurred at the same depth for both implants. The Ge PAI was carried out at 20 keV at 1×10^{15} Ge cm^{-2} dose, which gave an amorphous layer thickness of 40 nm, determined by Rutherford backscattering spectrometry and confirmed by XTEM and SIMS measurements.

The rapid thermal processing (RTP) was performed using a Process Products Corporation 18 Lamp annealer. Samples were cleaved into 1 cm x 1 cm squares and annealed using a 60 s isochronal scheme in the temperature range of 700 to 1000°C, and isothermally in the 750 to 900°C temperature range for times ranging from 3 to 10,000 seconds. For all anneals the ramp rate was kept constant at 50 °C/sec in N ambient. The Van der Pauw method was used to measure sheet resistance. The results for mobility and carrier concentration were obtained using Accent HL5500 Hall System.

RESULTS AND DISCUSSION

The electrical results after isochronal annealing studies are shown in Figures 1 through 3. Figure 1 shows Rs as a function of annealing temperature (time 60 s). Figure 2 shows areal "carrier concentration" over the same temperature range and Figure 3 the mobility. Looking first at Figure 1, there is a clear distinction between the B and BF_2 implanted samples, with both Si and SOI BF_2 implanted samples having lower Rs values after 800°C compared to corresponding B implanted samples. A similar trend is also observed in Figure 3, where the mobility of the samples implanted with BF_2 is consistently higher than those implanted with B.

In Figure 1 the B implanted samples show the classic deactivation reactivation behavior which is sometimes referred to as the "reverse annealing" effect. The "reverse annealing" is thought to be associated with the dissolution of the EOR defects and the release of silicon self interstitials which flow towards the surface, encountering the B profile and causing deactivation by clustering with the B atoms to form BICs. For the BF_2 implanted samples there is little evidence of the "reverse annealing" effect and Rs decreases with increasing annealing temperature. Thus one might speculate that the presence of F maybe inhibiting BIC formation.

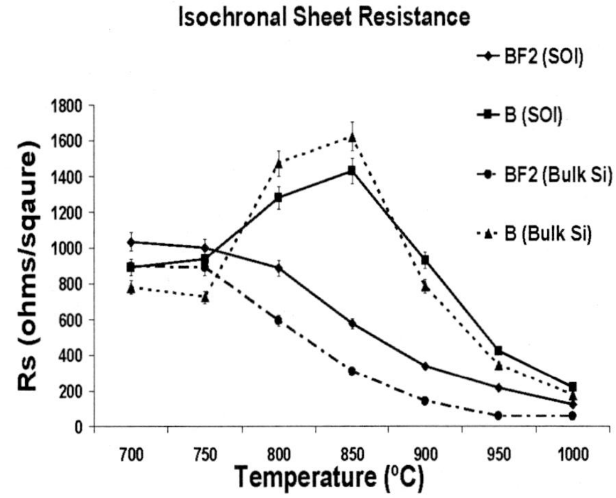

FIGURE 1. Van der Pauw resistivity measurements for samples implanted with a pe-amorphising implant of 20 keV Ge 1×10^{15} cm^{-2} in Si and SOI and then either 500 eV B 2×10^{15} cm^{-2} or 2.2keV BF_2 2×10^{15} cm^{-2}. Isochronal anneals were carried out for 60 s.

In Figure 1, for the B implanted samples the peak of the deactivation occurs at around 850°C, for both Si and SOI samples. Above 850°C, Rs decreases monotonically up to the highest annealing temperature of 1000°C. This trend corresponds to a dopant reactivation process which is thought to be due to the dissolution of Boron-Interstitial Clusters.

Comparing the Rs values for the B implant in SOI with those in Si one can see that the values of Rs in SOI at both 800°C and 850°C are lower than those in bulk Si. Figure 2 shows that this corresponds, as expected with a higher value of the areal "carrier concentration". These results agree with those of Hamilton *et al.*, [6] who showed that as the EOR defect band approaches the BOX silicon interface, this acts as an alternative sink in addition to silicon surface, for migrating interstitials. The results for B implanted samples in this study are in close agreement with those of Hamilton *et al.* [6] and therefore we might conclude that similar mechanisms are occurring.

Areal Carrier "concentration" values, shown in Figure 2, show higher values for the BF_2 implants above 800°C. It is also interesting to note that as the temperature is increased the values for the bulk silicon and SOI substrates diverge for the BF_2 implants. The values for the bulk substrates are consistently higher than those for the SOI in terms of areal "carrier concentration". This reflects the fact that in bulk Si

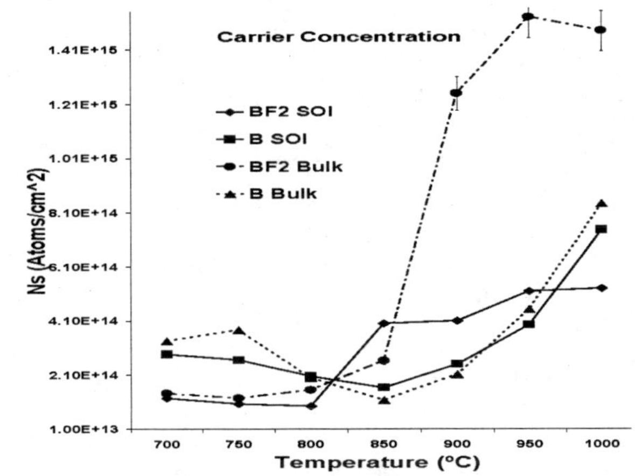

FIGURE 2. Areal carrier "concentration" versus temperature for a pe-amorphising implant of 20 keV Ge 1x 10^{15} cm^{-2} in Si and SOI and then either r 500 eV B 2 x 10^{15} cm^{-2} or 2.2 keV BF_2 2x 10^{15} cm^{-2}. Isochronal anneals were carried out for 60 s.

the B is free to diffuse while in SOI the BOX acts as an effective diffusion barrier and some B "pile up" is observed at the BOX interface.

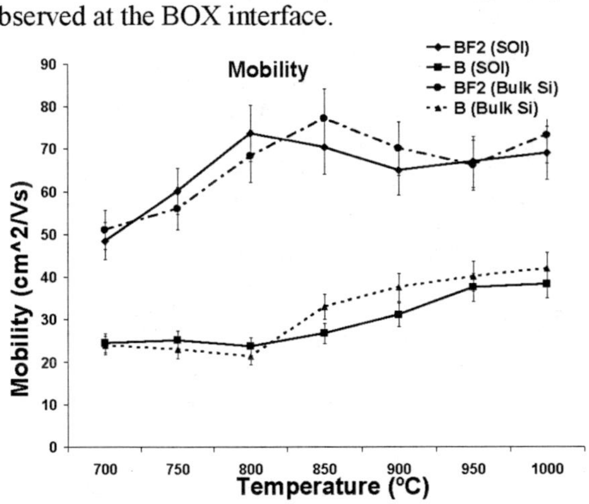

FIGURE 3. Hall Mobility versus temperature for a pre-amorphising implant of 20 keV Ge 1x 10^{15} cm^{-2} in Si and SOI and then either r 500 eV B 2 x 10^{15} cm^{-2} or 2.2 keV BF_2 2x 10^{15} cm^{-2}. Isochronal anneals were carried out for 60 s.

As has already been stated the most distinctive feature in the mobility results (Figure 3) is the difference between the Hall motilities for the B and BF_2 implants, with the BF_2 implant showing mobility values 2 to 3 times higher than those for the boron. The results so far show a snapshot, after 60 s, of how the EOR defects interact with the B and BF_2 implants in Si and SOI materials. To examine the influence of time, isothermal anneals were carried out at 800°C and 850°C for times between 10 and 1000 seconds. 800°C was chosen since this is the temperature where the "reverse annealing effect" is first observed for the isochronally annealed samples. Likewise 850°C was chosen as this corresponds to the peak of the deactivation. The results for isothermal anneal at 800°C and 850°C are shown in Figures 4 and 5 respectively.

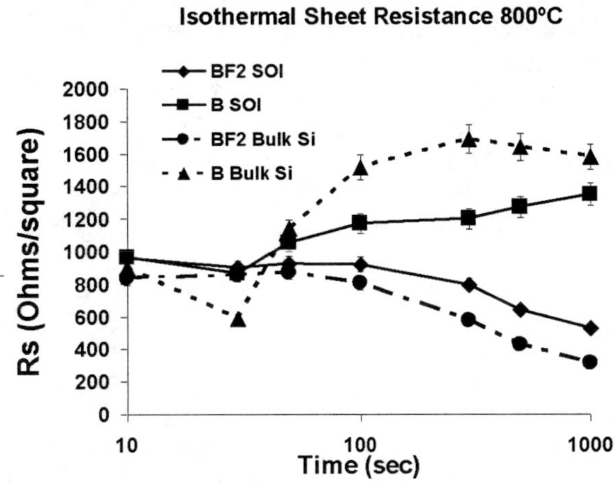

FIGURE 4. Sheet resistance versus annealing time for a pre-amorphising implant of 20 keV Ge 1x 10^{15} cm^{-2} in Si and SOI and then either r 500 eV B 2 x 10^{15} cm^{-2} or 2.2 keV BF_2 2x 10^{15} cm^{-2}. Isothermal anneals were carried out at 800°C.

In Figure 4, once again a clear difference is observed between the Rs values for the B and BF_2 implants. For the B implants the effects of the "reverse annealing" are more apparent for the Si than for the SOI and this may be because the BOX interface is acting as an alternative sink [6]. For the BF_2 implants the lowest value of Rs is observed in the bulk silicon,

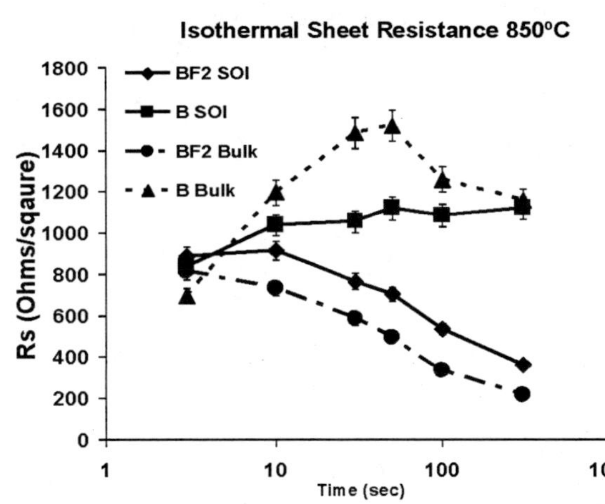

FIGURE 5. Sheet resistance versus annealing time for a pe-amorphising implant of 20keV Ge 1x $10^{15}$$cm^{-2}$ in Si and SOI and then either r 500 eV B 2 x 10^{15} cm^{-2} or 2.2keV BF_2 2x 10^{15} cm^{-2}. Isothermal anneals were carried out at 850°C.

especially at longer times. This ties in with the higher levels of dopant activation observed in Figure 2 and is probably a result of B diffusion and activation into the substrate.

The trends observed in Figure 4 become even more marked in Figure 5. Where the isothermal anneal was undertaken at 850°C to coincide with the peak in the "reverse annealing" profile (Figure 1). The peak deactivation for B implant occurs at shorter annealing times in this case.

The trends in Rs, between B and BF_2 in Figures 1, 4, and 5 closely resemble the results obtained by Cowern *et al.* [15], where a 30 keV Ge pre-amorphisation implant was used followed by 10 keV or and 22 keV fluorine implants, co-implanted with 500 eV B. Likewise the results for B in Si and SOI substrates agree with those reported by Hamilton *et al.* [6] for similar implantation conditions.

It can be postulated that the presence of the fluorine in the BF_2 implanted samples may inhibit BIC formation in a similar manner to that proposed in [15]. It has been proposed that the fluorine vacancy effectively traps the interstitials migrating from the end of range defects towards the silicon surface. Thus for the BF_2 implants the beneficial effects of the SOI observed for B only implants are lost due to the effects of the fluorine. Further experiments are required to determine the electrical leakage of such samples and to further investigate the role of F in these samples.

In summary, the BOX interface in SOI appears to have a beneficial effect for B only implants as it reduces the amount of deactivation observed during subsequent annealing. This is thought to be because the BOX interface provides an alternative sink for interstitials migrating from the EOR defect band. Likewise, the presence of fluorine is postulated to suppress BIC' formation by effectively trapping the silicon interstitials as they migrate towards the silicon surface.

ACKNOWLEDGEMENTS

M. Kah would like to acknowledge the support of the UK Engineering & Physical Sciences Research council for supporting his PhD studentship. The authors would also like to thank the IAEA for their support.

REFERENCES

1. International Technology Roadmap for Semiconductors ITRS Update Overview. 2006.
2. J. Liu and S. Mehta. *Mat. Res. Soc. Symp.* **610**. p B4.6.1.
3. A. Claverie, L. Laanab, C. Bonafos, C. A. Martinez, D. Mathiot. Nucl. Instr. *Nucl. Instr. and Meth. in Phys. Res. B* **96** (1995).
4. D. J. Eaglesham, P. A. Stolk, H.-J. Gossmann, and J. M. Poate. *Appl. Phys. Lett.* **65** (18), (1994).
5. L. Pelaz, G. H. Gilmer, H.-J. Gossmann, C. S. Rafferty, M. Jaraiz and J. Barbolla, , *Appl. Phys. Lett.*, **74**(24) p. 3657, (1999).
6. J. J. Hamilton N. E. B. Cowern, E. J. H. Collart, M. Bersani, D. Giubertoni, S. Gennaro K. J. Kirkby *Appl. Phys. Lett.* **91**, 092122 (2007).
7. J. M. Jacques K. S. Jones, L. S. Robertson, A. Li-Fatou, and C. M. Hazelton E. Napolitani, L. M. Rubin *J. Appl. Phys.* **98**, p (2005).
8. J. M. Jacques, L. S. Robertson, and K. S. Jones. *Mat. Res. Soc. Symp. Proc.* **717** p C4.6.1.
9. G. Impellizzeri, J. H. R. dos Santos, S. Mirabella, F. Priolo *Mat. Res. Soc. Symp. Proc.* **810** 2004.
10. R. G. Wilson, *J. Appl. Phys.*, vol. **54**,. p. 6879 (1983).
11. D. F. Downey, J. W. Chow, E. Ishida, and K. S. Jones, *Appl. Phys. Lett*, vol. **73**, pp. 1263-5, 1998.
12. J.A. Sharp , R.M. Gwilliam, B.J. Sealy, C. Jeynes, J.J. Hamilton, K.J. Kirkby. *Nucl. Instr. and Meth. in Phys. Res. B* **237** 93–97. 2005.
13. T.H. Huang, H. Kinoshita, D.L. Kwong, *Appl. Phys. Lett.* **65** (14) (1994) 1829.
14. SUSPRE software developed at the University of Surrey.
15. N. E. B. Cowern, B. Colombeau, J. Benson, and A. J. Smith, W. Lerch, S. Paul, and T. Graf, F. Cristiano and X. Hebras, D. Bolze. . *Appl. Phys. Lett.* **86**, 101905. (2005).

USJ Process Challenges for sub-45 nm CMOS

[a,1]Pankaj Kalra, [a,b]Prashant Majhi, [a]Hsing-Huang Tseng, [a]Larry Larson, [a,c]Raj Jammy, and [1]Tsu-Jae King Liu

[a]SEMATECH, Austin TX 78741
[1]University of California, Berkeley, CA 94720
[b]Intel, [c]IBM

Abstract. A review of the challenges associated with the formation of ultra-shallow junctions is presented. Flash annealing appears to be a promising candidate to replace conventional spike annealing for future CMOS generations. The process integration of high-κ/metal gate with flash annealing is also presented. It is shown that flash annealing retains high-κ/metal gate stack integrity, while achieving USJ benefits. However, it results in degraded interface quality which is improved by using a post-metallization anneal.

Keywords: CMOS scaling, ultra-shallow junctions, USJs, flash anneal, high-κ, metal gate
PACS: 85.30.Tv; 81.05.Cy; 85.40.Ry; 81.40.Ef; 85.50.-n; 81.05.By

INTRODUCTION

Transistor scaling has been the driving force for technology advancements in the semiconductor industry over the last few decades. In order to mitigate short channel effects, the gate-oxide thickness and source/drain junction depth have been scaled along with the gate length. Recently, however, gate-oxide-thickness scaling has slowed, as evidenced by the fact that an equivalent oxide thickness (EOT) of ~1 nm has been used for the past 2-3 generations of CMOS technology. Although significant progress has been made in the development of high-permittivity (high-κ) gate-dielectric materials and metal gate technology in recent years, it will be difficult to scale EOT well below 1 nm. This makes junction-depth scaling even more pressing for continued transistor scaling.

Future CMOS transistors (beyond the 45 nm technology node) should have ultra-shallow junctions (USJs) that are ~12-15 nm deep, with sheet resistance ~1000 Ω/□ in order to keep pace with historical advancements in high-performance logic devices [1]. These requirements are particularly difficult to meet for *p+/n* junctions (as compared to *n+/p* junctions). This paper therefore first discusses the formation of *p+* USJs.

Since high-κ/metal-gate stacks are already used in the most advanced 45 nm CMOS technology today, it is important to assess the compatibility of advanced annealing processes such as flash annealing with high-κ/metal-gate stacks. There are many reports on the impact of laser annealing or flash annealing for devices with poly-Si/silicon-oxide-based gate stacks, but relatively few reports for devices with high-κ/metal-gate stacks [2-7]. This paper therefore also discusses the impact of flash annealing on MOSFETs with Hf-based gate-dielectric and metal gate electrodes.

FORMATION OF P+ USJs

Blanket P+ USJ formation studies were performed using lightly doped (~10^{15} cm^{-3}) *n*-type Si (100) wafer substrates. Secondary Ion Mass Spectroscopy (SIMS) was used to measure the junction depth (X_j), taken to be the depth at which the dopant concentration falls to 5×10^{18} cm^{-3}. Sheet resistance (R_s) was measured by a non-contact method for improved accuracy [8]. The conventional method (four-point probe) yields erroneously low values of R_s due to leakage caused by probe-tip penetration to a depth lower than X_j. The error increases exponentially as X_j decreases, exceeding 1000 Ω/□ for X_j < 15 nm [9].

CP1066, *Ion Implantation Technology,* edited by E. G. Seebauer, S. B. Felch, A. Jain, and Y. V. Kondratenko

Experimental Results with Spike Anneal

Various combinations of low-energy boron implants (BF_2: 0.5-1.0 keV) and co-implants (N and F) were used to form *p+/n* junctions. A pre-amorphization Ge implant (PAI Ge) was done prior to boron implantation in order to eliminate channeling. The implant energies of these co-implanted species were chosen such that their projected ranges (R_p) are in-between the projected range of the implanted boron and the depth of the amorphous/crystalline interface. A spike anneal (1070°C) was used to activate the implanted dopants.

Figure 1 summarizes the experimental results obtained with spike annealing with and without co-implantation for *p+/n* junctions. These results indicate that it may be possible to achieve junction depths of ~20 nm, but this comes at the expense of higher sheet resistance. Furthermore, it will be increasingly difficult to meet the low sheet resistance requirement for X_j below 20 nm.

Experimental Results with Flash Anneal

A separate set of samples was used to form *p+/n* junctions by BF_2 implantation (0.8-1.0 keV, 5×10^{14}-2.5×10^{15} cm^{-2}, 7°). Again, PAI Ge was done to eliminate channeling. The implanted dopants were activated by flash annealing [10]. In a flash annealing process, the bulk of the wafer is first ramped up to an intermediate temperature (T_{int}). Then, only the device side of the wafer is heated to the peak anneal temperature (T_{peak}) by flash lamps; the bulk of the wafer acts as a thermal sink to rapidly cool the device-side surface.

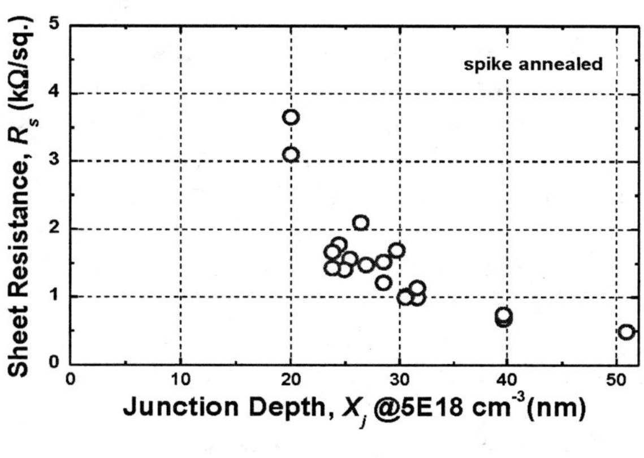

FIGURE 1. Sheet resistance (R_s) determined by non-contact method *vs.* junction depth (X_j) for *p+/n* junctions formed by spike annealing (1070°C).

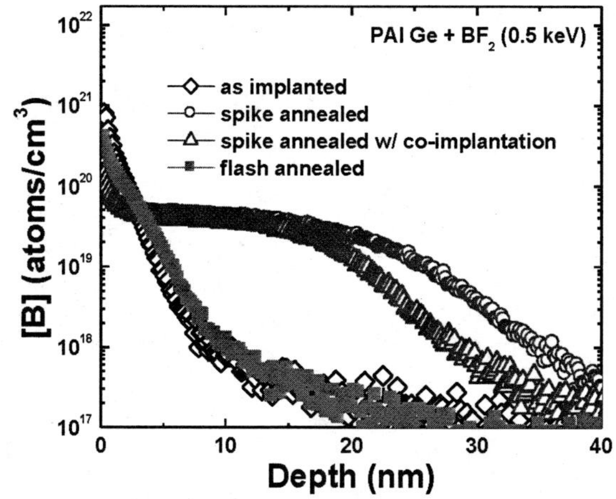

FIGURE 2. Boron profiles in samples which received a low-energy boron (0.5 keV) implant. Negligible diffusion is observed for the flash-annealed samples compared to the spike-annealed samples (with or without co-implantation).

SIMS analyses of boron (0.5 keV BF_2) implanted samples before and after flash annealing are shown in Figure 2. The profiles for spike-annealed samples are also plotted for reference. Negligible dopant movement is observed for the flash-annealed sample, as compared to the spike-annealed samples. Clearly, flash annealing reduces dopant diffusion as compared with spike annealing, and thus it is a promising technique for USJ formation for sub-45 nm nodes.

Various combinations of flash annealing parameters (T_{int}, T_{peak}, and n, where n is the number of flash shots) were also included in this study to elucidate interdependencies. The time in-between flash anneals was relatively long for the samples which received multiple flash anneals. The combination of T_{int} = 700°C and T_{peak} = 1300°C appears to be optimal for achieving shallow X_j with low R_s [9].

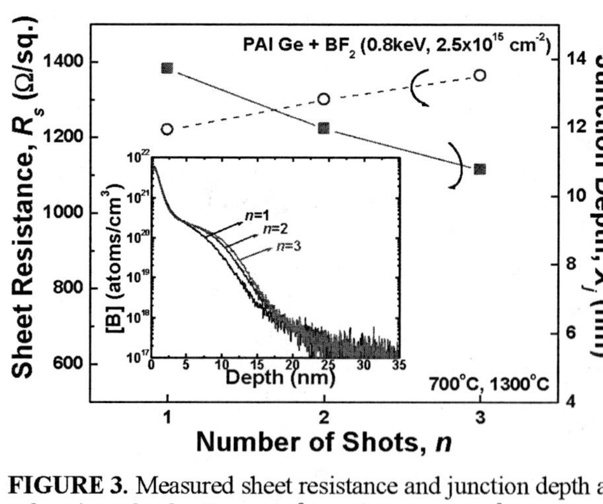

FIGURE 3. Measured sheet resistance and junction depth as a function of n for T_{int} = 700°C and T_{peak} = 1300°C. **Inset:** B profile evolution with the number of flash shots.

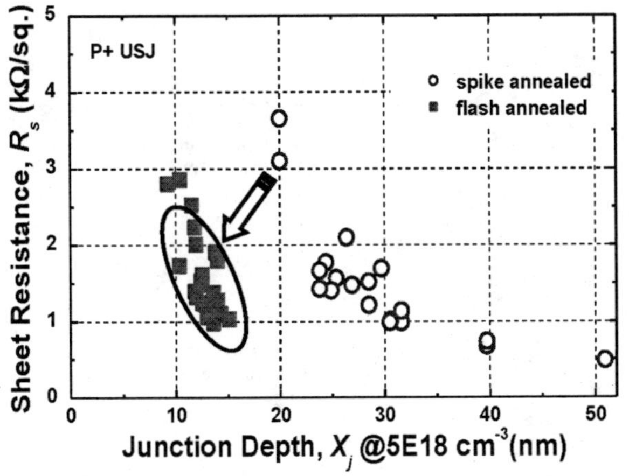

FIGURE 4. R_s-X_j summary for *p+/n* junctions formed by spike annealing (1070°C) or flash annealing. Flash annealing provides for less dopant diffusion and more dopant activation.

Figure 3 shows the impact of multiple shots of flash annealing. A 20% decrease in R_s is achieved as n is increased from 1 to 3. However, the junction depth is also increased, by ~1.5 nm. Therefore, multiple shots of flash annealing can be used to further reduce R_s with a slight trade-off in X_j.

Figure 4 summarizes the P+ USJ experimental results, which show that it is possible to achieve X_j in the range ~12-15 nm with R_s ~1kΩ/□ via flash annealing, which is adequate for 32 nm technology. Flash annealing not only reduces dopant diffusion but also improves dopant activation, as compared with spike annealing.

A challenge for advanced annealing techniques is annihilation of implantation-induced damage, because of their ultra-short anneal times. Figure 5 shows junction leakage measured by a non-contact method for various *p+/n* junctions that underwent either spike or flash annealing. (PAI Ge was used for all the junctions.) It can be seen that the flash-annealed junctions have higher leakage than the spike-annealed junctions, by ~100×. However, with F co-implantation, the junction leakage for the flash-annealed samples can be reduced significantly (by 5× to 135×, depending upon the initial implanted boron profile). The F implantation energy was selected so that most of the F dose lies in-between the boron and end-of-range (EOR) damage band interface. It is believed that F passivates the defects present in the EOR band.

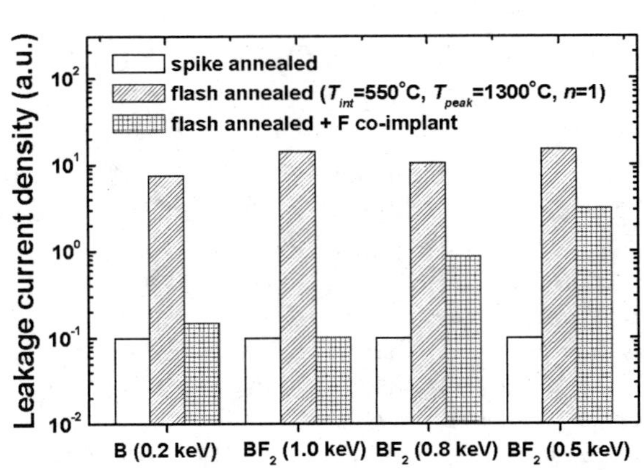

FIGURE 5. F co-implantation is effective for reducing the leakage of flash-annealed *p+/n* junctions.

Experimental Results with GCIB Doping

Using 1%B_2H_6 as the source gas, boron was introduced by a 5-kV or 3-kV gas-cluster ion beam (GCIB) doping process with a dose of 10^{14} cm^{-2} to form *p+/n* junctions. A summary of R_s *vs.* X_j for flash-annealed junctions formed by GCIB doping is shown in Figure 6 [11]. Clearly, 32-nm node USJ requirements (X_j ~ 12-15 nm, R_s ~ 1 kΩ/□) can be met with infusion doping and flash annealing.

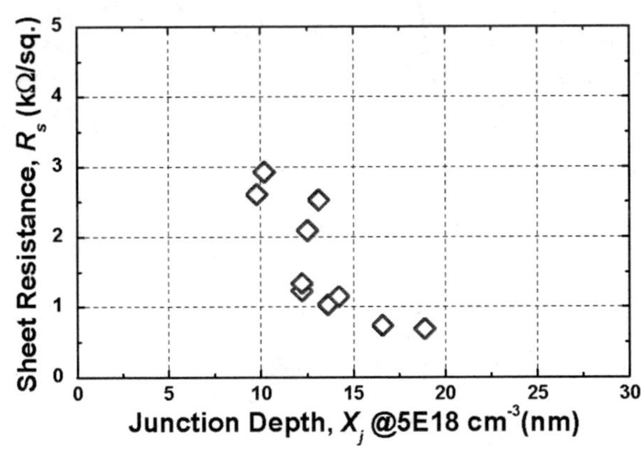

FIGURE 6. Sheet resistance (R_s) determined by non-contact method *vs.* junction depth (X_j) for *p+/n* junctions formed by GCIB doping and flash annealing.

PROCESS INTEGRATION OF HIGH-κ/ METAL-GATE WITH FLASH ANNEAL

The effects of flash annealing on the performance and reliability of MOSFETs with HfSiON/TiN gate stacks are discussed in this section [12]. A conventional gate-first CMOS process flow was used to fabricate devices with sub-100 nm gate length, as shown in Figure 7. After STI and well formation, the gate-dielectric stack was formed by depositing a 3 nm-thick HfSiO layer by ALD onto ~1 nm thick SiO_x, followed by a post-deposition anneal in NH_3. For the gate electrode, 10 nm of TiN was deposited by ALD and capped by 100 nm thick LPCVD poly-Si for ease of process integration. Source/drain-extension (SDE) and halo implants were performed next, followed by gate-sidewall spacer formation and deep-S/D implants. A spike anneal (1070°C) or flash anneal (T_{int}=800°C, T_{peak}=1300°C) was used to activate the implanted source/drain dopants.

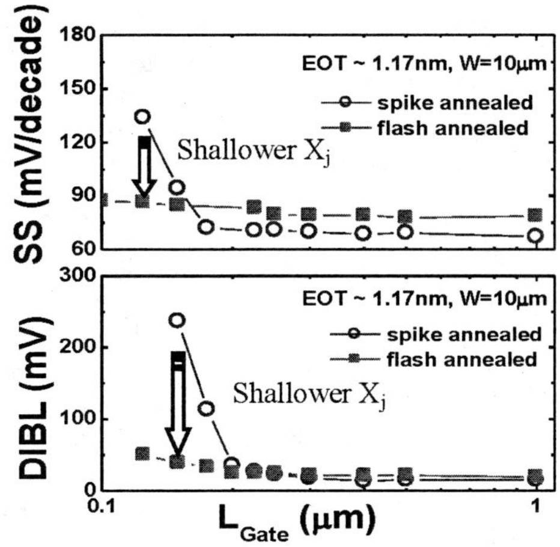

FIGURE 8. Measured DIBL and SS values for nMOSFETs fabricated using either spike or flash annealing to activate the implanted S/D dopants.

SCE Control

The sub-threshold swing (SS) and drain-induced barrier lowering (DIBL) of the transistors with different gate lengths are shown in Figure 8. It can be seen that short-channel effects (SCE) are well controlled down to 100 nm gate length for flash-annealed MOSFETs as compared with spike-annealed MOSFETs. The improved SCE can be attributed to shallower junction depth (X_j) in the flash annealed devices for identical SDE ion implantation conditions.

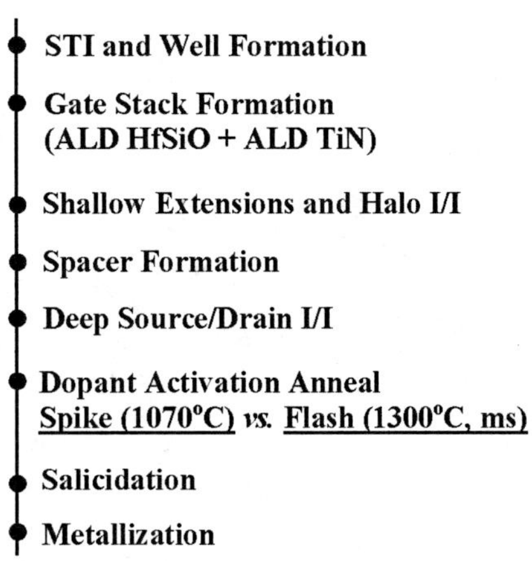

FIGURE 7. Gate-first CMOS process flow used to evaluate the impact of flash annealing on high-k/metal-gate stacks.

Gate-Stack Integrity

Figure 9(a) shows measured high-frequency (*f* = 100 kHz) *C-V* curve for nMOS capacitors. The EOT and the flat-band voltage (V_{FB}) values were extracted using the NCSU CVC program [13]. Comparable extracted values for EOT (~1.15 nm) and V_{FB} (~-0.52 V) are seen for the gate stacks exposed to flash *vs.* spike annealing. pMOS capacitors also exhibit similar features. The *C-V* curves for the flash annealed gate stacks show neither frequency dispersion nor hysteresis.

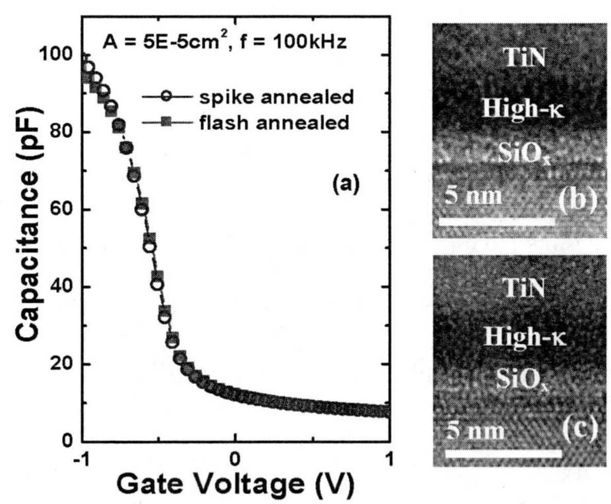

FIGURE 9. (a) Comparison of *C-V* curves for nMOS capacitors fabricated with spike or flash annealing; Cross-section HRTEM images — (b) spike annealed gate stack; (c) flash-annealed gate stack.

The physical structure of the gate stacks after S/D-activation annealing was analyzed using transmission electron spectroscopy (TEM). Figure 9 (b) & (c) show cross-sectional high resolution TEM (HRTEM) images for spike-annealed and flash-annealed gate stacks. These images indicate that there is negligible difference in the physical thicknesses of the HfSiON film and interfacial oxide layer for spike- *vs.* flash-annealed gate stacks. The measured thicknesses of the HfSiON and SiO_x layers are ~3 nm and ~1 nm, respectively.

Figure 10 shows the relationship between gate current density at $V_{GS} = V_{FB}{-}1$ V and EOT. For reference, the SiO_2/poly-Si gate leakage model is shown. The flash-annealed HfSiON/TiN gate stack data point falls is in agreement with the trendline for spike-annealed gate stacks, providing >500× leakage reduction at 1.15 nm as compared against a SiO_2/poly-Si gate stack.

Fast transient charging (FTC) in high-κ dielectrics is one of the major high-κ process integration issues. FTC has been shown to be a major contributor to low mobility, poor reliability, and V_{TH} instability [14-17]. In order to study the impact of flash annealing on charge trapping, single-pulse I_{DS}-V_{GS} measurements were performed. The results are shown in Figure 11. Minimal drop in I_{DS} over the entire pulse duration is observed for the flash-annealed device, indicating negligible charge trapping. This was further confirmed by long-time positive bias temperature instability (PBTI) testing.

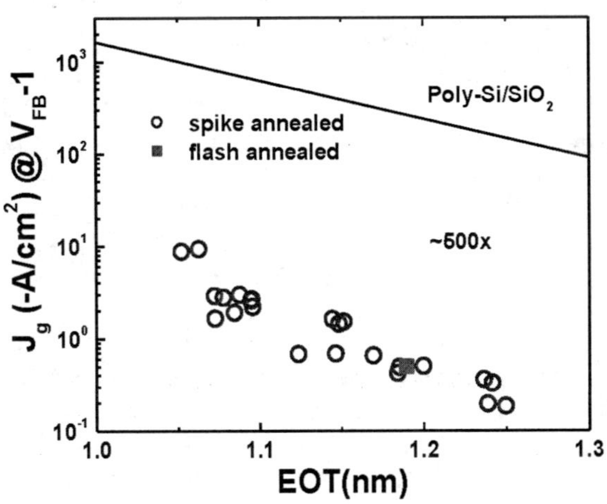

FIGURE 10. $J_{g,accumulation}$ *vs.* EOT for TiN/HfSiON capacitors fabricated with spike- or flash-annealing. Leakage current density is on par with the historical high-κ trend line.

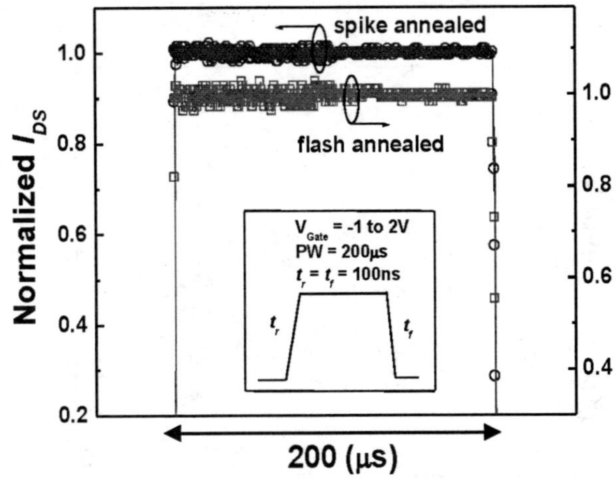

FIGURE 11. Single-pulse I_{DS}-V_{GS} measurements show insignificant ΔI_{DS} for both spike- and flash-annealed gate stacks, indicating negligible fast transient charge trapping. PW, t_r, and t_f correspond to the pulse width, pulse rise, and fall times, respectively.

Therefore, it can be concluded that the flash annealing process has minimal effect on gate-stack properties such as EOT, V_{FB}, and leakage, so it appears to be compatible with high-κ/metal-gate stacks. Furthermore, it is evident that bulk charge trapping in flash-annealed high-κ gate dielectrics is not an issue.

MOSFET Performance

Linear peak transconductance (G_m) as a function of gate length for spike- and flash-annealed nMOSFETs (W=10 µm) is shown in Figure 12. Significantly lower G_m is observed for flash-annealed devices as the gate length is scaled down to 100 nm; ~1.8× reduction is seen at L_g = 100 nm.

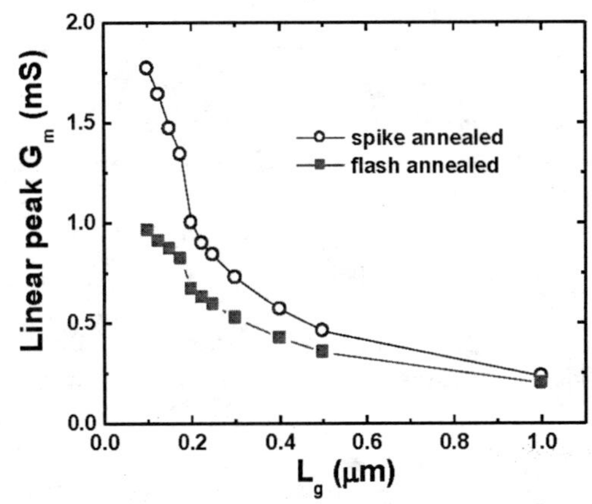

FIGURE 12. Effect of S/D activation anneal on G_m, plotted as a function of gate length, L_g. (W = 10 µm, V_{DS} = 50 mV).

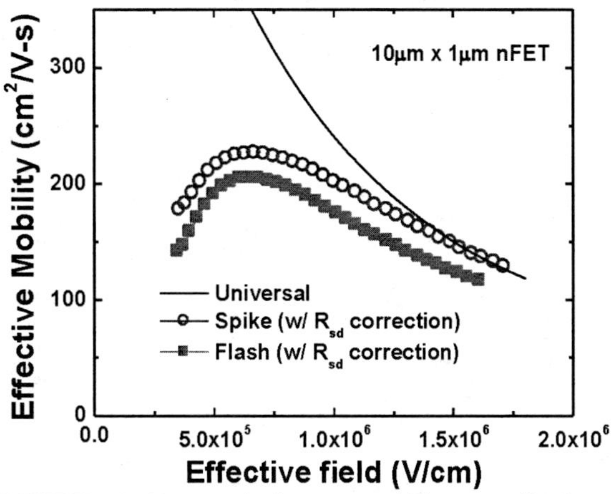

FIGURE 13. Extracted electron mobility *vs.* effective transverse electric field. Mobility is degraded for nMOSFET fabricated with flash annealing *vs.* spike annealing.

The effective electron mobility was extracted using the NCSU MOB2D model [18]. DC I_{DS}-V_{GS} measurements were done on 10×1 μm^2 nMOSFETs and were corrected for series resistance, R_{sd}. Both the peak and high-field mobility values are degraded for the flash-annealed device, as seen in Figure 13. For reference, the universal electron mobility model is also shown. Clearly, the performance of flash-annealed devices is degraded as compared with that of spike-annealed devices.

Fixed-amplitude charge-pumping (CP) measurements were done on 10×1 μm^2 nMOSFETs to understand the cause of mobility degradation. Figure 14 shows the measured interface trap density (N_{it}) *vs.* frequency. The density of traps at high frequencies, which correspond to traps located near to the substrate interface, is higher in the flash-annealed device. A larger slope of the N_{it} *vs.* frequency characteristic is also observed for the flash-annealed gate stack, indicating a higher density of traps away from the substrate interface (toward the HfSiON/SiO_x interface) in the flash-annealed device. Hence, the likely cause of the mobility loss seen in the flash-annealed device is higher interface state density.

Wafer-bow measurements performed before and after flash annealing help to explain the origin of the higher N_{it}. A large change in the wafer radius of curvature was observed, indicating wafer-level stress after flash annealing. This large change can be attributed to the significant thermal gradient through the thickness of the wafer during flash annealing, since only the device side of the wafer is heated to the peak temperature. Similar stress behavior has been reported for a pulsed laser annealing process [6].

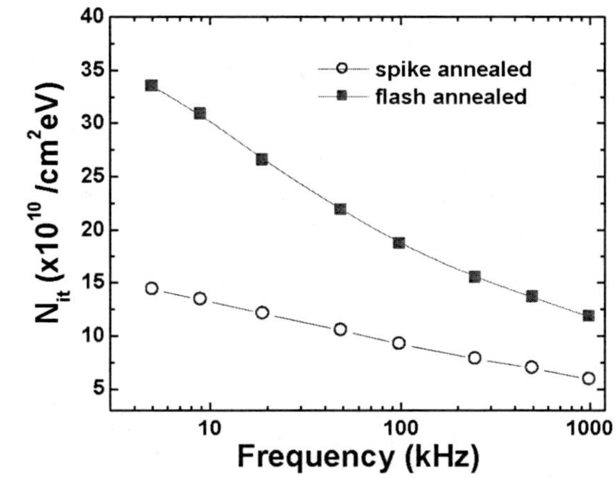

FIGURE 14. Charge pumping measurements of interface trap density (N_{it}) at various frequencies. Higher N_{it} and a significant difference in trap density profile are observed for the flash-annealed gate stacks as compared to the spike-annealed gate stacks.

At high frequencies, the observed N_{it} increase is probably caused by thermal-stress-induced defects (broken bonds) at the Si/SiO_x interface. The larger slope of the N_{it} *vs.* frequency characteristic at lower frequencies is indicative of a higher density of high-κ induced traps in SiO_x, which are likely to be oxygen vacancies [19].

Effects of Post-Metallization Anneal

As discussed above, flash annealing appears to be compatible with high-κ/metal-gate stacks, but a degraded Si interface is seen for flash-annealed FETs, resulting in mobility loss. This mobility degradation is a major roadblock for the adoption of flash annealing. In this section, the effects of a post-metallization annealing (PMA) process on the interface properties of flash-annealed MOSFETs are presented. The PMA was done after a conventional forming gas anneal.

Linear drain current (I_{DS}) and transconductance (G_m) versus gate voltage (V_{GS}) of a 10 μm ×1 μm nMOSFET before and after PMA are shown in Figure 15. Measurements were taken with V_{DS} = 50 mV. After the PMA, ~7-8% improvement is observed in I_{DS} and G_m. No discernible differences in EOT or V_{FB} are seen from the measured *C-V* curves before and after PMA.

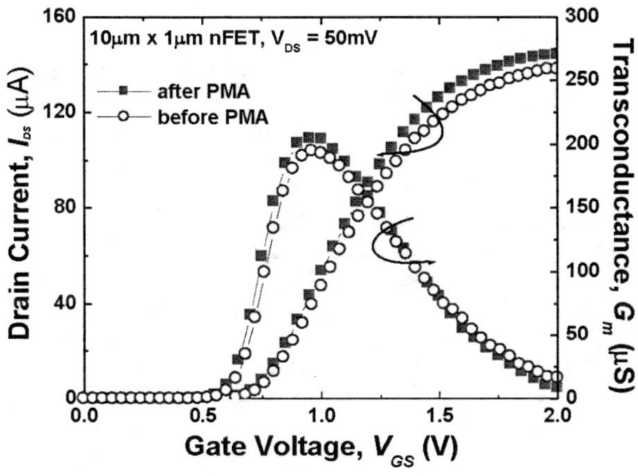

FIGURE 15. Samples annealed in the PMA show improved linear drain current and transconductance (V_{DS} = 50 mV).

The DC I_{DS}-V_{GS} measurements shown in Figure 15 were corrected for series resistance, then the effective electron mobility was extracted using the NCSU MOB2D model. It is found that PMA is effective for recovering mobility loss, as shown in Figure 16. Significant improvement is seen in the peak mobility, which is governed by interface quality. The improved interface quality is further evidenced by a lower sub-threshold swing for a 10×1 μm^2 nMOSFET (Figure 16 inset). The mobility recovery is attributed to a reduction in the interface trap density. However, for high transverse electric field, the gain in mobility diminishes, which indicates that the PMA only partially recovers mobility loss due to surface roughness scattering. Further optimization of the PMA conditions is necessary for further mobility gain.

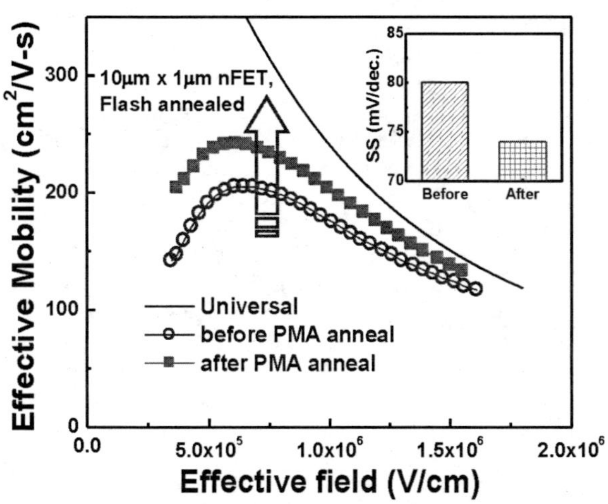

FIGURE 16. Mobility in a flash-annealed device is improved by PMA. Inset: The SS value is also lowered after PMA, for a 10 μm × 1 μm nMOSFET.

Challenges for Flash Annealing

As noted previously, the efficacy of flash annealing to annihilate implantation damage is a primary concern because of the ultra-short annealing time. Figure 17 shows the cumulative probability plot for *n+/p* junction leakage current density. It is seen that flash-annealed junctions leak more as compared to spike-annealed junctions, by ~10×. The leakage is a sensitive function of the location of residual effects with respect to the junction depletion region, which in turn depends on the S/D-extension and halo implant profiles. Therefore, careful optimization of the implantation conditions for the S/D extensions and halos are needed to reduce the junction leakage.

Figure 17 inset shows the gate-overlap capacitance (C_{ov}) distributions for flash- and spike-annealed devices. 20% reduction in C_{ov} is seen for flash-annealed devices. This reduction can be attributed to the shorter anneal time for flash annealing. It has been reported that a minimum length of overlap region (L_{ov}) is needed to avoid hot-carrier injection in the ungated region, which leads to reliability problems. Therefore, the small thermal budget of the flash annealing process may become an issue for highly-scaled CMOS devices, because it might not result in enough overlap between the gate electrode and underlying source/drain regions. To counter this problem, a low-temperature anneal can be used prior to flash annealing to diffuse the implanted S/D dopants to satisfy the L_{ov} requirement. The upper limit of the thermal budget for this anneal will be set by USJ requirements. This additional anneal can also help to annhiliate implantation damage, mitigating the increase in junction leakage.

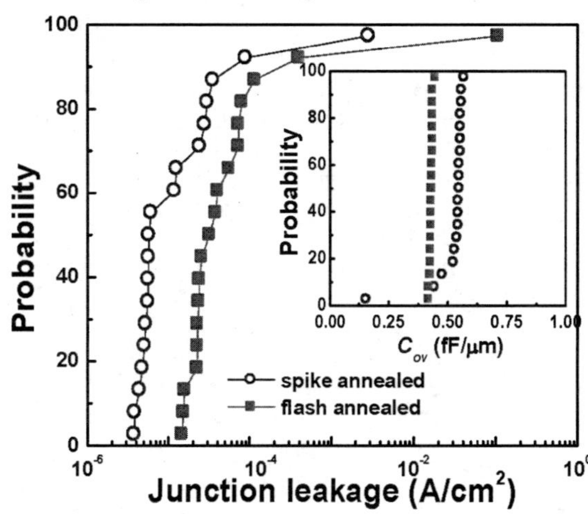

FIGURE 17. Junction leakage of *n+/p* diode after spike- or flash-annealing. **Inset:** Gate-overlap capacitance for flash- and spike-annealed MOSFETs.

CONCLUSIONS

This paper reviewed recent progress made towards the formation of USJs with the help of advanced annealing techniques, low-energy implants, and GCIB doping. Our experimental results obtained with flash annealing indicate that it is possible to achieve junction depths in the range of ~12-15 nm with sheet resistance of ~1 kΩ/□. This is considered to be adequate for the 32 nm technology node.

The process integration of high-κ/metal-gate stacks with flash annealing was discussed next. The flash annealing process was found to be compatible with the high-κ/metal-gate stack in terms of gate stack integrity, and to be effective for suppressing short-channel effects. Generation of O vacancies within the interfacial SiO_x layer due to thermal stress and high-temperature processing results in higher interface-state density and hence degraded mobility, however. A post-metallization annealing process is shown to be effective for recovering the mobility loss.

ACKNOWLEDGEMENTS

The authors would like to thank Mattson Technology for performing the flash annealing and Frontier Semiconductor for carrying out non-contact sheet resistance measurements.

REFERENCES

1. International Technology Roadmap for Semiconductors, 2006 Update.
2. T. Ito, K. Suguro, T. Itani, K. Nishinohm, K. Matsuo and T. Saito, *Symposium on VLSI Technology,* 53-54 (2003).
3. K. Adachi K. Ohuchi, N. Aoki, H. Tsujii, T. Ito, H. Itokawa, K. Matsuo, K. Suguro, Y. Honguh, N. Tamaoki, K. Ishimaru, and H. Ishiuchi, *Symposium on VLSI Technology*, 142-143 (2005).
4. A. Shima, Y. Wang, D. Upadhyaya, L. Feng, S. Talwar, and A. Hiraiwa, *Symposium on VLSI Technology*, 144-145 (2005).
5. A. Pouydebasque, B. Dumont, S. Denorme, F. Wacquant, M. Bidaud, C. Laviron, A. Halimaoui, C. Chaton, J. D. Chapon, P. Gouraud, F. Leverd, H. Bernard, S. Warrick, D. Delille, K. Romanjek, R. Gwoziecki, N. Planes, S. Vadot, I. Pouilloux, F. Arnaud, F. Boeuf and T. Skotnicki, *IEEE International Electron Devices Meeting,* 679-682 (2005).
6. S. Severi, E. Augendre, S. Thirupapuliyur, K. Ahmed, S. Felch, V. Parihar, F. Nouri, T. Hoffman, T. Noda, B. O'Sullivan, J. Ramos, E. San Andrés, L. Pantisano, A. De Keersgieter, R. Schreutelkamp, D. Jennings, S. Mahapatra, V. Moroz, K. De Meyer, P. Absil, M. Jurczak, and S. Biesemans., *IEEE International Electron Devices Meeting,* 859-862 (2006).
7. D. C. Gilmer J. K. Schaeffer, W. J. Taylor, G. Spencer, D. H. Triyoso, M. Raymond, D. Roan, J. Smith, C. Capasso, R. I. Hegde, and S. B. Samavedam, *Proc. European Solid-State Device Research Conference*, 351-354 (2006).
8. V. N. Faifer, M. I. Current, T. Nguyen, T. M. H. Wong, and V. V. Souchkov, *Ext. Abs. the Fifth International Workshop on Junction Technology*, 45-48 (2005).
9. P. Kalra, P. Majhi, H.-H. Tseng, R. Jammy and T.-J. King Liu, *MRS Spring Meeting, Symp. E*, (2008).
10. W. Lerch S. Paul, J. Niess, S. McCoy, T. Selinger, J. Gelpey, F. Cristiano, F. Severac, M. Gavelle, S. Boninelli, P. Pichler, D. Bolze, *Mater. Sci. Engg.*, **124-125**, 24-31 (2005).
11. P. Kalra, P. Majhi, H.-H. Tseng, R. Jammy and T.-J. King Liu, *The 17th International Conference on Ion Implantation Technology*, (2008).
12. P. Kalra, P. Majhi, D. Heh, G. Bersuker, C. Young, N. Vora, R. Harris, P. Kirsch, R. Choi, M. Chang, J. Lee, H. Hwang, H. -H. Tseng, R. Jammy, and T. -J. King Liu, *IEEE International Electron Devices Meeting,* 353-356 (2007).
13. J. R. Hauser and K. Ahmed, *Characterization and Metrology for ULSI Technology: 1998 International Conference* **449**, 235-239, (1998).
14. E. Gusev and C. P. D'Emic, *Appl. Phys. Lett.* **83,** 5223-5225 (2003).
15. S. Zafar, A. Callegari, E. Gusev and M. Fischetti, *J. Appl. Phys.* **93,** 9298-9303(2003).
16. L. Pantisano, E. Cartier, A. Kerber, R. Degraeve, M. Rosmeulen, G. Groseneken and H. E. Maes, *Symposium on VLSI Technology*, 163-164 (2003).
17. A. Kerber, E. Cartier, L. A. Ragnarsson, M. Rosmeulen, L. Pantisano, R. Degraeve, Y. Kim, and G. Groeseneken, *Symposium on VLSI Technology*, 159-160 (2003).
18. J. R. Hauser, *IEEE Trans. Electron Devices* **43**, 1981-1988 (1996).
19. G. Bersuker C. S. Park, J. Barnett, P. Lysaght, P. D. Kirsch, C. D. Young, R. Choi, B. H. Lee, B. Foran, K. v. Benthem, S. J. Pennycook, P. M. Lenahan, and J. T. Ryan, *J. Applied Physics*, **100**, 094108- 094108-6 (2006).

Comparison of B, BF_2 & $B_{18}H_{22}$ for Extension and BF_2, $B_{18}H_{22}$ & In HALO Implantation for 32nm Node Using Various Diffusion-less Annealing Techniques

John Borland[1] and Hiroki Kiyama[2]

[1]*J.O.B. Technologies, 98-1204 Kuawa St., Aiea, HI 96701*
[2]*Dai-Nippon Screen, 322, Furukawa-cho, Hazukashi, Fushimi-ku, Kyoto,612-8486, Japan*

Abstract. We compared various p-type dopant species for source drain extension (SDE) and pocket/HALO implantation when using diffusion-less annealing techniques for ultra-shallow junctions (USJ) at the 32nm node. For SDE, 200eVB, 900eVBF_2 and 3.8keV$B_{18}H_{22}$ with and without 5keVGe-PAI were studied and for HALO, 20keVBF_2, 45keVIn and 80keV$B_{18}H_{22}$. Diffusion-less Xe-lamp Flash anneal (FLA), 900°C spike anneal, 700°C SPE anneal and their combinations were examined for junction quality (dopant activation, junction leakage and implant damage annealing recovery) using RsL, photoluminescence (PLi) and thermal-wave (TW) rapid process optimization metrology techniques. Ge-PAI improved dopant activation (Rs) for the Flash and SPE anneals but not with spike anneal and BF_2 had the highest Rs values due to retained dose limits. Also Ge+BF_2 resulted in junction leakage degradation but the 900°C spike anneal resulted in stable defects and complete implant damage recovery. For HALO, the dopant activation is dose limited for boron dopant species while for indium dopant it is solid solubility limited (anneal temperature dependent) so indium had the highest Rs values but no degradation in junction leakage were detected.

Keywords: Ultra-shallow junction, Source drain extension, HALO, sheet resistance, junction leakage current, thermal-wave, photoluminescence, Flash annealing, diffusion-less annealing, spike anneal.
PACS: 81.70.Ha and 85.40.Ry

INTRODUCTION

For the 32nm node sub-10nm ultra-shallow junction (USJ) will be required. Usually Ge-PAI is used to achieve such shallow junctions with monomer B or BF_2 dopant species to eliminate boron channeling and enhance dopant activation with msec annealing (MSA) [1]. However, incomplete implant damage annealing can lead to unstable defects that can cause dopant deactivation and junction leakage degradation especially when using MSA [2] and by adding a low temperature anneal step results in stable defects [3-5]. Another concern is reducing device variation so detecting implantation and annealing micro-uniformity variation is also critical for rapid process optimization at the 32nm node [5-7].

EXPERIMENTATION

To continue the initial studies by Mineji [8] on various p-type dopants for SDE and HALO with diffusion-less annealing and Borland [5] on defect reduction and micro-uniformity improvement with MSA reported at IWJT 2007, in this study we used 300mm n-type wafers for all the p-type dopant implantations. For SDE implantation we compared: 1) B=200eV/1E15/cm^2, 2) BF_2=900eV/1E15/cm^2, 3) $B_{18}H_{22}$=3.8keV/5.56E13/cm^2, 4) Ge-PAI+B and 5) Ge-PAI+BF_2. A 5keV/5E14/cm^2 implant was used for Ge-PAI. For HALO implantation we used a 30 degree tilt and compared: 1) BF_2=20keV/3E13/cm^2, 2) Indium=45keV/3E13/cm^2 and 3) $B_{18}H_{22}$=80keV/1.67E12/cm^2.

For diffusion-less anneals we used the Dai-Nippon Screen (DNS) Xe-lamp Flash annealer and the Axcelis Summit RTA annealer comparing Flash, 700°C SPE, 900°C spike RTA, Flash+spike, spike+Flash, Flash+SPE and SPE+Flash annealing sequences. The wafers were then measured by RsL at Frontier Semiconductor for sheet resistance and junction leakage current, by PLi (photoluminescence) at Nanometrics and by TW (thermal-wave) at KLA-Tencor for implant annealing damage recovery and defect stability.

CP1066, *Ion Implantation Technology*, edited by E. G. Seebauer, S. B. Felch, A. Jain, and Y. V. Kondratenko

RESULTS

Source Drain Extension

The sheet resistance (Rs) results are shown in Figure 1 for the various diffusion-less annealing sequences. With Flash only Ge-PAI improved dopant activation and reduced Rs by about 50% for both B and BF_2. The lowest Rs was seen for Ge+B followed by B then Ge+BF_2 next $B_{18}H_{22}$ and finally BF_2 had the highest Rs value. The high Rs value seen with BF_2 compared to B for the 1E15/cm^2 dose, probably due to retained dose limit and surface sputtering effects reported by Borland for junctions <10nm deep and shown in Figure 2 [7].

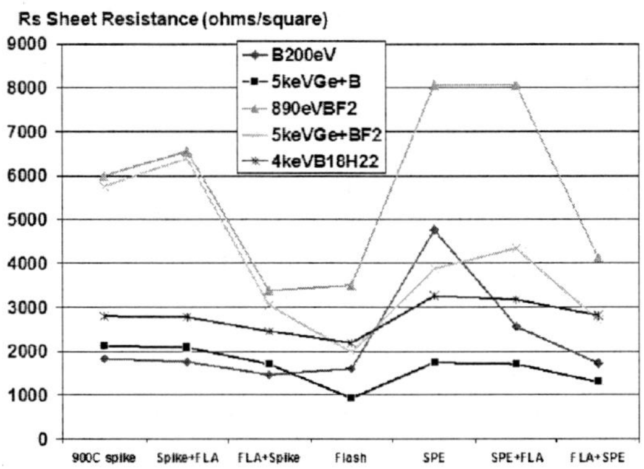

FIGURE 1. Rs sheet resistance results.

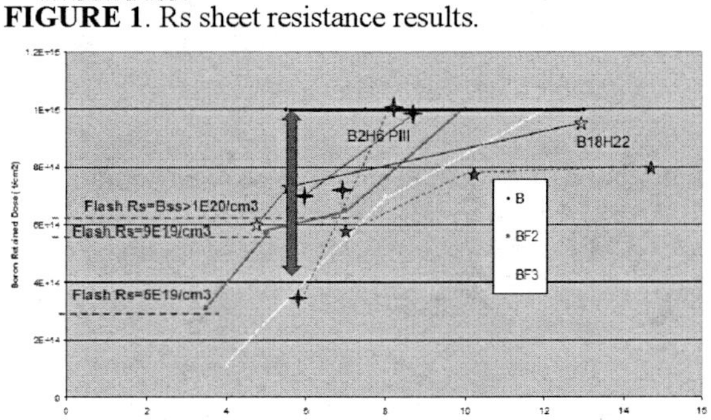

FIGURE 2. Boron retained dose limit issue for <10nm junctions with B, BF_2, BF_3, B_2H_6 & $B_{18}H_{22}$ [7].

Junction leakage results are shown in Figure 3 and with Flash anneal only the Ge+BF2 wafers showed junction leakage degradation of about 500x (>5E-5A/cm^2) compared to <1E-7A/cm^2 for the other wafers. These junction leakage results are better than those reported by Borland [5, 9, 10] using deeper Ge-PAI (10keV) implants with end of range (EOR) damage at 18nm depth. Mineji [11], Toda [4] and Camillo-Castillo [3] reported elimination of Ge-PAI EOR defects for energies <8keV but this was not the case with Ge+BF_2. The 900°C spike anneal combination recovered the leakage back to the very low level of <1E-7A/cm^2 for the Ge+BF_2 wafers suggesting complete implant damage anneal recovery. This is also verified by the TW results in Figure 4 and PLi results in Figure 5. Note that after Flash anneal the TW value for Ge+B is 167 and for Ge+BF_2 is 160 but with a Flash+spike or spike+Flash anneal sequence the TW values all drop to below 50. Similar effect of damage recovery was measured by PLi in Figure 5. With Flash only the Ge-PAI damage level is <300 PLi units so the residual implant damage from Ge+B must be shallower than the Ge+BF_2 case and the EOR damage within the junction resulting in low leakage <1E-7A/cm^2 for Ge+B compared to 5E-5A/cm^2 for Ge+BF_2 shown in Figure 3. With a spike+FLA or FLA+spike the PLi values increased to >600 for Ge+BF_2 and >1400 for Ge+B. So a spike anneal combination is critical for complete implant damage recovery. Another explanation for the leakage degradation with Ge+BF_2 is the F interaction with EOR defects. At IWJT 2008 Yamamoto[12] reported F co-implant does lead to leakage degradation. In an amorphous layer, F diffuses out of the silicon surface with spike annealing but remains substitutional in silicon and does not diffuse out with MSA. This high level of F from the BF2 implant could degrade junction leakage like C co-implant with MSA causing 3 orders of magnitude degradation in junction leakage as reported by Falepin [13] also at IWJT 2008. Comparison of junction leakage to TW is shown in Figure 6 for B, Ge+B and Ge+BF_2. For all the spike anneals and combinations with spike TW was <100 and leakage <1E-7A/cm^2. Note that without spike anneal high TW values between 120 to 180 were detected suggesting unstable residual implant damage but junction leakage for B and Ge+B were all below 1E-7A/cm^2 so these EOR defects must be shallow and within the junction. Comparing spike+FLA to FLA+spike no difference is seen from junction leakage and residual implant damage however, FLA first results in the lowest Rs value as shown in Figure 1. Besides complete implant damage recovery, the FLA+spike anneal sequence also eliminates the FLA Xe-lamp annealing signature shown in Figure 7 for PLi wafer image maps.

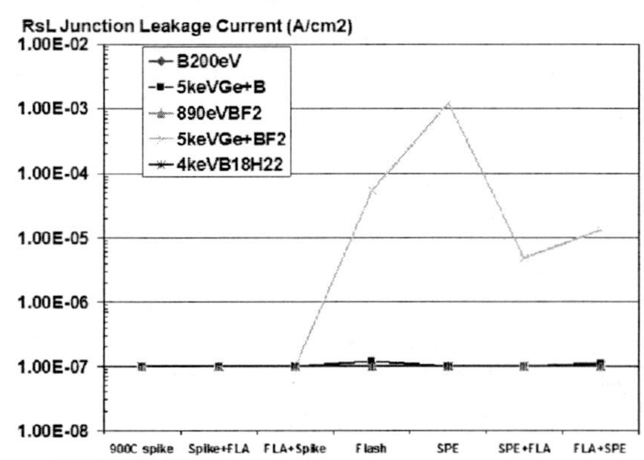

FIGURE 3. RsL junction leakage current results.

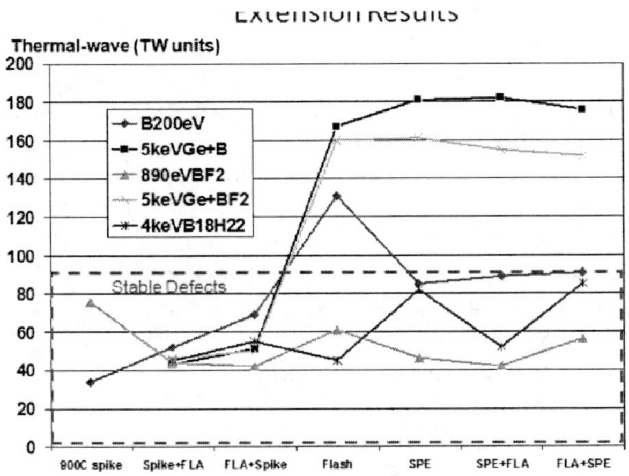

FIGURE 4. Thermal-wave results showing after anneal implant damage recovery.

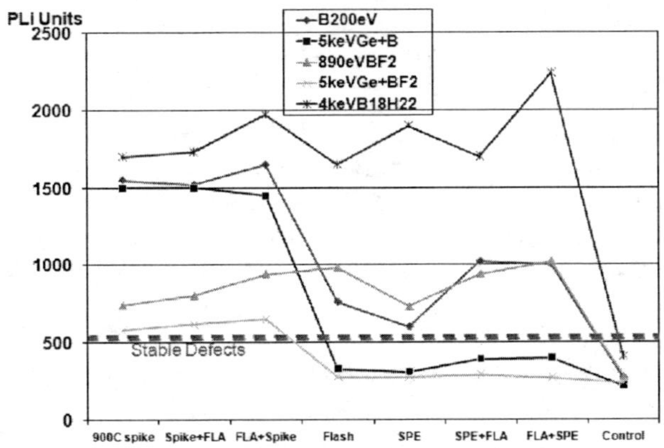

FIGURE 5. PLi results showing after anneal implant damage recovery.

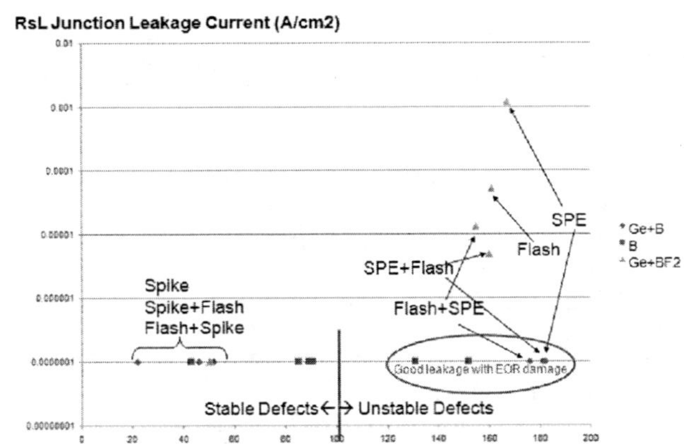

FIGURE 6. Correlation of junction leakage current to TW value.

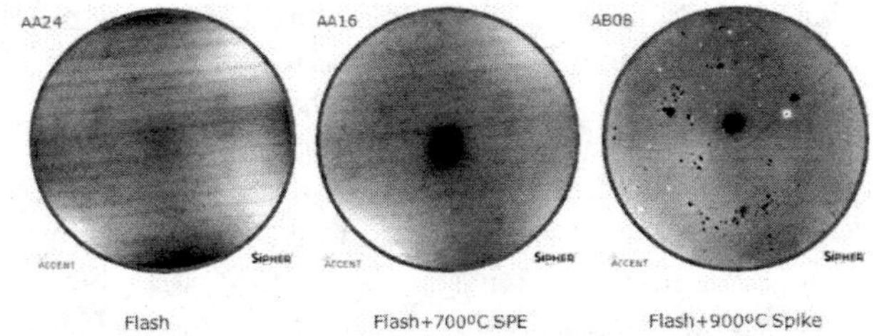

FIGURE 7. PLi wafer image showing elimination of Flash Xe-lamp anneal signature.

Halo

The Rs results are shown in Figure 8. For the boron doped HALOs there was no significant difference in Rs values for all the different anneals except for the SPE anneal because dopant activation for the 3E13/cm^2 dose is well below the boron solid solubility (Bss) limit so dose limited. For indium dopant activation was different and limited by the low indium solid solubility so with spike annealing and spike combination anneals the Rs value were all about 21,000 ohms/square while without spike anneal Rs was very high between 45,000 to 65,000 ohms/square. These Rs results for In are very different from those Mineji reported at IWJT 2007 [8], 4-5 times higher Rs while the B dopant Rs results are similar therefore this might be due to the different spike and Flash annealing equipment used as well as temperature calibration differences between these different annealers as reported by Borland [1] at IWJT 2008.

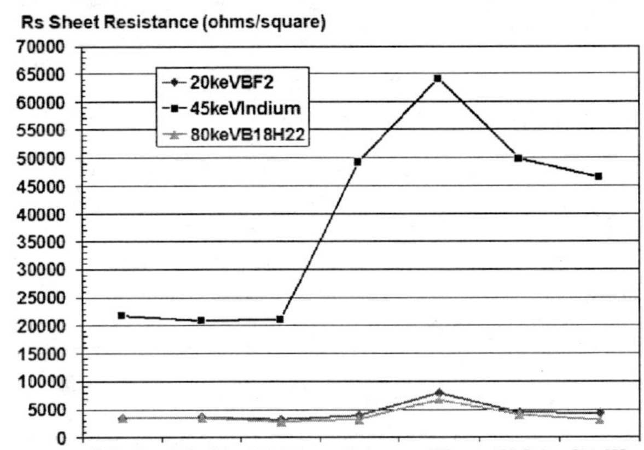

FIGURE 8. Rs sheet resistance results.

Junction leakage current results are shown in Figure 9 and all the results showed very good low leakage levels E-7A/cm^2 or below. Again these results were different from the In HALO leakage results reported by Mineji [8] were they observed E-5A/cm^2 leakage levels for indium HALO with Flash anneal and E-4A/cm^2 leakage levels for SPE annealing while in Figure 9 we report all <2E-7A/cm^2. Again, differences are annealing equipment for both Flash and spike anneals so absolute temperature differences are possible. The leaky junctions in the Mineji study may also explain the lower Rs values in their report. Defect detection using TW and PLi for after anneal residual implant damage are shown in Figures 10 and 11 respectively. No clear trends can be seen suggesting complete implant damage annealing for these 3E13/cm^2 dose dopant level.

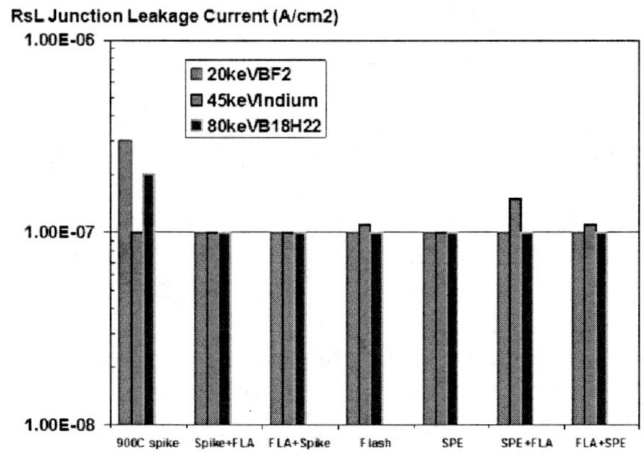

FIGURE 9. RsL junction leakage current results.

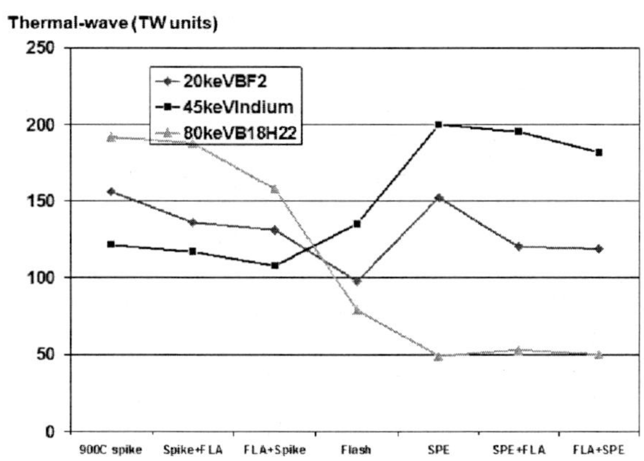

FIGURE 10. Thermal-wave after anneal implant damage recovery.

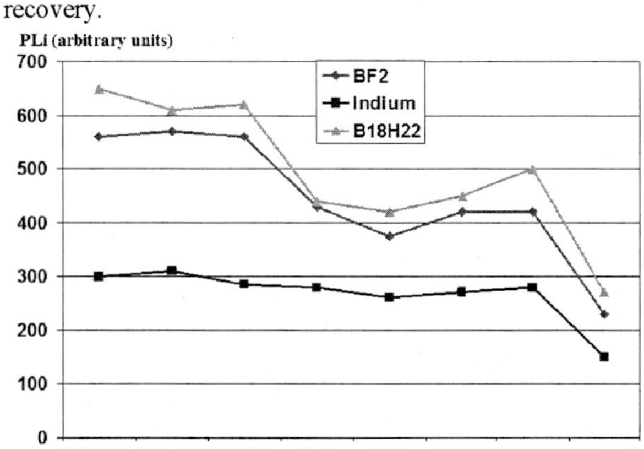

FIGURE 11. PLi after anneal implant damage recovery.

SUMMARY

32nm node p-type dopants were studied comparing B, BF_2, $B_{18}H_{22}$, Ge+B & Ge+BF_2 for extension and BF_2, $B_{18}H_{22}$ & In for HALO dopant implantation. Junction quality was determined using rapid process optimization metrology such as RsL for sheet resistance and junction leakage measurements while TW and PLi were used to determine implant damage recovery after annealing. With Flash annealing, only the Ge+BF_2 for SDE and In for HALO implants resulted in poor quality junctions. The Flash+spike annealing sequence resulted in high quality junctions for all the implants studied and also eliminated the Flash annealing signature micro-uniformity variation. Similar studies for n-type SDE and HALO dopant species are under investigation by many.

ACKNOWLEDGEMENTS

The authors are grateful to Mike Ameen of Axcelis Technologies for support with implants and spike anneals, Jeffri Halim of Frontier Semiconductor for the RsL sheet resistance and junction leakage measurements, Iad Mirshad and Brian Chung of KLA-Tencor for the thermal-wave measurements and to Andrzej Buczkowski of Nanometrics for the photoluminescence measurements.

REFERENCES

1. J. Borland *et al.*, IWJT 2008, paper 3.2, p. 68.
2. V. Moroz, MRS March 2008 Spring meeting, paper E6.6.
3. R. Camillo-Castillo *et al.*, MRS March 2008 Spring meeting, paper E6.9.
4. T. Noda *et al.*, SSDM 2007, A5-1, p.712.
5. J. Borland *et al.*, IWJT 2007, S4-7, p.69.
6. J. Borland, *Solid State Technology*, Jan. 2008, p.38.
7. J. Borland, Semiconductor International, Dec. 2006, p.49.
8. A. Mineji *et al.*, IWJT 2007, S4-8, p.73.
9. J. Borland *et al.*, IWJT 2006, paper 1.1, p.4.
10. J. Borland *et al.*, IIT 2006, paper P105.
11. A. Mineji *et al.*, SSDM 2006, P2-5, p. 412.
12. T. Yamamoto *et al.*, IWJT 2008, paper 3.4, p. 78.
13. A. Falepin *et al.*, IWJT 2008, paper 1.9, p. 46.

New Approaches for Characterization of Advanced Annealing Techniques for Ultra-Shallow Junction Formation

Paul Timans[1], Yao Zhi Hu[1], Jeff Gelpey[2], Steve McCoy[2], Wilfried Lerch[3], Silke Paul[3], Detlef Bolze[4], and Hamid Kheyrandish[5]

[1]*Mattson Technology, Inc., Fremont, California, USA (Paul.Timans@Mattson.com)*
[2]*Mattson Technology Canada, Inc., Vancouver, Canada*
[3]*Mattson Thermal Products GmbH, Dornstadt, Germany*
[4]*IHP, Frankfurt (Oder), Germany*
[5]*CSMA Ltd., Stoke-on-Trent, UK*

Abstract. Low thermal budget annealing approaches, such as millisecond annealing or solid-phase epitaxy (SPE), can electrically activate ultra-shallow junctions (USJ) without excessive diffusion, but they must also remove implant damage to minimize junction leakage. This paper presents results from annealing low-energy B implants into both crystalline and pre-amorphized silicon. Some wafers also received As implants for halo-style doping, and some halo-implanted wafers were pre-annealed at 1050°C before B-doping. The final anneal was either SPE at 650°C, spike annealing at 1050°C, or millisecond annealing with flash-assisted RTP™ (fRTP™) at temperatures between 1250°C and 1350°C. Electrical activation was assessed by sheet resistance (R_S) measurements with conventional four-point probing (4PP) as well as Hg-probe 4PP and a non-contact method. Residual damage was characterized by photoluminescence, thermal wave studies, optical reflectance and non-contact junction leakage current measurements. Damage from the heavy ions used for the halo and pre-amorphization implants dominates the annealing behaviour. Halo doping is the critical factor in determining junction leakage current. High temperature fRTP is shown to minimize residual damage.

Keywords: Ion implantation, doping, annealing, millisecond annealing, silicon, pn junctions, boron, CMOS, halo
PACS: 61.72.uf; 61.80.Jh; 66.30.J-; 66.70.Df; 72.40.+w; 73.40.Lq; 78.40.Fy; 78.55.Ap; 85.30.Kk; 85.40.Ry

INTRODUCTION

CMOS scaling typically increases channel doping to control short-channel effects, which reduces the width of junction depletion regions, raising the electric field and band-to-band tunneling leakage. Residual defects exacerbate the problem, e.g. through trap-assisted tunneling, hence annealing ion implantation damage is essential [1]. Very high temperature millisecond annealing electrically activates implanted dopants without excessive dopant diffusion, but some studies have suggested that conventional spike annealing is more effective in removing defects [2-6]. USJ can also be formed by low-temperature SPE, where dopants are activated by recrystallization of an amorphized silicon layer. However, SPE does not eliminate the end-of-range (EOR) defects found just beyond the initial amorphous/crystalline interface. These challenges make it essential to evaluate how well low-thermal budget annealing removes defects. Defect annealing can be studied by transmission electron microscopy or with test diodes, but these approaches can be time consuming and expensive. Hence there is much interest in non-invasive methods such as measurement of junction photo-voltages [7], photoluminescence [8] and thermal wave effects [9], which can rapidly characterize residual damage.

EXPERIMENT

200 mm diameter, 10 to 15 Ωcm n-type (100) Si wafers were cleaned in SC1 and SC2 solutions followed by an HF dip before ion implantation. 10^{15} B/cm^2 at 500 eV was implanted into crystalline Si (c-Si) or into amorphous Si (a-Si) layers formed by preamorphization implants (PAI) of 10^{15} Ge/cm^2 at 30 keV. Before the B or PAI implants, some wafers were implanted with 4×10^{13} As/cm^2 at 40 keV to simulate strong halo doping [10]. Some halo-doped wafers were annealed for 10 s at 1050°C before the PAI or B

CP1066, *Ion Implantation Technology,* edited by E. G. Seebauer, S. B. Felch, A. Jain, and Y. V. Kondratenko

implants. Spike anneals at 1050°C were performed in a Mattson 3000 RTP system, as were SPE anneals of 5 s at 650°C. Millisecond annealing was performed by fRTP, where the wafer was ramped at 150 K/s to 700°C or 750°C and then its front surface was heated to peak temperatures between 1250°C and 1350°C by a ~0.9 ms duration pulse of energy from a bank of flash-lamps [2]. The temperature of the top surface of the wafer was measured by an ultra-fast pyrometer and another pyrometer monitored the bottom surface.

Figure 1 shows typical B profiles after fRTP at 1300°C or spike annealing, obtained by secondary ion mass spectrometry with a Cameca IMS 7F instrument with 500 eV O_2 primary ions. The fRTP caused much less diffusion than was seen with spike annealing. Figure 1 includes a SRIM simulation of the As halo profile [11]. Pre-annealing the halo is expected to cause only ~ 6 nm of As diffusion [6]. High junction leakage is expected with halo doping because of the very high doping concentrations at the junction, > $6x10^{18}$ cm^{-3}. The As dose is also high enough to introduce very heavy lattice damage.

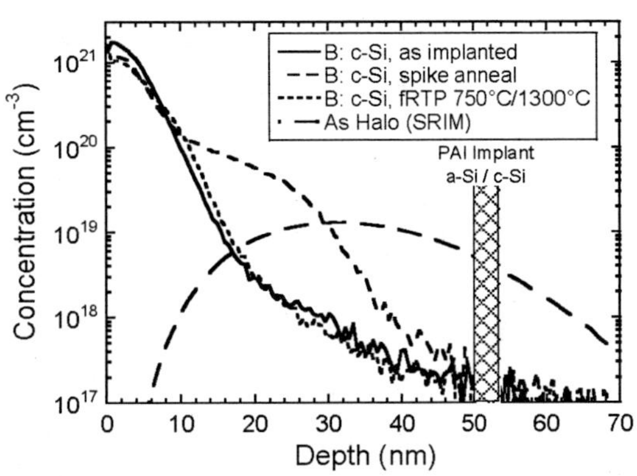

FIGURE 1. Doping profiles for the B doping and the halo-style As implant. EOR defects from the PAI are expected to form near the original location of the a-Si / c-Si interface.

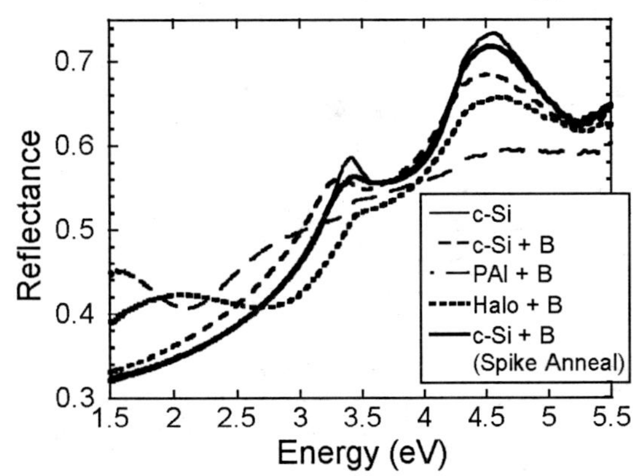

FIGURE 2. Reflectance spectra show the effects of ion-damage, amorphization and annealing.

Figure 2 shows reflection spectra of several samples, as measured with a KLA-Tencor UV1250SE system. The peaks near 3.4 and 4.6 eV are very sensitive to long-range crystalline order. The B implant reduces them, while surface amorphization eliminates them and introduces the reflectance oscillations from interference effects seen below 3 eV. The as-implanted PAI case is consistent with a 50 nm-thick a-Si-layer, and the halo-implanted case also shows the effects of heavy damage. Figure 1 shows the expected location of the EOR damage from the PAI. Annealing restores the shape of the spectra, although the reflectance peaks are slightly reduced in magnitude.

R_S was measured by conventional 4PP using a KLA-Tencor RS100 with type-D probes, 4PP with a Hg-probe from Four Dimensions, and the non-contact RsL[TM] method from Frontier Semiconductor. RsL uses dynamic junction photovoltage measurements to simultaneously determine R_S and junction leakage current density, J_L [7]. RsL gave realistic R_S values even with halo doping, unless very high junction leakage led to extremely small photovoltages. Figures 3(a) and (b) compare the RsL measurements with conventional 4PP and Hg-probe results, respectively.

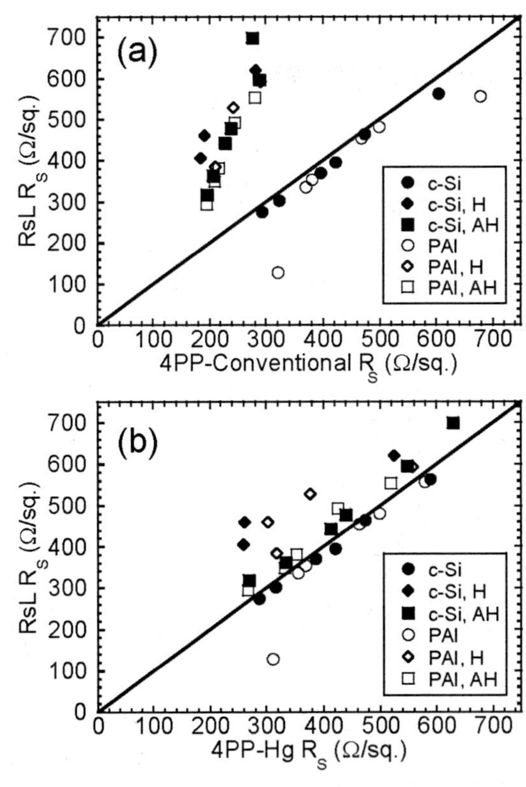

FIGURE 3. Comparison of sheet resistances from RsL and from (a) conventional 4PP or (b) Hg-probe 4PP. The conventional 4PP gives large errors with halo-doping (H), even if the halo is pre-annealed (AH). The Hg-probe 4PP and RsL results match for lightly-doped material, and also when the halo implant damage has been annealed.

Conventional 4PP gave unrealistically low R_S for all halo-doped cases but the results usually agreed with RsL for lightly-doped substrates. The Hg-probe results agreed with RsL for both lightly-doped substrates and pre-annealed halos. In a lightly-doped substrate a thick depletion region electrically isolates the surface layer from the substrate, and the junction forms beyond the EOR damage, so junction leakage effects are minimized. Halo doping leads to a very narrow depletion region that may contain residual defects, greatly increasing junction leakage. It seems that the conventional 4PP physically punched through the thin depletion region of the halo-doped cases, whereas this did not happen for the Hg-probe. However, the Hg-probing was affected by junction leakage if residual damage was present in halo-doped cases. Figure 4 shows R_S from the Hg-probe vs. junction depth results. fRTP gave superior dopant activation with reduced dopant diffusion as compared to the spike anneals. In this comparison, no benefit is seen from the use of PAI.

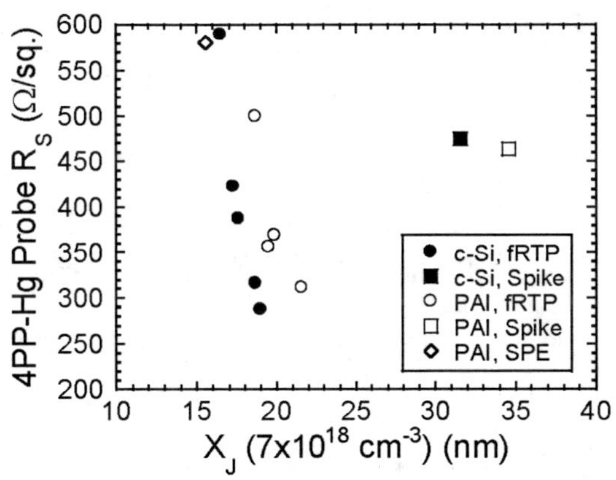

FIGURE 4. Sheet resistance versus junction depth for samples without halo doping.

Figure 5 shows J_L results from RsL for several implant and annealing conditions. B implants into c-Si gave J_L below the measurement limit of 0.1 μA/cm^2, whereas halo doping made J_L very sensitive to residual damage. Pre-annealing the halo implant greatly reduced J_L for implants into c-Si, but had less effect for PAI cases, because much of the halo-damaged region is subsequently amorphized by the PAI. SPE produces very leaky junctions, while spike anneals consistently give low leakage. PAI cases still showed significant leakage after fRTP at 1250°C, but 1300°C anneals were more effective. Optimized millisecond annealing significantly reduces the residual damage relative to that after low temperature SPE anneals.

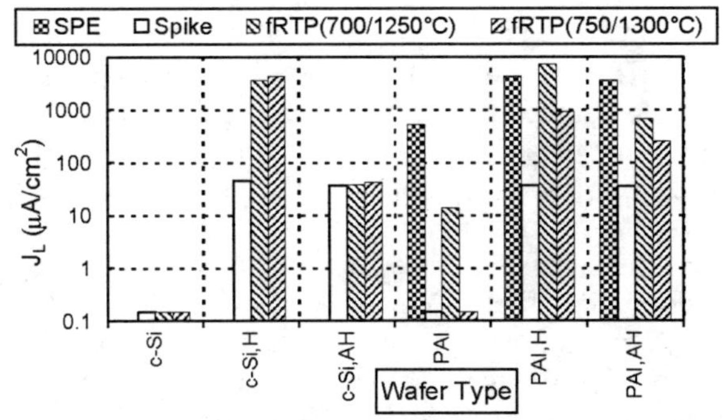

FIGURE 5. RsL leakage current for various anneals. Results are shown for implants into c-Si, including cases with halos (c-Si,H) and with pre-annealed halos (c-Si,AH). Analogous results are shown for PAI cases (PAI,H and PAI,AH).

Residual damage was also characterized by room temperature photoluminescence (PL) measurements performed using an instrument from Accent (now Nanometrics), where a detector collects light from radiative recombination of electron-hole pairs generated by a laser beam [8]. The PL signal is weak in silicon, but it is very strongly affected by defects. The signal is interpreted as a defect metric, with a higher number corresponding to a greater concentration of defects. Figure 6 shows this metric for the same wafers and anneals as in Figure 5. As-implanted samples show high defect levels that reduce with annealing. Implants into c-Si without the halo or with pre-annealed halos show very similar trends. The level is higher if the halo is not pre-annealed. In PAI cases, halo doping and pre-annealing have little effect because the PAI amorphizes most of the region affected by the halo. The PL trends show interesting similarities to those for J_L in Figure 5. For example, the spike anneal produces similar, low values in all cases. SPE anneals lead to high defect levels, while the 1250°C fRTP anneals slightly reduce them, and 1300°C anneals reduce them further.

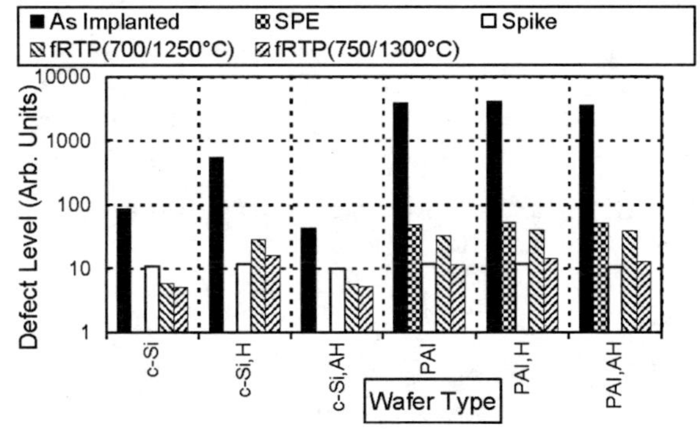

FIGURE 6. The defect level deduced from photoluminescence. Results are shown for the same implants and anneals as in Figure 5.

The wafers were also characterized by thermal wave (TW) measurements using a TP630XP system from KLA-Tencor. In these measurements, a modulated pump laser beam induces oscillations in the surface temperature and the distribution of charge carriers, and a probe laser beam measures corresponding reflectivity changes [9]. The magnitude of the TW signal is affected by implant damage and by the doping profile. Figures 7(a) and (b) show TW results for wafers without or with PAI implants, respectively. There is a general trend for the TW signal to decrease with increased thermal budget. However, there is also a strong effect from the halo-doping. Indeed, the results for cases with or without the pre-anneal of the halo are similar, suggesting that the electrical effect from the halo doping may even have more effect on the TW signal than the residual damage does.

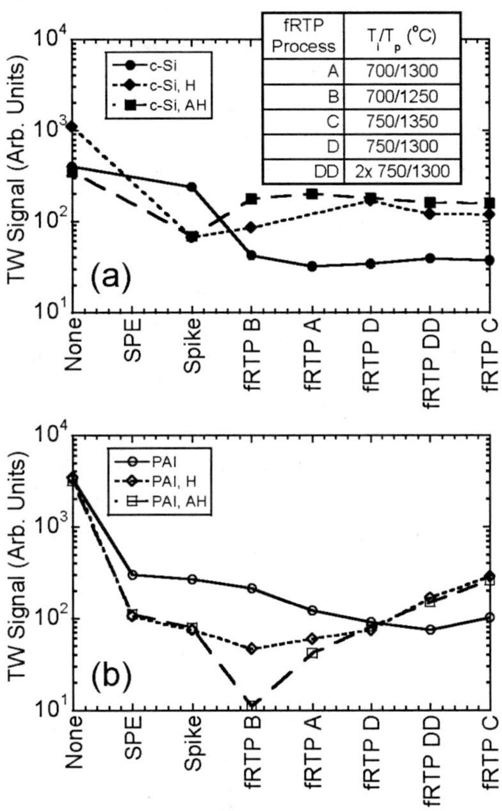

FIGURE 7. Thermal wave signals for implants into (a) c-Si, or (b) PAI wafers. Results are shown for various anneals, ordered with increasing peak temperature. The TW signal is strongly affected by the halo doping (halo-H, annealed halo-AH).

CONCLUSIONS

High temperature millisecond annealing activates dopants with minimal diffusion and also significantly reduces residual damage relative to that after SPE. However, the selection of the halo implant conditions is critical, and the damage from the halo implant must be considered in the optimization. In this study the halo dose was rather high, leading to very thin depletion regions and also introducing significant damage. Pre-annealing the halo implant damage reduced the junction leakage. PL, TW and RsL J_L measurements provide convenient and rapid assessment of residual damage and the effects of annealing.

ACKNOWLEDGEMENTS

The authors would like to acknowledge the help of H. Phan, G. Stuart and D. Camm of Mattson Technology; M. Current, T. Nguyen, J. Halim and V. Faifer of Frontier Semiconductor for the RsL results; J. Chen, T. Dimitrova, W. Liu, N. Jaeger and D. Dimitrov of Four Dimensions for the Hg-probe 4PP; A. Buczkowski, Z. Li and T. Walker of Accent (now Nanometrics) for the PL studies; and M. Bakshi, D. Shaughnessy and A. Salnik of KLA-Tencor for the TW results. This work was partially supported by the European Union (IST project 027152 ATOMICS) program.

REFERENCES

1. R. Duffy, A. Heringa, J. Loo, E. Augendre, S. Severi and G. Curtatola, *ECS Trans.* **3**(2) (2006) 19.
2. P. Timans, J. Gelpey, S. McCoy, W. Lerch and S. Paul, *Mater. Res. Soc. Symp. Proc.* 912, (2006) 3.
3. W. Lerch, S. Paul, J. Niess, S. McCoy, T. Selinger, J. Gelpey, F. Cristiano, F. Severac, M. Gavelle, S. Boninelli, P. Pichler and D. Bolze, *Mater. Sci. Eng. B* **124–125**, (2005) 24.
4. R. A. Camillo-Castillo, M. E. Law, K. S. Jones, L. Radic, R. Lindsay and S. McCoy, *Appl. Phys. Lett.* **88**, (2006) 232104.
5. T. Noda, S. Felch, V. Parihar, C. Vrancken, T. Janssens, H. Bender and W. Vandervorst, *Mater. Res. Soc. Symp. Proc.* **912**, (2006) 191.
6. P. Timans, Y. Z. Hu, J. Gelpey, S. McCoy, W. Lerch, S. Paul, D. Bolze and H. Kheyrandish, *Mater. Res. Soc. Symp. Proc.* (2008) in press.
7. V. N. Faifer, D. K. Schroder, M. I. Current, T. Clarysse, P. J. Timans, T. Zangerle, W. Vandervorst, T. M. H. Wong, A. Moussa, S. McCoy, J. Gelpey, W. Lerch, S. Paul, D. Bolze and J. Halim, *J. Vac. Sci. Technol. B* **25**, 1588 (2007).
8. A. Buczkowski, *ECS Trans.* **11**(3), 109 (2007).
9. W. L. Smith, A. Rosencwaig and D. L. Willenborg, *Appl. Phys. Lett.* **47**, (1985) 584.
10. S. Severi *et al.*, in *Technical Digest of the Int. Electron Devices Meeting 2006* (IEEE, Piscataway, 2006) p. 859.
11. The Stopping and Range of Ions in Matter, http://www.srim.org.

Electron Holography Characterization of Shallow Junction Fabricated By Diffusion-less Process for Sub-30nm Gate-Length MOSFETs

Nobuyuki Ikarashi[1], Kazuya Uejima[2], Takeo Ikezawa[3], Toshinori Fukai[2], Akio Toda[1], and Masami Hane[2]

[1]*Device Platforms Research Laboratories, NEC Corporation, 1120 Shimokuzawa, Sagamihara, 229-1198, Japan*
[2]*NEC Electronics Corporation 1120 Shimokuzawa, Sagamihara, 229-1198, Japan*
[3]*NEC Informatic Systems Corporation, 1120 Shimokuzawa, Sagamihara, 229-1198, Japan*

Abstract. Cross-sectional electron holography was used for two-dimensional potential mapping of source/drain extensions (SDEs) in scaled MOSFETs fabricated using state-of-the-art junction formation technology. First, we show that specimen-preparation artifacts, which have prevented detailed examinations of scaled SDEs, are significantly reduced by using low-energy backside Ar^+ milling. Next, we demonstrate that electron holography clearly reveals very shallow (10-nm-deep) SDE junction profiles. We also show that the experimental examinations, in conjunction with doping-process simulations, allow examinations of how dopant distribution, such as junction depth and abruptness, affect the potential distribution in planar-bulk MOSFETs approaching the scaling limit.

Keywords: Electron holography, Shallow junction, MOSFET
PACS: 61.05.jp, 85.40.Ry

INTRODUCTION

A critical issue in fabricating planar bulk MOSFETs approaching the scaling limit is suppressing the short-channel effect. We reported that elaborate control of the dopant distribution at channel/source-drain (SD) regions by using state-of-the-art junction formation technologies, such as cluster B implantation, millisecond annealing (MSA) and multi-halo implantation, significantly suppresses this effect [1].

Detailed experimental investigations of the dopant distribution in such MOSFETs are indispensable for further improving the junction formation technologies. A few experimental techniques, such as electron holography, can be used in examining the scaled MOSFETs (45-nanometer technology node and beyond) [2-4]. However, artifacts in their experimental results have not yet been clearly identified: for example, specimen-preparation artifacts, such as a surface damaged layer and thickness non-uniformity, have been problematic in electron holography examinations of scaled MOSFETs [4-7].

Accordingly we established a specimen-preparation technique that suppresses the artifacts and enables direct observations of the electrostatic potential in scaled MOSFETs. We first describe our specimen-preparation technique and characterize the artifacts. Second, we show that electron holography in conjunction with doping process simulations reveal the potential distributions in SD extension (SDE) junctions formed using a recent diffusion-less process.

EXPERIMENT

Electron Holography

In the absence of a dynamical diffraction effect and a magnetic field, the phase shift of an electron wave passing through a specimen is given by the expression

$$\phi(x,y) = C_E \cdot V(x,y) \cdot t(x,y) \qquad (1)$$

where C_E is an interaction constant [8] and $V(x, y)$ and $t(x, y)$ are the electrostatic potential and thickness of the specimen, respectively. The incident beam direction is set parallel to the z-direction, and the potential is assumed to be uniform along the z-direction. Hence, when specimen thickness $t(x, y)$ is uniform over the observation area, phase $\phi(x, y)$ changes linearly with electrostatic potential $V(x, y)$.

CP1066, *Ion Implantation Technology,* edited by E. G. Seebauer, S. B. Felch, A. Jain, and Y. V. Kondratenko

A JEOL 2010F transmission electron microscope and Gatan Holoworks software were used in the examinations. The spatial resolution, which was determined on the basis of the sampling frequency in the phase reconstructions, was 4 nm.

Samples

To examine the specimen-preparation artifacts in our electron holography technique, we prepared two samples; one was an nMOSFET sample fabricated using the standard production process, and the other was fabricated using the standard process but omitting the SD and SDE implantation processes (i.e., no pn junctions were formed). Hereafter, we refer to the latter sample as the sample without SD implantation.

We also prepared a sample with shallow SDE junctions by using $B_{18}H_{22}$ implantation, MSA, and multi-halo implantation [1]. The gate-length was 30 nm. A sample without the multi-halo implantation was prepared as a reference. Hereafter, the former and latter are referred to as multi-halo and single-halo samples, respectively.

Specimen Preparation

The two main problems in specimen preparation are surface damage and thickness non-uniformity [5-7]. The former produces an electrically inactive layer, or dead layer, and degrades experimental data. The latter locally changes phase distribution (see Eq. (1)) and has prevented detailed examinations of scaled MOSFETs.

Accordingly, we used low-energy backside Ar ion milling, instead of focused ion beam thinning (FIB), which is routinely used in preparing specimens from semiconductor devices (Figure 1). This is because the dead layer thickness can reportedly be reduced when a lower ion-beam-energy is used [9]: the accelerating energy was 3.0 to 3.5 keV in this experiment, while about 30 keV accelerating energy is usually used in conventional FIB.

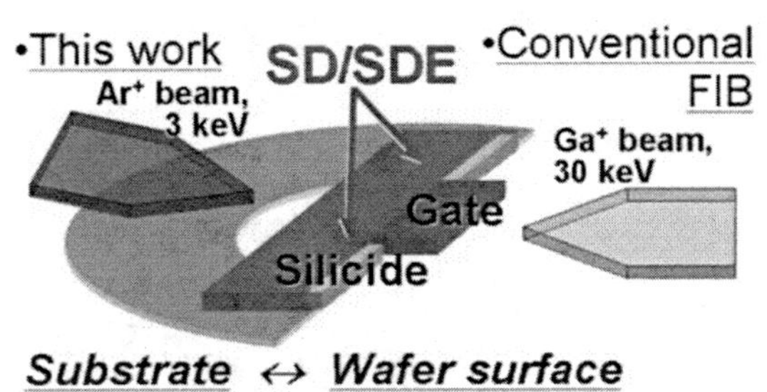

FIGURE 1. Schematic of low-energy backside Ar^+ milling in a specimen preparation.

In addition, we irradiated the ion beam only from the backside of the specimen (Figure 1). This is because thickness non-uniformity is mainly caused by ion beam irradiation from the surface side of the specimen due to differences in the sputtering yield of device structures such as the gate and the silicide. This non-uniformity can reportedly be reduced by backside ion beam irradiation in FIB [7].

RESULTS

Dead Layer Thickness

A schematic of a cross-section of a specimen is shown in Figure 2(a). We denoted the active layer thickness and the total specimen thickness as t_A, and t_S, respectively. We first determined t_A of a junction to estimate the dead layer thickness. We substituted the measured phase difference across the pn junction ϕ into Eq. (1), and set the potential difference V to 1.0 V [10] to calculate t_A of the junction. We then estimated t_S at the junction by using electron energy-loss spectroscopy [11].

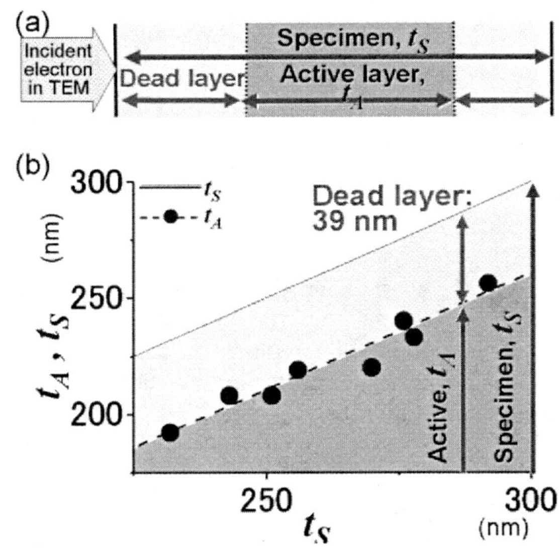

FIGURE 2. (a) Schematic of cross-section of specimen. (b) Plots of active layer thickness, t_A, vs. total specimen thickness, t_S. Thickness t_A changes almost linearly with t_S, and the discrepancy (indicated by red arrow), which shows dead layer thickness, is 39 nm.

Figure 2(b) shows plots of t_A vs. t_S. The diagram shows that t_A changes almost linearly with t_S, and the discrepancy between t_A and t_S was 39 nm. This discrepancy represents the dead layer thickness and is much thinner than that reported for the FIB prepared specimens, about 110 nm [5, 6].

Thickness Uniformity

Representative phase images of the nMOSFET sample and the sample without SD implantation are shown in Figures 3 (a) and (b), respectively. The phase distributions are indicated by the brightness distributions in the images. Clearly, in the nMOSFET sample, the SD regions appear brighter than the other regions. This indicates that the electron wave phases in these regions are different. In contrast, no such difference in brightness is visible in the sample without SD implantation, indicating that the electron wave phase is uniform over the observation area.

To quantitatively examine the phase distribution in the images, we measured the phase along the dashed lines in Figures 3(a) and (b). The results are shown in Figures 3(c) and (d), respectively. Evidently, the phase is uniform in the sample without SD implantation, but is different between the n-type and p-type regions in the nMOSFET sample. The fact that the observed phase is uniform in the sample without SD implantation, in which the $V(x, y)$ is uniform over the observation area, indicates that the thickness of the specimen is almost uniform (i.e., thickness non-uniformity is suppressed to a negligible level. See Eq. (1)). Therefore, we can conclude that the phase images of our specimens enable direct potential distribution analysis.

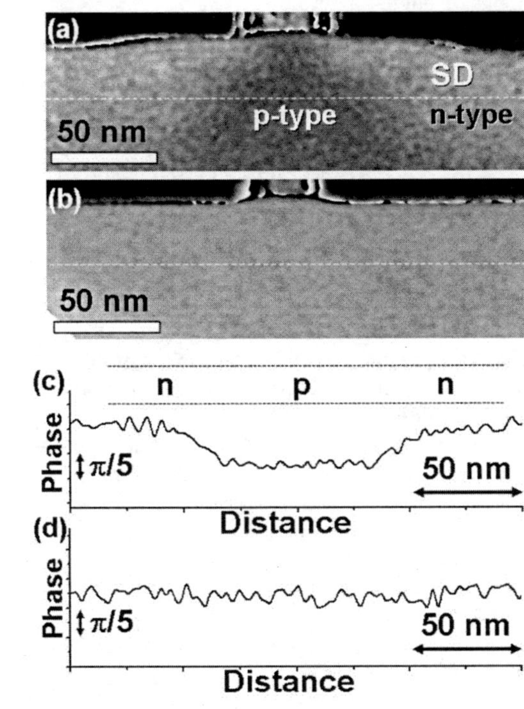

FIGURE 3. Phase maps of (a) nMOSFET, and (b) sample without SD implantation. (c) and (d) show line profiles of phases measured along dashed lines in (a) and (b), respectively. Uniform phase in (b) and (d) indicate that specimen thickness is uniform.

Analysis of SDE

Potential maps of the multi- and single-halo samples are shown in Figures 4 (a) and (c), respectively. The dark and bright areas correspond to the p+-type and n-type regions. We set the electrostatic potential of the p^+-type areas to -0.55 V and that of the n-type areas to +0.45 V [10], and indicated the 0.0-V equipotential lines by the solid lines. Figures 4(b) and (d) show the simulated dopant-concentration maps of the samples. The solid black lines indicate the metallurgical junction positions, at which the donor and the acceptor concentration are equal. The dashed lines indicate the 0.0-V equipotential lines calculated on the basis of the simulations. The calculated 0.0-V equipotential lines are superimposed on the experimental images, for comparing the experimental and simulated results: The dashed curves in Figure 4(a) and (c) show the calculated equipotential lines.

In each sample, the equipotential line obtained from the electron holography experiment agrees well in its position and shape with that obtained from the simulation. This agreement strongly suggests that the artifacts in this electron holography do not affect experimental results and that the modelings of the ion implantation and dopant diffusion/activation processes in this simulation are valid.

The potential maps (Figures 4(a) and (c)) indicate that, near the SDE junctions, the 0.0-V equipotential line positions are significantly different in the samples despite the similar metallurgical junction positions (Figure 4(b) and (d)): The distance between the equipotential line and the gate electrode was 10 nm in the multi-halo sample and 20 nm in the single-halo sample, while the junction depths in the samples were similar (10 nm in the former and 13 nm in the latter). This indicates that the dopant distribution does not directly show the resultant potential distribution, which determines performances of MOSFETs, such as short-channel characteristics. Consequently, electron holography directly reveals the potential distributions and provides useful information in MOSFET characterization.

In the deep SD regions, the simulated dopant distribution map of the single-halo sample, Figure 4(d), indicates that protuberances of the junction positions are formed in parallel with the <110> direction of the substrate. The most likely cause of this significant B diffusion is B channeling during B implantation because the MSA does not induce B transient enhanced diffusion (i.e. diffusion-less process) [10]. Evidently, such dopant distributions result in the gradual potential change at the SD junctions and narrow the distance between SD regions. In contrast, no such protuberances are visible in Figure 4(c). Therefore, these results indicate that B channeling play an important role in determining dopant distribution in the deep SD regions in the diffusion-less process and would affect the transistor performance, but such dopant distribution can be controlled by using multi-halo implantation.

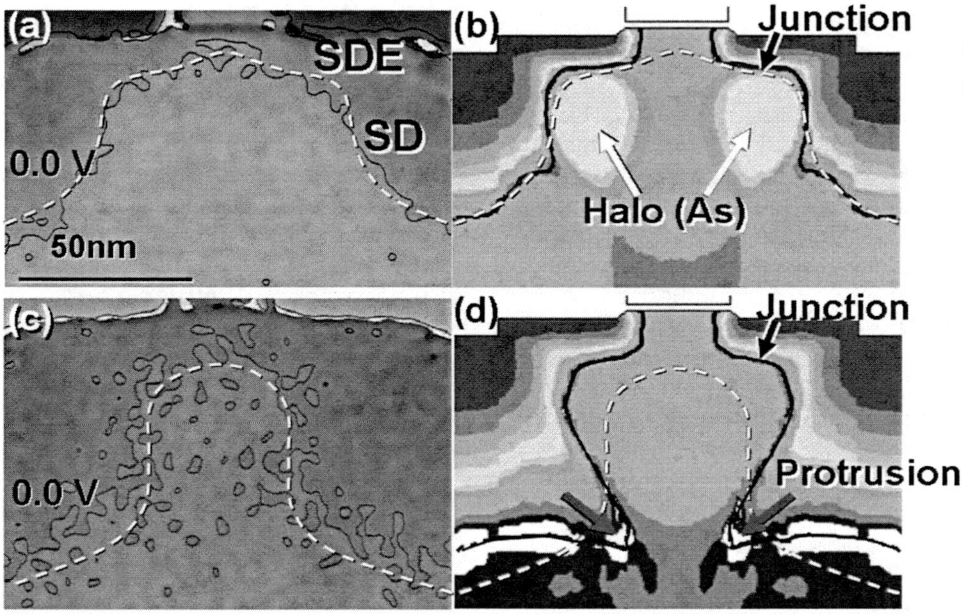

FIGURE 4. Potential maps of (a) multi-halo sample, and (c) single-halo sample. Dark and bright regions correspond to p^+- and n-type regions. Solid lines indicate 0.0-V equipotential lines. (b) and (d) show simulated dopant-distribution maps of multi-halo and single-halo samples, respectively. Black lines in simulated images show metallurgical junction positions. Dashed lines in (b) and (d) indicate 0.0-V equipotential lines determined from simulation and are superimposed on experimental images (dashed lines in (a) and (c)). Experimental and simulated 0.0-V equipotential lines agree with each other in each sample. Arrows in (d) indicate that protrusions of junction position are formed in deep SD regions of single-halo sample.

SUMMARY

Electron holography was used to reveal the electrostatic potentials in shallow junctions in a scaled (30-nm gate length) MOSFET. First, we demonstrated that the electron holography phase images directly reflected the potential distribution in the MOSFETs when specimen-preparation artifacts, such as the surface dead layer and thickness non-uniformity, were suppressed. We successfully reduced these artifacts using low-energy backside Ar ion milling. Second, we used electron holography to examine the potential distribution in SDEs fabricated by using a combination of $B_{18}H_{22}$ implantation, MSA, and multi-halo implantation. The analysis revealed that the SDE junctions were as shallow as 10 nm. Also, B channeling significantly affected the deep SD profiles, but the multi-halo implantation was able to control the profile.

ACKNOWLEDGEMENTS

We wish to thank Ms. M. Oshida for helping prepare specimens. We also thank people of Ultratech and Nissin Ion Equipment for processing the wafers.

REFERENCES

1. K. Uejima, K. Yako, N. Ikarashi, M. Narihiro, M. Tanaka, T. Nagumo, A. Mineji, S. Shishiguchi, and M. Hane, *Tech Dig. –Int. Electron Device Meet. 2007*, 151 (2007).
2. H. Fukutome, Y. Momiyama, E. Yoshida, M. Okuno, T. Itakura, and T. Aoyama, *Tech Dig. –Int. Electron Device Meet. 2005*, 4 (2005).
3. L. Zhang, K. Adachi, K. Tanimoto, A. Nishiyama, *Tech Dig. –Int. Reliability physics symp. 2007*, 516 (2007).
4. N. Ikarashi, M. Oshida, M. Miyamura, M. Saitoh A. Mineji, and S. Shishiguchi, to be published in *Jpn. J. Appl. Phys.* **47** (2008).
5. W. D. Rau, P. Schwander, F. H. Baumann, W. Hoppner, and A. Ourmazd, *Phys. Rev. Lett.* **82**, 2614 (1999).
6. A. C. Twitchett, R. E. Dunin-Borkowski, and P. A. Midgley, *Phys. Rev. Lett.* **88**, 238302 (2002).
7. R. E. Dunin-Borkowski, S. B. Newcomb, T. Kasama, M. R. McCartney, M. Wey-landa, and P. A. Midgley, *Ultramicroscopy* **103**, 67 (2005)
8. L. Reimer, *Transmission Electron Microscopy*, Berin: Springer-Verlag, 1982.
9. L. Xia, W. Wu, Y. Hao, Y. Wang, and J. Xu, *Appl. Phys. Lett.* **88**, 152108 (2006)
10. S. M. Sze, *Physics of Semiconductors, 2nd ed.* New York: John Wiley and sons, 1981.
11. D. Williams and C. Carter, *Transmission Electron Microscopy*, New York: Plenum Pub. Corp., 1996.

The Application of the Continuous Anodic Oxidation Technique for the Evaluation of State-of-the-Art Front-End Structures

S. Prussin[1], Shu Qin[2], Jason Reyes[1], Allen Mcteer[2]

[1] *Department of Electrical Engineering, University of California, Los Angeles, CA, 90095-1594, U.S.A*

[2] *Micron Technology Inc., Boise, ID 83707, U.S.A*

Abstract. The differential Hall effect CAOT permits rapid accurate measurement of mobility, resistivity, and carrier concentration. For single crystal structures with carrier concentrations less than 1E20 B cm^{-3} the resistivity-carrier concentration tracks the A.S.T.M. F-723 algorithm relationship. Above ~1E20 B cm^{-3}, a scattering defect mobility component is added to the phononic and coulombic components. This component is significantly greater for beamline implantation than for plasma doping for both single crystal junctions and polysilicon films.

Keywords: Differential Hall effect, mobility, plasma immersion ion implantation (PIII), continuous anodic oxidation technique, CAOT

PACS: 73.61.-r

INTRODUCTION

The CAOT represents an adaptation of the long established differential Hall Effect (DHE) technique which was first used as a means to evaluate ion implanted junctions in silicon [1] in 1970.

The CAOT is based on the recognition that the anodic oxide insulates the conductive silicon from the electrolyte. This allows us to carry out the entire series of anodic oxidations and electrical measurements with the silicon specimen continually in contact with the electrolyte and without removing the probes at the electrical contacts. This results in a number of outstanding features.

1. CAOT could be completely automated, allowing for push-button operation.
2. Time for a complete evaluation was greatly decreased, from hours to minutes.
3. Precision was increased, permitting atomic layer resolution, especially useful for surface evaluation.
4. Drift mobility profiles were obtainable.
5. Active dopant profiles can be obtained rather than the total dopant distribution obtained with SIMS.
6. The process is non-destructive with regard to the wafer, permitting CAOT to be used for production control.

A description of the CAOT together with examples of its application to USJ's and doped polysilicon gate electrode metrology was presented in 2007 [2].

Six electrical measurements are made at the end of each anodic oxidation interval, typically 15s. The depth of the silicon step is controlled by the anodic current density and time. For an anodic current density of 25mA/cm^2 for 15s, each step is ~15 Å; for an anodic current density of 5mA/cm^2 this is reduced to ~3 Å, or approximately one atomic layer of silicon. At each step a measurement is made of the sheet resistance and the Hall coefficient. These are mathematically converted into resistivity, mobility, and active dopant profiles.

A.S.T.M. ALGORITHM

The technique presently available for determining carrier concentration distribution is the spreading resistance probe (SRP). SRP measures a series of resistances, compares them to pre-measured resistivity standards, and yields a resistivity profile extending inward from the surface. The only available accurate correlation between resistivity and carrier concentration, the A.S.T.M. specification F-723, is based on an extensive study by the National Bureau of Standards for boron and phosphorus concentrations over the range of 1E14 cm^{-3} to 1E20 cm^{-3} [3]. Above 1E20 cm^{-3} the SRP uses an extrapolation of the A.S.T.M. algorithm, although there is a caveat in the A.S.T.M. specification F672-01 [4] with regard to using this resistivity-concentration correlation for concentrations exceeding 1E20 cm^{-3}.

With the D.H.E. evaluation we can generate the carrier concentration directly from our relationship:

$$n = \frac{1}{\rho \cdot \mu \cdot q} \text{ cm}^{-3}$$

This relationship gives us an accurate value of the carrier concentration whether the silicon is single crystal or polysilicon, perfect or damaged. We also find that our D.H.E. data closely tracks the A.S.T.M. algorithm up to a carrier concentration of ~1E20 cm^{-3} and no further.

DHE evaluations were carried out on ultra-shallow junctions and doped polysilicon films by beamline ion implantation and plasma immersion ion implantation (PIII), Table I.

Figure 1 represents the results of a D.H.E. evaluation of a beamline implant and of four specimens doped with PIII under different conditions. Also included is the A.S.T.M. algorithm which represents the carrier concentration which correlates with the resistivity profile measured with the SRP technique. All SRP data are found to lie on the A.S.T.M. curve.

For resistivities greater than 1.1E-4 Ω-cm (~1E20 B cm^{-3}) the D.H.E. and the SRP carrier concentration track one another for all specimens. For resistivities less than 1.1E-4 Ω-cm, we note that the curves begin to divert. This is attributed to the introduction of additional scattering mechanisms due to crystal defects, complexes, etc. reducing the hole mobility. Clearly, the result is most striking for the beamline implanted specimen M1. The surface concentration of M1, as determined by the D.H.E. technique, is 7.5E20 cm^{-3}. For an SRP evaluation, the identical surface resistivity would correlate with a concentration of 1.6E20 cm^{-3}.

CP1066, *Ion Implantation Technology,* edited by E. G. Seebauer, S. B. Felch, A. Jain, and Y. V. Kondratenko

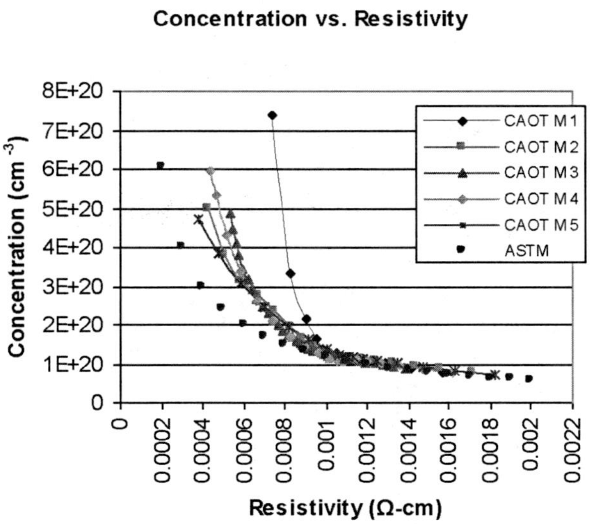

FIGURE 1. Comparison of mobility-carrier concentration profiles with A.S.T.M. F723 algorithm for beamline and PIII doped junctions.

TABLE I. List of Evaluated Specimens

Wafer	Energy	Dose	4PP	CAOT	Notes
			Rs	Rs	
#	(keV)	(cm^{-2})	(Ω/sq)	(Ω/sq)	
M1	2	5.00E+15	151.86	149.9	Beamline
M2	4.5	3.00E+16	159.97	158.3	PIII
M3	4.5	3.00E+16	162.37	161	PIII
M4	4.5	3.00E+16	165.7	165.4	PIII
M5	6	2.00E+16	130.12	136	PIII
M1C	6	2.00E+16	128.98	129.3	PIII + 2E13B @ 30keV
M2C	5	2.50E+16	148.13	145	PIII + 2E13B @ 30keV
M3C	5	2.50E+16	161.52	157.9	PIII
M5C	2	5.00E+15	143.49	145.6	Beamline + 2E13B @ 30keV
M6	4	1.50E+16	648.24	642.6	Beamline, 70nm Poly on 100nm SiO_2
M7	6	2.00E+16	1409.3	1422	PIII, 70nm Poly on 100nm SiO_2
M9	6	4.00E+16	1142.8	1145	PIII, 70nm Poly on 100nm SiO_2

DEFECT SCATTERING CONTRIBUTION TO MOBILITY

Since the N.B.S. study [5,6,7] was based on wafers cut from grown silicon boules, it is reasonable to assume that this data was accumulated from undamaged, single crystal silicon for concentrations below solid solubility. Under such conditions, we can assign two components to the hole and electron mobilities, the phononic, μ_{ph}, attributed to scattering by lattice vibration, and the coulombic, μ_{coul}, attributed to the scattering by doping ions in the lattice and therefore a function of the carrier concentration.

We can write:

$$\frac{1}{\mu_0} = \frac{1}{\mu_{ph}} + \frac{1}{\mu_{coul}} \qquad (1)$$

For carrier concentrations > ~1E20cm^{-3}, we can add the scattering contribution of crystal imperfections, and complexes, $\frac{1}{\mu_{def}}$.

$$\frac{1}{\mu} = \frac{1}{\mu_{ph}} + \frac{1}{\mu_{coul}} + \frac{1}{\mu_{def}} \qquad (2)$$

$$\frac{1}{\mu} = \frac{1}{\mu_0} + \frac{1}{\mu_{def}} \qquad (3)$$

$$\mu_{def} = \frac{\mu \bullet \mu_0}{\mu_0 - \mu} \qquad (4)$$

where μ = measured CAOT DHE mobility ($\frac{\mu_H}{r}$)

μ_0 = mobility calculated from the A.S.T.M. conversion relation.

μ_{def} = scattering component due to crystal imperfections and complexes

r = Hall scattering factor

Figure 2 illustrates the reciprocal mobility curves for two similarly implanted beamline shallow junctions and two similar PIII doped junctions. We note for all four that the defect mobility component becomes significant near the surface where we expect the highest concentrations and the heaviest damage and that as in Figure 1 the defect component for the beamline implantation is greater than that for the PIII doped specimens.

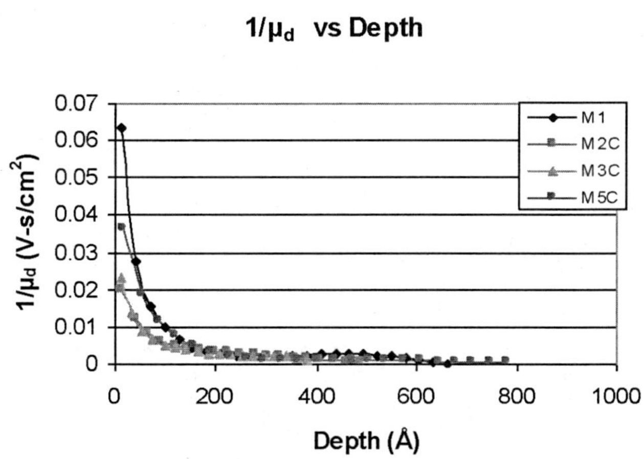

FIGURE 2. Comparison of reciprocal defect mobility components for beamline and plasma doped junctions.

HALL SCATTERING FACTOR

The Hall scattering factor, r, is defined as the ratio $\frac{\mu_H}{\mu_{drift}}$. For electron carrier concentrations, the value of r has been accepted as 1.0 [6]. For hole carriers the value of r has the range .7<r<.9 [8]. Both Thurber [5] and Masetti [9] chose values of 0.8.

Figure 3 represents the hole concentration distribution for specimen M5C, generated in two ways. One consists of using only the measured resistivity profile, converting this data into carrier concentrations using the A.S.T.M. algorithm. This represents the results which would be obtained using an SRP evaluation with a resistivity profile identical to the one measured by the CAOT.

The other two curves make use of the Hall mobility measurement for assumed r values of 1.0 and 0.8. For the limited number of specimens studied, a Hall scattering factor of 0.8 appeared to be appropriate.

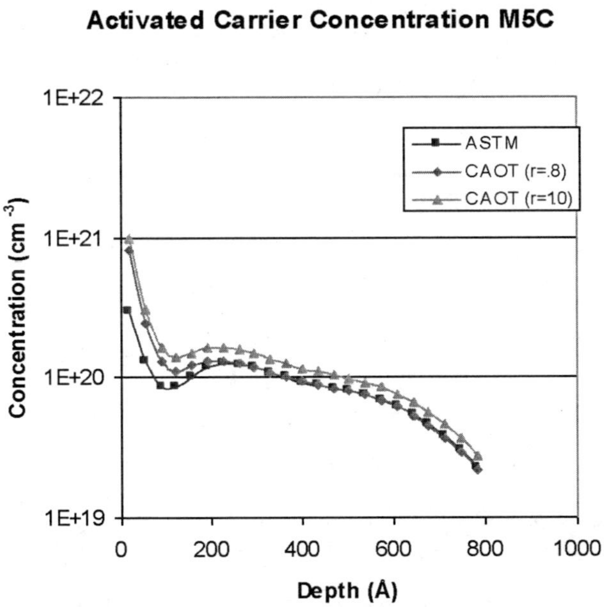

FIGURE 3. Comparison of carrier concentration profile from DHE (r=0.8) and DHE (r=1.0) with profile calculated from A.S.T.M. algorithm.

COMPARISON OF CAOT DHE WITH SIMS AND SRP EVALUATIONS

The validity of CAOT DHE evaluations is based on a comparison with the presently available techniques SIMS and SRP. The first was prepared by a SIMS/ARXPS technique [10]; the second was from an outside commercial source.

Figure 4 illustrates this comparison for specimen M5C. Above 1E20 cm^{-3}, we exceed the solid solubility of B and degrade the mobility, causing the resistivity values to correlate to higher hole concentrations as shown by the D.H.E.

Comparison of the D.H.E. carrier concentration distribution with the distribution calculated from the A.S.T.M. algorithm represents a means of determining the accuracy of the D.H.E. evaluation. Only if both the resistivity and mobility are accurately measured by the D.H.E. evaluation will the A.S.TM. and D.H.E. distributions track each other for concentrations less than 1E20 cm^{-3}.

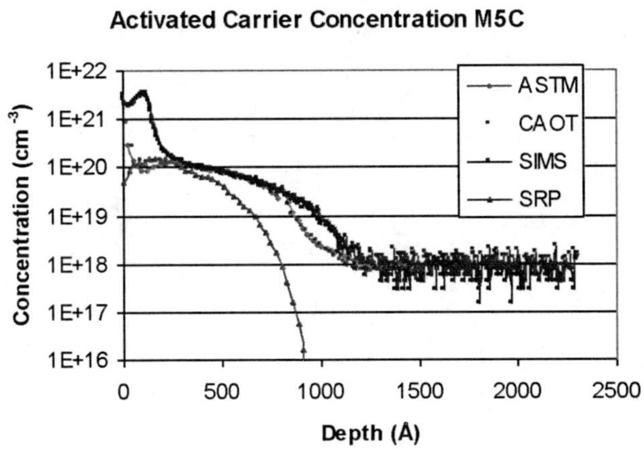

FIGURE 4. Comparison of SRP, SIMS, and DHE analyses with the A.S.T.M. algorithm

APPLICATION OF CAOT DHE FOR DOPED POLYSILICON FILMS

CAOT DHE studies were carried out on 70nm thick doped polysilicon films deposited on silicon wafers covered with 10nm and with 100nm thermal oxide. It was not possible at present to carry out successful CAOT DHE analyses on the wafers with the 10nm oxide film. It appears that the 10nm oxide is insufficiently thick to permit anodic oxidation to occur only at the electrolyte-specimen interface. This results in a change in anodic oxidation rate during the CAOT evaluation.

Evaluation of doped polysilicon films by SIMS leaves much to be desired. The total dopant distribution certainly fails to represent the electrically active carriers. SRP can contribute a useful resistivity profile, but the degradation of the drift mobility due to grain boundaries precludes the conversion of these values to carrier concentrations.

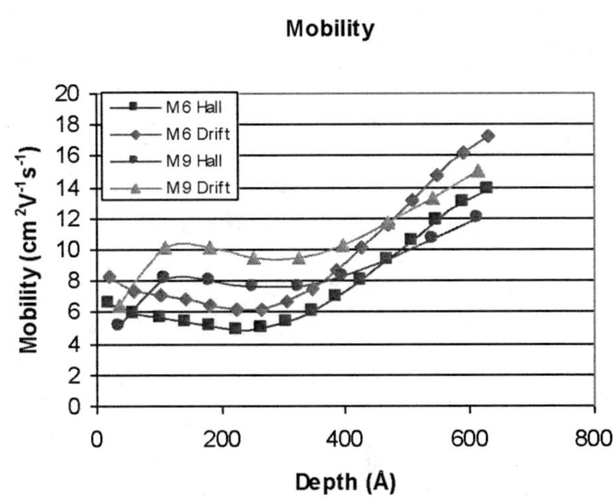

FIGURE 5. Comparison of mobility profiles for M6 (beamline doped) and M9 (PIII doped)

Figure 5 represents the mobility profiles for specimens M9 and M6. The Hall mobilities were measured and converted to drift mobilities by dividing by 0.8, the Hall scattering factor.

Figure 6 compares the SIMS results with those obtained by CAOT DHE for specimens M6 and M9. SIMS analysis provides us with the total B concentration. The CAOT DHE evaluation provides us with a layer by layer activated B distribution.

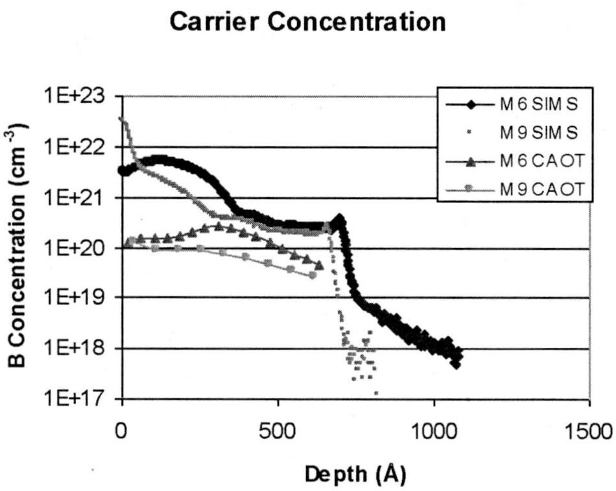

FIGURE 6. Comparison of SIMS with CAOT DHE for specimens M6 (beamline doped) and M9 (PIII doped).

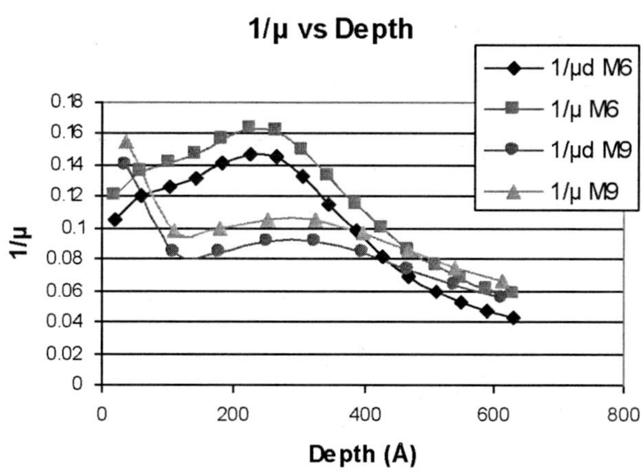

FIGURE 7. Reciprocal mobility defect component for M6 (beamline doped) and M9 (PIII doped)

The reciprocal mobility defect component dominates throughout the thickness of the doped polysilicon thickness, Figure 7. Since precipitates do not appear to affect the scattering phenomenon [11], the defect component must be attributed to the grain boundaries. Queirolo [12] has proposed a semi-quantitative relationship between the mean grain size and μ_d:

$$\mu_d = KD^{\alpha}$$

where K = 4.5E4 and α = 0.59 are constants and D (cm) = mean grain size.

From Figure 7, we note that the grain size tends to a minimum at the surface and a maximum at the interface of the film with the substrate. For a $\frac{1}{\mu_d}$ value of 0.16 at the surface, we would have a mean grain size of ~3μm and for a $\frac{1}{\mu_d}$ value of 0.04 at the interface, a mean grain size of ~30μm.

CONCLUSIONS

The CAOT applied to DHE permits us to accurately evaluate ultra-shallow junctions making it possible to also evaluate the higher carrier concentrations, greater than 1E20 cm^{-3}, found near the surface. Comparison of the carrier concentration distribution below 1E20 cm^{-3} with the A.S.T.M. algorithm permits us to evaluate the accuracy of our measurements. For doped polysilicon films, it is possible to determine the carrier concentration distribution as well as the mobility scattering component distribution due to grain boundaries.

REFERENCES

1. N.G.E. Johansson, J.W. Mayer, O.J. Marsh. *Solid-State Electronics*, **13**, 317-335, (1970)
2. S. Prussin, AIP Conference Proceedings, **931**, 275-278 (2007)
3. Standard Practice for Conversion Between Resistivity and Dopant Density for Boron Doped, Phosphorus-Doped, and Arsenic-Doped Silicon, A.S.T.M.Designation F723-99, Annual Book of A.S.T.M. Standards, 10.05 (2000)
4. Standard Test Method for Measuring ProfilesPerpendicular to the Surface of a Silicon Wafer using a Spreading Resistance Probe, A.S.T.M. Designation F672-01, Annual Book of AS.T.M. Standards, Vol 10.05 (2001)
5. W.R. Thurber, R.L. Mattis, Y.M. Liu, J.J. Filliben, *Journal of the Electrochemical Society*, **127**, 2291-2294 (1980)
6. W.R. Thurber, R.L. Mattis, Y.M. Liu, J.J. Filliben, *Journal of the Electrochemical Society*, **127**, 1807-1812 (1980)
7. W.R. Thurber, R.L. Mattis, Y.M. Liu, J.J. Filliben, N.B.S. Special Publication, 400-464 , May 1981
8. J.F. Lin, S.S. Li, L.C. Linares, K.W. Teng. *Solid-State Electronics*, **24**, 827-833 (1981)
9. Guido Masetti, Maurizio Severi, Sandro Solmi, IEEE Trans. Electron Devices, vol.ED-30, 292-295 (1982)
10. Shu Qin, Kent Zhuang, Shifeng Lu, Allen McTeer, Wendy Morinville, Kari Noehring. This Publication.
11. S. Solmi, M. Severi, R Angelucci, L. Baldi, R. Bilenchi. *Journal of the Electrochemical Society*, **129**, 1811-1818 (1982)
12. G. Querolo, E. Servida, L. Baldi, G. Pignatel, A. Armigliato, S. Frabboni, F. Corticelli. *Journal of the Electrochemical Society*, **137**, 967-971 (1990)

Ultra-Shallow P^+/N Junction Formation in Si Using Low Temperature Solid Phase Epitaxy Assisted with Laser Activation

Shuhei Hara[1], Yuki Tanaka[1], Takumi Fukaya[1], Satoru Matsumoto[1], Toshiharu Suzuki[2], Genshu Fuse[2], Toshio Kudo[3], and Susumu Sakuragi[3]

[1] *Keio University, 3-14-1 Hiyoshi, Kouhoku-ku, Yokohama, Kanagawa, 223-8522 Japan*
[2] *SEN Corporation, an SHI and Axcelis Company, SBS Tower 9F, 4-10-1 Yoga, Setagaya-ku, Tokyo, 158-0097 Japan*
[3] *Sumitomo Heavy Industries Ltd., 19 Natsushima-cho, Yokosuka, Kanagawa, 237-8555 Japan*

Abstract. A combination of Ge pre-amorphization implantation (Ge-PAI), low-energy B implantation and laser annealing is a promising method to form highly-activated, abrupt and ultra-shallow junctions (USJ). In our previous report of IIT 2006, we succeeded in forming pn junctions less than 10nm using non-melt double-pulsed green laser. However, a large leakage current under reverse bias was observed consequently due to residual defects in the implanted layer. In this study, a method to form USJ is proposed: a combination of low-temperature solid phase epitaxy and non-melt laser irradiation for B activation. Ge pre-amorphization implantation was performed at energy of 6keV with a dose of $3x10^{14}/cm^2$. Then B implantation was performed at energy of 0.2keV with a dose of $1.2x10^{15}/cm^2$. Samples were annealed at 400°C for 10h in nitrogen atmosphere. Subsequently, non-melt laser irradiation was performed at energy of $690mJ/cm^2$ and pulse duration of 100ns with intervals of 300ns. As a result, USJ around 10nm with better crystallinity was successfully formed. And the leakage current of pn diodes was reduced significantly. Moreover, it is proven from secondary ion mass spectroscopy (SIMS) analysis that transient enhanced diffusion (TED) of B is specifically suppressed.

Keywords: Ultra-shallow junction, low-temperature annealing, non-melt laser annealing
PACS: 71.55.Cn, 61.72.U-

INTRODUCTION

Miniaturization of MOSFETs by scaling-down of device dimensions requires the formation of ultra-shallow junction with a low sheet resistance [1, 2]. A lot of process techniques have been developed to meet the above requirements. Among these processes, a combination of Ge pre-amorphization implantation (Ge-PAI), low-energy B implantation and laser anneal (LA) is a promising method to form highly-activated, abrupt and ultra-shallow junctions [3, 4]. In the previous report of IIT 2006 [5], we succeeded in fabrication of pn junctions less than 10nm using non-melt double-pulsed laser. However, a large leakage current under reverse bias of pn diodes was observed consequently due to residual defects in implanted layers. On the basis of the previous results, in this study, the effect of low temperature annealing prior to non-melt laser anneal was introduced in order to improve crystallinity of implanted layers.

EXPERIMENTAL DETAILS

725μm thick, n-type Si (100) wafers with more than 10Ω · cm were used as substrates. First, Ge pre-amorphization implantation to suppress channeling of B ions was performed at energies of 3keV and 6keV with a dose of $3x10^{14}/cm^2$ and tilt angle of 7°. Then B low-energy implantation was performed at energy of 0.2keV with a dose of $1.2x10^{15}/cm^2$. Subsequently, samples were annealed in N_2 atmosphere at 400°C for 1.5h for Ge-PAI energy of 3keV and 10h for 6keV prior to the laser anneal. LA was carried out in air using a 527nm double-pulsed green laser with a pulse width of 100ns, oscillated at a frequency of 1kHz and delay time between the first and the second pulse of 300ns. The laser formed homogenized 0.25mm x 2.5mm line beam and samples were irradiated with overlap ratio of 90%. Laser energy densities were selected between $690\text{-}760mJ/cm^2$. Annealed samples

CP1066, *Ion Implantation Technology,* edited by E. G. Seebauer, S. B. Felch, A. Jain, and Y. V. Kondratenko

were analyzed with four-point probe method for the sheet resistance, secondary ion mass spectrometry (SIMS) for the dopant profiles and atomic force microscope (AFM) for surface morphology evaluation. For SIMS, O^{2+} was used as a primary ion at energy of 350eV. The junction depth was defined at B concentration of $1x10^{18}/cm^3$. The junction leakage current characteristics of pn diodes were measured. The crystallinity near the junction in implanted layers was analyzed by transmission electron microscope (TEM).

RESULTS AND DISCUSSION

AFM Analysis

Figure 1 shows the surface of samples with Ge-PAI energy of 6keV after laser irradiations with energy densities of (a) $700mJ/cm^2$ and (b) $730mJ/cm^2$ respectively.

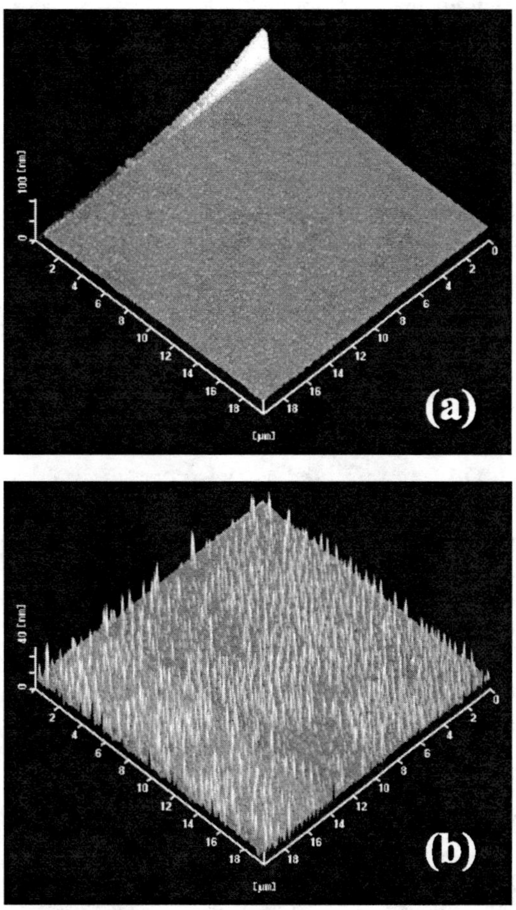

FIGURE 1. AFM images of annealed samples. (a) $700mJ/cm^2$, (b) $730mJ/cm^2$

The surface of the sample annealed at $700mJ/cm^2$ is smooth compared to the image of the sample annealed at $730mJ/cm^2$. It is considered that there is no melt of surface with the fluence of $700mJ/cm^2$. With energies higher than $710mJ/cm^2$, the surface starts getting rough. Therefore, it is expected that the threshold of laser energy density between a non-melting and a melting state is in-between 710-$720mJ/cm^2$ with the Ge-PAI energy of 6keV. Meanwhile, the threshold energy is in-between 720-$730mJ/cm^2$ with the Ge-PAI energy of 3keV. This difference indicates that there would be a larger amorphous region after pre-anneal for samples with Ge-PAI energy of 6keV. Since the absorption coefficient is much higher in amorphous silicon, laser energy is reserved more as heat in samples with Ge-PAI energy of 6keV.

Sheet Resistance Evaluation

Figure 2 shows the sheet resistances as a function of laser energy density after low-temperature pre-anneal (PA) and laser anneal for Ge-PAI energies of 3keV and 6keV. Firstly, the values of sheet resistance decrease greatly after pre-anneal for both samples. And it decreases gradually with higher energy density. It is thought that values decrease because of melt of the surface and sequential rapid B diffusion. Samples with Ge-PAI energy of 6keV show lower sheet resistance than 3keV, due to larger remaining amorphous region in implanted layer after pre-anneal. Also we can see the typical reverse-annealing phenomenon [6] just before melting energy density for each sample. As a result, the value of 714Ω/□ is obtained for the sample with Ge-PAI energy of 6keV at $690mJ/cm^2$.

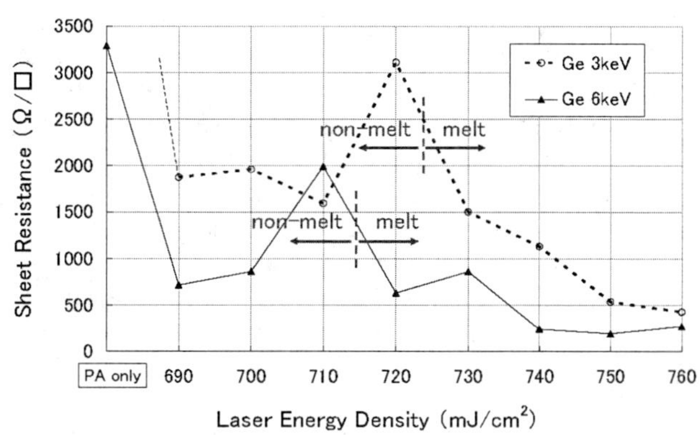

FIGURE 2. Sheet resistance vs. Laser energy density

SIMS Profiles

Figure 3 shows the result of SIMS profiles. Results of this study (with Ge-PAI energy of 6keV), as-implanted, and our previous data of 2007 [7] are included in the figure. It should be noted that the substrate of 2007 in Figure 3 had the same implantation condition except the Ge-PAI energy (3keV). In this case, samples were annealed at 650°C for 1h and were irradiated with 740mJ/cm^2 laser pulses.

We can see noticeable B diffusion over 7nm at the condition of year 2007. It is assumed that transient enhanced diffusion (TED) of B occurred in the pre-anneal process. On the other hand, the B diffusion is suppressed obviously with 400°C 10h pre-annealing. Annealing with a very low temperature may be capable of removing point defects in the implanted layer, which are responsible for TED. Though the boron profile shifts about 3nm deeper from that of as-implanted, the junction around 10nm with an abruptness of ~2nm/dec. is successfully formed with altered pre-anneal and laser anneal conditions.

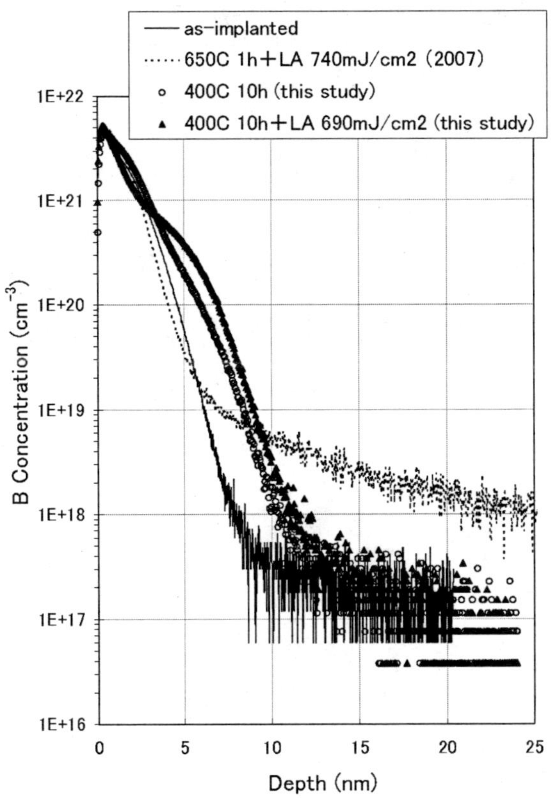

FIGURE 3. SIMS profiles of this study and the past years

Diode Leakage Current Evaluation

Figure 4 shows the *I-V* characteristics of pn diodes. The result what we reported at IIT 2006 [5] is also included. Sample conditions of 2006 are as follows: Ge-PAI energy of 3keV with a dose of 3x10^{14}/cm^2, B implantation energy of 0.2keV with a dose of 1.2x10^{15}/cm^2 and LA 760mJ/cm^2. There is a large leakage current under reverse bias for the sample of 2006 due to the residual defects in implanted layer. Meanwhile, for the sample of this study, a leakage current decreases and current density under forward bias becomes better at start. Besides following LA with 690mJ/cm^2 makes the reverse current much less (about 5 orders in magnitude) than that of year 2006. These results indicate that the combination of low temperature pre-anneal and non-melt LA improves the crystallinity of implanted layers and junction region by removing various defects.

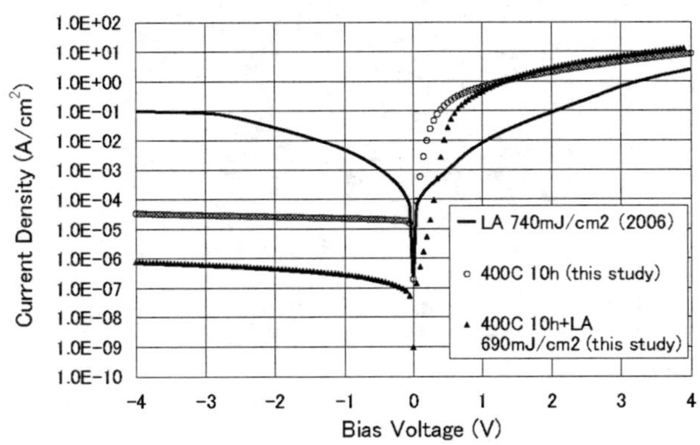

FIGURE 4. Diode characteristics

TEM Analysis

Finally, Figure 5 shows the cross sectional TEM (X-TEM) images near the junction in implanted layers of (a) after LA 760mJ/cm^2 and (b) after Pre-anneal 400°C 10h + LA 690mJ/cm^2. It is obvious that the surface region is destructed for the sample with only LA. This implies that the crystal regrowth was not enough with only LA and it contains lots of defects in the implanted layer still. On the other hand, the image of the sample with pre-anneal + LA shows the tidy lattice of Si suggesting that pre-anneal effectively restored the crystallinity of implanted layers.

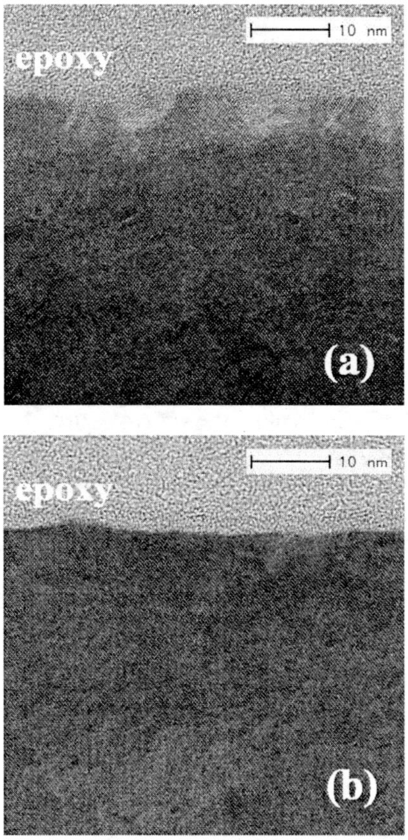

FIGURE 5. X-TEM image of implanted layers (a) with only LA 760mJ/cm^2, (b) with pre-anneal 400C 10h + LA 690mJ/cm^2

CONCLUSION

The methods fabricating ultra-shallow junction with the combination of low-temperature anneal and non-melt LA are shown in this study. It is found that low temperature anneal may be an effective process to reduce defects in implanted layers without large B diffusion. As a result, USJ of ~10nm with 714Ω/□ and better crystallinity is formed by using the combination method of Ge-PAI, low-temperature pre-anneal, non-melt LA. Since 10h-annealing is long for mass production, it is necessary to optimize the pre-anneal condition.

ACKNOWLEDGEMENTS

The authors would like to thank SEN Corporation, an SHI and Axcelis Company, Sumitomo Heavy Industries Ltd., MST and Sumika Chemical Analysis Service Ltd. for their technical support.

REFERENCES

1. International Technology Roadmap for Semiconductors (2006)
2. H. W. Kennel, P. H. Keys, M. Kiu, *Proc. of Workshop on USJT* (2005) 4
3. K. K. Ong, K. L. Pey, P. S. Lee, A. T. S. Wee, Y. F. Chong, K. L. Yeo and X. C. Wang, *Material Science and Engineering B*, **114-115** (2004) 25-28
4. M. Hernandez, J. Venturini, D. Zahorski *et al.*, *Applied Surface Science* **208-209** (2003) 345-351
5. Ryuta Yamada, Singo Seto, Soshi Sato, Yuki Tanaka, Satoru Matsumoto, Toshiharu Suzuki, Genshu Fuse, Toshio Kudo and Susumu Sakuragi, *Ion Implantation Technology*, American Institute of Physics 978-0-7354-0365-9/06 (2006)
6. Jian-Yue Jin, Jinning Liu, Ukyo Jeong, and Sandeep Metha, *J. Vac. Sci. Technol. B* **20**(1) (2002)
7. Takumi Fukaya, Ryuta Yamada, Yuki Tanaka, Satoru Matsumoto, Toshiharu Suzuki, Genshu Fuse, Toshio Kudo and Susumu Sakuragi, *15th IEEE International Conference on Advanced Thermal Processing of Semiconductors – RTP2007* (2007) 317-320

Local Resistance Profiling of Ultra-Shallow Junction with Spike Lamp and Laser Annealing Using Scanning Spreading Resistance Microscopy

Satoshi Abo[1], Yuji Tanaka[1], Kazuhisa Nishikawa[1], Fujio Wakaya[1], Toshiaki Iwamatsu[2], Hidekazu Oda[2] and Mikio Takai[1]

[1] *Center for Quantum Science and Technology under Extreme Conditions, Osaka University, 1-3 Machikaneyama, Toyonaka, Osaka, 560-8531 Japan*
[2] *Renesas Technology Corporation, 4-1Mizuhara, Itami, Hyogo, 664-0005 Japan*

Abstract. Local resistance profiles of ultra-shallow arsenic implanted into silicon with an energy of 3.5 keV and a dose of 1.2 x 10^{15} ions/cm^2 activated by conventional spike lamp and laser annealing were measured by SSRM in a nitrogen atmosphere with a depth resolution of less than 10 nm for investigating the combination of the conventional spike lamp and laser annealing. Spike lamp annealing at 1050°C followed by laser annealing at a power density of 0.42 kW/mm^2 was found to give the lowest sheet resistance. The resistance profiles obtained by SSRM also indicated the lowest resistance for the sample after spike lamp annealing at 1050°C followed by laser annealing with a power density of 0.42 kW/mm^2. Laser annealing alone with a power density of 0.42 kW/mm^2 resulted in the higher sheet resistance, though the shallower resistance profile could be obtained. Spike lamp annealing followed by laser annealing procedures are effective in activating shallow arsenic profiles.

Keywords: ultra-shallow junction, local resistance, laser annealing, Scanning Spreading Resistance Microscopy (SSRM)
PACS: 07.79.-v, 68.37.-d

INTRODUCTION

Shrinkage of ultra large scale integrations (ULSIs) is a trend for fabricating high speed and low power consumption devices. In the International Technology Roadmap for Semiconductors (ITRS) 2007 Edition, the depths of source and drain extension regions (X_j) for bulk micro processing unit (MPU) / application specific integrated circuit (ASIC) become 9, 8 and 7 nm in 2010, 2011 and 2012, respectively [1]. In our recent studies, the atomic depth profiles of ultra-shallow arsenic implanted into silicon were clarified by medium energy ion scattering (MEIS) with depth resolutions of 2.4 to 4.0 nm at depths of 10 to 20 nm [2, 3]. However, a ratio of the activated dopant cannot be observed by the MEIS measurement nor secondary ion mass spectroscopy (SIMS) method. Therefore, a high resolution electrically activated dopant profiling method is required, since conventional carrier profiling methods have poor depth resolutions.

In this study, the local resistance of ultra-shallow junction after spike lamp and laser annealing processes was measured by scanning spreading resistance microscopy (SSRM) in a nitrogen atmosphere with a depth resolution of less than 10 nm for investigating the combination of the spike lamp and laser annealing.

EXPERIMENTAL

Arsenic was implanted into silicon with an energy of 3.5 keV and a dose of 1.2 x 10^{15} atoms/cm^2. Implanted arsenic was activated by spike lamp annealing at 1000 and 1050°C and CO_2 laser annealing with power densities of 0.39 and 0.42 kW/mm^2. Ramp-up and -down rates of spike annealing were 200 and 80°C/sec, respectively. Expected peak temperatures of laser annealing were 1320 and 1390°C with power densities of 0.39 and 0.42 kW/mm^2, respectively, which are slightly less than the melting point of silicon (1414°C). These samples were investigated by SIMS for depth profiling of implanted atoms and by a four-point probe method for sheet resistances before SSRM measurements. In this study, two groups of the samples were measured. One was implanted arsenic activated with and without 0.42 kW/mm^2 laser annealing after 1050°C spike annealing.

CP1066, *Ion Implantation Technology,* edited by E. G. Seebauer, S. B. Felch, A. Jain, and Y. V. Kondratenko

This group was for observing the effects of additional laser annealing after spike annealing. The other was implanted arsenic activated with 0.42 kW/mm^2 laser annealing alone, 1050°C spike annealing alone, or 1000°C spike annealing followed by 0.39 kW/mm^2 laser annealing. These samples were almost the same sheet resistances of about 310 Ω/square. This group was for checking the inner resistance difference of the same sheet resistance samples. Samples were measured by SSRM for investigating the depth profiles of the local resistance.

SSRM is based on a conventional atomic force microscopy (AFM), where a probe tip for SSRM is an electrically conductive boron-doped diamond-coated probe. The local resistance is measured by the current through the probe, which is sensed by a logarithmic current amplifier, and calculated by the Ohm's law. The spatial resolution of the SSRM depends on a curvature radius of the probe and a contact force. The curvature radius of the probe in this study is 50 nm. The contact force in this study was 2.9 μN. The depth resolution in this study was less than 10 nm. The SSRM system cannot measure the local resistance along with the depth direction. Therefore, the samples were cleaved and polished to measure in-depth profiles. SSRM measurements were operated in a nitrogen atmosphere for suppressing the influence of surface oxidation and water adsorption.

RESULTS AND DISCUSSION

Table 1 shows the relationship between annealing conditions and sheet resistances obtained by a four-point probe method. The sheet resistances for samples with spike annealing alone at 1050°C, with laser annealing alone at a power density of 0.42 kW/mm^2, or spike annealing at 1000°C followed by laser annealing at a power density of 0.39 kW/mm^2 were almost the same at about 310 Ω/square. The sheet resistance for a sample with additional laser annealing at a power density of 0.42 kW/mm^2 after spike annealing at 1050°C was 20 % lower than that without laser annealing. Thus the spike annealing at 1050°C followed by 0.42 kW/mm^2 laser annealing results in the lowest sheet resistance.

TABLE 1. Annealing conditions and sheet resistances.

Annealing Condition	Sheet Resistance (Ω/square)
1050 °C Spike	3.01×10^2
1050 °C Spike + 0.42 kW/mm^2 Laser	2.42×10^2
0.42 kW/mm^2 Laser	3.20×10^2
1000 °C Spike + 0.39 kW/mm^2 Laser	3.15×10^2

Figure 1 shows the arsenic profiles obtained by SIMS for samples before and after spike annealing at 1050°C with and without additional laser annealing at 0.42 kW/mm^2. The arsenic profile of the as-implanted sample had a peak of 2×10^{21} atoms/cm^3 at a depth of 5 nm. On the contrary, the arsenic profiles with and without 0.42 kW/mm^2 laser annealing after 1050°C spike annealing were almost the same, which had a peak at the surface, decreased gradually from 10^{21} to 10^{20} atoms/cm^3 at depths from the surface to 20 nm and drastically decreased from 10^{20} to 10^{17} atoms/cm^3 at depths deeper than 20 nm. Thus the atomic arsenic profiles obtained by SIMS don't clarify the difference in sheet resistance in Table 1.

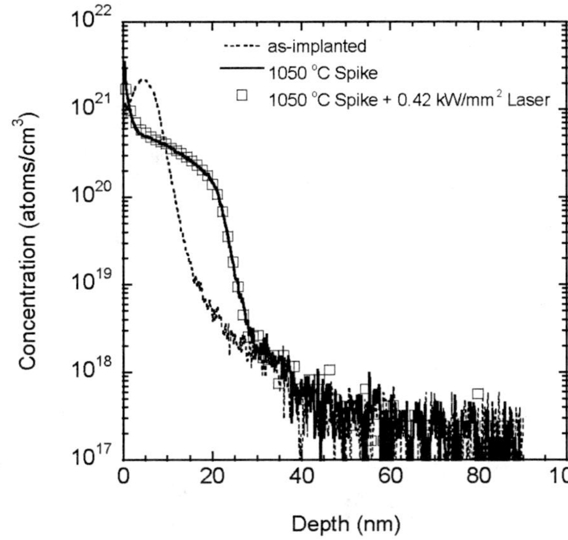

FIGURE 1. Arsenic profiles of as-implanted sample, samples with and without 0.42 kW/mm^2 laser annealing after 1050°C spike annealing measured by SIMS.

Figure 2 shows the local resistance profiles obtained by SSRM for arsenic implanted samples with and without 0.42 kW/mm^2 laser annealing after 1050°C spike annealing. The resistance profile for a non-doped sample was also shown for comparison. Both local resistances with and without laser annealing after spike annealing became the lowest at a depth of 20 nm and gradually recovered to the resistance of the non-doped sample (10^8 Ω), which indicates that implanted atoms close to the surface observed by SIMS were not activated by the spike lamp and laser annealing processes. The lowest resistance with additional laser annealing (2.2×10^3 Ω) was about fifteen times lower than that without laser annealing (3.1×10^4 Ω), which indicates that the laser annealing process after spike annealing can further activate the implanted atoms.

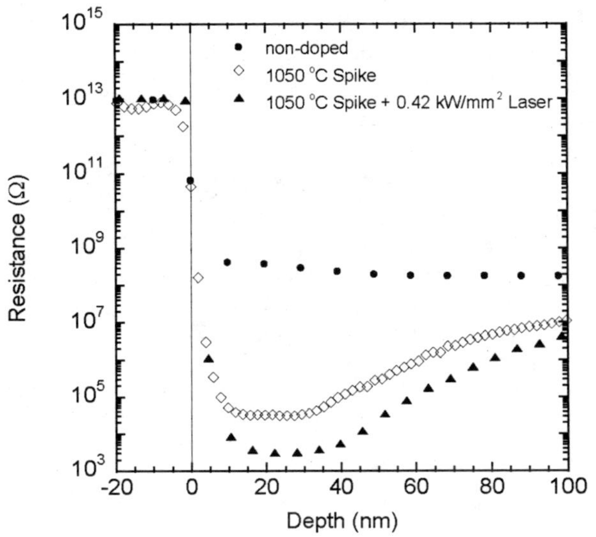

FIGURE 2. Local resistances of a non-doped sample, samples with and without 0.42 kW/mm^2 laser annealing after 1050°C spike annealing measured by SSRM.

Figure 3 shows the arsenic profiles obtained by SIMS for the as-implanted sample, samples with 0.42 kW/mm^2 laser annealing alone, with 1050°C spike annealing alone, and with 1000°C spike annealing followed by 0.39 kW/mm^2 laser annealing. Although the annealed samples showed almost the same sheet resistances of 310 Ω/square as shown in Table 1, the arsenic profiles with three different annealing processes differ at depths from 10 to 30 nm. In particular, the spike annealing at 1050°C enhanced arsenic diffusion at depths from 10 to 30 nm.

Figure 4 shows the local resistance profiles obtained by SSRM for arsenic implanted samples with 0.42 kW/mm^2 laser annealing alone, with 1050°C spike annealing alone, and with 1000°C spike annealing followed by 0.39 kW/mm^2 laser annealing. The resistance profile for a non-doped sample was also shown for comparison. The local resistance with 0.42 kW/mm^2 laser annealing alone became the lowest at depths from 10 to 30 nm with a resistance of 1.3 x 10^5 Ω and drastically recovered to 10^8Ω at 40 nm, which indicates that the laser annealing process can activate arsenic with less diffusion. However, the lowest local resistance with 0.42 kW/mm^2 laser annealing alone was slightly higher than those with 1050°C spike annealing alone, or 1000°C spike annealing followed by 0.39 kW/mm^2 laser annealing, which indicates that the laser annealing process alone was not enough to activate the implanted arsenic for realizing the lowest local resistance. The local resistance profiles for samples with 1050°C spike annealing, and 1000°C spike annealing followed by 0.39 kW/mm^2 laser annealing were almost the same, where the lowest local resistances were about 5 x 10^4 Ω at a depth of 20 nm and the local resistance gradually recovered to 10^8 Ω from 30 nm. This indicates that the spike annealing followed by laser annealing can realize a much higher activation of implanted arsenic.

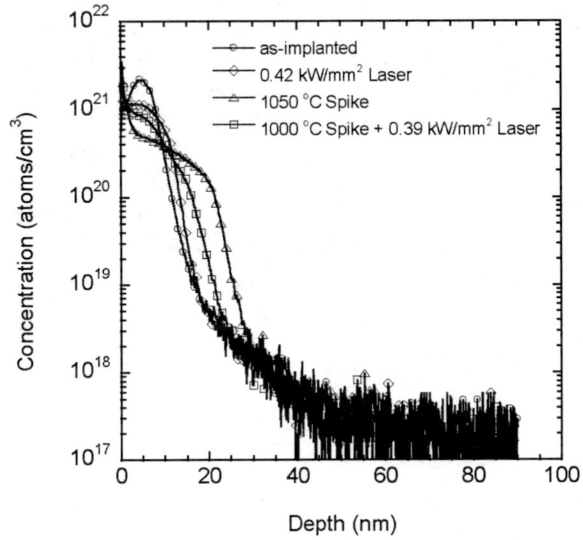

FIGURE 3. Arsenic profiles of an as-implanted sample, samples with 0.42 kW/mm^2 laser annealing, with 1050°C spike annealing, and with 1000°C spike annealing followed by 0.39 kW/mm^2 laser annealing measured by SIMS.

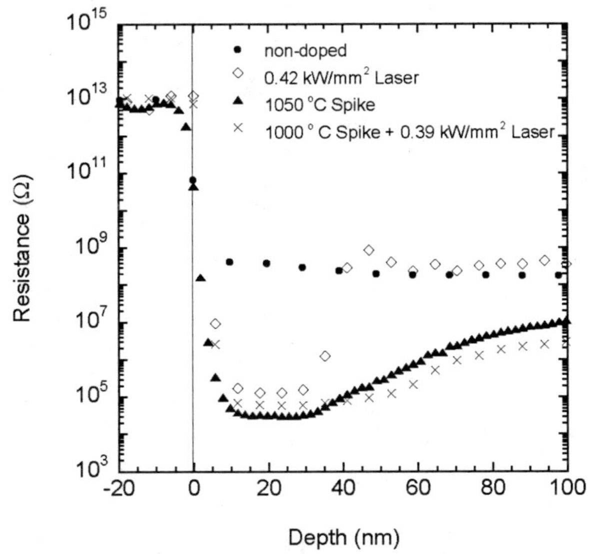

FIGURE 4. Local resistances of a non-doped sample, samples with 0.42 kW/mm^2 laser annealing, with 1050°C spike annealing, and with 1000°C spike annealing followed by 0.39 kW/mm^2 laser annealing measured by SSRM.

CONCLUSION

Local resistance profiles of ultra-shallow arsenic implanted into silicon with an energy of 3.5 keV and a dose of 1.2 x 10^{15} ions/cm^2 activated by conventional spike lamp and laser annealing were measured by SSRM in a nitrogen atmosphere with a depth resolution of less than 10 nm for investigating the combination of the conventional spike lamp and laser annealing.

Spike lamp annealing at 1050°C followed by laser annealing at a temperature close to the melting point of silicon (0.42 kW/mm^2) was found to give the lowest sheet resistance. The resistance profiles obtained by SSRM also indicated the lowest resistance for the sample after spike lamp annealing at 1050°C followed by laser annealing with a power density of 0.42 kW/mm^2. Laser annealing alone with a power density of 0.42 kW/mm^2 resulted in the higher sheet resistance, though the shallower resistance profile could be obtained.

Spike lamp annealing followed by laser annealing procedures are effective in activating shallow arsenic profiles.

REFERENCES

1. International Technology Roadmap for Semiconductors 2007 Edition, http://www.itrs.net/.
2. S. Ichihara, T. Nakagawa, M. Nitta, S. Abo, T. Lohner, C. Angelov, K. Ohta and M. Takai, *Nucl. Instr. and Meth. B* **219-220** p. 584 (2004).
3. S. Abo, S. Ichihara, T. Lohner, F. Wakaya, T. Eimori, Y. Inoue and M. Takai, *Nucl. Instr. and Meth. B* **237** p. 72 (2005).

A Study of Implanted BF_2 as a Function of Wafer Temperature During Implant

Tae-Hoon Huh[1], Byung-Jae Kang[2], Geum-Joo Ra[1], Kyung-Won Lee[1], Steve Kim[1], Ronald N. Reece[1], Leonard M. Rubin[1], Michael S. Ameen[1], Won-Min Moon[2], Min-Sung Lee[2], Young-Ho Lee[2], Jong-Oh Lee[2], Dong-Chul Park[2], Jung-Youn Lim[3], Youn-Soo Kim[3], Jae-Sang Ro[3]

[1]*Axcelis Technologies Inc., 108 Cherry Hill Drive, Beverly, MA 01915, USA*
[2] *Samsung Electronics Co., LTD., San#16 Banwol-Dong, Hwasung-City, Gyeonggi-Do, Korea*
[3]*Hong-Ik University, #72-1, Sang-Su Dong, Ma-Po Gu, Seoul City, Korea*

Abstract. The temperature effect for buried channel PMOS transistor characteristics was investigated. Generally, only dose, energy and implant angle have been considered as the major parameters for process matching between different high current implanters in transistor manufacturing. However, as the device is scaled down to sub-100nm size, additional parameters such as instantaneous dose rate and wafer temperature have become increasingly important for controlling the dopant profile by changing the annealing behavior of defects. The dose rate difference between ribbon and spot beam implanter was investigated through simulation and the wafer temperature difference was directly measured with special temperature measurement device. The peak height of both the boron and fluorine SIMS profiles corresponding to the location of the amorphous / crystalline (a/c) interface increased proportionally with increasing wafer temperature and to a lesser degree with increasing instantaneous dose rate. By increasing the wafer temperature, the threshold voltage (V_T) decreased dramatically while the active area sheet resistance remained constant. This higher secondary peak height resulted in enhanced lateral diffusion, shorter effective channel length, and therefore a lower threshold voltage (V_T).

Keywords: Implant, PMOS Transistor, Process Matching, Temperature Effect, BF_2, Fluorine Effect, Lateral Diffusion.
PACS: 52.77.Dq, 82.45.Vp

INTRODUCTION

As device geometries continue to shrink, the devices become more sensitive to material, surface, and especially process environments during ion implantation. Previously, only dose, energy, and beam angle have been the implanter parameters of concern in transistor matching [1,2]. However, as the device is scaled down to sub-100nm nodes, additional parameters such as instantaneous dose rate and wafer temperature during implant have become increasingly important because of their effect on the annealing behavior of defects [3-7]. For buried channel PMOS, the transistor characteristics are very sensitive to dose rate and wafer temperature during BF_2 implantation.

In semiconductor manufacturing, it is very important to match device parameters among different tools. This is especially true for implanters, because most transistor characteristics are set by the implanted dopant profiles. Therefore, it is crucial to understand machine to machine differences when matching device parameters. In this study, two different high current ion implanters, a ribbon beam implanter with a one dimensional mechanical scan (vertical scan only) system [8], and a spot beam implanter with a two dimensional mechanical scan (horizontal fast scan and vertical slow scan) [9] were compared to look for effects of dose rate and implant temperature on the resulting devices.

EXPERIMENTAL APPROACH

Bare wafer tests were initially performed to begin matching PMOS device parameters between the two implanters. Figure 1 shows SIMS data for the two different implanters. Both (a) as-implanted profiles and (b) post anneal profiles were exactly matched. On devices, however, the V_T was about 3% higher for the spot beam implanter, even though the active Rs was well matched (Table 1). To find the root cause of V_T difference, the machine difference was compared first.

CP1066, *Ion Implantation Technology,* edited by E. G. Seebauer, S. B. Felch, A. Jain, and Y. V. Kondratenko

Table 2 is the comparison table of two different types of implanters. Based on the comparison on Table 2, the instantaneous dose rate was calculated for 10 mA beam current. The dose rate of spot beam was one order higher than that of ribbon beam. For spot beam, it was 1.8×10^{16} ions/cm^2/s and for ribbon beam, it was 1.0×10^{15} ions/cm^2/s.

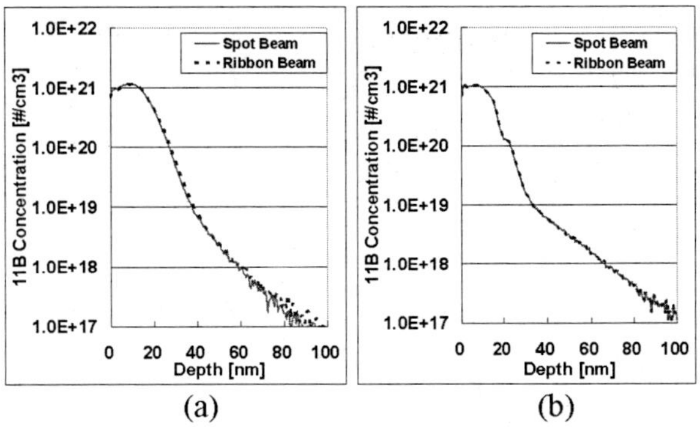

FIGURE 1. SIMS profiles of 11B for (a) as-implanted and (b) annealed samples. The profiles were well matched between spot and ribbon beam implanter. The implant was BF2, 10keV, 2×10^{15}cm-2 and the anneal was 850°C, 30 sec, in 100% nitrogen.

TABLE 1. First Device Test Results (Normalized).

	PMOS Vt	Active Rs
Ribbon Beam	100	100
Spot Beam	103	100

TABLE 2. Machine Comparison between Spot and Ribbon Beam.

Beam Type	Spot Beam	Ribbon Beam
Scan System	Mechanical X & Y	Mechanical Y Scan Only
Typical Scan Speed	X : 1~2 m/s Y : 2~8E-5 m/s	X : N/A Y : ~ 3.5E-4 m/s
Typical Beam Size	70 x 110 mm^2	330 x 50 mm^2
Cooling System	Water & Gas	Water & Gas

Wafer temperature during implant was also compared. Figure 2 showed the wafer temperature during implant which measured directly with a special measurement device. For spot beam, the wafer was repeatedly heated and cooled during implant. So the peak temperature was very slowly increased. However, for ribbon beam, the wafer temperature was monotonically increased during implant. So the maximum temperature during implant was higher than that of spot beam. Figure 2 showed that the maximum temperature of ribbon beam 680 W (40 keV, 17 mA) about 80°C, but that of spot beam 800 W (40 keV, 20 mA) was about 70 °C.

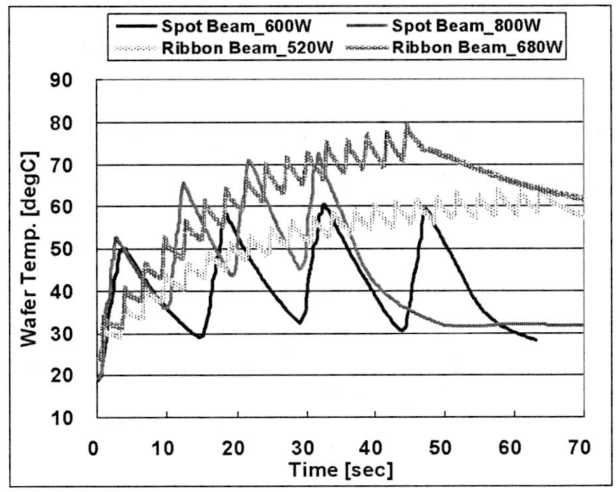

FIGURE 2. Wafer temperature during implant for both spot and ribbon beam. Implant energy was fixed at 40 keV, and the beam current was varied to change the total beam power.

The machine comparison suggested that spot beam has a higher dose rate and higher cooling efficiency. So, for device matching to ribbon beam, it is necessary to decrease the dose rate and lower the cooling efficiency. We varied the scan speed (instantaneous dose rate) and wafer temperature (by controlling the wafer chuck temperature) on the spot beam implanter. The test conditions are shown in Table 3. The implant for all tests was $^{49}BF_2^+$, 10keV, 2×10^{15}cm^{-2}, with a post-implant anneal of 850°C, 30 sec in a 100% nitrogen ambient.

TABLE 3. Implant Conditions Test in this Study

#	Fast Scan Speed (m/s)	Slow Scan Speed (mm/s)	Wafer Temp. (degC)	Remark
1	~ 1.0	~ 25	15	
2	~ 2.0	~ 45	15	**Normal Operation**
3	~ 3.0	~ 45	15	
4	~ 2.0	~ 45	45	

* Both fast scan and slow scan speed was varied by the beam current which was measured in real time during the implant.

RESULTS AND DISCUSSION

A. Bare Wafer Test

Figure 3 shows dopant profiles after annealing for both boron and fluorine for the "normal" mode of the spot beam implanter (#2 on Table 3). Even though the boron profile was exactly matched between the two implanters, the secondary fluorine peak is noticeably lower in concentration and slightly deeper in depth for the spot beam implanter. This indicates that the a/c interface is slightly deeper, and that the end-of-range defect density is lower for the spot beam implanter under these conditions. The lower defect density from the spot beam implanter leads to reduced gettering of fluorine at the former site of the a/c interface.

Despite the benefits of a lower defect density from the spot beam implanter, it was desired to match the device characteristics of the ribbon beam implanter, which was process of record. The dose rate was decreased by increasing both the fast and slow scan speeds. This increased the height of the fluorine secondary peak (Figure 4), but not by enough to match that of the ribbon beam implanter. The faster scan speed leads to a lower implant duty cycle. This likely results in more damage relaxation time [7], and thus a thinner amorphous layer thickness and more residual defects at the end of range after annealing.

The effect of wafer temperature during implant using the spot beam is shown in Figure 5. Increasing the wafer temperature during implant on the spot beam implanter resulted in higher secondary peak heights for both boron and fluorine. For the fluorine case, the secondary peak was higher than the ribbon beam reference. Implant damage relaxation is increased by the higher wafer temperature [6,10], leading to a thinner amorphous layer thickness. It is not clear at this time why the higher implant temperature increased the height of the secondary fluorine peak after anneal.

From these SIMS profiles we conclude that the spot beam implanter generated more primary defects than the ribbon beam implanter due to its higher instantaneous dose rate. However, fewer residual defects remained after post implant anneal as compared with the ribbon beam, indicating a better quality junction with the spot beam implanter. These extra residual defects from the ribbon beam are believed to cause faster lateral diffusion of boron and shorter effective channel length than the spot beam implanter. Even though it is not generally desirable to increase residual defect levels, increasing dose rate and/or the wafer temperature could be a solution for transistor matching between two different types of implanters.

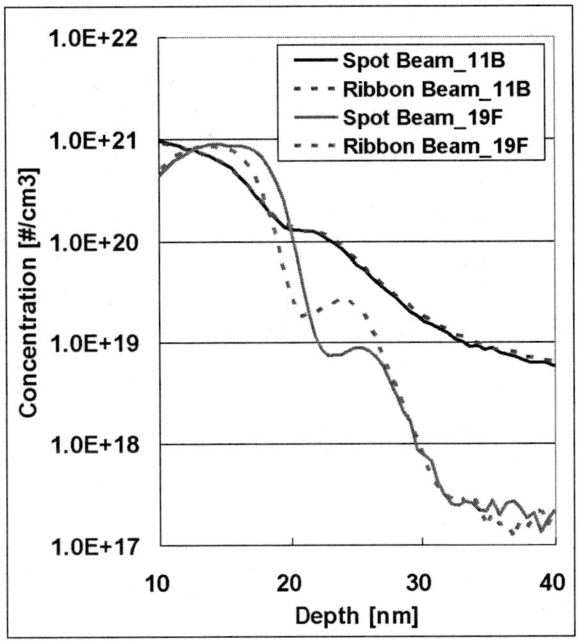

FIGURE 3. SIMS profiles of both ^{11}B and ^{19}F in the region of secondary peak. The implant and anneal conditions are same as in Figure 1.

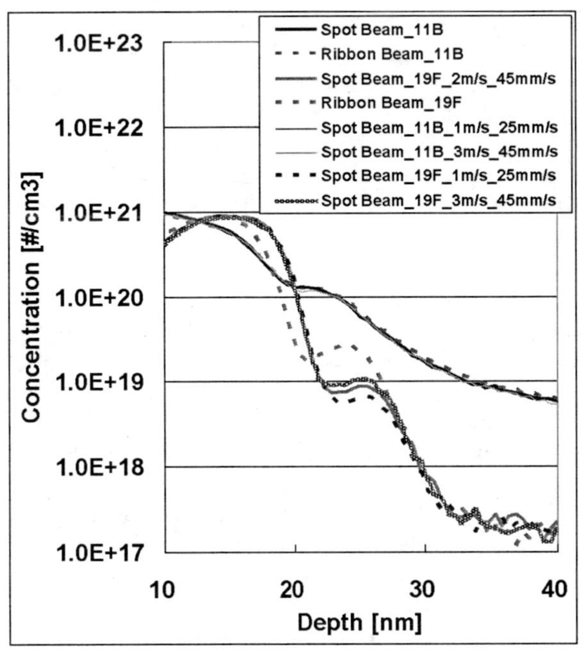

FIGURE 4. Effect of fast and slow scan speeds of a 2D mechanical scan system on the ^{11}B and ^{19}F profiles. Decreasing the dose rate by increasing both scan speeds increased the secondary fluorine peak height, but not to the level of the ribbon beam implanter.

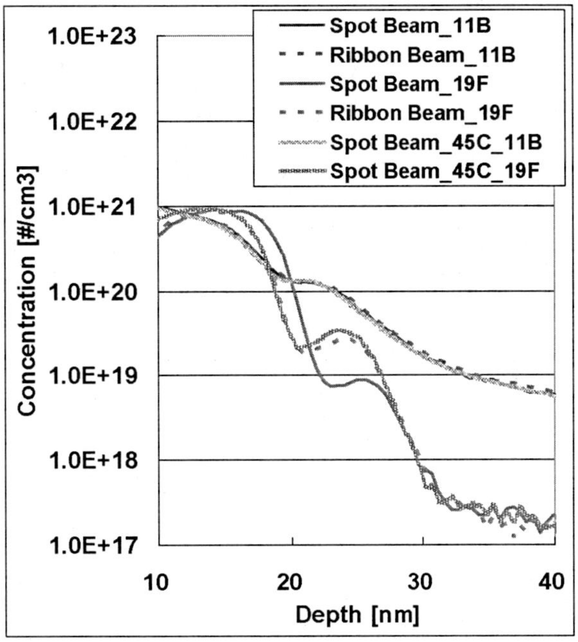

FIGURE 5. Effect of wafer temperature on profiles of both ^{11}B and ^{19}F. Increasing the wafer temperature up to 45°C increased the fluorine secondary peak to a level higher than that of the ribbon beam reference. The normal wafer temperature is 15°C.

B. Device Wafer Test

Based on the bare wafer test results, device wafers were implanted on the spot beam system at temperatures of 25, 30 and 35 °C. By matching the residual defect density through matching of the fluorine secondary peak, the threshold voltage could be also matched.

Figure 6 showed the device test results. The V_T difference between the ribbon and spot beam is less than 0.5% for all temperatures investigated. However, the overall deviation was reduced by increasing temperature. The devices implanted at 30°C were almost identical to the ribbon beam reference. The 3% V_T difference due to the difference in effective channel length was eliminated by the "warm" implant.

The active area Rs increased with increasing wafer temperature. The Rs sensitivity versus wafer temperature was about 2.0 ohm/sq per °C. The residual defects could be capturing both boron and fluorine. Because more defects would capture more boron, the higher implant temperature could be the cause of the increased active area Rs. The devices implanted at 30°C showed almost the same Rs value as the ribbon beam reference. The 30°C wafer chuck temperature on the spot beam implanter was determined to be the optimal condition for matching both PMOS V_T and active Rs.

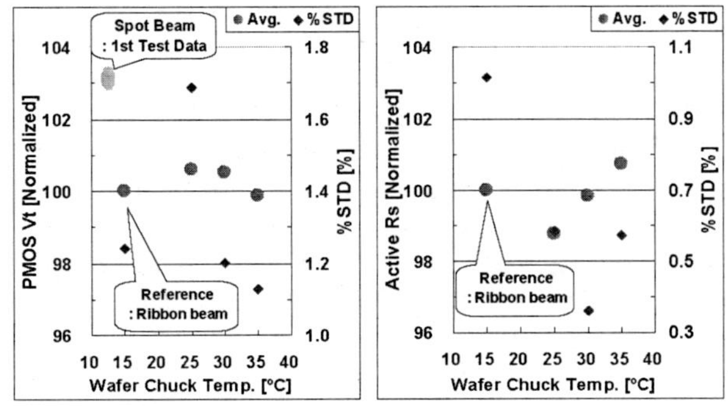

(a) (b)

FIGURE 6. Device test results for (a) PMOS V_T and (b) Active area Rs. The V_T difference between the ribbon and spot beam implanters was less than 0.5% for all temperatures investigated. The active Rs increased with increasing wafer temperature.

ACKNOWLEDGEMENTS

The authors would like to thank Axcelis Korea Field Service Team and Axcelis Product Support Team, especially YS Seo, SK Kim, TS Choi and Edgardo Deleon for reconfiguring the implant system for this study.

REFERENCES

1. A. Agarwal, Proceeding of the 13th International Conference on IIT 2000, p.293.
2. A. J. Murrell *et al.*, *J. Vac. Sci. and Technol.* **B 18**, 462 (2000).
3. K. S. Jones *et al.*, *J. Appl. Phys.* **63**, 1414 (1998).
4. R. D. Goldberg *et al.*, *Mater. Res. Soc. Symp. Proc.* **316**, 259 (1994).
5. R. D. Goldberg *et al.*, *Nucl. Instrum. Methods, Phys. Res.* **B 106**, 242 (1995).
6. J .A. Van den Berg *et al.*, *J. Vac. Sci. Technol.* **B 20**(3), 2002.
7. K. Shim *et al.*, *Proceedings of the 16th International Conference on IIT 2006*, p137.
8. C. Campbell, *et al.*, "Beam Angle Control on the VIISta 80 Ion Implanter", in *Proceedings of the 14th International Conference on IIT 2002*, p. 193.
9. P. Splinter, *et al.*, "Optima HD: Single Wafer Mechanical Scan Ion Implanter," in *Proceedings of the 16th International Conference on IIT 2006*, p. 601.
10. J .A. Van den Berg *et al.*, *Nucl. Instrum. Methods, Phys. Res.* **B 183**, 154 (2001).

The Effect of Amorphization Conditions on the Measured Activation of Source Drain Extension Implants

Jonathan England[1], Alexander Kontos[1], Anthony Renau[1], Russell Gwilliam[2], Andrew Smith[2], Andrew Knights[2] and Amitabh Jain[3]

[1]*Varian Semiconductor Equipment, 35 Dory Road, Gloucester, MA 01930, USA*
[2]*Surrey Ion Beam Centre, Nodus Laboratory, The University of Surrey, Guildford, Surrey. GU2 7XH, UK*
[3]*External Development & Manufacturing, Texas Instruments, 13121 TI Boulevard, MS 365, Dallas, TX 75243, USA*

Abstract. Un-patterned wafers were processed using low-dose Indium or medium-dose Germanium pre-amorphization implants (PAI) followed by p-type dopant implants of BF_2 or carborane (CBH). The wafers were then annealed by RTA (spike), laser anneal (LSA) or combination of LSA and spike. Active dopant distributions calculated from SIMS and sheet resistance measurements compared favorably with those determined by differential Hall, which is a challenging technique for shallow profiles. The trends in B diffusion behavior and activation are discussed in relation to the different implant damage budgets, damage evolution during the anneals and presence of fluorine. In particular, for low thermal budget LSA only anneals, CBH implants appear to give higher activation than BF_2 due to the absence of fluorine.

Keywords: Pre-amorphization, ion implantation, transient enhanced diffusion, activation, laser annealing
PACS: 61.72.uf; 85.40.Ry

INTRODUCTION

Many schemes have been proposed for the production worthy manufacture of shallow, high conductivity, low leakage 32nm transistor junctions. The most challenging step of these schemes often involves formation of the PMOS source drain extension. This study compared two PAI schemes previously used at larger design rules, BF_2 and ortho-carborane (CBH[1]) doping implants and various anneal strategies. Co-implants were not included. The use of the differential Hall (DH) technique for shallow junction metrology was also evaluated.

EXPERIMENTAL CONDITIONS

All implants were carried out into 200mm Si wafers at Varian using a VIISta HC implanter except for the Indium implants, which were carried out on a VIISta 810 medium current tool at Texas Instruments. PAI conditions of Ge/20keV/3.5E14/cm² or In/15keV/3E13/cm² were followed by dopant implants at 0° tilt of CBH/6.8keV/1.4E14/cm², or BF_2/2keV/1.4E15/cm². It was noted (see Figure 2) that the B energies were not exactly matched for these conditions. The B from the CBH had a nominal equivalent energy of 500eV whereas the B from the BF_2 was shallower, with an equivalent energy approximately 10% lower. The different combinations of implants were annealed with one of the following schemes: LSA at 1270°C for 0.8 ms; spike at 950°C for ~2 seconds; LSA followed by a spike (both with the same parameters as above). The sheet resistance, Rs, of each wafer was then measured by a KLA-Tencor RS100 4 point probe.

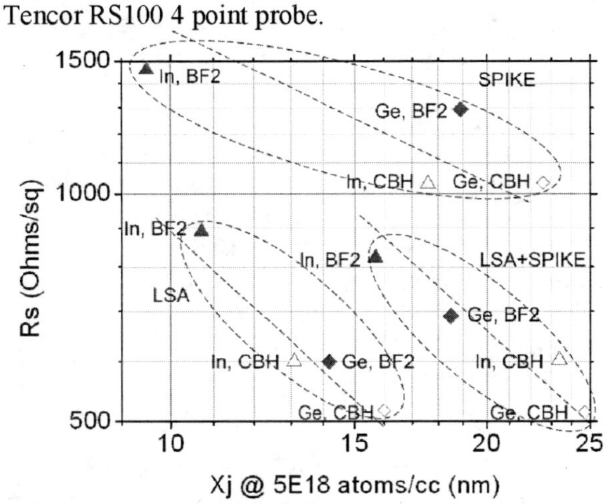

FIGURE 1. An Rs-Xj plot for the PAI, dopant combinations of this study. Ellipses enclose same anneal conditions; dotted lines are guides to the Rs-Xj trends.

CP1066, *Ion Implantation Technology,* edited by E. G. Seebauer, S. B. Felch, A. Jain, and Y. V. Kondratenko

TABLE 1. Comparison of CBH and BF_2 observations

Anneal	Activation	Diffusion	Mobility
Spike	Similar	CBH deeper	Similar
LSA	CBH higher	Similar	Similar
LSA+ Spike	Similar	CBH deeper	Similar

Next, pieces of each wafer were profiled by SIMS, which was carried out at Texas Instruments on a Cameca 6F with 500 eV O_2^+ ions and oxygen leak for boron. The B SIMS profiles for the CBH implants shows ^{10}B + ^{11}B, whilst the BF_2 shows ^{11}B only. In this study, the junction depth (Xj) was defined as the depth at which the B concentration was $5x10^{18}$ atoms/cm^3. The Rs–Xj plots for the various samples are shown in Figure 1 and the SIMS profiles are in Figure 2.

The SIMS and Rs data could be combined to estimate the maximum concentration of electrically active boron in the following way. It was assumed that at low concentrations, the activated profile was same as the SIMS profile. However, near the surface where the chemical concentration was high, the active boron concentration was assumed to reach a maximum value. Using the hole mobility as a function of active boron concentration given by Schroder[2], the contribution to the sheet resistance of each part of the boron depth profile could be calculated. The maximum active boron concentration was chosen so that the calculated value of the total sheet resistance matched the measured value. The effects of surface oxide were ignored. The estimated maximum activation concentrations and their associated hole mobilities are reported in Figure 3. The activation levels calculated from SIMS/Rs often agreed with another qualitative estimate of the active B concentration, which was indicated in the SIMS profile where the diffusion tail breaks from the immobile, surface peak[3].

Further pieces of each wafer had van der Pauw structures fabricated on them at The University of Surrey, and DH measurements were carried out[4]. This technique makes repeated Hall and resistivity measurements after thin layers of the material have been removed. In this way, the depth profiles of the electrically activated boron concentrations and mobility could be measured. The Hall scattering factor (the ratio of field-free mobility to that in a magnetic field) has been taken as unity. The technique is limited for shallow junctions by the accuracy of the material removal step. Also, depletion effects mean that the junction is distorted and can only be measured to a limited depth. For these reasons, the results are semi-quantitative but certainly indicative of trends.

Finally, junction leakage measurements were made on mesa diode structures of varying sizes fabricated on other pieces of each wafer at The University of Surrey. This is a more sensitive technique than assessing damage by cross sectional or planar TEM. The junction leakages for all the samples of this study were measured to be extremely small (~10^{-8} A/cm^2).

RESULTS AND DISCUSSION

The trends in the DH and SIMS/Rs results for activation and mobility (Figure 3) follow each other quite closely. Indeed, applying a reasonable Hall scattering factor of ~0.6 would bring the maximum activated B and mobilities into good agreement. The DH would still suggest higher activation (and lower mobility) for the LSA cases than the SIMS/Rs calculations. In most cases, the relative depths of the DH profiles for the various conditions in Figure 2 agree with the relative depths of the SIMS profiles. Hence DH was a useful technique for this study.

The total damage remaining at the completion of the implants, how this evolved during the various anneals and the presence of fluorine is an important framework within which to explain the observations of diffusion and activation behavior in this study.

When comparing the results for the two PAI conditions on the Rs-Xj plot (Figure 1), it can be seen that Ge always showed lower Rs and deeper Xj than In. TEM data confirms the TRIM prediction that the In PAI creates only a little damage, located close to the surface. The more complete and deeper damage created by Ge PAI released more interstitials from the EOR damage during the anneals. This may explain the observation of greater B diffusion after Ge, compared to In PAI. The activation measurements (Figure 3) suggest that the lower Rs values for Ge were a consequence of the deeper dopant profiles, rather than higher peak B activation.

TABLE 2. Comparison of observations for the various anneal conditions

Anneal	B Tail and F Out-Diffusion	EOR Decoration by F	Activation	Same Rs-Xj trend for In & Ge
Spike	High TED	No	Low	No
LSA	Low TED	Yes	High	Yes
LSA + Spike	Intermediate CED	No	High	Yes

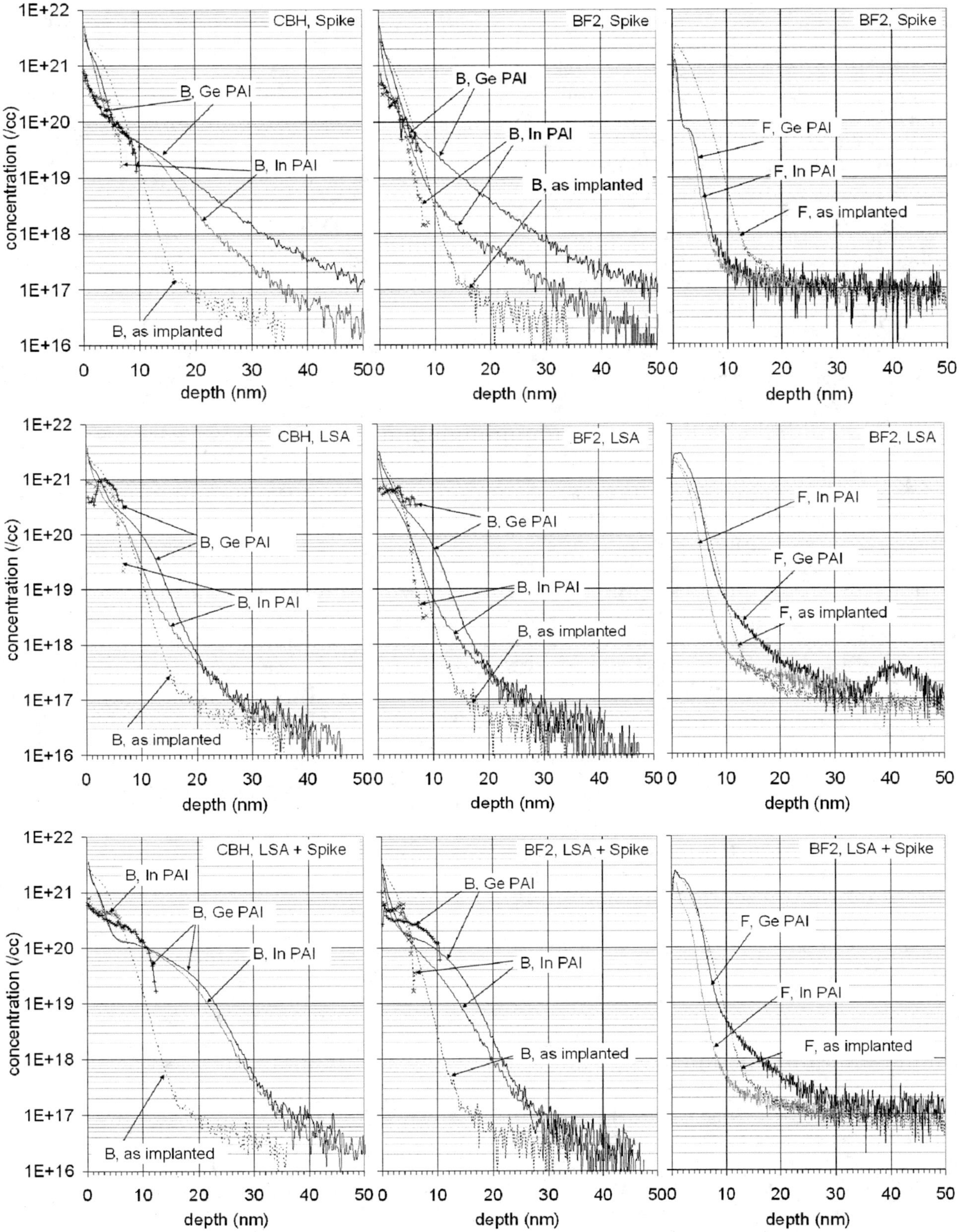

FIGURE 2. SIMS profiles (lines only) before and after annealing and differential Hall profiles (lines and symbols) after annealing for: B from CBH; B from BF_2; F from BF_2 implants.

The effects of the different dopant species are summarized in Table 1. The fact that different amounts of diffusion can be observed after BF_2 and CBH implants may be a consequence of the fluorine introduced by the BF_2. F is known to annihilate interstitials, which mediate B diffusion, through its ability to getter vacancies[3]. Other experiments suggest that the intrinsic C implanted as part of the CBH molecular implants has little effect on diffusion[1]. (In contrast, when the C is implanted separately, the deeper C impedes the EOR interstitial flux from interacting with the shallow B[5]). In addition, the CBH would be expected to retain more damage after implant than BF_2 due to the molecular effect[6] (and the CBH energy was also slightly higher). For the In PAI case, where the dopant damage was a significant addition to the total damage budget, this mechanism may account for the larger difference in Xj between CBH and BF_2, compared to the Ge case, where the dopant damage was swamped by that from the Ge.

The effects of the anneal conditions are summarized in Table 2. The observations of the SIMS profiles are consistent with the spike anneal promoting the greatest amount of damage evolution and interstitial release, leading to the deep tails indicative of transient enhanced diffusion (TED). In contrast the LSA evolved the damage such that few interstitials were released and the profiles remained shallow. The spike after the LSA appears to not have been able to release as many interstitials as the spike alone, as there is little further TED in the tail. However, the profiles become boxier as concentration enhanced diffusion (CED), driven by the boron activated during the LSA step, occurs.

The clearest trends in the activation measurements (Figure 3) are that the LSA gave much higher activation than either spike or LSA + spike, and that CBH showed higher activation than BF_2 after the LSA anneals. As the formation of Boron Interstitial Clusters (BICs) is an important mechanism in the de-activation of B, the first observation is consistent with the explanation again that the LSA released the lowest number of interstitials for BIC formation. The difference in activation for the dopant species after LSA may be explained by a lower competition for lattice sites between B and C after CBH implants compared to that between F and B after BF_2. There is a hint that CBH may have had slightly lower activation than BF_2 for the In PAI cases that were followed by spike and LSA + spike anneals. Similar to the diffusion argument, this may be explained by the dopant damage being a large fraction of the total damage budget (and the total damage was more surface peaked) for the In PAI so that the spike anneals may have allowed more extensive BIC formation after CBH than BF_2. After Ge PAI, the presence of F appears to be a more important factor.

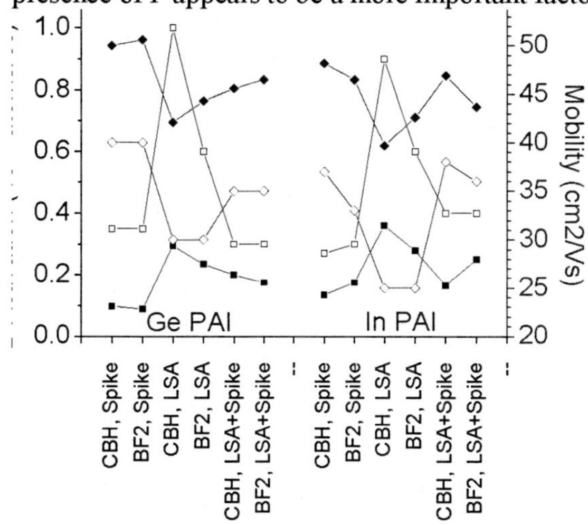

FIGURE 3. Boron activation: (■) calculated from Rs/SIMS; (□) measured by differential Hall; and mobility: (◆) from Rs/SIMS; (◇) differential Hall

CONCLUSIONS

Differential Hall has shown itself to be a useful, semi-quantitative technique for 32nm junction metrology. Many of the observations of PAI, dopant and anneal schemes could be explained by considering implant damage, thermal budgets and the presence of fluorine. In particular, the higher levels of boron activation after laser anneals for CBH compared to BF_2 implants, regardless of PAI scheme, appears to be attributed to the absence of fluorine, rather than damage considerations.

REFERENCES

1. A. Renau, *Extended Abstracts of the Seventh International Workshop on Junction Technology*, pp. 107-116 (2007)
1. D. Schroder, *"Semiconductor Materials and Device Characterization"* (2nd ed) John Wiley and Son, New York 1998, pp. 551-553
2. N. Cowern, B. Colombeau *et al., Appl. Phys. Lett.* **86**, 101905 (2005)
4. N. Bennett , A. Smith, B. Colombeau, R. Gwilliam, N. Cowern, B. Sealy, *Mat. Sci. and Eng.* **B 124–125**, 305–309 (2005)
5. E. Collart, D. Kirkwood *et al., Proceeding of IIT2004, Part II,* 11-14 (2005)
6. L. Perlaz, L.A. Marques and J. Barbolla, *J. Appl. Phys.* **96**, 5947-5976 (2004)

Photon Influence on P and B Diffusion

W. Aderhold[1], A. Hunter[1], S. B. Felch[2], and J. Ranish[1]

[1]*Applied Materials, 974 E. Arques Ave. Sunnyvale, CA 94085*
[2] *Spansion,* 915 Deguigne Drive, Sunnyvale, CA 94088

Abstract. We investigate photon effects for two thermal processes: implanted dopant activation and diffusion; and silicon oxidation. Because the Applied Materials Radiance*Plus* RTP system heats only one side of the wafer with lamps, the thermal and photon effects are separable by changing the side of the wafer that is irradiated. The difference in process results can then be interpreted in respect to the spectral difference of the lamp radiation and the grey body radiation emitted by the hot wafer. No significant effect due to the presence of high energy photons in either process was observed.

INTRODUCTION

Shallow junction formation with rapid thermal processing (RTP) requires high wafer irradiance for rapid heating in order to achieve high electrical activation (which has an Arrhenius temperature dependence) while minimizing diffusion of the implanted dopants. In the Applied Materials RTP systems the wafer requires irradiances >100W/cm^2 in order to heat the wafer at rates in excess of 300K/s. This paper is an investigation into the effect, if any, that the high photon flux has on the implant anneal.

Some authors have reported photon effects on diffusivity [1] and defect annihilation [2] in silicon. Lojek *et al.* reported improved defect annihilation with high energy photons for amorphizing implants [2] compared to lower energy photons in furnace-like anneals. In the more recent work [1] Jung and Seebauer observed diffusivity enhancements for Si self-interstitials in n-type Si, but not in p-type Si. As much of the diffusion of n-type and p-type source-drain regions is in the extrinsic concentration regime such change in the diffusivity of self interstitials should influence the dopant profiles of interstitially diffusing species (like B, P). The diffusivities of self interstitials as well as the dopant diffusivity are known to be dependent on the Fermi level [3,4]. A possible explanation for these phenomena is that the high energy photons create electron-hole pairs, increasing the minority carrier densities and thus shifting the Fermi levels. It is also speculated that the reported difference for n- and p-type diffusivities of self interstitials is due to the position of their ionization energies relative to the Fermi level [5]. As the work [1] was done with monochromatic 633nm laser with 1.6W/cm^2 [5] and at relatively low temperatures, we wanted to reproduce the results with a realistic spike anneal process. Because irradiances as low as 15W/cm^2 of lamp radiation will generate electron-hole pairs (due to photon energies greater than the band gap), similar enhanced diffusivity should be observed during RTP.

EXPERIMENTAL DETAILS

The Applied Materials Centura RTP chamber consists of three main parts: a light source, a wafer support mechanism and temperature measurement system. The light source is a honeycomb array of tungsten filament lamps. The wafer support design is such that lamp radiation is prevented from reaching the backside of the wafer, to prevent interference with the optical pyrometry system. The wafer is rotated during processing. Multiple optical pyrometers are situated at fixed radii beneath the wafer and measure the wafer temperature.

Because the underneath side of the wafer is kept isolated from direct lamp radiation, the Centura RTP tool is perfectly suited to conduct the experiments into photon effects by comparing the process results of wafers irradiated from the front side and the bottom side. The spectral irradiance of the RTP light source at various voltages is shown in Fig 1.

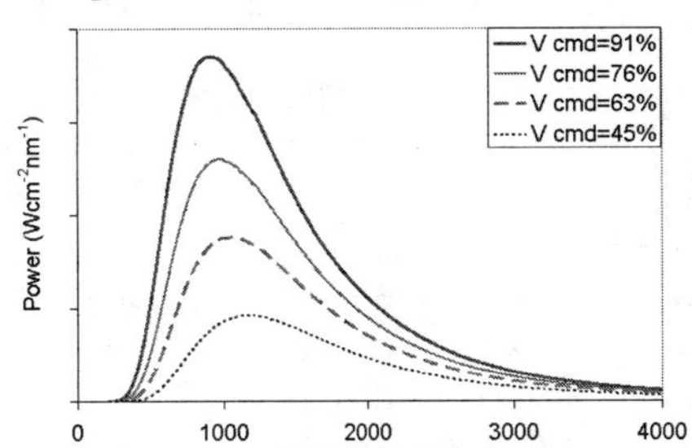

FIGURE 1. Spectral irradiance of RTP lamp source

CP1066, *Ion Implantation Technology,* edited by E. G. Seebauer, S. B. Felch, A. Jain, and Y. V. Kondratenko

To calculate the carrier density generated by the RTP tool we convert this spectrum into photon energy. Only photons with energies higher than the band gap $E_g(T)$ can generate electron hole pairs. Fig 2 shows the carrier density as a function of photon energy for two different wafer temperatures: 550°C (E_g=0.95eV) and 1000°C (E_g=0.77eV). To calculate the optical parameters of Si we used the RAD-PRO spreadsheet from Lee and Zhang [6].

The following are the experimental conditions.

1. 8 Ωcm n-type [100]-Si ,BF_2 ,3keV $1 \cdot 10^{15}$ cm^{-2} ,spike RTA to 950°C, 1000°C, and 1050°C
2. 8 Ωcm p-type [100]-Si , spike RTO 1000°C, 1050°C, and 1100°C
3. 8 Ωcm n-type [100]-Si , BF_2 3keV,$1 \cdot 10^{15}$ cm^{-2} , 15s soak RTA to 1040°C, and 1050°C
4. 8 Ωcm p-type [100]-Si , Ge PAI, B + C co-implants, spike RTA at 1050°C [7].
5. 8 Ωcm n-type [100]-Si , Ge PAI P + C co-implants spike RTA at 1050°C [8].

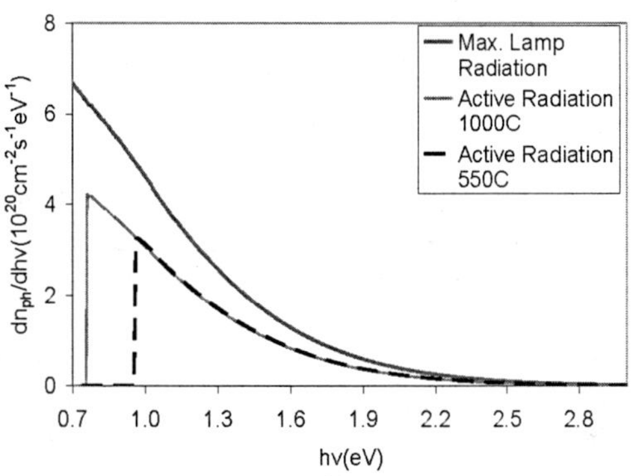

FIGURE 2. Carrier density vs photon energy generated by RTP lamps for two different wafer temperatures. "Active" here refers to the spectrum that results in carrier generation.

The temperature profiles and lamp power profiles for a typical spike process and a typical soak process are shown in Figures 3 and 4. The only difference between spike RTO and spike RTA is that RTO is done in an O_2 ambient and RTA is performed in N_2 ambient. Note that there are two distinct regimes for the RTP recipe: the preheat regime, typically 550°C, and the fast ramp regime where the wafer is rapidly heated to 1000°C. Therefore, we look at radiation effects for the two cases 1000°C and 550°C where the radiation effects are maximized. In contrast to the spike process the soak process uses much slower ramp rates, and corresponding lower lamp powers, but the lamps remain at intermediate power levels to keep the constant temperature during soak. The photon energies are much lower (spectrum shifts to longer wavelength) in a soak process as can be seen from figures 1 and 4.

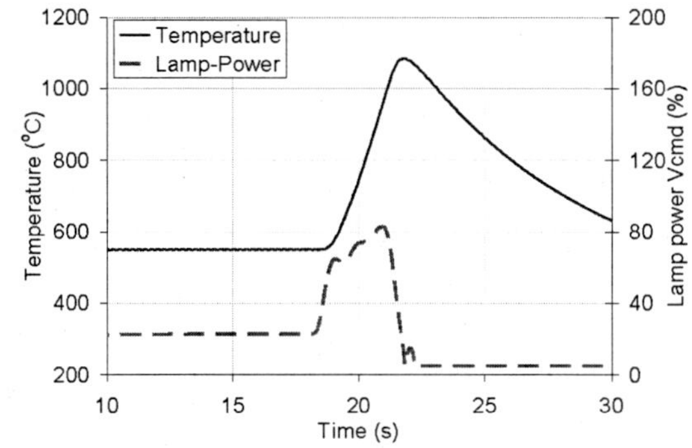

FIGURE 3. Spike temperature vs. time profile.

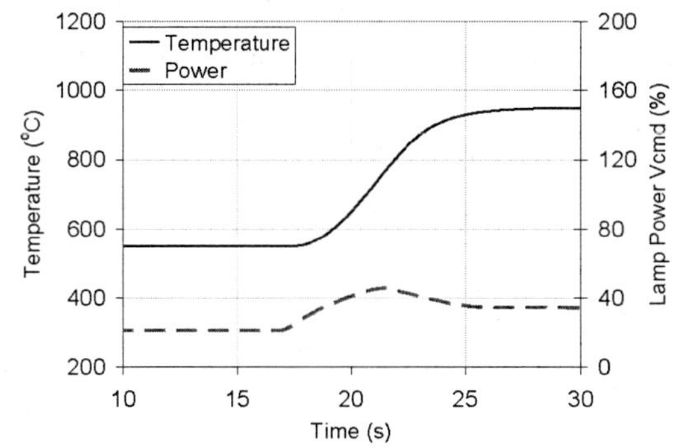

FIGURE 4. RTP soak temperature vs. time profile.

To verify that for the temperatures of interest no photons capable of generating electron-hole pairs can travel through the wafer thickness, we used the RAD-PRO spreadsheet. The result of this calculation is shown in Figure 5. and confirms this hypothesis even for the most intense lamp radiation used in the experiments. We can safely conclude that the underneath side of the wafer does not, in fact, receive any radiation from the lamps.

However, the side of the wafer facing the lamps does receive a high photon flux. Again using RAD-PRO we calculate the absorption in thin layers of silicon and obtain the minority carrier generation rate G per unit area through the depth of the wafer under the assumption that each photon with an energy E>Eg creates one carrier we get for x=Depth (nm):

$$G = nm/\tau = d(dnph/dhv)/dx \tag{1}$$

Again under assumptions for maximum irradiance we take the photon flux in Figure 2 for 1000°C and equation (1) we get Figure 6.

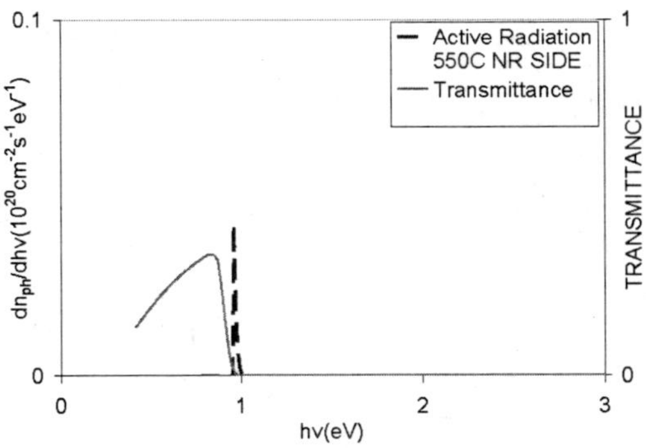

FIGURE 5. Active photon flux for 550°C on the NR side of an 8 Ωcm wafer with the transmittance from RAD-PRO calculation multiplied with the total active photon flux for 550°C from Figure 2.

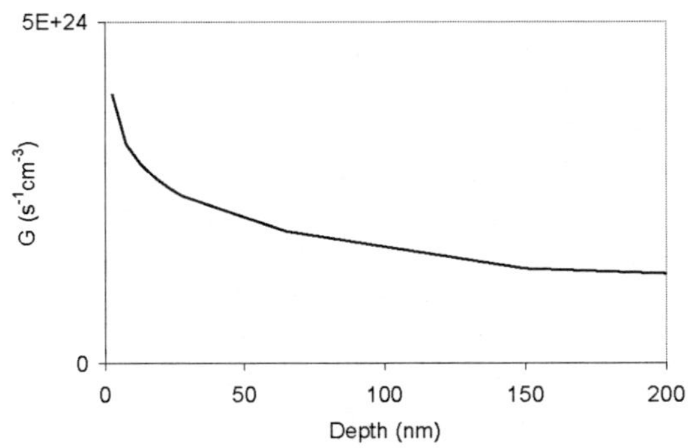

FIGURE 6. Worst case minority carrier generaton rate R side of an 8 Ωcm wafer at 1000°C

From the minority carrier lifetime, τ, we derive the minority carrier density. The carrier lifetime depends on a number of parameters:

$$\tau = 1/(Nt \,.\, vth \,.\, \sigma n) \tag{2}$$

Where, Nt is the density of traps, vth the thermal velocity and σn the capture cross section. These parameters are not well known especially for implanted substrates at process conditions. However, recent work by Wagner and Rau [9] extracted τ vs. temperature from experimental data on Si solar cells from which we extrapolate the value for 1000°C. The value for τ measured for implanted Si solar cells at room temperature is 20µs [10] from which we estimate the lifetime at 1000°C to be 0.2ms. From that value of τ the resulting surface concentration of minority carriers is 8E20 cm^{-3} for the carrier generation at 1000°C. Such high densities do not exist in real silicon as the carrier lifetime in the surface region is much lower due to recombination at defects, the surface, and due to diffusion in gradient direction. The diffusion length of carriers is given by:

$$L = (\tau \,.\, \mu \,.\, kT/q)1/2 \tag{3}$$

With the temperature dependence of the mobility µ ~ $T^{-3/2}$ [11] and the temperature dependence of τ the diffusion length increases by about a factor of 2 from room temperature to 1000°C. This yields a diffusion length at 1000°C to be approximately 10 µm. Therefore, a significant number of carriers generated at the surface cannot traverse the thickness of the wafer. The sheet resistance measurements for implant spike anneals were done with a type-H 4-point probe tool RS-100 from KLA-Tencor Inc. The oxide thickness was measured with the F5 ellipsometer from KLA-Tencor Inc. The SIMS profiles were measured at Evans Analytical Group

RESULTS AND DISCUSSIONS

The sheet resistance measurements across the wafer diameter of the BF2 implant anneals are shown in figure 7.

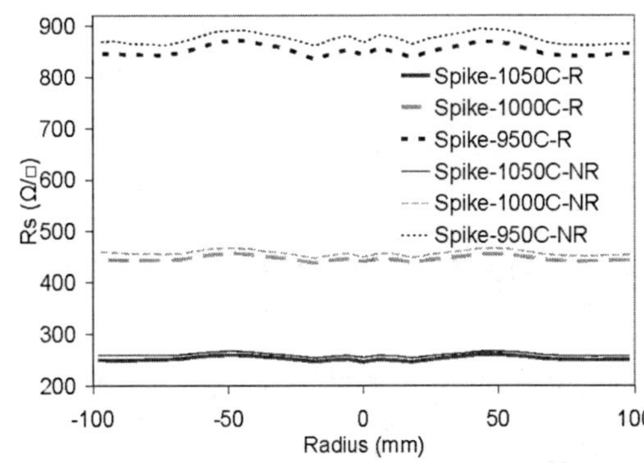

FIGURE 7. Sheet resistance for BF_2 implant (3KeV, $1*10^{15}cm^{-2}$) spike anneal Radiated (R) vs. Non-Radiated side (NR)

Wafers with the front side facing the lamps (R) and wafers with the front side facing down (NR) were processed with the same recipe temperatures of 950, 1000, and 1050°C, respectively. As can be seen from Figure 7, the wafers processed at the same temperatures measure slightly different sheet resistances for the R vs. the NR cases. This difference is consistent with a temperature shift of about 3 degrees from the front side of the wafer to the back. This is in accordance with results reported by Lojek [12] for temperature loss across the wafer with single sided heating. Importantly, the Arrhenius calculation yielded the same activation energy of 1.7eV for both the R and NR cases.

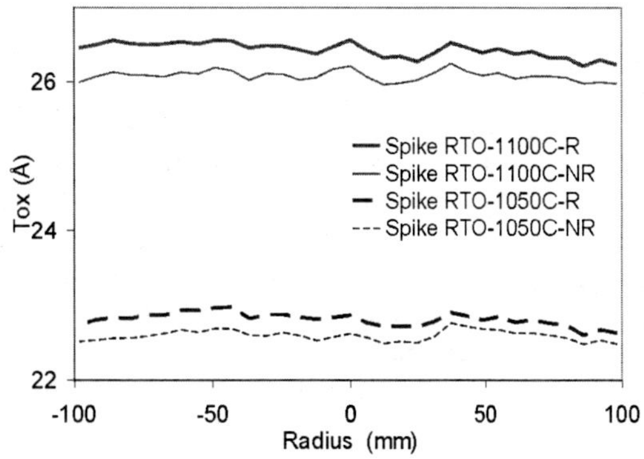

FIGURE 8. Oxide thickness for spike RTO Radiated (R) vs. Non-Radiated side (NR)

This result should be verifiable for different processes with the same thermal history. To verify this, we ran the spike RTO experiments. The results are shown in Figure 8 and confirm the same shift in temperature between the R and NR experiments. For the Arrhenius plot we used the grown oxide during spike, assuming a starting oxide of 14Å. The activation energy is smaller than for the spike anneal but is the same for both cases R and NR.

We also performed the same experiment for a short soak on the BF_2 implanted wafers. From an Arrhenius plot of the data we measure an activation energy of 1.21eV and get the same shift of 3K as in the previous experiments.

Finally, we adjusted the recipe temperatures to compensate for this 3K difference and annealed boron and phosphorous implated wafers. The SIMS profiles are shown in figures 9 and 10.

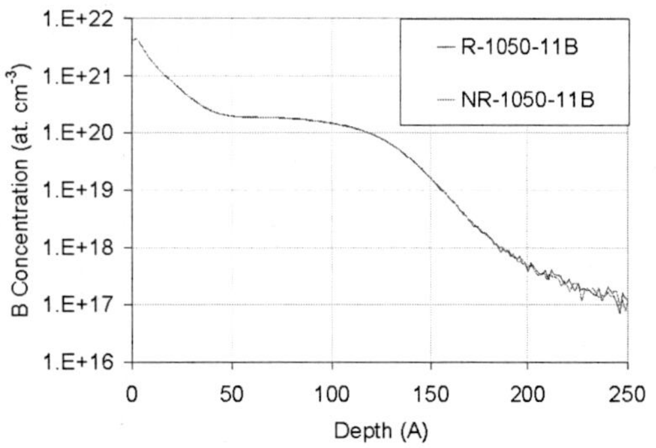

FIGURE 9. Boron SIMS profiles for 45nm P-MOS S/D after 1050°C spike anneal corrected for temperature difference across wafer radiated (R) vs. non-radiated (NR)

The SIMS profiles verify that the dopant diffusion is identical to the limits of the metrology between irradiating the front versus the back side of an implanted wafer.

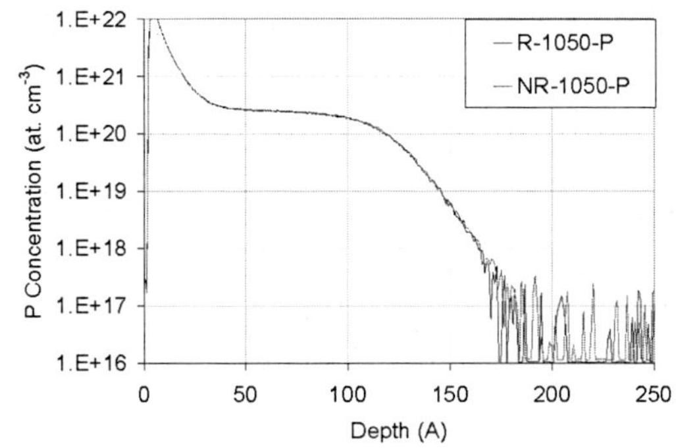

FIGURE 10. Phosphorus SIMS profiles for 45nm N-MOS S/D after 1050°C spike anneal corrected for temperature difference across wafer radiated (R) vs. non-radiated (NR)

CONCLUSIONS

We found no evidence for photon interaction in the activation of implanted anneals for either spike or soak processes. We find no evidence of photon interaction on oxidation for spike processes. We find no evidence of photon interaction in the diffusion of 45nm p- and n-type S/D dopants.

REFERENCES

1. M Jung, M.Y.L.; Seebauer, E.G.; 4th Int.Worksh. on Junct. Tech., IWJT '04. 15-16 March 2004 pp.87 – 89
2. Lojek, B.; Whiteman, M.; Ahrenkiel, R, -RTP 2001, 25-29 Sept. 2001 p.125 – 131
3. Pichler, P. "Intrinsic point defects Impurities and their diffusion in Si", Section 5.3.3 Wien, Springer, 2004
4. Pichler, P.; Ortiz, C.J.; Colombeau, B.; Cowern, N.E.B.; Lampin, E.; Claverie, A.; Cristiano, F.; Lerch, W.; Paul, S,. IEDM Tech. Dig. IEEE Int. 13-15 12.04 p.:967 - 970
5. Seebauer, E.G, (private communication)
6. Lee, B.J.; Zhang, Z. M., RTP-2005, p.275 – 281
7. Pawlak, B. J.; Janssens, T.; Brijis, B.; Vandervorst, W.; Collart, E. J. H.; Felch, S. B.; Cowern, N. E. B.;, *Appl. Phys. Lett.* **89**, 062110 2006
8. Pawlak, B. J.; Duffy, R.; Janssens, T.; Vandervorst, W.; Felch, S. B.; Collart, E. J. H.; Cowern, N. E. B, *Appl. Phys. Lett.* **89**, 062102 2006
9. Wagner, T.A.; Rau, U., in *Proc. of 3rd World Conf. on Ph. En. Conv., '3.* Vol. 2, 12-16 5'03 p. 1162 - 1165
10. Nubile, P.; Beloto, A.F.; Veissid, N.; Waldman, B.; Pereyra, I.; Andrade, A.M, 21st IEEE Ph. Spec. Conf., 1990., 21-25 May 1990 p.):295 - 298 vol.1
11. Sze, S. M., "Physics of Semiconductor Devices", 2nd Ed., John Wiley & Sons, New York, 1981, section 1
12. Lojek, B, , RTP-2005, 15-16 October 2005 Page: 289

PROCESS CONTROL AND YIELD

Threshold Voltage Mismatch Resulting from Mechanical Vibration

Janet M. Towner

ON Semiconductor, Pocatello, ID 83201

Abstract. A severe yield loss was experienced for several mixed-signal devices which relied on matched transistor pairs in their operation. The loss was caused by abrupt variations in threshold voltage over the short distance that separated the transistors. To determine if the problem was caused by nonuniform dosing, the threshold voltage implant was performed in five parts with the wafer repositioned each time for fivefold symmetry. A significant yield improvement resulted from this change. The pattern of loss was also reproduced on test wafers. Root cause of the micrononuniformity was traced to excessive mechanical vibration of the source rough pump; this pump was located inside the enclosure to isolate it from ground. The vibration was transmitted to the wafer in the direction of the mechanical scan.

Keywords: Ion Implantation, threshold voltage, microuniformity, vibration
PACS: 52.77Dq, 85.40.Ry.

INTRODUCTION

Mixed-signal devices offer advantages over multichip systems but are obviously more difficult to manufacture as they have additional sensitivities [1]. Transistor matching is an important parameter as it is the offset between these pairs that provide core design elements: temperature compensation and differential input circuits [2, 3].

Depressed sort yields were observed in several mixed-signal devices. The problem was intermittent and manifested in varying degrees from wafer-to-wafer within a lot. Over time the magnitude of the problem increased, eventually causing yield to drop by more than 70%. In order to meet delivery schedules, a substantial amount of additional material had to be manufactured to provide sufficient die. A short term solution was required to alleviate excessive costs and a long term solution, with root cause determination, was required to prevent future occurrences.

The primary products affected were those utilizing matched sets of transistors of relatively large physical size. The failure mode was determined to be a mismatch in threshold voltages (Vt). While transistors were functioning within specification, the offset between them caused functional failure. Typical deviations are less than 3mV; affected die showed mismatch far in excess of 25mV.

Surveying the yield losses it was apparent that there was a common, characteristic pattern (Figure 1). Sort wafers maps showed lines of evenly spaced, alternating functional and failing die with a fixed diagonal orientation. Such closely spaced striations were unlikely to result from variations in substrate doping, gate oxide thickness or dopant cross-contamination. For this reason, further scrutiny was placed on processes that can vary in this manner. This included ion implantation, particularly at the n-tub, p-tub and V_t adjust levels. Yields were analyzed according to process equipment. However, no consistent correlation was found. It was therefore necessary to run controlled experiments in order to assist in determining the underlying cause of the failures.

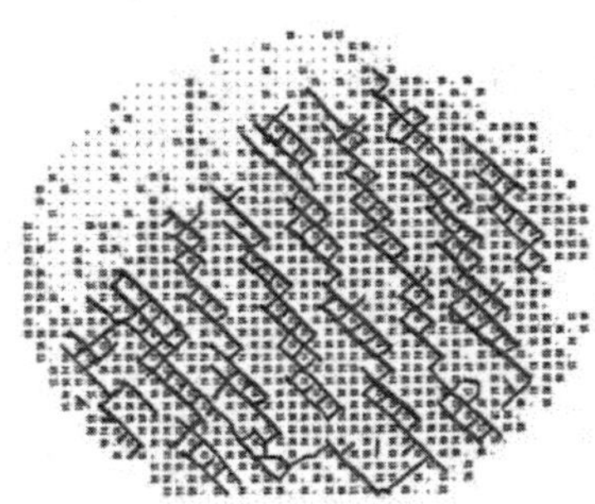

FIGURE 1. Poor sort pattern for Vt matching problem.

EXPERIMENTAL

Three types of implanters were used in manufacturing. One had an electrostatic scan in both

CP1066, *Ion Implantation Technology,* edited by E. G. Seebauer, S. B. Felch, A. Jain, and Y. V. Kondratenko

the vertical and horizontal directions, the second had mechanical scans in all directions and the third was equipped with a so-called hybrid system: wafers are mechanically moved through the beam vertically while being electrostatically scanned horizontally.

For this third type of implanter, the mechanical scan is at a rate of 254mm/sec. This is accomplished using a linear motor and air bearing assembly. In the horizontal direction, the beam is scanned at an average frequency of 1000Hz. This is accomplished with a scan generator and a 30kV amplifier. The system produced a saw tooth waveform that is applied across two conducting plates on either side of the beam path.

Test wafers consisted of silicon coated with 11nm of thermally grown oxide. The Vt adjust implant was performed at a dose of $2.1x10^{12}$ at/cm^2 with BF_2 at an energy of 100keV. A Thermawave was used to characterize the implants.

RESULTS AND DISCUSSION

The initial screening experiment was performed on the product most sensitive to this problem. The experimental design is shown in Table 1. This was a two-level, three-factor test. In this experiment, a direct comparison was made between all three types of implanters with 12 wafers split at each level.

TABLE 1. Average Net Die per Wafer (NDW) as a function of implanter and level

Level	Implanter	NDW	Stdev
N-TUB	2	588	405
	3	807	399
P-TUB	1	833	385
	2	562	400
Vt ADJUST	1	1075	91
	3	319	171

No interaction between implant levels was observed. While statistically significant differences were seen for all three implants (>99.8%), by far the largest difference in yield was at the Vt adjust level. The use of the third type of implanter at this level had the greatest negative effect on yield. In addition, the variation from wafer-to-wafer was increased substantially.

Both the Vtn and Vtp values were also significantly affected at this level. The Vtn increased for the third type of implanter by 8mV while the Vtp showed a 24mV increase. What was also significant was the increase in variation across the wafer, from 5 to 13mV for Vtn and from 3 to 5mV for Vtp. It is this implanter used at this level which will be discussed in the remainder of the paper.

Test wafers were implanted and measured to generate high resolution contour maps on the Thermawave. A remarkable similarity to the pattern of yield loss was seen (Figure 2), with similar angular orientation. Total Thermawave deviation corresponded to a dose of $7x10^{11}$ at/cm^2 or a 33% maximum-to-minimum variation. The similarity in patterns, along with the magnitude of the dose variation, gave confidence that the source of poor yield had been found.

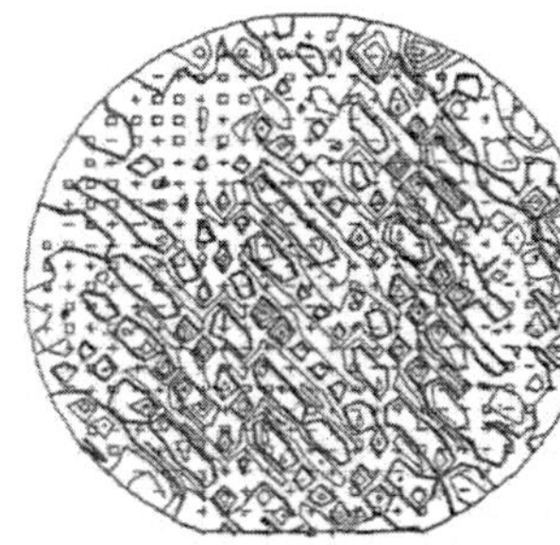

FIGURE 2: Thermawave contour map of Vt adjust implant from the third type of implanter. Average TW is 690.

To greater strengthen this conclusion, tests were next performed with the standard single step implant and one in which the implant was performed in 5 separate steps with the wafer rotated 72°, each step receiving 20% of the dose. A prime number of steps was selected so the pattern of nonuniformity would have the least probability of overlapping. Uniformity was expected to improve as the dose variation would be averaged out.

The corresponding contour map shown in Figure 3 shows significantly improvement compared with the standard wafer. The standard deviation decreased from 0.32% to 0.11%.

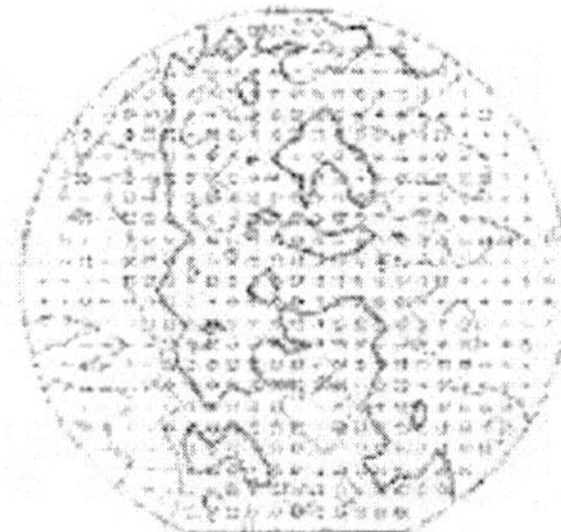

FIGURE 3. Thermawave contour map of Vt adjust implant performed in 5 segments. Average TW is 693.

This comparison was also made for 3 production lots; results are shown in Table 2. For two lots, there is a significant improvement in yield. For the third lot, no significant difference was observed. This result is not surprising, however, as this is the best yielding lot. As previously mentioned, the problem was intermittent and effected lots in varying degrees. Likewise, the poorest yielding lot showed the biggest improvement.

TABLE 2. NDW with standard and 5 step Vt adjust implant

Lot	Standard	5 step	Δ
1	682	979	297
2	1101	1263	162
3	1009	977	-32

In order to characterize the problem in more detail, wafers were implanted at a variety of orientations (single step). Results are shown in Figures 4 and 5. It appeared from these tests that wafer orientation influenced the periodicity and severity of the nonuniformity. In all cases the nonuniform bands are evenly spaced across the wafer. At 0° six bands are observed which are slightly skewed from the horizontal. At 122° the density of bands has greatly increased but the orientation changed by less than 45°.

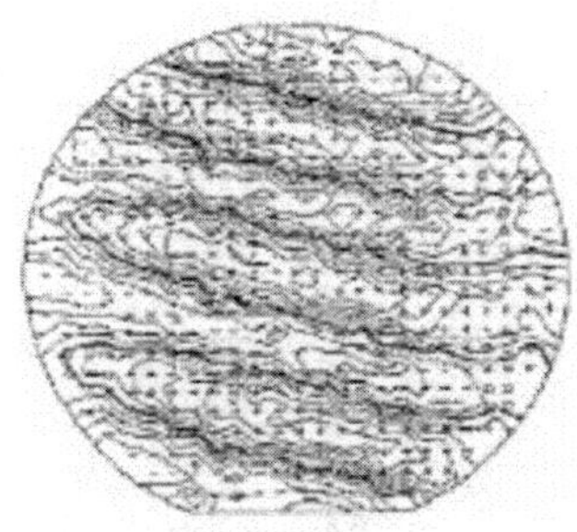

FIGURE 4. Thermawave contour map for the 0° implant orientation.

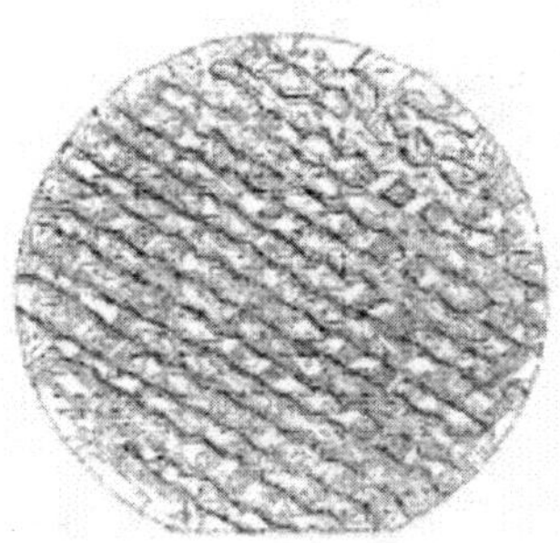

FIGURE 5. Thermawave contour map for the 122° implant orientation.

A possible explanation for the difference would be planar channeling [5]. If channeling was responsible the influence of the screen oxide should have an influence. However, tests with 0, 11 and 22nm oxide thickness showed no variation in the pattern only in the intensity of the Thermawave signal.

To gain better measurement resolution, horizontal and vertical line scans were made. By reducing the scan length to 100mm it allowed a measurement to be made every 1mm. In contrast the measurements were 4mm apart for a 100mm diameter contour scan. This change in metrology allowed the full signature of the problem to be captured. Results are shown in Figure 6 for wafers aligned at 0°, 30°, 60° and 90° orientations during implant. The periodicity and severity of the problem is seen to be independent of the wafer orientation. The different contour patterns are merely artifacts of limited resolution of the measurement tool.

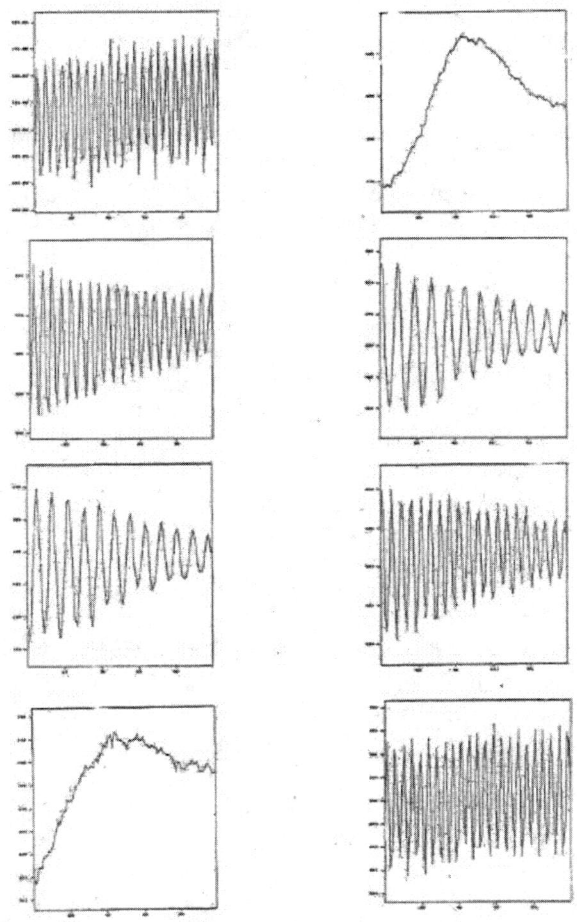

FIGURE 6. Vertical and horizontal Thermawave scans for wafers implanted at orientations of 0°, 30°, 60° and 90° orientations.

The vertical scan of the 0° implant and the horizontal scan of the 90° implant are identical, with the periodicity disappearing completely at an orthogonal orientation. The true periodicity is horizontal striping was 22.5 cycles over a 100mm length, equivalent to 4.44mm per cycle. The wafer moved in the vertical direction at the rate of 254mm/sec. This corresponds to a frequency of 57Hz for the 4.4mm period.

In the design of the implanter, great care had been taken to ensure any variation across the wafer from the electrostatic scan was corrected. Prior to implant, beam current is measured at 1000 points across the scan width. The dosimetry system compensates for variations in beam current by changing the slope of the scan waveform between each point. This is repeated iteratively until the desired uniformity, as specified in the recipe, is achieved.

A comparable cautious approach had apparently not been taken in the design of the mechanical direction of scan. The opening to the Faraday is 7mm wide and 35mm long, so it was not possible to measure variations in this direction in normal operation.

The periodicity of the wafer variation was confirmed with a y-axis displacement test. A method was developed where the majority of the beam was blocked in this direction and only the edge of the beam entered the Faraday using the wafer platen. The linear drive motor was disabled allowing the air bearing assembly to be disengaged. The platen was then manually positioned in the path of the beam. Measurement with an oscilloscope confirmed a 57Hz fluctuation was occurring.

To isolate the source of this noise, measurements were made on all power supplies, the analyzer magnet, the lens and quadropole magnets. No signal was observed at the signature frequency. The facility power lines were also eliminated as a possible cause of the noise. Excessive vibration was however noted in the extraction manipulator housing. Vibration measurements were made of the terminal source/backing pump. This pump is located inside the enclosure in order to isolate the high extraction voltages from ground. Data analysis confirmed rotational vibration at 3447rpm, the signature 57Hz frequency.

The pump was replaced and repeated testing showed excessive Thermawave variation was eliminated. As corrective actions, additional equipment and process monitors were implemented. The source pump was tested for vibration on a monthly basis and high resolution line scans were made on the weekly SPC test wafers.

CONCLUSIONS

A severe yield loss was experienced for several mixed-signal devices. The loss was caused by abrupt variations in threshold voltage over the short distance that separated the transistors. To determine if the problem was caused by nonuniform dosing, the threshold voltage implant was performed in five equal dosed segments with offsets in wafer orientation. A significant yield improvement resulted from this change. The pattern of loss was also reproduced on test wafers.

Root cause of the micrononuniformity was traced to excessive mechanical vibration of the source rough pump; this pump was located inside the enclosure to isolate it from ground. The vibration was transmitted to the wafer in the direction of the mechanical scan of the ion implanter.

ACKNOWLEDGEMENTS

The author would like to thank Dan Slusher, Don Vest and Chris Gibson.

REFERENCES

1. M. J. M Pelgrom, H. P. Tuinhout and M. Vertregt, *Proceedings IEEE IEDM*, 915-918 (1998).
2. M. J. M Pelgrom, A. C. J. Duinmaijer and A. G. P. Welbers, *IEEE J. Solid-State Circuits*, **24** 1433-1440 (1989).
3. S. J. Lovett, L. Wall, M. Welten, A. Mattewson and B. Mason, *IEE Colloquim on Improving the Efficiency of IC Manufacturing Technology*, 11/1-11/5 (1995).
4. A. Rosencwaig, J. Opsal, W. Smith and D. Willenborg, *Applied Physics Letters,* **46** 1013 (1985).
5. C. S. Murthy and G. R. Srinivasan, *IEEE Trans. Electron Devices*, **39** 264-274 (1992).

Using Ion Implantation to Streamline High Volt Processing in Standard CMOS

John J. Naughton and Janet M. Towner

ON Semiconductor, Pocatello, ID USA

Abstract. In a standard 0.5μm single well technology, a deep N-well is constructed using a 150keV P implant at a dose of 1e13 at/cm^2 and driven to a final depth of 4μm. To make this N-well suitable for automotive battery applications it is constructed with 2μm P-tub enclosure. Use of a high energy (780keV P+++) implant streamlined the requisite long diffusion cycle such that it could be processed in a standard diffusion drive. While this resulted in longer implant times, it reduced the overall cycle time for a low volume, high voltage custom devices.

Keywords: Ion Implantation, Diffusion, High Voltage, Automotive Electronics.
PACS: 52.77Dq, 61.72.u, 64.70.kg, 66.30.Xj, 85.40Ry

INTRODUCTION

Electronic features are becoming pervasive to automobiles. Applications range from fuel management, to safety systems, to chassis, and of course infotainment. According to industry analysts the average silicon content per vehicle will have almost doubled by 2010 from the previous ten years [1]. These features are also becoming a "selling point" in a consolidating automobile market. Complicating this growing field is that the development time for a new vehicle has been reduced from seven to as little as three years [2]. These factors, combined with cost and reliability, put significant pressure on automotive vendors including their custom IC suppliers [3]. In response to these demands we have developed high voltage features in a standard CMOS mixed signal technology to solve these requirements and constraints.

In a standard single well CMOS technology a deep N-well (DNW) is constructed using a 150keV P implant at a dose of 5e12 at/cm^2. It is then driven to a final depth of 4μm. To make this deep structure suitable for automotive applications it is constructed with a 2μm P-well block enclosure. This layout technique establishes a breakdown threshold in excess of 80V and the basis for 40V "tile" transistors. These devices are well suited for the automotive circuit protection requirements such as double battery and load dump. This type of technology provides the features, the reliability and the cost constraints attractive to automotive electronic suppliers need for power electronic systems [4].

BACKGROUND

Mixed-signal Application Specific Integrated Circuits (ASICs) are often produced in small lot sizes compared to mass produced standard products. This can complicate manufacturing lines when these custom processes require specific recipes that constrain certain Fab equipment or modules as in the case of a deep N-well drive. To alleviate this particular burden on our diffusion furnaces we had initially constructed our DNW module using standard drives by simply repeating the step three times. In this manner these low volume devices would be transparent to the manufacturing line, processing along with the standard CMOS products.

The purpose of this work was to evaluate process improvements by using a high energy (780keV P+++) implant for the DNW reducing the triple diffusion steps to a single standard drive. While this would result in longer setup and implant times, it would significantly reduce overall cycle time for the part. The obvious challenge would be to determine the energy and dose that could utilize an existing drive recipe to match the dopant profile. Complicating this exercise was the lack of a high energy implanter. However, each of the N-well drives is seven hours and forty five minutes. With an hour clean step of each, four steps could be eliminated for a total savings of seventeen

CP1066, *Ion Implantation Technology,* edited by E. G. Seebauer, S. B. Felch, A. Jain, and Y. V. Kondratenko

hours and thirty minutes. In addition, four queuing delays could be avoided.

In addition to this goal, an improvement in carrier mobility was desirable for the 40V PDMOS; its on state resistance (Ron) was greater than twice that of the N40V structure. Previous experiments manipulating dosage had shown a trade-off in Ron with breakdown voltage (BVDSS). We considered that better performance might be obtained using B+ rather than BF2+ for the p-channel resurface (PCR) implant considering the reduced damage to the substrate for implantation with elemental boron [5]. Since the PDMOS is most sensitive to the p-drift region [6] this might also show if there was an effect of the F+ with LOCOS formation in its extended drain.

The DNW is first established using a 150keV P implant at a dose of 5e12 prior to the standard N-well. Cross-contamination could be controlled because the implant was performed on a tool dedicated to P implants. Energy contamination was controlled by proper set up procedures and tight recipe limits [7].

Figures 1 and 2 are cross sectional representations of the devices. The 40V NMOS transistor is constructed in the standard cmos P-tub but using the DNW as an extended drain region underneath the field oxide. The extended drain region then transitions to standard N-well and then to N+ active for improved Ron, Figure 1. The 40V PMOS transistor is constructed inside the DNW for isolation. It uses a 40keV BF2 implant at a dose of 3e13 at/cm^2 for a PCR for the extended drain region in a standard P-tub, Figure 2.

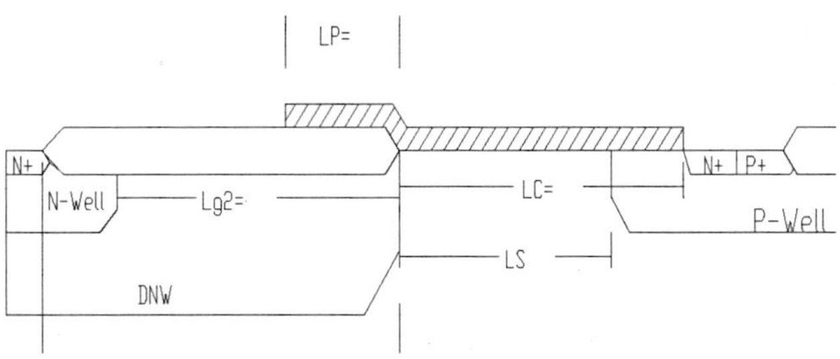

FIGURE 1. Cross sectional diagram of NMOS tile structure.

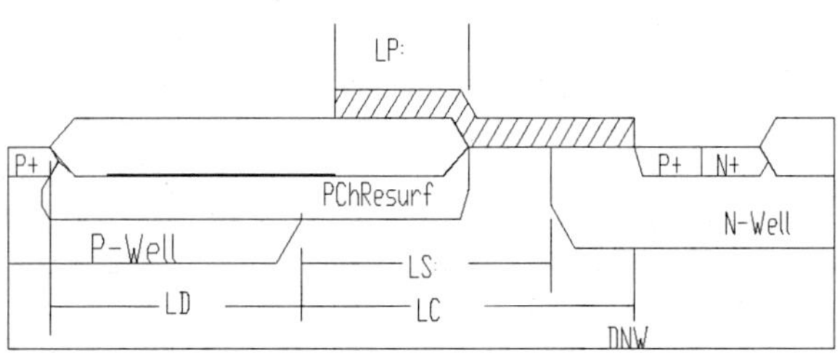

FIGURE 2. Cross sectional diagram of PMOS tile structure.

Figures 3 and 4 show a top view diagram of each of the tile devices. The center of the structure is the drain contact area that is surrounded by an extended drain region, the gate poly contacts, then the channel and then the source area contacts. This type of construction provides several advantages to a standard lateral DMOS; it is isolated on all sides minimizing its effect on any surrounding low volt circuitry, one only has to model a single structure, it simplifies layout such they can be "tiled" together with minimal consideration to surrounding circuitry [8].

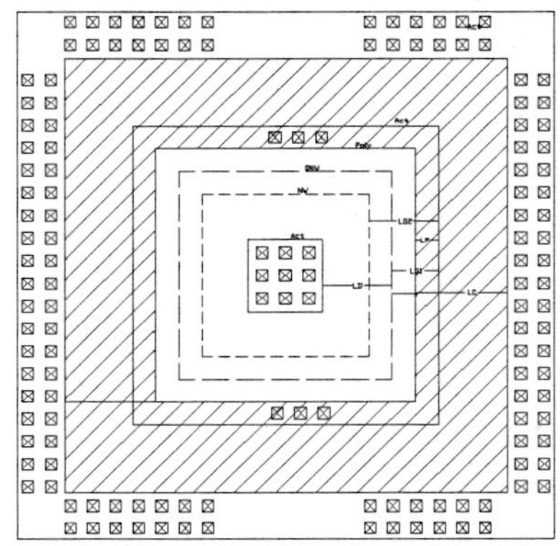

FIGURE 3. Top view diagram of NMOS tile structure.

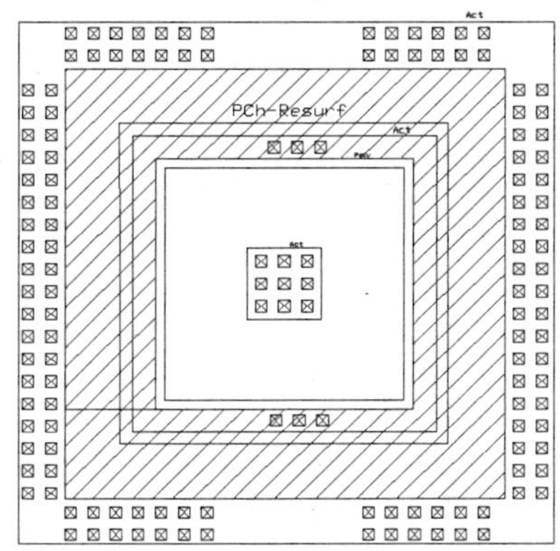

FIGURE 4. Top view diagram of PMOS tile structure.

EXPERIMENTAL

The energy targeted for the DNW was determined based on calculations using diffusion rates for a standard activation anneal. The one chosen grows 90nm of oxide in oxygen at 1050 C for 1hr then has a 2.25hr drive at 1150 C. This correlates to an equivalent implant energy of ~800keV. Fortunately this was close to the upper limit of 780keV P+++ for an EHP500i VSEA medium current implanter. Reproducible beam currents of 60μA could be achieved which only increased implant times to less than one minute.

TABLE 1. Design of experiment for Deep Nwell dose, energy and diffusion steps and P-channel resurf dose and energy.

Wafer	Deep N-well Dose	Deep N-well Energy	# Drives	P-Channel Resurf Dose	P-channel Resurf Energy
1	5e12 P+++	780keV	1	3e13 BF2+	40keV
2	5e12P+, 6.7e11 B+	150keV, 8keV	1	3e13 BF2+	40keV
3	5e12 P+++	780keV	1	3e13 BF2+	40keV
4	5e12 P+	150keV	3	3e13 B+	8keV
5	5e12 P+	150keV	3	3e13 BF2+, 4.5e12mP+	40keV, 10keV
6	5e12 P+	150keV	3	3e13 B+	8keV
7	5e12 P+	150keV	3	3e13 BF2+	40keV
8	5e12 P+	150keV	3	3e13 BF2+	40keV
9	5e12 P+	150keV	3	3e13 BF2+	40keV

Table 1 shows the design of experiment performed. Wafers 2 and 5 were also included to quantify the effect of phosphorus and boron cross-contamination at the DNW and PCR modules, respectively. Wafers 1, 3, 4, and 6 were replicates. Wafers 7 to 9 were the standard process flow.

RESULTS AND DISCUSSION

Prior to running split lot analysis test wafers were processed to compare depth and dopant concentrations by spreading resistance probe (SRP). Control wafers were run with the existing implant and triple diffusion module while other test wafers were processed with the high energy implant and single drive. Figure 5 shows the SRP results of the standard and Figure 6 of the high energy DNW. The results show that the depth and concentration of dopants were closely matched.

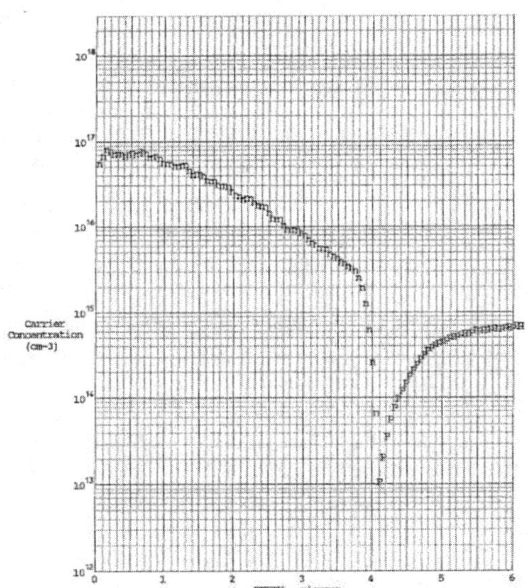

FIGURE 5. SRP results of Deep N-Well with standard implant and three diffusion cycles.

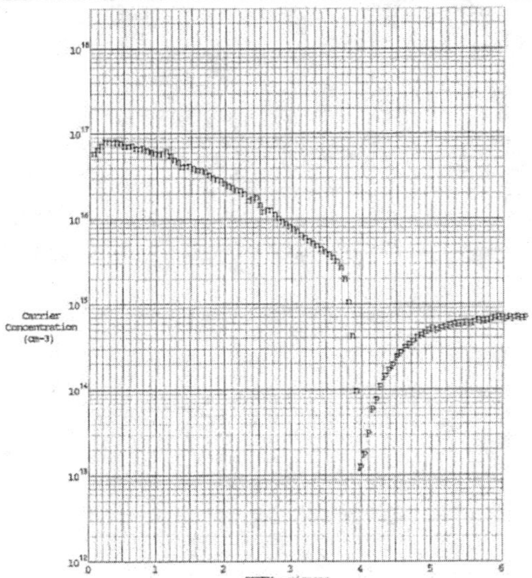

FIGURE 6. SRP results of Deep N-Well with high energy implant and one diffusion cycle.

The next analysis was based on the parametric performance of the N40V and P40V tile devices. These results were mixed but promising. Some of the critical electrical tests diverged based on the polarity of the structure. Figure 7 shows one example of this, Ron.

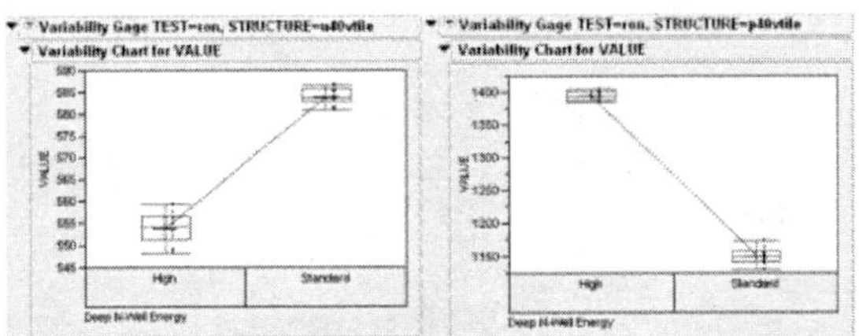

FIGURE 7. Divergent Ron results for the N40V structure (left) and P40V structure (right).

The most notable result was that the breakdown voltage (BVDSS) improved in both structures. Figure 8 illustrates these breakdown voltages.

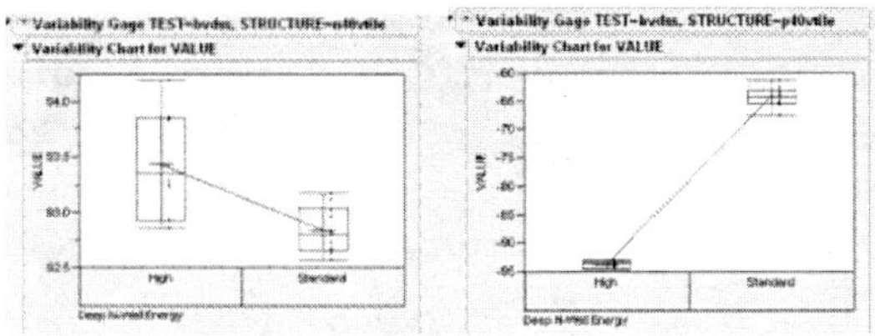

FIGURE 8. Improved BVDSS results the N40V structure (left) and P40V structure (right).

Table 2 shows a qualitative comparison of the observed electrical parametrics compared to the existing process module.

TABLE 2. Qualitative comparison of the electrical parametrics with the high energy implant.

Test	N40V Tile	P40V Tile
BVDSS	↑	↑
Gamma	↔	↔
Idsat	↑	↓
Isub	↓	↑
Kprime	↔	↔
Ron	↑	↓
Vtgm	↔	↔

Since the PCR implant is only used in the P40V tile device, the results of this experiment were based entirely on that devices parametric result. Essentially, the results of the experiment were indicative of an increase in dose. While saturation current, on state resistance and gain improved, it was at the expense of breakdown voltage. Table 3 summarizes these findings.

TABLE 3. Mean values of the P40V tile structure as a function of p-channel resurf species.

Test	B+	BF_2+	Result
BVDSS (V)	-64	-93	↓
Gamma (sqrtV)	0.48	0.44	↔
Idsat (A)	-0.01	-0.01	↔
Isub (A)	3.3e-5	2.7e-6	↔
Kprime ($\mu A/V^2$)	302.50	210.10	↑
Ron (Ohms)	1150	1400	↓
Vtgm (V)	-1.67	-1.65	↔

CONCLUSION

The ability to use a high energy implant to significantly reduce the cycle time of a high volt, deep N-well process module proved promising. Although there were mixed results, the improvement in BVDSS of both the N40V and P40V tiles indicated the process could be centered with a dose adjustment. The next step in the investigation would be a matrix to identify this dose.

The results of the species experiment to improve the P40V tile device performance were less encouraging. The reduction in BVDSS created almost a 30V mismatch between the P40 and N40 structures. However, the improvements in saturation current, on state resistance and gain were desirable. The next step in this investigation would be to look at another transition in the dopant gradient of the P40V structure using a lower dose at the junction in addition to the PCR to achieve the benefits

ACKNOWLEDGEMENTS

The authors would like to thank Muhammad Anser and Greg Scott for their technical consultation and development of the initial process technology. The authors would also like to thank Lynda Pierson for support with spread resistance probe data and Jeff Gerlach for his diffusion support.

REFERENCES

1. F.Marchio *et al.*, "A R-evolution in power electronics: from "intelligent" to "super smart power" in automotive", *Power Electronics in Transportation*, 2004, pp. 27-34.
2. F.Marchio, "Bridging the Gap between Silicon and System", 2007 EMC Compo Keynote address, Torino, It.
3. J.H. Renner, "An Integrated approach at Daimler-Chrysler" Integrated Reliability Workshop Final Report, 1999. IEEE International, 1999, pp. 152-153.
4. L. Lorenz, "Automotive Ind. & Multi-Market Marketing" Power Electronics and Motion Control Conference, 2006. IPEMC 2006 5th International. August 2006 Vol. 1 pp 1-7.
5. P.J Wrigh *et al.*, "The effect of fluorine on gate dielectric Properties" Electron Devices Meeting, 1987 International 1987, Volume: 33 pp. 574 - 577
6. P. Moens *et al.* "Development of an Optimized 40 V pDMOS device by use of a TCAD design of experiment methodology", 2000 International Conference on Simulation of Semiconductor Processes and Devices, pp. 276 – 279
7. P.F.H.M. Van der Meulen, "Energy Contamination Control in Multiple Charged Ion Implantations", 1992 IEEE/SEMI ASMC Conference Proceedings, pp. 96-99
8. Michael Riggs *et al.*, "High voltage Hall-Effect Sensor Interface in a Standard Digital CMOS Process" ESSCIRC 2006 Conference proceedings, pp. 584-587

EXLE-SIMS: Dramatically Enhanced Accuracy for Dose Loss Metrology.

W. Vandervorst[a], R. Vos[1], A. J. Salima[1], A. Merkulov[2], K. Nakajima[3] and K. Kimura[3]

[1]Imec, Kapeldreef 75 Belgium,
[a]also: Instituut for Kern- en Stralingsfysika, K.U.Leuven, Belgium
[2]Cameca, France
[3]Department of Micro Engineering, Kyoto University, Japan

Abstract. A detailed assessment of dose loss metrology for As implants in Si subjected to several cleaning steps is made by comparing SIMS profiles with high resolution RBS measurements. The results show that the normally used SIMS protocols overestimate the surface concentration significantly (~5x) and thus lead to a large overestimate of the (remaining) implant dose (~ 20%). We show that by going to EXLE-SIMS (extremely low energy SIMS) which uses bombardment energies as low as 150 eV (instead of the routinely used 500 eV), these surface artifacts can be dramatically reduced and close agreement with the RBS results can be obtained.

Keywords: Dose loss, SIMS, RBS, As, cleaning
PACS: 85.40.Ry, 81.70.Jb, 79.20.Rf

INTRODUCTION

Accurate dose metrology for ion implant processes is feasible with SIMS with a rather high reproducibility (~ 1%) [1]. The latter is achievable for relatively deep profiles where the SIMS surface transients are less important relative to the rest of the profile. When dealing with shallow profiles and multilayer structures, the situation is less favorable as SIMS is suffering from near-surface, associated atom migration, and ionization and sputter yield changes across the interface of multilayer structures. [2,3, 4,5].

In this paper we focus on the determination of the accurate profile for an As-implant in Si and in particular the determination of the amount of As lost when the sample is subjected to several cleaning steps. Since the Si-removal due to the cleaning steps is expected to be only on the order of 0.5 nm/step, the required accuracy and precision on the As-profile determination becomes 1-3 %. As the loss is confined to the surface, accurate dose loss metrology after cleaning requires the ability to probe very accurately the profile in the near-surface region including the profile across the native oxide. This requires analysis conditions which, at present, are not compatible with the conditions used for the highest profile precision. On the other hand the fact that the profile changes by the cleaning step are confined to the first nm also implies that the rest of the profile should be identical or even could be used in some kind of normalization procedure.

The artifacts related to SIMS profiles are primarily a less than adequate depth resolution causing unreal long tails on profiles which should be steeper and signal variations uncorrelated to the concentration variations of the species of interest. In particular these signal variations can be very pronounced at the start of a profile (near-surface transient) and at interfaces of multilayer structures (such as the SiO_2/Si transition) and (complex) correction algorithms have been reported in the literature to correct for them [6,7,8,9]. Rather than trying to optimize these correction algorithms, an exploration of alternative protocols inducing less variations and thus requiring smaller corrections would be very advantageous and ultimately lead to higher precision and accuracy.

Whereas routine SIMS conditions rely on ion beam energies around 0.5 keV, as this represents a reasonable compromise between erosion velocity, depth resolution and surface transients, recent studies with a new protocol, termed EXLE-SIMS for **Ex**tremely **L**ow **E**nergy SIMS, have demonstrated that by going to extremely low primary beam energies

CP1066, *Ion Implantation Technology,* edited by E. G. Seebauer, S. B. Felch, A. Jain, and Y. V. Kondratenko

(~ 100-150 eV) these interfacial variations can be suppressed significantly [4].

The objective of this paper is to assess in detail the accuracy of the standard approaches to this problem and at the same time to explore the EXLE-SIMS conditions when applied to dose loss metrology. In order to assess their performance also an alternative technique (high resolution Rutherford backscattering spectrometry, RBS) is used which is claimed to be quantitative and artifact free in multilayer analysis.

EXPERIMENTAL

The sample used is this study is an As-implant in Si (10 keV, 2 10^{15} at/cm^2) annealed at 1030°C (spike). Pieces of the sample are then subsequently exposed to different immersion times in a chemical solution (aerated 2% HF) to simulate consecutive cleaning steps in a normal process flow.

As–profiles are measured using SIMS (Cameca SC-Ultra, Cs) while monitoring As^-. The Cs-impact energy was varied between 150eV, 250 eV and 500 eV with a concurrent change in impact angle (32, 43 and 50°). Unless otherwise stated, no correction algorithm for surface transients nor the presence of the oxide was applied. One set of profiles was measured by a commercial laboratory using an exploratory protocol (500 eV Cs, 60°, $AsSi^-$) with a point-by–point data correction procedure (referenced further as PBCP) similar to PCOR [6,7]. This procedure considers the variation of the relative sensitivity factors as well as the erosion rates when going through the surface oxide. At least for a set of As-delta layers buried below a 5 nm oxide, this has led to reasonable results [5]. As a reference for assessing the accuracy of the SIMS profiles, RBS-profiles were collected using a system designed for high depth resolution measurements. High energy resolution is achieved with a magnetic sector analyzer (0.4 % or ~ 1.6 keV for a 400 keV He^+-beam) [10] which combined with an intermediate beam energy (400 keV He^+) to maximize the stopping power and with a tilted geometry (50° scattering angle and 54.7° incident angle) leads to a depth resolution (for As) of ~0.5 nm at the surface.

RESULTS AND DISCUSSION

Standard SIMS profiles

The SIMS profile, as measured by SIMS and treated with the PBPC-data algorithm is shown in Figure 1. for the reference As sample. As indicated by the O-signal, the surface is oxidized and an oxide ~ 0.8nm is present on the sample surface (when using 50% of the O-signal as an indicator for the oxide-silicon interface). At least based on the relative position of the As and O-signals, a pile up of the As at the oxide silicon interface is inferred.

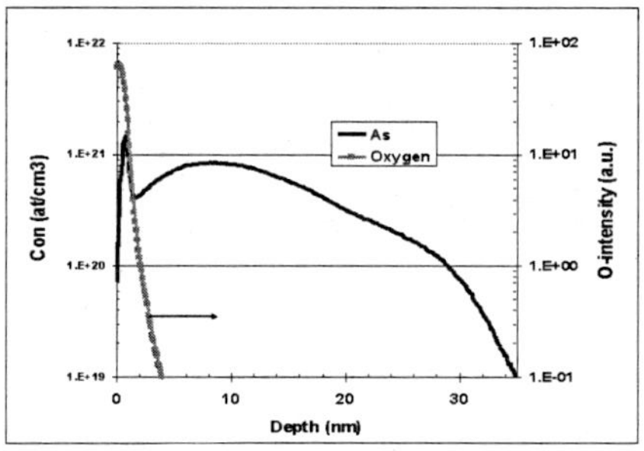

FIGURE 1. As- and O-profiles as determined with standard SIMS (500eV, Cs, 60°) and PBPC-data treatment.

When exposed to an etching solution, part of this profile will be removed and successive profiles after etching should be identical when disregarding the part contained in the removed Si. Hence we can estimate from the reference sample the precision required for the other samples by calculating the As-dose loss when x nm of Si are removed from this profile. As shown in Figure 2 for etching steps removing only 1 nm of Si this corresponds to a precision better than 3%.

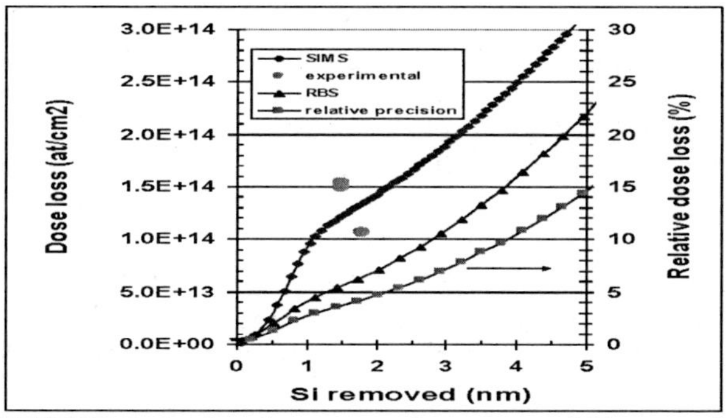

FIGURE 2. As dose loss versus the amount of Si removed. Calculated from Figure 1 (SIMS) and Figure 4 (RBS). Relative dose loss calculated relative to the dose of the SIMS profile. Experimental data points are calculated from the remaining dose (integrals of SIMS profiles, Figure 3) for the etched samples.

Crucial in this experiment is not only to provide information on the amount of As removed but also on the Si-loss. We can rely in our experiment on using the remainder of the entire As-profile as a depth indicator since this should not be influenced by the etching step. This procedure is illustrated in fig 3 (a,b) where the reference profile and the profiles after 10, 20 min etching are aligned in the deeper part of the As-profile. The shifts required to match these profiles correspond to the estimated Si-removal (1.5 and 1.8 nm respectively).

Integration of the different profiles provides the corresponding dose loss (cfr Figure 2) which is very reproducible (cfr the two measurements for 10 min etching) but at variance with the expected values based on the integration of the reference profile. In particular the lower dose loss for the longer etch time is disturbing and indicative of inconsistent results.

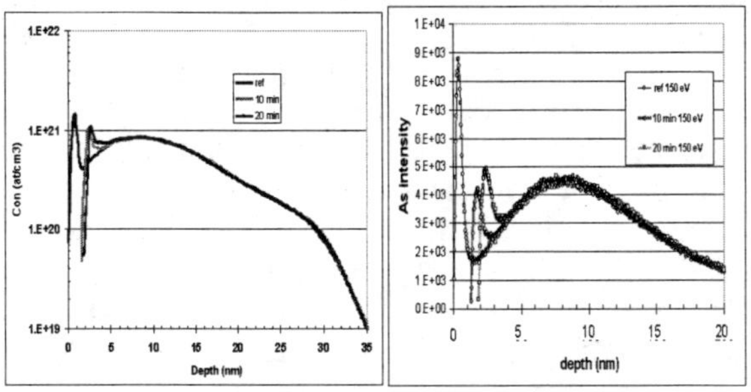

FIGURE 3. a) As (PBPC) SIMS profiles for 0, 10 and 20 min etching, aligned in the deeper part of the profile
b) similar (on a linear scale) for analysis with 150 eV.

The reason for the discrepancies lies in the details of the As-profiles (Figure 3) as they display very pronounced surface peaks for all the samples, even after prolonged etching. As the etching step is done at room temperature and cannot induce any further dopant migration, the observed profiles should be identical except for the part which is removed.

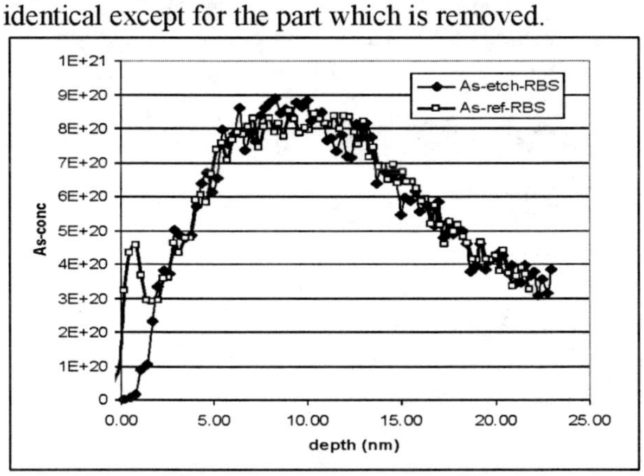

FIGURE 4. High resolution RBS As-profile of the sample after 0 and 10 min etching. The measured profile for the etched sample is shifted over 1 nm.

If this assumption is correct, the surface peaks must be interpreted as SIMS artifacts which distort the correct estimate of the dose loss. In order to clarify this, high resolution RBS was done on the reference sample and the one after 10 min etching (Figure 4). Both profiles overlay quite well after shifting the profile after etch over 1 nm, which corresponds to the Si-loss in this treatment. It is clear that the reference sample has an As-interfacial peak which is completely removed during the etching step. Based on the RBS-profiles we must conclude that the As-peaks observed in Figure 3 are pure SIMS artifacts. Apparently the PBPC algorithm (in combination with the 500 eV analysis) is not able to remove the artifacts completely. When performing the dose loss calculation using the RBS–profiles (cfr Figure 1) much smaller values are found as compared to the SIMS results, due to the much smaller As-interfacial peak.

EXLE-SIMS

The existence of these surface peaks is related to ionization yield changes when transitioning from the (native) oxide to the silicon substrate and reminiscent of earlier work on interfacial B/As profiling through thicker oxides [2,3,4,6,7]. In that case, the PBPC treatment (with 500 eV Cs) was successful in eliminating most of the transients and only suffered from a lack of depth resolution due to the high energy used. Despite this success, concerns were raised due to the very large amount of correction required. At the same time it was shown that when the primary beam energy was lowered substantially towards the conditions for EXLE-SIMS (150eV), a dramatic reduction in importance of the transients was observed and the raw data already reflected much closer the

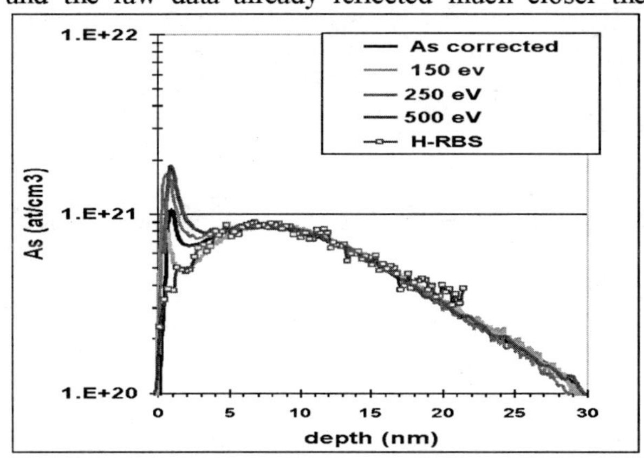

FIGURE 5. As-profiles (10 min etched) obtained at several beam energies (untreated data) versus PBPC-treated profile (500 eV) and RBS-profile.

actual concentration variations [5]. Hence by going to these conditions the magnitude of the corrections required could be drastically reduced and better accuracy with the PBPC algorithm could be expected.
Based on these observations we investigated the evolution of the surface peak when reducing the primary beam energy (Figure5). We compare here untreated SIMS profiles (Cameca SC-Ultra) with the PBPC profile as well as the RBS-profile. In all cases a substantial overestimate of the interfacial As-peak can

be seen which becomes dramatically smaller at 150 eV. The comparison untreated – PBPC profile also illustrates the improvement/limitation of the complex data analysis. Clearly at 500 eV impact energy (Figure 6) the surface peak is ~4x higher than the maximum of

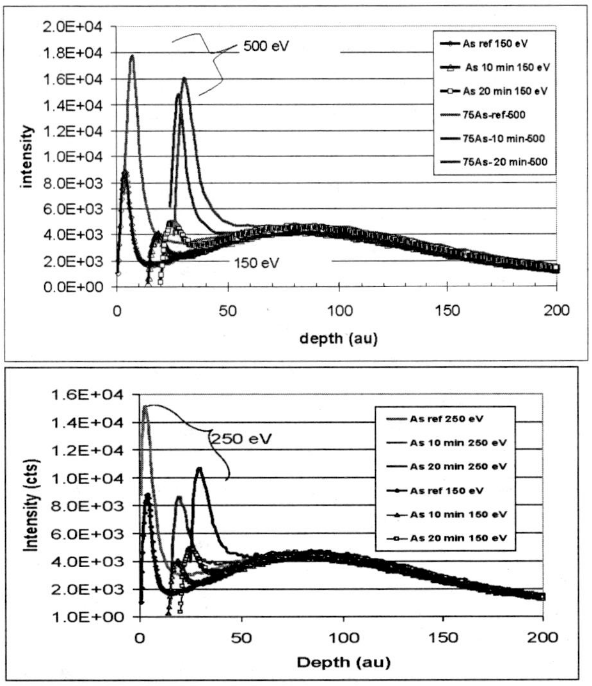

FIGURE 6. As-profiles after 0, 10 and 20 min etching. (a) comparison between the analysis using 500 and 150 eV, (b) using 250 and 150 eV. (no data treatment in all cases).

the As profile whereas this decreases significantly when going to 150 eV but still does not reach unity. As the RBS profiles show no peak at all a further reduction of the primary beam energy might be required. To assess this more quantitatively we have determined the ratio of this As-surface peak relative to

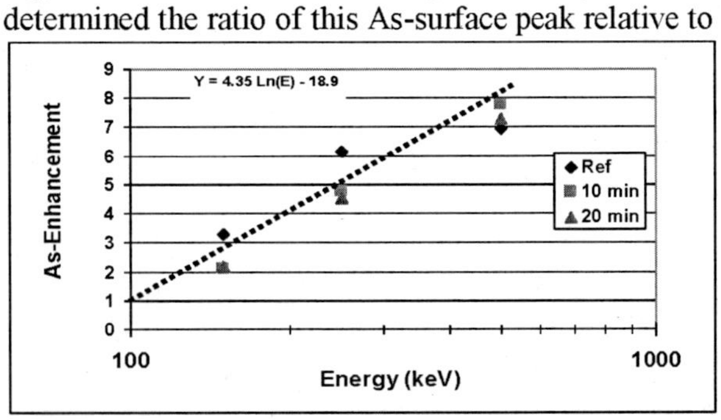

FIGURE 7. Enhancement of the surface As-peak as a function analysis energy. The trend line is a simple fit through the data points.

the (true) value taking from the RBS-profile (Figure 7). The amount of As-enhancement decreases drastically with bombardment energy (~ ln(E)), and its extrapolation indicates that profiles taken with 100 eV would be free of this artifact.

CONCLUSIONS

The analysis of As dose loss after a cleaning step requires a highly precise and accurate SIMS profile. Our results show that the normally used protocols always generate a strong As-surface peak even when a complex data correction approach is applied, which is not present in the corresponding RBS-profiles. This remaining surface peak then leads to a large overestimate of the (remaining) implant dose (~ 20%) and thus an incorrect assessment of the dose loss. The fundamental cause of the problem is a strong influence of the oxygen concentration on the As-ionization probability [12] which can however be reduced dramatically when using extremely low energy (< 150 eV, EXLE-SIMS) bombardment conditions. Application of a PBPC-like algorithm might lead to a complete removal of the surface peak when applied to these improved raw data where the amount of correction required becomes very limited.

REFERENCES

1. H.Maul, N.Loibl, U.Ehrke, A.Merkulov, P.Peres, M.Schuhmacher, AIP Conference Proceedings 866, American Institute of Physics, Melville, NY, 2006, pp. 594-598
2. W.Vandervorst, T.Janssens, B.Brijs, T.Conard, C.Huyghebaert, J.Fruhauf, A.Bergmaier, G.Dollinger, T.Buyuklimanli, J.A.Vandenberg and K.Kimura, *Applied Surface Science* **231-232** 618 (2004)
3. W.Vandervorst, T.Janssens, R.Loo, M.Caymax, .Peytier, R.Lindsay, J.Fruehauf, A.Bergmaier, G.Dollinger, *Applied Surface Science*, **203-204** 371 (2003).
4. W.Vandervorst_, (plenary lecture SIMS XV, Kanazawa, Japan), Applied Surface Science (2008) (accepted)
5. W.Vandervorst, AIP Conference Proceedings 931, American Institute of Physics, Melville, NY, 2007, pp. 233-245
6. C.W.Magee, R.S.Hockett, T.H.Buyuklimanli, I.Abdelrehim and J.W.Marino, AIP Conference Proceedings 931, American Institute of Physics, Melville, NY, 2007, pp. 142-145
7. T.H.Buyuklimanli , C.W.Magee, J.W.Marino and S.R.Walther, *J. Vac. Sci. Technol.* **B24**(1), 408 (2006)
8. T.Janssens and W.Vandervorst. Proc SIMS XII, ed. A.Benninghoven, P.Bertrand, H.N.Migeon and H.W.Werner, Elsevier, Amsterdam (1998) 151
9. D.Giubertoni, M.Bersani, M.Barozzi, S.Pederzoli, E.Iacob, J.A. van den Berg and M.Werner, *Appl. Surf. Sci.* **252**, 7214 (2006)
10. K.Kimura, S.Joumori, Y.Oota, K.Nakajima and M.Suzuki, *Nucl Instr. and Methods B* **219-220**, 351 (2004)
11. W.Vandervorst_, T.Janssens, C. Huyghebaert and B.Berghmans *Applied Surface Science* (2008) (accepted)
12. W.Vandervorst, to be published

Junction Photovoltage Metrology and High Resolution Mapping of Ion Implants Electrically Isolated from the Wafer Surface

Mr. Ferenc Korsós, Mr. Krisztian Kis-Szabo, Dr. Eric Don, Mr. Aron Pap, Dr. Tibor Pavelka, Mr. Cyrille Laviron, and Mr. Marcus Pfeffer

Semilab R.T. Budapest, Hungar.
STMicroelectronics, Crolles France
Fraunhofer IISB, Erlangen, Germany

Abstract. Application of the Junction Photovoltage (JPV) method for metrology of ultra-shallow implant (USJ) has been well documented. In this paper we report the application of the JPV method to control implant processes which create an electrically active layer isolated from the surface. In such structures four point probe (4PP) metrology is unreliable or impossible. This paper presents results obtained using a Junction Photo-Voltage (JPV) method optimized for characterization of these two processes, CMOS: Implantation through a screen oxide followed by anneal and also MeV p-type high energy implantation into n-type doped substrate followed by an anneal. Taking errors and necessary corrections into account the JPV method allows accurate high resolution mapping of such structures without the need for an extra process step. Results obtained from the JPV measurements were correlated to destructive off-line analytical measurement tools.

Keywords: JPV, Implant Dose, High Resolution. Wafer Map.
PACS: 02.60.Cb, 06.20.-f, 61.72.Vv, 73.50.Pz, 85.40.Ry

INTRODUCTION

The application of the Junction Photovoltage (JPV) method for metrology of ultra-shallow implants (USJ) has been well documented [1, 2]. The JPV method avoids damage to the junction by probe penetration and provides much faster high spatial resolution full wafer mapping.

Traditionally ion implants were made through a thin screen oxide as a safeguard against implant induced surface contamination [3]. In advanced CMOS processes the ion implant is still made through a screen oxide but more to improve implant spatial uniformity, or better control the depth profile [4]. The implant parameters are optimized taking account of the screen oxide, therefore monitor wafers used to check this process should have the same or very similar screen oxide. Conventionally, the implant uniformity across the implanted and annealed wafer is checked by measurement of sheet resistance by four point probe after removal of the insulating screen oxide. The non-contact JPV measurement method enables measurement of the sheet resistance in the top layer of a p-n junction under the screen oxide on a monitor wafer thus speeding process development and in-line control.

CMOS image sensor development has involved several generations of decreasing feature size to increase pixel count for consumer applications. In these imaging devices the normal CMOS deep retrograde well structure also takes on additional importance in helping to avoid photo-generated cross-talk between pixels [5]. Following such a high energy implant, typically above 750 keV, the type converted layer may be buried below a p-n junction which isolates the converted layer from the surface. In order to obtain a good four point probe sheet resistance measurement manufacturers traditionally employ a second lower energy “bridging” implant to connect the surface to the buried layer. In principle, the JPV measurement can be used to measure the sheet resistance of the buried implanted layer if the sheet resistance of the unconverted top layer is significantly higher than the sheet resistance of the buried layer.

CP1066, *Ion Implantation Technology,* edited by E. G. Seebauer, S. B. Felch, A. Jain, and Y. V. Kondratenko

Theory of the JPV Measurement.

In the simplest case, we illuminate an implanted wafer in its center in a round spot with a radius r_0. The used light is chopped by a frequency f and its fill factor is 0.5. It is also assumed that illumination is homogenous and its density is J (for the simplicity J means here part of whole illumination, which generates electron-hole pairs). Naturally, there is reflection from the surface of wafer R. Generated carriers charges the p-n junction, but not the all of them can reach the junction. To take account this effect an internal efficiency η is needed to be introduced, which gives the fraction of generated electron-hole pairs reaching the junction.

Assume that there is p-type Si substrate, which had been implanted by donor atoms (e.g. Phosporus), so we have an n-type layer on a p-type substrate. During illumitation generated electron-hole pairs move to the junction where they are being separated by an electric field. Electrons go to n-type layer and holes to p-type layer. There will be majority carriers in the layers and the capacitance of junction C_d is partly charged resulting in a photovoltage U. This voltage can be measured as an ac signal by capacitance electrodes. At the same time, carriers are spreading laterally with a current density I in the implanted layer, which has a sheet resistance R_s. There is also an ohmic leakage current through the junction characterized by its ohmic resistance R_d. At the edge of wafer there can be an additional leakage to the ground with an ohmic resistance R_e One can get equations for a homogenous case in center of wafer from the following idea:

Because of lateral spreading of carriers, there is a decrease in photovoltage proportional to the local sheet resistance:

$$dU = -IdR_s = -I\frac{R_s}{2\pi r}dr \, ,$$

where r is distance from the center of wafer. Because of capacitance and ohmic resistance of junction there is a change in current:

$$\boldsymbol{dI = -\frac{U}{dZ} = 2r\pi\left(J\Theta(r_0 - r) - \left(\frac{1}{R_d} + i\omega C_c\right)U\right)dr} \, ,$$

where $\omega=2\pi f$ and Θ(r) is the Heaviside-function. We get the following differential equations:

$$\frac{\partial U}{\partial r} = -IdR_s = -I\frac{R_s}{2\pi r},$$

$$\frac{\partial I}{\partial r} = 2r\pi\left(J\Theta(r_0 - r) - \left(\frac{1}{R_d} + i\omega C_d\right)U\right).$$

One can express I from the first equation and put into the second and after some manipulation:

$$r^2\frac{\partial^2 U}{\partial r^2} + r\frac{\partial U}{\partial r} - r^2\left(\frac{R_s}{R_d} + i\omega R_s C_d\right)U + J\Theta(r_0 - r) = 0.$$

Measuring signal through a range of frequencies this equation can be solved numerically and the three parameters, R_s, C_d, *and Rd* can be evaluated at the center of wafer. Since the center value of sheet resistance is known, an algorithm is used to determine sheet resistance on the whole wafer.

Results from Measurements

The following results were obtained on 200 mm wafers supplied from Fraunhofer IISB, Erlangen, Germany and University of Surrey, UK. In these wafers substrate carrier concentrations were in the range 1E15 to 1E16 cm^{-3}. The wafers were prepared specifically for this work.

Sheet Resistance Correlation Through Screen Oxide Implants

Four point probe measurement cannot be reliable through an oxide layer because contact between probe and implant layer is not good because the probe has to break through the oxide; in the worst case no contact is made. In contrast JPV measurement is non-contact via capacitive electrodes and so the presence of an insulating dielectric does not prevent measurement. However the JPV method does require a controlled photon flux and in the case of an oxidized wafer reflectance must be measured and used to correct the photon flux. We monitored a series of wafers of 70 keV Boron implants with and without oxide layer (See Table 1). In this study we were particularly interested in the low dose regime typical for critical threshold adjust measurements. A silicon oxide 20 nm thick was grown on all of the monitor wafers and used as a screen oxide. At each dose level two wafers were implanted. Following implant all the wafers were annealed the oxide being used in this case as a cap oxide.

Following annealing one wafer at each dose level was stripped of its oxide and the other was left intact. For this energy and dose simulation indicates around 230 nm ion penetration depth for Boron with a longitudinal straggling 70 nm. One can assume that after annealing junction depth is somewhere 300 to 400 nm. The dose range 1E13 1/cm^2 to 5E11 1/cm^2.

The sheet resistance was determined in center of wafer. Results can be seen in Table 1. The sheet resistance increases, as expected, with decreasing dose. The junctions were of good quality for dose above of 5E11 1/cm^2 whereas below that dose level the junctions the junctions exhibited high leakage.

TABLE 1. Sample conditions and sheet resistance in center of wafer.

Dose [1/cm^2]	Oxidized	R_s [Ω/sq]
5E11		97897
5E11	x	47562
1E12		33783
1E12	x	36540
5E12		5564
5E12	x	5259
1E13		3258
1E13	x	3025

The correlation between oxidized and stripped wafers is reasonably good except for very low dose levels below 1E12 cm^{-2}. Investigations are underway to determine if the errors in this ultra low dose region are due to measurement effects such as surface depletion modulation or if the error relates to our oxide stripping process, for example by removal dopant accumulated near the interface between the cap oxide and silicon during the annealing process.

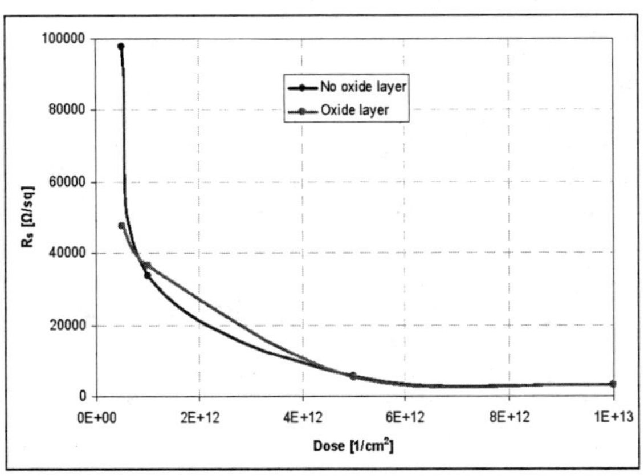

FIGURE 1. Sheet resistance versus implant dose for non-oxidized and oxidized samples.

In addition to the mean value correlation, it is important that the wafer map distribution is faithfully reproduced. In Fig 1 and 2 we show that the measurement through the oxide has little or no effect on the map distribution provided that suitable correction is made for any variation in reflectance across the wafer. However, some damage due to our oxide stripping process can be observed in Fig 2. We show here one dose level typical of all those in the study.

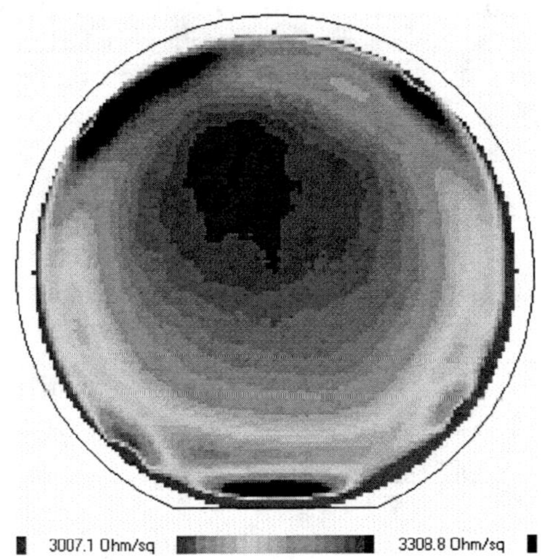

FIGURE 2. Wafer map of 1E13 dose implant with screen oxide stripped after implant and anneal.

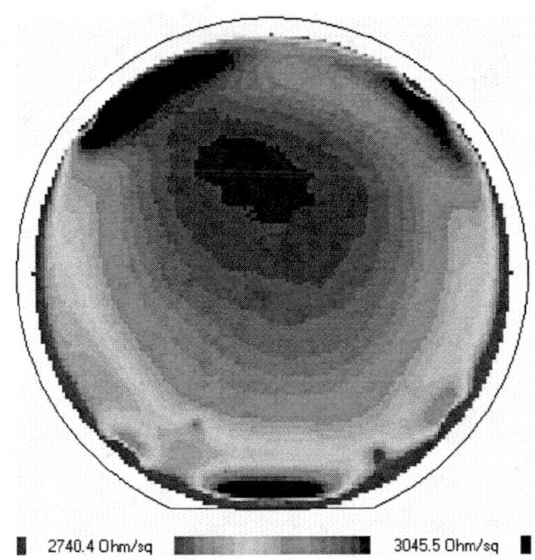

FIGURE 3. Wafer map of 1E13 dose implant with screen oxide still on wafer after implant and anneal.

Sheet Resistance Correlation for High Energy Implants

In order to illustrate the problem of monitoring high energy implants with conventional four point probe measurement, we produced a set of wafers all with As^+ implant dose of 3E13 cm^{-2} and energy in the range between 50 keV to 4 MeV into 10 ohm.cm boron doped substrates. The resulting implant structure was simulated using Synopsis TCAD (Version 2007.12). The results of this simulation are given in Table 2 and it can be seen that junction depth increases monotonically and sheet resistance decreases until the implant energy exceeds 600 keV, at which point a second junction is formed and the converted layer is effectively buried under the p-type surface. The width of this buried layer continues to increase with increasing energy and so we anticipate that the sheet resistance should continue to fall.

TABLE 2. Sample conditions simulated junction depths and sheet resistance values.

Dose [ion/cm^2]	3E+13					
Ion Species	As+		Substrate	Anneal		
twist	24°		10 ohmcm	30s/950°C		
tilt	7°					
Slot No	Energy [keV]	Deep juntion Depth	Shallow junction depth	Buried layer width	Rs of n Top Layer	Rs of p Top Layer
W2	100	0.55 µm	x	x	1.02E+03	x
W3	150	0.65 µm	x	x	8.96E+02	x
W4	200	0.75 µm	x	x	8.16E+02	x
W5	300	0.85 µm	x	x	7.20E+02	x
W6	600	1.05 µm	x	x	5.78E+02	x
W7	1200	1.45 µm	0.012 µm	1.44	x	1.21E+08
W8	2000	2.03 µm	0.118 µm	1.91	x	2.95E+06
W9	2800	2.5 µm	0.32 µm	2.18	x	6.40E+05
W11	4000	3.18 µm	0.65 µm	2.53	x	2.53E+05

In the range from 100 to 300 keV both the four point probe and JPV measurements correlate reasonably well with the theoretical sheet resistance up to 1Kohm/sq. The 4PP gives a large error at 600 keV.

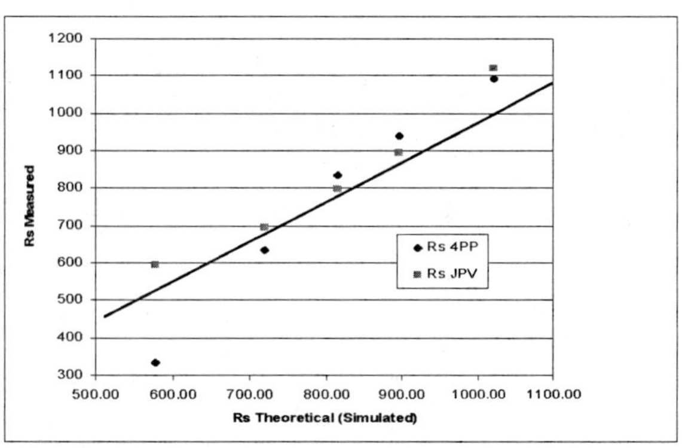

FIGURE 4. Measured sheet resistance versus calculated sheet resistance 100 to 300 keV.

However as the implant energy is increased to 600 keV and beyond, the simulation indicated that a buried layer isolated by a p-n junction is formed. The change of surface type was confirmed using a Semilab PN -100 type tester. In this region, the slope of measured sheet resistance verses implant energy is reduced but the JPV measurement retains higher sensitivity than the four point probe.

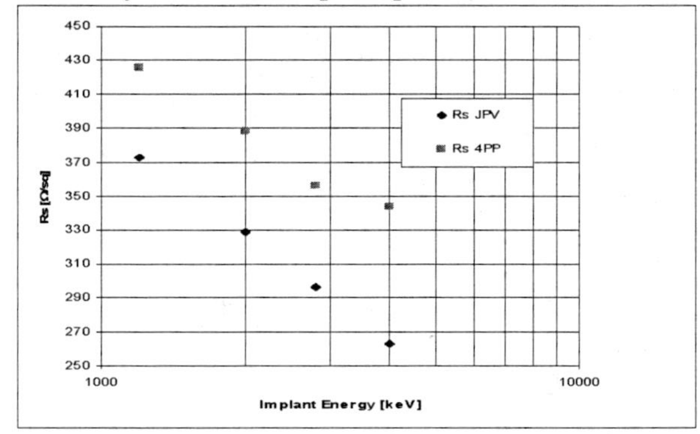

FIGURE 5. Measured sheet resistance versus ion implant energy for 4PP and JPV in the high energy range 1 to 4 MeV.

Results show that JPV gives accurate measurement in high Energy implant region where errors in 4PP measurement are caused by the isolating junction.

ACKNOWLEDGEMENTS

This work is part of the SEA-NET project with financial support of the European Community Framework 6 Program.

REFERENCES

1. E.Don *et al.*, "Metrology and High Resolution Mapping of Shallow Junctions formed by Low Energy Implant Processes" in. IIT 2006 11th – 16th June 2006. AIP Proc. Vol 886 p.534, (2006).
2. E.Don *et al.*, "Improvements in High Resolution Mapping of Sheet Resistance using Junction Photovoltage by Correcting for Wafer Edge and Compensating for Junction Leakage" in *International Workshop on Semiconductor Device Fabrication, Metrology and Modeling*, (Insight 2007) Napa, California (2007).
3. S. Mader and A.E. Michel, *J.Vac Sci Technol* **13** 391 (1976).
4. C.P. Change *et al.*, "Enabling shallow Trench Isolation for o.1micron Technologies and bejond" VLSI Symp. 1999 Section 11B-4 p. 161.
5. J. Boreland and N. Tokoru, "Consumer demands Make for a sizzling Imaging Market in Japan" *Solid State Technology* Nov. 2004 p S18.

Dynamics of Carrier Plasma and Thermal Waves in Ion Implanted Si with Varying Pump and Probe Wavelengths

Alex Salnik and Derrick Shaughnessy

Implant Metrology Division, KLA-Tencor Corporation
160 Rio Robles, San Jose, CA 95134 USA

Abstract. Dynamics of the carrier plasma- and thermal-waves in ion implanted Si is studied using the Modulated Optical Reflectance (MOR) technique with varying pump and probe beam wavelengths. It is shown that selection of the pump and probe beam wavelengths allows control of the MOR signal behavior on Si samples implanted with a wide range of implantation energies and doses.

Keywords: modulated, optical, reflectance, plasma, thermal, interference, ion implanted Si.
PACS: 78.20.Nv; 52.35.-g; 66.70.-f; 61.72.uf; 42.25.Gy.

INTRODUCTION

The Thermal Wave (TW) technology [1-3] based on the Modulated Optical Reflectance (MOR) effect [4] has long been used for the characterization of ion implantation processes in IC manufacturing. In our previous articles [5,6], we describe the use of a damage-based modeling to study the generation and detection of the carrier plasma- and thermal waves in discontinuously inhomogeneous semiconductors.

The MOR technique uses an intensity-modulated pump beam to produce carrier plasma, thermal, and elastic waves in the sample [1-3]. A non-modulated probe beam detects variations in the surface reflectance coefficient R introduced by the propagating and reflected waves in the sample. In the case of the ion implanted Si, the MOR signal is primarily driven by the plasma and thermal waves [5,6].

Characteristic experimental TW signal amplitude and phase dose dependencies obtained for As^+ implanted Si (100 keV) are presented in Figure 1 along with the results of the best fitting to the damage-based theoretical model [6]. In this example, the pump and probe wavelengths were 780 nm and 675 nm, respectively. As can be seen in this figure, the entire range of implantation doses can be divided into several characteristic regions where the TW signal amplitude and phase exhibit a clearly distinct behavior.

At low implantation doses (Region I), the TW signal is dominated by the plasma-wave component. The TW amplitude rapidly decreases with increasing implantation dose while the TW phase remains almost unchanged (Figure 1). In the Region II, the TW phase exhibits a sharp drop indicating a change in the dominant physical mechanism from plasma to thermal. At relatively high implantation doses (Regions III and IV), the TW signal is a result of several different physical effects including optical interference due to the presence of a distinct amorphous layer.

The dynamics of the plasma and thermal components of the TW response from ion implanted Si changes dramatically with the pump and probe beam wavelengths.

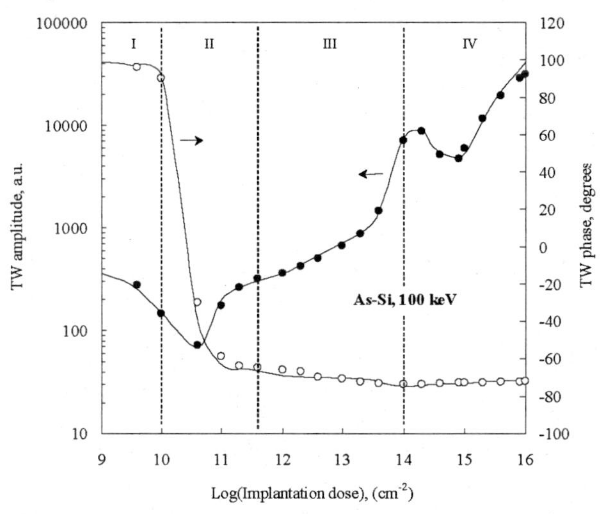

FIGURE 1. Experimental TW signal amplitude (left scale) and phase (right scale) dose dependencies obtained for As-implanted Si (100 keV) at 1 MHz modulation frequency (points). Solid lines are the best fit to a theoretical model [6].

CP1066, *Ion Implantation Technology,* edited by E. G. Seebauer, S. B. Felch, A. Jain, and Y. V. Kondratenko

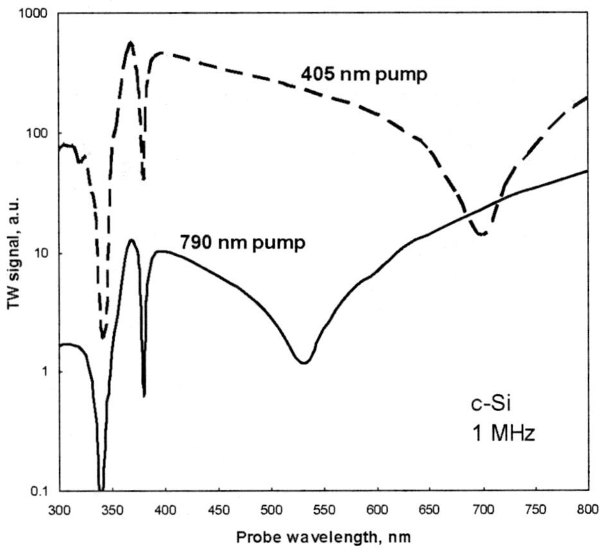

FIGURE 2. Calculated TW signal dependencies on the probe beam wavelength for c-Si with fixed pump wavelengths (405 nm and 790 nm).

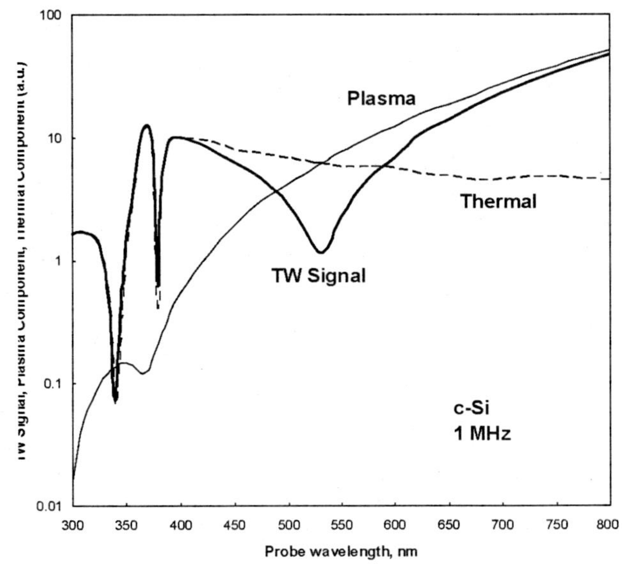

FIGURE 3. Calculated TW signal and amplitudes of the plasma and thermal components as a function of the probe beam wavelength for c-Si. Pump beam wavelength: 790 nm.

PLASMA AND THERMAL WAVES

Non-Implanted Si

The dynamics of the plasma and thermal waves in the MOR experiments with varying probe beam wavelength can be best appreciated in the case of non-implanted Si. Figure 2 shows the calculated TW signal response as a function of the probe beam wavelength for c-Si.

As shown in Figure 2 for both 790 nm and 405 nm pump beam wavelengths, the TW signal response in this case has several distinct features. In the near-UV probe beam wavelength range (< 400 nm), the TW amplitude varies significantly due to the strong optical absorption near the fundamental bands of Si E_1 and E_2.

With the probe beam wavelength increasing into the visible, the TW signal experiences a sharp drop due to the destructive plasma-thermal interference [5,6]. This effect is similar to the one we discussed earlier analyzing the TW signal response as a function of the implantation dose.

The position of the plasma-thermal interference depends on the thermal and electronic properties of c-Si and the pump beam wavelength. Therefore, the interference peaks for 405 nm and 790 nm pumps appear at different wavelengths of the probe beam (Figure 2). Note that the TW signal is much stronger for 405 nm pump beam wavelength because of the stronger optical absorption.

At the long probe beam wavelengths approaching IR, the TW signal amplitude is increasing monotonically for both pump beam wavelengths.

The "fine structure" of the TW signal response as a function of the probe beam wavelength is shown in Figure 3. The plasma wave component of the TW signal is increasing monotonically with increasing probe beam wavelength becoming the dominant physical mechanism at the long probe wavelengths near the IR region. The thermal component is decreasing steadily with increasing probe beam wavelength from its maximum at low probe wavelengths. At a certain critical value of the probe beam wavelength, the amplitude of the plasma wave component becomes commensurate with that of the thermal wave resulting in destructive interference (negative peak) in TW amplitude.

Implanted Si

To study the dynamics of the plasma and thermal waves in ion implanted Si, a series of the MOR experiments was conducted using different combination of the pump/probe beam wavelengths. Three pump beam wavelengths were used: 405 nm, 670 nm and 780 nm. Probe beam wavelengths used were 670 nm, 780 nm, and 1064 nm. In all experiments, the pump beam modulation frequency was kept constant at 1 MHz.

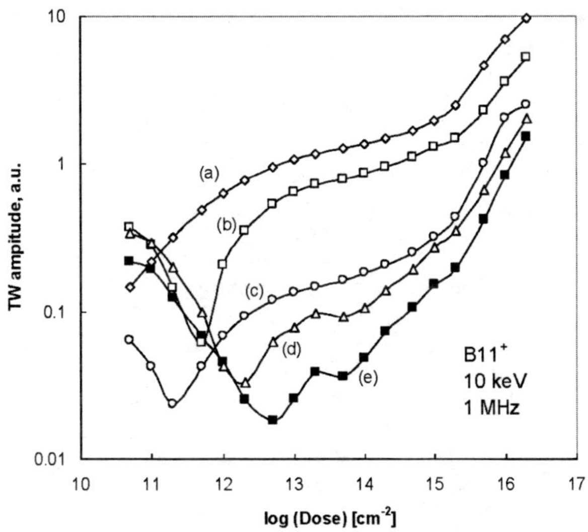

FIGURE 4. Experimental TW signal dose dependencies obtained from B^+-implanted Si (10 keV) with different pump/probe beam wavelength combinations: (a) 405/780 nm, (b) 405/1064 nm, (c) 780/670 nm, (d) 670/1064 nm, (e) 780/1064 nm.

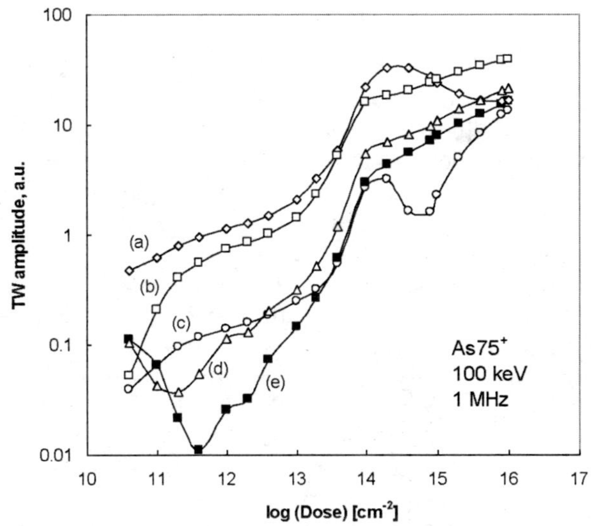

FIGURE 5. Experimental TW signal dose dependencies obtained from As^+-implanted Si (100 keV) with different pump/probe beam wavelength combinations. Notations are the same as in Figure 4.

Figure 4 shows experimental TW amplitude dose dependencies obtained from Si wafers implanted with B^+ ions at 10 keV. Several combinations of the pump/probe beam wavelengths were used: 405/780 nm, 405/1064 nm, 780/670 nm, 670/1064 nm, and 780/1064 nm marked (a) – (e) in Figure 4.

The effect of varying probe beam wavelength can be appreciated by comparison of curves (a) with (b) and (c) with (e) in Figure 4. Increasing the probe beam wavelength shifts the position of the plasma-thermal interference peak towards higher implantation doses and broadens the transition region. This behavior is consistent with our previous observations that longer probe wavelengths are more sensitive to the plasma waves in the samp aves [6].

The TW signal transition with varying pump beam wavelength can be analyzed using curves (b), (d), and (e) in Figure 4 that were obtained with the same probe beam wavelength of 1064 nm. The higher photon energy relative to the energy band gap and the stronger optical absorption for shorter pump wavelengths results in larger amplitudes of the thermal-wave component of the MOR signal. Also, examining the signals at the lowest doses where the material is approaching a crystalline structure it is evident that the amplitude of the plasma-wave contribution to the MOR signal is also higher for shorter pump wavelengths, a result consistent with the theoretical predictions for c-Si (Figure 2).

Increasing pump wavelength has a significantly stronger effect on the amplitude of the thermal-wave component relative to that from the plasma-wave resulting in a shift in the position of the plasma-thermal transition towards higher implantation doses and a broadening of the transition region.

Similar considerations can be made in the case of As^+ implantation shown in Figure 5. However, heavy As ions produce much more damage in Si than Boron ions resulting in a TW response with a much stronger thermal-wave component and consequently a thermally dominated signal at relatively low doses.

At high implantation doses when a distinct amorphous layer is formed, the TW signal is affected by the modulated interference effect [6-8] for pump/probe beam wavelength combinations that do not include 1064 nm probe wavelength (curves (a) and (c) in Figure 5). The implantation dose dependencies of the TW signal for pump/probe wavelength combinations with 1064 nm probe exhibit monotonic dependencies on dose in this region (curves (b), (d) and (e) in Figure 5).

We note the presence of an additional peak in the plasma-thermal transition region for curves (d) and (e) in Figure 4 and curve (e) in Figure 5. The nature of these peaks that are more pronounced for 780/1064 nm pump/probe beam wavelength combinations for both B and As implantation is currently under investigation.

CONCLUSION

The relative contribution of the plasma and thermal components of the MOR signal from ion implanted Si can be adjusted through judicious selection of the pump and probe wavelengths. While the weighting of the two contributions is affected by changing either the pump or probe wavelength, this sensitivity is caused by different physical mechanisms.

The pump beam wavelength defines the dynamics and relative magnitude of the plasma- and thermal-waves in the sample (shorter pump beam wavelengths result in relatively stronger thermal wave components) while the probe beam wavelength defines the relative sensitivity of the MOR signal to the plasma and thermal waves (longer probe beam wavelengths are more sensitive to plasma waves).

Therefore, by selecting the appropriate pump/probe beam wavelength combination it is possible to control the TW signal dependence on dose for ion implanted Si.

ACKNOWLEDGEMENTS

Authors would like to thank Dr. A. Kosterin for his help with experiments, and Drs. M. Bakshi and L. Nicolaides (both of KLA-Tencor Corp.) for helpful discussions.

REFERENCES

1. W. L. Smith, A. Rosencwaig, and D. Willenborg, *Appl. Phys. Lett.* **47**, 584 – 587 (1985).
2. W. L. Smith, A. Rosencwaig, D. L .Willenborg, J. Opsal, and W. Taylor, *Solid State Technol.* **1**, 85 – 90 (1986).
3. D. Guidotti and H. M. Van Driel, *Appl. Phys. Lett.* **47**, 1336 – 1339 (1985)
4. A. Rosencwaig, J. Opsal, W. L. Smith, and D. Willenborg, *Appl. Phys. Lett.* **46**, 1013 – 1016 (1985)
5. A. Salnik and J. Opsal, J. Appl. Phys. **91**, 2874 – 2882 (2002)
6. A. Salnik and J. Opsal, *Rev. Sci. Instrum.* **74**, 2144 – 2148 (2003)
7. J. Opsal, in *Review of Progress in Quantitative Nondestructive Evaluation*, edited by D. O. Thompson and D. E. Chimenti (Plenum, New York, 1987), **6B**, pp. 1339 – 1346.
8. J. Opsal, in *Review of Progress in Quantitative Nondestructive Evaluation*, edited by D. O. Thompson and D. E. Chimenti (Plenum, New York, 1989), **8B**, pp. 1241 – 1245.

Characterization of Charging Control of a Single Wafer High Current Spot Beam Implanter

Matthias Schmeide, Christoph Bukethal

Infineon Technologies Dresden GmbH & Co. OHG, Königsbrücker Str. 180, D-01099 Dresden, Germany
Matthias.Schmeide@Infineon.com

Abstract. This paper focuses on the characterization of charging control of an Axcelis Optima HD single wafer high current spot beam implanter using MOS capacitors with attached antennas of different size and shape. Resist patterns are implemented on Infineon Technologies own charging control wafers to investigate the influence of photo resist on charging damage.

Compared to batch high current implanters the design of the beamline and the beam shape are comparable to single wafer high current spot beam implanters, however due to the different scanning architecture the dose rate of the single wafer high current spot beam implanters is significantly higher compared to the batch tools. Therefore, the risk of charging damage will be higher.

The charging damage was studied as a function of the energy, the beam current and the most important plasma flood gun parameters. The results have shown that for very high antenna ratios the charging damage for single wafer implanters, even spot or ribbon beam implanters, is higher than for high current batch implanters.

Keywords: Ion implantation, high current single wafer implanter, charging neutralization, plasma flood gun.

PACS: 61.72.Tt, 85.40.Ry

INTRODUCTION

In the last 15 years, batch implanters were the most common tools for high current applications due to their straightforward construction, good cooling performance and above all the good cost of ownership. Batch implanters are still attractive for factories running low batch sizes, such as logic factories especially on 200mm wafers.

However, with reduction of the geometries below 200nm the structures on the wafer surface are very fragile and vulnerable to ballistic forces caused by particles emanating from the beam line in combination with the high spin speed of the disk. To solve this problem, the reduction of the spin speed of the disks was enabled in combination with the integration of an internal particle sensor located in the process chamber to detect particle induced structure damage.

These techniques resulted in a good correlation between defect density and yield [1].

Even with reduction of the spin speed the structure damage cannot be completely prevented at geometries well below 110nm without significant preventative maintenance. Thus, the consequence was to go to single wafer high current implanter platforms without any fast movement of the wafer.

Since about 2 years the Axcelis Optima HD, a single wafer high current implanter, was positioned on the market [2]. In contrast to the VSEA single wafer ribbon beam high current implanter platform VIISta, the Optima HD is able to reach acceptable beam currents without using deceleration for energies higher than 2keV in combination with a fast tuning time.

The design of the beamline and the beam shape of batch high current implanters and of single wafer high current spot beam implanters, such as the Optima HD, are comparable and as such one could expect similar charging behavior. However, due to the different scanning architecture together with a much lower mechanical scan speed, the dose rate of the single wafer high current spot beam implanters is significantly higher than that of the batch tools. Therefore, the risk of charging damage will be higher and thus, a new characterization and optimization of the plasma flood gun (PFG) is required.

This paper focuses on the characterization of the charging control of an Axcelis Optima HD single wafer high current spot beam implanter using Infineon Technologies own charging control wafers based on MOS capacitors with attached antennas of different size and shape.

The charging damage was studied as a function of the energy, the beam current and the most important plasma flood gun parameters. The results are discussed and compared to a batch implanter with a comparable beamline optic, the Axcelis ULTRA, and a ribbon beam system, the Varian VIISta80.

THEORY

Doping of substrate material using ion implantation generates a positive charge at the surface, which for instance

CP1066, *Ion Implantation Technology,* edited by E. G. Seebauer, S. B. Felch, A. Jain, and Y. V. Kondratenko

can generate damage of the gate dielectric material in CMOS devices, cause uniformity problems due to charge up of the beam itself, and photo resist break through and other process dependent problems. One of the important key factors is the beam potential itself, which is dependent on the mass, charge, energy and current density of the ion beam. In case of a circular beam, neglecting the space charge neutralization, the potential is given by eq. (1) [3], where I is the beam current, m the mass and q the charge of the ions, E is the beam energy, r is the beam radius and R is the distance to the nearest grounded surface.

$$V = \frac{I}{4\pi\varepsilon} \cdot \left(\frac{m}{2qE}\right)^{1/2} \cdot \left[1 + 2\ln\left(\frac{R}{r}\right)\right] \qquad (1)$$

This equation shows that for low energy beams with a high mass, like arsenic and a high beam current the potential will be increased.

To reach a space charge neutralization of the beam, in addition to the ionization of residual gas molecules, the attraction of electrons, and the generation of free electrons from striking and hitting the surfaces, a PFG is required to supply low energy electrons.

Even if the beam is neutralized, there will be some charging effects due to the typically broader spatial distribution of the electrons compared to the ions, resulting in a positive charge of the beam center surrounded by a negatively charged ring, cf. Figure1 [4].

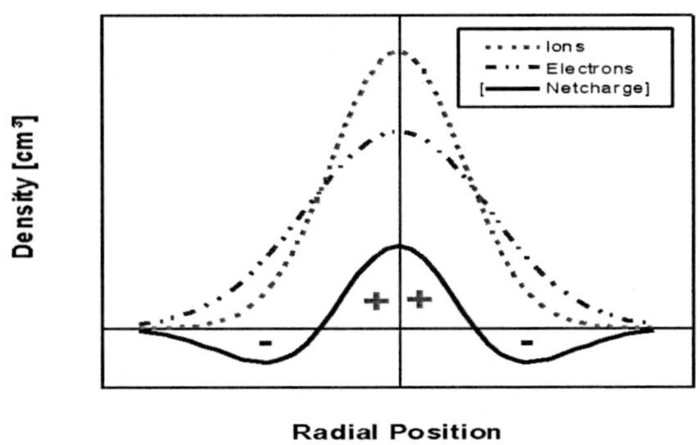

FIGURE 1. Approximate spatial distribution of ions and electrons in the ion beam and the net charge density.

The positive net charge in the beam center can generate charging problems and this will be dependent on the velocity of the scanning beam.

TEST STRUCTURES

Infineon Technologies' own charging monitoring test, described in detail in [5], was used. It is based on MOS capacitors of 10µm² active area, an oxide thickness of 4.5nm, and in situ phosphorus doped poly electrodes (antennas) with antenna ratios (AR) between 10^3 and 10^5.

The antenna ratio is defined as the ratio of poly area to the active area. To avoid the negative influence of the ion beam on the oxide, the polysilicon is overlapping the thin gate oxide. In addition to the antennas with a rectangular area (RECT), dense and wide fork types with small or thick fingers are used (F_DS, F_WS, F_DT, F_WT) (cf. Figure 2). The finger width of the small fingers is 0.25µm and of the thick fingers is 1µm. The space between the fingers is 0.4µm for the dense and 1.2µm for the wide fork antennas. Moreover, various resist patterns overlapping or surrounding the poly electrodes were realized as described in detail in [6].

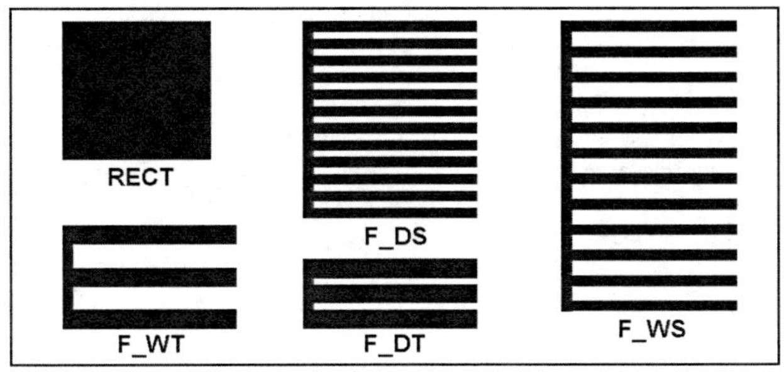

FIGURE 2. The most important antenna structures of the Infineon Technologies own charging test.

For the evaluation of charging damage, leakage current measurements under accumulation mode were performed at -3.5V. A yield criterion around 10^{-9}A was established to distinguish between damaged and undamaged devices. Measurements during the design period of the test have shown a good correlation between leakage current and charge-to-breakdown data [7].

EXPERIMENTAL

Based on eq. (1), single charged arsenic implants were performed with a dose of 5E15atoms/cm² to characterize the charging control of the Optima HD with energies of 5keV (which enables beam currents up to 15mA) and 40keV (with a maximum beam current of 25mA). To enable a comparison of the Optima HD to batch implanters and to single wafer ribbon beam implanters, the Axcelis ULTRA batch tool and the Varian VIISta80 single wafer tool were used applying different beam currents at 40keV arsenic with a dose of 5E15atoms/cm².

To study the influence of the most important PFG parameters on the charging performance, the PFG parameters arc current (1.5A - 4.5A), the Xe gas flow (0.5sccm - 1.0sccm) and the extraction voltage (2V - 6V) were varied for the 40keV arsenic implants.

After implantation of the antenna structured wafers, the resist, at the partial covered structures, was stripped. After thinning of the polysilicon, the leakage current of the wafer was measured and the yield was calculated.

RESULTS AND DISCUSSION

The results of the low energy performance on the Optima HD using an arsenic 5keV implant with a dose of 5E15atoms/cm² with 15mA are shown in Figure 3. It is obvious that the standard setting of the PFG (POR) with an arc current of 1.5A, an extraction voltage of 2V and 0.7sccm Xe flow were sufficient to reach a yield of 100% even for an AR of 10^5. Without using the PFG (OFF), the yield would decrease starting at an AR of 10^4 and would reach 0% at an AR of 10^5. Using a lower PFG setting (LOW), such as an arc current of 0.5A and a Xe flow of 0.2sccm, would not be sufficient for an AR of 10^5.

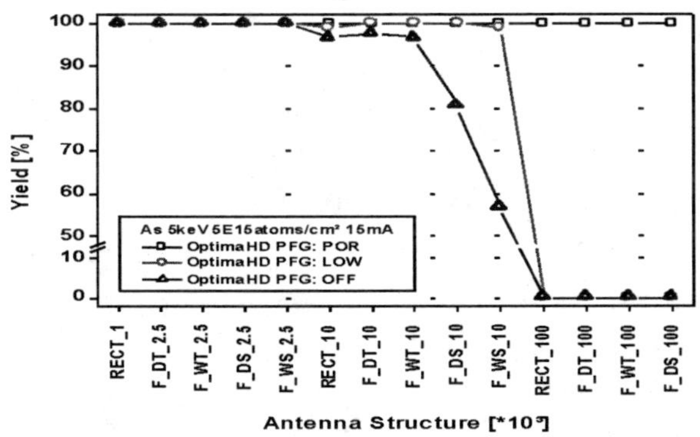

FIGURE 3. Yield of antenna structures implanted with As 5keV 5E15atoms/cm² 15mA on the Optima HD with different PFG settings.

Figure 4 shows, that the results of the 40keV test implants are more interesting than the 5keV test implants. It should be noted that during the first test sequence the backside grounding of the wafers on the electrostatic chuck was not working properly due to sticking grounding pins. These tests are marked with the term "Isol" (isolated) in Figure 4.

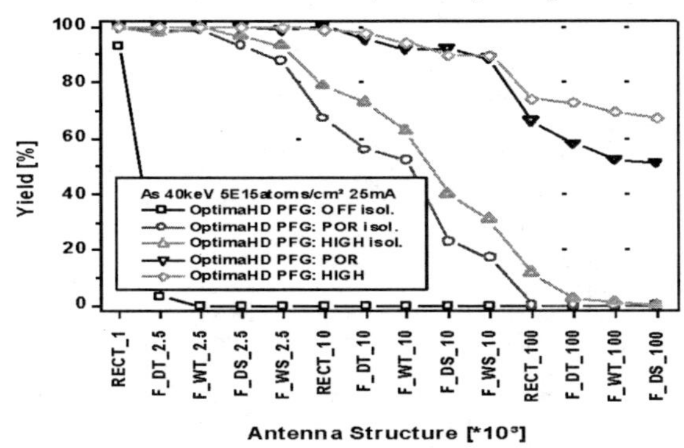

FIGURE 4. Yield of antenna structures implanted with As 40keV 5E15atoms/cm² 25mA on the Optima HD with different PFG settings with or without grounding of the wafers on the backside.

The test was repeated with a new electrostatic chuck and a new design of the grounding pins. The test was done using the standard settings of the PFG with an arc current of 1.5A and 0.9sccm Xe flow (POR) or with an arc current of 2.5A and 0.9sccm Xe flow (HIGH). The results show a dramatic decrease of the yield in the case where the grounding of the wafers on the backside did not work correctly.

However, with a correct backside grounding of the wafers and applying the standard PFG settings, the yield started to decrease at an AR of 10^4 and reached a yield of about 50% for an AR of 10^5 and 70% with the PFG setting with higher arc current values (HIGH).

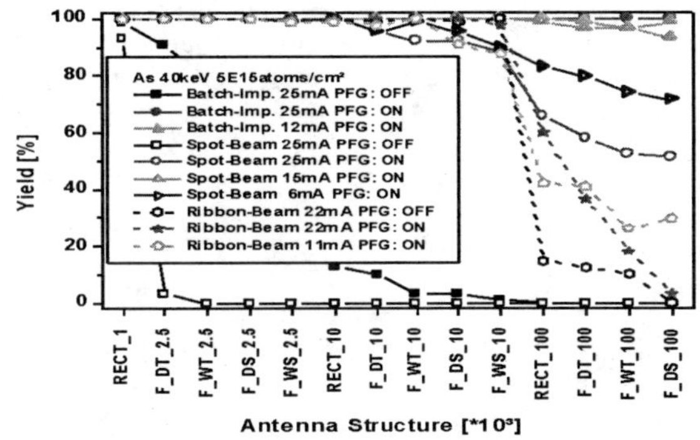

FIGURE 5. Yield comparison between batch implanter (ULTRA), single wafer spot beam implanter (Optima HD) and single wafer ribbon beam implanter (VIISta80) for As 40keV 5E15atoms/cm² with different beam currents and PFG settings.

Compared to the results of the Optima HD, the single wafer ribbon beam implanter VIISta80 shows a lower yield especially at an AR of 10^5. This could be due to the larger spatial area of the ribbon beam and the difficulty in delivering electrons across the beam compared to a spot beam. In contrast to this, the batch implanter Axcelis ULTRA shows a perfect yield also for the highest available AR (cf. Figure 5). This suggests that the duty cycle or dwell time of the beam could be an important factor.

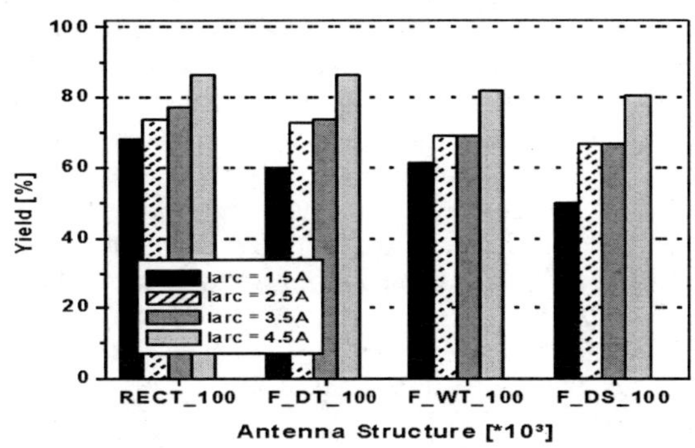

FIGURE 6. Influence of the ARC current of the PFG (Iarc) on the Optima HD at As 40keV 5E15atoms/cm² 25mA on the antenna structures.

Figure 5 shows, that increasing beam currents also decrease the yield if the shape of the beam is comparable. The beam current variation on the Optima HD has shown that the broader beam at 6mA induces a higher charging damage compared to the more focused beam at 15mA. The yield of the 6mA beam current is higher than that of the 25mA but much lower than that of the 15mA beam current, which nearly reached 90% yield also for the highest antenna ratio.

The variations of the PFG settings have shown that the influence of the extraction voltage can be neglected. Changing the gas flow between 0.5sccm and 0.9sccm increases the yield by only 5% at higher AR. The more important arc current of the PFG can increase the yield by up to 30% at the highest AR (10^5) as shown in Figure 6.

The influence of the scan speed on the charging performance of the gate oxide was demonstrated earlier using an Axcelis NV-GSD batch implanter with an 80keV arsenic beam with 12mA (cf. Figure 7). The variation of the PFG settings has not shown any influence on the yield results, but the reduction of the disk speed from 1210rpm to 409rpm results in a decreasing yield by about 10% for high AR.

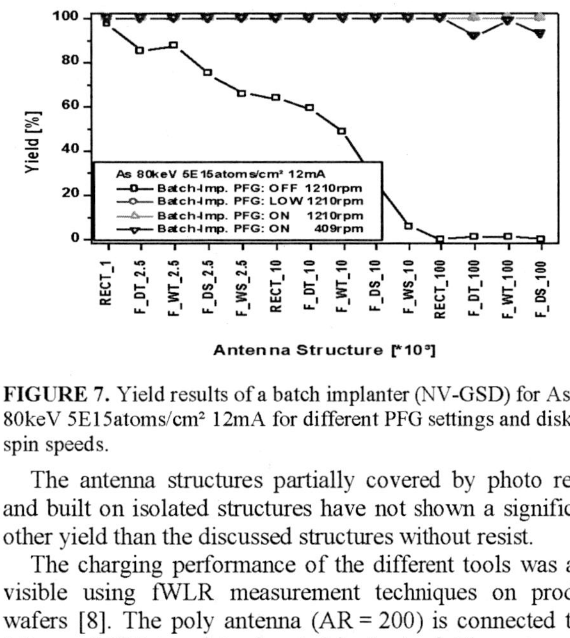

FIGURE 7. Yield results of a batch implanter (NV-GSD) for As 80keV 5E15atoms/cm² 12mA for different PFG settings and disk spin speeds.

The antenna structures partially covered by photo resist and built on isolated structures have not shown a significant other yield than the discussed structures without resist.

The charging performance of the different tools was also visible using fWLR measurement techniques on product wafers [8]. The poly antenna (AR = 200) is connected to a 5.2nm n-MOS transistor located in the kerf. The gate oxide leakage current and the V_{th} drift after revealing stress were evaluated. The V_{th} drift was significantly increased for wafers where the LDD and S/D implantation was done on the spot beam implanter compared to the batch implanter. There was no difference in the gate oxide leakage.

CONCLUSION

The results have shown that the single wafer spot beam implanter (Optima HD) had better yield results on MOS based charging monitor structures than ribbon beam single wafer implanters (without the PFG optimization described in [9]) even at higher beam currents. However, the performance of the batch implanter (Axcelis ULTRA) can not be reached. This comparison suggests that the duty cycle or dwell time on the wafer, and/or the spatial area of the ribbon beam make charging control more challenging. This hypothesis is supported by the observation that the main difference between the spot single, and spot batch system is the much lower scan speed of the single wafer spot beam implanter compared to the batch implanter, and the resulting duty cycle.

Finally, the PFG performance should be improved to reach 100% yield at an antenna ratio of 10^4 independent of the beam shape and beam current. This will increase the margin to the design rule which defines the maximum antenna ratio on the product.

ACKNOWLEDGEMENTS

We would like to acknowledge King Ang Ong from Infineon Kulim for processing the VIISta samples and Serguei Kontratenko and Bernhard Krimbacher from Axcelis for helpful discussions and advice.

REFERENCES

1. P. Splinter, M. Graf, C. Godfrey, Y. Huang, D. Polner, J. Danis, K. Ota, "Optima HD: Single Wafer Mechanical Scan Ion Implanter", in *Proceedings of the 16th International Conference on Ion Implantation Technology*, pp. 601-604 (2006).
2. M. Schmeide, M. Kokot, D.-W. Franke, B. Sauter, "Detection and Reduction of the Yield Impact of Particle Induced Structure Defects at Batch Ion Implanters", in *Proceedings of the 16th International Conference on Ion Implantation Technology*, pp. 562-565 (2006).
3. N. White, M.E. Mack, G. Ryding, D.H. Douglas-Hamilton, K. Steeples, M. Farley, V. Gillis, A. Wittkower, R. Lambracht, Solid State Technology 28, 151 (1985).
4. "Best methods and practices", Axcelis Technologies Inc., Beverly, MA (2005).
5. T. Dirnecker, "Untersuchungen von Aufladungseffekten bei der Ionenimplantation", Dissertation, Shaker Verlag Aachen (2006).
6. T. Dirnecker, L. Frey, A.J. Bauer, H. Ryssel, "Influence of Antenna Shape and Resist Patterns on Charging Damage During Ion Implantation", in *Proceedings of the 16th International Conference on Ion Implantation Technology*, pp. 291-294 (2002).
7. T. Dirnecker, A. Ruf, L. Frey, A.J. Bauer, D. Henke, H. Ryssel, "Influence of Photoresist Pattern on Charging Damage During High Current Ion Implantation", *7th International Symposium on Plasma- and Process-Induced Damage Proceedings*, pp. 106-110 (2002).
8. A. Martin, R. Vollertsen, "An Introduction to Fast Wafer Level Reliability Monitoring for Integrated Circuit Mass Production", Microelectronics Reliability 44, 1209-1231 (2004).
9. P.F. Kurunczi, A.S. Perel, E. Wright, S. Kikuchi, J.T. Scheuer, "Advanced Charge Control for Single Wafer Implanters", in *Proceedings of the 16th International Conference on Ion Implantation Technology*, pp. 445-448 (2006).
10. D.A. Kirkwood, T. Sakase, R. Miura, R.D. Goldberg, A.J. Murell, "Optimized Charging Performance on Quantum X Ion Implanters", in *Proceedings of the 16th International Conference on Ion Implantation Technology*, pp. 641-644 (2006).
11. S. Sakai, H. Fan, E. Chen, M. Tanjyo, "Charging Mechanism During Ion Implantation Without Charge Compensation", in *Proceedings of the 16th International Conference on Ion Implantation Technology*, pp. 460-463 (2006).
12. M.S. Ameen, M.E. Mack, "Wafer Cooling and Wafer Charging", Ed. J.F. Ziegler, in *Ion Implantation Science and Technology*, Ion Implantation Technology Co, Yorktown, NY (1996).

Reliability Tests of a Real Time Charge Monitor (RTCM)

G. Gammel, C. Biloiu, A. Perel, J. T. Scheuer, and A. Renau

Varian Semiconductor Equipment Associates, 35 Dory Road, Gloucester, MA 01930 USA

Abstract. Wafer charging effects that result from the ion implantation process are typically mitigated by employing a plasma flood gun to provide electrons to neutralize the positive ion beam as it strikes the wafer surface. As gate oxide thickness of semiconductor devices continues to shrink, control of the wafer charge-up during ion implantation becomes critical. Varian Semiconductor Equipment Associates (VSEA) has designed and built a real time charge monitor (RTCM) for *in situ* measurement of the beam-induced charging potential. Mounted close to the wafer and in the beam path, the RTCM continuously measures the deviation from quasi-neutrality as the wafer passes through the beam. In this paper we present reliability tests of this device for high current implanters over varied implantation conditions and failure modes. When correlated with the yield of antenna device test wafers, it was found that the RTCM signal is a sensitive and reliable tool to predict charging damage to the semiconductor devices.

Keywords: ion implantation, wafer charging, plasma flood gun, charge monitor.
PACS: 52.59.Sa, 52.77.Dq, 84.37.+q, 85.40.Ry

INTRODUCTION

Insufficient neutralization of charge build-up on the wafer surface during ion implantation can cause damage in two ways. First, it may induce micro-sparks due to charge build-up on photo-resist, which can result in catastrophic transistor damage. Second, in post-gate implants, it can cause excessive tunneling current to flow through the gate, which is one path electrons can take to neutralize the charge build-up, resulting in threshold voltage shifts or premature breakdown. The former is dependent on the instantaneous current only, while the latter is a function of both the instantaneous and the time-integrated current flows [1, 2].

In modern implanters the deleterious effect of positive wafer charging is counterbalanced by a plasma flood gun (PFG) whose role is to provide electrons for charge neutralization during implant. However, the PFG electron output might not be enough for complete charge neutralization or, conversely, excess PFG electron current can lead to negative charging damage. Therefore, during implant, the beam should be monitored *in situ* and in real time [3] to prevent the wafer surface from charging to unacceptable levels.

In the following, we present operating characteristics of a real time charge monitor (RTCM) designed and built to monitor the wafer charging levels during ion implantation on VSEA ribbon beam tools.

REAL TIME CHARGE MONITOR

In the VIISta [4] series of the VSEA high current ion implanters, the wafer to be implanted is loaded onto a platen, which is passed up and down by a grounded air bearing assembly through the fixed ribbon ion beam. The wafer sits on a thin layer of insulating material, and its electrical connection to the ground is made through the air bearing assembly by three spring-loaded ground pins, which make contact with the backside of the wafer. During implant, the current flow from floating front surfaces on the wafer to the ground can follow two paths: 1) through the PFG plasma and then to the grounded PFG arc chamber; 2) through the semiconductor structure on the wafer, then the substrate and ground pins. The most damaging path is when much of the current flows through the gate oxide of the semiconductor structure.

The importance of the plasma flood gun (PFG) electron emission current on the wafer charging level has been demonstrated in a test where the platen was isolated from the air bearing and connected to electrical ground through a resistor. It was found that the current flowing through this resistor during implantation decreased as PFG arc current increased. In other words, the amount of uncompensated positive charge on the wafer decreased as the amount of neutralizing negative charge, i.e., electrons emitted by the PFG (proportional to the PFG arc current) was increased.

CP1066, *Ion Implantation Technology,* edited by E. G. Seebauer, S. B. Felch, A. Jain, and Y. V. Kondratenko

RTCM Construction

In order to get reliable information on the wafer charging level, a charge monitor should be placed as close as possible to the wafer without interfering with the implantation process. Two RTCM sensors are placed 8" downstream of the wafer plane, symmetrically with respect to the wafer center, ½" beyond the wafer edge in the horizontal direction. These spatial locations ensure the sensors are always in the beam path, do not disturb the beam in the vicinity of the wafer, and provide information on charging state on each side of the wafer. All parts exposed to the beam are made of graphite (same material as beam dump, which sensors are just in front of) to avoid any possibility of metals contamination. Metals and particle performance are unaffected by the sensors.

RTCM Operation

The RTCM sensor is connected to the ground through a large resistor. The voltage developed across the resistor is a function of the impinging beam current as well as the local electron environment provided by the PFG. This voltage is a sensitive indicator of the uncompensated beam charge and implicitly of the wafer charging level. The measured signal is high if the PFG electron emission current is low, but it quickly drops as the PFG arc current is increased. At some point, the voltage levels off and further increase of PFG arc current has little effect on the measured RTCM voltage.

Figure 1 depicts time dependent PFG electron emission current and voltage detected by the RTCM sensor during implant for two levels of the PFG arc current. The initial voltage depends inversely on the initial PFG arc current. Then, during each pass of the wafer through the beam, the PFG emission current and the RTCM voltage rise as the wafer diverts the plasma surrounding the sensor, and then drop as the wafer exits the beam. The sensor voltage when the air bearing is at the top of the pass is generally higher than at the bottom of the pass because the air bearing itself still diverts some beam plasma. However, what is important from the wafer electrostatic charging perspective are the variations of the PFG electron emission current and RTCM voltage during scan. For the example shown in Figure 1, the larger PFG arc current results in a large variation in the PFG emission current of ~ 4 mA and a small variation in the RTCM sensor voltage of ~ 0.9 V (full lines). The lower PFG arc current results in a smaller variation of the emission current of ~2.2 mA and a larger variation of ~2.5 V of the sensor voltage (dashed lines).

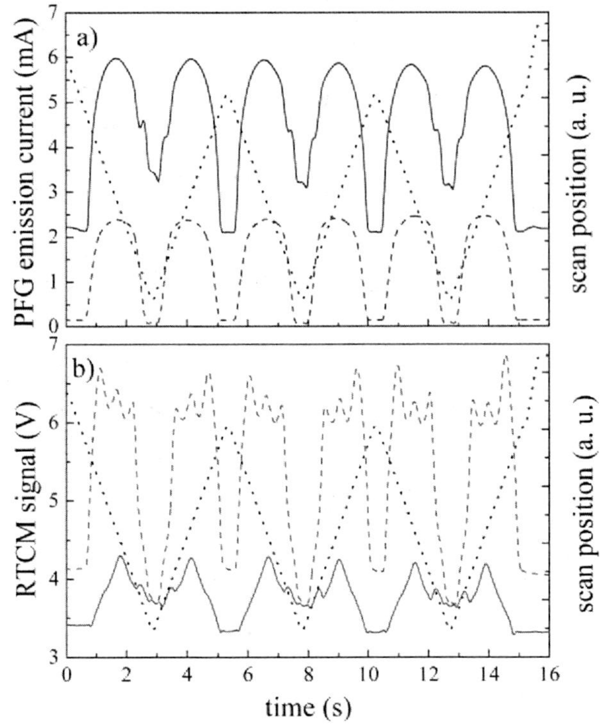

FIGURE 1. Typical signal waveforms obtained during implant: a) the PFG emission current; full and dashed lines correspond to high and low PFG electron emission current, respectively; b) the corresponding measured RTCM voltage. The triangular waveform (dotted line) is the position of the wafer relative to the beam; top and bottom positions – the wafer is below and above the beam, respectively.

The sensitivity of the RTCM to different implant conditions is shown in Figure 2 where a 2D map of the measured RTCM signal vs the arc and beam currents is given for a 40 kV As^+ implant.

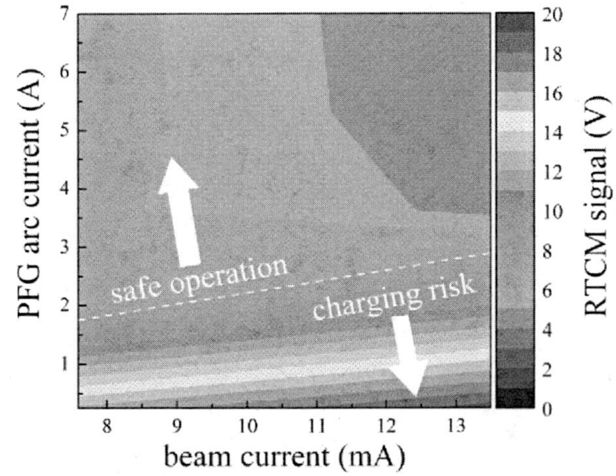

FIGURE 2. A 2D map of the RTCM voltage in the PFG arc current – beam current space. The white dashed line represents the boundary between the charging safe and charging risk operation modes.

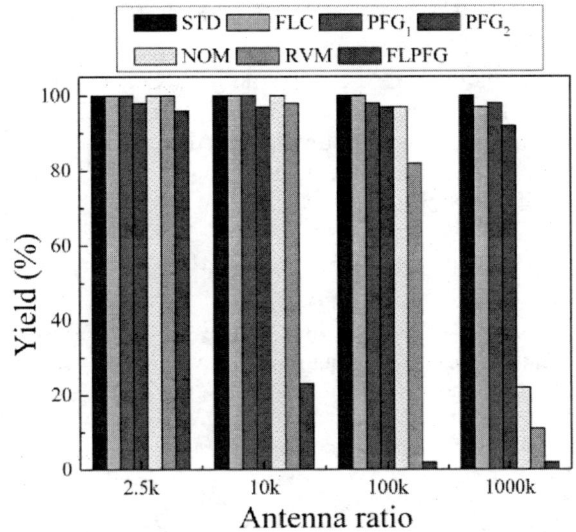

FIGURE 3. Philtech YST test device yield for different failure modes and antenna ratios.

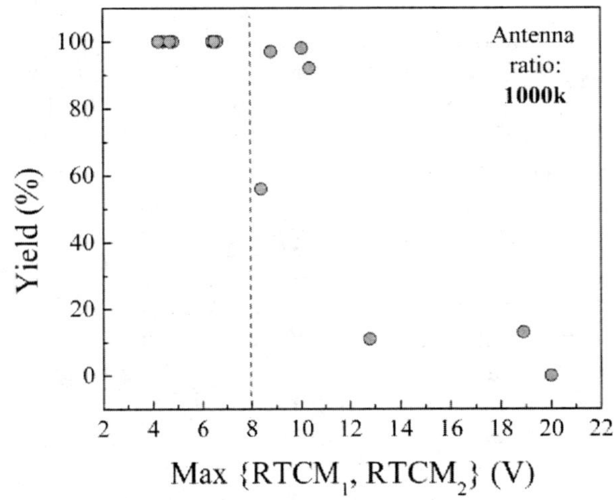

FIGURE 4. The correlation between wafer yield and the two RTCM sensors readings. Dashed line indicates the boundary above which the wafer yield decreased below 100%.

PFG arc current drives the PFG electron emission current, which in turn, changes the wafer charging level. High values of the beam current and low values of the PFG arc current (right bottom corner) lead to high voltages detected by the RTCM. Conversely, low values of the beam current and high values of the arc current (left top corner) lead to low voltages and implicitly low wafer-charging levels. Although the sensor voltage is not a direct measure of the voltage on the wafer, if the RTCM measured voltage exceeds a pre-determined empirical limit, the wafer is at risk of charge-up-damage. This limit is dependent on device layout, geometry, and specific implant conditions.

RELIABILITY TESTS

To validate the relationship between the RTCM detected voltage and the wafer yield, tests under various implantation conditions were performed using Philtech YST (yield safety test) wafers [5]. These wafers have box or comb antenna structures specially designed to study the gate damage due to tunneling through a 40 Å gate thickness. For these test wafers the antenna ratio (contact area above the gate to the gate area) spans from 1000k (10^6:1) to 2.5k (2.5×10^3:1). The bigger the ratio, the more sensitive the device is to tunneling damage. Since real devices have antenna ratios of ~ 10^3:1, these wafers are more sensitive to charge-up damage than most manufactured semiconductor devices.

The plasma flood gun (PFG) is a magnetically confined arc plasma source. Two such plasma guns placed in the beam vicinity pump neutralizing electron fluxes into a cage through which the ion beam passes. Since too few electrons emitted by the PFG is the most detrimental for the wafer charging-up, we focused our research on different PFG malfunctioning scenarios.

Shown in Figure 3 are yield data for the case of As^+ 20 kV energy, 10 mA beam current, 5×10^{15} ions/cm^2 dose implants obtained under various simulated failure conditions: only one plasma flood gun working ($PFG_{1,2}$), floating cage (FLC), floating PFG (FLPFG), no magnetic field confinement in the PFG (NOM), and reversed magnet polarity (RVM). For the sake of comparison the yields for standard operation mode (STD) are also provided. As can be seen, for the lowest antenna ratio all failure modes give good yield (>98%). For higher antenna ratio the yield decreased and, for the most sensitive case most devices failed (less than 20% yield) for scenarios involving reduced magnetic confinement and the lack of ground connection of the PFG.

Corroborating the results for different implantation conditions (species, energy, beam current, dose) and for all failure modes we were able to obtain a value of 8V (vertical dashed line in Figure 4) for the RTCM sensor signal for which all devices had 100% yield for the highest (1000k) antenna ratio. It would imply that, for real devices, less than 8V reading on the RTCM would ensure a charge-up-damage free implantation process. The difference in $RTCM_1$ and $RTCM_2$ voltages corresponded with asymmetry in beam current density (measured by a 2D profiler), and with Philtech yield differences across the wafer. The sensor reading the higher voltage corresponded to the higher density side of the beam, and reduced Philtech yield on that side of the wafer.

The sensor voltage reading repeatability was evaluated in house as well as at customer sites by tracking the RTCM signal under similar implanting conditions over three month operation time. The signal variability of <30% can be considered acceptable because no decrease in wafer yield was observed. Actually this apparently large figure proves the sensitivity of the RTCM: its indications tracks the gradual change of the implant conditions during such prolonged operation time.

CONCLUSIONS

Good sensitivity of the RTCM sensor against beam current, beam energy, and PFG arc current has been observed for a wide range of implant conditions. Excellent correlation was obtained between Philtech YST wafer yield and RTCM sensor voltage.

In tests with Philtech wafers, the yield was 100% even for the most sensitive antenna ratio, as long as RTCM voltage reading was less than 8V. For actual manufactured devices a sensor voltage less than this limit value will ensure a high implantation yield. Endurance tests showed that RTCM reading was reasonably stable over a typical implanter end-station maintenance interval (3 months).

ACKNOWLEDGEMENTS

The authors thank E. Cobb and L. Klos for their work on this project and for their helpful suggestions for the RTCM device improvement. Also, we thank R. Low for his encouragement and support in the early phases of this project.

REFERENCES

1. J. P. McVittie *Plasma Charging Damage: An Overview*- 1st International Symposium on Plasma Process-Induced Damage, Santa Clara, CA, USA, 1996, pp. 7-10
2. K. P. Cheung, *Plasma Charging Damage*, Springer-Verlag, New York LLC, 2001, pp. 41-161.
3. W. G. En and N. W. Cheung, IEEE Trans. Electr. Devices **44**, 379-387 (1997)
4. G. Mezack, T. Callahan, S. Metha, and U. Jeong *Advantages of the Varian VIISta Single Wafer High Current Ion Implanter for Advanced Device Fabrication*- Proceedings Internat'l Conference on Ion Implantation Technology, Alpbach, Austria, 2000, pp. 431-434
5. Philtech Inc., Tokyo, Japan, Philtech YST wafer unveils charge-up-damage for a better yield of processing- http://www.philtech.co.jp

Angle Effects in High Current Ion Implantation

Joseph C. Olson (joseph.olson@vsea.com), Christopher Campbell

Varian Semiconductor Equipment Associates, Inc., 35 Dory Road, Gloucester, MA 01930, United States

Kyoichi Suguro, Yoshimasa Kawase

Toshiba Corporation, Process and Manufacturing Engineering Center. 8, Shinsugita-cho, Isogo-ku, Yokohama 235-8522, Japan

Hiroyuki Ito

VSEKK, Yusen Shin-Yokohama Bldg., 3-17-2 Shin-Yokohama Kouhoku-ku, Yokohama-shi, Kanagawa 222-0033, Japan

Abstract. Previous studies conducted on batch high current implanters with 130 nm devices[1] have shown the importance of implant angle during source-drain (SD) and source-drain extension (SDE) implants. For these implants, errors in implant angle lead to device asymmetry and this device asymmetry has been cited as a reason for requiring single wafer high current implanters[2]. The sensitivity of device performance to angle is increasing as devices shrink. For current anneal technologies, angle effects are more importance in NMOS, due to the lower diffusion of arsenic. However, the importance of implant angle in PMOS is expected to increase as diffusionless anneals are adopted.

In this paper we report on angle effects in single wafer high current ion implantation, for the improvement of the characteristics of MOSFETs integrated into Systems-on-a-Chip (SoCs) of 65nm or beyond. The single wafer high current implanter and its angle measurement and control system will be described. A comparison of the implanter's angle measurements to device data will be presented.

Keywords: Ion implantation, implant angle control, semiconductor devices
PACS: 85.40.Ry, 85.30.De

INTRODUCTION

High current ion implantation for transistor doping has undergone a rapid transition from batch to single wafer equipment, driven by particle control and yield issues[3]. This transition also makes the superior angle control of single wafer equipment available for high dose implants. Previous work has shown the sensitivity of transistors to angle variation, in particular resulting in device asymmetry[1,2,4]. In this paper we describe device results for 65 nm transistors implanted on a single wafer high current tool, the VIISta HCS, as compared to reference devices processed on a batch implanter and as compared to the angle measurements collected by the VIISta HCS. The VIISta HCS is also described briefly.

DEVICE DESCRIPTION

Figure 1 shows TCAD simulation output for a 65 nm nMOS transistor and notes the source drain extension(SDE) region, which is typically doped on a high current implanter. Errors in implant angle alter the shadows cast by the gate and spacers, resulting in a change in the proximity of the source drain extensions to the gate. Previous reports[5] have noted that the transistor on-state current, I_{dsat}, degrades more rapidly when the source side of the transistor is shadowed

CP1066, *Ion Implantation Technology,* edited by E. G. Seebauer, S. B. Felch, A. Jain, and Y. V. Kondratenko

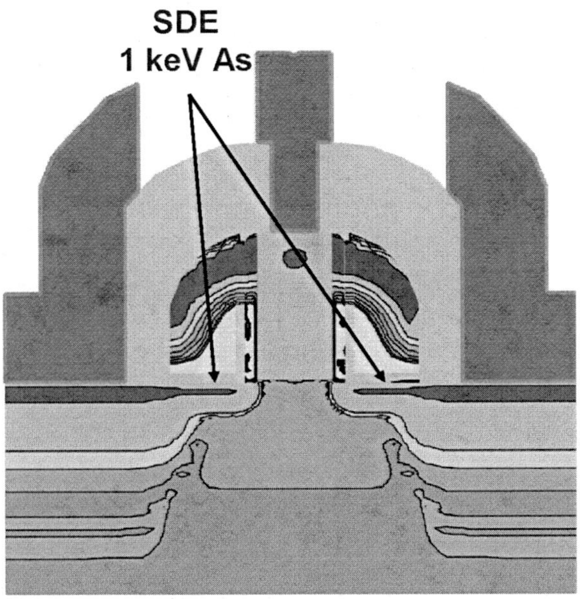

FIGURE 1. TCAD output from simulation of a 65 nm high performance logic transistor. The source drain extension region is highlighted.

than when the drain side is shadowed. TCAD results reported here show the same effect. In this paper, the asymmetry is exploited to examine implant angle integrity by looking at the skew (defined below) of device parameters. A skew measurement is made by comparing a device parameter, for instance I_{dsat}, to its value with the source and drain reversed, as shown in (1),

$$I_{dsat}skew = 2\left(\frac{I_{dsat} - I^{r}_{dsat}}{I_{dsat} + I^{r}_{dsat}}\right) \quad (1)$$

where I_{dsat} is the on-state current, and I^{r}_{dsat} is the on-state current with the source and drain reversed. Similar skew values can be computed for the threshold voltage and off-state leakage current.

The skew of I_{dsat} under SDE implant angle error, as predicted by the TCAD model, is shown in Figure 2. For the simulated device, the on-state current can be skewed by as much as 1% per degree of angle error. The chart shows that the sensitivity of on-state current to angle error decreases as the range of angles reaching individual transistors increases.

We will later describe the skew of the on-state current on actual devices. Figure 3 illustrates the tested geometry, utilizing a wafer with transistors oriented in two different directions: (a) gate vertical, and therefore sensitive to angle variation in the horizontal direction; and (b) gate horizontal, and therefore sensitive to angle variation in the vertical direction.

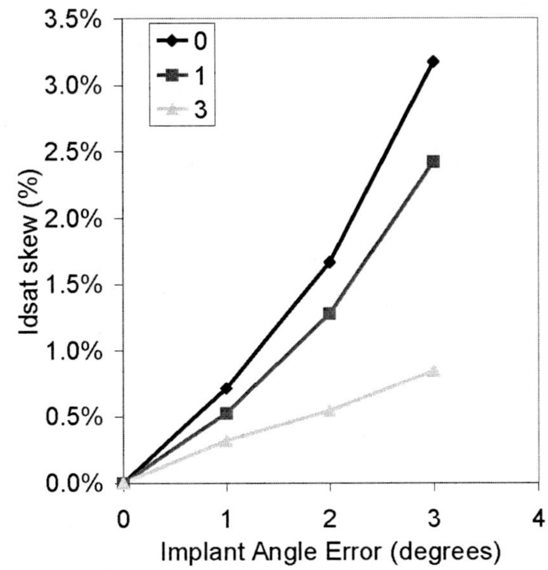

FIGURE 2. Skew of Idsat as predicted by TCAD for varying SDE implant angle errors. The skew is plotted for angle distributions with widths (1σ) of 0° (black diamonds), 1° (red squares) and 3° (grey triangles).

VIISTA HCS DESCRIPTION

Single wafer implants for this study were performed on the Varian VIISta HCS. The HCS system consists of a dual magnet ribbon beamline and a simple end station with mechanical wafer scanning in a direction perpendicular to the ribbon beam. In the HCS, the first magnet performs mass analysis on the ion beam and the second magnet is used to make the beam uniform and parallel. The HCS is shown in Figure 4.

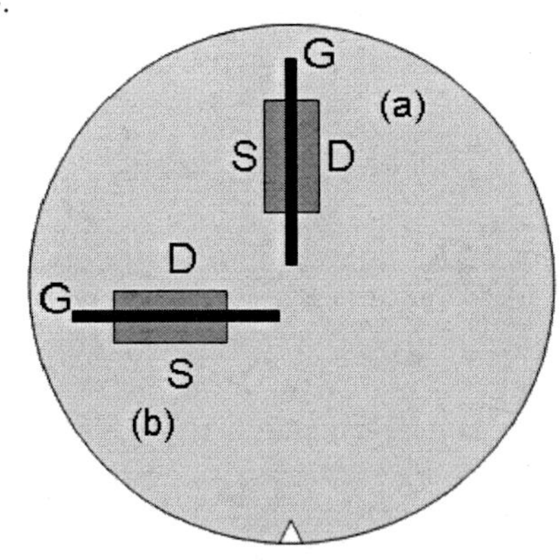

FIGURE 3. Layout of source (S), drain (D) and gate (G) for (a) vertical gate and (b) horizontal gate transistors.

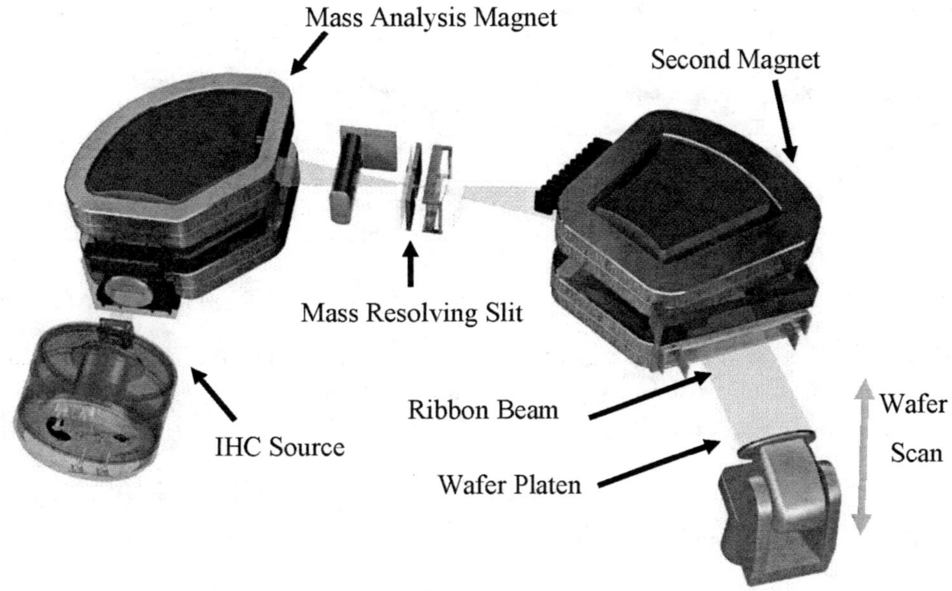

FIGURE 4. Schematic of the VIISta HCS high current implanter.

Beam angle measurement equipment is installed on the HCS used in this work. The angle measuring system consists of an array to sample the beam at multiple positions horizontally and vertically, and provides information on beam direction and within wafer variation of both horizontal and vertical angles. Varian angle measuring systems are described extensively elsewhere[6,7].

As part of every beam setup, the tool measures and reports the mean beam angle and the within wafer variation of mean angle with position (defined on the tool as beam angle spread(BAS)). Every implanted wafer is positioned to reduce to less than 0.1° the contribution of the mean beam angle to implant angle error. Within wafer angle variation is controlled by recipe setup and interlock. It is typical for established recipes to meet the recipe interlock with a high success rate. When this does not happen, an automated minimization of the variation is performed. In a test of sub 3 keV beams, the recovery routine was successful at restoring sub-interlock level angle spread in 23 out of 28 attempts, therefore preventing 82% of the failures. Figure 5 shows the result of an automated minimization for 1.5 keV BF_2.

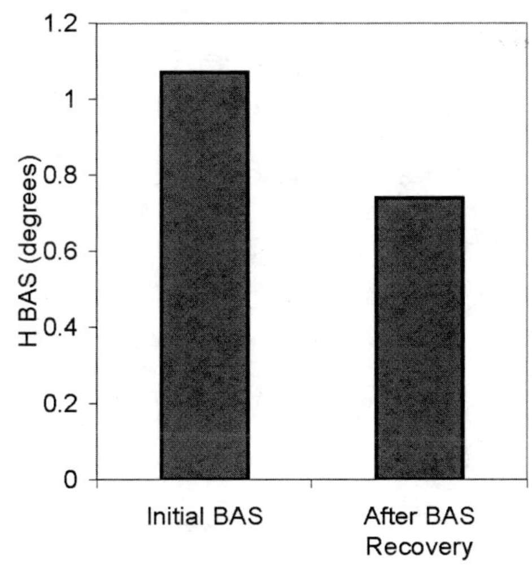

FIGURE 5. Results of automated angle recovery for 1.5 keV BF_2.

RESULTS

Table 1 shows a comparison of the I_{dsat} skew for NMOS transistors produced on the single wafer HCS and on a batch high current implanter. The single wafer architecture results in a reduction in the skew of the on-state current. Although the devices tested here are not the ones modeled by TCAD, the magnitude of the skew is consistent with the model.

TABLE 1. Comparison of NMOS Idsat skew between single wafer implanter (VIISta HCS) and batch implanter for single step (no wafer rotation implants). Both implanters produce <1% skew in quad mode (four rotation) implants.

	Single Wafer	Batch
Horizontal Gate	<1%	3.6%
Vertical Gate	<1%	<1%

Figure 6 shows the beam current, uniformity and BAS as measured on the VIISta HCS for 1 keV As, compiled from 291 beam setups. The data is organized by machine configuration and illustrates various productivity and process performance results.

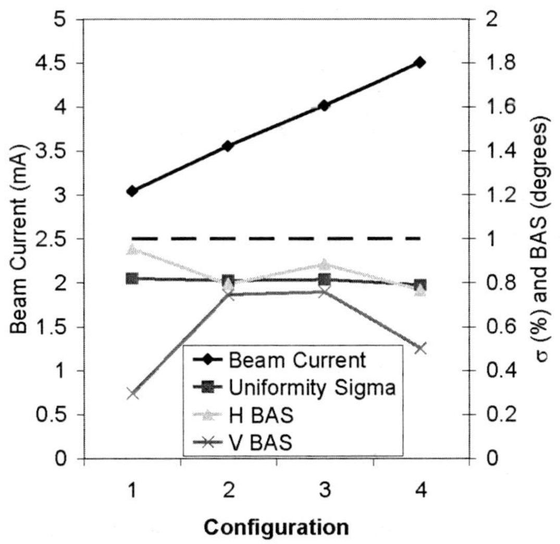

FIGURE 6. Measured beam current, uniformity and angle spread for 1 keV As for various configurations. The dashed line represents the process requirement for uniformity and angle spread.

DISCUSSION

In published simulations of device performance, angle variation in the NMOS source drain extension implant has been shown to be the single largest contributor of ion implantation to transistor variation[8]. The device data shown in Figure 3 and Table 1 demonstrate the value in eliminating the angle variation introduced from the cone angle of a batch machine by changing to a single wafer ion implanter.

In addition to the beam current, uniformity and angle data, Figure 6 also displays a process window which is determined by transistor performance requirements. A goal of the VIISta HCS development was to maintain or improve the margin between the process window and measured beam results with high productivity.

CONCLUSIONS

The introduction of single wafer implanters into high dose applications has resulted in improvements in transistor symmetry and performance. Attention by chipmakers to the key device requirements continues to drive ion implanter manufacturers to improve their equipment.

ACKNOWLEDGEMENTS

The authors gratefully acknowledge the assistance of Greg Redinbo for comments during the preparation of this manuscript.

REFERENCES

1. Yoneda, K. and Niwayama, M. "The drain current asymmetry of 130 nm MOSFETs due to extension implant shadowing originated by mechanical angle error in high current implanter", in *Extended Abstracts of International Workshop on Junction Technology*, 2002, pp. 19-22.
2. Suguro, K., "Prospects and Challenges in USJ Technology for Next Generation LSIs", vTech Symposium, San Francisco, 2005.
3. Pipes, Leonard *et al.*, "Characterization and reduction of a new particle defect mode in sub-0.25 μm semiconductor process flows" in *Proceedings of the 15th International Conference on Ion Implantation Technology*, edited by L. J. Chen, J. Poate and T.-F. Lei, Elsevier, Oxford, United Kingdom, 2005, pp. 330-335.
4. Jeong, Ukyo *et al.*, "Effects of Beam Incident Angle Control on NMOS Source/Drain Extension Applications" in *Proceedings of the 146th International Conference on Ion Implantation Technology*, edited by Bob Brown *et al.*, IEEE, Piscataway, New Jersey, 2003, pp. 64-68.
5. T. Ghani, K. Mistry, P. Packan, M. Armstrong, S. Thompson, S. Tyagi, M. Bohr, "Asymmetric Source/Drain Extension Transistor Structure for High Performance Sub-50nm Gate Length CMOS Devices," paper # T3-A1, 2001 Symp. VLSI Technol. Digest Tech. Papers, IEEE, New York, NY, US, 2001.
6. Campbell, Christopher *et al.*, "Beam Angle Control on the VIISta 80 Ion Implanter" in *Proceedings of the 146th International Conference on Ion Implantation Technology*, edited by Bob Brown *et al.*, IEEE, Piscataway, New Jersey, 2003, pp. 193-196.
7. Olson, Joseph C. *et al.*, "Metrology Requirements for Single Wafer Ion Implanters" in *Proceedings of the 16th International Conference on Ion Implantation Technology*, edited by Karen J. Kirkby *et al.*, American Institute of Physics, Melville, New York, 2006, pp. 538-541.
8. Gossmann, H.-J. *et al.*, *Solid State Technology* **50**(7) (July, 2007).

Key Technologies for Ultra High Dose CMOS Applications

Y. Jeon[1], I. Koo[1], J. Oh[2], SB. Lee[3], J. Butterbaugh[4], S. Jin[2], J.Lee[2], K. Rouh[2], M. Ju[2], S. Jeon[2], J. Ku[2], SW. Lee[3], MT. Ok[3], A. Lee[4], K. Kim[4], SW. Lee[5], KJ. Ju[5], JW Park[5] and V. Singh[1]

[1]*Varian Semiconductor Equipment Associates, Inc., 35 Dory Road, Gloucester, MA 01930, USA*
[2]*Hynix Semiconductor, Inc., Ichon-si, Kyoungki-do, 467-701, Korea*
[3]*Nanometrics, Inc., Milpitas, CA 95035, USA*
[4]*FSI International, Inc., Chaska, MN 55318, USA*
[5]*Varian Korea Ltd., Pyeongtaek-si Kyoungki-do 459-040 Korea*

Abstract: The trend towards shrinking advanced microelectronic Logic and DRAM devices will require ultra high dose implantation. One ultra high dose application in DRAM, being rapidly adopted in production is Dual Poly Gate (DPG). Three main challenges existed for the adoption of this high dose dual poly gate (DPG) doping applications: monitoring of high dose implantation, photoresist stripping and maintaining high throughput. In this paper we present how these challenges have been addressed. VSEA's plasma doping (PLAD) tool offers several unique advantages for DPG applications. When compared to conventional or molecular beam line implanters or other immersion techniques, PLAD delivers 3 to 7 times higher throughput (compared to traditional ion implanter) without dopant penetration through the thin doped polysilicon layer into the gate oxide. It also improves P+ poly silicon DPG device properties at superior throughput. In this work we demonstrate how hot spray photoresist strip processing eliminates the need for multiple-tools required for wet+ash+wet process. In addition to PLAD's patented in-situ dose control metrology we also demonstrate an ex-situ high dose implantation metrology using spectroscopic ellipsometer (SE) and spectroscopic reflectometer (SR). The technique shows good correlation ($R^2 \sim 0.99$) between implant dose and damaged layer thickness.

Keywords: Plasma doping, P^+ poly Si, ultra high dose implant metrology, high dose implant PR strip
PACS: 52.77.Dq

INTRODUCTION

The trend towards shrinking modern microelectronic devices requires new processing technologies for production worthy solutions. One new application that has emerged in DRAM process flow which requires low-energy high-dose implant is dual poly gate (DPG). In DRAM pMOSFETs applications B is implanted into Phosphorous doped poly silicon gate to achieve the required device performances - lower operating voltage, lower leakage current and high drive current. In this kind of ultra high implantation (> 3e16 dose), there are three key aspects for a production worthy implant solution - high throughput, photoresist strip and dose repeatability monitoring.

DPG doping requirements are very stringent, with the thin phosphorous doped poly silicon gate layers demanding energy contaminant-free low energy (≤ 5keV $^{11}B^+$) implants with doses 2 to 3E16 cm^{-2}. This combination of requirements has led to significantly reduced productivity for traditional beamline-based implant tools, due to the inefficiency of transport of low energy high space charge ion beams. Plasma doping (PLAD) avoids this limitation, as ions are transported across the plasma sheath across the entire wafer surface, which is a distance of about a centimeter as compared to the meter size scale of beamline ion transport [1, 2]. PLAD can also provide most precise energy control and process security because it has pulsed DC wafer bias power generator and real time in-situ dose control system.

Heavily implanted photoresist (PR) is especially challenging from strip perspective due to the dehydrogenated, amorphous carbon layer that forms on the surface [3]. The carbonized layer can be removed by plasma etching, or can be broken up by physical processes such as ion bombardment or the swelling of the underlying PR material. Physical processes, however, tend to leave residue where the carbonized resist contacts the wafer at the edges of features and particularly at the inner boundary of the wafer edge bead removal area. Plasma ashing processes used to strip implanted PR tend to oxidize the wafer surface and cause an unacceptable increase in Si material loss in subsequent processing steps. Interest in ash-free, all wet stripping processes is driven primarily by the desire to reduce surface damage and material loss, but also by a desire for a simplified stripping process for PLAD implants, and the elimination of a process step for all implants [4].

During the plasma doping process, dopant ions are implanted into the substrate wafer. The implanted profiles are surface peaked with dopant rich layer near the surface as shown in Figure 1. In general it is

CP1066, *Ion Implantation Technology*, edited by E. G. Seebauer, S. B. Felch, A. Jain, and Y. V. Kondratenko

challenging to monitor high dose, low energy implants, here we present on a multi-wavelength spectroscopic ellipsometer (SE) and spectroscopic reflectometer that can be used to determine both crystal damage layer in the silicon substrate caused by the implanted ions as well as the dopant rich layer near the surface. Additionally, this measurement method can be performed at any point on the wafer, hence providing dose uniformity.

In this work, we accomplished improved P+ poly silicon DPG device performance at superior throughput level without PR stripping issue. We utilized VIISta PLAD for high dose implant, ZETA® spray cleaning system (FSI) for photoresist removal and a combination of spectroscopic ellipsometer and reflectometer (Nanometrics Atlas system) for high dose monitoring.

EXPERIMENTAL RESULTS AND DISCUSSSIONS

In this work, we used a commercially available, 300 mm, FSI batch spray system to strip the PR after ultra high dose PLAD implantation. Because the H_2SO_4 is first heated to 95°C before mixing with H_2O_2, rapid radical production occurs while the fresh solution is in contact with the carbonized layer, so the radicals can react with the carbonized layer more efficiently before they decompose [4]. We analyzed the poly with SIMS and TEM, post implant, PR striping and cleaning process (ZETA®Zeta spray cleaning system).

Figure 1 shows the SIMS profiles as implanted and post PR strip. Because of the surface peaked profile, PLAD is very effective for low energy ultra high dose applications on thin doped poly silicon gate without dopant penetration. The PR stripping process has to be able to remove the boron oxide film and the dopant rich surface layer. We measured the dopant rich and damaged layer thickness before and after PR strip process using TEM, the dopant rich layer was removed effectively without PR residue, as well as the implant damaged layer was not etched (Figure 2).

During the BF_3 plasma doping process, dopant ions are implanted into the substrate wafer and dopant rich layer is formed on the surface simultaneously as shown in Figure 1. We have tested that the samples were implanted at 1e15 to 1.1e17 of nominal dose at 1 to 7kV bias voltage on n-type bare silicon wafers, those were prepared at two different process conditions that cause different thickness of dopant rich layers. To distinguish the dopant rich surface layer and implant damaged layer thickness, we used a multi-wavelength SE & SR system (Nanometrics's Atlas) within 193 to 780 nm. Figures 3 and 4 show the correlations of the layer thicknesses to implant dose and implant energy, respectively. Especially, the damaged layer thickness correlates to implant dose and the R square value of damaged layer thickness vs nominal dopant dose was 0.9912.

We have also compared the B implanted dopant depth at two different process conditions to decouple the dopant rich surface layer and damage layer thickness. We compared the damaged and the dopant rich surface layer thickness measured by TEM and ellipsometer in Figure 5. The ellipsometric damaged layer thickness was matched with TEM thickness very well, but the ellipsometric dopant rich surface layer thickness was not matched with TEM thickness very well because the optical material constants of the dopant rich layer are not easily accessible. If one measures the ellipsometric thickness after DI rinse, the error of the dopant rich and damaged layer thickness analysis can be minimized. From the results, this SE and SR monitoring method could be useful for the ultra high dose plasma doping applications as a nondestructive dose monitoring.

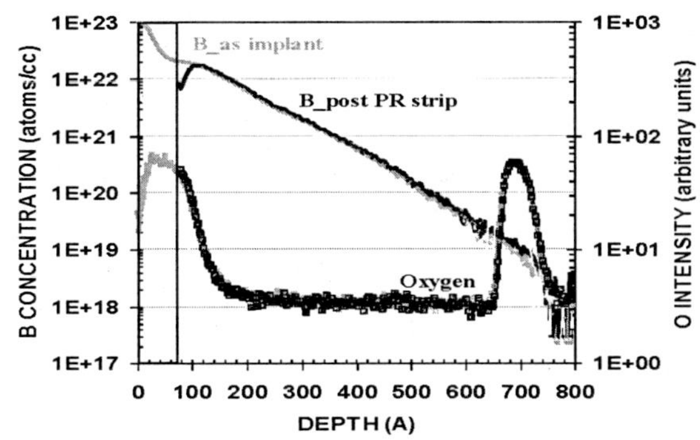

FIGURE 1. B SIMS profiles with BF_3 PLAD implant on poly silicon and post PR Strip and cleaning process.

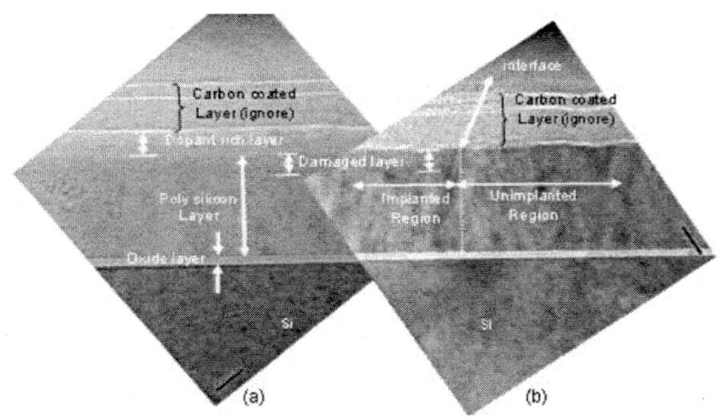

FIGURE 2. Cross-sectional TEM photographs implanted on poly silicon layers (a) with PLAD BF_3 implant and (b) post Zeta PR strip and cleaning process

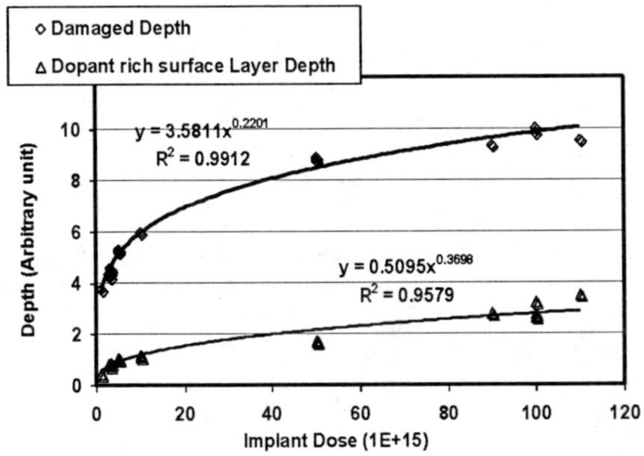

FIGURE 3. Correlation of implant dose vs. Dopant rich surface layer and damaged layer thickness analysis using SE analysis (implanted BF_3 at 7 kV)

For the p$^+$ poly-Si gate fabrication of 66 nm DPG DRAM devices, BF_3 plasma was utilized to counter dope (using PLAD) into phosphorous in-situ doped poly-Si layer after the plasma nitridation process of gate oxide. The batch spray system was used for PR striping and post cleaning process and a conventional poly implant annealing (PIA) was performed at 950°C or spike poly implant annealing at 1075°C with lamp-based equipment [5]. The electrical properties, such as the leakage current (I-V) and capacitance-voltage (C-V) characteristics of p+ poly-Si gate, were measured by 4155B and HP4284A, respectively.

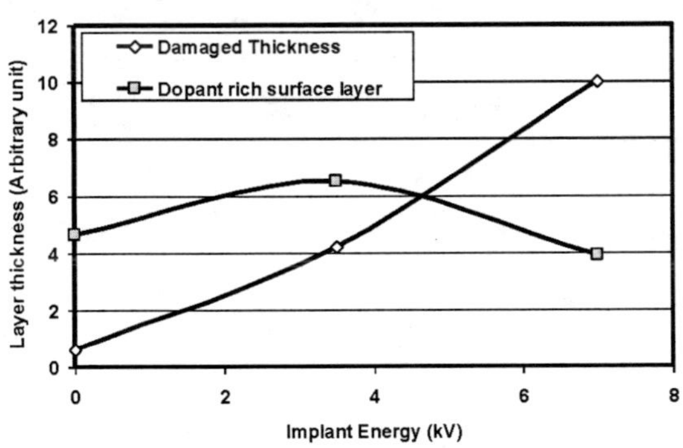

FIGURE 4. Energy effect analysis using ellipsometer (1e17 dose BF_3 implanted).

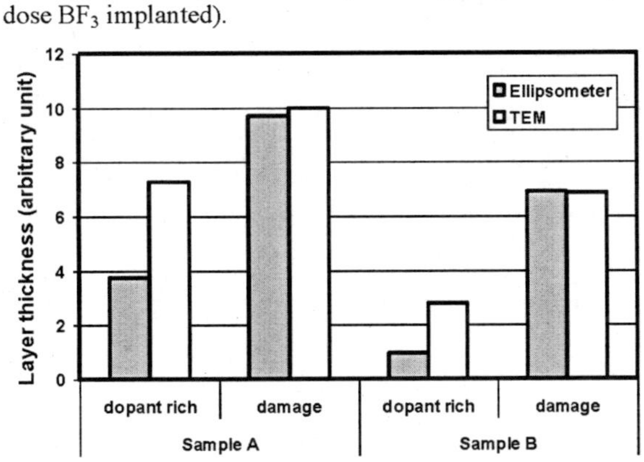

FIGURE 5. Comparisons of measured thickness by TEM and ellipsometer.

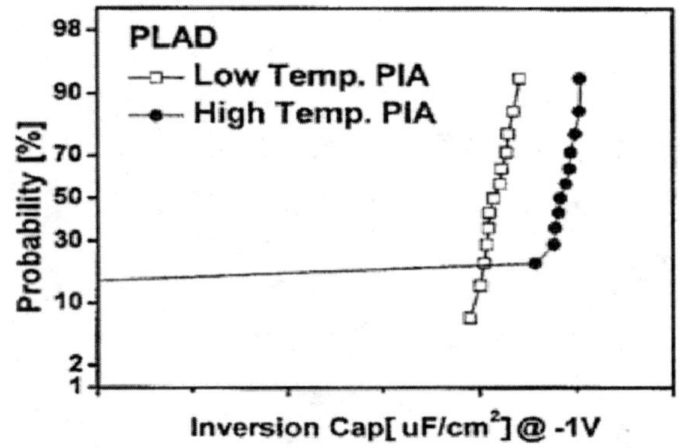

FIGURE 6. Inversion capacitances properties at different PIA temperatures.

The device performances such as I-V, C-V and inversion capacitance distributions were tested which the devices were BF_3 PLAD implanted and annealed at two different anneal temperatures. We matched the inversion capacitance and the poly depletion ratio (PDR) at seven times faster throughput using PLAD than conventional beam line implanter. The inversion capacitance and PDR can be improved by PIA temperature control, because the boron concentration can be increased at the interface of poly silicon and gate oxide layer by increasing the PIA temperature (Figure 6 and 7). PLAD provides wide process window (energy, dose and annealing temperature) to maximize PDR value without boron penetration for thin poly, due to unique energy contamination-free doping profile. The gate current properties are not sensitive to the PIA temperature conditions which are implanted by PLAD implant.

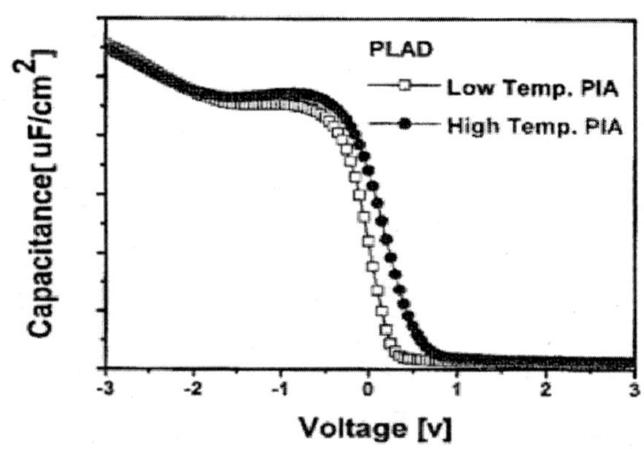

FIGURE 7. C-V characteristics of p+ poly-silicon gate

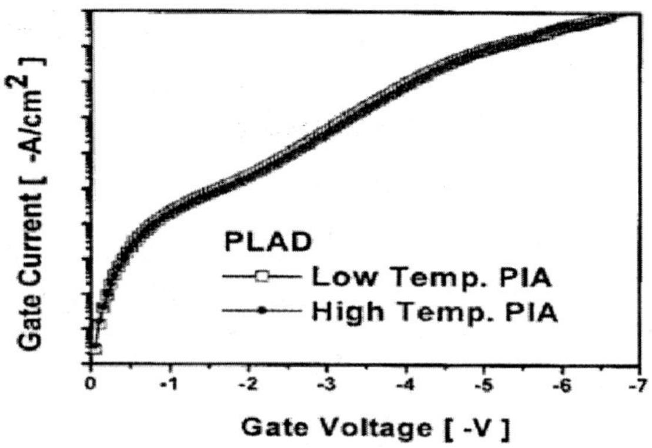

FIGURE 8. I-V characteristics of p+ poly-silicon gate

CONCLUSIONS

We demonstrated improved p+ Poly-Si Gated DPG device properties at superior throughput level (seven times faster than conventional beam line implant) without photoresist stripping issues using VIISta PLAD and a single hot spray photoresist strip process, eliminating the need for a multiple-tool, wet+ash+wet process. Also, we demonstrated a high dose metrology using a spectroscopic ellipsometer (SE) and a spectroscopic reflectometer (SR) which shows acceptable correlation ($R^2 \sim 0.99$) between implant dose and damaged layer thickness. This monitoring method could be useful for the ultra high dose plasma doping applications as a nondestructive dose monitoring.

REFERENCES

1. R Anthony Renau and Jay Scheuer, "Comparison of Plasma Doping and Beamline Technologies for Low Energy Ion Implantation", in *Proc. 2002 Int. Conf. Ion Implant. Tech.*, Taos, New Mexico USA, Sept. 2002.
2. Walther, S., Lenoble, D., Lallement, F., *Nuclear Instruments and Methods in Physics Research Section B*, **237**(1-2), p. 126-130.
3. T. Bausum, M. DeSarno and G. Dahrooge, Semi. Intl., Web Exculsive, June (2003).
4. K. Christenson, J. Butterbaugh, T Wagener, N. Lee, B. Schwab, M Fussy and J. Diedrrick, "All Wet Stripping of Implanted Photoresist," *Solid State Phenomena* **134**, pp 109-112(2008).
5. J.G. Oh, J.K. Lee, S.H. Hwang, H.J. Cho, Y.S. Sohn, D.S. Sheen, S.H. Pyi, S.W. Lee, S.H. Hahn, Y. Jeon, Z. Fang and V. Singh, IIT 2006 AIP Conference Proceedings, Volume 866, pp. 25-28 (2006).

vMask® Usage in Semiconductor Foundry Manufacturing

Ching I. Li, Hsien Hsiu Lai, Ron Liu, Chao Chun Chen, Y. R. Wang, Michael Chan, Chan Lon Yang, S. F. Tzou

United Microelectronics Corp., Central Research and Development Division, No. 18, Nan-Ke Rd. II, Science-Based Industrial Park, Shan-hua, Tainan County, Taiwan, R.O.C

Baonian Guo, Sungho Jo, Kyu-ha Shim, Youn Ki Kim, Todd Henry

Varian Semiconductor Equipment Associates, Inc., 35 Dory Road, Gloucester, MA 01930, US

Abstract. Time and cost of development are key factors for the success of advanced devices. Device development requires multiple iterations of carefully designed implant splits to optimize device performance and yield. Conventional whole wafer implant splits are both time consuming and costly, due in part to the need to gain adequate statistics with the relatively high wafer to wafer variability of early stage development devices. VMASK, which provides multiple implant conditions on a single wafer, can be applied for a variety of corner splits, such as well, halo, and LDD implants to achieve target device performance and optimize process flow. By reducing the impact of variability from other process steps and the direct reduction in the number of device wafers required by a factor of four, a significant reduction in both wafer cost and cycle time can be achieved. In this paper, the benefits of VMASK were discussed and evaluated in a high volume production foundry fab and applied in the process development.

Keywords: novel technique, ion implantation
PACS: 85.40.Ry, 61.72.Tt, 81.05.cY

INTRODUCTION

Reducing the cost and increasing the speed of process development continue to be major challenges for semiconductor fabs. Device development requires multiple iterations of carefully designed implant splits to optimize device performance and yield. Conventional whole wafer implant splits are both time consuming and costly because an entire wafer must be used to characterize one split condition. The poor utilization of silicon makes ion implantation the number one consumer of test wafers, consuming 60% of 45 nm R&D wafers [1].

Inadvertent statistical variation of processes from wafer to wafer, which is prevalent in the development phase of a technology node, is also a contributing factor to the increase in cost and time. These variations can be attributed to any of the numerous processes in IC processing which are often not mature during the technology development phase. This could misrepresent the results and frequently drive incorrect inferences of a designed experiment. In order to accelerate development and learning where the effort to reduce time-to-market is intense, minimizing these variations is critical for shortening the development time.

One approach is to commit additional lots to assess repeatability of a given result since the statistical performance within a designed split is often a gauge of the robustness of a chosen variable. Although helpful, this type of redundancy often increases cost and diverts valuable resources.

This cost impact is more pronounced with implant. When device development incorporates new processes or architecture changes, both drive and short-channel characteristics make it difficult to assess net benefits without making implant adjustments. Implant conditions are often adjusted to null out the parametric device shifts so that the impact of new processes can be evaluated and therefore requires more test wafers.

CP1066, *Ion Implantation Technology,* edited by E. G. Seebauer, S. B. Felch, A. Jain, and Y. V. Kondratenko

To address these issues, VMASK was developed for the Varian VIISta implanter platform. VMASK allows a single wafer to be implanted with four independent recipe conditions which include species, dose, energy, and tilt angle. With VMASK, four different implant corner tests can be performed on one test wafer. A detailed description of VMASK was reported previously [2] and VMASK is compatible with all VIISta high current, medium current or high energy implanters.

In addition to the savings in the number of wafers used for test, since all four implants are done on a single wafer, statistical variation contributed by other processing steps outside of the implant is eliminated.

In this paper, the benefits of VMASK in a high volume production foundry fab and application in process development were evaluated.

EVALUATION CRITERIA FOR HIGH VOLUME MANUFACTURING

In the previous paper, process characterizations relative to the baseline normal implants were investigated. They include metal contamination, cross contamination, and charging impact [2]. The additional evaluation was conducted to measure viability of VMASK in the high volume production fab: sustaining particle performance, sufficient usable area for various die sizes and process variation impact.

Particle

With VMASK operation, slightly elevated adders were expected in the high power, high dose implants. However, no appreciable increase in particles was detected for any implant condition. Figure 1 shows VMASK particle performance data (above 0.16μm) on a VIISta HCS in the production environment. VMASK split tests are labeled in the Figure 1. Particle performance with VMASK was similar to that during normal production operation.

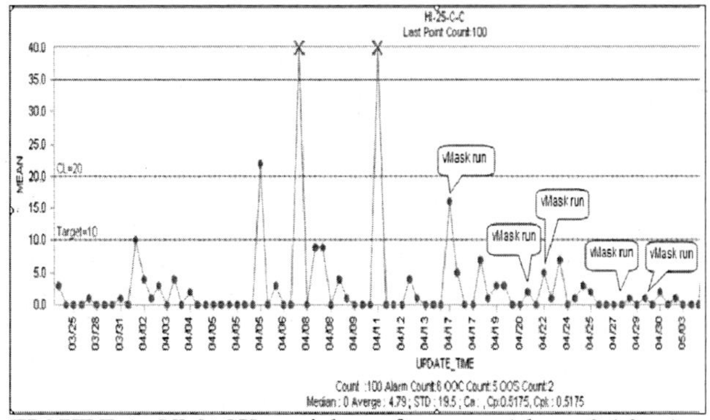

FIGURE 1. VMASK particle performance (above 0.16um) under production environment

Exclusion zone

In the high volume foundry fab where multiple devices and technology nodes are carried out simultaneously, die sizes could vary widely. For this purpose, the usable areas of VMASK implants (due to edge exclusion zone on VMASK) should be characterized. The edge exclusion zone is dependent on beam and implant tilt angles due to VMASK shadowing effects near the boundary. High tilt halo implants, As 30 keV, 7E13/cm^2, 4 steps and P 50 keV, 7E13/cm^2, 4 steps were selected to characterize the edge exclusion zones as a function of implant tilt angles at 0, 15, 30, 45° on a VIISta 900XP medium current tool. To characterize the edge exclusion zone, only one quadrant was implanted into one wafer. The quadrant implanted wafers were annealed and sheet resistivity was measured perpendicular to the quadrant edge at three different spots at 25, 75, and 125mm (Figure 2A). The step-like Rs shifts indicate consistent boundary exclusion zones for all three locations. Figure 2B shows the exclusion zone half width as characterized in the four edges. The exclusion zone half width (in mm) can be expressed as less than $5 + 12.7 \times \tan(\text{tilt})$.

Figure 2C shows a calculation of good die within 4 quadrants comparing to a full wafer at different die size. The die was assumed as a square with the same length and width. For 30mm by 30mm die size at 0° tilt implant, 44 good die within four quadrants out of 61 die per whole wafer were expected. Although this may not be adequate for statistical analysis for fully functional die, it is adequate for general process optimization based on test structures.

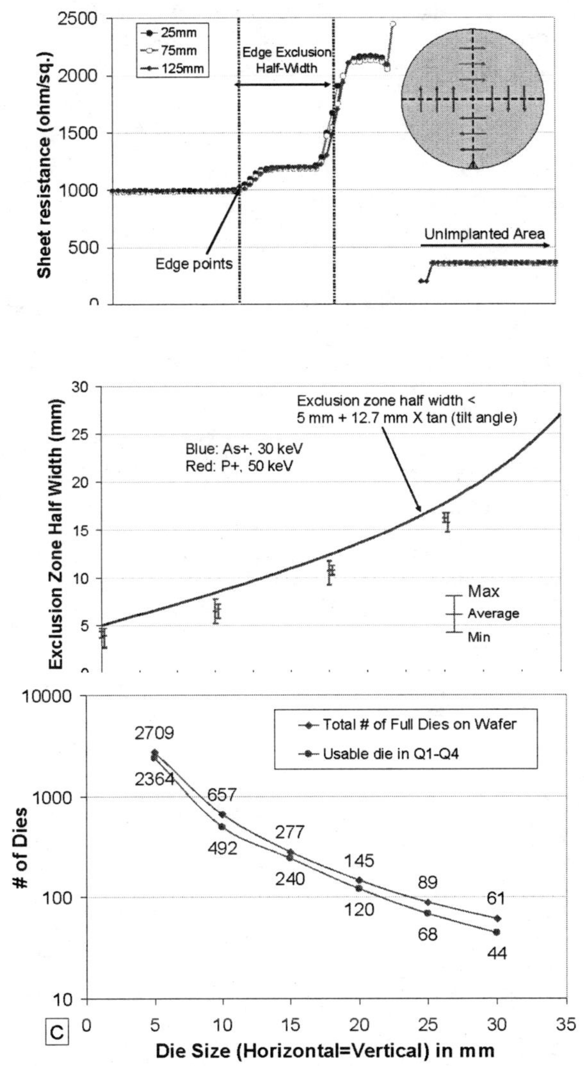

FIGURE 2. VMASK edge exclusion zone half width characterization as a function of tilt angle on a VIISta 900XP medium current tool. A) Step-like Rs line scans on one quadrant implanted wafer indicate exclusion zone (from implanted to native bulk silicon), the insert indicates 3 different Rs line scan positions, B) edge exclude zone half width as a function of tilt angles at 0, 15, 30, and 45° can be represented as 5 + 12.7 × tan (tilt) in mm, and C) total die number on a full wafer as a function of die size and useful die number using VMASK at 0° tilt.

ELIMINATION OF PROCESS VARIATIONS OUTSIDE OF IMPLANT

VMASK implant quadrants can be used for dose matching purpose to reduce test wafer usage by 75% as a new tool was being integrated into high volume production. Figiure 3 shows As 3keV NLDD implants done on a VIISta HCS for dose matching. Consistent results between data extracted from VMASK quadrants and full wafers demonstrated the reliable process result using VMASK.

Figure 4 is a study of amorphization thickness with Xe at different energy and dose for four quadrants using VMASK. The amphorization thickness was measured using FIB (focused ion beam, courtesy of FEI) by cutting a small piece in the center of each quadrant, and the PAI thickness was reported based on cross sectional Tecnai TF20 TEM (transmission electron microscopy). Samples were prepared using the FIB biopsy method that allows the TEM lamellae to be extracted without breaking the wafer [3]. These measurements can provide direct feedback on the energy and dose sensitivity for Xe PAI (pre-amphorization implant) thickness.

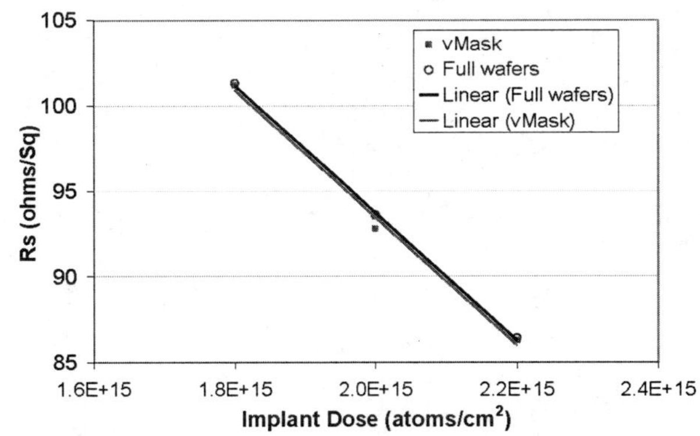

FIGURE 3 Dose matching and sensitivity on a VIISta HCS for As, 3keV NLDD. Consistent results can be achieved between data extracted from each quadrant using VMASK and full wafers.

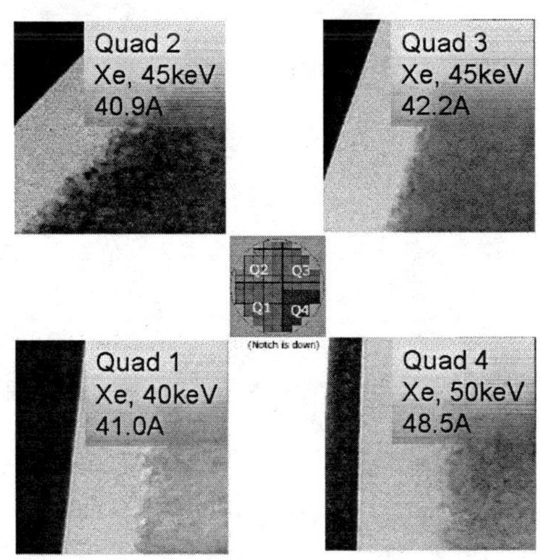

FIGURE 4. Xe PAI studie with Xe implants at various energy and dose using VMASK; TEM were prepared from a whole wafer using FIB extraction methods (courtesy FEI).

By reducing the impact of variability from other process steps, better splits using VMASK can be designed while saving test wafers. For instance, by processing three unique splits plus a reference split on one wafer, the wafer-to-wafer variability is eliminated within a given split. A reference quadrant within every wafer can be used as baseline reference and enables the monitoring of wafer to wafer variations at the same time. Thus the overall standard deviation within a given split can be significantly reduced as all candidates within a given split will be processed as a single wafer in every other part of the IC processing line. The significant reduction in within-split variations directly enables faster feedback and hence decreases development cycle time.

Figure 5 shows a 65nm P+ dose split active and poly Rs results using the VMASK with the same or different wafers along with one whole wafer reference. The active (Figure 5A) and poly Rs (Figure 5B) show similar trends for VMASK splits within wafers and wafer to wafer. The slight shift using VMASK relative to the whole wafer (noted by an oval) is due to variation within the lots (not shown).

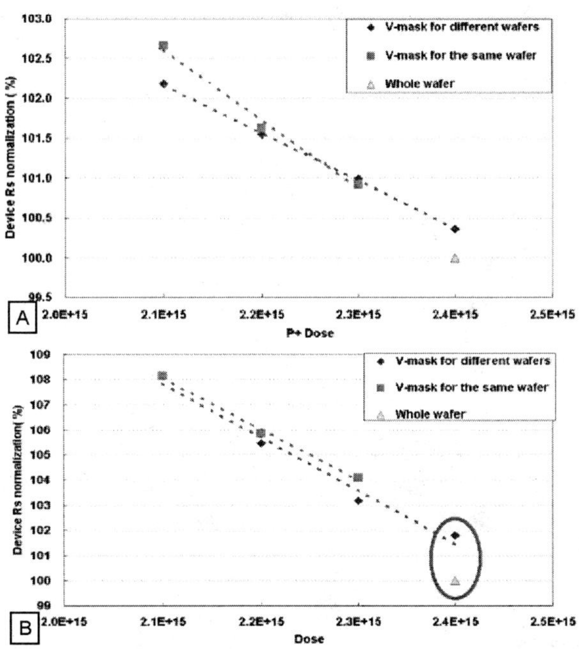

FIGURE 5. Normalized device dose split results for a 65nm P+ implant. Active Rs (A) and poly Rs (B) show similar trend for VMASK splits (wafer to wafer and within wafers).

CONCLUSIONS

Time and cost of development are both key factors for the success of advanced devices and semiconductor manufacturers. VMASK on VIISta platform implanters can be utilized to achieve as many as four different implant conditions on a single wafer. With VMASK, development wafer costs can be reduced up to 75%. Also, VMASK can improve experimental quality and efficiency, speed time to market and therefore enable users to be more competitive in the current marketplace.

ACKNOWLEDGMENTS

Guidance from D. Ramappa, T. Thanigaivelan, and T. Wu is greatly appreciated.

REFERENCES

1. Private communication on 45nm RD wafer consumption at a major US logic fab
2. S. R. Walther, S. Falk, S. Mehta, Y. Erokhin, and P. Nunan, "Using multiple implant regions to reduce development wafer usage," IIT2006, CP866, *Ion Implantation Technology*, edited by K. J. Kirkby, R. Gwilliam, A. Smith, and D. Chivers, pp. 409-412
3. N. Bicais *et al.*, "In-Line TEM Sample Preparation and Wafer Return Strategy for Rapid Yield Learning," *Proc. SPIE 6152*, 615217 (2006)

Accurate Dose Control with Pressure Compensation System on Single-Wafer Ion Implanters

Makoto Sano, Tatsuya Yamada, Fumiaki Sato, Mitsukuni Tsukihara and Michiro Sugitani

SEN Corporation, an SHI and Axcelis Company
1501 Imazaike, Saijo, Ehime, 799-1362 Japan
Phone: +81-50-3383-2060, FAX: +81-898-64-6152, E-mail: sano_mk@senova.co.jp

Abstract. On ion implanters, dosimetry is generally controlled by monitoring an electrical current detected with Faraday cup(s) instead of the actual number of dopants. In some cases, ions in a beam undergo charge exchange phenomenon, interacting with residual gas molecules in a beamline. This charge exchange of ions causes variations of the detected beam currents which result in inaccurate dosimetry of implanted dopants. We will introduce to you a system similar to Pressure Compensation of the batch-type ion implanters to single-wafer ion implanters, that is, the SHX high current ion implanter and the MC3-II medium current ion implanter. The system converts the beam current detected with Faraday cups to an adequate value for dose control as a function of beamline pressure and brings accurate dosimetry also in these single-wafer ion implanters.

Keywords: Dosimetry, Pressure increase, Outgassing, Charge exchange
PACS: 61.72.U-, 85.40.Ry

INTRODUCTION

On single-wafer high current ion implanters, the charge exchange phenomenon detracts accuracy of dosimetry in the same way as batch-type high current ion implanters. On single-wafer medium current ion implanters, influences of the charge exchange were not so serious in the past because the implanted beam power was relatively low and outgassing from photoresist on a wafer did not result in large problems. Using a high power beam, which is necessary for high throughput in a relatively high dose or high energy implantation, however, the beam current variations can cause serious problems even for medium current ion implanters. While one of the countermeasures for these problems on ion implanters is enhancement of vacuum pumping performance, the cost of equipments increases and its efficacy is rather limited in practice. Reduction of the beam current is another one, but it deteriorates the throughput considerably. To achieve high accuracy of dosimetry avoiding the influence of the charge exchange without both of the cost increase and the throughput deterioration, some batch-type ion implanters adopt Pressure Compensation system or Pcomp [1-4]. In the system, which has been being used in actual device manufacturing, a beam current detected with a Faraday cup is converted to a corresponding beam current without a charge exchange by a function of beamline pressure. A similar system to Pcomp of batch-type ion implanters was introduced to single-wafer ion implanters.

PCOMP PERFORMANCE

Formulas

In the Pcomp, beam current detected with Faraday cups is compensated as a function of beamline pressure as follows:

$$I_{00} = {}^{I_m}\!/_{f(P)} \tag{1}$$

where I_m is a beam current detected actually with Faraday cups, P is the beamline pressure, $f(P)$ is a function of P and I_{00} is a beam current expected to be detected in an ultimately high vacuum condition where $P = 0$. For the batch systems, eq.(1) is expressed as follows:

$$f(P) = \exp(-KP) \tag{2}$$

CP1066, *Ion Implantation Technology,* edited by E. G. Seebauer, S. B. Felch, A. Jain, and Y. V. Kondratenko

$$f(P) = 1 + \frac{K_2}{i}\left[1 - \exp(-K_1 P)\right] \quad (3)$$

where eq.(2) is applicable for high current implanters and eq.(3) is applicable for high energy implanters. K, K_1 and K_2 are coefficients which depend on ion species and implant energy, and i means an intended ion charge number. The eq.(3) is the more essential formula and the eq.(2) is obtained from eq.(3) by assuming that $K_1 = K$, $K_2 = -1$ and $i = 1$.

In the case of using a plasma shower assembly to mitigate wafer charging in ion implantation, plasma gas molecules coexist with the other gas molecules such as ones caused by outgassing out of photoresist coating on a wafer. We represent a pressure increment caused by the plasma gas molecules as P_A and modify the eq.(3) as follows:

$$f(P) = \left\{1 + \frac{K_{A2}}{i}\left[1 - \exp(-K_{A1} P_A)\right]\right\} \left\{1 + \frac{K_2}{i}\left[1 - \exp(-K_1 \langle P - P_A \rangle)\right]\right\} \quad (4)$$

where K_{A1} and K_{A2} are coefficients relevant to the plasma shower gas which depend on ion species and implant energy. P_A can be considered as an invariable for each implantation because the amount of gas flow is controlled to be constant all the while for an implantation period. $P\text{-}P_A$ is mainly regarded as a pressure increment induced by outgassing out of photoresist-coated on a wafer.

Beamline Effect

SEN's single-wafer ion implanters, that is, the SHX high current ion implanter and the MC3-II medium current ion implanter have the Angular Energy Filter or AEF just in upstream of a wafer process chamber. AEF deflects ion beams with an electric field and/or a magnetic field and separates off energy contamination components in the beam using an energy slit which is able to change its width for each implantation. When the charge exchange occurs in upstream of AEF, the dopants which undergo the phenomenon are eliminated by AEF and cannot reach to a wafer and as a result, the real number of dopants decreases (Figure1). The decrease of the real number of dopants due to AEF is called "beamline effect". The beam current detected with Faraday cups is influenced by both of the beamline effect and the charge exchange in downstream of AEF.

In consideration of the beamline effect, a beam current to be used for dosimetry, described as I_0, is represented as follows:

$$I_0 = I_{00} \times g(P) = I_m \times g(P) / f(P) \quad (5)$$

$$g(P) = \left\{1 + \frac{K_{A2}}{i}\left[1 - \exp(-\alpha K_{A1} P_A)\right]\right\} \left\{1 + \frac{K_2}{i}\left[1 - \exp(-\beta K_1 \langle P - P_A \rangle)\right]\right\} \quad (6)$$

where $g(P)$ is a function of P. The values of α and β are coefficients which reflect degree of the beamline effect and range from 0 to 1 in principle. Larger values of α and β mean more significant beamline effect. By assuming that K_2=K_{A2}=-1 and i=1, $g(P)/f(P)$ is modified for high current implanters, as follows:

$$g(P) / f(P) = \exp\left[(1-\alpha) K_{A1} P_A\right] \exp\left[(1-\beta) K_1 \langle P - P_A \rangle\right] \quad (7)$$

Eq.(7) means that the beamline effect influences dosimetry by using modified beam current in a fashion similar to decreasing K_{A1} and K_1.

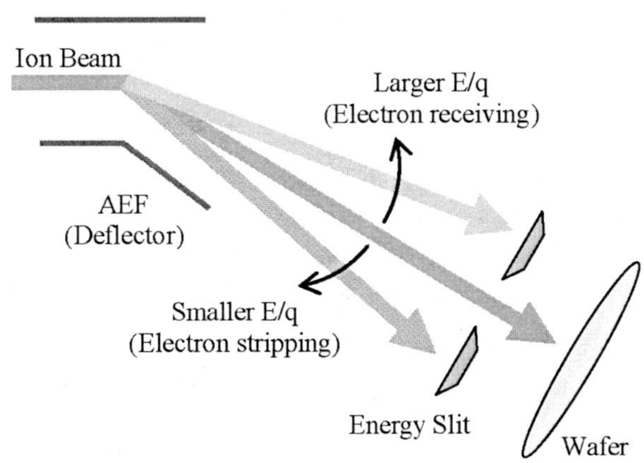

FIGURE 1. Schematic diagram of the beamline effect caused by charge exchanges in upstream of AEF, where E means ion energy and q means ion charge.

Variation of I_m and P

Figure 2 shows time dependency of I_m and P in (a), and a relation between I_m and P in (b), respectively measured in an implantation to a wafer coated with photoresist on the MC3-II, medium current ion implanter. The implant condition is B+/250keV/2800μA/2e14cm^{-2}/Tilt=7deg./Twist=23deg. I_m and P synchronously fluctuate periodically depending on a vertically up-and-down motion of the wafer. The beam current decreases by about 20% at maximum as the pressure increases. This may be interpreted that about 20% of B+ ions were neutralized by interaction with molecules spouted out from photoresist at the maximum pressure.

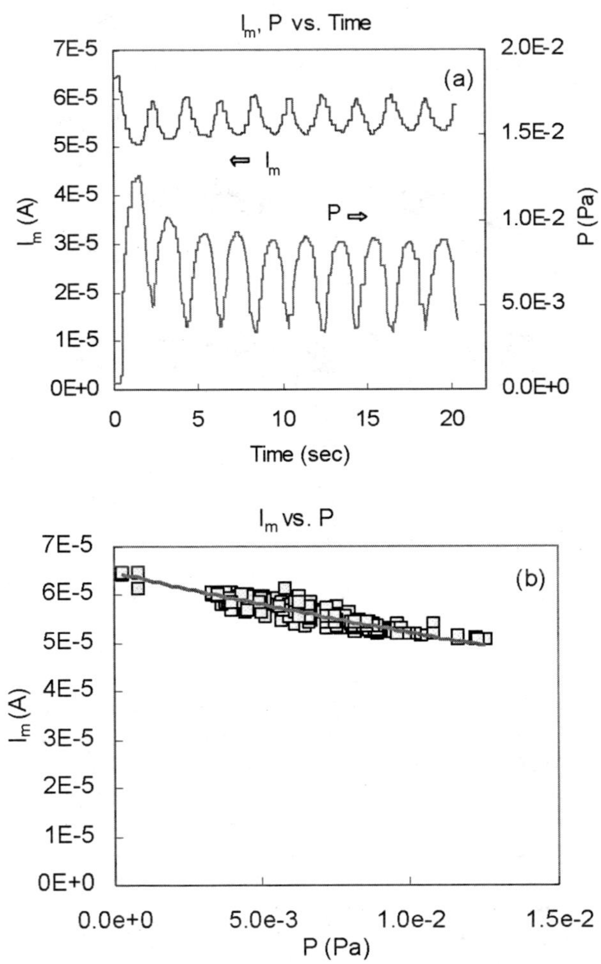

FIGURE 2. Variation of the beam current detected with Faraday cups and the beamline pressure detected with a hot cathode ion gauge in implantation of B+/250keV/2800uA/2e14cm^{-2}.

Measurement of Sheet Resistance

To measure the dose variation caused by photoresist outgassing, we prepared patterned wafers which were partially coated by photoresist and exposed Si surface as a straight line in a center region (Figure 3). After implantation, the coated photoresist was removed and the wafer was annealed by RTA process.

The results of sheet resistance measurement for B+/250keV/2800μA/2e14cm^{-2}/Tilt=7deg./Twist=23deg. are shown in Figure 3. Line scan measurements were done in the vertical direction (y) and the horizontal direction (x), respectively. Closed circles show the result for a bare wafer as a standard and closed squares show the result for wafers partially coated by photoresist without Pcomp. Comparing both of them, the latter wafers have lower values of sheet resistance which mean overdose induced by neutralization of B+ ions in high pressure conditions caused by photoresist outgassing. In the vertical direction, the sheet resistance profile for the implantation to the photoresist-coated wafer seemed to have an M-shaped curve which may indicate pressure variation depending on wafer position during implantation. On the other hand, in the horizontal direction, the sheet resistance profile for the implantation to the photoresist-coated wafer has a shape similar to the result for the bare wafer. The similarity of the both sheet resistance profiles in the horizontal direction is considered as invariance of horizontal beam profiles in both vacuum conditions. Open circles in Figure 3 show the values of sheet resistance for the photoresist-coated wafers in case of using Pcomp. The degree of overdose and the M-shaped curve in the vertical direction are improved adequately by using Pcomp.

Reduction in Impacts on Throughput

In some cases, increase of beamline pressure leads to large decrease of beam current detected with Faraday cups. The large decrease results in implant holds, which deteriorate throughput extremely, because there is concern that beam instability can cause inaccurate dosimetry. Figure 4 shows time dependency of I_m, I_{00} and P for an implantation of B+/320keV/2800μA/5e14cm^{-2}/Tilt=7deg./Twist=23deg. Red dot-lines represent interlock values of implant holds. Without Pcomp, the implant system monitors I_m as a beam current and implant holds occur frequently due to deviations of the beam current from a specified range. In contrast, Pcomp can not only improve dosimetry but also avoid this type of implant holds by monitoring the compensated beam current I_{00}. Trim factors can adjust dose amount in certain implant conditions but cannot enable to avoid implant holds triggered by beam instability judgment. As a result, Pcomp improves throughput significantly in some high pressure conditions.

CONCLUSIONS

On ion implantation, outgassing from photoresist coating wafers can induce significant dose shifts which can result in fluctuations of device characteristics within a wafer and/or in wafer-to-wafer. Pcomp, which is a system that compensates beam currents as a function of the beamline pressure, improves dose accuracy in implantation onto photoresist-coated wafers. Pcomp is effective for single-wafer ion implanters as well as for batch-type implanters. Pcomp is also inevitable to improve throughput in high pressure conditions by avoiding implant holds which occur by judging beam instability.

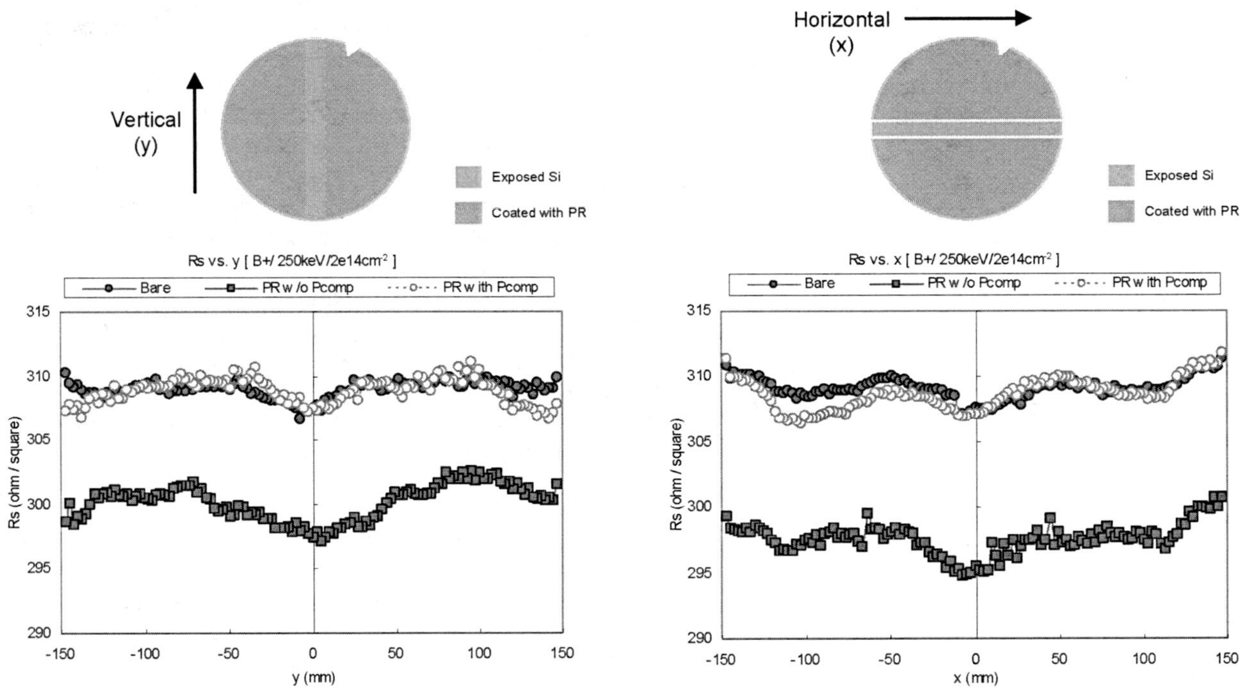

FIGURE 3. Sheet resistance variations in vertical and horizontal directions in implantation of B+/250keV/2800μA/2e14cm^{-2}.

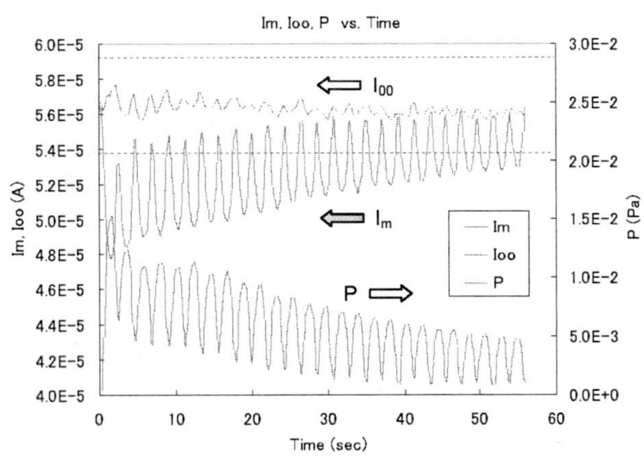

FIGURE 4. Avoidance of implant holds due to judging beam instability by monitoring compensated beam current in implantation of B+/320keV/2800μA/5e14cm^{-2}. Dot-lines show interlock limits.

REFERENCES

1. M. Sano, M. Kabasawa, H. Kariya and M. Sugitani, "Monitoring System of Reliability and Repeatability of Pressure Compensation," *Proceedings of 1998 International Conference on Ion Implantation Technology*, pp.322-325 (1998).
2. T. N. Horsky, "Photoresist Outgassing in High Energy and High Current Ion Implantation," *Proceedings of 1998 International Conference on Ion Implantation Technology*, pp.654-657 (1998).
3. H. Kariya, M. Sano, M. Tsukihara and M. Sugitani, "Automatic Generation of Pressure Compensation Factors," *Proceedings of 1998 International Conference on Ion Implantation Technology*, pp.562-565 (1998).
4. M. Sano, M. Kabasawa, M. Tsukihara and M. Sugitani, "Dose Control Accuracy in Pressure Compensation," *Proceedings of 2000 International Conference on Ion Implantation Technology*, pp.682-685 (2000).

Improved Beam Angle Control with SPV Metrology

K. Steeples[1], E. Tsidilkovski[1], A. Bertuch[1], E. Ishida[2], and A. Agarwal[3]

[1] *QC Solutions Inc., Billerica, MA, USA*

[2] *Spansion LLC, Sunnyvale CA, USA*

[3] *Spansion (currently at Twin Creeks Technologies, Santa Clara, CA)*

Abstract. A method of real-time monitoring of implant angle for state-of-the-art ion implant doping in integrated circuit manufacturing has been developed using Surface Photo Voltage measurements on conventional monitor wafers. Measurement results are analyzed and compared to other techniques.

Keywords: Metrology, channeling, implant, angle, defects.
PACS: 61.72.Tt, 61.80.-x, 85.40.Ry, 73.50.Pz, 73.50.Gr

INTRODUCTION

The importance of implant beam angle control particularly in single crystal materials has been well appreciated since the introduction of ion implant technology to the manufacture of electronic devices. In the early days of integrated circuit (IC) manufacturing, the design rules required angle control of just several degrees to reduce channeling to a manageable level, while analytical work required much tighter control, down to fractions of a degree. As IC density has increased, the requirements for implant angle control have tightened dramatically. Shadowing, asymmetry, and channeling on individual transistors and across the whole wafer, with low energy implants, have to be controlled to a fraction of a degree. These "scaled" transistors together with larger wafer sizes determine process capability and monitoring issues.

A conventional method of monitoring implant angle is to generate "U-curves", i.e. response curves as a function of beam angle [1]. This technique relies upon the selective implantation into a crystal channel (for example <100> or <112>). However, detection of channeling becomes more challenging as implant energy is reduced, and the channel acceptance angle increases. We show that conventional metrology loses sensitivity at about 15 keV.

We present a method and measurement results that extend the capabilities of conventional real-time implant monitoring for accurate implant angle control.

Surface Photo Voltage effect (SPV) is a well-established method [2] that is used for measuring various electrical parameters in single crystal semiconductors, such as minority carrier lifetime, doping density, interface defect density, as well as for processing equipment characteristics, such as furnace oxide quality and metal contamination. Practically, all the SPV parameters are not measured directly, but are calculated from the measured characteristics using appropriate theoretical models and comparing the results to some reference values. In general, the SPV phenomenon is based on two effects: 1) generation of electron hole pairs by light, 2) separation of the pairs by the surface electric field. In the depletion state, the minority carriers are collected at the surface, changing the surface potential and thus producing the SPV.

The implantation process introduces a large number of crystal defects N_d that act as recombination centers, reducing photo-carrier lifetime and thus affecting the SPV-measured dynamic charge [3]. SPV method is highly sensitive to implant parameters due to the strong dependence of the measured signal on the implant-induced defect density $\sim N_d^2$.

EXPERIMENTAL

N-type silicon wafer substrates, <100> crystal orientation, were implanted with boron with energies 5-80keV and doses 5e11 - 5e15 ions/cm^2. Implants were performed about the axial channel <112> at 35.3° tilt to investigate ion beam to wafer crystal

CP1066, *Ion Implantation Technology,* edited by E. G. Seebauer, S. B. Felch, A. Jain, and Y. V. Kondratenko

alignment. The implant angle was tilted +/-1 degree in relationship to the 35.3° crystal pole to determine sensitivity. Implanted boron channeling is maximized when proper ion beam to crystal alignment is achieved, i.e. along the <112> pole. The implanted wafers were measured as-implanted and following an RTA anneal. All wafer samples were 300mm in diameter and were implanted using either a high current or medium energy implanter. Measurements were performed using a small-signal SPV technique [3] for the post anneal wafers and using both SPV and the conventional damage metrology method for the as-implanted wafer samples.

The SPV measurements were performed on an ICT300 system manufactured by QC Solutions, Inc. (QCS). This measurement technique is a non-contact ac-SPV, utilizing a high frequency modulated light of low intensity. A blue light photo source was used for the measurements, and all SPV measurement values are reported as dynamic surface charge Q_d (C/m^3) with full wafer contour maps. The QCS sensing probe is capacitively coupled to the wafer and introduces light with the energy larger than the bandgap energy, creating electron hole pairs in the near-surface space charge region. Drift of the photo-carriers in the space charge region modulates the surface/interface potentials, determining the measured dynamic charge, Qd.

Same as-implanted wafers were measured with a conventional damage metrology technique to compare sensitivities of the methods to implant angle variation, particularly at low energies. Measurements were performed using 121 point pattern map with a reported signal in dimensionless units. Results of the measurements were normalized, which allows direct comparison between different methods.

RESULTS

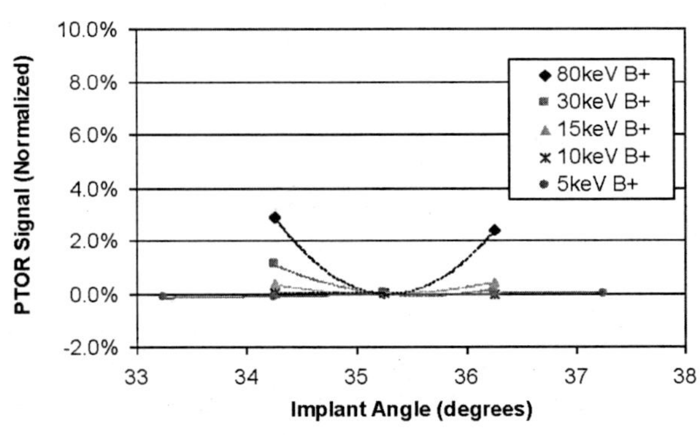

FIGURE 1. Energy Dependence of the Process Tool Of Record (PTOR): U-curve signal. B^+ 1e14cm^{-2} as-implanted.

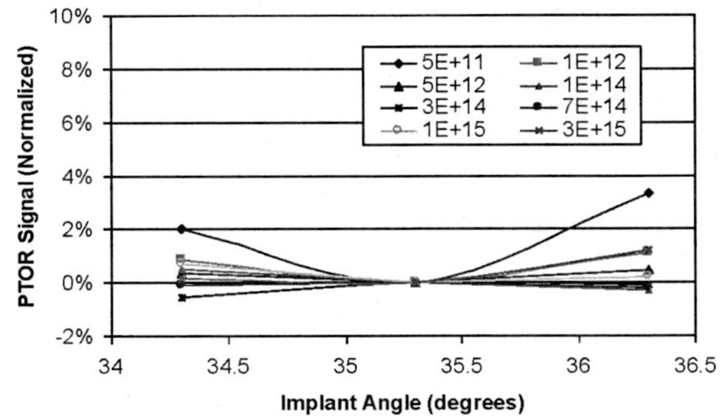

FIGURE 2. B^+ 10keV implant with various doses. PTOR dose response about the polar channel 35.3°.

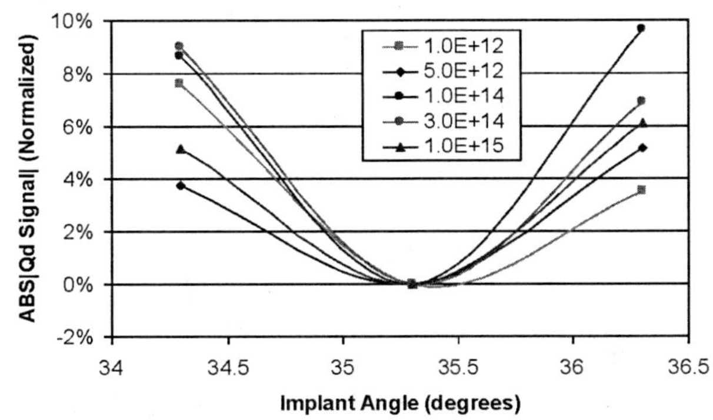

FIGURE 3. B^+ 10keV implant with various doses, un-annealed. SPV response about the polar channel 35.3°, <112> pole.

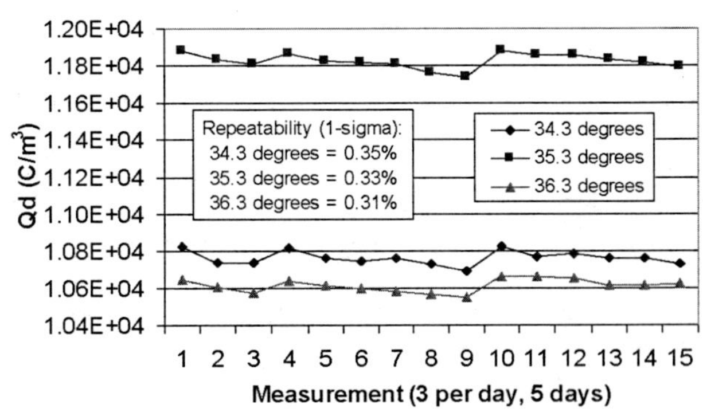

FIGURE 4. B^+ 10keV 1e14cm^{-2} implant monitor, 5-day repeatability with SPV technique.

DISCUSSION

Figure 1 shows U-curves for B 1e14cm^{-2} at different energies measured with PTOR technique. As the implant energy decreases the sensitivity of the technique is decreased, from ~ 3% per degree of tilt at 80keV to ~ 0 at 15keV.

Comparing the data presented in Figure 2 and Figure 3 we conclude that although both PTOR and SPV technique show typical U-shape pattern centered around 35.3 angle [4], the SPV-based method is clearly more sensitive to implant angle variation than the standard PTOR method. This difference is particularly evident at low implant energies below 15keV, where the PTOR method is essentially insensitive to beam angle.

Sensitivity of the SPV technique to implant angle is due to the same mechanism, which is responsible for the SPV dependence on implant energy [3]. Thus, for annealed wafers, a relationship between the implanted junction depth and light absorption length determines magnitude of SPV signal. For as-implanted wafers, dependence of the SPV signal on implant angle is based on the correlation between implant-induced damage profile and light absorption depth. When the implant depth is smaller than the absorption length, higher implant energy or more channeling angle results in larger dynamic charge Q_d value. This effect is primarily due to the quadratic dependence of Q_d on the effective width of an implanted area, which is proportional to implant depth. Therefore, even at low implant energies, where the channeling acceptance angle is greater and the random/de-channeled/channeled profiles are not resolved, SPV sensitivity to implant angle is strong due to the strong dependence of the total profile width on implant depth at low energies.

In case of deeper (higher energy) implants the dependence of Q_d on implant angle can change its sign due to the modulation of the resolved random and channeled profiles and associated depletion widths by trapped charge carriers in as-implanted wafers [3].

For annealed wafers, a similar effect of changing a sign has been observed. Although, in this case the mechanism of the effect is different: increase of the implant depth leads to the change of annealed profile shape, at certain energy, adding additional junction and therefore modifying the net SPV signal. This effect of reversed sign on annealed wafers was observed for B$^+$ 80keV.

A sensitivity to crystal ingot was noted for the SPV measurement of the wafers implanted with B 10keV 1e14cm^{-2}. No particular conclusion about this effect has been made at this time: possible factors affecting the measurement include crystal orientation cut, surface condition, oxygen content, etc.

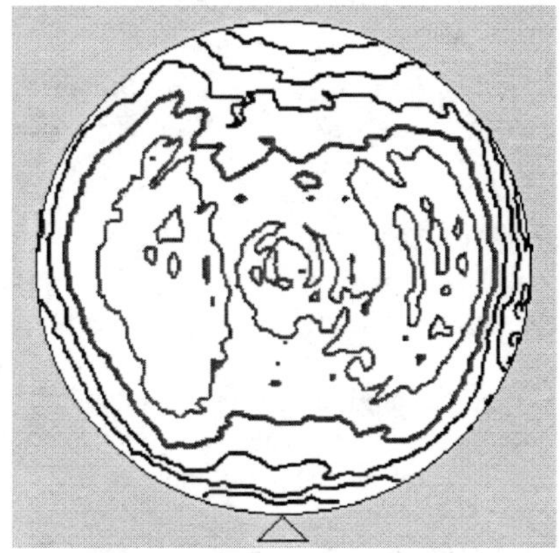

Qd C/m^3			
	X	1.064e+04	
	R	1.684e+03	15.8%
	S	2.929e+02	2.75%
	MIN	9.682e+03	91%
	MAX	1.137e+04	107%

FIGURE 5. Typical wafer map of as-implanted B$^+$ 10keV 1e14cm^{-2} uniformly implanted in n-type silicon substrate along the <112> axial channel. Contour interval = 2%.

SUMMARY

The channeling effect around <112> crystal pole in silicon was studied with standard implant metrology and SPV-based methods. The SPV technique clearly shows superior sensitivity to small implant angle variations for all investigated implant energies and doses, especially below 15keV. High sensitivity of the method combined with its capability to provide rapid full scale wafer mapping allows the use of SPV technique for improved implant angle control in real time manufacturing environment.

REFERENCES

1. R. B. Simonton *et al., Nucl. Instr. and Meth. B***55** (1991) p. 39-44
2. C. Krueger *et al., Ion Implantation Technology* **886** (2006) p. 531
3. K. Steeples and E. Tsidilkovski, *Ion Implantation Technology* **886** (2006) p. 558
4. E. Tsidilkovski, IIT 2008 Ion Implant School Course Handbook

Implant Angle Repeatability on Optima MD

Robert Rathmell, Jonathan David, and Mark Harris

Axcelis Technologies, Inc., 108 Cherry Hill Dr, Beverly, MA 01915 USA

Abstract. A very sensitive channeling implant has been used to quantify the repeatability of implant angles for the Optima MD over a period exceeding one month. Sensitivity of sheet resistance to variation in vertical and horizontal beam angles was measured for 500 keV P++ at a dose of 5E13 /cm^2 near orientations tilt/twist of 0°/0° and 35.26°/0° for (100) wafers. Sensitivity to tilt in the vertical plane at both conditions was at least a 10% change in Rs for a 0.5° change in tilt at a tilt of 0.75° away from the channel (i.e. at 0.75°/0° and 34.5°/0°), where the variation in Rs was linear with tilt. Horizontal angle sensitivity resulted in a 7.5% change in Rs for a 0.5° change in angle, but was less sensitive to smaller angle variation, since Rs varied roughly as a second order function near the channel. Sensitivity to tilt is reduced by about a factor of 3 for implants at 0°/0° compared to 0.75°/0° with a similar response at 35.26°/0° compared to 34.5°/0°. Implants were carried out using wafers from a single boule over a period exceeding one month at the orientation of 0.75°/0°. Angle control was enabled by in-situ measurement and correction of the beam angles in both the vertical and horizontal planes prior to implant. The repeatability of sheet resistance was used to infer the variation in angle. Attributing all Rs variation to be the result of beam angle variation showed that the standard deviation of beam angles for over 30 wafers is 0.03°.

Keywords: Implant angle control
PACS: 85.40.Ry, 81.05.Ea, 41.85.Qg

HARDWARE DESCRIPTION

Optima MD angle control hardware was described previously [1 - 3]. A brief summary of the techniques will be given here. The Optima MD is a medium dose, serial implant tool that uses a hybrid scanning system. The beam is scanned electrostatically in the horizontal plane and angles are "corrected" by a curved acceleration gap to form rays that are parallel to the central, unscanned ions. The wafer is scanned mechanically "vertically" in the tilted plane of the implant to pass through the ribbon-like beam to uniformly dose the entire wafer. Beam angles are measured in the horizontal and vertical planes prior to implant and automatically corrected to ensure the implant is performed at the angles specified in the recipe.

Horizontal Angle Measurement

Horizontal beam angles are measured by a moving profiler behind a mask with 7 slots. The angle corrections ensure that the 7 beamlets spanning the width of the wafer are parallel to each other and that the largest angle satisfies the recipe limit of Maximum Horizontal Beam Angle. Corrections are automatically made by steering the beam prior to the scan plates to correct the average angle or adjusting the parallelizing lens voltage, if necessary, to correct parallelism.

Vertical Angle Measurement

The beam is deflected vertically by nominally 15° by an electrostatic energy filter just before entering the end station. Measurement in the vertical plane of the angle of the scanned beam verifies that the actual Vertical Beam Angle, VBA, is known prior to implant. The measurement is made by the VBA monitor, which is a faraday with a thick mask with narrow slots that admits more current when it is aligned with the direction of the ions. The VBA monitor is mounted on the scan arm of the wafer so that it has a fixed offset from the wafer surface as shown in Figure 1. At the start of a process job, the scan arm tilts so that the VBA monitor collects current as it rotates from 5° above to 5° below the nominal angle and then again on the way back to generate two plots of current vs. angle. The two plots are compared to detect noise or beam drop-outs. The flux weighted average angle of the two data sets is the Vertical Beam Angle. The wafer angle is adjusted about an axis parallel to the scanned beam, if necessary.

Both measurement schemes are calibrated to crystal planes during machine setup to ensure that a 0° implant is normal to the surface of the wafer.

CP1066, *Ion Implantation Technology,* edited by E. G. Seebauer, S. B. Felch, A. Jain, and Y. V. Kondratenko

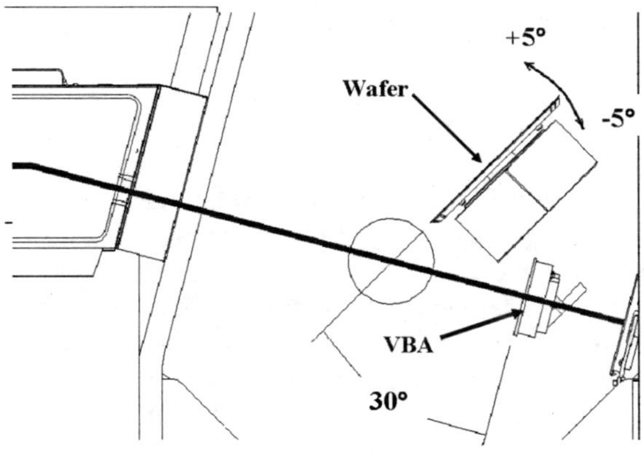

FIGURE 1. Scan arm is shown in the position to measure vertical beam angle. The VBA monitor is mounted to the scan arm with a fixed angle to the wafer.

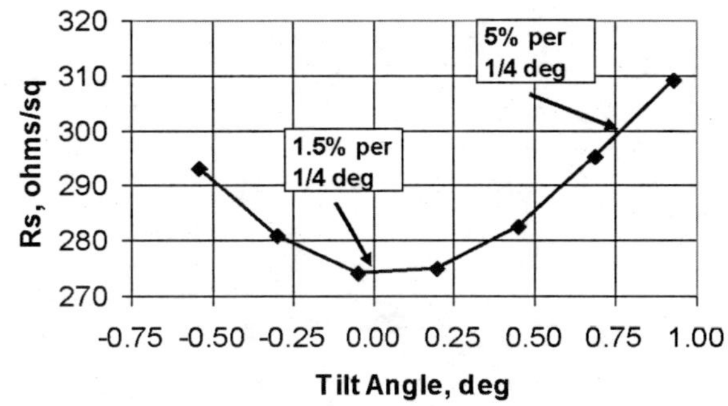

FIGURE 2. Average sheet resistance for a series of wafers implanted with 500 keV P++ at a dose of 5E13/cm^2 near the 0°/0°, <100> axis at a range of vertical tilt angles.

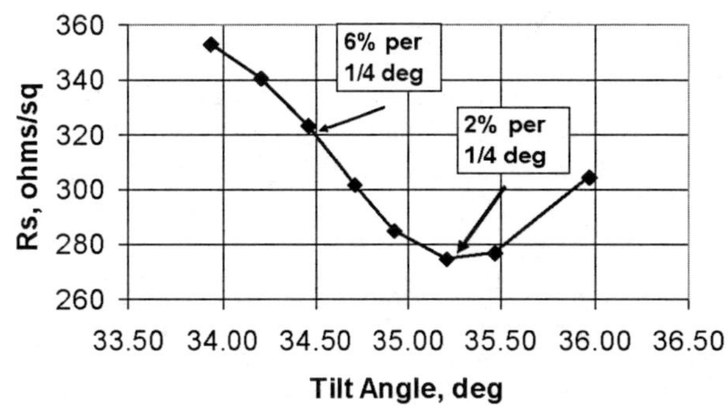

FIGURE 3. Average sheet resistance for a series of wafers implanted with 500 keV P++ at a dose of 5E13/cm^2 near the 35.26°/0°, <112> axis at a range of vertical tilt angles.

EXPERIMENT

Test Conditions

The primary goal of this experiment was to evaluate the ability of the angle control scheme of Optima MD to accurately and reproducibly measure and correct implant angles from batch to batch over a significant period of time. Prior experience has shown that high energy phosphorus implants at a dose of 5E13 atoms/cm^2 are very sensitive to channeling effects [3]. Two different crystal orientations were considered for the test. For (100) wafers, an implant at a tilt/twist of 0°/0° is aligned with the <100> axis, while an implant at 35.26°/0° is aligned with the <112> axial channel.

Vertical Angle Sensitivity

Implants were made at a range of tilt angles near these two axes to determine the sensitivity. The wafers were measured with a Therma-Wave Therma-Probe and a sheet resistance, Rs, probe. Results for Rs are shown in Figures 2 and 3. Therma Wave results (not shown) showed comparable sensitivity.

One can see that the sensitivity to small angle changes is greater at a tilt of about 0.75° away from the channeling axis and has a roughly linear slope. Sensitivity was comparable at both conditions and the 0.75°/0° orientation was selected for the repeatability test.

Horizontal Angle Sensitivity

In the horizontal plane on the Optima MD, one does not have to implant a series of wafers to produce the "V-Curve" of Rs vs. angle. One can implant a single wafer with the Parallelizing Lens off so that the scanned beam spans the wafer with angles that vary linearly by +/-1.2 degrees at this energy, as shown in Figure 4. The sensitivity of sheet resistance to angle is about the same for the <100> and <112> axes. The Rs increases about 7% at 0.5° from the channel, which, at this angle, is about 70% as sensitive as the vertical axis. Horizontal sensitivity was about the same at a tilt/twist of 0.75°/0° (not shown), because the channeling feature is largely planar. Horizontal sensitivity is less for small angle errors near 0°.

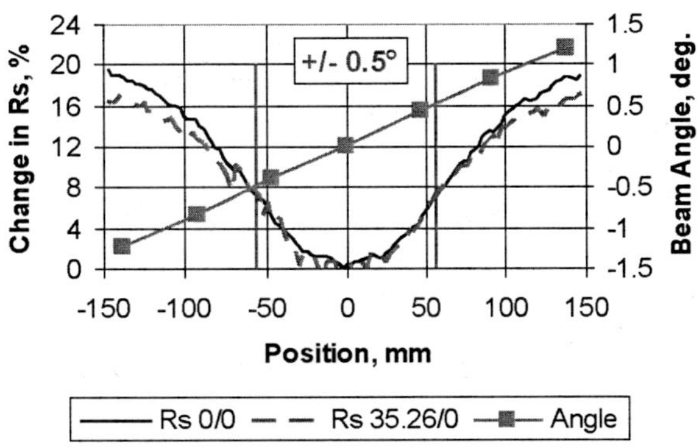

FIGURE 4. Variation of sheet resistance horizontally across the wafer implanted with 500 keV P+ at a dose of 5E13/cm^2 with the P-lens off so that angles vary linearly from one side to the other.

Test Angle Correction Accuracy

In order to confirm that angle corrections worked, the beam was tuned to cause a significant angle error, and wafers were implanted with and without automatic corrections enabled. Since the beam is deflected in the vertical plane by nominally 15° at the Angular Energy Filter of the Optima MD, the AEF can be adjusted to change the beam angle in the vertical plane. Wafers were implanted with the AEF adjusted to cause the beam to enter the end station at angles approximately +/-0.55° from the nominal 15° bend. In each case, one wafer was implanted with VBA corrections off, and one with corrections enabled. Figure 5 shows that the wafers implanted with corrections enabled fall within <0.1° to the expected Rs value for 0.75°/0° orientation.

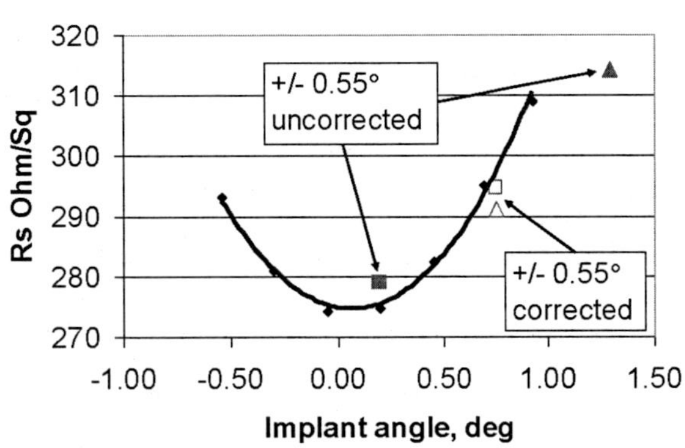

FIGURE 5. Wafers implanted at a recipe orientation of 0.75°/0° with the beam steered by the AEF to cause a vertical angle error of +/- 0.55° with automatic angle corrections on and off are compared with the V-curve.

Angle Repeatability Implants

Cassettes of wafers from the same boule used for the implants in Figures 2 to 5 were selected for a repeatability test. Using wafers from a single boule ensured that orientation of the crystal matched the sensitive region near a tilt of 0.75° shown in Figure 2 and minimized changes in the crystal cut error through the test. No attempt was made to compensate for crystal cut error in the selected material, although wafers with a reported crystal cut error of <0.1° were chosen, which is consistent with the curve in Figure 2.

Implants were performed daily for over a month using 500 µA of 500 keV P++, at a dose of 5E13 atoms/cm^2 and an orientation of 0.75°/0° tilt/twist. Recipe limits for horizontal beam angles were set to 0.1° to ensure that horizontal steering or P-lens voltage would be automatically adjusted if any of the seven measured peaks were >0.1°. Vertical beam angle auto-correction was enabled so that the wafer angle was adjusted about an axis parallel to the scanned beam to correct for variation in the VBA.

Results of the Angle Repeatability Implants

Sheet resistance repeatability of the implants is shown in Figure 6. Dashed lines above and below the data show the equivalent change in Rs corresponding to +/- 0.1° in angle variations. If one assumes that all of the variation is due to angle variation, the results show that the maximum angle variation was less than +/-0.1°. It is possible that some of the variation is caused by finite dose repeatability and metrology.

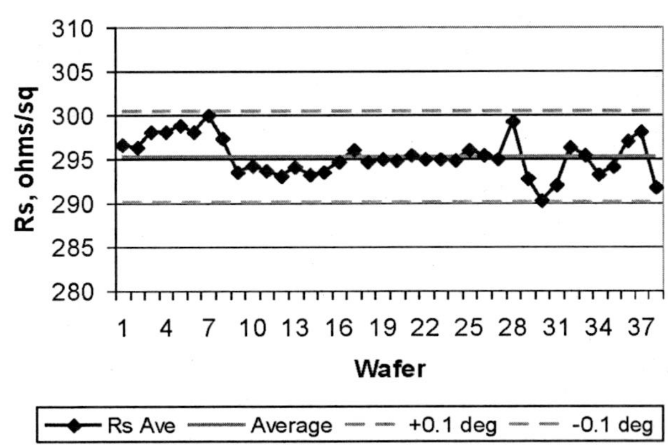

FIGURE 6. Sheet resistance results of daily implants of 500 keV P++ at 5E13 atoms/cm^2 at an orientation of 0.75°/0° tilt/twist. Upper and lower lines show the change in Rs equivalent to +/- 0.1 degree in tilt.

Implant Data Logs for these implants report the measured Horizontal Beam Angle, HBA, and VBA for these implants. Table 1 shows the average value and standard deviation of these angles taken from the IDL's for all implants. It also shows the average value and standard deviation of the sheet resistance. If the standard deviation of the sheet resistance is all attributed to angle variations, it would convert to a standard deviation of 0.032° for all of the implants.

TABLE 1. Data summary for repeatability implants

	Average value	Standard deviation	Angle standard deviation based on Rs
HBA	0.005°	0.019°	NA
VBA	0.76°	0.023°	NA
Rs	295.3	0.74%	0.032°

SUMMARY

Optima MD has the ability to measure and automatically correct implant angles in the horizontal scan plane and the vertical plane the start of a batch of wafers. A very sensitive channeling implant condition was selected to test repeatability of the implant angle using daily implants for over a month. The orientation of the implant was chosen as 0.75°/0° tilt/twist because of the sensitivity (5% change in Rs for 0.25° change in angle) and the near linearity of the response. It was also shown that comparable sensitivity could be obtained by implanting at 34.5°/0°. The variation in Rs was consistent with total angle variation of less than +/-0.1° with a standard deviation of 0.03° for all implants.

ACKNOWLEDGEMENTS

The authors would like to thank Fernando Silva and Khen Ung for running the implants.

REFERENCES

1. K. W. Wenzel, A. M. Ray, B. H. Vanderberg, and R. D. Rathmell, "Optima MD: Mid-Dose, Hybrid-Scan Ion Implanter", *Proc. of 16th Intl. Conf. on Ion Implantation Tech.*, Marseille, France, AIP Conf. Proc. 866, 2006, pp. 637-640.
2. R. D. Rathmell, B. H. Vanderberg, A. M. Ray, D. E. Kamenitsa, M. Harris, and K. Wu, "Implant Angle Control on the Optima MD", *Proc. of 16th Intl. Conf. on Ion Implantation Tech.*, Marseille, France, AIP Conf. Proc. 866, 2006, pp. 349-352.
3. R. D. Rathmell, D. E. Kamenitsa, M. L. King, and A. M. Ray, IEEE *Proc. of Intl. Conf. on Ion Implantation Tech.*, Kyoto, Japan, 1998, pp. 392-395

Characterization of a Small Form Factor Multipole RGA for Process Chamber Monitoring and for Reduction in Time to Complete Routine Preventive Maintenance

Ron Johnson, Brannon Winningham and John Schuur

INNOViON Corporation
4253 NE 189th Avenue
Gresham, OR 97230 US
rcjohnson@innovioncorp.com, bwinningham@innovioncorp.com, jschuur@innovioncorp.com

Abstract. A small form-factor, multipole Residual Gas Analyzer (RGA) has been used to study steady state and post-PM conditions in an Axcelis GSD ion implanter. The RGA properties and specifications are discussed and data is presented to illustrate N_2, O_2, and H_2O pump-down curve characteristics. Baseline performance is compared to performance following invasive activities to determine applicability for eliminating explicit He leak checking requirements and for the determination of how quickly a machine may be returned to production. The target ion implanter was an Axcelis GSD/200E.

Keywords: RGA, residual gas analyzer, multipole, leakcheck.
PACS: 07.30.Bx, 07.30.Hd

INTRODUCTION

The application of RGA systems to ion implanters is a natural fit. The high vacuum and simple nature of introduced species of an ion implanter is an environment that can be well studied by an RGA. Horiba/STEC manufacture a small form-factor, multipole RGA system which offers variations in performance that are dependent on the chosen sensor head. Intrusion of the sensor head into the vacuum chamber is minimal and less than 5 cm. Protrusion from the vacuum chamber is also much smaller than on conventional RGA systems and is less than 15 cm. The small size allows for novel sensor placement and easy portability between systems or vacuum sub-systems on a given machine. The penalty for the compact size of the head is in time and mass resolution and available mass range. The sensor used in this series of tests (MPA6-5-2/65K) had a maximum time resolution of 0.6 seconds/amu peak, a mass resolution of 0.8 amu and a mass range of 2-65 amu. The system was used primarily in a trend mode to evaluate vacuum response to a number of treatments. The ion implanter used for initial testing of the RGA was an Axcelis GSD 200E, installed in the Portland facility of Innovion Corporation.

INSTRUMENT DESCRIPTION

The Micropole RGA system consists of a sensor head, an electronics module "Spectrum Converter" that attaches to the head and via special cable to a serial interface box for communicating with a Windows-based computer. Finally the vendor supplies WinMPA software to control the sensor and acquire and display the data. Optional hardware for controlling vacuum components via user defined logic in response to sensor output is also available; this hardware was not used for this investigation. The software has a reasonable array of features for defining default and custom data acquisition profiles. The software and system were mostly straightforward to use and only required occasional references to the system manual. The sensor head ships with individualized calibration information which must be entered into the configuration data for the operating software. Since each sensor head is individually calibrated at the factory, no expensive calibration schemes are required by the end user. With a cost in the neighborhood of \$1.5k – \$2.0k, replacing the head on failure or when discernible drift/non-linearity is detected makes for reasonable Cost Of Ownership (COO).

There are a variety of scan modes for operating the sensor: Leak, Scan and Scan & Hold. The software

CP1066, *Ion Implantation Technology,* edited by E. G. Seebauer, S. B. Felch, A. Jain, and Y. V. Kondratenko

provides for several display formats and in Leak Mode will display a trend for Helium pressure as well as an analog 'bar' display similar to what is standard on most dedicated leak checkers. The mode most used in this investigation was a straightforward Scan which scans through the amu range from lowest to highest. There are also a handful of configurable parameters one can use to modify the Scan mode; the most useful of these is 'Fast Scan' which ignores intermediate peaks during Gas Trend measurement and so delivers an overall faster data rate. Finally, the Scan & Hold mode completes a single scan and then waits for user direction.

EXPERIMENTAL OUTLINE

The system was tested in He Leak Check mode and in Gas Trend mode using three different trend definition files. In the Gas Trend mode, specific amu peaks were chosen for each of the two trend definition files as being most likely show variation under the specific treatments used. Initially, a trend definition file was configured to detect ion implant species: ^{11}B, ^{19}F, $^{28}N_2$, ^{31}P, ^{40}Ar, along with $^{1}H_2$, ^{4}He and $^{18}H_2O$. A second trend definition file was created to optimize detection of atmospheric leaks into the process chamber and consisted of the following key peaks: $^{1}H_2$, ^{4}He, $^{18}H_2O$, $^{28}N_2$, $^{32}O_2$, ^{40}Ar, and amu 45 (reported to be the most dominant fraction for isopropyl alcohol (IPA)) [1]. Finally, for a scan of normal process operation where photoresist outgassing and argon from an active electron shower would be expected, but fast scans were desired, a 4 peak trend definition file was setup using the following amu: $^{1}H_2$, $^{18}H_2O$, $^{28}N_2$, ^{40}Ar.

It should be noted that the data presented here has not been corrected for ionization efficiency. The RGA sensor calibration uses nitrogen (N_2) as a standard. This investigation was primarily interested in qualitative results versus absolute pressure measurement.

Gas Trend Mode: Ion Beam Setups

Several different ion beam set-ups were completed to investigate whether the RGA would detect the individual specie components. A uniform set of ion source parameters was selected for all setups to minimize differences between gas species: Gas Flow = 3 sccm, Arc Current was 1 A, Source Magnet was 7 Amps, Arc Voltage was 60 Volts (except for BF_3, where Arc V was 90) and Extraction Voltage was 80 kV.

Ion beams were generated from each of the following source gases: arsine, phosphine, BF_3 and argon. Even though the particular RGA sensor selected for test is unable to detect amu 75, a 75As+ ion beam was set up to see whether any detectable species would be evident as one might expect if high current ion beams caused 'evaporation' of previously implanted and/or deposited contaminants. The actual sequence of beam tuning was: $^{75}As^+$ (15 mA), $^{1}H_2^+$(0.7 mA), $^{1}H^+$(0.6 mA), $^{31}P^+$(7.4 mA), $^{1}H_2^+$(0.5 mA), $^{1}H^+$(0.4 mA), $^{11}B^+$(1.6 mA), $^{49}BF_2^+$(3.2 mA), $^{19}F^+$(1.3 mA), $^{40}Ar^+$(4.3 mA). Perhaps due to the sensor head placement (lower portion of process chamber, behind the process disk, but near to the disk Faraday), no ion beam components were detected. Neither were there elevated levels of any of the other scanned peaks, except during use of phosphine gas: the H_2 level was elevated and averaged 4E-7 Torr versus 1E-7 Torr for other gas sources.

In a less controlled trial with the sensor head placed very near to the ion beam, expected species were difficult to detect. Most ion beam related species will be driven into machine surfaces with little back-scattering or surface sputtering and so will be unavailable for detection in the background gas of the implanter. It may also be that the ion beam is somewhat contained by the design of the machine.

Gas Trend Mode: Pumpdown Curves

For this phase of the investigation, no ion beams were present in the system, but a variety of activities were carried out to understand how the state of the vacuum chamber changed and to more fully test the capabilities of the RGA system. For this segment, baseline data were recorded along with data for vent recovery, vent recovery with IPA or H_2O cleaning and response to leak introduction. Note: although trends for several amu peaks were recorded, only those showing clear changes are being included in this report.

The first task was to look at baseline levels and understand how a simple vent would affect the vacuum system. Figure 1 illustrates a very fast recovery to baseline vacuum levels after an N_2 vent to atmosphere, but with the process chamber door (or 'clamshell') held closed. In the absence of exposure to atmospheric water, the vacuum system recovery is excellent.

Figure 2 illustrates how vacuum recovery is affected by full exposure of the vacuum system to room air for a 5 minute period. As expected, there is a marked increase in the time taken for the system to return to baseline levels, especially for the water peak.

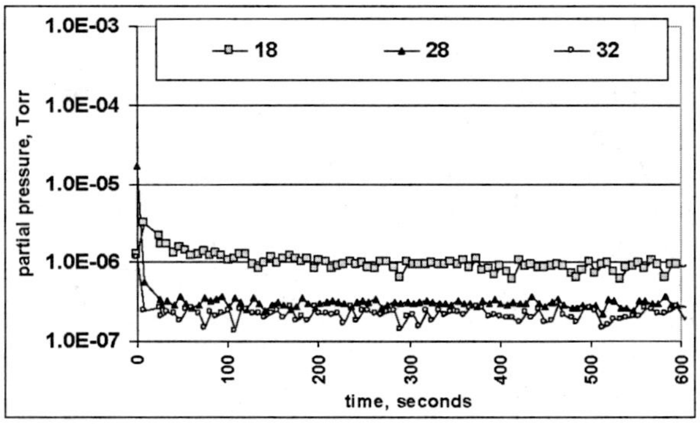

FIGURE 1. Recovery of vacuum state after N_2 vent; chamber door remained closed.

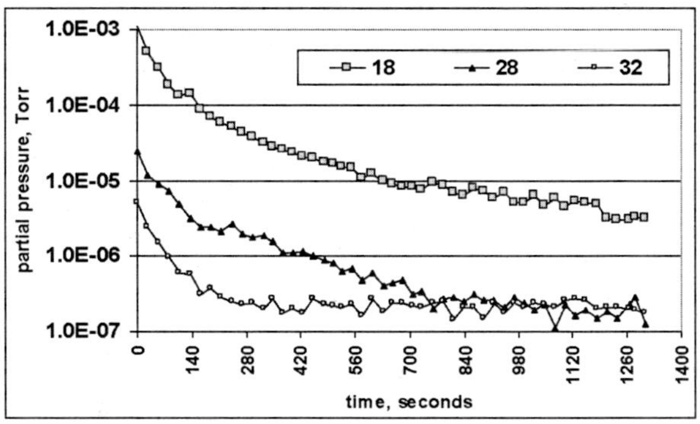

FIGURE 2. Recovery of vacuum state after N_2 vent; chamber door open to room air for 5 minutes.

Next a leak was introduced at the o-ring sealing surface of one of the ports on the vacuum system via inclusion of a foreign object with an approximate diameter of 50 microns. The data appear in Figure 3 which shows N_2 and O_2 reaching an equilibrium state after around 10 minutes while the H_2O peak continues to decline with time.

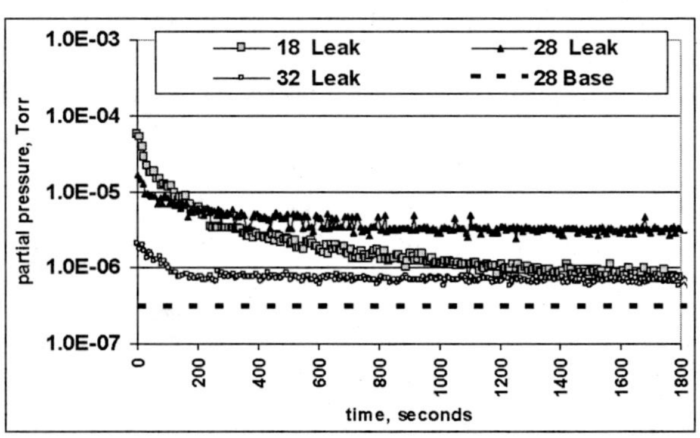

FIGURE 3. Response of system to introduced leak; atmospheric components remain well above baseline.

Finally, a variety of other test cases were investigated: wipe chamber with H_2O, wipe chamber with IPA, wipe chamber with H_2O and IPA, et cetera. In these several trials, very little new was discovered except that IPA, or its most common vacuum fraction amu 45, cleared the system very rapidly.

Gas Trend Mode: Wafers in Process

For this phase of the investigation, the RGA system was running with a 4 peak trend definition file and with Fast Scan enabled. Two types of wafer process were captured: 1)high current, single pass process with electron shower and 2)medium current, quad angle process without electron shower. Figure 4 illustrates the load/unload and implant sequence for 2 batches (1 lot) for the first case and Figure 5 illustrates the same total sequence for the second case.

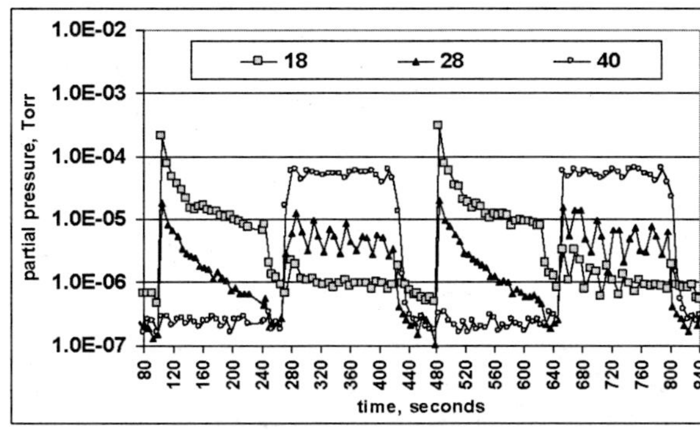

FIGURE 4. Two normal high current process cycles: load/unload and implant with photoresist and argon present. Two cycles.

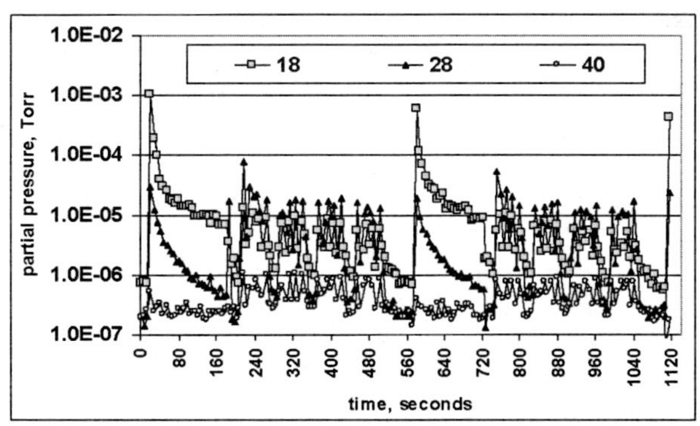

FIGURE 5. Two normal medium current, quad angle process cycles: load/unload and implant with photoresist present, but no argon.

The RGA captured the most relevant information for these processes. In the high current case, the argon is seen to switch on at the beginning of scanning which then coincides with increased detection of amu 28 (N_2, CO). It is interesting that the detected level of amu 18 (H_2O) is seen to decrease during scanning. It is not clear whether this is due to rapid outgassing induced by the ion beam or whether the RGA sensor was being somewhat overloaded by the elevated pressures. This feature bears additional investigation in future.

The quad angle implants are clearly seen in the four groupings of elevated N_2 levels and a modulation of the amu 18 levels. In these processes, without a high argon background, the amu 18 levels are more consistent with a simply decay over time (despite the obvious moudulation in level).

Helium Leak Check Mode

The Helium Leak check mode was tested only briefly for functionality. In this mode the system is configured to analyze only amu 4 and the software provides for two primary data displays: 1)trend display of helium pressure on log scale and 2)'analog' bar graph for rapid response. After creating a known leak at a particular fitting, helium was introduced to the general area and was immediately seen on the displays. The software also provides an audio output, but as it was running on a laptop computer, speaker volume was inadequate to overcome the background noise in the Fab. For this application, an external, amplified speaker set-up would be required. Overall function in He Leak Check mode was reasonable and the small size of the sensor makes it easy to move between various parts of an ion implanter. The system could reasonably be adapted to replace an external He Leak Check tool and its associated COO.

Vacuum Troubleshooting

Use of the RGA system to troubleshoot difficult to resolve issues related to process delivery is expected to be its most valuable application. For example, after a baseline for a particular machine is known, it would be simple to run an RGA scan during a process shutdown event and look for particular contaminant signatures. In any given year, there will be 1 event for every implanter or two that is not resolved by a standard He leak check or other common corrective maintenance approaches. The compact size of this multipole RGA would be a great asset in facilitating this type of problem solving. During the time under study, no such event was available to pursue and report on.

Another area of promising application is evaluation of high vacuum pump health. Both turbo and cryo pumps can fail slowly; this type of RGA system could enable simple, periodic checks to alert one to a pump whose performance is degrading.

CONCLUSION

The compact multipole RGA system was tested on an Axcelis GSD implanter and performed as expected for all phases of this investigation. It was helpful to have a visualization of the vacuum behavior of target species and their response to a variety of treatments.

This system has great potential for aiding in troubleshooting situations that do not yield good answers to conventional approaches and techniques. Further, the cost and COO of the system are such that it would be economical to permanently mount an individual sensor on each machine. The software allows for each sensor in a facility to have its own configuration file so the electronics can easily be moved from one system to another.

REFERENCES

1. INFICON Spectra Library Card, copyright 2002.

Correlation of Beam Transport Characteristics and Mass Resolving Power to Analyzing Magnet Pole Piece Wear in Axcelis HE Ion Implanters

Ron Johnson[1], John Schuur[1] and Randy Tysinger[2]

[1] INNOViON Corporation
4253 NE 189th Avenue
Gresham, OR 97230 US
rcjohnson@innovioncorp.com, jschuur@innovioncorp.com

[2] Maxim Integrated Products
14320 Jenkins Road
Beaverton, OR 97005 US
Randy_Tysinger@or.mxim.com

Abstract. An investigation into odd analyzing magnet tuning behavior on an Axcelis HE ion implanter revealed a correlation to excessive wear on the analyzer magnet pole pieces. Though the lifetime is generally quite good, the pole pieces are internal to the beamline design and are exposed to and sputtered by the ion beam during mass analysis. This erosion of the pole piece surfaces affects beam focus and transport, especially for lower current ion beams. Eventually, automatic beam set-up fails altogether as the system is unable to locate the correct amu peak. It is important to understand and correct any deficiencies in pole piece condition as long-term neglect will eventually lead to process errors. Data is presented to illustrate good and bad pole piece condition and a method is suggested for non-invasive testing.

Keywords: analyzing magnet, pole piece, mass resolution, amu, Axcelis HE.
PACS: 61.72U-, 07.75.+h, 07.77.Ka

INTRODUCTION

Odd analyzer tuning behavior seen in an Axcelis HE ion implanter was, at first, difficult to interpret as final implant data logs (IDLs) and the on-board tool SPC package were not giving indications of any problems with actual process delivery. End of line data from devices processed on the tool during this period of odd amu tuning also showed no significant trends, shifts or patterns in the data. No devices or wafers were ever the subject of scrap as a result of the observed anomaly. Specifically, during some Phosphorus ion beam set-ups, the Auto Beam software was consistently setting the amu to either 30 or 32 at the Injector Faraday (a beam current measuring element that precedes the Linac). Once the beam was routed through the Linac, subsequent 'AMU Tune' operations resulted in a final amu setting of 31, as expected. This paper presents the findings of an investigation to resolve the amu tuning problem.

PROBLEM DISCOVERY

Lengthy source set-up times for some Phosphorous processes alerted Operators to potential recipe problems. Observation of Auto Beam function for these processes was initially inconclusive as the ion source amu tune was inconsistent – on some trials the amu was low, on some trials it was high and for others it was nominal. This occasionally odd problem was not observed however once the ion beam had been accelerated through the Linac, that is to say, the amu was always correct for the final ion beam set-up (ion beam current was correct, etc). For the ion implanter in question, a daily sheet resistance monitor was

CP1066, *Ion Implantation Technology,* edited by E. G. Seebauer, S. B. Felch, A. Jain, and Y. V. Kondratenko

processed which also used Phosphorous; this monitor process had approximately 10x ion current v. the process with amu tuning problems. Eventually, consistent failure to low amu inspired more investigation, including on-board tool SPC, IDL review and review of amu spectrums. Figure 1 is analyzer spectrum data from an HE machine with the amu tuning problem. For both monoatomic and diatomic Phosphorous there is a 'horns' phenomena in the spectrum that has peaks above and below the target peak amu. (+/- 1 amu for $^{31}P^{+}$ and +/- 2 amu for $^{31}P_2^{+}$).

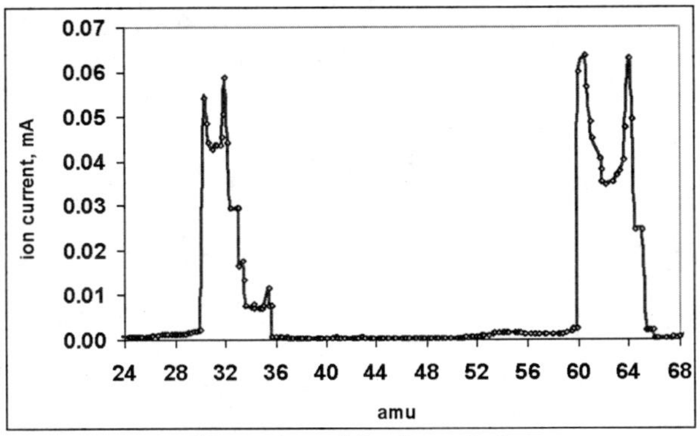

FIGURE 1. An illustration of the 'horns' phenomena observed for both $^{31}P^{+}$ and $^{31}P_2^{+}$ on an HE implanter.

PROBLEM IDENTIFICATION

Using the varying behavior of two well known processes with respect to the emergence of the amu tuning issue as a guide, a number of spectra were taken at different ion source Arc Current settings for both Phosphorous and Argon plasmas. The data is plotted in Figures 2 and 3, respectively. (Note: this data is taken from a different HE machine.)

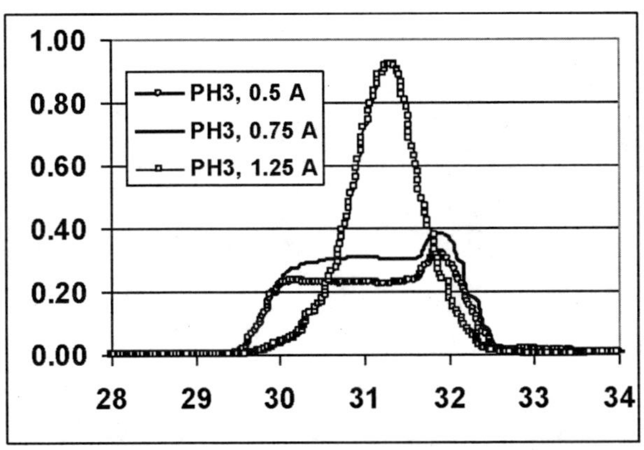

FIGURE 2. Change in PH3 amu spectrum for affected HE machine as a function of Arc Current.

Both Figures 2 and 3 show a clear dependence on Arc Current with respect to evolution of amu peak mismatch. It should be noted that this data was generated simply by varying Arc Current with all other source parameters, such as: Arc Volts, Gas Flow, Source Magnet, Suppression Voltage, etc. were held constant.

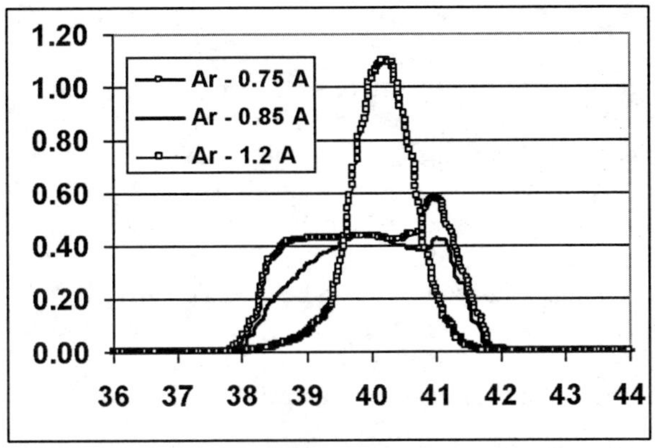

FIGURE 3. Change in Argon amu spectrum for affected HE machine as a function of Arc Current.

The data in Figures 2 and 3 inform us that the problem will be masked for many process set-ups – as beam current (Arc Current) increases, the problem with peak location is reduced.

PROBLEM RESOLUTION: MASS RESOLVING CAPABILITY

At this time, the source and beamline was inspected physically for abnormalities, and it was noted that the amu pole pieces, particularly those closest to the ion source and extraction assembly were severely eroded. As these are replaceable, it is a matter of correct Preventive Maintenance (PM) management to avoid issues with the pole pieces.

For the purposes of developing a diagnostic tool to improve the ability to identify and correct the problem in future operations, mass resolution of the analyzer magnet was measured before and after renewal of the analyzer pole pieces.

Per OEM specification [1], the expected mass resolution (M/ΔM) of an Axcelis HE machine is ≥ 50. The method for mass resolution measurement was to review a detailed amu spectrum and calculate the ration of the amu to the fill-width, half-maximum value of the amu peak.

The data in Table 1 clearly show a very poor mass resolution capability for the worn pole pieces which contrasts very strongly with the mass resolution values obtained for a newly installed set of analyzer pole pieces.

TABLE 1. Mass Resolution ($M/\Delta M$) of Axcelis HE with "worn" and "new" analyzer pole pieces

Species	"worn" = o; "new" = X	Ion current, uA	M/ΔM
$^{31}P^+$	o	320	13
$^{31}P^+$	**X**	**390**	**121**
$^{40}Ar^+$	o	360	13
$^{40}Ar^+$	o	1500	46
$^{40}Ar^+$	**X**	**660**	**120**

This exercise suggests a straightforward procedure for detecting amu pole piece problems. In the normal course of operations, the pole pieces will wear gradually over time; the data used in this paper were taken from machines which had been in continuous operation for > 5 years. So it is important to have a metric that lets you know when erosion of the pole pieces is causing an actual problem. The method of measuring mass resolution using a defined set of source parameters and comparing to an established standard (for example, mass resolution ≥ 50, 60, other).

CONCLUSION

Odd amu tuning behavior was noted in an Axcelis HE ion implanter for some processes. Investigation led to the understanding that the problem was limited to low current ion beam set-ups. A physical inspection of the beamline revealed worn analyzer pole pieces. The effect of this excessive wear on the pole pieces was confirmed by the measurement of poor mass resolution for ion beam set-ups of interest.

Replacement of the worn pole pieces resulted in an elimination of amu tuning issues for all processes. Subsequent measurement of very good mass resolution suggested a method for detection of the problem's recurrence.

REFERENCES

1. GSD/HE Specification # 58130, March 2003, Axcelis Corporation.

Simulation of Mass Interferences Considering Charge Exchange Events and Dissociation of Molecular Ions During Extraction

Volker Häublein[1], Lothar Frey[2], and Heiner Ryssel[1,3]

[1]*Fraunhofer Institute of Integrated Systems and Device Technology (IISB) Schottkystrasse 10, 91058 Erlangen, Germany*
[2]*Institut für Angewandte Physik, TU Bergakademie Freiberg, Leipziger Str.23, 09596 Freiberg*
[3]*Lehrstuhl für Elektronische Bauelemente, Universität Erlangen-Nürnberg, Cauerstrasse 6, 91058 Erlangen, Germany*

Abstract. Contamination due to mass interferences generally leads to severe problems in ion implantation. To be able to prevent this kind of contamination, the cause of the respective mass interference has to be known. Frequently, however, the transport mechanisms of the contamination are unknown and the search for potential transport mechanisms is troublesome. Simulation of mass interferences, therefore, is an excellent means to find potential transport mechanisms. In this paper, the simulation software ENCOTION (ENergetic COntamination simulaTION) is explained. In the latest version, an enhanced model was implemented in order to calculate the apparent mass of ions. The relevance of the new model is discussed by the example of tungsten contamination in the case of BF_2^+ implants. A new module that allows for the simulation of mass spectra was recently added to ENCOTION as a new feature.

Keywords: ion implantation, contamination, mass interferences, simulation
PACS: 41.85.Qg, 33.15.Ta, 85.40.Ry

INTRODUCTION

Experience shows that in spite of mass separation, undesired ions originating from the source chamber can be implanted energetically into a wafer. Charge exchange events and dissociation of molecular ions prior to mass analysis are the main reasons why contaminants are able to fulfill the magnetic rigidity requirements for transmission. Thereby, for mass analysis, it makes a difference whether those reactions occur during extraction or after extraction. In the first case, the so-called apparent mass depends not only on charge and mass states of the ions, but also on the extraction and the suppression voltage of the extraction unit [1].

In 2002, the software tool ENCOTION was presented for the first time [2]. It consisted mainly of a module that simulated potential transport mechanisms of undesired ions passing through the analyzer magnet. Since usually surprisingly large amounts of potential solutions are simulated, it quickly became obvious that finding contamination mechanisms without a simulation tool is not very promising. Later, a module was added to ENCOTION which simulated mass spectra of the potential transport mechanisms [3]. These *local* mass spectra around the peaks of the implanted ion species are especially useful to analyze the impact of mass resolution on the occurrence of contamination.

In this paper, a new module that simulates *global* mass spectra is introduced. This module is intended to provide a quick overview on what might be going on when certain elements are present in the ion source. As the simulation of transport mechanisms is concerned, ENCOTION's enhanced functionality is demonstrated for the case of W contamination during BF_2^+ implantation.

PHYSICAL RELATIONS

Mass interferences may occur since the analyzer of an implanter does not separate for the real mass of an ion, but for the so-called apparent mass. In the simplest case, when no charge exchange events or dissociation processes of the ion are involved, the apparent mass of the ions is calculates to

CP1066, *Ion Implantation Technology*, edited by E. G. Seebauer, S. B. Felch, A. Jain, and Y. V. Kondratenko

$$m_{\mathrm{app}} = \frac{m_{\mathrm{an}}}{q_{\mathrm{an}}}, \quad (1)$$

where m_{an} and q_{an} are mass and charge state of the analyzed ion, respectively. Equation (1) applies e. g. for Mo contamination in BF_2^+ implants, where $^{98}Mo^{++}$ ions have the same apparent mass.

The calculation of the apparent mass becomes more complex when charge exchange events and dissociation processes of the ions after extraction and prior to mass analysis have to be considered. In this case, the apparent mass is given by

$$m_{\mathrm{app}} = \frac{m_{\mathrm{an}}^2}{q_{\mathrm{an}}^2} \frac{q_{\mathrm{sc}}}{m_{\mathrm{sc}}}, \quad (1)$$

where m_{sc} and q_{sc} are mass and charge states of the extracted ion, respectively. Although this equation appears to be quite simple, it makes numerous contamination mechanisms possible, as was demonstrated by previous versions of ENCOTION [2]. An example where Eq. (2) applies is the Mo contamination during B^+ implants, where Mo^+ ions are stripped to Mo^{+++} ions prior to mass separation and, therefore, show the same apparent mass as $^{11}B^+$ ions [4].

When charge exchange or dissociation reactions happen during extraction, the electric field in the extraction unit (see Figure 1) takes effect on the energy of the ions. The apparent mass, therefore, depends on the extraction and suppression voltages. It was experimentally shown that those reactions are most likely to happen close to the extraction electrode [1]. The apparent mass, therefore, can be approximated by

$$m_{\mathrm{app}} = \frac{m_{\mathrm{an}}^2}{q_{\mathrm{an}}^2} \left[\frac{q_{\mathrm{sc}}}{m_{\mathrm{sc}}} + \frac{V_{\mathrm{sp}}}{V_{\mathrm{ex}}} \left(\frac{q_{\mathrm{sc}}}{m_{\mathrm{sc}}} - \frac{q_{\mathrm{an}}}{m_{\mathrm{an}}} \right) \right], \quad (1)$$

where V_{sp} and V_{ex} are the suppression and extraction voltage, respectively [1].

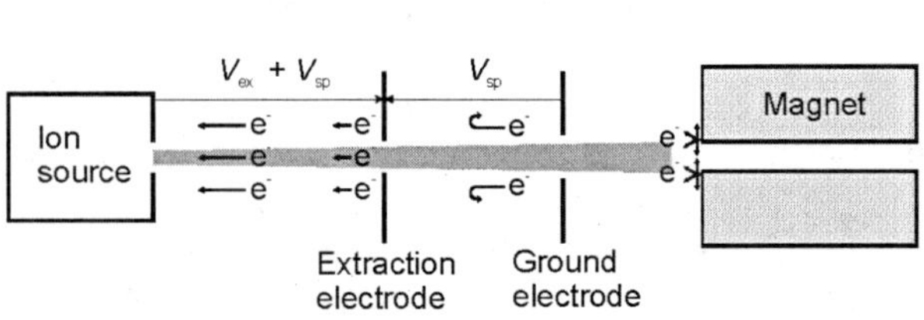

FIGURE 1. Schematic layout of the beamline from ion source to magnet. In order to suppress electrons, the extraction unit consists of an extraction and a ground electrode.

SIMULATION TOOL *ENCOTION*

ENCOTION is a user-friendly software based on Windows. It provides a database of all elements of the periodic table including all natural isotopes.

Features and Applications

For a simulation, only a few parameters have to be specified, such as the elements which are present in the ion source, the mass resolution of the implanter, the acceleration voltages, and several other parameters which set the complexity of the simulated ions and reactions [2]. The number of simulated transport mechanisms depends on the user's input (e. g., the number will be increased when the mass resolution is decreased). The simulated transport mechanisms are listed in a table, together with the corresponding apparent mass and its difference to the apparent mass of the implanted ion species. Furthermore, the effective energies of the contaminants are provided. This is especially useful when those energies can be compared to experimental data, since the number of potential transport mechanisms may be reduced significantly.

ENCOTION may not only be used when contamination just occurred. It is also an optimum tool in order to evaluate the contamination risks of new source materials [5] or source feed materials. Furthermore, it is extremely useful for identifying unknown peaks in mass spectra. By the simulation of mass spectra of the simulated transport mechanisms, it is possible to examine the correlation between contamination and the implanter's mass resolution and voltage settings [3].

Simulation of Mass Spectra

The functionality of the software was extended by its ability to simulate *global* mass spectra within a typical range, e. g. from 0-200 amu. In contrary to the *local* mass spectrum module that directly visualizes simulated transport mechanisms [3], this module does not consider changes in charge or mass state of ions, since this would cause an exuberantly growing number of peaks. The user may select up to five different elements which are present in the ion source. Moreover, the maximum charge state and the maximum number of atoms in a molecule have to be specified for each element. The elements can be weighed by a factor: the larger the factor, the higher the intensity of the corresponding peaks. According to the input, ENCOTION simulates multiply charged ions as well as molecules of the different elements. A useful feature is that any of the chosen elements can be activated or deactivated. The resulting spectrum can be compared directly to the

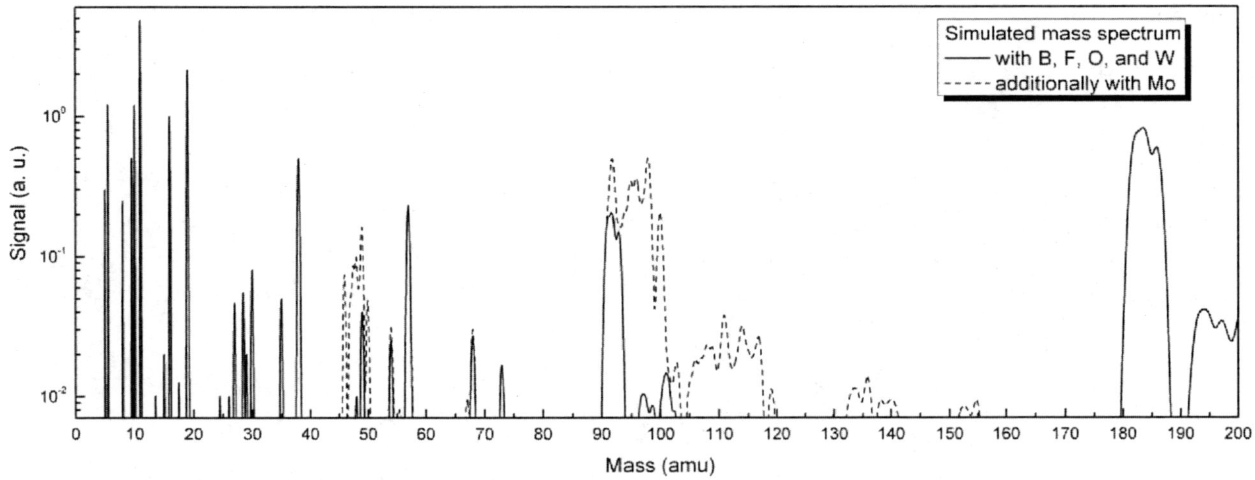

FIGURE 2. Simulated mass spectra from 0-200 amu.

previous one, making it possible to study the impact of a certain element on the occurrence of peaks. Figure 2 shows an example of a comparison of two simulated mass spectra. The simulated elements were B, F, N, O, and Mo. The same spectrum was simulated again without Mo.

W CONTAMINATION IN BF_2^+ IMPLANTS

In this section, the applications of ENCOTION are shown on a very relevant example. Due to the molybdenum contamination issue for BF_2^+ implants, molybdenum was replaced by tungsten as the standard ion source material, which raises the question about tungsten contamination. In fact, tungsten contamination, which could be attributed to the ion source, was reported early by Liebert *et al.* [6]. SIMS measurements revealed that tungsten gets implanted with an effective energy of about the extraction energy (when no post acceleration is applied). Therefore, mass interferences according to Eq. (1) can be excluded, since the necessary charge states of at least +4 would lead to much higher energies. A mass spectrum showed significant peaks of W^+ and WF_x^+ ions [6]. In the end, the following transport mechanisms were proposed, declaring the first one as the leading candidate [6]: (a) $W^+ \rightarrow W^{++}$, (b) $W_2^{++} \rightarrow W^{++}$, and (c) $WF^+ \rightarrow WF^{++}$.

Table 1 shows an excerpt of transport mechanisms that were simulated by ENCOTION under the assumption that changes in charge state and mass occur *after* extraction. It was supposed that only W, F, and B were present in the ion source. The above proposed transport mechanisms are also found by ENCOTION, although the reactions $W^+ \rightarrow W^{++}$ and $W_2^{++} \rightarrow W^{++}$, however, are only simulated when a poor mass resolution below 20 is set. The relation between mass resolu-

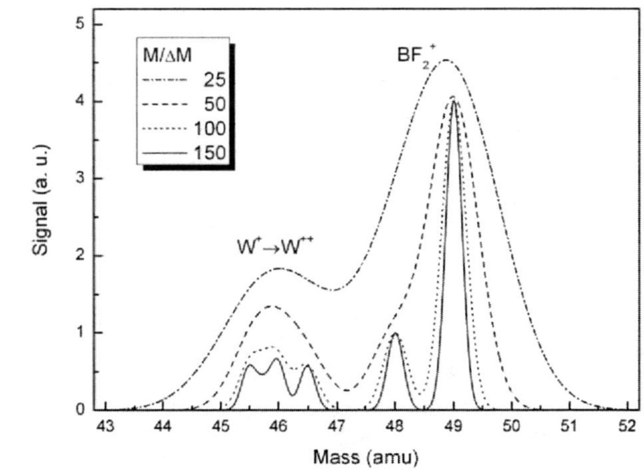

FIGURE 3. With ENCOTION simulated mass spectrum signals of the implanted ion species BF_2^+ and ions that undergo the reaction $W^+ \rightarrow W^{++}$ as a function of mass resolution.

tion and contamination is shown in Figure 3. Hence, for mass resolutions larger than 25, a contribution of this reaction seems to be unlikely. The reaction $^{186}WB^+ \rightarrow ^{186}WB^{++}$ may be excluded since no WB^+ peaks are present in the mass spectrum. Despite of its good agreement with the apparent mass of BF_2^+, the reaction $^{183}WF_2^{++} \rightarrow ^{183}WF_2^{+++}$ can be excluded due to the high effective energies. Because of suitable values for the apparent mass and the effective energies, the reaction $WF^+ \rightarrow WF^{++}$ may be considered a likely mechanism for the contamination with the isotope ^{180}W (natural abundance: 0.12 %), the contamination with heavier tungsten isotopes strongly depends on the implanter's mass resolution.

In the following, the transport mechanisms are considered for the case that changes in charge state or mass take place at the extraction electrode. Since all reactions of the transport mechanisms in Tab. 1 lead either to a charge increase or a mass decrease, the apparent mass will be according to Eq. (3) decreased. Table 2 shows the resulting apparent masses and effective energies (the suppression voltage V_{sp} is assumed to be 2 kV) for the reactions $W^+ \rightarrow W^{++}$ and $WF^+ \rightarrow WF^{++}$. Because of the lower apparent mass, the first reaction is even more unlikely than the same reaction taking place after extraction. The latter reaction, however, shifts the tungsten spectrum to the position of the BF_2^+ peak(s), making this reaction a major candidate for tungsten contamination in BF_2^+ implants (Figure 4).

TABLE 1. Simulated transport mechanisms for W contamination in BF_2^+ implants, when the reactions take place after extraction and prior to mass analysis. ΔM_{app} is the difference of the apparent masses of the respective mechanism and BF_2^+. $E_{eff,ex}$ and $E_{eff,tot}$ are the effective W energies after extraction and post acceleration, respectively (extraction and post acceleration voltages were set to $V_{ex} = V_{ac} = 50$ kV).

Transport Mechanism	M_{app} (amu)	ΔM_{app} (amu)	$E_{eff,ex}$ (keV)	$E_{eff,tot}$ (keV)
$^{186}W^+ \rightarrow ^{186}W^{++}$	46.49	-2.52	50.0	150.0
$^{186}W_2^{++} \rightarrow ^{186}W^{++}$	46.49	-2.52	50.0	150.0
$^{180}WF^+ \rightarrow ^{180}WF^{++}$	49.74	0.73	45.2	135.7
$^{182}WF^+ \rightarrow ^{182}WF^{++}$	50.24	1.23	45.3	135.8
$^{186}WB^+ \rightarrow ^{186}WB^{++}$	49.24	0.23	47.2	141.6
$^{183}WF_2^{++} \rightarrow ^{183}WF_2^{+++}$	49.10	0.09	82.8	207.0

TABLE 2. Simulated transport mechanisms for W contamination in BF_2^+ implants, when the reactions take place near the extraction electrode. $V_{ex} = V_{ac} = 50$ kV, $V_{sp} = 2$ kV.

Transport Mechanism	M_{app} (amu)	ΔM_{app} (amu)	$E_{eff,ex}$ (keV)	$E_{eff,tot}$ (keV)
$^{186}W^+ \rightarrow ^{186}W^{++}$	44.63	-4.38	48.0	148.0
$^{184}WF^+ \rightarrow ^{184}WF^{++}$	48.71	-0.30	43.5	134.1
$^{186}WF^+ \rightarrow ^{186}WF^{++}$	49.19	0.18	43.6	134.3

CONCLUSION

Simulation of mass interferences is a probate means to understand and solve contamination problems in ion implantation. The simulation software ENCOTION, with its modules to simulate potential transport mechanisms and visualize mass spectra, has become a powerful tool. Its functionality was demonstrated for tungsten contamination in BF_2^+ implants, where the reaction $WF^+ \rightarrow WF^{++}$, taking place during (and also after) extraction was found to be a likely transport mechanism. The functionality of ENCOTION will be increased continually. Current work deals with the simulation of cluster spectra.

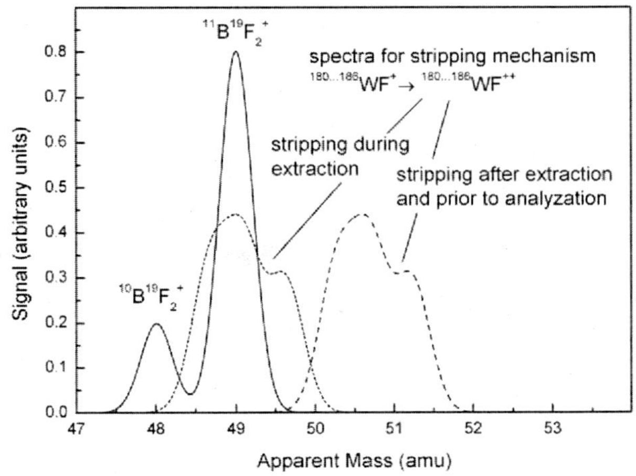

FIGURE 4. With ENCOTION simulated mass spectrum signals of BF_2^+ ions compared to signals of ions that undergo the stripping reaction $WF^+ \rightarrow WF^{++}$. When the stripping reaction occurs during extraction, W contamination for $^{11}BF_2^+$ implants is evident.

REFERENCES

1. Häublein, V., Frey, L., and Ryssel, H., *Nucl. Instr. And Meth. in Phys. Res. B* **237**, 346-350 (2005).
2. Häublein, V., Frey, L., and Ryssel, H., "ENCOTION-A new Simulation Tool for Energetic Contamination Analysis" in *Ion Implantation Technology 2002*, edited by B. Brown *et al.*, 2002 14th International Conference on Ion Implantation Technology Proceedings, Piscataway, NJ: IEEE, 2003, pp. 101-104.
3. Häublein, V., Frey, L., and Ryssel, H., "The Impact of Mass Resolution on Molybdenum Contamination for B, P, BF2, and As Implantations" in *Ion Implantation Technology-IIT 2006*, edited by K. Kirkby *et al.*, AIP Conference Proceedings 866, New York: American Institute of Physics, 2006, pp. 464-467.
4. Ryssel, H., Current, M., and Frey, L., "Contamination Control in Ion Implantation," in *Ion Implantation Science and Technology*, edited by J. F. Ziegler, Edgewater: Ion Implantation Technology Co., 2000, pp. 564-601.
5. Häublein, V., Walser, H., Frey, L., and Ryssel, H., "Investigation of Lanthanum Contamination from a Lanthanated Tungsten Ion Source" in *Ion Implantation Technology 2002*, edited by B. Brown *et al.*, 2002 14th International Conference on Ion Implantation Technology Proceedings, Piscataway, NJ: IEEE, 2003, pp. 346-349.
6. Liebert, R. B., Angel, G. C., and Kase, M., "Tungsten Contamination in BF2 Implants" in Ion Implantation Technology-96, edited by E. Ishida *et al.*, Proceedings of the Eleventh International Conference on Ion Implantation Technology, Piscataway, NJ: IEEE, 1997, pp. 135-138.

Non-Uniform Dose Mapping Controlled by Modulated Vertical and Horizontal Scans

S. Ninomiya, Y. Kimura, T. Kudo, A. Ochi, R. Toda, M. Tsukihara, F. Sato, G. Fuse, K. Ueno, and M. Sugitani

SEN Corporation, an SHI and Axcelis Company, 1501, Imazaike, Saijo, Ehime, 799-1362, Japan

Abstract. Since geometrics of semi-conductor devices continue to shrink, the requirement for each process becomes severer to keep uniformity of electrical parameters of the semi-conductor devices. A larger wafer also causes larger variations. Thus it has been strongly required for ion implantation process to compensate for the variations from other processes because of its good dose controllability. A newly developed mapping of intentional non-uniform dosage system, which is named "MIND system", is implemented in SEN's single-wafer-type implanters. The MIND system controls both horizontal and vertical scan speed simultaneously. Intentional two-dimensional non-uniform profiles of sheet resistance, such as concentric and eccentric profiles, are obtained only by single-step ion implantation.

Keywords: Control of scan speed, Compensation for other processes, Sheet resistance
PACS: 41.75.Ak, 61.72.Uf, 85.40.Ry,

INTRODUCTION

There are many processes to manufacture semiconductor devices. In order to obtain the same semiconductor devices in a wafer, it is quite important to keep uniformity within a wafer in a certain range. Since geometrics of semi-conductor devices have a tendency to shrink, uniformity requirements within a wafer for each manufacturing process become severer. In addition, since the size of a wafer becomes larger and larger, to keep uniformity in a wafer becomes harder and harder.

Previously, a great effort has been made to get good uniformity within a wafer in each process. For instance, electrical parameter variations for semiconductor devices are determined by the worst processes in all manufacturing processes, i.e., some "bottleneck processes" are presented. However, some processes are quite difficult to keep uniformity in a wafer within required ranges.

Consequently, in the future, a new concept to obtain good uniformity of semiconductor devices in a wafer is necessary. It should be pointed out that to keep good uniformity in each process is not indispensable. Electrical parameters should be the same for all manufactured semiconductor devices. In other words, it is practically enough to keep good "substantial" uniformity in several processes in a wafer. In this concept, all processes to manufacture semiconductor devices are classified into three types, that is, "bottleneck processes", "normal processes" and "support processes".

The "bottleneck processes" have essential difficulty to obtain good uniformity in a wafer. For example, dry etching process and plasma CVD process may be classified into these types of processes. The "normal processes" can obtain good uniformity by themselves. However, they can not bring intentional non-uniformity in a wafer because of their less controllability.

The "support processes" have ability not only to obtain its own good uniformity, but also to compensate for other processes by giving intentional non-uniform profile within a wafer. If you can get information on non-uniform profiles within a wafer of the "bottleneck processes", you can compensate for the variations by using subsequent "support processes". If you can predict non-uniform profiles at the "bottleneck processes", you can also use former "support processes" to compensate for the variations of subsequent "bottleneck processes".

Ion implantation processes are ones of the best candidates of the "support processes", because of their good dose controllability. That is, the dosage of the ion can principally give intentional non-uniformity within a wafer.

This paper reports on the newly developed Mapping of Intentional Non-uniform Dosage system,

CP1066, *Ion Implantation Technology,* edited by E. G. Seebauer, S. B. Felch, A. Jain, and Y. V. Kondratenko

which is named "MIND system", implemented in SEN's single-wafer type implanters [1]. The MIND system is found to be powerful tool to compensate for variations from other processes.

REQUIREMENTS FOR THE MIND SYSTEM

It is crucially important to know which types of non-uniformity are happened in "bottleneck processes". One of the types of the non-uniformity profiles in a wafer is an eccentric profile. For example, a gate critical dimension, gate spacer deposition thickness and gate spacer etch critical dimension are generally known to have variations of eccentric-profile, which cause the electrical parameter variations, such as Vt values and I_{on}/I_{off} [2].

It is possible for each ion implantation process to play a role in compensation for variations from other processes. Especially, when halo implantation processes are used to compensate, intentional eccentric dose mappings must be obtained only by a single-step implantation, because incident beam has finite tilt angle to the wafer plane for halo implantation. In other words, step-wise rotation implantation can not be used for halo implantation to form two-dimensional intentional non-uniform profile.

Considering halo implantations from the opposite side, another essential requirement is found. A center of the eccentric profile must be changed for each twist angle in order to implant the same dosage into both sides of halo positions in a single transistor. Since four-step implantations are naturally preferred to halo implantations, the implant system for compensation for variations should be suitable for these implantations.

Summarizing requirements to the MIND system, both 1) to obtain eccentric profile by a single-step and 2) to set a center of eccentric profile at an arbitrary position are essentially necessary. Technically, 3) good productivity is also required.

EXPERIMENTAL

SEN have two series of 300-mm single-wafer-type implanters; the MC3 series as medium-current implanters and the SHX series as high-current implanters. The both series have similar scan system, i.e., ion beam scan for the horizontal direction and a mechanical scan for the vertical direction. The MIND system has been developed to correspond to both series. Though results of the MC3 series are mainly reported in this paper, the similar results were obtained for the SHX series as well. The MIND system can be used in all implantation process which SEN's single-wafer-type implanters used.

In order to obtain two-dimensional non-uniform dose profiles, such as eccentric profiles, by a single-step ion implantation, it is necessary to control both vertical and horizontal scan speed simultaneously, which is carried out in the MIND system.

The MIND system has two correction functions. One is for the vertical scan and the other is for the horizontal scan. Such two correction functions can be generally selected independently and simultaneously. Only one of the correction functions also can be separately selected, if needed. By changing correction functions or their parameters, you can control the dose correction pattern easily and flexibly. Support software, which runs on an off-line PC, has been developed to generate correction functions and transfer them to our implanters.

The support software also simulates resultant non-uniform dose mapping and indicates as a correction pattern of a dosage. Figure 1 shows a predicted dose non-uniform map generated by the support software. In this case, four sets of different correction functions were used and four-step by 90 degrees step-wise rotation implantations were performed. The result of sheet resistance map of an actual implant is also shown in Figure 1. The prediction matches quite well even for complicated non-uniform dosage. This result means that the MIND system has high accuracy. One reason for such high accuracy is that SEN's single-wafer-type implanters have an electrical field ion beam scanning system. The MIND system is capable of controlling the ion dose mapping very precisely.

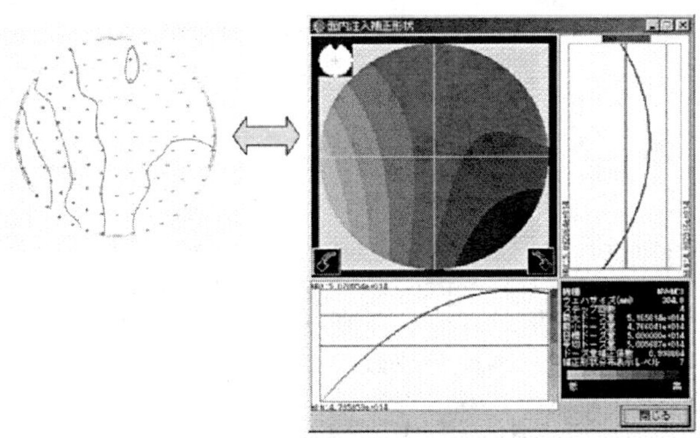

FIGURE 1. A predicted dose map generated by the support software and a result of sheet resistance. (B 20keV $1E15cm^{-2}$ on the SHX)

Correction functions are normally connected to an implantation recipe. An on-line treatment for the MIND system has been already developed. Any correction function is converted to seven-ordered rational function and characterized by the eight coefficients. In a on-line recipe, these eight values are used to designate correction functions.

For SEN's single-wafer-type implanters, beam angle is determined by focusing point of the parallel lens, which is independent from irrespectively ion dose. In other words, dose uniformity can be controlled by itself. The MIND system is designed not to disturb the superiority of the above concept, i.e., only ion dosage can be changed without any changing of the implant beam angles.

RESULTS AND DISCUSSIONS

Since the MIND system has high accuracy, the intentional non-uniform dose mapping can be just mathematically predicted by correction functions. Selecting proper correction functions both for horizontal direction and for vertical direction, a concentric profile of a sheet resistance map is obtained only by a single-step ion implantation, which is shown in Figure 2. Both hill-type non-uniform profile and cup-type profile can be obtained. Similar sheet resistance maps seem to be obtained by using step-wise rotation with one-dimensional modulations, but it is not true for considering halo implantation, which has been already pointed out above.

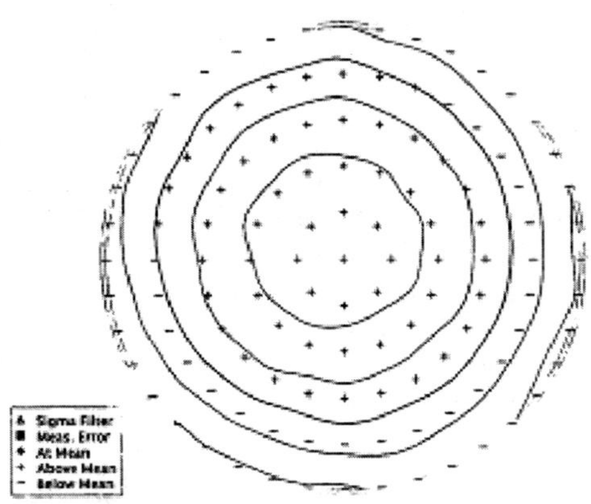

FIGURE 2. A concentric profile of a sheet resistance map by a single step ion implantation (B 100keV 5E14 cm^{-2} on the MC3-II)

Moving the center of the concentric profile toward a certain direction in a wafer plane, eccentric profile should be obtained. Simple mathematical transformation of coordinates to both horizontal correction functions and vertical ones allows us to obtain an intentional eccentric profile also by a single-step ion implantation shown in Figure 3. The center of the eccentric profile is set to the lower right of the wafer in this figure. This result means the MIND system can obtain eccentric profile by a single-step ion implantation, which is the first requirement we have pointed out in the above section.

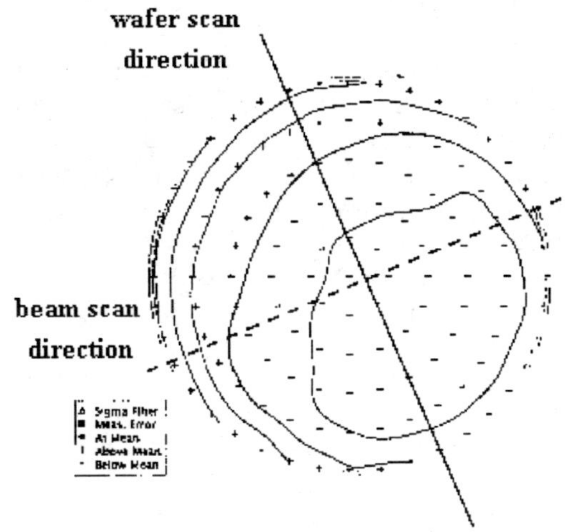

FIGURE 3. An eccentric profile of a sheet resistance map by a single-step ion implantation (B 100keV 5E14 cm^{-2} on the MC3-II)

Mathematical prediction shows that a center of an eccentric profile is independent from both the beam scan direction and the wafer scan direction. Figure 4 shows different kinds of the eccentric profiles of sheet resistance maps. The centers of the eccentric profiles are located onto beam scan direction, onto wafer scan direction at an upper right side and at a lower left side, which are completely predicted by the support software. This result means the MIND system can set a center of eccentric profile at an arbitrary position, which is the second requirement we have pointed out in the above section.

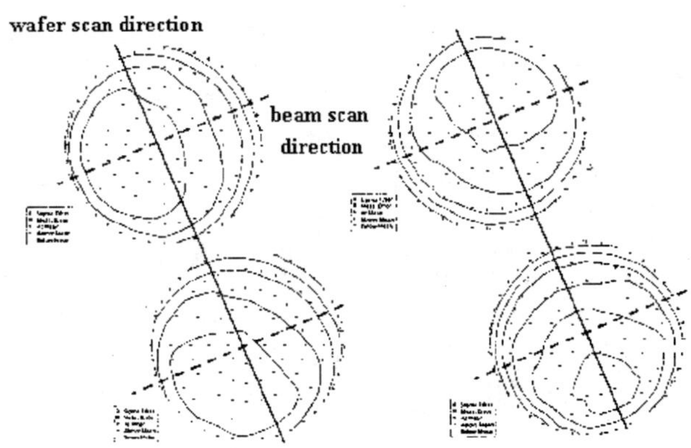

FIGURE 4. Eccentric profiles of sheet resistance maps with different center positions (B 100keV 5E14 cm^{-2} on the MC3-II)

As shown above, the MIND system generates both eccentric and concentric profiles for a single-step implantation. It is not necessary to use the multi-step implantation in which is a waste of time. In addition, the MIND system does not change auto-tuning sequence of the MC3 and SHX series, i.e., no additional sequences for corrections of the ion dose are added. Thus, productivity is quite the same as normal implantation. This fact means the MIND system has good productivity, which is the last requirement we have pointed out in the above section.

Since all requirements we have pointed out above are satisfied, the MIND system is found to have enough ability to compensate for variations from other processes.

There are three categories for single wafer processing, 1) scanned beam with one-dimensional wafer scan, 2) ribbon beam with one-dimensional wafer scan and 3) two-dimensional mechanical scan of the wafer. In order to obtain intentional non-uniform dose profile only by a single-step implantation, the second category, ribbon beam with one-dimensional wafer scan, is the most difficult. Two-dimensional mechanical scan can be easy to control non-uniform dosage in a wafer. Such scanning method has, however, comparatively bad productivity determined by limitation of mechanical scan speed. Therefore, the MIND system for the first category is the only way to form two-dimensional non-uniform dosage for high-tilt halo implantation with good productivity.

Even for zero-degree angle implantation, such as deep S/D implantation, the MIND system has a great merit in practice. Intentional concentric non-uniform profiles can also be obtained by using multi-step implantation with one-dimensional dose correction in this case. However, productivity for such implantations should become extremely worse in comparison to non-corrective implantation. First, it costs relatively long time for multi-step implantations themselves. In addition, it is necessary to divide ion dosage equally to the number of steps, which comes to reduce ion current. Worse productivity is not happened when the MIND system is used.

If more complicated patterns of the dosage profile for zero-degree angle implantation are required, one can combine the MIND system with step-wise rotation implantations. An example is shown in figure 5. An intentional concentric doughnut profile is obtained combining the MIND system with eight-step equiangular step-wise rotation implantations.

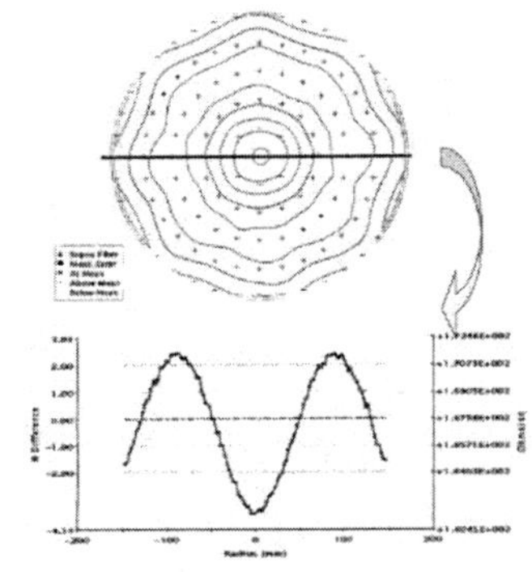

FIGURE 5. A concentric doughnut profile of a sheet resistance map (B 100keV 5E14cm^{-2} on the MC3-II)

Some other results of sheet resistance maps, such as a crescent profile and a checker profile, by using the MIND system with non-equiangular implantations have been already reported and discussed elsewhere. [1] It has been already pointed out that the MIND system with combining multi-step implantations reduces the number of split lot wafers for development of a new process.

CONCLUSIONS

A new intentional non-uniform dose mapping system, the MIND system, controlled by the both vertical and horizontal scan modulations has been developed. Two-dimensional eccentric profile of non-uniform dosage is obtained by a single-step ion implantation. The MIND system has enough ability to compensate for variations from other processes to manufacture semi-conductor devices. Especially, for compensation by halo implantations, the MIND system is the only way to compensate for variations from other processes with good productivity.

ACKNOWLEDGEMENTS

The authors thank Dr. Mitsuaki Kabasawa for fruitful discussions and Ms. Masako Kainou for paper presentation.

REFERENCES

1. S. Ninomiya *et al.*, *Proceedings of 8th Int. Workshop on Junction Tech.*, pp 25-28.
2. Min-Yong Lee *et al.*, *Proceedings of 16th Int. Conf. on Ion Implantation Tech.*, pp.62-64.

Micro Four-Point Probe with High Spatial Resolution for Ion Implantation and Ultra-Shallow Junction Characterization

Daniel Kjaer[1], Rong Lin[1], Dirch Hjorth Petersen[1, 2], Petros M. Kopalidis[3*], Ronald Eddy[4], David A. Walker[4], William F. Egelhoff, Jr.[5], Larry Pickert[6]

[1]*CAPRES A/S, Scion-DTU, Building 373, DK-2800 Kgs., Lyngby, Denmark*
[2]*DTU Nanotech - Dept. of Micro and Nanotechnology, Technical University of Denmark, B-345 East DK-2800 Kgs. Lyngby, Denmark*
[3]*Technological Educational Institute of Thessaloniki, Sindos 57400, Greece*
[4]*Innovion Corporation, Chandler, AZ 85226, USA*
[5]*National Institute of Standards & Technology, Gaithersburg, MD 20899, USA*
[6]*Microchip Technology Inc., Chandler, AZ 85224, USA*
* *kopalidis@urgrad.rochester.edu*

Abstract. The application of micro four-point probe technique in ion implantation non-uniformity mapping and analysis is demonstrated in this work. The technique uses micron-size probes with electrode pitch of 10 μm to achieve greatly enhanced spatial resolution of sheet resistance (Rs) measurements. Rs non-uniformities due to uneven dopant distribution or activation can be mapped with improved accuracy, making it easier to detect implanter scanning problems, dose and charge control malfunctions and annealer related non-uniformities. The technique's superior performance in spatial resolution over conventional four-point probe measurements is demonstrated by zero edge exclusion sheet resistance measurements at the wafer edge. In addition, the technique is used to investigate potential Rs variations between equivalent As+ and As_2+ implants with the same effective energy. Finally, repeatability and reproducibility are investigated by making multiple measurements on a selected ULE implanted and annealed wafer.

Keywords: Sheet resistance, four-point probe, ion implantation, process control, diagnostics
PACS: 07.10.Cm; 84.37.+q; 85.85.+j

INTRODUCTION

A major concern in ion implantation process qualification and control is to ensure consistent uniformity of dopant distribution over large numbers of processed wafers. Modern ion implanter design involves complex mechanisms of ion beam and/or wafer scanning as well as dose and charge control that, when deviating even slightly from their normal operation, can result in unacceptable non-uniformities and seriously affect implanter uptime and availability. Sheet resistance (Rs) measurements by conventional four-point probe (FPP) have long been used as a principal way of detecting and analyzing such wafer non-uniformities [1], however recent work has demonstrated that the technique sometimes becomes limited in its usefulness to characterize ultra-shallow implants, due to probe penetration [2] and leakage current [3], as well as non-uniformities, due to limited spatial resolution [4]. The micro four-point probe (M4PP) technique, utilizing micron sized and non-destructive probes, provides a straightforward solution to these difficulties.

In this work the main focus for Innovion is to determine the usefulness of the M4PP system, manufactured at Capres for process monitoring, process diagnostics and for wafer detail correlation to die problems.

EXPERIMENTAL

n- and p-type 200 mm wafers were implanted in a batch, high current ion implanter (Axcelis GSD-200E2) with energies and doses as shown in Table 1. The wafers were annealed in a rapid thermal annealer at either of two locations – Innovion and Microchip – using an Axcelis-Reliance 850 and a Mattson-AST SHS2800 respectively. The anneal recipe used with the Axcelis Reliance 850 rapid thermal annealer is 35 sec, 1100°C with 50% N_2 in air. A different anneal recipe is used with the Mattson-AST SHS2800, as follows: 18 sec, 1050°C with 10% O_2 in N_2 gas.

CP1066, *Ion Implantation Technology,* edited by E. G. Seebauer, S. B. Felch, A. Jain, and Y. V. Kondratenko

Test #	Species	Energy, keV	Dose, ions/cm^2	Rs, Ohm/sq. Capres (M4PP)	Rs, Ohm/sq. CDE (B-type FPP)	M4PP-FPP Diff., %	Notes
1	As+	2	1E15	163.64	146.87	10.80	Flood on
2	As+	2	1E15	149.25	141.82	5.11	Flood off for comparison
3	As_2+	2	1E15	152.85	148.94	2.59	Dimer; Particle energy and dose are same as in Run 1
4	B+	2	1E15	155.78	153.17	1.69	
5	BF_2+	3	1E15	273.52	268.94	1.69	Effective B energy = 0.67 KeV
6	As+	2	1.2E15	132.08	126.25	4.51	Dose Sensitivity vs. Test #1
7	As+	2	8E14	203.57	191.32	6.20	Dose Sensitivity vs. Test #1

TABLE 1. Experimental matrix of implanted species, energies and doses.

The FPP used at Innovion is a CDE ResMap Model 178 with a 1 mm pitch probe. The M4PP consists of an array of micro-machined, metal coated silicon cantilevers providing an extremely low contact force (~ 10^{-5} N) [4]. In this work M4PPs with 10 μm probe pitch were utilized.

RESULTS AND DISCUSSION

The average sheet resistances (Rs) measured with the CDE and the M4PP (see Table 1) indicate generally quite good agreement between the two instruments for all implanted wafers, with the differences in average Rs within 6.2 % with the exception of the first implant for which the difference is 10.8 %. It is not clear at the time of publication what caused the large difference in this implant. For all implants, the Capres M4PP gives slightly higher Rs values than the CDE. This is partially due to a difference in lab temperature, 19.5 °C for the CDE and 26.5 °C for the M4PP, and most likely partially due to leakage current on the CDE part [2,3]. The higher roughness of the CDE contour maps origins from unintended rotations of the sample during measurements as the wafer is not fixed in the tool.

Rs Comparison for Flood Off, Dimer and BF_2 Implants

The first three implants, cf. Table 1, involve a comparison of sheet resistance between FPP and M4PP for a baseline 2 keV, 1e15 As^+ implant, the same implant with the charge neutralization device of the implanter turned off and an equivalent As dimer implant. The Rs values for the baseline implant are 146.87 and 163.64 Ω/sq respectively for the 4PP and M4PP, a 10.80 % difference. Both probes show a drop in Rs and worsening uniformity when the charge neutralization is turned off, an expected consequence of the ion beam blowing up as it traverses the Si wafer from the aluminum disk. The Rs maps for the flood-off implants are shown in Figure 1.

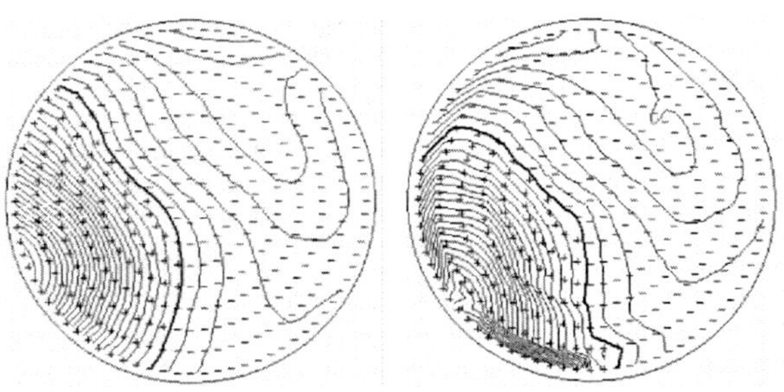

FIGURE 1. Rs contour maps from a) the M4PP and b) the CDE for a flood-off implanted wafer (notch down, 2 % contour intervals).

Compared to the baseline the Rs values from the As_2^+ implanted wafer differ by -1.4 % and 6.6 % for the FPP and the M4PP probes respectively. These variations are up to 3 times higher than previously published results [5], which could be due to the different type of implanter used in this study.

The next comparison involves a 3 keV, 1e15 BF_2 implant (annealed in the SHS2800) with an effective B energy of 0.67 keV. The Rs maps from the BF_2 implant are shown in Figure 2.

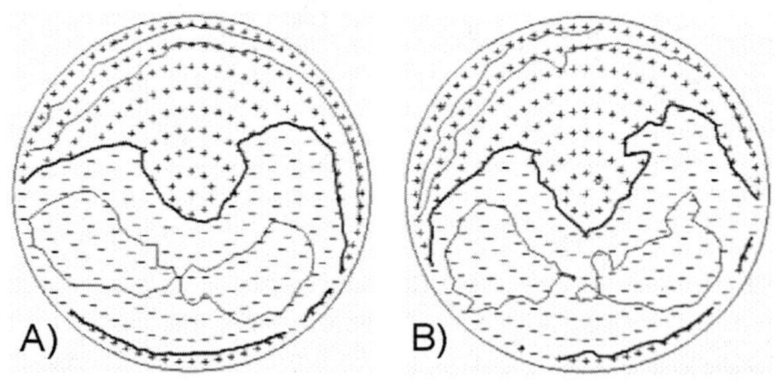

FIGURE 2. Rs contour maps from a) the M4PP and b) the CDE for a 3 keV, 1e15 BF_2 implanted wafer (notch down, 1 % contour intervals).

Reproducibility, Repeatability and Dose Sensitivity

Repeatability and reproducibility studies of the M4PP were performed on the BF_2 implant. The repeatability of the instrument was tested at five points on a 2×2 mm square. The sheet resistance at each position was measured repeatedly 20 times in a random manner. The mean Rs, the standard deviation (SD), and the relative SD calculated for the five points are given in Table 2.

X / µm	Y / µm	<Rs> / Ω/sq	SD / Ω/sq	SD / %
-1000	-1000	275.2	0.25	0.09%
-1000	1000	275.41	0.39	0.14%
0	0	275.24	0.24	0.09%
1000	-1000	275.15	0.25	0.09%
1000	1000	275.42	0.16	0.06%

TABLE 2. Repeability data for 5 locations measured 20 times each in a random manner.

The reproducibility of the M4PP was checked for four different probes: one used probe with more than 1000 touches and three new probes. A 2.5 mm line scan with 100 µm step size was repeatedly measured by these probes at the same locations. Table 3 summarizes the measurement results.

	Probe 1 (old)	Probe 2 (new)	Probe 3 (new)	Probe 4 (new)
<Rs> / Ω/sq	274.64	274.52	274.66	274.55
SD / Ω/sq	0.23	0.24	0.18	0.20
SD / %	0.08	0.09	0.06	0.07

TABLE 3. Reproducibility on a 2.5 mm line scan repeatedly measured by four M4PPs.

An additional long term reproducibility test was performed in a scan area of 5x5 points each separated by 10 mm. The scan was measured 3 times a day (with a time span of 4 hours) for 5 days over a period of 7 days and the wafer was transferred to and from the measurement system during the test. In Figure 3 the mean Rs values as well as the temperature for each measurement time are plotted.

The relative SD of the 15 scans was calculated to be 0.08 %. It is evident that the measurement deviations are correlated to and can be explained by the variation in lab temperature.

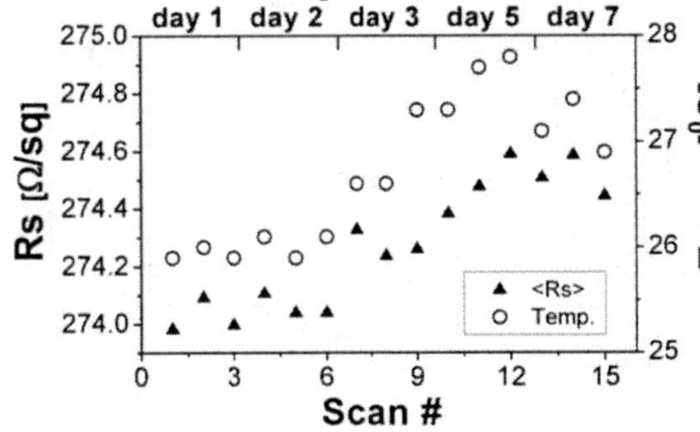

FIGURE 3. Rs and temperature for long term test.

The 2 keV, 1e15 As^+ implant (test # 1) was analyzed for dose sensitivity by running two additional implants with +/-20% dose difference (tests # 6 and 7). The dose sensitivity as described in reference [6] was 1.1 and 1.0 for the M4PP and the conventional 4PP respectively. This is in good agreement with the theoretical dose sensitivity which is 1.04 for the choice of doses while assuming constant mobility and 100% dose activation.

Edge Measurements

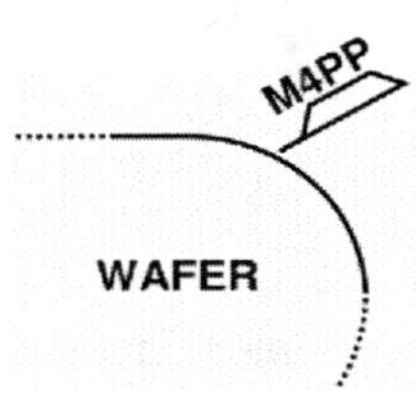

FIGURE 4. M4PP measuring beyond the wafer edge.

The Capres M4PP has due to the microscopic tip separation virtually no edge exclusion. This has been shown by performing a line scan all the way to the wafer edge and beyond (see Figure 4). The step size of the line scan was lowered to 5 µm as the probe approached the edge. For each measurement point an Rs value was read out together with the z-position of the probe thus effectively mapping the topology of the wafer edge. These results are plotted in Figure 5 and 6 together with information of the wafer slope, α and the measured sheet conductance, Gs. Beyond the wafer edge a significant increase in Rs is observed mainly due to the smaller projected area of this region with respect to the direction of the ion beam. The estimated sheet conductance based on the expression: $Gs \propto dose \times \cos(\arctan(\alpha))$, was found to deviate less than 2 % from the measured values for a slope angle up to 20° and the result was reproducible on different wafers.

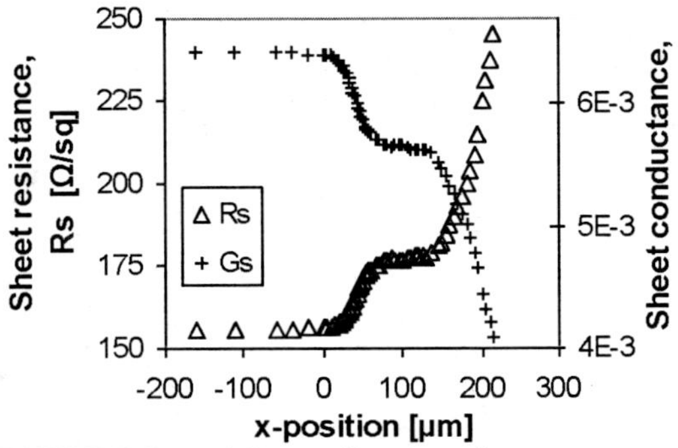

FIGURE 5. Rs and Gs as a function of x-pos.

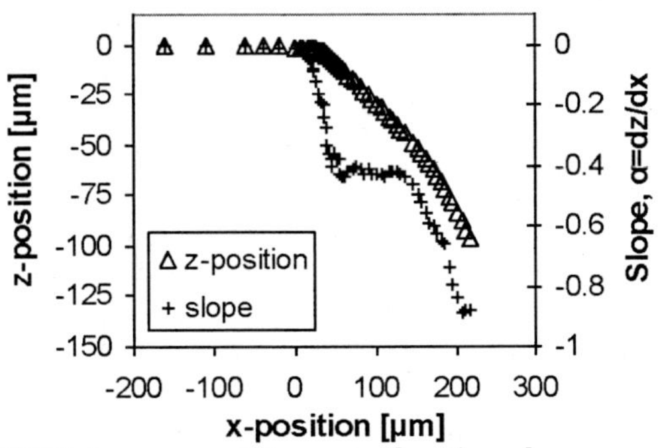

FIGURE 6. z-pos. and slope as a function of x-pos.

The M4PP can perform similar high resolution line/area scans anywhere on a wafer and thus the tool allows the user to map even very confined areas such as test pads in scribe lines, which can not be measured using a conventional FPP. With its micron-size probes and nanoscale positioning the M4PP is a natural choice for the assessment of the "critical diameter" that coincides with the corrected scan plane of all implanter types. This can be used to assess new recipe setups as well as to verify possible drift in that setup after longer wafer runs.

CONCLUSIONS

In this work we investigate the advantages of using M4PP technique for dose and uniformity control in comparison with the conventional FPP. In general we find that conventional FPP measures lower Rs values than M4PP which has previously been reported to be related to probe and substrate leakage current present for conventional FPP measurements. The M4PP technique utilizes micron-size probes and spacing between them, thus allowing for very localized measurements. This can give high resolution Rs information in tight areas of the wafer and can help to identify and trouble-shoot process problems related to implant and/or anneal. The zero probe penetration and micron-size probe pitch ensure more accurate and reliable measurements and overcome the shortcomings of a conventional FPP. Furthermore we demonstrate the repeatability and reproducibility of the Capres M4PP system to be below 0.1 %.

The identification of commercial instruments is to specify the experimental conditions and does not imply any NIST endorsement or recommendation that it is necessarily the best instrument for the purpose.

REFERENCES

1. C. Yarling, M. I. Current, "Ion Implantation Process Measurement, Characterization and Control", in *Ion Implantation Science and Technology*, J.F. Ziegler editor, Yorktown, NY, 692, 1996.
2. T. Clarysse, A. Moussa, F. Leys, R. Loo, W. Vandervorst, M. C. Benjamin, R. J. Hillard, V. N. Faifer, M. I. Current, R. Lin, and D. H. Petersen, Materials Research Society Symposium Proceedings 912, 197 (2006).
3. C. L. Petersen, R. Lin, D. H. Petersen, P. F. Nielsen, *Micro-Scale Sheet Resistance Measurements on Ultra-Shallow Junction* 14th IEEE Conference on Advanced Thermal Processing of Semiconductors Proceedings, IEEE, Piscataway, NJ, 153, 2006.
4. D. H. Petersen, R. Lin, T. M. Hansen, E. Rosseel, W. Vandervorst, C. Markvardsen, D. Kjær, and P. F. Nielsen, *Journal of Vacuum Science and Technology B: Microelectronics and Nanometer Structures* **26**, 362 (2008).
5. P. M. Kopalidis, B. S. Freer, M. Rathmell, *Journal of the Electrochemical Society* **152**, G623, (2005).
6. S. Falk, R. Callahan, P. Lunquist, "Accurate Dose Matching Measurements Between Different Ion Implanters", 11th International Conference on Ion Implantation Technology Proceedings, E. Ishida, S. Banerjee, S. Mehta, T. C. Smith, M. Current, A. Tasch editors, IEEE, Piscataway, NJ, 268, 1997.

Photothermal Investigation of Micro-Uniformity Problems Caused by Different Scan Systems

Hans Geiler[1], Klaus Brand[2], Hans-Joachim Selle[3]

[1] *JenaWavve GmbH, Konrad-Zuse-Str.5, D-07745 Jena, Germany*
[2] *RubitecGmbH, Universitätsstr.150, D-44801 Bochum, Germany*
[3] *Xfab Semiconductor Foundries AG, Grenzstrasse 28, D-01109 Dresden, Germany*

Abstract. To study beam scanning and beam profiling effects low energy implants of Boron (25 keV) and high energy implants of Helium (5.4 MeV) were carried out by use of different scanning systems including mechanical, electrostatic and hybrid scanning. The sensitivity of photothermal measurement by use of the excess carrier wave in the depth up to 50 µm is proved for buried damage detection and compared with the effect in shallow damage profiles. The micro-mapping capability of the photothermal techniques allows the detection of dose variations in a sub-mm-scale without Moiré effects from mapping steps. Conclusion for advanced dose monitoring by multi-frequency photothermal methods will be derived.

Keywords: dose control, photothermal metrology, scanning effects.
PACS: 61.72U-, 41.85Ew.

INTRODUCTION

The uniformity of the ion dose distributed over the whole wafer represents a basic requirement of any ion implanter and needs continuous progress in advanced control techniques. In the low-dose range the photothermal heterodyne technique allows the evaluation of damage profiles by the response of excess carrier and thermal waves immediately after ion implantation by wafer mapping with different modulation frequencies [1],[2]. The shrinking of the edge exclusion on the wafer and the scaling of the devices bring up some old problems of beam scanning related pattern formation [3]. Non-uniformities in cm-scale across the wafer or in the wafer edge region cause differences between the ICs and reduce the yield. Micro-non-uniformities in mm-scale and below cause parameter variations inside the chip and this may result in a total loss [2]. This paper expands the well known photothermal heterodyne techniques with the classical wafer mapping into the field of monitoring micro-non-uniformities by microscan-distances of about 50 µm (photothermal diffusion length limited) and avoids Moirè effects of mapping [2],[4]. With the two examples of a medium current low energy and a low current high energy implantation, respectively, the method of optimized control recipes will be demonstrated: (1) the parallel ribbon beam hybrid scanner with electrostatic scanned x-direction and mechanical scanned y-direction and (2) an electrostatic x-y-scanning machine for MeV-implantation. In case (1) the dose correction during scanning may cause stripe like dose variations on the wafer detectable by photothermal methods and by parameter variations of the sensitive devices in the test structures [5]. By photothermal micro-mapping in the mm-range with step widths of 50µm strong ripples in the micro-uniformity were found. In case (2) the application of a small beam diameter of about 3mm contains the danger of Lissajous pattern formation especially at small doses [3]. To exclude channeling effects different twist and tilt versions were realized and the revealed dose pattern is then compared with simulations.

Photothermal Heterodyne Monitoring by Full and Partial High Resolution Wafer Mapping

The photothermal heterodyne technique combines the complex response detection (real and imaginary part or phase and amplitude) versus modulation frequency with the mapping procedures for lateral dose control [4]. In the case of standard mapping point distances of 0.5 to 1 cm the danger arises to oversee micro-uniformity problems with lateral dimensions of 1mm (Moiré problem). Of course high resolution techniques like QC-mapping can be used on special prepared monitor wafers [6]. The advantage of photothermal measurement to be applicable directly on the product wafer immediately after ion implantation without additional preparation steps is somewhat compensated by the recording time of one map point of about 300 ms. To avoid throughput problems, a

CP1066, *Ion Implantation Technology,* edited by E. G. Seebauer, S. B. Felch, A. Jain, and Y. V. Kondratenko

combination of a full wafer map and micro-map in chip dimensions or in defined wafer areas can be applied. So a x-y-full wafer map with 200 map points and a 3 mm edge exclusion can be combined with a polar map in the 5 mm near the edge region running up to the facet boundary at the wafer edge or micro-mapping of test areas are used in product wafers to check scan errors or glitches in the scan control. For all investigations a standard TWIN-SC4 of the company PVA TePla AG was used, which supplies the real and imaginary part of the conversion coefficient K and the DC-reflectivity R. The local excitation is carried out by an intensity modulated laser beam (wavelength 785 nm, absorption depth about 8 μm in silicon) focused on the sample with a spot radius of 3 μm. The modulation frequencies vary between 150 kHz and 15 MHz to change the penetration depth of the excited carrier and thermal waves in the semiconductor. This special resolution limit can be shifted to about 10 to 20 μm, if modulation frequencies of 15 MHz and more will be used for the local excitation of the semiconductor. But, in the same moment the problem of sensitivity arises for high energy implants with a damage depth of about 30 μm, which could be overcome by use of modulation frequencies of about 100 kHz. As shown in [1] the deep profiles of MeV-implants need lower modulation frequencies less than 1 MHz to detect the buried radiation damage. The compromise in the current problem was found to measure the deep profile of He-implants (maximum in 27 μm depth) with 450 kHz and the shallow profiles of B (damage maximum 100 nm) with 1 MHz.

For case (1) for micro-uniformity tests with ribbon ion beams with hybrid scan techniques the needs for lateral resolution were found in the range of 100 μm which is non-critical and comprises in all cases the depth profile of damage. The second monitoring demand of case (2) arises in case of electrostatic dual scanning with small beam diameters especially in the low dose range. Deduced for the applied beam diameter the Lissajous pattern is expected to be revealed with a lateral resolution of 100 μm. Hence, the photothermal resolution could be guarantied with modulation frequency of 450 kHz.

EXPERIMENTAL CONDITIONS FOR MICRO-UNIFORMIT TESTING

(1) Boron Implantation With Ribbon Beam

The ribbon beam technique with electrostatic scanning in x-direction beam angle correction by an electrostatic lens needs beam profile controller for the dose correction during scanning. A channel B-implant of 25 keV (dose $5 \cdot 10^{11}$ cm^{-2}) was carried out with an Axcelis NV 8250 HT ribbon beam implanter in the high resolution scan mode. The feedback loop from the Faraday beam profiler to the correction of the scan velocity at the electrostatic deflection plates allows an in-situ dose correction and uniformity improvement. But, glitches in the control loop may cause systematic scan errors with resulting dose patter, which can be seen in the prober results (Figure 1).

To reveal such errors earlier and be able for precise adjustments of the control loop a photothermal dose monitoring is carried out. The Figure 2 shows the full wafer image together with a micro-map. This micro-map avoids misinterpretations by Moiré-effects caused by mapping point sampling and show in this case dose oscillations between $4.5 \cdot 10^{11}$ and $6 \cdot 10^{11}$ cm^{-2} with a lateral periodicity of about 10 mm. This pattern arises with low current of 3 μm and can be reduced in periodicity and amplitude with increasing beam current to 12 μm (reduction of scans). As can be seen from the geometrical dimension of the effects, the photothermal modulation frequencies in the range 1 to 3 MHz give enough lateral and vertical resolution to be sensitive for also small dose values.

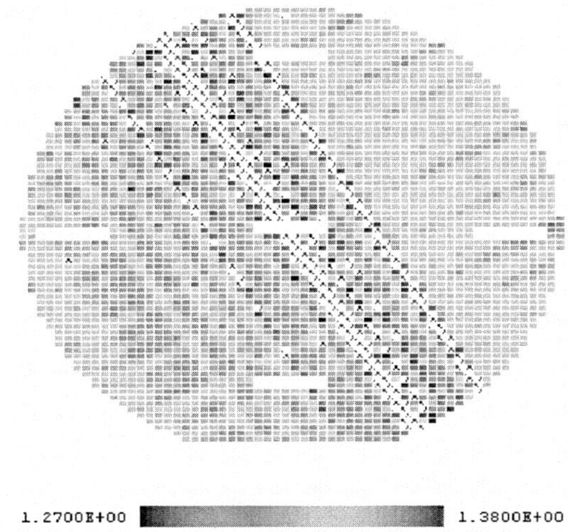

FIGURE 1. Test pattern of the final prober revealing systematic parameter deviations of the devices

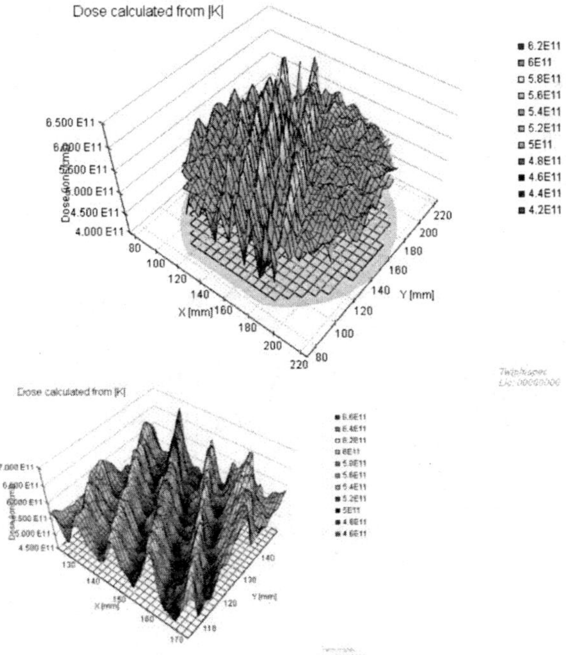

FIGURE 2. Dose monitoring by photothermal full wafer mapping (upper part and micro-mapping in a small area (lower part).

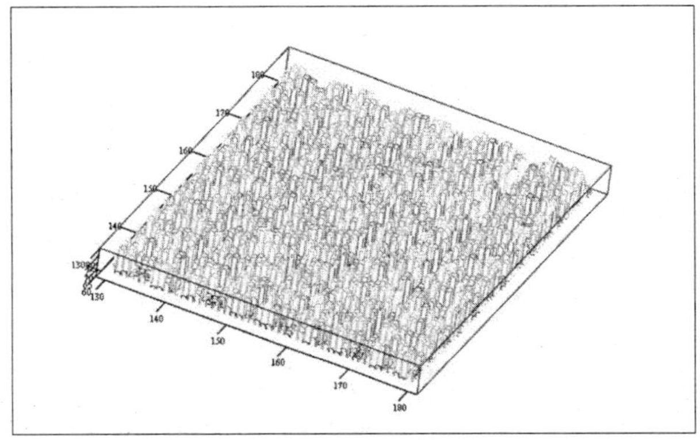

FIGURE 3. Simulated dose pattern in a micro-map with one strong periodicity in y-direction (fast scan) and a second periodic modulation with 37° inclination to the x-direction.

(2) MeV-He-Implantation with Electrostatic Dual Scan

The 5.4 MeV He^{+}-implantation was carried out with the Dynamitron tandem accelerator applying a dual electrostatic scan technique [7]. Taking into account the small mean beam diameter of 3 to 5 mm optimized ratios between the x- and-y-deflection frequencies of f_x=117 Hz and f_y=1019 Hz were selected to minimize scan pattern generation on the wafer. To avoid channeling effects the twist and tilt angles of 17° and 7°, respectively, were used. The expected micro-uniformity pattern simulated without dithering and a fixed phase of $\pi/8$ is shown in Figure 3.

This prediction was checked by 3 implants with twist angles +17°, 0° and -17°, respectively. In Figure 4 the micro-map with an alignment of the flat normal vector to the y-scan direction (fast scan) is shown. The pattern corresponds well with the simulated one in Figure 3. The cross check is realized by measuring the twisted wafers with the geometry shown in Figure 5. The pattern should rotate with the scan-directions if no crystallographic effects like channeling occur.

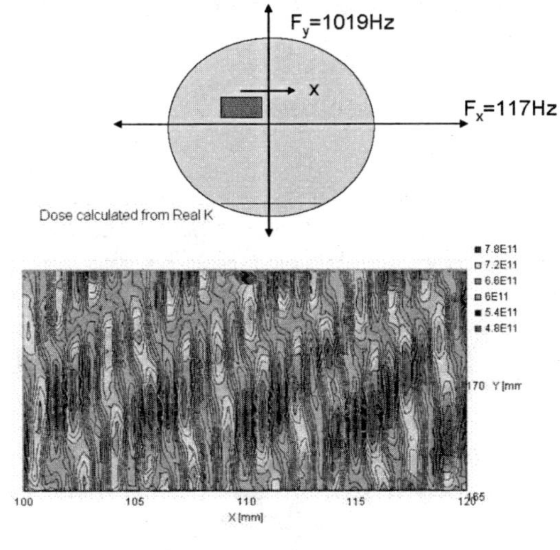

FIGURE 4. Dose image derived from the Re{K} by calibration in case of the upper scan geometry

SUMMARY

The micro-uniformity can be controlled without additional preparation by photothermal methods with the dual mapping method: standard wafermap with 1 cm mapping point distance and local micro-map with 100 μm mapping point distance. This avoids the Moiré effect of standard mapping and save time for improved throughput.

The micro-uniformity deviations in the low dose range of about 10% can be observed in cases of small beam diameters and also in case of feedback controlled ribbon beams. In the first case the pattern is

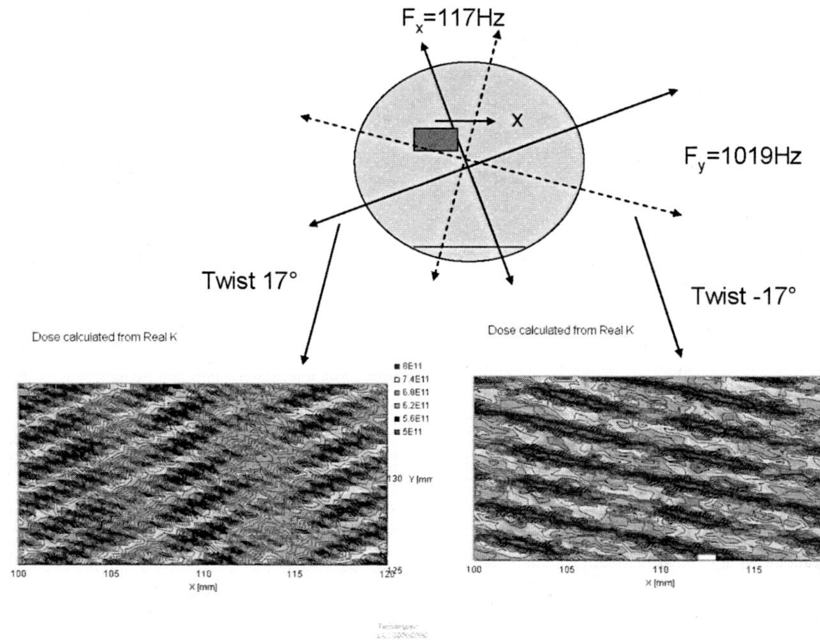

FIGURE 5. Twisted implant conditions for 5.4MeV He 6e11 cm^{-2} and the related micro-pattern of the dose.

caused by scanning effects like Lissajous pattern, because the scan pitch/beam width ratio cannot be unlimited reduced in low dose cases [2]. Beside the well selected frequency ratio [4] the phase dithering or interlacing with scan wave form shifts are recommended to avoid the pattern [8]. In the second case the dose pattern is caused by the beam profile scanning and glitches in the control loop. The dual mode dose monitoring by TWIN avoids such problems by early initiated fine tuning of the beam profile monitoring.

REFERENCES

1. K.Brand, H.Karge, H.D.Geiler; *Proc.Int. Conf. IIT-96 Austin;* Ed.by E.Ishida, L.Larson, A.Tasch etal.; Pub. IEEE Service Center 1997, p.218-221.
2. M.I.Current, N.Ohno, K.Hurley, W.A.Keenan, T.L.Quitner, Ch. Jeynes; *Solid State Technology*, **7** July 1993, p.111.
3. E.J.Rogers, NIM **189**(1981)305.
4. M.Wagner, S.Käpplinger, H.D.Geiler; *Proc.Int.Conf. Ion Implantation Technology IIT-94 in Catania*, Ed. by S.Coffa, G.Ferla, F.Priolo, E.Rimini; Elsevier 1995, p. 642-645.
5. U.Sobe, K-H.Rooch and D.Mörtl, "Simulation and Analysis of Analog Circuit and PCM (Process Control Monitor) Test Structures in Circuit Design", *Proc. DASS 2007*, Dresden, published in: Wissenschaftlichen Schriftenreihe "Eingebettete, Selbstorganisiernde Systeme", Band 5, p. 55 – 59.
6. C.Krueger; in *Ion Implantation Technology* (proc. of 16. conf IIT 2006 in Marseille), ed. By K.J.Kirkby, R.Gwilliam, A.Smith and D.Chivers, AIP, NY 2006, p.531.
7. K.Brand; *Proc.Int.Conf. Ion Implantation Technology IIT-94 in Catania*, Ed. by S.Coffa, G.Ferla, F.Priolo, E.Rimini; Elsevier 1995, p. 458-461.
8. S.R.Walter, B.Pedersen, P.Murphy, W.Goodenough; *Ion Implantation Technology IIT-2000 in Alpach,* Ed. by H.Ryssel, L. Frey, J. Gyulai and H.Glawischnig, IEEE 2000, p. 427-430.

High-Resolution Mapping of Low-Dose Implants

Jeffri Halim[1], Akira Mineji[2], Vladimir Faifer[1], Michael Current[3]

[1] *Frontier Semiconductor, 199 River Oaks Parkway, San Jose, CA 95134*
[2] *NEC Electronics Corp., 1120 Shimokuzawa, Sagamihara, Kanagawa 229-1198 Japan*
[3] *Current Scientific, 1729 Comstock Way, San Jose, CA 95124*

Abstract. High spatial resolution sheet resistance measurements, based on analysis of junction photo-voltage (JPV) signals, are used to reveal subtle dose uniformity variations for 40 and 90 keV Boron implants in the dose range of 10^{11} B/cm^2. The use of various combinations of electrostatic ion beam scanning with linear and rotational wafer motion resulted in dose variations with patterns characteristic to the particular combination of scanning methods. Detailed wafer maps, based on over 70,000 pixel sites per wafer, were obtained with JPV methods for sheet resistances approaching 1 M-Ohm/square, providing for evaluation of local and global dose uniformity performance.

Keywords: Dose uniformity, junction photo-voltage, sheet resistance.
PACS: 85.40.Ry, 72.40.+w

INTRODUCTION

The use of mapping of sheet resistance, Rs, to diagnose uniformity effects in implantation and annealing is now over 30 years old [1], beginning with 4-point probe measurements with ≈120 points per 100 mm wafer [2] to the present day use of non-contact junction photo-voltage (JPV) probes with over 70,000 measurement pixels per 300 mm wafer [3]. While difficulties with establishing Ohmic contacts for lightly doped surfaces has generally limited the routine use of contact 4-point probes to implant doses greater than 10^{13} ions/cm^2, non-contact JPV measurements of Rs, based on carrier spreading, do not have such limitations [4].

Control of low dose, below 10^{13} ions/cm^2, implants has long been important for the manufacture of high-precision resistors and setting threshold voltage in long-channel CMOS transistors. More recently applications for low dose implants have diversified to include formation of deep wells for high-voltage switches for automotive and other applications and optical imaging arrays.

The purpose of this work is to explore the impact of ion beam scanning and wafer motion on the dose uniformity of implants in the 10^{11} ions/cm^2 range with the use of high spatial resolution JPV measurements.

EXPERIMENTAL DETAILS

40 and 90 keV boron ions were implanted at doses from 10^{11} to $5x10^{13}$ B/cm^2 using two implanters, one with x-y electrostatic beam scanning, with an option for continuous wafer rotation during the implant, and one with x-scanning and y-direction linear motion (see Table 1). Lightly-doped, 150 and 200 mm n-Si(100) wafers, with resistiviy of ≈30 Ohm-cm, were used to avoid the formation of counter-doped surface junctions for all but the lowest dose implants (Figure 1).

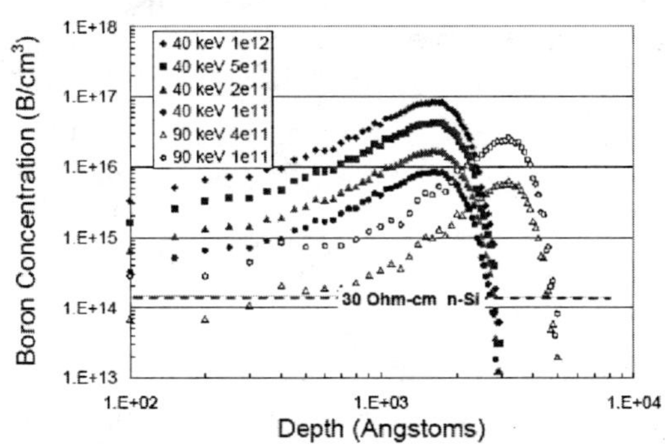

FIGURE 1. TRIM simulations of 40 and 90 keV B as-implanted profiles at various doses. Also shown is the estimated doping level for 30 Ohm-cm n-Si wafers.

CP1066, *Ion Implantation Technology*, edited by E. G. Seebauer, S. B. Felch, A. Jain, and Y. V. Kondratenko

TABLE 1. Implant conditions and RsL results

Diameter	E (keV)	Dose(B/cm^2)	Beam scanning	Wafer motion	Rs(Ohm/sq)	Sigma (%)
150 mm	90	4.10E11	x-y scan	no action	57,671.3	2.36
150 mm	90	4.10E11	x-y scan	continuous rotation	54,678.6	2.38
150 mm	90	4.10E11	x-y scan	no action	54,236.2	2.27
150 mm	90	4.10E11	x scan	y scan	53,653.5	1.45
150 mm	40	2.00E11	x-y scan	continuous rotation	JPV only	7.65
150 mm	40	2.00E11	x-y scan	no action	JPV only	5.84
150 mm	40	2.00E11	x scan	y scan	JPV only	8.74
150 mm	100	5.00E+13	x-y scan	no action	840.0	1.42
200 mm	90	1.E+11	x scan	y scan	NA	NA
200 mm	40	1.E+12	x scan	y scan	24,173.7	1.73
200 mm	40	5.E+11	x scan	y scan	66,774.7	1.50
200 mm	40	1.E+11	x scan	y scan	NA	NA

The wafers were annealed at 1060 C for 10 s in an RTP.

Measurement Details

Spreading of photo-induced carriers, measured with a dual electrode probe spaced ≈1 mm above the test wafers, was used to determine the sheet resistance of the p-n junctions [4-7]. This method measures junction Rs and carrier recombination and trap-assisted tunneling leakage currents, L, and is referred to as RsL measurements [6,8].

For these measurements a focused red laser-diode beam (660 nm wavelength, absorbtion depth in Si ≈7.8 um) with diameter <0.1 mm was used to obtain high spatial resolution Rs measurements. The spatial resolution was mainly determined by the diameter of light beam and the pixel-to-pixel step size. The wafers were measured using from 0.4 mm up to 1 mm step size, with pixel counts from 30,000 to 72,000 points per wafer. The time of measurement per wafer varied from ~4 minutes for 1,000 points, 30 minutes for 30,000 points to approximately 1 hour for a 72,000 point Rs map.

Wafer Maps

An beam scanning synchronization ("lock up") pattern is clearly visible in the 30k pixel map in Figure 2 and horizontal Rs line scan in Figure 3 for a 4x10^{11} B/cm^2 implant with an electrostatic x-y scanned beam although the overall Rs uniformity is modest, ≈2.4 %. The background concentric pattern is due to imperfections in the planarity of the stage rotation, slightly changing the wafer surface to probe electrode distances and the JPV signal used in the Rs analysis as it "wobbles".

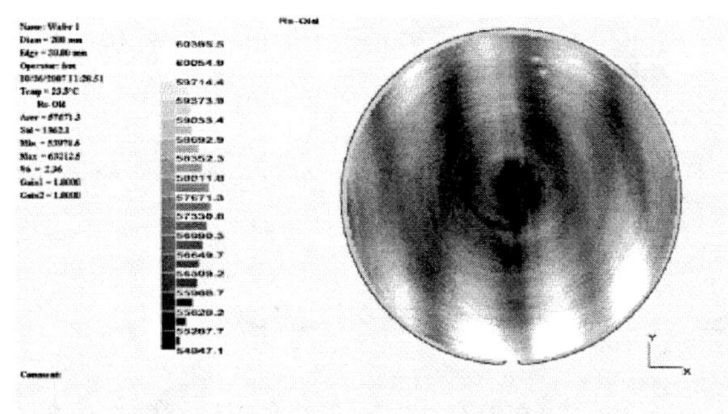

FIGURE 2. 30k pixel Rs map (0.6 mm^2 pixel area) of a 90 keV B, 4x10^{11} B/cm^2, 3.2 uA implanted 150 mm wafer with and x-y beam scan. Sheet resistance was 57,671.3 Ohms/sq. with a uniformity of 2.36%.

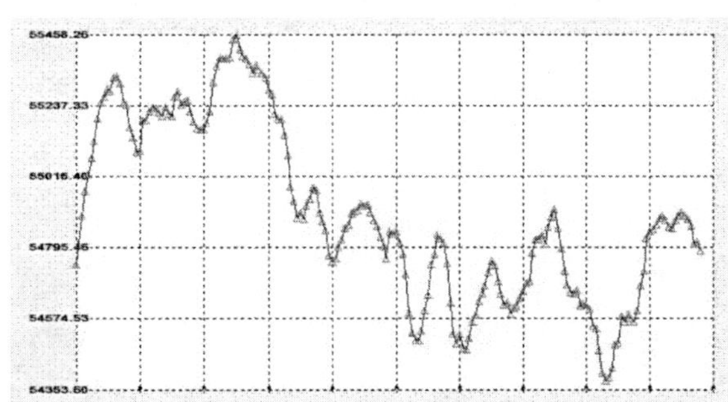

FIGURE 3. An Rs line scan across the center of the wafer shown in Figure 1. Rs =54,905.9 Ohms/sq., uniformity =0.5 %.

Using the same implant tool as the wafer shown in Figure 2, a lower dose implant with continuous wafer rotation during implant combined with an x-y beam scan resulted in a complex "standing wave" pattern (Figure 4). Similar patterns, with lower spatial frequencies, have been reported with the combination of beam and wafer scanning with wafer rotations [9].

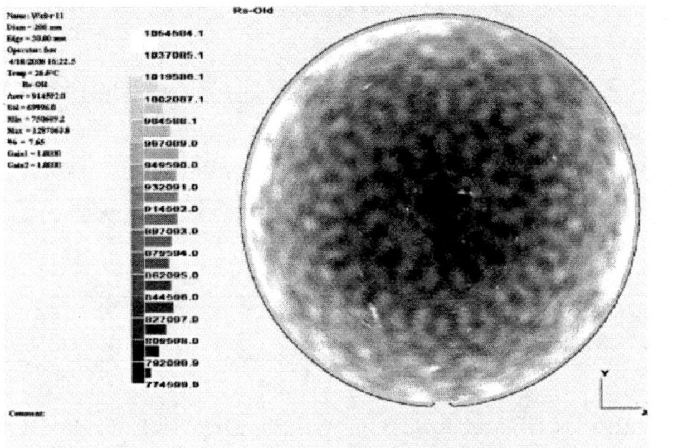

FIGURE 4. 72k pixel JPV map (0.24 mm^2 pixel size) of a 40 keV B, $2x10^{11}$ B/cm^2, 1.6 uA implanted 150 mm wafer with and x-y beam scan with continuous wafer rotation, with a uniformity of 7.65%.

When the continuous wafer motion is not used, just the x-y beam scanning, the "zig-zag" dose uniformity pattern, similar to the case in Figure 2, returns (Figure 5).

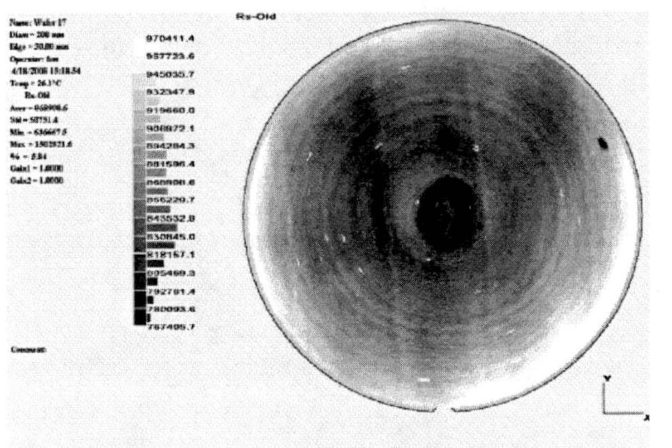

FIGURE 5. 72k pixel JPV map (0.24 mm^2 pixel size) of a 40 keV B, $2x10^{11}$ B/cm^2, 1.6 uA implanted 150 mm wafer with and x-y beam scan with no wafer rotation, with a uniformity of 5.84%.

Implants done at a dose of $4x10^{11}$ B/cm^2 with an implant tool with x-direction beam scanning and y (vertical) wafer motion showed clear horizontal stripping patterns suggesting variations in the beam current levels, beam size or beam sweep speeds between successive passes over the wafer (Figure 6). Horizontal dose variations have also been linked to non-steady wafer motion driven by vibration sources within an implanter [10].

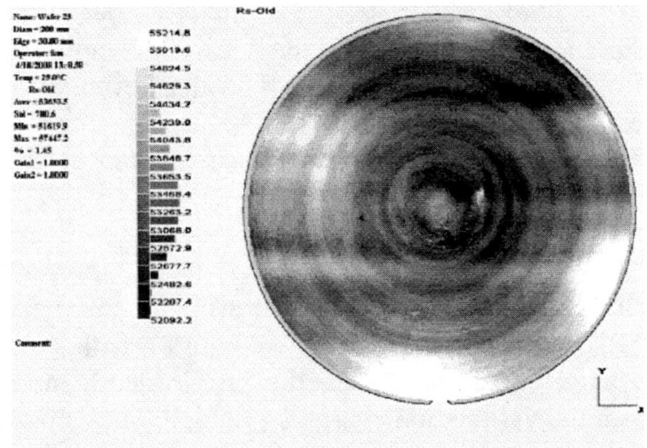

FIGURE 6. 72k pixel Rs map (0.24 mm^2 pixel size) of a 90 keV B, 4e11 B/cm^2, 6 uA implanted 150 mm wafer with an x beam scan with y wafer motion. Sheet resistance was 53,653.5 Ohms/sq. with a uniformity of 1.45 %.

A near repeat of the implant conditions shown in Figure 6, with a 200 mm wafer and minor adjustments to the scanning conditions, showed no evidence of the horizontal Rs bands (Figure 7). The remaining features in the Rs map suggested the effects of "water spots" from post-implant cleaning.

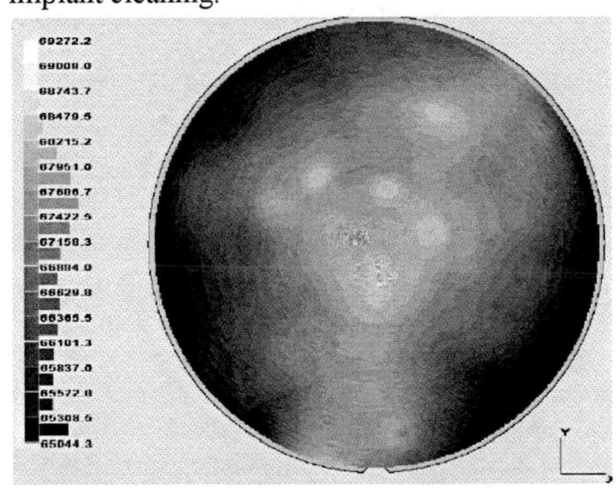

FIGURE 7. 72k pixel Rs map (0.42 mm^2 pixel size) of a 40 keV B, 5e11 B/cm^2 implanted 200 mm wafer with an x beam scan and y wafer motion. Sheet resistance was 66,774.7 Ohm/sq. with a uniformity of 1.5 % (compare to Figure 6)

DISCUSSION

The implants at doses of less than $4x10^{11}$ B/cm^2 did not give measurable Rs values. The high sheet resistance for $2x10^{11}$ B/cm^2 and less, greater than 300 k-Ohm/square, was beyond the dynamic range of the dual electrode probe and amplifiers (strong JPV signals were measured from the central electrode but no JPV signal above noise levels for the outer electrode). However, a stable JPV signal was measurable on the central RsL electrode directly over the illumination spot. Since the local JPV signal is directly proportional to the Rs value, even these very low dose cases still can be used to provide dose uniformity values and maps of scan effects (Figures 4 & 5).

The Rs values in the $4x10^{11}$ to $5x10^{13}$ ions/cm^2 range vary with implant dose with a sensitivity, (ΔRs/ΔDose), close to unity (Figure 8), with some detailed variations due to the profile depth range from the various ion energies (40 to 100 keV) used in this study (see Figure 1). So the dose variations observed in these implants are similar to the measured Rs and JPV variations, in a range from about 1.5 to 9 %.

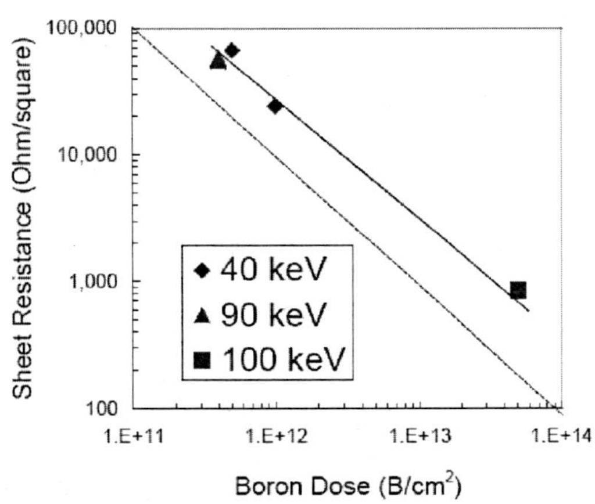

FIGURE 8. Rs values vs dose for a variety of Boron energies. Note sensitivity of about 1.

The scan synchronization effects seen here are characteristic of scanned ion beam systems. Even when the drive frequencies for the scanners are chosen to be non-rational multiples of each other [11], to avoid direct Lissajous pattern overwriting, the ion beam, being a lossy plasma, responds with a range of frequencies. So there is always some level of beam overlap occurring, depending on the frequency response of the ion beam and the drive frequencies used in the scanners. These residual scanning patterns resulting from dynamic effects in the ion beam can be compensated to some degree by various "target repositioning" techniques [12]. However when these "quad implant" methods are extended to the use of continuous rotation during the implant, resonances between the ion scan and wafer rotation frequencies result in additional, more complex, dose variation patterns (Figure 4) [9].

SUMMARY

Dose uniformity patterns arising from ion beam scanning and wafer motion synchronization, although modest, are now measurable with non-contact JPV techniques for doses in the 10^{11} ions/cm^2 range and sheet resistances approaching the 100 k-Ohm/square regime.

High spatial resolution Rs and JPV mapping, with pixel sizes of less than 1 mm and pixel counts of >70,000 per wafer, provides detailed evaluation of both local and global dopant activation uniformity arising from implant and anneal conditions. This provides a new and useful means for improvement in low-dose uniformity and dose controls.

ACKNOWLEDGEMENTS

We acknowledge the valuable contributions of our co-workers, Dr. Seiichi Shishiguchi and Mr. Norihiko Matsuzaka of NEC and Dr. Naotsugu Ohno of FSM to this work.

REFERENCES

1. D. S. Perloff, *J. Electrochem. Soc.* **123** (1976) 1745.
2. M. I. Current, D.S. Perloff, L.S. Gutai, in Ion Implantation Techniques, eds. H. Ryssel, H. Glawischnig, Springer (1982) 235-254.
3. V. Faifer, M. Current, N. Ohno, J. Halim, J. Lu, ECS Trans. 13-1 (2008) 37-43.
4. M. I. Current *et al.*, a tutorial chapter on "New Metrologies for Implant Process Controls", in Ion Implantation Science and Technology, ed. J.F. Ziegler, IIT Press (2008).
5. V. N. Faifer *et al.*, Mat. Res. Soc. Symp. Proc. Vol.810 (2004) C11.9.
6. V. N. Faifer *et al., J. Vac. Sci. Technol.* **B25** (2007) 1588-1592.
7. T. Clarysse *et al., J. Vac. Sci. Technol.* **B26** (2008) 420-424.
8. V. N. Faifer, M.I. Current, D.K. Schroder, Appl. Phys. Lett. 89 (2006) 151123.
9. N. Nagai, *Nuc. Intsr. Meth.* **B55** (1991) 253-256.
10. J. M. Towner, D. C. Slusher, these Proceedings.
11. N. Turner, in Ion Implantation: Equipment and Techniques, eds. H. Ryssel, H. Glawischnig, Springer (1982) 126-142.
12. N. Turner, *Nuc. Instr. Meth.* **B6** (1985) 224-227.

Improved R_s Monitoring for Robust Process Control of High Energy Well Implants

J. H. Kim[1], S. Kim[1], G. J. Ra[1], K.W. Lee[1], R. N. Reece[1], and S. Y. Bae[2]

[1]*Axcelis Technologies, Inc., 108 Cherry Hill Drive, Beverly, MA 01915, USA*
[2] *Hynix – ST Semiconductor Limited., K7 Plot, Export Processing Zone, ZiangSu, Wuxi, China*

Abstract. We describe a robust method of improving precision in monitoring high energy ion implantation processes. Ion implant energy accuracy was measured using an unpatterned implanted layer on an intrinsic *p*-type silicon wafer. To increase Rs sensitivity to energy at the well implant process, a p-n junction structure was formed by p-well and deep n-well implants into the *p*-type Si wafer. It was observed that the depletion layer formed by the p-n junction was very sensitive to the energy variation of the well implant. Consequently, it can be recommended to monitor well implant processes using the Rs measurement method described herein.

Keywords: Sheet resistance (Rs), PN junction, HE implanter, LINAC, FEM calibration

INTRODUCTION

As devices are scaled down to sub-micron dimensions, strict control of the implant process is needed to support yield improvement and to respond quickly to yield excursions. Process control and device matching as well as device yield depends on the accurate control of implant dose and energy at the retrograde well implant step. To meet this requirement, implant process monitoring is traditionally carried out by several methods including Therma Probe® (TW), measure of silicon lattice damage, Sheet Resistance (Rs), a measure of the resistance of a doped and activated silicon layer, and Secondary Ion Mass Spectroscopy (SIMS), a profiling technique of dopant concentration vs. depth. Therma Probe measurement is non-destructive and needs no additional process steps after implantation. While TW is easier to perform than sheet resistance (Rs) techniques, Rs measurement is the dominant implant monitoring technique since Rs is directly related to the electrical characteristics of devices. All three techniques remain useful and popular, each with specific and different sensitivities and limitations.

Well implant processes in sub-micron devices can be particularly sensitive to energy variation, whether from voltage offsets in conventional DC high energy implanters, or from final energy magnet (FEM) calibration on RF linear accelerator systems [1]. Energy variations of well implants cannot be easily monitored by conventional measuring methods since energy sensitivity is quite a bit lower than dose sensitivity [2]. In this paper, we introduced a P-Well /Deep N-Well diode junction structure not only to detect such energy variations of implanters, but also to match N or P-well Rs values in production devices.

EXPERIMENT

Boron and Phosphorus species at each P-well and Deep N-well energy were implanted into crystalline Cz, *p*-type bare Si (100) using an implanter with an RF linear accelerator. The post implant anneal (PIA) was carried out at 1000°C in a rapid thermal annealer that is standard to conventional CMOS processing. All anneals were performed in an N_2 ambient. Sheet resistance values were measured using a four-point probe.

Rs Comparison and Monitoring With Bare Si Wafers

We compared the P-well Rs trend in Deep N-well of ~ 90nm DRAM devices manufactured with three similar high energy implanters. The P-well Rs of devices implanted on implanter A shows higher Rs than the other two systems, B and C.

Comparative data is shown in Figure 1. Implanter A shows Rs approximately 5% higher for the same recipe, and had been matched using standard Rs techniques using just one implant step. This result reveals that conventional Rs matching by a single implant dose split test on Si wafers is not always an accurate method for matching between implanter machines.

CP1066, *Ion Implantation Technology*, edited by E. G. Seebauer, S. B. Felch, A. Jain, and Y. V. Kondratenko

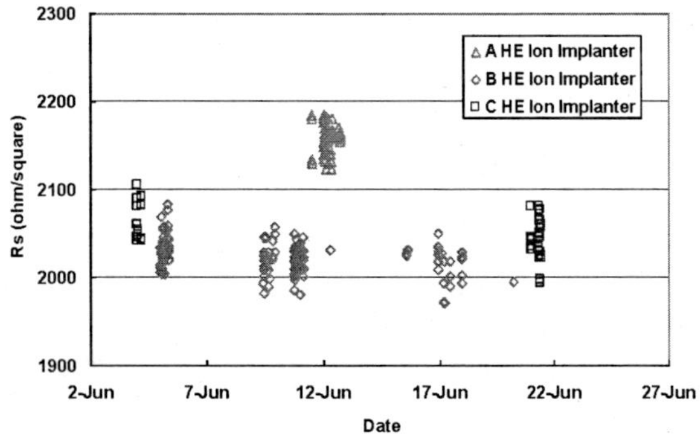

FIGURE 1. P-well Rs trend of device lots from each high energy implanter.

The results indicated that these unmatched Rs characteristics appeared at only the triple well structure, which couples the P-well and deep N-well. Figure 2 shows the overlapped SIMS profiles of implant 400keV $^{11}B^{+}$ ions, $1.0x10^{13}$ ions/cm^2 by implanters A, B, & C. The profiles do not reveal any significant differences in energy (depth or Rp) and dose among three systems by SIMS analysis.

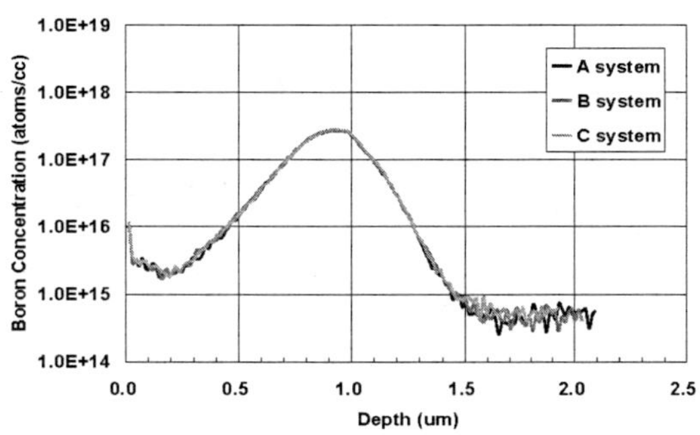

FIGURE 2. SIMS comparison of P-well implants of 400keV $11B^{+}$ ions to a dose of $1.0x10^{13}$ ions/cm^2 on the bare test wafers.

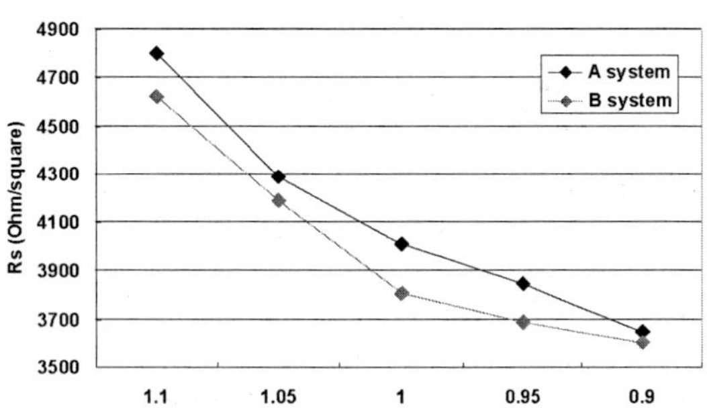

FIGURE 3. Rs comparison by measuring PN junction structure (P-well/deep N-well) with energy split tests between A and B implanters. Annealing perfomred at 1000℃ 10sec N2 condition

Finally we designed a junction diode structure in the bare Si wafers which coupled P-Well implants 400keV $^{11}B^{+}$ ions at a dose of $1.0x10^{13}$ ions/cm^2 to a deep N-well implant of 1.2 MeV $^{31}P^{+}$ ions to a dose of $1.0x10^{13}$ ions/cm^2. We compared Rs values of the energy split test and RTA at 1000°C and 10s, as shown in Figure 3.

We demonstrated that the Rs value increased with increasing P-well implant energy, and Rs decreased with decreasing P-Well energy. It indicated that the Rs in P-Well region could be precisely controlled by P-type dopant concentration except that which was overlapped and compensated with N-type dopant in deep N-well region, by controlling the energy of P-well implant.

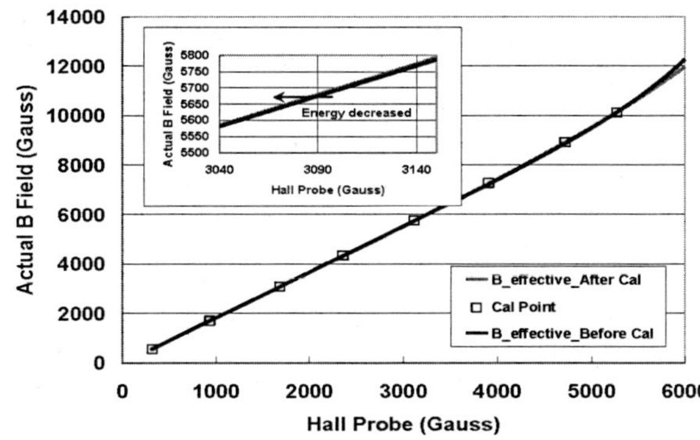

FIGURE 4. P-well energy was decreased by FEM (final energy magnet) calibration for implanter A. The inset shows that hall probe gauss measurement decreased the P-well energy up to 10 keV.

In the RF LINAC based high energy system, final energy is determined by FEM calibration performed on each implanter. Calibration is carried out by using a DC ion beam of 80keV Ar^{+}, 10keV B^{+}, 270keV Xe^{+++}, and 210keV Xe^{+++} [1]. Figure 5 shows the P-well Rs matching after reduction of 10keV on system A by such a FEM calibration.

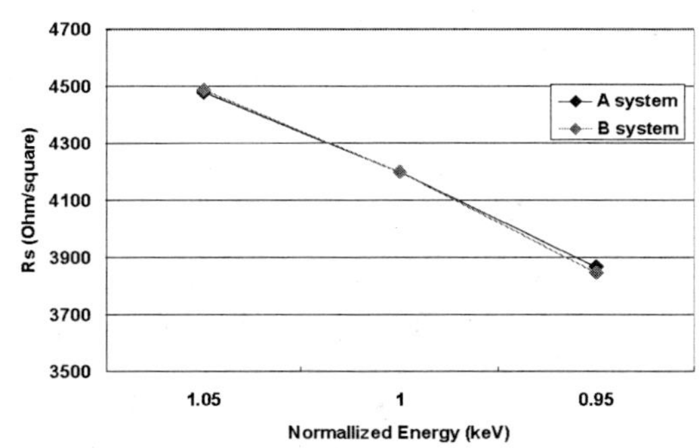

FIGURE 5. Rs monitoring results after FEM calibration (energy reduced about 10 keV on implanter A). The anneal was 1000℃ 10sec N_2 after P-well/deep n-well implant performed on the bare test wafers.

The Rs trend of 90 nm DRAM devices after FEM calibration shows the energy is well matched between the three HE implanters as shown in Figure 6.

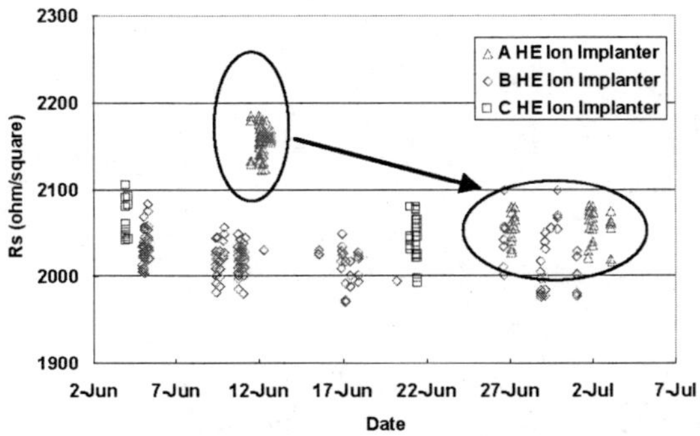

FIGURE 6. Rs trend comparison of device lots for each high energy implanter. The arrow indicates Rs trend is successfully changed by the FEM calibration of an implanter.

CONCLUSIONS

We conclude that an Rs monitoring method by a p-n junction structure on a bare Si wafer is a more sensitive tool to detect energy variation and to match multiple implanters in a high volume manufacturing fab.

REFERENCES

1. Noriyuki Suetsugu, Hiroyuki Kariya, Mitsuaki Kabasawa and Michiro Sugitani, in *Proceedings of the International Conference on Ion Implantation Technology* 2000, pp. 448-451, Sep. 2000.
2. Dennis Kamenitsa and Peter Lillian, IBID, pp. 674-677.

Evaluation of Laser Scattering Technique and Spectroscopic Ellipsometry for In-Line Ion Implantation Characterization

F. Milesi[a], E. Nolot[a], Z. Mehrez[a], F. Mazen[a], and S. Favier[b]

[a] *CEA-LETI MINATEC, 17 rue des Martyrs, 38054 Grenoble Cedex 9, France*
[b] *ST MICROELECTRONICS, 850 rue Jean Monnet, 38920 Crolles Cedex, France*

Abstract. Today, most semi-conductor fabrication units are looking for fast and cheap in-line characterization tools to develop and monitor each process step. In this study, we evaluated new ways to characterize as-implanted process by ion implantation with equipment typically dedicated for other applications. First, we will demonstrate how sensitive spectroscopic ellipsometry and laser scattering technique can be to slight changes in ion implantion parameters (i.e. energy, ion dose, tilt, etc.). Then, these results will be compared to classical in-line ion implantation doping characterization technique (i.e. Sheet resistance measurement). In this study, ion implantation has been realized on medium current ion implanters (E500 from Varian and NV-8200P from Axcelis). Two different species have been implanted (one heavy ion: Arsenic and one light ion: Hydrogen) into silicon with various energy, ion dose, tilt and beam uniformity. In-line characterization was performed on as-implanted wafers, using two differents means: i) a Woollam M2000 spectroscopic ellipsometer, working in the 0.7-6.5eV range and ii) the low frequency part of the scattered light (called "haze") of KLA-Tencor SP1^{DLS} particle counter. First promising results indicate that these techniques could be used to characterize in-line low-doping and no-doping species.

Keywords: Ion implantation, In-line characterization, spectroscopic ellipsometer, scattering light haze measurements.
PACS: 06.20.-f; 42.25.Fx; 81.20.-n; 85.40.-e

INTRODUCTION

Today, there are only few means of ion implantation process characterizations. These means are off-line techniques like SIMS1. or in-line techniques like sheet resistance measurement (i.e. four-point probe2. or surface photo voltage3. methods) but this last technique is limited to doping ions and requires a post anneal in order to form an electric junction. The only in-line technique to characterize as-implanted processes is Therma wave based technique (i.e. Therma-Probe4. from KLA-Tencor or TWIN5. from PVA Tepla).

Our work proposes to evaluate the response to ion implantation of various in-line characterization techniques already used for other applications. In this study, laser scattering technique6. and spectroscopic ellipsometry7. response to various implantation conditions (species, dose, energy, etc.) have been compared with sheet resistance technique.

The interest of these two techniques is available in all fabrication units and full-map of 200/300mm wafer in less 1min.

EXPERIMENT

200mm single crystal, (100)-oriented silicon was implanted with either doping species using an Axcelis NV-8200P medium current ion implanter and Hydrogen using a Varian E500. The data reported here concerns two ions (one light ion: Hydrogen and one heavy ion: Arsenic), implanted with various energy, dose and beam uniformity.

Two different techniques have been evaluated as ion implantation characterization method:

- the low frequency part of the scattered light, called "haze", from KLA-Tencor SP1^{DLS} particle counter,
- a Woollam M2000 spectroscopic ellipsometer.

For doping processes, all wafers have been annealed with a Axcelis RTP850 furnance at 1000°C, and the sheet resistance has been measured with a Napson WS3000, in order to compare sheet resistance measurements and haze results.

CP1066, *Ion Implantation Technology,* edited by E. G. Seebauer, S. B. Felch, A. Jain, and Y. V. Kondratenko

Laser Scattering Technique

The wafer is illuminated (Figure 1) by a laser (wavelength = 488nm), with a beam either normal to the surface or oblique (~70° to the normal). The scattered light, coming from different defects and the background signal of the surface is collected by two separate channels (toward some photomultipliers), which differ in position and size: the wide collector (elliptic mirror with an angle of between 25° and 70° to the normal) and the narrow collector (lens with a angle of between 5° and 25° to the normal) channels.

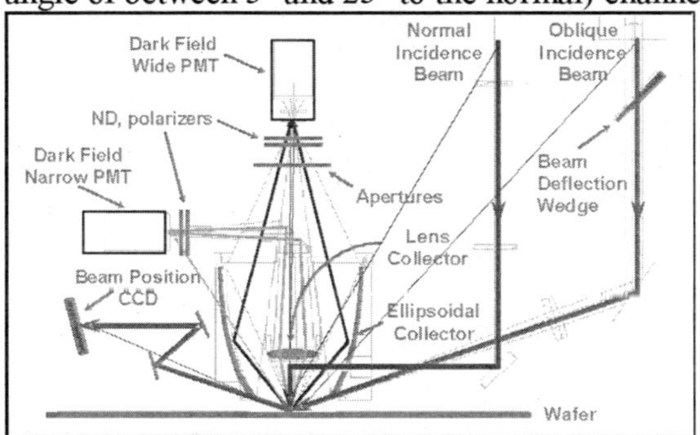

FIGURE 1. Principle of KLA-Tencor SP1^{DLS}

$$\text{Haze} = \frac{\text{diffuse beam intensity by the surface} \times 10^6}{\text{incidence beam intensity}} \text{[ppm]}$$

Haze signal is very sensitive to surface conditions (roughness, thickness, optical index of the damage zone, ...). For our study, we used either Median or Average (because Median and Average data are identical except in the case of the no-uniformity beam) and Standard Deviation parameters. The best sensitivity of the haze measurements is obtained with the oblique beam and the wide collector channel except for low energy implantation (<10keV).

The figure 2 shows the same tendencies between a sheet resistance mapping and haze mapping (Figure 2).

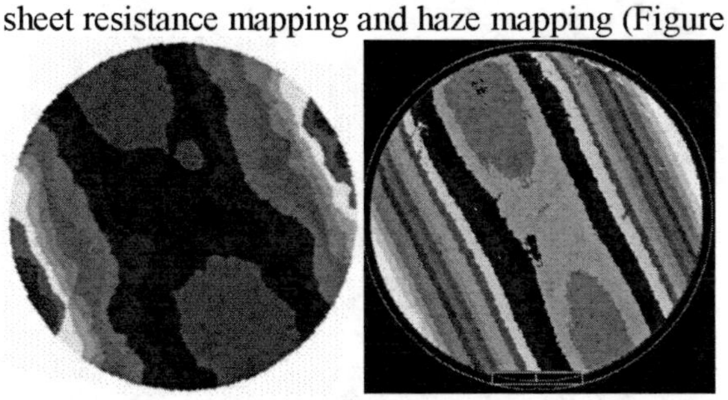

FIGURE 2. Full wafer sheet resistance and haze mapping for As+ 200keV 9E13at/cm² ion implantation process with controlled beam uniformity.

Dose Characterization

First, the beam uniformity correlation versus standard deviation of the sheet resistance technique and the haze measurements has been studied (Tab. 1). The dose uniformity has been calculated by standard calculation of the sheet resistance variation (a narrow dose variation corresponds to a linear sheet resistance variation). Results show that there is a good correlation between haze measurements and beam uniformity for As+ 200keV 9E13at/cm² process. The sensitivity of the haze measurements is better than sheet resistance technique (sheet resistance: 0,5 and haze: 3).

Then, with the same process (As+ 200keV), we have looked at the haze variation (compared to sheet resistance variation) with a wide range of dose (from 9E12 to 9E14at/cm²). The figure 3 shows that haze is a function of the ion dose implantation as sheet resistance measurements. On the contrary with H+ 76keV process with a wide range of dose (from 1E15 to 1E17at/cm²), haze measurements are also a function of the dose but reverse compared to As+ 200keV process and the sensitivity of the measurements is too low (Figure 3). Nevertheless, at only high dose (>1E16at/cm²), the twist rotation (i.e. 22°) and five of the six mechanical wafer clamps (Figure 4) are revealed by full wafer haze mapping.

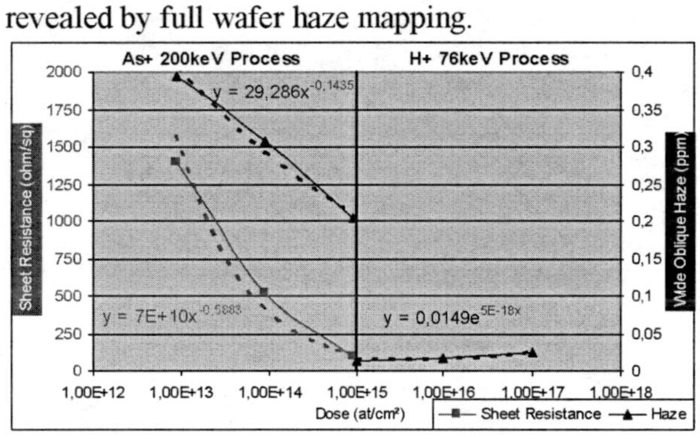

FIGURE 3. Haze and sheet resistance variation as functions of dose for two different processes

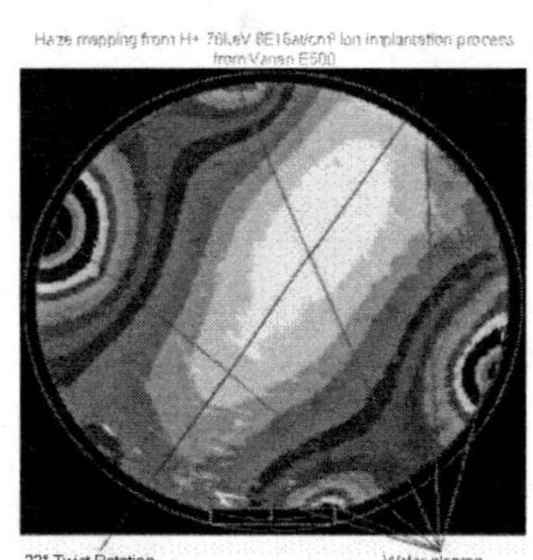

FIGURE 4. Full wafer haze mapping (wide – oblique) for H+ high dose process from Varian E500.

Energy Characterization

Secondly, we studied the As+ 1E14at/cm² process with a wide range of energy (from 5 to 250keV). Haze and sheet resistance show an evolution with energy increase, on the contrary H+ 1E14at/cm² process is not sensitive (Figure 5).

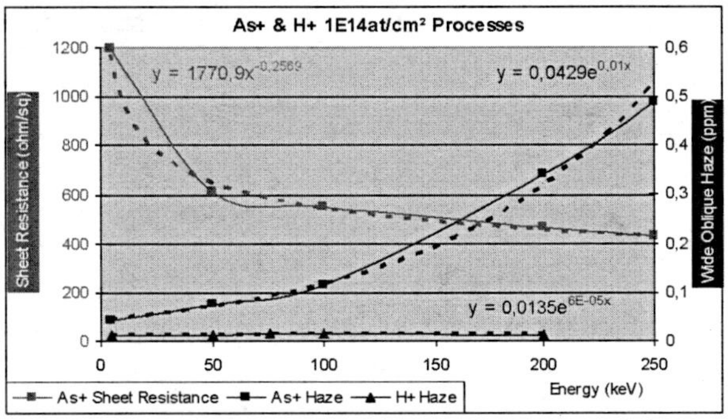

FIGURE 5. Haze and Sheet resistance variation as functions of energy for the As+ & H+ 1E14at/cm² ion implantation processes.

Species Characterization

Third, haze evolution as functions of species (implantation of four different species: Hydrogen, Boron, Phosphorous and Arsenic at 200keV and at the same Rp) has been studied and as dose and energy characterizations, the laser scattering technique is sensitive to ion implantation species (Figure 6).

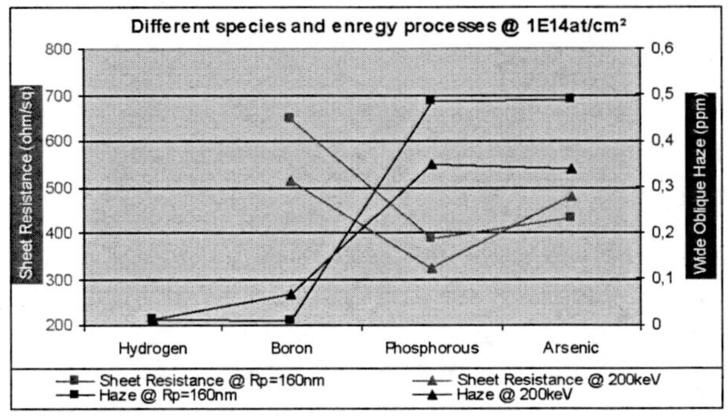

FIGURE 6. Haze and Sheet resistance variation as functions of species and energy for 1E14at/cm² ion implantation processes.

Signal of laser scattering tools enables a characterization about as-implanted processes as functions of ion implantation parameters (dose, energy or species) but only with "aggressive" conditions (heavy ions, high doses or high energies) in order to have a good sensitivity of the haze measurement. It would reveal the ion implantation damages into the crystal silicon because we have the same haze mapping before and after desoxidization but not the same measurements; and it doesn't run more after a thermal anneal.

Moreover, this technique could be a new tool to calibrate ion implanters, since mechanical twist rotation or beam centering could be accurately evaluated through haze mappings.

TABLE 1. As+ 200keV 9E13at/cm² process trough 30nm of oxide with controlled beam uniformity

Beam uniformity (%)	Standard deviation (%)					
	Haze (Oblique Wide Collector & Mediam Data)				Sheet Resistance	
	As-implanted with 30nm oxide	Dose uniformity	As-implanted after desoxidation	Dose uniformity	As-implanted after desoxidation	Dose uniformity
Blank Wafer	2.7		1.7		0.3	
0.4	3.2	**1.1**	3.1	**1.1**	1.0	**1.8**
3.5	11.6	**3.9**	10.4	**3.5**	2.3	**3.9**
10.5	34.9	**11.6**	32.3	**10.8**	5.9	**10.1**

Spectroscopic Ellipsometry

Spectroscopic ellipsometry (SE) is a non destructive and fast optical technique. The polarization state of the reflected light is often presented in terms of $\tan\Psi$ and $\cos\Delta$ which are the amplitude and projected phase of the complex reflectance ratio:

$$\rho = Rp/Rs = \tan\Psi \, e^{i\Delta}$$

Rp and Rs are the complex reflection coefficients of light polarized parallel to (p) or normal to (s) the plain of incidence. SE is intrinsically sensitive to these film optical properties. To increase the precision, the measurements were performed with the full spectral range (190-1700nm) and at 4 angles of incidence (65-70-75-80°).

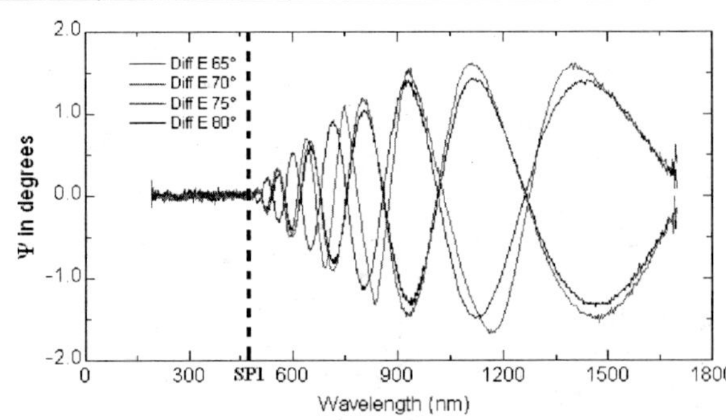

FIGURE 7. Phase difference between two Hydrogen processes (H+ 69keV and H+ 76keV) measured by spectroscopic ellipsometry.

The Figure 7 shows $\Delta\Psi = 0$ at 488nm (SP1 wavelength), so it is impossible to characterize the difference betwen of two Hydrogen processes with the

SP1. For a wavelength higher than 600nm, there is a phase difference between the both processes and it would be possible to characterize Hydrogen ion implantation.

The dielectric function of single crystalline silicon (c-Si) differs significantly from that of amorphous silicon (a-Si). The ion implantation caused disorder can be described as a graded Effective Medium Approximation8. (EMA) by mixing c-Si and a-Si with controlled distribution profile. The agreement between the measured and the model-generated data is quantified via the Mean Squared Error (MSE). The subscript "i" identifies each unique wavelength and angle of incidence, σ is the standard deviation, N is the total number of (ψ, Δ) pairs and M is the number of fit parameters.

$$MSE = \sqrt{\frac{1}{2N-M}\sum_{i=1}^{N}\left[\left(\frac{\psi_i^{mod}-\psi_i^{exp}}{\sigma_{\psi,i}^{exp}}\right)^2+\left(\frac{\Delta_i^{mod}-\Delta_i^{exp}}{\sigma_{\psi,i}^{exp}}\right)^2\right]}$$

TABLE 2. EMA grated layer model

4	srough	1.208nm
3	Graded (EMA)	178.042nm
2	EMA (a-Si)/80% Si-jaw	0nm
1	a-Si	0nm
0	Si-jaw	1nm

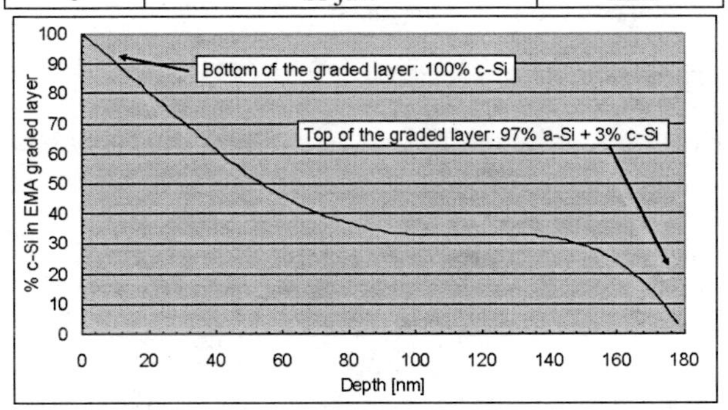

FIGURE 8. EMA grated layer from As+ 220kev 9E13at/cm² process

For the Arsenic 220keV 9E13at/cm² process, an EMA grated layer model (Table 2 and Figure 8) has been done, enabling us to know the thickness of the grated layer by ion implantation and also to have an idea of the amorphized layer thickness. Thanks to these measurements, it is possible to know the thickness and the crystal structure of the grated layer as functions of energy and dose, and the repeatability of the model has been realized with three different wafers for various energies and doses. The figure 9 shows that the grated layer is a function of energy and also the same conclusion is true for the dose.

CONCLUSION

Damages generated into the crystal silicon by ion implantation are revealed by the laser scattering technique. For ion implantations with "aggressive" conditions (i.e. heavy ion, high energy or high dose) the sensitivity of the laser scattering measurements permits us to characterize as-implanted processes and the full wafer haze mapping enables us to have a fast and in-line idea of the ion implantation uniformity. A calibration must be done for each process in order to have a standard measurement. Besides with spectroscopic ellipsometry, the thickness of the grated layer is known. So, both these techniques are new ways to characterize surface and depth as-implanted processes.

In order to have a characterization tool for all as-implanted processes, manufacturers would improve the sensitivity of the haze measurements by using laser scattering with wavelength higher than 600nm.

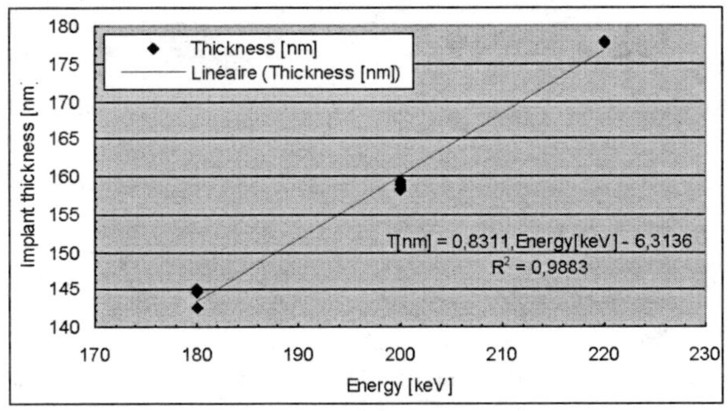

FIGURE 9. Thickness of the grated layer as functions of energy from Arsenic 9E13at/cm² process.

REFERENCES

1. A. Benninghoven, F.G. Rüdenauer, H.W. Werner, "Secondary Ion Mass Spectrometry: Basic concepts, instrumental aspects, applications, and trends", Wiley, New York, 1987.
2. F.M. Smits, "Measurements of sheet resistivity with the four-point probe", in *BSTJ*, **37**, 1958, pp371
3. E. Tsidilkovski, K. Crocker, K. Steeples, "Ion implant process monitoring with a dynamic surface phot-charge technique", in *15th Annual IEEE/SEMI ASMC Conference Proceedings*, Boston MA, 2004, pp181-186.
4. N. Pearce, Li Zhou, P. Graves, H. Takeo, T. Romig, "Production implant monitoring using the Therma-Probe 500", in *Ion Implantation Technology,* 1996, pp206-209.
5. See www.pvateplaamerica.com/semi_implantdosage.php
6. E. Nolot, B. Arrazat, S. Favier, Y. Borde, J. F. Damlencourt, B. Vincent, O. Kermarrec, V. Carron, F. Nemouchi, P. Vandelle, Y. Bogumilowicz, A. Danel, "Laser Scattering: a Fast, Sensitive, In-Line Technique for Advanced Process Development and Monitoring", in *AIP Conf. Proc*, 2007, Vol. 931, pp116-120.
7. J. Hilfiker, J. Woollam, G. Mowry, P. Chow, J. ELman, "Automated Spectroscopic Ellipsometry", *The Industrial Physicist*, March 1996, pp. 30–34.
8. W. R. Tinga, W. A. G. Voss, D. F. Blossey, "Generalized approach to multiphase dielectric mixture theory", *J. Appl. Phys.*, **44**(9) pp 3897-3902, (1973).

Carrier Density Profiling of Ultra-Shallow Junction Layers Through Corrected C-V Plotting

James Chen[1], Dimitar Dimitrov[1], Tatiana Dimitrova[1], Paul Timans[2], Jeff Gelpey[3], Steve McCoy[3], Wilfried Lerch[4], Silke Paul[4], and Detlef Bolze[5]

[1]*Four Dimensions, Inc., 3140 Diablo Ave, Hayward, California, 94545, USA*
[2]*Mattson Technology, Inc. Fremont, California, USA*
[3]*Mattson Technology Canada, Inc., Vancouver, Canada*
[4]*Mattson Thermal Products GmbH, Dornstadt, Germany*
[5]*IHP, Frankfurt (Oder), Germany*
Tel.: (510) 782 1843, Fax.: (510) 786 9321, E-mail: tcchen@4dimensions.com

Abstract. The aim of this report is to present and justify a new approach for carrier density profiling in ultra-shallow junction (USJ) layer. This new approach is based on a capacitance measurement model, which takes series impedance, shunt resistance and the presence of a boron skin on the USJ layer into account. It allows us to extract the depletion layer capacitances in the USJ layer from C-V plotting more accurately and hence to obtain better carrier density profiles. Based on this new approach the carrier density profiles of different USJ layers with and without halo-style implants are obtained and discussed.

Keywords: USJ, Annealing, Carrier density profile, Mercury probe, Fine sheet resistivity mapping
PACS: 84.37+q, 68.55, 81.70Jb.

INTRODUCTION

Profiling of carrier density in a heavily doped ion implanted layer has rarely been performed, mainly because C-V plotting with a metal contact to such a layer is usually very difficult because the contact shunt resistance is too low for the plotting to be able to cover a reasonable depth. However, it has been found that the situation can be different for at least some ultra-shallow junctions (USJ). Here the implantation dose is so heavy and yet its depth is so shallow that profiling of a USJ layer's carrier density to a meaningful depth becomes possible. This may be useful for checking the effects of annealing and ion implantation processes. We can obtain such profiles because:

(1) At the surface of the USJ layer (p type), the implanted boron density is so high and the annealing time is so short that the boron activation ratio is low enough to create a very thin boron skin at the surface as shown schematically in Figure 1. This boron skin appears to prevent the metal contact to the USJ layer from being too leaky to allow good C-V measurements.

(2) The carriers under the boron skin in the USJ layer are so close to the surface that the depletion layer can reach a significant portion of the depth of the USJ layer without requiring excessive bias in the C-V plotting for the profiling.

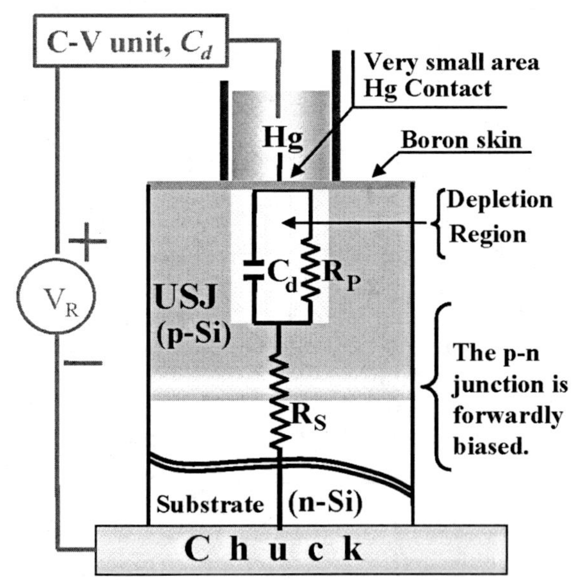

FIGURE 1. Carrier density profiling of USJ layers (the dimensions are not proportional for clarity).

Although the impurity under the boron skin is not fully activated, the carrier density under the boron skin is still very high and the boron skin is extremely

CP1066, *Ion Implantation Technology,* edited by E. G. Seebauer, S. B. Felch, A. Jain, and Y. V. Kondratenko

thin. This means that the area for the C-V measurements should be kept very small in order to make good measurements. Also, the measurement frequency should be high enough to reduce the effect of shunt resistance on the measurement. Furthermore, the measured capacitance still can be corrected for the error caused by parasitic factors such as series impedance, parallel capacitance and shunt resistance. Carrier density profiles of USJ samples with various implantations and annealing processes have been obtained using the method based on the above consideration and are presented in this report.

EXPERIMENTAL DETAILS

A. Experimental Technique

Figure 2 shows an equivalent circuit for the depletion layer capacitance measurement in the case of a metal contact to the surface of a wafer for C-V measurement. In the figure, C_d is the depletion layer capacitance of the Schottky contact under reverse bias, R_p is the equivalent resistance in parallel to C_d, R_s is the equivalent resistance in series with the C_d, L_S is the equivalent series inductance in the measurement circuit.

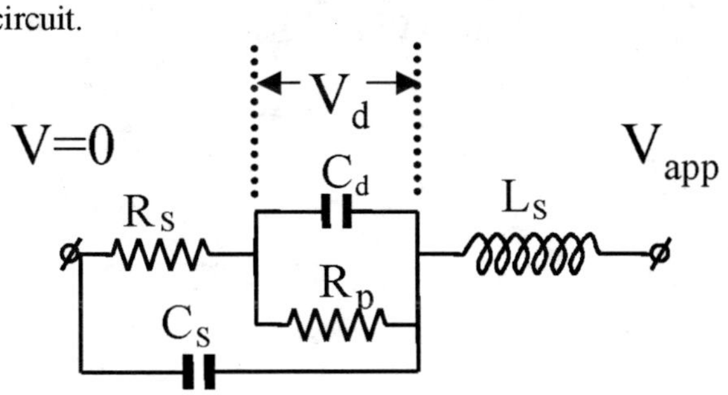

FIGURE 2. Small Signal equivalent circuit of a Schottky contact to USJ layer for C-V measurements.

Using an impedance meter to make measurements to such a structure, one can obtain a complex impedance, ***Z***, at each bias with [1]:

$$\mathrm{Re}(Z) = R_S + \frac{R_p}{1+\omega^2 R_p^2 C_d^2} \quad (1)$$

$$\mathrm{Im}(Z) = \omega L_S - \frac{\omega R_p^2 C_d}{1+\omega^2 R_p^2 C_d^2} \quad (2)$$

Thus one may determine C_d by measuring ***Z*** at three frequencies, $\omega = 0$, ω_1, and ω_2 then use Eqs.(1) and (2) to solve for C_d. Many samples of USJ, with and without halo, were measured in this way for extracting their carrier density profilings.

B. Sample Preparation

The wafer substrates used in this study are (100) oriented n-type Si with resistivity of 10-15 Ω.cm. The USJ layers were implanted with tilt/twist low energy (500 eV) B+ with dose $1\text{x}10^{15}$ B/cm^2 . Some of these wafers were also implanted with a halo-style implant of $4\text{x}10^{13}$ As/cm^2 at an energy of 40 keV, and some also with a pre-amorphization implant of $1\text{x}10^{15}$ Ge/cm^2 at 30 keV.

TABLE 2. Flash Assisted Rapid Thermal Annealing.

Process	Pre-heat [°C]	Time [mSec]	Temperature [°C]
A	700	1	1300
B	700	1	1250
C	750	1	1350
D	750	1	1300
2xD	750	2	1300

Doping activation of all wafers was carried out by flash-assisted rapid thermal processing (fRTPT™) as shown in Table 1. In this approach, wafers are pre-heated to a temperature of 700 ^{0}C or 7500 ^{0}C, and then the implanted surface is heated to a high temperature by a powerful pulse of energy from a bank of flash-lamps [2,3].

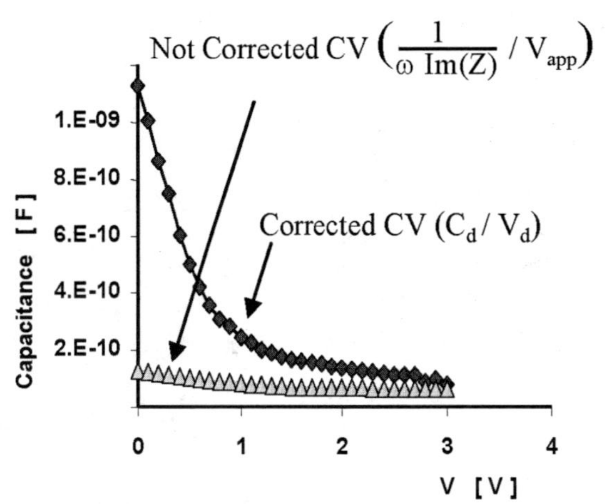

FIGURE 3. Correction of the C-V plot of an USJ wafer.

EXPERIMENTAL RESULTS AND DISCUSSIONS

The typical correction of the C-V plot and the corresponding carrier density profile are illustrated in Figure 3 and Figure 4.

Measurements were made with CV map 3093, an automatic mercury probe system, [4], in conjunction with an Agilent 4285A precision LCR meter. The mercury contact area is $9.0x10^{-5}$ cm^2. In this setup, the impedance due to wire inductance L_S is low enough to be negligible at 1 MHz and lower frequencies. Therefore, the measurements are necessary only at two frequencies, f = 0 and 1MHz, for determining C_d, R_S and R_p. Very small mercury contact area is necessary for maximizing the shunt resistance, R_p and minimizing the influence of the inductance, L_S.

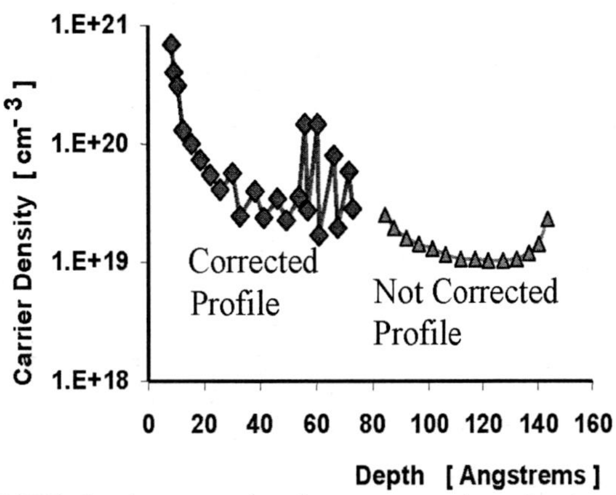

FIGURE 4. A corrected and not corrected carrier density profile for the C-V plots shown in Figure 3.

The corrected C-V, (C_d/V_d), plot was then used to calculate the majority carrier density, p(W), and the width of the depleted area W was obtained from the well known equations [5]:

$$W = A.\varepsilon_S\varepsilon_0 \frac{1}{C_d}, \qquad (3)$$

$$p(W) = \frac{C_d^3}{2A^2\varepsilon_0\varepsilon_S \frac{dC_d}{dV_d}}, \qquad (4)$$

where V_d is the voltage drop across C_d obtained by subtracting the voltage due to the leakage current through the series resistance, R_S, from the voltage between the mercury probe and the chuck.

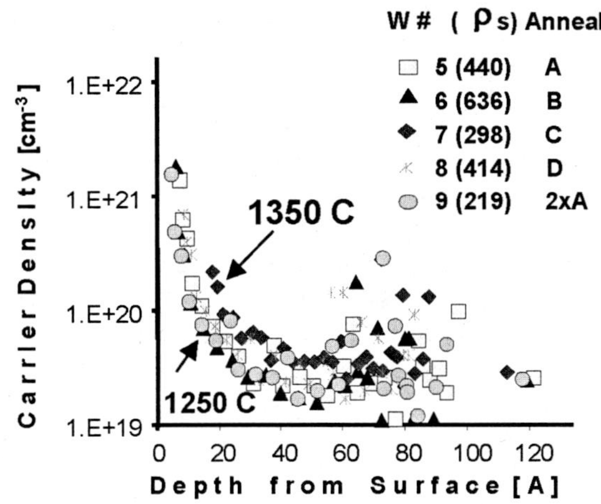

FIGURE 5. Comparison of the corrected carrier density profiles of USJ wafers without Halo implantation, flash-assisted annealed at different temperatures and time as shown in Table 1.

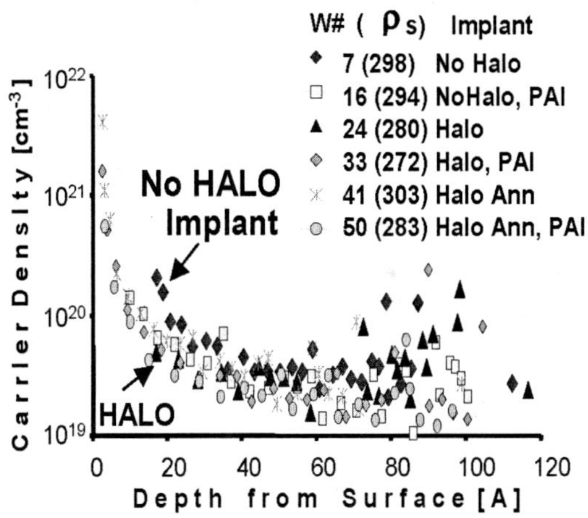

FIGURE 6. Comparison of corrected carrier density profiles of USJ layers with different implantations. All wafers have the same annealing for electrical activation of the implantation – 1 mSec at peak temperature 1350°C (process C in Table 1).

The corrected carrier density profiles of USJ layers without halo implantation are compared in Figure 5 for different annealing cycles as described in Table 1. For depths beyond 70Å even the corrected carrier density profiles become noisy. The main reason for the noise observed at higher depth is the increase of the leakage current at higher d.c. bias which significantly reduces the shunt resistance. So, to compare the effect of different millisecond flash-assisted annealing, we should use the corrected profiles up to approximately 70Å. The plots show that the wafer annealed at highest peak temperature of 1350°C has the highest carrier density, and wafer annealed at the lowest peak temperature of 1250°C has the lowest carrier density. The observation that at a higher temperature more implanted atoms are electrically activated, is also supported by independent four-point probe measurements of the sheet resistance of the wafers. The reliability of the four-point probe measurements of USJ implanted wafers was discussed previously [6].

Figure 6 shows a comparison of corrected carrier density profiles for different implantation processes combined with annealing at 1350°C. The highest carrier density profile, at about $9.5x10^{19}$ cm^{-3} at 20Å depth, is observed for the wafer without halo implantation (black diamonds in Figure 6). Halo implanted wafers have lower densities of electrically activated carriers – at about $6x10^{19}$ cm^{-3} at 20Å depth. This difference can be expected because the halo implantation partially compensates the USJ low energy boron implantation. Here we need to mention that in this comparison between wafers with and without halo implants, the carrier density at 20Å from the surface is not inversely proportional to the sheet resistance measured by the four-point probe, which is shown in brackets in the legend in Figure 6. This may be a consequence of the effect of junction leakage on the four-point probe measurements, when heavy halo doping is used. It is expected, however, that mapping of integration of carrier density over a range of depth would be useful for monitoring the uniformity of implantation and annealing. The advantages of using this mapping method are high geometric resolution (~ 100 μm) and no interference from bad contacts or junction leakage.

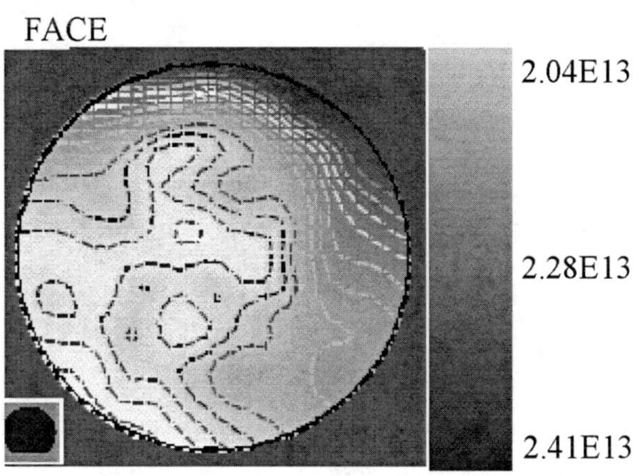

FIGURE 7. Contour map of the carrier density integration for USJ on a wafer. The contour map is based on 81 site corrected CV plots, measured and mapped automatically by CVmap 3093 system. See ref. [4].

Figure 7 shows a contour map of such integration of carrier density for a wafer with USJ on halo and PAI and annealed through process C as shown in Table 2. Comparing it with a corresponding map by four-point probe, the map is much less noisy. Diameter scan by the same technique of laser annealed USJ wafer from a separate source clearly reveals the stitches of laser scan.

CONCLUSION

A technology for USJ carrier density profiling has been developed and used to monitor and analyze the effects of different annealing and implantation processes for formation of USJ layers. By taking into account the series resistance, R_S, and the shunt resistance, R_d, we are able to improve the accuracy of measurement of the depletion capacitance, C_d, and hence obtain more realistic carrier density profiles. It is quite possible that, using an automatic mercury probe, this technique can provide us with better wafer maps than that of four-point probe and its very fine geometric resolution can reveal laser annealing stripes.

REFERENCES

1. A. Straub, R. Gerbs, *et al.*, *J Appl Physics,* **97** (2005) p.083703-1.
2. P. Timans, J. Gelpey, S. McCoy, W. Lerch and S. Paul, *Mater. Res. Soc. Symp. Proc.* 912, (2006).
3. W. Lerch, S. Paul, J. Niess, S. McCoy, T. Selinger, J. Gelpey, F. Cristiano, F. Severac, M. Gavelle, S. Boninelli, P. Pichler and D. Bolze, *Mater. Sci. Eng. B* **124–125**, (2005) 24.
4. CV map 3093 is a product of Four Dimensions, Inc
5. D. K. Schroeder, "Carrier and Doping Density," in *Semiconductor Material and Device Characterization*, 3d Edition, Hoboken, New Jersey: John Wiley & Sons, Inc., 2006, pp. 61-126.
6. James Chen, Dimitar B Dimitrov, Chi-Yiu So, "Measuring USJ Junction Conductivity and Metal Film to Semiconductor Contact Resistivity with a Modified Four Point Probe Setup" in *Proc. of Intern. Workshop on Junction Technol.* (Shanghai, China, 2007) p. 81.

Evaluation of Pre-Amorphous Layer by Spectroscopic Ellipsometry

Satoshi Shibata[1], Fumitoshi Kawase[1], Akihiko Kitada[2] and Takashi Kouzaki[2]

[1] *Tonami plant, Semiconductor Company, Matsushita Electric Industrial Co., Ltd.*
[2] *Matsushita Electric Industrial Co., Ltd., R&D,*
271 Higashikaihotsu, Tonami Toyama, 939-1312, Japan

Abstract. We have employed Spectroscopic Ellipsometry (SE) as a measurement technology to non-destructively and simply measure pre-amorphous layers made by ion implantation. In this paper, we examine the differences between amorphous film thickness measured by cross-section-TEM and SE, aiming at a highly precise evaluation of pre-amorphous layers created during shallow junction formation at energies below 10 keV. Our data indicates that the amorphous layer thickness measured with SE also includes a portion of the heavily-damaged crystal near the amorphous-crystal interface, where EOR defects are located after annealing. For example, thickness of amorphous layers formed by helium in plasma doping tools can be measured with this technology.

Keywords: Pre-amorphous, Ion implantation, Spectroscopic ellipsometry, Plasma doping, cross-section-TEM
PACS: 68.37.Lp, 68.35Dv

INTRODUCTION

In shallow junction formation technology, control of pre-amorphous layers is an important element and it is indispensable to know the layer thickness for control of device manufacturing. Cross-sectional transmission electron microscope (X-TEM) and Rutherford back scattering spectroscopy (RBS) have been used to measure amorphous layer thickness, however these methods require preparation of measurement samples by destruction of wafers into smaller segments and are not suitable for in-line process controls. Spectroscopic ellipsometry (SE) is a local (≈30 um spot size), non-destructive technique which has been used for measurements of amorphous layer thickness for implant energies greater than 20 keV [1] and has been applied for measurements of annealing temperatures [2-4].

In this work, we apply SE and X-TEM methods for the measurement of amorphous layer thickness for lower energy (<20 keV) and lower dose ($5x10^{14}$ ions/cm^2) implants with Ge and As ions. We also apply the SE method to amorphous layers formed by He^+ implantation with plasma immersion methods [8,9].

EXPERIMENTAL METHODS

a) Experimental methods

Starting materials were p–Si(100) 200 mm wafer with resistivity of 10-15 Ohm-cm. The amorphous layers were formed by implanting As or Ge ions using a batch-type high current ion implantation machine at energies from 4 to 20 keV with doses from $2x10^{14}$ to $5x10^{14}$ ions/cm^2. SE measurements were taken with a Xe-lamp as a light source over a wavelength band of 250 to 850 nm, with a spot size of about 28 micrometers in diameter. X-TEM measurements were made with a maker instrument with a magnification calibration by Si lattice imaging.

b) Amorphous layer thickness

Figures 1a and 1b show the spectrum of cos[delta] and tan[psi] measured by SE for As ion implantation at energies of 4-50 keV, at a dose of $3x10^{14}$ As/cm^2. It can be see that their spectra shift to higher wavelengths as As energy increases.

Figure 2 shows the implanting energy dependence of amorphous film thickness. The wafer surface was dealt with as amorphous this time. In addition, the harmonic oscillator model [5-7] is used for calculation of the film thickness in the SE measurement. The X-TEM film thickness used a numerical conversion of the image contrast by image processing, defined as the depth where the contrast difference between each layer becomes one half. Amorphous film thickness increases with energy for both the SE and X-TEM measurements with the SE results ≈2 nm thicker than the TEM values.

CP1066, *Ion Implantation Technology,* edited by E. G. Seebauer, S. B. Felch, A. Jain, and Y. V. Kondratenko

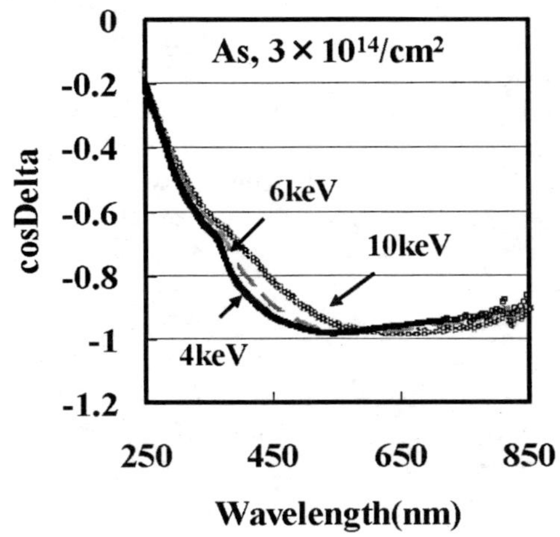

(a) The spectrum cos [delta]

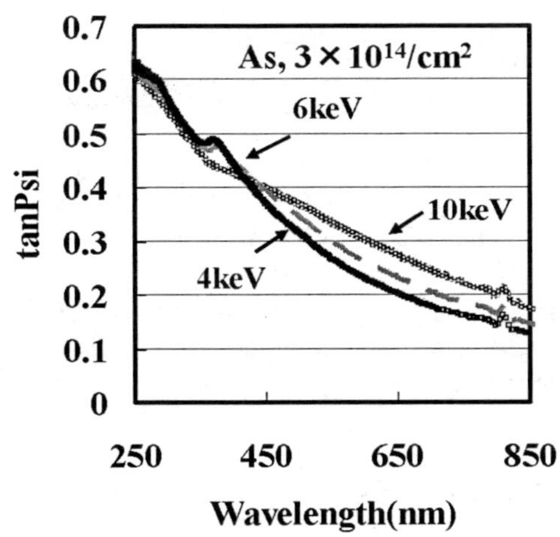

(b) the spectrum tan[psi]

FIGURE 1a, 1b. The spectrum of cos[delta] and tan[psi] measured by SE for As ion implantation.

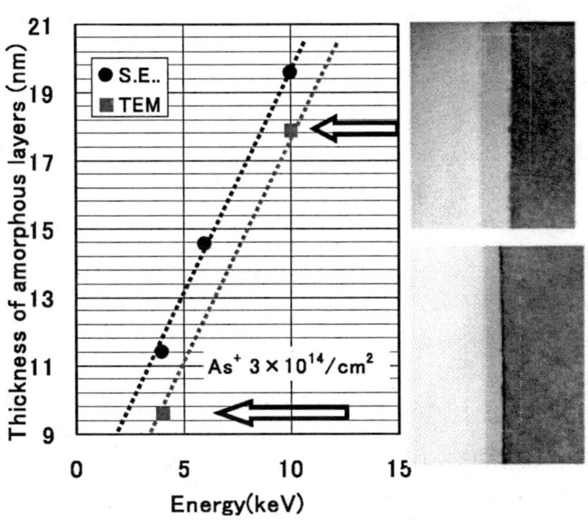

FIGURE 2. The implanting energy dependence of amorphous film thickness.

Amorphous film thickness is increasing similarly both the result in SE, and the result by TEM with the increase in energy. Between the two measurement methods, a ≈2 nm film thickness difference exists. The SE measurements are an average from 49 locations over the wafer surface, while the TEM samples were taken from the wafer center.

So, we performed a comparison with the value of the SE measurement on graph which carried out numerical conversion of the TEM image contrast, in order to clarify the observed thickness difference. The results are shown in Figure 3a and 3b. Along our definition, the amorphous layer thickness, measured by X-TEM, measures from surface to the center section of the domain where numerical value of contrast shifts rapidly. On the other hand, the result of SE shows even the place where the numerical value of contrast had fallen completely. That is to say, our data indicates that the amorphous layer thickness measured with SE also includes a portion of the heavily-damaged crystal near the amorphous-crystal interface.

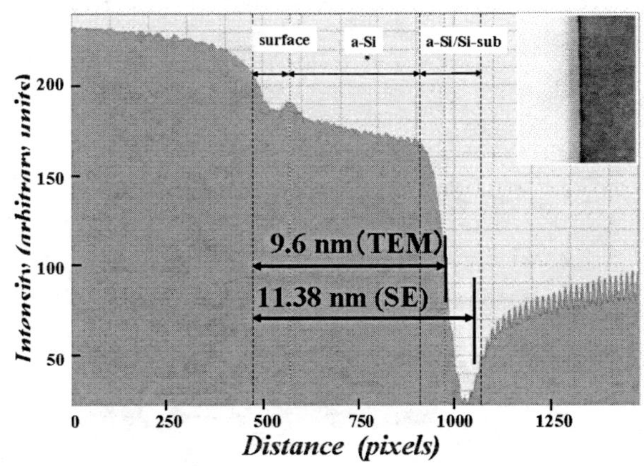

(a) As 4 keV $3x10^{14}$ As/cm^2

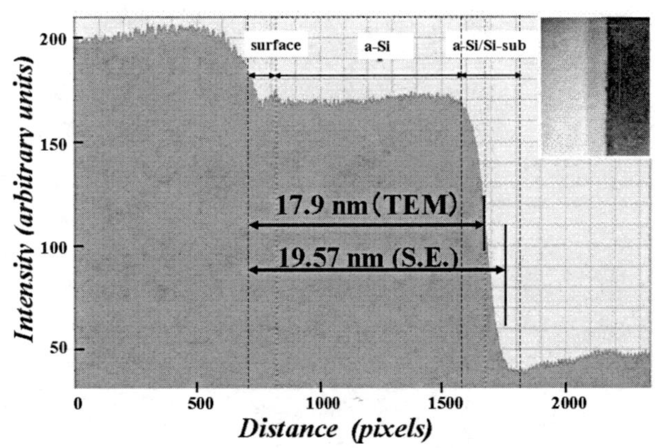

(b) As 10keV $3x10^{14}$ As/cm^2

FIGURE 3a, 3b. The comparison with the value of the SE measurement on graph which carried out numerical conversion of the TEM image contrast.

APPLICATIONS

a) The pre-amorphous layer formed of germanium ion implantation

Figure 4 shows the measurement by SE of the amorphous layer thickness formed by Ge ion implantation. The amorphous layer thickness formed by ^{74}As ion implantation is also shown for comparison. The amorphous layers formed by these similar-mass (^{75}As and ^{72}Ge) ions increase in thickness with energy along the same trend.

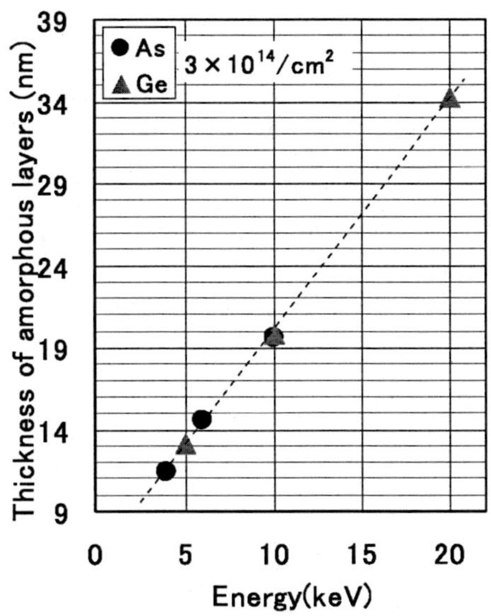

FIGURE 4. The measurement by SE of the amorphous layer thickness formed by Ge ion implantation.

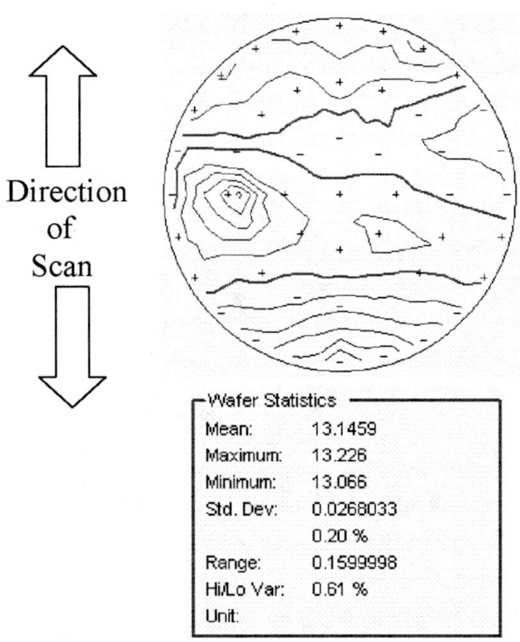

FIGURE 5. The 49 points cross-wafer uniformity map measured by SE of a pre-amorphous layer formed by Ge implantation

Figure 5 shows a 49 points cross-wafer uniformity map measured by SE of a pre-amorphous layer formed by Ge implanted with a batch type ion implantation machine. The Ge implant condition was a dose of $3x10^{14}$ Ge/cm^2 at an energy of 5 keV. Differences of the amorphous layer thickness over the wafer is 0.2 nm or less. However, the signature of the beam scanning can still be seen in the amorphous layer thickness map.

b) The amorphous layer formed by helium plasma in plasma doping

Table 1 shows the measurement result in the SE of the amorphous layer thickness formed by helium ions in a RF plasma doping tool.

TABLE1. The amorphous layer thickness for helium plasma immersion implant

Bias Power (peak to peak voltage)	100W (100 V)	400 W (400V)
Thickness (nm)	6.39	17.15
Uniformity (%)	1.83	0.80

Figure6 shows amorphous layer thickness for He$^+$ PIII implants measured by TEM (Sasaki04) and SE (this work). Actually, it shows that both results are not based on other conditions (plasma density, He ion dose, pulse repetition rate and wafer temperature during implant. etc.), but coincide a dependence on bias voltage.

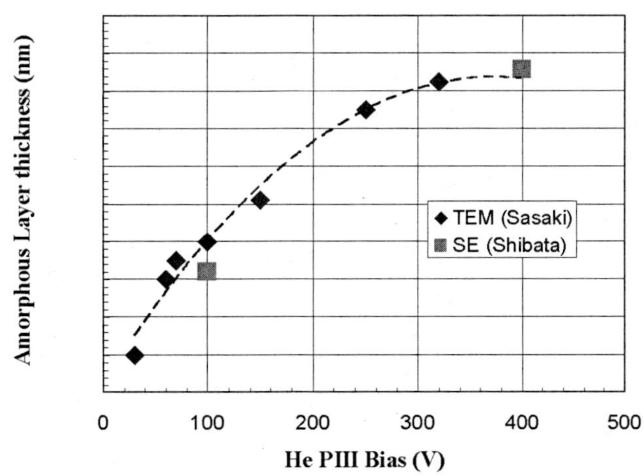

FIGURE 6. The amorphous layer thickness for He+ PIII implants measured by TEM (Sasaki04) and SE (this work).

Figure 7 shows the thickness uniformity mapping by SE of 121 points at the bias power of 100W. The thickness uniformity is ≈2% or less, and the possibility that the uniformity of amorphous layers formed by He plasmas may improve is high.

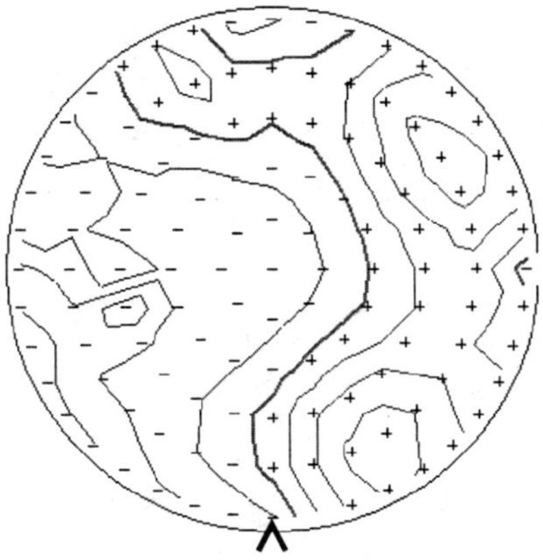

Wafer Statistics	
Mean:	6.3866
Maximum:	6.647
Minimum:	6.136
Std. Dev:	0.1165962
	1.83 %
Range:	0.5109997
Hi/Lo Var:	4.00 %
Unit:	

FIGURE 7. The thickness uniformity mapping by SE of 121 points at the bias power of 100W.

CONCLUSION

We have measured the pre-amorphous layer thickness formed of ion implantation using SE and X-TEM methods. The amorphous layer thickness from SE appears to measure from the wafer surface into the heavily damaged region below the amorphous layer imaged by TEM. SE methods provide an accurate measure (within 2 nm) of amorphous layer thickness by a non-destructive, non-contact method, enabling rapid, multi-point mapping of thickness over a wafer.

ACKNOWLEDGEMENTS

The authors would like to thank Y. Baba, H. Kamiyanagi, and E. Kanazaki for cooperation of this work, and to thank H. Fukumoto, and H. Kawahara and H. Kuroda for their generous support of this work, to thank Dr. M. I. Current, Dr. N. Ohno, Dr. G. Fuse, and Prof. Kitamura for fruitful discussion.

REFERENCES

1. S. Shibata, Y. Nambu, R. Etou, and G Fuse, "Evaluation of High Dose Ion Implantation by Spectroscopic Ellipsometry", IIT-98 IEEE Proc. 98X144 (1999) 465-467.
2. S. Shibata, Y.Nambu, H. Izutani , T. Morita, K. Yamazawa, and M. Arai, "Real Exposed Temperature Evaluation Using Reordering of Implanted Amorphous Si Layers", ISSM2003.
3. Y. Nambu, S. Shibata, K. Itonaga, S. Hashimoto, and G Fuse, "Reordering of Implanted Amorphous Si Layers with Low Temperature RTA", IIT-98 IEEE Proc. 98X144 (1999) 642-645.
4. K. Yamazawa, M. Arai, S. Shibata, Y.Nambu, H. Izutani, and T. Morita, "A wafer surface temperature measurement method utilizing the reordering phenomena of amorphous silicon (The REAL method)" SICE-ICASE International Joint Conference 2006
5. F. Wooten, "Optical Properties of Solids", Academic Press, New York 1972, P42.
6. L. Ward, "The Optical Constants of Bulk Materials and Films", Adam Hilger, Philadelphia, 1988, P14
7. J. Shin Chen, S. Chao, J. Shium, Kao H. Niw, and C. Hsin Chen, "Appl. Opt. 35", 90, 1996.
8. Y. Sasaki, C. G. Jin, H. Tamura, B. Mizuno, R. Higaki, T. Satoh, K. Majima, H. Sauddin, K. Takagi, S. Ohmi, K. Tsutsui, and H. Iwai, "B2H6 Plasma Doping with "*In-situ* He Pre-amorphization."," Symposia on VLSI Technology 2004, pp. 180-181.
9. Y. Sasaki, C. G. Jin, K. Okashita, H. Tamura, H. Ito, B. Mizuno,H. Sauddin, R. Higaki, T. Sato, K. Majima, Y. Fukagawa, K. Takagi, I. Aiba, S. Ohmi, K. Tsutsui, and H. Iwai, "New method of plasma doping with In-situ Helium pre-amorphization,"IIT2004.

Quantitative Evaluation of Ion-implanted Arsenic in Silicon by Instrumental Neutron Activation Analysis

Toshiko Takatsuka,[1] Kouichi Hirata,[1] Yoshinori Kobayashi,[1] Takayoshi Kuroiwa,[1] Tsutomu Miura,[1] and Hideaki Matsue[2]

[1]*National Metrology Institute of Japan, National Institute of Advanced Industrial Science and Technology, 1-1-1 Higashi/Umezono, Tsukuba, Ibaraki 305-8565/8563, Japan*
[2]*Quantum Beam Science Directorate, Japan Atomic Energy Agency, 2-4 Shirakata-Shirane, Tokai-mura, Naka-gun, Ibaraki 319-1195, Japan*

Abstract. Certified reference materials (CRMs) of shallow arsenic implants in silicon are now under development at the National Metrology Institute of Japan (NMIJ). The amount of ion-implanted arsenic atoms is quantified by Instrumental Neutron Activation Analysis (INAA) using research reactor JRR-3 in Japan Atomic Energy Agency (JAEA). It is found that this method can evaluate arsenic amounts of 10^{15} atoms/cm^2 with small uncertainties, and is adaptable to shallower dopants. The estimated uncertainties can satisfy the industrial demands for reference materials to calibrate the implanted dose of arsenic at shallow junctions.

Keywords: Instrumental Neutron Activation Analysis, arsenic implants, Certified Reference Material, uncertainty
PACS: 82.80.Jp, 85.30.Tv, 85.40.Ry

INTRODUCTION

For the miniaturization of silicon semiconductor devices, dopants are ion–implanted at lower energies to make shallower electric junctions. According to the international technology roadmap for semiconductors (ITRS) 2007 [1], dopant concentration should be controlled in 4% precision at present and in 2% after the year 2010 as indicated in Table 1. Several techniques, such as secondary ion mass spectrometry (SIMS) and Rutherford backscattering spectrometry (RBS), are widely used for characterizing the dopant concentration versus depth. In order to quantify the dopant concentration by these techniques, reference materials are required for calibration. Although standard reference materials of ion-implanted arsenic in silicon are available [2], the depth of arsenic is rather deep compared to that of recent silicon devices. Therefore, new reference materials with shallower arsenic implants are required.

Certified reference materials (CRMs) of shallow arsenic implants in silicon are now under development at the National Metrology Institute of Japan (NMIJ). Although isotope dilution (ID-) ICP-MS is widely used as one of primary methods for quantitative analyses of elements, it is impossible to quantify the amount of arsenic without any stable isotopes by ID-ICP-MS. We are quantifying the amounts of ion-implanted arsenic atoms by instrumental neutron activation analysis (INAA), which is adaptable to the analyses of mono-isotopic elements.

TABLE 1. Requirements for dopant metrology [1]

Year	2007	2008	2010	2012	2014
DRAM 1/2 pitch (nm)	65	57	45	36	28
At-line dopant concentration precision	4 %	4 %	2 %	2 %	2 %
Metrology for junction depth (nm)	12.5	11	9	7	-

CP1066, *Ion Implantation Technology,* edited by E. G. Seebauer, S. B. Felch, A. Jain, and Y. V. Kondratenko

EXPERIMENTAL

Arsenic ions were implanted into 300 mm-Si(001) wafers at a nominal dose of 3×10^{15} /cm^2 and 10 keV in energy. The wafer was diced into 15×15 mm^2 pieces by a dicing saw and eight of them were randomly selected as samples to quantify arsenic amount and to test the homogeneity in a wafer. Surface areas of samples were measured by taking images using a commercial optical scanner with a high resolution. From the images, the lengths of four sides were measured to calculate the surface areas. Length calibration was performed by comparing length of each side of samples to that of a reference sample. All sides of the reference sample were measured by a SI calibrated metrology system.

For INAA, all samples were cleaned in isopropyl alcohol with ultrasonic and rinsed in ultrapure water, subsequently cleaned ultrasonically in high-purity nitric acid, and then rinsed in ultrapure water. After the cleaning, all samples were dried for a day in a clean room of class 10000.

For the production of standards to calibrate arsenic amounts, a working standard solution was prepared by diluting SRM 3103a [3] (commercially available standard solution from NIST) gravimetrically. Portions of the solution were dropped from a polyethylene pipette onto the cleaned blank silicon pieces which were prepared from another 300 mm wafer without ion implantation, followed by weighing every portion to calculate the arsenic content. Conventionally a standard solution being dropped on pieces of filter paper to keep arsenic inside, we utilized pieces of a silicon wafer instead of filter paper in this study for the reasons stated later in the Results section. After each piece of silicon had been dried in a desiccating box, the surface was covered with a silicon chip cut from another blank wafer.

All samples and standards covered with silicon chips were put in folded aluminum foil and stacked in two irradiation containers I and II: four standards (Std. A-D), three samples (S1-3), and a blank (B1) in container I, three standards (Std. E-G), five samples (S4-8), and another blank (B2) in container II. The neutron irradiation of each container I or II was performed separately for four hours at 9.6×10^{13} thermal fluence rate of research reactor JRR-3 in Japan Atomic Energy Agency (JAEA) [4]. When neutrons are irradiated to samples, some elements change into radio-isotopes. Concerning arsenic atoms, ^{75}As changes into ^{76}As from which gamma rays are emitted, while for silicon matrix, ^{31}Si with a half-life of 2.62 hours arising from ^{30}Si shows rather strong activity of beta-rays. After the beta-rays from silicon matrix had substantially decayed, all samples were taken out of the containers. Every sample with a cover silicon chip was heat-sealed in a clean polyethylene envelope for gamma-ray counting.

Gamma-ray activity of each sample was measured by a high-purity germanium detector with an energy resolution around 1.8 keV and relative efficiency 20% at 1333 keV. Every sample was set on a plastic plate positioned at an appropriate distance from the detector to obtain suitable counting rate. If the counting rate is too large, pulse pileup occurs frequently and results in counting loss even with pileup correction. For too small the rate, it takes long time to acquire enough signal intensity. Therefore the distance was set to 11 cm from the detector face, i.e., 14.3 cm from the active area of the detector (calculated from another set of experiments). Gamma-ray measurements were performed at least twice per sample and the order was randomized to eliminate the counting time difference. Counting times were determined so that more than 150,000 total counts could be accumulated in the characteristic arsenic peak from the samples.

Because the energy of gamma-ray depends on the element and the intensity does on the amount of the element, the amounts of arsenic can be evaluated from gamma-ray spectra. In case of arsenic analysis, the doublet peaked at 559 and 563 keV in each spectrum was integrated to obtain the intensity, and then was compared between samples and standards to determine arsenic amount.

RESULTS AND DISCUSSION

Figure 1 shows a gamma-ray spectrum, which is obtained from a silicon sample with arsenic ion implantation. Several sharp peaks are clearly observable. Most of the peaks are from ^{76}As and some are from ^{24}Na. The largest peak at around 560 keV is picked up and the counts are integrated over the peak for every sample and standard. Concerning silicon blank chips, it is confirmed that no arsenic count is observed. In the detailed spectrum for a sample plotted with a solid line in Figure 2, the doublet peak at 559 + 563 keV can be clearly seen and no other peak exists in this range. For comparison, spectra obtained from standards with silicon and filter paper are plotted with black and gray closed circles, respectively. The spectrum for the standard with silicon shows almost the same characteristics as that for the sample. For the standard using filter paper, in contrast, an additional small peak is observed near the arsenic peak at 559 keV as indicated with an arrow. This gamma-ray arises from ^{82}Br generated from ^{81}Br in commercial filter paper even after chemical cleaning, which may affect the arsenic peak intensity in counting. To remove the additional peak, we employ the pieces of silicon wafer instead of filter paper for substrates. Another reason for using silicon substrates concerns absorbance differences between samples and standards due to matrix elements. Gamma-ray self-absorption is small but exists even in less than 1 mm-thick material. The difference of the absorption can be eliminated or made very small when silicon is used for the matrix of both samples and standards. The last reason is to make standards nearly identical to the samples in size and shape. Counting efficiency depends on the counting geometry, which can be defined by the distance between samples/standards and the detector, areas and thickness of samples/standards, and arsenic distribution in samples/standards, and so on. Therefore, it is efficient to apply silicon substrate for a reduction of differences between samples and standards.

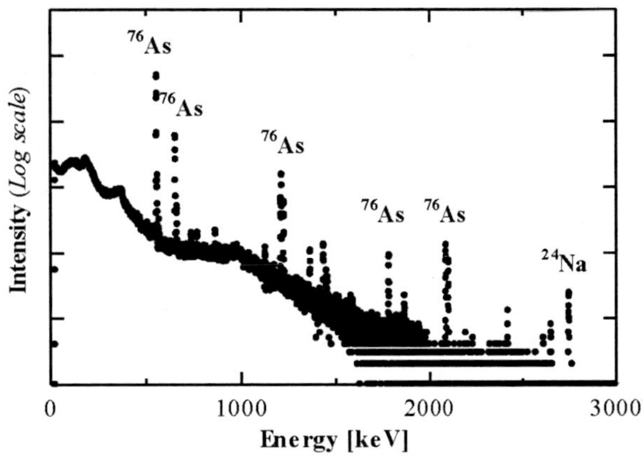

FIGURE 1. Gamma-ray spectrum of arsenic in silicon.

The results from the measurements are summarized in Table 2. The integrated peak intensity from each spectrum is corrected into the intensity at the irradiation time using a conventional radioactive decay equation [5], and indicated as count rates in Table 2.

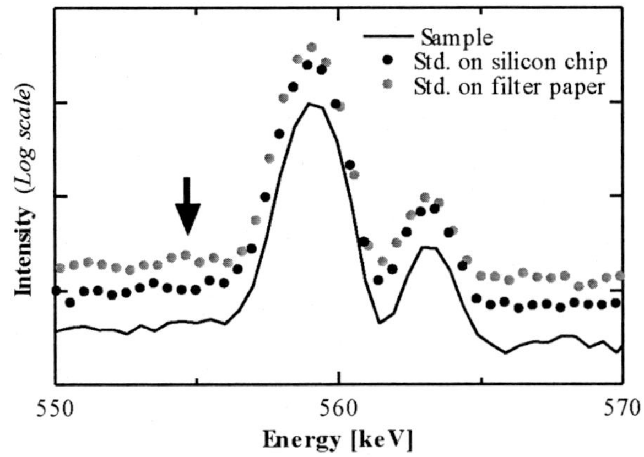

FIGURE 2. Gamma-ray spectra of standards with silicon and filter paper.

Arsenic amount in each standard is calculated from the weight of standard solution deposited onto the blank silicon. Arsenic amounts in the samples are calculated by comparing the intensities between the samples and the standards, both of which were in the same container when irradiated by neutron. The obtained arsenic amounts in samples range between 864 ng and 876 ng, and the variations could include measurement errors, distribution of arsenic contents in a wafer, the differences in area among the sample pieces, and so on. Dividing the amounts by areas, arsenic concentrations are calculated to be in the range from 3.03 to 3.09 × 10^{15} atoms/cm^2 with a relative standard deviation of 0.35%. The averaged concentration becomes 3.05 × 10^{15} atoms/cm^2, which is almost equivalent to the nominal implanted dose of 3 × 10^{15} atoms/cm^2.

It is necessary to estimate the uncertainties for the measurements by INAA in order to evaluate the arsenic concentration in reference materials [6]. Some of the uncertainties are picked up and discussed here.

TABLE 2. Results summary

Sample	Count rate (cps)	As amount (ng)	Area (cm^2)	As concentration (atoms/cm^2)
S1	313.6	868.7	228.3	3.059×10^{15}
S2	312.2	864.2	228.1	3.045×10^{15}
S3	312.1	863.9	228.9	3.033×10^{15}
S4	309.1	863.3	228.8	3.033×10^{15}
S5	310.7	868.7	229.3	3.046×10^{15}
S6	312.4	874.1	228.9	3.069×10^{15}
S7	313.0	876.0	228.2	3.085×10^{15}
S8	310.2	867.0	229.1	3.041×10^{15}
Average	-	**868.2**	**228.7**	$\mathbf{3.051 \times 10^{15}}$

Five kinds of the uncertainties are listed below and summarized in Table 3. All uncertainties below are interpreted as relative standard uncertainties, which are often called one sigma.

Preparation of standards: Initial standard solution was SRM 3130a with a relative standard uncertainty of 0.14%. The diluted solution of 30 ppm was used as a working standard solution. The weights of the solution deposited for the preparation of standards are about 30 mg per sample. Taking account of the uncertainty for the weighing 0.05 mg, relative standard deviation for the solution weight is 0.05%. The total standard uncertainty would be 0.15% for seven standards.

Calibration of arsenic amounts: Arsenic amounts in samples were calculated from the count rate of gamma-rays using a linear calibration curve. The calibration curve was obtained from the count rates and arsenic amounts of standards by the method of least squares. The overall uncertainty for the calibrating arsenic amounts can be predicted to be 0.35% in relative standard uncertainty for the eight samples. This uncertainty includes measurement precision of the repetition and sample-to-sample variation for both samples and standards, errors in the calibration curve, and gamma-ray counting statistics for all the samples and standards.

Geometry: Silicon substrates of samples and standards are nearly identical to each other in area and thickness, but arsenic distribution is slightly different between them. Samples have an arsenic distribution from the near surface to 20 nm deep in silicon, while arsenic atoms exist on the surface of silicon substrate for standards. Considering the distance between the detector and sample surface of 14.3 cm, it is estimated that the distance between the detector and averaged arsenic position of samples is 14.3001 cm, and that of standards is 14.3 cm. Estimating a 0.01 cm uncertainty for the distance, the uncertainty would be 0.04%.

Areas: Sample areas are calculated from the length of each side of the rectangles that was measured by using an optical scanner with a resolution of 6400 dpi. Estimating a 10 dots' uncertainty for the length in scanned images, uncertainty for the area could be calculated to be 0.31 %. The calibration from "dot" into "meter" is done by using a reference sample with an uncertainty of side length less than 0.002%. As for area difference between samples, the observed relative standard deviation of the mean value is 0.19% for eight samples, so that the uncertainty due to measurement precision is 0.068% (0.19 divided by $8^{1/2}$). Therefore total uncertainty of sample area is estimated to be 0.31%.

Dead time correction: During the total measurement time of 2.5 days, gamma-ray intensity from ^{76}As decreases due to the decay with a half-life of 26.3 hours. All the samples and standards were counted over the same fractional dead time of the detector between 0.4 and 1.6%, and decrease of the dead time in a measurement was less than 0.04%. Correction of the dead time during a measurement can be done as [observed count rate] / (1 - [dead time]). Estimating the uncertainty of 0.02% in measurements at 1% dead time, uncertainty for the dead time error would be 0.02%.

These results lead to the conclusion that the combined standard uncertainty is 0.49% on this set of experiments and the expanded uncertainty as defined by ISO becomes 0.98% in a coverage factor of 2.04, which gives an approximate confidence of 95%. Since precisions of device process parameters and metrology are widely described in 3 sigma, the obtained uncertainty can be expressed as 1.47% in 3 sigma. The estimated uncertainty is small enough compared to the process range of 4% required at present for the dopant concentration precision, and the uncertainty might be acceptable for the precision of 2% in 2010 or later.

TABLE 3. Uncertainties

Source	Uncertainty
Preparation of standards	0.15 %
Calibration of arsenic amounts	0.35 %
Geometry	0.04 %
Areas	0.31 %
Dead time correction	0.02 %
Combined uncertainty	**0.49 %**

SUMMARY

Arsenic implants in silicon were quantified by INAA and the uncertainties were estimated. The reduction of uncertainties was achieved using the standards with silicon substrate. It is found that this method can evaluate arsenic amounts of 10^{15} atoms/cm^2 with small uncertainties, and is adaptable to shallower dopants. The estimated uncertainties can satisfy the industrial demands for reference materials to calibrate the implanted dose of arsenic at shallow junctions.

REFERENCES

1. International Roadmap for Semiconductors 2007.
2. Certificate of SRM 2134, NIST(2000).
3. Certificate of SRM 2133, NIST(2004).
4. http://www.jaea.go.jp/04/qubs/english/neutron_e.html.
5. G. Gilmore and J. D. Hemingway, *Practical gamma-ray spectrometry*, Chichester: John Wiley & Sons, 1995.
6. R.R. Greenberg, R.M. Lindstrom, and D.S. Simons, *Journal of Radioanalytical and Nuclear Chemistry* **245**, 57-63 (2000).

Correlation of Measured Surface Contaminants as a Function of Ion Beam Current in GSD Ion Implanters

Ron Johnson, Ron Eddy and John Schuur

INNOViON Corporation, 4253 NE 189th Avenue, Gresham, OR 97230 USA
rcjohnson@innovioncorp.com, reddy@innovioncorp.com, jcshuur@innovioncorp.com

Abstract. An investigation into measured differences in surface contaminant levels when evaluating several ion implant tools and/or tool types. For a given species and energy, beam current appears to have a strong effect on final contaminant levels. Ideally one evaluates process integrity at machine set-up conditions that approximate or exceed the worst case conditions required for regular, volume wafer production. Some contaminants may exhibit a 'threshold' effect where they are easily observed with sufficient beam current, but not present at all for lower beam currents. Knowing the characteristics of the measurement process is essential in obtaining reliable results with a clear interpretation and better facilitates cross-site or cross-platform comparisons.

Keywords: surface contamination, TOF-SIMS, Al, Fe, Mg, aluminum disk, Axcelis GSD.
PACS: 81.05.Cy, 81.65.-b,

INTRODUCTION

Analysis of surface contaminants for most semiconductor manufacturing equipment is a mandatory activity to ensure on-going process integrity. In ion implantation, the nature of the interaction of the ion beam with the materials of construction of the machine is of primary concern. In addition to run to run variability of internal cleanliness of a particular machine, there can be run to run variability in ion source and beamline settings that cause variations ion beam current density, et cetera.

In the course of conducting a variety of process qualifications, surface contamination data, usually TXRF or TOF-SIMS and sometimes VP-ICPMS, is routinely requested. The qualification activity can be either between sites within an organization or between different organizations. A normal protocol would be for identical processes to be run on a qualified machine and a target machine for best comparison (reduced variability) using single-sourced wafers from the same batch. It does happen that identical processes are not always executed leading to uncertainty in explaining deviations in the final data. This investigation sought to provide a measure of quantification of the variability as a function of beam current (or beam current density).

EXPERIMENT DESCRIPTION

The standard process for Innovion is a 5.0E15 at/cm^3 arsenic process at nearly maximum beam current capability. For the GSD machine, this beam current has been set to 15 mA. An energy of 80 keV was chosen for the process as it is the energy most used by our daily equipment monitor recipes and is also a reasonable reflection of average customer conditions. An additional factor for our operation is that the energy be something that can be compared very directly across multiple tool platforms. Since there is a significant implanter in our inventory that is limited to 40 kV Extraction voltage (Varian E220), the Metals Survey process also limits Extraction voltage to 40 kV even though our GSD tools are capable of 90 or 80 kV.

An experiment of this sort required some consideration for implant sequence and tool 'conditioning' to ensure a reasonable outcome (and therefore justification of the laboratory fees!). The concept is that each test wafer needs to have a reasonably identical process environment. We elected to process dummy wafers through an Argon conditioning implant prior to the beginning of the test sequence and in-between each of the test wafer runs. In all, four test wafer runs were completed at beam

CP1066, *Ion Implantation Technology,* edited by E. G. Seebauer, S. B. Felch, A. Jain, and Y. V. Kondratenko

currents of 1 mA, 5 mA, 10 mA and 15 mA. In addition, 1 unimplanted wafer was analyzed as a control. It is important to note that the test implanter is SMIF-equipped, which means process wafers are not exposed to Fab air, though they are exposed to HEPA-filtered air in the end-station.

Finally, the analysis method used was TOF-SIMS as offered by EAG, Sunnyvale, CA. This method has been used for the prior two years and offers a good survey of the surface, including aluminum detection. The primary trade-off between TOF-SIMS and TXRF (or VP-ICPMS) is one of small analysis spot size versus large spot (or whole wafer). Overall, the technique provides good value for the cost and has comparable or better detection limits compared to TXRF.

Process Sequence; additional operating conditions

The final process sequence is listed in Table 1.

TABLE 1.
TOF-SIMS Test Wafer Implant Sequence

Process Parameters	# Batches / Net Dose
1. 75As+, 5E15, 80 keV, 15 mA	2 / 1E16 at/cm^3
2a. 40Ar+, 5E15, 80 keV, 10 mA	2 / 1E16 at/cm^3
2b. 75As+, 5E15, 80 keV, **1 mA**	1 / 5E15 at/cm^3
3a. 40Ar+, 5E15, 80 keV, 10 mA	2 / 1E16 at/cm^3
3b. 75As+, 5E15, 80 keV, **5 mA**	1 / 5E15 at/cm^3
4a. 40Ar+, 5E15, 80 keV, 10 mA	2 / 1E16 at/cm^3
4b. 75As+, 5E15, 80 keV, **10 mA**	1 / 5E15 at/cm^3
5a. 40Ar+, 5E15, 80 keV, 10 mA	2 / 1E16 at/cm^3
5b. 75As+, 5E15, 80 keV, **15 mA**	1 / 5E15 at/cm^3

The other important implant conditions to state are for the configuration of the Secondary Electron Flood system. The SEF system is enabled, i.e., ON, for all high dose processes on the GSD. This system will add to the vacuum load (and interactions) in the process chamber and will also contribute some particles/contamination via the action of the filament. For the TOF-SIMS experiment, the SEF gas flow was set to 3 sccm (Argon), and the Primary Emission Current was set to a multiple of 15x beam current which is a fairly standard level for this system.

EXPERIMENTAL RESULTS

The primary data are reported in multiples of E10 at/cm^2. See Table 2 for a complete listing of species and data. What stands out about these results is how clean the test implanter is, independent of beam current. Most elements in the survey are below the detection level of the measurement technique. Only Al, B, Mg and Zn show any levels above 1E10 at/cm^2. All of these species, with the exception of Zn do show a correlation to beam current. Among the other elements in the survey, only Ca shows a beam current dependence. The details of the data will be discussed below.

Aluminum and Boron Trends

The Boron level decreased with increasing beam current and was highest in the Control wafer. See Figure 1 for a plot of the Aluminum and Boron levels. The Al level trended with beam current, but reached a plateau at 10 mA. In this type of implanter with a non-Si coated disk, these levels are not unusual. Figure 2 is a plot comparing reduced Boron level with increasing beam current density.

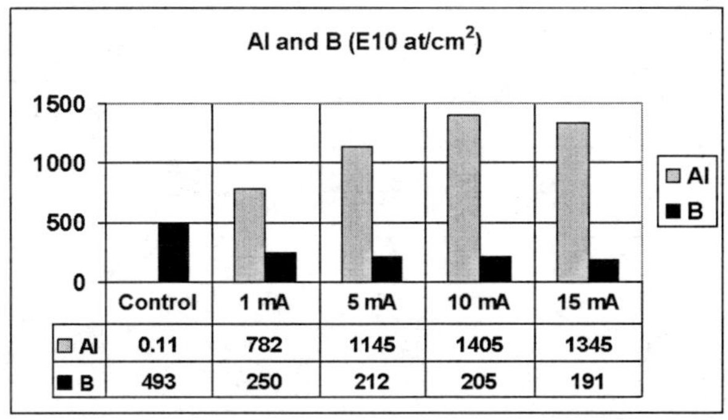

FIGURE 1. Plot of Al and B concentration v. beam current.

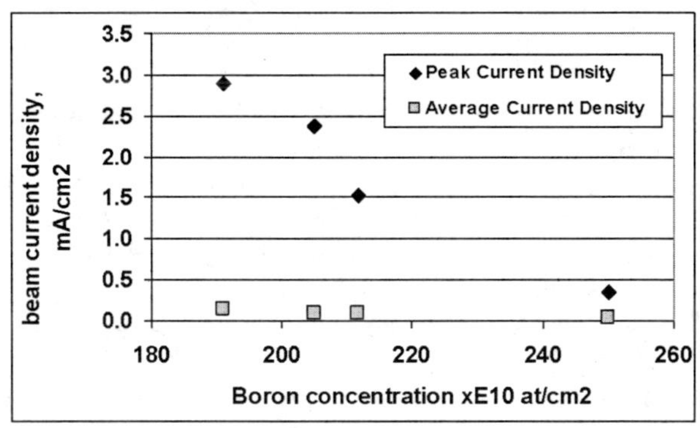

FIGURE 2. Plot of Boron concentration v. beam current density; highest density corresponds to 15 mA condition.

Boron is usually understood to be sourced from the materials used for HEPA filter construction and other environmental sources. In the layout of the Fab, the wafers only see HEPA filters briefly during wafer loading and the Control wafer, with the highest Boron level did not leave its original cassette. This result can be explained by the action of the beam having a 'cleaning' effect on the B levels that correlates to beam current density. As beam current density increased, perhaps due to local heating, the B levels were reduced on the wafer. This result may explain why B levels can sometimes be variable in these types of tests.

Magnesium and Iron Trends

The level of Mg had a trend similar to Al for the four test conditions while Fe was constant (approximately 0.4). An interesting feature of this data is the ratio of Mg/Al that was measured: 91:1, 88:1, 90:1 and 90:1 for the four conditions in increasing beam current. This ratio reflects the expected composition of Al 6061 alloy[1] that is typically used for implanter construction. Expected level of Magnesium is between 0.8% and 1.2% (80:1 and 120:1, respectively).

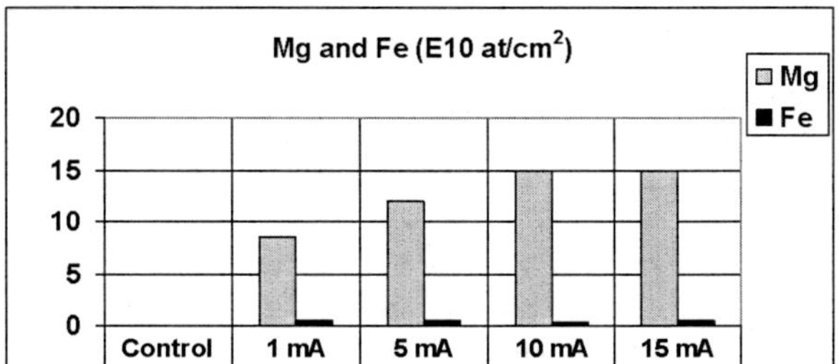

	Control	1 mA	5 mA	10 mA	15 mA
□ Mg	0.034	8.6	12	15	15
■ Fe	0.077	0.41	0.42	0.35	0.42

FIGURE 3. Plot of Mg and Fe concentration.

CONCLUSION

The test results suggest the implanter under test was in good operating condition. Only a handful of elements exceeded a level of 1E10 at/cm^2, and with the exception of Zn, all are well understood. There was good correlation of beam current to Al levels and a counter trend of Boron reduction correlated to increasing beam current density. This work suggests low risk in using higher beam currents for production processes.

The one anomaly was the Zn level for the 10 mA condition. This test had neither the highest beam current nor the highest beam current density, yet a large Zn response was seen. In some cases, a high Zn level is seen for this type of test, though it is usually associated with a newer disk and sometimes a conditioning protocol is used to reduce the Zn level prior to introducing product wafers. The disk in these tests had been in operation for about one year, so another mechanism is needed to explain the result. It is possible that the small analysis area contributed an exaggerated response to the presence of some Zn on the wafer surface. Also, disks under extended usage may exhibit behaviors where the Zn segregates in the Al and migrates to the surface. To sort this out would require a more ambitious study.

ACKNOWLEDGEMENTS

The authors would like to acknowledge the assistance of Jim Norberg of EAG in interpretation of some of the results. We would like to thank Peter Kopalidis for useful discussions on Zn in Aluminum implant disks. We would also like to thank Mike Murawski for analysis of essential beam characteristics.

REFERENCES

1. International Alloy Designations and Chemical Composition Limits for Wrought Aluminum and Wrought Aluminum Alloys (Revised 2001)
2. Private communication with Peter Kopalidis, 6/2/08, Technological Educational Institute of Thessalonica, Greece.

Table 2. TOF-SIMS Analysis Survey; all values x E10 at/cm^2. '<' denotes Detection Limit										
Beam current, mA	**Control**		**1 mA**		**5 mA**		**10 mA**		**15 mA**	
specie	*Spot 1*	*Spot 2*	*Spot 1*	*Spot 2*	*Spot 1*	*Spot 2*	*Spot 1*	*Spot 2*	*Spot 1*	*Spot 2*
Li	<0.017	<0.017	<0.010	<0.010	<0.012	<0.011	<0.012	<0.012	<0.012	<0.012
B	504	481	253	247	210	194	210	199	193	190
Na	0.12	0.2	0.04	0.041	0.11	0.11	0.16	0.15	0.16	0.14
Mg	<0.034	<0.034	8.7	8.5	13	12	16	15	15	14
Al	0.11	0.11	792	772	1150	1040	1450	1360	1360	1330
K	0.0057	0.56	0.012	0.011	0.0087	0.022	0.0082	0.012	0.0099	0.012
Ca	<0.022	<0.022	0.044	0.045	0.062	0.11	0.11	0.14	0.1	0.1
Ti	<0.57	<0.57	<0.34	<0.34	<0.40	<0.36	<0.41	<0.40	<0.41	<0.40
V	<0.037	<0.037	<0.026	<0.036	<0.026	<0.023	<0.026	<0.026	<0.029	<0.026
Cr	<0.044	<0.044	0.18	0.16	0.19	0.19	0.21	0.22	0.21	0.2
Mn	<0.049	<0.045	<0.14	<0.12	<0.13	<0.12	<0.17	<0.15	<0.18	<0.15
Fe	<0.077	<0.077	0.41	0.41	0.39	0.43	0.35	0.35	0.38	0.46
Ni	<0.15	<0.15	<0.092	<0.090	<0.11	<0.098	<0.11	<0.11	<0.11	<0.11
Co	<0.096	<0.097	<0.057	<0.057	<0.067	<0.062	<0.069	<0.067	<0.070	<0.068
Cu	<0.33	<0.33	<0.22	<0.28	<0.23	<0.23	<0.25	<0.28	<0.33	<0.29
Zn	<0.63	<0.63	<0.52	<0.49	<0.43	<0.40	3.6	2.2	<0.47	<0.46
Sn	0.53	0.67	nm	nm	nm	nm	nm	nm	Nm	nm

MATERIALS SCIENCE

An Atomistic Model of Stressed Si Solid-Phase Epitaxy

N. G. Rudawski,[1] K. S. Jones,[1] and R. Gwilliam[2]

[1]*Department of Materials Science and Engineering, University of Florida, Gainesville, FL 32611-6400*
[2]*Nodus Accelerator Laboratory, Advanced Technology Institute, Surrey Ion Beam Centre, Guildford, Surrey GU2 7XH*

Abstract. An atomistic model of stressed solid-phase epitaxial growth of ion-implanted Si is presented. Macroscopic growth kinetics are modeled as the result of crystal island nucleation and subsequent island ledge migration in the growth interface between amorphous and crystalline phases with a rate-limiting approach taken regarding the influence of each process on the macroscopic growth velocity. Due to differences in the volumetric deformation associated with the transition state of each process, the application of stress alters nucleation and migration kinetics differently thus making the macroscopic growth interface velocity a complex function of in-plane uniaxial stress.

Keywords: Solid-phase epitaxial growth, stress, strain, ion-implantation
PACS: 64.60.Cn, 81.15.Np, 61.72.uf, 64.70.kg

INTRODUCTION

Solid-phase epitaxial growth (SPEG) of ion-implanted Si has long been a topic of great technological importance [1]. In particular, stress-influenced SPEG has become of greater interest in recent years. The earliest work on stressed-SPEG revealed the macroscopic velocity, v, of an advancing growth front between amorphous (α) and crystalline phases to be exponentially-enhanced with a state of pure hydrostatic compressive stress [2]. Subsequent work by Aziz *et al.* studied the relationship between v and uniaxial stress in the plane of the growth interface (σ_{11}) and suggested that v may be slightly enhanced (retarded) by in-plane tension (compression) [3]. In conjunction with the work from pure hydrostatic compression experiments, this led to the advancement of the activation strain tensor, $\Delta V_{ij}^* = kT\partial\ln(v)/\partial\sigma_{ij}$, as a means to describe the influence of a generalized stress state, σ_{ij}, on v where kT has the usual meaning and positive (negative) elements of σ_{ij} are tensile (compressive) [4]. By convention, 1 and 2 are the in-plane axes and the 3 axis is aligned with the growth direction. Thus, v with σ_{ij} is given by

$$v = v(0)\exp\left(\frac{\Delta V_{ij}^*\sigma_{ij}}{kT}\right), \qquad (1)$$

where $v(0)$ is the $\sigma_{ij} = 0$ growth velocity. The basis for using ΔV_{ij}^* to explain stress-influenced SPEG stems from the well-known observation that $v(0) = v_0\exp(-\Delta G^*/kT)$ over a wide temperature range where v_0 is a temperature-independent pre-factor and $\Delta G^* = 2.7$ eV is the activation energy[1] for macroscopic growth as given by transition state theory (TST).

Epitaxial growth results from the nucleation of crystal islands in the growth interface with subsequent in-plane migration of island ledges. In the case of liquid-phase epitaxy, nucleation and migration are both mediated by the flux of atoms to the growth interface and thus the growth rate is proportional only to diffusivity in the liquid phase [4]. However, the case of SPEG is inherently different since the atoms used for growth are already attached to the growth interface. Thus, following crystal island nucleation, only the atoms in contact with the island ledges can be used for migration. Hence, macroscopic SPEG is controlled by the kinetics of island nucleation as well as the kinetics of ledge migration. Although, the strong Arrhenius-type behavior of $v(0)$ suggests that one of these processes is inherently much slower than the other such that SPEG can be approximated as resulting from a single atomistic process.

Here, using new experimental results of stressed-SPEG of (001) Si, an atomistic model is advanced which addresses the sequential nature of nucleation and migration processes in macroscopic growth kinetics and isolates the stress dependence of each process.

CP1066, *Ion Implantation Technology,* edited by E. G. Seebauer, S. B. Felch, A. Jain, and Y. V. Kondratenko

EXPERIMENTAL

A polished 50 μm-thick (001) Si wafer was Si^+-implanted at 50, 100, and 200 keV to doses of 1×10^{15}, 1×10^{15}, and 3×10^{15} cm^{-2}. Subsequently, samples were cleaved along <110> directions into ~0.2×1.8 cm^2 strips (with 1 and 2 axes taken to be [110] and [$1\bar{1}0$] crystal directions) and uniaxially-stressed up to magnitude of 1.3±0.1 GPa along [110] (σ_{11}) as presented elsewhere [5]. Stress-free, tensilely-stressed, and compressively-stressed strips were annealed simultaneously at 525 °C in N_2 ambient for 1.0 – 4.0 h. No detectable stress relaxation occurred during annealing. Growth was examined using weak-beam dark-field (WBDF) cross-sectional transmission electron microscopy (XTEM). XTEM specimens ~10 μm long were prepared via focused ion beam (FIB) milling within a lateral distance of ±3 mm from the strip centers to minimize the presence of any thermal gradient. Due to the very small specimen length to strip length ratio, it is reasonably assumed no intra-specimen stress gradients existed.

RESULTS

Figure 1(a) presents a WBDF-XTEM image of the as-implanted structure indicating an initial α-Si layer 350 ± 5 nm thick continuous to the surface. Following annealing for 4.0 h with σ_{11}= 0, shown in Figure 1(e), 138 ± 5 nm of growth occurred. The error in all growth measurements is given as the root-mean-squared roughness of the α/crystalline interface. End of range (EOR) defects were observed near the initial α/crystalline interface for all specimens [6]. As reported by others, these defects do not interact with the growth interface and do not influence SPEG kinetics [6]. In the cases of σ_{11} = −0.5, −1.0, and −1.3 GPa, shown in Figures 1(b) – (d), 77 ± 9, 72 ± 8, and 68 ± 9 nm of growth occurred which is less than the σ_{11} = 0 case. Growth interface roughening was observed with σ_{11} < 0, consistent with prior reports [7]. For cases of σ_{11} = 0.5, 1.0, and 1.3 GPa, shown in Figures 1(f) – (g), 138 ± 6 nm of growth occurred, similarly to the σ_{11} = 0 case. The surface-EOR distance showed no detectable

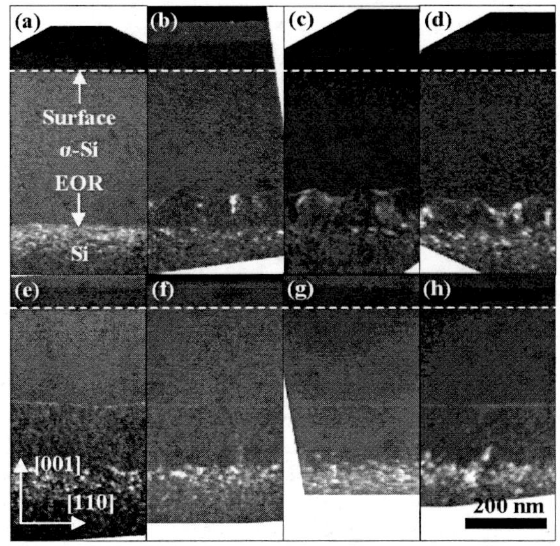

FIGURE 1. a) WBDF-XTEM image of the as-implanted structure. WBDF-XTEM images of specimens after annealing for 4.0 h with σ_{11}= b) −0.5, c) −1.0, d) −1.3, e) 0, f) 0.5, g) 1.0, and h) 1.3 GPa.

variation between as-implanted and annealed specimens indicating no detectable flow of α-Si [8]. Annealing of specimens for 1.0, 2.0, and 3.0 h was performed for all values of σ_{11} and the resulting amounts of growth were measured (not presented). From this data, v versus σ_{11} was calculated as presented in Figure 2. No detectable difference was observed in v for $0 \le \sigma_{11} \le 1.3$ GPa with v = 34 ± 2 nm/h. In compression, v was retarded to a limiting rate of 17 ± 2 nm/h for $-1.3 \le \sigma_{11} \le -0.5$, while σ_{11} = −0.25 GPa retarded v to 22 ± 2 nm/h. The behavior of v versus σ_{11} predicted by Eq. (1) (Aziz *et al.* [3]) is provided for reference. No predicted increases in v were observed for $0 < \sigma_{11}$ while retardation for $\sigma_{11} < 0$ was consistent with prior work [3]. However, in compression, the retardation appears to reach a limit and the reductions are greater than predicted.

DISCUSSION

It appears that the presented results cannot be reasonably explained by the activation strain tensor [3] thus it becomes necessary to consider a new model to explain the data. In particular, as was first suggested by Williams *et al.* for Si SPEG, the sequential nature of nucleation and migration processes must be addressed [9]. It must first be recognized that a growth interface has an equilibrium planar density of nuclei, N_D^*, which is a function of T [10]. Of course, time is required to reach this equilibrium planar concentration. Hence, the equation governing planar density, N_D, as a function of time, t, is given by

$$J = \frac{N_D^* - N_D(t)}{\tau_n}, \quad (1)$$

where J is the rate of island nucleation per unit area and τ_n is the characteristic timescale require to reach equilibrium. At t = 0, N_D = 0 and thus Eq. (2) is approximated as $J \sim N_D^* / \tau_n$ which is time-independent. The contribution from J to the overall growth velocity is given by JhA, where h = 0.14 nm and A are the monolayer height and plan-view area of an individual nuclei.

Next, migration is addressed. Without stress, the rate of ledge migration, $\tau_{m,ij}^{-1}$ for the α-Si/(001) Si interface is a second-order tensor analogous to diffusivity or mobility given by

$$\tau_{m,ij}^{-1} = \begin{pmatrix} \tau_{m,11}^{-1} & 0 \\ 0 & \tau_{m,11}^{-1} \end{pmatrix}. \quad (1)$$

The reciprocals of the elements of Eq. (3) may be considered as timescales associated with the migration process. Thus, the velocity of ledge migration is given as $v_{m,i} = \tau_{m,ij}^{-1}\Delta x_j$ where Δx_j is the ledge migration vector dependent on the crystallography of the interface and the coordinate frame of reference. In the presented case, $\Delta x_1 = \Delta x_2 = 0.19$ nm.

Both $v_{m,1}$ and $v_{m,2}$ contribute to the overall growth velocity and are presumably independent of one another. In other words, if migration along one direction became infinitely slow, macroscopic growth could still occur from migration along the other direction since the nuclei possess finite size upon formation.

Accounting for the sequential, rate-limiting nature of nucleation and migration and the independence of migration directions on macroscopic growth kinetics, v is given by

$$v = \frac{AhJv_{m,1}}{AhJ + v_{m,1}} + \frac{AhJv_{m,2}}{AhJ + v_{m,2}}. \quad (1)$$

Next, it is assumed $AN_D^* \sim 1$ since approximately 1 island will be found in area A on a surface with planar density N_D^*. Thus, making the necessary substitutions, Eq. (4) in final form is given by

$$v = \frac{h}{\tau_n + \tau_{m,11}(h/\Delta x_1)} + \frac{h}{\tau_n + \tau_{m,11}(h/\Delta x_2)}. \quad (1)$$

Therefore, v is independent of N_D^* and depends only on the timescales associated with nucleation and migration.

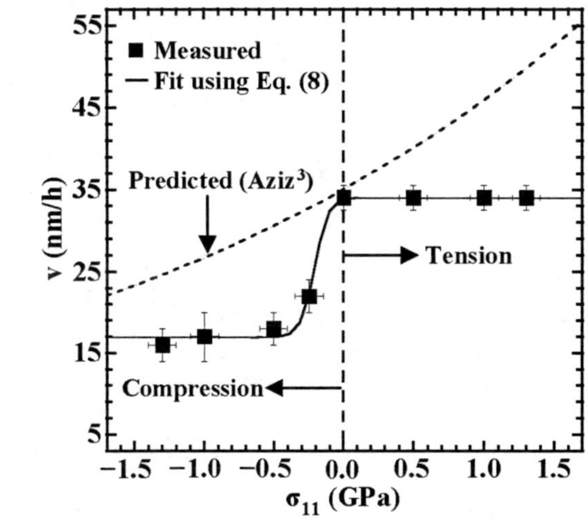

FIGURE 2. Plot of measured v versus σ_{11} behavior with the behavior predicted by Aziz *et al.* [3] and the behavior as fit using Eq. (8).

Presumably, τ_n^{-1} and $\tau_{m,11}^{-1}$ are Arrhenius in nature and thus each may be modified via application of σ_{ij} as given by TST. In terms of nucleation, the nucleation activation strain tensor $\Delta V_{ij}^n = kT\partial\ln(\tau_n^{-1})/\partial\sigma_{ij}$, describes the change in τ_n^{-1} with respect to σ_{ij}. For the α-Si/(001) Si interface,

$$\Delta V_{ij}^n = \begin{pmatrix} \Delta V_{11}^n & 0 & 0 \\ 0 & \Delta V_{11}^n & 0 \\ 0 & 0 & \Delta V_{33}^n \end{pmatrix}. \quad (1)$$

The nucleation of a crystal island causes volumetric deformation primarily in the growth direction, thus implying $\Delta V_{11}^n \sim 0$. Thus, σ_{11} does not alter nucleation kinetics and $\tau_n^{-1} \sim \tau_n(0)^{-1}$ where $\tau_n(0)^{-1}$ is the stress-free nucleation rate [11].

The volumetric deformation associated with the migration activation state is likely quite different from the activation state for nucleation. The migration activation strain tensor, $\Delta V_{kl}^{m,ij} = kT\partial\ln(\tau_{m,ij}^{-1})/\partial\sigma_{kl}$, describes the change in $\tau_{m,ij}^{-1}$ with respect to σ_{kl}.[11] Thus,

$$\tau_{m,ij}^{-1} = \tau_{m,ij}(0)^{-1}\exp\left(\frac{\Delta V_{kl}^{m,ij}\sigma_{kl}}{kT}\right), \quad (1)$$

where $\tau_{m,ij}(0)^{-1}$ is the stress-free migration rate tensor. Since the only term in σ_{ij} is σ_{11}, only $\Delta V_{11}^{m,11}$ and $\Delta V_{11}^{m,22}$ are of relevance in $\Delta V_{kl}^{m,ij}$. The greatest in-plane volume change presumably occurs parallel to, rather than orthogonal to a given migration direction, thus implying $|\Delta V_{11}^{m,22}| \ll |\Delta V_{11}^{m,11}|$. Thus, making the necessary substitutions, v with σ_{11} is given by

$$v = \frac{h}{\tau_n(0) + 2^{-3/2}\tau_{m,11}(0)\exp\left(\frac{-\Delta V_{11}^{m,11}\sigma_{11}}{kT}\right)} + \frac{h}{\tau_n(0) + 2^{-3/2}\tau_{m,11}(0)} \quad . \quad (1)$$

Equation (8) was fit to the data in Figure 2 producing $\tau_n(0) = 29 \pm 1$ s, $\tau_{m,11}(0) = 1.00 \pm 0.01$ s, and $\Delta V_{11}^{m,11} = (12 \pm 1)\ \Omega$ where Ω is the atomic volume of Si. The nucleation timescale is much greater than the migration timescale which confirms prior suggestions from Williams *et al.* [9]. The positive value of $\Delta V_{11}^{m,11}$ suggests in-plane expansion associated with ledge migration.

In the case of $0 \leq \sigma_{11}$, Eq. (8) reduces to the tensile saturation velocity, $v_t = 2h/\tau_n(0)$, while in the case of $\sigma_{11} << 0$, Eq. (8) reduces to the compressive velocity limit, $v_c = h/\tau_n(0)$. Thus, in both cases, SPEG is limited by nucleation and $v_t/v_c \sim 2$ independently of growth temperature, which has been experimentally observed [12].

Hence, for island ledge migration with $\sigma_{11} = 0$, crystal ledges migrate rapidly and evenly in both in-plane directions since $\tau_{m,ij}(0)^{-1}$ is isotropic. In the case of $\sigma_{11} < 0$, ledge migration along 1 is enhanced relative to 2. Thus, nucleation is the limiting step with $0 \leq \sigma_{11}$ since migration is rapid in both directions thus explaining why v does not change with $0 \leq \sigma_{11}$. With $\sigma_{11} < 0$, migration along 1 is retarded but unchanged along 2. Thus, the growth rate is effectively halved since only migration along 2 effectively contributes to the growth rate.

Importantly, the large value of $\Delta V_{11}^{m,11}$ suggests migration may involve the coordinated motion of several atoms along a growing island ledge rather than individual atomic motion. Analogous coordination is observed in the deformation of metals where the associated activation volume can be on the order of ~100 times the atomic volume of the material [13]. Thus, $\Delta V_{11}^{m,11} = (12 \pm 1)\Omega$ is reasonable for the presented case if migration involves such coordination.

Also, the morphological instability observed with $\sigma_{11} < 0$ [Figures 1(b) – (d)] can be readily predicted as a result of Eq. (8). A growth interface with perturbations under in-plane macroscopic compression, shown in Figure 3(a) has concentrated compressive stress (σ_{11}^{local}) in the troughs of the interface and tensile stress at the peaks since the Young's modulus of α-Si approximately half that of the crystalline counterpart.[8] Thus, as per Eq. (8), the peaks grow faster than the troughs and interface roughens. In the case of a growth interface with perturbations under macroscopic tension, shown in Figure 3(b), concentrated tension exists in the troughs with

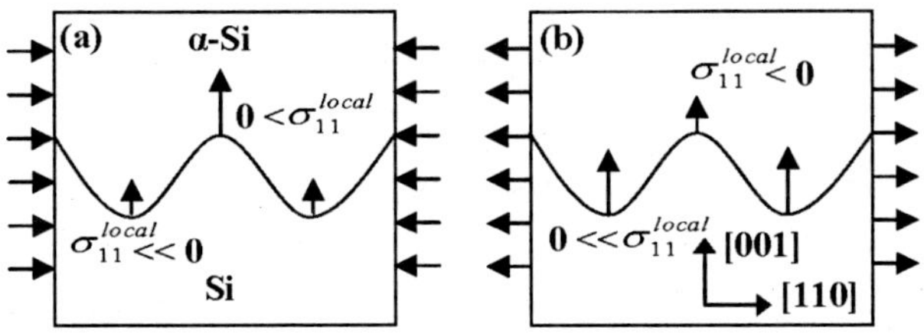

FIGURE 3. Schematic representations of growth interface morphological (in)stability with macroscopic in-plane a) compression and b) tension.

compression in the peaks. Thus, as per Eq. (8), the troughs grow faster than the peaks and the interface planarizes.

CONCLUSIONS

An atomistic model of stressed-SPEG was presented which advanced a rate-limiting approach between nucleation and migration processes on growth kinetics. It was advanced that applied stress alters the kinetics of each process differently thus making the growth velocity a complex function of in-plane uniaxial stress.

ACKNOWLEDGMENTS

The authors acknowledge the Semiconductor Research Corporation (Task ID 1372.003) for funding this research.

REFERENCES

1. G. L. Olsen and J. A. Roth, *Mater. Sci. Rep.* **3**, 1 (1988).
2. G.-Q. Lu, E. Nygren, and M. J. Aziz, *J. Appl. Phys.* **70**(10), 5323 (1991).
3. M. J. Aziz, P. C. Sabin, and G.-Q. Lu, *Phys. Rev. B* **44**(18), 9812 (1991).
4. H. A. Wilson, *Phil. Mag.* **50**, 238 (1900).
5. N. G. Rudawski, K. S. Jones, and R. Gwilliam, *Appl. Phys. Lett.* **91**(17), 172103 (2007).
6. K. S. Jones, S. Prussin, and E. R. Webber, *Appl. Phys. A* **45**(1), 1 (1988).
7. W. Barvosa-Carter, M.J. Aziz, L.J. Gray, T. Kaplan, *Phys. Rev. Lett.* **81**(7) (1998) 1445.
8. A. Witvrouw and F. Spaepen, *J. Appl. Phys.* **74**(12), 7154 (1993).
9. J. S. Williams, R. G. Elliman, W. L. Brown, and T. E. Seidel, *Phys. Rev. Lett.* **55**(14), 1482 (1985).
10. B. Lewis, *J. Cryst. Growth* **21**, 29 (1974).
11. N. G. Rudawski, K. S. Jones, and R. Gwilliam, *Phys. Rev. Lett.* **100**(16), 165501 (2008).
12. N. G. Rudawski, K. S. Jones, and R. Gwilliam, *Mater. Sci. Eng. R – Rep.* **61**, 40 (2008).
13. R. J. Asaro and S. Suresh, *Acta Mater.* **53**, 3369 (2005).

The Role of Implanter Parameters on Implant Damage Generation: an Atomistic Simulation Study

J. Singer[1,3,5], M. Jaraíz[2], P. Castrillo[2], C. Laviron[3], N. Cagnat[4], F. Wacquant[4], O. Cueto[3], A. Poncet[5]

[1]*NXP Semiconductors, 17 rue des Martyrs, Grenoble, 38054, France. julien.singer@cea.fr*
[2]*University of Valladolid, Valladolid, 47011, Spain*
[3]*CEA-Léti, MINATEC, 17 rue des Martyrs, Grenoble, 38054, France*
[4]*STMicroelectronics, 850 rue Jean Monnet, Crolles, 38926, France*
[5]*Lyon Nanotechnologies Institute (UMR-CNRS 5270), Villeurbanne, 69621, France*

Abstract. In this article we use atomistic process simulation to study the effect of some implanter parameters on damage accumulation for two types of implanters: a batch tool with a spot ion beam and a single-wafer tool with a ribbon beam. The studied parameters are the scanning speed, the wheel rotation speed and the beam diameter for the former, and the scanning speed and the beam width for the latter. We kept constant not only the species, dose and energy, but also the average dose rate, managed by the beam current, the wafer temperature (T). In such conditions damage accumulation is expected to be constant. However we show that beam focalization has a strong impact, because it affects the instantaneous dose rate. We define the instantaneous dose rate as the dose rate seen by a point of the wafer while it passes through the beam. Its effects on damage accumulation are comparable to those of wafer T and average dose rate.

Keywords: Ion implantation, Si, atomistic simulation, damage accumulation, beam focalization, instantaneous dose rate.
PACS: 61.72.uf, 61.80.Az, 61.80.Jh, 85.40.Ry

INTRODUCTION

Ion implantation generates cascades of damage within the silicon crystal. Interstitial-Vacancy (IV) Frenkel pairs accumulate and form highly damaged (but still crystalline) regions, conventionally called amorphous pockets (AP's) [1]. When the defects concentration exceeds a threshold, the silicon becomes amorphous. The amorphization by ion implantation depends directly on species, energy, dose, tilt and rotation. But other parameters affect it: for instance, the effects of implant T and average dose rate (DR_{avg}) are both significant [2,3]. In the present study, the effect of other parameters specific to two different ion implanter types is also investigated. The batch (BH) tool processes 13 wafers at once, is placed on a wheel which spins around, and is scanned over the spot ion beam. Scanning speed $V_{scan(BH)}$, wheel rotation speed $V_{rot(BH)}$ and beam diameter $D_{beam(BH)}$ (focalization) can vary. The single-wafer (SW) tool scans the wafer over a rectangular, ribbon ion beam. Scanning speed $V_{scan(SW)}$ and beam width $W_{beam(SW)}$ (focalization) will be studied. Figure 1 explains schematically the beam movements on the wafer for both implanter types.

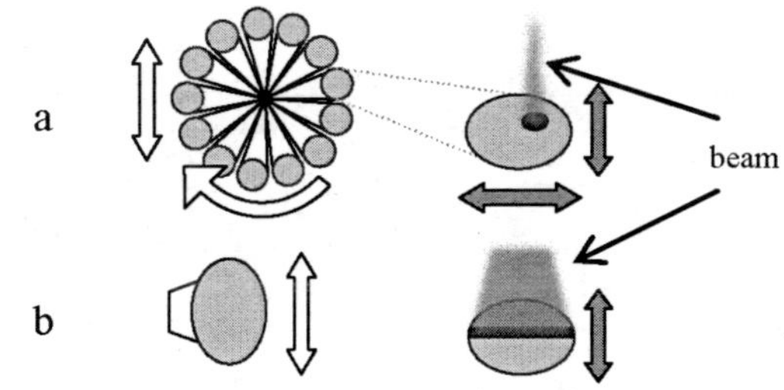

FIGURE 1. Schematic representation of a BH implanter and its beam on the wafer (a) and of a SW tool and its beam (b). Open arrows represent the wafer(s) movements in the implanter (the beam is fixed) and filled arrows show the resulting movements of the beam with respect to the wafer.

THEORY AND MODELING

Let's consider a point on the wafer. Each time it passes through the ion beam, there is a pulse of implantation. For the SW tool, the number of pulses coincides with the number of passes (Figure 2 (a)). For the BH tool, the implantation consists of a series of bursts of short pulses (Figure 2 (b)). Each wheel revolution produces one pulse and each scan leads to one burst. Figure 2 (c) schematically explains for the

CP1066, *Ion Implantation Technology,* edited by E. G. Seebauer, S. B. Felch, A. Jain, and Y. V. Kondratenko

BH tool that due to $D_{beam(BH)}$, any point on the wafer is exposed several times during a scan, which results in bursts of pulses.

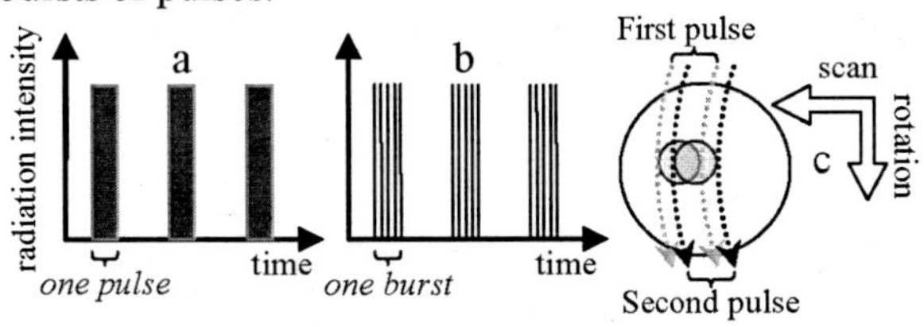

FIGURE 2. Implantation modes for (a) SW and (b) BH implanters. (c) The BH tool, for each wheel revolution a pulse of implantation is performed on the wafer.

The relevant implanter characteristics affecting the characteristics of pulses are presented in table 1 for the BH tool and in table 2 for the SW tool.

TABLE 1. Implanter parameters of the BH tool.

Name	Parameter description	Reference value
$D_{wheel(BH)}$	wheel diameter	130 cm
$V_{rot(BH)}$	rotation speed	$15\ rnd\cdot s^{-1}$
$D_{scan(BH)}$	scanning amplitude	35 cm
$I_{beam(BH)}$	beam current	5 mA
$D_{beam(BH)}$	beam diameter	5 cm
$V_{scan(BH)}$	scanning speed	$5\ cm\cdot s^{-1}$

TABLE 2. Implanter parameters of the SW tool.

Name	Parameter description	Reference value
$D_{scan(SW)}$	scanning amplitude	35 cm
$I_{beam(SW)}$	beam current	5 mA
$L_{beam(SW)}$	beam length	30 cm
$W_{beam(SW)}$	beam width	3 cm
$V_{scan(SW)}$	scanning speed	$20\ cm\cdot s^{-1}$

DR_{avg} is usually considered. It is calculated as:

$$DR_{avg} = \frac{Dose_{tot}}{Time_{tot}}, \tag{1}$$

and is proportional to the beam current I_{beam}. However we can also define the instantaneous dose rate DR_{inst}. This is the dose rate during a short time, for instance within a pulse:

$$DR_{inst} = \frac{\delta(dose)}{\delta(time)}, \tag{2}$$

where δ(dose) is the dose implanted during the short time δ(time).

In this study we use the atomistic process simulation tool DADOS. It has been proven to be reliable due to physical models implementation [4]. In particular, ion implantation is modeled in two stages. First the Marlowe code generates a cascade of IV pairs for each implanted ion using the Binary Collision Approximation [5]. The cascades are then implanted one after the other. The number of cascades N_{casc} that are simulated depends on the simulated surface $Surf_{simul}$:

$$N_{casc} = Dose_{tot} \cdot Surf_{simul}. \tag{3}$$

The time Δt_{casc} between two cascades depends on DR_{inst}:

$$\Delta t_{casc} = \frac{1}{DR_{inst} \cdot Surf_{simul}}. \tag{4}$$

During this relaxation time the diffusion of interstitials (I's) and vacancies (V's) and their recombination are simulated at the implant T, which is considered to be constant. DADOS allows the simulation of ion implantation as a series of short pulses. The input parameters are described in table 3. They are enough to define the implantation profiles of Figure 2. The calculus of these parameters is based on the implanter parameters presented in table 1 and table 2.

TABLE 3. Parameters used as inputs in DADOS.

Abbreviation	Parameter description
$N_{casc/pulse}$	number of cascades per pulse
t_{pulse}	pulse duration
Δt_{pulse}	time between two pulses
$N_{pulses/burst}$	number of pulses per burst
Δt_{burst}	time between two bursts

It is also possible to simulate a time-uniform implantation. In that case DR_{inst} equals DR_{avg} and Δt_{casc} is then:

$$\Delta t_{casc} = \frac{1}{DR_{avg} \cdot Surf_{simul}}. \tag{5}$$

Finally at any time of the process DADOS allows access to atomistic data, that are not experimentally available.

RESULTS

For both implanters I_{beam} is kept constant in order not to modify the DR_{avg}. Moreover D_{scan} is taken bigger than the wafer diameter (300 mm), which is in turn the value retained for $L_{beam(SW)}$ of the SW tool. Implant conditions were chosen close to amorphization limit in order to highlight the effect of all these parameters. If any, the transition values presented here can not be generalized. For each implant condition a specific study should be performed in order to see if a variation would be critical or not.

For the batch tool the realistic range of values for $V_{scan(BH)}$ is between 1 and 30 $cm\cdot s^{-1}$. For a too low $V_{scan(BH)}$, $N_{pulses/burst}$ becomes too high, and it is better for uniformity to do several scans instead of only one or two. The results of $V_{scan(BH)}$ variation on damage accumulation are depicted in Figure 3 (a). Damage accumulation is given as a volume fraction, with respect to the reference value. The trend we observe is that there is not an abrupt transition between an amorphizing and a non-amorphizing range.

Although $V_{rot(BH)}$ is usually fixed on a given machine, we wondered how its variation would affect damage accumulation. For uniformity reasons, the value of $V_{rot(BH)}$ would not go down under 1 $rnd\cdot s^{-1}$.

The upper value of 35 rnd·s^{-1} will probably not be exceeded. In this range no significant variations of amorphization is observed, as it is shown in Figure 3 (b).

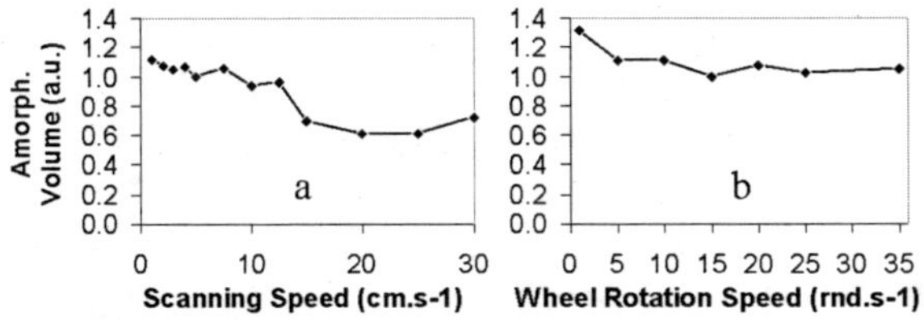

FIGURE 3. Effect of (a) $V_{scan(BH)}$ and (b) $V_{rot(BH)}$ on damage accumulation.

Finally, the variation of $D_{beam(BH)}$ results from focalization. As $D_{beam(BH)}$ decreases (resp. increases), pulses are shorter (resp. longer) but more (resp. less) "intense" – bigger (resp. smaller) $N_{casc/pulse}$. Thus DR_{inst} is directly increased (resp. decreased). Realistic values for $D_{beam(BH)}$ variations are between 1 and 15 cm. The results are represented in Figure4. We observe that $D_{beam(BH)}$ variations can result in drastic changes on damage accumulation during implantation.

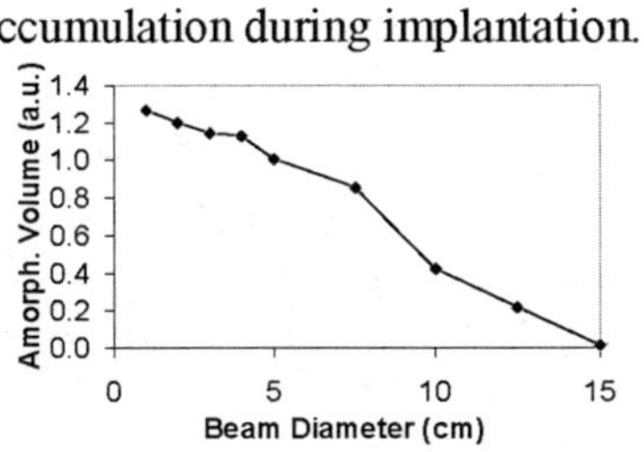

FIGURE 4. Effect of $D_{beam(BH)}$ (focalization) on damage accumulation.

Now we analyze the results for the SW tool. Plausible values for $V_{scan(SW)}$ are contained within the 10 – 40 cm·s^{-1} range. Figure 5 (a) shows that the amorphization of silicon is not affected by $V_{scan(SW)}$. $L_{beam(SW)}$ being equal to the wafer diameter, only $W_{beam(SW)}$ is affected by the focalization of the rectangular, ribbon ion beam of the SW implanter.

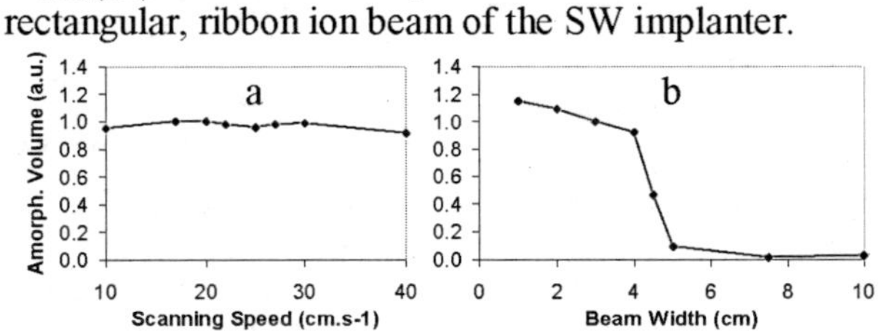

FIGURE 5. Effect of (a) $V_{scan(SW)}$ and (b) $W_{beam(SW)}$ (focalization) on damage accumulation.

The effect of the beam focalization on damage accumulation is quite remarkable, as seen in Figure 5 (b). There exists a threshold value (dependent on the other parameters and on the implant conditions) over which the amorphization volume decreases suddenly.

DISCUSSION

We brought to the fore that the beam focalization has a strong impact on damage accumulation, while V_{scan} and–for the BH tool–$V_{rot(BH)}$ do not affect it. The reason can be found by examining the DR_{inst} evolution. In Figure 6 the evolution of DR_{inst} for a BH tool is plotted as a function of the implanter parameters. It appears that DR_{inst} only depends on $D_{beam(BH)}$. Indeed the same number of cascades is implanted within a shorter pulse.

For the SW tool the assessment is the same. As it is shown in Figure 7, DR_{inst} only depends on $W_{beam(SW)}$ and not on $V_{scan(SW)}$.

And indeed, for both implanter types, plotting the damage accumulation as a function of DR_{inst} (Figure 8) leads to a threshold effect, similar to the behavior observed with DR_{avg} variations during time-uniform implants [2].

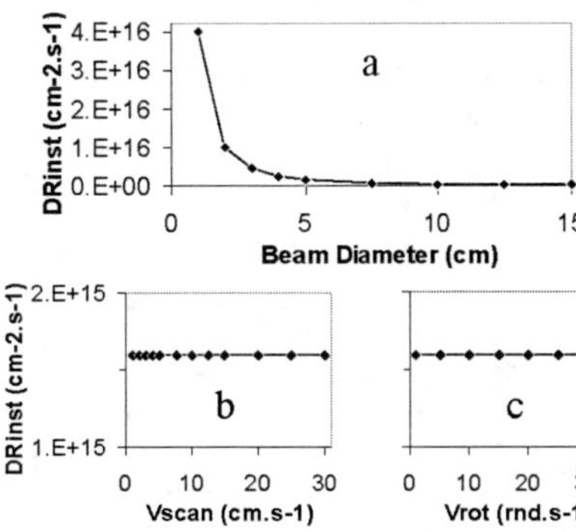

FIGURE 6. Evolution of DR_{inst} with (a) $D_{beam(BH)}$, (b) $V_{scan(BH)}$ and (c) $V_{rot(BH)}$, for the BH tool.

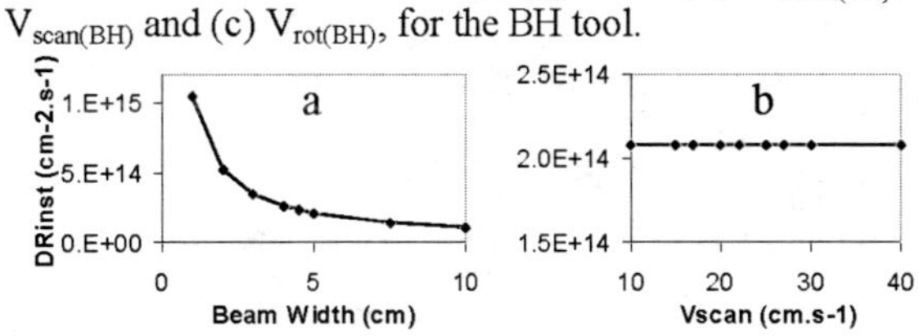

FIGURE 7. Evolution of DR_{inst} with (a) $W_{beam(SW)}$ and (b) $V_{scan(SW)}$, for the SW tool.

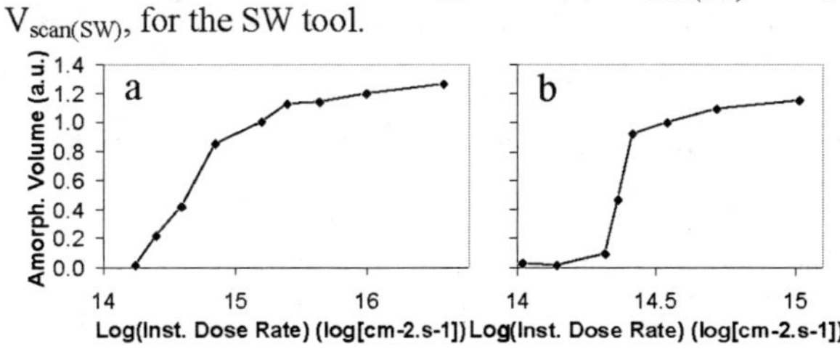

FIGURE 8. Damage accumulation as a function of DR_{inst} for the BH tool (a) and for the SW tool (b).

The DR_{inst} parameter has a drastic effect on damage accumulation because it affects the size of AP's after a pulse. In order to facilitate the analysis we will now stay with a damage below the amorphization threshold and look at the number of I's that are present within AP's. Big AP's ($N_{I's} + N_{V's} \geq 40$) form during implantation by the overlap of several cascades, as shown in Figure 9. As we saw earlier, IV pairs recombination rate within AP's depends on their size. Bigger AP's are more stable, i.e. recombination is less efficient [6]. Figure 10 (a) shows that recombination in AP's resulting from a higher number of cascades hardly occurs in comparison with dynamic recombination of few cascades. Figure 10 (b) highlights that these are the biggest AP's which remain while the smallest underwent recombination.

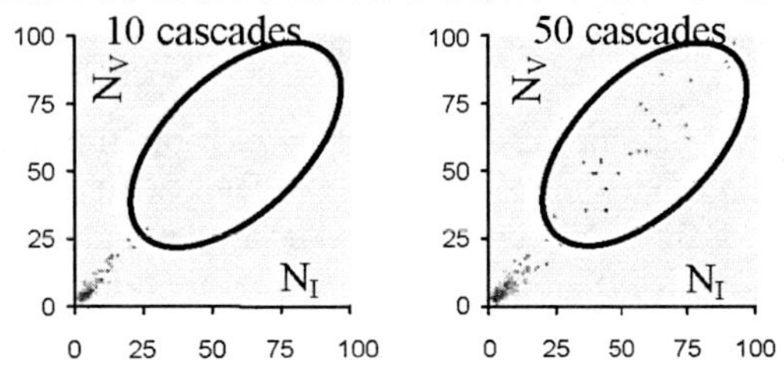

FIGURE 9. This graph plots the AP's composition histogram. The X and Y axes represent the number of I's and V's, respectively, in AP's. Each point represents a type of AP, containing X I's and Y V's. The grey level is representative of the concentration of (I's+V's). More big AP's are present when several cascades overlap.

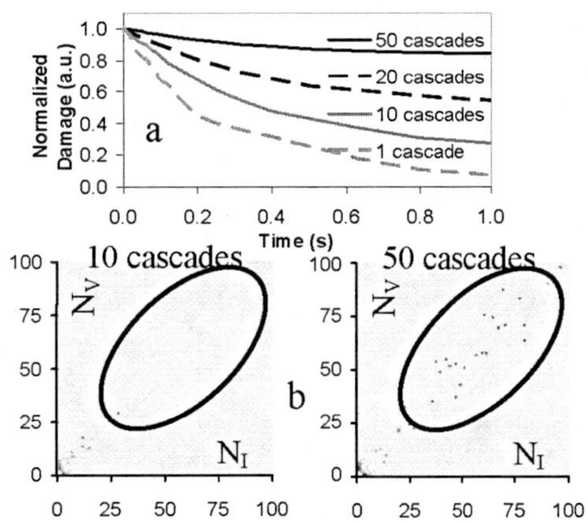

FIGURE 10. (a) Normalized I's dose in AP's as a function of relaxation time at room T, for different initial defect densities. (b) Comparison with Figure 9 shows that unlike small AP's, the big ones did not undergo recombination after 1s relaxation at room-T.

Thus at the end of a pulse, more AP's remain that are bigger if DR_{inst} is increased. This is plotted in Figure 11. In this condition, even if relaxation happens between each pulse, damage accumulation will be more important with a higher DR_{inst}.

CONCLUSION

Our atomistic process simulations concerned two types of implanters, a BH tool with a spot ion beam and a SW tool with a ribbon ion beam. We studied the effect of some implanter parameters on damage

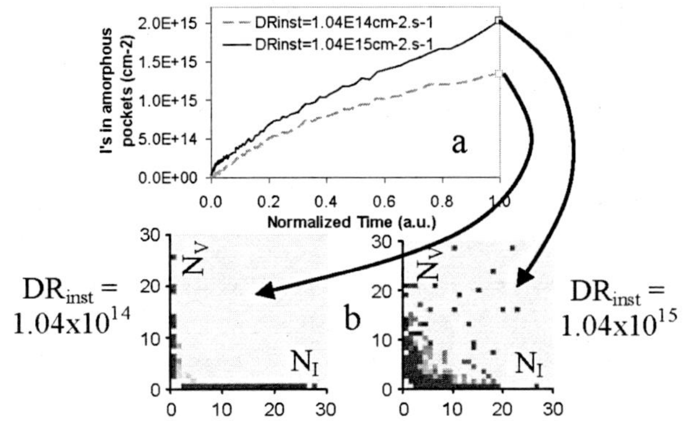

FIGURE 11. (a) The curves represent the evolution of the number of interstitials present in AP's as a function of normalized time (0 is the beginning of the pulse and 1 is its end). (b) Moreover AP's histograms show that with the lower DR_{inst} (larger $W_{beam(SW)}$) there are less AP's containing both I's and V's. This is evidence that recombination was more efficient.

accumulation during the implantation. The studied parameters are $V_{scan(BH)}$, $V_{rot(BH)}$ and $D_{beam(BH)}$ for the BH tool, and $V_{scan(SW)}$ and $W_{beam(SW)}$ for the SW tool. We kept constant not only the implanted species, dose and energy, but also the DR_{avg} (and thus I_{beam}) and the wafer T. In these conditions our simulations showed that only beam focalization ($D_{beam(BH)}$ and $W_{beam(SW)}$) did affect damage accumulation. We introduced the DR_{inst}, which is the dose rate seen by a point of the wafer while it passes through the beam. It is directly related to beam focalization. DR_{inst} has a strong impact on damage accumulation, at the same level than the DR_{avg} or the wafer T. Though beam focalization depends on the energy, it may vary from one implanter to the other, the user having no control on it. However it is also important to consider this parameter for experiments concerning amorphization and implantation defects studies.

REFERENCES

1. Caturla, M.-J., Díaz de la Rubia, T., Marqués, L. A., and Gilmer, G. H., *Phys. Rev. B* **54**, 16683-16695 (1996).
2. Schultz, P. J., Jagadish, C., Ridgway, M. C., Elliman, and R. G., Willimas, J. S., *Phys. Rev. B* **44**, 9118-9121 (1991).
3. Goldberg, R. D., Williams, J. S., and Elliman, R. G., *Nucl. Instr. Meth. in Phys. Res. B* **106**, 242-247 (1995).
4. Jaraíz, M., Gilmer, G. H., Poate, J. M., and Díaz de la Rubia, T., *Appl. Phys. Letters* **68**, 409-411 (1996).
5. Robinson, M. T., and Torrens, I. M., *Phys. Rev. B* **9**, 5008-5024 (1974).
6. Mok, K. R. C., Jaraíz, M., Martin-Bragado, I., Rubio, J. E., Castrillo, P., Pinacho, R., Barbolla, J., and Srinivasan, M. P., *J.Appl. Phys.* **98**, 046104 (2005).

Importance of the Carbon Kick-out Mechanism in Reducing Transient Enhanced Diffusion

Aaron Vanderpool

Intel Corporation 4500 S. Dobson Rd. Chandler, AZ 85248

Abstract. Co-implants have proven to be an effective means of reducing Transient Enhanced Diffusion (TED). Combinations of co-implants can significantly reduce TED but they are largely dependent upon mechanism employed. The effect of Fluorine, Carbon and Germanium combinations on Boron and Phosphorus diffusion found that the triple combination is slightly less effective than specific double combinations. A Ge and C combination work best for Boron Ultra-Shallow Junctions (USJ). F co-implants slow Solid Phase Epitaxial re-growth rates and as a result may be reducing C incorporation into the lattice reducing the C Kick-out mechanism. A F and C combination produced better results than the triple combination in Phosphorus USJ. Perhaps adding Ge stabilizes the substitutional C again reducing C Kick-out and the consumption of the Interstitials that drive TED. It is concluded that the Carbon Kick-out mechanism is the dominate mechanism in controlling TED and should be considered to achieve optimal USJ results.

Keywords: Co-implant, Boron, Phosphorus, Arsenic, Vacancy, Interstitialcy, Dissociative, Ultra-Shallow Junction.
PACS: 66.30.-h, 41.75.Ak, 81.05.Cy, 61.72.U–

INTRODUCTION

The formation of high quality Ultra-Shallow Junctions (USJ) is critical to semiconductor device performance. The Transient Enhanced Diffusion (TED) of electrical dopants in Silicon must be overcome to produce high quality USJ. Co-implants have been shown to be effective in reducing TED by various mechanisms and in some case have been combined for greater effect. However, not all co-implant combinations will complement each other to reduce TED. Successful co-implant combinations depend upon the mechanisms they employ.

TED CONTROL MECHANISMS

Electrical dopants that diffuse along interstitial pathways cluster with a Silicon self-interstitial (I) via the Interstitalcy mechanism [1,2] shown in FIGURE 1. Ion implantation damage super-saturates the lattice with interstitials significantly elevating diffusion until the damage is annealed out.

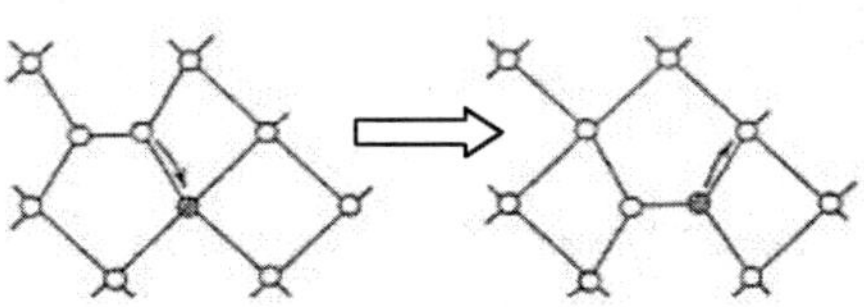

FIGURE 1. Schematic representation of the Interstitalcy mechanism [1].

To reduce TED the formation of the interstitial, clusters must be inhibited or the interstitial concentrations reduced. The most straight forward method to reduce the interstitial population is to amorphize the lattice and re-grow it using Solid Phase Epitaxy (SPE) during the activation anneal. SPE eliminates the super-saturation in the re-grown region but not below it in the implant generated End of Range (EOR) damage.

Low energy Germanium co-implants have been used to amorphize the lattice to just below the depth of the USJ in order to minimize the EOR damage. EOR damage is reduced because the large atomic mass of Ge, and the low implant energy, allow the ion to stop quickly in the lattice mostly containing the EOR damage in the amorphous layer. In addition, upon re-growth Ge adds local stain to the lattice where the USJ dopant is located reducing TED [3,4] by inhibiting interstitial cluster formation.

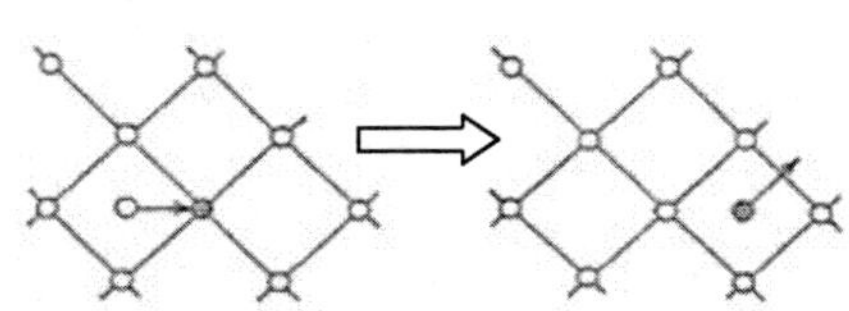

FIGURE 2. Schematic representation of the Kick-out mechanism [1].

Carbon co-implants can be added to an amorphous layer to reduce TED. Upon SPE re-growth a fraction of the C becomes substitutional. As interstitials diffuse towards the surface from the EOR damage they Kick-out the substitutional C and take their place while the C becomes interstitial. This mechanism is schematically pictured in FIGURE 2 and was well characterized by Rucker [5,6].

The oldest co-implant used to reduce TED is Fluorine. Recent calculations by Diebel [7], Shano [8] and others indicate that Fluorine is most likely to cluster with Vacancies (F-V). Interstitials diffusing

CP1066, *Ion Implantation Technology,* edited by E. G. Seebauer, S. B. Felch, A. Jain, and Y. V. Kondratenko

from the EOR damage will annihilate the cluster releasing the Fluorine and consuming the interstitial. Analytical evidence of these clusters has been found using Positron Annihilation Spectroscopy [9]. Experiments to increase F-V cluster concentrations by increasing the vacancy concentration have been able to reduce Boron TED [10] as well.

CO-IMPLANT COMBINATIONS

At first glance the mechanisms associated with these three co-implants appear to be independent and capable of all being utilized to control TED in the same USJ. In fact, some combinations of these co-implants have been used to complement each other [10,11,12]. However, not every combination is complementary.

Control of Boron TED

Previous Soak anneal work [10] has found that the combination of Fluorine and Carbon co-implants were much more effective at reducing Boron TED than either of them alone. In FIGURE 3 the Spike annealed F and C combination (#1) was compared with the more common Ge co-implant (#2) technique.

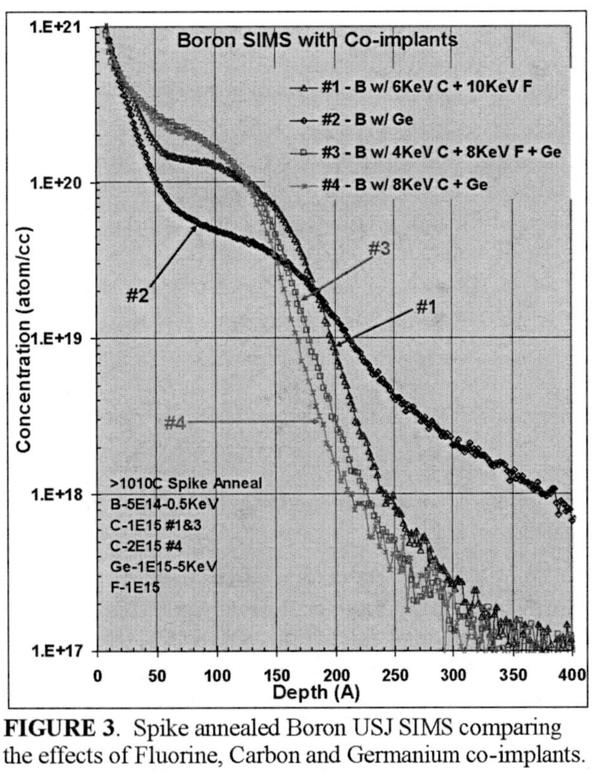

FIGURE 3. Spike annealed Boron USJ SIMS comparing the effects of Fluorine, Carbon and Germanium co-implants.

The F and C combination is much more effective at reducing Boron TED than only the Ge co-implant as the interstitials are consumed by more than one mechanism. However, when Ge is paired with C (#4) B TED is reduced even further. This must indicate that either F is reducing the effectiveness of the C mechanism to some extent or the Ge mechanism is more effective than the F mechanism. While a comparison of the Ge profile (#2) with pervious work[10] shows that Ge is more effective than F it is also true that F is reducing the effectiveness of the C mechanism. The F, C and Ge combination (#3) in FIGURE 3 provides evidence of this. When F is added to the best case Ge and C combination the profile degrades slightly indicating that the F mechanism is not helping to reduce B TED. It is well documented that F slows SPE re-growth in USJ [13]. Perhaps when the SPE re-growth rate is reduced, less C is incorporated into substitutional sites reducing the C Kick-out consumption of interstitials. In any case, the F and Ge mechanism are only working to compliment the C mechanism which is the common factor in achieving high levels of TED reduction.

Control of Phosphorus TED

In FIGURE 4 the individual effect of Fluorine (#5) and Carbon (#6) co-implants on Phosphorus TED demonstrate that the C mechanism is stronger in reducing TED than the F mechanism.

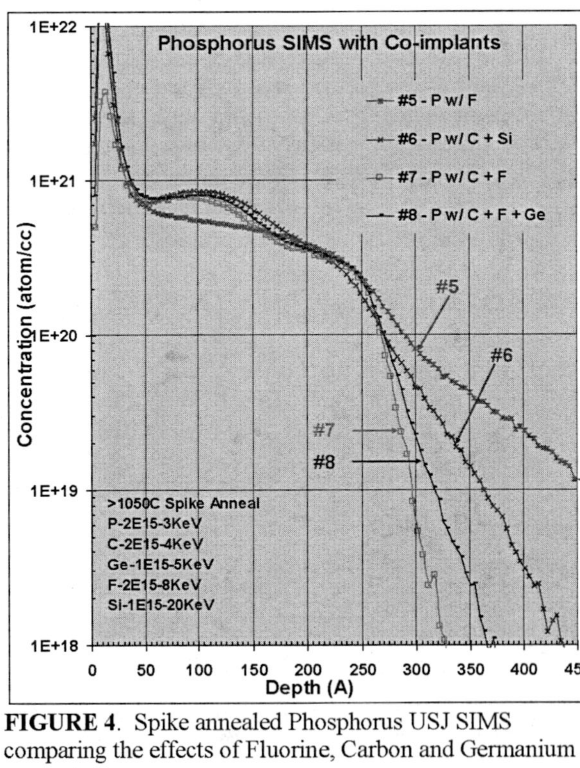

FIGURE 4. Spike annealed Phosphorus USJ SIMS comparing the effects of Fluorine, Carbon and Germanium co-implants.

The combination of F and C (#7) produce the best profile in the figure where the separate mechanisms work together to reduce TED. However, when Geranium is added to the F and C combination (#8) Phosphorus TED increases. In this profile Ge must either be negatively affecting the F mechanism or the C mechanism. This means either the Ge is reducing the formation of F-V clusters (or interstitial annihilation of the F-V cluster) or it is stabilizing the substitutional C reducing C Kick-out. Because of the local strain Ge adds to the lattice F-V clusters should be enhanced by the addition of Ge since the vacancy will relieve some of the Ge strain. By the same token, retention of substitutional C should help to relieve the local strain generated by the Ge. Therefore, it is likely that the Ge in the F, C and Ge profile (#8) is reducing C kick-out because the diffusion increased rather then decreased because F-V mechanism was amplified. This means that minimizing Phosphorus TED requires co-implant combinations with control mechanisms that interfere with the C the least. It also means the C mechanism is the most important of the three.

Differences Between Boron and Phosphorus TED

There is one significant difference between the TED control findings in FIGURE 3 and FIGURE 4. The F and C combination (#1) for the Boron USJ is worse than the F, C and Ge combination (#3) whereas the F and C combination (#7) for the Phosphorus USJ is better than the F, C and Ge combination (#8). Even though the difference is small it must be due to the effect Ge has on the electrical dopant in question. Because of its smaller size in the lattice, substitutional Boron acts to relieve some of the local strain Ge places on the lattice. As discussed, this inhibits the formation of Boron Interstitial clusters reducing TED. However, Phosphorus is much closer in size to Silicon and so does not relieve some of the local strain caused by Ge. This causes the F mechanism to be the second strongest in reducing Phosphorus TED and is one of the reasons why Phosphorus TED is harder to control.

Arsenic Diffusion

TED control in Arsenic USJ has never been necessary because of the excellent profiles it produces with no help. This is probably becasue 40% of its diffusion is believed to be along interstitial pathways [1,14] while the other 60% is believed to diffuse via a vacancy mechanism. This raises the question of what affect TED control co-implants will have on an As USJ. The result of a Carbon co-implant is displayed in FIGURE 5. Arsenic with no co-implants (#10) produces a excellent profile. The solubility of the bulk of this USJ is higher than a Boron USJ with a steep roll-off of 3.5 nm/decade and a junction depth of 24 nm at a concentration of 1.0E19 cm^{-3}. This profile does have an interstitial diffusion tail below 1.0E18 cm^{-3}. When a C co-implant (#9) is used with As the diffusion tail is almost entirely suppressed however the junction depth increases to 26 nm and the profile gets steeper to 2.2 nm/decade. The containment of the interstitial tail is consistent with consumption of Interstitials by the Kick-out mechanism observed with Boron and Phosphorus C co-implants.

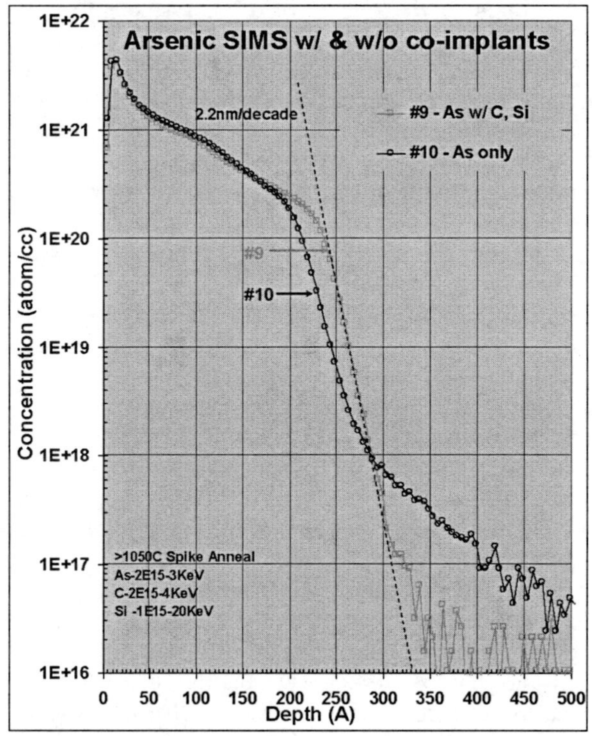

FIGURE 5. Spike annealed Arsenic USJ SIMS evaluating the effects of Carbon co-implantation.

The increase in junction depth must be a result of the Kick-out and Dissociative mechanism. The increase in steepness and junction depth is the result of increasing the vacancy diffusion of As. Carbon co-implants facilitate this because substitutional C undergoes a Dissociative reaction pictured in FIGURE 6 that elevates the vacancy concentration. This elevation causes many of the interstitial diffusing As atoms in the tail to diffuse via the Vacancy mechanism in FIGURE 6 in the bulk of the USJ leading to a steeper but slightly deeper profile. This is a higher quality USJ.

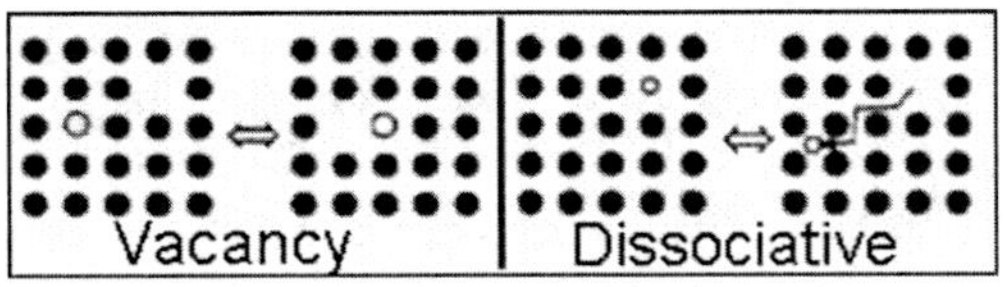

FIGURE 6. Schematic representation of the Vacancy and Dissociative diffusion mechanism [1].

CONCLUSIONS

To improve the quality of USJ it is possible to combine some common co-implants but not all combinations complement each other. The mechanisms that generate a reduction in TED must be carefully considered so that second order effects like SPE growth rate or strain compensation do not degrade the effectiveness of more important mechanisms like Carbon Kick-out. In addition, the USJ species type affects the best methods to control TED. Boron's small size in the Silicon lattice allows it to be more easily controlled than Phosphorus which will likely determine the ultimate extent to which Phosphorus TED can be controlled. It is extremely convenient that SiGe is used to increase hole mobility in B USJ since Ge has the unique size advantage needed to control Boron. In general however, Carbon co-implants do the most to control TED in conjunction with another co-implant. One of the common features with all the plots that have C co-implants is an elevated solubility throughout the bulk of the USJ. The C co-implant in the As plot (FIGURE 5) was able to convert most of the interstitial diffusion to vacancy diffusion. The As profile changes this generated are reflected to some extent in all of the Boron and Phosphorus profiles with Carbon co-implants. In general, among the TED control mechanisms used in these plots for Boron and Phosphorus the mechanism that interferes the least with the Carbon Kick-out mechanism achieved the largest reduction in TED. Therefore, it is reasonable to conclude that the Carbon Kick-out mechanism is the dominate control mechanism among these co-implants. In the design of USJ, integrators will be able to produce the highest performance junctions if the strength of the Carbon Kick-out mechanism is utilized.

ACKNOWLEDGEMENTS

I am very grateful for the consultation and excellent SIMS analysis of Andre Budrevich in the Intel Portland Technology Development lab. Without his expert technique the subtle differences in these SIMS profiles would not have been visible and these conclusions could not have been drawn.

REFERENCES

1. P. M. Fahey, P. B. Griffin, and J.D. Plummer. *Rev. Mod. Phys.* **61** 289 (1989).
2. B. Sadigh, T. J. Lenosky, S. K. Theiss, M.-J. Caturla, *et al,. Phys. Rev. Lett.* **83** 4341–4344 (1999).
3. N. Moriya, L. C. Feldman, H. S. Luftman, C. A. King, J. Bevk, B. Freer. *Phys. Rev. Lett.* **71** 883 (1993).
4. P. Kuo, J. L. Hoyt, J. F. Gibbons, J. E. Turner, D. Lefforge. *Appl. Phys. Lett.* **66** 580 (1995).
5. H. Rucker, B. Heinemann, D. Bolze, R. Kurps, D. Kruger, *et al., Appl. Phys. Lett.* **74** 3377 (1999).
6. H. Rucker, B. Heinemann, D. Bolze, D. Knoll, D. Kruger, R. Kurps, *et al.* IEDM Technical Digest. International, an IEEE conference; 14.4 (1999).
7. M. Diebel and S. T. Dunham. Silicon Front-End Junction Formation Technologies, MRS **717** C4.5 (2002).
8. T. Shano, R. Kim, T. Hirose, Y. Furuta, H. Tsuji, M. Furuhashi, K. Taniguchi. IEDM Technical Digest. International, an IEEE conference; 37.4 (2001).
9. X. D. Pi, C. P. Burrows, and P. G. Coleman. *Phys. Rev. Lett.* **90** 155901 (2003).
10. A. Vanderpool and M. Taylor, *Nucl. Instr. and Meth. in Phys. Res. B* **237** 142 (2005).
11. B.J. Pawlak, R. Lindsay, R. Surdeanu, P. Stolk, K. Maex, X. Pages. 14th International Conference on Ion Implantation Technology (2002).
12. A. Vanderpool, A. Budrevich and M. Taylor. *AIP Conf. Proc. for IIT2006* **886** 41 (2006).
13. J.M. Jacques, L.S. Robertson, K.S. Jones, J. Bennett, and M. Rendon. Silicon Front-End Junction Formation Technologies, MRS **717** C4.6 (2002).
14. S. C. Jain, W. Schoenmaker, R. Lindsay, P. A. Stolk, S. Decoutere, *et al., J. Appl. Phys.* **91** 8919 (2002).

Fracture in Hydrogen-Implanted Germanium

F. Mazen[1], A. Tauzin[1], L. Sanchez[1], F. Chieux[1], C. Deguet[1], E. Augendre[1], T. Akatsu[2], C. Richtarch[2], and L. Clavelier[1]

[1] *CEA-LETI-MINATEC – 17, rue des Martyrs, 38 054 GRENOBLE CEDEX 9 – FRANCE*
[2] *SOITEC S.A., Parc Technologique des Fontaines, 38190 BERNIN – France*

Abstract. We have studied the mechanism of fracture in hydrogen-implanted Ge. First, the as-implanted Ge state and its evolution during subsequent annealing were characterized via TEM and FTIR-MIR spectroscopy. Results showed that the extended defects formation and growth follow the same basic mechanism in Ge as in Si, which is the reference material. Nevertheless, the global damage level in the implanted Ge layer is higher compared to Si. Second, the fracture step was studied via the fracture kinetics analysis, SIMS and AFM on the transferred layer. An activation energy comparable to the reported data from blistering studies was obtained. Just like in Si, the Cmax of H in Ge measured via SIMS was found to decrease during the fracture anneal. This decrease is associated with the formation of gaseous H_2 that pressurizes the internal cavities and then contributes to the fracture. Finally, a high roughness of the Ge transferred layer was measured, which results from the large thickness of the implantation damaged zone.

Keywords: Germanium, Germanium On Insulator, Fracture, Hydrogen.
PACS: 68.55.Ln, 78.-30.-J,68.37.Lp,68.35.Ja

INTRODUCTION

Germanium is now considered a material of interest for various applications in both micro- and opto-electronics fields. For example, in the CMOS field, for sub-22 nm node, the theoretical high carrier mobility and compatibility with high k materials are at the basis of the renewed interest in Ge [1]. Also, a good match with GaAs for both its lattice parameter and its coefficient of thermal expansion has been another reason for interest in Germanium in the field of high-efficiency multijunction solar cells [2]. Recent developments on thin Germanium-On-Insulator (GeOI) substrates have given rise to a new generation of innovative germanium templates.

The feasibility of GeOI substrates using the Smart Cut™ technology has been demonstrated in previous publications [3] [4]. In this study, we focus on a key point of this technology, i.e. the fracture mechanism in hydrogen implanted Ge. Indeed, since the Smart Cut ™ discovery [5] and its application to Silicon-On-Insulator (SOI) manufacturing, many studies have been performed to characterize and understand the fracture mechanism in silicon; see for example the review [6]. By comparison, only a few papers [7-9] deal with hydrogen behavior in Ge for implant conditions compatible with layer transfer applications. These papers are limited to blistering studies or fundamental basic characterizations of hydrogen implanted Ge.

In this work, the first part is concerned with the characterization of germanium after the hydrogen implantation step in our experimental conditions. The second part deals with the fracture step (i.e. the transfer of the implanted Ge layer onto a foreign substrate) of the Smart Cut™ process for GeOI fabrication.

EXPERIMENTAL

Four-inch diameter (100) bulk Ge substrates were covered with a 250 nm thick SiO_2 layer via PECVD. The SiO_2 was densified at 600°C. Samples were then implanted with H^+ ions at 7 X 10^{16} at/cm² and energy of 80 keV without active control of sample temperature during implantation. Implanted wafers were then molecularly bonded to Si or Ge substrates covered with a thin SiO_2 layer [4]. The microstructure of the as-implanted and annealed (but not fractured) samples was analyzed via Transmission Electron Microscopy (TEM). Multiple Internal Reflexion Fourier Transform Infrared Spectroscopy (FTIR-MIR) experiments were performed following the procedure detailed in [10]. For splitting kinetics studies, the bonded samples were submitted to isothermal annealing in a furnace up to the occurrence of fracture.

CP1066, *Ion Implantation Technology,* edited by E. G. Seebauer, S. B. Felch, A. Jain, and Y. V. Kondratenko

Secondary Ion Mass Spectroscopy (SIMS) was performed on as-implanted and on transferred Ge layers. The morphology of the transferred Ge layer was characterized via Atomic Force Microscopy (AFM).

GERMANIUM IMPLANTATION

Figure 1 shows TEM images of an as-implanted Ge sample. 1a) is a large scale micrograph of the damaged zone. Most of the defects are situated in a large zone wich has a thickness of around 400 nm. The most defective zone is situated in the last 150 nm at the bottom of the damaged zone. In this region, small microcracks parallel to the sample surface can be distinguished.

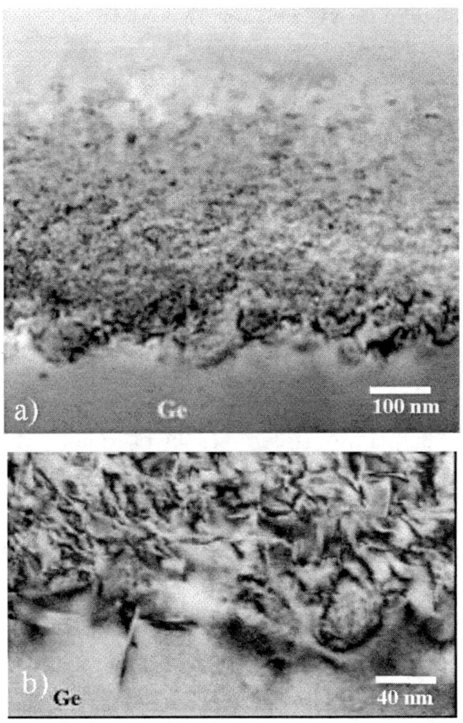

FIGURE 1. TEM images of as-implanted Ge at $7X10^{E16}$ H+/cm² at 80 keV. a) Large scale – the image is defocalized to distinguish the microcracks b) zoom on a detail of the most damaged zone.

The large-scale damage level in the hydrogen implanted layer is a major difference between the Ge and the Si reference case. Indeed, for the same implantation dose and energy range, the reported defective zone in Si is at least two times thinner [11].

Image 1b) shows a detail of the most damaged zone of the image 1a). A high density of (100) platelets, parallel to the Ge surface, is observed. Their length varies between 10 and 30 nm. The same kind of defects has been identified in (100) Si as the fracture precursor. We confirm the presence of vertical (100) platelets already observed in Ge, but we were not able to detect (111) oriented platelets and nanocavities [8]. It may be due to the high damage level of the Ge layer for the conditions studied here.

Figure 2 shows the evolution of the H implanted Ge microstructure after a thermal treatment (not sufficient to reach the Ge fracture). The large scale damage remains very high. In the most heavily damaged zone, both lateral and vertical dimensions of the small microcracks that were identified in Figure 1a, have increased. Their lateral propagation has led to the formation of microcraks of length longer than 150 nm. Due to the large thickness of the damaged zone, microcracks propagate in various planes at different depths. We will see later that this point has consequences for the transferred layer topology.

In conclusion of the TEM study, the global mechanism of extended defects (i.e., platelets and microcracks) formation and growth in Ge seems to be in agreement with the Si case. Platelets are present after implantation. They are the precursor of microcracks that grow and propagate during the susbequent annealing.

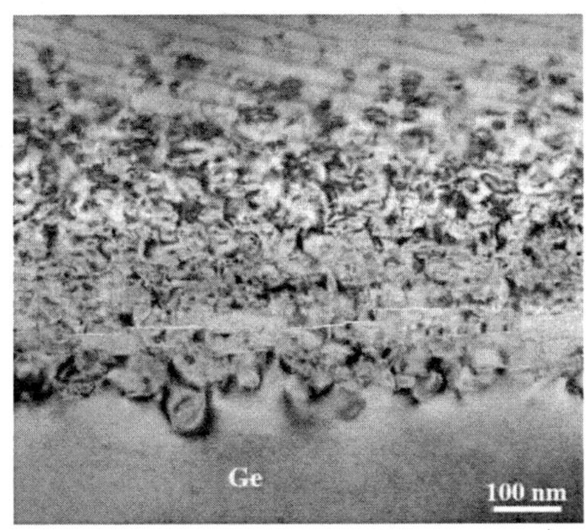

FIGURE 2. TEM images of Ge implanted at $7 X 10^{16}$ H+/cm² at 80 keV and annealed for 10 minutes at 325°C.

Infrared spectroscopy allowed us to study the chemical configurations of H atoms implanted in Ge. Figure 3 shows FTIR-MIR spectra of as-implanted and annealed Ge samples. The annealing conditions were chosen to get close to the fracture but without significant appearance of blisters at the Ge surface. Indeed, too much surface relief could perturb the FTIR-MIR measurement.

The peaks related to H in Ge bonds are located between 1900 and 2100 cm^{-1}. The contribution to the spectra of the different configurations of H in Ge is not yet fully understood. After implantation, the principal absorption peak is centered at 2032 cm-1. It could be attributed to hydrogenated divacancy V_2H_6 [9]. After annealing, the intensity of this peak slightly decreases and the global spectra shift towards a lower wavenumber region. Compared to the as-implanted spectrum, the contribution of three peaks increases : at 2011cm^{-1}, 1992 cm^{-1} and 1976 cm^{-1}. In our case (thermal treatment up to the occurrence of blistering), the peaks at 1976 cm^{-1} and 1992 cm^{-1} can be attributed to Ge monohydrides on the (111) and (100) internal surfaces, respectively. The peak at 2011 cm^{-1} is strongly magnified after the annealing and has a major impact on the IR spectrum. It has been recently attributed to Ge dihydrides on internal surfaces [9].

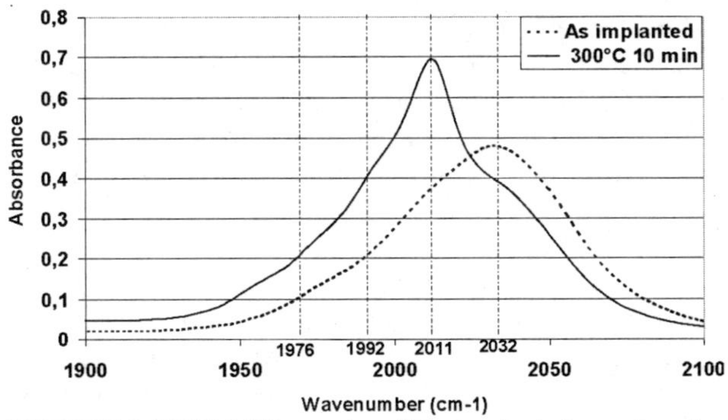

FIGURE 3. FTIR-MIR spectra of Ge implanted samples. As implanted 7X10^{16} H+/cm² at 80 keV and annealed at 300°C during 10 min.

All these peak attributions and their evolution are consistent with observed growth and propagation of microcraks after annealing. Nevertheless, the evolution of the IR spectra with temperature is different from comparable experiment in Si [12]. In silicon, the anneal leads to a strong decrease of the intensity of the peaks associated with hydrogenated multivacancy and Si point defects. Contrary to the Ge case, this decrease is not associated with an increase of the peaks related to internal surfaces. Indeed their intensity slightly decreases during annealing. It is supposed that the strong decrease of the intensity of the global peak related to H bonded in Si after annealing is due to the formation of gaseous H_2 molecules which are not detectable via IR spectroscopy. In our study, we obtained a global increase of the intensity of the peaks related to bonded hydrogen in Ge after annealing. This point needs to be further investigated since SIMS analysis (see Figure 5) has shown that gaseous H_2 is, like in Si, produced during the annealing of hydrogen implanted Ge.

GERMANIUM FRACTURE

Figure 4 shows a splitting kinetics of implanted and bonded Ge samples. The data were obtained by measuring the necessary time to reach the fracture (i.e. to transfer the implanted Ge layer onto the foreign substrate) at different temperatures. It has been demonstrated that layer transfer (or blistering) is thermally activated [11]. Thus, the activation energy (Ea) associated with the fracture mechanism can be extracted from an Arrhénuis plot of the measured time versus temperature.

We obtained Ea=1.8 eV in our experimental conditions. This value is in agreement with the result reported in [7] about the blistering kinetics of hydrogen implanted Ge. In Si, it has been proposed that the activation energy Ea for fracture corresponds to the sum of the activation energy of free hydrogen diffusion in silicon and hydrogen bond energy in implanted silicon [13]. The same mechanism can then be proposed for fracture in Ge, with 0.38 eV [14] as the energy for free H diffusion in Ge and ~1.4 eV for the bond energy of H in the implanted Ge matrix.

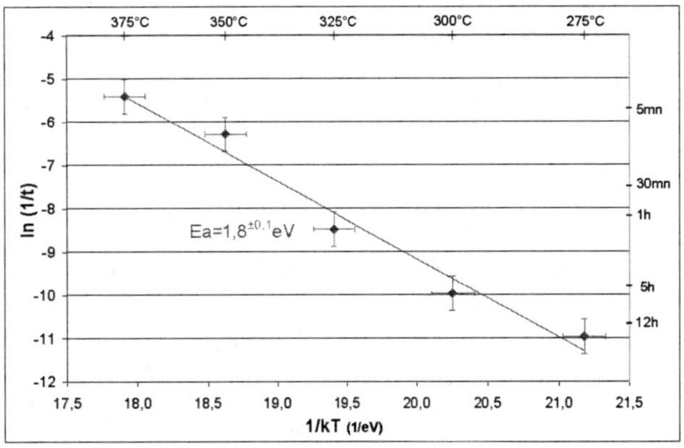

FIGURE 4. Fracture kinetic of Ge implanted with 7X10^{16} H+/cm² at 80 keV and bonded to a foreign substrate.

Figure 5 shows the comparison of SIMS profiles of the as-implanted Ge and the transferred Ge layer. The profile after splitting is adjusted to fit with the as implanted one. Then, the maximum depth of "after split" profile corresponds to the surface of the transferred Ge layer. A reference profile of Ge before implantation shows that the strong H peak at Ge/SiO2 interface is not due to the implantation.

First, the results show that the Cmax decreases from 3 X 10^{21} at/cm^3 after implant to 1.6 X 10^{21} at/cm^3 after splitting. Comparable results have been obtained for the Si case and have been attributed to the formation of gaseous H_2 molecules during the annealing. Since H_2 is not as well detected via SIMS as bonded H, that leads to a decrease of the apparent H concentration in the SIMS profile [12].

Secondly, the depth where fracture occurs (i.e the surface of the transferred layer) is shifted by around 100 nm deeper from the as implanted Cmax. This shift is too high to be attributed to experimental variations

(SIMS precision or non uniformity of the transferred layer thickness). For the Si case, it has been reported that for standard process conditions the fracture occurs at (or very close to) the Cmax depth [6]. Nevertheless, a shift between Cmax and fracture depth was observed in Si implanted with a very high dose [15]. It was attributed to the inhibition of the fracture propagation in a too highly damaged matrix. Since we observed a very high level of damage in the implanted Ge layer, the shift of the fracture line from Cmax can be attributed to the same causes.

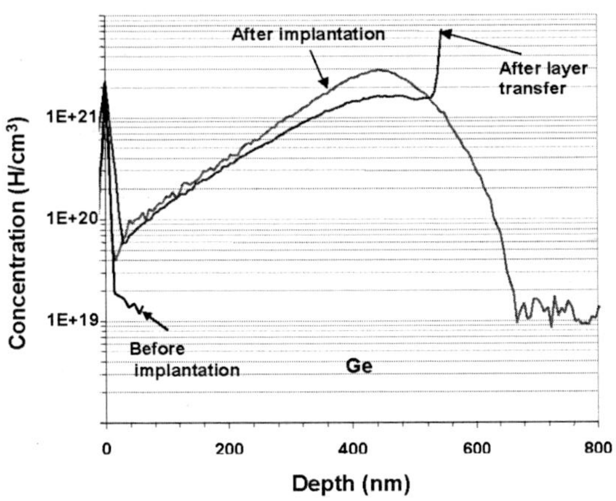

FIGURE 5. SIMS profiles of Ge substrate, Ge substrate after implantation (7 X 10^{16} at/cm² at 80keV) and transferred layer after splitting. The profile after splitting is adjusted to fit with the as implanted one.

Figure 6 shows an AFM image of the transferred Ge layer surface. We measured a roughness of 16.5 nm RMS. This value is roughly double of a Si layer transferred with similar implant conditions [11]. This is consistent with TEM results after implant, which showed that the damaged layer thickness is also double that in the Si case. We can then suppose that the high roughness of the Ge transferred layer is directly due to the as implanted state microstructure: during Ge fracture, microcracks can propagate in a heavily damaged layer that is thicker than for the Si case.

FIGURE 6. AFM image of transferred Ge layer.

CONCLUSION

We have studied the fracture in H implanted Ge. The global mechanism seems to be the same as the reference case, fracture in Si with similar implant conditions. 1) Platelets are the precursors of the fracture. They are already present just after implantation. After annealing, microcracks grow and propagate until fracture occurs. 2) After annealing, hydrogen is mainly bonded to internal surfaces. 3) H_2 gas is formed during annealing and may contribute to fracture by pressurizing existing cavities. 4) Ge fracture is a thermally activated mechanism that can be linked to H-diffusion in the presence of defects.

Some particularities of the Ge fracture can be highlighted: 1) the as-implanted layer is highly damaged. That leads to a high roughness of the transferred layer. 2) after annealing, the population of bonded hydrogen seems to increase and there is no evidence of H_2 formation on IR spectra.acknowledgements

The authors aknowledge A.M Papon for TEM analysis, F. Laugier for SIMS analysis and N. Rochat for IR spectrscopy. The authors acknowledge all the people from LETI and SOITEC who contribute to this project.

REFERENCES

1. C.Chui *et al.*, IEDM Proceedings, p. 437 (2002).
2. J.M. Zahler *et al.*, *Thin Solid Films*, **403**, p. 558 (2002).
3. F. Letertre *et al.*, *MRS proceedings* 809, p. 153 (2004)
4. C. Deguet *et al.*, Electronics Letters **42** p. 415 (2006)
5. M.Bruel et.al, *Electronics Letters* **31** p. 1201 (1995)
6. K.K. Bourdelle, in these Proceedings.
7. S. Bedell *et al.*, *Journal of Applied Physics* **90**, p 1138 (2001).
8. T. Akatsu *et al.*, *Applied Physics Letters* **86**, 181910 (2005)
9. J.M. Zahler *et al.*, *Physical Review B* **75**, 035309 (2007)
10. N.Rochat. Technological diploma, INPG France (2000)
11. Aspar et.al, *Journal of Electronic Material* **30**, 834 (2000)
12. S. Personnic *et al.*, *Journal of Applied Physics* 103, 023508 (2008).
13. M. Bruel *et al.*, *MRS Bulletin 23*, p35 (1998)
14. Van Wieringen *et al.*, Physica XXII, p 849, (1956)
15. T. Hochbauer *et al.*, *Journal of Applied Physics* **92**, p. 2335 (2002)

Ion Beam Layer Separation of Cadmium Zinc Telluride

Rabi S. Bhattacharya[1], P. He[1], Y. Xu[1] and M. Goorsky[2]

[1] UES, Inc.4401 Dayton-Xenia Road, Dayton, OH 45432
[2] University of California at Los Angeles, 10920 Wilshire Blvd, Suite 107, Los Angeles CA 90024

Abstract. We have investigated the approach of ion induced layer separation process for layer splitting from Cadmium Zinc Telluride (CZT) bulk single crystal and transferring and bonding the separated layers with Silicon (Si) wafers. Layer separation experiments have been carried out at UES using 1 MeV H^+ ions from the high energy accelerator (1.7 MV Tandetron). Ion dose and annealing temperature for complete separation of 1 cm x 1 cm size layers have been optimized. Bonding of CZT with Si was accomplished using various IR transmitting chalcogenide glasses. Cracking of separated CZT films was occurring for chalcogenide glass bonded films. Optimization of thermal treatment has led to the minimization of such cracks. Detailed characterizations of the separated films will be presented.

Keywords: Ion Implantation, Smart-Cut, Layer Separation, Cadmium Zinc Telluride, Infrared Detector, Chalcogenide
PACS: 81.05.Dz, 81.20.-n.

INTRODUCTION

Mercury cadmium telluride (HCT) is the material of choice for detectors in various applications, e.g., thermal imaging, weapons guidance systems, satellite surveillance, night vision, nuclear radiation and gamma ray detectors, nuclear medicine and related applications [1]. It is a ternary semiconductor compound, which exhibits a wavelength cutoff proportional to the alloy composition. The actual detector in the photoconducting mode of operation is composed of a thin layer (10 to 20 µm) of HCT with metallized contact pads defining the active area. Photons with energy greater than the semiconductor band-gap energy excite electrons into the conduction band, thereby increasing the conductivity of the material. The wavelength of peak response depends on the material's band-gap energy and can easily be varied by changing the alloy composition. In order to sense the change in conductivity, a bias current or voltage is required. Another mode of detection is a p-n junction that operates in a photo-voltaic configuration and there is no constant voltage applied. This is the preferred mode in the modern world of focal plane arrays (FPA).

The growth of HCT has been traditionally done on CdTe substrates. Due to the lattice mismatch between HgCdTe and CdTe, $\Delta a/a \sim 3\times10^{-3}$, there are a great number of misfit dislocations that are generated at the interface and reduce the quantum efficiency [1]. The role of lattice matching in determining the quantum efficiency of IR devices has been shown to exist in a similar material system, PbSnTe [2], and shown to apply for the HCT system as well. The state-of-the-art for the growth of $Hg_{1-x}Cd_xTe$, with x~0.2, is to use $Cd_{1-y}Zn_yTe$, with y~0.045, that matches, to within a value of $\Delta a/a \sim 2\times10^{-4}$, the lattice constant of HCT. It is only for a perfect lattice matched system that one can expect to have BLIP (background limited IR performance) conditions. The use of CdZnTe (CZT) is obviously the only way to go today.

The growth of high quality CZT substrates is not an easy task, as the distribution parameter of the Zn component from the melt into the solid is larger than 1, and therefore the y value changes rapidly along the growing boule. Only a small part of the whole crystal is suitable for matching the HCT film's lattice constant with the right value. The main supplier of today's needs is one company and the price of 5x5 cm^2 wafers is very high, above $10,000.00. Thus, there is a need to find a better way of manufacturing those CZT substrates to meet the needs of the IR industry. And the needs will grow immensely in the coming decade as the US military will have to replenish the thousands of systems with IR modules used during the Iraqi Freedom War including all the missiles used and the systems on armored vehicles that got destroyed.

The objective of this work was to develop a cost effective process of preparing large area CZT wafers for the epitaxial growth of HCT and have it ready on a Si wafer in order to facilitate the electronic readout circuitry. The method we proposed is known as the

CP1066, *Ion Implantation Technology,* edited by E. G. Seebauer, S. B. Felch, A. Jain, and Y. V. Kondratenko

"Smart-Cut" technique, invented in France, and applied already to prepare thin films of Si, GaAs, InP, InSb, $LiNbO_3$, and SiC [3-6].

EXPERIMENTAL DESCRIPTION

Ion implantation has been carried out at UES by using the 1.7 MV Tandetron accelerator. Hydrogen ions of energy in the range 250 keV to 3.0 MeV can be easily produced in this machine. The UES machine is equipped with a 4" wafer handler that can accommodate 25 wafers for implantation. The beam is scanned to provide a very uniform implant over the whole wafer. Ions are implanted nearly normal to the surface of the CZT wafer without masking. A small offset angle is used to prevent channeling. The energy of implantation can be adjusted to select a desired depth of penetration and therefore a pre-determined film thickness. The substrate can be heated or cooled to a desired temperature during implantation.

Implanted CZT is subjected to thermal annealing at various temperatures. Implanted crystals were transferred onto clean Si substrates and placed in the furnace at room temperature. The temperature is then gradually increased at 5-10°C/min ramp rates to desired temperature and kept constant over a certain period of time (typically >6 hours), and then cooled down to room temperature.

CZT single crystal substrates of size 1 cm x 1 cm were purchased from Nikko Materials Co., Ltd. CZT (211)A surfaces were implanted with 1 MeV H^+ ions at various doses. Characterizations were done by standard X-ray diffraction (XRD) $\theta/2\theta$, high resolution XRD, atomic force microscopy (AFM), optical and transmission electron microscopy (TEM), and Fourier Transformed Infrared (FTIR) Spectroscopy.

RESULTS

Initial experiments of ion implantation with substrate at room temperature even at very high ion doses, > $3x10^{17}$ cm^{-2}, did not result to any bubble formation or separation. Ion implantation experiments were then carried out with substrate mounted on a cold finger (cooled with liquid nitrogen). Figure 1 shows the bubble formation on the surface of CZT after 1 MeV H^+ ion implantation at $1.5x10^{17}$ cm^{-2} and annealing at 180°C for 1 hour. Figure 1d shows the section of the sample where layer exfoliation has taken place. Figure 2a shows the cross-section TEM (XTEM) micrograph of the as-implanted CZT and Figure 2b shows the cross-section optical micrograph of the separated layer after annealing. The thickness of the separated layer matches quite well with the depth of the damage band as measured by XTEM. XRD $\theta/2\theta$ spectra of the separated CZT film on Si and the bulk CZT are shown in Figure 3. Overlapping CZT peaks showed good crystalline quality of the separated film that is comparable to that of the substrate. Figure 4 shows the AFM image of the surface of the CZT substrate after layer separations. A similar image was observed for the surface of the separated film. The rms roughness of the separated surfaces of film and substrate is about 16 nm for both.

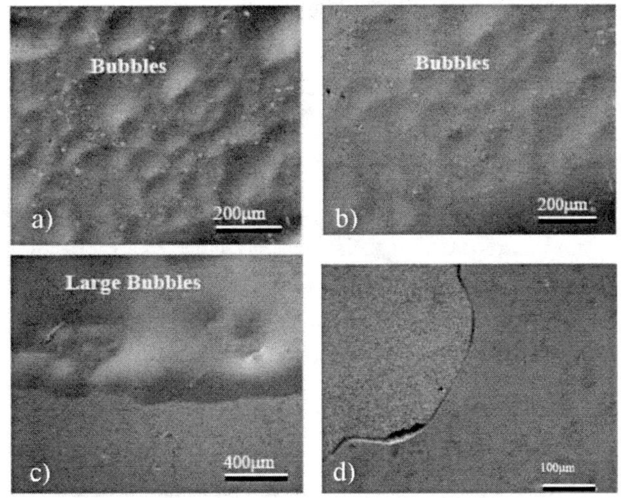

FIGURE 1. Optical micrograph of 1 MeV H^+, $1.5x10^{17}$ cm^{-2} implanted and 180°C, 1 hour annealed CZT.

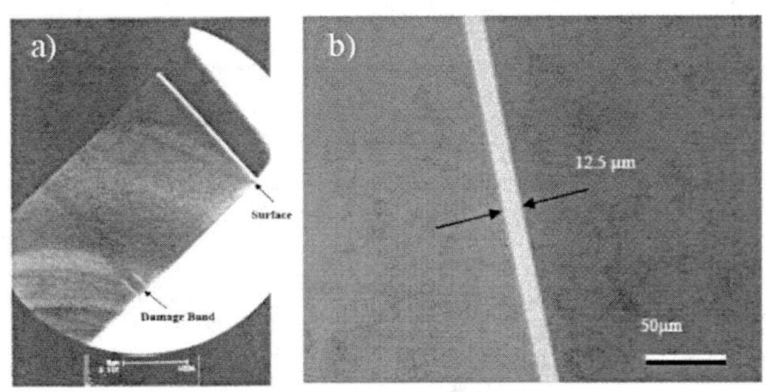

FIGURE 2. (a) Cross-section TEM of the 1 MeV H^+, $1.5x10^{17}$ cm^{-2} Implanted and (b) cross-section optical micrograph of the separated CZT layer.

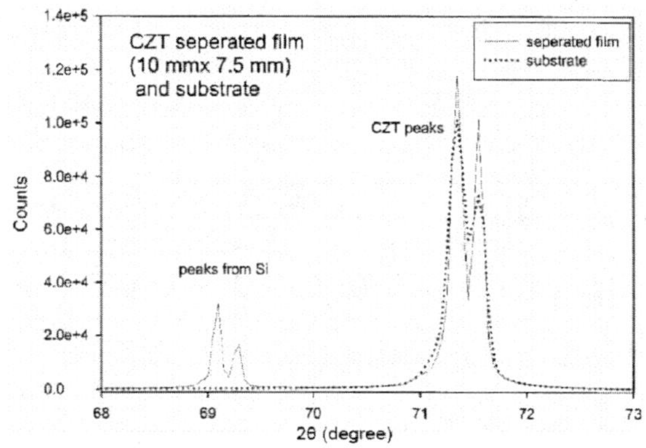

FIGURE 3. XRD Spectra of Separated CdZnTe Film and the Substrate.

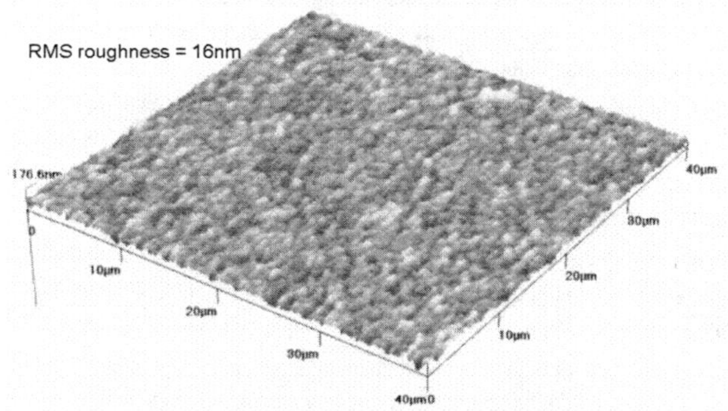

FIGURE 4. AFM Image of CZT Substrate.

In order to separate a large area film and transfer the layer on to Si wafer, bonding of the implanted surface with the Si wafer prior to annealing is required. As shown by previous authors in developing the smart-cut technology that bonding to a handle wafer provides the necessary restoring force to drive the crack propagation parallel to the surface by suppressing the blistering. Most of the semiconductor materials studied so far for layer separation could be attached to Si by direct bonding. This was commonly achieved by depositing a thin layer of Si_3N_4 by plasma enhanced chemical vapor deposition (PECVD) on the highly polished and flat implanted semiconductor materials and the handle Si wafer. Thermal oxidation of Si to form SiO_2 was also used for bonding. These approaches did not work for CZT for three reasons: first, the quality of surface finish of the CZT was not as good as most of the optical device quality semiconductor materials; second, the surface of CZT was not very stable under high energy H^+ ion bombardment and created some roughness; and third, the diffusivity of hydrogen in this material is very high, even at room temperature. The evidence of the high diffusivity was evident from the fact that upon warming up the implanted sample to room temperature, the bonding had to be accomplished within a few hours or else no separation took place. The instability of the CZT surface made it impossible to implant through a deposited layer of Si_3N_4 on CZT. Thus, we looked for other approaches of bonding. Using a third body to glue the two surfaces appeared attractive because it can accommodate the surface roughness. The bonding material should preferably be IR transparent since the light for detection needs to transmit through the substrate, at least in the most popular device configuration [1]. Chalcogenide glasses are IR transparent materials and also melt at low temperatures. We have used Ge(30)As(13)Se(32)Te(25), known as IG3 from Vitron, Inc., for bonding because of its low transition temperature around 275°C and broad IR transmittivity. The material was evaporated by electron beam on Si wafer to a thickness of about 5 microns. Figure 5 shows the bonded and separated CZT film of size 1 cm x 1 cm from the same size substrate. Although the full size layer separation and transfer onto Si was accomplished, problems remain because of the presence of cracks due to large thermal expansion coefficient of IG3.

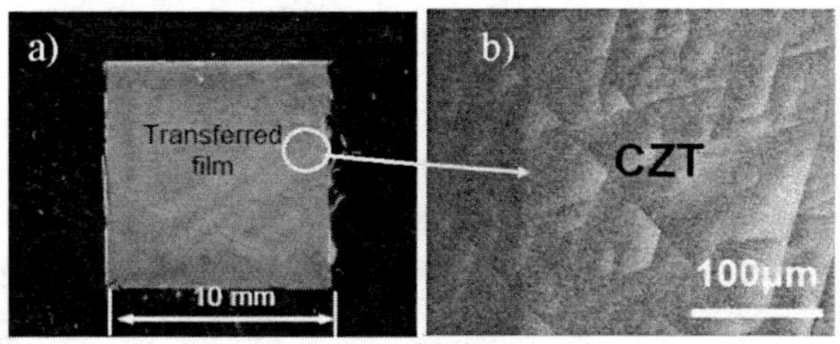

FIGURE 5. Optical Photo/Micrograph of Bonded and Separated CZT Film of Size 1 cm x 1 cm from the Same Size Substrate.

The exfoliated CZT layer was examined by high resolution XRD. Figure 6a shows a triple axis diffraction scan including the Si (004) reflection and the CZT (224) reflection. This ω-2θ scan plots the relative lattice parameters of the two materials. The Si reflection is centered at zero arcsec and the CZT peak is 4112 arcsec from the Si substrate peak; this is consistent with a fully relaxed CZT lattice constant, indicating that the layer is not under measurable strain in this arrangement. Figure 6b shows a full reciprocal space map that includes both the Si (004) and the CZT (224) reciprocal lattice points. The scan from Figure 6a is a line scan through this map vertically along Q_x=0. Again, the silicon lattice point is centered at 0,0 . The CZT peak maximum is tilted to the right of the Si peak. This is a typical result for wafer bonded structures where the lattice planes of the transferred layer do not necessarily align parallel with the substrate lattice planes.

a)

b)

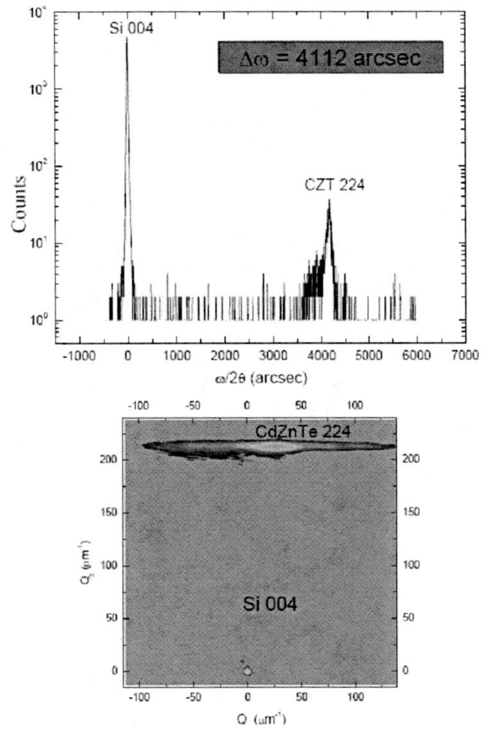

FIGURE 6. a) Triple Axis Diffraction Scan Including the Si (004) Reflection and the CZT (224) Reflection of the Exfoliated CZT Layer on Si. b) Reciprocal Space Map that Includes Si (004) and CZT (224) Reciprocal Lattice Points.

In order to determine the role of hydrogen in H-induced exfoliation, we have performed Fourier Transformed Infrared (FTIR) spectroscopy of implanted and virgin CZT. Figure 7 shows the transmission FTIR spectra of virgin and 1 MeV H^+, 1.5×10^{17} cm^{-2} implanted CZT.

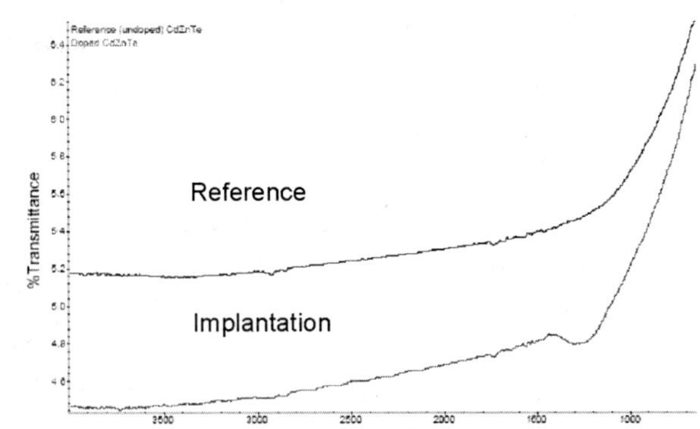

FIGURE 7. Transmission FTIR Spectra of Virgin and 1 MeV H^+, 1.5×10^{17} cm^{-2} Implanted CZT.

The small bands near 2925, 2850 and 1740 cm^{-1} are possibly due to adsorbed aliphatic ester. There is a broad band at 1200-1300 cm^1 visible only for implanted sample. The origin of this broad band remains unidentified. In Si, SiH band is found at approximately 2000 to 2200 cm^{-1} [6]. One would expect the position of two atom vibration for CdH, ZnH and TeH near that of SiH assuming that the force constants are similar. This result indicates that hydrogen atoms are not bonded with CZT, indirectly supporting the fact that they may be forming H_2 molecules that are very mobile and diffuse out at ambient temperature in a short period of time.

CONCLUSION

In conclusion, we have shown that large area layer separation of single crystal CZT using MeV H^+ ions is possible under suitable implantation conditions and bonding with carrier Si wafer. The bonding was accomplished with a chalcogenide glass layer deposited on Si. The separated CZT layers are of good crystalline quality. The rms roughness of the separated layer and the bulk crystal surface is about 16 nm. FTIR results show that the implanted hydrogen atoms do not form bonds with metallic elements of CZT. More work is needed to eliminate the cracks in the bonded film. However, results show promise for the use of transferred CZT layer as templates for the epitaxial growth of HCT for IR detector applications.

ACKNOWLEDGEMENTS

This work was supported by the National Science Foundation, Small Business Innovative Research Program, Grant No. DMI-0522039. The authors acknowledge Howard Evans for ion implantation and Evans Analytical Group for FTIR measurements.

REFERENCES

1. P. Norton, *Opto-Electronics Review* **10**(3), 159 (2002).
2. S. Rotter, D. Kasemset and C. G. Fonstad , "The Role of Lattice Matching in Improving the Performance of PbSnTe IR Photo-Diodes", *Electron Device Letters*. EDL-3, **66** (1982)
3. M. Bruel, *Nucl. Instr. Meth. In Phys. Res., Section B* **108** (1996) 313.
4. M. Levy, R. M. Osgood, Jr., R. Liu, L. E. Cross, G. S. Cargill III, A. Kumar and H. Bakhru, *Appl. Phys. Lett.* **73** (1998) 2293.
5. Q. Y. Tong and U. M. Gosele, *Advanced Materials* **11** (1409) 1999.
6. S. Hayashi, D.Bruno and M. S. Goorsky, *Appl. Phys. Lett.* **85** (2004) 236.
7. M. K. Weldon, V. E. marsico, Y. J. Chabal, A. Agarwal, D. J. Eaglesham, J. Sapjeta, W. L. Brown, D. C. Jacobson, Y. Caudano, S. B. Christman, and E. E. Chaban, *J. Vac. Sci. Technol.B* **15** (1997) 1065.

Boron Diffusion in Amorphous Germanium

L. A. Edelman[1], K. S. Jones[1], R. G. Elliman[2], and L. M. Rubin[3]

[1] *Department of Materials Science & Engineering, University of Florida, Gainesville, FL 32611-6400*
[2] *Department of Electronic Materials Engineering, Research School of Physical Sciences and Engineering, Australian National University, Canberra ACT 0200, Australia*
[3] *Axcelis Technologies, Inc., Beverly, MA 01915*

Abstract. Rapid boron diffusion in amorphous germanium is observed during solid phase epitaxial regrowth. Boron diffusion in amorphous germanium is seen to increase approximately 4 orders of magnitude relative to amorphous silicon. Weak bonding between germanium and boron coupled with a distinct dopant profile indicate a different diffusion mechanism than that of boron in amorphous silicon.

Keywords: Boron, Germanium, Amorphous, SPER, and Diffusion.
PACS: 66.30.J-, 81.15.Np, 82.80.Ms, 81.40.Jj, 81.05.Gc, 61.43.Dq

INTRODUCTION

Material alternatives to Si have gained serious interest as potential solutions to continue Moore's law. Silicon-germanium (SiGe) alloys and Ge have higher carrier mobilities than Si while maintaining basic group IV characteristics [1]. Current devices depend on the use of a pre-amorphizing implant (PAI) in Si prior to dopant implantation during the fabrication of ultra-shallow junctions. PAI in Si has several advantages including suppression of ion channeling and increased dopant activation due to dopant trapping during low-temperature solid phase epitaxial regrowth (SPER) of the amorphous (*a*-) layer [2]. PAI in Ge also results in high levels of active B following annealing and reduces dopant channeling in Ge, although the channeling tail appears to be exponential rather than Gaussian [3,4]. However, prior to completion of SPER, the implanted ions can diffuse in the amorphous phase. In the case of ultra-low energy boron in *a*-Si, the dopant profile for B in *a*-Si exhibits a unique shape with a characteristic "knee" that is consistent with the presence of an immobile fraction above concentrations of approximately $2x10^{20}$ cm^{-3} and a second mobile fraction below this value [5]. For *a*-SiGe alloys with Ge concentrations above 12 at%, the immobile B fraction breakpoint is maintained while the diffusion-length of the mobile B fraction is reduced after low temperature thermal processing [5]. In the case of PAI Ge, negligible diffusion was observed following low temperature annealing although the solubility limit for B was increased [6,7]. While recent work has focused on dopant diffusion in thin layers of *a*-Ge, there has been no attempt to report on B diffusion in thick layers of *a*-Ge [7].

EXPERIMENTAL PARAMETERS

Ge and Si wafers of (001) orientation were preamorphized from the surface to a depth of approximately 850 nm, by cooling them to approximately 77K and implanting with Si^+ ions ($5x10^{15}$ Si^+/cm^2 at 500 keV, plus $1x10^{15}$ Si^-/cm^2 at 95 keV). The samples then received a B implant of $1x10^{15}$ cm^{-2} at 500 eV. Post-implant anneals were conducted either in a quartz tube furnace under N_2 ambient or a Rapid Thermal Anneal (RTA) for *a*-Ge anneals at temperatures above 400 °C. Isothermal anneals were determined based on extrapolated SPER velocities for *a*-Ge and *a*-Si with anneal times chosen in order to limit the diffusion measurements to only the amorphous phase (i.e. prior to interaction with the advancing amorphous/crystalline interface) [8].

Cross-sectional transmission electron microscopy was used to measure *a*-layer depths before and after annealing and to confirm that B diffusion was restricted to the amorphous phase. Secondary ion mass spectroscopy (SIMS) profiles were collected using an O_2^+ beam with net impact energy of 1 kV and a beam current of 25 nA. The *a*-Ge data was not normalized

CP1066, *Ion Implantation Technology,* edited by E. G. Seebauer, S. B. Felch, A. Jain, and Y. V. Kondratenko

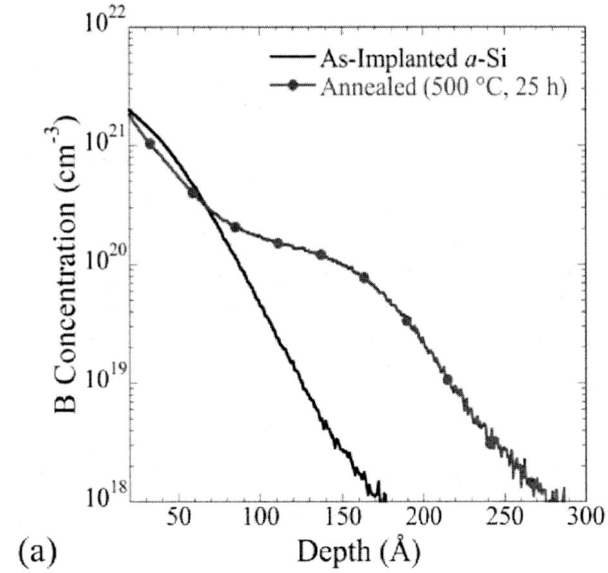

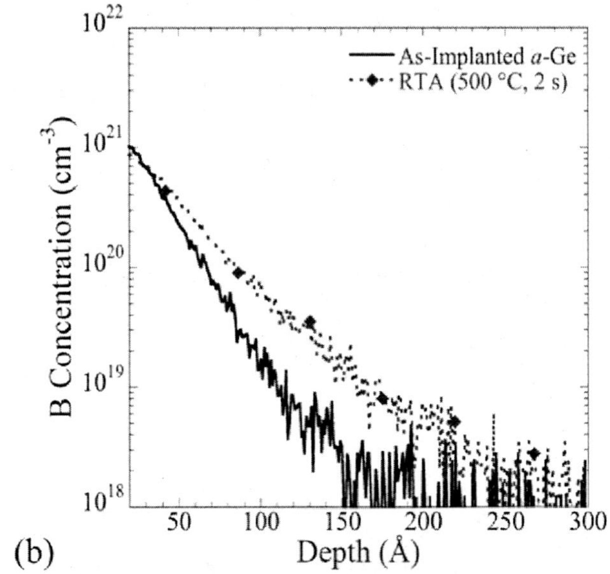

FIGURE 1. SIMS profiles of as-implanted B and after annealing at 500 °C in (a) *a*-Si for 25 h and (b) *a*-Ge for 2 s. Data for *a*-Si was normalized to the dose while *a*-Ge data was not normalized.

due to SIMS sensitivity, while the *a*-Si was normalized to the dose. In order to calculate B diffusivity, the diffusion length was measured at a B concentration of $3x10^{19}$ cm^{-3} and equated to $2\sqrt{(D_B t)}$.

RESULTS AND DISCUSSION

Figure 1 shows the B diffusion profiles following implantation and after annealing at 500 °C for 25 h in (a) *a*-Si and 2 s in (b) *a*-Ge. B in *a*-Si exhibits an immobile fraction breakpoint at $2x10^{20}$ cm^{-3} and the mobile fraction below this breakpoint moves approximately 10 nm. However B in *a*-Ge shows no clustering and B diffusion length is over 3 nm.

Figure 2 shows the B diffusion profiles in *a*-Ge after implantation and following RTA at 450 °C for 8 and 16 s. After 16 s, B motion has a diffusion length of approximately 2 nm at this temperature and B exhibits concentration dependent diffusion. Assuming steady-state diffusion behavior, the diffusivity can be calculated from the measured diffusion length determined at a B concentration equal to the solubility limit divided by *e*, or $3x10^{19}$ cm^{-2}. B diffusivities determined in this way are approximately constant with time and exhibit no transient behavior. At 450 °C, the time-averaged diffusivity of B in *a*-Ge is $5.6x10^{-16}$ cm^2/s. Compared to the extrapolated B diffusivity in *a*-Si at 450 °C ($1.1x10^{-19}$ cm^2/s), B motion in *a*-Ge is considerably faster [5]. And at 500 °C, the diffusivity of B in *a*-Ge is $1.7x10^{-14}$ cm^2/s, and in *a*-Si it is $2.2x10^{-18}$ cm^2/s. B is not constrained in the same way in *a*-Ge as it is in *a*-Si, and B diffusion in *a*-Ge cannot be explained by simple chemical interactions. B diffusion in *a*-Ge proceeds more rapidly than in *a*-Si at comparable temperatures.

Previous work by Satta *et al.* reported no B diffusion after annealing a thin (100 nm) *a*-Ge layer at temperatures up to 600 °C [6]. At 600 °C, SPER in *a*-Ge proceeds at a velocity near $1x10^3$ nm/s, making B diffusion in the amorphous phase difficult to measure. The activation energy for regrowth in *a*-Ge is 2.02 eV [8]. This value is considerably lower than the expected activation energy for B diffusion in *a*-Ge. By comparison, during low temperature thermal processing in *a*-Si, the activation energy for SPER is

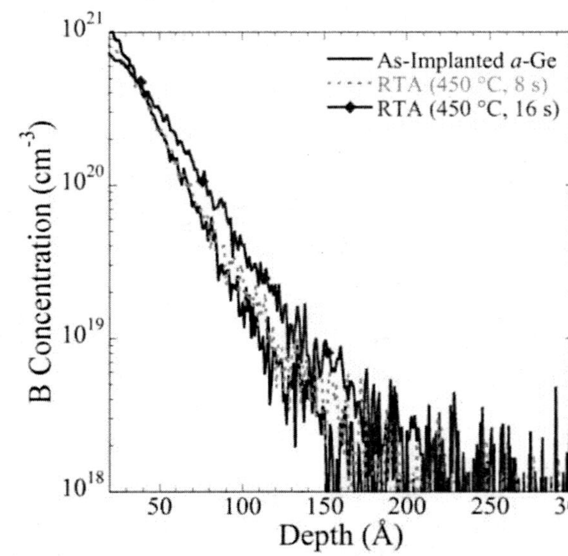

FIGURE 2. SIMS profile of as-implanted B in *a*-Ge and after RTA at 450 °C for 8 and 16 s. Data shown is not normalized.

approximately equal to that for B diffusion (2.6 and 2.8 eV respectively) [8,9]. It is possible that in *a*-Ge, the mechanisms for SPER and B diffusion directly compete, with diffusion dominating at lower temperatures and regrowth dominating higher temperatures. However at the lower temperatures where SPER can be effectively controlled, the resulting B diffusion is expected to proceed very slowly and outside the detection limit of SIMS. This study was able to isolate a small temperature region where B diffusion is measurable and SPER can be controlled.

While most Group III dopants diffuse via a vacancy mechanism in *c*-Ge, work by Uppal *et al.* determined that B diffusion is mediated by an interstitial mechanism due to its high activation energy (4.65 eV) and prohibitively low diffusion coefficient [10,11]. Janke *et al.* found a high formation barrier for B-I defects and a prohibitively high diffusion coefficient associated with a barrier for dissociation of the B-I defect in *c*-Ge [12]. Additional modeling found the migration length of B-I complex in *c*-Ge to be independent of temperature with a solubility similar to that of B in *c*-Si [13]. Although the activation energy for B diffusion in *c*-Ge is higher than that in *c*-Si, the diffusion mechanisms are expected to be similar. The same cannot be said for B diffusion in *a*-Si and *a*-Ge.

Recent work by Mirabella *et al.* proposed an activation energy for B diffusion *a*-Si dependent on the capture and release of dangling bonds [14]. Other molecular dynamics simulation results suggest an additive kinetic barrier for B diffusion composed of a formation energy for mobile B and a smaller migration energy for B in *a*-Si [15].

The bonding between Ge and B is prohibitively low for Ge to effectively trap B with second nearest neighbor B-Ge pairs having a bonding energy of 40 meV and first and third nearest neighbor pairs having a bonding energy of 10 meV [16]. In *c*-Ge, the solid solubility of B is 5.5×10^{18} cm^{-3}, while in *a*-Ge it is comparable to that of B in *a*-Si despite the absence of an immobile B fraction [17]. The absence of an immobile B fraction and high B diffusivity values at low temperatures indicates a kinetic barrier to diffusion, specifically the migration energy for mobile B.

CONCLUSIONS

In conclusion, B was found to readily diffuse in *a*-Ge with a distinct diffusion profile. For the time intervals studied, no transient diffusion was observed. The diffusivity of B in *a*-Ge is approximately four orders of magnitude faster than that in *a*-Si at the same temperature.

ACKNOWLEDGEMENTS

The authors would like to thank the Major Analytical Instrumentation Center at the University of Florida for use of the Focused Ion Beam and Transmission Electron Microscope and Mihkail Klimov at the University of Central Florida for the SIMS analysis. This research is supported by the Semiconductor Research Corporation.

REFERENCES

1. L. L. Berger, *CRC Handbook of Chemistry and Physics*, edited by David R. Lide (Taylor & Francis, London, 2005) pp. 12-82.
2. M. Aboy, L. Pelaz, J. Barbolla, R. Duffy, and V. C. Venezia, *Mater. Sci. and Eng. B* **124-125,** 205-209 (2005).
3. Y.-L. Chao, S. Prussin, J. C. S. Woo, and R. Scholz, *Appl. Phys. Lett.* **87**, 142102 (2005).
4. K. Suzuki, K. Ikeda, Y. Yamashita, and S. Takagi., *Jap. J. of Appl. Phys.* **46,** 926 (2007).
5. L. A. Edelman, M. S. Phen, K. S. Jones, R. G. Elliman, and L. M. Rubin, *Appl. Phys. Lett.* **92,** 2108-2110 (2008).
6. A. Satta, E. Simoen, T. Clarysse, T. Janssens, A. Benedetti, B. De Jaeger, M. Meuris, and W. Vandervorst, *Appl. Phys. Lett.* **87,** 172109 (2005).
7. A. Satta, E. Simoen, T. Janssens, T. Clarysse, B. De Jaeger, A. Benedetti, I. Hoflijk, B. Brijs, M. Meuris, and W. Vandervorst, *J. of the Electrochem. Soc.* **153,** G229-G233 (2006).
8. P. Kringhoj and R. G. Elliman, *Phys. Rev. Lett.* **73,** 858-861 (1994).
9. L. A. Edelman, R. G. Elliman, L. Rubin, L. Washington, and K. S. Jones, *J. Vac. Sci. Technol.* B **26,** 333-337 (2008).
10. S. Uppal, A. F. W. Willoughby, J. M. Bonar, N. E. B. Cowern, T. Grasby, R. J. H. Morris, and M. G. Dowsett, *J. Appl. Phys.* **96,** 1376 (2004).
11. S. Uppal, A. F. W. Willoughby, J. M. Bonar, A. G. R. Evans, N. E. B. Cowern, R. J. H. Morris, and M. G. Dowsett, *J. Appl. Phys.* **90,** 4293 (2001).
12. C. Janke, R Jones, S. Oberg, and P. R. Briddon, *J. Mater. Sci: Mater. Electron* **18,** 775 (2007).
13. P. Delugas and V. Fiorentini, *Phys. Rev. B* **69,** 085203 (2004).
14. S. Mirabella, D. De Salvador, E. Bruno, E. Napolitani, E. F. Pecora, S. Boninelli, and F. Priolo, *Phys. Rev. Lett.* (2008).
15. I. Santos, W. Windl,. L. Pelaz, and L. A. Marques: First Prinicples Study of Boron in Amorphous Silicon in *Doping Engineering for Front-End Processing*, edited by B.J. Pawlak, M. Law, K. Suguro, M.L. Pelaz (Mater. Res. Soc. Symp. Proc. **1070**, Warrendale, PA, 2008), E05-07.
16. J. Bang, J. Kang, W.-J. Lee, and K. J. Chang, *Phys. Rev. B* **76,** 064118(7) (2007).
17. F. A. Trumbore, *Bell Syst. Tech. J.* **39,** 205-233 (1960).

Optically Stimulated Diffusion in Ultra-Shallow Junction Formation

Y. Kondratenko, C. T. M. Kwok, R. Vaidyanathan, and E. G. Seebauer*

Department of Chemical & Biomolecular Engineering, University of Illinois, Urbana, IL 61802 USA
* *eseebaue@illinois.edu*

Abstract. Formation of pn junctions in advanced Si-based transistors employs rapid thermal annealing after ion-implantation in order to increase the activation of dopants. There has long been suspicion that the strong lamp illumination required for this procedure may nonthermally influence the diffusion of dopants. Identification of such effects is difficult in conventional RTA geometries because lamps provide both heating and photostimulation, and because the interpretation of conventional dopant diffusion experiments is impeded by complex dopant-defect interactions. We have circumvented these problems with a new experimental design in which heating and illumination can be decoupled. Data interpretation has been simplified by examining the motion of isotopically labelled ^{30}Si tracer in an epitaxial ^{28}Si matrix using SIMS depth profiling, and more recently with dopants such as arsenic in standard ion-implanted Si. Results for self-diffusion show that for n-type Si, self-diffusion rates are increased nonthermally by more than an order of magnitude for modest illumination intensities of roughly 1 W/cm^2. Diffusion rates for silicon self-diffusion are shown to change by factors of up to 25 in response to optical fluxes on the order of 1 W/cm^2. Results depend on doping type; the rates of both interstitial formation and migration are affected in the case of n-type material. There is no comparable effect for p-type material, however. Significant effects appear for boron as well in standard implantation experiments.

Keywords: Photostimulated diffusion, Illumination effects, Dopant diffusion enhancement, Annealing, Ultra-Shallow Junctions.
PACS: 66.30.Lw, 85.40.Ry.

INTRODUCTION

Formation of pn junctions in advanced Si-based transistors employs rapid thermal annealing (RTA) of ion-implanted regions in order to increase the activation of dopants. The need for ever-shallower junctions with low sheet resistances has driven a shift to "spike" RTA with exceptionally high heating rates near 400 °C/s. There has been some evidence that the strong lamp illumination required for this procedure may nonthermally influence the diffusion of dopants in Si. For example, for phosphorous diffusion, Noel and coworkers [1,2] found that continuous wave illumination with ultraviolet light at 300W enhances diffusion. Along similar lines, for arsenic diffusion, Fair *et al.* [3] found that RTA with broadband incandescent lamps with λ_{peak}= 0.8 μm enhanced diffusion over furnace heating with λ_{peak}= 2.0 μm. More recently, Lojek *et al.* [4] have reported evidence for athermal motion of boron and arsenic. Other examples exist as well [5-7].

In all these cases, however, heating by the probe light or changes in heating configuration as probe intensity varied cast at least some doubt on the interpretation of the results. More definitive results have awaited experimental configurations in which heating and illumination could be decoupled.There is certainly physical reason to suppose that nonthermal illumination can affect bulk diffusion in Si. For example, a standard diffusion mechanism involves spontaneous creation of interstitials and vacancies from a single lattice site. These two defects are then relatively free to move until they are annihilated. Both vacancies and interstitials (of Si and dopants) are capable of existing in multiple charge states [8]. The formation energy and entropy [9] for the forward reaction and the migration energies [10] of the point defects thus formed can vary with charge state. Since illumination affects the availability of charge carriers and therefore the most favored charge state of the defects, illumination can in principle change both the formation and migration energies (and entropies) of point defects [11].

CP1066, *Ion Implantation Technology,* edited by E. G. Seebauer, S. B. Felch, A. Jain, and Y. V. Kondratenko

Such effects have already been quantified in the analogous case of surface diffusion, where low-level optical illumination < 2 W/cm^2 was shown to either enhance or inhibit diffusion of indium, germanium and antimony on Si by close to an order of magnitude, depending on the doping type (n or p) of the underlying substrate [12, 13]. In that case, the effects seemed to arise primarily from illumination-induced changes in the formation energy for surface vacancies and mobile atoms.

Both formation and migration mechanisms involve changes in the average electrical charge state of the relevant species. Identification of such effects is difficult in conventional RTA geometries because lamps provide both heating and photostimulation, and because the interpretation of conventional dopant diffusion experiments is impeded by complex dopant-defect interactions and the presence of built-in electric fields in typical test structures. In this paper we present unambiguous experimental measurement of nonthermal illumination-enhanced diffusion in unimplanted silicon. Additionally, this effect has been qualitatively demonstrated for boron diffusion in ion implanted silicon. In case of boron the illumination also affected dopant activation

EXPERIMENTAL

Experiments examined the decay of a step concentration profile of ^{30}Si tracer atoms in an epitaxial ^{28}Si matrix using specimens akin to those employed in studies of thermal diffusion [14]. We used a novel experimental design employing a new heating configuration in which heating and illumination were largely decoupled. Experiments examined the decay of a step concentration profile of ^{30}Si tracer atoms in an epitaxial ^{28}Si matrix. Natural Si contains 3.10% of the mass 30 isotope. A 0.5 μm-thick layer of Si depleted in this isotope (to 0.002%) was deposited (by Isonics Corp. [15]) using low-pressure chemical vapor deposition. Dopants of B or As were incorporated into the grown layer at a concentration of $\sim 1\times 10^{19}$ cm^{-3}, matched in type and amount to that of the substrate.

Temperature uniformity on the illuminated and dark sides was ensured through a structure consisting of two single-sided Si samples sandwiched together with Ta foil. A chromel-alumel thermocouple was spot-welded on the top of the foil (Figure 1), with the junction wedged against the sample sandwich. Heating above 600°C led to silicidation at the interfaces of Si/Ta/Si sandwich structure, thereby helping to ensure uniform temperature throughout. Optical view factors for the two sides were also matched. Experiments were performed in an ultrahigh vacuum chamber to minimize surface oxidation and corresponding OED effects. To remove native oxide and other contaminants, specimens were cleaned with an HF-based etching solution immediately prior to mounting in vacuum. Illumination was accomplished with a He-Ne laser operating at 633 nm, and isotopic profiles on the illuminated and dark sides were compared with SIMS after annealing.

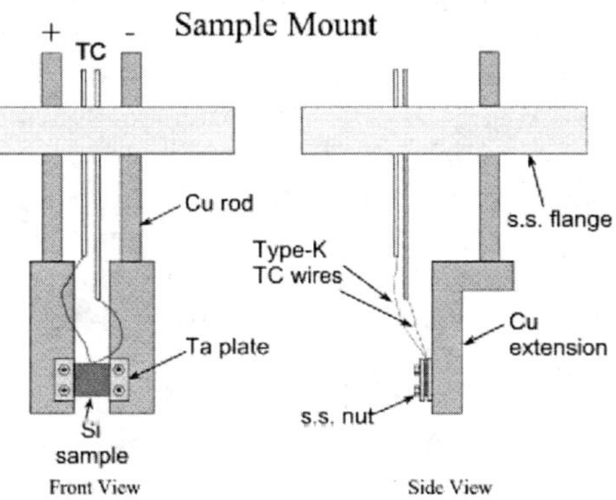

FIGURE 1. Detail of the sample mount. Optical illumination impinged from the left in the side-view figure.

For boron diffusion experiments the samples were supplied by Applied Materials (B implanted in Si (100) wafer at 3 keV and dose of 2 x 10^{15} cm^{-2}). Implanted samples were coupled with the backing sample (As doped with a resistance of 0.0015 Ω cm.) in order to allow resistive heating. The samples were subjected to the same cleaning procedure and placed in the same sample mount. Temperature was monitored via K type thermocouple which was pressed against the surface of implanted sample. A 75 W air-cooled Xe arc lamp was used for sample illumination. A Macor mask was used to mask half of the sample area. Optical glass filters (UG-5 and KG-2), in conjunction with the glass window, confined the spectrum of the incident light to wavelengths of 350–400 nm, which helped to eliminate heating by blocking unnecessary photons. The maximum intensity was 0.5 W/cm^2 over a 6mm diameter spot. Control experiments that measured the sample temperature showed no significant rise in temperature during the span of typical annealing time. The samples were annealed for 1 hr at temperatures ranging from 700°C to 750°C.

After annealing, SIMS was performed on both the dark and illuminated regions of the sample. Cameca 5f SIMS instrument was used for boron samples. An O_2 (+) primary beam with an energy of 7.5 kV was used. The negative sample voltage was 4.5 kV resulting in a net impact energy of 3 keV. Due to the shallow implantation, it was important to use a low energy primary beam to eliminate any spurious SIMS artifacts. The ion current was about 100nA with a 300 μm raster size.

RESULTS AND DISCUSSION

Figure 2 shows that the diffused profiles do not fit the Gaussian form expected for concentration-independent diffusion of a single species. Instead, the profile looks closer to linear on a logarithmic scale, signifying an exponential decay. Such behavior has been observed previously in the case of boron diffusion in Si [16, 17], and offers a signature of diffusion by a quick-moving intermediate species such as an interstitial [15, 16]. Ural *et al.* did not observe this state, probably because their diffusion times were sufficiently long for a more typical error-function solution to develop. Indeed, profiles in the present experiments exhibited error-function shapes at long times, and exhibited faster self-diffusion in n-type material compared to p-type, in accord with the literature [18].

Figure 3 compares diffusion profiles for illuminated and unilluminated specimens of n-type material. Illumination increases the diffusivity substantially. *By contrast, no such effect could be observed for p-type material* in the temperature range 600-900°C for intensities up to 2 W/cm^2. The difference in behavior between n and p-type material gives strong evidence that the observed enhancement in n-Si is genuine, and not an artifact of some unknown heating or similar spurious effect.

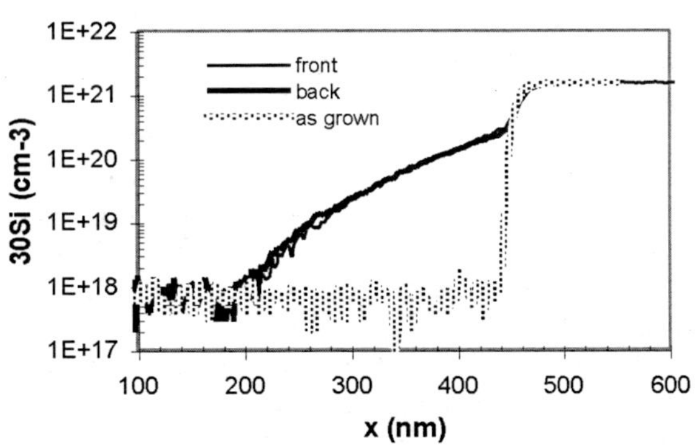

FIGURE 2. Diffusion profiles of ^{30}Si in control experiments with no illumination. Material was As-doped to 10^{19} cm^{-3}. Anneal was 1 hr at 950°C.

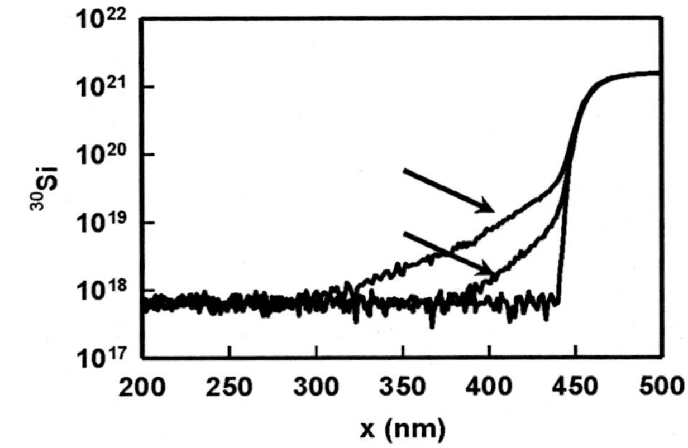

FIGURE 3. SIMS profile of 30Si showing illumination enhancement at 800°C for 1-hr (n-type).

These data can be rationalized in terms of changes in self-interstitial charge state induced by photogenerated charge carriers. Although self-diffusion within Si can be mediated by both vacancies and self-interstitials, there exists considerable evidence [19-21] that interstitials dominate the mass transport at the temperatures of this experiment. Both interstitials [18,19] and vacancies [22] are well known to assume a variety of charge states depending on the type and concentration of the dominant charge carrier (electrons or holes). The most likely charge states for Si interstitials in n-type Si are not definitively established from the literature. The primary candidates are Si_i^{++}, Si_i^{o}, and Si_i^{-}; Si_i^{+} is destabilized by negative-U behavior [18]. Charged defects have formation energies that vary with charge state.

Migration energies can also vary, although the situation is complicated for Si_i, which may change charge state during a diffusional hop [18]. In the present experiments, interstitials probably form mainly via a Frenkel pair mechanism in which creation of one interstitial also creates one vacancy. Thus, changes in the primary charge state of either the interstitial or vacancy can affect the apparent interstitial formation energy.

Experiments with dopant diffusion in silicon show that photostimulation can affect diffusion and activation of dopants in a way that involves more than just heating. Figure 4 illustrates that boron diffusion can be either enhanced or inhibited depending upon temperature. Soak annealing at temperatures below 730°C the diffusion was inhibited, while above this “crossover” temperature diffusion was enhanced. Illumination also affected dopant activation. Sheet resistance was higher in illuminated samples annealed below 730°C, while higher temperature produced samples with lower sheet resistance in illuminated samples. These effects are complicated, but they have precedent as described in the Introduction [11, 12].

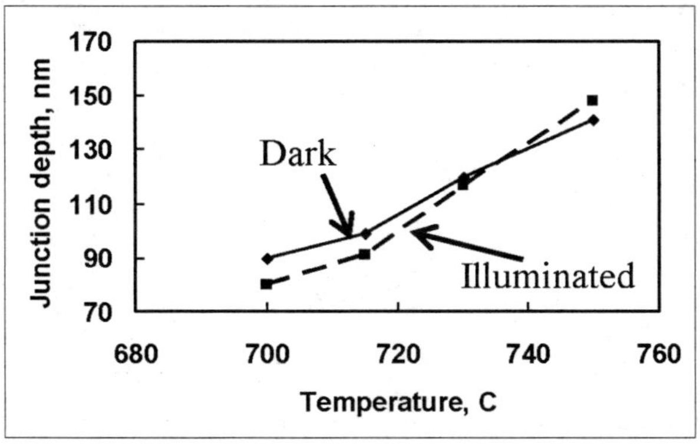

FIGURE 4. Differences in junction depth ([B]=5×10^{18} atoms/cm^3) between illuminated and non-illuminated samples after soak annealing at 700-750°C.

The observed changes from photostimulated diffusion inhibition to enhancement hints that several opposing kinetic steps may be involved in dopant diffusion and activation. Illumination could potentially affect any one or all of them in ways that are hard to predict *a priori* without reliable modeling. It is necessary to point out that illumination intensities used in these experiments are orders of magnitude lower than those encountered in industrial processes. Judicious experimental design coupled with modeling will be necessary in order to predict and exploit this nonthermal phenomenon at higher temperatures and light intensities.

CONCLUSIONS

Strong evidence for nonthermal illumination effects on diffusion in bulk Si at low light levels has been presented. The same effect was observed in boron diffusion. Furthermore, illumination also affected dopant activation. These results show unambiguous experimental evidence for nonthermal illumination effects in implanted silicon.

ACKNOWLEDGEMENTS

The authors would like to acknowledge Applied Materials (S. Felch, M. Foad, H. Graoui, and S. Moffatt) for supporting two authors with the Graduate Fellowship (Y. Kondratenko and R. Vaidyanathan), providing samples and SIMS characterization of boron implanted samples. This work was supported by NSF (CTS 98-06329 and 02-03237). SIMS was performed at the Center for Microanalysis of Materials at the F. Seitz Materials Research Laboratory at UIUC, which is partially supported by the U.S. Department of Energy under grant DEFG02-91-ER45439..

REFERENCES

1. S. Noel, L. Ventura, A. Slaoui, J. C. Muller, B. Groh, R. Schindler, B. Froeschle, and T. Theiler, *J. Electron. Mat.* **27** (1998) 1315.
2. R. B. Fair and S. Li, *J. Appl. Phys.* **83** (1998) 4081.
3. B. Lojek, R. Whiteman and R. Ahrenkiel, Proc. RTP 2001 (RTP2001 and IEEE, 2001) 125.
4. E. Wieser, H. Syhre, F. G. Rudenauer, and W. Steiger, *Phys. Stat. Sol. A* **81** (1984) 247.
5. Y. Ishikawa and M. Maruyama, Jpn. *J. Appl. Phys.* **36**, (1997) 7433.
6. J. Gyulai, H. Ryssel, L. P. Biro, L. Frey, A. Kuki, T. Kormany, G. Serfozo, and N. Q. Khanh, *Radiation Effects and Defects in Solids*, **127** (1994) 397.
7. J. A. Van Vechten, "A Simple "Man's View of the Thermochemistry of Semiconductors," in Handbook on Semiconductors, ed. T. S. Moss, p. 1, North-Holland, New York (1980).
8. J. A. Van Vechten and C. D. Thurmond, *Phys. Rev. B*, **14** (1976) 3539, 3551.
9. R. B. Fair, "Physics and Chemistry of Impurity Diffusion and Oxidation of Silicon," in Silicon Integrated Circuits, Part B, ed. D. Kahng, p. 1, Academic Press, New York (1981).
10. M. Y. L. Jung, R. Gunawan, R. D. Braatz and E. G. Seebauer, Rapid Thermal and Other Short-Time Processing Technologies II (ECS Vol. 2000-9, 2000) 15.
11. R. Ditchfield, D. Llera-Rodriguez and E.G. Seebauer, *Phys. Rev. Lett.* **81** (1998) 1259.
12. R. Ditchfield, D. Llera-Rodriguez and E.G. Seebauer, *Phys. Rev. B* **61** (2000) 13710.
13. Ural, P. B. Griffin, and J. D. Plummer, Phys. Rev. Lett. 83 (1999) 3454; A.Ural, P. B. Griffin, and J. D. Plummer, *Physica B* **273/274** (1999) 512.
14. Isonics Corporation, 5906 McIntyre Street, Golden, CO 80403, Attn: Steve Burdin (303) 279-7900.
15. N.E.B. Cowern, Phys. Rev. Lett. 65 (1990) 2434; N. E. B. Cowern, G. F. A. van del Walle, P. C. Zalm, and D. J. Oostra, *Phys. Rev. Lett.* **69** (1992) 116.
16. N. E. B. Cowern, G. F. A. van de Walle, D. J. Gravesteijn, and C. J. Vriezema, *Phys. Rev. Lett.* **67** (1991) 212.
17. G. E. Murch and A. S. Nowich, Diffusion in Crystalline Solids, (Orlando, FL, Academic Press, 1984).
18. Y. Bar-Yam and J. D. Joannopoulos, *Phys. Rev. B*, **30** (1984) 2216.
19. P. E. Blochl, E. Smargiassi, R. Car, D. B. Laks, W. Andreoni and S. T. Pantelides, *Phys. Rev. Lett.*, **70**, (1993) 2435.
20. W. A. Harrison, *Phys. Rev. B* **57** (1998) 9727.
21. J. A. Van Vechten, *Phys. Rev. B* **33** (1986) 2674.

Numerical Solution of Scattering Integral for Universal Interatomic Potential

Gholamhosain Haidari

Physics Department, Malayer University, Malayer, Hammedan, Iran
E-mail: moh1135@gmail.com

Abstract. The solution of Scattering Integral is needed for Monte Carlo simulation of ion implantation. Universal interatomic potential is used to make scattering integral. This integral doesn't have an analytical solution. Different methods for numerical solution were studied for different values of related variables. The results reveal that the numerical solution obtained by the Midpoint method has enough accuracy for different values of related parameters.

Keywords: Scattering Integral, Universal Interatomic Potential, Numerical Integration, Ion Implantation.
PACS: 02.60.Jh, 02.70.Uu, 34-50.-s, 34.50.Bw

INTRODUCTION

Monte Carlo methods are a widely used class of computational algorithms for simulating the behavior of various physical and mathematical systems and for other computations. The Monte Carlo method has been used to predict the transport of energetic ions in solids [1]. This numerical simulation method requires repeated calculation with the lattice. Numerical evaluation of the scattering integral can be difficult because the integrand is weakly singular at the distance of closest approach. There are different algorithms for numerical integration [2]. In this essay, several algorithms for obtaining a numerical solution are presented. Comparison between them and their precisions are discussed. The interaction of a moving charged particle with nuclei of a solid can be described by a series of classical elastic scattering processes. It is assumed by the Monte-Carlo method that only two particles (the ion and one target nucleus) are considerably involved in a scattering process. This is called the binary collision approximation, which reduces the multi particle problem to a classical elastic two particle problem which is characterized by the conservation of energy and momentum. In a two body collision influenced by a central potential field $V(r)$, the scattering angle is given by [1].

$$\theta(p,E_c)=\pi-2\int_{r_{\min}}^{\infty}\frac{p.dr}{r^2\sqrt{1-\frac{V(r)}{E_c}-\frac{p^2}{r^2}}} \qquad (1)$$

Where θ is the scattering angle, p is the impact parameter, $r_{\min}$ is the distance of closest approach, E_c is the particle energy, which is measured in central of mass coordinates, i.e.:

$$E_c=\frac{M_2}{M_1+M_2}E_1 \qquad (2)$$

Where M_1 and M_2 are the masses of the projectile and target particle, respectively, and E_1 is the projectile energy in the lab coordinates. The distance of closest approach, which is the point at which the projectile's radial velocity is zero, is determined from the following equation

$$1-\frac{\Phi(r_{\min})}{E_c}-\frac{\beta^2}{r_{\min}^2}=0 \qquad (3)$$

To non-dimensionalize the integral, the following substitutions are made

$$x=\frac{r}{a_u},\ \beta=\frac{P}{a_u},\ \varepsilon=\frac{aE_c}{Z_1Z_2e^2},\text{and}$$

$$v(x)=\frac{Z_1Z_2e^2}{ax}\Phi(x) \qquad (4)$$

CP1066, *Ion Implantation Technology,* edited by E. G. Seebauer, S. B. Felch, A. Jain, and Y. V. Kondratenko

Where $\Phi(x)$ is the screening function and a_u is the screening length, which depends on the Bohr radius and the atomic numbers of the two particles. Ziegler, Biersack and Littmark have performed potential calculations for a wide range of particle pairs. On the basis of their calculations they could extract an approximate analytical expression for the dimensionless screening potential, the universal screening potential (USP). This potential should be suitable for arbitrary atom pairs:

$$\Phi(x) = 0.1818e^{-3.2x} + 0.5099e^{-09423x} + 0.2802e^{-0.4028x} + 0.02817e^{-0.2016x} \quad (5)$$

The universal screening potential is a function of the dimensionless reduced radius x which is related to the real radius by the scaling length a_u suggested by Ziegler, Biersack and Littmark [1]:

$$a_u = 0.8854\frac{a_\circ}{Z_1^{0.23} + Z_2^{0.23}} \quad (6)$$

($a_\circ = 0.529A^\circ$ is the Bohr radius)

The resulting integral is:

$$\theta(p, E_c) = \pi - 2\int_\eta^\infty \frac{\beta.dx}{x^2\sqrt{1 - \frac{\Phi(x)}{\varepsilon} - \frac{\beta^2}{x^2}}} \quad (7)$$

The dimensionless distance of closest approach η is found from

$$1 - \frac{\Phi(\eta)}{\varepsilon} - \frac{\beta^2}{\eta^2} = 0 \quad (8)$$

Numerical solution

As seen from Eq. (8), the integrand in Eq. (7) is singular at x=η. In general, closed-form solutions for the integration are not available. The approximate analytical integration scheme is difficult and limited [3]. Hence numerical integration is advantageous, particularly for an up to date computer. To facilitate this integration, an additional substitution is introduced:

$$I = \int_0^{1/\eta} (1 - \frac{\beta^2}{\eta^2} - \frac{\Phi(\frac{\eta}{1-\eta\omega})}{\varepsilon\eta} + \frac{\Phi(\frac{\eta}{1-\eta\omega})\omega}{\varepsilon} + \frac{2\beta^2\omega}{\eta} - \beta^2\omega^2)^{-0.5} d\omega$$

$$\omega = \frac{1}{\eta} - \frac{1}{x}, \quad \theta = \pi - 2\beta I \quad (9)$$

The new integrand is singular at $\omega = 0$. For numerical integration, some methods such as the first four closed Newton-Cotes quadrature formulas (Trapezoidal Rule, Simpson's Rule, Simpson 3/8 Rule, and Boole's Rule) where use $f(\omega = 0)$, are unsuitable. Midpoint, Gaussian Quadrature, and Monte Carlo methods [4] were used for numerical integration. The results for these methods are shown in Table1.

As illustrated in Table 1 and Figure 1, for the Gaussian Quadrature method, the integration was more accurate by increasing the number of points or nodes (1 up to 500), as was expected. But for larger amounts of ε, the accuracy went up slowly and wasn't sufficient. For example, for N=500, the error was about 0.07% for $\varepsilon = 2*10^{-5}$ and more than 100 % for $\varepsilon = 2$. The Mathematica Software values were chosen for calculation of error.

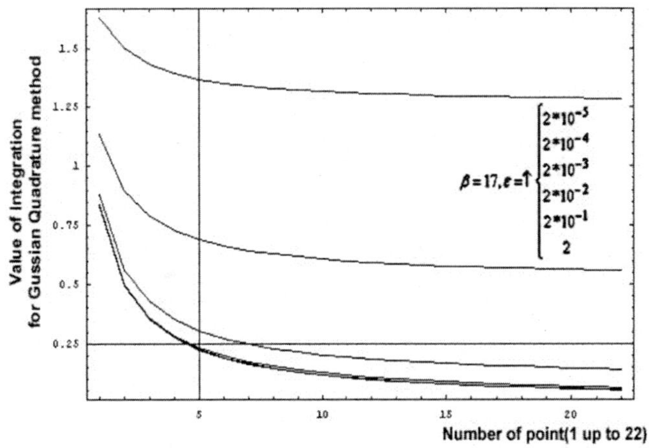

FIGURE 1. Scattering angle for specified values of ε and β that computed with various points of Gaussian quadrature method (1 $\rightarrow$ 22).

The results for the Midpoint method are summarized in Table1 and Table2 for different value of partition number (N). The integration is more accurate by increasing N; moreover the runtime is proportional to N. We have to choose between accuracy and runtime. In Table 2, the error for $\varepsilon = 2$ is still large and needs a greater N. Hence, what number for N to select? Figure 2 illustrates the percent of errors for different values of N for the case ($\varepsilon = 2, \beta = 17$) that has the largest error in Table 2.

TABLE 1. Scattering angle (in radians)

Solved by Mathematica Software	Monte Carlo method N=100000	Mid point method N=100000	Gaussian quadrature method			ε	β
			500 points	20 points	15 points		
3.02817	3.02814	3.02825	3.0283	3.02982	3.03035	2×10^{-5}	1
2.34945	2.34742	2.34997	2.35032	2.3611	2.36485	2×10^{-5}	7
1.25888	1.26069	1.26022	1.26114	1.28859	1.29816	2×10^{-5}	17
2.96084	2.96076	2.99073	2.96105	2.96362	2.96452	2×10^{-4}	1
1.89873	1.89997	1.89962	1.90024	1.91838	1.92471	2×10^{-4}	7
0.517232	0.478553	0.51932	0.520868	0.563605	0.578559	2×10^{-4}	17
2.8192	2.82096	2.81944	2.81962	2.82455	2.82628	2×10^{-3}	1
1.12534	1.13174	1.12692	1.12808	1.16027	1.17153	2×10^{-3}	7
0.0896378	0.0884909	0.0922294	0.0942219	0.14718	0.165739	2×10^{-3}	17
2.44935	2.44869	2.44991	2.45035	2.4618	2.46582	2×10^{-2}	1
0.301181	0.306444	0.30357	0.305402	0.354221	0.371328	2×10^{-2}	7
0.00979359	0.00847814	0.0124862	0.0145693	0.0695785	0.0888622	2×10^{-2}	17
1.42737	1.4267	1.42885	1.43002	1.46031	1.47093	2×10^{-1}	1
0.0376828	0.040494	0.0403479	0.0424089	0.096856	0.115942	2×10^{-1}	7
0.000988744	0.00414503	0.00369266	0.00578583	0.0610243	0.0803889	2×10^{-1}	17
0.277102	0.268072	0.279585	0.281517	0.33223	0.350011	2	1
0.00386989	0.0063966	0.00657094	0.00866182	0.063842	0.0831861	2	7
0.0000989712	0.00375567	3.02825	0.00489823	3.02982	0.0795326	2	17

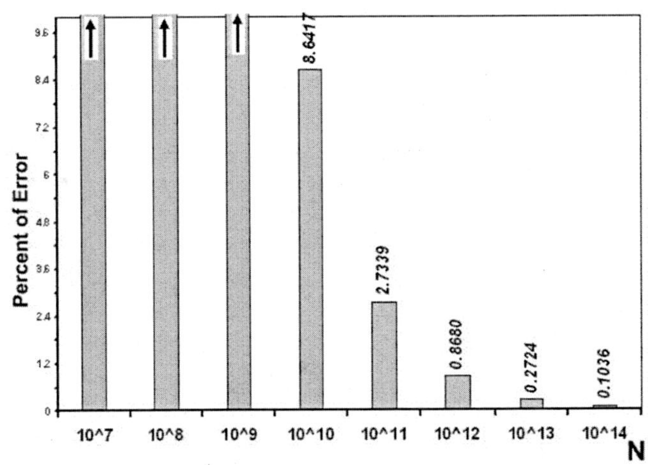

FIGURE 2. Percent of Error for the case ($\varepsilon = 2, \beta = 17$) for different value of N.

TABLE 2. Comparison scattering angle between Midpoint and Monte Carlo method.

IMSL Library Re.[3]	Mid point method		ε	β
	N=10^7	N=10^6		
3.03	3.02818	3.0282	2×10^{-5}	1
2.35	2.3495	2.34961	2×10^{-5}	7
1.26	1.25902	1.25931	2×10^{-5}	17
2.96	2.96085	2.96088	2×10^{-4}	1
1.90	1.89882	1.89901	2×10^{-4}	7
0.517	0.51744	0.517891	2×10^{-4}	17
2.82	2.81922	2.81927	2×10^{-3}	1
1.13	1.1255	1.12584	2×10^{-3}	7
0.0896	0.0898969	0.0904573	2×10^{-3}	17
2.45	2.44941	2.44941	2×10^{-2}	1
0.301	0.30142	0.301937	2×10^{-2}	7
0.00982	0.0100629	0.0106451	2×10^{-2}	17
1.43	1.42752	1.42752	2×10^{-1}	1
0.0377	0.0379493	0.0385256	2×10^{-1}	7
0.00102	0.00125913	0.00184379	2×10^{-1}	17
0.277	0.27735	0.277887	2	1
0.00388	0.00413998	0.00472402	2	7
0.000131	0.000369473	0.000954384	2	17

However, the Monte Carlo integration isn't suitable for this one variable integration; it was used and a comparison between these methods was made. As illustrated in Table1, the error of Monte Carlo method is unsuitable and there isn't a satisfactory result even for a large number of N. Figure 3 illustrates the scattering angle for ($\varepsilon = 2, \beta = 1$). That program was run 100 times.

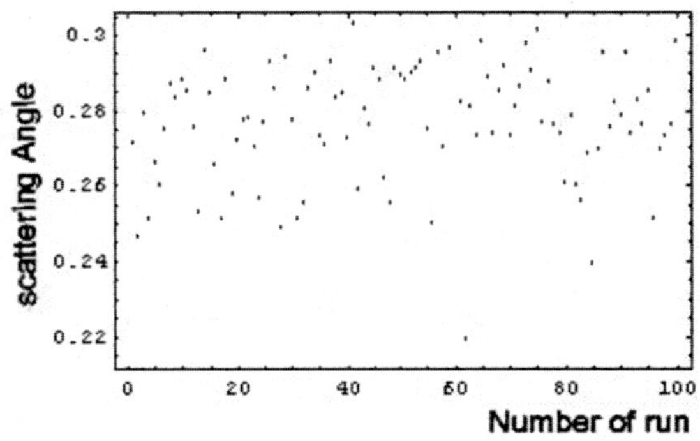

FIGURE 3. Scattering angle for $\varepsilon = 2, \beta = 1$ was calculated 100 times by Monte Carlo integration

CONCLUSION

The above numerical results for Scattering Integral show that the Gaussian Quadrature method does not have enough accuracy in some cases even for 500 points. Also, the Monte Carlo method is not suitable for such integration. The Midpoint method with suitable N can have adequate accuracy.

ACKNOWLEDGEMENTS

This work is supported by Malayer University.

REFERENCES

1. J.F. Ziegler, J.P. Biersack, and U. Littmark, *The Stopping and Range of Ions in Solids* vol.1, Pergamon Press, New York, 1985.
2. Josef Stoer and Roland Bulirsch. "Introduction to Numerical Analysis." New York: Springer-Verlag, 1980.
3. J.P.Blanchard *et al.* An approximate solution to the scattering integral for generating interatomic potentials, *J. Appl. Phys.* **61** (8), 15 April 1987.
4. R. E. Caflisch, "Monte Carlo and quasi-Monte Carlo methods", *Acta Numerica* 7, Cambridge University Press, 1998, pp 1-49.

Angular Distributions of Sputtered Atoms from Semiconductor Targets at Grazing Ion Beam Incidence Angles

M. Sekowski[1], A. Burenkov[2], J. Hernández-Mangas[3], A. Martinez-Limia[2], and H. Ryssel[1, 2]

[1] *Lehrstuhl für Elektronische Bauelemente, Cauerstrasse 6, 91058 Erlangen, Germany*
[2] *Fraunhofer Institut für Integrierte Systeme und Bauelementetechnologie (IISB), Schottkystrasse 10, 91058 Erlangen, Germany*
[3] *Departamento Electricidad y Eletrónica, Universidad de Valladolid, 47011 Valladolid, Spain*

Abstract. Angular distributions of ion sputtered germanium and silicon atoms are investigated within this work. Experiments are performed for the case of grazing ion incidence angles, where the resulting angular distributions are asymmetrical with respect to the polar angle of the sputtered atoms. The performed experiments are compared to Monte-Carlo simulations from different programs. We show here an improved model for the angular distribution, which has an additional dependence of the ion incidence angle.

Keywords: Sputtering, Germanium, Silicon, Grazing ion incidence, Angular distribution
PACS: 81.65.Cf, 79.20.Rf

INTRODUCTION

Focused ion beam systems (FIB systems) are used for surface structuring of semiconductors like germanium and silicon [1]. In order to understand the ion-etch process effects like sputtering, ion reflection, swelling, etc., have to be well known. As a consequence of sputtering, the redeposition of atoms occurs. Especially in etching of small sized structures, this effect plays an important part. Ion etch simulations consider the redeposition by assuming a symmetrical angular distribution which is described by a cosine of the polar emission angle ϑ to the power of n, where $1 < n < 3$ [1]. This assumption is sufficiently accurate for normal incidence ions. But for grazing ion incidence, as it occurs on the sidewalls of the produced structures, the angular distribution is asymmetrical with respect to the target normal. We measured, therefore, the distributions of sputtered germanium and silicon using two different experimental setups. Additionally, we made Monte-Carlo simulations of the sputtering process using different Monte-Carlo programs [2,3,4]. The observed differences between measurements and simulations are discussed. Finally, the improved models for the angular distribution of sputtered atoms are shown in the last chapter.

EXPERIMENTAL SETUP

Two setups were realized for obtaining angular distributions of germanium and silicon:

a) A small piece of a germanium wafer was used as a target for argon ions with a kinetic energy of 20 keV. This setup was placed in a large vacuum chamber of an ion implantation system, which is usually used for tribology research (Figure 1). The sputtered germanium is collected on a 300 mm silicon wafer with a distance to the target of 20 cm. The implantation angle was varied from 60° to 85° by tilting the germanium target with respect to the ion beam. As the collector sample is a standard silicon wafer, we were able to analyze its surface contamination by a total-reflection x-ray fluorescence system (TXRF). Measuring under total reflection conditions results in a very low detection limit of 1×10^{11} at. cm^{-2} to 1×10^{12} at. cm^{-2}, depending on the analyzed atom species. The TXRF system Atomika 8300W, which has a spot size of 0.8 cm, was used for this investigation. This setup has, therefore, an angle resolution of about 5°, which is adequate to monitor local inhomogeneities.

b) The setup for the silicon sputtering is realized in a vacuum chamber of a FIB system. As the chamber

CP1066, *Ion Implantation Technology,* edited by E. G. Seebauer, S. B. Felch, A. Jain, and Y. V. Kondratenko

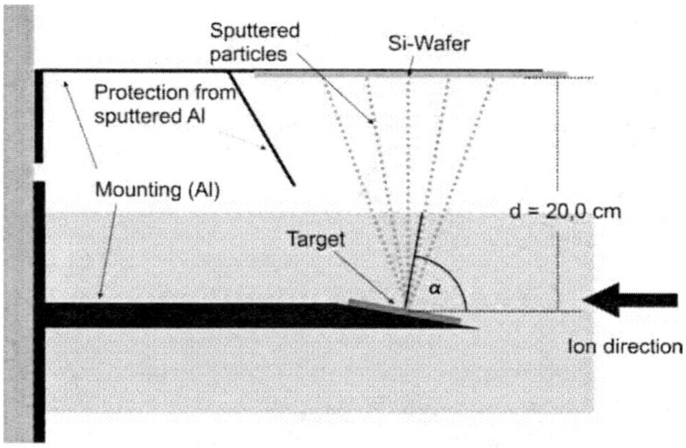

FIGURE 1. Experimental setup a) for the Ar^+:Ge sputtering experiments. The ion incidence angle α was varied from 60° to 85°.

has much smaller dimensions compared to the tribology vacuum chamber in a), the distance between the target and the collector was only 5.5 mm. Gallium ions with kinetic energies of 30 keV were sputtering the silicon atoms with angles of incidence of 70° and 80°. The collector was a piece of a silicon wafer with a 3 µm polymer coating on it. The coating was used for collecting the sputtered silicon atoms, and the silicon wafer at the bottom served for mechanical stability and also as heat sink, since the collectors were analyzed by Rutherford Backscattering Spectroscopy (RBS). The RBS measurements were performed at the University of Minsk (Belarus) with spatial resolutions of 300 µm. This results in an angle resolution of 6°. In difference to a) (Figure 1) the target and the collector were arranged parallel in b), which has to be considered in the evaluation of the collected material.

Table 1 shows the measured and simulated values for the total sputter yield. The Monte-Carlo simulations were performed with the program MC_Sim [2]. At very high values of the angle of incidence, the measured sputter yield is much lower than expected from the simulation. This was the reason why the angular distributions for the Ar^+:Ge, 85° experiment were no more measurable due to too low surface concentrations of germanium on the collector wafer.

RESULTS

Experiments for Ar^+:Ge

For the Ar^+:Ge experiments, 300 mm silicon wafers were used as collectors. The silicon wafers were analyzed spatially resolved by TXRF. From Monte-Carlo simulations it is known that the sputtered atoms have only a few eV of kinetic energy, thus their penetration depth into the collector is only a few nanometers. Analyzing just the collector surface with the TXRF gives us therefore the desired information about the sputtered atoms in a particular direction. Ion implantation dose for all Ar^+:Ge experiments were on the order of 1 x 10^{15} cm^{-2}. This dose was chosen for two reasons: first, a lower dose would lead to crystal orientation effects in the angular distribution [5]. Second, a higher ion doses leads to a notable material removal. Under these conditions ion etching is known to produce local structures and rippling on the target surface [6]. This roughness influences the sputtering and the obtained angular distributions. Figure 2 shows the measured germanium distribution on the silicon wafer in the ion incidence plane. ϑ is the polar angle of the sputtered atoms with respect to the target normal. Backward emission is indicated by negative ϑ.

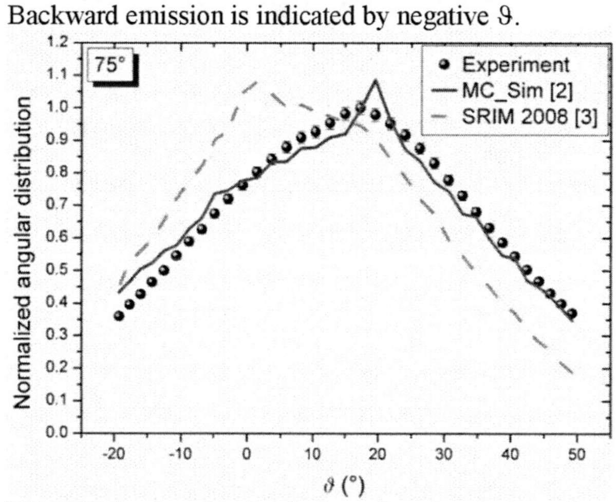

FIGURE 2. Angular distribution of sputtered germanium. 20 keV argon ions with an angle of incidence of 75° were used for sputtering. ϑ is the polar angle with respect to the target normal. The continuous lines represent Monte-Carlo simulations.

TABLE 1. Experimental and simulated values for the sputter yield for the experiments Ar^+:Ge and Ga^+:Si.

Parameter \ Experiment	Ar^+:Ge, 60°	Ar^+:Ge, 70°	Ar^+:Ge, 75°	Ar^+:Ge, 80°	Ar^+:Ge, 82.5°	Ar^+:Ge, 85°	Ga^+:Si, 70°	Ga^+:Si, 80°
Simulated sputter yield (MC_Sim)	7	8	10	11	10	8	10	11
Experimental sputter yield	10.8	11.4	10.3	11.4	1.3	0.02	13	16

In general, the observed distributions show a local maximum at a direction in a right angle to the primary ion incidence beam (15° in Figure 2). This maximum is related to primary knock-on atoms (PKA).

In addition to the experiments, Monte-Carlo simulations were performed and compared to the experimental values. For the simulations, different Monte-Carlo codes [2,3] were used. In order to be able to consider a realistic angle resolution as in the experiments, the sputtered atoms in the simulation were collected on a virtual plane with a distance according to the experimental setup. In Figure 2, the Monte-Carlo simulations are depicted by continuous lines. The position of the PKA peak is reproduced well by all simulations, but the magnitude varies. The differences between the simulation codes are primarily due to different treatments of the target surface.

Experiments for Ga^+:Si

The parallel-aligned target-collector combination was placed in a vacuum chamber of a FIB system. The disadvantage of measuring the collector with RBS, compared to the TXRF, is the high detection limit of about 1×10^{15} cm^{-2}, which requires a very large amount of sputtered silicon atoms. In consequence, the implantation doses for the experiments were on the order of 1×10^{18} cm^{-2}, which leads to etching processes of the silicon surface of several micrometers. The created groove influences the angular distribution, as sputtering from the edges additionally occurs. Figure 3 contains the angular distribution for silicon atoms sputtered with gallium ions with an angle of incidence of 80°.

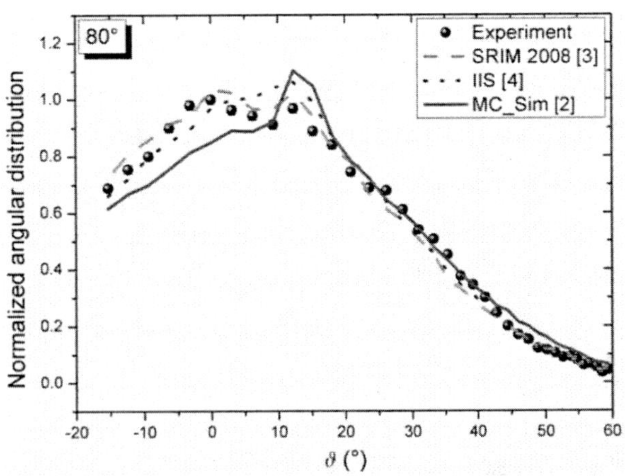

FIGURE 3. Angular distribution of sputtered silicon. 30 keV gallium ions with an angle of incidence of 80° were used for sputtering. The continuous lines represent Monte-Carlo simulations.

It should be noted that the angular distributions in Figure 2 and 3 have not been corrected to the different collector-target alignments.

Like already mentioned in the Ar^+:Ge experiments, the Monte-Carlo simulations reproduce the position of the PKA peak quit well, but the magnitude is overestimated.

ANALYTICAL MODELS

The experimental results in the last chapter can be used to establish analytical models for the angular distributions. Up to now, reduced models assuming a cosine distribution for the angular distributions are used in etch simulations [1]. In order to be able to consider the mentioned PKA peak in the angular distributions, we added another cosine to the model, which depends on the angle of ion incidence:

$$Y_S = b\cos^n(\vartheta) + c\cos^m(\vartheta - (90 - \alpha')). \quad (1)$$

Y_S is the angular distribution per solid angle, b, c, n and m are parameters depending on the primary beam and the target material, and

$$\alpha' = \alpha - \delta \quad (2)$$

corresponds to the primary ion beam incidence angle α reduced by δ. δ has a value of 2° for Ar^+:Ge and 3° for Ga^+:Si, and originates from the surface refraction due to the surface binding energy [7]. As we used plane collectors in our experiments, Eq. (1) has to be extended by a factor κ [8] to obtain the angular distribution $f(\vartheta)$ measured on the collector:

$$f(\vartheta) = \kappa(\vartheta) \cdot Y_S(\vartheta). \quad (3)$$

For the Ar^+:Ge setup, κ depends on the angle of ion incidence

$$\kappa(\vartheta) = \cos^3(\vartheta - (90° - \alpha)), \quad (4)$$

whereas for the parallel alignment in the Ga^+:Si experiment applies:

$$\kappa(\vartheta) = \cos^3(\vartheta). \quad (5)$$

The parameters b, c, n, and m were obtained by backward engineering of the experimental results (see Figure 4).

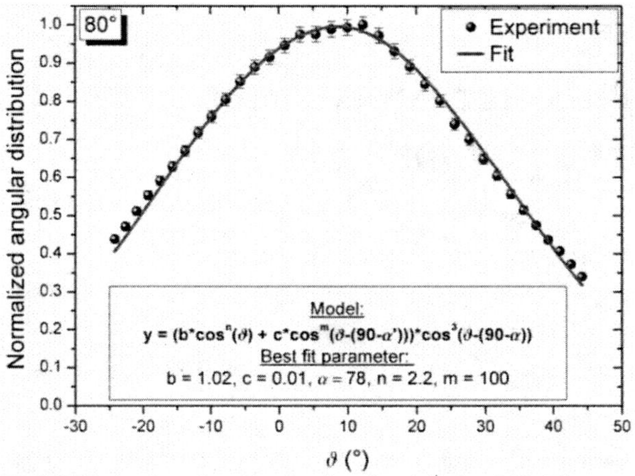

FIGURE 4. Angular distribution for Ar^+:Ge, 80°. The analytical model (Eq. (3)) is adapted by minimizing χ^2.

treatment of the target surface in the Monte-Carlo simulations.

A more precise analytical model is adapted to the experiments. The accordance in the case of Ar^+:Ge is very good, for Ga^+:Si the noisy distribution due to high implantation doses derogates the results.

ACKNOWLEDGEMENTS

The authors would like to thank E. Birnbaum for the help with the ion implantation system and F. Komarov from the University of Minsk (Belarus) for the RBS measurements. This work was supported by the DFG (Deutsche Forschungsgemeinschaft).

TABLE 2. Model parameters (Eq. (3)) for the experiments Ar^+: Ge and Ga^+:Si.

Parameter \ Experiment	Ar^+:Ge, 60°	Ar^+:Ge, 70°	Ar^+:Ge, 75°	Ar^+:Ge, 80°	Ar^+:Ge, 82.5°	Ga^+:Si, 70°	Ga^+:Si, 80°
b	1.2	1.08	0.86	1.02	0.99	0.3	0.28
c	0.06	0.04	0.15	0.01	0.01	0.7	0.73
n	1.73	1.99	1.5	2.20	2.36	1.0	2.0
m	120	60	10	100	100	0.9	2.8
α	58°	68°	73°	78°	80.5°	67°	77°
χ^2	$4\cdot10^{-5}$	$5\cdot10^{-5}$	$1\cdot10^{-3}$	$2\cdot10^{-4}$	$4\cdot10^{-4}$	$2\cdot10^{-3}$	$6\cdot10^{-4}$

Table 2 shows the results for all experimental data. The mean reduced standard deviation χ^2 was calculated for all angular distributions to indicate the agreement of the model to the experimental values. The Ar^+:Ge experiments could be modeled very well, the obtained model parameters are all in the same range. For Ga^+:Si the high implantation dose produced additional noise to the angular distributions, which leads to a quite high χ^2 value. Nevertheless, the obtained parameters for the analytical model are nearly the same for both implantation angles

CONCLUSION

Experiments, Monte-Carlo simulations and analytical models for angular distributions of ion sputtered germanium and silicon are reported within this work. For the experiments the collector-technique was used, where a plane sample collects the sputtered material and is investigated afterwards. According to the experiments, Monte-Carlo simulations were performed to compare the results. The PKA peak due to primary knock-on atoms is reproduced in both cases, but with a different magnitude which results from an inaccurate

REFERENCES

1. S. Beuer, M. Rommel, C. Lehrer, E. Platzgummer, S. Kvasnica, A.J. Bauer and H. Ryssel, *Microelectron. Eng.* **84**, 810-813 (2007).
2. M. Ullrich, A. Burenkov and H. Ryssel, *Nucl. Instrum. Methods Phys. Res. B* **228**, 373-377 (2005).
3. http://www.srim.org.
4. J.M. Hernández-Mangas, J. Arias, L. Bailón, M. Jaraíz and J. Barbolla, *J. Appl. Phys.* **91,** 658-667 (2002).
5. H. Sommerfeldt, E.S. Mashkova and V.A. Molchanov, *Physics Letters* **38A**, 237 – 238 (1972).
6. A. Brown, J. Erlebacher, W.L. Chan and E. Chason, *Phys. Rev. Lett.*, **95**, 056101 (2005).
7. J.P. Biersack and W. Eckstein, *Appl. Phys. A* **34**, 73 – 94 (1984).
8. A.I. Dodonov, S.D. Fedorovich, E.A. Krylova, E.S. Mashkova and V.A. Molchanov, *Radiation Effects* **107**, 15 – 21 (1988).

Surface Modification of Polymer Substrates by Oxygen Ion Irradiation

G. H. Takaoka, H. Ryuto, R. Araki and T. Yakushiji

Photonics and Electronics Science and Engineering Center, Kyoto University, Nishikyo, Kyoto 615-8510, Japan

Abstract. Oxygen cluster ions and/or monomer ions were used for the sputtering and the surface modification of polymers such as polycarbonate (PC) and polyethylene terephthalate (PET). For the case of oxygen cluster ion irradiation, the sputtered depth increased with increase of the acceleration voltage, and the sputtering yield was much larger than that by the monomer ion irradiation. The sputtered particles represented the polymer structure, which indicated that the bond scission by the cluster ion irradiation resulted in an ejection of monomer molecule through the intermolecular collision. On the other hand, for the oxygen monomer ion irradiation, the implanted depth increased with increase of the acceleration voltage, and the bond scission occurred at the deep region through the binary collision with the high energetic ions. Therefore, the sputtering yield for the polymer surfaces decreased, and the sputtering effect became very small. Furthermore, the simultaneous use of oxygen cluster and monomer ions was more effective for oxidation of the PET surfaces rather than the monomer ion irradiation or the cluster ion irradiation. As a result, the contact angle measurement showed that the wettability of the PET surfaces irradiated by the simultaneous use of oxygen cluster and monomer ions was much enhanced.

Keywords: cluster ion beam, oxygen ion, sputtering, polymer, PET, wettability
PACS: 36.20.Hb, 36.40.Wa, 61.80.Jh, 68.47.Mn, 68.49.Sf, 82.80.Pv

INTRODUCTION

Polymers such as polycarbonate (PC) and polyethylene terephthalate (PET) have light weight and desirable mechanical properties as well as various chemical properties, which have attracted much interest in engineering applications such as electronics, optics, sensor and biomimetics[1]. For example, artificial ligaments such as PET fabrics have been used for reconstruction of damaged ligaments, in which they are fixed to bone by using plug and staples[2,3]. However, PET does not bond to living bone directly, and the surface modification such as wettability is required in order to achieve bone-bonding ability, i.e., bioactivity. In another application of polymers to the biosensor [4] such as DNA chip, surface etching and patterning have an important role, and high-rate etching and smooth surface formation of polymer substrates have attracted much interest.

The ion beam process including ion implantation technology has a high potential for material modification of polymers. The advantage of the ion beam process is based on the ability of controlling the kinetic energy precisely by the acceleration voltage. Another advantage is that various kinds of ion species such as atomic, molecular and cluster ions are available, and the interaction of these ions with the solid surface is different depending on the ion species used. In particular, the impact process of cluster ions on the solid surface is much different from the single ion impact, and represents unique irradiation effects such as the high-density irradiation, the multiple collision and the low-energy irradiation effects[5,6]. Furthermore, equivalently low-energy and high-current ion beams can be realized using cluster ion beams. On the other hand, the monomer ion beams with high incident energy produce free radicals and bond scission on the polymer surface. The irradiation effect of monomer ion beams is effective for cross-linking and/or carbonization, which results in the surface modification of the polymer substrate. In this article, the sputtering effects of polymer substrates such as PC and PET substrates are investigated by using oxygen (O_2) cluster and/or monomer ion beams, and the sputtering process by the cluster and monomer ion irradiation is discussed. Furthermore, the wettability of the PET surfaces is investigated by contact angle measurement.

EXPERIMENTS

The detail of the experimental apparatus was described elsewhere [7]. Briefly, neutral beams of O_2 clusters were produced by adiabatic expansion of a high-pressure gas such as 3500 Torr through a glass nozzle into a vacuum region. The average size (peak size) was approximately 2000 molecules per cluster. The cluster beams formed were ionized by electron

CP1066, *Ion Implantation Technology,* edited by E. G. Seebauer, S. B. Felch, A. Jain, and Y. V. Kondratenko

bombardment. An electron voltage for ionization (Ve) was 300 V, and an electron current for ionization (Ie) was adjusted between 100 mA and 500 mA. The ion beams contained both the cluster ions and monomer ions, and the ratio of the monomer ions to the cluster ions was about five. These ion beams were irradiated on PC and PET substrates. The molecule of PC and PET consists of benzene ring as well as hydrogen (H), carbon (C) and oxygen (O) atoms, as shown in Figure 1. The substrate temperature was room temperature. When only the cluster ion beams were irradiated on the substrate surfaces, the retarding method was used for the separation of cluster ion beams from monomer ion beams. The cluster size used was larger than 100 molecules for O_2 cluster ion beams. When the ion irradiation was performed by using only the monomer ion beams, the inlet pressure decreased to 1000 Torr, and the O_2 monomer ion beams were obtained. The cluster ions as well as the monomer ions were accelerated by applying the acceleration voltage between 0 kV and 10 kV. The dose was changed ranging from 1.0×10^{13} to 1.0×10^{15} ions/cm^2. The background pressure around the substrate was 1.0×10^{-6} Torr, which was attained using a turbo-molecular pump.

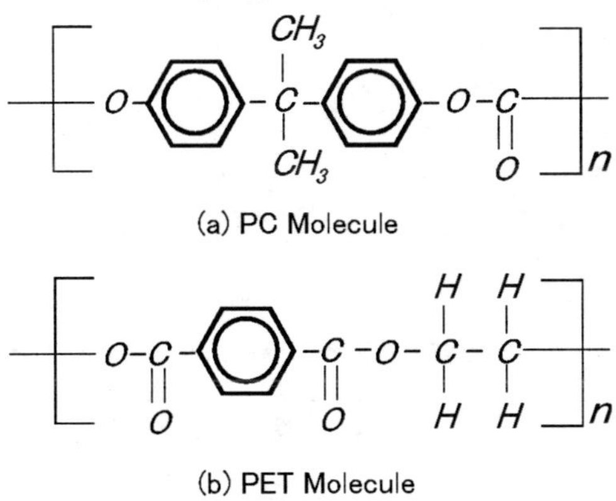

FIGURE 1. Molecular structure of (a) PC substrate and (b) PET substrate.

RESULTS AND DISCUSSION

In order to clarify the sputtering effects by O_2 ion irradiation on the polymer surfaces, the sputtered depth was measured by the step profiler (Veeco Instruments: DEKTAK-3173933). Figure 2 shows the dependence of sputtered depth for PC and PET substrates on acceleration voltage for O_2 cluster and monomer ions. The ion dose was 1.0×10^{15} ions/cm^2. As shown in the figure, the sputtered depth by the cluster ion irradiation increases with increase of the acceleration voltage, and it is approximately 60 nm for the PC and PET substrates at an acceleration voltage of 10 kV. Taking account of the sputtered depth and the ion dose, the sputtering yield can be calculated by estimating the density of PC and PET such as 1.2 g/cm^3 and 1.39 g/cm^3, respectively. The molecular weight is 254 for PC and 192 for PET, respectively. The sputtering yield at an acceleration voltage of 10 kV is 17 molecules-per-ion for the PC surface and 26 molecules-per-ion for the PET surface, respectively. On the other hand, for the O_2 monomer ion irradiation, the sputtered depth for the PC and PET substrates is almost zero, and the sputtering does not occur at an ion dose of 1.0×10^{15} ions/cm^2. This is ascribed to the difference in the bond scission process between the cluster and monomer ion irradiation. For the case of the O_2 monomer ion irradiation, the implanted depth increased with increase of the acceleration voltage, and the bond scission occurred at the deep region through the binary collision with the high energetic ions. In contrast, for the case of the cluster ion irradiation, the high density irradiation and the multiple collisions occur near the surface, and the sputtering yield is much higher than that by the monomer ion irradiation.

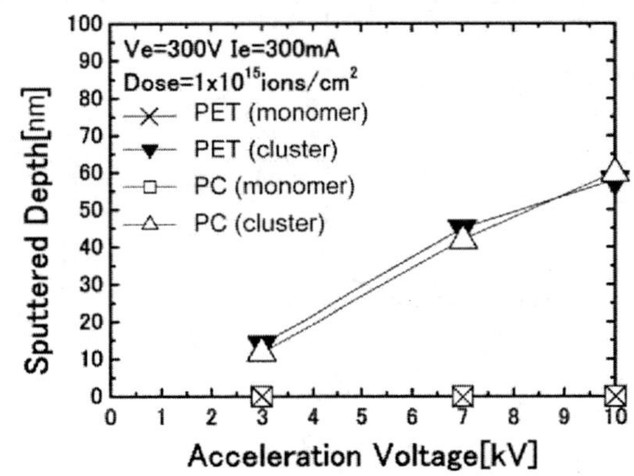

FIGURE 2. Dependence of sputtered depth for PC and PET substrates on acceleration voltage for O_2 cluster and monomer ions.

The sputtered particles from the PC and PET substrates by O_2 cluster ion irradiation were collected on the Si surfaces, and these samples were investigated with X-ray photoelectron spectroscopy (XPS). Figure 3 shows Cis XPS spectra of (a) PC unirradiated substrate and sputtered particle as well as (b) PET unirradiated substrate and sputtered particle, respectively. The acceleration voltage and ion dose were 7 kV and 1.0×10^{15} ions/cm^2, respectively. As shown in Figure 3(a), the PC unirradiated substrate exhibits peaks assigned to the benzene ring at 285 eV, (-C-O-) around 286.8 eV and –O-(C=O)-O around

290.9 eV, respectively. For the sputtered particles, a new peak assigned to O=C-O around 289.1 eV appears, although the peak assigned to –O-(C=O)-O around 290.9 eV becomes very weak. The PC molecule connects each other through the C-O bonding, which corresponds to a bond with the –O-(C=O) in the adjacent molecule. The XPS measurement indicates that the C-O bonding is broken by the cluster ion irradiation, and the sputtered particles are ejected while remaining the monomer structure. On the other hand, for the case of the PET substrate, the unirradiated substrate exhibits peaks assigned to the benzene ring or ethylene bond at 285 eV, (-C-O-) around 286.8 eV and O=C-O around 289.1 eV, respectively. The sputtered particles exhibit the same peaks as the unirradiated substrate, as shown in Figure 3(b). This indicates that the sputtering of PET substrates occurs through ejection of the sputtered particles as a unit of monomer by the O_2 cluster ion irradiation, which results in the high rate sputtering.

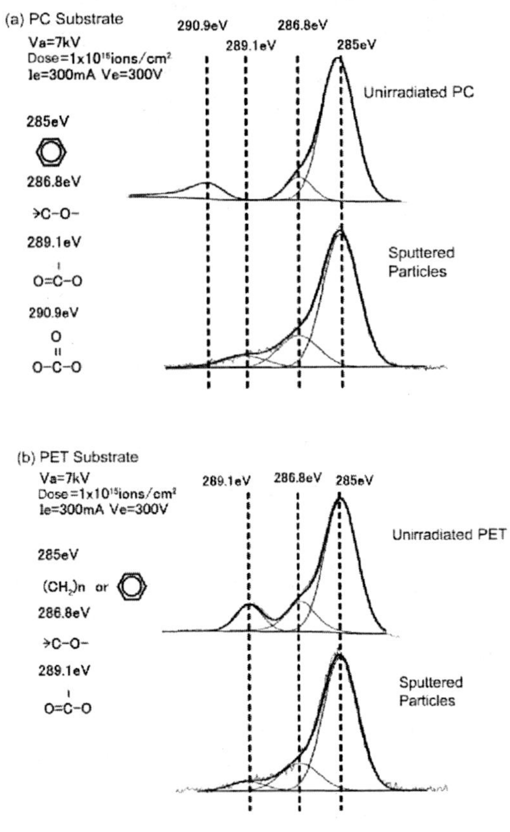

FIGURE 3. Cis XPS spectra of (a) PC unirradiated substrate and sputtered particle as well as (b) PET unirradiated substrate and sputtered particle.

Figure 4 shows Cis XPS spectra of PET substrates irradiated with (a) O_2 cluster ion beams, (b) O_2 monomer ion beams and (c) the simultaneous use of O_2 cluster and monomer ion beams. The acceleration voltage and ion dose were 7 kV and 1.0 x 10^{15} ions/cm^2, respectively. Compared with unirradiated surface, which was shown in Figure 3(b), the PET surface irradiated with O_2 cluster ion beams shows the weaker peak assigned to O=C-O around 289.1 eV. This is due to the bond scission, which results in that the monomers of PET are ejected from the surface irradiated by the O_2 cluster ion irradiation. On the other hand, for the O_2 monomer ion irradiation, the PET surface exhibits a few additional peaks such as –O-(C=O)-O besides those of the unirradiated surface. This occurrence is explained in terms of the bond scission through the binary collision with the high energetic ions. Furthermore, the PET surface irradiated by the simultaneous use of O_2 cluster and monomer ion beams exhibits a higher intensity of the (–C-O) peak than those irradiated by the O_2 cluster ion beams or O_2 monomer ion beams. Also, the full width at half maximum of the peak becomes larger for the simultaneous use of O_2 cluster and monomer ion

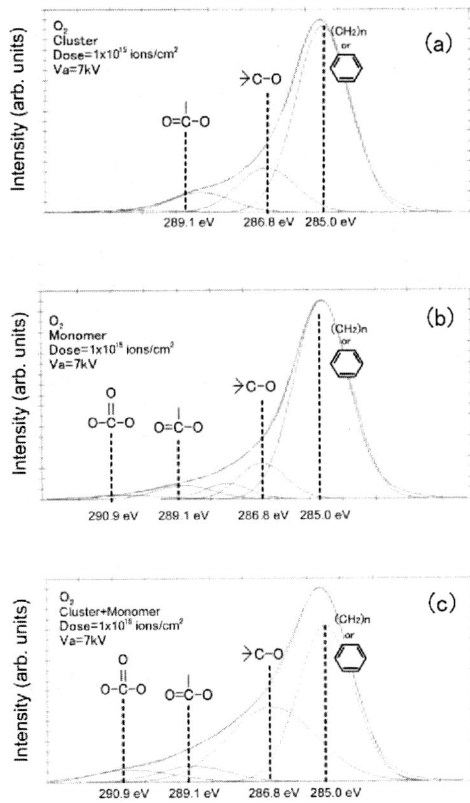

FIGURE 4. Cis XPS spectra of PET substrates irradiated with (a) O_2 cluster ion beams, (b) O_2 monomer ion beams and (c) the simultaneous use of O_2 cluster and monomer ion beams.

beams. The change of the surface structure is ascribed to the occurrence that the chemical bonds of PET are broken by O_2 monomer ion beams to form dangling bonds, and at the same time, a large number of oxygen atoms are supplied to the dangling bonds by O_2 cluster ion beams to form hydrophilic groups such as (–C-OH)[8]. With regard to the PET surface state, the atomic force microscope (AFM) measurement showed that the surface roughness was 0.3 nm for the unirradiated surface. Even after the O_2 cluster ion irradiation, the surface roughness was less than 1 nm.

The oxidation of the PET surfaces influenced the wettability. For the evaluation of the wettability, the contact angles were measured for the water droplet, which was put on the PET surfaces just after their removal from the vacuum chamber. Figure 5 shows the dependence of the contact angle on the acceleration voltage for the PET surfaces irradiated by the O_2 cluster and/or monomer ion beams. The ion dose was 1.0 x 10^{15} ions/cm^2. As shown in the figure, the contact angle is 80 deg for the unirradiated surface. For the case of O_2 ion beam irradiation, the contact angle decreases with increase of the acceleration voltage up to 10 kV. In particular, for the simultaneous use of the O_2 cluster and monomer ion beams, the contact angle is approximately 1 degree, and the super-hydrophilic property of the PET surface is obtained. This is due to the occurrence that the sufficient amount of dangling bonds are produced by the O_2 monomer ion irradiation and also that the sufficient supply of oxygen atoms is achieved by the O_2 cluster ion irradiation. The formation of the hydrophilic groups might be responsible for the low contact angle of the irradiated PET substrate.

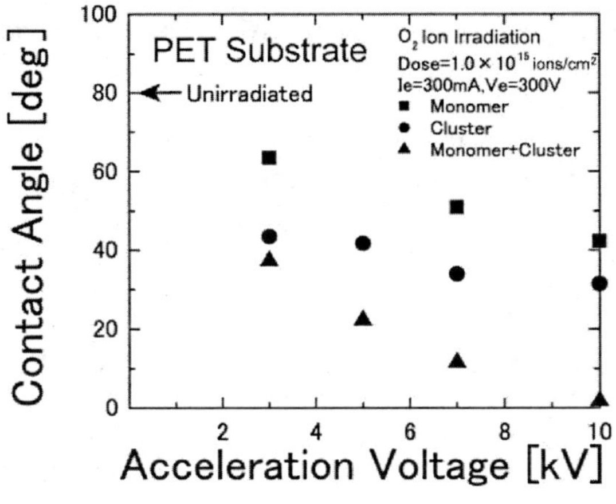

FIGURE 5. Dependence of the contact angle on the acceleration voltage for the PET substrates irradiated by the O_2 cluster and/or monomer ion beams.

CONCLUSION

Oxygen cluster ions and/or monomer ions were used for the surface modification of polymers such as polycarbonate (PC) and polyethylene terephthalate (PET). The irradiation effects of these ions on the surfaces were investigated by changing the acceleration voltage between 0 kV and 10 kV. For the case of oxygen cluster ion irradiation, the sputtered depth increased with increase of the acceleration voltage, and the high rate sputtering was performed due to the high density irradiation and the multiple collision effects. Furthermore, the XPS measurement showed that the sputtered particles were ejected while remaining the monomer structure. On the other hand, for the oxygen monomer ion irradiation, the sputtering did not occur at an ion dose of 1.0 x 10^{15} ions/cm^2. Furthermore, for the simultaneous use of oxygen cluster and monomer ions, the wettability of the PET surfaces was enhanced. In this method, high-energetic monomer ion beams had an important role in bond scissor, and oxygen cluster ion beams could provide much oxygen molecules to the polymer surfaces, which resulted in enhancement of the surface oxidation.

ACKNOWLEDGEMENTS

A part of this work was supported by "Nanotechnology Support Project" of the Ministry of Education, Culture, Sports, Science and Technology (MEXT), Japan.

REFERENCES

1. Y. Bar-Cohen and Q. Zhang, *Mater. Res. Bull.* **33** 173 (2008).
2. T. Kokubo, M. Kawai, M. Kawashita, K.Yamamoto and T. Nakamura, Key. *Engin. Mater.* **284-286** 775 (2005).
3. T. Kokubo, T. Ueda, M. Kawashita, Y. Ikuhara, G.H. Takaoka and T. Nakamura, *J. Mater. Sci.: Mater. Med.* **19** 695 (2008).
4. R. Langer, *Mater. Res. Bull.* **31** 477 (2006).
5 I.Yamada and G.H.Takaoka, *Jpn. J. Appl. Phys.* **32** 2121 (1993).
6. I.Yamada, *Eur. Phys. J. D.*, **9** 55 (1999).
7. J. Matsuo, H. Abe, G.H. Takaoka and I. Yamada, *Nucl. Instrum. Methods, B* **99**, 244 (1995).
8. D.T. Clark, B.J. Cromarty and A. Dilks, *J. Poly. Sci. Part A: Poly. Chem.* **16** 3173 (1978).

How Vacancies Assist in the Formation of Antisite-Complex Centers in 3C-SiC during Ion Implantation

S. T. Nakagawa[1], A. Okamoto[1], M. Ohishi[1], H. Saito[1], H. Hashimoto[1], and G. Betz[2]

[1] *Material Science, Graduate School of Science, Okayama University of Science, Okayama 700-0005, Japan*
[2] *Inst. f. Allgemeine Physik, Technische Universität Wien, Wiedner Hauptstraße 8-10, A-1040, Wien, Austria*

Abstract. Ion-beam irradiation onto a crystal often causes defects that cannot be completely annealed out. We had found a new candidate of such a thermally persistent defect in 3C-SiC using an empirical Molecular Dynamics simulation at a high temperature of 2000 K. Although it emerged in minor cases (~13 %), it was a peculiar complex defect made of antisites (C_{Si} $(Si_C)_4$) with a tetrahedral structure. Making use of a crystallographic analysis called the Pixel Mapping method, we studied the intermediate processes that lead to clustering. Whenever the tetrahedral clusters noticeably emerged, a cooperative motion of vacancies occurred remarkably in the long-range-order (LRO) of the crystal. A sudden change in the LRO profile linked with a burst of antisites, since then the new peculiar clusters was appreciably produced. The number of silicon antisite (Si_C) was much more than the other one (C_{Si}). From the same moment, the number of "*antisite-vacancy complex (Si_C V_{Si})*" also increased in tandem. We have confirmed that a cooperative motion of vacancies could trigger forming the peculiar type of complex defects.

Keywords: long-range-order, defect analysis, vacancy, antisite, 3C-SiC.
PACS: 61.50.Ah, 61.72-y, 61.72jd.

INTRODUCTION

For innovations in nanotechnology, it is very important to identify the details of intermediate processes, because a small change in an environmental condition may improve the succeeding synthetic processes. One of the most valuable roles of numerical computation is to find a key to alter the expected scenario, in addition to predict unknown phenomena. Molecular Dynamic (MD) simulation is often adopted as a conventional atomic-scale methodology for this purpose.

The ion implantation is an indispensable technique to introduce dopants into semiconductor crystal, which is a desired defect. Nevertheless the violent doping inevitably produces undesired defects. Some of them are thermally persistent defects remaining even above annealing temperature. In order to reduce the thermally persistent defects [1, 2], it is worthwhile to identify the structure of those defects and the cause of appearance or disappearance. They are thought to correlate closely with antisite defects in the case of compound semiconductors [2].

The compound semiconductors have been intensively investigated because of its superior high performance compared to Si. If an interstitial improperly recombines with another type of vacancy, antisite replacement occurs. 3C-SiC is one of the most promising compound semiconductors [3].

Several types of antisite-related complex defects due to ion irradiation have been characterized in 3C-SiC, e.g. D_I center [4] where Si_C and C_{Si} align in the direction of <111> and D_{II} center [5] where dicarbon antisite $(C_2)_{Si}$ aligns in the direction of <100>. Moreover, Rauls *et al.* [1] proposed a tetrahedral cluster of $Si_C(C_{Si})_4$ with the assist of antisite-vacancy (AV) complex as Si_C-V_{Si} or C_{Si}-V_C. This static model was based on the fact of low formation energy of C_{Si} rather than Si_C [6].

We have performed MD simulations to study the irradiation effect on the formation of defect complexes at high temperatures in 3C-SiC. With 5 keV boron entering (001) surface, we did not observe the expected cluster of $Si_C(C_{Si})_4$ [1]. We found a new type of tetrahedral cluster made of four antisites ($C_{Si}(Si_C)_4$), although the occurrence probability was as low as about 13 %.

Here we examine the vacancy contribution to the cluster formation, from the viewpoint of the long-range-order (LRO) in a crystal because AV complexes were thought to be the pivot in the Raul's model.

CP1066, *Ion Implantation Technology,* edited by E. G. Seebauer, S. B. Felch, A. Jain, and Y. V. Kondratenko

MOLECULAR DYNAMICS (MD) CALCULATIONS AND DEFECT IDENTIFICATION

The details of an MD simulation are given elsewhere [7]. Boron ions impact on (001) 3C-SiC with energies of 5 keV, where a MD super cell contains 8000 atoms (10 × 10 × 10 unit cells). One impact corresponds to the ion fluence of 5.273×10^{12} ions/cm^2. Every impact waits 10 ps after the previous impact for thermalization, where the heat-bath temperature is set at 2000 K. The projectile ion is B, not Si nor C, because the variation in the stoichiometry causes significant influence on the formation energy of antisite related defects in 3C-SiC. No agglomeration among B ions was observed with this number of ion impacts.

In order to analyze atomistic defects including structure-transitions caused by an external or an internal disturbance in a crystal, we have proposed a numerical methodology termed Pixel Mapping (PM) on the basis of crystallography. This method was first designed for a diamond-type crystal [8], then for 24 prototype cubic crystals [9].

The essential differences of the PM method from other analyses are in the "*prototype classification*" and "*coarse-grained viewpoint*". It identifies point defects, linear defects, planar defects (including antiphase domains in compound crystals), also structure change (including amorphization).

The idea of the "*coarse-grained viewpoint*" is that PM adopts certain margin to recognize an atom in a crystalline space. The coordinates of an atom are all expressed by the set of integers (*l, m, n*) by segmenting the crystalline space of a cubic crystal into small pixels [8]. The specific lattice sites are known from the integer addresses of relevant atoms. Table 1 shows the case of 3C-SiC. If a pixel (1) or (2) that should contain the lattice site is vacant, it indicates a vacancy is present. If another type of atom occupies there, it indicates an antisite defect is present [9].

TABLE 1. The classification of pixels from their integer ad- resses in 3C-SiC. The pixels labeled (1) and (2) show normal lat- tice sites for Si and C, respectively. The symbol "(non)" indicates the impossible address. The last column shows interstitials, see Figure 3.

(*l, m, n*)	all even	all odd	otherwise
l+m+n =4k	(1) Si_{Si} or C_{Si}	(non)	(non)
l+m+n=4k+1	(non)	{T}	{T}
l+m+n=4k+2	{H}	(non)	(non)
l+m+n=4k+3	(non)	(2) C_C or Si_C	(2) C_C or Si_C

As far as concerned tetrahedral-type clusters, an order parameter "*q*" was introduced [10]. Let us assume that "*i-th*" atom forms "*i-th*" tetrahedral cluster with four neighbor atoms, the q_i is defined as follows,

$$q_i = 1 - \frac{3}{8}\sum_{j=1}^{3} \sum_{k=j+1}^{4} \left(\cos \vartheta_{jik} + \frac{1}{3} \right)^2 \quad (1)$$

where θ_{jik} is the included angle formed by the central (*i*-th) atom and two (*j*-th and *k*-th) surrounding atoms. For a regular tetrahedron, $q_i = 1$. If surrounding atoms distribute completely random, $q_i \sim 0$.

Without this Eq. (1), for a tetrahedral cluster $A(B)_4$ one examines the dispersion of included angles ∠BAB and bonding lengths AB. In a regular tetra-hedron, the included angle "∠BAB" is 109.5 deg and AB has a fixed length.

With these two analyses, we cannot identify the orientation of the tetrahedral cluster in a crystal. Sometimes the identification of orientations is very substantial to constructing a theoretical model. It is usually impossible to determine the orientation of bonding with the nearest neighbor atom using low index as <111>. In PM, however, two atoms are expressed by integer addresses of atom A(*l, m, n*) and atom B (*l', m', n'*), thus the orientation of 'A --> B' is simply given with low index (*l'-l, m'-m, n'-n*). This way of determination is simple if two atoms are already selected. Otherwise, PM can determine even the chain's direction making use of Miller index determined by PM [11, 12].

Usually a LRO parameter is a single variable and is used to evaluate the degree of perfection of a crystal in order to verify the amorphization. In the PM method, however, we have defined a set of several components for the LRO parameter for respective prototypes. All those sets are summarized in one table called "PM Table" [9,13]. The purpose was to examine the significance of intermediate process when the disturbance to a crystal causes change from one prototype to another, not only to confirm amorphization. The structure change from one prototype to another is automatically known by comparing LRO components concerned in the PM Table, without modeling to guess the resultant new prototype.

In PM, a cubic crystalline space is segmented into small pixels that are further grouped into several groups depending on the type of Bravais lattice [9]. Each LRO component means the occupied fraction of pixels in each kind of pixels. A 3C-SiC crystal has a zincblende structure that is represented by ZnS and belongs to the group No.1 in PM Table [9]. A crystal in the group No.1 has fcc Bravais lattice. Except for two superimposed fcc crystals (β-Cristobalite and Laves), the number of the LRO components is eight for "*fcc-family crystals*" that have commonly fcc-type primitive vectors. Therefore each LRO components of 3C-SiC is defined as follows for i =1, 8 (see Table 1).

$$N_i = \frac{\text{total number of atoms staying duly at i - th kind of pixels}}{\text{total number of atoms}} \quad (2)$$

From any change in the LRO components, the structure change is well known. Table 2 shows the *ideal CA transition* in 3C-SiC in terms of LRO components. Note that when the LRO components are "$N_{(1)}$ = 1/2, $N_{(2)}$ = 1/2 and all others N_i = 0", it means that the 3C-SiC is a perfect crystal.

TABLE 2. An ideal CA transition (perfect crystal → amorphous- state) in terms of LRO components, N_i. Other notations are same to Table 1 [9].

Pixel (*l,m,n*) *l+m +n*=	Lattice sites (1) or (2)	(metastable) {T} or {H}	Interstitials <1'>, <2'>, <T'>, <H'>
4k	1/2→1/16: (1)	(non)	0 → 3/16 <1'>
4k+1	(non)	0→1/16 {T}	0 → 3/16 <T'>
4k+2	(non)	0→1/16 {H}	0 → 3/16 <H'>
4k+3	1/2 →1/16: (2)	(non)	0 → 3/16 <2'>

RESULTS

In order to simulate a stochastic phenomenon as ion irradiation, we adopt many different initial conditions to MD simulation by varying seed numbers. Although it happened in minority cases (~ 13 %), a new type of cluster $C_{Si}(Si_C)_4$ remarkably emerged [14], where no $Si_C(C_{Si})_4$ emerged as was proposed [1]. Figure 1 shows (a) a typical geometry of it and (b) spatial distribution of them. Using PM, it is known that all bonding lobes orient along <111> in a tetrahedron.

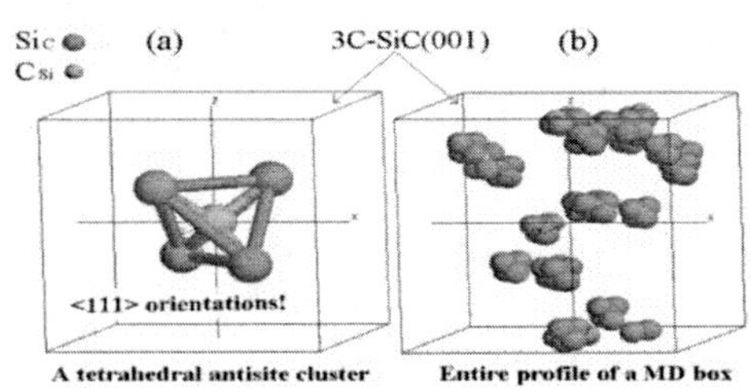

FIGURE 1. The antisite cluster $C_{Si}(Si_C)_4$ produced as a result of 5 keV boron ion impact at 2000 K. A typical shape is shown in the left figure (a), where all bonding orientations towards surrounding antisites from the central antisite are <111>. It was taken by zooming in from the entire distribution of those clusters within a MD super cell as shown in the right figure (b).

According to conventional analyses, the tetrahedral order parameter $\langle q \rangle$ shown in Eq. (1) was 0.96 as an average, which is close enough to unity and clusters are regarded almost regular tetrahedrons. On the other hand, the distribution of included angles "∠ Si_C - C_{Si} - Si_C" and bonding lengths "C_{Si} - Si_C" are as follows. The included angle was 109.9 deg in average with standard deviation of σ_θ = 7.7 deg. The bonding length was 0. 1932 nm in average with standard deviation of σ_r = 0.01268 nm. These conventional analyses also prove that the new type of cluster is close to the regular tetrahedron. However, no analyses can identify the crystallographic orientation of bonding as <111>.

Figure 2 shows correlation between AV complexes (in the left ordinate with triangles) and antisite clusters $C_{Si}(Si_C)_4$ (in the right ordinate with open circles) as ion irradiation continued. Both kinds of AV complexes "Si_C -V_{Si}" and "C_{Si} –V_C" slowly increased in tandem. In major cases (a), neither prominent increase of any AV complexes nor appreciable antisite-clusters was observed. In minor cases (b), prominent increase appeared cooperatively both in antisite clusters and AV complexes, as is indicated by a black arrow after certain number of ion impacts had been stabilized. Whenever the $C_{Si}(Si_C)_4$ cluster appeared remarkably, AV(Si_C -V_{Si}) overwhelmed AV (C_{Si} –V_C) in number as shown in Figure 2b. The assistance of AV for antisite clusters was already proposed [1], thus we examine the reason what may cause it.

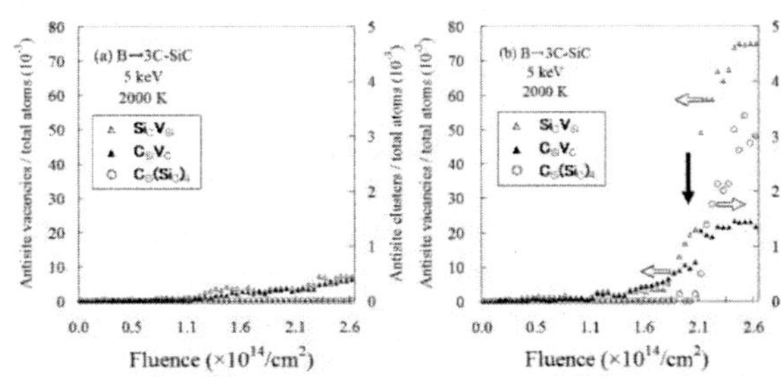

FIGURE 2. The growing profile of AV complexes as a function of boron fluence in a major case (a) and a minor case (b). The black arrow indicates the moment when clusters $C_{Si}(Si_C)_4$ appeared suddenly. The white arrows indicate the respective ordinates.

Figure 3 shows the irradiation effect on atomic relocation, in terms of the number of occupied pixels, in the same cases as shown in Figure 2. The abscissa exactly corresponds to that used in Figure 2. That is, total number of ion impact of 50 corresponds to fluence of 2.637×10^{14} ions/cm^2, as two black arrows in Figure 2b and Figure 3b correspond to each other.

First four symbols indicate number of interstitial atoms located at respective interstitial pixels (Table 1). Therefore, if multiplied the values of the ordinate by 1/8000, Figure 3 turns to show partly the profile of LRO components. The latter four symbols (▲ , △ , ○ , ●) distinguish the kind of atoms at lattice site (pixel), i.e., they are (Si_{Si}, C_C, Si_C, C_{Si}), respectively.

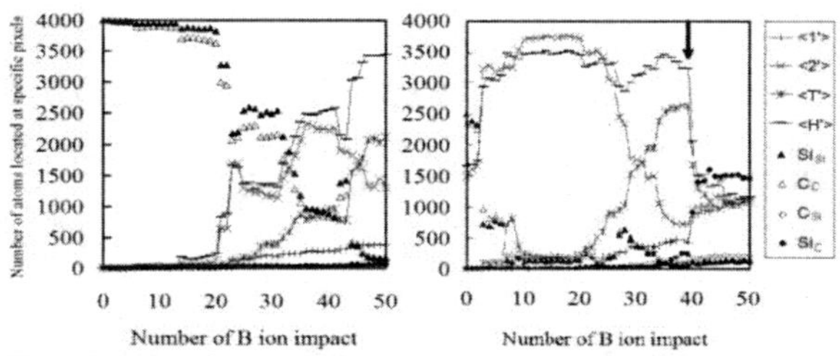

FIGURE 3. Figure 1 The redistribution of 8000 atoms in a crystal following ion impacts, in terms of the number of occupied pixels. See Table 1 for the notation of interstitial pixels. Triangles (▲,△) and circles (○, ●) indicate the proper and improper occupation of Si and C atoms, respectively. The arrow indicates the same moment that was indicated in Figure 2b.

The most striking difference between two figures is the cooperative and preferential production of Si_C, after the moment shown by arrow in Figure 3b. Whenever the peculiar type of cluster emerged, as the black arrow indicates that moment in Figure 3b, somewhat cooperative recombination occurred after the LRO was disturbed due to successive ion bombardment.

In an ordinary case (Figure 3a), a crystal is collapsed by ion irradiation and many Frenkel pairs are produced. As irradiation continued, interstitials just move around, and no specific trend is observed to lead cluster formation. On the other hand, in some particular case (Figure 3b), many Frenkel pairs are also produced. After leaving their original lattice sites, atoms preferably moved into the two kinds of interstitial sites, <2'> and <H'>. A similar change of preferable occupation was also observed in the case of low energy ion impact of c-Si [8]. All of sudden many vacancies started to recombine with interstitials but leading to improper recombination. Most changes were that i) vacancies cooperatively started to recombine with interstitials as if one kind of LRO might have exerted, ii) most of those recombinations were improper, and iii) much larger Si_C than C_{Si} resulted, which formed $Si_C(C_{Si})_4$.

CONCLUSIONS

We have examined the contribution of vacancies for a complex defect formation in 3C-SiC, making use of an empirical molecular dynamic (MD) simulation. For the defect analysis to identify various defects we used the pixel mapping (PM) method for cubic crystals. Previously, at a high temperature of 2000 K, we had observed the appreciable growth of a new type of antisite clusters of $C_{Si}(Si_C)_4$, although with low probability (13%). Whenever the clusters were formed, we observed a drastic change in long-range-order (LRO) parameter. Many vacancies were produced first. After a while, all of sudden major part of them started to recombine improperly with interstitials, which produced many antisites and also the antisite-vacancy (AV) complexes. Then, those clusters were formed. Furthermore, it was exclusively confirmed that bonding with surrounding antisites all oriented toward <111>, and the antisite clusters were almost regular tetrahedrons. Therefore, it can be hypothesized that tetrahedral antisite cluster of $Si_C(C_{Si})_4$ is formed after the cooperative but improper recombination of vacancies, due to somewhat LRO interaction. It caused a burst of both antisites and AV complexes. The role of AV complex can be to assist the diffusion of antisites and gather them at a lattice site at high temperatures.

REFERENCES

1. E. Rauls, Th. Frauenheim, A. Gali, and P. Deák, *Phys. Rev. B* **68** 155208 (2003).
2. M. Bockstedte, A. Mattausch, and O. Pankratov, *Phys. Rev. B* **68,** 205201 (2003).
3. *SiC Materials and devices* (Academic Press, San Diego,1998) ed. Y. S. Park.
4. A. Gali, P. Deák, E. Rauls, N. T. Son, I. G. Ivanov, F. H. C. Carlsson, E. Janzén, and W. J. Choyke, *Phys. Rev. B* **67,** 155203 (2003).
5. A. Mattausch, M. Bockstedte, and O. Pankratov, Phys. Rev. B **69,** 045322 (2004).
6. L. Torpo, S. Pöykkö, R. M. Nieminen, *Phys. Rev. B* **57,** 6243(1998).
7. S. T. Nakagawa, K. Hashimoto, and G. Betz, *Nucl. Instrum. Methods Phys. Res., Sect. B.* **258,**194 (2007).
8. S. T. Nakagawa, *Phys. Rev. B* **66,** 094103 (2002).
9. S. T. Nakagawa, *J. Phys. Soc. Jpn.* **76,** 034603 (2007).
10. J. R. Errington, P. G. Debenedetti, and S. Torquato, *Phys. Rev. Lett.* **89,** 215503 (2002).
11. S. T. Nakagawa, H. J. Whitlow, and G. Betz, *Surf. and Coat. Tech.* **201,** 8393 (2007).
12. S. T. Nakagawa and G. Betz, *Nucl. Instrum. Methods Phys. Res., Sect. B* **230,** 210 (2005).
13. S. T. Nakagawa: in *Ion Beams in Nanoscience and Technology*, ed. R. Hellborg (Springer-Verlag, Berlin, in press) Chap. 5.
14. S. T. Nakagawa, A. Okamoto, and G. Betz, to be published in *Nucl. Instrum.Methods, Phys. Res., Sect. B* (2008).

Ion Implanted Left-Handed Tunable Metamaterial for Microwave Circuit Applications

H. Yang[1], A. Kafaratzis[1], Z. Hu[1], N. Peng[2] and R. Gwilliam[2]

[1] *School of Electrical and Electronic Engineering, The University of Manchester, PO BOX 88, Manchester, M60 1QD, United Kingdom*
[2] *Surrey Ion Beam Centre, Advanced Technology Institute, Surrey University, Guildford, GU2 7XH, United Kingdom*
E-mail: z.hu@manchester.ac.uk

Abstract. Metamaterials are artificial structures that can be designed to exhibit specific electromagnetic properties not commonly found in nature. In recent years metamaterials with simultaneously negative permeability and permittivity otherwise known as left handed materials (LHM) or double negative materials, have attracted the attention of scientific and engineering communities. This paper reports active reconfigurable ion implanted left-handed metamaterials for radio frequency and microwave nonlinear circuit applications, demonstrating the advantage of the ion implantation technology in making low-cost pulse shaping, harmonic generation and tunable filters for wireless communication systems.

Keywords: Left handed materials, varactor, left-handed soliton generator, left-handed harmonic generator, left-handed filter.
PACS: 41.20.-q, 41.20.Jb

INTRODUCTION

The concept of left handed materials (LHMs) was first theoretically studied by Veselago [1], he concluded that the direction of the Poynting vector of a plane wave is opposite to that of its phase velocity in a substance with simultaneously negative values of permittivity and permeability, leading to the medium supporting backward wave propagation. Such a special material, however, was not available at the time and little attention was given to these interesting properties. More than thirty years later, Smith *et al.* [2] inspired by Pendry *et al.* [3, 4] demonstrated the existence of negative refraction when split ring resonators and thin metallic wires are arranged in an array. This breakthrough has brought a great interest to scientific and engineering societies. A wave of developing metamaterials with new physically realizable response functions that do not occur, or may not be readily available, in nature and their applications in physics and engineering is currently under way.

Nonlinearity of LHMs has shown a great potential in circuit applications such as pulse generation, harmonic generation, and dynamic tuning [5]. This paper reports, for the first time, nonlinear LHMs based on Si varactors fabricated by ion implantation technology (IIT) for radio frequency and microwave circuit applications. The obvious advantage in employing Si device by IIT is cost-effective and easy to be integrated with other circuit functional blocks. Ion implanted left-handed (LH) soliton generator, LH harmonic generator and LH filter dynamic tuning property were investigated in this report.

FABRICATION OF SILICON VARACTOR BY ION IMPLANTATION

Modelling of Si varactor fabricated by IIT was carried out by using SILVACO software. Designated IV and CV characteristics could be achieved by sequential ion implantations of 1×10^{13} ions/cm^2 of P at 5 keV, 2.5×10^{11} ions/cm^2 of B at 20 keV, 7.5×10^{13} ions/cm^2 of P at 600 keV, 5×10^{13} ions/cm^2 of B at 280 keV and 3×10^{14} ions/cm^2 of P at 1300 keV. High resistive FZ Si wafer was implanted using both DF implanter and 2MV implanter at Surrey Ion Beam Centre. The as implanted wafer was annealed at 1000 °C in flowing ULSI grade nitrogen gas atmosphere in order to activate all dopants to form Schottky diodes. Both IV and CV characteristics were monitored.

CP1066, *Ion Implantation Technology,* edited by E. G. Seebauer, S. B. Felch, A. Jain, and Y. V. Kondratenko

SOLITON PROPAGATION IN LEFT-HANDED ION IMPLANTED NONLINEAR TRANSMISSION LINES

Soliton waves have been an integral part of nonlinear sciences for more than forty years. Its theory has been used to develop pulse compression transmission lines and harmonic generators [6]. Potential application of solitons for pulse shaping for ultra-wideband (UWB) is studied in this section.

Solitons can be described as wave pulses that can propagate through a medium without distortion. A distinct property of soliton waves is the fact that they can preserve their amplitude and shape after collision with other solitons. The presence of such waves in a medium is a consequence of the balance between dispersion and nonlinearity. Consequently, a soliton can travel through a nonlinear dispersive medium without altering its shape, as if it were travelling through a linear dispersive medium.

A LH nonlinear transmission line that can support soliton propagation can be formed by connecting ion implanted varactor diodes in series and inductors in shunt as depicted in Figure 1.

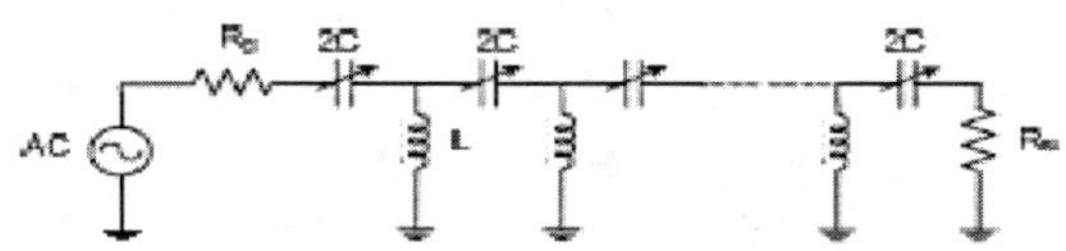

FIGURE 1. Circuit diagram of the left-handed transmission line for soliton propagation.

The propagation factor of the line is given by

$$\beta(V) = -\frac{1}{\omega\sqrt{LC(V)}} \tag{1}$$

where ω is the angular frequency, L is the inductance and C(V) denotes the voltage-dependent capacitance of the ion implanted varactor diode.

Figure 2 shows the soliton propagation along the nonlinear transmission line. The input signal is a Gaussian pulse modulated by a 3 GHz sine-wave.

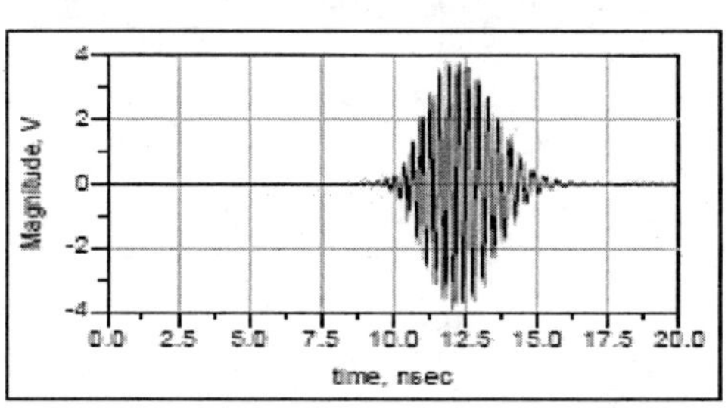

FIGURE 2. Soliton propagation along the left-handed nonlinear transmission line.

HARMONIC GENERATION BASED ON ION IMPLANTED NONLINEAR LEFT-HANDED TRANSMISSION LINE

Nonlinear transmission lines are also used extensively for harmonic generation. In this section we investigate harmonic generation based on left-handed nonlinear transmission line. The nonlinear elements in our investigation are zero-biased back-to-back ion implanted varactor diodes. The schematic circuit is shown in Figure 3.

The formula for the variable capacitance as a function of voltage is given as:

$$C(V) = \frac{C_{jo}}{\left(1 + |V / V_{bi}|\right)^{M}} \tag{2}$$

where C_{jo} is the zero-bias diode capacitance, V_{bi} the built-in voltage and M the grading coefficient.

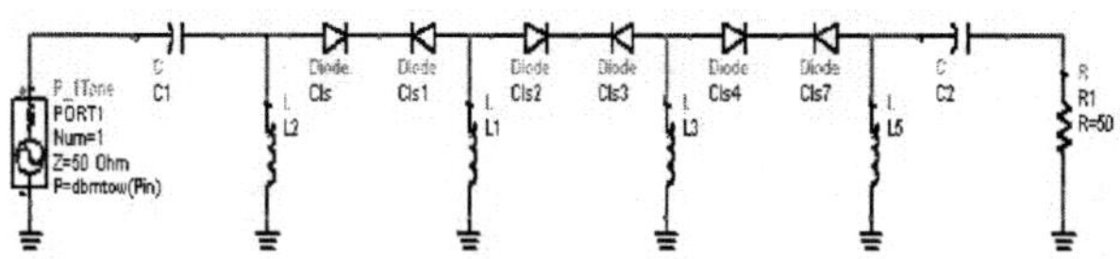

FIGURE 3. Circuit diagram of the left-handed transmission line for harmonic generation.

The third harmonic conversion efficiency of left-handed and conventional transmission lines is illustrated in Figure 4. This is simulated results based on the extracted ion implanted varactor diodes. The aim is to investigate the possibility of ion implanted metamaterials for lower THz applications.

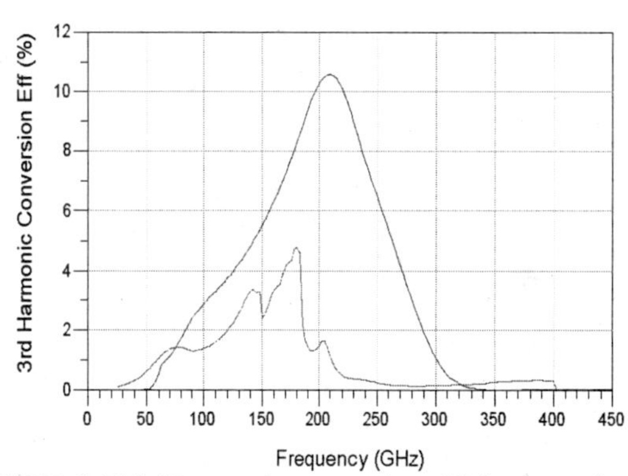

FIGURE 4. Third harmonic conversion efficiency against fundamental frequency for left-handed transmission line (blue) and conventional transmission line (red).

TUNABLE FILTERS USING ION IMPLANTED VARACTOR DIODES

The tunable bandpass filter was proposed by using implanted diodes as varactors. The schematic layout of the filter is show in Figure 5. The X band tunable filter has a constant tunable bandwidth of 570 MHz with return loss better than 10 dB. The transmission coefficient and return loss with various bias illustrated in Figure 6, clearly showing tuning capability of the varactor diodes.

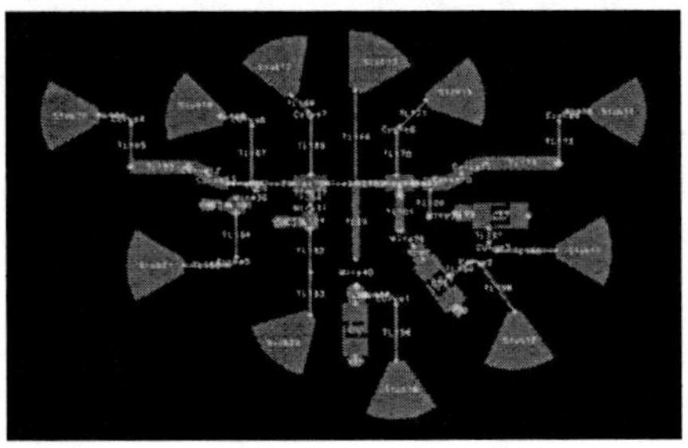

FIGURE 5. Circuit layout of the left-handed tunable bandpass filter.

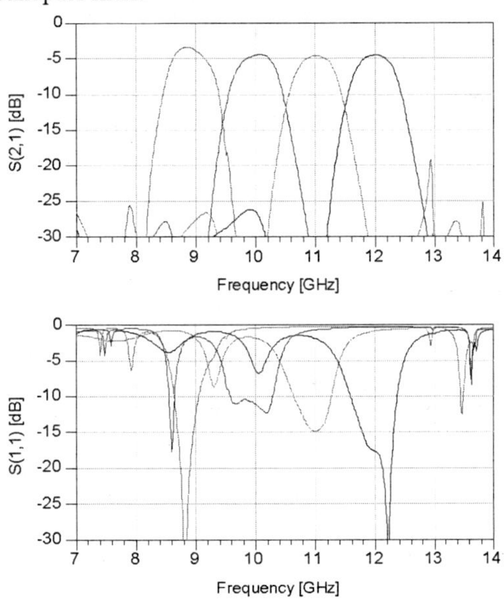

FIGURE 6. Transmission and reflection coefficients of the LF tunable filter consisting of ion implanted varactor diodes.

CONCLUSION

The preliminary results for nonlinear wave propagation, harmonic generation and active filter tuning based on Si ion implanted varactors are all very promising. Fine tuning of Schottky diode fabrication processes, in particular the post implant thermal annealing, is still under way. It is clear that ion implantation technology is a powerful tool in realizing compact novel active nonlinear left-handed meta-circuits for RF/microwave circuit applications at low cost.

ACKNOWLEDGEMENTS

The authors wish to acknowledge the EPSRC for supporting this work under grants EP/C015339 and EP/D032210. Thanks to Dr. Ahmed Neijm of SILVACO for his help on running varactor modeling by using SILVACO software.

REFERENCES

1. V. G. Veselago, "The electrodynamics of substances with simultaneously negative values of ε and μ", *Soviet Phys. Usp*, **10**, 509-514 (1968).
2. D. R. Smith *et al.*, "Composite Medium with Simultaneously Negative Permeability and Permittivity", *Phys. Rev. Lett.*, **84**, 4184-4187 (2000).
3. J. B. Pendry *et al.*, "Low frequency plasmons in thin-wire structures", *J. Phys. C.*, **10**, 4785-4809 (1998).
4. J. B. Pendry *et al.*, "Magnetism from Conductors and Enhanced Nonlinear Phenomena", *IEEE Trans. Microwave Theory Tech.*, **47**, 2075-2084 (1999).
5. I. V. Shadrivov *et al.*, "Nonlinear Tunable Microwave Metamaterials", *NATO Advanced Research Workshop, META '08 Metamaterials for Secure Information and Communication Technologies*, 7-10 May, 2008, Marrakesh, Morocco
6. M. J. W. Rodwell, *et al.*, "Active and nonlinear wave propagation devices in ultrafast electronics and optoelectronics", *Proceedings of the IEEE*, **82**, 1037-1059 (1994).

Platinum vs. Palladium in Catalyst-Based Hydrogen Sensors Used for Wide Temperature Range Hydrazine Leak Detection

Claudiu Muntele and Daryush Ila

Center for Irradiation of Materials, Department of Physics, Alabama A&M University, Normal, AL 35762, USA

Abstract. Here we are addressing possible solutions to challenges associated with using palladium and platinum as active agents in reverse-biased p-n structures and linear (resistive) structures for hydrazine and hydrogen detection schemes to operate in a temperature range from cryogenic to ambient (room temperature). Preliminary results at room temperature on devices fabricated using high-temperature sensor technology show only a limited response to hydrogen as opposed to dry air.

Keywords: Sensors, p-n structures, ion implantation, catalyst, hydrogen, hydrazine.
PACS: Replace this text with PACS numbers; choose from this list: 81.05.Hd, 82.65.+r, 89.60.Ec

INTRODUCTION

Silicon carbide based non-linear electronics devices (MOSFET, metal-semiconductor, or p-n junctions) are promising candidates for hydrogen detection schemes if used in conjunction with a catalyst such as platinum or palladium [1, 2]. For the past decade, the emphasis was mostly on high temperature applications in the automotive industry (for hydrogen-fueled engines) and aerospace industry (for jet engines), but now the focus is broadening to include auxiliary systems such as storage tanks, fuel lines, fuel production systems, all operating in a wide range of temperatures, from ambiental (RT – room temperature) to cryogenic. Hydrazine is a particular hydrogen-containing chemical of interest, as is widely used as a fuel in rocket propulsion systems, fuel cells, pesticides, dyes etc. It is also a neurotoxin, causing damage to most organs in the human body. Therefore sensors able to detect hydrazine leaks at sub-ppm level at RT (anhydrous hydrazine melts at 2°C and evaporates at 113.5°C) and in a wide humidity range (hydrazine is fully miscible in water) are highly desirable. Sensitive analytical methods have been developed for the determination of hydrazine in air, water, food, drugs, and cigarette smoke. Minimum detection limits reported [3] are 2 $\mu g/m^3$ (1.5 ppb) air (gas chromatography), 5 µg/liter (5 ppb) water (colorimetry), 1 µg/liter (~1 ppb) blood, plasma, or urine (gas chromatography/mass spectrometry), and 3000 µg/kg (3 ppm) drug (gas chromatography). However, all these methods involve complicated analytical instrumentation generally available only in a specialized laboratory environment. From the variety of detection schemes available for portable devices, two seem of being the most commonly used: color-changing paper readers (e. g. MDA 7100 and newer) and fuel cell-based detectors (e.g. products of PureAire Monitoring Systems, Inc.).

Our group's focus and past experience is on using either platinum or palladium as active detection agents implanted into silicon carbide substrates to fabricate p-n structures able to react to either molecular hydrogen or hydrogen-based chemicals in hot environments (300 to 800°C range). While high temperature operation is completely characterized for such catalysts, low temperature operation of these schemes presents catalyst-related challenges (such as catalyst "poisoning" due to surface passivation), some of them little understood, some well characterized, but without a functional solution.

EXPERIMENTAL WORK

We used 6H n-type silicon carbide for producing non-linear (p-n junctions, n-p-n stack) devices, and 6H semi-insulating silicon carbide for producing linear (resistive) devices. In both contexts we introduced Pd or Pt distributions in concentrations of no more than 10 at% under the surface of the silicon carbide host

CP1066, *Ion Implantation Technology,* edited by E. G. Seebauer, S. B. Felch, A. Jain, and Y. V. Kondratenko

using low energy ion implantation. Such devices give a good current response (nA to mA, for 1V bias) to hydrogen at temperatures from 300 to 800°C, with sensitivities better than 1000 ppm [4]. Here we tested similar devices implanted with 1×10^{15} cm^{-2} Pd of 35 keV and 1×10^{16} cm^{-2} Pt of 13 keV at 7 deg off the normal in order to avoid channeling effects. Both Pd and Pt implantations were carried out at a substrate temperature of 500°C in order to minimize the implantation damage. Electrical measurements were carried out at RT in a low-noise environment (Faraday cage) using a system of Keithley instruments interfaced with a computer for data collection. Current vs. voltage measurements were carried out in both dry air and forming gas (argon with 4% molecular hydrogen, labeled as FG in this article).

RESULTS AND DISCUSSIONS

In Figure 1 we show the current vs. voltage plot for an n-p-n stack of junctions formed by implanting Pd into n-type silicon carbide. Pd acts as a p-type dopant in this material. The plot is typical for a system of back-to-back diodes, in which the current flow through the stack is always controlled by the reverse-biased junction. We carried out the measurements from both negative to positive and positive to negative directions for each situation (air or FG) to observe whether any hysteresis is present. While no significant hysteresis is observed, no significant difference between the current in air and in FG is observed either.

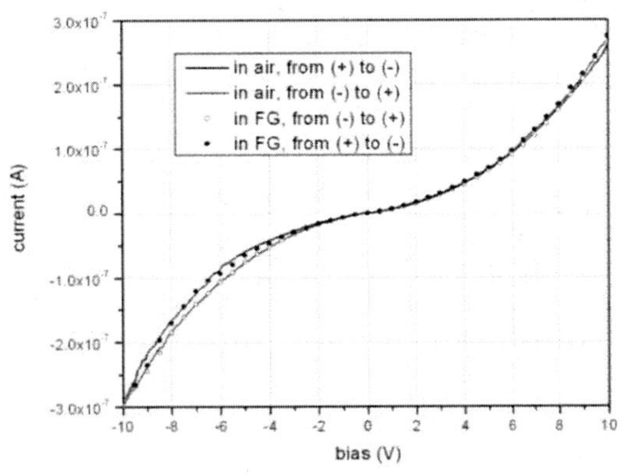

FIGURE 1. Current vs. voltage (at RT) for Pd-implanted silicon carbide in both air and forming gas.

The results of the measurements on resistive-type devices on semi-insulating silicon carbide implanted with Pt are not much more promising for the same voltage range (Figure 2). We carried out these measurements in the same manner as for the previous case, in both positive and negative directions, for both, air and FG, situations. Here, a large hysteresis is observed, due to a significant capacitive behavior. The same capacitive behavior is responsible for the fact that the current values are positive for the (-) to (+) travel, and negative for the opposite direction. The relationship between the voltage and current governing this behavior is described in Eq. 1:

$$I = \frac{dQ}{dt} = C\frac{dV}{dt} \tag{1}$$

When the voltage travels from (-) to (+), dV is always positive, hence the positive values for the current. Similarly, when the voltage travels from (+) to (-), dV is always negative, hence the negative values for the current. In our measurements, we had voltage step of 0.1 V every 1 s, therefore the average value of the capacitance becomes ~1.4 pF (using an average value of 1.4×10^{-13} A in either direction, from the plots).

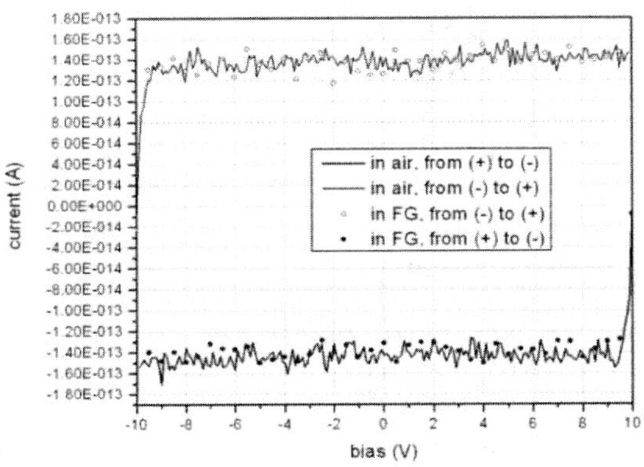

FIGURE 2. Current vs. voltage (at RT) for Pt-implanted silicon carbide in both air and forming gas.

Using the same resistive device, exploratory measurements were carried out in an extended voltage range, from 0 to 100 V, with a larger time delay between the voltage step and the moment the measurement was taken (5 s), in order to allow for parasitic capacitances to relax. In these conditions, the currents measured were of the same magnitude as the sensitivity limit of the instrument (1×10^{-15} A), and therefore largely affected by noise (see Figure 3). However, a trend seems to become visible: while the current through the device while in air has an increasing tendency, the current seems not to change while in FG, symptomatic of hydrogen incorporation (possibly as metal hydrides), and therefore increased resistivity.

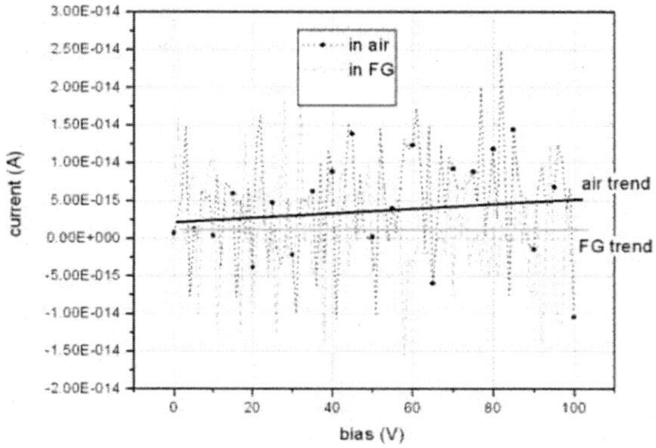

FIGURE 3. Current vs. voltage (at RT) for Pd-implanted silicon carbide in both air and forming gas.

CONCLUSIONS

Preliminary measurements on devices fabricated using directly adopted technology developed for high temperature chemical sensors show that some sensing is still available at RT. While the response is very weak, the basic mechanism seem to function. In order to improve the sensing, one direction would be to bring the catalyst distribution closer to the surface in a higher concentration (overdoping), at least for the linear device configuration. Also, while the hydrogen reaction happens without any significant energy released, the breakdown of hydrazine (experimental setup currently under construction) is strongly exoenergetic, therefore heating up the sensitive area of the device and augmenting the catalytic effect. Furthermore, a FET configuration might prove more sensitive than a p-n junction or a resistor type of device.

ACKNOWLEDGEMENTS

This research is sponsored by the Center for Irradiation of Materials, Alabama A&M University and by the AAMURI under the National Science Foundation Grant number EPS-0447675.

REFERENCES

1. S. M. Savage, A. Konstantinov, A. M. Saroukhan, and C. I. Harris, *Mat. Sci. Forum* **338-342**, 1431-1434 (2000).
2. G. W. Hunter, P. G. Neudeck, M. Gray, D. Androjna, L. Y. Chen, R. W. Hoffman Jr., C. C. Liu, and Q. H. Wu, *Mat. Sci. Forum* **338-342**, 1439-1442 (2000).
3. International Programme on Chemical Safety, Environmental Health Criteria for Hydrazine, 1987.
4. C. I. Muntele, D. Ila, D. J. Larkin, D. B. Poker, D. K. Hensley, *Mat. Res. Soc. Symp. Proc.* **647**, (2001).

IMPLANT TECHNOLOGY

Vertical Beam Angle Control: an Advancement/Requirement in Modern Ion Implant Manufacturing

Christian Krueger[1], Robert Rathmell[2], Dennis Kamenitsa[2], and Bernhard Krimbacher[3]

[1]*Fab36, AMD, Dresden, Germany,*
[2]*Axcelis Technologies Ltd., Beverly, Massachusetts,*
[3]*Axcelis Technologies Ltd., Dresden Germany*

Abstract. As the industry moves to the new technology nodes of 45 nm and 32 nm devices, implant angle control becomes even more crucial for consistent device performance. Commercial single wafer ion implanters are able to measure and correct the horizontal incident angle of the ion beam. But the vertical beam angle (VBA) control has become a very important parameter as well. In this work the authors demonstrate the impact of a tilt variation for a 65 nm and a 45 nm MOS transistor generated by different beam setups on one machine. Comparisons are made for each technology with a controlled angle variation of ±4°. The Vt-distribution should be reduced with better incident angle control allowing faster development of a new transistor node base line using the new VBA control technique from Axcelis.

Keywords: Angle control, process control, threshold voltage
PACS: 85.40.Ry, 81.05.Ea, 41.85.Qg

INTRODUCTION AND THEORY OF OPERATION

Manufacturing integrated devices at 45 nm and 32 nm technologies and beyond becomes more complicated due to the tighter restrictions on transistor parameter variation. To fabricate a symmetric or an asymmetric transistor requires exacting process conditions and results for every manufactured wafer. Tight control of threshold voltage (V_t) or Miller-Capacitance (C_M) is needed to improve the performance of certain devices and achieve the transistor node requirements with stable results. The new medium current implanter Optima MD™ manufactured by Axcelis ® is using a new technique for the vertical beam angle (VBA) measurement in order to compensate for a vertical beam angle variation from the expected or desired beam angle. The actual vertical tilt angle between the plane of the scanned beam and the plane of the wafer is measured via a profile mask attached to the linear portion of the mechanical scan arm supporting the wafer. Any offset in the vertical beam angle is checked against the recipe parameters and will be corrected automatically.

VBA HARDWARE

The vertical beam angle monitor, as shown in Figure 1, is rigidly mounted on the arm that supports the electrostatic chuck (1) such that there is a fixed offset between the angle of the surface of the VBA monitor and the surface of the wafer of about 30°. The VBA monitor is a faraday (2) with a thick slotted mask that limits the current transmitted to the collector as the slots in the mask are tilted relative to the beam angle [1]. This will produce a current as a function of the tilt angle of the VBA monitor that peaks when the slots are aligned with the beam and that allows accurate determination of the "center of mass" or flux weighted vertical angle of the beam. The fixed offset between the VBA monitor and the wafer surface is calibrated with a sensitive channeling implant and allows one to set the chuck tilt relative to the actual beam angle to the value specified in the recipe.

The vertical beam angle is measured at the start of a processing job by tilting the chuck so that the VBA monitor is in line with the ion beam, as shown in Figure 1a. Data is collected at +/-5° about this point as the chuck moves down and up through this range of angles in about 4 seconds. This provides two sets of data that can be checked for consistency and beam dropouts. At these tilt angles, the wafer is safely above the ion beam. After the VBA measurement, the chuck moves to the implant position, as shown in Figure 1b, and passes through the scanned beam at the desired tilt angle relative to the measured beam angle, while the VBA monitor remains fixed to the tilt mechanism and is safely out of the beam. The chuck and wafer are tilted about an axis parallel to the plane

CP1066, *Ion Implantation Technology,* edited by E. G. Seebauer, S. B. Felch, A. Jain, and Y. V. Kondratenko

of the scanned beam in order to achieve the desired tilt.

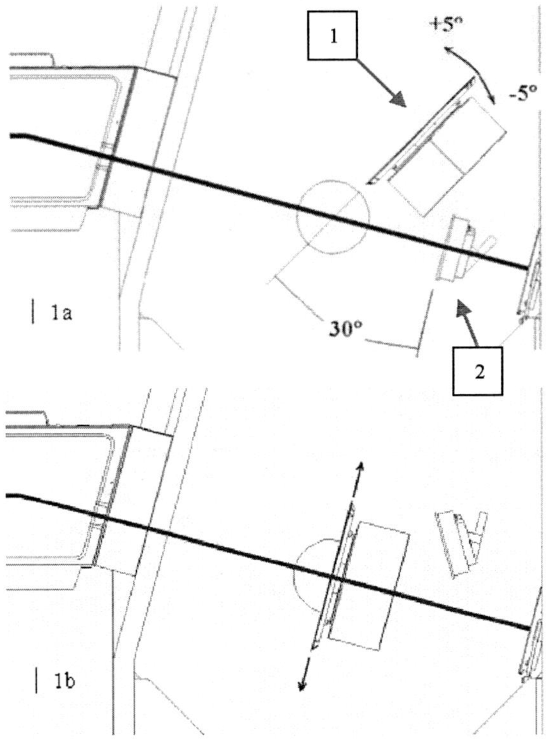

FIGURE 1. a) Scan arm at tilt angle to measure vertical beam angle of the accelerated ion beam. (1) electrostatic chuck with wafer (2) VBA-Faraday-Cup
b) Scan arm at an implant position with VBA out of the beam (Figure by Axcelis)

STARTING CONDITIONS

Figure 2 shows a full range-median-average plot of VBA measurements for 12 different recipes on one of the Optima MD^{TM} implant tools when the VBA was measured, but not corrected. The actual VBA varies somewhat from setup to setup and the amount of variation depends on the recipe conditions. Naturally, low energy recipes trend to show higher variations compared to recipes with higher energies due to a somewhat larger angle spread and magnification of small angle errors in deceleration and focusing elements.

The maximum variation for a 10 keV Boron (11) recipe is ~ 0.5° compared to a very low variation of less than 0.1° from setup to setup for Arsenic (2 & 3) or Phosphorus (1 & 8) recipes with energies between 50 keV and 250 keV. However, the data also show that there can be over 1° variation from one recipe to another if no correction for VBA is used. In addition, the setup for a given recipe may be different on every tool as well. Therefore, such factors could lead to significant angle variation and make it very difficult to

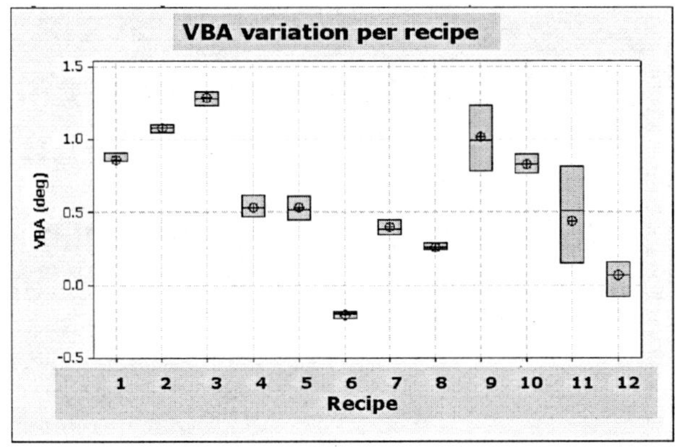

FIGURE 2. VBA variation on one implanter for different recipes over a period of one month (Graph by AMD)

reduce the transistor parameter variation for a stable transistor node development.

The horizontal beam angles are independently measured and automatically corrected at seven points across the wafer [1] to a recipe limit, which in this study was set to <0.5° (0.3° was typical).

ACTIVE INCIDENT ANGLE CORRECTION

Active correction of the vertical tilt of the wafer by adjusting the wafer orientation about an axis parallel the scanned beam plane in order to compensate for the measured VBA offset could be one of the key factors to get the transistor parameter distribution better under control. The new capabilities with the Optima MD^{TM} are able to do this very well. Figure 3 shows a graph for a number of different beam setups. The data show

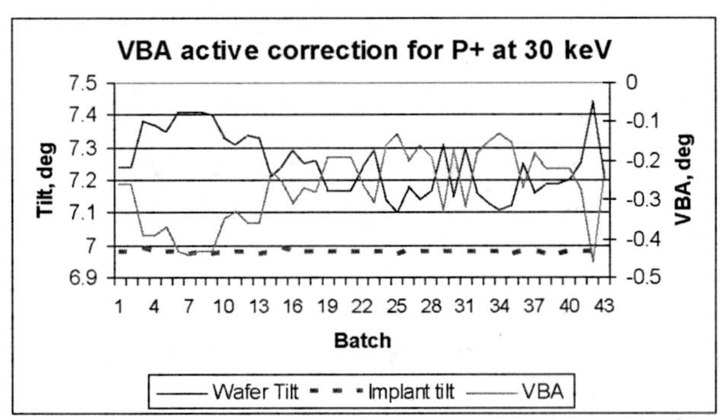

FIGURE 3. VBA active correction for a 30 keV Phosphorus recipe with 7° Tilt requirement (Graph by Axcelis)

the measured Vertical Beam Angle and the corresponding mechanical wafer tilt. The sum of these is the actual angle between the beam and the surface of the wafer and is shown to be very constant for all processing jobs.

A separate paper at this conference uses a sensitive channeling condition to show that the angle measurement of the Optima MD™ is repeatable relative to the planes of the crystal from setup to setup. [2].

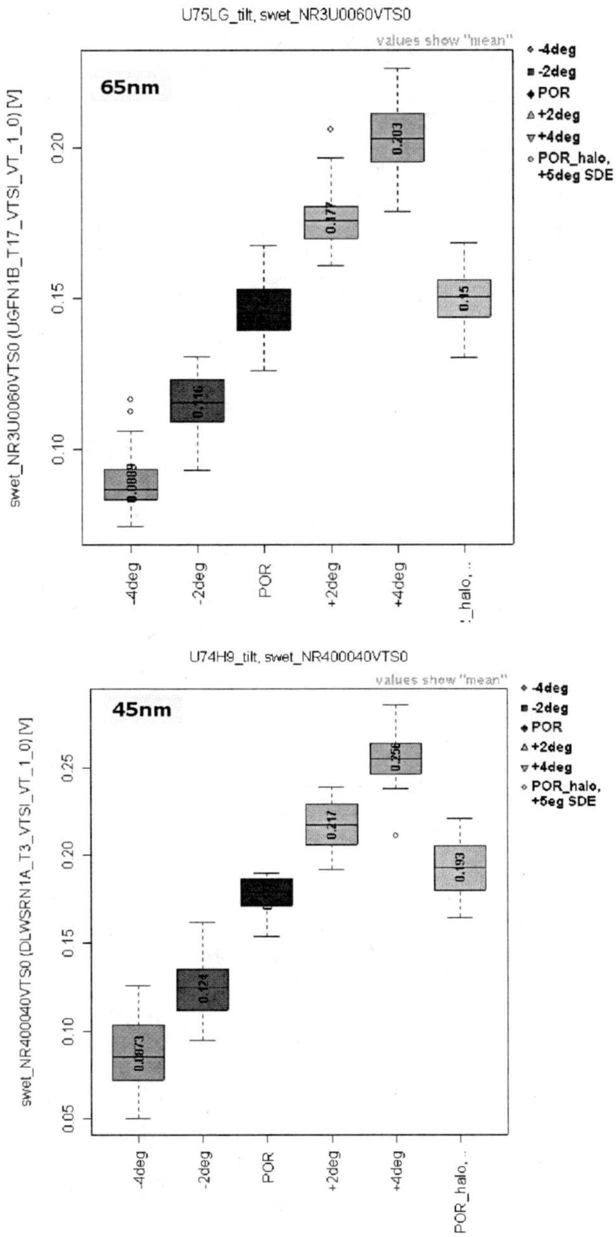

FIGURE 4. V_t-sensitivity for a NMOS transistor at different technology levels (Graph by AMD)

It was expected that the angle compensation would reduce the transistor parameter distribution for a certain device in a long term trend.

DEVICE DATA

In order to determine the sensitivity of device parameters, V_t and C_M, to wafer tilt, a wafer split experiment with a relatively high tilt angle variation was carried out using a Halo-and a Source Drain Extension, SDE-implant. The tilt of the wafer (the orientation of the wafer about an axis parallel to the plane of the scanned beam) was purposely varied ± 4° for the Halo implant, compared to the Process of Record, POR. A separate implant using a +5° offset in the SDE tilt was done with the Halo implant at the normal tilt angle. The resultant device data show a clear V_t-dependency on the incidence angle for the Halo implants. Surprisingly, variations of the tilt angle of the Halo implant influence the results much more than the Extension implant tilt variation. Figure 4 shows the influence the tilt angle changes for 65 nm and 45 nm technologies. The V_t-sensitivity of that particular NMOS transistor is different for each technology level, yielding 14 mV/deg and 21 mV/deg, respectively, with the higher sensitivity for the smaller node, as expected. Theoretically this effect would influence the transistor node development significantly if there are variations larger than ~0.5° at the 45 nm node and would make a clear interpretation of "controlled" experiments very difficult.

In reality there is a relatively small reduction in the V_t parameter distribution on long term trends comparing data before active correction of VBA to those after active correction, as shown in Figure 5 for 4 months of data. The red bar shows when VBA auto-correction was turned on. That means that this parameter distribution is more influenced by other parameters like perhaps a variation in gate length, C_M, or some other implant condition than by the implant incident angle itself. It is also possible that the implant angle was not ranging very far even before turning on active correction as was seen for some recipes in Figure 2.

While the effect on the long term distribution is relatively small, Figure 4 shows that there is a high sensitivity to beam angle for halo implants, and it is more and more important to minimize all relevant sources of variation for an effective device development.

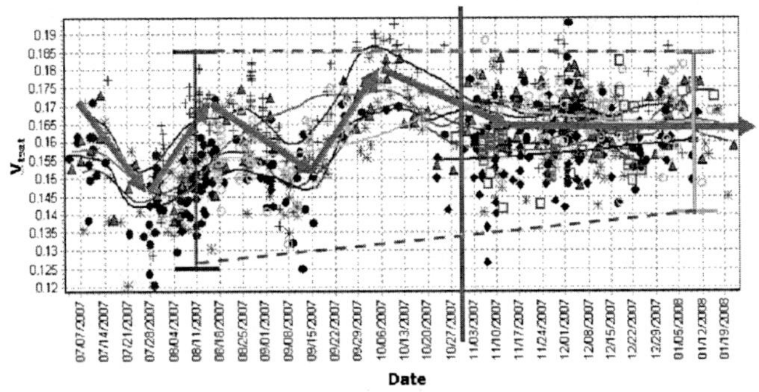

FIGURE 5: Long term V_t behavior prior to and after the use of the VBA unit to correct angles on all Optima MD™ (Graph by AMD)

CONCLUSION

The Optima MD Vertical Beam Angle measurement allows precise control of the orientation of the wafer about an axis parallel to the plane of the scanned beam during implant. Some devices studied show a high sensitivity such that a variation of ~0.5° could result in significant changes in V_t. As expected, the sensitivity is increasing as the node size decreases. The fact that controlling the VBA very tightly did not change the observed variation in device electrical parameters of real product so much suggests that either there was not much angle variation in the critical implants prior to implementing active control or parameters other than beam implant angle are causing lot-to-lot variations that may be seen.

REFERENCES

1. R. D. Rathmell, *et al.*, "Implant Angle Control on Optima MD", in *Proceedings of the 16th International Conference on Ion Implantation Technology.*, Marseille, France 349-352 (2006)
2. R. D. Rathmell and Jonathan David, "Implant Angle Repeatability on Optima MD", in these Proceedings.

Benefits of Zero Degree Single Wafer High Energy Implants for Advanced Semiconductor Device Fabrication

Woojin Lee, Thirumal Thanigaivelan, Hans-Joachim Gossmann, Russell Low, Benjamin Colombeau, Kerry Lacey, Mark Merrill, Anthony Renau

Varian Semiconductor Equipment Associates
35 Dory Road, Gloucester, MA 01930

Abstract. High energy well implants for retrograde well formation are usually tilted to avoid channeling. However, this can cause the well profile under the STI (Shallow Trench Isolation) to be skewed or asymmetric due to shadowing by the resist. This results in poor inter-well isolation and increased leakage currents. This problem becomes more pronounced below 65 nm design rules. In this paper, using experimental data and TCAD simulations we demonstrate the improvement in inter well isolation and junction characteristics that can be achieved with true-zero well implants. Finally, we briefly discuss the corresponding die shrinkage that can be expected.

Keywords: Zero degree implant, Shadowing effect, Inter-well isolation
PACS: 85.40.Ry, 85.30.De

INTRODUCTION

In order to limit uncontrollable channeling variations across the wafer, implants are usually performed with up to 7° of tilt. Although this may mask the effect of an implant angle that may be varying by a few degrees, it causes another problem which can degrade device performance. This is because high energy implants require a thick photo-resist mask, which will shadow significantly any implant that is not perpendicular to the wafer (i.e. not zero degree) [1]. Figure 1 illustrates the shadowing that occurs with a non-zero degree implant. Such a non-zero implant may lead to shifts in well boundaries [2], and poor and varying inter-well isolation.

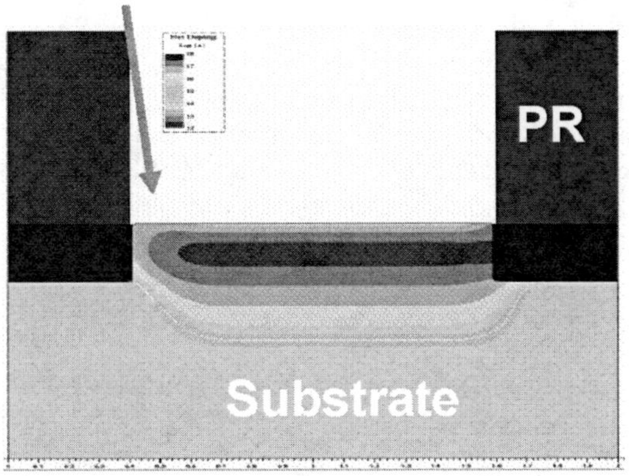

FIGURE 1. Illustration of the shadowing that occurs with non-zero degree implants. This example shows the asymmetric doping that occurs when 500 keV P^+ is implanted at 2° tilt.

The shadowing problem is compounded in advanced devices by scaling trends. Lateral device dimensions have reduced faster than the highest implant energies. Photo-resist thickness has therefore not scaled with device size and the problems due to shadowing have increased, particularly beyond the 65 nm node. To avoid these problems, high energy implants need to be done at a true zero degree angle. However, it has been seen that a true zero implant cannot be done on a batch tool due to the cone angle effects [3]. This has driven the development of the single wafer high energy ion implanter. In addition, it has been recently discussed that zero degree tilted implant may improve junction leakage [4].

In this paper, we present experimental and TCAD simulation data to demonstrate how leakage currents can be improved with true zero degree implant. Finally, we discuss the impact of implant tilt on die size reduction.

WELL IMPLANT PROFILE

Key device parameters such as drive and leakage currents, threshold voltage and well isolation are affected by the distribution of dopant and defects in

CP1066, *Ion Implantation Technology,* edited by E. G. Seebauer, S. B. Felch, A. Jain, and Y. V. Kondratenko

and below the channel region. In order to optimize these distributions, three implants are often used to dope the channel and provide isolation. These implants are generally done as a sequence from highest energy to lowest energy. The resulting dopant distribution is illustrated in Figure 2, which shows a typical ^{11}B triple well SIMS profile implanted at tilt angles of 7° and 0°.

The profile has been broken into three regions in order to explain the impact on device parameters. Region 1 is the channel region and the dopant concentration here will affect the threshold voltage (Vt), the drive current (Ion), the gate induced drain leakage (GIDL) and channel resistance. Dopant and defect concentrations in region 2 affect Source/Drain (S/D) punch through as well as junction capacitance. The deepest portion of the well, region 3, is doped by the highest energy implants and these affect isolation parameters like n+/pwell and p+/nwell breakdown.

As shown in Figure 2, the 0° implant allows ions to be channeled. It therefore gives a deeper implant. In the case of the zero degree implant, ions travel down the channels between the lattice atoms and the reduced scattering causes the ions not only to travel farther but also to cause less damage to the lattice. Zero degree implants therefore result in less residual defects.

There is another aspect of channeled well profiles that is worth mentioning. As Figure 2 illustrates, the third peak is much broader for the zero degree implant (since it is composed of two peaks, representing the channeled and the un-channeled ranges).

Figure 3 compares the leakage currents of 70 nm DRAM devices made with well implants using 5° tilt to those made with 0° tilt. It can be seen that, even at 70nm, 0° wells show lower leakage currents.

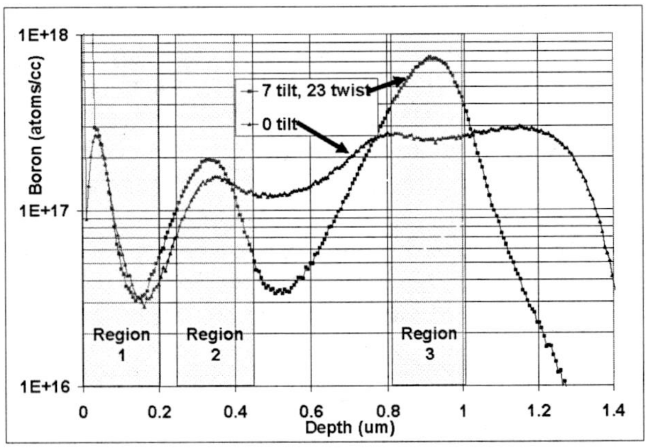

FIGURE 2. SIMS profiles for a typical triple p-well showing as-implanted boron depth distributions for 0° and 7° implants through a 50A screen oxide. The 3 implants for the 7° sequence were: 400 keV, 2e13 then 105 keV, 4e12 then 10 keV, 2e12. For the 0° sequence they were: 300 keV, 2e13 then 95 keV, 6e12 then 10 keV, 2e12.

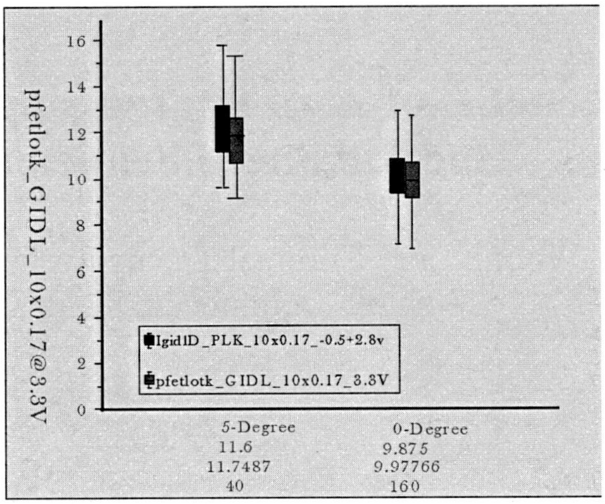

FIGURE 3. Comparison of 70 nm DRAM leakage currents for with well implants performed at 0 and 5 degrees.

Effect of Well Implant Profile on Device Characteristics

As discussed earlier, a typical retrograde well implant process employs a series of well implants in the sequence of decreasing implant energies ending with the lowest energy implant which is a Vt-like implant. The channeled implants provide deeper and more gradual junctions for the same energy and dose as compared to the non channeling, 7° implants. The effect of the profile difference on the device characteristics is discussed below. Figure 2 compares channeling 0° implants with non-channeling 7° implants.

Region 1: For the same energy and dose the channeled implants, 0°, have low dopant concentration at the peak range, Rp, and a more gradual profile. Also the channeling implant has less Si damage due to fewer collisions with the lattice resulting in fewer Si defects. The advantage of the reduced peak dose and reduced Si defects is believed to be lower leakage as shown in Figure 3.

Table 1 summarizes the experimental data comparing a 5° and 0° implant process for a 90nm technology node device.

TABLE 1. Device parametric response to 0° implants compared to POR 5° process

Device Parameter	With 0° well implant
GIDL	15 – 20% reduction
Junction Capacitance	~ 20% reduction

The S/D punch through and the S/D to well junction leakage is controlled by the dopant concentration and profile in the region 2. With the channeled dopant profile for the 0° well implant condition the Si defects are reduced.

This results in lower junction capacitance and lower junction leakage as shown in Table 1. The lower junction capacitance is due to the reduction of the impurities in the junction region (2) and the lower leakage is due to the reduction of electric field at the junction.

Region3: The isolation performance is determined by the distance between the S/D to well region, which includes the STI width, and the N and P well doping concentration and boundary. Higher angle implants result in skewed well boundaries between the nWell and pWell. The nWell implant undercuts the resist in the nFET side to a certain degree and vice versa.

A tilt of 7° corresponds to a rotation of the incident beam away from the surface normal in a counter-clockwise direction. As expected the 7° tilted implants create a shadow and shift the boundary by $S = t \times tan(7°)$, where S is the width of the shadow region and t is the resist thickness. For a resist thickness of 2.5 μm, the shadow width is 307 nm. The parameter S is called the resist setback required to offset for the shadowing. In a typical isolation scheme with a Shallow Trench Isolation process, the well spacing must be at least twice this resist setback.

Isolation Improvement with 0° Well Implants

TCAD simulations were performed to gain insight into the impact of tilt angle and STI dimenions. on electrical device parameters. They utilize integrated 2D-implant [5] and process/device [6] simulators [7]. In the following we will restrict ourselves to a discussion of isolation performance, specifically to leakage across the STI.

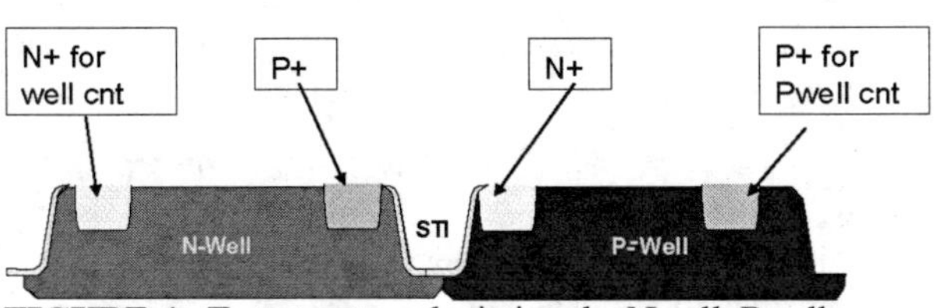

FIGURE 4. Test structure depicting the Nwell, Pwell, N+SD and P+SD contacts.

The test structure consists of a pFET (on the left in Figure 4) and a nFET (on the right), separated by STI of a depth of ≈ 400 nm. A process flow including dopant activation representative of an 80 nm DRAM process is implemented. The wells are implanted while protecting the well of the respective opposite polarity by resist of 2.5 μm thickness. Tilt angles ranging from 0° to 7° were used, where a positive tilt angle implies a direction of the ion beam from the upper left corner to the lower right corner in Figure 5. The STI length was varied between 50 nm and 500 nm. n+ and p+ contacts to the nWell on the left and the pWell on the right, respectively, are made, as are appropriate n+ and p+ SD contacts. Charge distribution contours are shown in the device cross-sections of Figure 5 for the resulting test structure, for a STI length of 500 nm (top) and 50 nm (bottom). Red colors denote n-type charge, blue colors p-type.

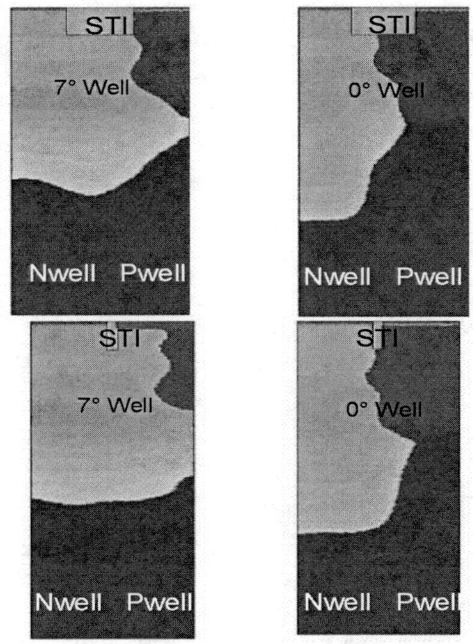

FIGURE 5. Contour plot of net doping concentration for the 7° and 0° well process. The distance is the same in x- and y-axis.

The 7° implant noticeably shifts the well boundary to the right, with the 500 nm STI length barely providing isolation, while the nWell encroaches deep into the pWell for the 50 nm STI length.

To quantify leakage across the STI, voltage is applied between the nSD contacts of the nFET (right) and the nWell well-tie of the pFET (left) Between the right electrode, of the nSD contact, and the left electrode, of the nWell contact, a npn bipolar transistor with its base in the bulk Si is formed. Applying a negative voltage to the n+ electrode while keeping the nWell contact at ground potential makes the right electrode the emitter and the left one the collector, creating an emitter-coupled circuit with an essentially grounded base. This causes the npn transistor to turn on very quickly. Note that this is not due to leakage since it occurs even for the wide 500 nm STI. Neither is it due to breakdown since impact ionization and tunneling have been turned off in the simulation. Of course, for large encroachments of the nWell into the pWell a short results between n+ and nWell regions and the bipolar transistor action is not necessary for a large current.

Applying a positive bias to the right electrode while keeping the left one grounded reverses the roles; the right electrode is now the collector and the left one the emitter. In this case the npn transistor is off until sufficient voltage drop develops between the base and the emitter. This situation is similar to latch up, except that in the present case removal of the bias will also turn off the current flow, since there is only one bipolar transistor and therefore nothing "to latch".

We define a n +-nWell turn-on voltage TV as that positive bias on the nFET nSD that gives rise to a current of 1x10-9A/µm. It is shown in Figure 6 as a function of STI length and for a variety of tilt angles.

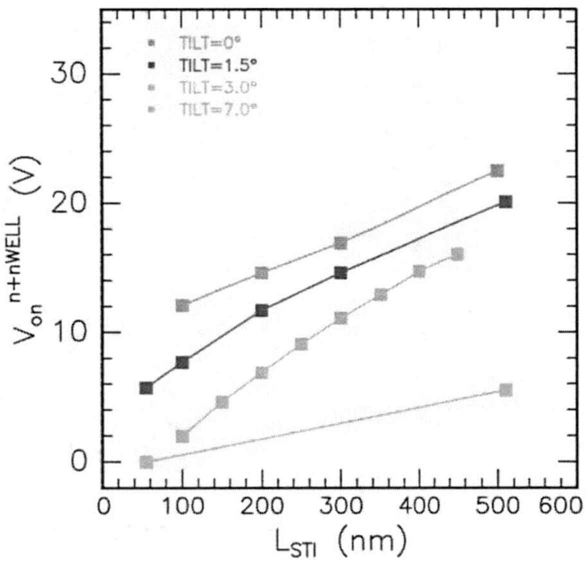

FIGURE 6. n+ - Nwell turn on voltage as a function of STI widths, for various incident angles.

The larger the angle, the smaller the TV, the steeper the rise with L_{STI} (with the exception of 7° incidence). A 0° process yields the highest TV.

Die Size Reduction

Finally, we comment on the minimum width/length of the STI trench for the True Zero™ process. The resist will not cast a shadow for a 0° well implant and in the absence of variations does not need a resist setback. For a process with finite variations the resist set-back has to be at least equal to the total tolerance.

Table 2 lists the estimated STI width required as a function of well implant angle and the potential die size reduction as a function of the percentage of transistors that can be shrunk. Clearly, a true zero implant allows better integration.

TABLE 2. Die size reduction achievable with a 0° process, assuming a transistor with $W=4L_g$ and the listed fractional area of the chip that is occupied by core transistors.

Tilt (°)	L_{STI} (nm)	ΔA			
		5%	10%	30%	40%
1.5	150	-2%	-5%	-14%	-18%
3.0	250	-3%	-6%	-18%	-24%
7.0	500	-4%	-8%	-24%	-32%

CONCLUSIONS

We have demonstrated the device benefits of 0° well implants by analyzing the different regions of the well profile, with the help of modeling and device parametrics. By adopting a 0° well implant the junction leakage and junction capacitance can be reduced by 15 to 20% and the packing density increased by decreasing the die size .

ACKNOWLEDGEMENTS

The authors would like to thank Mr. C.N. Hwang, Mr. J.H. Yang, and Mr. Y.C. Chuang at Winbond for sharing valuable experimental data for this research.

REFERENCES

1. K. Kapila, A. Jain, M. Nandakumar, S. Ashburn, K. Vasanth, S. Stridhar, "The Effect of Deterministic Spatial Variations in Retrograde Well Implants on Shallow Trench Isolations for sub-0.18um CMOS Technology" IEEE Trans. Manu.,.**12**(4).
2. T. Yamashita, M. Kitazanwa, Y. Kawasaki, H. Takashino, T. Kuroi, Y. Inoue, M. Inuishi, "Advanced Retrograde Well Technology for 90nm-node Embedded SRAM by High-Energy Parallel Beam" *Japanese Journal of Applied Physics, Part 1* **41**(4B) P.2399-403.
3. Y. Hai, E.N. Shauly, "Influence of Batch-to-Batch Substrate variation and Cone Effect on High Energy Implant Distribution profile", 14th International Conference on Ion Implantation Technology Proceedings, 2002.
4. M. Kitwaza, T. Yamashita, Y. Kawasaki, T. Kuroi, T. Eimori, M. Inuishi and Y. Ohji, "Novel Substrate Engineering for Hogh Performance CMOSFETs using Channeling Ion Implantation" ,SSDM, 0728, 2003
5. IMSIL-2003, Gerhard Hobler, TU Vienna, Vienna, Austria 2003.
6. M. R. Pinto, D. M. Boulin, C. S. Rafferty, R. K. Smith, W. M. Coughran, Jr., I. C. Kizilyalli, M. J. Thoma, "Three-dimensional characterization of bipolar transistors in a submicron BiCMOS technology using integrated process and device simulation", Proc IEDM-1992, 923-6 (1992).
7. SENTAURUS 2007.3, Synopsys.

Neutralization of Space Charge Effects for Low Energy Ion Beams Using Field Emitters

D. Nicolaescu[1], S. Sakai[1], K. Matsuda[1], Y. Gotoh[2] and J. Ishikawa[2]

[1]*Nissin Ion Equipment Co., Ltd., 575 Kuze Tonoshiro-cho, Minami-ku, Kyoto 601-8205, Japan*
[2]*Department of Electronic Science and Engineering, Kyoto University*
Kyotodaigaku-Katsura, Nishikyo-ku, Kyoto 615-8510, Japan

Abstract. The paper presents models and computations for neutralization of space charge effects using electrons provided by field emitter arrays. Different ion species ($^{11}B^+$,$^{31}P^+$,$^{75}As^+$) with energy in the range E_{ion}=200 eV-1 keV have been considered. The ion beam divergence is studied as a function of electron beam geometry and physical parameters (electron and ion energy, electron/ion current ratio I_{el}/I_{ion}). The electron beam geometry takes into account electron source positions and initial launching angles. It is shown that optimal ion beam neutralization occurs for low energy electrons emitted parallel to the ion beam.

Keywords: ion beam, ion beam, space charge, field emitters.
PACS: 41.85.Ja,41.75.Ak,85.45.Db.

INTRODUCTION

Shrinking of semiconductor the device in size requires implantation performed with ion beams that have high current density, low energy and precise parallelism. However, it is difficult to transport to target such a beam of particles because space charge effects make the beam diverge. Current systems use plasma flood guns [1] as electron sources to control wafer charging and neutralize space charge effects in the region nearby the wafer. A new method for neutralization of space charge effects in high current ion implantation systems has been recently proposed [2], where electrons are provided by field emitter arrays. Using field emitters in implantation systems imposes special requirements, such as non-contamination of the target wafer with metallic species, low energy and narrow energy distribution for emitted electrons, and long lifetime operation of the array in relatively poor vacuum [3, 4]. Another problem, which is the subject of this paper, is related to the emitter array positioning and electron beam properties for optimal ion beam neutralization. Models and computation results for neutralization of space charge effects using electrons provided by field emitter arrays are presented in this article. It is shown that optimal ion beam neutralization occurs in the case of low energy electrons that are emitted along (or parallel to) the ion beam direction [5].

MODELING FRAMEWORK

Neutralization of space charge effects for the ion beam is obtained allowing electrons to flow together with ions. Space charge neutralization occurs for ion and electron current densities j_i and j_e that are correlated with their velocities v_i and v_e by [6]:

$$j_e = j_i \left| v_e / v_i \right| \quad (1)$$

Electrons are generally faster than ions, so for neutralization of space charge effects, the electron current needs to have higher density than the ion current. Writing energy instead of velocity in the previous equation and considering a laminar flow model, it can be concluded qualitatively that for space charge neutralization, lower electron/ion current ratios are required in the cases when the ions have higher energy and lower mass and the electrons have lower energy. Actual flow of particles is complex and elaborate numerical computations are required for quantitative estimations.

In ion implantation systems such as Exceed 3000, the region that is comprised between the collimator magnet and target (wafer) is basically field free. However, ions and supplied electrons, if it is the case, are subject to non-negligible forces of mutual interaction that are associated with space charge effects. In this modeling framework, local fields and charged particle trajectories have been computed with

CP1066, *Ion Implantation Technology,* edited by E. G. Seebauer, S. B. Felch, A. Jain, and Y. V. Kondratenko

the commercial program Opera3D/Scala from Vector Fields [7]. A stand alone user developed program coded in Mathcad [8] has been used as the mandatory input/output (I/O) interface and for information processing. The interface program writes information about initial conditions for charged particles in the system (mass, energy, direction of velocity, position and number of particles), as well as other global information (ion and electron currents). Opera3D/Scala computations use this information about initial conditions and compute local electric fields and charged particle trajectories. This last information is written as a "tracks" type named file. The file is then read by the I/O interface and used for further processing. Extracted information includes, among others, the position and impact angle of ions at target. The number of particles in the system has been usually chosen as 1000, a number high enough to obtain beam statistics within a reasonable computation time interval (such as a half-hour or less). However, assessment of computational precision has been carried on for ion beams comprising up to 20,000 charged particles, showing that no significant change in results for characterization of space charge effects appears. The computation unit offers flexibility to compare and assess which is the arrangement of field emitters arrays optimal for ion beam neutralization.

INITIAL CONDITIONS

Different ion species ($^{11}B^+$,$^{31}P^+$,$^{75}As^+$) with energy in the range E_{ion}=200 eV-1 keV have been considered. The field free region modeled is 34 cm wide along X direction.

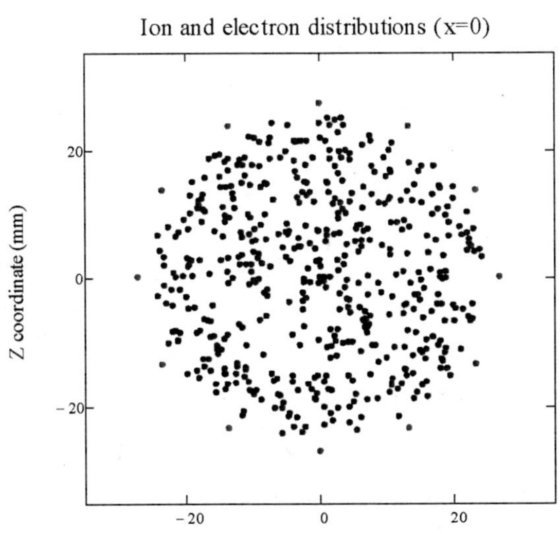

FIGURE 1. Model showing ions (blue dots) that have random initial position within a circle and field emitter arrays (red dots) that are symmetrically placed around the ion beam.

For simulation results reported in this article, ions have initial positions randomly chosen within a circle of radius R_{ion}=25 mm in the initial YZ plane x=0. As concerns the initial conditions for electrons supplied in the system by field emitter arrays, several situations have been considered. A first approach was to consider laterally injected electrons as described in Ref. [9], although without any additional electrodes. The problem symmetry, as well as detailed simulations, suggested that achieving uniform electron and ion mixing is the key element for better ion beam neutralization. An "ideal" reference configuration is obtained for ions and electrons having random positions and uniform inter-mixing. A possible physical configuration is shown in Figure 1, with several electron sources (in this case, 12) symmetrically distributed around the ion beam that has N_{ion}=500 ions.

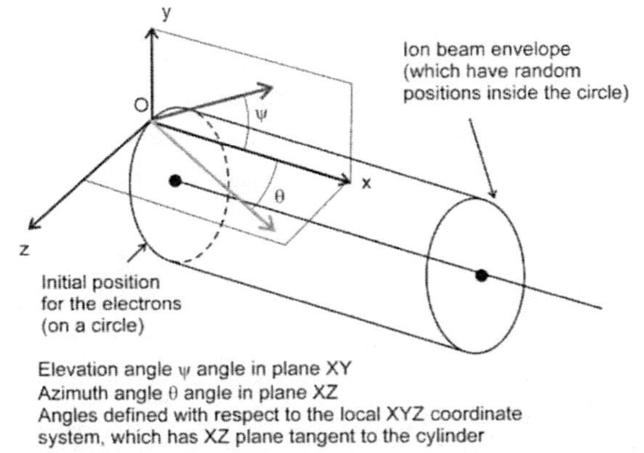

FIGURE 2. Definition of initial conditions for electron launching for azimuth and elevation angles.

Field emitters are devices usually micromachined from silicon having a sharp tip (with radius in the nm range) which is surrounded by a circular gate electrode (with radius 1 μm or less) placed in its vicinity [2]. A group of several emitters (hundreds or thousands) form a small emitter array that may be mm sized. Very high electric fields (greater than 1 V/nm) appear even for reasonable gate potentials, forcing the electrons from the emitter to tunnel in vacuum. Usually, the electrons are emitted within a cone with a half-angle opening around 15 deg or more. A second gate electrode may be used for precise electron beam focusing [10,11]. Different elevation ψ and azimuth θ angles, defined as shown in Figure 2, have been considered as initial condition for electron emission, in order to describe electron emission within the mentioned cone region and also different orientations of the emitter array with respect to the ion beam.

The current that is associated with the ion beam is I_{ion}=125 μA. Field emitted electrons have low energy and a relatively narrow energy distribution [2]. For most simulations, if not otherwise mentioned, electron energy is E_{el}=10 eV and electron/ion current ratio is I_{el}/I_{ion}=24 or 34.

MODELING RESULTS

In Figure 3, an example is given for ion trajectories that are curved solely due to space charge effects when there are no field emitted electrons. Different types of ions encounter different space charge effects. For fixed ion energy, higher mass ions have slower velocity and more important beam deflection due to space charge effects. This situation is outlined in Figure 4. Simple models [12] outline that the beam radius at target increases as $R_{ion} \sim \sqrt{M_{ion}}$.

The ion beam radius at target has been used to quantitatively assess the ion beam neutralization process. In Figure 5, an example is given for ion beam neutralization that occurs when electrons are emitted. Patterns appear for the electron flow because precise similar conditions are considered for electrons. The ion beam has greatly reduced diameter at target compared to the previous case. In Figures 6 and 7, ion beam neutralization is discussed as a function of (elevation and azimuth) launching angles for the electrons. In both cases, optimum ion beam neutralization is obtained when the electrons have initial velocity direction parallel to the ion beam [5]. In Figures 8 and 9, ion beam neutralization is discussed as a function of electron/ion I_{el}/I_{ion} current ratio and electron energy. Lower electron energies allow lower current ratios I_{el}/I_{ion} to be used for effective for ion beam neutralization. For low electron energy, the curve in Figure 9 is steep. This result shows that efficient ion beam neutralization requires not only lower electron energy, but also requires narrow distribution for the electron energy.

CONCLUSIONS

Based on presented and other results as well, the following conclusions can be drawn. Field emitter arrays (FEAs) are electron sources with better controlled parameters such as lower energy, narrow energy distribution and electron emission direction, and therefore can be effective for ion beam neutralization. Optimal ion beam neutralization is obtained for FEAs symmetrically placed nearby the ion beam and launching the electrons parallel with it, possibly as focused beams. Distributing the electron sources along the ion beam direction is also beneficial.

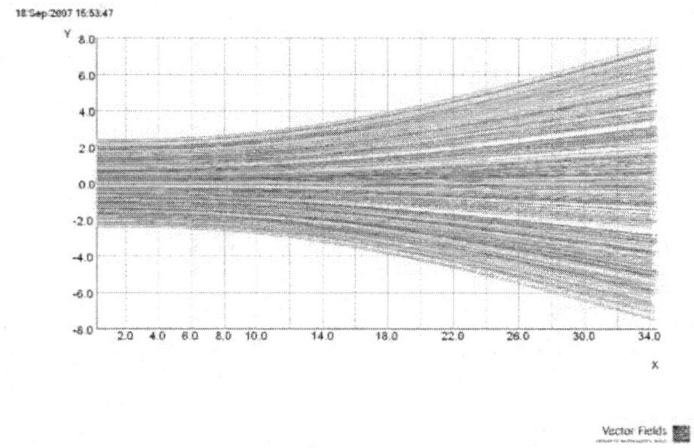

FIGURE 3. Example of ion trajectories curved as a result of space charge effects. No electrons are considered in the system. $^{31}P^{+}$, E_{ion}=500 eV.

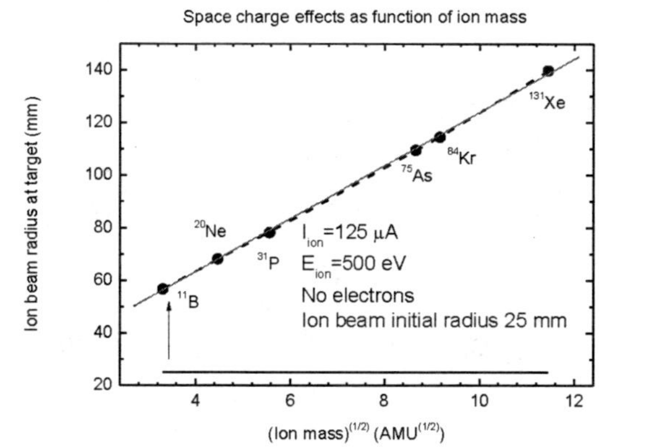

FIGURE 4. Simulation results showing that, for given ion energy, space charge effects are more important for ions with higher mass. The beam radius at target increases as $R_{ion} \sim \sqrt{M_{ion}}$.

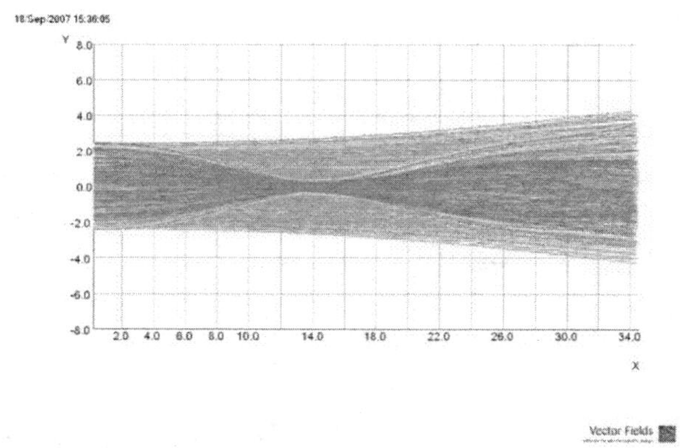

FIGURE 5. Example of ion beam neutralization occurring in the presence of emitted electrons. $^{31}P^{+}$, E_{ion}=500 eV, E_{el}=10 eV, I_{el}/I_{ion}=34.

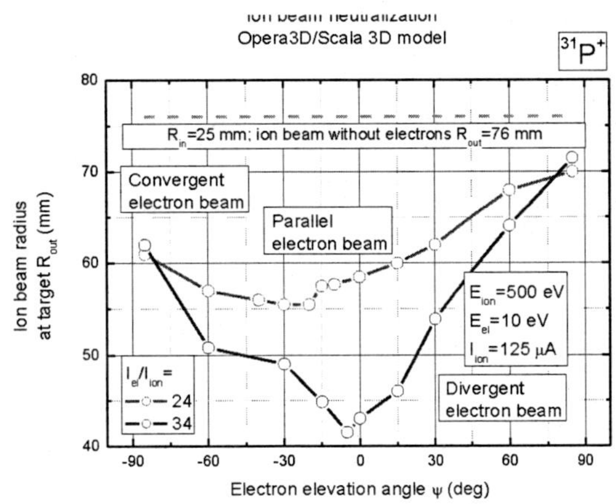

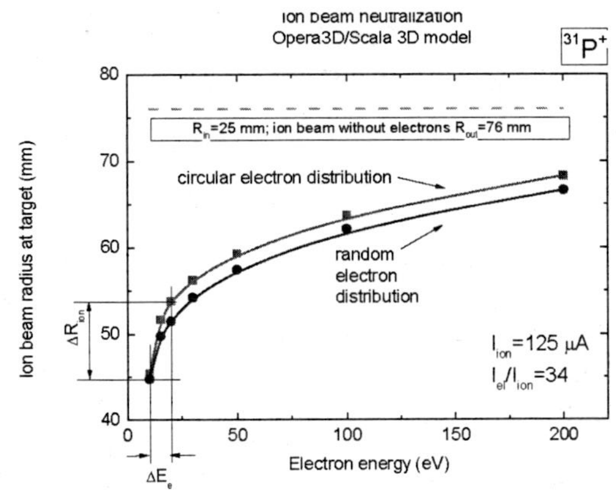

FIGURE 6. Ion beam radius at target as a function of electron elevation angle ψ.

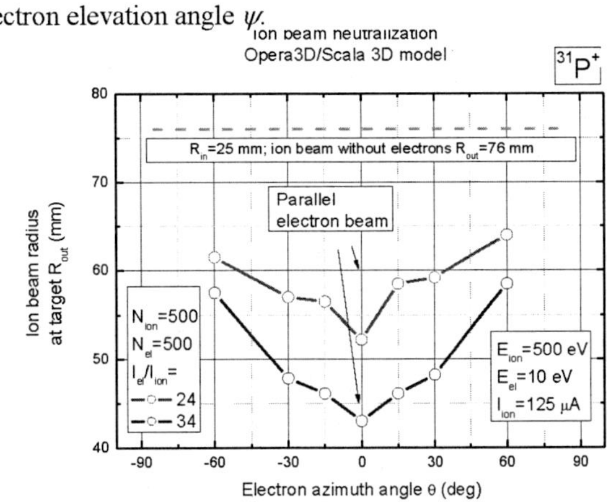

FIGURE 7. Ion beam radius at target as a function of electron azimuth angle θ.

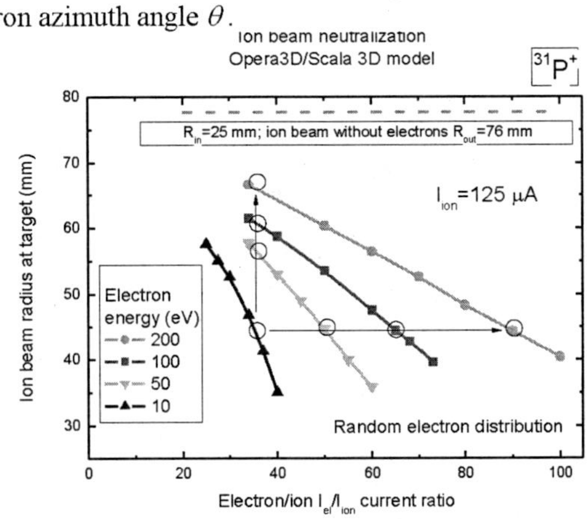

FIGURE 8 Ion beam neutralization as a function of electron/ion current ratio.

FIGURE 9. Ion beam neutralization as a function of electron energy.

ACKNOWLEDGEMENTS

This research has been carried out as the project entitled "Development of non-diverging transport and irradiation techniques by low-energy ion beams for advanced semiconductor device manufacturing process" under the contract of Innovation plaza Kyoto, Japan Science and Technology Agency (JST)."

REFERENCES

1. A.C. Wooding, D.G. Armour, J.A. van den Berg,, A.J.T. Holmes, C. Burgess and R.D. Goldberg, *Nuclear Instruments and Methods in Physics Research B* **237** 273-277 (2005).
2. J. Ishikawa, Y. Gotoh, K. Nakamura, T. Kojima, H. Tsuji, T. Ikejiri, S. Sakai, S. Umisedo, N. Nagai, M. Nagao and S. Kanemaru, *Nuclear Instruments and Methods in Physics Research B* **237** 390-394 (2005).
3. A. Oowada, M. Takeuchi, Y. Sakai, Y. Gotoh, M. Nagao, H. Tsuji, J. Ishikawa, S. Sakai and T. Kimoto, *J. Vac. Sci. Technol. B* **26**, 876-879 (2008).
4. M. Takeuchi, Y. Gotoh, H. Tsuji, J. Ishikawa and S. Sakai, *Rev. Sci. Instrum*. **79**, 02B712 (2008).
5. J. Ishikawa, D. Nicolaescu, Y. Gotoh and S. Sakai, Patent WO 145355 A2 (21.12.2007).
6. H. Riege, *Nuclear Instruments and Methods in Physics Research A* **451** 394-405 (2000).
7. Opera3D/Scala from Vector fields, Ltd., http://www.vectorfields.com.
8. Mathcad 14, http://www.ptc.com.
9. J. Ishikawa, Y. Gotoh, K. Matsuda and S. Sakai, Patent Abstracts of Japan, 2007035370 A (08.02.2007).
10. D. Nicolaescu, V. Filip, J. Itoh, *J. Vac. Sci. Technol. B* **19**, 892-896 (2001).
11. D. Nicolaescu, V. Filip, J. Itoh, Jpn. *J. Appl. Phys*. **40** (Part 1), 83-86 (2001).
12. R. G. Wilson and G. R. Brewer, *Ion Beams with Applications to Ion Implantation* (Robert E. Krieger, Malabar, 1979), Chap. 3, p. 132.

Introduction of the MC3-II/WR System, an Extended Energy Medium Current Ion Implanter

Michiro Sugitani, Fumiaki Sato, Masazumi Koike, Makoto Sano, and Kazuyoshi Ueno

SEN Corporation – an SHI and Axcelis company, 1501 Imazaike, SaiJo, Ehime, 799-1362, Japan

Abstract. The MC3-II/WR is a medium-current ion implanter, newly developed by SEN Corporation. The most significant change from the original MC3-II is an expansion of its energy coverage with an extended terminal voltage from 260 kV to 320 kV. This expansion takes in a large portion of high energy applications and results in significant cost reduction of device production. The MC3-II system was developed to meet requirements of advanced LSI's, especially requirements for implant accuracy through its controllability of beam quality, keeping productivity high or. The MC3-II/WR inherits the MC3-II's advantages and enhances its capability in the energy region and mechanical throughput.

Keywords: Ion implanter system, Medium current
PACS: 41.75.Ak, 61.72.U-, 85.40.Ry

INTRODUCTION

The MC3-II, SEN's state-of-the-art medium-current ion implanter, was developed as a successor of the MC3 [1]. The original MC3 has two significant features. One is a superior beam line which can afford ideal parallelism of beam angles and uniform beam angle divergence throughout the beam scan range. The other is a unique mechanical scan mechanism which ensures the unmoving implant position, no matter where the mechanical scan position is. A combination of the two features results in an important implant characteristics, that is, uniform implant angle and angle divergence in a whole area of a wafer even when an implant tilt angle is not zero. This capability recently becomes more important because shrinkage of devices makes MOSFETs very sensitive for avoiding asymmetric electrical characteristics [2].

The MC3-II also has such capability as well as more productivity than the original MC3. The productivity enhancements of the MC3-II are on a mechanical throughput which is enhanced to 450 WPH and the beam currents are also increased, especially in an energy region as low as 3 keV, and beam auto-setup time which is reduced down to less than three minutes. Reliability in volume production is also improved.

Having the same superiority in beam quality and the same productivity as the MC3-II, the MC3-II/WR is designed to have an extended terminal voltage up to 320 kV, where the original MC3-II has 260 kV of a terminal voltage at maximum. This energy extension provides several benefits to users. Because the beam line of the MC3-II/WR is exactly as the same as the MC3-II's, the low energy capability, i.e., beam currents, beam parallelism and angle divergences are not sacrificed and only high energy capability is simply added.

BACKGROUND

From the very beginning, medium-current implanters have capability to perform high energy applications. Use for such applications, however, has been very limited because of several reasons.

1) Energy contamination
2) Severe dose shift in high beam power operation
3) Poor dose uniformity in high beam power operation
4) Poor beam current
5) Short source lifetime in multiply-charged ion generation

High energy ion implanters, such as the NV-GSD-HE3 (HE3) [3], have been considered the most efficient implanters for the high energy applications of the energy range of higher than 300 keV because of many concerns of medium-current implanters as above and of unique capability of chain-recipe implants [4] which increase productivity of the HE3 dramatically. On the other hand, medium-current implanters have been being improved to overcome the concerns for these ten years.

For example, indirectly heated cathode ion sources (IHC) broke down the issues (4) and (5). Usual lifetime of the IHC for multiply-charged ions is about 5 to 10 times longer than that of the Bernas type ion sources at the maximum beam current. Of course, the

CP1066, *Ion Implantation Technology,* edited by E. G. Seebauer, S. B. Felch, A. Jain, and Y. V. Kondratenko

maximum beam current of the IHC is about 3 to 10 times higher than that of the Bernas type. The energy filters after final acceleration or deceleration are another improvement to break down the issue (1).

Then, the issues (3) and (4) are the final obstacles to use the medium ion implanters for the high energy applications. These issues are caused by one single reason. Out-gassing from a wafer generated during an implantation is a root cause of the dose shift and poor uniformity. High power beams induce more out-gas, and then more dose shifts and poorer uniformity. The pressure compensation [5, 6] is the answer from SEN to overcome the remaining issues and it works successfully to use the maximum beam current from the IHC even in the high energy region. Thus, a throughput for some high energy applications of the MC3-II/WR becomes as the same as the HE3's, sometimes, higher when an implant dose is smaller than 5e12/cm^2 because of the difference of mechanical throughputs. It can be indicated that the MC3-II/WR, which has a simpler system structure than the HE3, shows better cost performance than the HE3. Needless to say, the HE3 still has significance of existence because of its coverage of higher energy region up to 3 to 4 MeV.

There is a proactive reason to use medium-current implanters for the high energy applications. As well known, the multi-wafer implanters like the HE3 have implant angle deviation which is the worst at zero degree implants. The angle deviation induces a different channeling effect among positions in a wafer [7]. Well tuned beam of the MC3-II/WR can provide almost a half of angle deviation of the HE3 even which utilizes a reduced cone angle disk. The channeling difference in a whole wafer implanted by the MC3-II/WR is quite small compared to that of the HE3.

SYSTEM CONCEPT

Components of the beam line of the MC3-II/WR are conceptually the same as those of the original MC3. To facilitate understanding of the system, the beam line layout is shown in Figure 1. The ion source is of an indirectly heated cathode type as mentioned above and the beam scanning is an electrostatic one. The beam parallelizing system is also of an electrostatic basis. The energy filtering system also uses an electrostatic one and is functionally independent from the parallelizing system. These features make it possible to precisely control the dose and angles.

PERFORMANCE

Energy ranges of the MC3-II/WR for single-charge, double-charge and triple-charge ions are extended from 260 keV to 320 keV, from 520 keV to 640 keV and from 780 keV to 960 keV, respectively from the original MC3-II. Sometimes these extended energies include significant applications.

In order to extend the terminal voltage, what are changed from the original MC3-II are very few points. The first is extension of the extract voltage from the ion source. Consequently the maximum voltage of the parallel lens becomes higher. Next, the post acceleration voltage is slightly higher. Finally, air gaps around the terminal are set to larger values. These changes have very small risks for a stable operation. In fact, after confirmation of the sufficient high voltage isolation, the system is working almost at the same reliability rate as the original MC3-II in volume production in the actual field. Even though the MC3-II/WR extended its energy range from the original MC3-II, power consumption increases very little and is still mush less than that of other medium-current implanters with a magnetic parallel lens and the linac-type high-energy implanters.

As mentioned in the previous section, the pressure compensation must be used to obtain the maximized throughput. It is also already mentioned that the pressure compensation improves the dose uniformity, but it has a confinement. The uniformity

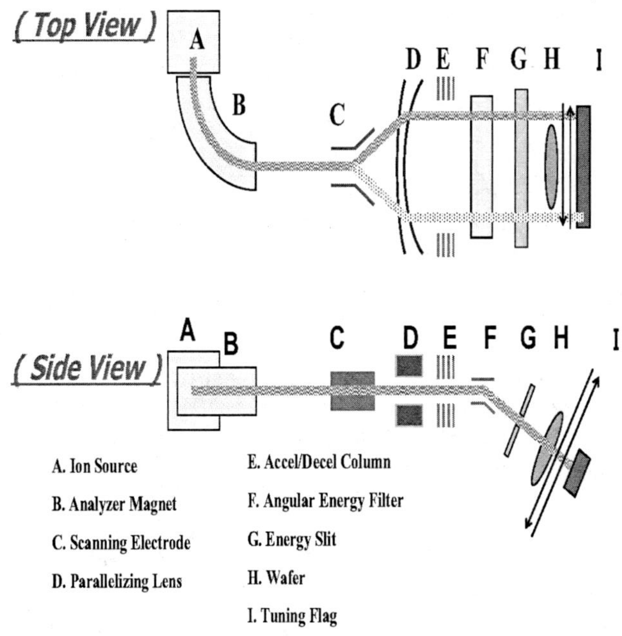

FIGURE 1. Beam line components of the MC3-II/WR.

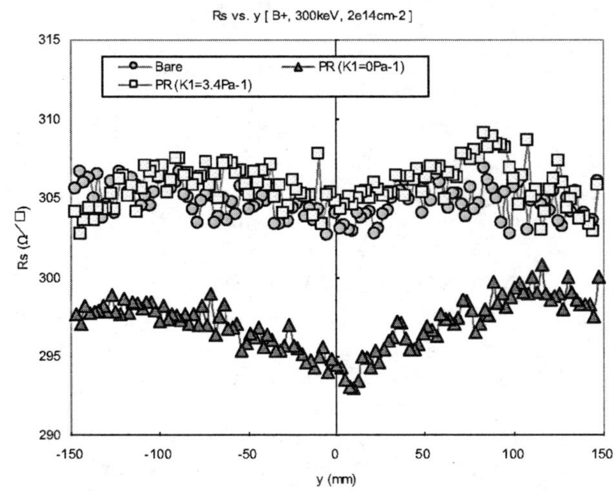

FIGURE 2. Comparison of sheet resistance for a bare wafer (diamond; blue) and PR coated wafer (triangle; purple). So significant over-dose is observed for the PR coated wafer due to the beam neutralization that the pressure compensation must be used. Then, if a compensated beam current used, the final dose for a PR coated wafer (rectangle; yellow) is almost the same as that for a bare wafer.

improvement works only for the mechanical scan direction. If the dose non-uniformity occurs in the beam scan direction due to out-gassing from the wafer, the pressure compensation can not improve it because the non-uniformity in the beam scan direction is induced not due to the pressure change but to the path length difference. One of the reasons is a fact that the pressure change is much slower than the beam scan speed. Fortunately and definitely, because the path

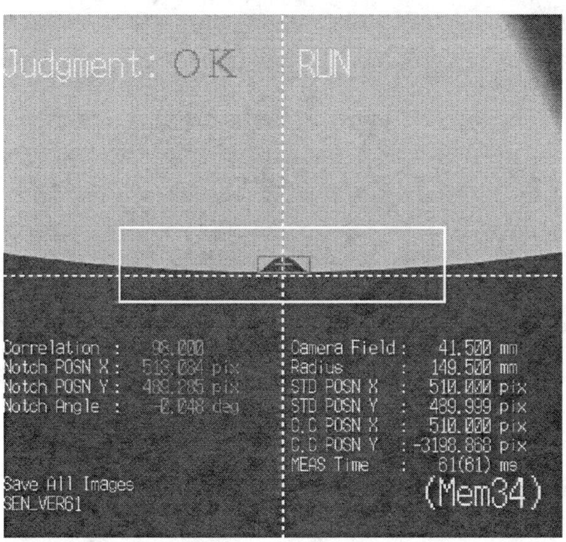

FIGURE 3. Notch image of a wafer on the platen detected by a CCD camera .

length in the pressure change region of the MC3-II and WR (MC3-II series) is same at any beam scan position, the dose uniformity in the beam scan direction does not change even at the high pressure condition as shown in Figure 2. The implant condition is of B^+, 300 keV and 2E14/cm^2. The dose shift occurs equally at any beam scan position and the pressure compensation works well to reproduce the bare wafer data correctly as shown in Figure 2.

The MC3-II series has inherited the precise dose and angle control of the original MC3 and features a new angle control system and a vertical beam profile detection system to achieve more precise implantation [8]. The angle control system is a new notch aligning system using a 2-D camera. Usually, notch aligning is done before a wafer is transferred onto the implant platen. In this case, there remains a possibility that the notch position moves slightly due to a mechanical handling. This system observes the notch position at the platen and no wafer transfer after aligning. The camera for notch sensing is set just above the load/unload position of the platen. The camera works at the time when the platen holds a new wafer. Precision of the notch detection is less than 0.1 degree and accuracy of the notch alignment is ±0.2 degrees. This system makes the MC3-II series more reliable and accurate. Figure 3 shows a measurement of the notch position and twist angle.

The vertical profile monitor system is working as a tilt angle monitor. The monitor itself measures a beam density in the vertical direction as shown in Figure 4. The system derives an incident beam angle against the

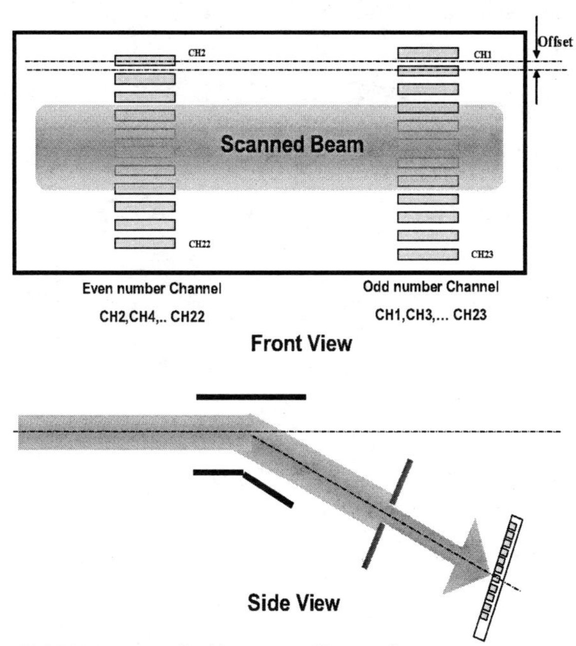

FIGURE 4. Vertical beam profile monitor.

wafer surface plane. A combination of the notch aligner and the vertical profile monitor ensures both the twist angle and tilt angle.

The MC3-II series also has a new feature of dose control. That is "Dose modulated implantation" [9, 10]. This function works to compensate non-uniformity imposed by other processes like etching and annealing. This feature is based on an assumption that if the non-uniformity results in deviation of device characteristics and the characteristics can be modified by implant dose control, intentional non-uniform implant dose can suppress the deviation. There must be many non-uniform shapes and the dose modulation system is desired to be flexible enough to compensate as many patterns as possible. Capability of precise dose control of the MC3-II series makes it possible to control modified dose in the both mechanical scan and beam scan directions and the flexibility to control the direction independently makes it possible to generate various kinds of two-dimensional patterns.

Figure 5 shows a sample of the system with an actual non-uniform implant of an eccentric profile in contradistinction to the predicted dose map generated by the support software.

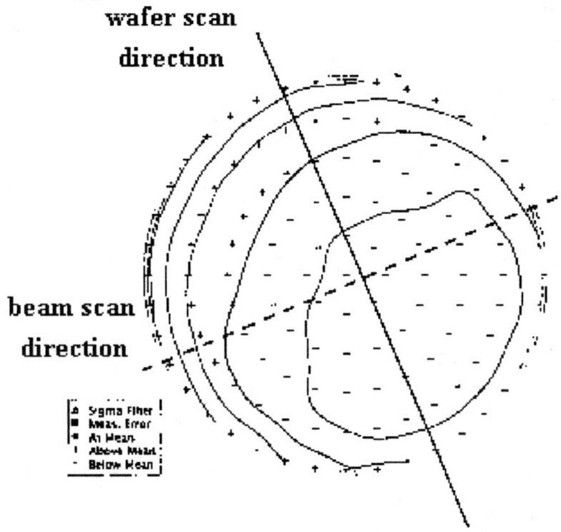

FIGURE 5. An eccentric profile of a sheet resistance map by a single-step ion implantation in a condition of B^+, 100keV, 5E14 cm^{-2}.

CONCLUSIONS

The MC3-II/WR was designed as one type of the MC3-II series and featured an extended terminal voltage of 320 kV to cover a higher energy region with more cost effectiveness than high energy implanters, keeping superior performance as a medium-current implanter. The maximum voltage is generated successfully and stably in use at the maximum beam current. It also features a more accurate twist angle control system and a flexible intentional non-uniform dose control system.

The new medium-current ion implanter with precise implant control and high productivity as well as low power consumption can contribute any kind of device manufacturers both in quality and cost performance, appealing an ecological point which will be more and more important in the near future.

ACKNOWLEDGEMENTS

The authors thank Dr. Fuse and Dr. Kabasawa for fruitful discussions.

REFERENCES

1. O. Campbell *et al.*, IIT 1998, p154.
2. G. Fuse *et al.*, IWJT 2007 S7-4, p121.
3. M. Sugitani *et al.*, IIT 1998, p192 .
4. S. F. Nitodas, IIT 2000, p173.
5. M. Sano *et al.*, IWJT 2007, p127.
6. M. Sano *et al.*, IIT2008 to be published.
7. L. Rubin, IIT 2002, p17.
8. M. Sato at al., IIT 2008 to be published.
9. S. Ninomiya *et al.*, IWJT 2008 S1-4, p25.
10. S. Ninomiya *et al.*, IIT2008 to be published.

Optima XE Single Wafer High Energy Ion Implanter

Shu Satoh*, Joseph Ferrara, Edward Bell, Shital Patel and Manny Sieradzki

Axcelis Technologies, Inc. 108 Cherry Hill Drive, Beverly, MA 01915, USA
** shu.satoh@axcelis.com*

Abstract. The Optima XE is the first production worthy single wafer high energy implanter. The new system combines a state-of-art single wafer endstation capable of throughputs in excess of 400 wafers/hour with a production-proven RF linear accelerator technology. Axcelis has been evolving and refining RF Linac technology since the introduction of the NV1000 in 1986. The Optima XE provides production worthy beam currents up to energies of 1.2 MeV for P^+, 2.9 MeV for P^{++}, and 1.5 MeV for B^+. Energies as low as 10 keV and tilt angles as high as 45 degrees are also available., allowing the implanter to be used for a wide variety of traditional medium current implants to ensure high equipment utilization. The single wafer endstation provides precise implant angle control across wafer and wafer to wafer. In addition, Optima XE's unique dose control system allows compensation of photoresist outgassing effects without relying on traditional pressure-based methods. We describe the specific features, angle control and dosimetry of the Optima XE and their applications in addressing the ever-tightening demands for more precise process controls and higher productivity.

Keywords: Ion Implantation, Implant Systems, Single Wafer, High Energy
PACS: 85.40.Ry,41.85.-p

INTRODUCTION

Device scaling and the recent interest in channeled implants in the MeV energy range has imposed very tight process control requirements including implant angle control and contamination. At the same time, there is a need for higher throughput and small lot sizes management. The Optima XE was developed specifically to meet these manufacturing needs. The Optima XE is the first production single wafer high energy implanter using RF linear accelerator technology, while also maintaining strict controls on traditional implant qualities such as dose, implant uniformity, contamination control, and high beam current capability. This paper describes overall system capability while highlighting unique features and experimental results as well as a description of a new high productivity endstation providing wafer throughput in excess of 400 wafers per hour.

SYSTEM OVERVIEW

Beamline

The layout of Optima XE is shown in Figure 1. The ion beam is generated by a field-proven "Eterna" ELS (Extended Life Source) ion source and is extracted at up to 90 kV through an electrode with three independent axes of motion.

After extraction and formation, the beam is transported through a 70 degree triple-indexed mass analyzer magnet. The beam is then accelerated by a 13.56 MHz twelve resonator RF linear accelerator. Beam focusing is provided by electrostatic quadrupole lenses incorporated in the linac. The accelerator in the present system is an evolution of the RF linear accelerator utilized by the longtime industry-standard Paradigm XE [1] high energy ion implanter. Axcelis has been refining linear accelerator technology for over twenty years, and there are hundreds of implanters using linacs in the field. The most significant advances in accelerator design for the Optima XE include improved vacuum performance and easier maintenance.

The system is capable of accelerating singly-charged phosphorus ions up to 1.2 MeV and singly-charged boron ions up to 1.5 MeV. Higher charge state ions are employed to reach higher energies. The ability to utilize singly-charged ions for many high energy implants adds to high productivity, high reliability, and long ion source life. Production worthy beam currents are achievable for energies ranging from 10 keV up to 4 MeV. For low energy beams, the

CP1066, *Ion Implantation Technology,* edited by E. G. Seebauer, S. B. Felch, A. Jain, and Y. V. Kondratenko

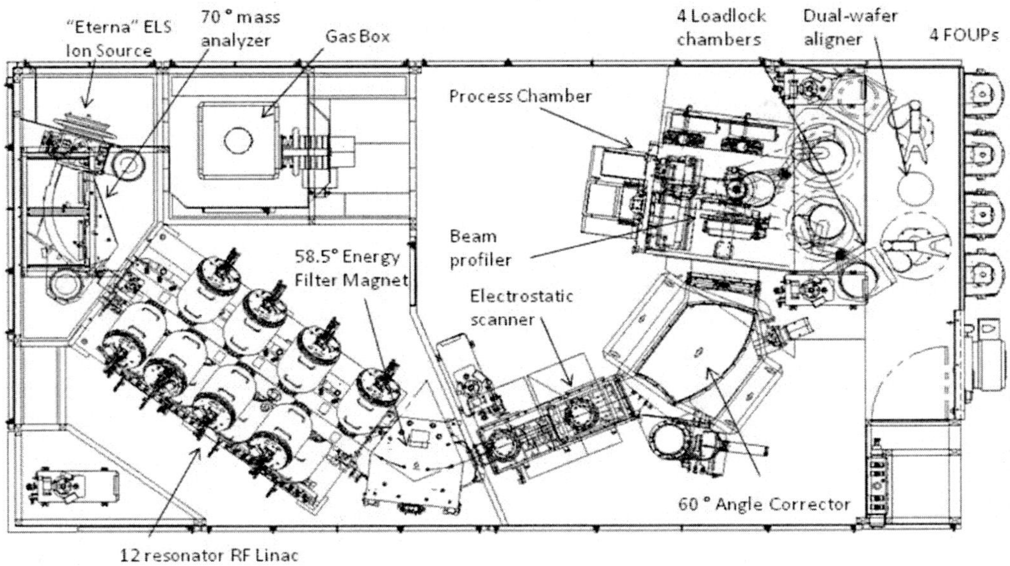

FIGURE 1. The OptimaXE beamline layout.

resonators on the linac are turned off and focusing in the accelerator region is provided by the quadrupoles.

For customers with lower maximum energy requirements, there is a version of the implanter, the Optima HE, incorporating a six resonator linac that can accelerate singly-charged phosphorus ions up to 600 keV.

Upon exiting the linac, the beam passes through a 58.5 degree energy filter magnet that selects only those ions that have been accelerated to the desired energy. The combination of the analyzing magnet and the linac provides a high degree of resolving power.

After passing through the energy resolving aperture, the accelerated and filtered beam is focused by an electrostatic quadrupole doublet and then scanned horizontally at 1 kHz. The scanner consists of two scan plates each controlled by a bipolar power supply with the control voltage driven by an arbitrary function generator.

A 60 degree dipole-field angle corrector magnet focuses the diverging beam from the scanner to produce a uniform, parallel ion beam at the wafer plane with no line of sight from the scanner region to the wafer plane. The horizontal angle can be checked at any time by the use of an *in situ* beam parallelism check. Vertical angles can be controlled by detecting the vertical angle of the beam using the Axcelis VBA [2] and adjusting the steering of the beam upstream of the wafer. The angle control system is discussed in detail elsewhere in these proceedings [3].

Horizontal beam uniformity is controlled by a sequential loop of measuring the beam profile in the wafer plane with a traveling beam profile cup. Following the measurement, a new scan waveform with correction in instantaneous dV/dt is created to drive the scan amplifiers.

Vertical dose uniformity is controlled by continuously monitoring and controlling the wafer mechanical scan speed based on the output of the "PR cups" which monitor the beam current for drift, photoresist outgassing and beam glitches.

Endstation

The newest endstations for low and medium dose single wafer ion implantation equipment must be able to achieve mechanical throughputs in excess of 400 wafers per hour with highly reliable wafer handling and superb levels of cleanliness. Design considerations also include precise alignment of the wafer to the ion beam and minimizing product at risk at any single station.

The endstation is shown in Figure 2, and innovations in the design include a unique, high-speed wafer notch aligner; high levels of parallel wafer handling, sensors for monitoring each wafer's location continuously; and centralized wafer handling controls.

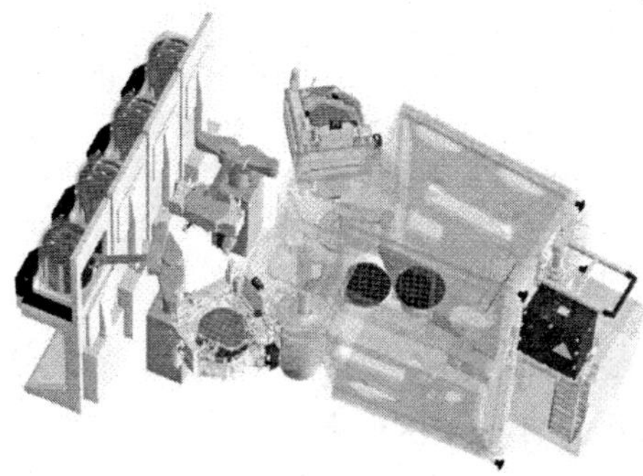

FIGURE 2. OptimaXE high throughput endstation.

All robot motions are accomplished with smooth, blended moves and very low wafer accelerations and decelerations. All wafer handling is performed in the horizontal plane. Four loadlocks are employed to allow for long venting and roughing cycles. A centralized control system is used for all wafer handling, loadlock, wafer scanning, and dose control functions which results in virtually no software latency typical in systems employing multiple control computers controlling different parts of the endstation.

During implant, the wafer is scanned con-focally so that the wafer motion is always parallel to the wafer surface, regardless of tilt angle. Wafers may be implanted at up to a 45 degree tilt angle, allowing the implanter to be used for channel engineering implants as well as traditional high energy implants.

Real time trending capability is provided, including the ability to record machine parameters at a high rate. Data are archived and are available for inspection and analysis at any time.

MACHINE PERFORMANCE

The Optima XE has been characterized across the full range of operation. The following sections highlight some of the results achieved as well as some of the important innovations embodied in the design.

Beam Performance

The Optima XE produces fully auto-tuned ion beams of high currents to enable high wafer throughputs.

Table 1 lists the maximum energy and beam currents available for selected ion species.

TABLE 1. Maximum Energy and Beam Current

Species	Maximum Energy (keV)	Maximum Beam Current (puA)
B^+	1500	1000
B^{++}	2500	100
P^+	1200	800
P^{++}	2900	500
As^{++}	800	250

Photoresist Outgassing Control

One of the challenges in high energy implantation is heavy outgassing from the photoresist layer under bombardment from high energy ions [4]. Endstation pressure from outgassing can reach 1 x 10^{-4} Torr at its peak, increasing the chance of charge exchange reactions between ions and residual gas. This effect not only alters the assumed linear relationship between electrically measured ion current and actual particle dose on wafer, but it also changes the actual beam uniformity on wafer. If not properly compensated, the outcome of heavy photoresist outgassing would be manifested by a significant dose shift and uniformity deterioration.

On the Optima XE, a new continuous beam current monitoring system, PR cup dose control, was developed to eliminate the need for traditional pressure compensation. A pair of Faraday cups, one on each side of the scanned beam, are located in a remote area upstream of the wafer where electrical current closely approximates actual flux on the wafer. The current readings from these cups are applied to a servo control system for mechanical wafer scan speed in order to maintain constant dosing of the wafer. Figure 3 demonstrates the effectiveness of the technique. The figure shows the results of vertical and horizontal diameter scans of a PR wafer implanted with a 2 MeV P^{++} 480 puA beam. Despite a pressure rise to over 1 x 10^{-4} Torr due to PR outgassing, the diameter scans in both vertical and horizontal show the correct dose with non-uniformities < 0.5%.

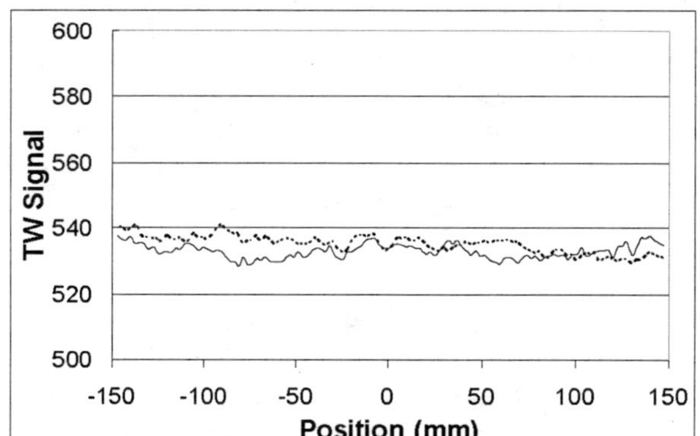

FIGURE 3. Figure showing vertical (solid line) and horizontal (dashed line) diameter scans for a 2 MeV P^{++} 480 puA PR implant.

A large process chamber with the fortified pumping system, the PR cup dose system has successfully demonstrated the ability to maintain dose and uniformity integrity under extremely heavy outgas loads exceeding 1 x 10^{-4} Torr in the process chamber. Figure 4 shows Rs shift in three trials of implanting bare wafers and PR wafers with high energy beams. The columns in the figure represent results from a 2 MeV P^{++} 480 puA 1.98 x 10^{13} cm^{-2} implant, a 1.2 MeV P^{+} 640 uA 1 x 10^{13} cm^{-2} implant, and a 300 keV B^{+} 400 uA 2 x 10^{13} cm^{-2} implant, respectively. The shift in measured Rs value for all nine wafers is under one percent.

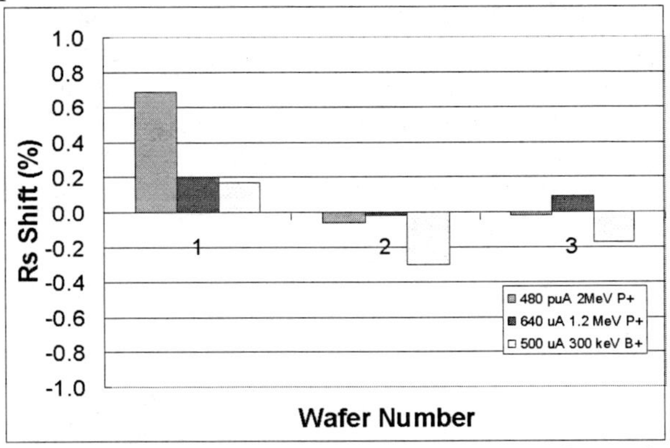

FIGURE 4. Rs shift between bare and PR wafers under various implant conditions (see text).

Horizontal Implant Angle Control

There are two aspects in horizontal beam angle control, one is the angle of incidence onto the wafer surface, and the other is parallelism across the wafer. In a beam parallelizing scheme using a magnetic angle corrector, the two properties are interrelated and we cannot change one property independently without affecting the other. This can be a problem if the optics of the corrector magnet varies with beam rigidity. However, the magnetic field profile in the Optima XE corrector magnet does not vary significantly due to the finite permeability of the steel. In this system, once implant angle and parallelism are verified for one beam energy, they are automatically satisfied for all energies. There is no need to introduce another optical adjustment. Angle variation between low rigidity beams and high rigidity beams has been shown to be well below +/-0.1 degrees [3].

Beam Energy Control

The energy filter magnet on the Optima XE is a dipole magnet with a parallel-to-point focal design. The ion beam must go into the filter as a parallel beam in order to form a sharp image at the resolving slit downstream of the filter. The magnetic field value of the energy filter magnet is calibrated carefully against DC beams of various known magnetic rigidity values, so that the energy of beam through the magnet is accurately known from its magnetic field value.

In an RF linac, final beam energy depends on many parameters. The energy filter magnet is used as a standard to which all the parameters in the linac are tuned, and this function as an energy standard is the most important function of the filter magnet, although the filter aspect is certainly necessary.

One important factor in achieving energy accuracy is maintaining beam injection angle into the filter. In order to ensure high repeatability of the energy setting, a Faraday cup has been positioned at the entrance of the energy magnet to which a newly added tuning process carefully centers the beam. An independent *in situ* monitoring scheme for beam energy is in development.

Contamination Control

The Optima XE has been carefully designed to minimize the deposition of contamination on the wafers from both the beamline and the endstation. Energetic particles are prevented from reaching the wafer by the high resolving power of the combination of optical elements in the beamline.

Trace metal analysis for thirty contaminate elements after ion implant has been performed. All heavy metals levels were below 1 x 10^{10} cm^{-2}. Levels for aluminum were below 3 x 10^{10} cm^{-2}. In addition, particulate analysis resulted in <100 adders at 65 nm bin size.

SUMMARY

The design of the Optima XE ion implanter has been guided by recent trends in ion implantation. The characterization of the Optima XE has demonstrated with excellent results that it can meet and exceed these requirements.

REFERENCES

1. D. A. Marshall, C. Dolan, and M. A. Harris, "Introducing the Axcelis Paradigm and Paradigm XE 300mm Multi-Wafer Ion Implanters for Low-Tilt Well and Channel Applications" in *2004 International Conference on Ion Implantation Technology Proceedings* (2004).
2. R. Rathmell, J. David, "Implant Angle Repeatability on Optima MD", in these Proceedings.
3. Edward Bell and Shu Satoh, "Angle Control on the Optima HE/XE Ion Implanter", in these Proceedings.
4. T.C. Smith, "Photoresist and Particulate Problems", in *Ion Implantation, Science and Technology, 6th Ed,.* (1996), p. 629.

Radiation Issues Surrounding Very High Energy Ion Implantation

Nicholas R. White[a], Nobuhiro Tokoro[b], Edward Bell[c]

[a]Albion Systems, 9 Spy Rock Hill, Manchester, MA 01944, USA
[b]Genesis Magtech Inc., 6 Centennial Drive, Peabody, MA 01960, USA.
[c]Axcelis Technologies Inc., 108 Cherry Hill Drive, Beverly, MA 01915, USA.

Abstract. The requirements for doping of semiconductor device layers have pushed the boundaries of implanter energy, for example, with the use of >5 MeV energies for fabrication of optical sensory arrays. With the higher energy comes the need to understand associated radiation risks. This paper presents original measured data quantifying the range in which safe commercial implantation can be accomplished without radiation hazards.

Keywords: ion implantation, x-ray generation, neutron generation.
PACS: 25.45-z, 25.70-z, 27.10, 27.20, 29.20.Ba, 85.40

INTRODUCTION

The implantation of light dopants (B, P) into Si at energies >500 keV has been widely used since the mid-80's for formation of retrograde profile wells and buried layers intended to reduce latch-up in CMOS devices and sensitivity to radiation-induced soft-errors in DRAMs. Throughout this period, a small number of companies used MeV-range implanters to create deep dopant profiles for light-sensitive CCD and CMOS imagers. The recent expansion of the use of digital imagers in cell phones, web-cameras and other surveillance applications has greatly widened the use of high-energy implantation. The desire to increase the sensitivity of semiconductor-based imagers for military, security, medical and space imaging applications made it desirable to extend use of higher energy dopant beams beyond 3 MeV to the area of 6 to 12 MeV.

The work described here was performed as preparation for specifying a high energy implanter system, which was subsequently built and used in volume production. The work involved determining the highest energy at which conventional dopant ions could be implanted into silicon in a normal manufacturing environment, and in considering the feasibility of extending the energy range still higher. Although this work is several years old, it was not previously brought to the attention of the implant community. The data for neutron production by boron ions on carbon is particularly important.

Background

We were concerned to understand the dominant risks of producing neutron and gamma radiation with dopant beam in the rarely used energy range from 6 MeV to about 12 MeV. We knew pure deuterium beams to be particularly problematic, with $D(d,n)^3He$ reactions between the incident beam and previously implanted D in wafers, beam stops and apertures occurring at energies as low as 20 keV[1]. For this reason, most commercial ion implantation equipment contains software interlocks to prevent the selection of AMU = 2 beams, with strong warnings against the use of D_2^+ beams, while still allowing the use of $^4He^+$ ions.

Energetic proton beams implanted with machines which are also used for doping operations in Si can produce strong gamma reactions through ^{11}B(p, gamma)^{12}C reactions above a resonance energy of 163 keV with gamma energies at 4.44, 11.52, and 15.96 MeV. The B atoms are present in all electrodes, apertures, and beam stops that were exposed to B beams during doping operations prior to the use of the proton beam. Such proton implants could be used to create deep (≈100 nm) H-rich cleave planes, at doses in the mid-10^{16} H/cm^2 range, for layer splitting operations (Smartcut™ and H-cut) used in combination with wafer bonding for formation of laminated heterogeneous layered structures. At energies above 400 keV, proton gamma production, with 1.94 MeV gammas, can occur with ion beam

CP1066, *Ion Implantation Technology,* edited by E. G. Seebauer, S. B. Felch, A. Jain, and Y. V. Kondratenko

exposure to graphite apertures and beam stops through the $^{12}C(p, gamma)^{13}N$ reaction [1].

Data from the University of Oxford [2] revealed the likelihood that fusion reactions involving the boron ions themselves could in fact be the dominant source of neutrons, and we therefore undertook to quantify this problem, and determine safe limits and methods of operation.

The authors solicited expert help from a number of sources, described below. Having obtained this information, its validity was confirmed by building and testing complete implanter systems; furthermore the information was applied to the careful design, building and testing of SEMI S2 compliant radiation interlocking and failsafe exposure protection systems, some aspects of which are worth discussion.

NEUTRON RADIATION ISSUES WITH MEV IMPLANTERS

The control of x-radiation from the accelerator is accomplished relatively easily, by a combination of internal suppression of electrons to limit their acceleration energies, and external lead shielding on the accelerator tank, which, once installed, is non-intrusive and is never to be removed. However, the potential hazard of neutron radiation requires far more care and study.

For many light ion species, exothermic nuclear reactions are possible but are not commonly observed, because the beam ions have insufficient energy to overcome the coulomb repulsion between their nuclei and those of the target atoms. As the ion energy increases, the probability of overcoming this 'Coulomb barrier' [3] between the ion and target nuclei and enabling nuclear reactions increases, potentially generating fast neutrons and gamma rays. The Coulomb barrier, expressed in terms of the energy of projectile ions in the target frame of reference for head-on collisions, can be estimated [4] to within ~1 MeV using the following:

$$CB = \frac{e^2 Z_p Z_t}{4\pi\varepsilon_0 R_0\left[A_p^{\frac{1}{3}} + A_t^{\frac{1}{3}}\right]}\,\frac{A_p + A_t}{A_t} \qquad (1)$$

where e is the electronic charge, Z and A mean atomic number and mass number respectively, the suffix p refers to the projectile, t to the target, and. R_0 $=\sim 1.5 \times 10^{-15}$m. For ^{11}B ions, exothermic reactions can occur with many target atoms with mass greater than 12, as well as any ^2D in the target (which is occasionally used to enhance the properties of photoresists).

Exothermic reactions can also occur at energies less than the Coulomb barrier, due to the finite probability of quantum mechanical tunneling of the incoming ion into the target nucleus as the Coulomb barrier energy is approached. The probability falls exponentially as the energy is reduced. The probability is related to the time the two nuclei spend in close proximity, and to their smallest separation, and for exothermic reactions this can be significant even at energies several MeV less than the Coulomb barrier.

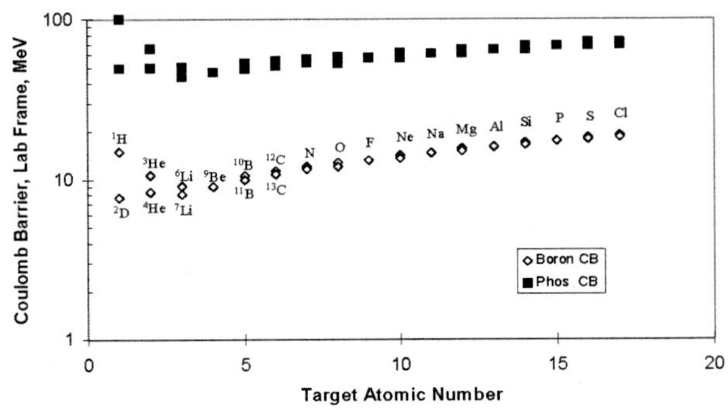

FIGURE 1. Coulomb barrier estimated energies for ^{11}B and ^{31}P ions for target elements from H to Cl.

The Coulomb barrier ion energy is shown in Figure 1 for ^{11}B and ^{31}P ions. The electrostatic repulsion barrier for incident ^{31}P ions is much higher than the ion energies considered here, ≈10 MeV. However ^{11}B can undergo a number of reactions in this energy range, especially for light target elements targets. The relative importance of various reactions can be estimated by combining data for neutron production for B ions at 12MeV and above [2] with a model of the relative abundances of target elements for a photoresist (containing H and O) masked Si wafer in an Al holder.

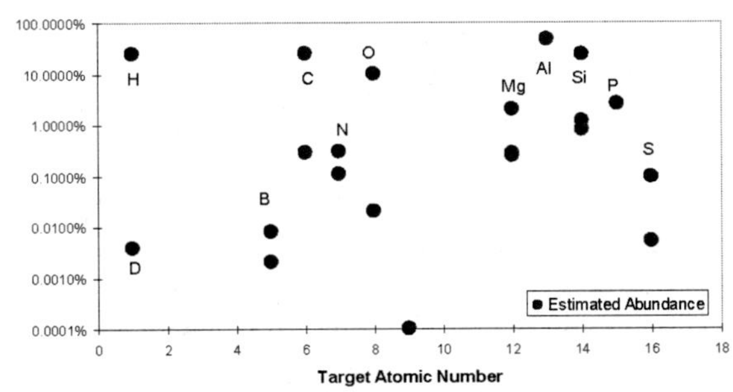

FIGURE 2. Estimated relative abundance of different elements in an implanter endstation struck by an ion beam.

From this estimate, Figure 3 shows that the B ion reaction with C in the photoresist mask is the leading producer of exothermic reactions generating neutrons. Note that in this model target, the natural abundance of ^2D in photoresist is assumed. If deuterium-enhanced photoresist were used, then neutron production rates increase accordingly to high levels.

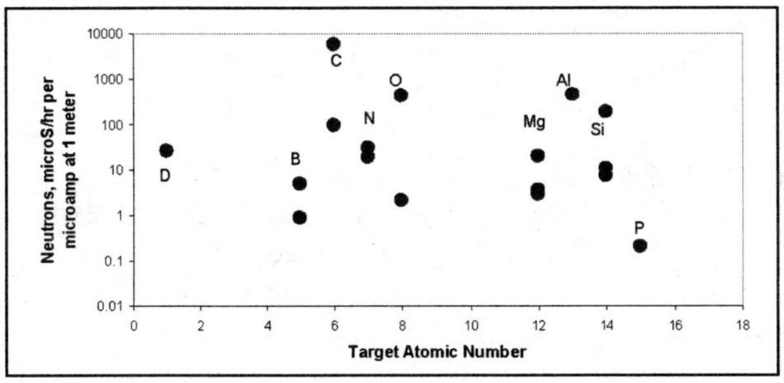

FIGURE 3. Estimated neutron production from 12 MeV boron, by target species, assuming the abundances shown in Figure 2.

The $^{12}C(^{11}B,n)^{22}Na$ reaction produces copious energetic (5 to 8 MeV) neutrons at a B ion energy of 12 MeV. ^{22}Na further decays to ^{22}Ne with a half-life of 2.6 years emitting 1.2745 MeV gammas, as well as by proton emission. A preliminary experiment was performed for us by Litherland and colleagues at the University of Toronto [5] and confirmed the nuclear reaction involved. This experiment implanted a boron target with carbon-12. They then looked for activity from Sodium-22 which is one byproduct of the neutron-generating reaction. This gave us the ability to measure the reaction at a rate down to 1% of the permissible limit in the implanter. The energies run were 10 MeV and 8 MeV carbon ions on boron, and were equivalent to 9.17 and 7.33 MeV boron-11 on carbon-12. These established the slope of the yield curve. The measurement did not allow us to specify accurately the boron current which is permissible into a pure carbon target at 7.3 MeV, which was one of our goals. This was because the ^{22}Na produced at these energies may also decay by proton emission, which we were unable to detect.

In order to extend the measurements of Guo *et al.* [2] at 12 MeV, neutron generation data over a range of energies was solicited and collected from the well-calibrated facilities at University of Oxford, IBM at Yorktown Heights, the Ion Engineering Center at Hirakata, Japan and the Institute of Ion Beam Physics and Materials Research at Rossendorf (Figure 4). Neutron counts were measured (or converted) to the forward direction, where the flux is peaked by approximately a factor of 2. A major uncertainty in the quantitative comparisons is the range of weighting factors for the conversion between deposited energy, Gray, Gy (J/kg), and the biological dose equivalent, Sievert, Sv (J/kg), for neutrons. The neutron weighting factor, q or RBE, can vary from 5 to more than 15 depending on the neutron energy spectrum. For the data shown in Figure 5, a biological weighting factor of 10 was used.

To convert this data to machine operational conditions consistent with safe operations, defined to be an exposure rate of less than 0.6 μSv/hr at any location where a human operator can be located during continuous and unrestricted machine operation, a model wafer target was used with 50% photoresist coverage on Si and a beam-on-target ratio of 50%, accounting for typical beam scanning and duty cycle. Safe Boron beam currents under these model conditions, based on the data shown in Figure 4, are shown in Figure 5. Also shown is an example of higher beam currents allowable with the exclusion of human

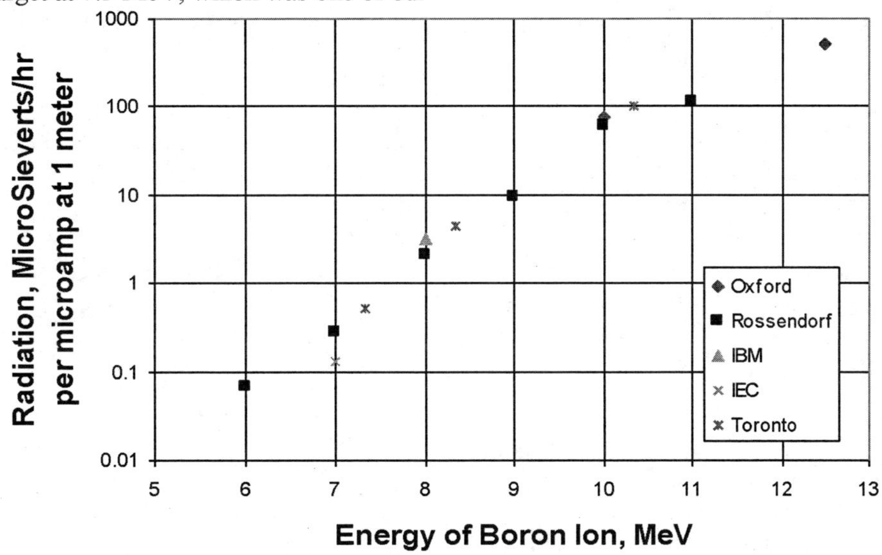

FIGURE 4. Radiation in the forward direction produced by boron on pure carbon, as a function of boron ion energy.

operators from the vicinity of the machine during operation and a radiation license for the tool site. A beam current of 0.1 particle μA at a B beam energy of 8 MeV (projected range ≈10 μm), corresponding to operations at less than 0.6 μSv/hr exposures for all human operator locations, is sufficient for a wafer throughput of 20 300 mm wafers per hour for a dose of 10^{11} ions/cm^2, which is adequate for deep profile applications for high-sensitivity optical sensors.

Interlock Implementation

A number of special beamline components, neutron radiation sensors and interlocks were implemented for high energy (>3 MeV) operation for B beams. It is undesirable to intercept high energy boron beams on graphite, yet graphite is generally preferred because it is relatively benign as a sputtered contaminant. Heavy metals will not produce neutrons and have a low sputter yield, particularly for light ions at such high energies, so heavy metal (e.g. tantalum) beam stops are advisable.

Where the risk of sputter contamination seems too great, aluminum may be used; its Z is high enough to suppress neutron production, but its sputter yield is higher.

A 2 MV tandem accelerator was used to produce the desired 8 MeV B^{3+} ions. It will also produce a weak beam of 10 MeV B^{4+}, which is magnetically separated. The higher energy beam is intercepted with a tantalum beam stop to suppress additional neutron generation.

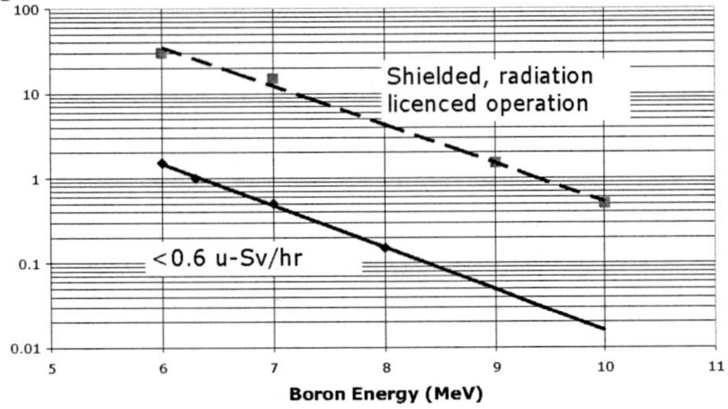

FIGURE 5. Permissible beam currents of boron in μA. Solid: unshielded and unlicensed continuous use; dashed: with special radiation precautions.

A Ludlum neutron monitor with Bonner sphere moderator was provided between the implant location and the operator's station to provide a rapid warning of excessive neutron levels. It was linked to a multi-level hardware alert and hard system shutdown interlock. The neutron level hardware was designed to be fail-safe against loss of system power, disconnections of components and short-circuits to ground of any signal lead. A signal rate-dependent time averaging system was implemented to allow operations close to the neutron exposure limit of 0.6 μSv/hr (or less if desired) without allowing statistical fluctuations to cause false alarm shutdowns. The sampling time was adjusted with the local dose rate so that a shutdown action was commanded only when the average dose rate was measured to exceed the design limits with a precision of 2 standard deviations. Door interlocks (hardware) prevented access within 1 meter of the implant location or of the beam dump.

A hardware interlock prevented the acceleration of deuterium or hydrogen. This interlock prevented the beam from being accelerated unless the injector magnet was energized at a field sufficient to prevent deuterium from being injected into the accelerator. The interlock comprised a simple reed-switch magnetic field threshold in the analyzer magnet.

Beam charge-state, energy, and species were always known and displayed without ambiguity with an accuracy of +/- 0.5 AMU (i.e. the species was known with effective certainty) and within 10 keV. The terminal voltage of the accelerator was redundantly measured by both a resistor chain and by a generating voltmeter. Software was used to provide real-time monitoring and information, and to automate the operation while avoiding forbidden operating regimes, but fail-safe hardware interlocks provided the ultimate protection.

ACKNOWLEDGEMENTS

All work unless otherwise stated was carried out at Diamond Semiconductor Group LLC. Work on interlocks for high-energy B beams was carried out by P. Harvey. We gratefully acknowledge useful discussions, sharing of data, and some original experiments with T. Zabel at IBM, E. Litherland at the University of Toronto, H.R.McK. Hyder (ex Oxford and Yale Universities), and several others.

REFERENCES

1. K. Sadaatmand, E. McIntyre, S. Roberge, Z. Wan, K. Wenzel, R. Rathmell and J. Dykstra, *IIT 98*, IEEE 98EX144, 1999, pp. 292-295.
2. Z. Y Guo, P. T. Allen, G. Doucas, H. R. McK. Hyder, "Thick Target Fast Neutron Yields", *Nucl. Intr. and Meth.* **B29** (1987) p. 500.
3. H. A. Enge, Introduction to Nuclear Physics, Addison-Wesley (1986) pp 5-9.
4. ibid, pp 286-290.
5. A. E. Litherland, private communication.

In-Situ Ion Source Cleaning: Review of Chemical Mechanisms and Evaluation Data at Production Fabs

R. Kaim[1], S. Bishop[1], O. Byl[1], D. Eldridge[1], P. Marganski[1], J. Mayer[1], J. Sweeney[1], S. Yedave[1], D. Fuchs[2], S. Spreitzer[2], J. Vogel[2], J. Dunn[3], P.Lundquist[3], J. Rolland[3], T. Romig[4], D. Newman[4], M. Mitchell[4], K. Ditzler[4]

[1] *ATMI Inc., 7 Commerce Drive, Danbury, CT 06810, USA*
[2] *Infineon Technologies AG., Wernerwerkstr.2 , 93049 Regensburg, Germany*
[3] *Atmel Corporation, Colorado Springs, Colorado, 80906, USA*
[4] *Texas Instruments Inc., Dallas, Texas 75243*

Abstract. Since the concept of chemical *in-situ* ion implanter cleaning was introduced at IIT2006 [1], evaluations of the XeF_2 cleaning technology have taken place or are ongoing at more than 40 production fabs worldwide. Testing has been focused on assessing effects of cleaning in the source arc chamber and extraction regions. In this paper we describe use of the cleaning technology in a production environment and summarize evaluation data showing advantages of the technology for improving ion source life, reducing glitching, improving beam auto-tuning and avoiding species cross-contamination. More details of the evaluations are given in several separate papers submitted to this Conference.

We have supplemented the fab production data with laboratory experiments designed to investigate the reactivity of XeF_2 and fundamental aspects of the source deposition and cleaning processes. These experiments are summarized here, and more details can be found in separate papers submitted to this Conference

Keywords: ion implant, ion source, AutoClean, xenon difluoride
PACS: 61.72.uf, 85.40.Ry, 07.77.Ka.

INTRODUCTION

Xenon Difluoride (XeF_2) is a white/colorless crystalline powder with a vapor pressure of 4.5 torr (0.6 kPa) at room temperature. The Xe-F bond energy is the lowest of any of the fluorides, and as such the material has been widely evaluated as an *in-situ* chemical cleaning reagent for ion implanters. The material is delivered in a customized cylinder [2] which may be installed in the gas box of any production ion implanter.

In this paper, we will review data from production fabs which have been evaluating the XeF_2 *in-situ* cleaning process. We will also describe laboratory studies of XeF_2 reactivity, as well as experiments to evaluate source deposition and etching mechanisms.

DATA FROM PRODUCTION FABS

In order to implement *in-situ* cleaning in production, a cylinder of XeF_2 is installed in the implanter gas box and a "cleaning recipe" is developed. Details of the recipe depend on the implant process and the composition, location and deposition rate of the material to be cleaned. However, in most cases *in-situ* cleaning is performed twice daily for about 10 – 15 minutes.

Below we will review evaluation and qualification data from three different fabs, with different implant processes, ion source types and implanter models.

Ion Source Life

Since it usually determines the frequency of implanter shutdowns for maintenance, improvement of ion source lifetime is often the most important motivation for implementing *in-situ* cleaning.

Figure 1 shows a compilation of ion source lifetime data [3] with two different in-situ cleaning recipes. With a 15 minute clean every 12 hours, the ion source life increased about 24% from an average of 168 hours to an average of 208 hours. However, it was observed that the arc chamber walls were noticeably rougher after using the 15 minute recipe, indicating possible over-cleaning. The clean time was reduced to 10 minutes, and an average source life time of about 340 hours was achieved, an improvement of about 100%.

CP1066, *Ion Implantation Technology,* edited by E. G. Seebauer, S. B. Felch, A. Jain, and Y. V. Kondratenko

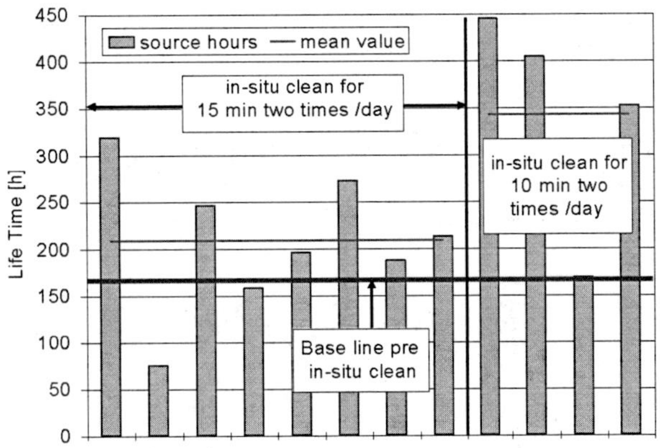

FIGURE 1. Source life times with different cleaning times, compared to the base line without *in-situ* cleaning [3].

Figure 2 shows data from 20 weeks of testing [4] of the *in-situ* cleaning process, in which an important success factor was achieving a two week PM target, requiring consistency and repeatability of the source lifetime. Prior to testing, the average source life was determined to be 10 days (240 hours). During the 20 weeks of testing, the source life consistently met the two week PM target, with no premature failures due to ion source causes. There were 25-30% fewer source changes compared to an equivalent implanter not running *in-situ* clean.

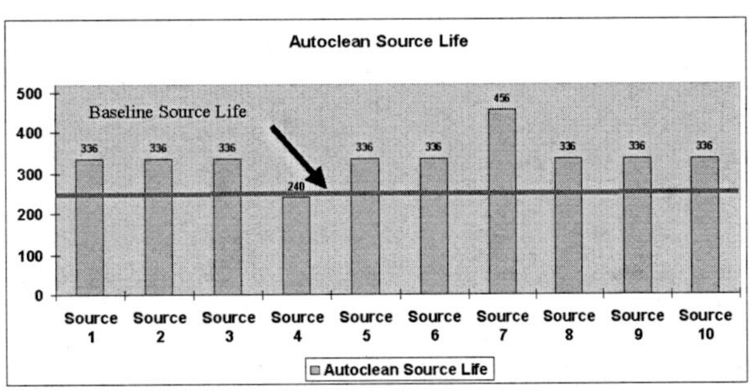

FIGURE 2. Source life improvement with *in-situ* clean [4].

To date, tests of source life improvement with *in-situ* cleaning have involved more than 40 fabs, with numerous different tool types and implant processes. Average source life prior to testing has ranged from as little as 50 hrs to as much as 360 hrs. In most cases, use of *in-situ* cleaning has resulted in increased source life, indicating that source life in production ion sources is usually driven by failures related to deposited materials which are capable of being removed by chemical cleaning.

Glitching and Instability

Beam glitching and instability are important consequences of materials deposited on insulators. Figure 3 shows reduction of glitching in a side-by-side comparison of two implanters, one with and one without use of *in-situ* cleaning [4].

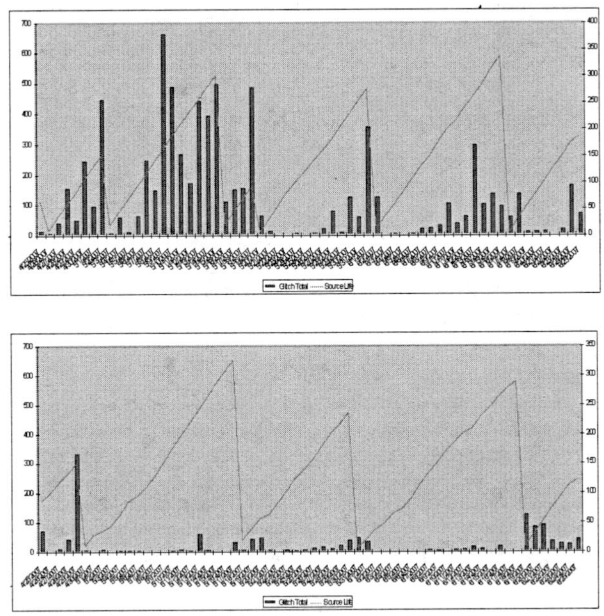

FIGURE 3. Total glitches (bars), compared to source life (line), for implanters without (upper figure) and with (lower figure) *in-situ* cleaning [4].

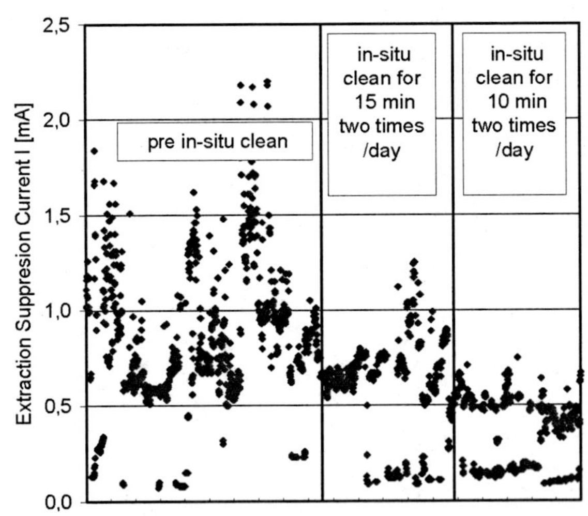

FIGURE 4. Effect of *in-situ* clean on extraction suppression current [3].

Coating of suppression electrodes with conductive deposits often causes beam instability and may require interrupting production to perform maintenance. Figure 4 shows measurements [3] of the suppression current without cleaning and with the 10 minute and 15 minute *in-situ* cleaning recipes referred to in the previous section.

Beam Auto Setup

Beam glitching is often the root cause of difficulties in automatic beam setup, resulting in lengthy setups or setup failures. We have therefore studied the effect of *in-situ* cleaning on the auto setup time and the auto setup failure rate, both of which can be important factors in tool productivity.

Figure 5 shows the effect of *in-situ* cleaning on auto setup interruption events [3]. The reduction in auto setup interruptions with the 10 minute cleaning recipe resulted in a 2.5% increase in productive time for the high current implanter.

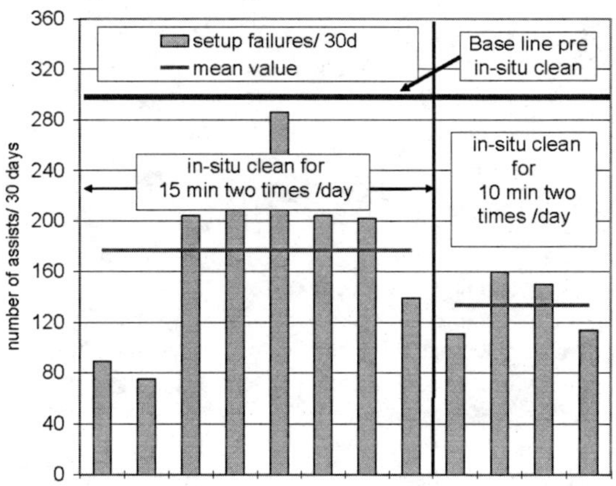

FIGURE 5. Effect of *in-situ* cleaning on auto tune interruption events [3].

Romig *et al.* [4], using a different implanter system, have found that use of in-situ cleaning resulted in a 16% reduction in average auto setup time for production recipes.

Species Cross-Contamination

Deposits inside the arc chamber are responsible for the well-known "memory effect" in which beam currents from species attributable to the previous feed gas can be extracted long after the source feed gas has been changed. The presence of beams from a previous dopant causes risk of species cross-contamination if there are overlapping masses in the beams. The best known example is contamination of $^{49}BF_2^+$ implants by $^{50}PF^+$ arising from phosphorus deposits from previous operation with PH_3.

Figure 6 shows measurements [5] of the P^+ beam current observed during BF_3 operation, beginning immediately after switching from 24 hours of operation with PH_3. The figure compares cleaning effectiveness of Argon gas and XeF_2. The authors also showed that the P^+ beam current measurements correlate with SIMS measurements of P (derived from PF) in wafers implanted with BF_2. The improved effectiveness of XeF_2 compared to Argon supports previous work [6] in which XeF_2 was found to be more effective than BF_3 in reducing cross-contamination.

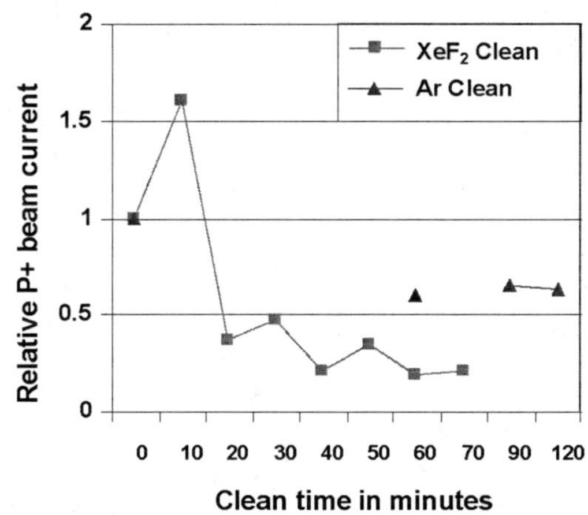

FIGURE 6. Comparison of argon and XeF_2 clean chemistries for reducing residual P+ beam current while running BF_3

LABORATORY EXPERIMENTS

Laboratory testing has been carried out to determine the reactivity of XeF_2 with dopant materials and with implanter materials of construction. XeF_2 readily etches boron, phosphorus, arsenic, tungsten and molybdenum. On the other hand, XeF_2 has been found to be inert with aluminum, graphite, Viton, Vespel and boron nitride. XeF_2 is therefore a suitable in-situ cleaning agent, etching dopant deposits as well as tungsten or molybdenum deposits created in the source arc chamber, while being inert to most other materials inside the implanter vacuum system.

We have studied the problem of tungsten transport and deposition inside the arc chamber when using fluoride feed gases in a tungsten ion source [7]. This problem occurs to some extent with all fluorides but is particularly severe with GeF_4. Figure 7 shows a comparison of mass spectra of various fluorides running under similar conditions in a tungsten ion source. XeF_2 produces the largest quantity of tungsten and its fluorides, and in particular it is the only fluoride which produces significant amounts of WF_5^+. GeF_4 also produces significant amounts of tungsten and its fluorides, while BF_3 and SiF_4 produce much smaller amounts.

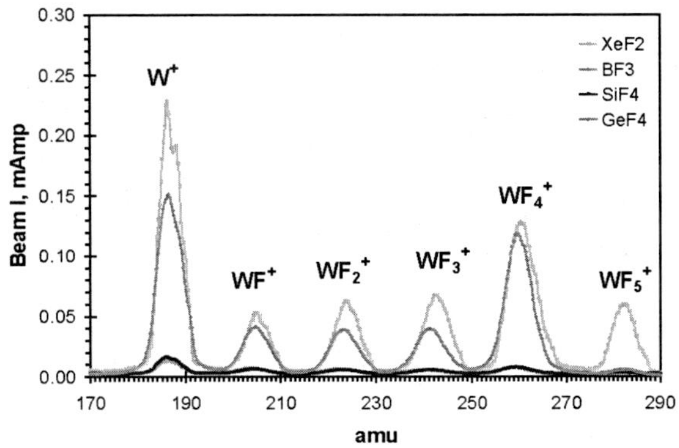

FIGURE 7. Mass spectra from plasma of different fluorides running under similar conditions in a tungsten Bernas source [7].

Since XeF_2 is capable of producing high concentrations of tungsten fluorides in the arc chamber, it is important to determine under which conditions tungsten etching will occur, as opposed to conditions in which there is deposition of tungsten. Figure 8 shows the important effect of temperature, as measured by the filament current of a Bernas source in which XeF_2 is flowing. For lower temperatures (filament current below about 120A), the filament is etched and there is weight loss. At higher temperature (including with plasma), deposition of tungsten occurs. This behavior can be explained in terms of the W-F thermodynamic equilibrium as a function of temperature.

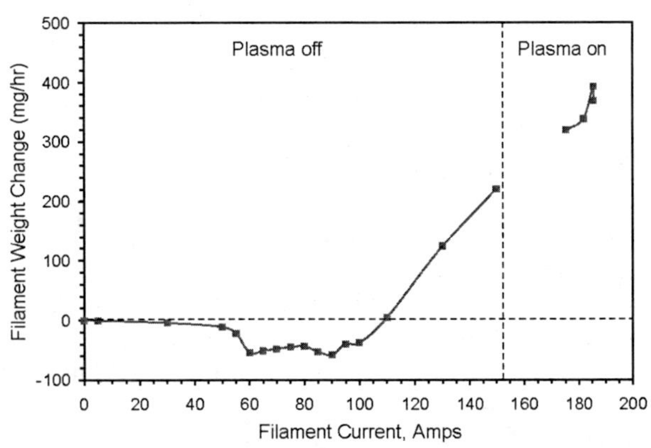

FIGURE 8. Filament weight change *vs.* filament current in a Bernas ion source with 2.2sccm. XeF_2 flow [7].

FUTURE DEVELOPMENTS

ATMI's ion source test stand is being used to develop "static" *in-situ* cleaning, in which XeF_2 vapor is introduced into an entire vacuum chamber at a pressure of 1-3 torr. Unlike the presently used *in-situ* cleaning, the "static" method enables cleaning of components such as vacuum chamber walls, liners, bushings and electrodes. EDS measurements on test coupons [9], and *in-situ* FTIR measurements [8] have demonstrated 6x reduction in phosphorus deposits from PH_3, and 5x reduction in arsenic deposits from AsH_3 operation.

CONCLUSIONS

Data from production fabs has shown that *in-situ* cleaning can increase source life, reduce glitching, improve auto setup performance and avoid species cross-contamination. The reactivity of XeF_2 with various implanter materials has been investigated, and the mechanism of tungsten etching and deposition by fluorides in ion sources has been studied.

REFERENCES

1. S. Bishop, R. Kaim, S. Yedave, J. Arno, F. DiMeo Jr, and M. Wodjenski, "The Development of In-Situ Ion Implant Cleaning Processes", in *Proceedings of the 16th International Conference on Ion Implantation Technology 2006*, Marseille, France
2. J. Sweeney, P. Marganski, R. Kaim, M. Wodjenski, J. Gregg, S. Yedave, S. Sergi, S. Bishop, and D. Eldridge"Optimization of Xenon Difluoride Vapor Delivery", in these Proceedings.
3. D. Fuchs, S. Spreitzer, J. Vogl , S. Bishop, D. Eldridge, R. Kaim, "Decreasing Beam Auto Tuning Interruption Events with In-Situ Chemical Cleaning on Axcelis GSD", in these Proceedings.
4. T. Romig, D. Newman, M. Mitchell, K. Ditzler, D. Eldridge, S. Bishop, J, Mayer, "Implanter Source Life And Stability Improvement Using In-Situ Chemical Cleaning", in these Proceedings.
5. J. Dunn, J. Rolland, P. Lundquist, S. Bishop, "Manufacturing Assessment of an XeF_2 In Situ Clean Process for Mitigation of Species Cross Contamination", in these Proceedings.
6. J. Dunn, P. Lundquist, J. Rolland, S. Bishop, D. Eldridge, R. Kaim, "In situ chemical cleaning for improved ion implanter utilization", *Solid State Technology*, March 2008.
7. O. Byl, S. Yedave, S. Sergi, J. Sweeney, S. Bishop, R. Kaim, D. Eldridge, "Tungsten Transport in an Ion Source", in these Proceedings.
8. S. Yedave, J. Sweeney, O. Byl, S. Letaj, M. Wodjenski, M. Hilgarth, P. Marganski, S. Bishop, D. Eldridge, R. Kaim, "Development of "Static" In-Situ Implanter Chamber Cleaning", in these Proceedings.
9. M. Stawasz, S. Yedave, L. Hiscock, J. Sweeney, R. Kaim, "Surface Analytics In Support Of The Development Of Static AutoClean – An In-Situ Cleaning Process For Ion Implanters", in these Proceedings.

Layer Transfer with Implant-Induced Defects: a Path to Advanced Engineered Substrates

Konstantin K. Bourdelle

SOITEC, Parc Technologique des Fontaines, 38190 Bernin, France

Abstract. Ion implantation has been long used as the predominant method for introducing doping impurities into semiconductor devices. It involves delivering the ions into precisely defined locations and engineering the dopant profiles in the semiconductor matrix through a proper selection of implant species, implantation energies, and most importantly, suitable annealing steps. The Smart Cut™ technology utilizes ion implantation for an entirely different purpose. Light ions, most often hydrogen and/or helium, are used to engineer the amount and nature of the lattice damage that under the right conditions leads to the detachment of a thin layer of material and its transfer to a new substrate. The Smart Cut technology is used to fabricate Silicon On Insulator (SOI), strained SOI and hybrid orientation wafers and to transfer layers of Ge, GaN, SiC, and other materials for high-end CMOS and photonics applications. The layer transfer mechanism is determined by the introduction of a variety of point and extended defects during implantation. The nature and morphology of the implant-induced defects depend strongly on the implant and substrate parameters. Understanding of physics behind these phenomena and engineering of implant-induced defects to achieve optimal parameters of the transferred layers represent a significant technical challenge. The corresponding issues will be described and discussed.

Keywords: Hydrogen, Implantation, Hydrogenated Defects, Platelets, Microcracks
PACS: 6172Cc, 6172Qq, 6180-x.

INTRODUCTION

The Smart Cut technology has developed into the mainstream industrial way to produce high quality Silicon On Insulator (SOI) wafers [1]. Millions of SOI wafers for a variety of electronic applications have been made by this process, in which a layer of single crystalline Si is removed from a donor wafer and transferred to a handle wafer, with a buried layer of SiO_2 providing electrical isolation between the two [2]. The technology, in most instances, is based on three main steps: ion implantation, wafer bonding, and separation and fracture to prepare a thin layer. By varying thickness, composition, and crystal orientation of the top layer as well as composition and orientation of the handle wafer, the Smart Cut technology enables fabrication of highly sophisticated engineered substrates for the semiconductor industry: SOI for HOT (Hybrid Orientation Technology) [3]; Direct Silicon Bonding (DSB) wafers [4], silicon on quartz (SOQ) [5]; SiGe on insulator (SGOI) and strained Si on insulator (sSOI) [6]; and Ge on insulator (GeOI) [7]. The technology can be used for transfer of SiC, GaAs, InP, GaN, and thin layers of other materials used in high-end microelectronic, optoelectronic and photonics applications (for recent reviews see [8,9]).

Blistering phenomena caused by surface bombardment with hydrogen or inert gases have been observed by many researches long before the Smart Cut invention. Hydrogen ions, when implanted to a dose of > 1×10^{16} cm^{-2}, produce fine planar defects, platelets, in the Si lattice. Depending on the dose, implant temperatures and other parameters the platelets evolve into microcracks or microcavities. Some hydrogen ions attach themselves to the dangling Si bonds in the microcracks, while other fill these voids. If such an ion-implanted wafer is heated up, typically to 400-500°C for a sufficient period of time, more hydrogen segregates into the microcracks in the form of molecular hydrogen, H_2, the pressure builds up to a point of fracture, and the surface of Si becomes pockmarked with blisters. This is clearly an undesirable effect of ion implantation: for recent review of blistering phenomena see, e.g. [10].

Michel Bruel discovered that the "undesirable" blistering effect, could be used to produce very well controlled transfer of thin layers from one semiconductor substrate to another. The most important ingredient of Bruel's method was a stiffener

CP1066, *Ion Implantation Technology,* edited by E. G. Seebauer, S. B. Felch, A. Jain, and Y. V. Kondratenko

placed on top of the implanted wafer. This thick and stiff layer prevents blistering and redirects the pressure that builds up in microcracks in a lateral direction. After the stiffener is attached to the suitably implanted wafer, some form of force is applied to eventually split the crystal. Heating of the wafer builds up pressure within the weakened plane that can cause a split [11]. Alternatively, application of a mechanical stress will also cause cleavage of the crystal.

Despite significant progress achieved during the past years the details of fundamental phenomena that drive the fracture and determine the quality of the transferred layers are still under intensive investigation.

In this paper the author will review the results of the recent studies related to the Smart Cut layer transfer in different materials. The paper will focus on: (1) group IV materials: Si of different crystal orientation, SiGe alloys and Ge; (2) SiC and III-V semiconductors.

FUNDAMENTALS OF LAYER TRANSFER IN GROUP IV MATERIALS

Smart Cut™: industrial technology

In commercial applications the layer transfer is usually done with the total implant dose exceeding 5×10^{15} cm^{-2}. That practice dictates the use of high current implanters for industrial applications of Smart Cut to maximize the throughput and minimize the cost of the final product. The thickness of the transferred layer is largely determined by the energy of implanted ions. The industrial implant equipment available on the market determines the practical energy range from 10 to 200 keV. Therefore accurate simulations of implant distributions in this energy range are of practical importance.

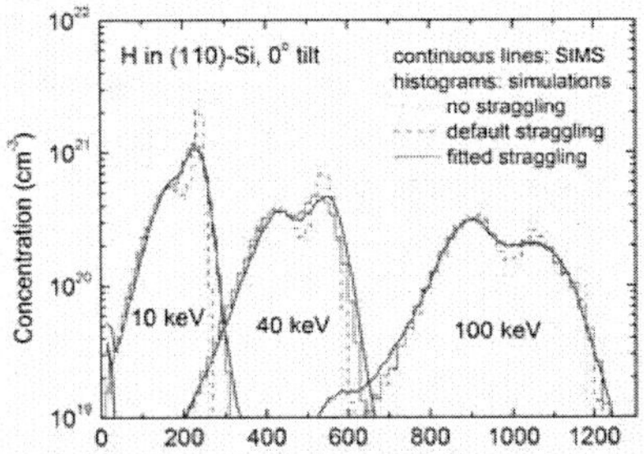

FIGURE 1. Comparison of simulated depth distributions of H in Si (histograms) and SIMS profiles (continuous lines). Implants have been done into (110) Si at 0° tilt angle (from [12]).

Commercial implant simulators work quite well for dopants, but, e.g., for H in Si the experimental data on electronic stopping are widely scattered at energies around and below the stopping power maximum. To address this issue a binary collision code IMSIL has been calibrated for accurate simulations of the implant profiles of light ions. The code calibration for H ions in (100) and (110) Si is described in [12]. An example given in Figure 1 demonstrates an excellent agreement between experimental and simulated profiles for H implanted into (110) Si in channeling direction.

H layer transfer in (100) Si layers

Better understanding of hydrogen-induced fracture phenomena in Si has been developed at LETI and Soitec in late 90s [11]. A recent study by S. Personnic *et al.* [13] has established a broad and quantitative picture of fracture in H implanted (100) Si. In this work Si substrates were implanted with H at an energy of about 40 keV and a dose of a few 1×10^{16} cm^{-2} through a 145 nm thick thermally grown oxide. After implantation and hydrophilic bonding to a base substrate, the structures were cut into sample pieces. These samples underwent annealing at temperatures between 350 and 500 °C for different times, t_a, ranging from 30 seconds to 51 hours. To compare results obtained at different temperatures, the annealing time t_a can be expressed as fraction or percentage of the time needed to obtain the full layer transfer at a given temperature: the splitting time, $t_s(T)$.

Figure 2 shows the evolution of the H depth distribution during annealing. Two main features are observed. First, there is a "transient" signature: changes in the depth profile occur only during initial stage of the anneal, i.e. for $t_a < t_s(T)/10$. The redistribution shows the evidence of an "uphill diffusion": the concentration gradients at the slopes of the profile increase after annealing while a plateau is observed in the R_p region. These are classical features observed when the precipitation of a second phase from a supersaturated matrix occurs since the precipitates tend to absorb the free mobile species around them. Second, there is an apparent H "dose loss": reduction in the integrated area under the profile after anneal. That is not consistent with the small redistribution of H in the surface region and might be due to H precipitation in a form that is not or only partially detected by SIMS.

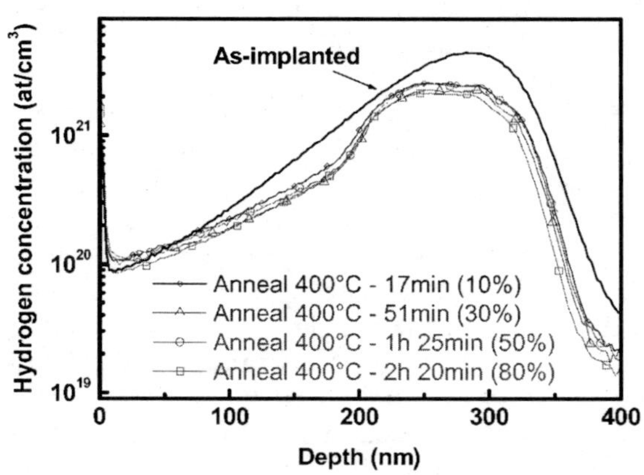

FIGURE 2. Hydrogen SIMS profiles before and after isothermal annealing at 400 °C (from [13]).

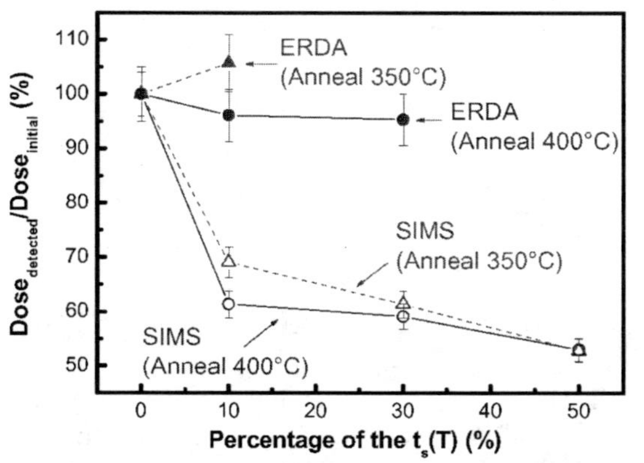

FIGURE 3. Normalized hydrogen dose measured by SIMS and ERDA as a function of percentage of $t_s(T)$ for isothermal anneals at 350 °C and 400 °C (from [13]).

To verify this assumption additional experiments with the Elastic Recoil Detection Analysis (ERDA) technique sensitive to the total hydrogen content in the sample regardless of the chemical state have been done. The normalized amounts of H determined by SIMS and ERDA are plotted in Figure 3 as a function of percentage of $t_s(T)$ for two temperatures. It is seen that the amount of hydrogen detected by SIMS strongly decreases during the first stage of annealing, whereas the ERDA signal does not change significantly with time. This seems to be a confirmation that the decrease of the SIMS signal seen in Figure 2 is not due to the out-diffusion of H from the implanted region but to some rearrangement of the H atoms in a form not detectable by SIMS. The Fourier Transform Infrared Spectroscopy (FTIR) results (not shown here) showed that the fast rearrangement of the hydrogenated defects also occurs during the initial stage of annealing.

Cross Sectional Transmission Electron Microscopy (XTEM) was used to quantitatively follow the evolution of the characteristics of the populations of platelets during annealing of the same set of samples. A typical damage zone was observed between 170 and 370 nm below Si surface and the fracture plane was at a depth of about 300 nm. Two families of platelets were found with habit planes either lying on the (001) plane parallel to the wafer surface or on two sets of (111) plane. The (111) platelets were located deeper in the damage zone and were detected in considerably smaller densities than the (001) platelets. Using the method described in details in Ref. 14 a statistical analysis of the characteristics of the platelet populations was found after each annealing step.

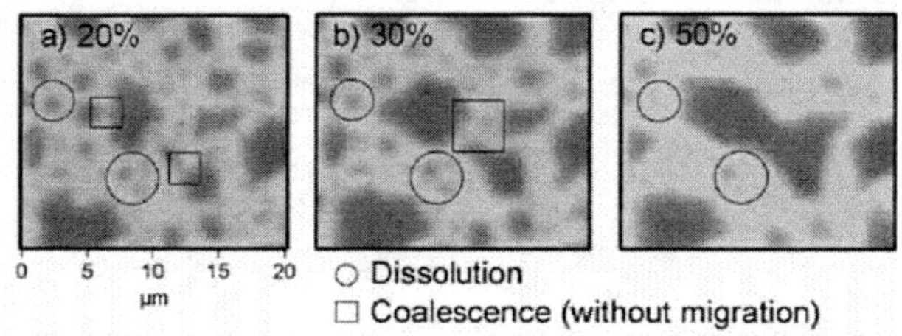

FIGURE 4. Optical profiler images of buried microcracks after an annealing at 350 °C for 20% (a), 30% (b), and 50% (c) of $t_s(T)$. The images illustrate the Oswald ripening and the coalescence mechanisms (from [13]).

The characterization of annealed samples by XTEM is restricted to times shorter than 50% of $t_s(T)$ for which the maximum platelet diameter is about 40 nm. The optical characterization of the microcrack evolution at ~ 300 nm below the Si surface was therefore performed using phase shifting interferometry technique [13]. Figure 4 shows the isothermal evolution of the microcracks at 350 °C. Two distinct phenomena are observed. First, the smallest microcracks dissolve (circles in Figure 4) while the largest grow. This behavior suggests that microcracks are also involved in Ostwald ripening, i.e. interchange the vacancies and hydrogen atoms they are composed of. The growth of the largest microcracks, however, occurs also in part through the coalescence between neighboring, already large, microcracks (squares in Figure 4). This phenomenon does not involve the motion of these defects but instead appears as the result of the sudden overlapping of their strain fields as they grow by the regular Ostwald ripening mechanism. Such a geometric coalescence leads to the random formation of large microcracks with the size up to tens of micrometers after annealing for 50% of $t_s(T)$.

The following conclusions have been reached by Personnic *et al.* [13]. Growth kinetics of hydrogenated extended defects can be divided into three stages.

During the first 10% of the splitting time, the transient process of H_2 formation induces the development of two distinct hydrogenated extended defects (HED) populations: platelet type defects and microcracks of several microns in diameter. The fracture plane therefore is already defined during the first stage of annealing. For longer anneals, between 10 and 80% of the $t_s(T)$, the smaller HEDs, platelets, grow via an Ostwald ripening mechanism. For larger HEDs, microcracks, the growth mechanism is more complex involving, besides Ostwald ripening, crack propagation and coalescence. It is driven by the continuous H_2 formation. In the last stage of anneal, the larger microcracks under high hydrogen gas pressure have strong mechanical interactions by the sudden overlapping of their strain fields leading to a predominant geometric coalescence process. This "catastrophic" mechanism results in the full layer transfer of Si.

H layer transfer in Si layers with other crystal orientations

Mechanism of layer transfer was intensively studied in (001) Si whereas other surface orientations, i.e. (110) and (111), received little attention. The corresponding data are described in Refs. [3, 15, 16]. The XTEM analysis of the damage zone in the as-implanted (110) wafer revealed the following HEDs: (100) and (111) platelets. The platelet density was, however, lower in (110) than in (100) substrate. Interestingly, in (110) wafers the (100) platelets perpendicular to the surface were observed and (111) platelets have the different angle relative to the surface, i.e. 35.3° vs. 54.7° for (100) Si. In the (111) substrates (111) platelets parallel to the surface with the size of 5 nm were found. In addition a lower density of {1-11} platelets with an angle of 70° relative to the surface was detected.

The XRD study of out-of-plane strain [15] revealed that the strain is strongest in (111) substrates, followed by (110) and (100) substrates. Atomic displacements, as obtained from the Debye-Waller parameter, follow the same trend. After annealing at the temperatures between 350 and 450 °C, a strong decrease in maximal deformation for all the substrates has been observed. Triple axis rocking scans showed the broadening of the curves attributed to the to mosaic tilt due to platelet and microcrack nucleation.

Raman measurements of the dose dependence of compressive biaxial in-plane strain induced by H implants in (100) and (110) Si [16] revealed that the absolute value of strain increases linearly with the defect density. For H implants the slope depends on the substrate orientation: the compressive strain increases more rapidly with the calculated average defect density (or dose) in Si (110) than in Si (100). The orientation dependence of strain might be linked to the differences in platelet orientation relative to the wafer surface. In Si (100) and Si (110) two platelet variants are found: {001} and {111}. However, in Si (110), unlike in (100) Si, there are no platelets parallel to the surface. These differences in platelet morphology may be the cause of the orientation dependence of H-induced strain.

The (100) wafers demonstrate the fastest splitting kinetics and (110) substrates have the slowest one. The activation energy to achieve splitting is smaller for (100) Si as compared to other orientations. The kinetics dependence may be due to different areal atomic densities (highest for (110) orientation) and correspondingly to the different density of bonds to be broken during microcrack propagation. The difference in platelet size and density may also influence the kinetics behavior at lower temperatures. No significant differences in absolute values of post-split roughness or in surface morphology for the wafers of different orientations have been found.

He and H co-implantation into (100) Si

Both H and He implants are known to induce blistering and exfoliation in different materials. A detailed XTEM study of the extended defects formed after ether H or He implantation into (100) Si in the low dose, 1×10^{16} cm^{-2}, regime was performed in Ref. [17]. The majority of He-induced defects in the as-implanted state of Si were nanometer size bubbles filled with He, whereas platelets are formed after H implants. Figure 5 shows different type of extended defects formed after short, 3 min, anneal at 350 °C. Evolution of He-induced defects during annealing results in dramatic spatial rearrangement and significant reduction in the density of the small nano-bubbles. A short anneal (see Figure 5a)) leads to the transformation from the homogeneous three-dimensional distribution of the nano-bubbles (as-implanted state) to a "preferential" three-dimensional distribution around the large platelets. A longer annealing leads to the formation of two-dimensional ring-like structure in the plane of the platelet around its edges and significant reduction in the density of the platelets perpendicular to the surface. In contrast, the annealing of H implanted samples does not change significantly the nature of the extended defects: all the platelets types found after a shorter anneal are also observed in the sample after longer anneal. These defects are: (100)-type (parallel and perpendicular to the surface) and {111} platelets (see Figure 5b)).

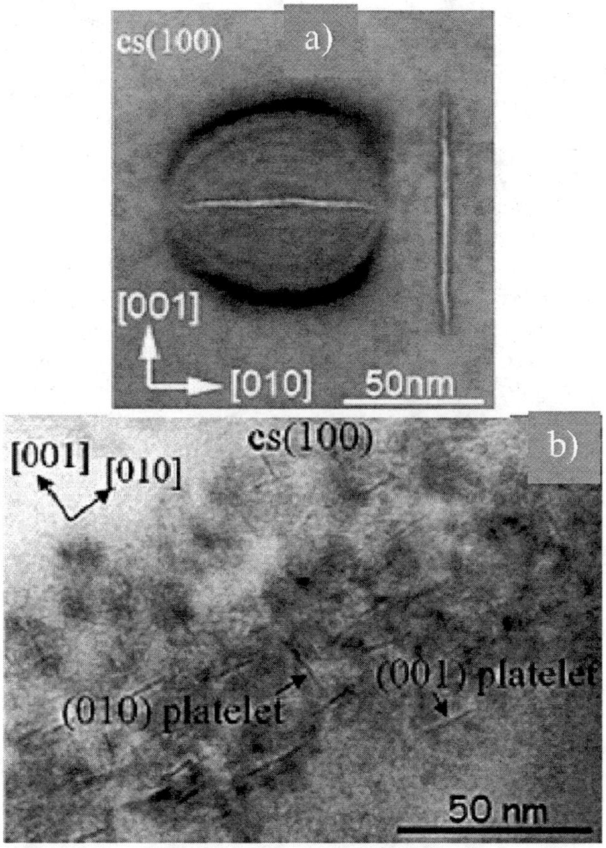

FIGURE 5. XTEM micrographs showing different types of defects in He- (a) and H- (b) implanted Si after anneal at 350 °C for 3 min (from [17]).

H implant at doses in the range of 1×10^{16} - 1×10^{17} cm^{-2} is typically necessary to obtain fracture. Therefore dose reduction may be helpful for more economical fabrication of the SOI wafers. It was shown that H and He co-implantation leads to a significant decrease of the total implant dose necessary to achieve Si layer transfer [18]. Fracture enhancement due to He co-implant was attributed to the rapid He diffusion to H-filled platelets enhancing the pressure that causes layer detachment. A more recent report [19] on co-implantation showed an improvement in the quality of transferred Si layers and more favorable splitting kinetics as compared to H implant only case.

In addition, it has been found that increase in total co-implant dose above 2.5×10^{16} cm^{-2} leads to thermal splitting with a significant reduction of thermal budget. That is of special importance in the fabrication of some heterostructures, e.g. SOQ and GeOI.

It was shown [20] that under certain He and H sequential implant conditions, a "swiss-cheese-like" structure with highly damaged but still crystalline pockets embedded into amorphous material is formed. This partial amorphization significantly affects the damage morphology and the thickness of the transferred layer. The later study [21] has identified the infrared absorption modes corresponding to the formation of the partially amorphized layer.

Layer transfer in SiGe and Ge

The study of H implantation into relaxed SiGe layers [22] epitaxially grown on (100) Si substrates reveals similar defect structure to that observed in Si. The majority of defects in as-implanted samples are (100) platelets parallel to the surface. Like in Si in $Si_{0.8}Ge_{0.2}$ the small microcracks do not always lie exactly on (100) planes. In SiGe very often {311} defects are found to connect two (100) platelets. In SiGe samples {311} defects can be easily imaged in the area below the damage zone.

The splitting kinetics results obtained for 40 keV H implants into Si, $Si_{0.8}Ge_{0.2}$ and $Si_{0.7}Ge_{0.3}$ at a dose of 6×10^{16} cm^{-2} revealed an Arrhenius-type behavior for all materials [22]. First, at temperatures below 550°C, the time to achieve the splitting at a given temperature of isochronal anneal increases with the increase in Ge concentration. Second, the activation energy to achieve splitting increases from 1.5 eV to 2.0 eV while Ge concentration increases from 0 to 30%. The AFM analysis showed that for the same H implant conditions the surface of transferred SiGe layers is rougher than that for Si after splitting at 500°C. Since the density and size of (100) platelets and {311} defects are significantly different for Si and SiGe that can explain the slower splitting kinetics and higher post-split roughness observed in SiGe samples.

The results of H and He co-implantation into relaxed $Si_{0.8}Ge_{0.2}$ layer presented in Ref. [22] indicate that it is possible, like in Si, to decrease the total implant dose necessary to achieve good splitting quality by factor three. An XTEM analysis of the co-implanted $Si_{0.8}Ge_{0.2}$ sample in the as-split state showed that the depth of the damage zone significantly decreases as compared to H-only case.

In H implanted Ge wafers the majority of the extended defects were identified as platelet-like structures lying on {100} and {111} planes [23]. The (111) oriented platelets were observed in the deeper region of H implant distribution. In Ge wafers, a high density of {311} defects and nanobubbles with the average size of 2 nm are observed. Some of them were connected to (100) platelets. Interestingly, in Si it was not possible to clearly observe the {311} defects. In Ge the {311} defects can be imaged better, because of the larger atomic mass and better contrast (see Figure

6). In addition, more efficient damage production in Ge than in Si at the similar implant conditions may lead to significantly higher density of {311} defects in Ge.

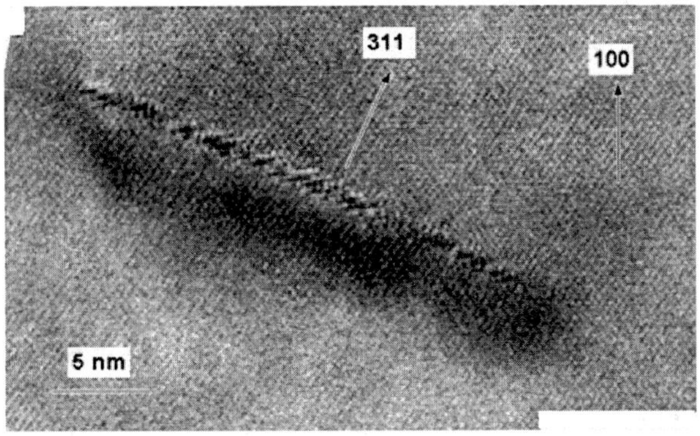

FIGURE 6. High-resolution TEM micrograph of a {311} defect in Ge (from [23]).

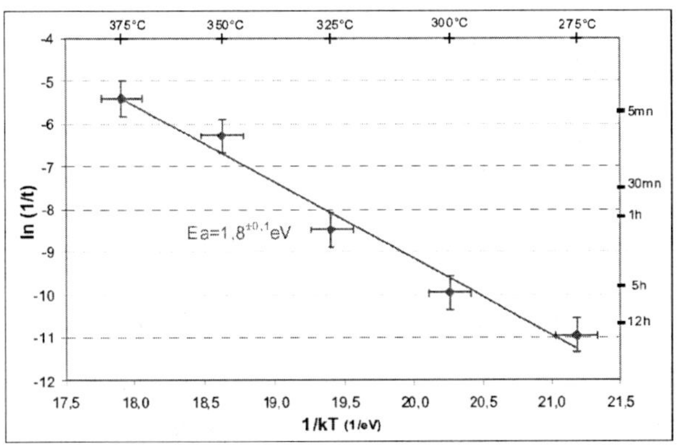

FIGURE 7. Splitting kinetics of Ge implanted with 80 keV H at a dose of 7×10^{16} cm^{-2} (from [24]).

The splitting kinetics for Ge layer transfer is shown in Figure 7. The recent results on fracture in Ge are presented by Mazen *et al.* [24].

SMART CUT™ IN OTHER MATERIALS

The Smart Cut technology is a powerful tool applicable to many materials. That opens a path to the formation of SiC or III-V based engineered substrates by layer transfer of materials like SiC [25], GaAs [27], InP [28], GaN [29] on a silicon, poly-SiC, sapphire, ceramic, or metal substrates.

Layer transfer in SiC

SiCOI engineered substrate was demonstrated by Di Cioccio *et al.* [25]. Hydrogen implantation conditions were studied extensively to determine splitting process window with different SiC poly-types (6H, 4H, 3C). Dose capable of resulting in layer transfer in SiC was found in the same range as for silicon: between 5×10^{16} and 1×10^{17} cm^{-2} depending on the implant energy. Splitting behavior does not depend significantly on the polytype and doping level. Grisolia *et al.* [26] have studied defects and microcracks created by hydrogen in SiC and their subsequent evolution with annealing. It has been observed that platelet nucleation and growth follow an Ostwald ripening law before leading to microcracks formation and ultimately to crystal splitting. This behavior is similar to what has been observed and recently quantified in the Si [13]. Figure 8 shows an XTEM picture of the transferred SiC layer.

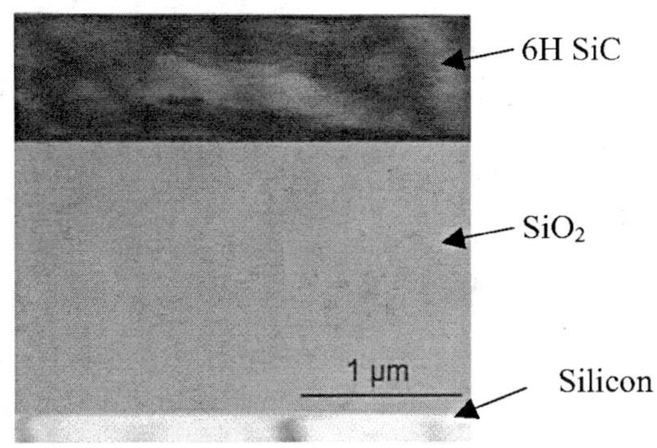

FIGURE 8. XTEM picture of transferred SiC layer (from [9]).

Layer transfer of III-V semiconductors

The first transfer of semi-insulating GaAs layer on silicon was reported by Jalaguier *et al.* [27]. Hydrogen implantation led to a splitting at the temperatures between 400°C and 700°C. Implantation was performed through SiO_2 deposited layer at a dose in the range of 5×10^{16} and 1×10^{17} cm^{-2} at 100 keV.

Based on optimized H implantation conditions and bonding through insulating layers (PECVD SiO_2 layers on InP and silicon), InP transfer on silicon was demonstrated by Jalaguier *et al.* [28]. The 3" InP wafers were semi-insulating. The InP layer transfer on silicon was performed with a good splitting quality using anneal in 250°C – 400°C temperature range. Almost no visible macroscopic defects have been observed. XTEM analysis of the transferred InP layer showed a thin damaged layer close the transferred surface.

First Smart Cut™ layer transfer of GaN has been reported by Tauzin *et al.* [29]. Starting GaN material was 4 μm thick GaN MOCVD epilayer grown on sapphire, either not intentionally doped or n-type (Si doped). Corresponding blistering study showed an activation energy of 2 eV between 300 °C and 700 °C, with a minimal dose of 1×10^{17} cm^{-2} to form blisters. This activation energy is close to the diffusion barrier for hydrogen in n-type GaN, estimated around 1.99 – 2.17 eV. XTEM analysis of as-implanted GaN sample (60 keV energy at a dose of 3×10^{17} cm^{-2}) showed damage zone located from 240 to 590 nm away from the surface and approximately 350 nm wide. SEM analysis showed that GaN transferred layer thickness was around 400 nm. Implantation conditions can be adjusted to modify this thickness to between 0.2 and 0.8μm. XRD analysis showed a FWHM of 390 arc-sec in the rocking curve mode, which is close to the original GaN epilayer value at 310 arc-sec. Post-split AFM analysis provided an RMS roughness value of about 5 nm and a peak-to-valley value around 46 nm for 5x5 μm^2 scan.

CONCLUSIONS

A review of the Smart Cut Technology for different substrates has been done. The point and secondary defects leading to the implant-induced separation and eventual layer transfer have been described. The ways to choose the optimal implant conditions for fabrication of high quality engineered substrates have been presented and discussed.

ACKNOWLEDGEMENTS

The author thanks S. Personnic, P. Nguyen, X. Hebras, A. Boussagol, T. Akatsu, N. Ben Mohamed, N. Sousbie, C. Richtarch, B. Faure, C. Aulnette, N. Daval, E. Guiot, A. Tauzin, F. Mazen, C. Deguet, N. Rochat, M. Kennard, C. Lagahe, I. Cayrefourcq, F. Letertre, C. Mazuré, H. Cerva, G. Hobler, N. Cherkashin, A. Claverie, C. Villeneuve, and V. Paillard for technical support and pertinent discussions.

REFERENCES

1. M. Bruel, *Electron. Lett.* **31**, 1201 (1995).
2. C. Mazuré, *ECS Trans.* **6** (4), 3 (2007).
3. K. K. Bourdelle *et al.*, 2004 IEEE Int. SOI Conf. Proc., p. 98 (2004).
4. K. K. Bourdelle *et al.*, *ECS Trans.* **3** (4), 409 (2006).
5. F. Brunier *et al.*, Proc. 2003 IEEE Int. SOI Conf., 59 (2003).
6. B. Ghyselen *et al.*, *Solid-State Electron.* **48**, 1285 (2004).
7. F. Letertre *et al.*, Mater. Res. Symp. Proc. **809**, B4.4.1 (2004).
8. B. Faure, *ECS Trans.* **3** (6), 293 (2006).
9. F. Letertre, Mater. Res. Symp. Proc., in press (2008).
10. B. Terreault, *Phys. Status Solidi A* **204**, 2121 (2007).
11. B. Aspar *et al.*, *J. Electron. Mater.* **30**, 834 (2001).
12. G. Hobler, K. K. Bourdelle, and T. Akatsu, *Nucl. Instrum. Methods Phys. Res.* B **242**, 623 (2006).
13. S. Personnic, K. K. Bourdelle, F. Letertre, A. Tauzin, N. Cherkashin, A. Claverie, R. Fortunier, and H. Klocker, *J. Appl. Phys.* **103**, 023508 (2008).
14. J. Grisolia, G. Ben Assayag, A. Claverie, B. Aspar, C. Lagahe, and L. Laanab, Appl. Phys. Lett. 76, 852 (2000).
15. N. Sousbie, L. Capello, J. Eymery, F. Rieutord, and C. Lagahe, *J. Appl. Phys.* **99**, 103509 (2006).
16. C. Villeneuve, K. K. Bourdelle, V. Paillard, X. Hebras, and M. Kennard, *J. Appl. Phys.* **102**, 094905 (2007).
17. X. Hebras, P. Nguyen, K. K. Bourdelle, F. Letertre, N. Cherkashin, and A. Claverie, *Nucl. Instrum. Methods Phys. Res.* B **262**, 24 (2007).
18. A. Agarwal, T. E. Haynes, V. C. Venezia, O. W. Holland, and D. J. Eaglesham, *Appl. Phys. Lett.* **72**, 1086 (1998).
19. C. Lagahe-Blanchard *et al.*, Electrochem. Soc. Symp. Proc. PV2003-19, p. 346 (2003).
20. P. Nguyen *et al.*, *J. Appl. Phys.* **97**, 083527 (2005).
21. P. Nguyen *et al.*, *J. Appl. Phys.* **101**, 033506 (2007).
22. P. Nguyen *et al.*, Electrochem. Soc. Symp. Proc. PV2005-03, 185 (2005).
23. T. Akatsu, K. K. Bourdelle, C. Richtarch, B. Faure, and F. Letertre, *Appl. Phys. Lett.* **86**, 181910 (2005).
24. F. Mazen *et al.*, these proceedings.
25. L. Di Cioccio *et al.*, *Mat. Sci. and Eng. B* **46**, 349 (1997).
26. J. Grisolia *et al.*, *J. Appl. Phys.* **87**, 8415 (2000).
27. E. Jalaguier *et al.*, *Electron Lett.* **34**, 408–409 (1998).
28. E. Jalaguier *et al.*, Proc. 11th Intern. Conf. on InP and Related Materials, Davos, Switzerland, PDB 4, pp 26–27 (1999).
29. A. Tauzin *et al.*, Proc.-Electrochem. Soc. PV 2005-02, 119 (2005).

Introduction of the SHX-III System, a Single-Wafer High-Current Ion Implanter

Michiro Sugitani, Mitsukuni Tsukihara, Mitsuaki Kabasawa, Koji Ishikawa, Hiroki Murooka, and Kazuyoshi Ueno

SEN Corporation, an SHI and Axcelis Company
1501 Imazaike, SaiJo, Ehime, 799-1362, Japan

Abstract. The SHX-III system, categorized as a single-wafer high-current ion implanter, has been developed by SEN Corporation in order to meet all the requirements for high dose and relatively high mid-dose applications, including high-tilted multi-step implantation. Recently the three major advanced device types, namely logic devices, memory and imagers, started to require high-current ion implanters in diverse ways. The SHX-III is designed to fulfill such a variety of requirements in one system. The SHX-III has the same end station as the MC3-II/WR, SEN's latest medium current implanter, which has a mechanical throughput of 450 WPH. This capability and precise dose control system of the SHX-III causes dramatic productivity enhancement for application of mid-high dose, ranged between 5E13 to 2E14 atoms/cm2, usually performed by medium current ion implanters. In this paper the concept and performance of the SHX-III will be described, concerning influence of device characteristics. A concept and performance data of the SHX figure that this system can provide implant quality and productivity as far as the 32 nm node.

Keywords: single-wafer, high-current implanter
PACS: 41.75.Ak, 61.72.U-, 85.40.Ry

INTRODUCTION

Recently the three major advanced device types, namely logic devices, memory chips, and imagers, started to require high-current ion implanters in diverse ways. For the latest high-performance, low-power and low-standby-power logic devices, accuracy of extremely low-energy implant, such as dose uniformity, less energy contamination (hopefully no contamination), beam parallelism, and uniform beam divergence within a wafer, are to be satisfied simultaneously at a high level, providing high productivity as well. For memory production, high productivity for poly-silicon gate implant must be guaranteed without energy contamination even using a high deceleration ratio for the sake of well designed energy filter. For advanced imagers, extremely low level of metal contamination capability is the key issue.

MOS-FETs have been requiring shallower junctions to enhance the electric characteristics and reduce power consumption, suppressing the short channel effect. In order to form shallow junction as shallow as or shallower than 20 nm, very low energy implant must be introduced. In the case of P-channel transistors, boron ion energy of sub-keV is required. Due to the nature of ion beam dynamics, such a low energy ion beam tends to expand and results in significant beam current loss.

Fighting for productivity of low energy implant for shallow junctions is a long term issues on the recent high current ion implanters. There are many proposals as countermeasures. The most well-known way is using deceleration [1-2]. Cluster or high mass molecule ions are other candidates and several molecules [3-5] and gas cluster [6] ions are proposed. Plasma doping is another candidate to enhance a throughput of ultra-shallow junction doping [7].

At the moment, deceleration is the only way in the actual manufacturing lines. Sometimes the deceleration ratio is limited at a very low value due to low acceptable levels of energy contamination. In this case, available beam current becomes very low and the productivity of source-drain extensions also becomes low.

The SHX series, which SEN Corporation – an SHI and Axcelis company (SEN) developed as a single-wafer high-current implanter, furnishes a unique answer to this issue. The SHX series has an energy filter after deceleration. This system can theoretically eliminate energy contamination and can use very large deceleration without concern about energy contamination.

At the same time, as the junction depth becomes shallower, regions of halo implantations become shallower. Historically, these implantations have been performed by medium-current implanters due to poor dose and angle controllability of high-current implanters. But, the productivity for halo implantations is getting lower and lower due to very

CP1066, *Ion Implantation Technology,* edited by E. G. Seebauer, S. B. Felch, A. Jain, and Y. V. Kondratenko

poor low energy capability of medium-current implanters.

Although the original SHX series has sufficient dose and angle controllability and sufficient beam current to perform halo implantations, the productivity is limited not by the beam current but by its mechanical throughput. Therefore, the SHX-III, a third generation of the series, has been developed to add more capability to the original SHX-II, especially enhancing the mechanical throughput.

BASIC CONCEPT

Utilizing the same concept as the SHX-II, the SHX-III has additional functionality. At first, let us explain the major concept of the SHX-II, which the SHX-III also has. The beam line components of the both systems are exactly same. This means that beam quality and quantity of the both system are exactly same. The conceptual beam line is described in Figure 1.

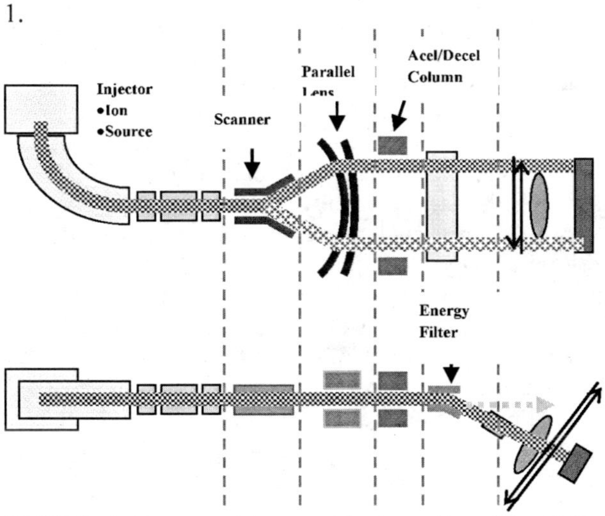

FIGURE 1. Conceptual layout of the beam line of the SHX series.

The most discriminative feature of the SHX series is an energy filter located after the parallel lens and the final acceleration and deceleration column. In medium-current implanters like the MC3 series of SEN [8, 9], an electrostatic energy filter is used. In high-current implanters, however, because low energy and high current beam must be transported efficiently, the SHX series utilizes a magnetic energy filter additionally. As mentioned in the previous section, this energy filtering system makes it possible to operate at a very high deceleration ratio. In Figures 2 and 3, typical SIMS profiles of the SHX series are shown, comparing those of a drift mode on the LEX series [10], SEN's multi-wafer high-current implanters.

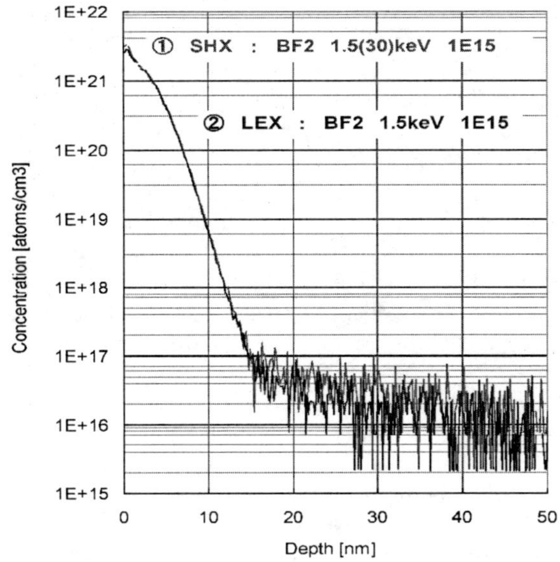

FIGURE 2. Comparison of SIMS profiles by different two systems, the SHX and LEX series. The implant conditions are same at BF_2^+, 1.5 keV and $10^{15}/cm^2$. The LEX (blue line) is operated by use of the drift mode which ensures no energy contamination as a reference. The SHX (red line) is operated by use of the deceleration mode from 30 keV down to 1.5 keV. The deceleration ratio is 20 and no significant energy contamination is observed in a concentration range above $10^{17}/cm^3$.

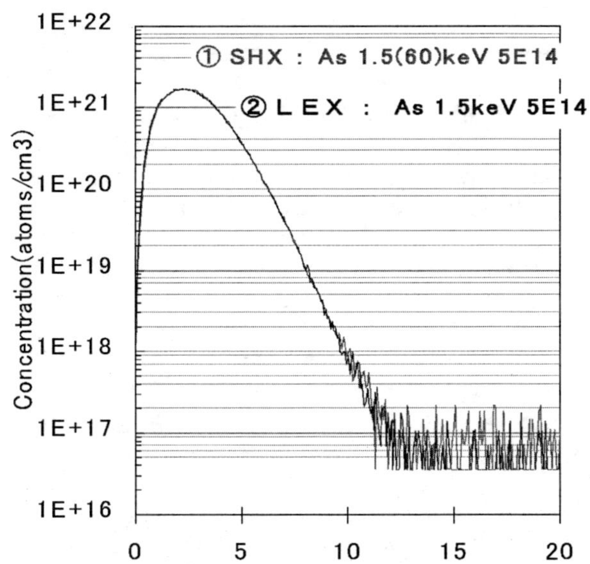

FIGURE 3. Comparison of SIMS profiles by different two systems, the SHX and LEX series. The implant conditions are same at As^+, 1.5 keV and $5x10^{14}/cm^2$. The LEX (blue line) is operated by use of the drift mode which ensures no energy contamination as a reference. The SHX (red line) is operated by use of the deceleration mode from 60 keV down to 1.5 keV. The deceleration ratio is 40 and no significant energy contamination is observed in a concentration range above $10^{17}/cm^3$.

Both sets of data show a superior energy filtering capability of the SHX series. Even with a deceleration ration of 40 as shown in Figure 3, the energy contamination level looks negligible. Strictly speaking, the contamination level is less than 0.03 % and the level is almost as same as accuracy of the SIMS analysis.

The second unique feature is a combination of the electrostatic beam scanning and the electrostatic deceleration parallel lens. This system ensures not only a beam parallelism but also other beam qualities like angle divergence and dose uniformity within a wafer.

Thirdly, the SHX series has a new type of ion source, named D-Source [11]. The D-Source of a kind of indirectly heated cathode types was originally developed for the LEX series, SEN's multi-wafer high-current implanters, in order to increase beam currents and to extend a source lifetime. The D-Source on the SHX series can provide sufficient beam currents especially in the energy region above 1 keV. In conjunction with its energy contamination-free capability, the productivity for poly-silicon gate doping is higher than conventional single-wafer high-current implanters.

NEW CONCEPT

The SHX-III is newly featured with a high speed end station implemented from the MC3 series. The mechanical throughput is enhanced from 220 wafers per hour (WPH) of the SHX-II to 400 WPH or more for 1-step implants. For 4-step implants, it is enhanced to 200 WPH from 70 WHP of the SHX-II.

In the case of halo implantations, high-tilt and multi-step implantations are usually used. Even with the multi-step implantation, each step requires sufficient dose uniformity. This is one of the reasons and the biggest reason why the existing single-wafer high current implanters can not be used for this purpose. As shown in Figure 4, the SHX series has very good capability in terms of uniformity to perform the high tilt implant of relatively low dose for the high- current implanters and relatively high dose for the medium-current implanters. In the case of the MC3 series even with the high speed end station of 450 WPH, an actual throughput for an implant at B+, 3 keV, 4-step 10^{14} ions/cm^2 (2.5x10^{13} ions/cm^2 for each step) is less than 25 WPH because of a low beam current at such a low energy region. In the case of the SHX-II, the throughput increases to 70 WPH, limited by the mechanical throughput.

Finally, in the case of the SHX-III, the throughput becomes 200 WPH or more because of abundant beam current and the high speed end station. Of course, energy contamination is eliminated in this condition as well.

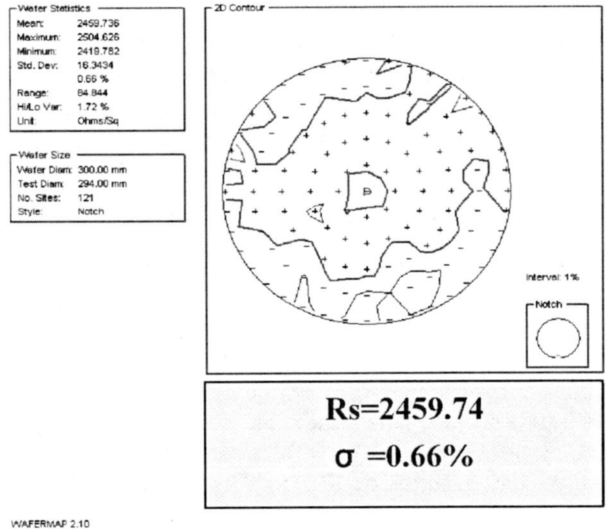

FIGURE 4. Sheet resistance map for a high-tilt application. The implant condition is at B^+, 3keV, 2.5x10^{13} ions/cm^2. The tilt/twist angles are 25°/0°. The beam current is 1 mA. Only one scan implant was performed.

From the basic concept throughout the SHX series can provide the essentially precise implantation in terms of energy contamination, angle uniformity and dose uniformity. Additionally, the SHX-III has a high speed end station and expands its process coverage from the conventional high dose applications, including the source/drain extensions, to the halo implantations and other relatively high-dose applications, which are currently covered by medium-current implanters, with an extremely high throughput and adequate precision.

OTHER FEATURES

Inheriting the end station features of the MC3-II, the SHX-III also has a precise notch alignment system [12]. The precision of notch detection is ±0.1° and alignment accuracy is ±0.2°. This capability enhances implant angle controllability of the SHX series.

As a member of the single-wafer implanters of SEN, the SHX-III can support an intentional non-uniform implant system, the MIND system [13, 14]. The system has two purposes. One is to compensate for process deviations brought on by other processes. Usually, they come from etching, CVD and/or annealing processes. If the deviation has a concentric profile and can be compensated by a single-step implant, one-dimensional scan modulation with multi-step implantations can generate a specified profile of a dose map [11].

However, if even one of the two above conditions is not satisfied, two-dimensional scan modulation is required in the case of a hybrid scan system with a fast beam scan and a slow mechanical scan. The MIND system can provide a various types of profiles using a support software which can generate a machine precognitive data using a user friendly GUI. A sample of very complicate profile pattern is demonstrated in Figure 5. If the multi-step implant is used with the two-dimensional scan modulation, much more complicated profiles can be achieved as shown in Figure 6. The MIND system gives a wide flexibility for a process integration and can contribute yield improvement.

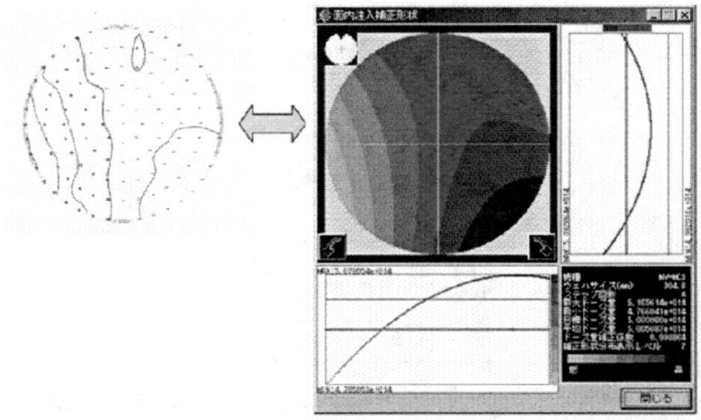

FIGURE 5. A predicted dose map generated by the support software and a result of sheet resistance. The sheet resistance map on the left hand side is a result of an implant condition at B^+, 20keV, $1x10^{15}$ atoms•cm^{-2}.

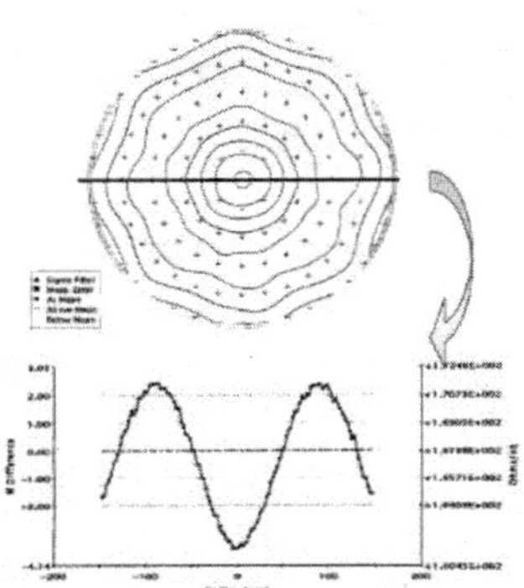

FIGURE 6. A concentric doughnut profile of a sheet resistance map at an implant condition of B+, 100 keV, and 5x1014• ionscm^{-2} on the MC3-II.

CONCLUSIONS

The original SHX-II provides sufficient energy contamination-free beam current at a very low energy region and precise control of dose uniformity and implant angles. A newly developed single wafer high-current implanter, the SHX-III, keeps superior characteristics of the SHX-II and has an enhanced mechanical throughput. A combination of high quality implant and high mechanical throughput on a high-current implanter develops a new process coverage.

The SHX-III appears to be a very productive high-current implanter for the three major types of the LSI manufacturing in terms both of quality and productivity for the current and future generations.

ACKNOWLEDGEMENTS

The authors thank Dr. Genshu Fuse and Mr. Masaki Ito for fruitful discussions.

REFERENCES

1. G. Angel *et al.*, Proceedings of the 12th International Conference on Ion Implantation Technology, p188, 1998.
2. H. Murooka *et al.*, Proceedings of the 13th International Conference on Ion Implantation Technology, p699, 2000.
3. K. Goto *et al.*, IEDM-96, IEEE(1997), p768.
4. D.C. Jacobson *et al.*, Proceedings of the 13th International Conference on Ion Implantation Technology, p300, 2000.
5. A. Renau, IWJT 2007, S7-1, p107.
6. M. E. Mack, Proceedings of the 15th International Conference on Ion Implantation Technology, p235, 2004.
7. S. B. Felch *et al.*, Proceedings of the 10th International Conference on Ion Implantation Technology, p81, 1994.
8. O. F. Campbell *et al.*, Proceedings of the 12th International Conference on Ion Implantation Technology, p154, 1998.
9. M. Sugitani *et al.*, in these Proceedings.
10. M. Tsukihara *et al.*, Proceedings of the 14th International Conference on Ion Implantation Technology, p373, 2002.
11. H. Izutani *et al.*, IWJT 2008, S1.5, p29.
12. F. Sato at al., in these Proceedings.
13. S. Ninomiya *et al.*, IWJT 2008, S1-4, p25.
14. S. Ninomiya *et al.*, in these Proceedings.

Nissin's New Cluster Implanter: CLARIS

S. Umisedo[1*], N. Hamamoto[1], Y. Koga[1], H. Une[1], N. Maehara[1], T. Nagayama[1], M. Tanjyo[1], N. Nagai[1], T. N. Horsky[2], R. D. Goldberg[2], S. K. Hahto[2], D. G. Bilbrough[2], D. C. Jacobson[2], H. F. Glavish[2]

[1] *Nissin Ion Equipment Co., Ltd, 575 Kuze Tonoshiro-cho, Minami-ku, Kyoto, Japan, 601-8205*
[2] *SemEquip Inc., 34 Sullivan Road, Suite 18, Billerica, MA 01862, USA*
**e-mail: Umisedo_Sei@nissin.co.jp*

Abstract. A new implanter for cluster ion implantation has been developed for production use. Some key performances will be described to show that equivalent beam current enhancement by using cluster implantation supported by high precision beam and dose control together with good tool stability enables a production worthy high throughput implantation for advanced device fabrication.

Keywords: Cluster, Implantation, Shallow Junction, Equipment
PACS: 37.20.+j, 85.40.Ry

INTRODUCTION

Cluster ion implantation technology has been revealed to have many advantages for fabrication of advanced semiconductor devices [1,2]. Ultra low energy implantation, self amorphization, and low leakage current are the typical characteristics of boron cluster implants such as $B_{10}H_{14}$ and $B_{18}H_{22}$. Also, carbon cluster implants such as $C_{14}H_{14}$ or C_7H_7 are being studied for the use in diffusion control or for Si stress engineering [3].

For a sufficient utilization of such an advantageous technology, a new cluster implanter has been developed as a production machine through a joint development program between Nissin Ion Equipment Co., Ltd. and SemEquip, Inc [4,5].

In this paper, many of its tool performance features are described to understand how the tool serves the cluster implant technology with production quality and reliability.

BEAM PERFORMANCE

Beam Current

First of all, the triggering feature of the cluster ion implantation, i.e., enhancement of ultra low energy beam current is shown in Figure 1. Cluster ion beam is extracted from SemEquip's 350 ClusterIon® source [6,7]. The beam is then mass analyzed, drifted or decelerated, X-Y focus adjusted, scanned in the horizontal plane and then collimated before being implanted into a wafer [4]. The beam current at the target at equivalent energies of 2 to 3 keV exceed 20 pmA in drift mode operation which is free of energy contamination. Since the punch through of energy contaminant for poly gate implant causes damage on the device, a high current but energy contaminant free beam serves as a strong tool for poly gate implantation.

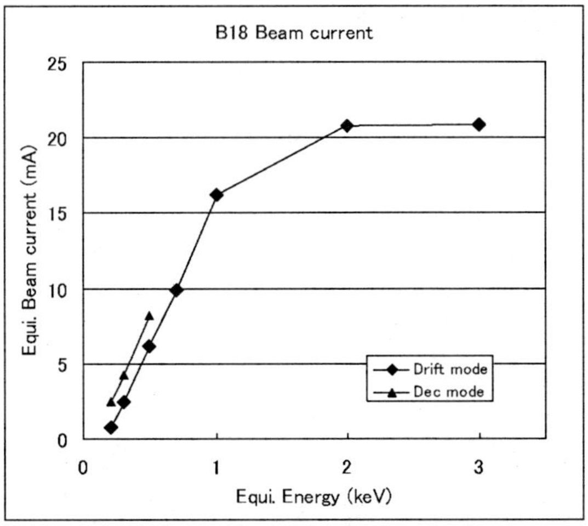

FIGURE 1. B_{18} beam current with drift mode operation plotted as equivalent particle mA vs equivalent B energy.

It should also be noted that even for the energies as low as 500 eV, the equivalent beam current with drift mode is more than 6 mA, which clearly shows the

CP1066, *Ion Implantation Technology*, edited by E. G. Seebauer, S. B. Felch, A. Jain, and Y. V. Kondratenko

advantage for USJ formation such as SDE, while deceleration mode is also available when even higher beam currents are needed for lower energies.

Beam Angles

Figure 2 shows the beam parallelism and divergence angle for FWHM beam size in the horizontal plane of B_{18} beam plotted against equivalent B energy. It is noted that beam parallelism is controlled to less than 0.3 degree for all the available energy range while the divergence angle is also suppressed mostly below 0.5 degree.

It should be noted that these data are obtained for large beam currents at low energies with drift mode, i.e. not using deceleration just prior to the wafer, and also for the beams that contain 15 mass peaks, which shows the success of this beam line design for cluster ion beam transportation.

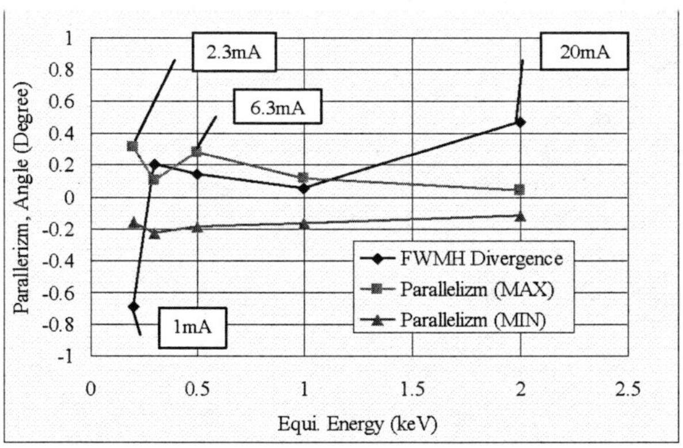

FIGURE 2. Beam parallelism and divergence in horizontal plane plotted against equivalent B energy.

Dose Uniformity Within a Wafer and Repeatability Within a Lot

With the above confirmed angle precision, the beam is scanned in the horizontal plane and dose uniformity is controlled by this scanning velocity, which can precisely control the implant dose as is well established by the Nissin Exceed series. In Figure 3, a dose uniformity measured with T.W. is shown for equivalent 300 eV, 2.1 pmA drift mode implantation. It should be noted that even at such a low energy, the implant dose uniformity is precisely controlled to 0.23%.

The implant uniformity was not only confirmed by T.W. measurements but also by as implanted SIMS profiles at three different positions: the center of the wafer, and the left and right edge of the wafer in the beam sweep direction. In Figure 4, only the ^{11}B component is plotted for demonstration purposes, but from this plot it is clear that all three profiles are identical, which shows the implant profile uniformity in addition to implant dose uniformity.

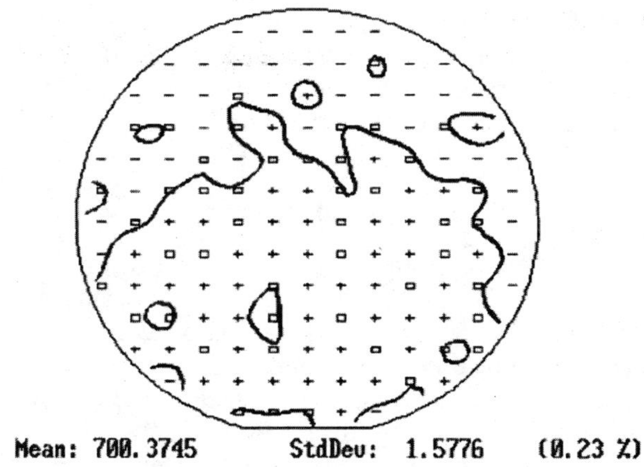

FIGURE 3. Dose uniformity within a wafer measured with TW after implanted with equivalent 300 eV.

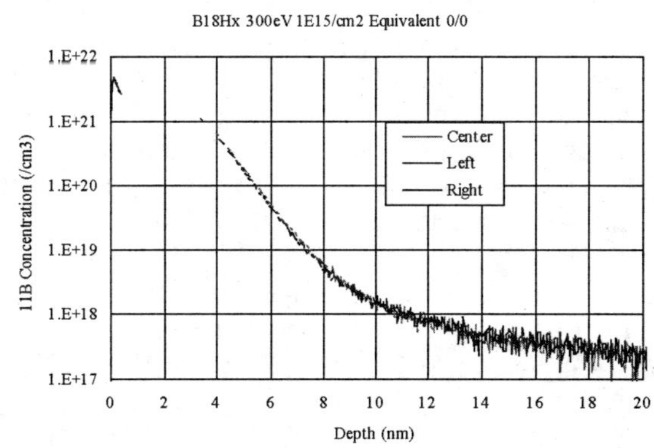

FIGURE 4. As implanted SIMS profile of B_{18} implanted wafer measured at left, center and right position on the wafer in beam sweep direction.

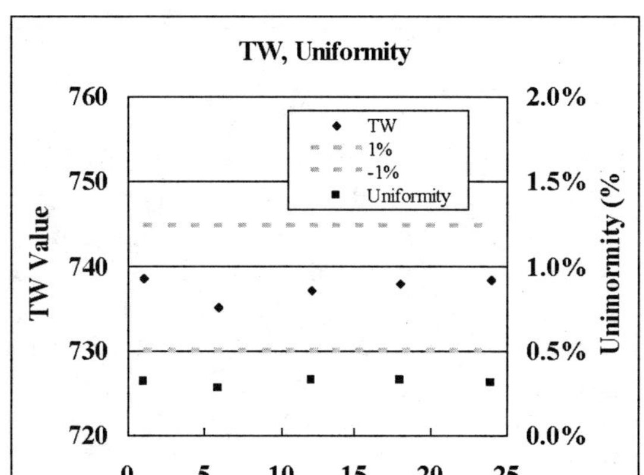

FIGURE 5. Dose repeatability within a lot.

Moreover, the repeatability data within a lot is shown in Figure 5. As can be seen from this data, the dose variation is well controlled within 1%, the standard deviation being 0.19% in this case.

Metal Contamination

It was also confirmed that although a beam with such a massive ion could have a large sputtering effects on the beam line walls, it did not increase any metal contamination to the wafer. In addition, since, as will be described later, the system needs to perform in-situ deposition cleaning, there was also concern as to whether any etched out component materials might somehow be carried in the beam to the wafer and increase the contamination. However, the data on metal contamination before and after in-situ cleaning after 100 hour beam operation in Table 1 show that such an effect is completely negligible. In Table 1, the results are shown in units of 1E10/cm2 for a wafer which was implanted with 500 eV 1E15 B/cm2 equivalent dose. All the measured elements were found to be less than 10 ppm which shows very clean implantation was achieved.

TABLE 1. Metal contamination on B_{18} implanted wafer before and after in-situ deposition cleaning.

Element	Al	Cr	Cu	Fe	Mg	Mo	Na	Ni	W	Zn
100H,Before FCL	0.44	—	0.80	0.098	0.19	—	1.3	0.078	—	0.38
100H,After FCL	—	—	0.70	—	0.19	—	0.60	0.078	—	0.28
Det. Limit	0.14	0.070	0.057	0.065	0.15	0.038	0.16	0.062	0.020	0.056

Wafer Throughput

It was also confirmed that even with such a precise and clean ion beam, the wafers can actually be processed with production worthy throughputs due to its large equivalent particle beam current as explained previously. Figure 6 shows the time chart of an equivalent 2 keV, 5E15 B/cm^2 implant with drift mode operation. The data show that such an implant can process 30 wafers per an hour, which is a good number for production use. Also, it was confirmed that 300 eV 1E15 B/cm^2 implants can be done at 32 wph throughput by using deceleration mode and a gas bleed into the beam line.

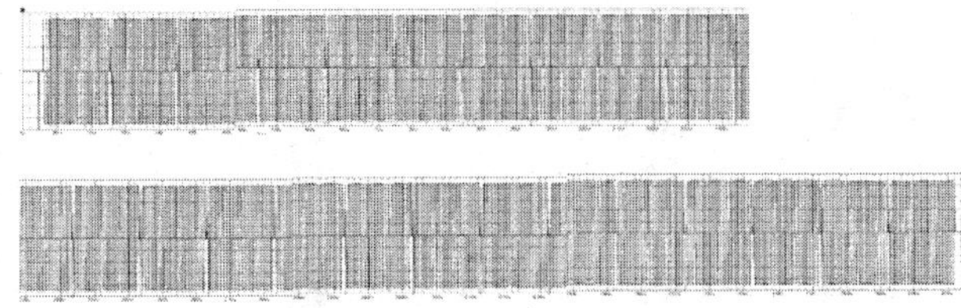

FIGURE 6. Equivalent 2 keV 5E15 implant throughput data recorded by mechanical scan velocity of the wafer.

TOOL STABILITY

In-situ Cleaning System

One of the key technologies for utilizing borohydride cluster implantation for production use is the solution for a large amount of insulating deposition inside the source. As a successful solution to this issue, an *in-situ* deposition cleaning system was developed [8].

In the new implanter system as a production tool, an advanced delivery system is equipped to supply the cleaning material gas from the ground potential either from the user's facility or from the cylinder cabinet placed in the beam line enclosure. Such a system enhances the advantage of the cleaning system by eliminating the tool down time for the replacement of material gas cylinder.

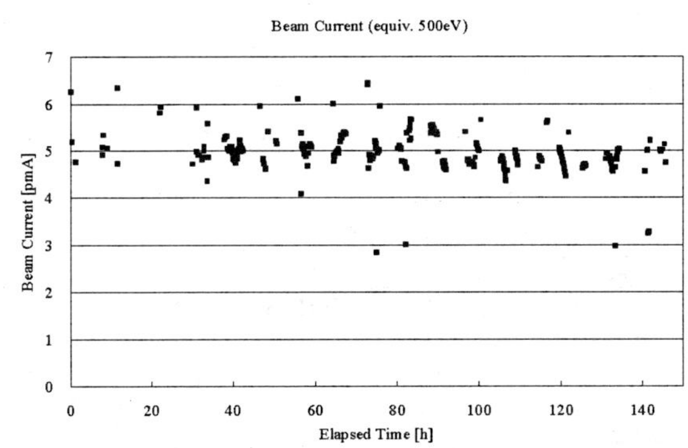

FIGURE 7. Equivalent 500 eV beam reproducibility over 100 h source operation including in-situ cleaning cycles.

Source Life

Figure 7 shows beam current data over continuous operation of the tool with implantation and in-situ cleaning alternatively cycled. In this run, several implant recipes have been used, but only the beam current of the equivalent 500 eV implant is plotted here. As can be seen from this plot, the ion source could deliver a cluster beam for over 150 h by periodically performing the *in-situ* cleaning.

Dose Repeatability Between Lots

While running the tool for 150 h, the dose repeatability was also examined between lots by T.W. measurements. Figure 8 shows the results. Although the mean value fluctuates more than in Figure 5, the standard deviation in this data was calculated as 0.67% which shows a good reliability as a production tool.

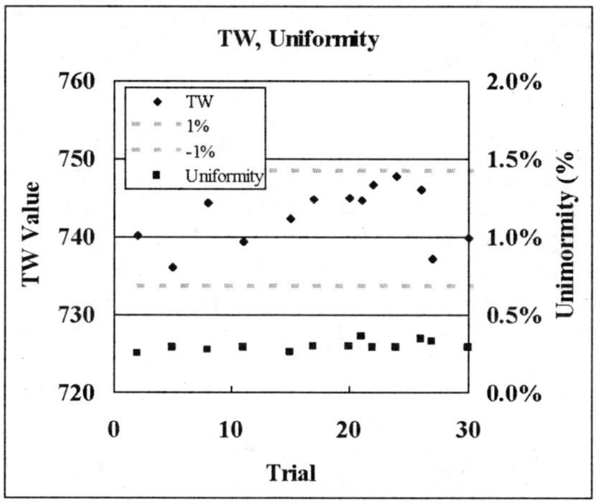

FIGURE 8. Equivalent 300 eV dose repeatability between Lots over 100 h source operation including in-situ cleaning cycles.

Particles

Some repeated data for particles both before and after the *in-situ* cleaning have been measured during this run. In Figure 9 data show that any particle increase is well below the tool specifications of 30 pcs.

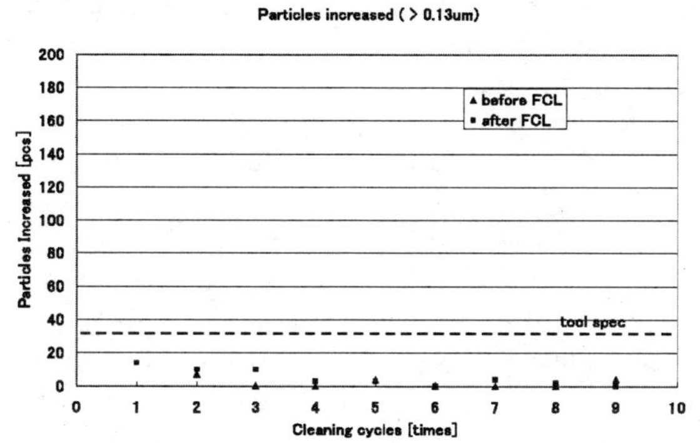

FIGURE 9. Added particles after B_{18} implant before and after in-situ cleaning cycles.

Though the available data for this measurement was restricted to the first few cleaning cycles due to measurement problems later on, the presented data imply that there is no significant trend of increasing particles that will be carried or generated by the beam while continuously implanting wafers, which is evidence for the reliability of the long term use as a production machine for advanced device fabrication.

CONCLUSION

A new cluster implanter has been realized through a Nissin-SemEquip joint development program. Its beam current and beam quality were confirmed to have desirable performance for implantations to form ultra-shallow junctions as well as poly gates with high throughput and good quality. The tool stability is also being continuously improved and the reliability for production use is now being rigorously evaluated. We will continue to investigate and validate its capability as a production machine for advanced semiconductor devices.

ACKNOWLEDGEMENTS

The authors would like to thank all the members of Nissin and SemEquip who joined in this project.

REFERENCES

1. T. Aoyama *et al.*, *Decaborane Ion Implantation for Sub-40-nm Gate-Length PMOSFETs to Enable Formation of Steep Ultra-Shallow Junction and Small Threshold Voltage Fluctuation* , IWJT2005. pp. 27.
2. N. Hamamoto *et al.*, *Decaborane implantation with the medium current implanter*, IIT2004 pp. 95.
3. W. Krull, , INSIGHT2007. pp. 189.
4. H. F. Glavish *et al.*, *A Beam Line System for a Commercial borohydride Ion Implanter,*, IIT2006. pp. 167.
5. N. Hamamoto *et al.* , *Development of Nissin new boron cluster ion implanter*, IWJT2007, pp. 125.
6. T. N. Horsky *et al.*, *Universal Ion Source*^{TM} *for Cluster and Monomer Implantation* , IIT2006, pp. 159.
7. D. Adams *et al.*, *A Vaporizer forDecaborane and Octadecaborane* , IIT2006, pp. 178.
8. T. N. Horsky *et al.*, *Boron Beam Performance and in-situ Cleaning of the ClusterIon*^{R} *Source*, IIT2006, pp. 198.

Implant Angle Monitor System of MC3-II

Fumiaki Sato, Makoto Sano, Hiroaki Nakaoka, Yoshito Fujii, Tetuya Kudo, Makoto Nakanishi, Masazumi Koike and Yasushi Fujino

SEN Corporation, an SHI and Axcelis Company
1501 Imazaike, SaiJo, Ehime, 799-1362 Japan

Abstract. Precise implant angle control is required for the latest generation of ion implanters to meet further shrink semiconductor device requirements. Especially, the highest angle accuracy is required for Halo implant process of Logic devices. The Halo implant angle affects the device performance, because slight differences of beam divergence change the overlap profile towards the extension. Additionally, twist angle accuracy is demanded in case of channeling angle implant. Therefore monitoring beam angles and wafer twist angles is important. A new monitoring system for the MC3-II, SEN Corp.'s single wafer type medium current implanter has been developed. This paper describes the angle control performance and monitoring system of the MC3-II. For the twist angle control, we developed a wafer notch angle monitor. The system monitors the wafer notch image on the platen. And the notch angle variation is calculated by using image processing method. It is also able to adjust the notch angle according to the angle error. For the tilt angle control, we developed a vertical beam profile monitor. The monitor system can detect beam profile of vertical directions with horizontally scanning beam. It also measures beam angles of a tilt direction to a wafer. The system configuration and sample beam data are presented.

Keywords: Angle Control, Ion Implantation
PACS: 61.72.uf, 06.30.Bp

INTRODUCTION

Precise implant angle control is required for the latest generation of ion implanters to meet further shrink semiconductor device requirements [1]. Especially, the highest angle accuracy is required for halo implant process of Logic devices [2]. Therefore, monitoring the beam angle is important in order to obtain better uniformity of beam characteristics within a whole wafer surface. Higher twist angle accuracy is also demanded in the case of implants with channeling angles implants. In addition, tilt angle accuracy has importance for halo implant. The tilt angle variation affects devices performances. Moreover the slight difference of beam divergence results in deviation of an overlap profile for the extension. Thus the beam profile monitor in the vertical direction is extremely important. This paper describes capability of a new monitoring system and its angle control performance.

TWIST ANGLE MONITOR

Figure 1 shows wafer handling sequence of the MC3-II. Wafers are aligned at the notch aligner system in the in-air module. After alignment, the wafers are carried into a load lock chamber and transferred to the electric-static-clamp (ESC). The wafers are chucked on the platen and rotated to the targeted twist angle that was indicated in a recipe.

Even if the notch is located by the aligner, there is small angle variation because of in-vacuum handling and chucking after alignment. Some devices need precise twist angle control, so we have developed the correction system on the platen. The wafer notch image is monitored by the camera that is located outside of the process chamber. The wafer notch position is monitored on the ESC after chucking.

System Description

The twist angle monitor system configuration is shown in Figure 2. The camera and the lighting system are mounted on the process chamber. The process chamber vacuum is isolated by the clear window on the chamber. The notch image is taken by the camera through the view port. The angle variation is calculated by the image processing unit using pattern matching method. The system also can compensate the angle variation. If a user sets some tolerance, the system compensates the angle variation by adjusting

CP1066, *Ion Implantation Technology,* edited by E. G. Seebauer, S. B. Felch, A. Jain, and Y. V. Kondratenko

the rotation axis of the ESC. After adjusting, the system rechecks the notch position to confirm the angle variation is within the tolerance.

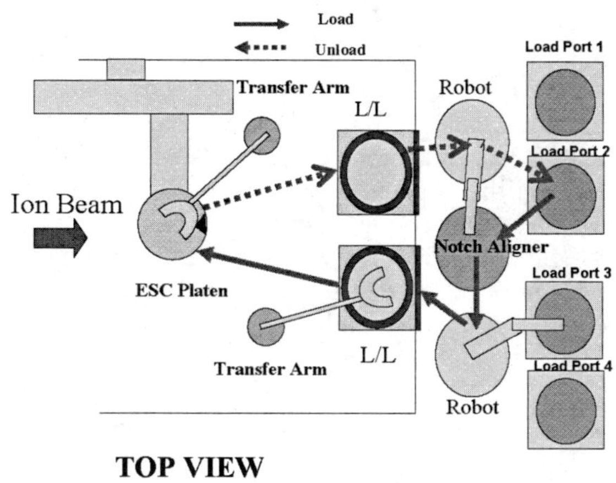

FIGURE 1. MC3-II wafer handling sequence.

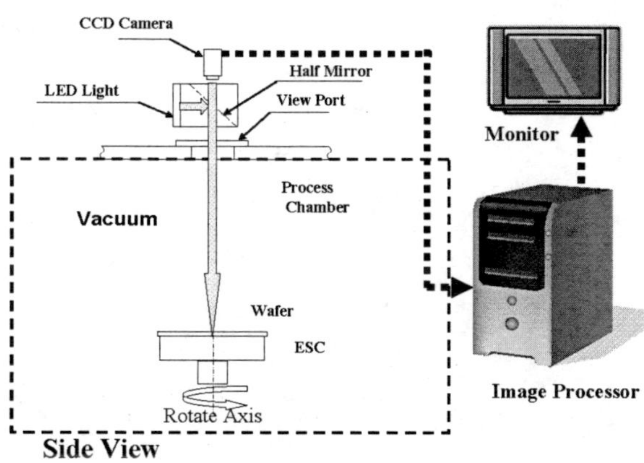

FIGURE 2. The twist angle monitor system configuration

Monitor Performance

The monitor system accuracy is evaluated by a special reference wafer. Figure 3 is the reference wafer image taken by the monitor system. The reference wafer has equally spaced machining holes. Each hole is positioned accurately. Each hole's image is compared to the center hole's image as a reference. And the angle difference values are calculated. The data of the result is shown in the Figure 4. It shows good linearity of this monitor system.

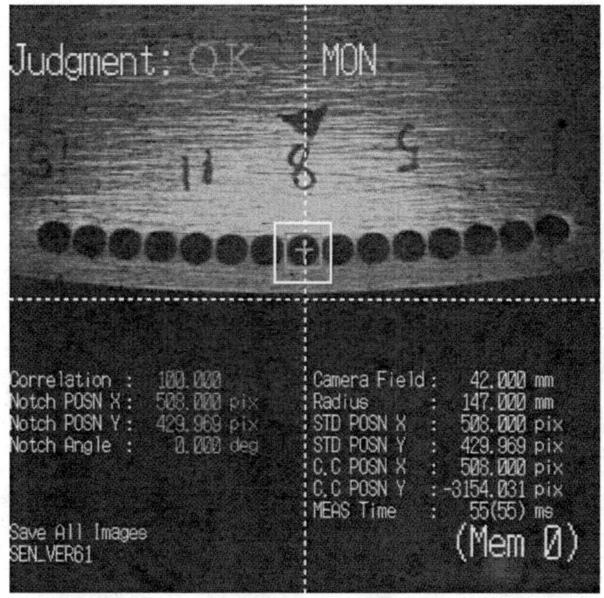

FIGURE 3. A monitor image of the reference wafer. Each hole is located at 1 degree offset.

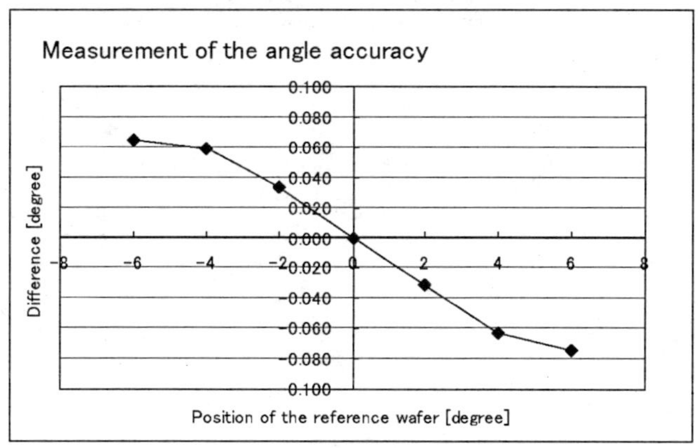

Difference= Measurement Value (N) – (2 x N)
N=-3, -2, -1, 1 2, 3

FIGURE 4. Twist monitor system angle detection accuracy .

Angle Correction Performance

The twist angle monitor system also can be used to compensate the twist angle variation. MC3-II can set tolerance of the notch variation. The ESC platen has the rotation axis to set the twist angle. The angle error is corrected by the rotation. After the correction, the monitor system rechecks the notch position.

Figure 5 is an example of the results of wafer image processes. The monitor displays the notch image, correlation value, and the angle variation.

We evaluated the correction performance of the twist angle monitor system. Firstly, we operated 25 wafers without compensation. Secondly, we operated the same wafer with correction of ± 0.2 degree tolerance. Figure 6 shows the result of the angle variation data. The data show that the twist angle monitor system can compensate the variation within ± 0.2degree.

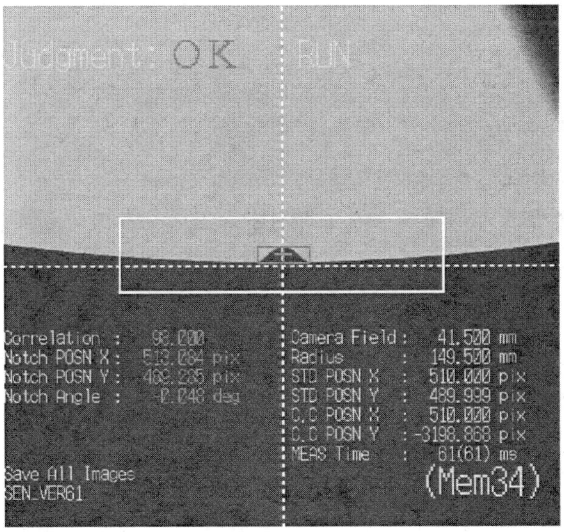

FIGURE 5. Sample of a wafer notch image

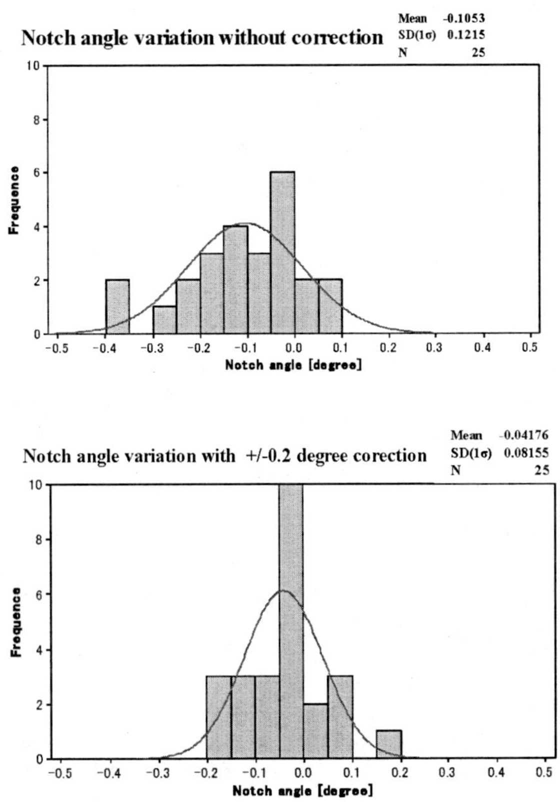

FIGURE 6. Sample data of notch angle correction.

TILT ANGLE MONITOR

Figure 7 shows the beam line of the MC3-II. The ion beam is extracted from an ion source (A). The ion beam is scanned by a scanning electrode (C) horizontally after mass analysis with an analyzer magnet (B). The beam angle is corrected at Parallelizing lens (D). And then the beam is accelerated/decelerated at the column (E). The beam is filtered with the angular energy filter (F) and energy slit (G). The wafer (H) is scanned mechanically in the vertical direction. The MC3-II has a flag faraday to measure ion beam current for beam tuning. We have replaced this unit with a multi channel faraday in order to detect vertical beam profiles.

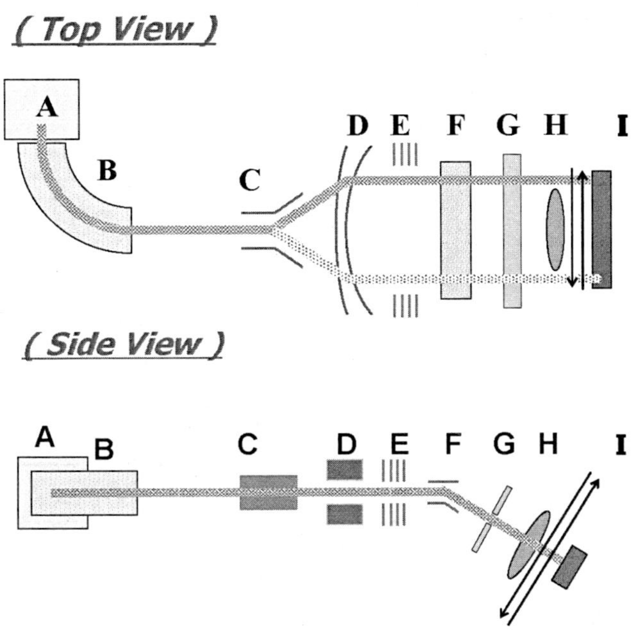

A. Ion Source
B. Analyzer Magnet
C. Scanning Electrode
D. Parallelizing Lens
E. Accel/Decel Column
F. Angular Energy Filter
G. Energy Slit
H. Wafer
I. Tuning Flag

FIGURE 7. MC3-II beamline

System Description

Figure 8 shows a configuration of the Vertical beam profile detector. There are 21 small graphite electrodes. Each strip is isolated electrically. The left side block and the right side block have a half pitch shift of the small electrode. Therefore the beam detection resolution in the vertical direction is half pitch of the electrode.

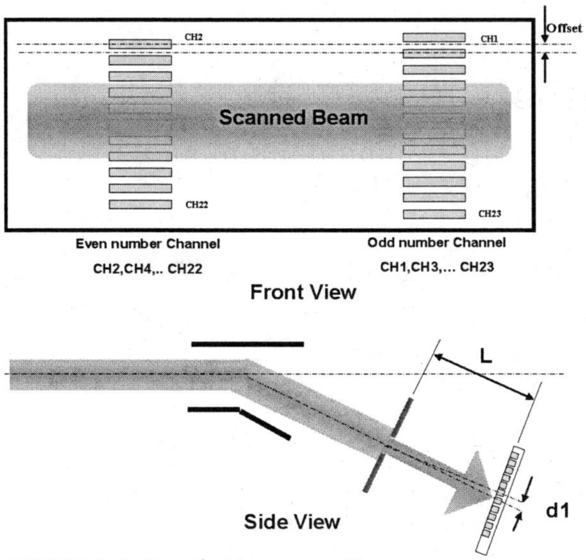

FIGURE 8. Vertical beam profiler

The scanned beam is detected by each electrode. The detected beam current is converted to the digital value by the analog/digital converter shown in figure 9. Then the MC3-II makes a beam profile in the vertical direction.

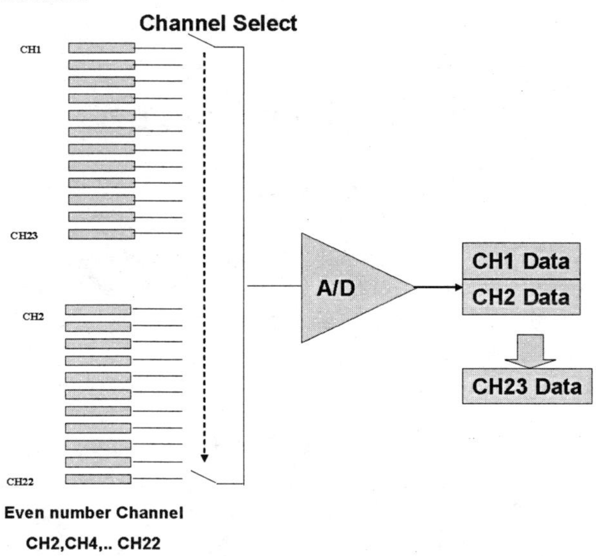

FIGURE 9. Circuit diagram of the vertical beam profiler

The monitor system itself is a beam profile monitor in the vertical axis of the beam scanning plane. The system calculates beam angele from this profile data. The beam angle is estimated from the following equation:

$$beam\_angle = \arctan\left(\frac{d1}{L}\right) \qquad (1)$$

where d1 is the difference between the peak position of the beam profile and the virtual center position of the detector. L is a distance between the energy slit and the vertical beam profiler. The twist angle monitor system has been developed. The twist angle is monitored by the camera and the variation can be adjusted within ±0.2 degree. The vertical beam profile monitor has also been developed.

The monitored profile data can be used for derivation of tilt angle and its divergence of the tilt angle direction:

$$beam\_divergence = \arctan\left(\frac{d2}{L}\right) \qquad (2)$$

where d2 is standard deviation of the beam width. Typical beam profile data is shown in figure 10.

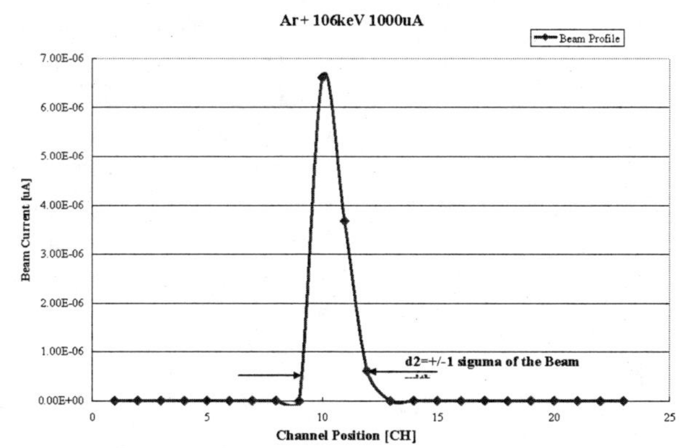

Conditions:
Ar+ 106keV Ib = 1000uA
Measurement result
Beam angle - 0.44 degree
Beam divergence 0. 14 degree

FIGURE 10. MC3-II beam profile data

CONCLUSION

The twist angle monitor system has been developed. The twist angle is monitored by the camera and the variation can be adjusted within ±0.2 degree.

The vertical beam profile monitor has also been developed. The monitored profile data can be used for derivation of tilt angle and its divergence of the tilt angle direction.

REFERENCES

1. Y. Kikuchi *et al.* "Profile and Angle Measurement System of SHX" in *Proceedings of the 16th International Conference on Ion Implantation Technology* 2006 pp.393-396
2. G. Fuse *et al.*, IWJT2007 S7-4 p121

Development of a Co-Axial Hot Cathode for Magnetized Ion Source Plasma

N. Miyamoto [a], N. Hamamoto [a], S. Imakita [b], A.G. Mendenilla [b] and M. Wada [b]

[a] *Nissin Ion Equipment Co. Ltd. (Minami-ku, Kyoto, 601-8205 Japan)*
[b] *Graduate School of Engineering, Doshisha University (Kyotanabe, Kyoto, 610-0321 Japan)*

Abstract. Directly heated high temperature cathodes of refractory metals such as tungsten run electric current of more than several tens of amperes. The electric current makes magnetic field around the cathode wire, and the magnetic field causes inhomogeneous emission of electrons from the cathode. To solve this problem we have designed the cathode having a co-axial heater current flow structure, and mounted it in a Bernas-type ion source. A plasma produced by co-axial hot cathode showed a clearer column along the external magnetic field and less displacement in the direction perpendicular to the field than that produced by a hair-pin filament. Stable discharge current as high as 5000 mA was obtained for Ar and BF_3 gases with the co-axial cathode. Boron and phosphorus ion beams were extracted from the source on an actual ion implanter. The ion beam currents were 1.5 times as large as those obtained with a hair-pin filament.

Keywords: ion source, hot cathode, plasma instability.
PACS: 29.25.Ni, 41.75.Ak

INTRODUCTION

Life of plasma producing electrodes often limits continuous operation time of an ion implantation system. Thus, efforts are being made to realize longer life ion sources with different kinds of plasma excitation processes [1]. High temperature cathodes exhibit relatively short life time. A heat load and ion sputtering determine the lifetime of the cathode.

A high temperature cathode having a co-axial heating electrical current path is proposed with the structure adaptive to a conventional Bernas-type ion source [2]. A co-axial current flow structure does not produce significant magnetic field at the surface of the cathode, so that an external magnetic field which confines a source plasma is not disturbed by the operation of the cathode. As a result, a beam current can be increased with keeping an arc discharge current and a heat load of the cathode. In other words an arc discharge and a heat load can be decrease with keeping a beam current. That means the co-axial cathode is expected to have a long life. In this paper, a characteristic of arc discharge and results of beam extraction test with a co-axial cathode are described.

EXPERIMENTAL SETUP

Ion Source

A high temperature cathode having a co-axial heating electrical current path is produced with the structure adaptive to a Nissin's BEAR (Bernas-type Electron Active Reflection) source [2]. A schematic diagram of the co-axial cathode is shown in Figure 1. The co-axial cathode consists of a tungsten centre wire, a thin-wall tantalum tube, and a tantalum ring connecting the centre wire and the tube at the cathode end. The connecting part of the cathode end ring is welded to ensure electrical contact. At the other end, the cathode terminals are connected to molybdenum

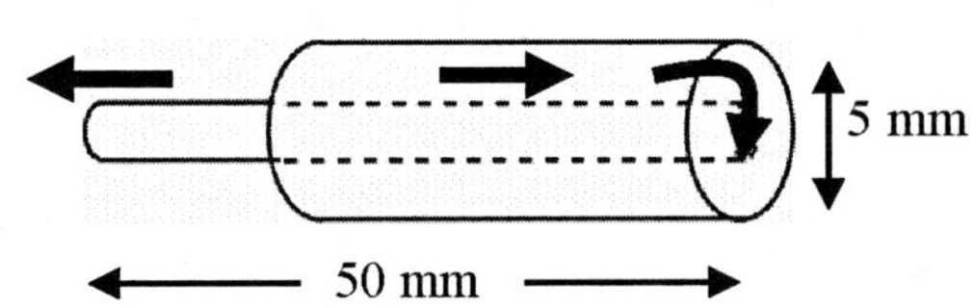

FIGURE 1. Schematic illustration of a co-axial cathode.

CP1066, *Ion Implantation Technology,* edited by E. G. Seebauer, S. B. Felch, A. Jain, and Y. V. Kondratenko

current feed-throughs inserted from the end flange supporting the arc chamber of the ion source. The outer diameter of the cathode tube is 5 mm, and the length of the cathode is about 50 mm.

A conventional BEAR source uses a hair-pin shaped 2 mm diameter tungsten filament as the cathode. According to previous research, thermionic electron emission is localized near the region of the negative end of the filament [3, 4]. The thermal electron was trapped near the cathode by magnetic field which is produced by the heating current, which should cause a wide energy distribution function of the emitted electrons. A co-axial cathode structure producing small magnetic field at the surface of the external conductor can emit electrons straightly from tip of the co-axial cathode.

The arc chamber is made of molybdenum with same dimensions as the BEAR source used in Nissin's EXCEED series. Two electron reflectors biased at several tens of volts negative with respect to the anode are located at the both ends of the chamber inside. These are biased at different potentials to enhance the plasma density.

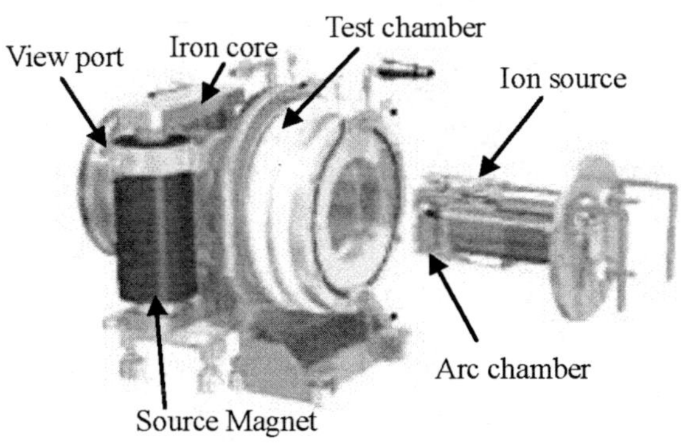

FIGURE 2. Experimental setup of test chamber.

Ion Source Housing Chamber

Figure 2 shows a chamber housing the test ion source. The chamber is identical to the Nissin's medium ion implanter except the absence of the extraction electrode. At the opposite side of a flange mounting the ion source, a CCD camera observes the arc chamber directly through a viewing port. A 230 l/s turbo molecular pump backed up by an oil rotary pump evacuates the test ion source. The gas pressure is measured by a vacuum ionization gage.

EXCEED 3000 Medium Ion Implanter

A beam extraction test was carried out at Nissin's EXCEED 3000 medium current ion implanter [5]. The beam formation components are an ion source, a mass separation magnet, an acceleration column, a final energy magnet, a beam sweep magnet and a collimator magnet. EXCEED 3000 has a pair of multiple Faraday cup systems. The front Faraday cups and the back Faraday cups are located about 300 mm upstream and down stream of the platen, respectively. The front Faraday cup system has 16 Faraday cups and the Back Faraday cup system has 11 Faraday cups. By measuring the beam current signals from the front and the back Faraday cup systems the beam alignment or the spatial variation of the beam incident angle to the wafer along the beam scanning direction (horizontal direction) is calculated and these angle errors are tuned within an acceptable value by adjusting the beam line parameter.

RESULTS AND DISCUSSIONS

Discharge Characteristics

Arc current plotted as a function of the heating current to a co-axial cathode in 0.6 Pa BF_3 gas is shown in Figure 3. The arc current increased with increasing heating current, and the arc current reached 5000 mA at 165 A of heater current. The 5000 mA discharge current was also confirmed with Ar gas.

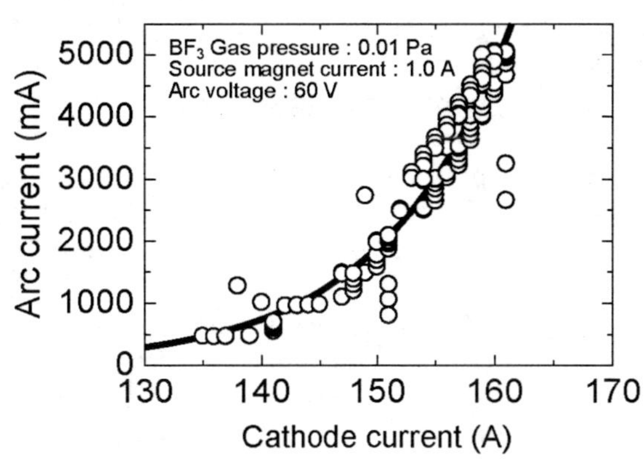

FIGURE 3. Arc current characteristics as a function of cathode current at BF_3 gas discharge

The shape of the produced plasma in the arc chamber was observed through a viewing window by opening the front plate of the chamber. Figure 4 shows the shape of an Ar plasma column recorded with a CCD camera coupled to a 10 nm bandwidth optical interference filter at 488 nm center wavelength. The external magnetic field is along the perpendicular direction in the figures. The plasma produced with a hair-pin filament expanded in the direction perpendicular to the field as shown in the left panel of the figure, while that produced with the co-axial cathode in the right panel exhibited a sharp cylindrical shape. A small shift toward left in the figure was probably the bending of the external magnetic field. From these results, it is understood that the larger displacement of the arc for a hair-pin filament was not due to the error magnetic field, but was caused by the disturbed the magnetic field by the filament heating current.

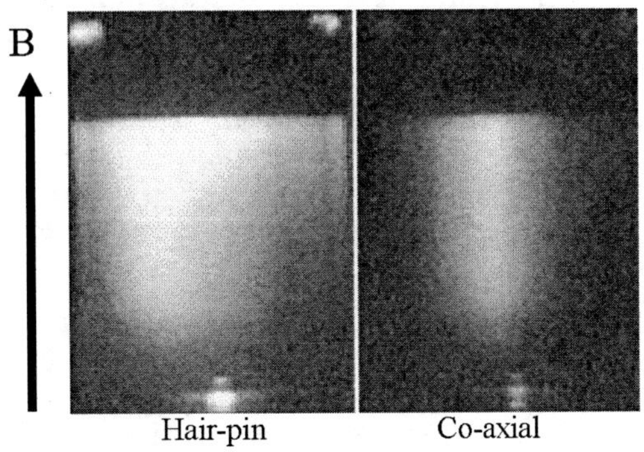

FIGURE 4. Typical arc discharge in the arc chamber various type of hot cathode.

Beam Extraction Test

Ion beam extraction test with co-axial cathode was carried out at the Nissin's EXCEED 3000 medium current ion implanter.[5] A BEAR ion source with co-axial cathode was installed on the EXCEED 3000 instead of normal BEAR source. Power supplies necessary to produce a source plasma were kept unchanged. The mass-separated ion beam current transported down stream of the wafer position were measured by the Back Faraday cup array.

The $^{11}B^+$ ion beam current is plotted as a function of the arc current in Figure 5. These data are obtained at 100 keV beam energy, 40 kV extraction potential, 1.6 A of source magnet current corresponding to 100 G linear magnetic field, 0.01 Pa BF_3 gas pressure out side of the arc chamber, and 60 V arc voltage. The $^{11}B^+$ ion beam current produced by a co-axial cathode was about 2.8 times as high as that of the hair-pin cathode at the same arc current. The extraction current was also increased so that the reason of the beam enhancement is assumed to the ion production increasing. As shown above, the co-axial cathode can produce more ion beam current at the same arc current with compared to the hair-pin filament configuration.

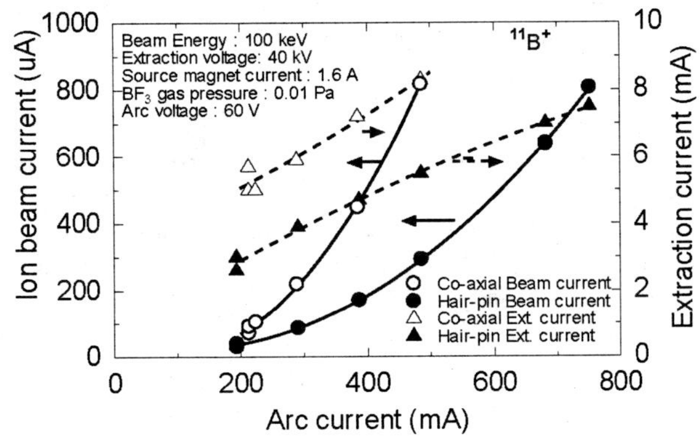

FIGURE 5. $^{11}B^+$ beam currents plotted as functions of arc current for two types of cathode.

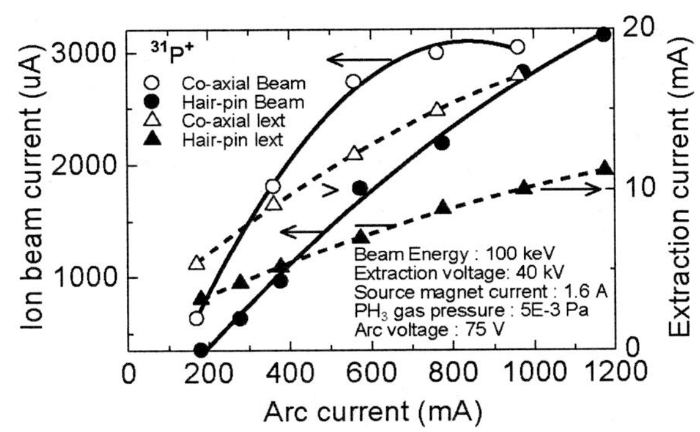

FIGURE 6. $^{31}P^+$ beam current and extraction current as functions of arc current for two types of cathode structure.

The dependence upon the arc current of $^{31}P^+$ ion beam current and that of extraction current are shown in Figure 6. The experimental conditions are the same, with those for $^{11}B^+$ ion beam extraction, except 0.05 Pa of PH_3 gas pressure, and 75 V of arc voltage. The $^{31}P^+$ ion beam current and the extraction current also increased with increasing arc current. The beam current produced by the co-axial cathode saturated above 750 mA arc current, while the extraction current did not saturate against the increase of arc current. The reason of this beam current saturation is possibly caused by mis-match of source extraction optics. However it is also confirmed that the beam current produced by a co-axial cathode at 750 mA was 1.4 times higher than that produced by a hair-pin cathode.

Beam currents of multiply charged phosphorus ions were measured to see the effectiveness of the co-axial cathode if it produces higher charge state ions. The results are shown in Figure 7 and 8 for doubly charged and triply charged phosphorus ions, respectively.

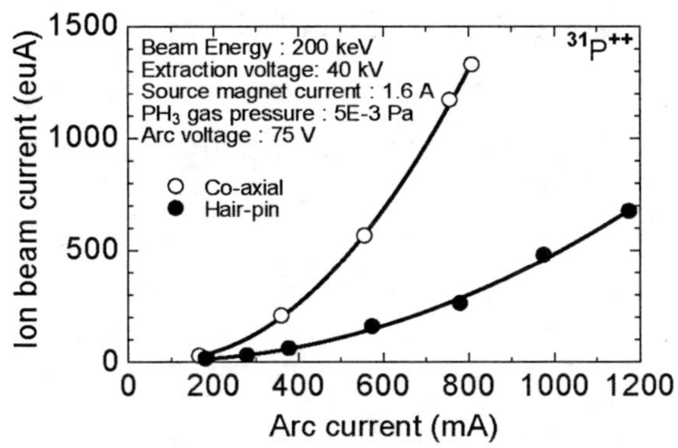

FIGURE 7. $^{31}P^{++}$ beam currents plotted as functions of the arc current for different types of cathode structure.

The co-axial cathode produced both ion species more efficiently than a hair-pin filament cathode as shown in the figures. The co-axial cathode produced nearly 5.1 times as high as the doubly charged phosphorus ion beam current produced by a hair-pin filament, and 5.6 times as high as the triply charged phosphorus ion beam by a hair-pin filament. The observed highly concentrated arc discharge can benefit the efficient production of the highly charged ions.

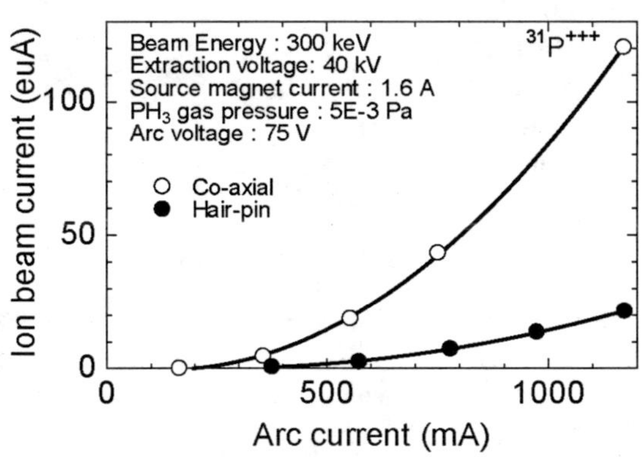

FIGURE 8. $^{31}P^{+++}$ beam currents plotted as functions of the arc current for different types of cathode structure.

The efficiency showed a dependence upon both the intensity of the external magnetic field and source gas pressure. Further optimization of operation for larger multiply charged ion beam seems possible with the co-axial cathode structure, and the research is being made to clarify the mechanism of enhancing the production of multiply ions with a co-axial cathode.

SUMMARY

A co-axial cathode adaptive to a Bernas type ion source for ion implanter was developed. The cathode produced little magnetic field by heating current and does not affect the magnetic field structure to confine the source plasma. The plasma column produced by the co-axial cathode was more concentrated than that produced by a hair-pin filament cathode.

The boron and phosphorus ion beam extraction experiments were carried out at an actual ion implanter. Boron and phosphorus ion beam were produced by using a co-axial cathode. The ion beam current has increased in agreement with the result of the discharge test. The extracted currents of singly charged boron ion beam were enhanced by a factor of 2.5. The phosphorus ion beams were enhanced by factors 1.4 and 5.6 for singly charged ions and triply charged ions, respectively. The co-axial cathode can enhance the ion beam current at the same arc condition with compared to the hair-pin filament.

This nature of the co-axial cathode can supposed to be extended the operation time of the ion source.

ACKNOWLEDGEMENTS

The authors thank K. Miyabayashi and S. Sakai for their technical advice on the experiment. Valuable discussions and comments from the colleagues of the system development group of Nissin Ion Equipment Co. are greatly acknowledged.

REFERENCES

1. R. Hollinger, M. Galonska, B. Gutermuth, F. Heymach, H. Krichbaum, K.-D. Leible, K. Ochs, P. Schaffer, S. Schaffer, P. Spadtke, M. Stork, and A. Wesp, *Rev. Sci. Instrum.* **79**, 02C703 (2008).
2. N. Miyamoto, K. Miyabayashi, T. Yamashita, H. Fujisawa, *Rev. Sci. Instrum.* **73**, 819 (2003).
3. K. W. Ehlers and K. N. Leung, *Rev. Sci. Instrum.* **50**, 356 (1979).
4. M. E. Arciaga, T. Kasuya, A. G. Mendenilla, H. J. Ramos, M. Wada, The Science and Engineering Review of Doshisha University, Vol. **43** (2002).
5. S. Sakai, M. Tanjyo, N. Hamamoto, S Umisedo, T. Kobayashi, T. Yamashita, T. Matsumoto, T. Ikejiri, K. Tanaka, Y. Koga, S. Yuasa, M. Naito, N. Nagai, in *Proceedings of 16th International Conference on Ion Implantation Technology*, 2006, 605.

Advanced Middle-High Energy Range Medium Current Implanter EXCEED9600A

Takashi Nogami, Masayoshi Hino, Tomoaki Kobayashi, Hideki Fujita, Takao Matsumoto

Nissin Ion Equipment Co., LTD. (Kyoto, Japan)

Abstract. EXCEED9600A was developed not only as medium current implanter, but also as implanter for the high energy range. The energy domain that can be used is 3 to 320 keV in single ion operation. In comparison with EXCEED3000AH implanter (3 to 250 keV), the upper energy limit has been increased by 70 keV. Comparing with EXCEED3000AH, most of the system components are the same, except for the high voltage cabinet and acceleration tube, which have increased volume and length, respectively. Several productivity improvement functions are added, for example, throughput has been improved by reducing auto set-up time. In this paper, the detailed machine performance is reviewed and the improved productivity and the reliability characteristics are also precisely explained.

Keywords: Ion implantation
PACS: 85.40.Ry

INTRODUCTION

An explanation about the relationship between the MOS energy and the dose energy for our implanters is needed first. The maximum energy for EXCEED3000AH implanter is 750 keV. The maximum energy of the EXCEED9600A implanter reported in this article is 960 keV. The medium current implanter is currently used in new market applications being operated at energies higher than 750 keV. For this purpose, customers use a medium current implanter as backup unit of the high energy implanter. The development of the new EXCEED9600A fulfills this demand from our customers.

FEATURE

Next we describe the main characteristic of EXCEED9600A. The injection energy is 320 keV. This breaks down to the extend voltage of 40 keV and acceleration voltage of 280 keV. EXCEED9600A carries IHC-R ion source as standard equipment and secured the current expensive beam value in the polyvalent ion. In addition, as high-productivity item, we take in beam current –UP, reduction of implant loss time.

SPECIFICATIONS OF THE PROCESS

The main characteristic included in Process Specifications is enabling the ion energy of 960 keV as upper limit. In addition, the implantation current for EXCEED9600A is increased compared to EXCEED3000AH case. Two examples are given, B+, 300 keV, 1600 uA (EXCEED3000AH : B++, 300 keV, 500 euA), P++, 600 keV, 1350 euA (EXCEED3000A; P+++, 600 keV, 260 euA). Another EXCEED9600A specification is related to improvement of effective-throughput (wafers/hours) as example of High-productivity.

Case of B, 300keV, 2E13,

EXCEED3000AH : 74(wafers/hours),

EXCEED9600A : 200(wafers/hours), about 2.7 times.

Please refer to figure 1 for the details.

In addition, case of P, 600keV, 2E13,

EXCEED3000AH : 32(wafers/hours),

EXCEED9600A : 148(wafers/hours), about 4.6 times.

Please prefer to figure 2 for the details.

Operation of EXCEED9600A at lower ion energies is similar with EXCEED3000AH (for lower energies, the two implanters are equivalent). As a general feature, it can be said that EXCEED9600A is an implantation system that contributes to productivity improvement by using high energy implantation.

CP1066, *Ion Implantation Technology,* edited by E. G. Seebauer, S. B. Felch, A. Jain, and Y. V. Kondratenko

For recipe < 11B 300keV 2e13 >

	Chrge	No. of Scans	Cycle Time (sec)	Imp.THP (wfs/h)	Effective THP (wfs/h)
EXCEED3000AH	2	21	39.83	90.4	74
EXCEED9600A	1	2	9.59	375.3	200

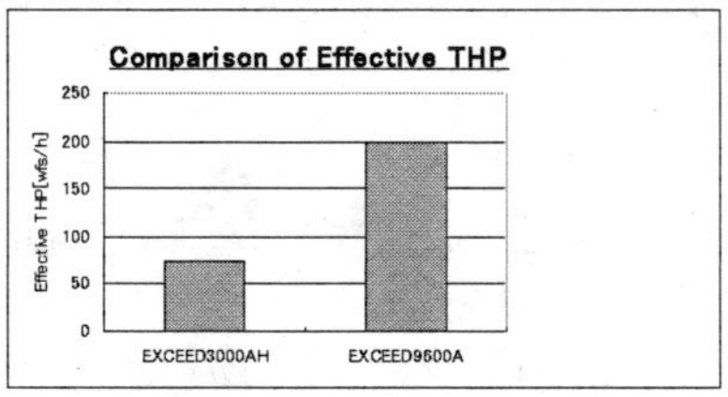

FIGURE 1. High Productivity for HE Recipe(1)

For recipe < 31P 600keV 2e13 >

	Chrge	No. of Scans	Cycle Time (sec)	Imp.THP (wfs/h)	Effective THP (wfs/h)
EXCEED3000AH	3	62	102.19	35.2	32
EXCEED9600A	2	2	15.8	227.8	148

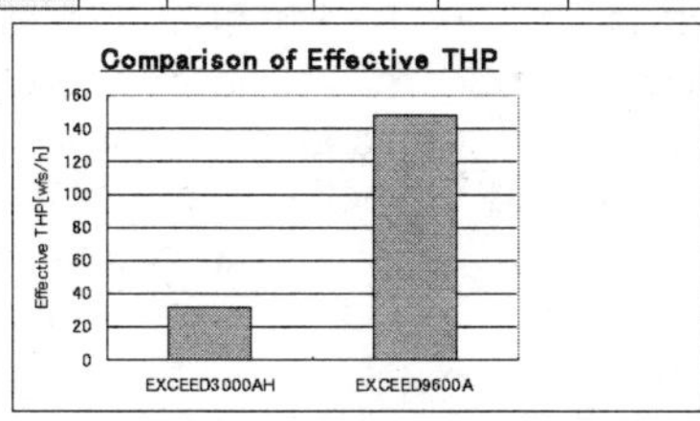

FIGURE 2. High Productivity for HE Recipe(2)

As far as Uniformity Check, even in the high energy domain it is within 0.5% which is a characteristic of EXCEED9600A (Figure 3,4).

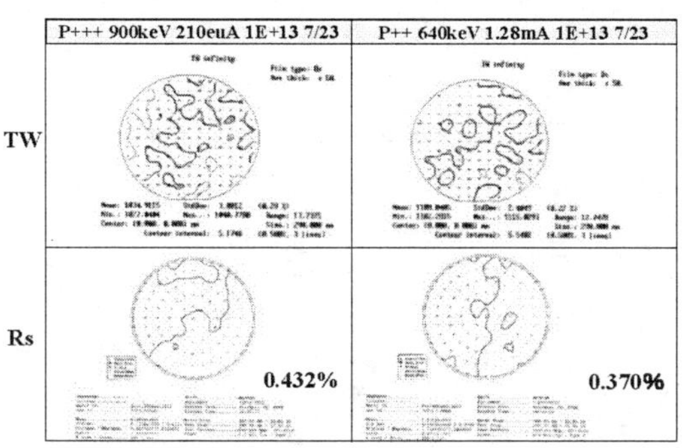

FIGURE 3. Uniformity Check(1)

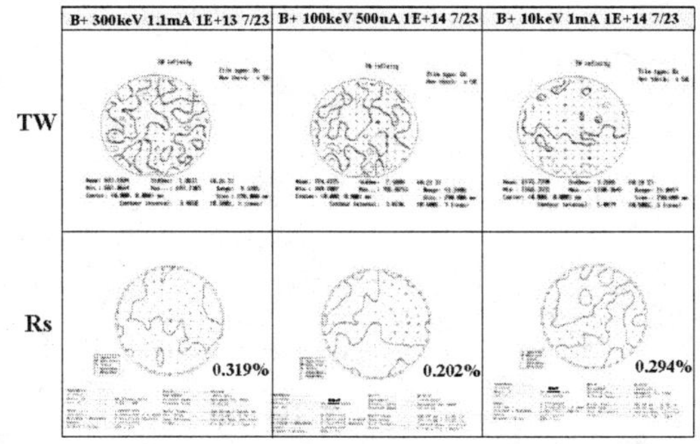

FIGURE 4. Uniformity Check(2)

As far as cooling capability data, even operation at high energy (480W) confirms that we do not have any problem. The wafer cooling system is the same as EXCEED3000AH. The end station succeeds to a type same as EXCEED3000AH-G2. Therefore, it can be said that the end station of EXCEED9600A is the machine which had high reliability and high throughput with the results in EXCEED3000AH.

We understand that we don't have any problem from figure 5 about SIMS Analysis data either.

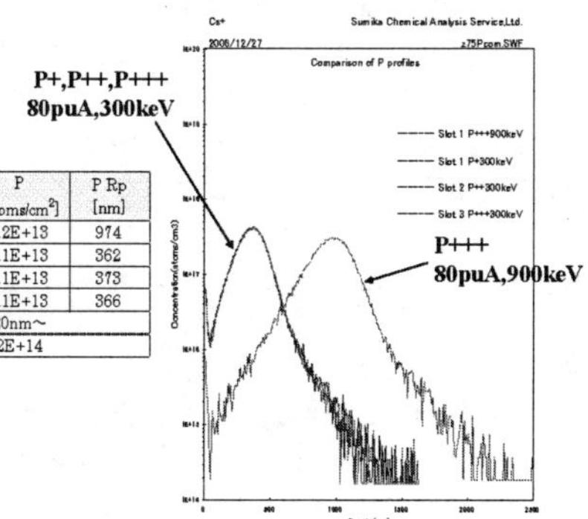

Data

Sample name	P [atoms/cm^2]	P Rp [nm]
Slot 1 P+++900keV	1.2E+13	974
Slot 1 P+300keV	1.1E+13	362
Slot 2 P++300keV	1.1E+13	373
Slot 3 P+++300keV	1.1E+13	366
Range of calculation (nm)	20nm~	
B. G. Level (atoms/cm3)	2E+14	

FIGURE 5. SIMS Analysis Data

SPECIFICATIONS OF HARDWARE

Layout

Please refer to Figure 6. It is the comparison of the layout EXCEED9600A and EXCEED3000AH.

The differences are as follows.

1. The dimensions of the high voltage cabinet
2. Acceleration Tube (Addition of Second Focus electrode)
3. Insulation Transformer. Exchange from Oil transformer (220 kV) to Dry (280 kV)
4. Acceleration power supply (Exchange from 220 kV P/S to 280 kV)

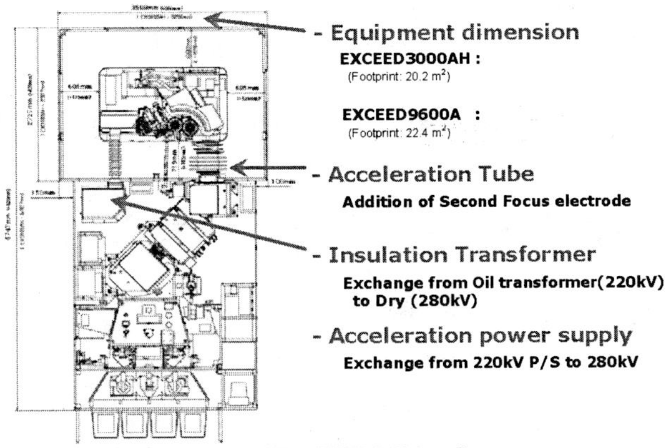

FIGURE 6. EXCEDD9600A Layout

Risk Assessment

First, the acceleration mode voltage has been improved, the voltage stablity is given. We clamp the voltage of the insulation part which measurement we lose prior confirmation and we did electrolysis simulations and the security of the safety factor of the insulation design As a result the safety of high voltage terminal is definitely incrased.

In addition, there is a problem of the power supply protection at the time of the electric discharge because the acceleration mode voltage improved. For prevention, we carry out a power supply protection circuit by electric discharge GAP and the reinforcement of the anti-electric discharge measure.

Increase in acceleration voltage causes a problem with the insulation up to the dry process transformer. This problem increased, we examined it and confirmed successful elimination. We proceeded with the development of this 280 kV breakdown voltage by using the transformer of 220 kV breakdown voltage which had necessary specifications with the results in EXCEED3000AH and developed the experience for the cause and succeeded in manufacturing.

Because the insulation distance became long, the beam line height increased. As a result, the FOUP setting height is changed too, but the standard top of SEMI-S2 does not have any problem. We were able to successfully complete the development of software for EXCEED9600A based on the previous version software developed for EXCEED3000AH. Changes introduced associated with increasing the upper limit of energy and adding E-Lense for the new acceleration tube.

Implant Angle Accuracy

To improve precision of the injection angle, we changed a definition of a parallel degree of a beam as described below.

Before, Parallelism +/-0.5 deg. (Average across the wafer)
After; Parallelism +/-0.1 deg (Average across the wafer)
This machine has an adjustment function to satisfy the above specifications. We enable correspondence to a more high-definition device injection by using this function.

OPTIONAL FUNCTIONS

Beam X-Y Monitor

Beam X-Y Monitor is a system monitoring the shape of the beam independently in an X direction and a Y direction. We made it possible to monitor the beam shape of the Y direction this time. The Faraday cup can move in the Y direction and intercept the beam, and it is a method to monitor a beam. The shape of a monitored beam is displayed by man machine interface (MMI) .The information displayed by MMI is the width of the Y direction of the beam and the width of the X direction and an incidence angle and an emission corner of the beams. Please refer to figures 7, 8, and 9 for the details. It continues about the automatic control of the emission corner of the beam as the next step and is carrying out development now.

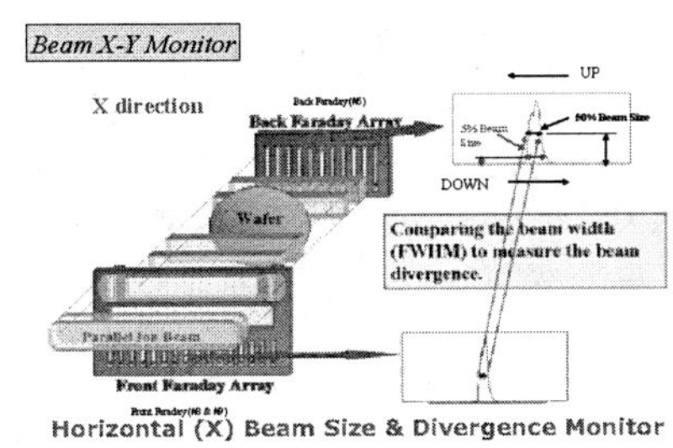

FIGURE 7. Beam X-Y Monitor(1)

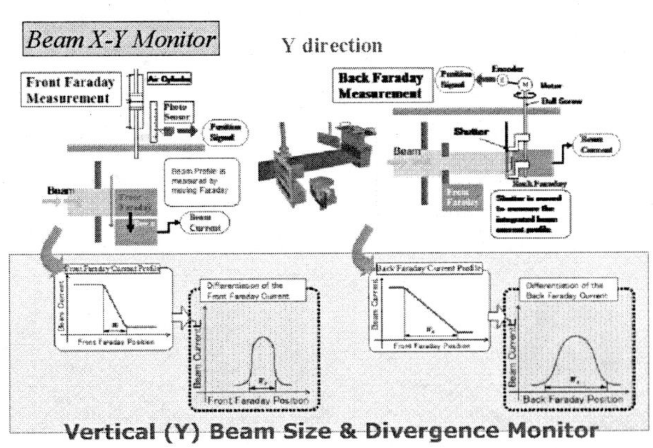

FIGURE 8. Beam X-Y Monitor(2)

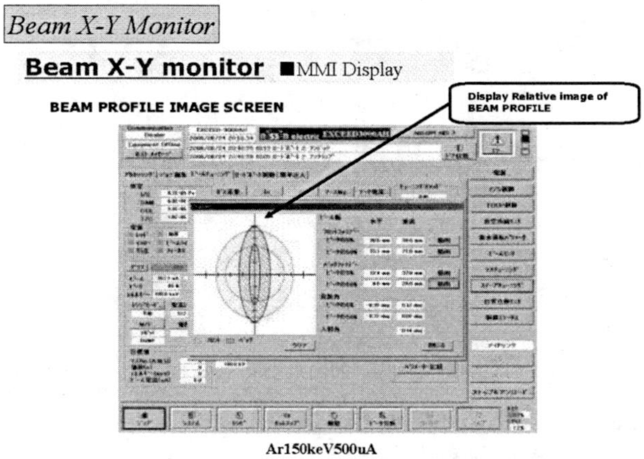

FIGURE 9. Beam X-Y Monitor(3)

High Productivity: Shortening of the Implant Time

The injection condition more than twice reduces the number of scans by lowering the scan speed and reduces loss time with the label top and bottom ends of the scan. As a result, we shorten the injection time and contribute to productivity improvement. In addition, the number of scans changes it to an even number in the case of the odd number and reduces loss time for scan and can shorten the injection time.

For example, if the dose is : 5.5E12/cm2 ~ 2.75E13/cm2 in a range of 5.5E12/cm2~2.75E13/cm2, Two times of number of scans was above and we did all the scans twice and compared throughput when we lowered the scan speed in the case of speed uniformity (320 mm/sec) As a result, about 40 wfs/h increase in throughput was observed on average,. In addition, on average of the odd number of scans, each about 50 wfs/h increases on about 30 wfs/h, the average of the even number of scans. We can plan throughput improvement by reducing the number of scans in this way and contribute to productivity improvement.

Low Energy Beam Current Up

We describe quantity of beam up by the low energy. For measures, we add a permanent magnet for convergence to the entrance of the COL magnet and let you converge the Y direction of the beam. By this, we raise transportation efficiency of the ion beam, and quantity of beam plans improving. For an effect, about 1.5 times increase in a domain less than energy 15 keV.

CONCLUSIONS

We can become the opposition of the H/E machine by handling a part of the high energy domain with this machine. Because this machine is the device which assumed EXCEED3000AH as a basic concept, it has the same high reliability as in EXCEED3000AH, a device comprising the stocktaking. This machine has a many functions aimed for productivity improvement.

REFERENCES

1. S. Sakai, *et al.*, in *Proceedings of the 16th International Conference on Ion Implantation Technology 2006*, pp.605-608.

High Productivity Implantation "PARTIAL IMPLANT"

Masayoshi Hino, Naoki Miyamoto, Shigeki Sakai, Takao Matsumoto

Nissin Ion Equipment Co., LTD. (Kyoto, Japan)

Abstract. The patterned ion implantation "PARTIAL IMPLANT" has been developed as a productivity improvement tool. The Partial Implant can form several different ion dose areas on the wafer surface by controlling the speed of wafer moving and the stepwise rotation of twist axis. The Partial Implant system contains two implant methods. One method is "DIVIDE PARTIAL IMPLANT", that is aimed at reducing the consumption of the wafer. The Divide Partial Implant evenly divides dose area on one wafer surface into two or three different dose part. Any dose can be selected in each area. So the consumption of the wafer for experimental implantation can be reduced. The second method is "RING PARTIAL IMPLANT" that is aimed at improving yield by correcting electrical characteristic of devices. The Ring Partial Implant can form concentric ion dose areas. The dose of wafer external area can be selected to be within plus or minus 30% of dose of wafer central area. So the electrical characteristic of devices can be corrected by controlling dose at edge side on the wafer.

Keywords: Ion implantation
PACS: 85.40.Ry

INTRODUCTION

Cost reduction and productivity improvement are important subjects at the production field. Reducing consumption of wafers is required as part of the cost reduction at the semiconductor development field. Experimental ion implantation needs many wafers to be implanted under various conditions. In particular, dose matching implantation needs many wafers to fix the dose rate to standard implanter. Consumption of wafers can be reduced if several different dose areas can be formed on one wafer surface.

On the other hand, production yield drops due to variations of device electrical characteristics may become an issue at the production field. Inhomogeneous resist thickness for central and external areas of the wafer cause inhomogeneous dopant depth of the implantation process. As a result, variations of the electrical characteristics of devices on the wafer appear. These variations may be compensated if using different implantation dose rates. As a result, production yield drop (especially related to external areas of the wafer) are diminished or eliminated, leading to productivity improvement.

The patterned ion implantation system "PARTIAL IMPLANT (PI)" has been developed for the medium current ion implanter to solve the problem. A performance test of the Partial Implant System is carried out at Nissin EXCEED3000AH. In this paper, the procedure of the Partial Implant and the result of the implantation profile are presented.

IMPLANTATION SYSTEM

EXCEED3000AH has a pair of multiple Faraday systems, the Front Faraday and the Back Faraday, located at about 300 mm upstream and down stream of the platen, respectively [1]. Implant system adopts Hybrid Scan System that uses mechanical vertical wafer scanning and electromagnetic horizontal beam sweep. Ion dose uniformity is provided by controlling the wafer scan speed and the beam sweep speed.

The Partial Implant System can form several different dose areas by intentionally controlling the wafer scan speed. The Partial Implant System contains two implant methods. One is "DIVIDE PARTIAL IMPLANT (Divide PI)" that is aimed at reducing the consumption of the wafer. Another is "RING PARTIAL IMPLANT (Ring PI)" that is aimed at improving yield by correcting electrical characteristics.

CP1066, *Ion Implantation Technology,* edited by E. G. Seebauer, S. B. Felch, A. Jain, and Y. V. Kondratenko

PARTIAL IMPLANTATION SYSTEM

Divide Partial Implant

Figure 1 shows the schematic diagram and implant procedure of the Divide PI. The Divide PI evenly divides the dose area on one wafer surface into two or three different dose areas by controlling vertical mechanical scan speed.

A transition area that varies from one dose to another dose appears at the Partial Implant. Size of the transition area depends on the vertical beam size. A beam which has small vertical size is required to produce a narrow transition area. Mask slits for the Partial Implant (PI Slit) are used in order to limit the vertical beam size.

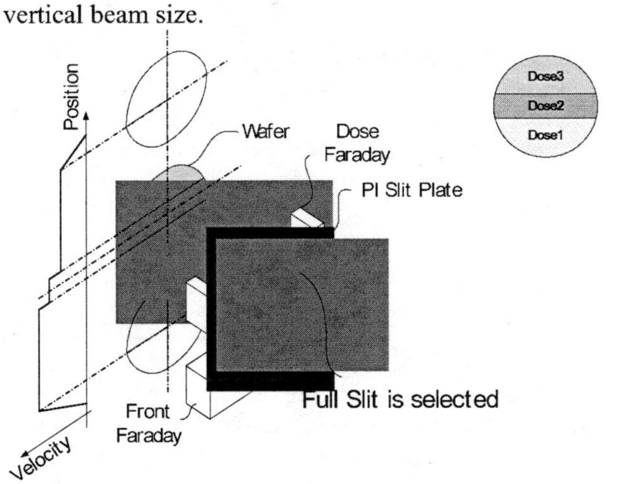

FIGURE 1. Schematic diagram of the Divide Partial Implant.

The PI Slit is 11 mm in height and 380 mm in width. The PI Slit is located on a PI Slit Plate which made of carbon. The PI Slit plate is located at upstream side of the Front Faraday. The PI Slit plate is controlled individually from the Front Faraday. When the Divide PI was carried out, the PI slit plate was moved to select the PI Slit. Then the wafer scan speed was controlled to produce the two or three different dosage area. Each divided areas by the 2 or 3-division Divide PI are numbered from lower area of stood wafer.

Stepwise Rotation Implant of Single, Bi and Quad Mode are adopted to the Divide PI. The Bi and Quad Mode are used when ion should be implanted from 2 or 4 directions to devices on the wafer.

Ring Partial Implant

Figure 2 shows a schematic diagram and implant procedure of the Ring Partial Implant. The Ring PI forms two or three concentric dose areas by controlling both the vertical mechanical scan speed and the stepwise rotation of wafer twist axis. The Ring PI compensates for the difference of the electrical characteristics of device on the wafer by controlling the dose rate concentrically on the wafer surface. The Ring PI that forms two concentric dose areas is "Single Ring Partial Implant (Single Ring PI)". A central circle and one outer ring dose areas are formed. The Ring PI that forms three concentric dose areas is "Double Ring Partial Implant (Double Ring PI)". A central circle and two outer ring dose areas are formed. Each concentric dose areas and these radiuses are numbered from inner area.

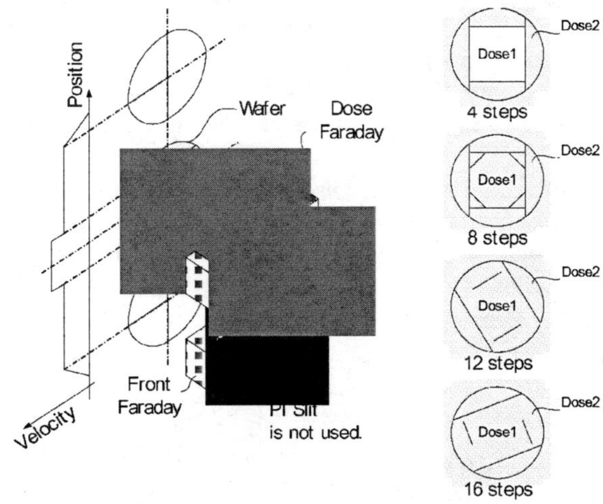

FIGURE 2. Schematic diagram of the Ring Partial Implant.

In case of usual Ring PI, the PI slit is not used. Because the required gradual dose was assumed not as rapid as needed the PI Slit. However, when the Ring PI is required to narrow ion beam, the PI Slit can used.

Number of the Stepwise Rotation for the Ring PI can be selected from 4, 8, 12 and 16 steps. The dose area form is more circular by increasing stepwise of rotation. However, throughput wafer out process is decreased with increasing stepwise of rotation. The number of steps for the Ring PI should be selected to keep the yield of the device each wafer or to keep the throughput of wafer out.

IMPLANT RESULT

Divide Partial Implant

The 3-division Divide PI by $^{11}B^{+}$ has been carried out with the condition or following table. Two reference implantations by the normal implant system with the conditions that 3.0E13/cm^2 and 5.0E13/cm^2 of dose are also carried out to compare these results.

Sheet resistance map (anneal condition: 900°C for 30sec) is shown in Figure 3.

TABLE 1. Condition of 3-division Partial Implant.

Energy	Beam Current	Tilt Angle	Twist Angle
100 keV	200 uA	7 degrees	23 degrees
Dose1	**Dose2**	**Dose3**	**Stepwise Rotation**
3.0E13/cm^2	5.0E13/cm^2	3.0E13/cm^2	Single

Sheet resistance that is measured along the direction which makes 23 degrees with vertical line is shown in Figure 4. Two solid lines show the reference sheet resistance value. The plotted value by open circle is result of 3-division Divide Partial Implant. The sheet resistance at each divided dose area with one Divide PI has been equivalent to two reference value of sheet resistance at each area. Moreover the difference of the sheet resistance to the reference value is less than 0.4% at each dose area. This graph shows that ion dose is exactly implanted at each area by 3-division Divide PI and that the width of the transition area is enough lower about 15 mm width.

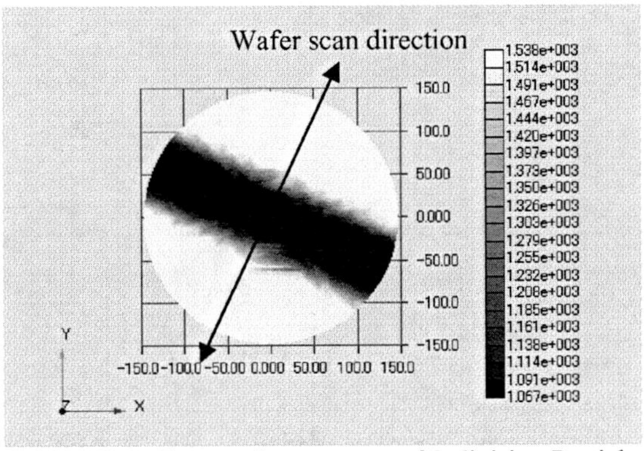

FIGURE 3. Sheet resistance map of 3-division Partial Implant.

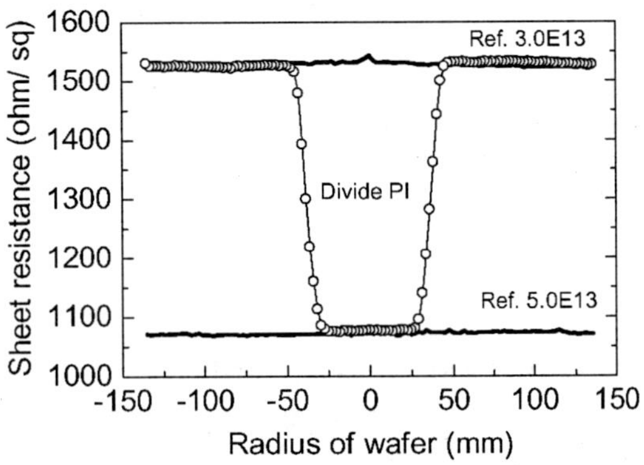

FIGURE 4. Sheet resistance variation along 23 degree direction.

Ring Partial Implant

The Single Ring Partial Implant by $^{11}B^{+}$ has been carried out with parameters shown in the following table. Two reference implantations by the normal implant with the conditions that 3.0E13 /cm^2 and 2.6 E 13 /cm^2 of dose are also carried out to compare these results. This Single Ring PI has been carried without the PI Slit.

TABLE 2. Condition of Single Ring Partial Implant.

Energy	Beam Current	Tilt Angle	Twist Angle
12 keV	200 uA	7 degrees	10 degrees
Dose1	**Dose2**	**Radius of central area**	**Stepwise Rotation**
3.0E13/cm^2	2.6E13/cm^2	120 mm	16 steps

Sheet resistance map (anneal condition: 900 degrees of temperature for 30sec) is shown in Figure 5. Center areas are 3.0E13 /cm^2 of Dose1. Gray area is transition area to 2.6E13 /cm^2 of Dose2. This map shows that 2 concentric dose areas have been able to be formed by the Single Ring PI. Sheet resistance that is measured along 0 degree direction is shown in Figure 6. Two solid lines show the reference sheet resistance value in this graph. The plotted value by open circle is result of Single Ring Partial Implant. This graph shows that ion dose is exactly implanted at each area by Single Ring Partial Implant.

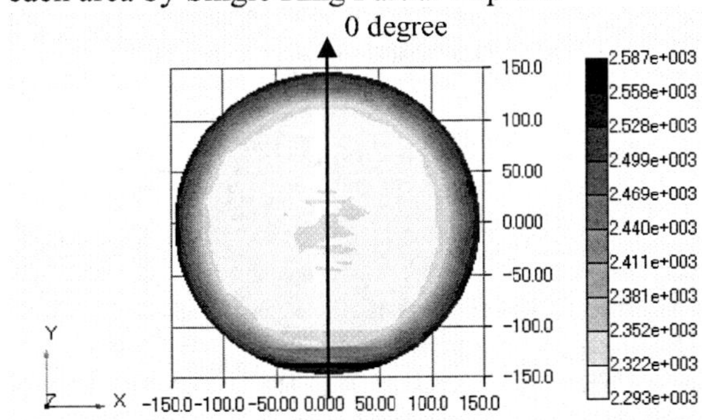

FIGURE 5. Sheet resistance map of Ring Partial Implant.

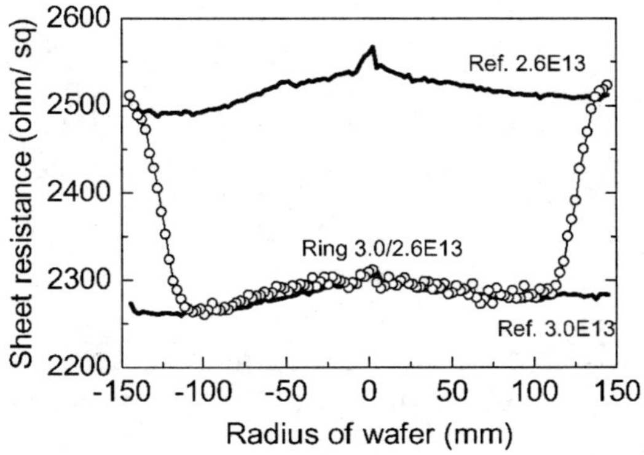

FIGURE 6. Sheet resistance variation along 0 degree direction.

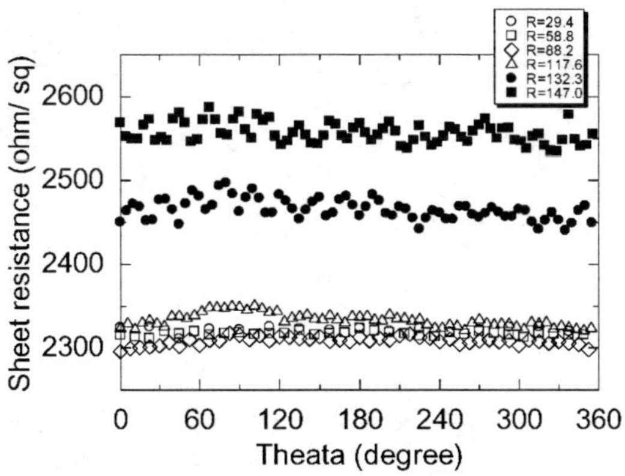

FIGURE 7. Sheet resistance variation along circumference direction.

Figure 7 shows sheet resistance value that is plotted along circumference direction at the 29. 4mm, 58.8 mm, 88.2 mm, 117.6 mm, 132.3 mm and 147.0 mm of radiuses. The plotted value by open symbols is sheet resistance at the Dose1 area. The plotted value by close symbols is sheet resistance at the Dose 2 area.

TABLE 3. Ion dose uniformity along circumference direction.

Radius (mm)	Average (ohm/sq)	Standard Deviation (ohm/sq)	Uniformity (%)
29.4	2321.89	3.49	0.15
58.8	2316.88	2.47	0.11
88.2	2307.93	4.75	0.21
117.6	2333.06	8.14	0.35
132.3	2465.27	12.23	0.50
147.0	2556.94	12.07	0.47

Ion dose uniformities along circumference direction at each radius that are calculated from raw sheet resistance data are shown in Table 3. Standard deviation of the sheet resistance at the Dose2 area (radius: 132.3 mm and 147.0 mm) is larger than the Dose1 area (radius: 88.2 mm and less). Ion dose non-uniformity at the outer ring dose area tends to increase. However, these values are acceptable low.

CONCLUSION

The Partial Implant System has been developed with purpose of being a productivity improvement tool.

The obtained sheet resistance data of the Divide PI shows that the ion dose area on the wafer surface can be divided evenly into two or three difference ion dose areas by the Divide PI. The ions can be implanted precisely into each divided dose areas. The transition area between different dose areas can be reduced with using the PI Slit. The Divide Partial Implant can contribute to reduce consumption of wafers.

The obtained sheet resistance data of the Ring PI shows that the concentric ion dose area can be formed by the Ring PI. The ions can be implanted with precise dose at both central and outer dose areas. Precision of the ion dose uniformity along the circumference at outer ring dose area is satisfactory when the Ring PI with 16 times of stepwise rotation implant is carried out. The Ring Partial Implant can contribute to the correcting the electrical characteristic of devices on the wafer.

ACKNOWLEDGEMENTS

The authors would like to thank Y.Shinomiya for technical assistance. They would also like to thank their colleagues of the system development group for their valuable discussions and comments.

REFERENCES

1. S. Sakai, *et al.*, in *Proceedings of the 16th International Conference on Ion Implantation Technology 2006*, pp.605-608

Increase of Boron Ion Beam Current Extracted from a Multi-Cusp Ion Source in an Ion Doping System with Mass Separation

Yutaka Inouchi, Shojiro Dohi, Masahiro Tanii, Junichi Tatemichi, Masashi Konishi, Masaaki Nukayama, Kazuhiro Nakao, Koichi Orihira, and Masao Naito

Nissin Ion Equipment Co., Ltd., 29, Hinokigaoka, Minakuchi-cho, Koka, Shiga, 528-0068 Japan

Abstract. A multi-cusp ion source with large area extraction area has a capacity to extract high current ion beams. However it is difficult to obtain high current boron ion beams from boron trifluoride plasma as input power density is low and BF_2^+ is a main ion species. It was found that fluoride ion currents were selectively decreased by applying positive potential to a plasma electrode with respect to a plasma chamber. Using this method to a multi-cusp ion source with adequate magnet configuration, high current boron ion beams of 500 μA/cm were obtained.

Keywords: Ion Source, Multi-cusp, Ion implanter, Ion doping, Boron trifluoride, Langmuir probe, Flat-panel-display, Ribbon beam, long pole gap
PACS: 61. 72. Tt; 68. 55. Ln; 85. 40. Ry

INTRODUCTION

Recently, for advanced applications of Low Temperature Poly-crystalline Silicon Thin Film Transistor (LTPS-TFT) such as system-on-glass devices and OLED displays, precise and stable doping processes have been required. To satisfy their requirements, implanter type with mass analysis has replaced shower type as doping equipments. Recently, implanter types are used for S/D processes as well as channel and LDD processes.

Nissin developed an ion doping system with mass separation "ion doping iG4" referred as iG4 for such various processes [1-2]. The original system could not give so high performance for P type S/D doping processes because of poor current of B^+. B^+ ions were produced in BF_3^+ plasma. B^+ fraction is usually small, while the fraction of BF_2^+ is large. Especially, a multi-cusp ion source used in iG4 could not be expected to obtain large B^+ fraction, as input power density was not so high in spite of fine plasma confinement. Thus, to obtain high B^+ ion currents, wastefully large extraction currents were necessary. In this paper, a new method to obtain large fraction B^+ ion currents is proposed, which increases B^+ beam currents in small extraction currents on a multi-cusp ion source with large extraction area.

FEATURE OF ION DOPING iG4

Beam line of iG4 has a simple concept of keeping wide ribbon beam shape from an ion source to a target. This concept allows a high mass resolution and an easy control of a beam profile at a target by ion source parameters. Figure 1 shows the schematic diagram of iG4 beam line which consists of a mass-analyzing magnet and an analyzing slit.

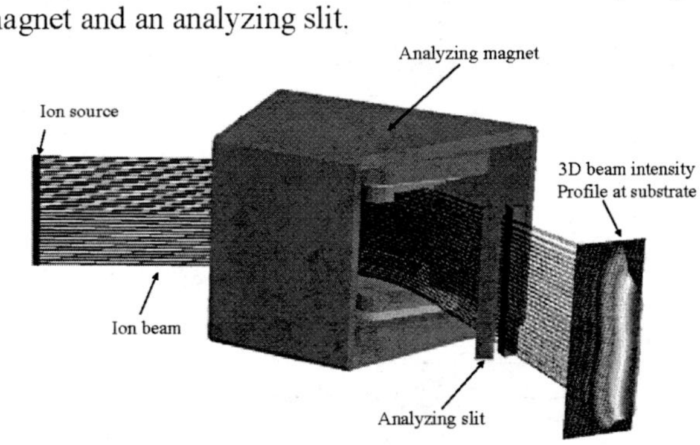

FIGURE 1. Schematic diagram of Ion Doping iG4 beam line.

This figure shows that the ribbon beams are bended and transported horizontally through a long pole gap in the analyzing magnet keeping almost the same shape. The bending angle of 70 degrees and an adjustable

CP1066, *Ion Implantation Technology,* edited by E. G. Seebauer, S. B. Felch, A. Jain, and Y. V. Kondratenko

analyzing slit width can control a mass resolution of M/ΔM from 4 to 10.

To extract wide ribbon beams, a multi-cusp ion source which has a large internal volume of 95cm x 20 cm x 20 cm is used in iG4. Figure 2 shows the schematic diagram of the ion source.

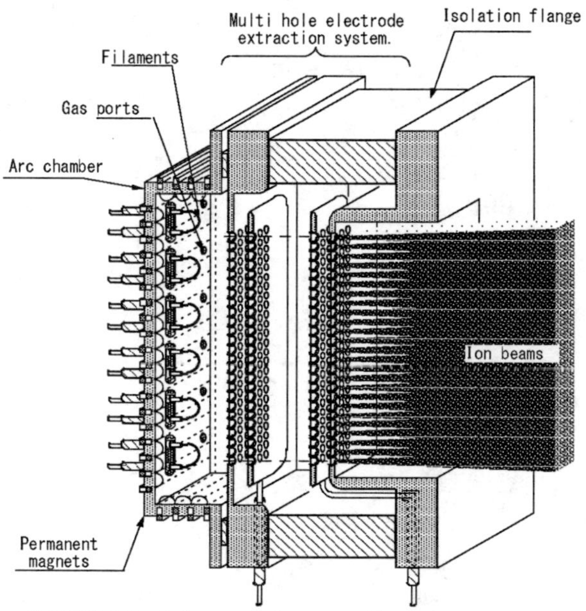

FIGURE 2. Schematic diagram of ion source.

An arc discharge with six filaments produces a plasma, and ions are extracted and accelerated in a electrode system. We use multi-hole or multi-slit type electrodes to extract wide ribbon beams which size is 80cm x 3 to 5cm. Divergence of ion beams extracted from each hole or slit is small enough to control a target beam profile by controlling a plasma density profile with the use of the six filaments. Target beam profiles are measured by multiple faraday cups placed at the back of the analyzing slit.

BORON FRACTION IMPROVEMENT

On multi-cusp ion sources, potential of a plasma electrode is usually cathodic or floating to keep small plasma loss area as it is difficult to confine plasma magnetically on its surface. If potential of the plasma electrode is positive with respect to an anode plasma chamber, total extraction currents may be poor. However, we were interested in a fact as follows;

A. Total extraction ion currents decreased sharply as the plasma electrode potential increased.
B. But B^+ ion currents did not so decrease.

This fact implies that positive potential of the plasma electrode selectively decreased ion currents other than B^+, that is, F^+, BF^+, BF_2^+ ion currents. Thus B^+ ion fraction in extraction currents must be improved. To maintain B^+ ion currents as high as possible with the positive potential plasma electrode, we tested some magnet configurations of the plasma chamber and selected one which showed the highest performance of B^+ ion extraction.

To confirm and study the effect of positive potential applied to the plasma electrode, we measured target currents for each ion species at several positive potentials, and plasma parameters were also measured by a single probe. In these experiments, one of filament ports was used for the probe, thus only five filaments were used for a plasma production. Figure 3 shows the schematic diagram of the experimental set-up with a plasma electrode power supply and a single probe. In the experiments, potential of the plasma electrode was changed by the plasma electrode power supply and each potential was measured with respect to the chamber walls.

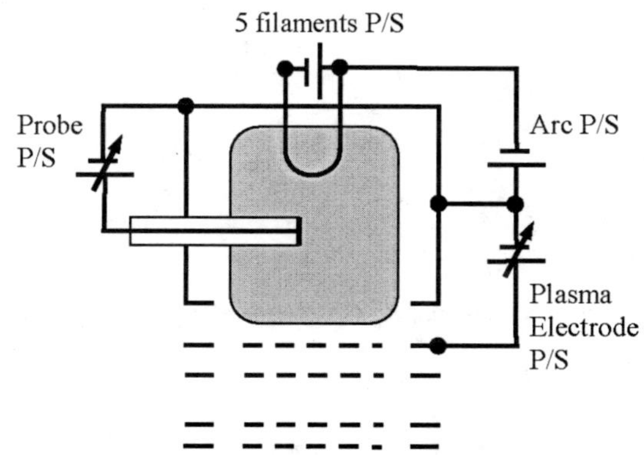

FIGURE 3. Schematic diagram of experimental set-up with a plasma electrode power supply and a single probe.

Ion Fraction Measurement

B^+ ion currents were adjusted to 300 μA/cm at the energy of 30 keV with several potentials of the plasma electrode, and mass spectra were measured. As the potential was increased, the extraction currents from the ion source decreased and F^+, BF^+, BF_2^+ ion currents also decreased. Figure 4 shows the dependence of these ion currents on the plasma electrode potential. Each current was normalized by B^+ current which was adjusted at 300 μA/cm, and arc currents and plasma electrode currents which were measured by a power supply were plotted in Figure 5.

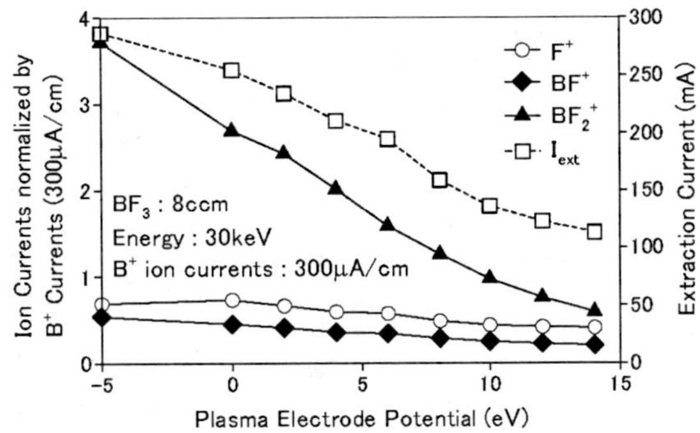

FIGURE 4. Dependence of F^+, BF^+, BF_2^+ ion currents on potentials of the plasma electrode with respect to the plasma chamber. Extraction currents were also plotted.

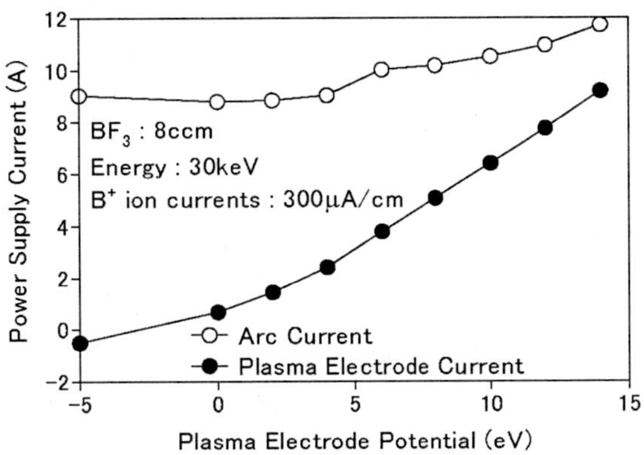

FIGURE 5. Dependence of arc and plasma electrode currents on potentials of the plasma electrode with respect to the plasma chamber.

Single Probe Measurement

Single probe measurements were carried out at several plasma electrode potentials. A plane probe made of a 5 mmϕ molybdenum rod which side face was covered by an aluminum tube was placed at the center of the plasma chamber. Langmuir probe characteristics are shown in Figure 6. There was an obvious difference between positive and negative potentials for their ion saturation currents.

Several plasma parameters were evaluated by probe characteristics. Figure 7 shows the dependence of plasma potential on plasma electrode potentials. Figure 8 shows the electron density and the electron temperature. Electron density had a similar dependence on the plasma electrode potential with arc current in Figure 5.

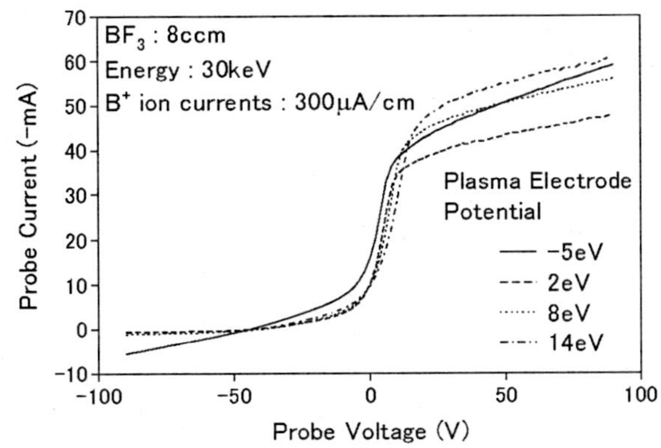

FIGURE 6. Langmuir probe characteristics at several plasma electrode potentials. Probe voltage was changed with respect to the plasma chamber potential.

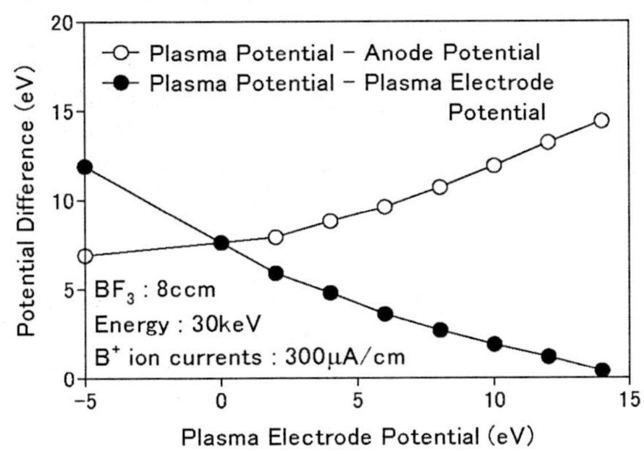

FIGURE 7. Dependence of plasma potential on plasma electrode potentials.

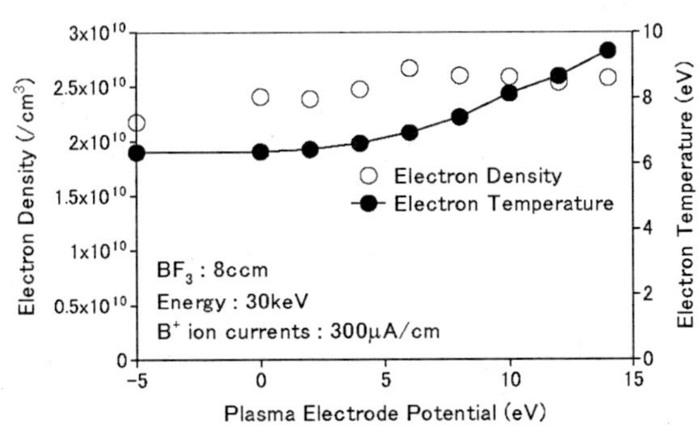

FIGURE 8. Dependence of electron density and electron temperature on plasma electrode potentials.

DISCUSSIONS

Several experiments were performed with a multi-cusp ion source which magnets were aligned adequately for B^+ ion extraction.

Our concern was what caused the selective decrease of F^+, BF^+, BF_2^+ ions except B^+ ions in the extraction currents.

Figure 7 shows that the plasma potential increased as the plasma electrode potential increased, while the difference from the plasma electrode decreased. At positive potential of the plasma electrode, the electrode was worked as anode as well as walls of plasma chamber. On the other hand, the large potential difference between plasma and the chamber walls produced the ion sheath close to the chamber walls. The ion loss at the chamber walls was affected by magnetic field. BF_2^+ ions which mass is about 5 times heavier than that of B^+ ions may escape to the walls easier across magnetic field. This expectation agrees with the result in Figure 4. As the plasma potential increased from -5 eV to 14 eV, BF_2^+, BF^+ and F^+ ion currents decreased to 1/6.3, 1/2.7 and 1/1.7 respectively.

It is expected that the recombination of F^+, BF^+, BF_2^+ ions by electrons or negative ions occurs more often than that of B^+ ions. But now, we do not have a definite opinion that the selective decrease has a close relation to the recombination.

Difference of sputtering yield between B^+ and BF_2^+ can explain the difference of probe currents in ion saturation region of probe characteristics in Figure 6.

Increase in Boron Currents

It was confirmed that the positive potentials improved the boron fraction in the extraction currents. We measured boron beam currents at several arc currents. In this experiment, the plasma electrode potential was fixed at 8 eV, and only 5 filaments were used as other experiments. Figure 9 shows the dependence of beam current density and extraction current on arc current. They were proportionate to arc current. Boron beam current of 500 μA/cm (above 40 mA) was obtained at arc current of 15 A and extraction current of 230 mA.

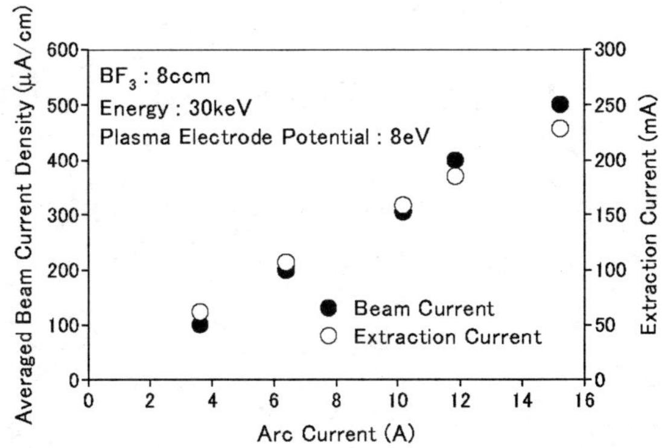

FIGURE 9. Dependence of averaged beam current density and extraction current on arc current. The plasma electrode potential was 8eV with respect to the plasma chamber.

CONCLUSIONS

In a multi-cusp ion source, positively applied potential to the plasma electrode with respect to the arc chamber decreased extraction currents. But we were interested in the fact that decrease of B^+ ion currents was much smaller than that of F^+, BF^+, BF_2^+ ion currents. Single probe measurements were carried out to study how positive potential had an influence on plasma parameters. Some discussions were made about the results. Positively applied potential adds the function as main ion loss area to the chamber walls. We think that ion loss rate at the chamber walls is affected by the magnetic field and it is dependent on its mass.

The positive potential of plasma electrode and the adequate magnets configuration of a multi-cusp ion source improved the fraction of B^+ ion currents in the total extraction currents. High current B^+ ion beams were obtained in proportion to input arc power at considerably small extraction current. B^+ beam current of 500 μA/cm (above 40 mA) was obtained at arc current of 15 A and extraction current of 230 mA.

REFERENCES

1. Y. Inouchi *et al.* IDW/AD'05 pp. 297-300.
2. S. Dohi *et al.*, Proceedings of the 16th International Conference on Ion Implantation Technology , edited by K. J. Kirkby *et al.*, al., AIP Conference Proceedings 866, American Institute of Physics, Melville, NY, 2006, pp. 417-420.

Beam Uniformity Controllable Ion Source with a Long Slit

T. Ikejiri, W. Zhao, K. Tanaka, T. Igo, T. Kinoyama, S. Tamura, and T. Yamashita

Nissin Ion Equipment Co., LTD
575 Kuze-Tonoshiro-Cho, Minami-Ku, Kyoto, 601-8205 Japan

Abstract. With scaling up of a substrate of LCD and silicon, an ion source with a longer extraction area than the substrate has been used for an ion implantation system. Wide sheet beam emitted from a long extraction area is effective to decrease space charge effect of a high current ion beam. An ion source which has this structure has been applied for FPD. This ion source controls the source plasma partially. This function is one of the methods to adjust the uniformity of the ion sheet beam. However, there are technical difficulties for this ion source. Therefore, we have developed three types of ion sources which have a longer extracting slit than the substrate and different direction of the source magnetic field. This paper shows the structures and the experimental result of these ion sources.

Keywords: ion source, uniformity control, sheet beam
PACS: 07.77.Ka

INTRODUCTION

With the increase of LCD and silicon substrate size, the width of the ion beam has been increased in the ion implantation system. It is necessary for keeping the productivity with growing up the substrate size to increase beam current. However increase of beam current causes the ion beam to spread by the build up space charge effect and the beam transportation to be difficult.

To solve this problem, the ion source that has longer extracting area than the substrate size is proposed and has already been applied to an ion implantation system for LCD [1]. The other feature of this ion source is partial control of the plasma in the arc chamber that is one of the methods to adjust the uniformity of the ion sheet beam.

However, it is technically difficult for this ion source to generate high current and fine uniform ion beam.

Therefore, we have developed three types of ion sources which have different direction of source magnetic field and an extracting slit longer than the substrate.

EXPERIMENT

All of the developed ion sources have a long slit in vertical direction. The long ion beam is extracted from the long slit perpendicular to the face of the slit and is bended by an analyzer magnet without focus or defocus in vertical direction of the beam.

In this paper, the ion beam traveling direction is defined as Z-direction, the direction of the long side of the slit is Y-direction and the rectangular direction for Z and Y, transverse direction, is X-direction. The coordinate system and measurement system are shown in Figure 1.

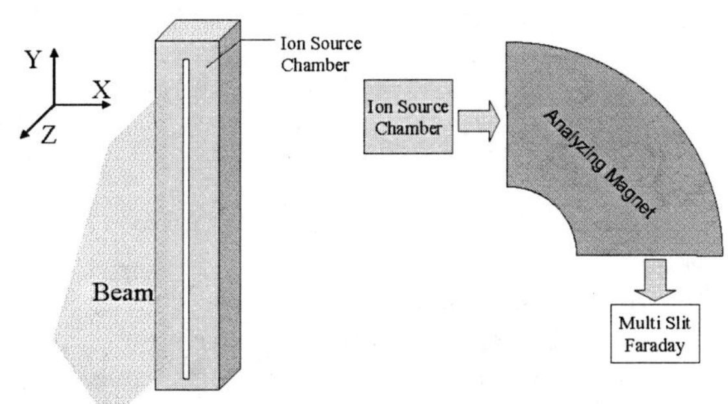

FIGURE 1. Coordinate system and measurement system

The extracted ion beam is bended 90 degree with the analyzer magnet and measured by a faraday with multi slits lined in Y-direction. The ion beam is estimated by the Y-direction current line density profile calculated using the measured data.

Three deferent ion sources with change of the direction of the magnetic field generated with a source magnet have been made and tested. The specifications of these ion sources are listed in TABLE 1.

CP1066, *Ion Implantation Technology,* edited by E. G. Seebauer, S. B. Felch, A. Jain, and Y. V. Kondratenko

TABLE 1. Tested ion source types

Source magnet field direction	Cathode type	Beam uniformity control method	Ion source chamber size
X	Filament 7pc	7 filaments current control	(approx.) W70mm D70mm H450mm
Y	IHC* 2pc	4 source gas lines controlled independently	
Z	Filament 7pc	7 filaments current control	Slit length 400mm

* IHC (Indirectly heated cathode)

The details and experimental results of these ion sources are described below.

X-directional Magnetic Field Ion Source

The ion source with X-directional magnetic field is shown in this chapter. This ion source is able to control the plasma density locally for Y-direction by limiting the arc discharge in X-direction.

Structure

Figure 2 shows the over view of this ion source and Figure 3 shows the top cross section view. The seven filaments are aligned in Y-direction on both X-sides alternately. This arc chamber is divided to seven parts by six plates that are between the filaments. The reason why these plates are set up is that the previous experiment without these dividing plates showed very low controllability of the beam current density.

To control the current of these filaments controls the plasma density in the ion source arc chamber locally and controls the beam current line density.

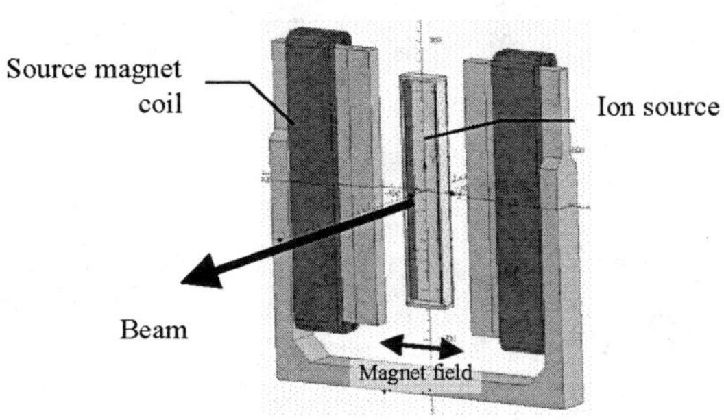

FIGURE 2. Overview of X-directional magnetic field ion source

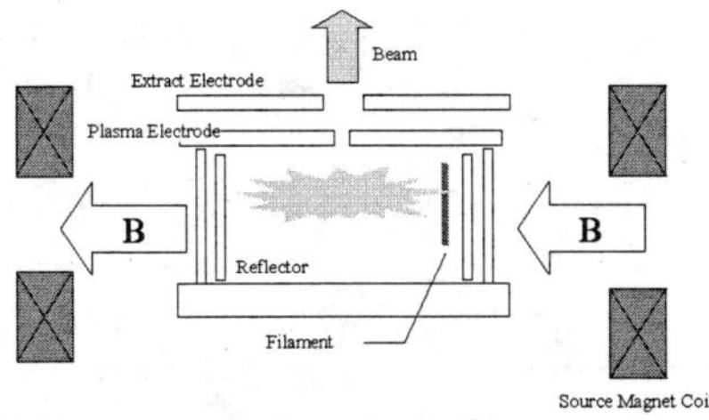

FIGURE 3. Top cross section view of ion source

Experimental Result

Figure 4 shows Y-direction beam current line density profile when the one filament of upper (2nd filament), center (4th filament) and lower (6th filament) filament each was drove. This proves that the beam current density at the filament driving point was intensified locally.

Figure 5 shows the line density profile of the uniform beam controlled by each filament current. For Ar^+ 15 keV beam, the beam current 5.55 mA and the uniformity $\sigma/X=1.81\%$ was obtained.

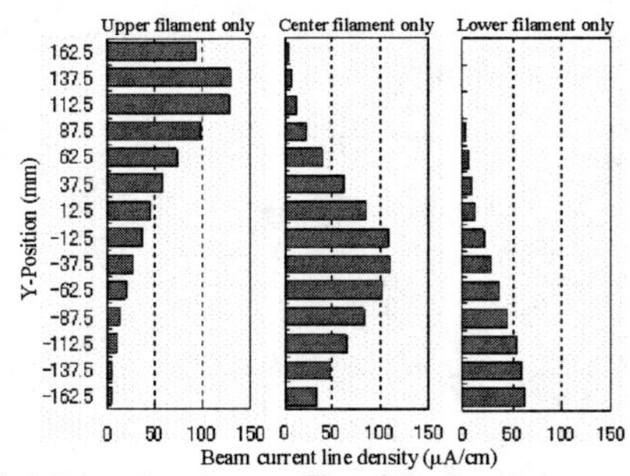

FIGURE 4. Line density profile of the beam by one filament (Ar^+ 15 keV/2mA (approx.))

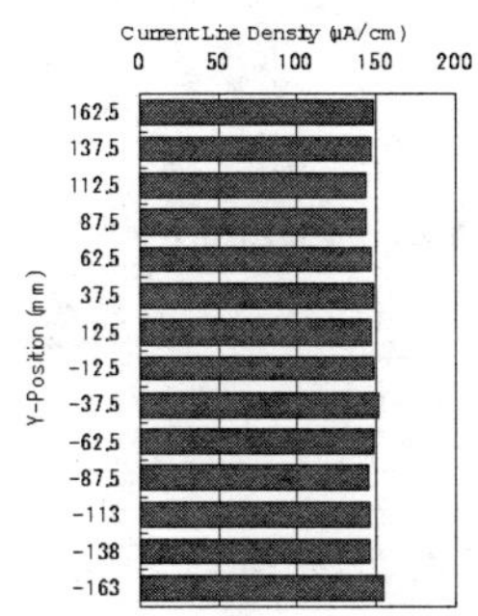

FIGURE 5. Line density profile of the uniformity controlled beam. Ar^+ 15keV 5.55mA $\sigma/X=1.81\%$

It was difficult to obtain high current ion beam from this ion source. This reason is likely that the beam extracted area was only a part of the arc discharge area which spreads in X-Y plane.

In addition, the extracted ion beam was bended 1 to 2 degrees to Y-direction by the X-direction source magnetic field. Therefore, it is necessary for getting parallel beam at the target to adjust the extraction angle or re-bending in the beam line after the extraction.

Y-directional Magnet Field Ion Source

In the first experiment, we tested multi cathodes to control the plasma density partially in the ion source under Y-directional source magnetic field. But the result was not effective. It seems likely that the effect of the plasma density adjusted locally by the cathode was transmitted in the arc chamber in a moment. Consequently we designed the ion source with four source gas lines controlled independently and attempted to adjust the local gas pressure in the arc chamber to control the beam uniformity.

Structure

The over view of the Y-directional source magnetic field ion source and the cross section view are shown in Figure 6. The window flame type magnet with two coils apply Y-directional magnetic field to the ion source. The ion source is equipped with two IHC (Indirectly heated cathode) one above the other and four source gas lines that are able to be controlled independently.

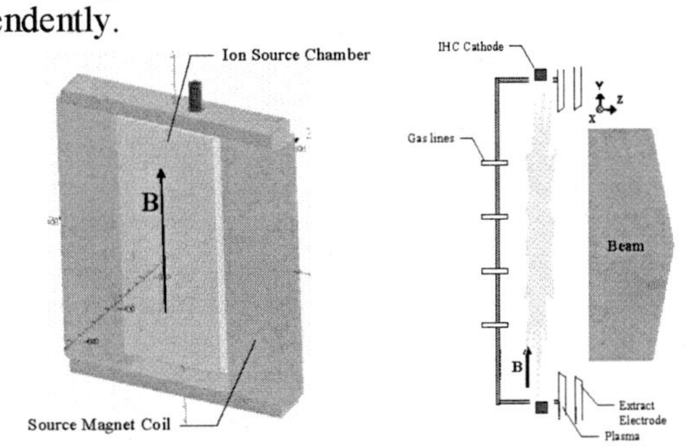

FIGURE 6. Overview and side cross section view of Y-directional magnetic field ion source

Experimental Result

The variation of the beam current density when 5 ccm of BF_3 gas was injected from one line each is shown in Figure 7. Upper side line is “line 1” and lower side line is “line 4”.

It shows that the density near the gas injected position increased.

Figure 8 shows the current density profile uniformed by the gas control. The beam current 8.15 mA and uniformity σ /X=1.49% at B^+ 20keV was obtained.

The effect of the control by the gas flow was mild compared with the other types of the source magnetic field ion sources, and the sharp ununiformity was not generated.

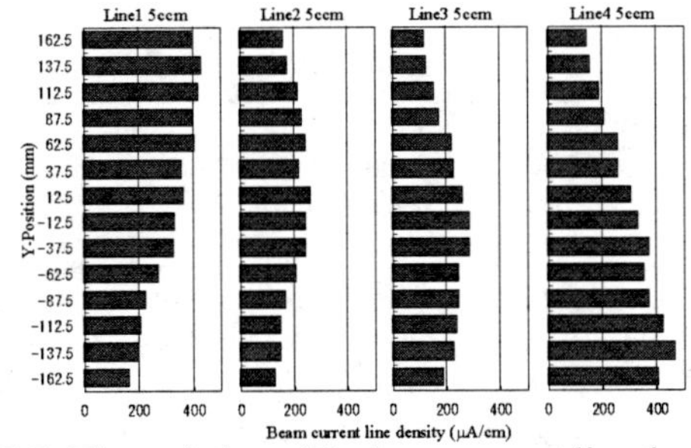

FIGURE 7. The variation of the beam current line density when BF_3 gas injected from one line each (B^+ 25keV 6~10mA)

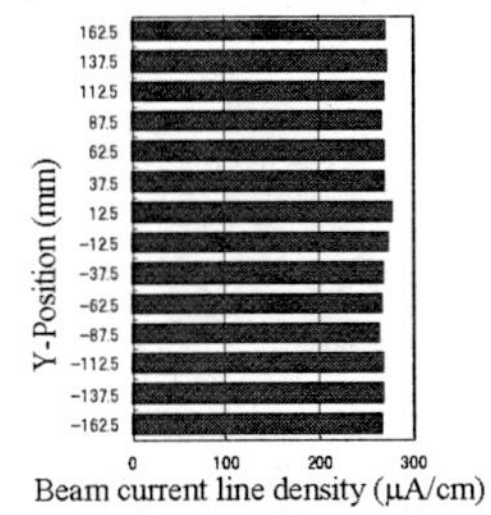

FIGURE 8. The line density profile of the uniformity controlled beam . (B^+ 20keV 8.15mA σ/X=1.49%)

On the other hand, the extracted ion beam was bended 1 to 2 degrees to X-direction by the Y-directional source magnetic field. This phenomenon was corrected by dividing the extract electrode into right and left and applying difference potential of 1 kV to them.

Z-directional Magnetic Field Ion Source

Finally, an ion source with Z-directional magnetic field is given. In this ion source, similarly as the X-directional ion source, the intensity of the plasma can be controlled partially by the arc discharge limited in Z-direction. And the beam is not bended by the source magnet because the direction of the extracted beam is identical with the direction of the magnetic field of the source magnet.

Structure

The overview and the top cross section view of the Z-directional magnetic field ion source are shown in Figure 8. A magnetic field is generated on Z-direction by two air core coils. Seven filaments are placed on the opposite face from the plasma electrode, and an anode electrode is set up between the filaments and the suppression electrode. One of the suppression electrodes is combined with the plasma electrode to extract the ion beam.

Generally, in ion sources with a plasma electrode combined with the reflector electrode, the sheath voltage between the reflector electrode and the plasma is generated comparatively large, as the result of this the ions are accelerated previously in the ion source and extracted efficiently. This is effective to obtain high current density ion beam.

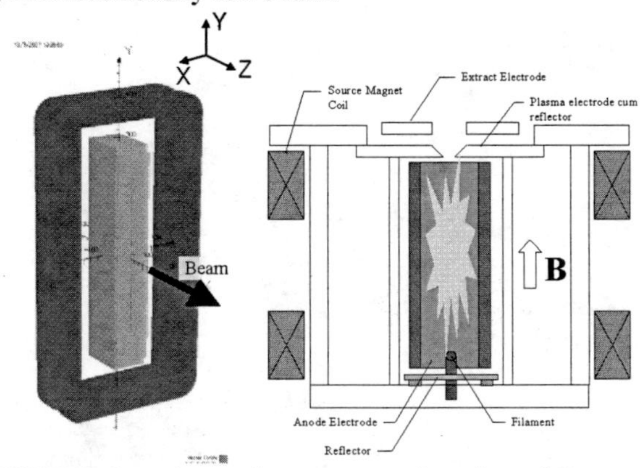

FIGURE 8. Overview and top cross section view of Z-directional magnetic field ion source.

Experimental Result

Figure 9 shows beam profiles when one filament of the seven filaments each is driven. The filament placed the top of the chamber is called 1st filament and the bottom one is called 7th filament. As shown in Figure 9, the intensity of the beam can be controlled partially by each filament.

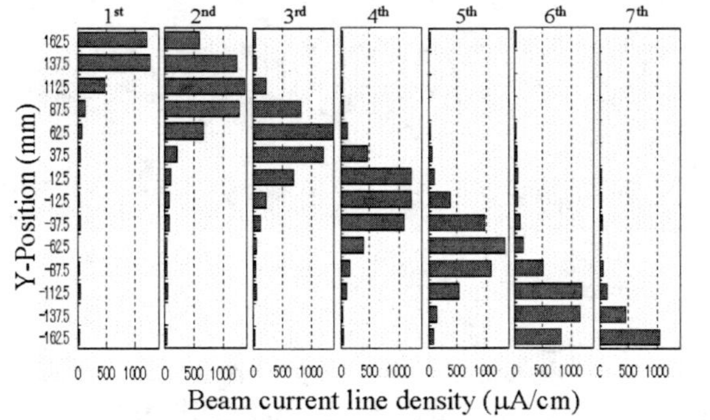

FIGURE 9. Line density profile of the beam by one filament (Ar+ 20keV)

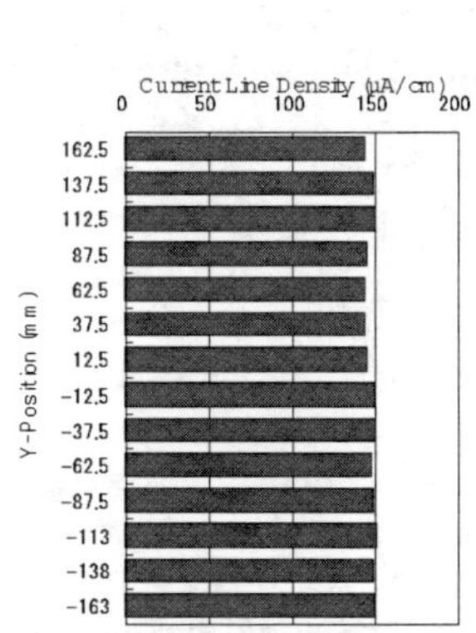

FIGURE 10. Uniformity controlled profile (B^+25keV 4.45mA σ/X=1.8%)

Using this ion source, B^+ beam have been obtained with the energy 25 keV, the beam current 4.45 mA and the uniformity 1.8 % STD.

This ion source can control the partially current line density of the ion beam dynamically, but sometimes the total beam uniformity is decreased caused by the strong localization of the beam density. Therefore this localization should be smoothed by optimization of the filament position or increase of the filament number or other.

CONCLUSIONS

We have developed three types of ions sources with a long slit that are able to control the uniformity of the ion beam and obtained the result shown in TABLE 2.

TABLE 2 Summary of the result

Source magnet field direction	X	Y	Z
Cathode	7 pc filament	2pc IHC	7pc filament
Gas Control	One line	4 line independent control	One line
Uniformity controllability (achieved uniformity)	Generate strong localization (1.81%STD)	mild response (1.49%STD)	Generate strong localization (1.80%STD)
Max Beam current	Low ~8mA(Ar^+)	High ~25mA(B^+)	Mid ~15mA(P^+)
Lifetime and maintenance	Not good many filaments	Better IHC, Gas control	Not good many filaments
Beam bend caused by source magnetic field	Y-direction Difficult to correct	X-direction Possible to correct by extract electric field	Good Not bend

This table suggests that Y type source has high possibility to obtain the fine uniformity of high current ion beam. X and Z type ion source obtained under 2% STD of uniformity, but points where beam density was locally strong was sometimes generated. The point like this was not generated for Y type ion source.

ACKNOWLEDGEMENTS

We would like to thank the members of the Tokin Machinery Corporation for their constant suggestions and support.

REFERENCES

1. Shojiro DOHI *et al.* "Increase of Beam Current Mass-Separated by Long Gap Dipole Sector Magnet for S/D Process in FPD Manufacturing" in *Proceedings of the 16th International Conference on Ion Implantation Technology 2006*, Marseille, France, 11-16 June 2006

Next Generation Angle Control for Medium Current and High Energy Implanters

Joseph C. Olson*, Atul Gupta, Hans-Joachim L. Gossmann, Dennis Rodier

Varian Semiconductor Equipment Associates, Inc., 35 Dory Road, Gloucester, MA 01930, USA
**joseph.olson@vsea.com*

Abstract. Angle control continues to increase in importance with device scaling. For instance, threshold voltage and on-state current in advanced logic devices depend critically on the angle accuracy of medium current halo implants. TCAD simulation results, showing that on-state current changes of greater than 1% per degree of implant angle are possible, are presented. We report here on improvements to angle performance and control in both the horizontal and vertical directions. Beam data covering the full operating space of the VIISta 900XP is presented, demonstrating total angle control of better than ±0.2°. The data set is chosen to emphasize performance of typical halo implants. Single wafer high energy angle control data is also presented.

Keywords: ion implantation, implant angle control
PACS: 85.40.Ry

INTRODUCTION

Control of implant angle is a central concern in all segments of the ion implant market. Source drain extension implants in the high current segment, halo implants in the medium current segment, and well implants in the high energy segment [1] all depend on precise and accurate implant angle control to achieve optimal results. In this paper we report on progress in three areas of ion beam angle control, specifically: (1) TCAD simulation of advanced device geometries showing the sensitivity of device parameters to ion beam angle; (2) performance of a new vertical angle detector; and (3) measured beam angles on the VIISta 900XP medium current and single wafer high energy ion implanters.

SIMULATION RESULTS

TCAD was used to simulate 32 nm half pitch high performance logic transistors, in order to evaluate the absolute and relative effects of various types of process errors on the resulting electrical parameters of devices. The simulation and results are described more completely elsewhere [2]. As a summary, Figure 1 shows selected process variables and their impact on the on-state current I_{dsat} of the nFET. The simulation predicts that, for a halo implant, an error of 1° in mean beam angle will result in a change in I_{dsat} of 1%, the most significant consequence of medium current process errors.

VERTICAL ANGLE DETECTOR PERFORMANCE

Varian scanned beam implanters have both measured and optimized horizontal beam angles for some time, utilizing techniques that have been described previously [3,4]. Horizontal, in this case, refers to angles along the scanned direction of the beam. As devices shrink and angle requirements tighten, a more complete picture of the beam is required. To that end, a vertical (perpendicular to the direction of beam scanning, parallel to the direction of mechanical wafer scanning) angle detector has been developed to provide this information.

In this paper we report on the performance of the new detector, rather than describe its design or operation. To characterize this detector, we placed it in a VIISta 900XP medium current implanter and oriented the detector to measure horizontal angles. For a set of test beams, the beamline controls were used to alter the beam angles and in each case the horizontal beam angles were measured by both the new detector and the implanter's existing system. Figure 2 shows a comparison of the measured beam angles, confirming that the new vertical angle detector provides an accurate measurement of beam direction. Beams

CP1066, *Ion Implantation Technology*, edited by E. G. Seebauer, S. B. Felch, A. Jain, and Y. V. Kondratenko

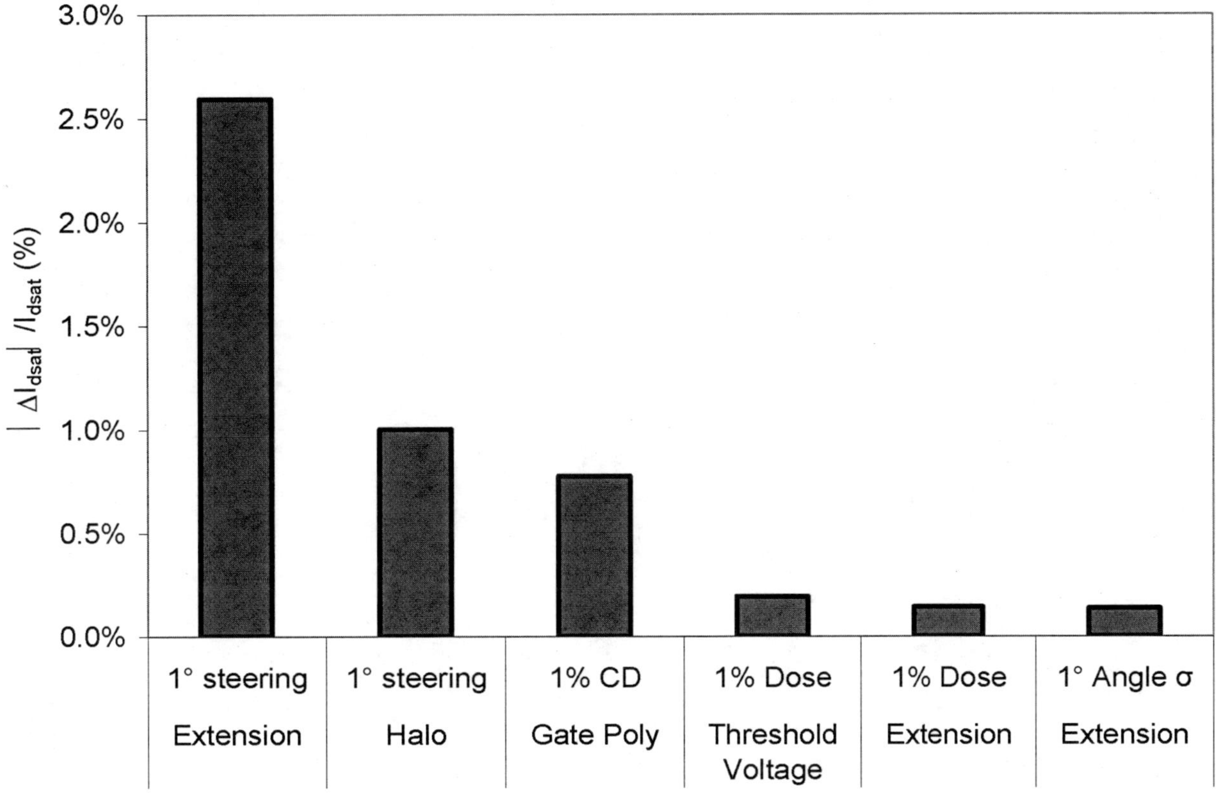

FIGURE 1. Impact of selected process variables on on-state current.

studied during this test consisted of 200 keV phosphorus, and boron at 5 keV, 35 keV, and 540 keV.

Offset between the two detectors was removed by a calibration procedure. Once the calibration procedure was completed, the detectors tracked each other to within better than 0.02° on average over the tested range of approximately ±5°. In production operation, any measured vertical beam angle would be accounted for by tilting the wafer.

VIISTA 900XP ANGLE PERFORMANCE

The horizontal angle characteristics of the VIISta 900XP were evaluated over its entire operating space. Detailed descriptions of the VIISta 810 medium current scanned beam ion implanter have been published elsewhere [5,6]. The VIISta 900XP is an evolution of the basic 810 design.

In Figure 3, the average within wafer horizontal angle variation of mean beam direction for beams covering the entire operating range of the tool is plotted. The data set consists of over 1600 setups collected during a four month period. Repeat setups were averaged and plotted as a single point. For all energies, species and currents tested, the average within wafer angle variation was less than 0.06°.

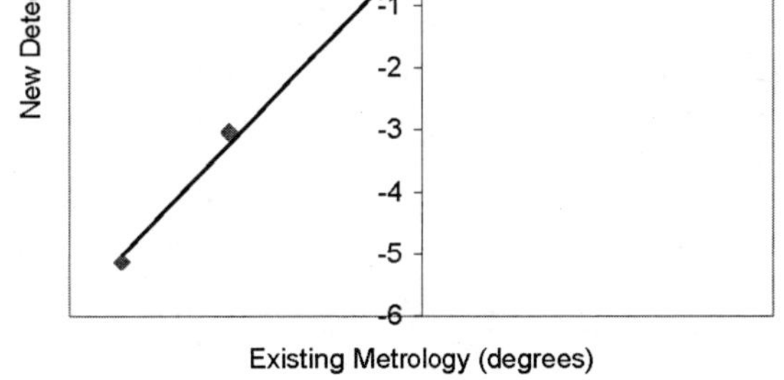

FIGURE 2. Comparison of beam angle measurements made by the existing VIISta 810 angle metrology and the new VIISta 900XP vertical angle detector. Both detectors were oriented to measure horizontal angles in this test.

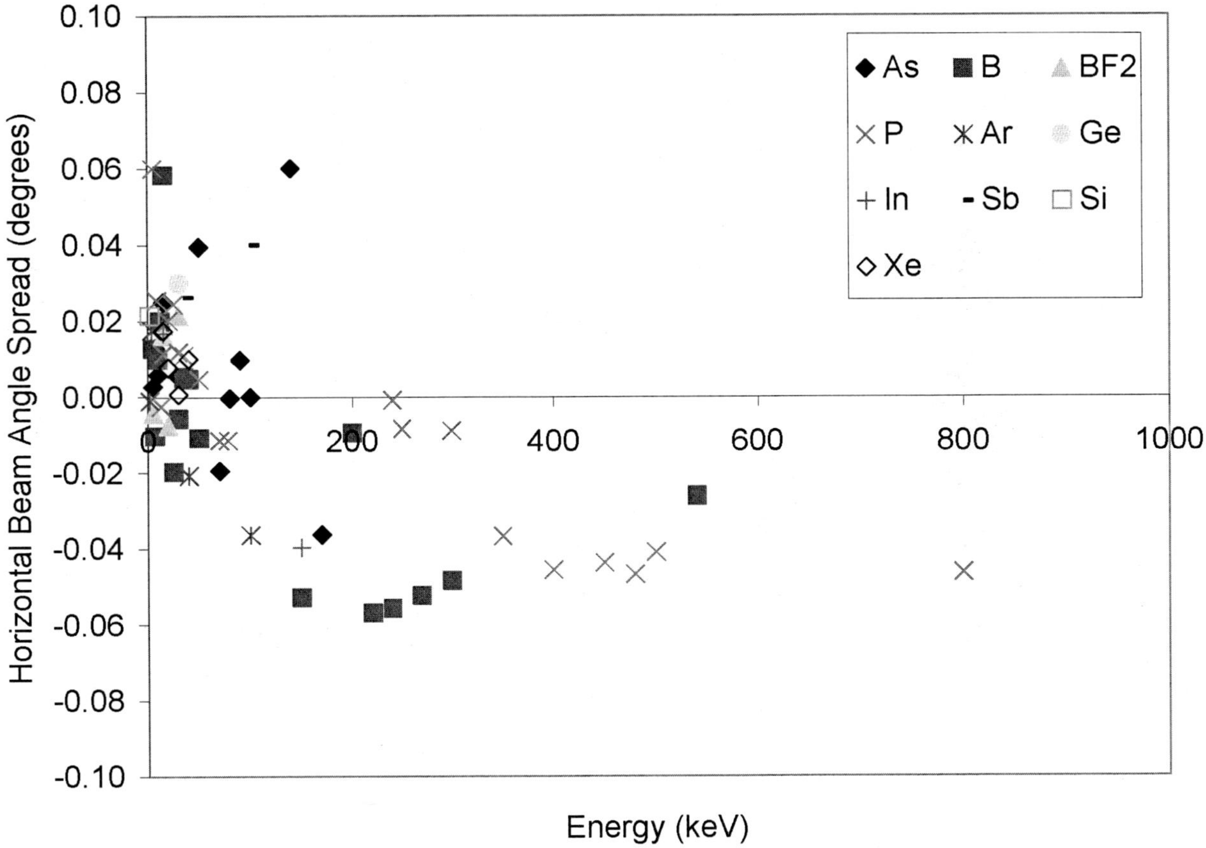

FIGURE 3. Measured average across wafer horizontal angle variation for 1600 setups over a four month period on a VIISta 900XP medium current implanter.

SINGLE WAFER HIGH ENERGY ANGLE PERFORMANCE

The horizontal angle characteristics of a scanned beam single wafer high energy implanter were also evaluated over its entire operating space. In Figure 4, the average within wafer horizontal angle variation of mean beam direction for beams covering the entire operating range of the tool is plotted. The data set consists of over 2500 setups collected during a four month period. Repeat setups were averaged and plotted as a single point. Again, for all energies, species and currents tested, the within wafer angle variation was less than 0.06°.

CONCLUSIONS

Real time measurement and correction of beam angles enables the scanned beam VIISta tools described here to meet the most stringent requirements for angle control in advanced applications. The measurement and control mechanisms for angles are effective over the entire range of beam species and energies, demonstrating horizontal angle variation across wafer of less than 0.1°. A new vertical angle detector has been designed to extend that performance.

ACKNOWLEDGEMENTS

The authors gratefully acknowledge the contributions of Todd MacEachern and James Carroll in fielding and testing the vertical angle detector.

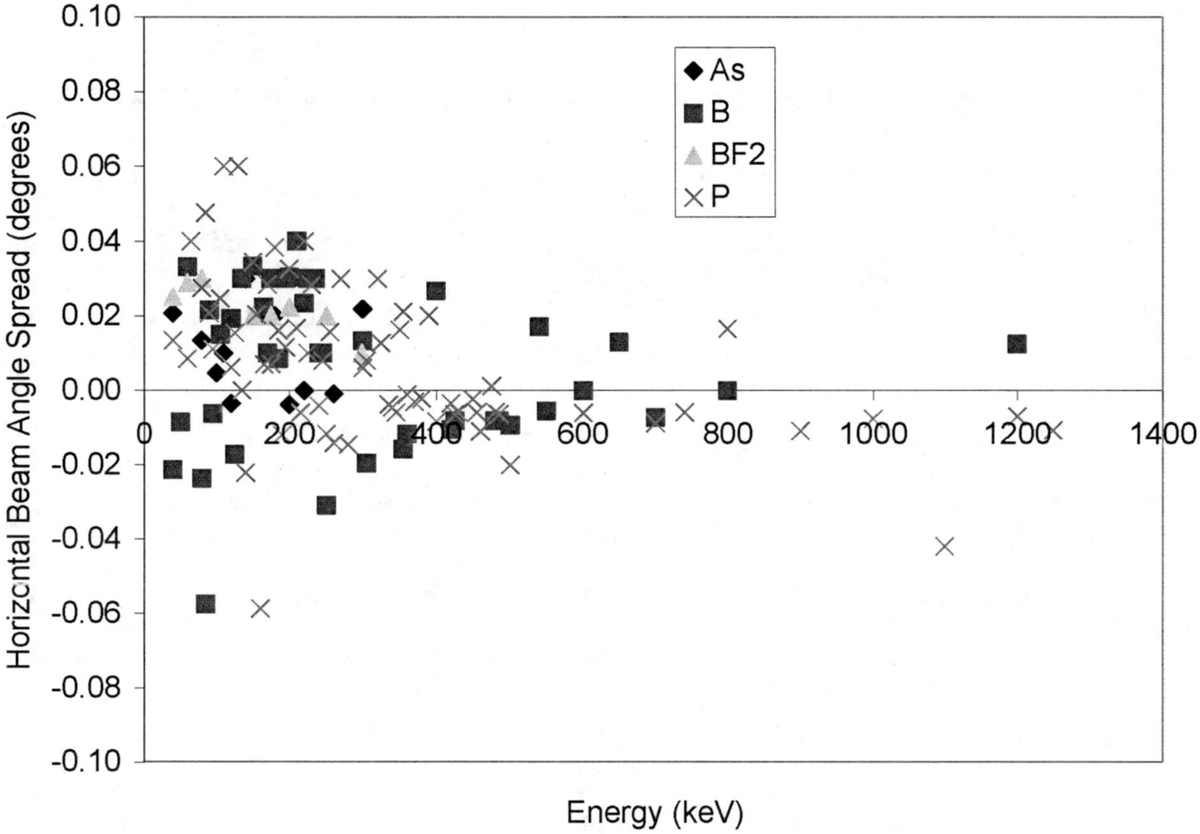

FIGURE 4. Measured average across wafer horizontal angle variation for 2500 setups over a four month period on a single wafer high energy implanter.

REFERENCES

1. Lee, W. *et al.*, "Benefits of Zero Degree Implantation Utilizing Single Wafer High Energy Implanter for Advanced Semiconductor Devices" in *Proceedings of the 17th International Conference on Ion Implantation Technology, 2008.*
2. Gossmann, H.-J. *et al.*, *Solid State Technology* **50**(7) (July, 2007).
3. Olson, J.C., Renau, A. and Buff, J., "Scanned Beam Uniformity Control in the VIISta 810 Ion Implanter" in *1998 International Conference on Ion Implantation Technology Proceedings, Kyoto, Japan*, edited by J. Matsuo, G Takaoka, and I. Yamada, Piscataway, NJ: IEEE, 1999, pp. 169-172.
4. Olson, J.C. and Renau, A., "Control of Channeling Uniformity for Advanced Applications" in *2000 International Conference on Ion Implantation Proceedings*, edited by Heiner Ryssel *et al.*, Piscataway, NJ: IEEE, 2000, pp. 670-673.
5. Renau, A. and Hacker, D., "The VIISta 810 300mm Medium Current Impalnter" in *1998 International Conference on Ion Implantation Technology Proceedings, Kyoto, Japan*, edited by J. Matsuo, G Takaoka, and I. Yamada, Piscataway, NJ: IEEE, 1999, pp. 158-161.
6. Renau, A., "The Beam Line Architecture of the VIISta 810 Medium Current Implanter" in *1998 International Conference on Ion Implantation Technology Proceedings, Kyoto, Japan*, edited by J. Matsuo, G Takaoka, and I. Yamada, Piscataway, NJ: IEEE, 1999, pp. 162-165.

Sources for Low Energy Extreme of Ion Implantation

A. Hershcovitch[1], V. A. Batalin[2], A. S. Bugaev[3], V. I. Gushenets[3], B. M. Johnson[1], A. A. Kolomiets[2], G. N. Kropachev[2], R. P. Kuibeda[2], T. V. Kulevoy[2], E. S. Masunov[4], E. M. Oks[3], V. I. Pershin[2], S. V. Petrenko[2], S. M. Polozov[4], H. J. Poole[5], I. Rudskoy[2], D. N. Seleznev[2], P. A. Storozhenko[6], A. Ya. Svarovski[7], and G. Yu. Yushkov[3]

[1] *Brookhaven National Laboratory, Upton, New York 11973, USA*
[2] *Institute for Theoretical and Experimental Physics, Moscow, Russia*
[3] *High Current Electronics Institute Russian Academy of Sciences, Tomsk, 634055 Russia*
[4] *Moscow Engineering Physics Institute, Kashirskoe sh. 31, Moscow, 115409, Russia*
[5] *PVI, Oxnard, California 93031-5023, USA*
[6] *State Research Institute for Chemistry and Technology of Organoelement Compounds 38, sh. Entuziastov, Moscow, 111123, Russia*
[7] *Siberian Divisions of Russian National Research Center "A.A. Bochvara Scientific Research Institute for Inorganic Materials," Seversk, 636070 Russia*

Abstract. A joint research and development effort focusing on the design of steady state, intense ion sources has been in progress for the past four and a half years. The ultimate goal is to meet the two, energy extreme range needs of mega-electron-volt and 100's of electron-volt ion implanters. This endeavor has resulted in record steady state output currents of higher charge state Antimony and Phosphorous ions: P^{2+} (8.6 pmA), P^{3+} (1.9 pmA), and P^{4+} (0.12 pmA) and 16.2, 7.6, 3.3, and 2.2 pmA of Sb^{3+} Sb^{4+}, Sb^{5+}, and Sb^{6+} respectively. During the past year the effort was channeled towards low energy implantation, for which the effort involved molecular ions and a novel plasmaless/gasless deceleration method. To date, 3 emA of positive Decaborane ions were extracted at 14 keV and a smaller current of negative Decaborane ions were also extracted. Additionally, a Boron fraction of over 70% was extracted from a Bernas-Calutron ion source.

Keywords: ion sources for ion implantation
PACS: 41.75Ak, 41.75Cc, 81.15Jj, 52.77Dq.

INTRODUCTION

Various types of ions are implanted, over a wide range of energies into some of the materials used in the construction of semiconductors. These energies range from as low as approximately 100 eV for shallow surface implantations, to as high as multi-MeV for deep implantation into the substrate. State of the art ion sources meet industry needs for the energy range of about 10 keV to about 300 keV. But at the two extremes (100's of eV and at multi-MeV) of the energy range, there is a lot of room for improvement due to space charge limitations at the low energy range and due to inefficiency in acceleration at the higher energy range. This paper is a synopsis of an extensive ion source R&D program designed to address industry needs.

Originally, the collaboration started to develop pulsed metal vapor ion sources with enhanced charge states. We utilized an external electron beam in two ion sources provisionally dubbed E-MEVVA. Lead and Bismuth, which previously achieved doubly charged ions, were ionized to ion charge states of Pb^{+7} and Bi^{+8} with ion currents exceeding 200 mA [1,2].

The natural next step was to adapt these charge enhancement characteristics to ion sources that generate steady state multi-charged B, P, As, and Sb ions. These technical enhancements can be adapted to DC ion implanters [3] in order to improve upon present day high-energy ion implanters that use rf accelerators. Progress in generating higher charge state B, P, and Sb ion beams is briefly reported in section I.

However, we soon realized the semiconductor industry has greater needs in the area of low energy (100's of eV) ion implantation, where space charge problems associated with lower energy ion beams limit implanter ion currents, thus leading to low production rates. To tackle the space charge problem, two approaches were followed: using molecular ions and ion beam deceleration with space charge compensation. Recent results from a Decaborane ion source are described in section II; additionally a novel gasless/plasmaless ion beam deceleration method is also mentioned in this section.

CP1066, *Ion Implantation Technology*, edited by E. G. Seebauer, S. B. Felch, A. Jain, and Y. V. Kondratenko

Finally, a spin-off result of an ion source, where over 70% of the extracted ion beam consists of singly charged boron, is briefly described in section III.

I. HIGH CHARGE STATE ION SOURCES

Ion beams containing record high charge states of Phosphorous and Antimony have been extracted from ion sources located at HCEI and at ITEP respectively. For some of the higher charge states, the improvement was greater than an order of magnitude over existing technologies.

At HCEI the ion source is a modified Bernas-Calutron ion source with 1 mm by 40 mm aperture. The source employs a design similar to that of the Russian ion implanter "Vesuvius" [4] which can generate record high charge states of Phosphorous ion beams. This kind of ion beam generator could be considered as a combination of Bernas ion source [5] and Calutron ion source [6]. In this ion source a conventional gas delivery system was replaced by an oven. After optimizing all ion source operating parameters: power, magnetic field and oven temperature, record yields of P^{2+} (8.6 pmA), P^{3+} (1.9 pmA), and P^{4+} (0.12 pmA) were extracted from the modified Bernas-Calutron ion source [7]. It is significant to observe that the previous best results [8,9] were P^{2+} (3 pmA), P^{3+} (0.2 pmA), and only a miniscule P^{4+} output. Further details and experimental results can be found in reference 7. Since the ion source contains a magnetic field perpendicular to ion beam extraction, charge state distribution measurements are being repeated with magnetic separation for confirmation.

Additionally, from this ion source (when operating with Boron), close to 1 emA of B^{+2} ions were extracted.

Record enhancement of Antimony charge states were obtained in an ITEP Bernas ion source in which a staggered, oscillating electron beam was generated [10]. Current levels reaching a Faraday cup after magnetic separation were 16.2, 7.6, 3.3, and 2.2 pmA of Sb^{3+}, Sb^{4+}, Sb^{5+}, and Sb^{6+} respectively. Additional results as well as a detailed investigation can be found in reference 10. Ion source extraction area is 20 mm^2.

II. IONS FOR SHALLOW IMPLANTATIONS

Since the invention of the transistor, the trend has been to miniaturize semiconductor devices. This has resulted in the need to decrease ion implantation energy, since shallow profile implantation is desired. But, due to space charge (intra-ion repulsion) effects, low energy ion beams are characterized by low current. Neutralizing plasmas, utilized in today's implanters, to reduce space charge offer only a partial solution and often result in implanting undesirable impurities. Therefore, low energy ion implanters have low production rates. Consequently, increasing the current of pure, low energy ion beams is of paramount importance to the semiconductor industry.

To mitigate the contamination problem, our collaboration is involved in two projects: molecular ions and a beam decelerator that compensates for space charge effects without gas or plasma. The latter is a highly proprietary novel technique for a low energy high current ion beam propagator. This technology produces ion implantations that are contaminant free! A record of invention has been filed by the collaborators.

Decaborane ($B_{10}H_{14}$) was introduced into the ITEP Bernas source and the spectrum shown in figure 1 was obtained [11]. A Decaborane current of 1 emA was extracted. Later, a somewhat smaller current of negative Decaborane was also obtained [12]. This significance result opens the possibility of merging negative and positive Decaborane beams, while slowing them down to further reducing the space charge problem.

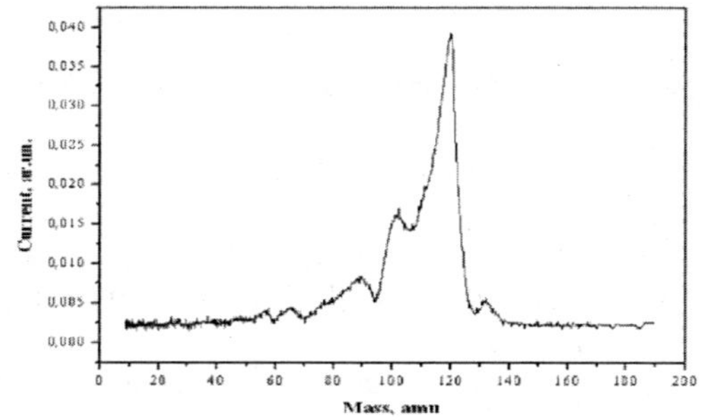

FIGURE 1. Original spectrum of Decaborane ion beam.

Recently, significant improvements were made to the ion source and the diagnostics. Three emA of decaborane ions were extracted from a 1.5 mm by 20 mm slit at 14 kV. Discharge current was only 85 mA. These results exceed any previously obtained results anywhere.

As it can be seen from figures 2 and 3, the decaborane spectrum was measured with unprecedented resolution. Figure 2 shows the measured decaborane spectrum, while figure 3 displays the theoretically predicted spectrum based on the isotopic abundance of the various boron isotopes.

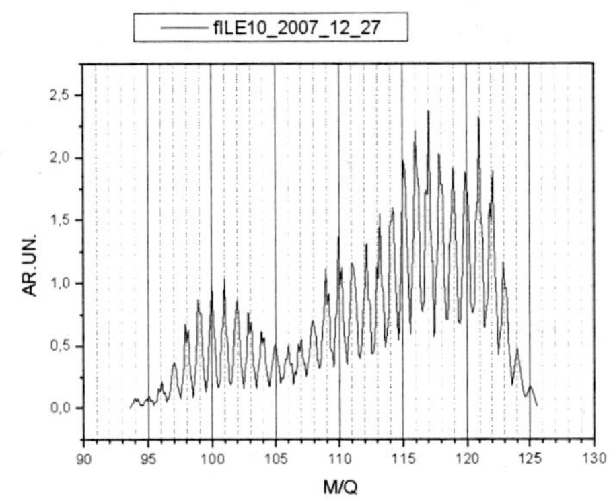

FIGURE 2. Measured Decaborane spectrum.

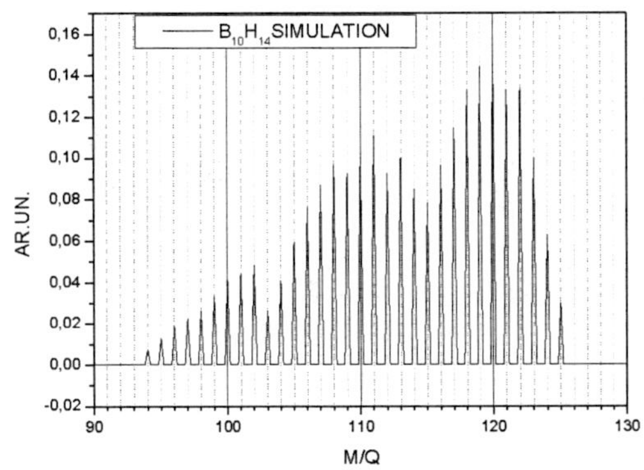

FIGURE 3. Simulation of expected Decaborane spectrum.

One puzzle remains: the heaviest molecular ion has mass 125 instead of the expected 124 (the experimental setup was even tested with xenon).

There is an inherent difficulty in transporting various species from Bernas type ion sources, since they contain magnetic fields perpendicular to the extraction direction that bend and separate ion species during extraction. The challenge is to find a transport system where electro-static forces compensate for this effect of the magnetic field. Such a system was successfully set up by adding a focusing element followed by a deflection element (to compensate for source magnetic field bending) and another focusing element (before the bending magnet). Simulation of extraction and transport of a ribbon beam over a wide range of masses from Boron to Decaborane was performed [13]. Results of simulations [12] that were performed with a modified (ITEP) version of KOBRA had excellent agreement with experimental results.

Presently, we think that Decaborane current is limited by bending magnet acceptance. Enlarging source aperture size by a factor of six might result in a proportional increase in Decaborane ion beam current.

III. HIGH FRACTION BORON YIELD

Final results to report are that intense beams of boron ions were extracted from the HCEI modified Bernas-Calutron ion source. The anticathode was placed inside a discharge chamber and instead of using the conventional boron-trifluoride (BF_3) gas; a solid lithium-boron-tetrafluoride ($LiBF_4$) compound was heated to release boron. For optimal ion source parameters beams of up to 41 mA were extracted. Singly charged boron made up over 70% of the total ion beam [14].

By comparison, no more than 25% of the extracted beam from conventional ion sources is Boron. Additionally, BF_3 is extremely toxic, while $LiBF_4$ is a safe compound.

Ion source performance is optimized by adjusting the temperature of the oven containing the solid $LiBF_4$ compound. Once steady optimal boron vapor flow into the ion source is achieved, discharge current and voltage is adjusted by increasing hot cathode emission current. Table 1 shows source performance (from low top row, to high, bottom row). More details can be found in reference 14.

IV. ADDITONAL INFORMATION

Substantial R&D was done [15-19] in addition to the results presented here. Some of the results include diagnostics [16] or ion source simulations [18], which may aid in further R&D, or disappointing results [15,17] of a duoPIGatron ion source.

TABLE 1. Boron charge-state fractions for some operating parameters of the Bernas-Calutron ion source.

Source Current & Voltage	Total Extracted Ion Beam	$^{11}B^{+}$ Fraction	$^{11}B^{++}$ Fraction
2 Amp, 250 Volt	25 mA	72 %	-
3.6 Amp, 190 Volt	34 mA	71 %	0.6 %
4 Amp, 110 Volt	41 mA	68 %	1.1%

V. PRESENT R&D PROGRAM

Given the focus of the ion implantation field on ultra-shallow junction formation, our program has changed course again. For the next three years, the ITEP group will focus on generation of octadecaborane ($B_{18}H_{22}$) with isotopically pure boron 11 (^{11}B), while the HCEI will focus on plasma immersion operating without working gases and eventually implanting isotopically pure ions.

DISCUSSION

The main objective of our program has been to develop commercial ion implantation sources for the semiconductor industry. Over the past four and a half years substantial results were obtained by our collaboration in spite of the fact that many changes in the research program have been introduced during this relatively short period. Our near term plan is to develop octadecaborane and plasma immersion sources with desirable properties that commercially available sources do not have.

ACKNOWLEDGEMENTS

One of the authors (A.H) is grateful to V. Benveniste, K. Saadatmand, B. Vanderberg, M. Graf, J. Poate W. DiVeriglio, R. Rathmell, P. Splinter, And M. Sieradzki, and A. Perel for many stimulating and very useful discussions. Work is supported by research contracts between BNL and HCEI with ITEP under the IPP Trust-2 program and by the U.S. Department of Energy.

REFERENCES

1. A.S. Bugaev, V.I Gushenets, E.M. Oks, G.Yu Yushkov, T.V. Kulevoy, A. Hershcovitch and B.M Johnson, *Appl. Phys. Lett.*, **79,** 919 (2001).
2. V.A. Batalin, A.S. Bugaev, V.I. Gushenets, A. Hershcovitch, B.M. Johnson, A.A. Kolomiets, R.P. Kuibeda, T.V. Kulevoy, E.M. Oks, V.I. Pershin, S.V. Petrenko, D.N. Seleznev, and G.Yu. Yushkov, *Journal of Applied Physics*, **92,** 2884 (2002).
3. V.A. Batalin, A.S. Bugaev, V.I. Gushenets, A. Hershcovitch, B.M. Johnson, A.A. Kolomiets, R.P. Kuibeda, B.K. Kondratiev, T.V. Kulevoy, I.V. Litovko, E.M. Oks, V.I. Pershin, H.J. Poole, S.V. Petrenko, D.N. Seleznev, A. Ya. Svarovski, V.I. Turchin, and G.Yu. Yushkov, *Rev. Sci. Instrum.* **75**, 1900 (2004).
4. V.V. Simonov, L.A. Kornilov, A.V. Shashelev, E.V. Shokin, *Ion Implantation Equipment* (Radio i Svjaz', Moscow, 1988, in Russian).
5. I. Chavet and R. Bernas, *Nucl. Instrum. Methods* **51**, 7 (1967).
6. H. Rassel and H. Glawischnig, Eds., *Ion Implantation Techniques* (Springer, New York, 1982), p.39.
7. V.I. Gushenets, A.S. Bugaev, E.M. Oks, A. Hershcovitch, B.M. Johnson, T.V. Kulevoy, H.J. Poole, and A.Ya. Svarovski, *Rev. Sci. Instrum.* **76**, 083301 (2005).
8. T.N. Horsky, *Rev. Sci. Instrum.* **69**, 840 (1998).
9. K. Saadatmand, *Rev. Sci. Instrum.* **69**, 859 (1998).
10. T.V. Kulevoy, R.P. Kuibeda, S.V. Petrenko, V.A. Batalin, V.I. Pershin, N. Kropachev, A. Hershcovitch and B.M. Johnson, V.I. Gushenets and E.M. Oks, and H.J. Poole, *Rev. Sci. Instrum.* 77, 03C110 (2006).
11. T.V. Kulevoy, S.V. Petrenko, R.P. Kuibeda, V.A. Batalin, V.I. Pershin, A.V. Koslov, and Yu. B. Stasevich, A. Hershcovitch and B.M. Johnson, E.M. Oks and V.I. Gushenets, H.J. Poole, P.A. Storozhenko, E.L. Gurkova, and O.V. Alexeyenko, *Rev. Sci. Instrum.* **77**, 03C102 (2006).
12. A. Hershcovitch, V. A. Batalin' A. S. Bugaev, V. I. Gushenets, B. M. Johnson, A. A. Kolomiets, G.N. Kropachev, R. P. Kuibeda, T. V. Kulevoy, I. V. Litovko, E.S.Masunov, E. M. Oks, V. I. Pershin, S. V. Petrenko, S.M.Polozov, H. J. Poole, I. Rudskoy, D. N. Seleznev, P. A. Storozhenko, A. Ya. Svarovski, G. Yu. Yushkov, *Rev. Sci. Instrum.* **77**, 03B510 (2006).
13. S.V. Petrenko, G.N. Kropachev, R.P. Kuibeda, T.V. Kulevoy, and V.I. Pershin, E.S. Masunov and S.M. Polozov, A. Hershcovitch and B.M. Johnson, H.J. Poole, *Rev. Sci. Instrum.* **77**, 03C112 (2006).
14. V.I. Gushenets, A.S. Bugaev, E.M. Oks, A. Hershcovitch, B.M. Johnson, T.V. Kulevoy, H.J. Poole, A. Ya. Svarovski, *Rev. Sci. Instrum.* **77**, 03C109 (2006).
15. A. Hershcovitch, B. M. Johnson, V. A. Batalin, G.N. Kropachev, R. P. Kuibeda, T. V. Kulevoy, A. A. Kolomiets, V. I. Pershin, S. V. Petrenko, I. Rudskoy, D. N. Seleznev, A.S. Bugaev, V. I. Gushenets, I. V. Litovko, E. M. Oks, G. Yu. Yushkov E.S.Masunov, S.M.Polozov, H. J. Poole, P. A. Storozhenko, A. Ya. Svarovski, *Rev. Sci. Instrum.* **79**, 02C507 (2008).
16. V.I. Gushenets, A.S. Bugaev, E.M. Oks, A. Hershcovitch, T.V. Kulevoy, I.G. Brown, *Rev. Sci. Instrum.* **79**, 02B701 (2008).
17. V.I. Gushenets, E.M. Oks, A. Hershcovitch, B.M. Johnson, *Rev. Sci. Instrum.* **79**, 02B307 (2008).
18. I. Rudskoy, T.V. Kulevoy, S.V. Petrenko, R.P. Kuibeda, D.N. Seleznev, V.I. Pershin, A. Hershcovitch, B.M. Johnson, V.I. Gushenets, E.M. Oks, H.P. Poole, *Rev. Sci. Instrum.* **79**, 02B313 (2008).
19. T.V. Kulevoy, S.V. Petrenko, R.P. Kuibeda, D.N. Seleznev, V.A. Koshelev, A.V. Kozlov, Yu.B. Stasevich, V.I. Pershin, A. Hershcovitch, B.M. Johnson, V.I. Gushenets, E.M. Oks, H.P. Poole, *Rev. Sci. Instrum.* **79**, 02B501 (2008).

Integration of an Axcelis Optima HD Single Wafer High Current Implanter for p- and n-S/D Implants in an Existing Batch Implanter Production Line

Matthias Schmeide[1], Serguei Kontratenko[2], Ralf Peter Müller[3], Bernhard Krimbacher[2]

[1] *Infineon Technologies Dresden GmbH & Co. OHG, Königsbrücker Str. 180, D-01099 Dresden, Germany*
Matthias.Schmeide@Infineon.com
[2] *Axcelis Technologies, Inc., 108 Cherry Hill Dr., Beverly, MA, USA*
[3] *Qimonda Dresden GmbH & Co. OHG, Königsbrücker Str. 180, D-01099 Dresden, Germany*

Abstract. This paper is focused on the integration and qualification of an Axcelis Optima HD single wafer high current spot beam implanter in an existing 200mm production line with different types of Axcelis batch implanters for high current applications. Both the design of the beamline and the beam shape are comparable between single wafer and batch high current spot beam implanters. In contrast to the single wafer high current ribbon beam implanter, energy contamination is not a concern for the considered spot beam tool because the drift mode can be used down to energies in the 2keV region. The most important difference between single wafer and batch high current implanters is the significantly higher dose rate and, therefore, the higher damage rate for the single wafer tool due to the different scanning architecture. The results of the integration of high dose implantations, mainly for p- and n-S/D formation, for DRAM 110nm without pre-amorphization implantation (PAI), CMOS Logic from around 250nm down to 90nm without and with PAI, are presented and discussed. Dopant concentration profile analysis using SIMS was performed for different technologies and implantation conditions. The impurity activation was measured using sheet resistance and in some cases spreading resistance technique was applied. The amorphous layer thickness was measured using TEM. Finally, device data are presented in combination with dose, energy and beam current variations. The results have shown that the integration of implantation processes into crystalline structure without PAI is more complex and time consuming than implantations into amorphous layer where the damage difference due to the different dose rates is negligible.

Keywords: Ion implantation, high current single wafer implanter, high current batch implanter, spot ion beam.
PACS: 52.77.Dq, 81.05.Bx.

INTRODUCTION

With increasing wafer size from 200mm to 300mm in the late 1990s the first single wafer high current implanter platforms were introduced into the market. Their major differences compared to the batch implanters, which were the most common tools for high current applications over the last 20 years, are the lower mechanical scan speed and the better angle control.

With reduction of the geometries below 200nm the structures on the wafer surface become very fragile and vulnerable to ballistic forces caused by particles emanating from the beam line in combination with the high spin speed of the disk of the batch implanter. To solve this problem, the spin speed of the disks was reduced and an internal particle sensor was integrated to detect particle induced structure damage [1]. Even with reduction of the spin speed the structure damage cannot be completely eliminated at geometries well below 110nm. Thus, the consequence was to go to single wafer high current implanter platforms with different scanning architectures.

Since then, about 2 years ago the Axcelis Optima HD, a single wafer high current spot beam implanter, was positioned on the market [2]. Due to using a comparable beamline and a similar shape of the spot beam as for the batch high current implanters, the Optima HD is able to generate acceptable high beam currents without using deceleration mode for the beam energies of ~2keV and above in combination with a fast tuning time to provide high throughput.

The most important difference between single wafer and batch high current implanters is the significantly higher dose rate and therefore, the higher damage rate for the single wafer tool due to the different scanning architecture [3].

This paper is focused on the integration and qualification of an Axcelis Optima HD single wafer high current spot beam implanter into an existing 200mm production line with Axcelis GSD ULTRA batch implanters for p- and n-S/D formation, for different technologies, like DRAM 110nm, CMOS Logic from around 250nm down to 90nm.

EXPERIMENTAL

The basic dose matching of the Optima HD to the installed Axcelis batch implanter base was performed on bare monitor wafers with and without oxide on top. The wafers were

CP1066, *Ion Implantation Technology,* edited by E. G. Seebauer, S. B. Felch, A. Jain, and Y. V. Kondratenko

implanted with B, BF_2 and As ions. After implantation the wafers were annealed with Rapid Thermal Processing (RTP) at 1120°C for 10s in an N_2 ambient. Then, the oxide was removed and sheet resistance was measured on KLA Tencor RS100 equipment. Based on the results, the dose trim factor was used to adjust the dose of the production recipes.

To perform device parameters matching for p- and n-S/D processes a Design Of Experiment (DOE) was run on split lots including a dose and energy variation. In addition, the beam current and the beam shape were varied at the technologies without PAI, like the 110nm DRAM and the 250nm Logic. The wafers were processed and probed using the standard process and the results were compared. To verify the results and study dose rate effects additional SIMS, TEM, SRP, and sheet resistance (RS) measurements were done for some conditions including beam current variation.

RESULTS AND DISCUSSION

The tool dose matching, based on the RS measurements of implanted bare wafers, has shown comparable results between the Optima HD and the ULTRA. The RS results have shown that the most recipes in the dose range lower than $1x10^{15}$ ions/cm^2 are matching well. Due to the comparable beamline optics, beam divergence and beam shape differences can be neglected. Furthermore, in these experiments drift mode was used for all beam energies higher than 1.8keV and we do not consider energy contamination effects in contrast to the matching between a batch implanter and a single wafer ribbon beam implanter, as discussed in [4].

However, due to a significant difference in dose rate and beam duty cycle on the wafer it was expected that for some process recipes additional implant parameter optimization is required to get exact matching of the device characteristics. Besides dose we used ion energy as an additional matching parameter.

A. Overall matching results

The overview of the process matching results for source drain (S/D) processes, shown in Table 1, demonstrate that for all processes ion dose adjustment based on the electrical parameters is required. For technologies without PAI an additional implant parameter tuning is necessary, such as the energy to compensate the effects of different amorphization behavior due to the higher beam current density and the different duty cycle or dwell time of the beam on the wafer. To confirm the major contribution of the dose rate effects on the device parameters it was demonstrated that the dose rate adjustment by a significant decrease of the ion beam current by about 10 times on the single wafer implanter provides comparable device results with the batch system without dose and energy adjustment.

However, to keep a high throughput for high dose implantations the maximum beam current has to be used. It was found that with adjustment of the dose and energy it is possible to match the different important device parameters well, even though the impurity distribution profile after implantation and RTP will not be exactly the same. For some processes related to BF_2^+ implantation it was found that the exact integration cannot be achieved with energy and dose adjustment and additional optimization of the other processes, like halo and Vt implantation, is required. Exemplary, the DRAM 110nm and the p-S/D of the 250nm Logic will be discussed below.

TABLE 1. Overview of the required adjustments for different technologies and process steps, with and without pre-amorphization implants (PAI).

Technology, Process step	Species	PAI	Energy Adjust.	Dose Adjust.
DRAM 110nm p-S/D	BF_2	No	-	-
DRAM 110nm n-S/D	As	No	Yes	Yes
DRAM 110nm array	As	No	Yes	Yes
Logic 250nm p-S/D	B	No	Yes	Yes
Logic 130nm p-S/D	B	Yes	No	Yes
Logic 130nm n-S/D	P	Yes	No	Yes
Logic 90nm p-S/D	B	Yes	No	Yes
Logic 90nm n-S/D	As	Yes	No	Yes

B. 110nm DRAM p-S/D matching

The results of the DOE for the Optima HD are shown in Table 2. Compared to the ULTRA, the electrical parameters for overlap capacitance (OVL cap.) and diffusion resistance (Diff. Res.) were lower and the absolute value of threshold voltage (Vt) was about 20% higher. These data suggest that the higher dose rate effects during S/D implantation resulted in a "cooler" transistor with a lower lateral diffusion and longer channel.

TABLE 2. Electrical parameters on product for 110nm DRAM p-S/D (20keV BF_2 3E15cm^{-2}).

Implant tool	OVL Cap. fF/µm (a.u.)	Diff. Res. Ohms/sq. (a.u.)	Vt. mV (a.u.)
ULTRA reference	100	100	100
HD original settings	98	92	118
HD OVL optimized	100	88	100
HD RS optimized	90	100	138

From the DOE, it is seen that by increasing the energy by 8% (HD OVL optimized) we are able to match OVL cap. and Vt, but, at the same time the Diff. Res. goes further to lower values. The only acceptable way to match the Diff. Res. is to decrease the energy by 8% and the dose by 10% (HD RS optimized). Doing this, the halo and Vt adjust implantations have to be optimized to match OVL cap. and Vt, which requires two additional work flows. The same problem was already found during the integration of a batch implanter at Infineon's 300mm production line equipped with single wafer ribbon beam implanters at the previous 140nm DRAM process [5].

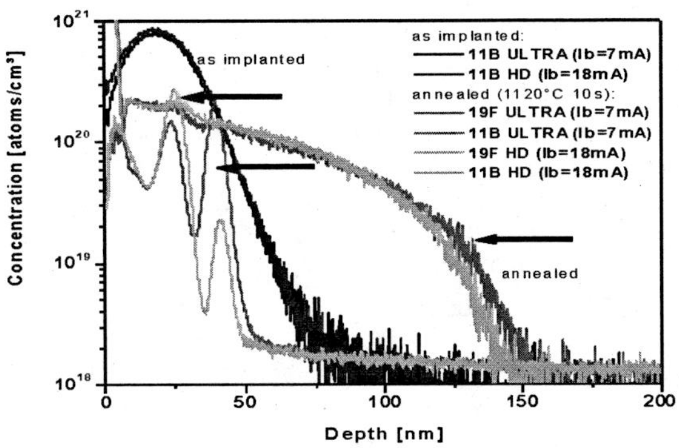

FIGURE 1. SIMS profile of B and F for 20keV BF_2 3E15cm^{-2} on Optima HD (HD) and ULTRA as implanted (B only) and after anneal.

Figure 1 shows boron and fluorine distribution profiles measured by SIMS for the considered BF_2^+ implantation. As-implanted boron and fluorine (not shown) profiles are comparable for both single wafer and batch implanter. During RTP impurity cluster formation and accumulation occurs at the projected range (Rp) and end of range (EOR) defect areas [6]. The position of the second peak (EOR), which defines the amorphous layer thickness, is deeper for the wafer implanted on the Optima HD with the higher dose rate compared to the ULTRA batch tool.

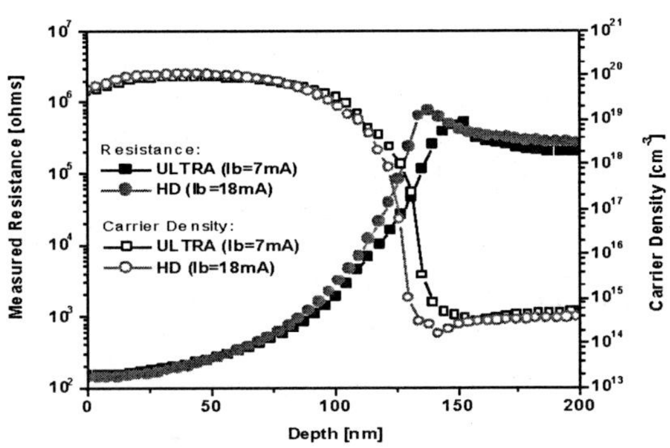

FIGURE 2. SRP results of 20keV BF_2 of 3E15cm^{-2} on Optima HD (HD) and ULTRA using the standard anneal.

Study of the amorphous layer thickness using TEM confirmed that after implantation it is deeper for the high dose rate process compared to the standard one, 40nm and 38nm, respectively. Different amorphous structure resulted to shallower boron profile and higher electrical activation after high temperature RTP at 1120°C for the Optima HD compared to ULTRA implanter. SRP measurement results presented in Figure 2 are well agreed with SIMS and RS data. Analytical measurement results show a good correlation with the observed device electrical parameters.

C. 110nm DRAM n-S/D and array matching

For the n-S/D and for the array contact implantation, arsenic ions with energies of 20keV and 10keV and a dose accordingly of $3x10^{15}$ and $1.3x10^{15}$ ions/cm^2 were used.

The DOE results for the n-S/D are shown in Table 3 and for the array contact implant in Table 4.

For n-S/D with the original HD settings based on the dose matching a similar trend with lower overlap capacitance, lower diffusion resistance, and higher Vt value was observed as we saw for BF_2 p-S/D process (paragraph IIIB) even though the difference from the reference is smaller. This suggests that we have got again a longer n-channel transistor with lower lateral S/D diffusion using the high dose rate process. It was found that OVL Cap. and Vt matching to the reference can be achieved by increasing the arsenic energy from 20keV to 22keV (HD OVL optimized).

TABLE 3. Electrical parameters on product for 110nm DRAM n-S/D (20keV As 3E15cm^{-2}).

Implant tool	OVL Cap. fF/µm (a.u.)	Diff. Res. Ohms/sq. (a.u.)	Vt. mV (a.u.)
ULTRA reference	100	100	100
HD original settings	98	95	108
HD OVL optimized	100	95	101

As-implanted 20keV arsenic SIMS profiles, presented in Figure 3, show a slightly lower tail for the Optima HD compared to the ULTRA, which is a result of the different dose rate and dwell time. Similar profile tails difference and slightly shallower profile depth was observed for the Optima HD samples after RTP. Assuming the importance of particularly the lateral diffusion on the transistor parameters an additional study is required.

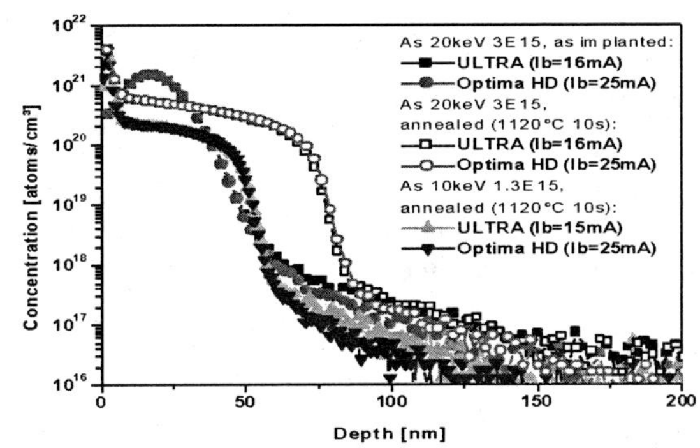

FIGURE 3. SIMS profile of As 10keV 1.3E15cm^{-2} and As 20keV 3.0E15cm^{-2} on Optima HD and ULTRA as implanted and after anneal.

Compared to the results of the n-S/D process, different optimization is required for the array contact implant (Table 4).

TABLE 4. Electrical parameters on product for 110nm DRAM array n-S/D (10keV As 1.3E15cm^{-2}).

Implant tool	OVL Cap. fF/µm (a.u.)	Diff. Res. Ohms/sq. (a.u.)	ICELL mV (a.u.)
ULTRA reference	100	100	100
HD original settings	101	100	113
HD ICELL optimized	100	103	100

Due to the presence of carbon in array device for TED reduction, the OVL Cap. reduction is less pronounced (even increases, which is not understood so far). Instead of an increasing energy to compensate for the higher amorphization rate, the energy has to be decreased in this case from 10keV to 8.5keV in combination with a lower dose of 5% (HD ICELL (cell leakage) optimized).

D. 250nm Logic p-S/D and matching

For the p-S/D formation of the 250nm Logic, a 7keV boron implantation with a dose of $4.5x10^{15}$ ions/cm^2 is used, which is commonly applied as well for poly-Si structures doping to build resistors for oscillators, so that the poly-Si resistance directly correlates to the yield.

In order to be closer to the production flow of this device, for the SIMS samples preparation we used n-type wafers with 10nm screen oxide in combination with the production RTP annealing recipe (1040°C at 9s). To do a comparison of dopant profiles for both high and low level rates on Optima HD two beam currents were used, 15mA and 2mA. The SIMS measurement results (cf. Figure 4) show comparable profiles after RTP between the ULTRA and the Optima HD with a lower beam current of 2mA. Boron profile implanted on the Optima HD with 15mA beam current is shallower after RTP and the accumulated boron peak position is deeper.

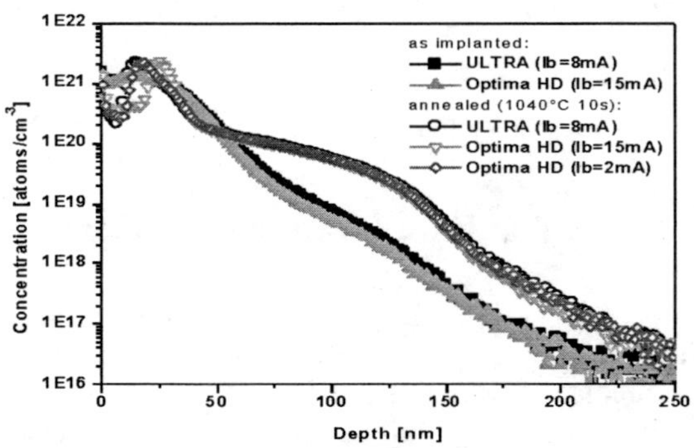

FIGURE 4. SIMS profile of B for 7keV boron 4.5E15cm^{-2} on Optima HD and ULTRA as implanted and after anneal.

A spreading resistance profiling (SRP) was done for the same wafers to compare electrically activated profiles, Figure 5. SIMS and SRP profiles comparison shows that boron peak accumulated during RTP is electrically inactive and related to boron cluster formation with silicon atoms [7]. Profiles implanted on the Optima HD with a high beam current/dose rate are shallower and electrical activation is higher compared to the ULTRA tool. The profile from Optima HD implanted with lower beam current/dose rate is closely matched to the profile from the ULTRA.

The best device electrical parameters matching and comparable yield, based on the DOE results, to the target was achieved in this case by decreasing the energy and increasing the implanted dose.

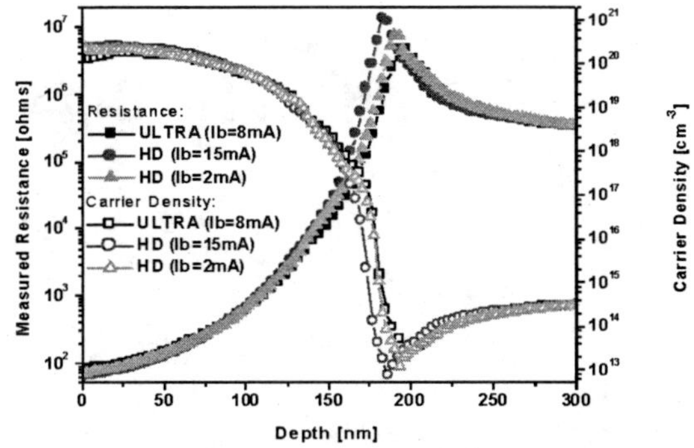

FIGURE 5. SRP results of 7keV B of 4.5E15cm^{-2} on Optima HD (HD) and ULTRA using the standard production anneal.

CONCLUSION

The presented results show that the single wafer spot beam high dose implanter (Optima HD) can be matched to an existing production line using batch implanters (ULTRA) for the different technologies with and without pre-amorphization implantation. However, the integration of different processes is more complicated compared to a simple dose matching showing the importance of the implant parameters such as implant dose rate and wafer temperature in addition to the energy and the dose. These parameters play an important role in defect accumulation during implantation and following dopant redistribution and activation during RTP. The influence of these effects is more pronounced for BF_2 ions, which may require other implantation process optimization in addition. Furthermore, the matching could be product dependent including the device design, the temperature budget and the device parameters importance priority.

It was demonstrated that the existing combination of a higher dose rate and the wafer temperature on Optima HD used for the S/D formation results in a "cooler" transistor performance with lower lateral diffusion and longer channel. This could be a potential advantage of the single wafer spot beam tool for new device applications.

REFERENCES

1. M. Schmeide, M. Kokot, D.-W. Franke, B. Sauter, *Proceedings of the 16th Int. Conf. on Ion Implantation Technology*, pp. 562-565 (2006).
2. P. Splinter, M. Graf, C. Godfrey, Y. Huang, D. Polner, J. Danis, K. Ota, *Proceedings of the 16th Int. Conf. on Ion Implantation Technology*, pp. 601-604 (2006).
3. M. Ameen, M. Harris, R. Reece, these proceedings.
4. V. Kaeppelin, Z. Chalupa, L. Frioulaud, S. Mehta, *Proceedings of the 16th Int. Conf. on Ion Implantation Technology*, pp. 353-356 (2006).
5. D.-W. Franke, M. Schmeide, *et al.*, *Proceedings of the 15th Int. Conf. on Ion Implantation Technology*, Part II, pp. 102-106 (2004).
6. N. Cagnat, C. Laviron, N. Auriac, J. Liu, *Proceedings of the 16th Int. Conf. on Ion Implantation Technology*, pp. 133-136 (2006).
7. P. Kopalidis, S. Kontratenko, *J. Electrochem. Soc.*, **V152**(5), pp. 375-377 (2005).

Optima HD Imax: Molecular Implant

D.R. Tieger, P.R. Splinter, T.J. Hsieh, W.P. Reynolds

Axcelis Technologies Inc., 108 Cherry Hill Drive, Beverly, Ma. 01915, USA

Abstract. Molecular implantation offers semiconductor device manufacturers multiple advantages over traditional high current ion implanters. The dose multiplication due to implanting more than one atom per molecule and the transport of beams at higher energies relative to the effective particle energies result in significant throughput enhancements without risk of energy contamination. The Optima HD Imax is introduced with molecular implant capability and the ability to reach up to 4.2 keV effective ^{11}B from octadecaborane ($B_{18}H_{22}$). The ion source and beamline are optimized for molecular species ionization and transport. The beamline is coupled to the Optima HD mechanically scanned endstation. The use of spot beam technology with ionized molecules maximizes the throughput potential and produces uniform implants with fast setup time and with superior angle control. The implanter architecture is designed to run multiple molecular species; for example, in addition to $B_{18}H_{22}$ the system is capable of implanting carbon molecules for strain engineering and shallow junction engineering. Source lifetime data and typical operating conditions are described both for high dose, memory applications such as dual poly gate as well as lower energy implants for source drain extension and contact implants. Throughputs have been achieved in excess of 50 wafers per hour at doses up to $1x10^{16}$ ions/cm^2 and for energies as low as 1keV.

Keywords: Ion implantation, high current implanter, wafer scan
PACS: 85.40.Ry

INTRODUCTION

After more than a decade of research and development [1,2], ion implantation with molecular ions is becoming accepted as a solution to semiconductor manufacturing productivity bottlenecks in low energy implantation and for enhanced device performance through material modifications. To this end, production equipment is being designed to optimize performance for molecular ionization and transport [3,4]. The Optima HD Imax is a production ion implanter which is designed to maximize the productivity offered by molecular implant and subsequently minimize the Cost per Wafer manufacturing of the end user. The implanter is a mass analyzed beamline with the exact mechanics of a traditional high current machine. These mechanics are ionization, beam extraction, beam analysis and beam transport. While all the components are familiar to the experienced user of ion implant equipment, the incorporation of molecular species allows the tool to be productive at both high doses (>$1x10^{16}$ ions/cm^2) and at the lowest energies of interest (<500eV).

Molecular Implantation Overview

Molecular implant for p-type dopants is being realized with the use of octadecaborane ($B_{18}H_{22}$). The implant of 18 atoms per ion charge enables the high dose capability with demonstrated performance on the Optima HD Imax of greater than 3mA electrical current or 54mA of effective ^{11}B current (see equation 1).

$$I_{11B} = 18.0 \times I_{B18} \quad (1)$$

As depicted in Figure 1, the ionization process of the $B_{18}H_{22}$ molecule results in a left skew spectrum spread of molecules starting with $B_{18}H_x$ and following with decreasing boron atoms (i.e. $B_{17}Hx$, $B_{16}H_x$, etc.). Depending upon the mass resolution, this results in a molecular weight of approximately 210. When compared to elemental boron, this average molecular weight translates to an energy ratio of 19.1:1 (see equation 2).

$$E_{11B} = 0.0524 \times E_{B18} \quad (2)$$

As can be concluded, the heavy molecular ion limits the practical maximum energy range of the

CP1066, *Ion Implantation Technology,* edited by E. G. Seebauer, S. B. Felch, A. Jain, and Y. V. Kondratenko

machine by requiring a very high magnet field in the mass analysis electromagnet. The Optima HD Imax has been designed to allow for a maximum ^{11}B effective of 4.2keV at 80 keV extraction.

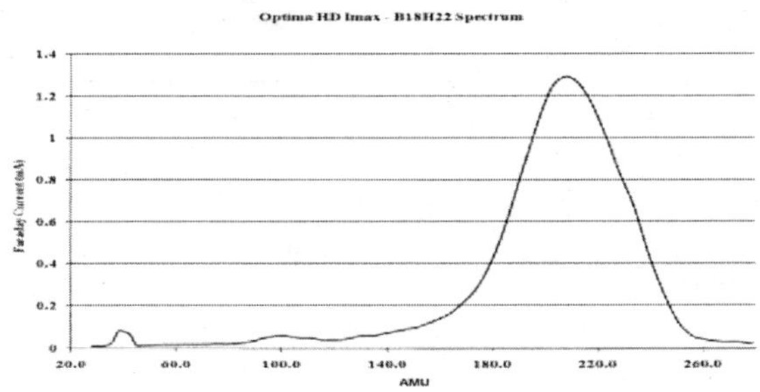

FIGURE 1. $B_{18}H_{22}$ Beam Spectrum

OPTIMA HD IMAX ARCHITECTURE

The Optima HD Imax shares a design architecture with the Optima HD which is considered a traditional high current ion implanter [5]. Wherever possible, design choices were made in such a way as to maximize commonality with the standard Optima HD but not if such choices would compromise the performance of the molecular beams that the Optima HD Imax delivers. The ion source, extraction system, aspects of the analyzing magnet and resolving slit have all been modified for the molecular implant version but the process chamber and two dimensional mechanical scan end-station are the same between the two Optima HD models. Shown in Figure 2 is the Optima HD Imax beamline.

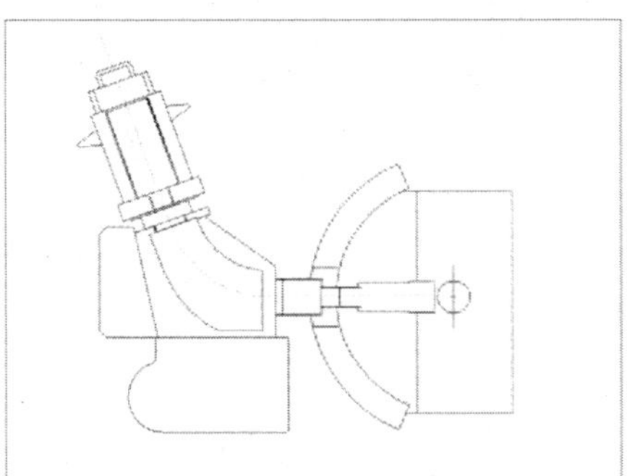

FIGURE 2. Optima HD Imax Beamline

Ion Source System

The ion source of the Optima HD Imax is the ClusterIon® Model 350 from SemEquip, Inc. [6]. The ClusterIon® source is an electron impact source which consists of an emitter and a 90 degree bend of the resultant electron beam which is then directed into the ionization chamber creating a stream of electrons parallel to the chamber extraction slit. The source optics are designed with an extended extraction slot to increase extraction area relative to standard high current tools. The molecular species used on the Optima HD Imax are distributed in solid form, therefore a vaporizer system has been developed by SemEquip and ATMI [7]. Due to the vapor pressure of $B_{18}H_{22}$, deposits readily form in the ionization chamber, electrode and surrounding vacuum housings. To prevent degradation of source performance and source lifetime, these deposits are automatically cleaned at regular intervals. To enable the cleaning, the source system encompasses the hardware needed to perform in-situ cleaning by reaction with atomic fluorine. The gas cabinet has the capability to house both argon (for both process and warm-up of the source) and NF_3 cleaning gases. To prevent condensation and clogging, the delivery path of the vaporized $B_{18}H_{22}$ to the ionization chamber is heated and kept in the temperature range of 90 -130 °C.

Extraction Region

The extraction electrode consists of a typical ground and suppression aperture pair. Unique to this design, the plates of the electrodes sandwich a heating coil which is elevated to approximately 150 °C to eliminate condensation of the vaporized $B_{18}H_{22}$.

The ground and suppression apertures feature an extended extraction area geometry that is matched to the source ionization chamber slit. Due to the wide electrical operating range of 2kV to 80kV, care has been taken to design aperture geometries that are effective at all energies. In addition to the aperture optimization of size and curvature, the electrode manipulator must allow for a sufficient gap axis travel to match the operating range [8].

Analyzer Magnet Design

The analyzer electromagnet shares many features with that of the Optima HD. For example, the magnet yoke and coils are identical. However, to allow for the large magnetic rigidity that is required to bend mass 220 at 80 keV (17,600 AMU-keV mass x energy product) with a field of ~ 1 Tesla, the bend angle has been decreased to 70 degrees and the radius has been enlarged. In addition, a pole index has been applied and the gap modified but the resultant ion beam trajectory is very similar for the two machines. Finally, it should be noted that the optical path length from ion source to wafer of the

Optima HD Imax, 1.7 m, is approximately the same as the standard Optima HD.

Resolving Aperture

To maximize beam current for applications where throughput is extremely important, such as in the dual poly gate (DPG) process, a relatively large resolving aperture is employed. However, for applications that require more stringent mass resolution to limit the contribution from the $B_{17}H_x$, a smaller resolving aperture is desirable [9]. In fact, the Optima HD Imax system has been designed to allow for multiple resolving apertures such that there are three, linearly actuated apertures: a large and small process aperture and a very small (3 mm) aperture for optical setup and AMU calibration.

Scanning System

The implanter beamline generates and transports a fixed spot beam which is coupled to a two dimensionally scanned mechanical endstation. This two dimensional scanning results in uniform exposure of the fixed ion beam over the wafer as every point on the wafer is translated through every point of the beam over multiple scans. During processing, the wafer is translated through the ion beam with a pendulum. The pendulum is capable of effective linear speeds of up to 3m/s while a second stage perpendicular to the pendulum moves the entire assembly through a slow scan of approximately 0.1m/s. With proper choice of scan parameters (number of slow scans and fast scan speed) the typical uniformities are < 1% (sigma). In addition the use of beam profiling has demonstrated that the horizontal angle can be known to better than 0.5 degrees [5].

MULTIPLE SPECIES

The ability to run multiple species on an ion implanter is key to its versatility. In addition to supporting octadecaborane, the Optima HD Imax machine is capable of running multiple forms of carbon molecules. Two process applications have emerged for carbon, the first being the traditional use of carbon for controlling boron diffusion and the second is to induce strain in the n-type channel. Research and development has been conducted using two molecules, ionized C_7H_7 as resolved from $C_{14}H_{14}$ and ionized $C_{16}H_{10}$. While both molecules can potentially be used for boron diffusion retardation and to produce tensile strain, one may be selected over the other due to the required energy, dose or resultant damage. As stated, the starting material for the C_7 implants is $C_{14}H_{14}$. A typical spectrum of carbon ionization for $C_{14}H_{14}$ on the Optima HD Imax is shown in Figure 3. The maximum effective carbon energy for C_{16} (mass 202) is ~4keV while the top energy for C_7 (mass 91) is ~10keV.

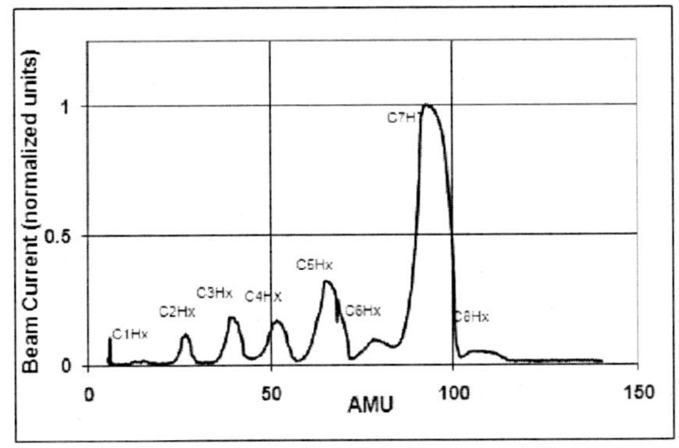

FIGURE 3. $C_{14}H_{14}$ Beam Spectrum

SYSTEM PERFORMANCE

The beam current multiplication of Equation 1 means, of course, that the electrical doses for an Optima HD Imax are much lower than a standard high current tools: Even ultra high, $2x10^{16}$ atom/cm^2 effective doses are only ~$1x10^{15}$ atom/cm^2 electrical and the majority of implant applications are in the 10^{13}-10^{14} range. This leads to high throughputs relative to standard high current machines. The overall productivity gain of the Optima HD Imax incurs a reduction of 12% due to the requirement for periodic in-situ cleaning, however, the 3x to 4x beam current premium over traditional High Current machines more than makes up for the lost availability due to the cleaning step. For example, as can be seen in Figure 4, effective beam currents in excess of 45ma can be sustained for effective ^{11}B energies of 4keV. In comparison, traditional high current ion implanters provide beam currents in the 10mA range for the analogous ^{11}B monomer implant.

Source Lifetime

The Optima HD Imax has been operating at customer sites for almost two years and are now qualified for mass production. In that time, multiple source life marathons have proven that the technology is robust during manufacturing operating conditions. Shown in Figure 4 is such a source life demonstration of 200 hours of continuous operation. During this marathon the average time between cleans was 7.17 hours and the average time of clean (beam-off to beam-on) was 59 minutes. Finally, the average beam current for this marathon was 45.7 mA effective boron current.

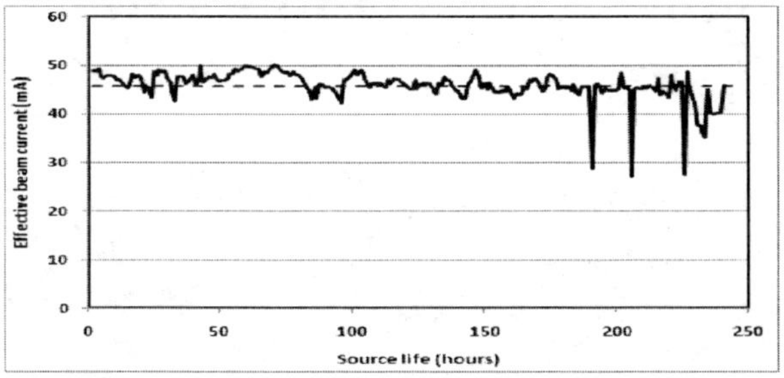

FIGURE 4. Beam current during a 200 hour marathon at a customer site demonstrating effective average ^{11}B current 45.7ma at 4keV.

Operating Conditions

The time duration between cleans is a direct contributor to the effective duty factor of the Optima HD Imax tool. Additionally, the beam current stability will also contribute directly to the productivity equation. Shown in Figure 5 is one such period of 8 hours from late in the same customer source life marathon and where the beam current remained constant to 94% of the starting value.

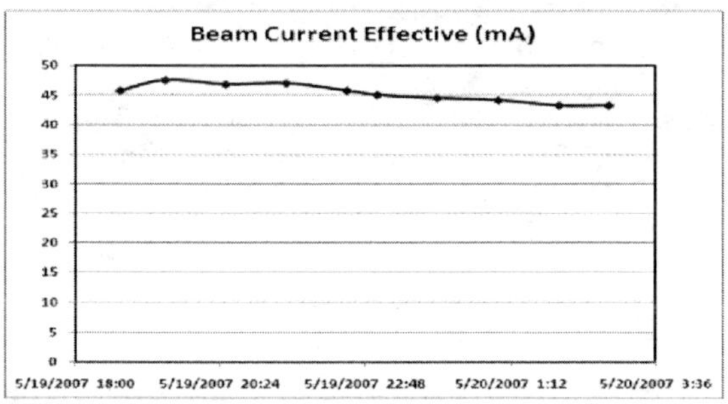

FIGURE 5. Beam current through an 8 hour cycle between cleans during a 200 hour marathon at a customer site.

Throughput Results

The Optima HD Imax can be compared to a standard high current tools' throughput over a variety of applications. Shown in Table 1 are the following: a comparison for 4.2 keV ^{11}B monomer equivalent DPG running at one customer site, a comparison of 2 keV ^{11}B monomer equivalent S/D, a comparison for 10 keV BF_2 replacement, and a contact implant being run at another customer site. In all cases the throughput from the Optima HD Imax, even with the downtime for cleaning and a trim correction for dose retention [10], is far superior to the throughput from conventional high current tools. This throughput advantage translates into lower Cost per Wafer [8] and demonstrates that the Optima HD Imax is a superior technological advance for high dose implantation.

Process	Tool	Energy (keV) - species	Species	Effective Dose (atoms/cm^{2})	Raw Throughput (wph)	Net Imax enhancement
DPG	STD tool	4.5	B	6.00E+15	35	
DPG	Imax	80	B_{18}	5.50E+15	85	2.1
S/D	STD tool	2	B	4.40E+15	28	
S/D	Imax	38.2	B_{18}	4.40E+15	109	3.4
P+CON	STD tool	10	BF2	5.00E+15	40	
P+CON	Imax	34	B_{18}	5.00E+15	71	1.6

TABLE 1. Throughput comparisons between STD high current tool and Optima HD Imax for three different processes.

SUMMARY

The Optima HD Imax is a two dimensional mechanical scan machine with molecular implant capability. The dose multiplication of the boro-hydride molecules allows for superior beam current and throughput compared to traditional high current implanters. Field operating results have shown that molecular implantation is a viable and productive solution for device manufacturing.

REFERENCES

1. D. Takeuchi, N. Toyoda, *et al.*, "Shallow Junction Formation by Polyatomic Cluster Ion Implanation", Technical Report of IEIC, SDX 1995, pp 83-89.
2. A. Perel, *et al.*, "Decaborane Ion Implanation", proceedings, IIT 2000, pp 304 – 307.
3. H. F. Glavish, T.N. Horsky, *et al.*, "A Beam Line System for a Commercial Borohydride Ion Implanter", Proceedings, IIT 2006, pp 167 – 170.
4. A. Renau, "A Better Approach to Molecular Ion Implantation", 7th International Workshop on Junction Technology, IEEE 2007. pp 107 – 112.
5. P. Splinter, M. Graf, *et al.*, "Optima HD: Single Wafer Mechanical Scan Ion Implanter", Proceedings, IIT 2006, pp 601-604.
6. T. Horsky *et al.*, "Boron Beam Performance and in-situ Cleaning of the ClusterIon® Source", Proceedings, IIT 2006, pp 198 – 201.
7. D. Adams *et al.*, "A Vaporizer for Decaborane and Octadecaborane", Proceedings, IIT 2006, pp 178 – 181.
8. D.R. Tieger *et al.*, "Beam Current Improvements on the Axcelis Optima HD Imax Implanter", this Proceedings.
9. D.R. Tieger *et al.*, "ClusterBoron™ Implants on a High Current Implanter", Proceedings, IIT 2006, pp 206 – 209.
10. M.A. Harris *et al.*, "Dose Retention Effects in Atomic Boron ClusterBoron™ (B18H22) Implant Processes", Proceedings, IIT 2006, pp 155 – 158.

Beam Current Improvements on the Axcelis Optima HD Imax Implanter

D. R. Tieger, P. R. Splinter, T. J. Hsieh, W. P. Reynolds

Axcelis Technologies Inc., 108 Cherry Hill Drive, Beverly, Ma. 01915, USA

R. D. Goldberg, D. C. Jacobson, E. K. McIntyre, S.K. Hahto, D. G. Bilbrough, D. Oved, T. N. Horsky

SemEquip Incorporated, 34 Sullivan Road, North Billerica, Ma. 01862, USA

Abstract. The ability to use large molecular species such as octadecaborane ($B_{18}H_{22}$) has been investigated at semiconductor device manufacturers as a way to significantly increase wafer throughput relative to standard high current ion implanters. Over the past two years, improvements in the machine design to support the use of $B_{18}H_{22}$ have led to beam current increases of 50% from 30pmA to 45pmA of boron at an equivalent boron energy of 4keV. This boron energy is required by p+ doping of dual poly gate (DPG) structures in DRAM. Beam current has also been significantly improved at the low equivalent boron energies anticipated to be required by 32nm processes for PMOS source/drain extensions (SDE). For example, at 250eV equivalent beam energy, a 100% increase in cluster boron beam current has been attained.

This paper describes the techniques by which these beam current improvements were accomplished, primarily through the refinement of ion beam optics. Other techniques for increasing overall tool productivity are also described, such as beam utilization and overall operational improvements. Wafer throughputs are presented for critical p+ implant processes such as dual poly gate (DPG), source/drain (S/D), source drain extension (SDE) and S/D contact. The higher throughputs resulting from these changes are translated into a cost per wafer (CPW) model and it is demonstrated that an increase in average beam current is the largest contributor to a reduction in CPW.

Keywords: Ion implantation, high current implanter, wafer scan
PACS: 85.40.Ry

INTRODUCTION

Over the past few years, the use of large molecular species in ion implantation for the production of semiconductor devices has advanced as the productivity and process benefits are further demonstrated. The productivity opportunity is a direct result of the relatively large effective beam current applied during the doping process. Equipment designers are faced with new challenges to optimize the transport of these large molecules through beamlines that are similar or even directly compatible with those designed for monomer specie transport. Optimization of source and extraction optics, mass analysis and drift regions of the beamline are necessitated to maximize the resultant beam flux available for processing. Traditional ion transport modeling software is utilized for 1st order refinement of these optics; however, the designer must also resort to empirical methods for final conclusions on the aperture geometries.

While beam current improvements offer the largest sensitivity to productivity, other factors such as the ion beam size, ease of setup, beam stability and beam energy also affect the maximum throughput. Further, the beam utilization as determined by the combination of the tools scanning system, coupled with beam size must be considered. Lastly, the use of large molecules and associated vapor pressures require frequent automated cleans which increase the final Cost per Wafer (CPW). Although the cleaning has an impact on the tool utilization, the CPW benefit is positive as compared to conventional implant for the majority of process steps tested including source/drains (S/D), dual poly

CP1066, *Ion Implantation Technology,* edited by E. G. Seebauer, S. B. Felch, A. Jain, and Y. V. Kondratenko

gate (DPG), source drain extensions (SDE) and contact implants.

Optima HD Imax Architecture

The Optima HD Imax has much in common with the Optima HD, which is the traditional high current ion implanter of the Optima platform [1]. The beamline generates and transports a fixed spot beam which is then coupled to a two dimensionally scanned mechanical endstation. The wafer scanning is accomplished with a fast scan (1– 3m/s) moving the wafer with a horizontal pendulum motion through the beam while a second vertical slow scan (0.1m/s) motion is superimposed on the fast scan. This scanning combination sweeps every point on the wafer through the beam multiple times in order to achieve uniform implants.

The elements of the beamline are shown in Figure 1 and consist of a SemEquip Incorporated 350 molecular source [2], heated extraction electrode, 70 degree analyzing magnet, plasma electron flood and a magnetically suppressed faraday.

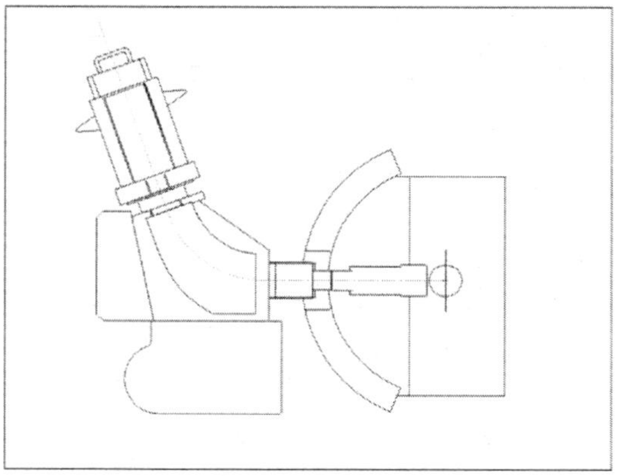

FIGURE 1. Optima HD Imax Beamline

The ionization is provided by an electron impact source [3, 4] producing low ion temperatures thus avoiding excessive disassociation of the large molecule. A schematic representation of the source is shown in figure 2. Electrons, emitted from a heated filament, are deflected through 90° into the ionization chamber, where they are confined by a permanent magnetic field. Once inside the chamber, the electrons ionize the source feed material and the resultant ions are extracted and accelerated by the electric field defined by the ion source and the suppression and ground electrodes attached to a 2-axis manipulator.

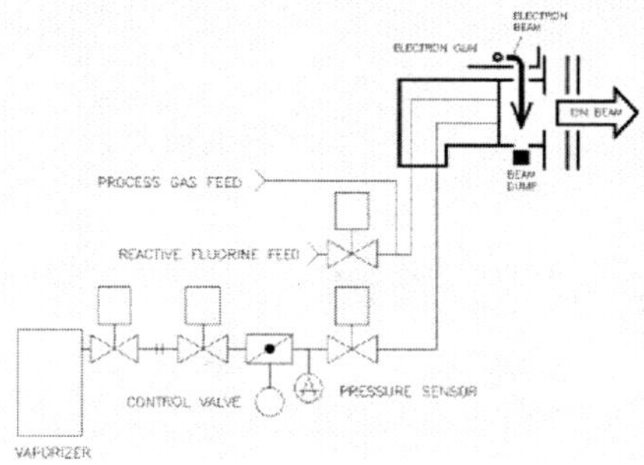

FIGURE 2: A schematic representation of the ClusterIon® ion source

The electromagnet for mass analysis is capable of resolving a mass x energy product of 17,600 AMU-keV. Specifically, $B_{18}H_{22}$ can be transported at 80keV with an effective ^{11}B energy of 4.2keV. The transport design supports boron hydride species as well as molecular carbon implants of $C_{16}H_{10}$ and C_7H_7.

TECHNIQUES FOR BEAM CURRENT IMPROVEMENT

The paths of the electrons injected into the ionization chamber are controlled by the magnetic fields located around the electron gun and ionization chamber. The second of these fields contains an inflexion point near the entrance aperture which can reflect electrons whose trajectories do not fall within the required acceptance window. By making modifications to the field, the electron trajectories were tailored to optimize the number of electrons injected into the ionization chamber. Additionally, the position of the ion column produced along the path of the electrons was altered to better match a newly modified source aperture.

The new source aperture was developed as part of a program to increase the efficiency of the extraction system. Optical modeling was used to design the extraction optics and match the resultant beam emittance to the acceptance of the Optima HD Imax beamline. Calculations of the ion trajectories were performed using 3-D Lorentz code [5] for a range of extraction, suppression and ground electrode geometries. The output from a typical calculation is displayed in Figure 3.

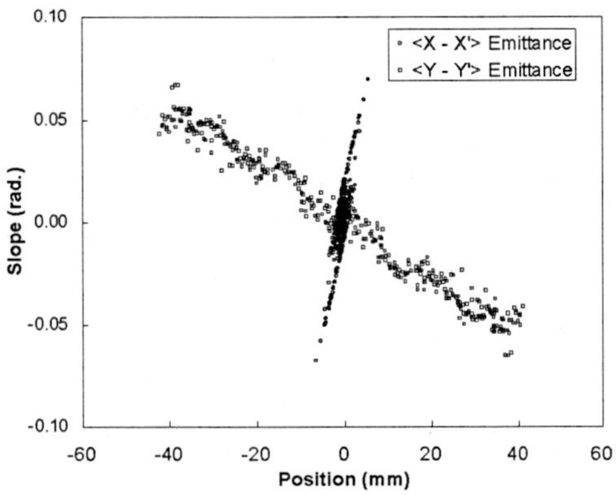

FIGURE 3: Modeled emittance plot of the extracted $B_{18}H_{22}$ ion beam

Information from the beam emittance models was subsequently used to represent the starting conditions for a second series of calculations, performed using Transport [6], to determine the transmission efficiency of the ion beam through the beamline. This procedure was conducted for each set of candidate ion optics. Empirical testing was used to fine tune the final electrode designs, which underwent marathon testing under production like conditions in a customer fab. Ultimately different extraction areas were found to be optimal for different extraction energies so that depending on the application, different choices of aperture sizes could be made.

Beam Utilization

The vertical beam size at the wafer can be adjusted by varying the curvature of the electrodes in the extraction triode (extraction aperture, suppression and ground aperture). Horizontal size at the wafer comes from a combination of horizontal size of the extraction triode and the opening of the resolving aperture. Also, there is an energy dependence to the beam sizes coming from the different magnifications caused by the extraction to suppression gap axis settings. Beam size will also be affected by space charge effects but much less than for standard high current implanters. Utilizing both first order modeling and empirical testing led to the final extraction sizes and curvatures.

The scan optimization of the Optima HD Imax takes into account beam size, fast scan speed and number of slow scans in order to most efficiently implant the desired dose. By adjusting all of these inputs the most efficient beam utilization and thus the highest possible wafer throughput can be obtained.

RESULTS OF BEAM CURRENT IMPROVEMENTS

Results from extended tests at customer sites indicated that the modifications made to the system allowed octadecaborane ion beams equivalent to atomic boron beams of 45mA at 4keV to be sustained over a 200hr source life [7]. The high beam current at high energy (> 1.5keV) was achieved in part by an increase in the overall transmission ratio of the boron hydride ion beam to 41%.

On the low energy (< 1.5keV) side improvements in ion optics and efficiency allowed for doubling of the beam current down to 250eV effective energies. Shown in Figure 4 are the beam current specifications of the Optima HD Imax, although it should be noted that beam currents in excess of 54 mA (effective) have been demonstrated at effective energies of 4keV. Despite the energy multiplication of the molecular species, which allows for transport at higher energies by a factor of roughly 20, transporting the lowest energies is still the largest challenge to the beamline design.

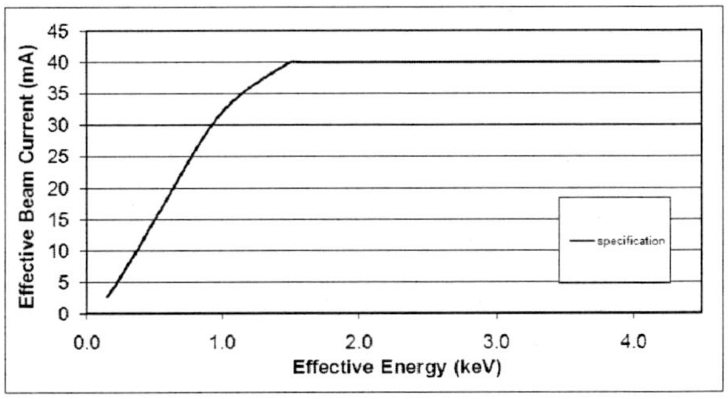

FIGURE 4: Beam Current Specification of the Optima HD Imax.

WAFER THROUGHPUTS

For energies over 1.5keV, the throughput advantage relative to a traditional high current tool is substantial. In Figure 5 is plotted the effective Optima HD Imax throughput vs. dose assuming a beam current of 40mA. Also shown in the same figure is the throughput enhancement relative to a standard high current tool. The standard high current tool was assigned a benchmark beam current of 9ma at 2keV which currently is achieved through the use of deceleration architectures and a comparable drift beam current would be lower.

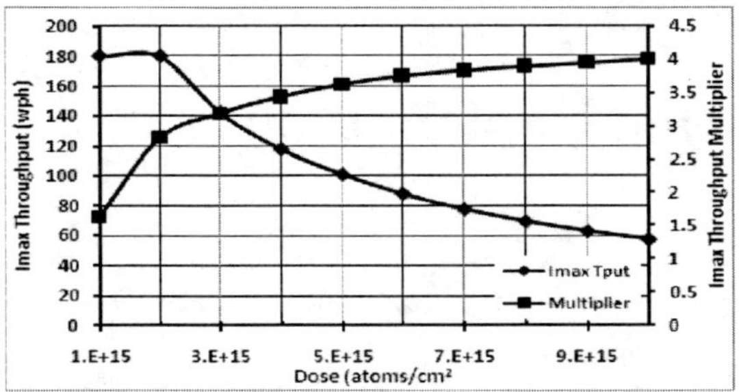

FIGURE 5: Optima HD Imax Throughput (wph) vs. Dose and comparison to throughput for standard high current implanter at 2 keV.

COST OF OWNERSHIP MODELS

The throughput advantages of the Optima HD Imax tool can be translated into cost per wafer production advantages. Shown in Figure 6 is a CPW ratio (Optima HD Imax to standard high current) as a function of beam current. For baseline assumptions the dose is $5x10^{15}$ atoms/cm^2 and the standard tool's beam current is again assumed to be 9mA, appropriate for 2keV. In this comparison, the time for cleaning has been incorporated into the analysis and is assumed to be 12%.

Finally, shown in Figure 7 are the sensitivities to a) the cleaning duration and b) time between cleans. Clearly, improved beam current and not down-time due to either of those is the most sensitive knob for CPW.

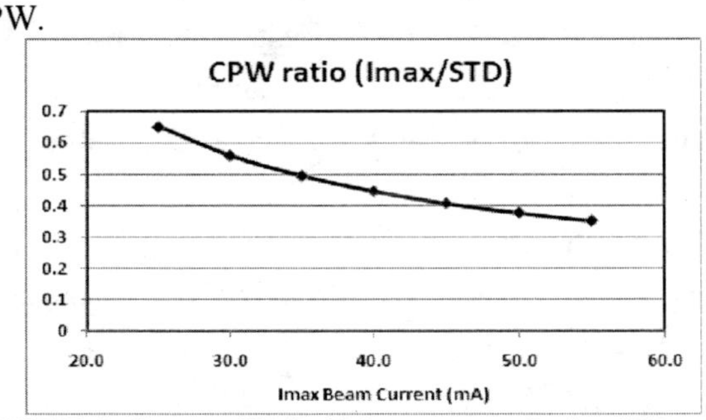

FIGURE 6: Cost per Wafer Enhancement: Optima HD Imax relative to standard high current implanter at 2 keV, 5e15 dose, as a function of Optima HD Imax beam current.

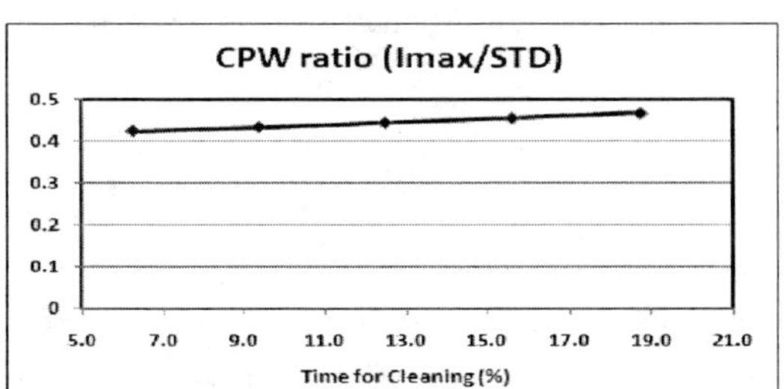

FIGURE 7a): Cost per Wafer Enhancement: Optima HD Imax relative to standard high current implanter at 2 keV, 5e15 dose, as a function of time for source cleaning.

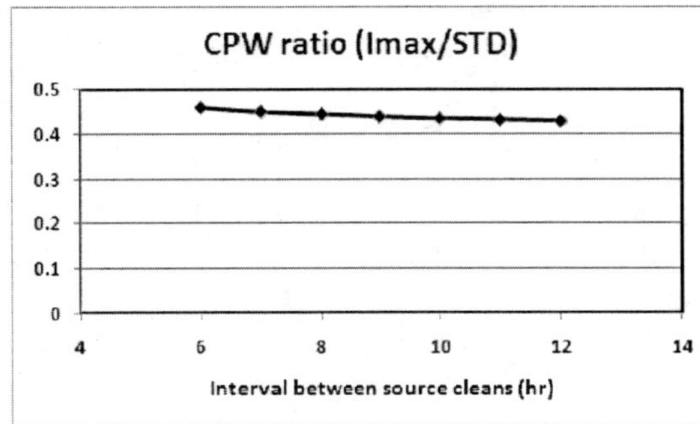

FIGURE 7b): Cost per Wafer Enhancement: Optima HD Imax relative to standard high current implanter at 2 keV, 5e15 dose, as a function of interval between source cleans.

SUMMARY

The Optima HD Imax is a two dimensional mechanical scan machine with molecular implant capability. The ion optics have been optimized as to support the generation and transport of molecular species for boron hydrides and carbon. Together with the beam current improvements, consideration for Cost per Wafer sensitivities such as utilization and operational considerations were provided. The productivity of the tool provides for significant benefits to the semiconductor manufacturer's cost of ownership equation as compared to traditional ion implant.

REFERENCES

1. P. Splinter, M. Graf, *et al.*, "Optima HD: Single Wafer Mechanical Scan Ion Implanter", in *Proceedings of the 16th International Conference on Ion Implantation Technology* 2006, pp 601-604.
2. T. N. Horsky, *et al.*, "Boron Beam Performance and in-situ Cleaning of the ClusterIon® Source, in *Proceedings of the 16th International Conference on Ion Implantation Technology* 2006, pp. 198-201.
3. T. Horsky, D. Jacobson, *et al.*, "Performance of the SemEquip Ion Source", in *Proceedings of the 14th International Conference on Ion Implantation Technology* 2002, pp.
4. T. Horsky, "ClusterIon Source for Cluster Implantation", in *Proceedings of the 14th International Conference on Ion Implantation Technology* 2004, pp.
5. A. Asi, "Boundary Element Method (BEM) for Charged Particle Optics," *Proceedings of 46th SPIE Annual Meeting*, San Diego, California, 2001
6. K.L. Brown, D.C. Carey, Ch. Iselin and F. Rothacker: Transport, a Computer Program for Designing Charged Particle Beam Transport Systems. See yellow reports CERN 73-16 (1973) & CERN 80-04 (1980).
7. D.R. Tieger *et al.*, "Optima HD Imax: Molecular Implant", in these Proceedings.

Metal Ion Sources for Ion Beam Implantation

W.J. Zhao, Z.Q. Zhao, and X.T. Ren

Institute of Heavy Ion Physics & State Key Labarotary in Nuclear Physics and Nuclear Technology, Peking University, Beijing 100871, P.R. China

Abstract. In this paper a theme touched upon the progress of metal ion sources devoted to metal ion beam implantation (MIBI) will be reviewed. A special emphasis will be given to some kinds of ion sources such as ECR, MEVVA and Cluster ion sources. A novel dual hollow cathode metal ion source named DUHOCAMIS will be introduced and discussed.

Keywords: Metal ion sources, Metallic ion beam implantation, Micro-metallurgy, Nano-clusters.

INTRODUCTION

The metallic ion beam implantation (MIBI) has been attracting much more attention since their potential applications in the area of surface modification of materials [1,2,3] and in semiconductors and insulators applied to Diluted Magnetic Semiconductor (DMS) [4,5], Electroluminescent (EL), Giant Magnetic Resistance (GMR), which are attractive for the developing field of spintronics and photonics.

Various methods to produce ions from metallic materials have been developed [6-11], including evaporation with different oven-technologies, metallic ions from volatile compound (MIVOC), cathode sputtering of metals, and ablation by vacuum arc or laser beam. However, different method should be combined with different kind of ion source to satisfy different applications.

For metal ion sources devoted to ion beam implantation, it should meet various requirements, such as: universal availability to produce a variety of ion species (specially for rare earth and transition metals), high average current intensity (from hundreds of microampere to several Milliamperes), good beam optics behavior (easy to handle and control the beam transport), low consumption rates of metals, long stable running time and long lifetime.

In this paper, the progress of metal ion sources devoted to MIBI will be reviewed. A special emphasis will be given to some kinds of ion sources such as ECR, MEVVA and Cluster ion sources. A novel dual hollow cathode metal ion source named DUHOCAMIS will be introduced and discussed.

ECR ION SOURCES

Compared with hot filament sources such as Freeman source, Electron Cyclotron Resonance (ECR) ion sources [12] have no hot filament, and can run with aggressive gases as well as long stable operation time and long lifetime. For the RF frequencies higher than 2.45 GHz, ECR sources can deliver high ion beam current with high charge states. When the RF frequencies are equal to or lower than 2.45GHz, this kind of ECR source is also named Microwave (MW) ion source [13], in particular, they are favored to feed corrosive elements or compounds such as oxygen, or halides for metallic ion beam production. It is one kind of ion source equipped for semiconductor ion implanter. It has been shown that various metallic ion beams of milliampere class have been obtained such as Al^+,Ti^+,Hf^+,Sc^+,etc, by using halides [14]. The method of MIVOC is better to be used for MW producing metallic ions, but evaporation would be not recommended because of metallic film deposition on the microwave window and other insulator parts.

CP1066, *Ion Implantation Technology,* edited by E. G. Seebauer, S. B. Felch, A. Jain, and Y. V. Kondratenko

Recently, high charge state-metallic ion beam production is increasing in interest to meet requirements of physical experiments, all methods to feed metal elements in to discharge have been used and developed, e.g. the inductively and resistively heated ovens [7,9], plasma heated rod [8], sputter [9] and MIVOC method [10] as well as ablation method with vacuum arc plasma mini-generator [15]. Some of metallic ion beams obtained from the ECR ion sources at IMP, CHINA are given in Table 1 [16]. These studies led to possibility of multiply charged metal ion beams which made it easy to reach high energy at low extraction voltage, but the average particle beam current is far from application of ion beam implantation now.

TABLE 1. Some of metallic ion beams at IMP, China. Unit: euA

	^{40}Ca	^{26}Mg	^{64}Zn	^{63}Cu	^{58}Ni	^{58}Ni*	Fe	Fe*
7^+		50						
8^+		70						
10^+				26		64	74	
11^+	132[287]		26	42			30	210
12^+	76[258]				30	74	25	175
13^+	208		50	39		57	20.5	142
14^+				34	40			
15^+			30	10	59	31	19	
16^+	[75]				78	17		25

Note: The data in [] were obtained by the big inner micro-oven, the data signed "*" by MOVIC.

VACUUM ARC ION SOURCES

The metal vacuum arc ion source (MEVVA /VARIS) is one kind of advanced ion source to produce high current, multiply charged and metallic ion beams with a large area [17]. This kind of source can provide ions of nearly all the metal elements in the periodic table, with a pulsed beam current of over one A for most heavy metallic ions. The average beam current of the MEVVA100, the biggest MEVVA in China, can reach 100 mA and the beam spot can be 800 mm in diameter [18]. Owing to macro particles and noisy discharge, it has now more applications to surface modification of materials with non-mass/charge analyzed ion beam implantation, plasma immersion ion implantation and metallic film deposition.

To solve the "noise problem" of the extracted ion beams, the GSI-MEVVA-4 and the LBNL-MEVVA-5 have been modified for many years, the successful experiment with a setup of GSI-VARIS for accelerator injection of U^{4+} beam has been made [19,20]. However, for metal ion beam implantation, to use ion beams with ion mass/charge analysis and definite ion energy as well as easy to control the beam transport, there are still a number of research work that should be made for this kind of ion source.

CLUSTER ION SOURCES

With growing interest in using cluster ion beams as alternative monomer-ion beams, various cluster ion sources have been developed, for example, Pulsed Arc Cluster Ion Source, Hollow Cathode Ion Source [21]. In our lab, a Magnetic Sputtering and Evaporating Cluster Source has been created [22,23]. Using this cluster ion source, we have prepared the embedded cluster insulator films such as Cu:CaF_2, Ti:CaF_2 and Al:CaF_2 films onto silica substrate, in which embedded metal can be prevented from oxidizing.

According to specific phenomena of cluster-surface interaction, the energy per atom of the cluster is of paramount importance [24-26]. Therefore, an ideal cluster ion source should generate ionized clusters with the definite element or composition, as well as the controlled size, flux and energy in a suitable range to meet different applications.

DUHOCAMIS—A DUAL HOLLOW CATHODE METALLIC ION SOURCE

A schematic diagram of DUHOCAMIS with a bottle-magnetic field structure is shown in Figure1 [27]. A tubular hollow sputter cathode is equipped with one anode and one cathode on each side. The ion beam is radially extracted through a slit in the tubular sputter electrode. The discharge mode of the source could be selected as dual hollow cathode discharge (DHCD) or PIG by regulating the electrical potential of the hollow sputter electrode.

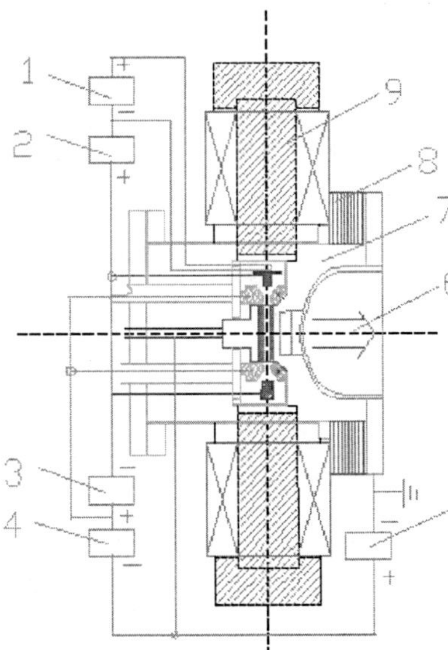

FIGURE 1. Schematic diagram of DUHOCAMIS

As a metal ion source, the notable characteristic for DUHOCAMIS is that it has a large internal sputtered surface of a long tube filled with very stable plasma, which is focused by a bottle-magnetic field; consequently it much enhances metallic ion yields and utilization ratio of sputtered materials, also ensures high stability and reproducibility for discharge parameters and ion beam parameters, as well as long lifetime. Some data of extracted ion beams are listed in Table 2. For high current ion beam implantation, it should be further enhanced the average beam current. It means that the source should be operated at the high duty cycle mode and improved the cooling condition.

TABLE 2. DHCD Ion Current Yields.

DHCD Ion Current Yields after mass analysis:
C-PIG-Magnet (bottle field 0.15T). Pulse (100 ms. 1ms)

Element Ion	Discharge Power	Ion Current	Extraction Voltage
27Al 1+	8 kW	45 mA	23kV
Ti 1+	22	24	25
Ti 2+	13	15	15
51 V 1+	14	13	13
Cu 1+	4	16	18
Mo 2+	7	10	22
Pb 2+	.8	2	12

CONCLUSION

Development of metallic ion sources, such as ECR, MEVVA, CLUSTER and DUHOCAMIS ion sources to generate high current ions for transition metals and rare earth metals, would promote the research and the applicAtion of spintronics and photonices, besides traditional micro-metallurgy. The requirements of metallic ion beam production for this aim can be considered similarly to the requirements for the semiconductor ion implantation; in that case controllable ion beams with mass analysis and definite particle energy are necessary. In the meantime, multiply charged ions and cluster ions can be available for ions with high energy and low energy, respectively.

ACKNOWLEDGEMENTS

This work was supported partly by NSFC under contracts of 19975003.

REFERENCES

1. N.J. Mikkelsen,J.Pedersen, C.A.Straede, *Surf. Coat. Technol.* **158-159** (2002) 42-47.
2. R.J.Rodriguez,A.Medrano, M.Rico, *et al.*, *Surf. Coat. Technol.* **158-159** (2002) 48-53.
3. W.J.Zhao,X.T.Ren,H.W.Zhao, *Rev. Scien. Instrum.* **77**, 03C113 (2006).
4. K.Potzger,S.Q.Zhou,H.Reuth,*et al.*, *Appl.Phys.Lett.***88**, 052508 (2006).
5. X.M.Lu, L.Shao, X.M.Wang,*et al.*, *Vaccu. Sci. Tech.* **B.20**,922 (2002).
6. B.H.Wolf, "Production of ions From Nongaseous Materials" in *Handbook of ion sources,* P331, B.H.Wolf, Ed. CRC Press, Inc.,(1995).
7. H.Koivisto,P.Frondelius,T.Koponen,*et al.*, *High Ener. Phys. and Nucl. Phys.* **31**, Supp.I (2007) 41.
8 M.Kiderra,T.Nakagawa,T.Kageyama,etal., *High Ener. Phys. and Nucl. Phys.* **31**, Supp.I (2007) 97.
9 R.Vondrasek, R.Scott,R.Pardo, High *Ener. Phys. and Nucl.Phys.* **31**, Supp. I(2007)101.
10. H.Koiviso,*et al.*, NIM, B94 (1994) 291-296.
11. P. Leherissier,C.Barue,C.Canet,*et al.*, *High Ener.Phys. and Nucl.Phys.* **31**, Supp.I (2007)85
12. R.Geller *et al.*, Proc.1st Int. Conf.Ion Sources, Saclay, 537 (1969).
13. N.Sakuda *et al.*, *Rev. Sci. Instrum.*, **48** (1977)762.
14. N.Sakuda, NIM, **B 21**, (1987)168.
15. A.V. Vodopyanov, *et al.*, *High Ener. Phys. and Nucl. Phys.*.**31**, Supp. I, 159 (2007).
16. Cao Yun, *et al.*, *High Ener. Phys. and Nucl. Phys.* **31**, Supp.I,. 70 (2007).
17. I.Brown, "The Metal vapor Vacuum Arc Ion Source" in *The Physics and Technology of Ion Sources,* P331, I.Brown, Ed. Wiley,New York, (1988).
18. A.D.Liu,*et al.*, *Surf. Coat. Technol.* 65-69, (2005) 193.
19. E. Oks,*et al.*, *Rev. Sci. Instrum.*, 73 (2002) 735.
20. R.Hollinger, et.al., *Rev. Sci. Instrum.*, 75 (2004) 1595.
21. http://www.physik.uni-rostock.de/cluster/
22. M. Habberland,M.Karris, in Averback Robert S, Ed *MRS Symposium Proceedings,* **206**, (1991): 291.
23. Zhao Zi-Qiang,Wei Lun-Cun,Wang Hao,*et al.*, *Nucl.Sci.Tech.*,7(1996)240.
24. Ziqiang Zhao, *Nucl. Tech.***25**(9) (2002) 699.
25. V.N.Popok, *et al.*, *Vacuum,* **76** (2004) 265.
26. M.Moseler, *et al.*, *NIM*,B 164-165 (2000)522.
27. W.J.Zhao, M.W.O.Mueller, J.Janik, K.X.Liu and X.T.Ren, *Rev. Sci. Instrum.* 79, 02B315 (2008).

Broad Ion Beam Extraction from Large Bore ECR Ion Source with Cylindrically Comb-Shaped Magnetic Fields Configuration by Feeding Simultaneously 11 to 13 GHz and 2.45 GHz Microwaves

Y. Kato[1], T. Satani[1], Y. Matsui[1], T. Watanabe[1], M. Muramatsu[2], K. Tanaka[3], T. Asaji[3], A. Kitagawa[2], F. Sato[1], T. Iida[1]

[1]*Osaka Univ., 2-1 Yamada-oka, Suita, Osaka 565-0871, Japan.*
[2]*NIRS, 4-9-1 Anagawa, Inage-ku, Chiba 263-8555, Japan*
[3]*Tateyama Machine Co. Ltd., 30 Shimonoban, Toyama, Toyama 930-1305, Japan*

Abstract. We tried to enlarge the operation window of an electron cyclotron resonance (ECR) ion source for producing the ECR plasma confined by cylindrically comb-shaped magnetic field, and for extracting the broad ion beam under the low pressures and low microwave powers. The magnetic field by permanent magnets constructs ECR zones at different positions for 2.45 GHz and 11 to 13 GHz microwaves, respectively. According to probe measurements, profiles of plasma density and temperature are different for using each single microwave. We conduct production of ECR plasma by launching simultaneously these two frequency microwaves, and obtain flat profiles of the electron density and the electron temperature. These profiles are not achieved by feeding single frequency microwave. It is found that plasma can be controllable on spatial profiles beyond wide operation window of plasma parameters. We conducted preliminary extracting and forming large bore ion beam from this source. We will make this source a part of tandem type ion source for the first stage. We investigated feasibility and hope to realize the device which has wide range operation window in a single device to produce many kinds of ion beams as like to universal source based on ECR ion source.

Keywords: Large bore ECR ion source, Two frequencies microwaves, Cylindrically comb shaped magnetic field.
PACS: 29.25Ni., 37.20+j, 52.50.Dg, 52.50.Sw.

INTRODUCTION

An electron cyclotron resonance ion source (ECRIS) has developed with the times [1,2]. Nowadays the ECRIS is used with not only semi-conductors but also with a wide variety of applications. For example, they are applied to heavy ion cancer therapy [3], investigation for discovery of new elements, and super-heavy nucleus science [4], and so on. Recently, the ion sources are not only used in producing multi-charged ions, but also cluster ions, iron encapsulated fullerenes which are important in the field of bio-nano technology [5]. A new type of ion source with a wide operation window is expected for various uses. We are planning to build the device which gives these required performance features in a single apparatus.

We developed a novel magnetic field configuration ECR ion source. The magnetic field configuration is constructed by a pair of comb-shaped magnetic field by all permanent magnets and has opposite polarity each other with ring-magnets. The comb-shaped magnet cylindrically surrounds the plasma chamber. This magnetic configuration suppresses the loss due to $\boldsymbol{E}\times\boldsymbol{B}$ drift, and then plasma confinement is enhanced. We tried many experiments by this ECRIS [6-8]. We will make this source a part of tandem type ion source for the first stage. We conduct preliminary extracting and forming large bore ion beam from this source [9]. Extraction broad-ion-beam from the first stage plasma and transfer like a shower to plasma generated by second stage ECRIS. We hope to realize a device which has a very wide range operation window in a single device to produce many kind of ion beams, *e.g.*, from multi-charged ion to cluster ion, nano-tube, fullerenes, including impurities trapping, etc, as a universal source.

In this paper, we try to control plasma parameters in an unused way, *i.e.*, by feeding simultaneously-2.45

CP1066, *Ion Implantation Technology,* edited by E. G. Seebauer, S. B. Felch, A. Jain, and Y. V. Kondratenko

GHz and 11 to 13 GHz frequency microwaves for broad-ion-beam extraction at the first stage.

EXPERIMENTAL

Outline of Experimental Device

The top view and the side view of the large bore ECRIS as the first stage of tandem ECRIS is shown in Figure 1. The base pressure is about 10^{-5} Pa. Two frequencies microwaves are supplied to the plasma chamber (200 mm in diameter and 320 mm in length) with the cylindrically comb-shaped magnetic fields. 2.45 GHz microwaves generated by magnetron (max. power: 1.3 kW) are transformed from the waveguide mode to the coaxial mode, and launched to the chamber along *z*-axis by L-shaped antenna. 11 to 13 GHz microwaves amplified by traveling-wave transformer amplifier (TWTA, max. power: 350 W) from the synthesizer are directly fed to the chamber beyond quartz window along to *z*-axis. Incident and reflected microwaves are tuned by the stainless steel meshed plate for the 11 to 13 GHz microwaves. The 2.45 GHz microwaves are tuned by the three stub tuner. The operating pressure is about 10^{-2} Pa. The plasma parameters are measured by Langmuir probe. The measurement position ranges vertically from the chamber wall (*y*=100mm) to the center of the chamber (*y*=0mm). Ion beam is preliminarily extracted from this source by the large bore extractor consisted of three electrode plates with multiply holes, and measured by the faraday cup (20 mm in diameter and 37 mm in length) located at *z*=360 mm, *i.e.* 263 mm apart from the electrode. The position and gap length of these electrodes have not yet optimized, and are now under investigation [9].

Configuration of Magnetic Field

Figure 2(a) shows side view of the cylindrically comb-shaped magnets surrounding the chamber, *i.e.* octupole magnets with a pair of ring magnets which polarity is opposite each other. Figure 2 (b) and (c) show contour plot and line forces in the cross section A-A' and B-B' indicated in Figure 2(a). The upper figures of Figure 2(d) and (e) depict contour plots and field lines in the radial and *z* direction cross section at the positions indicated by C and D angular cross section in Figure 2(b). The lower figures denotes field strength along field line labeled by (1)~(11) corresponding to the upper field lines. The positions of resonance zones for 11 GHz microwaves are formed around peripheral region nearby the chamber wall. Those for 2.45 GHz are formed around the center of the chamber, but vanish at the center cross section. The probe measurement conducted along to D angular cross section indicated in Figure 2 (b).

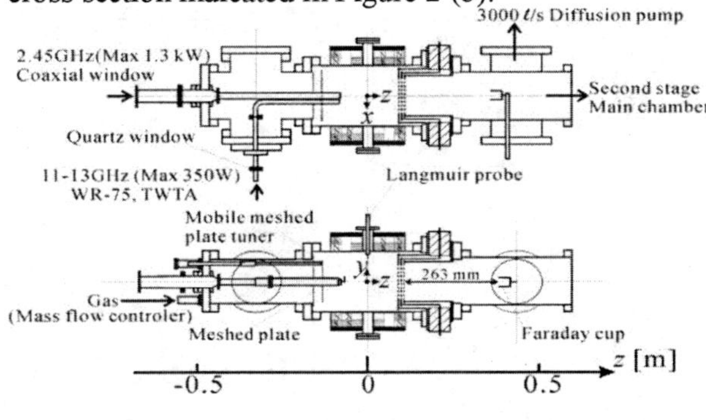

FIGURE 1. The large bore ECRIS as the first stage of tandem ECRIS.

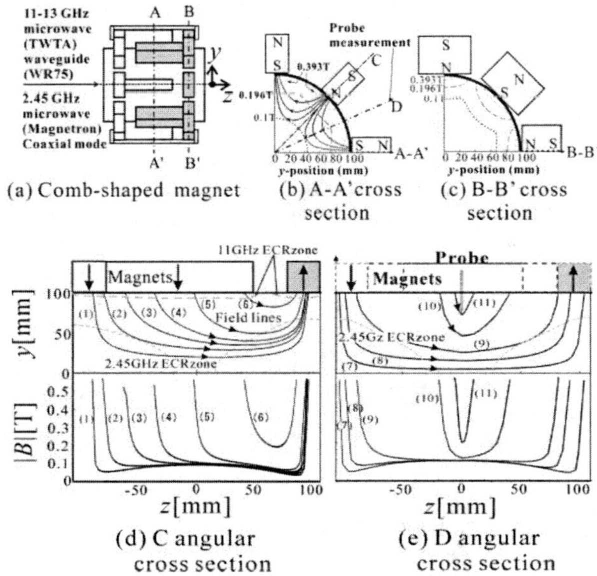

FIGURE 2. Side view of the cylindrically comb-shaped magnets (a). Contour plots of field strength and field lines in A-A' (b) and B-B' cross sections (c), respectively. Those in radial-z direction at C (d) and D angular (e).

RESULTS AND DISCUSSION

By using 2.45 and 11 to 13 GHz microwave individually, we can conduct experiments for production of ECR plasma and extraction of ion beams. We first investigate profiles of plasma parameters and characteristics in feeding each single frequency microwave to the ECRIS. And then we introduce simultaneously two frequency microwaves, measure the profiles, survey the experimental conditions, and confirm the effect of simultaneously feeding microwaves on the ECRIS plasma. We conduct preliminary ion beam extraction.

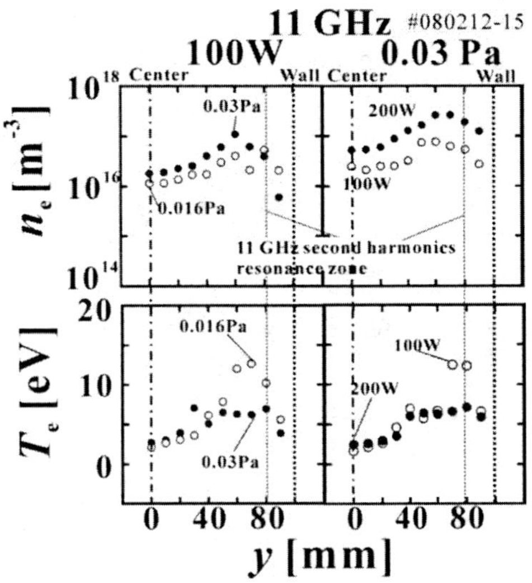

FIGURE 3. Profiles of n_e and T_e at the center in the cylindrically comb-shaped magnetic field ECRIS in the case of microwave frequency 11 to 13 GHz.

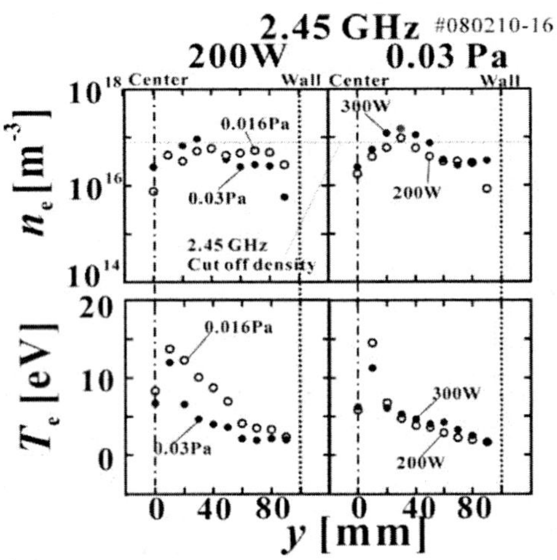

FIGURE 4. Profiles of n_e and T_e at the center in the cylindrically comb-shaped magnetic field ECRIS in the case of microwave frequency 11 to 13 GHz

ECR Ion Source Plasma Characteristics in Single Frequency Microwave

11 to 13 GHz Frequency Microwaves

Figure 3 shows the experimental results obtained by 11 GHz frequency microwaves. The upper and lower figures depict profiles the electron density n_e and the electron temperature T_e. The left and right side figures are the cases of microwave power 100 W at 0.03 Pa and 0.016 Pa, and those of pressure 0.03 Pa at microwave powers 100 and 200 W, respectively. Each horizontal axes indicate y directions, where y=0 and y=100 mm means the center and the chamber wall. The peak positions of the n_e and the T_e are observed around the second harmonics resonance zone at y=80 mm [8]. These tendencies, *i.e.* peaking in the peripheral region are caused from the peripheral formations of the fundamental and the second harmonic resonance zones, and are largely-unaltered by variation of microwave powers and pressures.

2.45 GHz Frequency Microwaves

Figure 4 shows the experimental results obtained by 2.45 GHz frequency microwaves. The upper and lower figures depict profiles of the n_e and the T_e. The left and right side figures are the cases of microwave power 200 W at 0.03 Pa and 0.016 Pa, and those of pressure 0.03 Pa at microwave powers 200 and 300 W, respectively. Horizontal axes are the same of Figure 3. It is found that the n_e attains and crosses over the cutoff density depicted by dotted lines in Figure 4 for 2.45 GHz under low microwave powers. The peak positions are observed around the center at y=0 mm in contradiction to the case of 11 GHz microwaves. While these facts are resulted by the construction of the ECR zone for 2.45 GHz frequency microwaves around the center, these tendencies are also largely-unaltered by variation of microwave powers and pressures. It is considered that the profile controlling by single microwave frequency in fixed magnetic field consisted by only permanent magnets are limited within operating pressure and microwave powers.

Profile Controlling by Two Frequencies of Microwaves

Next we try to control the n_e and the T_e profiles by feeding simultaneous two frequencies of microwaves, *i.e.* 11 to 13 GHz and 2.45 GHz. Our experimental procedure is the following: First we launch the 2.45GHz microwaves into the chamber flowing Ar operating gas, ignite, and then sustain the ECR discharge. We superimpose 11 to 13 GHz microwaves to this initial ECR plasma, *i.e.* simultaneous two frequencies microwaves operations. The appearance from the viewing port obviously changes to be more luminous in contrast to the case of single 2.45 GHz frequency microwaves. Finally we turn off 2.45 GHz microwave and hold the ECR plasma by only 11 to 13 GHz microwaves. We recognize the difference in plasma emissions in each case of single 2.45 GHz and 11 to 13 GHz. We conduct measurements of the n_e and the T_e profiles in these steps.

Figure 5 shows the typical n_e and T_e profiles obtained by these experimental results. The operating pressure is 0.03Pa. The microwave powers are 200 W and 100 W for 2.45 GHz and 11 GHz, respectively.

Figure 5 (a) shows the peaking profiles around the center and periphery regions for the 2.45 GHz (open circles) and the 11 GHz (open squares) lined by gray

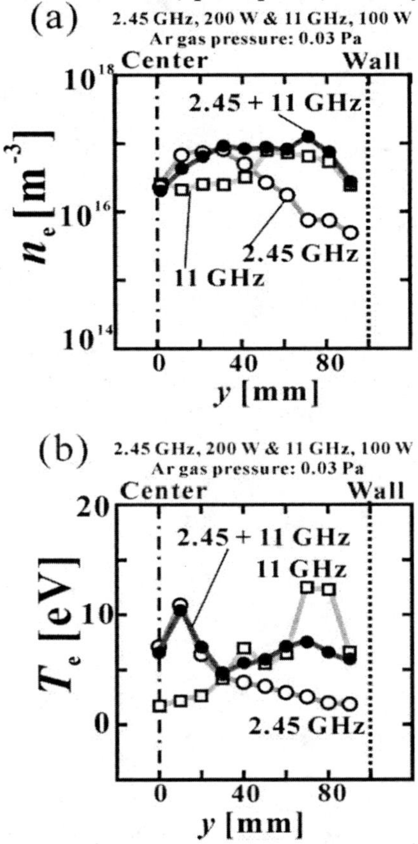

FIGURE 5. Profiles of n_e and T_e at the center in comb-shaped ECRIS in the case of two frequency microwaves.

lines, respectively. On the other hand, we obtain the flat profiles by launching simultaneous two frequencies microwaves as shown by closed circles and solid lines in Figure 5(a). The n_e profile can be controlled by varying each frequency microwave power. Figure 5(b) shows the T_e profiles corresponding to the n_e profiles as shown in Figure 5(a). The T_e profiles has basically the similar tendency to the case of the n_e, but flatting the T_e peaking is a little harder than that of the n_e at the low pressure. The T_e profile can be also controlled by varying each frequency microwave power.

The cylindrically comb-shaped magnetic field configurations are easy to scale up to more large size by selecting the number of multipole and the strength of the ring like magnets. The disadvantages of fixed magnetic configuration of ECR zone, operation and controllability of the profiles are redeemed by frequency-controllable TWTA, and moreover simultaneously feeding large different frequency microwaves as like our experimental results.

Preliminary Ion Beam Extractions

We conduct preliminary ion beam extraction from the cylindrically comb-shaped magnetic field configuration ECR ion source [9]. Now we are under optimizing extraction conditions of ion beams, *i.e.* the position, the gap length, the shape of the multi-hole extractor. We are investigating the effect of two frequencies of microwaves to the ion beam. At the present we obtain only slightly difference between two frequencies of microwaves, because of the luck of optimizing the extraction of ion beam, which is now under investigation.

ACKNOWLEDGEMENTS

The authors wish to thank Dr. A. G. Drentje, KVI, Professor Y. Yoshida, Toyo Univ., and Dr. S. Biri, ATOMKI for helpful discussions.

REFERENCES

1. R. Geller, *Electron Cyclotron Resonance Ion Sources and ECR Plasma*, IOP, Bristol, 1996.
2. A. G. Drentje, *Rev. Sci. Instrum*. **74**, 2631-2645(2003).
3. M. Muramatsu, *et.al*, contribution in IIT2008; A. Kitagawa, M. Muramatsu, M. Sasaki, S. Yamada, T. Sakuma, N. Sasaki, H. Takahashi, W. Takasugi, M. Yamamoto, S. Biri, K. Sudlitz, and A.G. Drentje; *Rev. Sci. Instrum*. **75**, 1476-1478(2004).
4. K. Morita, *et.al*., J. *Phys. Soc. Jpn*. **73**, 2593-2596(2004).
5. F. Éva, S. Biri, I. Iván, , *Fullerenes, Nanotubes and Carbon Nanostructures*, **15**, 249-256(2007); S. Biri, A.Valek, Kenez, A. Janossy, A. Kitagawa, , *Rev. Sci. Instrum*., **73** , 881-883(2002); K. Tanaka, *et.al*, contribution in IIT2008
6. T. Asaji, Y. Kato, F.Sato, T. Iida and J. Saito, , *Rev. Sci. Instrum*., **77**, 113503 1-6(2006).
7. Y. Kato, H. Sasaki, T. Asaji, T. Kubo, F. Sato, and T. Iida, , *AIP Conference Proceedings* **866**, 373-376(2006).
8. Y. Kato, T. Satani, T. Asaji, F. Sato, and T. Iida, , *Rev. Sci. Instrum*., **79**, 02A323 1-4(2008)
9. Y. Watanabe, *et.al*, contribution in IIT2008.

Optimization of Xenon Difluoride Vapor Delivery

Joseph Sweeney, Paul Marganski, Robert Kaim, Mike Wodjenski, John Gregg, Sharad Yedave, Steve Sergi, Steve Bishop, David Eldridge, and Peng Zou

ATMI, Inc., Danbury, Connecticut, USA 06810

Abstract. Xenon difluoride (XeF_2) has been shown to provide many process benefits when used as a daily maintenance recipe for ion implant. Regularly flowing XeF_2 into the ion source cleans the deposits generated by ion source operation. As a result, significant increases in productivity have been demonstrated. However, XeF_2 is a toxic oxidizer that must be handled appropriately. Furthermore, it is a low vapor pressure solid under standard conditions (~4.5 torr at 25°C). These aspects present unique challenges for designing a package for delivering the chemistry to an ion implanter. To address these challenges, ATMI designed a high-performance, re-usable cylinder for dispensing XeF_2 in an efficient and reliable manner. Data are presented showing specific attributes of the cylinder, such as the importance of internal heat transfer media and the cylinder valve size. The impact of mass flow controller (MFC) selection and ion source tube design on the flow rate of XeF_2 are also discussed. Finally, cylinder release rate data are provided.

Keywords: Xenon Difluoride, XeF_2, Ion Implant, Vapor Delivery, Chemical Etching
PACS: 47.60.Dx, 47.85.Dc, 44.20.

INTRODUCTION

Xenon difluoride (XeF_2) has recently been shown to be an effective in-situ cleaning agent for ion implant tools, resulting in increased source lives, reduced glitch rates, and mitigation of cross-contamination risks [1,2]. While the low vapor pressure of XeF_2 is a strongly desirable feature from a safety perspective, it poses challenges to the delivery of the vapor. Factors such as tubing and valve conductance, flow rate control, cylinder heat transfer (internal and external), and avoidance of condensation must be well understood in order to provide adequate and reliable vapor flow to the ion implant tool.

Two methods of cleaning have been under development. In the first method, termed "Dynamic" in-situ cleaning, XeF_2 flows at a constant rate directly into the arc chamber via a standard SDS® low-pressure drop gas stick. The primary purpose of this method is to clean deposits that may have formed within the arc chamber during the implantation process. In the second method, termed "Static" in situ cleaning, XeF_2 is dosed directly into the source or analyzer chamber. The flow rate is much greater than in "Dynamic" mode, with the intent of cleaning deposits that may be present within the broader source housing and analyzer chamber.

Solutions to the vapor delivery challenges associated with each cleaning process are described.

PHYSICAL PROPERTIES OF XENON DIFLUORIDE

Xenon difluoride is a white-to-colorless crystalline solid at room temperature with a relatively low vapor pressure (~4.5 torr at 25°C), and a packaging density of approximately 2 g/mL. It has a pungent, ozone-like odor, and, as an inorganic fluoride, OSHA and ACGIH each list the exposure limit at 2.5 mg/m^3.

Especially important for developing a vapor delivery system are the enthalpy of sublimation (55.2 kJ/mol) and the vapor pressure as a function of temperature, described by equation (1) and shown graphically in Figure 5 [3]. As XeF_2 is vaporized, the heat of sublimation results in a decrease in temperature of the solid. This manifests itself in a reduction in the XeF_2 vapor pressure, which can affect the maximum vapor flow rate of the material.

$$\log P_{(torr)} = \frac{-3057.67}{T} - 1.235 \log T + 13.97 \qquad (1)$$

VAPOR DELIVERY

To ensure reliable and sufficient XeF_2 vapor delivery, it is necessary to focus on the entire delivery system, including the cylinder design, delivery tubing

CP1066, *Ion Implantation Technology*, edited by E. G. Seebauer, S. B. Felch, A. Jain, and Y. V. Kondratenko

length and diameter, and mass flow controller, as each component represents a potential resistance to vapor flow. Additionally, it is important to understand the pressure-temperature profile along the delivery lines in order to avoid condensation, and thereby clogging.

Cylinder Design

Vaporization of solid XeF_2 is governed by heat and momentum transfer phenomena. A simplified model of the processes is shown in Figure 1. First, heat must transfer to the cylinder wall from the ambient air or from an external means of heating, such as a heating blanket. For the purposes of developing the XeF_2 cylinder, however, the intent was to not rely upon such an external heating means. Therefore the external heat transfer rate, Q_{cv}, is estimated from the convection heat transfer formula:

$$Q_{CV} = hA\left(T_{amb} - T_{w,o}\right) \tag{2}$$

where A is the cylinder surface area, T_{amb} is the ambient temperature, $T_{w,o}$ is the temperature of the exterior cylinder surface, and h is the heat transfer coefficient, which is obtained from the natural convection correlation presented by Cebeci [4]:

$$Nu = \left[1 + 1.446\left(Gr^{-0.25}\left(\frac{L}{D}\right)\right)^{0.904}\right] Nu_{FP} \tag{3}$$

Here, Nu is the Nusselt number defined as hL/k_{air}, Gr is the Grasholf number ($\beta\Delta TgL^3/\nu^2$), L is the cylinder, height, D is the cylinder diameter, k_{air} is the thermal conductivity of air, and Nu_{FP} is the Nusselt number for a vertical flat plate, details of which, along with the Grasholf number, can be found in any general text on heat transfer, such as that by Incropera and DeWitt [5].

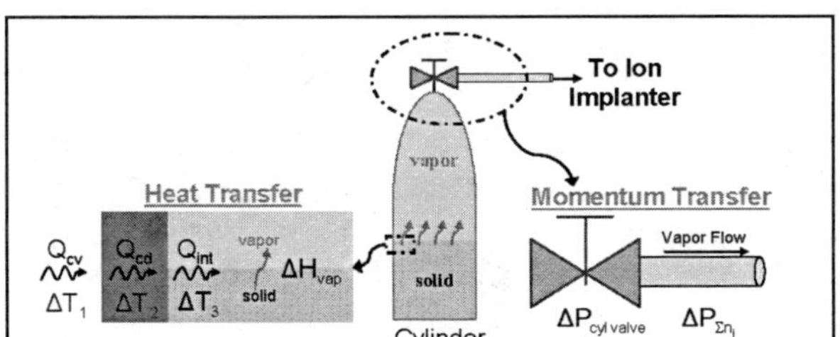

FIGURE 1. Heat and momentum transfer for the XeF_2 cylinder.

Second, heat must transfer via conduction through the cylinder wall. From an energy balance, this rate, Q_{cd}, must equal Q_{cv} at steady state. It can be expressed analytically by solving the radial form of the steady state heat equation with zero energy generation:

$$Q_{cd} = \frac{2\pi H k_{cyl}\left(T_{w,o} - T_{w,i}\right)}{\ln\left(r_2/r_1\right)} \tag{4}$$

where k_{cyl} is the thermal conductivity of the cylinder wall material, H is the cylinder height, $T_{w,i}$ is the inside wall temperature, and r_2/r_1 is the ratio between the outer and inner cylinder radii.

Finally, heat is transferred into the cylinder interior. By the same argument made earlier, this rate (at the inner cylinder surface) must be equal to Q_{cv}, and Q_{cd}, and it must be balanced exactly by the heat loss due to sublimation:

$$Q_{loss} = \dot{n}\Delta H_{vap} \tag{5}$$

where $\dot{n}$ is the molar sublimation rate of XeF_2. The radial temperature distribution within the cylinder is obtained by again solving the steady state heat equation, this time including an energy loss term [5].

$$T(r) = \frac{Q_{loss} r_1^2}{4Vk_{eff}}\left(1 - r^2/r_1^2\right) + T_{w,i} \tag{6}$$

Here, V is the volume of XeF_2 (solid phase), k_{eff} is the effective internal thermal conductivity of the cylinder, and Q_{loss} is taken to be negative. For a particular sublimation rate, equations 2-6 are used to determine the overall temperature distribution. This profile is then used to estimate the vapor pressure of XeF_2 in the cylinder (via equation 1). Once the vapor pressure is known, the pressure drop across the cylinder valve and additional delivery tubing and components can be determined. For the cylinder valve (or any downstream valve), expressions for choked and non-choked flow are taken from Swagelok [6]. For the delivery tubing, the following compressible flow expression is used [7]:

$$p_1^2 - p_2^2 = G^2 \frac{RT}{M_w}\left[\frac{4fL}{D} + 2\ln\left(\frac{p_1}{p_2}\right)\right] \tag{7}$$

Where p_1 is the upstream pressure, p_2 is the downstream pressure, G is the mass flux of XeF_2, R is the universal gas constant, T is the gas temperature, M_w is the molecular weight of XeF_2, L is the tube length, D is the tube diameter, and f is the friction factor. The total pressure drop that is calculated across the cylinder valve and subsequent delivery tubing and components must be equivalent to the difference between the vapor pressure of XeF_2 and the pressure of the chamber into which the vapor is being dosed (such as the arc chamber or source housing). Typically, an iterative process is required to determine the XeF_2 vapor flow rate that ensures this condition is met.

Simulations can be performed using equations 1-7 in order to evaluate the effect of various parameters. Of particular interest is to understand if an internal heat transfer medium is beneficial and to estimate the required cylinder valve size. For the purposes of the simulation, a cylinder of height 12" and diameter 4" is assumed – nearly identical to an SDS® JY cylinder. Additionally, the delivery tubing is assumed to have length of 15 feet and diameter of ¾ inches. Note that while most gas box tubing is no larger than 3/8", it is expected that in the "Static" in-situ cleaning application, XeF_2 will be delivered to the implanter via tubing separate from that of the gas box, and therefore a larger tubing could be used which would enhance the delivery rate.

In Figure 2, the theoretical steady state flow rate of XeF_2 is shown as a function of the internal effective thermal conductivity. While the authors are not aware of published data for the thermal conductivity of XeF_2 in granular powder form under vacuum conditions, it is likely to be small. For perspective, silica powder has a thermal conductivity between 0.05-0.1 W/(m·K) and evacuated silica powder has a conductivity almost 2 orders of magnitude smaller still [5]. Assuming the conductivity of XeF_2 is in a similar range, Figure 2 indicates that improving the internal heat transfer via the incorporation of internal conducting media will improve the vapor delivery rate significantly. However, once the effective internal conductivity is higher than about 4 W/(m·k), the additional benefit in flow rate is minimal because the external heat transfer rate is limiting.

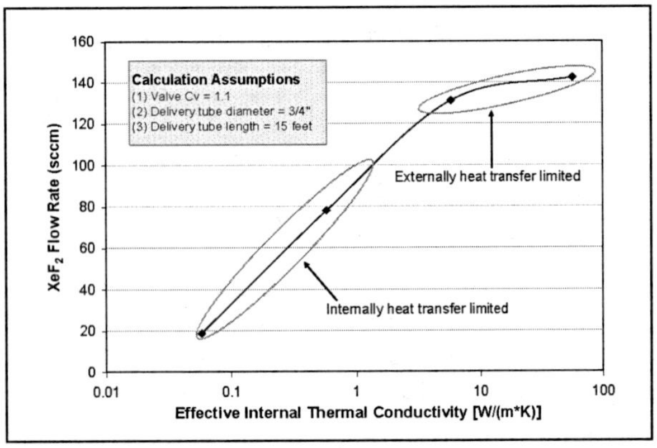

FIGURE 2. Effect of internal conductivity on XeF_2 flow.

The results of a second simulation are shown in Figure 3. In this case, the bulk internal thermal conductivity is taken to be approximately 5 W/(m·K). The steady state flow rate of XeF_2 is plotted against the cylinder valve C_v, which is an indicator of its conductance. For small valves, the XeF_2 flow is limited by the valve itself. As the valve size is increased (greater than about C_v = 0.8), the flow rate becomes limited by the delivery tubing.

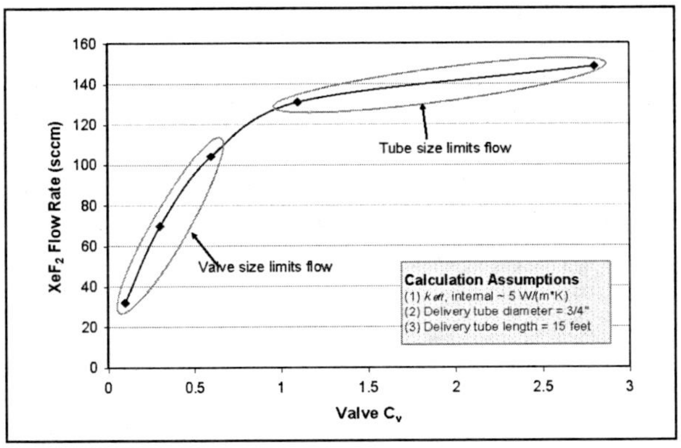

FIGURE 3. Effect of cylinder valve size on XeF_2 flow.

For cylinder design purposes, the intent is to ensure that the features of the cylinder – the internal thermal conductivity and the valve conductance – are not limiting factors of the XeF_2 flow rate for the majority of situations. Therefore, the cylinder is equipped with internal heat transfer media with a bulk conductivity of about 5 W/(m·K), and a cylinder valve with a C_v of about 1. Experimental results (not shown) have been within about 10-20% of the simulation results.

Vapor Delivery from the Gas Box

While the cylinder has been designed to deliver substantial flow rates of XeF_2 (important for the "Static" cleaning application), different delivery challenges arise in the "Dynamic" cleaning application in which the cylinder is installed within the gas box.

First, because the vapor pressure of XeF_2 is only about 4 torr, it is critical to use a low pressure drop mass flow controller such as those used on SDS® gas sticks. We have found that the newest low pressure drop MFC's – those specified to work down to 4 torr inlet pressure – can enable flow rates over 10 sccm in the absence of additional conductance limitations. This is compared to 1-2 sccm for standard low pressure drop MFC's.

A second vapor delivery challenge is on implanters that have 1/8" diameter tubing within the source housing. As shown in Figure 4, the maximum flow rates are limited to less than 2 sccm – data are from calculations using equation (7). When considering additional components that limit flow, such as control valves, and an MFC, the flow rates generally drop to less than 1 sccm. Field results have shown that for installations having 1/8" source tubing, flow rates can be as small as 0.2 sccm. For this reason, it is important to use ¼" diameter source tubing, which can enable flows of 1-4 sccm or higher depending on the MFC and stick design.

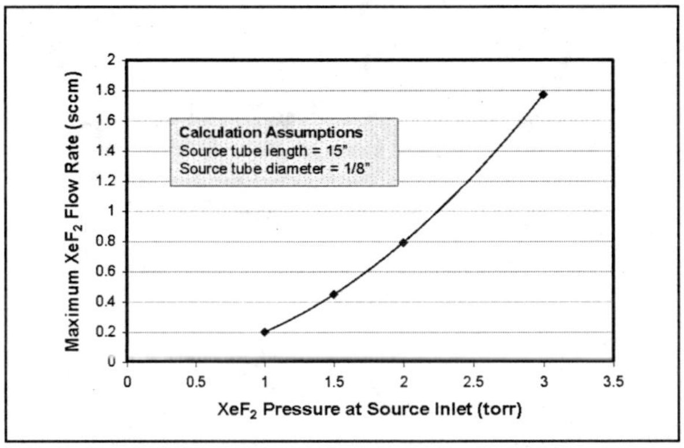

FIGURE 4. XeF_2 flow rate through a short 1/8" tube.

Another reason to replace 1/8" source tubes is to reduce the risk of condensation. Because XeF_2 is a solid at room temperature, if at any point in the delivery lines the temperature and pressure are such that their intersection is above the solid-vapor equilibrium curve shown in Figure 5, condensation can result. In one field installation, this was observed at the source flange which was actively cooled. Replacement of the 1/8" tubing with ¼" tubing reduced the XeF_2 pressure at the source flange and eliminated the condensation.

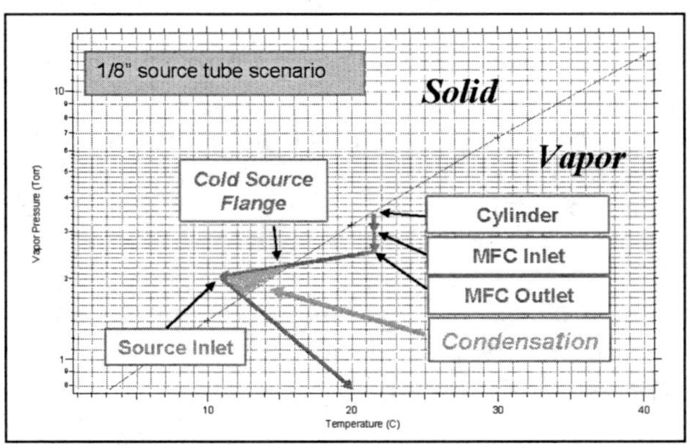

FIGURE 5. Vapor pressure curve of XeF_2 showing how condensation can result with 1/8" source tubes.

CYLINDER RELEASE RATE

A cylinder containing XeF_2 was opened to air in order to measure the release rate. Cylinder mass and HF concentration at the cylinder outlet were monitored with time (Figure 6). The results showed only a small mass increase (due to the mass of air rushing into the cylinder), and a very small HF concentration (~1 ppm) was detected at the cylinder outlet.

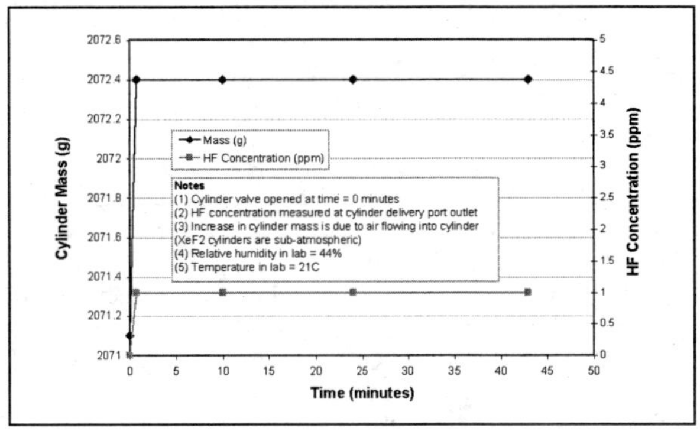

FIGURE 6. Release rate data for XeF_2.

CONCLUSION

A cylinder has been designed to maximize the vapor delivery rate of XeF_2 without the need for active heating. This has been accomplished through the use of internal heat transfer media and a relatively high conductance cylinder valve. Additionally, for gas box installations, it is important to use a low pressure drop MFC, as well as a ¼" diameter source tube. Finally, the XeF_2 release rate from a cylinder has been experimentally shown to be extremely small.

REFERENCES

1. J. Rolland, P. Lundquist, J. Dunn, S. Bishop, D. Eldridge, R. Kaim, "Insitu Chemical Cleaning for Improved Ion Implanter Utilization", *Solid State Technology*, p. 22, March 2008.
2. R. Kaim, S. Bishop, D. Eldridge, J. Mayer, J. Sweeney, O. Byl, S. Yedave, and P. Marganski, "In-Situ Ion Source Cleaning: Review of Chemical Mechanisms and Evaluation Data at Production Fabs", These proceedings
3. F. Shreiner, G. McDonald, and C. Chernick, "The Vapor Pressure and Melting Points of Xenon Difluoride and Xenon Tetrafluoride", *Phys. Chem.*, vol 72, no 4, 1162-1166 (1968).
4. T. Cebeci, "Laminar-Free-Convective-Heat Transfer from the Outer Surface of a Vertical Slender Circular Cylinder", in *Proceedings of the 5th International Heat Transfer Conference*, Section NC 1.4, Tokyo, vol. 3, (1974), pp. 15-19.
5. F. Incropera and D. DeWitt, *Fundamentals of Heat and Mass Transfer*, 5th ed., Wiley, New York, (2002), Chapters 3 and 9.
6. Technical Bulletin No. 8, August, 1994 – The Swagelok Company
7. R. Perry and D. Green, *Perry's Chemical Engineers' Handbook,* 7th ed., McGraw-Hill, (1997).

Tungsten Transport in an Ion Source

O. Byl, S. Yedave, S. Sergi, J. Sweeney, S. Bishop, R. Kaim, D. Eldridge

ATMI, 7 Commerce Drive, Danbury, Connecticut, USA

Abstract. We present a study that gives a deeper insight into the failure mechanism of tungsten ion sources used for ion implantation in the semiconductor industry. Our experimental data, corroborated by literature thermodynamics reports, indicates that tungsten migration is the primary cause of failure for processes utilizing fluorine-containing feed gases. It is shown that tungsten is the predominant constituent of most deposits, and that volatile tungsten fluorides are the transport species responsible for metal redistribution in ion sources. It was also demonstrated that AutoClean^{TM}, a new product of ATMI, can be used for filament growth control.

Keywords: Tungsten transport, ion source failure, ion implantation, AutoClean, xenon difluoride.
PACS: 85.40.Ry; 55.77.Dq.

INTRODUCTION

Tungsten has one of the highest melting points, lowest vapor pressures and lowest coefficient of thermal expansion among metals. Due to these properties it has become a preferred material for construction of arc chambers in ion sources used for ion implantation in the semiconductor industry. Ion implantation process involves ionization and fragmentation of a feed gas in a plasma constrained in the arc chamber, extraction and separation of implant species, and their implantation into silicon wafers. Due to the energetic nature of this process, ion source performance deteriorates with time, requiring laborious maintenance procedures to ensure proper operation. In many cases ion source failure involves insulator plating, filament breakage or growth, flaking and shorting of electric components, filamentary growth at the aperture, etc. The extent and domination of any of these failures depend upon the type of feed gas and tool operation conditions. Operation of an ion source with fluorine containing feed gases (BF_3, GeF_4, SiF_4) usually requires more frequent preventive maintenance interventions due to extensive tungsten deposition onto the arc chamber components. The problem has been known to result from chemical transport of tungsten within the arc chamber [1] and is normally addressed by balancing fluoride and hydride recipes. However, no reliable in-situ solution has been reported to-date. In this article we report the result of a study of tungsten transport in the arc chamber and demonstrate a method of filament growth control by application of AutoClean, a new product of ATMI.

MATERIALS AND METHODS

AutoClean Technology

AutoClean is an innovative combination of materials, packaging and process optimization developed by ATMI for improvement of ion source lifetime [2, 3] AutoClean technology is based on regular application of XeF_2 to maintain cleanliness of ion sources by residue removal due to chemical reaction of XeF_2 with deposits. The reaction products are predominantly volatile fluorides that can be removed by pumping.

Ion Source Test Stand

The Ion Source Test Stand (Test Stand hereafter) is a state of art tool that enables investigation of the performance of many materials used in ion implantation as well as the behavior of different parts of the ion implanter. In simple terms the Test Stand is a fully functional ion implantation tool without a wafer handling system. More details about the Test Stand can be found elsewhere [4].

XeF_2 *Ex-situ* Cleaning Tests

Ex-situ cleaning of contaminated arc chamber parts was carried out to test the ability of XeF_2 to clean specific parts as well as to find out the chemical

CP1066, *Ion Implantation Technology,* edited by E. G. Seebauer, S. B. Felch, A. Jain, and Y. V. Kondratenko

composition of effluent gases. The experiments were performed in a stainless steel test chamber equipped with a load lock, pressure transducers, vacuum generation system and gas dosing manifold. Progress of the reaction could be monitored with a residual gas analyzer by continuously leaking the test chamber gases into the RGA compartment via a leak valve. Additionally, gas phase composition could be probed with an in-line FTIR spectrometer.

Typically, a test consisted of pre-test X-ray dispersion (EDX) analysis of the deposits. Then, after weighing the part, it was loaded into the test chamber and cycle-purged for moisture removal. The part was then exposed to XeF_2 for some time with FTIR spectroscopy analysis of the gas phase products. The cycle was repeated several times until no more reaction products were detected. Upon cleaning the part was weighed and EDX analyzed again.

RESULTS

Parts *Ex-situ* Cleaning with XeF_2

Ex-situ cleaning of arc chamber parts has been carried out to determine the composition of the residues. In this study we used an arc chamber equipped with alumina insulators and a graphite repeller in a 52 hour process with 5.5-6.0 mA B^+ beam current at 40 keV extraction voltage with BF_3 as a feed gas. The insulators and the repeller were observed to be substantially metalized upon post-run examination of the ion source. The results of EDX analysis of the deposits pointed out that tungsten was the main component of the residue.

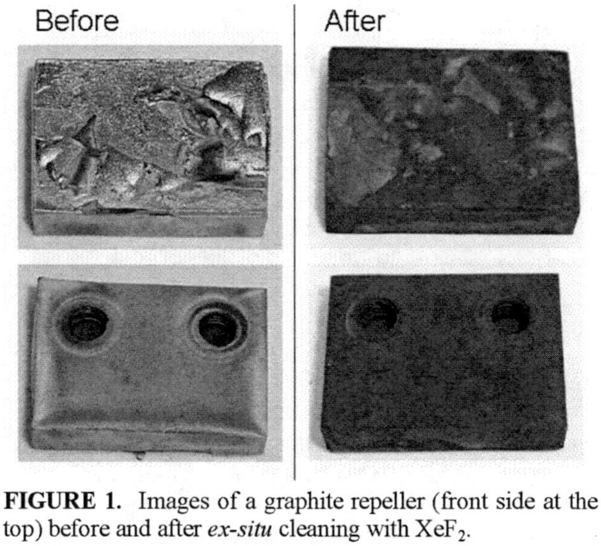

FIGURE 1. Images of a graphite repeller (front side at the top) before and after *ex-situ* cleaning with XeF_2.

Detection of boron by EDX analysis was not possible due to detector limitations, and to compensate this, XeF_2 *ex-situ* cleaning of all parts was carried out. Our previous tests indicated that XeF_2 reacts with boron and tungsten at room temperature generating only BF_3 and WF_6. Since the arc chamber walls and the filament were made of tungsten, only results from cleaning of the insulators and the repeller could be used for determining the composition of the deposits. Moreover, cleaning of the repeller gave the most accurate results because of the large amount of the deposit.

Figure 1 shows repeller images before and after cleaning. As one can see, all the metal deposits were removed. Only BF_3 and WF_6 reaction products were detected by IR spectroscopy, and therefore the cleaning end-point was determined by the absence of BF_3 and WF_6 and presence of XeF_2 in the gas phase. Since the repeller weight loss due to the cleaning reactions, $\Delta m(sample)$, and the total weight of XeF_2 used $m(XeF_2)$ were known, the weight of boron $m(B)$ and tungsten $m(W)$ could be determined by solving a system of mass balance equations (1), where $M(B)$, $M(W)$, $M(XeF_2)$ are atomic or molecular weights of boron, tungsten and xenon difluoride respectively.

$$\begin{cases} m(B)+m(W)=\Delta m(sample) \\ 3\dfrac{m(B)}{M(B)}+6\dfrac{m(W)}{M(W)}=2\dfrac{m(XeF_2)}{M(XeF_2)} \end{cases} \quad (1)$$

The solution of (1) indicates that boron constitutes less than 1% (at. %) of the deposit.

Filament Growth Controlled by AutoClean Process

After determining that tungsten is the major constituent of the deposits and learning from the literature that the direction of W transport can be controlled by changing the temperature of the part, we attempted to control filament weight gain or loss by application of *in-situ* AutoClean and by controlling filament temperature in a Bernas ion source. Since it was difficult to measure the filament temperature directly (especially in plasma conditions), the filament current was used as a metric for temperature. Figure 2 shows the rate of filament weight change relative to its current. Filament weight change rate was determined with and without plasma. In plasma-off conditions one can see three regions with respect to the current: 1) 0 – 50 Amps with small weight loss, 2) 60 – 100 Amps with 54 to 38 mg per hour weight loss, and 3) above 120 Amps filament weight was increasing with current. In plasma-on conditions we observed only filament growth.

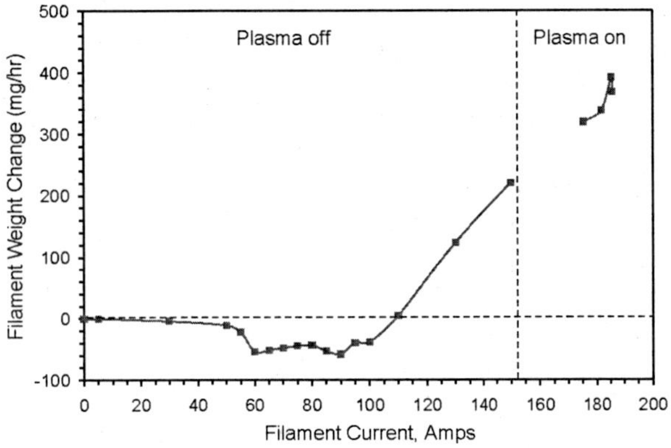

FIGURE 2. Filament weight change with filament current. Bernas ion source. XeF_2 flow: 2.2sccm.

WF_x Generation by Fluorine-Containing Implant Gases

It is known that tungsten fluorides (WF_x, x = 1,2,.6) are the carriers of tungsten within arc chambers. To determine the difference between XeF_2 and the three major fluorides (BF_3, SiF_4 and GeF_4) used for ion implantation, their mass spectra were measured under the same source plasma conditions.

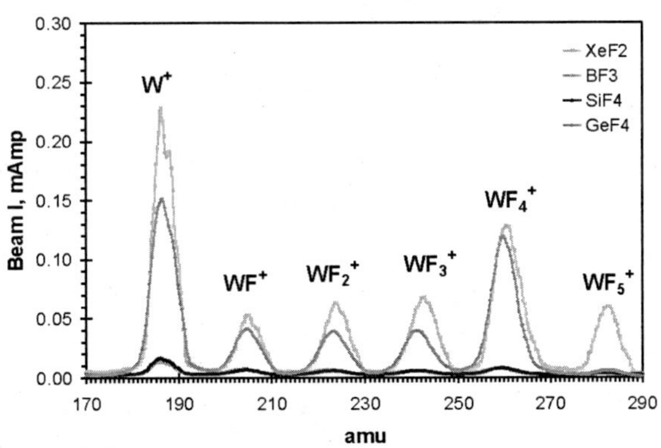

FIGURE 3. (in color) Mass spectra of XeF_2, BF_3, SiF_4 and GeF_4 – WF_x^+ region. Gas flow: 1 sccm, arc voltage: 90-100 V, arc current: 0.5 Amps, extraction voltage: 20 keV.

Figure 3 shows the 170 – 290 amu region where W^+ and WF_x^+ exhibit their peaks. XeF_2 generates the highest amount of tungsten and tungsten fluorides, which is expected because of the high reactivity of XeF_2 with tungsten. GeF_4 generates slightly lesser amounts of WF_x^+ than XeF_2. Both BF_3 and SiF_4 generate the smallest amounts of W^+ and WF_x^+ relative to XeF_2 and GeF_4. When normalized with respect to WF^+, (not shown here) the profiles of GeF_4, SiF_4 and BF_3 are quite similar in the 170 – 290 amu region with the exception of GeF_4 in the 250 – 270 amu region where it exhibits 2x intensity for the WF_4^+ peak. It should be pointed out that only XeF_2 generates a substantial amount of WF_5^+, which is believed to come from WF_6 fragmentation.

DISCUSSION

Tungsten Transport with Plasma Off

Without the influence of plasma, tungsten transport in an arc chamber is defined by the fluorine concentration and temperature distribution within the arc chamber. A simplified mechanism of XeF_2 interaction with tungsten can be described as follows: once XeF_2 is introduced into the arc chamber it spontaneously reacts with tungsten surfaces, releasing xenon as a gas and generating surface fluorides that quickly desorb as WF_x with x = 6 for temperatures below 1200K [5, 6]. The reaction rate increases with temperature [7] which results in unequal etching in a system with temperature gradients. In the case of a Bernas arc chamber the filament is the source of the heat and at low filament currents the temperature is not high enough to compensate for its small surface area and therefore most of the etching occurs on the arc chamber walls. As the filament temperature increases the etching of the filament becomes substantial, giving rise to the observed negative weight change in Figure 2 (60 – 100 Amps region).

As the filament temperature increases, the weight loss turns into a gain due to tungsten transport from the arc chamber walls to the filament. To explain the tungsten transport phenomenon one should consider the following chemical equations:

Etching	$W(s) + XeF_2(g) \rightarrow WF_x(g) + Xe$
Deposition I	$WF_x(g) \leftrightarrow W(s) + WF_y(g)$ $(y<x)$
Deposition II	$WF_x(g) \leftrightarrow W(s) + F_z(g)$
W vaporization	$W(s) \leftrightarrow W(g)$

The etching reaction generates WF_x with x decreasing with temperature. As WF_x readsorbs on tungsten surface it may decompose leaving tungsten behind and desorb as a lower fluoride (Deposition I). This process starts to dominate at higher temperatures and results in tungsten deposition onto the filament. At very high temperatures tungsten fluorides decompose according to the Deposition II reaction, further enhancing tungsten deposition. At extremely high temperatures (~3000 K) tungsten starts to vaporize. Since temperatures are not this high, vaporization does not contribute greatly to tungsten transport in the arc chamber.

In a hypothetical closed system with tungsten walls and some initial fluorine concentration at uniform temperature, a state of equilibrium is reached with a certain gas phase composition. The equilibrium may be characterized by the total amount of tungsten in the gas phase relative to the total amount of fluorine, $n(W)/n(F)$. This is a function of temperature with a maximum value dependent on temperature and the amount of fluorine [8]. At an initial theoretical fluorine pressure of ~10 mTorr the $n(W)/n(F)$ ratio exhibits a maximum between 1500 and 2200 K. This becomes important when a system has temperature gradients, as in the case of the arc chamber with hot filament and at constant flow of XeF_2. At steady state conditions, local equilibria will be established, resulting in gradients of gas phase tungsten concentrations and causing tungsten transport from the locations with high $n(W)/n(F)$ values to those with low $n(W)/n(F)$.

The filament growth above 100 Amps (Figure 2) can be explained in terms of low $n(W)/n(F)$ value over the filament relative to that over the other parts of the arc chamber.

Tungsten Transport with Plasma On

With plasma on, tungsten transport becomes perturbed with the following processes: 1) ionization and fragmentation of gas phase species, 2) ion sputtering, and 3) ion extraction into the beam. Even though these processes add substantial amount of complexity still the concept of tungsten transport due to the difference in local concentration of tungsten in the gas phase may be very useful for explaining some problems encountered during ion implantation. As an example, we would like to interpret the following problem: frequently IHC sources develop a cone-shaped deposit on the anticathode with no deposition on the cathode or minor signs of cathode etching in processes that use fluoride gases. Since the cathode and anticathode are at the same potentials and have roughly the same areas, they both experience similar sputtering effects due to influx of WF_x^+ ions and WF_x neutrals. However, the cathode is at higher temperature with high enough local $n(W)/n(F)$ value to compensate the total influx of tungsten whereas the anticathode being at lower temperature has lower $n(W)/n(F)$ that results in anticathode growth. The difference between filament growth for Bernas ion sources and no growth for the IHC cathode in this instance may result from lower temperature of the IHC cathode cup relative to the filament in the Bernas source. It also should be mentioned that in this case the attraction of WF_x^+ ions toward the cathode and anticathode plays an important role as it does not allow tungsten transport toward the arc chamber walls.

Anticathode growth is a problem for GeF_4 process. Figure 3 shows that GeF_4 generates nearly the same amount of tungsten fluorides as XeF_2 and about an order of magnitude higher than BF_3 and SiF_4. Moreover, because of the low intensity of WF_5^+ for GeF_4 relative to that of XeF_2 it is possible that GeF_4 plasma is even richer in gas phase tungsten than XeF_2 if corrected for the same amount of fluorine delivered into the arc chamber. The exact nature of this phenomenon is under investigation now, but it is obvious that the weakness of Ge-F bond relative to B-F or Si-F contributes to elevated levels of fluorine in the arc chamber for GeF_4 processes, dramatically enhancing the effects of tungsten transport and shortening ion source lifetime.

CONCLUSIONS

In this study we demonstrate the usefulness of the concept of local concentration of gas phase tungsten derived from tungsten transport phenomenon in explaining more complex problems related to tungsten redistribution in arc chambers of ion sources. We also have shown that AutoClean can be successfully used to mitigate some of these problems.

REFERENCES

1. A. Renau, "Ion Sources," in *Ion Implantation Science and Technology,* edited by J.F.Ziegler, Elgewater, MD: Ion Implantation Technology Co., 2006, p. 13-33.
2. R. Kaim, S. Bishop, D. Eldridge, J. Mayer, J. Sweeney, O. Byl, S. Yedave, P. Marganski, *These Proceedings.*
3. J. Dunn *et al.*, in these Proceedings.
4. S. Yedave, J. Arno, S. Bishop, F. Dimeo Jr., R. Kaim, L. Wang, "ATMI's Ion Implant Process Efficiency Research Laboratory (IIPERL)" in *Proceedings of the 16th International Conference on Ion Implantation Technology 2006*, edited by K. J. Kirkby *et al.*, AIP Conference Proceedings 866, American Institute of Physics, Melville, NY, 2006, pp. 651-654.
5. H. F. Winters, *J. Vac. Sci. Technol. A* **3**, 700-704 (1985).
6. W. Chen and J. T. Roberts, *Surf. Sci.* **324**, 169-184 (1995).
7. A. Bensaoula, J. A. Strozier, A. Ignatiev, J. Yu, J. C. Wolfe, *J. Vac. Sci. Technol. A* 5, 1921-1924 (1987).
8. G. M. Neumann, *J. Fluorine Chem.* **1**, 473-486 (1971/72).

Decreasing Beam Auto Tuning Interruption Events with *In-Situ* Chemical Cleaning on Axcelis GSD

Dieter Fuchs[1], Stefan Spreitzer[1], Josef Vogl[1], Steve Bishop[2], David Eldridge[2], Robert Kaim[2]

[1] *Infineon Technologies AG., Wernerwerkstr.2 , 93049 Regensburg, Germany*
[2] *ATMI Inc., 7 Commerce Drive, Danbury, CT 06810, USA*

Abstract. Ion beam auto tuning time and success rate are often major factors in the utilization and productivity of ion implanters. Tuning software frequently fails to meet specified setup times or recipe parameters, causing production stoppages and requiring manual intervention. Build-up of conductive deposits in the arc chamber and extraction gap can be one of the main causes of auto tuning problems. The deposits cause glitching and ion beam instabilities, which lead to errors in the software optimization routines. Infineon Regensburg has been testing use of XeF_2, an in-situ chemical cleaning reagent, with positive results in reducing auto tuning interruption events.

Keywords: Process stability, In-situ cleaning, Chemical cleaning, Beam auto tuning, Ion source life time.
PACS: 85.40.Ry; 81.65.Cf; 52.77.Bn

INTRODUCTION

Before the introduction of indirectly heated cathode ion sources the main reason for change of the ion source was simply a broken filament. With the introduction of indirectly heated cathode ion sources it turns out that sources have to be changed more often due to shorts caused by deposits on the insulators or build up of flakes inside the ion source arc chamber and the extraction/suppression electrodes in front of the source. To minimize source failures due to deposits and to extend source life, an in-situ cleaning process based on XeF_2 as reagent was developed [1].

But before an ion source fails, the conductive deposits influence the stability of the ion beam as well as the success rate and the duration of the auto tune sequences of an ion implanter. Both the source life and the auto tuning success rate have an impact on the overall utilization efficiency of production ion implanters.

The positive effect on ion source life in ion implanters of the newer generation has already been reported [2, 3]. We have tested the chemical cleaning on an Axcelis NV-GSDE200 high current implanter using an ELS source with a tungsten arc chamber. In addition to the ion source life, we monitored the auto tune interruption events and alarms and the duration of these events. We only took into account events and alarms which occurred when the tool was set to "productive" mode by the fab automation system. Standby or down periods were not regarded for the auto tune behaviour and alarms.

EXPERIMENTAL

The etch gas was introduced into the ion source like a standard dopant gas. The XeF_2 gas bottle has to be installed in an SDS gas string, due to the low vapor pressure of XeF_2 (about 4 Torr at room temperature). As there was no spare gas string in the tool, we used the phosphorous SDS gas string to install the bottle. For the cleaning sequences we ionized the XeF_2 and were running Xe^+ beams for 15 min every 12h. The gas flow was 2 sccm, resulting in a beam current of about 350 µA.

The implanter we used was running less than 5% phosphorous beams before we installed the xenon difluoride and about 60% boron beams. This ensures that the load of the ion source was identical in the chemical cleaning test phase and the reference period before.

At the beginning the Xenon beam was started and stopped manually by the operator, but this does not ensure a well defined cleaning time of 15 min. Therefore we implemented the clean as an implant on dummy wafer with a certain dose. The dedicated clean time comprises beam tune time, implant time and

CP1066, *Ion Implantation Technology,* edited by E. G. Seebauer, S. B. Felch, A. Jain, and Y. V. Kondratenko

unload time. After unloading the dummy wafers, the fab automation started the next production run.

Interruption events and alarms were monitored via APC software. Figure 1 shows the histogram of the regarded events during a source life. The duration of interruptions and additional retunes were extracted from the log files of the tool.

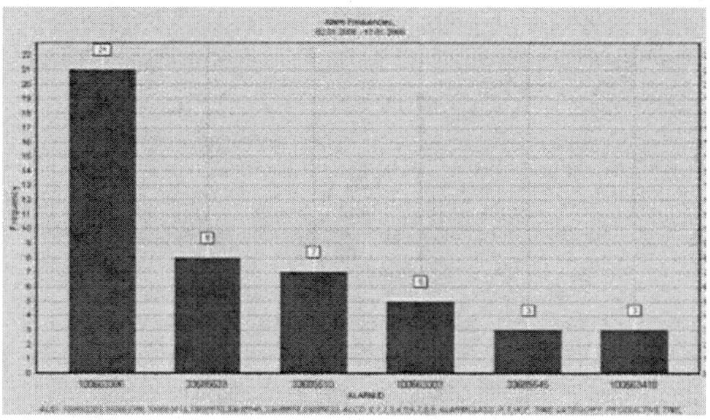

FIGURE 1. Example of a histogram of interruption events occur during an ion source life (productive time).

RESULTS AND DISCUSSION

Starting the tests with the in-situ clean for ion sources, first we checked if we could see any influence on the defect density at the wafer level. Particle tests directly after the clean cycle and within the next ten runs did not result in a different particle level. Regarding the implemented particle tests, we did not observe any difference between the periods with and without in-situ clean (see Figure 2).

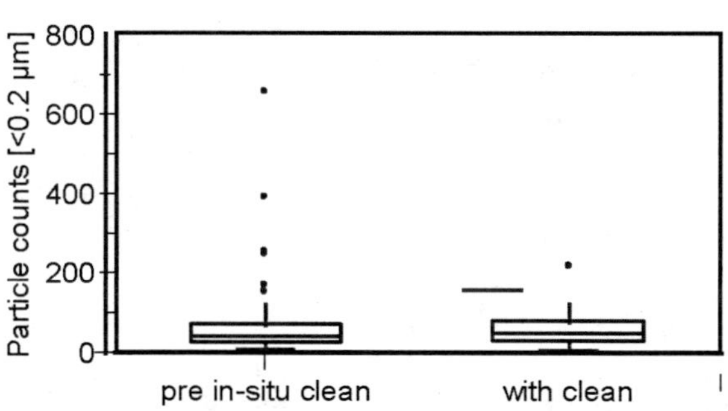

FIGURE 2. Comparison of the defect density values prior the in-situ clean and within the test period.

A) Ion Source Life

Starting with the clean cycles we observed a significant increase of the source life, but not of the magnitude reported in other publications [2, 3]. The ion source life increased about 24% from an average of 168 h to an average of 208 h.

In addition we noticed that the arc chamber walls got a very rough surface by using the in-situ clean.

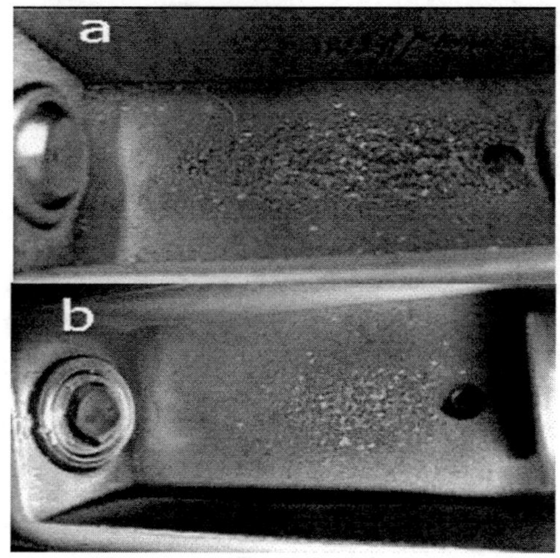

FIGURE 3. Roughness of the arc chamber walls a) with the use of in-situ clean (15 min every 12 h) and b) before the implementation of in-situ clean.

To reduce the etching effect at the arc chamber walls we reduced the duration of the clean from 15 min to 10 min per cycle. The shorter clean time was still effective enough to keep the insulators of the ion source clean.

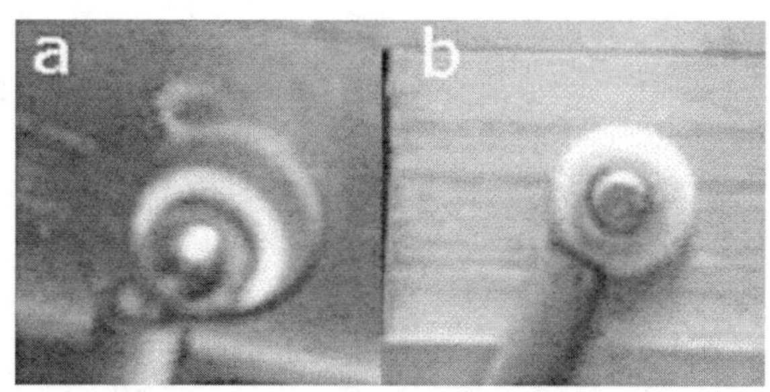

FIGURE 4. Cleanness of insulators of used ion sources a) without in-situ clean, b) with reduced clean time of 10 min per cycle.

The reduction of the clean time not only shows up with a smoother surface of the arc chamber walls, it results as well with an additional increase of the ion source life time. With the shorter clean time we achieved an average source life time of about 340 h;

this indicates an improvement of about 100%. Detailed ion source lifetime data are shown in figure 5.

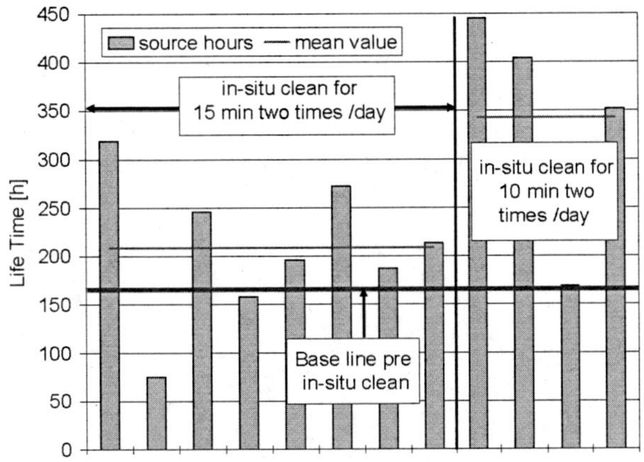

FIGURE 5. Source life times achieved with different cleaning times compared to the base line without in-situ cleaning.

B) Beam Auto Tuning Interruption Events

A cleaner ion source and high voltage apertures should result in more stable ion beam and better auto tuning behavior of the beam. Since the Axcelis GSD 200 high current implanter does not monitor glitching events, we chose the following interruption events and set up failures for monitoring the stability of the beam:

Auto beam recovery event
Excessive beam noise
Unstable beam
Maximum beam below required beam
Auto tune fails, iteration limits exceeded

These events were monitored for each source life when the tool was set to the "productive" status by the fab automation, and the data was then summed over consecutive 30 day periods of productive time.

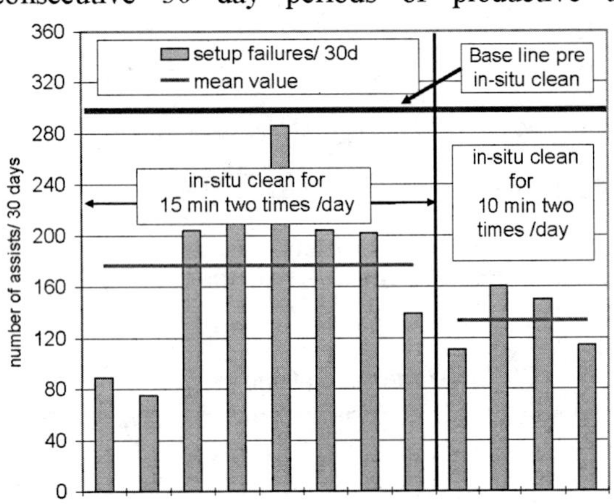

FIGURE 6. Auto tune interruption events with in-situ chemical cleaning.

The interruption events were cut in half with the shorter cleaning time. In the first test period the events dropped from an average of 312 to 177, a reduction of 43%. With the optimized 10 min clean, the number was lowered to an average of 134, a reduction of about 57%.

The better beam stability can also be documented by the lower leakage current at the suppression electrode of the extraction assembly. This indicates that the suppression electrode and its insulator were also effectively cleaned with running the Xenon beam. Even for this parameter the shorter clean shows lower values.

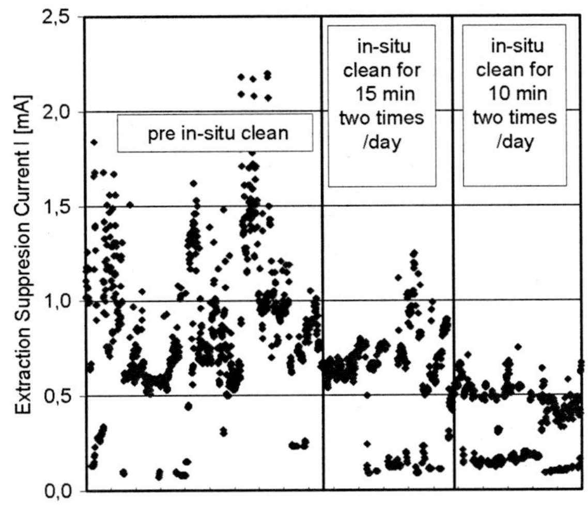

FIGURE 7. Extraction suppression current change with in-situ clean.

The lowering of the interruption events point to more stability for the ion beam, to more stable processing of the wafers and therefore to higher process quality.

C) Productive Time Loss Due to Tuning Interruption Events

But the lower number of interruption events has to lead to additional available production time. We used the time stamps for the occurrence and the clearance of the events out of the log files to determine the time loss by the interruption events. The time for recovering from a short interruption is not monitored as tool down time in the fab automation and is therefore a hidden loss of productivity.

The average daily time loss for beam failure recovery and additional retune was 91 min in the time before the in-situ clean. This loss was lowered to 46 min with the 15 min clean and to 38 min with the 10 min clean procedure. But you have to take the additional time for the cleaning into account and therefore we have gained 15 min per day with the 15 min clean and about 33 min with the 10 min clean. The Axcelis GSD200 high current implanters are running with an uptime of about 85%. The additional 33 min represent additional 2.5 % productive time for your high current tool.

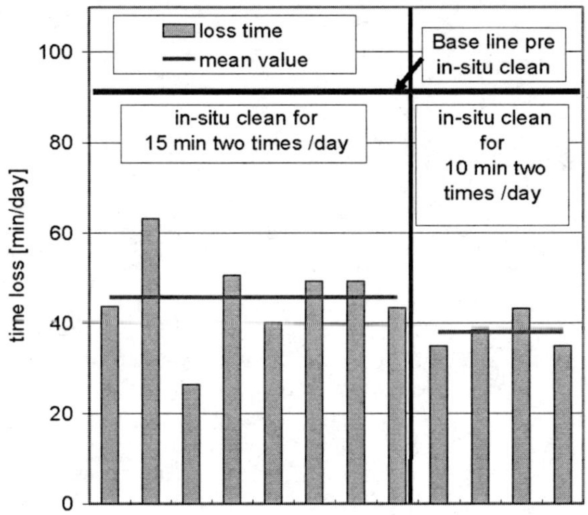

FIGURE 8. Reduced time loss per day with in-situ clean. Additional clean time not included in the figure.

CONCLUSION

The testing of in-situ cleaning with XeF_2 on an Axcelis GSDE200 high current implanter has shown that implementation of the clean doubles the source life time and cuts the interruption events in half. In addition, the suppression current of the extraction was lowered significantly. This results in a more stable ion beam and therefore in a higher quality implant process.

On the other hand, the longer source life and the lower number of interruption events result in higher availability of the production tool. For the examined tool the availability increased by about 3.5%, of which 2.5% is due to fewer interruptions and about 1% is from saved ion source changes.

REFERENCES

1. S. Bishop, R. Kaim, S. Yedave, J. Arnó, F. DiMeo, M. Wodjenski, "The Development of In-Situ Ion Implant cleaning Processes", *Proceedings of the XVI International Conference of Ion Implantation Technology*, 2006, pp. 477-480.
2. T. Romig, D. Eldridge, S. Bishop, *ISMI Symposium,* Austin, Texas Oct. 24th (2007).
3. R. Kaim, D. Eldridge, S. Bishop, P. Lundquist, J. Rolland, J. Dunn, "In-situ chemical cleaning for improved ion implanter utilization" *Solid State Technology,* **51**(3), March 2008

Implanter Source Life and Stability Improvement Using *In-Situ* Chemical Cleaning

Terry Romig, Doug Newman, Mike Mitchell, Kevin Ditzler

Texas Instruments Inc., Dallas, Texas 75243

David Eldridge, Steve Bishop, James Mayer

ATMI, Danbury, Connecticut 06810

Abstract. Current 300mm fabs rely heavily on automation to provide manufacturing efficiency. While implant processes and equipment follow this trend, tool availability and maintenance cycles are often driven by the ion source and adjacent areas which suffer from premature failures due to unwanted material deposits. While working in a high volume production environment, side by side comparative data has been collected on two tools running similar processes, with one tool having integrated in-situ cleaning cycles and the other with no in-situ cleaning. This paper will discuss significant improvements achieved in beam stability, glitch rate, ion source lifetime and maintenance cycles which were achieved on the tool with integrated in-situ cleaning. A program was established at Texas Instruments' DMOS6 wafer fab in early 2007 to explore and document process and equipment performance. Other critical areas, such as particle and metals contamination will be discussed with inferences as to potential yield improvements.

Keywords: Ion Implant, 300mm, AutoClean, source life, In-Situ clean

INTRODUCTION

The high current Ion Implanter is used extensively in the modern 300mm production environment, and its performance and availability is critical for maintaining a smooth manufacturing flow and device quality. At Texas Instruments' DMOS6 facility in Dallas, Texas, the Varian VIISta HCP high current implanters were having inconsistent ion source operation, causing shortened source life, having instability during operation causing production delays, and generating high cost of consumables for these source failures.

Texas Instruments' engineers wanted to find a way to perform biweekly maintenance on their high current implanters without unscheduled source failures. Sources were only lasting 10 days on average. The Implanter required other maintenance in the source area to be performed on a 2 week (14-15 day) interval, thus requiring additional source changes, and/or shorter PM intervals for the other source chamber components. In addition, beam uniformity of the "ribbon" beam in the VIISta HCP makes stability, and initial beam setup uniformity, critical to operational throughput and within wafer device parametric quality.

ATMI's recently introduced In-Situ chemical cleaning process or "AutoClean[tm]" process [1] was proposed to address these issues on the high current implanters.

This paper reports the results of this In-Situ cleaning method on the performance of a Varian VIISta HCP Implanter in high volume production.

IN-SITU SOURCE CLEANING

The In-Situ source cleaning process (AutoClean) consists of a gas cylinder installed in the implanter gas box, upgraded MFC for improved flow, and process recipes

The testing and evaluation was performed on a Varian VIISta HCP ion implanter installed in TI's DMOS6 production wafer fab in Dallas, Texas.

The etch gas is introduced into the ion source arc chamber in the same manner as a standard dopant gas. Ionized into a plasma, or simply introduced into the source without power, the etch gas removes dopant deposits such as As, P, B, and unwanted Tungsten [2] and Molybdenum that form on the arc chamber slits and cathode during BF_3 and GeF_4 operations.

CP1066, *Ion Implantation Technology,* edited by E. G. Seebauer, S. B. Felch, A. Jain, and Y. V. Kondratenko

Implanter Test Procedure

A standard recipe was created on the test implanter, using low total source arc power. The AutoClean gas flow (XeF_2) was set to a standard 1 sccm flow.

Cleaning frequency was scheduled for 15 minutes of clean time every 12 hours of production operation. Fab automation was programmed to execute an automatic clean routine on a set time basis, thus automating the In-Situ cleaning ensuring consistency with minimal manufacturing intervention.

Data was collected for 5 months, logging performance parameters including source life, glitch rate, beam uniformity, and recipe setup time.

PROJECT RESULTS

Source Life

The first critical performance data collected was overall source life using the In-Situ cleaning process. Prior to beginning the testing, data was collected and the average source life without using AutoClean was determined to be 10 days (240 hours). During the 20 weeks of testing AutoClean, the source life consistently met the 2 week PM target, with no source failing prematurely for ion source causes. In Figure 1, source #4 was removed early due to a filament power supply failure. Source #7 was run to end of life failure and lasted 456 hours until failure. This source life test represented a potential extension of 90%.

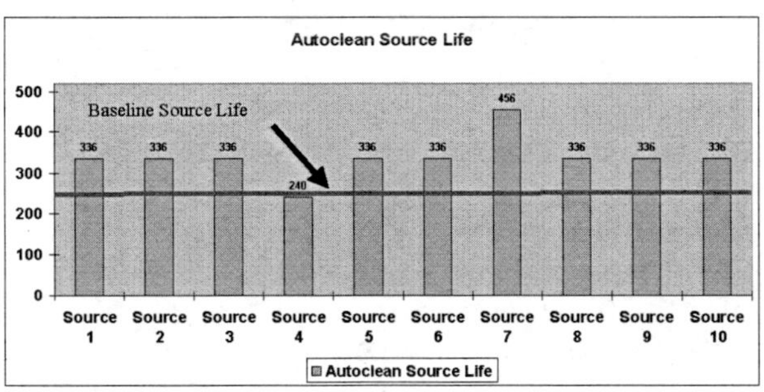

FIGURE 1. Source life results running AutoClean on a Varian ViiSta HCP

More importantly, no premature failures (less than 2 weeks) were recorded during testing, resulting in 25-30% fewer source changes compared to an equivalent implanter not running AutoClean. In addition, PM activity was planned and predictable resulting in improved labor utilization and reduction in production interruptions.

Glitch Rate

The second significant performance parameter measured during testing was ion beam glitch rate during production wafer implanting. In side by side comparisons between 2 systems running the same process recipe mix, the system running AutoClean was consistently more stable over the life of the ion source. Total ion beam glitches recorded for a 2 month period during testing showed the system running AutoClean recorded 1490 total glitches (figure 3), compared to 7720 glitches (figure 2) recorded for the standard system.

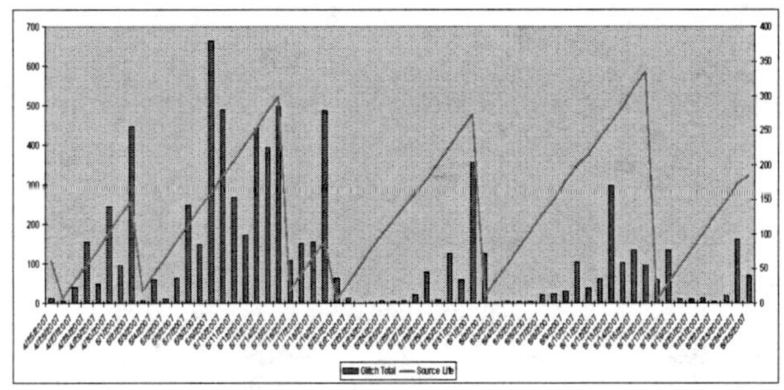

FIGURE 2. Total Glitches (7720), compared to source life, and recorded daily for system running standard processing with no In-Situ cleaning

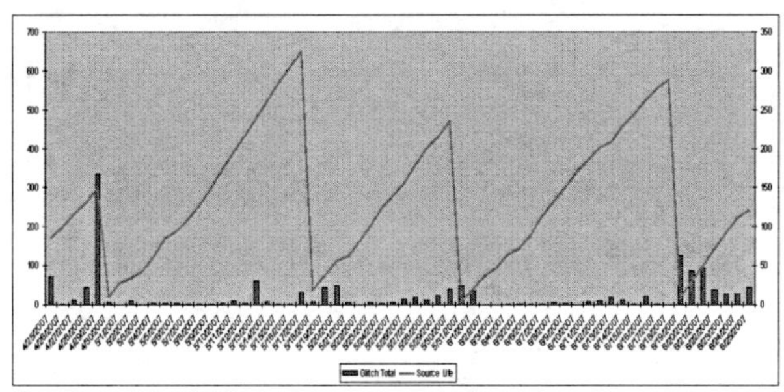

FIGURE 3. Total Glitches (1490), compared to source life, and recorded daily for system running In-Situ cleaning process.

The data in figures 2 and 3 show clearly that the In-Situ cleaning process significantly reduced the glitch rate as the source ages, producing glitch rates at the end of the source life that are comparable to new source performance.

When glitch rate data was sorted by individual wafers (figure 4), the In-Situ clean system showed a measurable reduction in the number of average glitches recorded for specific process recipes.

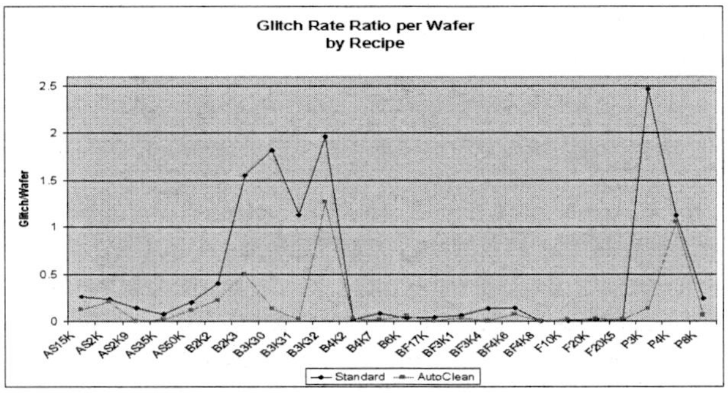

FIGURE 4. Reduction in glitches per wafer sorted by recipe

This illustrates the potential for individual wafer quality improvement, and the ability to improve overall device performance.

Beam Uniformity

A third important implanter performance metric recorded was overall average ion beam uniformity as measured by the implanter's profile system. As measured by overall percent of beam uniformity, sorted by specific process recipe, the In-Situ clean system reported an average improvement of 32% in beam setup uniformity before implant start.

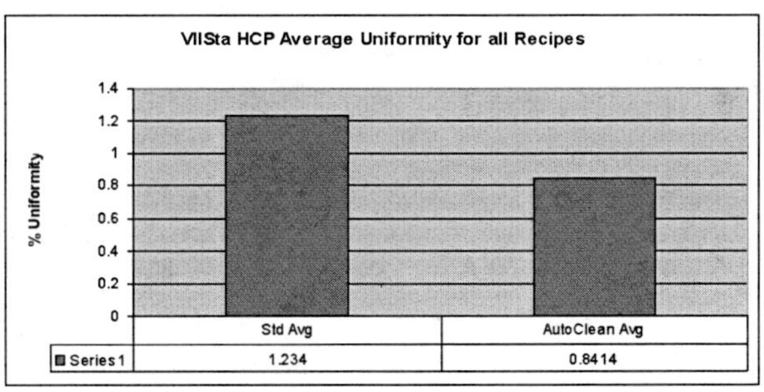

FIGURE 5. Average Ion Beam uniformity improved by 32% for the In-Situ clean system.

Recipe Setup Time

During analysis of the average recipe setup time, the In-Situ clean system experienced a 16% reduction in beam setup time. Several factors can contribute to the average setup time, including ion beam stability, beam uniformity, and beam optics. Figure 6 shows the average setup time by recipe for the In-Situ clean system and a standard system.

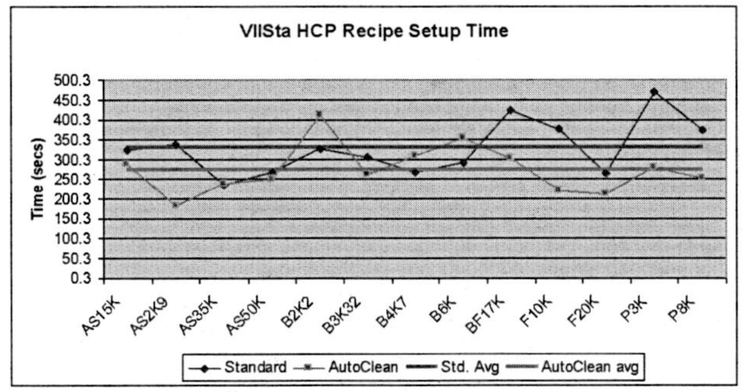

FIGURE 6. Average recipe setup time reduction for In-Situ clean system is 16%

Cost of Ownership

As a result of performance improvements, the reduction in number of source changes, and additional production time returned to manufacturing, the overall cost of ownership for the HCP high current system running In-Situ cleaning process has shown to be very positive.

When source life is extended to 2 weeks (336 hours) consistently, it significantly reduces the cost of consumables and labor required to maintain the implanter in production operation. This reduction can be modeled, and results in a savings in direct parts and labor of 3% per tool per year.

In addition, throughput modeling has estimated additional wafer output of 2000 additional wafers per year, per tool, can be achieved as a result of reduced downtime, and increased availability of the implanter.

CONCLUSIONS

The overall performance benefits and cost savings of the In-Situ cleaning process (AutoClean) offered by ATMI have shown significant improvements in manufacturing costs and production time resulting in the following benefits: 1) extended source life, with estimated 10+ fewer source changes per year needed, resulting in reduced cost of consumables and labor; 2) reduced labor costs through avoidance of manual cleaning and manual interventions; 3) 80 hours per year reduction in downtime caused by beam glitching, extraction arcing, and beam uniformity failures; 4) increased fab operational efficiency with more effective recovery from source instability, zero unexpected source failures; 5) parts are cleaner during PMs, resulting in enhanced safety with reduction in personnel exposure to hazardous materials & process byproducts.

CONTINUING DEVELOPMENT

Texas Instruments will be continuing to pursue further development of the In-Situ cleaning process, and expanding the installed base in volume manufacturing. Plans are in place to develop procedures for cleaning sources immediately after installation. Results of glitch rates in figure 3 indicate that In-Situ cleaning of a new source can reduce glitching and improve stability for the first 24 hours after a source is installed.

In addition, a further study will be done to determine if overall device performance is improved, including yield, due to the improved stability and uniformity of the implanted beams on production wafers.

Texas Instruments will work with implant vendors to create one-button AutoClean routines in system software to further automate and improve the in-situ cleaning process.

REFERENCES

1. S. Bishop, R. Kaim, S. Yedave, J. Arno, F. DiMeo Jr, and M. Wodjenski, "The Development of In-Situ Ion Implant Cleaning Processes", in Proceedings of the 16th International Conference on Ion Implantation Technology 2006, Marseille, France
2. S. Yedave, O. Byl, S. Sergi, J. Sweeney, R. Kaim, and D. Eldridge, "Tungsten Transport in Ion Source", in these Proceedings 2008, Monterrey, CA

Manufacturing Assessment of an XeF_2 *In-Situ* Clean Process for Mitigation of Species Cross Contamination

James P. Dunn[1], James L. Rolland[1], Paul Lundquist[1], Steve Bishop[2]

[1] *Atmel Corporation, Colorado Springs, Colorado, 80906, USA*
[2] *ATMI, Danbury Connecticut, 06810, USA*

Abstract. The residue buildup in an ion source can be energetically transported down the ion implanter beamline delivering unwanted dopants to the target area. Of particular concern are contaminants within one AMU of the target species such as $^{50}PF^+$ in a $^{49}BF_2^+$ beam causing counter-doping dose errors. Significant beam currents attributable to memory effect have been observed more than an hour after the species gas has been changed. This paper shows how Atmel Corporation, Colorado Springs, Colorado evaluated a new XeF_2 *in-situ* clean process as an effective means of reducing the risk of species cross contamination.

Keywords: ion implant, implanter, source,
PACS: 41.75.Ak; 61.72.up; 07.77.Ka

INTRODUCTION

A beta level evaluation of a new *in-situ* XeF_2 clean chemistry for ion implantation was conducted by Atmel Corporation, Colorado Springs, Colorado. The evaluation included spectrum analysis, profile SIMS and sheet resistance evaluation of the effectiveness of the in-situ clean process for reducing cross-contamination risks in high current ion implanters.

The purpose of the beta level evaluation was to determine the feasibility of performing non-invasive maintenance on the ion implant tools to address the risk of species cross contamination, enhance the source productive life and improve ion source stability over the useful life of the source. The results of a successful evaluation are evident in reduced operating costs, demonstrated consistent performance of the implant tools and the achievement of dependable operation.

The most common configuration of ion sources in use are the hot filament plasma type, which are the largest single cause of ion implanter downtime, both planned and unplanned. Depending on the implant species and beam current mix, ion sources may have a useful life of as much as several hundred hours, or as little as 50 hours. When the source fails, the implanter must be shut down, vented for source replacement and cleaning, then pumped down and re-qualified for production.

Surprisingly, the most common cause for ion source end of life is not failure of the hot filament. In most cases, failure is caused by buildup of solid deposits inside the ion source arc chamber and on the high voltage electrodes in the neighborhood of the source. Deposits inside the arc chamber are generally composed of arc chamber construction materials (e.g. W, Mo), with significant amounts of dopant materials. The buildup of dopant material, such as phosphorus or arsenic within the ionization chamber is a ready source of energetic contaminants and must be accounted for.

The least invasive solution has been to run a warm-up beam to ionize the contaminants and slowly extract the undesirable dopant until it has declined to an acceptable level. This can be done with a process gas or an inert such as argon until a clean spectrum is obtained. Additionally, the deposits build up over time to form flakes or films which are sufficiently electrically conductive to cause leakage currents across insulators, instability of the source plasma or high suppression currents. The XeF_2 clean chemistry was evaluated as a way to maintain the source integrity and beam purity while minimizing the need to intrude on the vacuum system.

THE CLEAN PROCESS

The XeF_2 *in-situ* clean process is a chemical process to convert contaminants into volatile gases that can be removed from the tool source area by the pumping system. A powerful fluorinating agent (XeF_2) is introduced into the source area where it readily disassociates to react with the dopant residues in and around the source area. The additional aspect of tungsten transport can also be managed to directly affect the source filament characteristics. A hot filament clean process encourages re-deposition of arc chamber tungsten

CP1066, *Ion Implantation Technology,* edited by E. G. Seebauer, S. B. Felch, A. Jain, and Y. V. Kondratenko

onto the filament where appropriate. A cold filament clean process avoids this same deposition, which is preferred in a tool that already runs large amounts of fluorine based dopants, such as BF_3.

EXPERIMENT

One deleterious effect of the deposits in and around the ion source is the so-called "implanter memory effect". It is well known that when changing from one dopant source gas to another, ions from elements of the first dopant can be extracted from the ion source plasma long after flow of the first dopant gas has been terminated. At Atmel, we have seen several milliamps of beam current persist long after the source gas has been turned off. In certain cases, this effect may cause serious contamination of the desired ion beam and consequent contamination of the implant process. Atmel has studied the well-documented case of P contamination in a BF_2 implant – the yield consequences of this contamination are so serious that some fabs will avoid scheduling phosphorus and BF_2 implants on the same tool, resulting in loss of flexibility in scheduling the implant workload. The P / BF_2 contamination is caused by phosphorus depositing in the ion source arc chamber when implanting with PH_3. After changing to BF_3 gas for the BF_2^+ implant, some of the fluorine reacts to form $^{31}P^{19}F^+$ whose mass of 50 is close enough to the desired mass of 49 ($^{11}B^{19}F_2$) to be co-implanted with the BF_2^+ ions. For this reason, Atmel has restricted the use of BF_2^+ implants on some high current systems with marginal mass resolution capability

We have previously tested the effects of introducing a highly reactive fluorinating vapor into production tools [1]. The testing demonstrated a significant increase in source life and RGA analysis showed no adverse effects, so we were comfortable in implementing the in-situ clean process to test implanter memory effect mitigation. The test protocol was designed to evaluate the effects of the in-situ clean process. SIMS was used to measure the amount of P^+ co-implanted with the BF_2^+. Spectrum analysis showed the change in ratio of the ionized products. R_s measurements clarified the effects on wafer resistivity.

The reagent was delivered from a cylinder with dimensions approximating those of an SDS JY, and was easily adapted to the argon slot in the tool gas box. Installation involved minor changes to the gas pigtail and an improved conductance for the source internal gas feed line. Heating of the gas line to avoid the possibility of product condensation was not found to be necessary. The tool argon slot was used and the argon MFC was deemed adequate since the XeF2 vapor pressure was sufficient to provide adequate flow.

The testing was done without plasma using a warm filament idled back to about 90 Amperes, with a manual introduction of the clean gas. The clean gas was introduced with the MFC set to 5sccm for 30 seconds. The MFC set point was then increased to 10 sccm and allowed to stabilize at the XeF_2 vapor pressure of about 4 torr in the cylinder. The XeF_2 flowed at vapor pressure with evaluation wafers run for at 10 minute intervals.

The test tool was run almost exclusively on phosphorus for 24 hours after a fresh source had been installed. A spectrum was run for comparison purposes. The tool was then switched to BF_3 gas and two N-type bare silicon wafers and one P-type bare silicon wafer were implanted with a high dose (5 x 10^{15} ions/cm^2) of BF_2^+ for SIMS and R_s testing. Spectrums were then obtained at 100% and 2% mass resolving slit settings. At this point the in-situ clean gas was introduced for 10 minutes before switching back to the BF_3^+ gas to repeat the above testing regimen. The spectrum data shows the changing ratio of the PF^+ ions as compared to the BF_2^+ and for the P^+ ions as compared to the BF^+.

The experimental parameters were selected based on their significance in proving a reduction in implanter memory effect beam current and the risk of cross contamination.

Qualification Parameters

Profile SIMS

Co-implantation was evaluated by use of a SIMS analysis. A BF_2^+ implant was run with a 5 x 10^{15} dose and evaluated for P^+ in the resulting test wafers. The in-situ clean was run for 10 minutes between test wafers. What we saw was an initial increase in the level of P^+ as the contaminants were volatilized, and a subsequent decrease in the level of the P^+ contaminants as the by-products were extracted from the source.

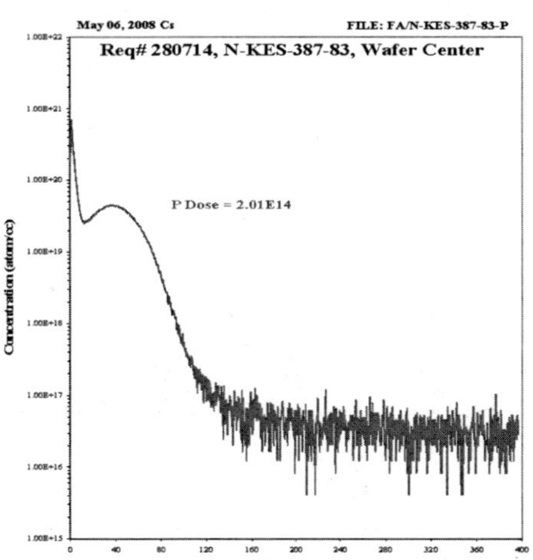

FIGURE 1. SIMS analysis of the P^+ contamination in a BF_2^+ implant after 24 hours of a phosphorus beam.

The initial SIMS check (Figure 1) was made after running the tool with a P^+ beam for 24 hours. The BF_2^+ implant showed a significant level of P+ co-implantation

After two 10 minute clean cycles were run, the SIMS test wafer (Figure 2) shows a 50% increase in the co-implanted P^+ as the phosphorus residues in and around the source area were volatilized.

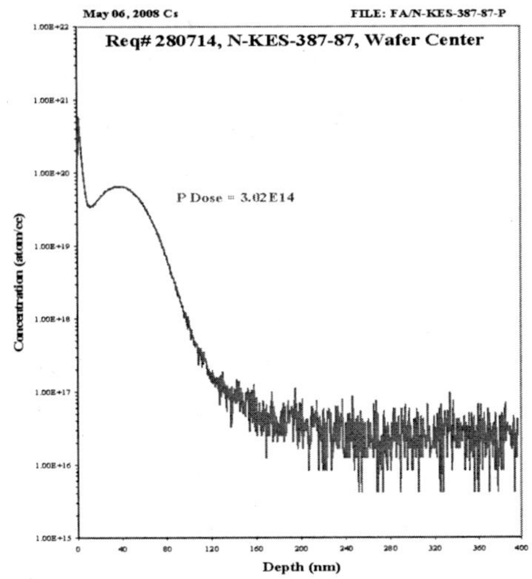

FIGURE 2. SIMS analysis of the P^+ contamination in a BF_2^+ implant after 20 minutes of XeF_2 in-situ cleaning.

After running the XeF_2 in-situ clean for another 10 minutes, a SIMS wafer was implanted (Figure 3). The test wafer shows that the 30 minute cumulative effect was a reduction of the P^+ contamination by a factor of three.

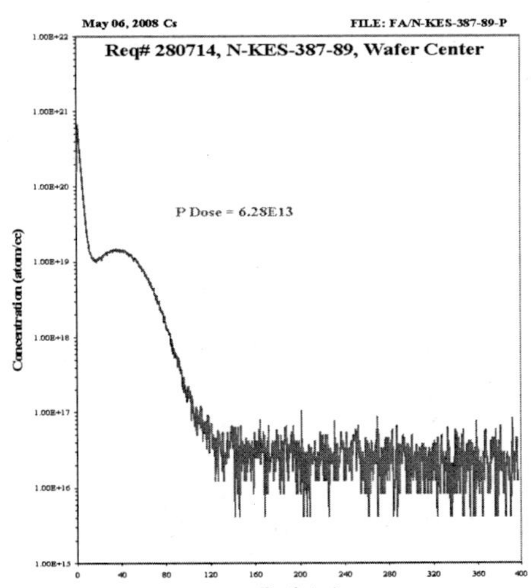

FIGURE 3. SIMS analysis of the P^+ contamination in a BF_2^+ implant after 30 cumulative minutes of XeF_2 in-situ cleaning.

Spectrums

The criterion for spectrum analysis was an evaluation of historic ion peak data for comparable or reduced ratios of the undesired energetics. This was a quick and easy test for assessment of the energetic component of material contaminants.

The effects of the clean process were most clearly seen in the ratio of $^{30}BF^+$ to $^{31}P^+$. The spectrums were run with tight defining apertures for mass resolution. The 2% open aperture necessary for AMU resolution resulted in a 39% reduction of available beam current compared to full current at 35% open. The initial spectrum showed almost equal beam currents for $^{30}BF^+$ and $^{31}P^+$ (Figure 4). At the 2% MRS setting, resolution between the PF^+ and $BF2^+$ peaks still was not as clearly defined as the resolution between the BF^+ and P^+ peaks. This is consistent with the species cross contamination risk seen with product.

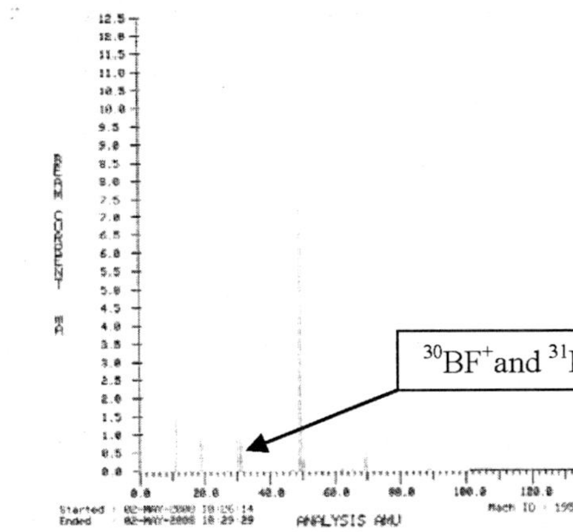

FIGURE 4. Spectrum analysis before the start of the XeF_2 in-situ clean

The spectrum check after 10 minutes of XeF_2 flow (Figure 5) showed a 2X increase in the P+ beam current due to the cleaning effect of the XeF_2.

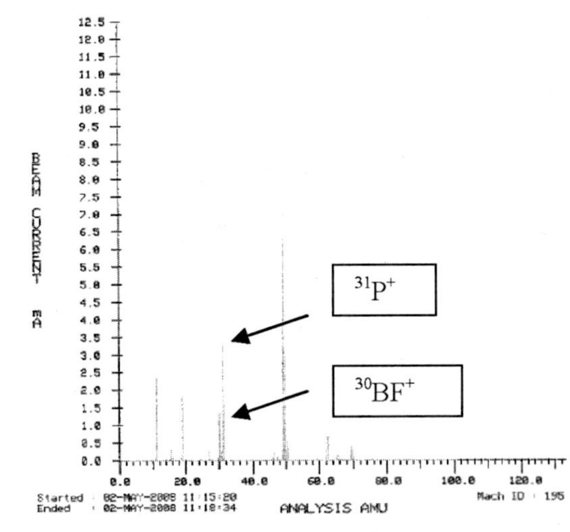

FIGURE 5. Spectrum analysis after running the XeF_2 in-situ clean process for 10 minutes

After the cumulative 40 minute clean, a SIMS wafer (Figure 6) showed the $^{31}P^+$ beam current was now only a third of the $^{30}BF^+$ beam current. The in-situ use of XeF_2 was demonstrably more effective than the typical argon clean (Figure 7).

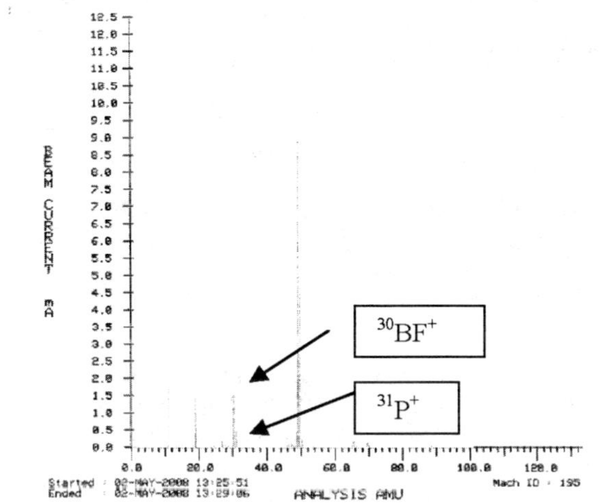

FIGURE 6. Spectrum analysis after running the XeF_2 in-situ clean process for 40 cumulative minutes

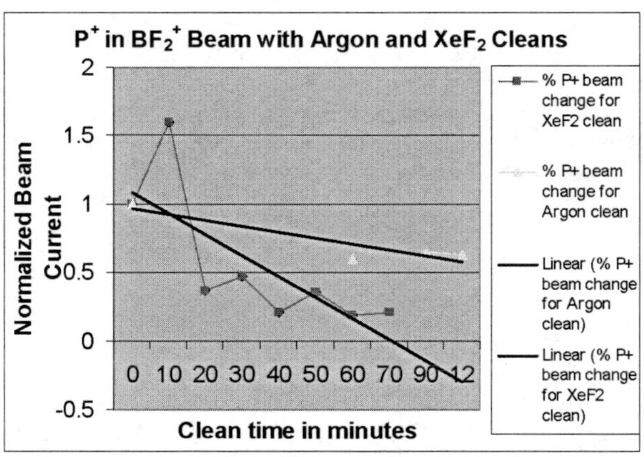

FIGURE 7. A comparison of argon and XeF_2 clean chemistries for reducing residual phosphorus contamination

R_s

Sheet Rho test wafers were also implanted with the BF_2^+ beam. The criterion was a demonstrated ability to minimize the effects of counter-doping on the resistivity test wafers.

The SIMS dose of 5 x 10^{15} was used though it was well outside the normal dose used for resistivity checks. The purpose was to look for relative changes in the dosing. The Rs wafers showed the identical trend that was seen in the SIMS analysis wafers and the spectrums (Table 1).

The wafer resistance increased for the first 20 minutes, then dropped as the source area was cleared of the counter-doping P^+. This was consistent with the other test data and reinforced the conclusions reached.

TABLE 1. Sheet Resistance (R_s) checks

XeF2 Flow duration	Resistance Average	Non-uniformity
0 minutes	53.924 Ω/sq	0.524 %
10 minutes	57.253 Ω/sq	0.509 %
20 minutes	57.220 Ω/sq	0.487 %
30 minutes	50.436 Ω/sq	0.543 %

CONCLUSIONS

The typical method of addressing the species cross-contamination risk consists of running a species warm-up or an argon beam until a spectrum shows an acceptable decrease in current for the unwanted species. Prior testing on a high current tool clearly showed the two standard methods to be functionally ineffectual as compared to the in-situ use of XeF_2 . Though the typical solutions will diminish the "implanter memory effect" over time, the in-situ clean was a much more efficient means of addressing the "implanter memory effect" problem. The reduction in beam current from residual by-products was accomplished in a very timely fashion without breaching the source vacuum. The SIMS wafers, spectral analysis and Sheet Rho resistivity tests all clearly indicated similar results, so further testing can be based on any of the above methods. The volatilized residues showed an increased incidence in less than 20 minutes and a marked decrease from the staring levels at 30 minutes. These results were compatible with earlier testing [2]. This was in marked contrast to the normal clean-up methods which encourage compromising on the allowed level of the undesired dopant or a source change to maintain production requirements.

The in-situ clean process has the potential to increase production's versatility by making the use of BF_2 much more accessible by reducing the risks historically associated with its use on some tools.

Further testing with an ionized clean process is planned for investigation of the tungsten transport effect on source filament life. The ability to improve ion source life via cleaning of the source and surrounding area in conjunction with increased filament life could well make this an integral part of regular tool maintenance.

REFERENCES

1. S. Bishop, R. Kaim, S. Yedave, J. Arno, F. DiMeo, M. Wodjenski, "The Development of *In-Situ* Ion Implant Cleaning Processes", in *Proceedings of 16th International Conference on Ion Implantation Technology*, American Institute of Physics, 2006.
2. J.Rolland, P. Lundquist, J. Dunn, S. Bishop, D. Eldridge, R. Kaim, "Insitu Chemical Cleaning for Improved Ion Implanter Utilization", *Solid State Technology*, March 2008, pp 22-24.

Surface Analytics in Support of the Development of Static AutoClean^{TM} – an *In-Situ* Cleaning Process for Ion Implanters

Michele Stawasz, Sharad Yedave, Laura Hiscock, Joseph Sweeney, Robert Kaim

ATMI, Inc., Danbury, Connecticut, USA

Abstract. Static AutoClean™ is a new *in-situ* cleaning strategy in development at ATMI® that enables increased process efficiency and safety in the ion implantation process. Like the Dynamic *in-situ* AutoClean technology previously introduced and released by ATMI, Static AutoClean utilizes XeF_2 chemistry for *in-situ* cleaning of hazardous contaminants and deposits. Static AutoClean, however, is targeted towards cleaning areas of the beam-line (like electrode insulators or source bushings) where cleaning efforts using Dynamic AutoClean may not be sufficient. An explanation of this cleaning strategy and results showing its effectiveness will be presented in a separate paper at this conference (S. Yedave *et al.*).

This paper presents the surface analytical data and methods used to understand and evaluate the effectiveness of Static AutoClean in removing contaminants from surfaces within the source vacuum chamber. Energy Dispersive Spectroscopy (EDS) was used to track the magnitude and spatial distribution of contaminants present on the surfaces within various regions in the source chamber space of an implanter following ion source operation with a commonly used dopant gas. After *in-situ* cleaning, these same components and surfaces were re-evaluated by EDS to quantitatively determine the reduction in surface contaminants present within the chamber. Learnings regarding the distribution of implant process contamination within the source chamber as well as Static AutoClean's effectiveness in removing it will be presented.

Keywords: Static AutoClean, AutoClean, In-situ Cleaning, Analytics
PACS: 06.20._f, 07.78._s, 07.81._a, 68.37.Hk, 78.70.En, 85.40.Ry

INTRODUCTION

Deposition of implant process products on surfaces within the arc chamber and ion source is a key limitation of ion source lifetime and a source of particulate contamination on implanted wafers. Reducing the severity of this deposition is the goal of AutoClean. This is accomplished through the reaction of gaseous XeF_2 with the process residues present to produce volatile reaction products which are removed by vacuum. AutoClean Static is performed by utilizing alternating cycles of XeF_2 gas introduction and evacuation. During gas introduction, XeF_2 is allowed to diffuse throughout the implanter area of interest to enable cleaning of areas and parts not directly exposed to the ion beam.

ASSESSMENT OF CLEANING EFFICIENCY

In the development of AutoClean Static, several model experiments were performed on an implant test stand to evaluate the ability of the product to remove implant process residues from the ion source vacuum chamber [1]. To evaluate the distribution of implant process residue on surfaces around the chamber, clean unused coupons composed of 316 stainless steel (SS) were placed in various locations (see Figure 1). Experiments were performed in two stages: (1) dosing of the implanter source chamber surfaces with implant residues formed from the use of common dopant gases (PH_3 shown here), and (2) cleaning of the source chamber using AutoClean Static. The details of the dosing and cleaning stages are presented elsewhere in these proceedings [1]. *In-situ* evaluation of the gaseous products of the reaction between XeF_2 and chamber residues was accomplished with analytics such as FTIR and RGA. Such analytics provided feedback on the progression of the cleaning reaction but provided no spatial information on cleaning efficacy in different locations within the chamber or of specific parts. *Ex-situ* surface analytical methods were necessary to evaluate the composition and distribution of residues on the chamber walls and surfaces of implanter parts.

CP1066, *Ion Implantation Technology,* edited by E. G. Seebauer, S. B. Felch, A. Jain, and Y. V. Kondratenko

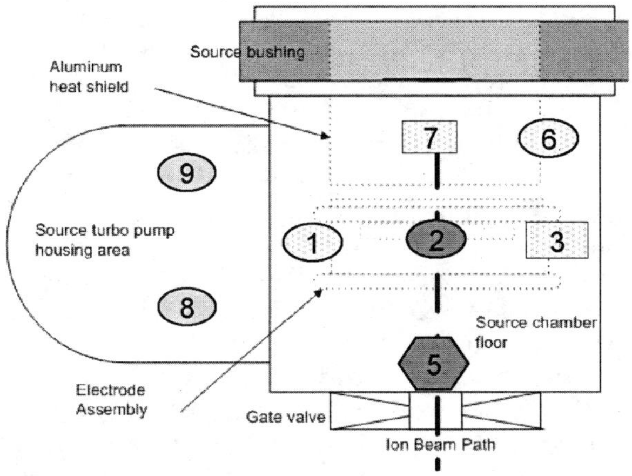

FIGURE 1. Top view of ion source chamber showing placement of numbered SS test coupons. Shading of coupons shown above represents the relative amount of phosphorous implant residue deposited for each coupon, with increasing shading darkness corresponding increasing magnitude of residue deposition. Shapes of the coupon positions represent the amount of cleaning achieved by Static AutoClean. Positions represented by ovals showed > 80% P removal, rectangles < 80%, and hexagons < 50% removal.

Surface Analytics

Scanning Electron Microscopy (SEM) and Energy Dispersive Spectroscopy (EDS) were used to evaluate process residue morphology, composition, magnitude, and distribution on SS coupon as well as implanter part surfaces following dopant gas dosing and XeF_2 cleaning. In addition, to obtain a baseline for coupon and part morphology and composition, characterization of the coupon and part surfaces occurred prior to their installation/placement in the source vacuum chamber.

EDS was utilized for compositional analysis of the coupon and part surfaces for several reasons. EDS is coupled with SEM, utilizing the electron beam to excite atoms in the sample surface to depths of several microns (depth dependent on the elemental composition of the material, its density, and the accelerating voltage utilized). In addition to providing compositional analysis with spatially-localized micron-level lateral resolution coupled with morphological analysis, the relatively large interaction depth of the electron beam into the sample and detection of subsequent emitted X-Rays emanating from sub-surface atoms allows for simultaneous detection of both substrate element peaks and deposited films/contaminants at its surface. As long as the surface residue film does not exceed a thickness approximated by equation 1, comparative analysis of the k-ratio of a predominant substrate peak to the k-ratio of the predominant residue peak for the analyzed surfaces of interest both following residue dosing and residue cleaning allows a determination of the % of residue cleaning achieved by AutoClean Static.

$$R = \frac{0.064\left(E_0^{1.68} - E_C^{1.68}\right)}{d} \tag{1}$$

Where E_0 is the accelerating voltage used (in keV), E_C is the absorption edge energy for the material probed (keV), and d is the density of the material probed (in g/cc) [1,2].

The k-ratio is the ratio of the net integral of the element of interest in the residue or substrate to the net integral of the pure element from a calculated spectrum. This offers an advantage over other surface analytical techniques such as XPS, Auger, and SIMS, which only probe the first few Å or nm of the surface, giving no indication of relative amount of residue removed without destructive depth profiling. The use of EDS allowed the repeated nondestructive *ex-situ* analysis of the same location on each sample following various process steps. In addition, the detection sensitivity of EDS (0.1 wt%) is appropriate for this application, which does not require the high detection sensitivities of XPS and Auger.

Implant Residue Dosing and Cleaning

Figure 2 shows optical and SEM images of SS coupon 1 (a) before installation into the implanter test stand, (b) after exposure to a P implant process (PH_3 source gas), and (c) after cleaning with AutoClean Static. Locations where SEM and EDS analysis were performed on the coupon are indicated. P implant process residue deposition was observable by eye as intensely-colored blue/black, red/violet, and yellow regions in image (shown in b) which were not present prior to the implant process (a). Such deposition was non-uniform in spatial distribution both across each of the coupons studied and from coupon to coupon.

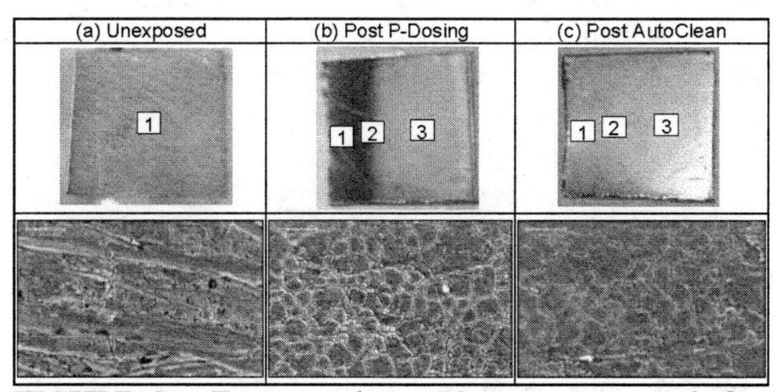

FIGURE 2. Top row shows SS test coupon 1 before installation into the test stand (unexposed), after exposure to a P implant process, and after cleaning with AutoClean Static. Positions of SEM/EDS analysis are indicated for each stage. Bottom row shows SEM images acquired at 2000X of position 1 for the same stages.

Numbering of spots probed on each of the coupons post P residue deposition and post-cleaning used the following scheme: spot #1 corresponds to the most highly colored location on the coupon, spot #2 – intermediate, and spot #3 – lightest coloration. A single spot was probed in the center of the coupon prior to installation into the implanter test stand.

To ensure that EDS spectra were quantitatively comparable from position to position, stage to stage in the experiment, and from coupon to coupon, strict protocol was followed during EDS spectral acquisition. Spectral acquisition parameters such as accelerating voltage, acquisition time, count rate, beam current, dead time and magnification were all kept constant. Likewise, sample/detector geometry parameters such as working distance, sample orientation, sample tilt, and take off angle were all kept constant. Composition of the implant process residue films was determined using the Phi-rho-Z method [2,4]. The quantity of the residue films was determined using the expression shown in equation 2.

$$MeasuredP = \frac{k-ratioP}{k-ratioX} \quad (2)$$

For the implant residue films, the k-ratio was derived from the Kα peak of P in the acquired spectra and the same peak calculated for pure P. Likewise, for the SS coupon substrates, the k-ratio was derived from the measured Fe Kα peak and the same peak calculated for pure Fe. Residue on graphite electrodes (C Kα), alumina insulators (Al Kα), and tungsten arc chamber parts (W Mα) were also evaluated in this manner, but are not presented here. This standardless method has been found to be quite effective [2,5]. Since the electron beam used to produce the detected X-rays interacted with, not only the residue film, but with the underlying substrate, this "Measured P" is analogous to an expression of residue film thickness and/or density. The % cleaning of the residue achieved by the AutoClean Static process was determined using the expression shown in equation 3.

$$\%Cleaning = \left[\frac{\left(\frac{k-ratioP}{k-ratioX}\right)_{dosed} - \left(\frac{k-ratioP}{k-ratioX}\right)_{cleaned}}{\left(\frac{k-ratioP}{k-ratioX}\right)_{dosed}} \right] * 100\% \quad (3)$$

Films deposited during the implantation process of P were found to contain P with no O present. Small percentages of W were also observed in some coupons, presumably originating from the W arc chamber. Figure 3 shows the EDS spectra for coupon 1 spot 1 obtained before installation on the test stand, after implant residue deposition, and after AutoClean Static (indicated by arrows). Since film residue varied in coloration significantly and with small length scale (~1mm) across the exposed face of the coupon, XRD could not be used to characterize the identity of the film due to the large spot size of the in-house XRD. It is likely that the film is composed of P_4, which has several different allotropes varying in coloration and density [6]. Interestingly, the color of several of the P_4 allotropes (black, violet, red, and yellow) corresponds with colors observed in the deposited films created here. In addition, the densities of each of the colored allotropes shows the same trend as the average measured P for each color observed and probed with EDS (Figure 4). It is also possible, however, that film coloration is due to differing film thickness of a single P_4 allotrope. A similar spot size limitation prevented determination of the film thickness of a single colored area of the observed residues using spectroscopic ellipsometry in-house. Since thickness determination is dependent on density, the presence of allotropes within the probed area would result in error in the measured thickness. Further work will include a small

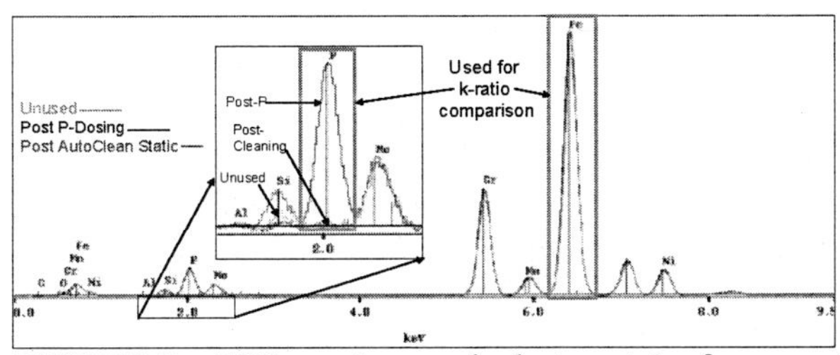

FIGURE 3. EDS spectra acquired at spot 1 of coupon 1 before exposure to P implant process, after exposure, and after AutoClean Static (inset, indicated by arrows). The peaks used to determine the k-ratios for P (residue) and Fe (SS coupon), which were then used to calculate measured P (equation 2) and % cleaning (equation 3), are shown boxed.

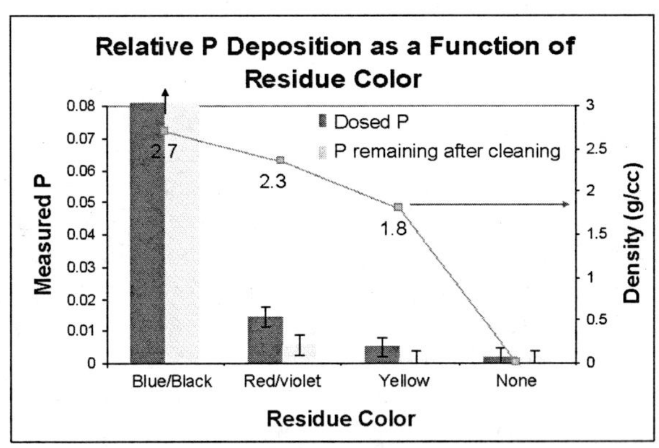

FIGURE 4. Plot showing a clear correlation between observed coupon film color and measured P. Dark columns indicate the average measured P in colored areas following dosing of the coupons; light- after applying AutoClean Static. The density of each of the suspected colored P_4 allotropes is noted and indicated by the data points corresponding to the right axis.

spot-size XRD and ellipsometry evaluation of each of the differently-colored areas of the deposited films to confirm the allotrope(s) present and the thickness of each colored area of the film.

Figure 5 shows the measured P probed after dosing with implant residue and following reaction with AutoClean Static as a function of coupon and spot#. Coupons showing the largest levels of measured P were located closest to the beam line and in the source turbo pump housing area. The coupon with the lowest amount of residue dosing was located inside the heat shield, which acted as a barrier to implant process product diffusion. The % cleaning achieved by AutoClean Static was over 80% for the majority of spots probed for the coupons investigated. Many of the spots which showed poor cleaning percentages also showed levels of P residue close to the detection limit of the technique, possibly leading to a false indication of poor cleaning. However, one coupon (5) clearly showed poor cleaning despite high levels of dosed P residue. This coupon was located along the beam line at the exit of the source chamber. We propose two hypotheses for the poor cleaning observed at this position: (1) the quantity of XeF_2 that diffused to this position within the chamber was inadequate to remove the large quantity and/or possibly high density of P_4 residue present on the coupon surface; (2) the location of this coupon along the beam line and near the exit of the chamber caused it to be exposed to stray implant ions and experience implantation. Implanted ions would be inaccessible to surface reaction with XeF_2 and thus would not be cleaned.

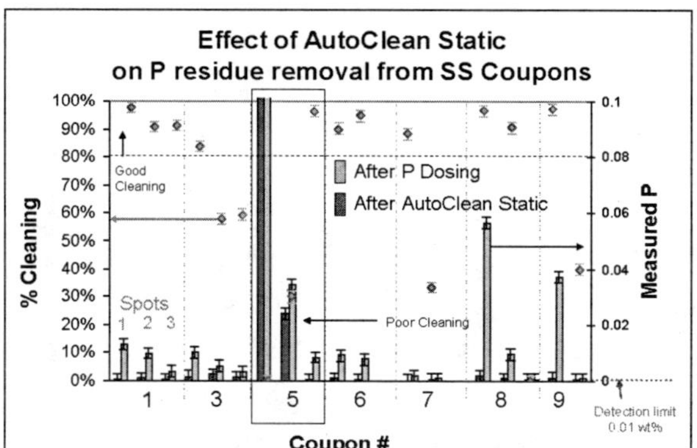

FIGURE 5. %Cleaning (left axis) of P implant residue by AutoClean Static as a function of coupon (and spot); and Measured P residue (right axis) following residue dosing (light bars) and cleaning (dark bars). "Good" cleaning is defined as diamond points located above the horizontal dashed line inside the plotted area. Coupons with "Poor" cleaning are indicated by the box. The horizontal dotted line at bottom right indicates the detection limit of EDS.

CONCLUSIONS

Regular use of Static AutoClean can prevent buildup of implant process deposition in an ion implanter. Static AutoClean was shown to remove P deposition from coupons located in various positions within an implant source chamber test stand. The majority of areas showed >80% cleaning. Areas that showed poor % cleaning also showed either large amounts of P deposition, possibly due to P implantation, or amounts of P deposition close to the detection limit of EDS, potentially leading to false indications of poor cleaning. Additional Static AutoClean cycles may be needed to remove heavy deposition, however. Similar analyses (not shown here) utilizing SEM/EDS have been applied with success by ATMI to implanter parts such as arc chamber apertures, filaments, insulators, base and side plates, electrodes, slits, and shields in various locations within the implanter.

EDS is a valuable, non-destructive analytical technique for measuring % cleaning of implant residue levels in the same locations before and after processing. The large interaction volume of EDS may allow detection of both substrate and surface residue films. Detection of residue and substrate signatures simultaneously allows comparative determination of amount of residue present on surface.

ACKNOWLEDGEMENTS

The authors would like to thank the Analytical Solutions Group at ATMI for performing the SEM and EDS analyses described here.

REFERENCES

1. Yedave, S., Sweeney, J., Letaj, S., Wodjenski, M., Hilharth, M., Marganski, P., Bishop, S., Eldridge, D., Kaim, R. "Development of "Static" *In-Situ* Implanter Chamber Cleaning" in *Proceedings of the 17th International Conference on Ion Implantation Technology 2008*.
2. Friel, J.J. 1995. *X-Ray and Image Analysis in Electron Microscopy*. Princeton Gamma-Tech Inc., Princeton, NJ.
3. Anderson, C.A., Hasler, M.F. (1966) Proc. *4th Intl. Conf. on X-Ray Optics and Microanalysis*, Hermann, Paris, 310.
4. Williams, D.B., Carter, C.B. 1996. *Transmission Electron Microscopy, vol IV: Spectrometry*. Plenum Press, New York.
5. ASTM E 1508 (1993), *ASTM Annual Book of Standards* Vol. 03.01, American Society for Testing and Materials, Philadelphia.
6. Weast, *R. C. CRC Handbook of Chemistry and Physics, 1st Student Edition 2000*. CRC Press, Boca Raton.

Development of "Static" *In-Situ* Implanter Chamber Cleaning

Sharad Yedave, Joe Sweeney, Oleg Byl, Shkelqim Letaj, Mike Wodjenski, Monica Hilgarth, Paul Marganski, Steve Bishop, David Eldridge, Robert Kaim

ATMI, Inc., Danbury, Connecticut, USA 06810

Abstract. Since the introduction of XeF_2 *in-situ* cleaning, its use in production implanters has been mainly focused on cleaning ion sources by flowing the cleaning vapor through the source arc chamber. This has been called "Dynamic" *in-situ* cleaning. "Static" *in-situ* cleaning is a different method under development at ATMI which allows an entire vacuum chamber and its contents to be cleaned. The chamber is filled to a pressure of 1 – 3 Torr of XeF_2 vapor, which reacts with deposited material on all internal surfaces, and the reaction by-products are then pumped away. When applied to the source vacuum chamber, the Static cleaning method allows cleaning vapor to contact components, such as the HV bushing and the manipulator assembly, which may not be adequately cleaned with the Dynamic method.

Recently, ATMI has installed a prototype Static *in-situ* cleaning system on an in-house Ion Source Test Stand in Danbury, CT. This paper will describe the prototype cleaning system and process and its applicability to production implant systems. We will also present experimental data showing removal of various dopant residues and the cleaning effectiveness for different components and surfaces inside the source vacuum chamber.

Keywords: Xenon Difluoride, XeF_2, Ion Implant, Vapor Delivery, Chemical Etching
PACS: 85.40.Ry, 61.72.U-, 81.65.Cf, 52.77.Dq, 47.70.Fw.

INTRODUCTION

Recent data have shown the effectiveness of using Xenon Difluoride (XeF_2) to clean the arc chambers of ion implant tools, resulting in a number of important production benefits [1,2]. In this process, called "Dynamic" in-situ AutoClean™, XeF_2 vapor flows from the gas box directly into the arc chamber where it reacts with deposits, forming volatile byproducts that are then pumped away (Figure 1). Due to the high reactivity of XeF_2, very little cleaning occurs outside of the arc chamber in this process. In order to complement the "Dynamic" process, ATMI has been developing a "Static" in-situ cleaning system and method that will target the entire vacuum chamber and its contents, such as source chamber surfaces, the high voltage bushing, the manipulator assembly, etc. (Figure 2). This process does not require plasma, and it is expected to improve the implant tool utilization by removing deposits before they accumulate to levels that can cause problems, such as arcing. It is also expected that in-situ cleaning may reduce the potential for personnel exposure to toxic materials that may occur during routine preventive maintenance operations.

Presented here are development details and performance data of the "Static" cleaning system and process.

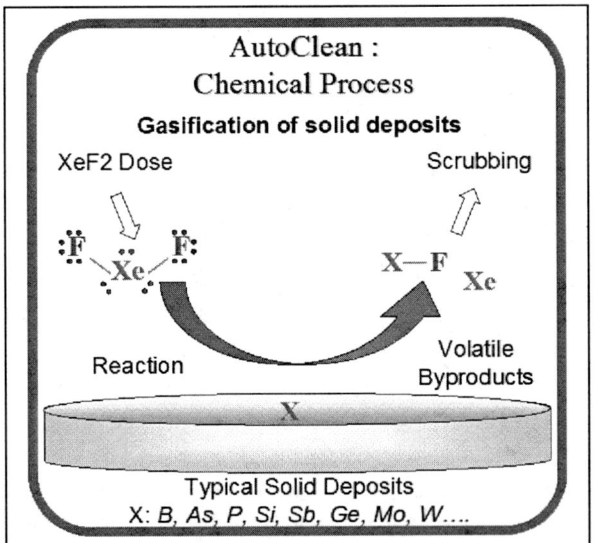

FIGURE 1. XeF_2 cleaning mechanism

CP1066, *Ion Implantation Technology,* edited by E. G. Seebauer, S. B. Felch, A. Jain, and Y. V. Kondratenko

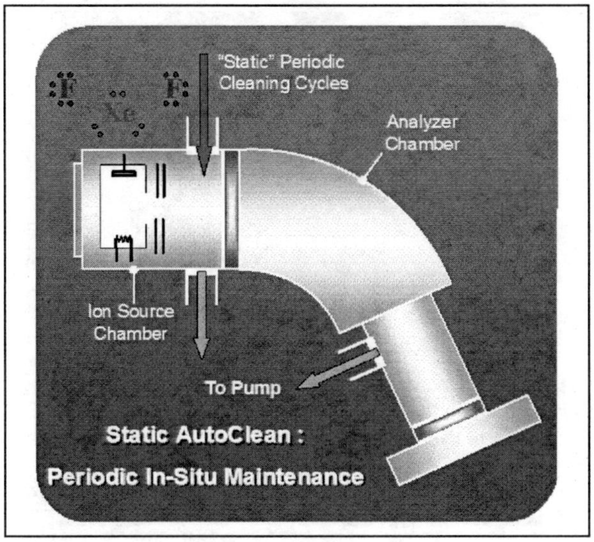

FIGURE 2. "Static" cleaning concept.

DEVELOPMENT

A prototype "Static" cleaning system has been installed on ATMI's ion source test stand, which is illustrated in Figure 3. The test stand has been outfitted with RGA and FTIR analytical capabilities for the purpose of monitoring chemical reactions. Specific details of the test stand have been previously reported [3].

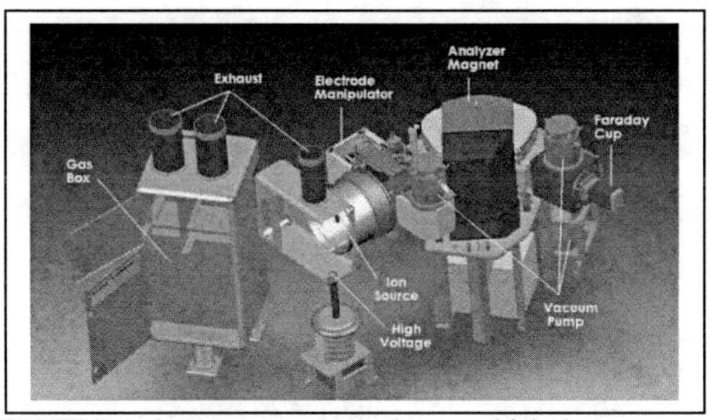

FIGURE 3. Ion source test stand onto which the prototype "Static" system has been installed.

The "Static" prototype system has the following features:

- XeF_2 cylinder capable of delivering high vapor flow rates [4] to a spare port on the test stand,
- master control hardware and PC (Figure 4), interlocked with source test stand hardware,
- LabView program code enabling automated vapor delivery and evacuation of source chamber, with programmable number of cycles, and soak time.

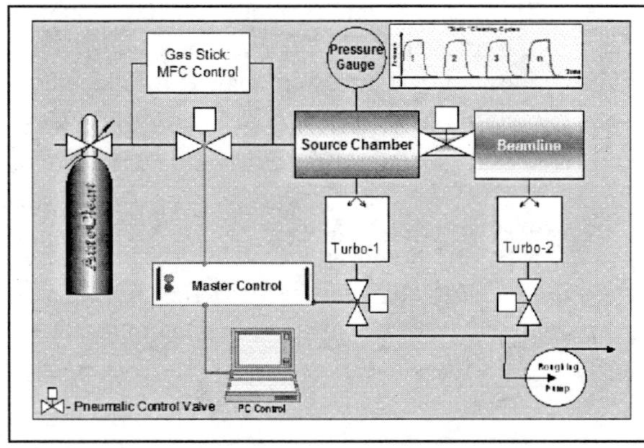

FIGURE 4. Installation schematic of prototype system

Experiment Methodology

In order to determine the effectiveness of the "Static" cleaning process, it was first necessary to deliberately contaminate the source test stand by running dopant beams of arsenic and phosphorus for specified lengths of time. Stainless steel coupons were placed at various locations within the source housing so that the dopant deposition might be quantified as a function of location (Figure 5).

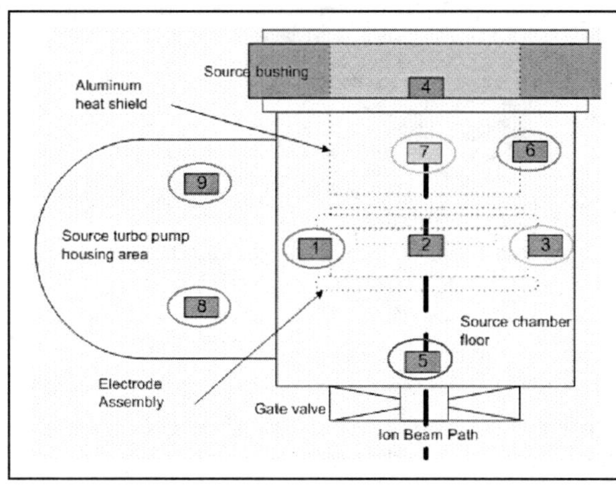

FIGURE 5. Placement of coupons in source housing during cleaning experiments.

After flowing arsine or phosphine and running beam for the specified time, the coupons were removed from the source housing and analyzed with Energy Dispersive Spectroscopy (EDS) to determine the surface residue content (see Stawasz *et al.* for additional details [5]).

The coupons were then placed back into the source housing and "Static" cleaning was performed. The cleaning process was comprised of the following steps (shown graphically in Figure 6):

- filling the source housing with XeF_2 vapor to a pressure of 1-3 Torr,
- allowing the vapor to soak for a selected length of time, and monitoring the reaction with FTIR, RGA, and chamber pressure,
- evacuating the source housing with the roughing pump,
- repeating the process until the byproduct concentrations had been significantly reduced.

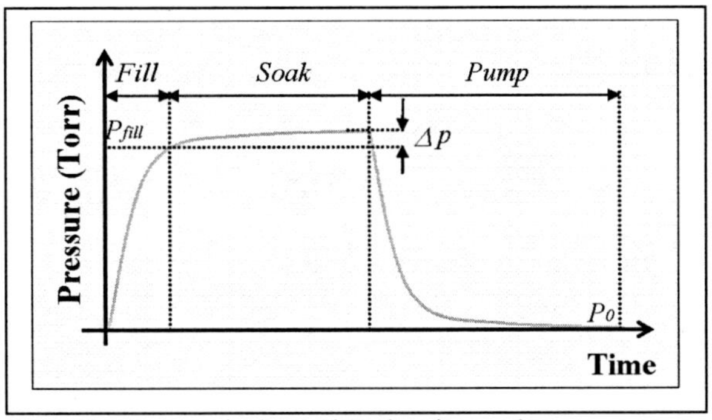

FIGURE 6. A typical "Static" cleaning cycle

After the completion of the cleaning cycles, the coupons were again removed from the source housing and analyzed with EDS.

RESULTS AND DISCUSSION

Separate experiments were performed to evaluate the capability of the "Static" cleaning process to remove arsenic and phosphorus deposits.

Arsenic

An arsenic beam was run for 24 hours under the following process conditions:

- AsH_3 Flow = 1 SCCM
- Beam Current = 2-5 mA
- Extraction Voltage = 20 kV

The results of the cleaning cycles are shown in Figures 7 and 8. EDS measurements showed a reduction of arsenic content on a test coupon from 73 atomic % to less than 1%. FTIR data indicated a 5x reduction in the concentration of AsF_3 – a byproduct of the cleaning reaction – after about 10 cycles. The FTIR data also show the presence of BF_3. This is a result of residual boron in the source housing.

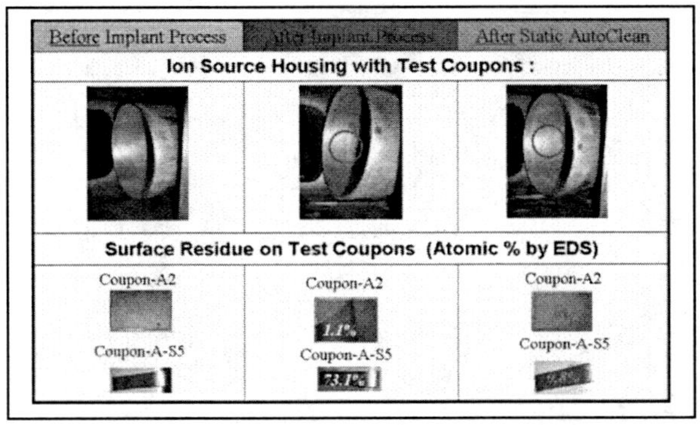

FIGURE 7. Pictures of arsenic contamination on coupons and source chamber surfaces. Cleaning cycles significantly reduced the arsenic levels as shown visually as well as by EDS.

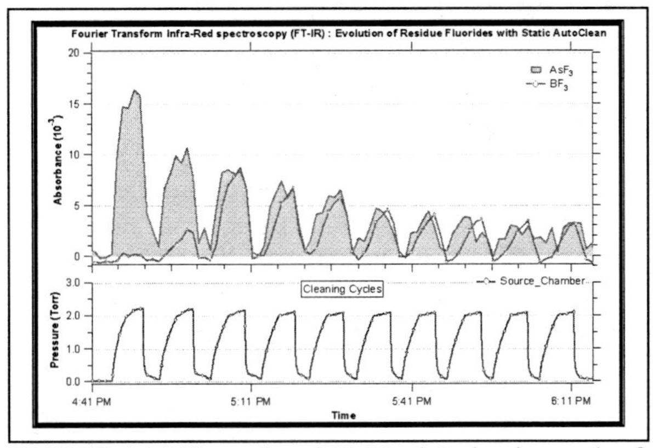

FIGURE 8. FTIR results of 10 "Static" cleaning cycles for arsenic

Phosphorus

Experimental conditions for the phosphorus test were the same as those for arsenic:

- PH_3 Flow = 1 SCCM
- Beam Current = 2-5 mA
- Extraction Voltage = 20 kV
- Beam Time = 24 hours

The results of the cleaning cycles are shown in Figures 9 and 10. EDS measurement of a test coupon showed a reduction of phosphorus from 42 atomic % to less than 1%. RGA data indicated nearly a 6x reduction in the intensity of phosphorus fluorides after only 3 cleaning cycles. Note that PF_3 and PF_5 are byproducts of the cleaning reaction with phosphorus.

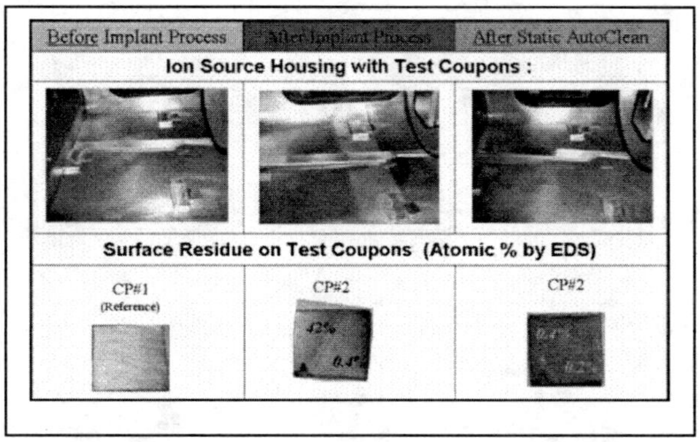

FIGURE 9. Pictures of phosphorus contamination on coupons and source chamber surfaces. Cleaning cycles significantly reduced the phosphorus levels as shown visually as well as by EDS.

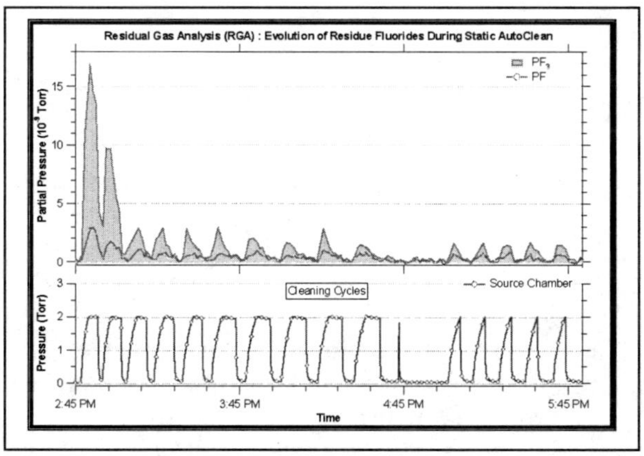

FIGURE 10. FTIR results of numerous "Static" cleaning cycles for phosphorus

CONCLUSION

The "Static" in-situ cleaning process has been shown to effectively clean dopant residues (P, As) on internal source chamber surfaces and test coupons, while not requiring plasma. Cleaning performance has been verified by RGA, FTIR, and EDS analytical techniques, as well as visually. The Static cleaning process complements the "Dynamic" AutoClean process which is already widely used in production ion implanters.

REFERENCES

1. J. Rolland, P. Lundquist, J. Dunn, S. Bishop, D. Eldridge, R. Kaim, "Insitu Chemical Cleaning for Improved Ion Implanter Utilization", *Solid State Technology,* Page 22, March 2008.
2. R. Kaim, S. Bishop, D. Eldridge, J. Mayer, J. Sweeney, O. Byl, S. Yedave, and P. Marganski, "In-Situ Ion Source Cleaning: Review of Chemical Mechanisms and Evaluation Data at Production Fabs", in these Proceedings.
3. S. Yedave, J. Arno, S. Bishop, F. DiMeo, R. Kaim, and L. Wang, "ATMI's Ion Implant Process Efficiency Research Laboratory (IIPERL)", in *Proceedings of 16th International Conference on Ion Implantation Technology,* American Institute of Physics, 2006.
4. J. Sweeney, P. Marganski, R. Kaim, M. Wodjenski, J. Gregg, S. Yedave, S. Sergi, S. Bishop, D. Eldridge, and P. Zou, "Optimization of Xenon Difluoride Vapor Delivery", in these Proceedings.
5. M. Stawasz, S. Yedave, L. Hiscock, J. Sweeney, and R. Kaim, "Surface Analytics in the Support of the Development of Static AutoClean – An In-Situ Cleaning Process for Ion Implanters", in these Proceedings.

Angle Control on the Optima HE/XE Ion Implanter

Edward Bell and Shu Satoh

Axcelis Technologies, Inc
108 Cherry Hill Drive, Beverly, MA 01915, USA
edward.bell@axcelis.com

Abstract. The Optima HE/XE is the latest generation of high energy ion implanter from Axcelis, combining proven RF linear accelerator technology with new single wafer processing. The architecture of the implanter is designed to provide a parallel beam at the wafer plane over the full range of implant energies and beam currents. One of the advantages of this system is the ability to control both the horizontal and vertical implant angles for each implant. Included in the design is the ability to perform *in situ* measurements of the horizontal and vertical angles of the beam in real time. The method of the horizontal and vertical angle measurements is described in this paper.

Keywords: Angle Control, Ion Implantation, Single Wafer.
PACS: 41.85.-p, 85.40-Ry

INTRODUCTION

As more zero degree implants are required for high energy implantation, control of the implant angles across the wafer becomes of critical importance[1,2]. Implanter design has adapted in response to this trend. The Optima XE is a hybrid scan single wafer ion implanter the controls angles to better than +/-0.1 degrees. The implanter combines proven RF linear accelerator technology for ion acceleration with a high throughput endstation employing single wafer processing. A detailed overview of the implanter is presented in another paper in these proceedings[3]. The emphasis of this paper is on beam angle measurement and control.

SOURCES OF ANGLE VARIATION

Sources of angle variation in an ion implanter that must be accounted for in order to achieve a uniform implant over the entire wafer include mechanical alignment, beam steering, beam parallelism, and wafer crystal cut.

The first component of variation includes errors in mechanical alignment of the beamline components that affect the angular relationship between the ion beam and the wafer. This can be mitigated by proper alignment of the endstation during machine installation. Fine tuning of the angles is performed based on the results of wafer implants as described below.

A second source of angular error includes the residual horizontal and vertical angles between the ion beam and the wafer after each beam tuning. These angles can be accounted for by proper beam tuning during recipe setup.

A third source of variation concerns the horizontal beam parallelism at the wafer plane. Since the Optima XE is a hybrid scan machine, optical errors in the design of the scanner or corrector magnet can cause the beam parallelism to vary across the wafer. Additionally, saturation in the steel of the corrector magnet can cause a situation where the parallelism is satisfactory for some beam rigidities, but not for others.

A final source of error is wafer crystal cut angle. This variation must be included in any angle variation analysis, but is outside the control of the implanter designer, although a scheme has been devised to address crystal cut errors[4].

ANGLE CONTROL ON THE OPTIMA XE

The layout of the Optima XE is shown in Figure 1. For the purposes of angle control, the important sections of the implanter are those after the beam is accelerated by the RF linac. The accelerated beam is scanned horizontally at 1 kHz by a scanner to an angle of approximately +/- 9 degrees. The scanned beam is then made parallel by passing the beam through a 60

CP1066, *Ion Implantation Technology,* edited by E. G. Seebauer, S. B. Felch, A. Jain, and Y. V. Kondratenko

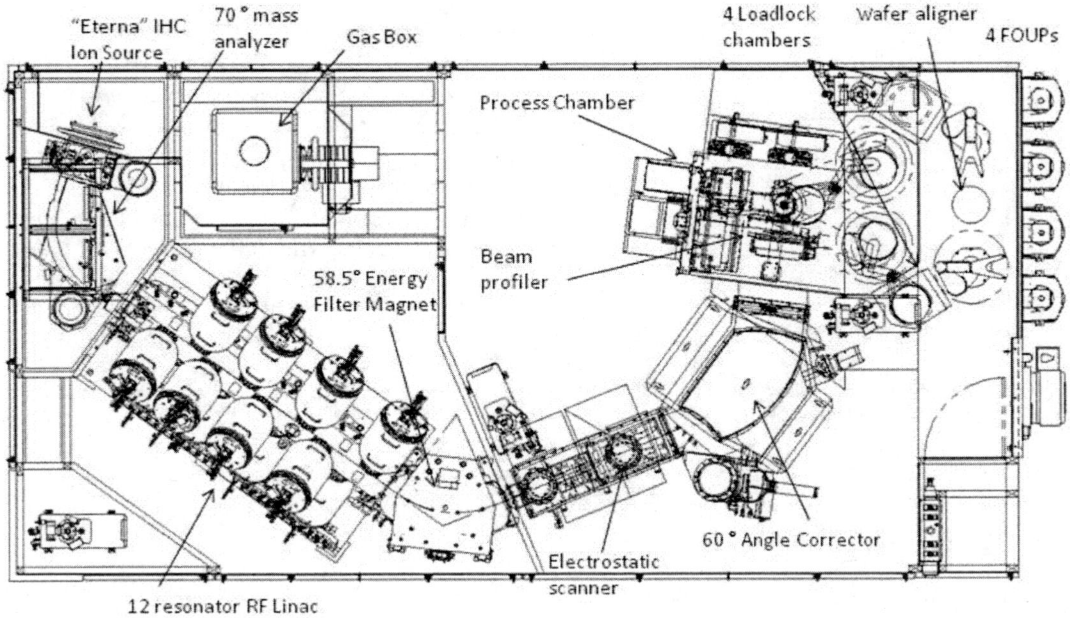

FIGURE 1. Figure showing the layout of the Optima XE

degree corrector magnet designed to produce a parallel beam upon exiting the corrector magnet. The ability to produce a parallel beam at the wafer plane depends on the design of the scanner and corrector magnet. The following sections detail the methods for measuring and correcting angle variation on the Optima XE.

Horizontal Angles

There are two aspects of beam angle control in the horizontal plane. The first aspect is the angle of incidence of the beam onto the wafer surface, represented here by an angle of beam on center. The other important parameter is the parallelism of the beam, meaning the variation of the beam angle in the scanned beam across the wafer.

In a beam parallelizing scheme using a magnetic angle corrector, the two properties of implant angle and parallelism are interrelated and one cannot change one property independently without affecting the other. However, with careful machine design, a solution that satisfies both requirements for all beams can be found.

Beam Center

The first aspect of angle control in the horizontal plane concerns the angle of the unscanned beam to the wafer plane. Contributions to this angle include components due to endstation positioning and beam center steering. The correct angle of the beam center to the wafer is determined during machine installation by implanting a series of wafers with varying horizontal implant angles on each wafer. The different angles are produced by changing the angle corrector magnet current for each wafer, and the required profiler position is recorded for each case. After probing the wafers, the optimum setting of the profiler for the unscanned beam is determined by curve fitting and the value is entered into the machine configuration file. Once this value is set, there is no need to change it unless the machine undergoes a procedure that may change the relative positions of the beamline components.

An example of the calibration process is shown in Figure 2. For this case five wafers were implanted and the minimum signal, corresponding to the maximum amount of channeling, can be determined to be at -0.09 degrees. The position of the profiler that produces this angle is saved in the machine configuration file. During beam tuning, the profiler is positioned at the configuration value and the corrector magnet is adjusted until the beam is maximized in the Faraday. This procedure ensures that the beam center is perpendicular to the wafer for each beam setup.

Beam Parallelism

The parallelism of the ion beam across the wafer, and the ability to quantify the parallelism is also very important. On the Optima XE, the parallelism of the

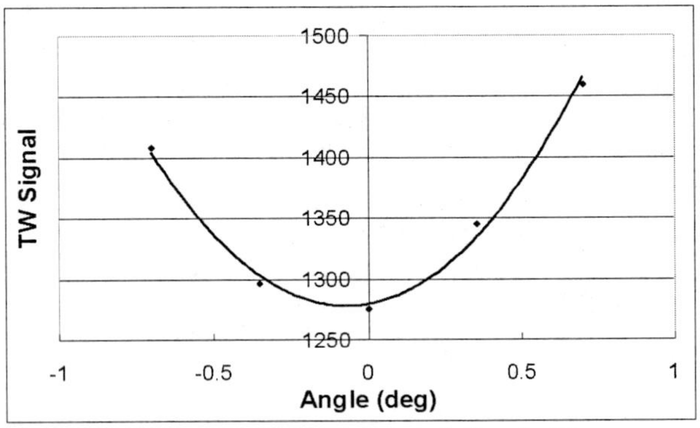

FIGURE 2. Figure showing the variation in Thermawave signal for various horizontal implant angles. The minimum occurs when the beam is perpendicular to the wafer.

ion beam can be measured automatically at anytime, using an *in situ* device. The technique is shown schematically in Figure 3. After a beam is tuned, a mask with seven vertical slit apertures is inserted into the beam, allowing the parallelism to be measured at seven positions across the scanned beam. The mask is located after the final optical element in the beamline and is well upstream of the wafer plane. The slits all have well known dimensions and positions and are evenly spaced in the horizontal plane at 50 mm intervals. Next, the profiler, located at the wafer plane, is scanned through the beam and the positions of the seven beamlets are determined. Using the distance between the aperture plate and the wafer plane, one can determine the range of horizontal angles present in the beam.

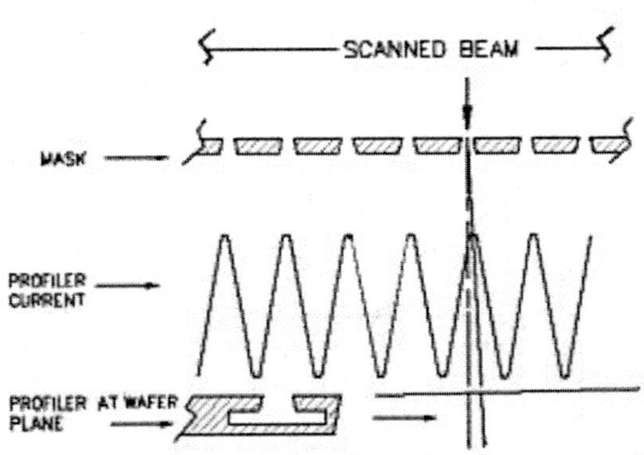

FIGURE 3 Figure showing the beam parallelism measurement technique (see text)

On the Optima XE, the parallelism of the beam has been measured for a wide range of beam stiffness as well as on several different machines. An example of the parallelism for various beam rigidities is shown in Figure 4. In the figure, the parallelism data are plotted for a low stiffness beam (shown as the solid points) and a high stiffness beam (shown as the open points). As can be seen, the measured angles are well under +/- 0.1 degrees.

The fact that the parallelism does not change significantly is due to the design of the corrector magnet. Magnet steel is subject to saturation when driven to high fields, and this change in properties can cause the field profile to change. This, in turn, causes the focal properties of the magnet to change for different ion beam rigidities.

In the Optima XE, the corrector magnet was designed for a maximum field of 1.0 T in the pole gap, and there is sufficient steel in the flux return path such that the field profile undergoes minimal changes between low rigidity beams and high rigidity beams.

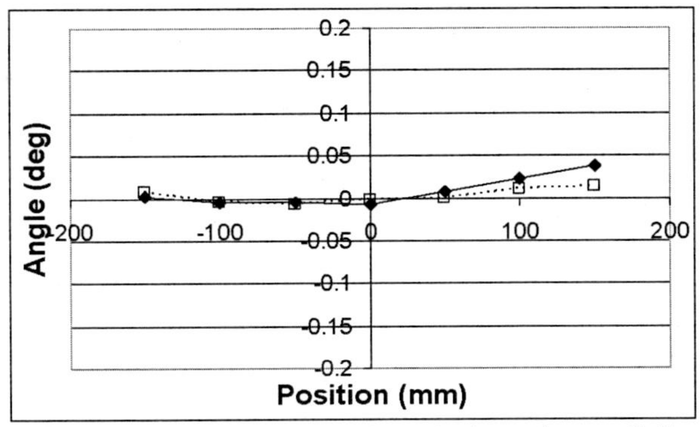

FIGURE 4. Figure showing horizontal parallelism measurements for a low stiffness beam (solid points) and high stiffness beam (open points).

Vertical Angles

Similar to the case for angles in the horizontal plane, the proper tilt angle for perpendicularity between the beam and the wafer in the vertical plane is determined during machine installation by implanting several wafers at various wafer tilt angles and determining the exact tilt that corresponds to wafer perpendicularity. The corresponding angle is saved in the machine configuration file.

In the vertical plane, the angle of the beam can be measured at any time utilizing the Axcelis VBA [5] vertical beam angle measuring device. The VBA is a deep Faraday cup with a high aspect ratio slotted cover which is mounted on the wafer scan arm. To determine the angle of the beam in the vertical plane, the scan arm is tilted about an axis parallel to the plane of the scanned beam until the maximum beam current is read by the Faraday. Using the known angular relationship between the VBA and the wafer plane, the angle of the beam in the vertical plane is determined.

For vertical angle correction of the ion beam during a recipe tune, the VBA can first be used to determine the starting angle of the beam. Next, the beam can be deflected vertically upstream of the wafer, and the vertical angle of the beam can be measured again.

RESULTS

Wafers implanted with high energy beams at a zero degree implant angle can be a rigorous test of beam parallelism, since the amount of channeling varies strongly with even small changes in implant angle. Figure 5 shows a Thermawave wafer map for a 900 keV P^+ implant on the Optima XE. There is no significant difference in the TW signal across the wafer.

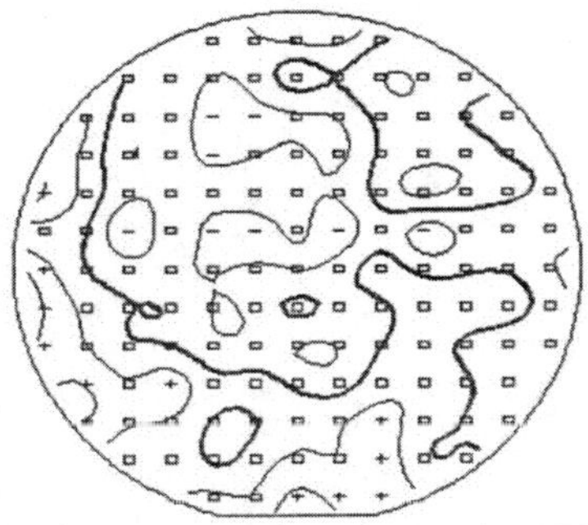

FIGURE 5. Thermawave plot of a 900 keV P^+ zero degree implant.

As another example, a wafer was implanted with a 615 keV B^+ beam. SIMS analysis for the implant condition shows no variation in implant profile across the wafer. Figure 6 shows a comparison of the four edges of the wafer with the center and shows that there is no significant variation from left to right or top to bottom.

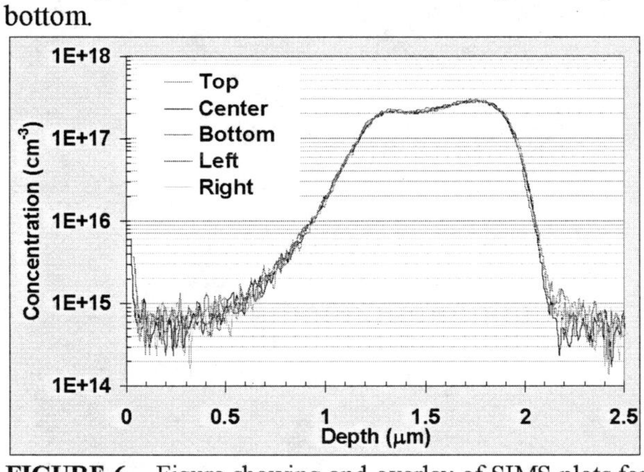

FIGURE 6. Figure showing and overlay of SIMS plots for five positions on the wafer surface for a 615 keV B^+ zero degree implant.

SUMMARY

The measurement and control methodology regarding beam angles on the Optima XE has been described. Procedures for minimizing angle variation arising from beam center steering, beam parallelism, and endstation positioning have been detailed and results of testing shown.

REFERENCES

1. Leonard M. Rubin, Wesley Morris, and Craig Jasper, "Process Control Issues for Retrograde Well Implants for Narrow n+/p+ Isolation in CMOS", Conference Proceedings, 14th International Conference on Ion Implantation Technology (2002), pp 17-20.
2. R. D. Rathmell, B. Vanderberg, A. M. Ray, D. E. Kamenitsa, M. Harris, and K. Wu, "Implant Angle Control on Opitma MD", Conference proceedings, 16th International Conference on Ion Implantation Technology (2006), pp. 349-352.
3. Shu Satoh, Joseph Ferrara, Edward Bell, Shital Patel, and Manny Sieradzki, "Optima XE Single Wafer High Energy Ion Implanter", these proceedings.
4. US Patent No. 6,881,967, Ray, Andrew M., Method of Correction for Wafer Crystal Cut Error in Semiconductor Processing (April 19, 2005).
5. R. Rathmell, J. David, "Implant Angle Repeatability on Optima MD", these proceedings.

MOLECULAR AND CLUSTER ION BEAMS

ClusterBoron™ Implant Alternative to BF_2 PMOS SDE

Thomas Feudel[1], Ralf Illgen[1], Christian Krüger[1], Marek Braun[1]
Karuppanan Sekar[2], David Lee[2] and Wade Krull[2]

[1] *AMD Saxony, LLC & Co. KG, Wilschdorfer Landstr 101, D01109 Dresden, GERMANY.*
[2] *SemEquip, Inc., 34 Sullivan Road, North Billerica, Massachusetts 01862, USA*

Abstract. As device geometries scale, the formation of the SDE becomes increasingly difficult and increasingly important. For advanced technologies, new methods such as ultra-low-energy boron implantation and millisecond annealing (flash or laser) are necessary to achieve the required junction characteristics. In addition, these processes must be compatible with the remainder of the process flow, which might include advanced dielectrics, stress technologies, SOI, etc. The emergence of ClusterBoron as a high productivity alternative for the low energy implant creates interest in device performance possible in a realistic process flow.

This paper will present an evaluation of the use of ClusterBoron for the PMOS SDE in an advanced 65nm logic process which includes laser annealing, e-SiGe stress layers and SOI. The conventional process uses a BF2 SDE process. Complete device characteristics will be shown comparing the ClusterBoron SDE to the conventional BF2. It will be shown that the ClusterBoron process achieves better boron activation, leading to enhanced transistor drive current. It will also be shown that the ClusterBoron is compatible with the SiGe stress layers and SOI structure. In summary, ClusterBoron presents an attractive alternative to the conventional BF2 process for advanced PMOS SDE.

INTRODUCTION

One of the issues with continued scaling of transistor structures is the shallow junction required for the PMOS SDE. Scaling requires that the junction depth is reduced with each generation to maintain control of short channel effects. For the 45nm generation, a PMOS SDE junction depth of less than 20nm is required, while the 32 nm generation will require junction depths around 15nm. At the same time, it is desired to keep the sheet resistance of this junction around 1000 ohm per square to minimize impact on the drive current. This set of requirements makes low energy and high dose boron implants critical to successfully forming an appropriate junction. Boron implants of less than 400eV will be required to meet the 32 nm requirements, and conventional implant tools have very low throughput for this type of recipe, especially when the sensitivity to energy contamination is considered. Most organizations use a BF2 implant to increase the productivity of the production process. The recent emergence of ClusterBoron implant as an alternative low energy boron process offers a high productivity solution to these issues. However, it must be shown that the ClusterBoron implant is compatible with the other process modules already incorporated into a modern CMOS flow. In particular, this study addresses the compatibility of the ClusterBoron implant with e-SiGe stressors. The concern is that the self-amorphizing nature of the ClusterBoron implant may allow the relaxation of the SiGe stress. Our experiment looks at the use of ClusterBoron for both the SDE and SD on a 65nm SiGe/SOI technology.

EXPERIMENT

This experiment uses a straightforward split lot approach to evaluating the use of ClusterBoron implant for the PMOS SDE or SD junctions. The appropriate dose and energy of the alternative implant recipes cannot be calculated directly because the ClusterBoron implant and BF2 implants behave differently during annealing, due to the fluorine inhibition of boron diffusion. Rather, an empirical approach with splits on ClusterBoron implant energy and dose was used. The matrix included three doses and two energies for the SDE and two energies and two doses for the SD. In all cases, only one ClusterBoron implant was included on any split; the Process-of-record (POR) BF2 recipes were used for the other implants.

CP1066, *Ion Implantation Technology,* edited by E. G. Seebauer, S. B. Felch, A. Jain, and Y. V. Kondratenko

RESULTS

In general, the devices with ClusterBoron implant behaved appropriately and parametric trends were observed. In Figure 1, we show the unsilicided active resistance. It is seen that the ClusterBoron splits produced lower active resistance, except for the one with the lowest energy and dose. As ClusterBoron dose and energy are increased, the active resistance is reduced. It is also seen that the ClusterBoron splits for the SD implant also match the POR for unsilicided active resistance. These results show that the ClusterBoron implant activates as well or better than the BF2 POR with the production flow's annealing sequence.

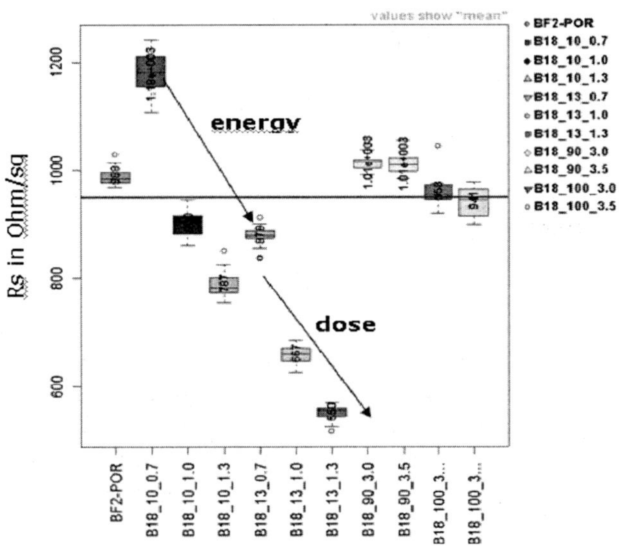

FIGURE 1. Unsilicided Active Resistance

The next parameter investigated is the threshold voltage. The data for V_{tsat} is shown in Figure 2. It is seen that the 10keV ClusterBoron SDE shows normal variation, with a sensitivity of 10mV/1E12. The POR is matched with a 10keV, 1.0E15 ClusterBoron implant. On the other hand, the 13keV ClusterBoron SDE show much higher sensitivity with 20mV/1E12. All of the 13keV splits gave V_{tsat} values that were higher than the POR. The V_{tsat} shows very low sensitivity to the SD implant splits and all of the ClusterBoron SD implants produced V_{tsat} values close to the POR.

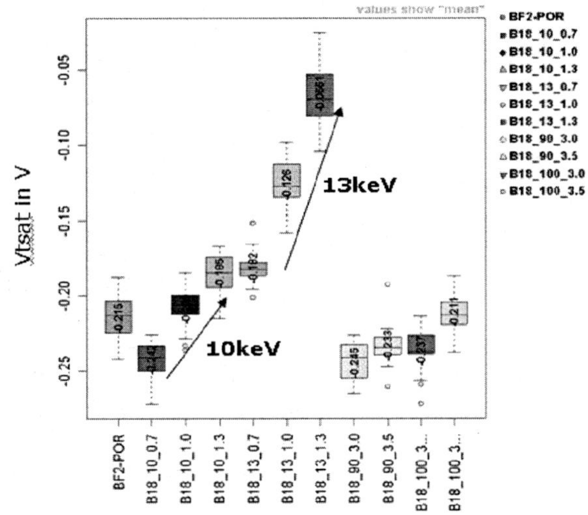

FIGURE 2. V_{tsat} (nominal devices).

The results for junction capacitance are shown in Figure 3. It is seen that the ClusterBoron SDE produces a junction capacitance similar to the POR. All of the 10keV ClusterBoron SDE splits have junction capacitance values that are close to the POR, with the higher dose splits having slightly lower junction capacitance. Similarly, the 13keV ClusterBoron SDE splits show decreasing junction capacitance as the dose is increased. The ClusterBoron SD splits show that the lower dose matches the POR regardless of energy, while the higher dose gives a lower junction capacitance.

The most important results of this work are shown in Figure 4, where we show the universal curve for the SDE implant splits. It is seen that all of the ClusterBoron SDE splits show better performance when compared to the POR. The improvement of performance is approximately 13% for all ClusterBoron SDE recipes. The 10keV ClusterBoron splits are shown in Figure 4a while the 13keV ClusterBoron SDEs are shown in Figure 4b. For the 10keV SDE, it is seen that the behavior is essentially independent of dose, with all the data falling about the same line. In addition to the higher drive current, the slope of the universal curve is also improved and the off state distribution is similar to the POR. For the 13keV SDEs shown in Fig 4b, the drive current is enhanced but the slope is the same as the POR and the off-state currents are slightly elevated compared to the POR.

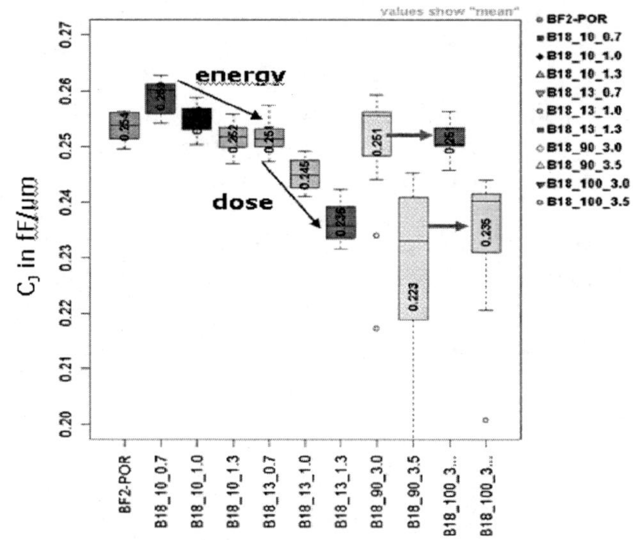

FIGURE 3. Junction Capacitance

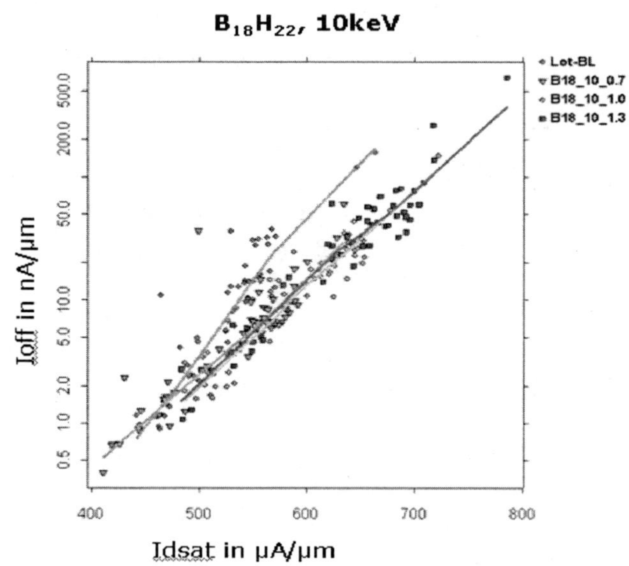

FIGURE 4a. Universal Curve for SDE implants with ClusterBoron energy of 10keV.

In Figure 5 we show the universal curve for the splits with ClusterBoron for the SD. It is seen that all the data fits within the same distribution. There is no performance gain observed by the use of ClusterBoron for the SD implant. However, it is still significant that the results are the same. The deep SD implant occupies the same volume as the e-SiGe which enhances the device performance by placing stress on the channel. By producing the same universal curve, the ClusterBoron SD implants must also be producing the same stress effect, that is, the ClusterBoron self-amorphization does not degrade the stress enhancement from the e-SiGe.

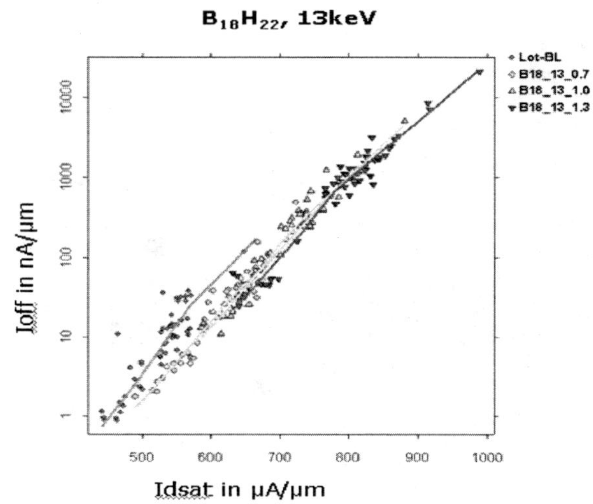

FIGURE 4b. Universal curve for ClusterBoron SDE implants with 13keV energy.

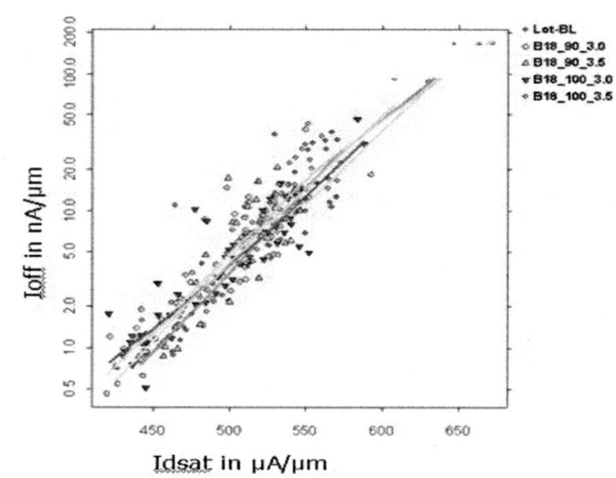

FIGURE 5. Universal curve for ClusterBoron SD implants.

The junction leakage characteristics are shown in Figure 6. While most of the data show leakage that is very similar to the POR, there are slight differences. The behavior of the two energies used for SDE are slightly different, with the 13keV SDE showing lower leakage values. For the ClusterBoron SD splits, all data show junction leakage in the same range as the POR.

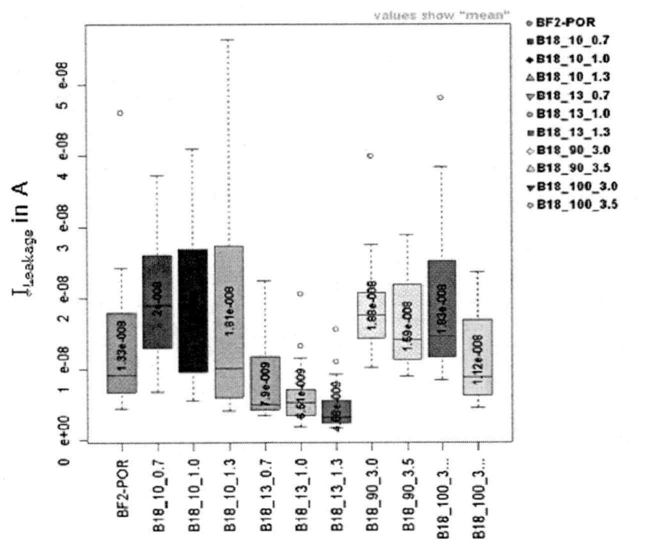

FIGURE 6. Junction leakage data.

SUMMARY

The results of this work show that the ClusterBoron implant can be used as an alternative to the conventional BF2 SDE without any degradation of performance. In addition, ClusterBoron can be used to replace the SD implant also without performance degradation. In fact, the experiment shown here produced a performance enhancement of >10% when ClusterBoron was used for the SDE. There results show definitively that the unique characteristics of the ClusterBoron implant, such as self-amorphization, due not degrade the stress enhancement produced by incorporating e-SiGe structures in the PMOS SD regions. Thus, the high productivity offered by the ClusterBoron implant can be realized without device degradation.

REFERENCES

1. Wade Krull, "Simplifying the 45nm SDE Process with Cluster Boron and Cluster Carbon Implantation", in *Proceedings of the 16th International Conference on Ion Implantation Technology 2006*, Marseille, France o1204
2. Tsutomu Nagayama, "Implantation characteristics by Boron Cluster ion implantation", in *Proceedings of the 16th International Conference on Ion Implantation Technology 2006* Marseill, France p123
3. Oh Jaegeun, "Investigation of covered p+Poly-Si Gate formed by B18Hx+ Cluster ion implantation", in *Proceedings of the 16th International Conference on Ion Implantation Technology 2006*, Marseille, France o301

$B_{18}H_{22}$ Implantation for the Source-Drain Extension in 45nm-node PMOSFETs and Beyond

Y. Kawasaki, S. Endo, M. Kitazawa, Y. Maruyama, T. Yamashita, T. Kuroi, H. Yoshimura and M. Kojima

Renesas Technology Corp., 4-1 Mizuhara, Itami, Hyogo, 664-8641, Japan

Abstract. We investigated the application properties of cluster implantation for MSA processing. We successfully replaced BF_2 with $B_{18}H_{22}$ to SDE implantation in pMOSFETs with the activation processing of MSA and spike-RTA. In the activation processing with only MSA, we found from blank wafer that $B_{18}H_{22}$ implant is more advantageous from the viewpoint of R_s, X_j and leakage current, compared to B, BF_2 and B with Ge-PAI.

Keywords: Semiconductors, $B_{18}H_{22}$, cluster implantation, millisecond annealing
PACS: 61.82.Fk

INTRODUCTION

In order to form the source-drain extension (SDE) in 45nm-node pMOSFETs and beyond, both shallower boron-doped layers and higher boron-activation are required. In dopant processing, the cluster implantation of $B_{18}H_{22}$ or $B_{10}H_{14}$ is noticed concerning the shallower profile such as little channeling or non-energy contamination [1]. These ion species can give not only a shallower profile but also superior beam characteristics in ultra low energy. In activation processing, a novel annealing technique such as flash lamp or laser is already established in the development of device. This equipment can supply the thermal budget at the temperature higher than 1200°C for extremely short time in the millisecond range. Currently the millisecond annealing (MSA) is utilized in combination with the conventional spike-RTA, which assists that SDE has the overlap with gate electrode due to boron diffusion.

In this work, we investigated the application properties of $B_{18}H_{22}$ implantation for MSA processing. First, we tried to replace the typical BF_2 with $B_{18}H_{22}$ in SDE implantation of 45nm-node pMOSFETs. In this case, boron is activated with the combination of first MSA and the following spike-RTA. It was demonstrated that such thermal sequence is effective for transistor characteristics [2]. Next, we focused on the properties of $B_{18}H_{22}$-implanted layer activated with only MSA, which will be necessary for the SDE formation beyond 45nm. We evaluated the fundamental characteristics of $B_{18}H_{22}$ layer after MSA in a blank wafer and compared to those of B, BF_2 implants and B implant with pre-amorphization implant of Ge (Ge-PAI).

EXPERIMENTAL

Figure1 shows part of the process steps in pMOSFET fabrication. After the gate electrode formation, either BF_2 or $B_{18}H_{X\ (X \leq 22)}$ was implanted into SDE region with 1.5 keV or 6 keV. These correspond to equivalent boron energy (EBE) of 0.3 keV calculated from the mass number. The ion dose of $B_{18}H_X$ implant was programmed to be equal to boron concentration of $5E14/cm^2$. Subsequently, MSA was performed at 1220°C utilizing a CO_2 laser. The temperature was controlled with laser power and was calibrated using the power of silicon melt as a reference. After the B implant into deep source-drain (SD) region, spike-RTA of 1050°C was done in the lamp-based equipment.

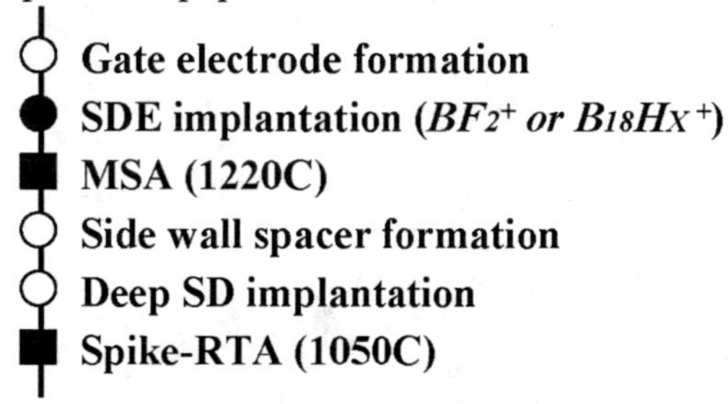

FIGURE 1. Process flow of pMOSFET.

CP1066, *Ion Implantation Technology*, edited by E. G. Seebauer, S. B. Felch, A. Jain, and Y. V. Kondratenko

In the experiment for only MSA processing, some blank wafers were prepared. The $B_{18}H_X$ was implanted at various EBEs of ≤0.5 keV and the dose from 3E14 B/cm^2 to 1E15 B/cm^2. Also B or BF_2 was implanted at equivalent energies and doses to $B_{18}H_X$ implant. Some of conditions with B implants were provided with Ge-PAI, which was carried out at energies of 5, 10 or 20 keV with the dose of 5E14/cm^2. The MSA temperature ranges from 1220°C to 1350°C. As the fundamental characteristics, sheet resistance (R_s) and profiles of boron and hydrogen were examined. The R_s and the profile were measured with four-point probe and SIMS respectively. Additionally, the leakage current in p-n junctions was measured for blank wafer in new type of tool, RsL.

RESULTS AND DISCUSSIONS

Transistor Characteristics in MSA and Spike-RTA

Figure 2 shows roll-off characteristics of V_{th} in the saturation region of pMOSFETs formed with implantations of either BF_2 or $B_{18}H_X$. Both roll-off characteristics are the same each other. Figure 3 shows I_{on}-I_{off} characteristics for pMOSFETs with same conditions as Figure 2. The I_{on}-I_{off} curve of $B_{18}H_X$ implant has the same characteristics as that of BF_2 implant.

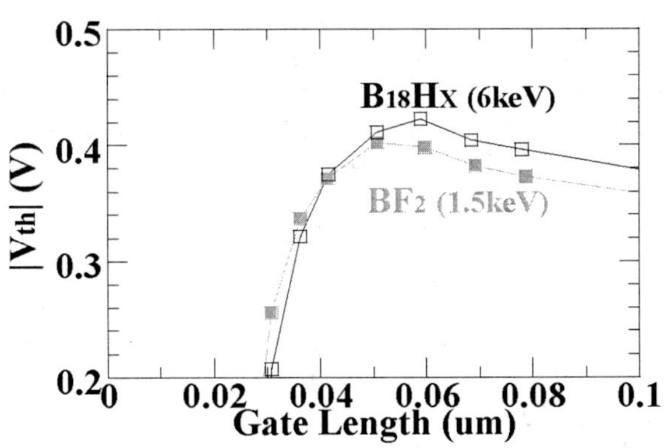

FIGURE 2. The dependence of $V_{th(sat.)}$ on gate length.

In SDE region, a large number of hydrogen atoms are induced with boron atoms as $B_{18}H_X^+$. Because they might become the precursor for fixed charges at the Si/SiO_2 interface, the negative bias temperature instability (NBTI) degradation was evaluated for pMOSFETs. Figure 4 shows NBTI induced V_{th} shift. Both curves are almost the same each other. Thus it was verified that there is no degradation of reliability also in pMOSFET with MSA and spike-RTA, as it is known in the case of only spike-RTA [3]. Figure 2, 3 and 4 promise that it is possible to replace BF_2 implant with $B_{18}H_X$ implant.

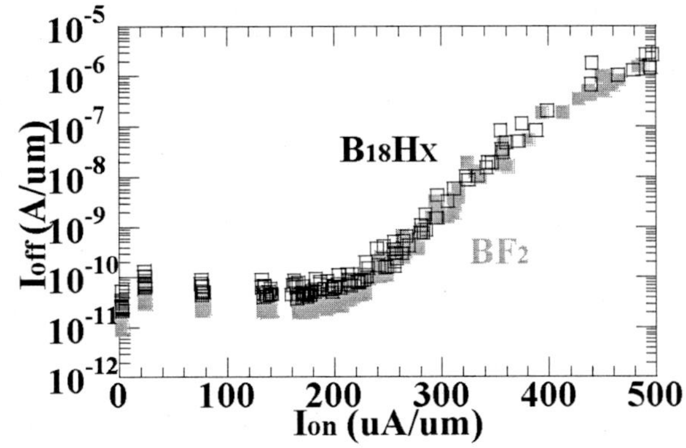

FIGURE 3. I_{on}-I_{off} characteristics.

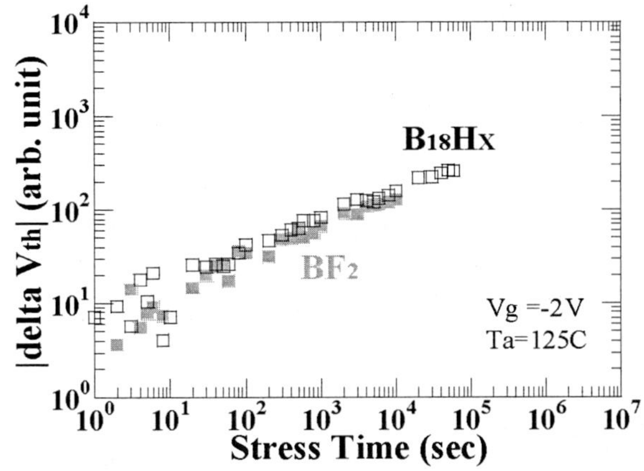

FIGURE 4. NBTI induced V_{th} shift.

Fundamental Characteristics in only MSA

Figure 5 shows the dependence of R_s on MSA temperature in B, BF_2 and $B_{18}H_X$ implants, which were performed at EBE of 0.5 keV, with dose of 5E14 B/cm^2. The R_s of $B_{18}H_X$ implant shows much lower value than those of B and BF_2 implants at the MSA temperature of 1220°C and the R_s-curve of $B_{18}H_X$ implant has a smaller gradient. The gradient of R_s-curve means the sensitivity of R_s to temperature. Therefore $B_{18}H_X$ implant gives rise to lower and more stable R_s than those of B and BF_2 implants at low temperature.

In order to determine such unique characteristic of $B_{18}H_X$ implant, the dependence on temperature was examined for B implant with Ge-PAI. It is already reported that the R_s of B implant with Ge-PAI becomes much lower than that without Ge-PAI at low power in flash lamp [4]. Also in this work with laser, same tendency is observed between the R_s-curves plotted in Figure 5. It is speculated that the re-alignment of silicon in the amorphous layer during MSA enhances boron-activation. The R_s-curves of B implant with Ge-PAI wholly shift to lower side with increase of Ge energy or amorphous thickness, because more boron atoms locate within amorphous layer.

Unlike the B implant, both the BF_2 implant and the $B_{18}H_X$ implant generate an amorphous layer. The thicker amorphous layer of the $B_{18}H_X$ implant contributes lower value and smaller gradient to the R_s-curve, compared to BF_2 implant. The R_s-curve of $B_{18}H_X$ implant approximates to that of B implant with Ge-PAI of 5 keV. Therefore $B_{18}H_X$ implant with EBE of 0.5 keV is worth B implant with Ge-PAI of less than 5 keV.

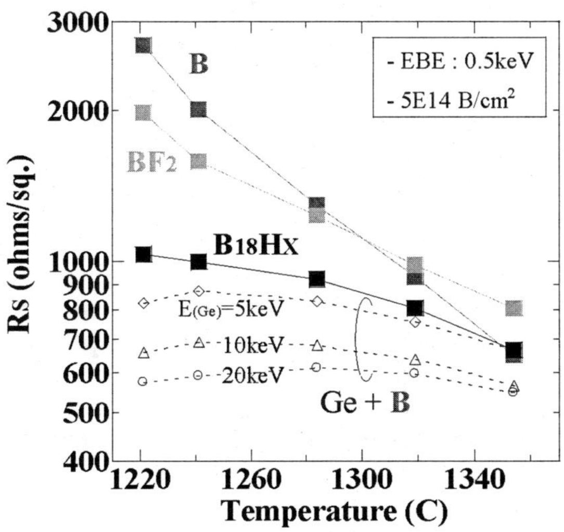

FIGURE 5. The dependence of R_s on MSA temperature in B, BF_2 and $B_{18}H_X$ implants and B implant with Ge-PAI.

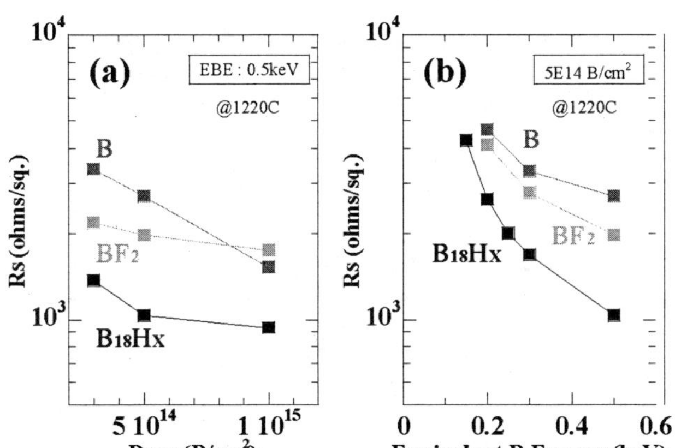

FIGURE 6. The dependence on implant parameters in three types of ion species at 1220℃. (a) dose under EBE of 0.5keV and (b) energy under the dose of 5E14 B/cm^2.

Figure 6 shows the dependence of R_s on dose and EBE in three types of ion species at 1220°C where the R_s specifically varies with the mass number of ion species in Figure 5. Both curves of $B_{18}H_X$ implants extensively show much lower value than those of B and BF_2 implants in Figure 6. At 3E14 B/cm^2 in Figure 6(a), the R_s of $B_{18}H_X$ implant becomes less than half of one of B implant. At 0.3 keV in Figure 6(b), the R_s of $B_{18}H_X$ implant becomes less than half of one of BF_2 implant.

Figures 7 and 8 show boron profiles for the samples compared in Figure 6 (a) and (b) respectively. Both profiles of $B_{18}H_X$ implants are steeper and shallower than those of B and BF_2 implants with equivalent condition. It is thought that this is because as-implanted profiles are retained due to the MSA of 1220°C which can minimize the boron diffusion.

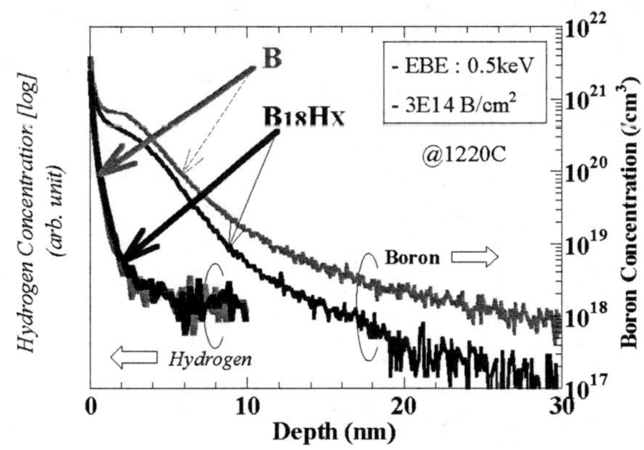

FIGURE 7. SIMS profiles of boron and hydrogen after MSA.

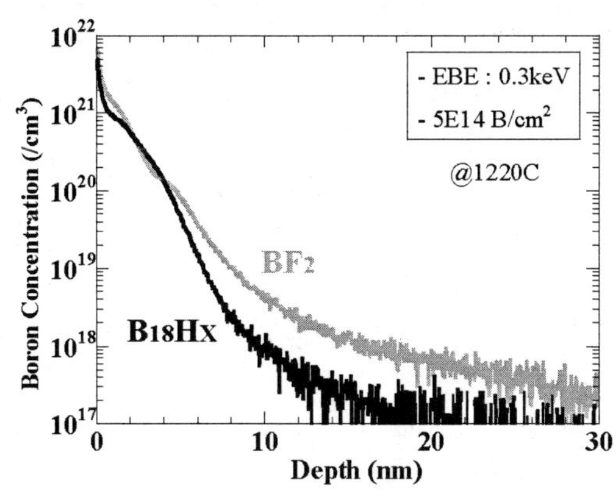

FIGURE 8. SIMS profiles of boron after MSA.

For the samples in Figure 7, also hydrogen profiles in activation layers were examined. The hydrogen profile of $B_{18}H_X$ implant plotted in same figure, perfectly fits that of B implant which obviously induces no hydrogen. Therefore the hydrogen in $B_{18}H_{22}$-implanted layer out-diffuses even for extreme short time in MSA processing. This fact ensures the result of reliability in Figure 4, because the hydrogen in SDE region disappears in the step of MSA in Figure 1.

The boron profiles in Figure 7 and 8 were summarized with a R_s-X_j plot in Figure 9. The X_j is defined as the distance below the surface where the concentration of boron is 1E18 /cm^3. It can be seen that two kinds of R_s-X_j data of $B_{18}H_X$ implants are greatly enhanced, compared to those of B and BF_2 implants with equivalent condition. It is guessed that the carrier concentration in activation layers of $B_{18}H_{22}$ implants becomes higher than those of B and BF_2 implants with equivalent condition.

In this figure, also the R_s-X_j data in the same conditions as SDE in Figure 1 are plotted. It means that the spike-RTA of 1050°C is added for the above activation layers, which are formed with the implants at EBE of 0.3 keV and the MSA at 1220°C. Contrary to the above result, the difference between BF_2 data and $B_{18}H_X$ data become much smaller and the relationship between R_s and X_j in them becomes a trade-off in small region. This indicates that the carrier concentration in activation layer of $B_{18}H_X$ implant is nearly identical to that of BF_2 implant with equivalent condition. This result proves that BF_2 and $B_{18}H_X$ show almost the same characteristics in I_{on}-I_{off} and roll-off of pMOSFETs.

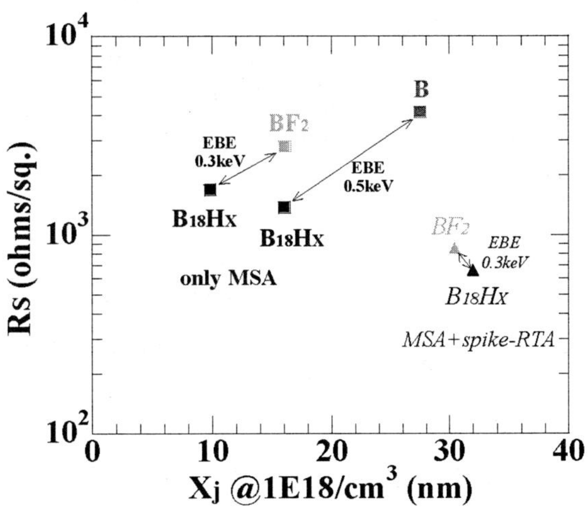

FIGURE 9. R_s-X_j plot in three types of ion species with activation sequence of only MSA and MSA+spike-RTA. The conditions with the same EBE are connected by an arrowed line.

Figure 10 shows the leakage current measured for the samples of Figure 5 using RsL tool, which has non-contact probe as an adaptable concept and utilizes the junction photo-voltage [5]. The conditions of B implant with Ge-PAI result larger leakage current with increase of Ge energy, because secondary defects locate at the deeper position than junction depth. On the other hand, all the curves in three types of ion species show the lower value. Especially, both the curves of $B_{18}H_X$ implant and BF_2 implant show lower detection limit independently on the temperature.

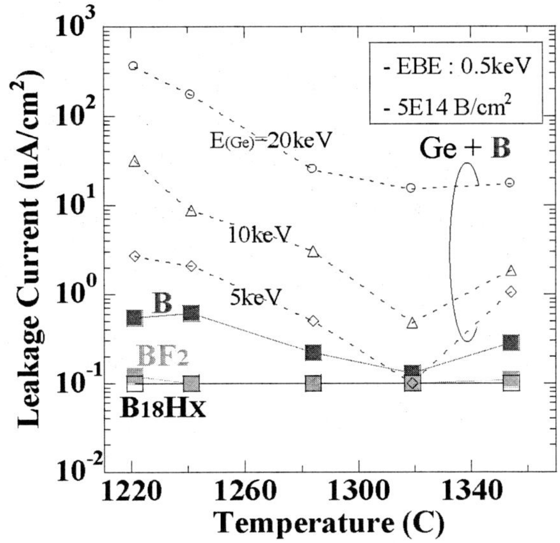

FIGURE 10. The dependence of leakage current on MSA temperature in B, BF_2 and $B_{18}H_X$ implants and B implant with Ge-PAI.

CONCLUSION

We successfully replaced BF_2 with $B_{18}H_{22}$ to SDE implantation in 45nm-node pMOSFETs with the activation processing of MSA and spike-RTA. In the activation processing with only MSA, we found from blank wafer that $B_{18}H_{22}$ implant is more advantageous from the viewpoint of R_s, X_j and leakage current, compared to B, BF_2 and B with Ge-PAI.

ACKNOWLEDGEMENTS

The author would thank to NISSIN's engineers and SemEquip's engineers for the supports of $B_{18}H_{22}$ implant and to Dr. J. Halim of FSM for measurement with RsL.

REFERENCES

1. Y. Kawasaki *et al.*, *Nucl. Inst. And Meth.* **B237** (2005) 25.
2. S. Endo *et al.*, Extended abstracts of the 7th International Workshop on Junction Technology, 2007, pp. 135-136.
3. M. Ishibashi *et al.*, Extended abstracts of the 5th International Workshop on Junction Technology, 2005, pp. 31-34.
4. S. Kato *et al.*, Extended abstracts of the 7th International Workshop on Junction Technology, 2007, pp. 143-146.
5. M. Current *et al.*, Extended abstracts of the 7th International Workshop on Junction Technology, 2007, pp. 43-46.

Improved Re-Crystallization of p^+ Poly-Si Gates with Molecular Ion Implantation

Jin-Ku Lee[1], Min-Ae Ju[1], Jae-Geun Oh[1], Sun-Hwan Hwang[1], Seung-Joon Jeon[1], Ja-Chun Ku[1], Sungki Park[1], Kyung-Won Lee[2], Steve Kim[2], Geum-Joo Ra[2], Ron Reece[2], Leonard M. Rubin[2], W. A. Krull[3], and H. T. Cho[3]

[1]Hynix Semiconductor Inc., San 136-1 Ami-ri, Bubal-eub, Ichon-si, Kyoungki-do, 467-701, Korea
[2]Axcelis Technologies Inc., 108 Cherry Hill Drive, Beverly, MA 01915, USA
[3]SemEquip Inc. 34 Sullivan Road, N.Billerica, MA 01862, USA

Abstract. Implantation of $B_{18}H_{22}$ molecules at 80 keV and doses up to 4x10^{16}cm^{-2} were evaluated for the application of p-type counterdoping of *in situ* n-type doped polysilicon gates. Compared to conventional B implants, molecular implantation provides greatly improved throughput without the risk of energy contamination. Implants at these high doses resulted in poor re-crystallization of the polysilicon layer due to the formation of excessive cluster-type defects. Subjecting the polysilicon to either UV-curing or low temperature soak annealing prior to dopant activation was not effective in improving the re-crystallization process. However, breaking the dose into two portions at two different energies was shown to significantly improve re-crystallization of the polysilicon layer. Improved dopant activation was confirmed by a >90% reduction in ring oscillator delay time on a 60 nm PMOSFET.

Keywords: Octadecaborane, Ion Implantation, Polysilicon Doping
PACS: 61.72.-y, 61.72.U-, 68.55.Ln, 85.40.Ry

INTRODUCTION

Modern DRAM fabrication requires very high dose boron doping to form the p+ poly gate by counterdoping of an *in situ* n-type doped poly layer [1,2]. High doping is also required for low resistance shallow junction formation and bit line contact resistance improvement. However, traditional beam line implantation suffers from either low beam current in drift mode or energy contamination in deceleration mode. For very high dose implants (more than 4x10^{16}cm^{-2}), boron forms clusters near the projected range during annealing. These clusters prevent the damaged layer from achieving complete re-crystallization, and cannot be eliminated by additional thermal processing. The result is poor device electrical characteristics, especially ring oscillator (R/O) delay.

In order to improve the re-crystallization of the implanted layer, we studied $B_{18}H_{22}$ molecular ion implantation process. For the same equivalent B energy, $B_{18}H_{22}$ has the same projected range (Rp) but better dopant activation. This improved activation is demonstrated on a 60 nm PMOS device.

EXPERIMENTAL

Initial work to measure the degree of re-crystallization was done on bare silicon wafers. We implanted B at energy of 5 keV and $B_{18}H_{22}$ at 30 to 80 keV (1.6 to 4.2 keV B equivalent) and doses of 1x10^{15}cm^{-2} to 5x10^{16}cm^{-2}. After annealing, various thermal treatments were investigated. These include poly implant annealing (PIA) for 10 to 30 seconds at temperatures of 700 to 1000°C, low temperature annealing for 10 to 20 minutes at 500 to 600°C, UV curing at 400°C for 10 to 15 minutes, and PIA at 950 and 1000°C. Breaking the 4x10^{16} cm^{-2} dose into two implants, one at 80 keV and another at 30 to 50 keV was also investigated. The doping characteristics and surface morphologies were analyzed with secondary ion mass spectrometry (SIMS), transmission electron microscopy (TEM), and a conventional four point probe.

We then implanted B or $B_{18}H_{22}$ at 30 to 80 KeV, 4x10^{16} cm^{-2}, into a 60 nm p+ poly gated device and measured the electrical characteristics including ring oscillator delay.

CP1066, *Ion Implantation Technology,* edited by E. G. Seebauer, S. B. Felch, A. Jain, and Y. V. Kondratenko

Figure 1 shows the processing sequence for these devices.

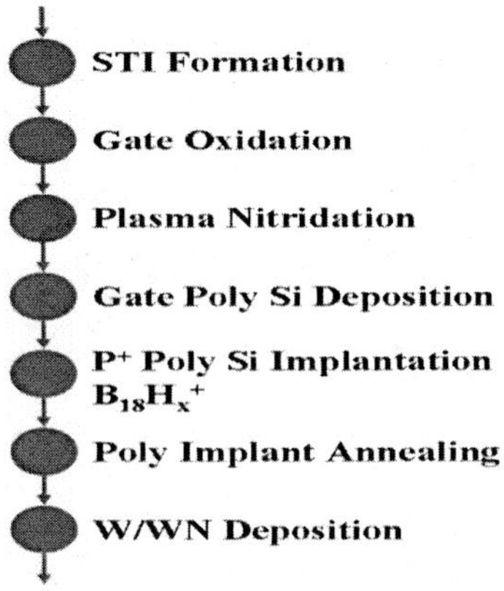

FIGURE 1. Process sequence for p+ poly-Si gate fabrication.

RESULTS AND DISCUSSION

Figure 2 shows the doping profile and cross sectional TEM of a boron, 5 keV, $4x10^{16}cm^{-2}$ implant annealed at 950°C. The boron has piled up at the projected range (~130Å) after annealing, and the damaged layer could not be re-crystallized due to the heavily piled up boron dopants seen on the TEM image. This damage takes the form of boron-interstitial clusters [3].

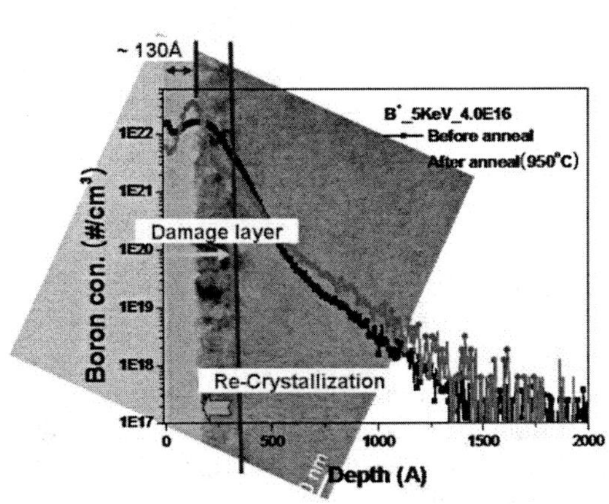

FIGURE 2. SIMS and cross sectional TEM for a B, 5 keV, $4x10^{16}$ cm^{-2} implant annealed at 950°C.

Figure 3 shows boron profiles for 80 keV $B_{18}H_x^+$ (4.2 keV B equivalent) at various doses before and after annealing at 950°C. The boron piled up at the projected range during the thermal treatment for all doses $\geq 2x10^{16}$ cm^{-2}.

Figure 4 shows TEM images for $2\text{-}4x10^{16}$ cm^{-2} doses before and after annealing. While the damaged layer at $2x10^{16}$ cm^{-2} has been re-crystallized well, the damaged layer (Figure 3, ~110A) with $4x10^{16}$ cm^{-2} couldn't be re-crystallized after annealing. This is similar to monomer boron implantation.

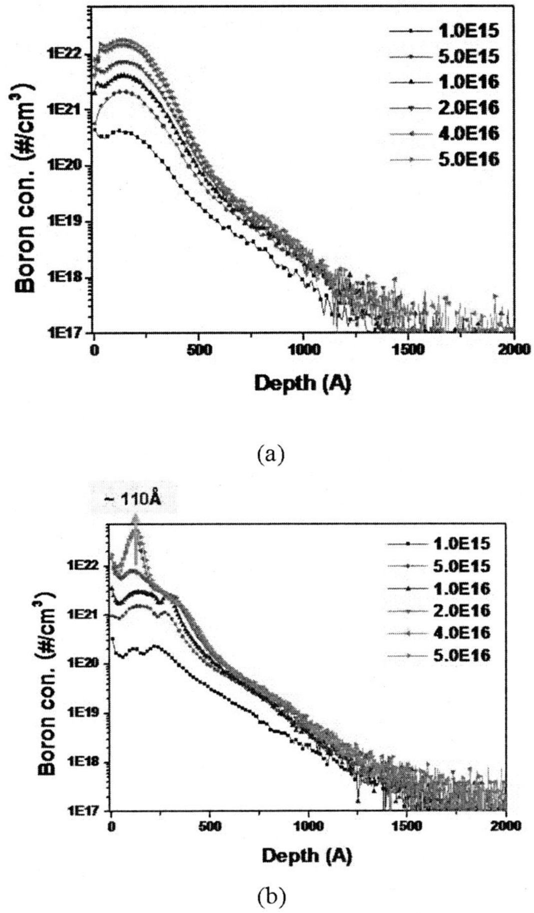

FIGURE 3. B profiles for 80 keV $B_{18}H_x^+$ at various doses before (a) and after (b) annealing at 950°C.

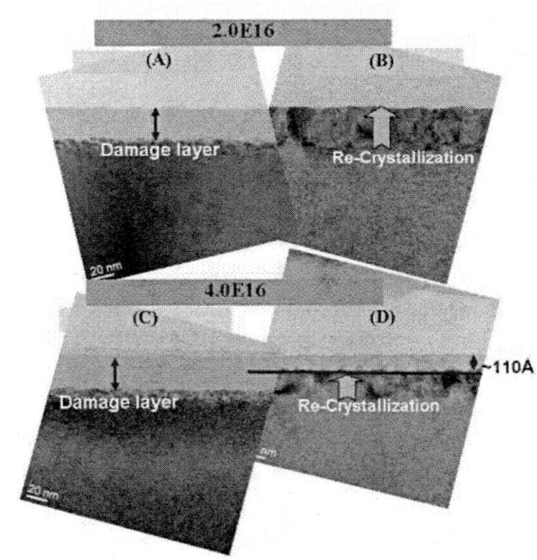

FIGURE 4. TEM images before and after annealing of (a,b) $2x10^{16}$ cm^{-2} (c,d) $4x10^{16}$ cm^{-2} dose.

Figure 5 shows SIMS and TEM data for 80 keV, $4x10^{16}$ cm^{-2} $B_{18}H_x^+$ implants with various annealing conditions. The anneal temperatures are 700, 950, and 1000°C for 20 seconds. Additionally, there were anneals of 10 and 30 seconds at 1000°C. There were no indications of dopant movement or damage layer evolution for 700°C anneals. This temperature was too low to activate the dopants. All samples show boron pile up at Rp for anneals ≥950°C, with no re-crystallization of the damage layer visible by TEM

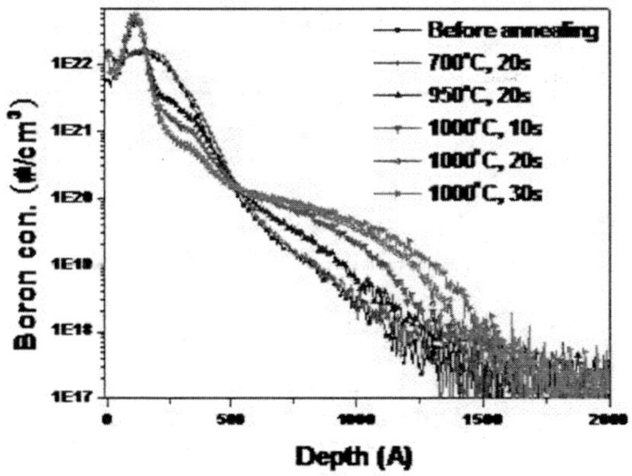

(a)

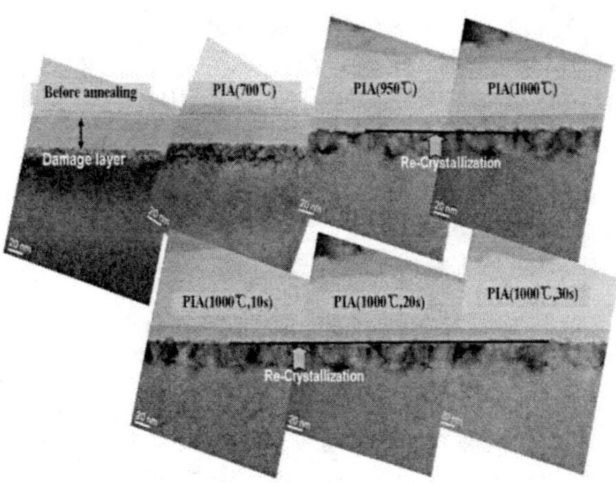

(b)

FIGURE 5. (a) SIMS profiles and (b) TEM images for 80 keV, $4x10^{16}cm^{-2}$ $B_{18}H_x^+$ implants with various annealing conditions.

Figure 6 shows SIMS profiles and TEM images for 80 keV, $4x10^{16}$ cm^{-2} $B_{18}H_x^+$ implants with UV curing and low temperature annealing before PIA. Pre-anneal treatments like UV-curing (Figure 6a) and low temperature annealing (Figure 6b) did not significantly affect the amount of boron pile up or the re-crystallization of the damage layer as indicated by TEM (Figure 6c).

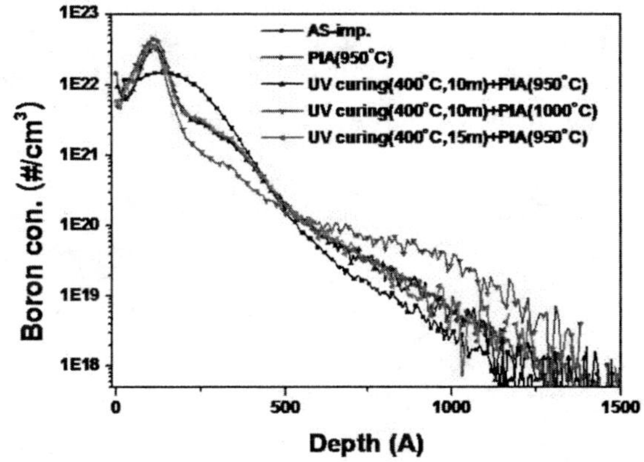

(a)

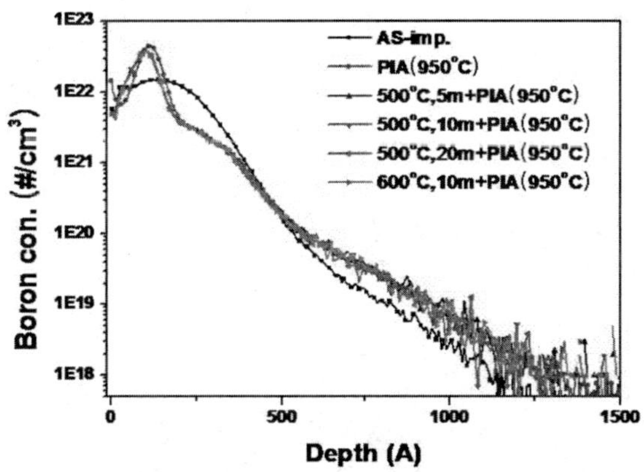

(b)

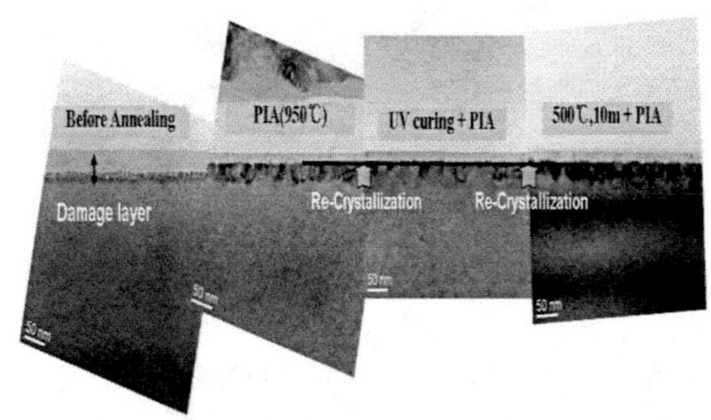

(c)

FIGURE 6. SIMS profiles and TEM images for 80 keV, $4x10^{16}$ cm^{-2} $B_{18}H_x^+$ implants with UV curing and low temperature annealing before PIA.

Figure 7 shows SIMS data for single step implants and when the dose is implanted at two different energies. The total dose was $4x10^{16}$ cm^{-2} and the post-implant annealing was at 1000°C. The dopant profiles are different after annealing as the peak locations have shifted significantly. The amount of boron at the surface decreased for the two step implant process.

Figure 8 compares the damage layer before and after annealing for single implant and two energy implants. The damage layer from the two step implant showed better recovery after annealing than the single implant step. This is attributed to reduced boron pileup at the projected range (Figure 7).

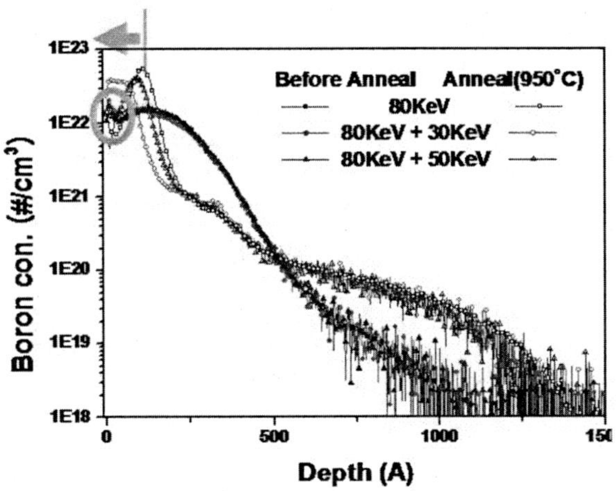

FIGURE 7. SIMS profiles of single energy and two energy implants. The total dose was $4x10^{16}$ cm^{-2} in all cases.

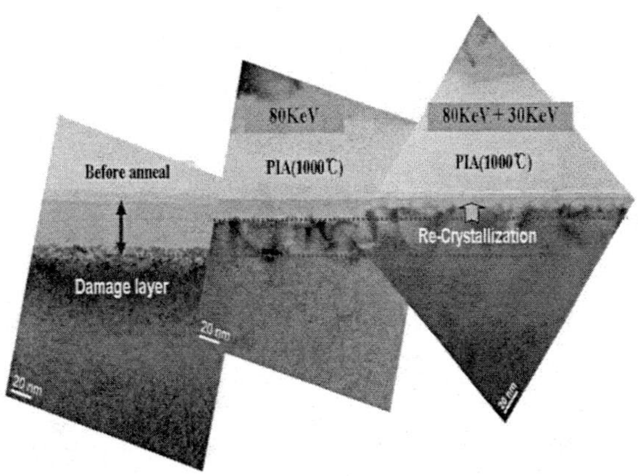

FIGURE 8. Comparison of implant damage layer before and after annealing for single implant and two energy implants.

Figure 9 shows the R/O delay results from a 60 nm PMOS device with the two step implant process. Breaking the dose into two different energies improved the R/O delay by over 90% as compared with a single high dose implantation.

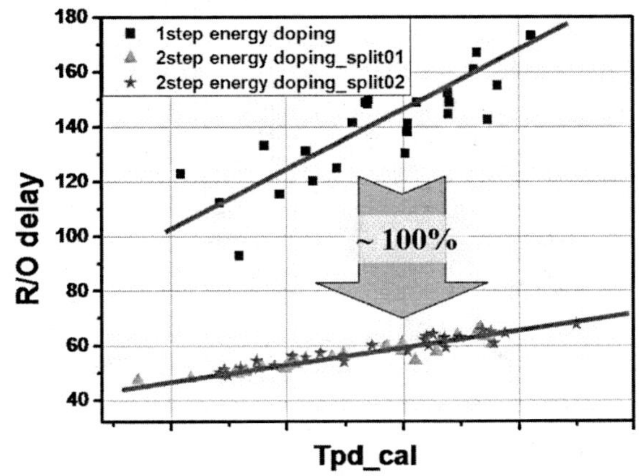

FIGURE 9. Ring oscillator delay results from 60 nm PMOS device.

CONCLUSIONS

We improved the re-crystallization of the damage layer from a $B_{18}H_x$ implant by breaking the high dose ($\sim 4x10^{16}cm^{-2}$) into two implants at two different energies. The two step implant was shown to move the dopant pile up region closer to the surface without significantly changing the rest of the dopant profile. In addition to improved damage recovery in the near surface region, R/O delay was reduced by almost 100% in a 60 nm MOSFET.

REFERENCES

1. S. H. Hwang, *et al.*, "Investigation of Converted p+ Poly-Si Gate Formed by $B_{18}H_X$ Cluster Ion Implantation", in *Proceedings of the International Conference on Ion Implantation Technology*, 2006, pp. 163-166.
2. D. Henke, *et al.*, "P-Type Gate Electrode Formation Using $B_{18}H_{22}$ Ion Implantation", in *Proceedings of the International Conference on Ion Implantation Technology* 2006, pp. 202-205.
3. S. Solmi, *et al.*, "Boron-interstitial silicon clusters and their effects on transient enhanced diffusion of boron in silicon", *J. Appl. Phys.* **88**, 4547 (2000).

Approaches to USJ Formation Beyond Molecular Implantation

C. Hatem, A. Renau, L. Godet, A. Kontos, G. Papasouliotis, J. England, and E. Arevalo

Varian Semiconductor Equipment Associates, 35 Dory Road, Gloucester, MA 01930, USA

Abstract. As junction depth requirements approach sub 10nm and the sensitivity to residual implant damage continues to increase, the capability to produce abrupt, shallow profiles while maintaining low residual damage becomes a difficult challenge. Implantation induced amorphization has been widely applied to reduce channeling tails of implanted dopant profiles for integrated circuit manufacturing. This has been required to meet aggressive junction depth targets. The problem, however, is that pre-amorphization creates high defect densities that remain near the former amorphous-crystalline interface post anneal. These end of range (EOR) defects become of greater concern as the industry begins to move towards millisecond anneal technologies. Millisecond anneal, while capable of close to diffusionless activation and abrupt junctions, has caused concern for its inability to fully repair these EOR defects. There has been a recent focus on removing traditional PAI through molecular implantation with limited success. Towards this end we have investigated alternative techniques to reduce EOR damage while maintaining the junction depth, sheet resistance and abruptness. Here we describe the results of two of these techniques. The subsequent reduction in EOR through the use of each process and the resultant Rs, junction depth and abruptness are detailed.

INTRODUCTION

As junction depth requirements approach 10nm for future low power logic device nodes, the sensitivity to residual damage and the resultant junction leakage has significantly increased. As a result, the leading cause of residual damage, the traditional germanium pre-amorphizing implant (PAI), has been removed for certain applications. The challenge has now become meeting the junction depth requirements of 32nm without PAI.

In parallel, efforts to enhance activation are driving millisecond annealing technologies (ms) to be incorporated into the process flow. However, the thermal budget of the millisecond anneal may not be sufficient to remove the residual damage from the implant step. Thus, the activation process now has two steps: a short RTP anneal in the 1000°C range followed by a ms anneal step where the peak temperature is in the range of 1200-1330°C. Additionally memory manufacturers wish to mitigate residual damage resulting from implant to reduce leakage and increase retention times.

Here we present two novel damage engineering techniques, that result in little to no residual damage and may enable the use of a single ms anneal activation step.

METHODS AND EXPERIMENT

Two novel methods for damage engineering have been developed:

1. Supplemental Amorphization Technique – This PLAD implant technique replaces the traditional PAI step with an inert specie. The result is the prevention of subsequent dopant implant channeling and little to no residual damage. Initial evidence also suggests the ability to tailor the junction depth independently of activation. This will be referred to as PR1 for the remainder of this paper.
2. Enhanced Amorphization Technique – This technique is a treatment to the substrate during implantation that lowers the amorphization threshold, resulting in a more abrupt implant profile, an enhanced amorphization by the dopant implant, and a subsequent lack of residual damage. This will be referred to as PR2 for the remainder of this paper.

TEM analysis and the use of fluorine decoration were used as metrics to measure amorphization and residual damage of substrates treated with PR1 and

CP1066, *Ion Implantation Technology,* edited by E. G. Seebauer, S. B. Felch, A. Jain, and Y. V. Kondratenko

PR2. The use of fluorine decoration has been previously used to indicate changes in end of range damage (EOR) post anneal. This has been reported and correlated to leakage in memory cells [1].

PR2 was used to demonstrate the ability to enhance amorphization by carbon compared to recent published literature on molecular carbon.

The formation of ultra-shallow junctions using PR1 and PR2 with ms anneal only is also reported and the resultant Rs and Xj suggest a clear advantage in achievable activation compared to traditional implant processes. The experiment included the use of carborane (CBH, $C_2B_{10}H_{12}$) as a source feed material for molecular implantation [2].

EXPERIMENTAL RESULTS

PR2 was applied to a silicon wafer implanted with 4mA of 5keV Carbon 1E15 cm^{-2}. A control wafer, without PR2 was also implanted with the same conditions. The samples were then sent for xTEM analysis for amorphization measurements. From the xTEM results in Figure 1, a very shallow amorphous layer can be seen near the surface in the baseline sample (A); below this region is approximately 25nm of extended damage before transition to the crystalline substrate. In sample B the application of PR2 to the same implant results in a clean amorphous layer approximately 22nm in depth with a very short transition to the crystalline region. Each sample then underwent a 1025°C spike anneal in 100% nitrogen and were analyzed for residual damage. In sample (C) there is clear evidence of large defects areas while the addition of PR2 (D) resulted in clean, full recrystallization.

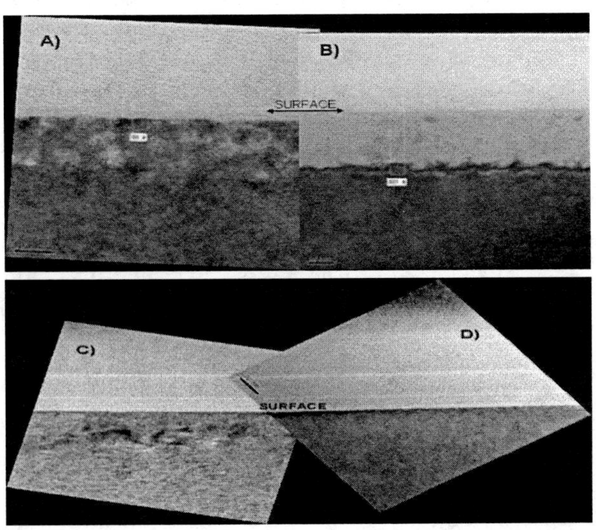

FIGURE 1. xTEM (scale 5nm) showing the amorphization of 5keV C 1E15 alone (A) and with PR2 (B) and the corresponding residual damage in (C) and (D).

A more detailed study was completed using carbon and PR2 to compare to recently reported molecular carbon amorphization results [3]. Figure 2 summarizes the effect of PR2, as measured by xTEM, with atomic carbon as compared to molecular carbon. As seen in the results, with the addition of PR2, atomic carbon implantation exceeds the amorphization of the molecular carbon at all experimental conditions.

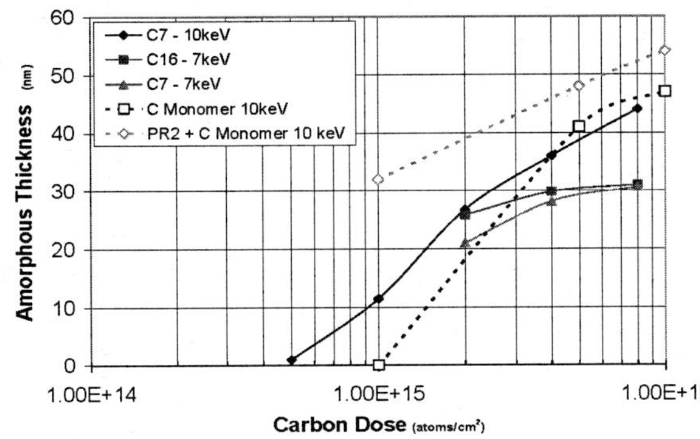

FIGURE 2. Resultant amorphization layer thicknesses of molecular carbon and monomer carbon with PR2.

PR1 was also evaluated by TEM analysis for amorphization. Figure 3 shows that a clean amorphous region is created using PR1.

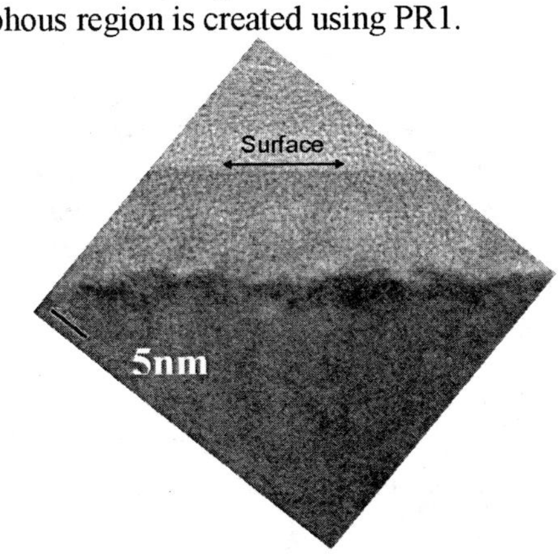

FIGURE 3. TEM result confirming the ability of a light ion specie to fully amorphize during PR1

In addition to xTEM analysis another method to characterize residual damage is the ability of Fluorine to decorate EOR. Previous reports [1] have indicated that the second fluorine peak correlates to residual damage and leakage in a device. SIMS analysis was used to quantify the reduction of the second peak for two sample sets evaluating the application of PR2.

The first sample was a 15keV BF_2 1E15 cm^{-2} implant. This sample and a control underwent an RTP anneal at 900°C for 120 sec. The second sample was a 15keV As 1E15 implant. A second fluorine implant was used as a marker and the sample and a control then underwent a flash anneal where an intermediate temperature of 750°C was held for 10 seconds and then a peak temperature of 1230°C was achieved. In Figure 4, EOR for each sample set has been completely removed as indicated by the absence of the second Fluorine peak.

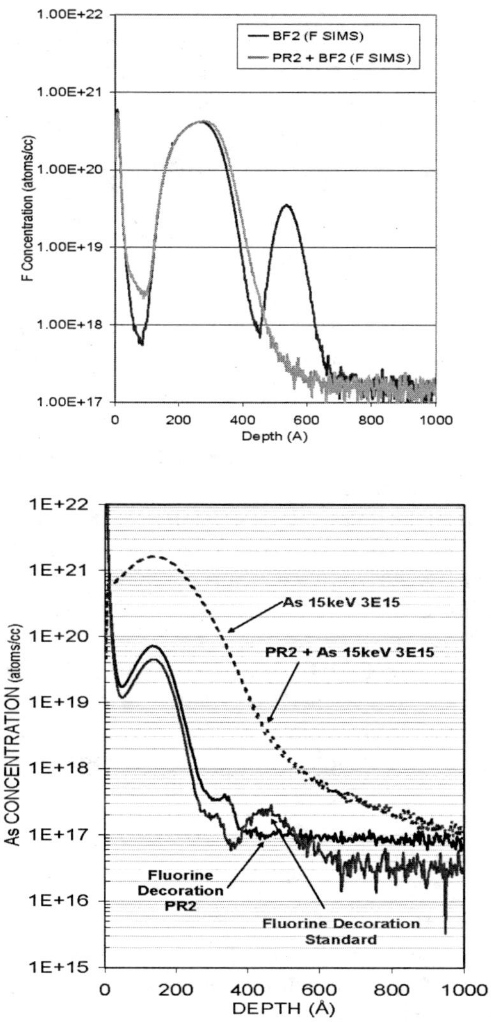

FIGURE 4. SIMS fluorine profile showing the use of PR2 to eliminate the 2nd peak associated with EOR and leakage.

A reduction in leakage was seen with the application of PR1 to USJ formation. Wafers were fabricated with diode structures, implanted with PR1 and BF2 or CBH, and then underwent a flash anneal where an intermediate temperature of 750°C was held for 10 seconds and then a peak temperature of 1230°C was achieved. A control sample with a traditional Ge PAI (12keV Ge 3E14 cm^{-2}) step was used for comparison. The samples were then sent for leakage measurements using two different methods: FSM using a JPV RsL Leakage method and Surrey University using a fabricated diode to measure leakage. Figure 5 shows a significant reduction in junction leakage for the PR1 sample when compared to a traditional Ge PAI. A similar result was shown with non melt LSA, peak temperature of 1330C.

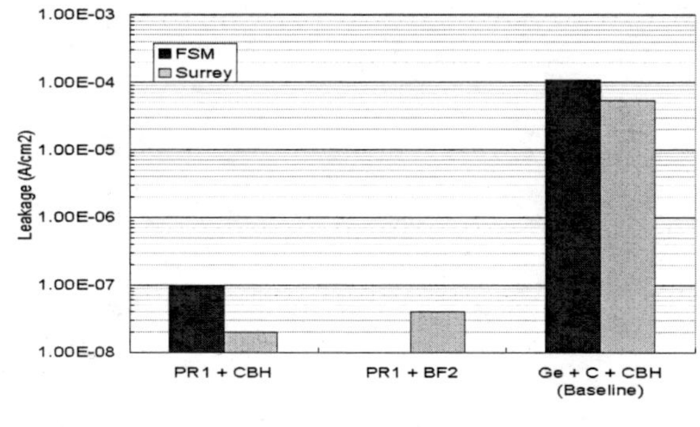

FIGURE 5. Leakage results from FSM and University of Surrey.

These leakage results show the combination of PR1 and ms-anneal only result in levels below the detection limit for FSM samples and extremely low for the results from the University of Surrey.

In an exciting discovery, PR1 has shown the ability to precisely control the Xj independently of activation (Rs). In Figure 6A several samples were treated with PR1. Each sample was then followed by an increasing implant dose of CBH. As seen in the chart the diffusion of boron dopant stops abruptly at 13nm regardless of the implanted dose. This allows the junction depth to be set by PR1 and the target activation to be dialed in with CBH implant dose. Figure 6B further verifies the ability to tune the junction depth independent of activation. In this chart BF2 implant was held constant and PR1 was modulated.

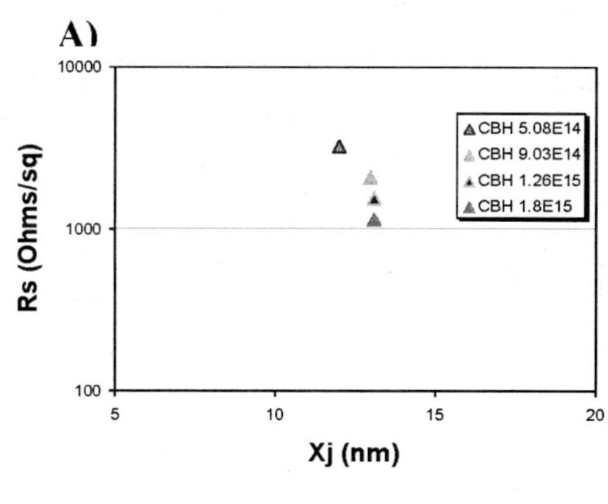

B)

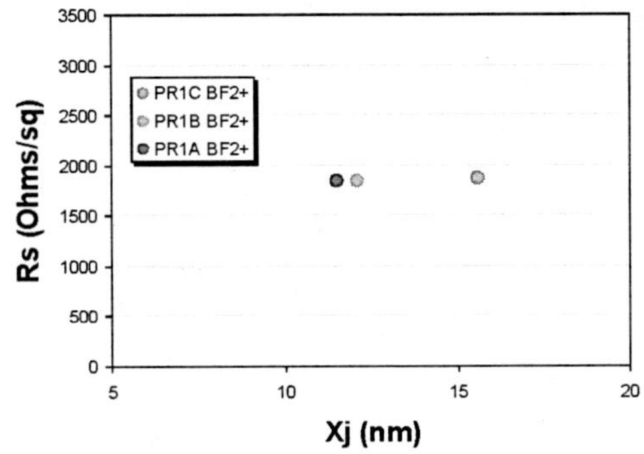

FIGURE 6A AND 6B. Rs vs. Xj plot showing the diffusion stopping power of PR1.

A summary of these recent results for PR1 and PR2 using LSA or Flash anneal can be found in Figure 7. These are some of the most advanced data published. The PR1 + CBH point corresponds to the previous leakage data in Figure 5. This is only initial data and there is room for further optimization.

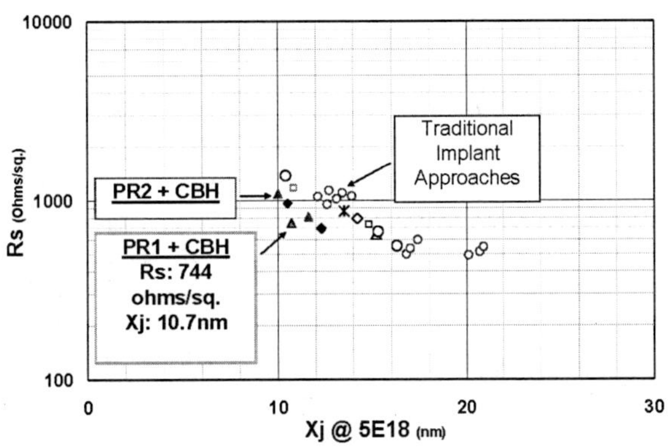

FIGURE 7. Initial Rs/Xj results for PR1 and PR2

The ability of PR2 to enhance amorphization and the impressive results with atomic carbon previously mentioned have led to the use of PR2 and a 5keV carbon $1E15/cm^2$ implant as the amorphization step, followed by a CBH implant equivalent to 350eV $1E15/cm^2$. The sample then underwent a flash anneal with a peak temperature of 1230°C. The resultant SIMS profile can be seen in figure 8 and is compared to a sample preamorphized with 12keV Ge $6e14/cm^2$.

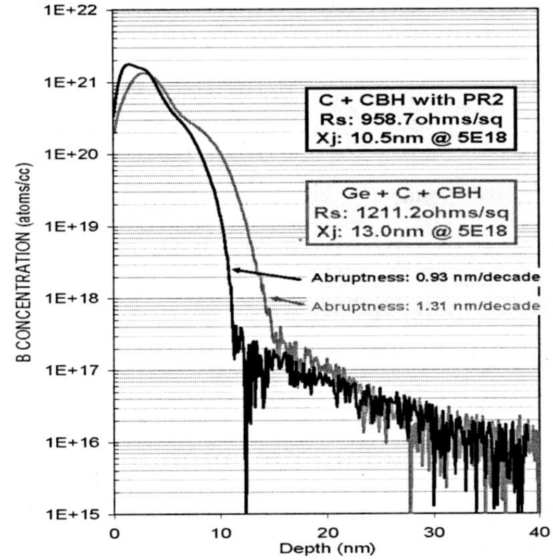

FIGURE 8. Initial Rs/Xj results for PR2 and CBH

CONCLUSIONS

In this paper we have demonstrated two unique approaches to damage engineering that result in little to no residual damage and a major reduction in junction leakage with a single ms anneal approach. The implementation of these processes could lead to a significant improvement in activation, junction depth, and junction leakage.

Furthermore PR1 may enable independent control of junction depth and activation while reducing residual damage. While PR2 has shown promise for significant reduction in EOR and leakage; specific applications to memory devices are also being explored.

ACKNOWLEDGEMENTS

The authors would like to thank Jay Scheuer, Chris Rowland, Greg Redinbo, Naushad Variam, Ethan Wright, Jay Johnson, and Roger Hyde for their contribution to this work.

REFERENCES

1. S. Bae, *et al.*, "P+ Source/Drain Ion Implantation Yield Improvement Through Fluorine Second Peak Reduction", *Proceedings IIT 2004.*
2. T. Renau, *et al.*, "A Better Approach to Molecular Implant", *Proceedings IWJT 2007.*
3. K. Sekar, *et al.*, "Self Amorphization with Cluster Implantation", *Proceedings Insight 2003*

Beam Angular Divergence Effects in Ion Implantation

T.N. Horsky[1], M. I. Current[2], S. K. Hahto[1], D. G. Bilbrough[1], D. C. Jacobson[1], W. A. Krull[1], R. D. Goldberg[1], N. Hamamoto[3], and S. Umisedo[3]

[1]*SemEquip Inc., 34 Sullivan Road, Suite 18, Billerica, MA 01862, USA*
[2]*Current Scientific, 1729 Comstock Way, San Jose, CA, 95124 USA*
[3]*Nissin Ion Equipment Co., 575 Kuze Tonoshiro-cho, Minami-ku, Kyoto, Japan, 601-8205*
e-mail: thorsky@semequip.com

Abstract. An important difference between monomer ion beams and heavy molecular beams is a significant reduction in beam angular divergence and increased on-wafer angular accuracy for molecular beams. This advantage in beam quality stems from a reduction in space-charge effects within the beam. Such improved angular accuracy has been shown to have a significant impact on the quality and yield of transistor devices [1,12]. In this study, $B_{18}H_x^+$ beam current and angular divergence data collected on a hybrid scanned beam line that magnetically scans the beam across the wafer is presented. Angular divergence is kept below 0.5 deg from an effective boron energy of 200 eV to 3000 eV. Under these conditions, the beam current is shown analytically to be limited by space charge below about 1 keV, but by the matching of the beam emittance to the acceptance of the beam line above 1 keV. In addition, results of a beam transport model which includes variable space charge compensation are presented, in which a drift mode $B_{18}H_x^+$ beam is compared to an otherwise identical boron beam after deceleration. Deceleration is shown to introduce significant space-charge blow up resulting in a large on-wafer angular divergence. The divergence effects introduced by wafer charging are also discussed.

Keywords: molecular implantation, ion beam, angular divergence
PACS: 61.72Tt

INTRODUCTION

The need for near-"diffusion-less" anneal conditions to achieve ≈10 nm junction depths for CMOS source/drain extensions has placed stringent requirements on the implantation process for the vertical and lateral placement of dopant and other ions in close matching to the needs of the transistor design. Considerable effort and ingenuity has been incorporated in the design of ion beam lines and wafer scanning systems to provide consistent control of beam incidence angle over the wafer surface to within a small fraction of a degree [1]. Next, control of the optical properties of the ion beam itself, including the beam divergence, is an increasingly relevant issue.

The impact of highly divergent ion beams, such as those created by accel-decel optics, on local dopant uniformity near gate and mask edges has been noted since the early days of sub-keV high current implanters [2]. These dopant uniformity effects included under-dosing due to beam shadowing and local over-dosing due to ion scattering from gate and mask edges. Indications of additional effects due to local disruptions of space charge balances by device-scale structures were seen in the current-voltage characteristics for 0.5 keV B^+ beams implanted into sub-micron mask openings [3].

The recent introduction of heavy molecular ion beams has demonstrated several advantages for high dose, low energy doping of semiconductors, including higher throughput for p^+ doping, enabling of defect-free junctions with reduced junction leakage, improved activation and contact resistance, as well as providing self-amorphizing implants for simplified and lower cost process flows [4,5,6]. The additional advantages of decreased beam divergence for molecular ion beams has been noted in the increased control of threshold voltage distributions and improved drive current symmetry for 40 nm CMOS transistors [7].

Space Charge Limits on Beam Transport

The space charge-limited charge density at ion extraction, the so-called Child-Langmuir relation, can be expressed as:

$$J_{max} = 1.72\,(Q/m)^{1/2}\,V^{3/2}\,d^{-2}, \qquad (1)$$

Where J_{max} is in mA/cm^2, Q is the ion charge state, m is the ion mass in AMU, V is the extraction voltage in kV, and d is the gap width in cm. While derived for a planar diode, this relation is also applicable to beam transport, particularly for an uncompensated beam. In practice, beam ionization of residual gases provides some charge compensation through low-energy electron capture by the beam potential, although the ionization cross sections fall steeply below about 10 keV. Plasma floods also provide useful compensation, although they tend to provide their primary benefit at their point of use, their thermal electrons being localized by fields present in the beam line. Wafer charging is a continuing yield concern for advanced devices, and is typically controlled by the incorporation of a plasma flood near the wafer surface.

CP1066, *Ion Implantation Technology,* edited by E. G. Seebauer, S. B. Felch, A. Jain, and Y. V. Kondratenko

The cluster implant process requires only one electrical charge per cluster of n atoms, rather than requiring every dopant atom to carry one electrical charge, largely ameliorating wafer charging for large n. The transport efficiency (beam transmission) can also be improved by using ionized clusters, since the dispersive Coulomb force is reduced with a reduction in charge density. Implanting a cluster of n like atoms provides the equivalent of a much lower energy monatomic implant since each atom of the cluster is implanted with the ion energy divided by n, whereas the implanter is also operated at an extraction voltage about n times higher than the required monomer implant energy.

By extension of equation (1), the following figure of merit, Δ, can be defined to quantify the increase in throughput, or implanted dose rate, for a cluster ion implant relative to monomer implantation:

$$\Delta = n\,(U_n / U_1)^{3/2}\,(m_n / m_1)^{-1/2}. \qquad (2)$$

Here, Δ is the maximum relative improvement in dose rate (atoms/sec) achieved by implanting a cluster with n atoms of the dopant of interest at an energy U_n relative to the single atom implant of an atom of mass m_1 at energy U_1, where $U_i = eV_i$. In the case where U_n is adjusted to give the same dopant implantation depth as the monatomic (n=1) case, equation (2) reduces to:

$$\Delta = n^2. \qquad (3)$$

Thus, the implantation of a cluster of n dopant atoms has the potential to provide a dose rate n^2 higher than the conventional implant of single atoms.

EXPERIMENTAL METHOD AND DATA

Ion beam characteristics were measured on a magnetically scanned implanter whose design has been previously disclosed [8]. As shown in Figure 1, the system consists of ClusterIon® source, 120° analyzer magnet, quadrupole triplet, scanner, and collimator. Beam diagnostics located behind the wafer plane to measure spatial uniformity, size, parallelism, and current, have also been described previously [9].

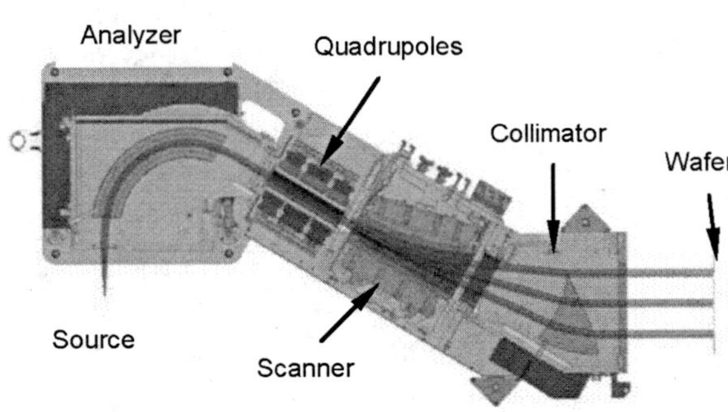

FIGURE 1. Beam line test system. Diagnostics are directly behind the wafer position.

Beam angle measurements were carried out with multiple Faraday cup systems (front & back Faraday arrays) located about 300 mm upstream and down stream of the wafer. The beam parallelism (beam incident angle across the wafer) was calculated by comparing the relationship of the beam position vs. time between the front and back Faraday arrays. In addition, as the beam profile can be measured at each Faraday, the nominal beam divergence α_p was determined by comparing the average beam diameter at FWHM between these Faraday arrays.

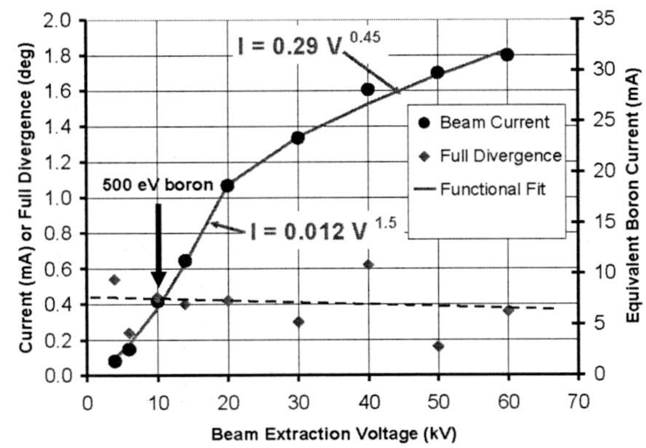

FIGURE 2. On-wafer currents and full divergence angles for scanned $B_{18}H_x^+$ beams as a function of energy.

Figure 2 shows scanned $B_{18}H_x^+$ beam current (circles) and full divergence angle (diamonds) as measured by the system of Figure 1. The divergence data and related dashed trend line show that beam divergence remained relatively constant over the energy range 4 keV to 60 keV, centered about 0.4 degrees. We define the full beam divergence φ as:

$$\varphi = 2\,|\alpha_p| \qquad (4)$$

Where α_p is the half-pencil angle about the beam axis. Since divergence is small throughout the energy range (corresponding to an equivalent boron implant energy range of 200eV to 3keV), the principal effect of energy on beam transport is an increase in beam current transmission with increasing energy. We have therefore fitted the beam current data of Figure 2 with the following binary function:

$$I = 0.012\,V^{1.5}, \quad 4\text{ kV} \le V \le 20\text{ kV}; \qquad (5)$$
$$I = 0.29\,V^{0.45}, \quad 20\text{ kV} < V \le 60\text{ kV}. \qquad (6)$$

This function suggests that in the low energy regime (below about 1 keV effective boron energy), transport is limited principally by space charge, as expressed by Eq. (1), whereas at higher energies, transmission is limited principally by the acceptance of the beam line. The fundamental premise underlying latter conclusion can be seen by examining the 2D beam emittance, ε_{2D}, according to Eq. (7):

$$\varepsilon_{2D} = (\pi/2)\, D\, (\upsilon_T/\upsilon_{//}), \qquad (7)$$

Where D is the beam diameter, and υ_T and $\upsilon_{//}$ are the ion velocities transverse to and along the beam direction, respectively, in the dispersive plane. Since $(\upsilon_T/\upsilon_{//})$ is clearly proportional to $V^{-½}$, we conclude that transmission at higher energies is principally determined by matching of the beam emittance to the acceptance of the beam line.

Beam Transport Modeling

To better understand the effects of space charge on the average angle the beam makes with the wafer, we modeled two ion beams, the first carrying 7.5 mA of boron at 500 eV, the second carrying 0.42 mA of $B_{18}H_x^+$ beam at 10 keV. In practice, these beams would deliver the same dose rate and implant energy to the wafer, the first presumably after deceleration, and the second in drift. We assumed the same source emittance and beam divergence after mass analysis and before deceleration, about 0.5 m before the wafer.

Lorentz 2D V6.4 from Integrated Engineering Software, Inc. was used to generate the model. Rays were launched from identical initial positions to create a Gaussian current density. This density was scaled according to the particular ion current being launched. The ions were launched from a surface to create a divergent beam, and then accelerated over a very short gap to the final required energy before allowing the ion to drift for 0.5 m and impact on the wafer. Space charge was calculated and stored in a mesh as described in the Lorentz 2D V6.4 user's manual [10]. Output generated included trajectory directional vectors that were used to determine the beam angle and pencil angle at the wafer.

The divergence of the decelerated boron beam will be a function of three terms: the emittance of the beam, lens aberrations introduced by the decel stage, and space-charge blowup of the decelerated beam. Lens aberrations were not included in this model. The divergence increase of the decelerated boron beam due to a fixed beam emittance was derived from equations (4) and (7), *i.e.*, $\varphi_{boron} \sim (10\text{kV}/0.5\text{kV})^{½}\, \varphi_{B18} \approx 4.5$ times the $B_{18}H_x^+$ divergence, or about 1.9 deg. Figure 3 shows the output for a 10 keV, 0.42 mA $B_{18}H_x^+$ beam with 43% space charge compensation. The divergence at the wafer predicted by the model for these conditions is 2 degrees. Figure 4 shows the output for a 500 eV, 7.5 mA $^{11}B^+$ beam with 98% space charge compensation. The model predicts a divergence of about 23 degrees at the wafer. It was necessary to provide nearly full space charge compensation to the ^{11}B beam due to severe blowup over its 0.5 m path length when lower compensation values were used. When the model of Figure 3 was run for $B_{18}H_x^+$ with 98% space charge compensation, the angular divergence was 0.47 degrees, in good agreement with the measured data of Figure 2, suggesting that 98% compensation is a realistic value for this 2D model.

FIGURE 3. 10 keV, 0.42 mA $B_{18}H_x^+$ beam terminating at the wafer, assuming 43% space charge compensation. Half the beam is shown about the positive Z-axis. The drift length is 0.5 m.

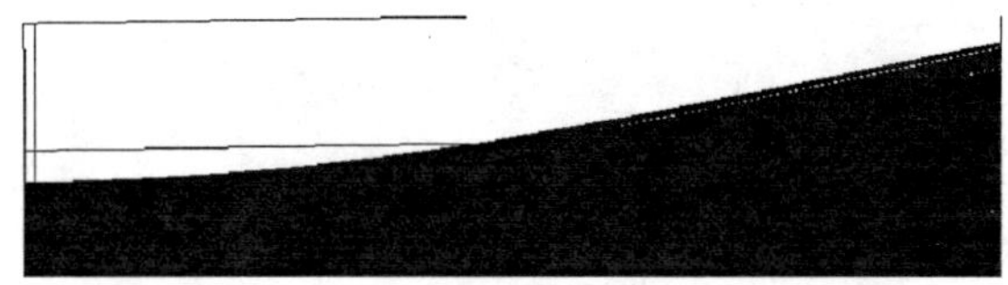

FIGURE 4. 500eV, 7.5 mA boron beam terminating at the wafer, assuming 98% space charge compensation. Half the beam is shown about the positive Z-axis. The drift length is 0.5 m.

Wafer charging can also affect incident beam angles. We added to the models used to generate Figure 3 and Figure 4 a photoresist-covered surface charged to 50V, which showed that the $^{11}B^+$ beam and $B_{18}H_x^+$ beams respond differently to the surface fields. Due to the higher stiffness and reduced charge density of the molecular beam, there is a negligible increase in divergence, while the lighter boron beam is deflected by the charged surface to a greater degree, increasing divergence by 9% from 22.9 degrees to 25 degrees. Estimates of the plasma density needed for a complete suppression of low energy beam self-repulsion for B^+ beams, so that the plasma Debye length is less than the ion spacing, leads to densities of $>10^{12}$ ions/cm^3 for 1 keV B^+ beams with modest electron temperatures, >3 eV [11]. These plasmas would be 2 orders of magnitude denser than typical support plasmas used to suppress charging effects on device wafer surfaces.

Given the angular divergence results, we can deduce that the beam transmission through a patterned feature, for example through a via for the S/D contact plug (which can include aspect ratios as high as 100:1), will be significantly reduced for large-divergence beams. These implants tend to be low energy and high dose, e.g., as low as 1 keV at >1E16 dose for 32 nm devices. A much higher total dose would be required for large-divergence beams relative to well-collimated beams, reducing wafer throughput.

SUMMARY

Beam current and full divergence angle has been measured over the effective energy range 200eV—3keV using $B_{18}H_{22}$ feed material in a hybrid magnetically scanned beam line. The full beam divergence was kept below about 0.5 deg over the energy range, resulting in reduced beam current at low energies. Fitting of the data showed that below about 1 keV boron energy (20 keV $B_{18}H_x^+$ energy), the beam current was proportional to $V^{1.5}$, indicating that beam transmission was limited by space charge. Above 1 keV, beam current was found to be proportional to $V^{0.45}$, indicating that beam transmission was limited by matching of the beam emittance to the acceptance of the beam line. Transport models were constructed of two ion beams, the first carrying 7.5 mA of $^{11}B^+$ at 500 eV, the second carrying 0.42 mA of $B_{18}H_x^+$ beam at 10 keV. The degree of space charge compensation was varied from about 43% to 98%. The 98% case yielded significant differences between the monomer and cluster beams, the full divergence angles at the wafer being about 23 degrees for the former, and about 0.5 degrees for the latter. The model was extended to include a charged wafer surface, which resulted in a 9% increase in the divergence of the monomer beam, and a negligible increase in divergence of the cluster beam.

REFERENCES

1. H-J. Gossmann, G. Redindo, Y. Eurokin, T. Romig, K. Elshot, J-J. Xu, J. McComb, *Solid State Technology*, July 2007, 71-76.
2. M.I. Current, D. Lopes, M.A. Foad, J.G. England, C. Jones, D. Su, *J. Vac. Sci. Technol.* **B16**(1) (1998) 327-333.
3. W. Lucaszek, M.I. Current, S. Daryanani, L. Larson, T. Rhoad, J. Shields, M. Vella, D. Wagner, Proc. of Plasma and Process-Induced Damage, P2ID03, (2003).
4. K. Yako *et al.*, 68th JSAP, Hokkaido, JP, 2007.
5. K. Uejima *et al.*, Proc. IEDM07, 2007, 151-154.
6. J. Borland *et al.*, Proc. IWJT, May, 2006
7. T. Aoyama, M. Fukuda, Y. Nara, S. Umisedo, N. Hamamoto, M. Tanjo, T. Nagayama, Internat. Workshop on Junction Technol., IWJT-05 (2005).
8. H. Glavish *et al.*, Proc. IIT2006, Marseilles, France (2006) pp. 167-170.
9. Nariaki Hamamoto *et al.*, 7th Int.Workshop on Junction Technol., June 8—9, 2007, Kyoto, Japan.
10. Lorentz User's Guide Ch. 4, Integrated Engineering Software, Inc.
11. M.I. Current, W. Lucaszek, M.C Vella, Proc. of 5th Inter. Symp. on Plasma Process-induced Damage, P2ID-2000, pp 137-140 (2000).
12. K. Okabe *et al.*, these proceedings, and references contained therein; also Fujitsu, IWJT 2005.

Infusion Doping for Sub-45nm CMOS Technology Nodes

[a,1]Pankaj Kalra, [a,b]Prashant Majhi, [a]Hsing-Huang Tseng, [a,c]Raj Jammy, and [1]Tsu-Jae King Liu

[a]*SEMATECH, Austin, TX 78741*
[1]*University of California, Berkeley, CA 94720*
[b]*Intel,* [c]*IBM*

Abstract: The requirements for ultra-shallow junctions will be difficult to meet for sub-45 nm CMOS technology nodes with conventional low-energy ion implantation and fast-ramp spike annealing. This paper explores the possibility of using infusion doping in conjunction with flash annealing to achieve shallower junctions with lower sheet resistance. No evidence of channeling in (100) and (110) silicon substrates is seen for infusion doping. A detailed investigation is carried out to study the effects of flash anneal parameters. Experimental results indicate that it is possible to achieve X_j in the range ~12-15 nm with R_s ~1000 Ω/□, which is adequate for the 32 nm technology generation.

Keywords: ultra-shallow junctions, USJs, infusion doping, GCIB, flash anneal
PACS: 41.75.-i; 36.40.-c; 85.30.Tv; 81.05.Cy; 81.40.Ef

INTRODUCTION

The steady miniaturization of the metal-oxide-semiconductor field-effect transistor (MOSFET) has driven the growth of the semiconductor industry over the past 40 years. This is because transistor scaling has provided for dramatic improvements in performance and reductions in cost per function. The biggest challenge for continued transistor scaling is to adequately suppress the short-channel effect (SCE) and drain-induced barrier lowering (DIBL), which degrade performance. SCE and DIBL are dependent on the MOSFET design parameters according to following equations [1]:

$$SCE = \frac{\varepsilon_{Si}}{\varepsilon_{ox}}\left(1+\frac{X_j^2}{L_{el}^2}\right)\frac{T_{ox}}{L_{el}}\frac{T_{dep}}{L_{el}}V_{bi}$$

$$DIBL = \frac{\varepsilon_{Si}}{\varepsilon_{ox}}\left(1+\frac{X_j^2}{L_{el}^2}\right)\frac{T_{ox}}{L_{el}}\frac{T_{dep}}{L_{el}}V_{d}$$

where L_{el} is the electrical channel length, X_j is the source/drain-extension junction depth, T_{ox} is the gate-dielectric thickness, T_{dep} is the channel depletion depth, ε_{Si} is the dielectric constant of the semiconductor material (silicon), ε_{ox} is the dielectric constant of the gate dielectric, V_{bi} is the pn-junction built-in potential, and V_d is the drain bias voltage. As the scaling of T_{ox} and T_{dep} slows down due to the approaching of fundamental limits, aggressive X_j scaling will become imperative to control SCE and DIBL. Future generations of CMOS technology will require ultra-shallow junctions (USJs) with sub-15nm X_j and sub-1000Ω/□ sheet resistance [2].

These USJ requirements are especially difficult to meet for *p+/n* junctions formed using conventional doping (low-energy ion implantation) and annealing (rapid thermal processing) techniques, due to boron transient enhanced diffusion (TED). Infusion doping utilizing a gas-cluster ion beam (GCIB) is an alternative doping technique that has been proposed to overcome the limitations of beam-line implantation [3]. Flash annealing is a candidate technique to replace conventional spike annealing. In a flash annealing process, the wafer temperature is first ramped up to an intermediate temperature (T_{int}); then the device side of the wafer is rapidly heated to the peak temperature (T_{peak}) using a flash lamp, to achieve high dopant activation (due to the very high T_{peak}) while minimizing dopant diffusion (due to the ~ms annealing time at T_{peak}).

EXPERIMENTAL DETAILS

USJ experiments were performed on lightly doped (8-24 Ω-cm) *n*-type (100) and (110) silicon wafers. Using 1%B_2H_6/Ar as the source gas, boron was introduced by a 5kV or 3kV GCIB doping process to form *p+/n* junctions. The GCIB dose was adjusted

CP1066, *Ion Implantation Technology,* edited by E. G. Seebauer, S. B. Felch, A. Jain, and Y. V. Kondratenko

to give an equivalent 10^{15} cm^{-2} boron dose. The cluster energy is transferred to the substrate surface upon impact, causing rapid heating and dissociation of the molecule into its constituent atoms. The Ar atoms leave the surface, while the boron atoms intermix with, or are infused into, the substrate. The kinetic energy of each boron atom is very low, and hence intermixing occurs within a very shallow depth. A spike anneal (1070°C) or flash anneal was performed to activate the dopants. Flash anneal parameters (T_{int}, T_{peak}, and n, the number of flash anneals) were varied to investigate their impact for the infusion doping process. The time between flash anneals was relatively long for samples which received multiple flash anneals.

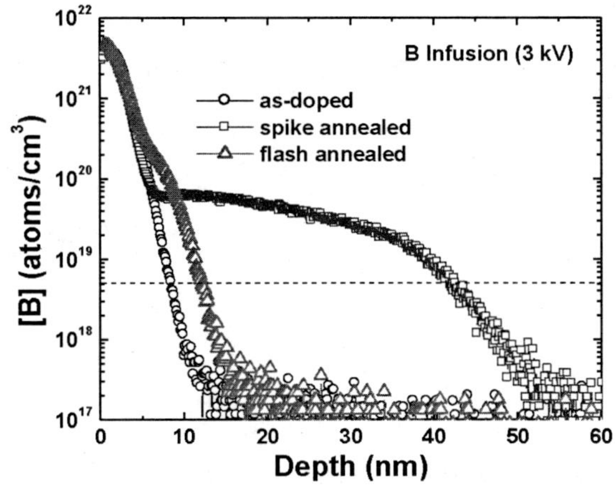

FIGURE 2. SIMS B profiles for samples doped with the B Infusion process. Negligible diffusion is observed for the flash-annealed sample.

RESULTS AND DISCUSSION

Secondary Ion Mass Spectroscopy (SIMS) was used to determine X_j, which is taken to be the depth at which the dopant concentration falls to 5×10^{18} cm^{-3}. Sheet resistance (R_s) was measured by a non-contact method for improved accuracy [4].

SIMS Analyses

Figure 1 shows the B profiles of samples as implanted with BF_2 (0.8keV, 2.5×10^{15} cm^{-2}, 7°) or as-infused with B (3kV). No pre-amorphization implant was done for the sample with BF_2 implantation. The as-infused profile shows no evidence of channeling for the (100) silicon substrate, in contrast to the implanted profile which shows a channeling tail starting at ~10^{19} cm^{-3}.

Figure 2 shows spike- and flash-annealed profiles for a 3kV B infusion process. The as-infused profile is also shown for comparison, with the 5×10^{18} cm^{-3} level indicated as a reference to measure junction depth. Spike annealing results in much deeper junctions (>40nm), whereas flash annealing causes little dopant diffusion.

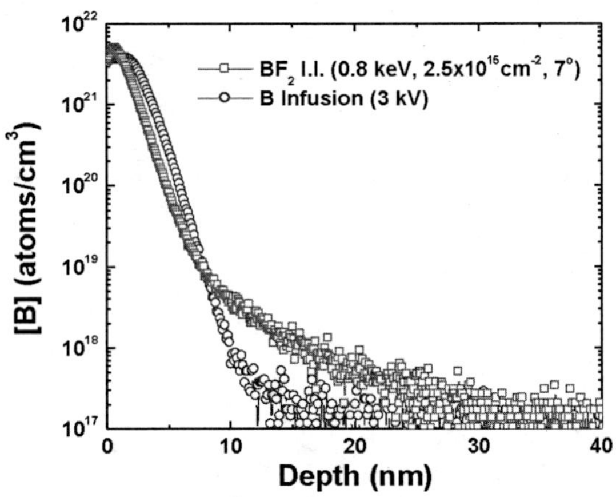

FIGURE 1. SIMS B profiles, for samples doped with BF_2 implantation or B infusion processes before activation anneal. No PAI was done prior to BF_2 implantation.

Impact of Flash Anneal Parameters

Figure 3 shows the effect of T_{int}, for two different infusion process conditions. R_s is reduced considerably when T_{int} is changed from 800°C to 700°C, for both 1250°C and 1300°C peak temperature. A small change in X_j is also observed, however this difference is within the depth resolution of SIMS. Thus, 700°C is preferred over 800°C for the intermediate temperature to achieve higher dopant activation.

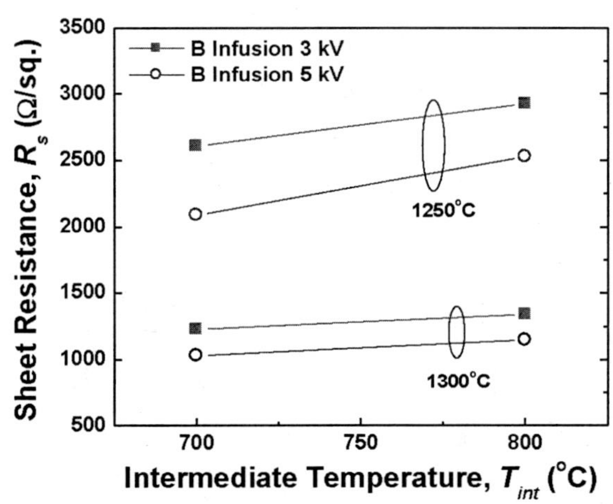

FIGURE 3. Measured sheet resistance as a function of T_{int}, for two different peak temperatures.

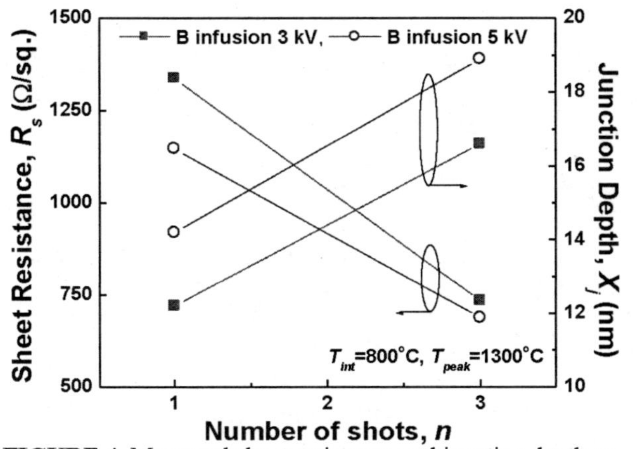

FIGURE 4. Measured sheet resistance and junction depth as a function of n for $T_{\rm int}$ = 800°C and $T_{\rm peak}$ = 1300°C.

Furthermore, lower R_s is achieved for T_{peak}=1300°C as seen in Figure 3, with minimal increase in X_j (ΔX_j ~0.5nm, close to the depth resolution limit of SIMS). The combination of 700°C intermediate temperature and 1300°C peak temperature is found to be optimal in this study. Figure 4 shows the impact of multiple shots of flash annealing. 40-45% decrease in R_s is observed as the number of shots, n, is increased from 1 to 3. However, X_j is also increased (ΔX_j ~4.5nm).

In order to elucidate the effects of the flash anneal parameters on dopant activation, a simple model is used to estimate the dopant activation efficiency, η [4]:

$$\eta \equiv \frac{R_{SIMS}}{R_{meas}}$$

where $R_{\rm SIMS}$ is the sheet resistance calculated from the SIMS profile assuming 100% dopant activation, and $R_{\rm meas}$ is the measured sheet resistance of the same junction. $R_{\rm SIMS}$ is calculated by dividing the doped

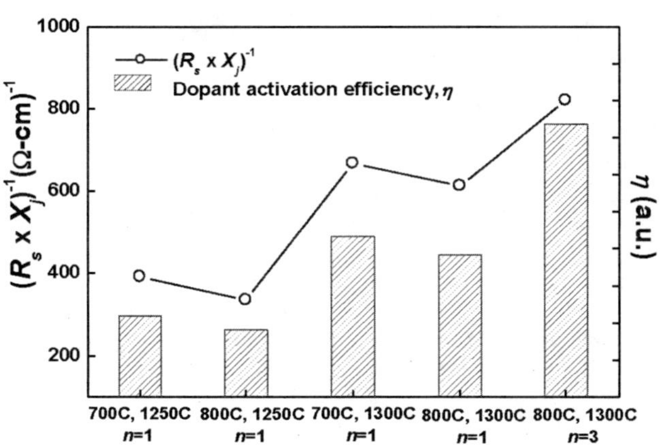

FIGURE 5. $(R_s \times X_j)^{-1}$ and dopant activation efficiency for various flash anneal conditions.

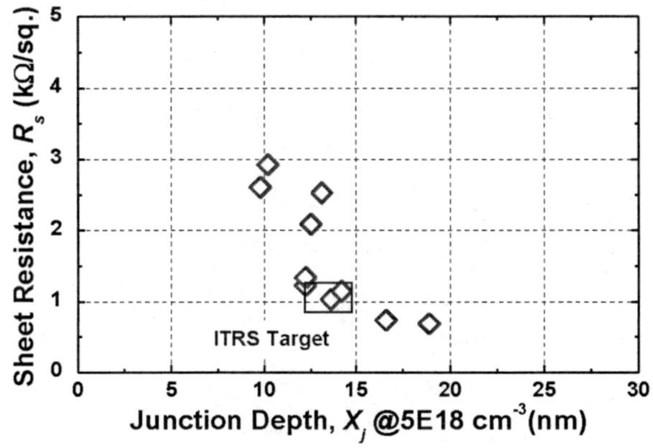

FIGURE 6. Sheet resistance (R_s) determined by non-contact method *vs.* junction depth (X_j) for $p+/n$ junctions formed by infusion doping and flash annealing.

semiconductor into multiple parallel slices. The calculated η values for different flash anneal conditions are summarized in Figure 5.

It is observed that η improves when $T_{\rm int}$ is reduced from 800°C to 700°C, for $T_{\rm peak}$ = 1250°C and n = 1. For a single-shot flash anneal, η is improved when $T_{\rm peak}$ is increased from 1250°C to 1300°C. The highest η is achieved with 3 flash anneal shots. These findings are consistent with conclusions drawn from a comparison of $(R_s \times X_j)^{-1}$, as shown in Figure 5.

The R_s-X_j summary plot for flash-annealed $p+/n$ junctions formed by infusion doping is shown in Figure 6. Clearly, 32nm-node USJ requirements (X_j~12-15nm, R_s~1kΩ/□) can be met by infusion doped, flash annealed junctions.

GCIB Doping for (110) Substrates

To keep pace with high transistor drive current requirements for future generations of CMOS technology, it has been proposed to fabricate pMOSFETs on (110) Si to achieve higher hole mobility than on (100) Si [6]. Because of the wider channels in (110)-oriented Si as illustrated in Figure 7, increased ion channeling is of concern, as it would result in larger X_j and thereby make it difficult to suppress short channel effects in (110) pMOSFETs.

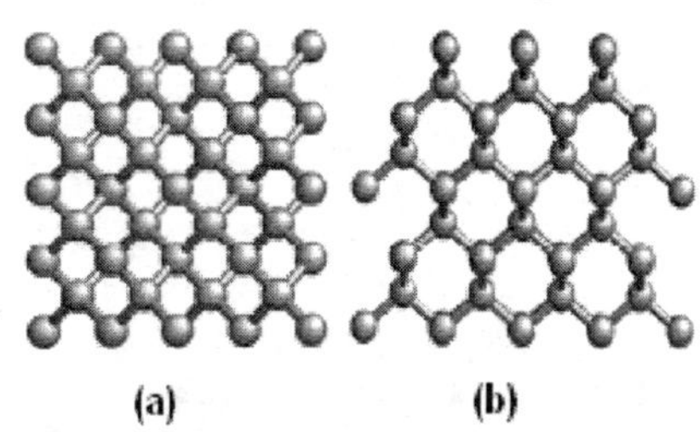

FIGURE 7. Atomic lattice of Si as viewed in the **(a)** <100> and **(b)** <110> directions.

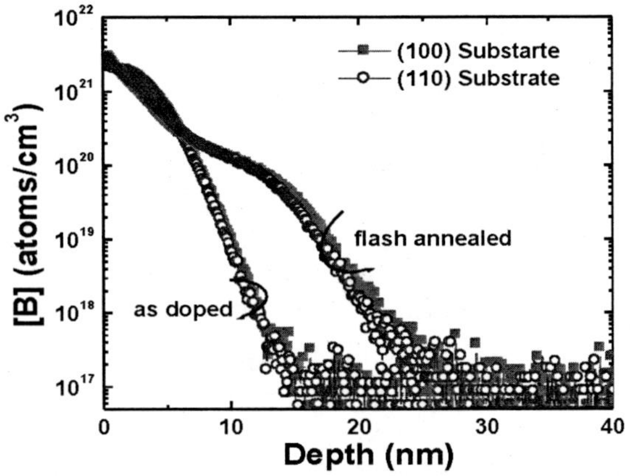

FIGURE 8. SIMS B profiles for unannealed and flash-annealed (T_{int} = 800°C, T_{peak} = 1300°C, n=3) samples doped with a boron infusion process (5kV), for (100) and (110) substrate crystalline orientations.

Infusion doping does not suffer from the problem of ion channeling, as shown earlier in Figure 1. This feature of Infusion doping can be used to overcome a limitation of the ion implantation technique. Figure 8 shows SIMS results for infusion doped (110)- and (100)-oriented silicon wafer substrates, before and after flash annealing.

There appears to be a very small difference in the boron profiles for the two substrate orientations, but this difference is within the depth resolution of SIMS. These results indicate that infusion doping is insensitive to substrate crystalline orientation and thus is attractive for fabrication of ultra-shallow junctions for (110) pMOSFETs, if channeling is the limiting factor.

CONCLUSIONS

GCIB doping is an attractive technique for USJ formation. In contrast to ion implantation, it shows no evidence of channeling for either (100)- or (110)-oriented Si substrates. Flash annealing is a promising candidate to replace conventional spike annealing. The best results are obtained with T_{int} = 700°C and T_{peak} = 1300°C. Multiple shots of flash annealing further lower R_s, with a slight trade-off in X_j. Sub-45 nm node USJ requirements can be met with infusion doped, flash annealed junctions.

ACKNOWLEDGEMENTS

The authors would like to thank Epion (now TEL Epion) for the GCIB doping, Mattson Technology for flash annealing, and Frontier Semiconductor for non-contact sheet resistance measurements.

REFERENCES

1. T. Skotnicki, *Proc. European Solid-State Device Research Conference*, 19-33 (2000).
2. International Technology Roadmap for Semiconductors, 2006 Update.
3. J. Hautala, J. Borland, M. Tabat and W. Skinner, *The Fourth International Workshop on Junction Technology*, 50-53 (2004).
4. V. N. Faifer, M. I. Current, T. Nguyen, T. M. H. Wong, and V. V. Souchkov, *Ext. Abs. the Fifth International Workshop on Junction Technology*, 45-48 (2005).
5. P. Kalra, P. Majhi, H.-H. Tseng, R. Jammy and T.-J. King Liu, *MRS Spring Meeting, Symp. E*, (2008).
6. M. Yang, M. Ieong, L. Shi, K. Chan, V. Chan, A. Chou, E. Gusev, K. Jenkins, D. Boyd, Y. Ninomiya, D. Pendleton, Y. Surpris, D. Heenan, J. Ott, K. Guarini, C. D'Emic, M. Cobb, P. Mooney, B. To, N. Rovedo, J. Benedict, R. Mo, H. Ng, *IEEE International Electron Devices Meeting*, 453-456 (2003).

Dopant Activation and Defect Analysis of Ultra-Shallow Junctions Made by Gas Cluster Ion Beams

Yan Shao[1*], John Hautala[1], Larry Larson[2**], Amitabh Jain[3]

[1] *TEL Epion, 37 Manning Rd., Billerica, MA 01821*
**Yan.shao@us.tel.com*

[2] *FEP Division, SEMATECH, 2706 Montopolis Drive, Austin TX 78741.*
***Larry.Larson@SEMATECH.org*

[3] *External Development and Manufacturing, Texas Instruments, 13121 TI Boulevard, MS 365 Dallas, TX 75243.*

Abstract. A long-standing problem in the fabrication of junctions by ion implantation is the enhanced diffusion caused by interstitial agglomeration and dissolution on annealing. Another side effect is residual damage that leads to junction leakage. The gas cluster ion beam technique (GCIB) offers a means of introducing dopant atoms without the creation of interstitials due to the low energy non-ballistic nature of the individual atoms and the localized thermal spike associated with the collision event between the large cluster and the silicon surface.

We have employed cluster ion beams to infuse B into Si substrates to fabricate ultra-shallow junctions. Various annealing techniques were used to investigate GCIB B activation. We find that annealing at temperatures around 950°C results in profiles that do not exhibit the extended tail characteristic of enhanced diffusion. This is in contrast to profiles in ion implanted samples, which inevitably result in enhanced diffusion. For laser and flash anneal, because the time scale of process is short, multiple anneal passes are needed in order to achieve low sheet resistance with little diffusion. We have shown that control of interfacial and surface smoothness is critical in improving junction characteristics.

Keywords: Gas Cluster Ion Beams, doping, transient enhanced diffusion, activation, laser annealing
PACS: 61.72.uf; 85.40.Ry

INTRODUCTION

As device dimensions decrease, the demand for ultra-shallow source drain extension (SDE) formation has been rapidly increasing. For 45-nm node and beyond, a SDE junction in the range of 10 nm to 20 nm wih sheet resistance below 1000 ohm/sq is needed depending on the device application and trade-offs between dopant activation, junction depth (Xj) and junction quality. Formation of ultra-shallow P-type junctions using B dopant is even more challenging due to the low energy requirement on doping and the rapid diffusion of B during thermal annealing. Therefore the doping and annealing techniques are key technologies for shallow junction formation.

Regular ion implantation needs a pre-amorphorization step, usually Ge or Si, to avoid B channeling and B acting with Si self interstitials generated by the implant. But the pre-amorphorization implant also introduces end of range damage (EOR), which releases interstitials during anneal, causing interstitial enhanced diffusion [1,2]. Gas Cluster Ion Beam (GCIB) infusion is a promising doping technology that provides shallow doping with high dopant concentration gradients [3,4]. During infusion processing, clusters, with large numbers of atoms or molecules (n>5000), are delivered at high energy (1–60 keV) to the substrate surface. Since the kinetic energy is shared with thousands of atoms in the cluster, the average energy of an individual atom or molecule in GCIB processing is small, typically < 10eV. The impact of a cluster on the substrate surface produces a highly localized transient thermal and pressure spike which lasts only a few picoseconds [5]. Dopant species, which are carried in the cluster, are infused into the substrate during this rapid dissipation of energy and pressure. This doping mechanism is fundamentally different from ion implantation. The GCIB infusion doping is a self-

CP1066, *Ion Implantation Technology,* edited by E. G. Seebauer, S. B. Felch, A. Jain, and Y. V. Kondratenko

amorphizing process. Therefore, there is no need for the pre-amorphizing implant step and there is no EOR damage.

In this study, B was infused into Si substrate by GCIB using either 5%B_2H_6/He or 1%B_2H_6/Ar gas to fabricate ultra-shallow junctions. Various annealing techniques were used to investigate GCIB B activation. The result of the activation was evaluated by sheet resistance, dopant profile and junction depth, leakage current and TEM for defect analysis.

GCIB INFUSION AND ACTIVATION

Boron infusion was carried out on 200 mm N-type substrates in the energy range of 3 kV to 10 kV with process gas of either 1%B_2H_6/Ar or 5%B_2H_6/He using a TEL Epion GCIB system. All wafers received an HF last cleaning step before the GCIB process. The as-infused B profile in this energy range has a junction depth of 100 Å ~ 120 Å at the 1E18 atom/cm^2 level, equivalent to a 500 eV ion implant with Ge pre-amorphorization. Dopant activation was achieved using spike annealing from 900 °C to 1050 °C and laser anneal at 1270 °C. Two step activation using 450 °C very low temperature anneal (VLTA) for 1 hour in a tube furnace followed by a second step high temperature activation was also studied. This low temperature annealing step was used in an effort to improve amorphous/crystalline interface roughness and aid in recrystallization/dopant activation in the second step. To compare with the GCIB process, B ion implantation was also prepared at 2 kV 1.4E15 atoms/cm^2 using BF_2, with Ge PAI at 20 keV 3.5E14/cm^2 dose and a C ion implant at 3 keV 5E14/cm^2 dose. After processing, the B profiles were measured by secondary ion mass spectroscopy (SIMS). The B SIMS profiles plotted are the sum of ^{10}B and ^{11}B; this is particularly important in the case of GCIB as both isotopes are allowed in the process. The junction depth (xj) was defined as the depth at which the B concentration was 1E18 atoms/cm^3. Sheet resistance was measured on a KLA-Tencor RS100 4 point probe with type H probe. In order to investigate the residual damage, cross section and plan view TEM were also used.

RESULTS AND DISCUSSION

Figure 1 shows the sheet resistance as a function of junction depth for various the annealing schemes tested, together with the theoretical dopant activation concentration of 3E19 and 1E20 calculated by assuming a box like B profile. For all the activation schemes tested, the activation level was in the range of 3E19 to 5E19 B solid solubility in terms of a boxed shape profile. In general, the low sheet resistance associated with ms spike anneals is accompanied with a substantial amount of B diffusion, while the μs laser anneal give limited B diffusion due to the short time scale of the annealing process. The resulting sheet resistance is also higher.

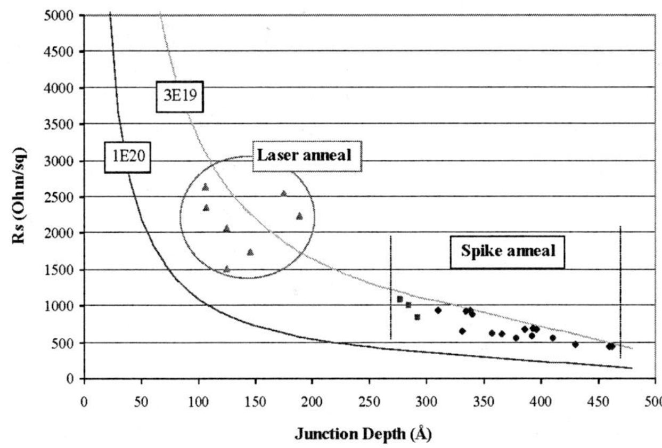

FIGURE 1. Activation results for infusion doped wafers after various spike anneals and laser anneals.

The B diffusion during spike anneal is dominated by concentration enhanced diffusion. This is different from the interstitial enhanced diffusion in ion implanted samples. Figure 2 compares B profiles of GCIB process and ion implantation before and after a 950°C spike anneal. This anneal condition is known to lead to the formation of a 'TED tail' as a manifestation of EOR disorder.

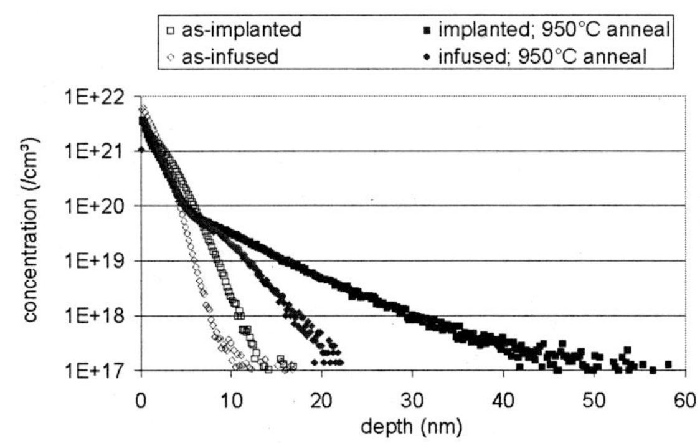

FIGURE 2. Comparison between the diffusion behavior of infused and ion implanted samples.

The GCIB process used 5%B_2H_6/He gas at 7 kV with 5E14 B dose. The ion implantation was done with 20 kV Ge implant at dose of 3.5E14 followed by 2 kV 1.5E15 BF_2 implant. The as-infused sample has a slightly shallower junction depth than the implanted sample. After a 950 °C spike anneal, B diffuses into the Si substrate with a slight degradation of the concentration slope with respect to the as-infused profile, indicating that the diffusion is mostly concentration enhanced diffusion of B. On the other hand, an extended tail is observed in the implanted sample. This enhanced B diffusion is caused by self intersitial flow from the EOR region to the surface B rich region, interacting with the implanted B causing transient enhanced diffusion.

To improve the B activation with limited diffusion, a combination of annealing techniques, such as multiple laser cycles and very low temperature (<450 °C) furnace anneal (VLTA) followed by laser anneal are tested. In this paper, we will be focusing on the effect of the VLTA on activation.

The low temperature anneal step has been shown to reduce the amorphous-crystalline interface roughness in the case of ion implanted samples with benefits on subsequent laser annealing. The interface roughness could affect the density of regrowth related defects in the recrystallized layer [6]. With a smoother amorphous-crystalline interface, the second step recrystallization and dopant activation anneal can be more effective. It was proposed [7] that the point defect relaxation during the low temperature annealing process initiates a reversed phase transition, i.e. a crystalline to amorphous phase transition, which results in the dissolution of detached microcrystalline, and crystalline material protrudes into the amorphous layer at the amorphous-crystalline interface. The cross section TEM clearly shows the expected result.

(a) (b)

FIGURE 3. The effect of very low temperature anneal on interface and surface roughness. (a) as-infused and after low temperature anneal. (b) after 1270 °C laser anneal with and without VLTA.

Figure 3a is the comparison of a-c interface roughness before and after the low temperature anneal. The sample had a 5 kV 1%B_2H_6/Ar GCIB process with equivalent B dose of 1E15. The as-infused sample has an amorphous layer 43-46Å thick with an amorphous/crystalline peak to valley interface roughness of ~25Å. After 400°C 60 min. anneal the amorphous layer thickness is slightly reduced to ~ 40 Å. The interface roughness dropped to ~ 5Å. Figure 3b compares the surface roughness of samples with and without the low temperature anneal after second step laser anneal. This time the sample had 10 kV 5%B_2H_6/He GCIB process and the laser anneal was 1270 °C. Laser annealing was performed in two consecutive cycles based on learning from previous work. The amorphous layer was fully recrystalized with no observable defects. Furthermore, the sample which received the low temperature anneal step showed a smoother surface.

Figure 4 shows the SIMS profiles of the two samples in Figure 3b together with a sample that received Ge and C implants prior to the infusion step. The pre-amorphous implant was done on standard implanter using 20 kV 3.5E14 Ge implant and 3 kV 5E14 C implant. All three samples received the same GCIB process and the same 1270 °C laser anneal. Boron depth profiles are shown for the unannealed conditions and after laser anneal. It can be seen that two samples with and without the VLTA process show exactly the same SIMS profiles. The sample which received ion implantation has a greater amount of diffusion. The implantation is expected to produce amorphousness to a depth of approximately 300 Å. It has been shown that boron undergoes enhanced diffusion in the amorphous phase. The greater depth of the amorphous layer relative to the samples with no implantation means that boron has longer to diffuse before the amorphous layer recrystallizes.

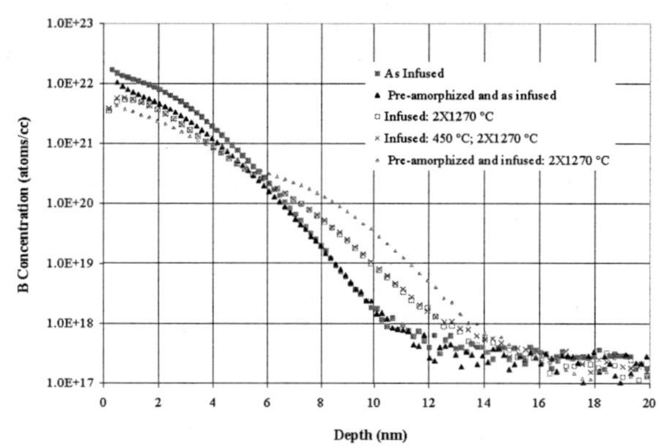

FIGURE 4. The effect of 450°C VLTA or pre-amorphization on the diffusion of infused boron.

The Rs-xj data for the three annealed profiles of figure 4 is shown in Figure 5. A dramatic drop in sheet resistance is seen as a result of the VLTA cycle. This brings the activation level to above 5E19/cc level in terms of a boxed shape profile. The sheet resistance measurements were confirmed by an independent non-contact RsL measurement in order to verify the effect of the VLTA cycle, table 1. Further work is planed to understand the mechanism in greater detail. The sample amorphized by ion implantation also shows a low sheet resistance. This is consistent with the VLTA result; since the amorphization depth is deeper than GCIB process, the a-c interface roughness is set by the implantation rather than the infusion step.

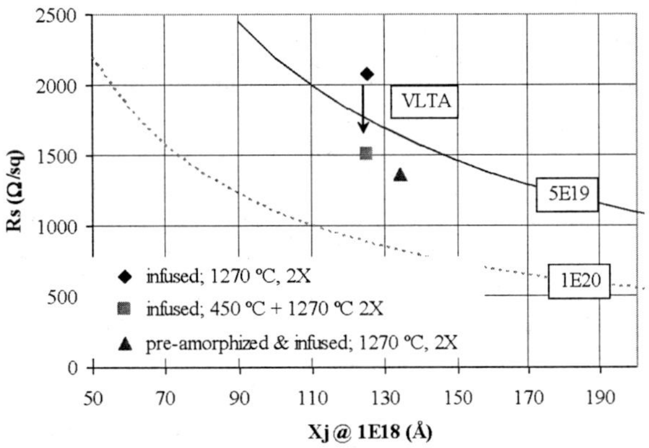

FIGURE 5. Improving dopant activation with combination of annealing techniques.

Finally, the same three samples were measured for Rs and leakage current using Frontier Semiconductor's non-contact RsL measurement [8]. Results are tabulated in table 1. The sheet resistance from RsL measurement agrees well with the 4 point probe measurement. The fact that lower leakage current is achieved for GCIB only process indicates that GCIB is effective in suppressing Si self interstitials generated during infusion.

TABLE 1. Compare sheet resistance measurement using 4PP and none contact RsL

#	Anneal	Rs Measurement (Ω/sq)		Leakage I (uA/cm^2)
		4 PP	RsL	
Infusion	laser	2076	2103	0.1
Infusion	VLTA + laser	1513	1547	0.437
Pre-amorphize + infusion	laser	1365	1299	1.475

SUMMARY

This study has demonstrated the basic difference between the GCIB process and ion implantation in terms of avoiding end-of-range disorder and its deleterious effects. A two step anneal scheme with a very low temperature anneal followed by laser anneal was investigated on GCIB infused samples. It was found that the VLTA helps to improve the interface roughness between crystalline and amorphous Si. This interfacial smoothening is necessary in order to obtain high quality junctions by GCIB.

ACKNOWLEDGEMENTS

The authors would like to thank Andrei Li-Fatou for his help with SIMS analysis and Frontier Semiconductor for none contact RsL sheet resistance and leakage current measurement.

REFERENCES

1. C. Bonafos, M. Omri, B. de Mauduit, G. BenAssayag, A. Claverie, D. Alquier, A. Martinez, and D. Mathiot, *J. Appl. Phys.* **82**, 2855 (1997)
2. B. J. Pawl, R. Surdeanu, B. Colombeau, A.J. Smith, N.E.B. Cowern, R. Lindsay, W. Vandervorst, B. Brijs, O. Richard, and F. Cristinao, *Appl. Phys. Lett.* **84**, 2055 (2004)
3. Ho Lee, Hwa Sung Rhee, *et al.*, IWJT 2006
4. T. Yamashita, T. Hayashi, Y. Nishida, Y. Kawasaki, T. Kuroi, H. Oda, T. Eimori and Y. Ohji, IWJT2005, Kyoto Japan
5. I. Yamada, *et al.*, *Materials Science and Engineering Reports,* **34**(6), 231-295 (2001).
6. Heather Banisaukas, Kevin S. Jones, Somit Talwar, Scott Falk, Dan F. Downey, Mat. Res. Soc. Symp. Vol. 610, B10.3.1 – 6, (2000)
7. G.A. Rozgonyi, E. Myers, and D.K. Sadana, ECS (1986)
8. V.N. Faifer and M.I. Current, *Appl. Phys. Lett.* **89**, 151123 (2006)

Summary of Industry-Academia Collaboration Projects on Cluster Ion Beam Process Technology

Isao Yamada[a*], Jiro Matsuo[b] and Noriaki Toyoda[a]

[a]*Graduate School of Engineering, University of Hyogo 2167 Shosha, Himeji, Hyogo 671-2280 Japan*
[b]*Quantum Science and Engineering Center, Kyoto University, Gokasyo, Uji, Kyoto 611-0011 Japan*

Abstract. Processes employing clusters of ions comprised of a few hundred to many thousand atoms are now being developed into a new field of ion beam technology. Cluster-surface collisions produce important non-linear effects which are being applied to shallow junction formation, to etching and smoothing of semiconductors, metals, and dielectrics, to assisted formation of thin films with nano-scale accuracy, and to other surface modification applications. In 2000, a four year R&D project for development of industrial technology began in Japan under funding from the New Energy and Industrial Technology Development Organization (NEDO). Subjects of the projects are in areas of equipment development, semiconductor surface processing, high accuracy surface processing and high-quality film formation. In 2002, another major cluster ion beam project which emphasized nano-technology applications has started under a contract from the Ministry of Economy and Technology for Industry (METI). This METI project involved development related to size-selected cluster ion beam equipment and processes, and development of GCIB processes for very high rate etching and for zero damage etching of magnetic materials and compound semiconductor materials. This paper describes summery of the results.

Keywords: Cluster ion beam, Cluster size, Shallow implantation, Surface smoothing, Surface damage.
PACS: 61.72T;79.20R;81.15J;3640

INTRODUCTION

With the progress of electronic devices to below submicron level (<100nm), the fabrication processes for fine structure become more important. In addition to the electronic devices, optical and magnetic devices, or sensors and mechanical devices mounted on IC chips, also require nano-scale fabrication techniques. The fabrication process of the nano-structure is termed "nano-technology". The ion beam process is one of the basic technologies in nano-structure fabrications such as etching or implantation of impurities, and a low energy ion beam process is required to be compatible with the shrinking device scale. However, it is difficult in principle to attain both low energy and high ion current with an atomic ion beam because of divergence by the space charge effects. The gas cluster ion beam process offers solutions to these problems which are facing conventional ion beam processing. Several possible application areas and the equipment development studied by industry-academia collaboration projects are shown here.

INDUSTRIAL EQUIPMENT DEVELOPMENT

Cluster ion currents in early equipment were very low, sometimes only of the order of a few nanoamperes, and the factors responsible for beam intensity limitations were not well understood at the time. In order to generate and transport even the very low cluster ion currents which could be produced in early systems, gas cluster ion beam (GCIB) equipment typically employed costly high-capacity vacuum pumps. By 1995, it was recognized that it would not be practical to use substantially greater gas flows in order to increase cluster ion beam currents to the levels to be required for production processors. No other groups or institutes in the world had yet paid attention to the concept of cluster ion beam equipment and its possible use for surface processing.

During collaboration with Kyoto University in support of work sponsored by the Japan Science and Technology Agency, Epion corporation began development of commercial GCIB equipment in 1995.

CP1066, *Ion Implantation Technology,* edited by E. G. Seebauer, S. B. Felch, A. Jain, and Y. V. Kondratenko

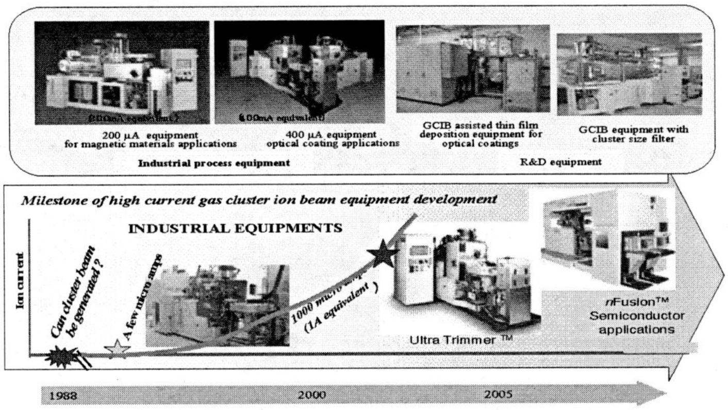

FIGURE 1. Milestones of GCIB equipment development.

Efforts made by Epion Corporation to improve generation and effective use of neutral clusters and to increase efficiency of cluster ionization and subsequent transport and use of the ionized cluster beams, without increasing gas consumption and pumping requirements, were successful. Cluster ion beam currents of several hundred microamperes became possible with source gas flows of less than 500 sccm. In 1997, Epion delivered the first GCIB R&D system to Kyoto University. At present beam current levels, many production applications of GCIB are now economically practical. Other applications will require beam currents of 1000 μA or more. In 1999, Epion Corporation began to manufacture the first production GCIB systems. These were GCIB Ultra-Smoother™ machines designed to produce ultra-smooth surfaces on critical layers used in read-write heads for computer hard disks. In 2001, Epion Corporation in the U.S. initiated GCIB semiconductor applications development. One application for which the fundamental properties of GCIB are ideally suited is in ultra-shallow junction formation. Shallow doping applications of GCIB, which are referred to as "infusion" processes, employ Ar gas clusters as hosts into which fractional amounts of dopant-containing gases such as B_2H_6, PH_3 or AsH_3 are added. Present shallow junction formation throughput of the equipment exceeds twenty 300 mm wafers per hour. Figure 1 shows the milestones of the equipment development.

PROCESS DEVELOPMENTS

In 2000, a four year R&D project for development of industrial technology began in Japan under funding from the New Energy and Industrial Technology Development Organization (NEDO). Subjects of the project are (1) high performance thin film development (*Adachi New Industrial Co., Ltd.*), (2) equipment development (*Epion Japan Corporation*), (3) semiconductor device development by shallow implantation (*Fujitsu Ltd.*), (4) measurement and evaluation technology (*Ion Engineering Res. Inst.*), (5) SR mirror development (*Nippon Pillar Packing Co., Ltd.*), (6) SR X-ray lithography C*VD diamond mask (*Mitsubishi Materials Corp.), (7) high hardness DLC film development (*Nomura Plating Co., Ltd*), and (8) thin film deposition equipment development (*Shimadzu Corporation*).

In a 5 year project which was started in 2002, several nano-processing technologies have been developed by Japanese companies under a Japanese nanotechnology program called "Advanced Nano-Fabrication Process Technology Using Quantum Beams". The project was fully supported by NEDO /METI (New Energy and Industrial Technology Development Organization /the Ministry of Economy and Technology Industry).

The R&D topics of the project involve surface smoothing processes and equipment development. These include (1) non-damage surface smoothing of magnetic materials surfaces (damage depth <1 nm and surface roughness Ra<3nm) in order to satisfy requirements for next generation HDD sensor head (*Hitachi Ltd.*), (2) non-damage surface smoothing (Ra<0.5 nm without scratch traces) for poly-crystalline SiC wafers which are used as monitor wafers for CVD processes (*Mitsui Engineering & Shipbuilding Co., Ltd.*), (3) high speed surface smoothing of laser annealed poly-Si surfaces (<5nm) which are used for FET of FPD (*Mitsubishi Electric Corp.*) and high speed surface smoothing (Ra<1nm) of Si nano-structure surfaces for photonics materials (*Japan Aviation Electronics Industry, Ltd.*). Equipment development has also been done by (5) large cluster size selection system development at the accuracy of $\Delta E = \pm 1$ eV /atom (*Kawasaki heavy Industries, Ltd.*) and (6) stable and reliable GCIB equipment development (*Epion Corporation*).

Hitachi Ltd. has studied non-damaging processing for magnetic materials. Increasing of areal density in hard disk drives (HDDs) requires decreasing the flying height of the magnetic recording head over the disk. In order to make further reduction of the flying height, smoothing technology for magnetic materials surface is extremely important because roughness and damage on the surface decrease the sensitivity of sensors. GCIB processing has been studied for surface smoothing under extremely low damage conditions on a very thin PtMn films. Typical thickness of the PtMn films was 80 nm. The film surfaces before irradiation exhibited hillocks with diameters of about 10 nm. Figure 2 shows the results of GCIB processing. After irradiation by GCIB at 20 kV (ion dose: 3×10^{15} cm^{-2}) from an incident angle of 80°, the hillocks disappeared and a smooth surface with an average roughness of 0.5 nm was obtained.

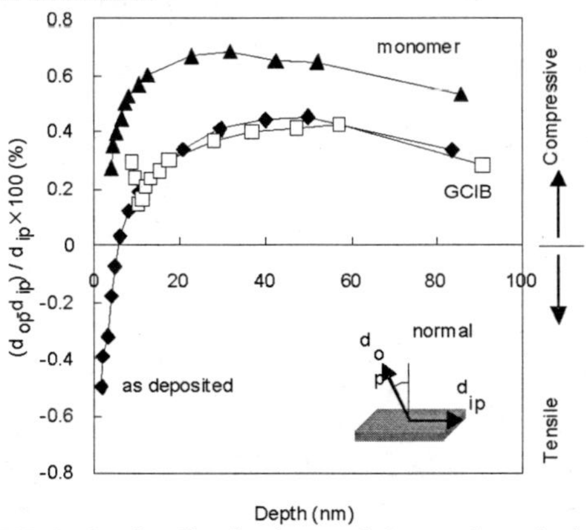

FIGURE 2. Strain distributions of the as-deposited, GCIB and monomer ion beams irradiated PtMn films.

Surface damage has been evaluated by using X-ray reflectivity (XRR) and grazing incidence X-ray diffractometry (XRD). For the XRR measurements, a laboratory X-ray source with $CuK_{\alpha 1}$ radiation at a wavelength of 0.15405 nm was used and for the grazing incidence X-ray diffractometry (XRD) measurements, synchrotron radiation at a wavelength of 0.15418 nm was used. The strain is defined as $\{ [(d_{op}) - (d_{ip})] / (d_{ip}) \}$x100 (%) by the in-plane (d_{ip}) and out-of-plane (d_{op}) lattice space of fcc (111) . Strain depth distributions for as-deposited, Ar ion irradiated and GCIB irradiated films are shown in Figure 2. The as-deposited PtMn films show compressive and tensile stresses in the bulk and surface regions, respectively. On the other hand, the films irradiated by GCIB and Ar ion beam exhibit compressive stress throughout the films. Monomer ion bombardment was made with an incident angle of 80° at the acceleration voltage of 100 V. The strain profile of the GCIB irradiated film is similar to that of the as-deposited film. In contrast, the strain distribution of the Ar ion beam irradiated film does not correspond with that of the as-deposited film throughout the film thickness. The results show that by GCIB processing, a damaged layer thickness of less than 1.5 nm can been achieved.

Mitsui Engineering & Shipbuilding Co., Ltd has studied poly-SiC surface smoothing. The developed SiC wafer is expected to be used for a monitor wafer in semiconductor device fabrication processes. Silicon carbide is a hard covalently bonded material. Mechanical polishing of SiC is typically done with diamond based slurries, where the abrasive size is successively reduced and eventually ending with a sub-micron slurry to achieve the desired roughness. However scratches left after the mechanical polishing is a problem issue for the applications. To demonstrate the possibility for removing scratches produced on poly-SiC wafer surfaces, GCIB processes have been studied. The SiC wafers were irradiated by Ar cluster ion beams. The result shows that the scratches induced by mechanical polishing on the initial SiC wafer were removed and a smooth surface of average roughness 0.7 nm was obtained.

Mitsubishi Electric Corp. has studied surface smoothing of laser-crystallized polycrystalline silicon thin films. A laser crystallization technique is a useful for obtaining high quality poly-Si films. The laser process however results in formation of many hillocks near the grain boundaries of the poly-Si films. This result makes it difficult to realize high-performance poly-Si TFT. The thickness of the laser crystallized poly-Si film is typically about 50 nm, and the created hillock height is almost the same as the film thickness. Surface smoothing without thinning the film under damageless conditions are important factors for downscaling of the TFTs. By selecting cluster source materials, irradiation energy, combinations of different source materials, and glancing angle irradiation processing, surface smoothing by GCIB has effectively eliminated the hillocks on poly-Si films. Damage caused by GCIB irradiation has also been investigated. By using glancing angle bombardment of GCIB irradiation, C-V measurements of MIS capacitors have shown that electrical-activated damage generated by GCIB irradiation can be suppressed. The total etching depth of poly-Si depends on the cluster size of GCIB. By controlling mean cluster size of SF_6-GCIB to 3000 molecules with acceleration voltage of 20 kV, a flat surface with low etching depth of less than 2 nm has been successfully obtained.

Japan Aviation Electronics Industry, Ltd. has studied surface smoothing of the side walls of high aspect ratio Si pillar structures for photonic crystal DWDM devices. Figure 3 shows (a) structure of photonic device and (b) the surfaces pre- and post-GCIB treatments. The successive etching and deposition by inductively coupled plasma-reactive ion etching (ICP-RIE) are usually used to fabricate the tall pillar structures. These processes result in excessively rough side walls. Polishing of sidewall surfaces has been done by GCIB. Surface protuberances with depths of 40 nm are irradiated by 20 kV GCIB at 83 degrees from the sidewall. The surface protuberances are removed selectively with an etching depth of only 50 nm. If ultra-smooth sidewalls can be fabricated by sidewall polishing to 0.3 nm in average roughness, corresponding transmission loss would be only 0.12 dB/mm. This estimated transmission loss is lower than current results obtained on photonic crystal structures.

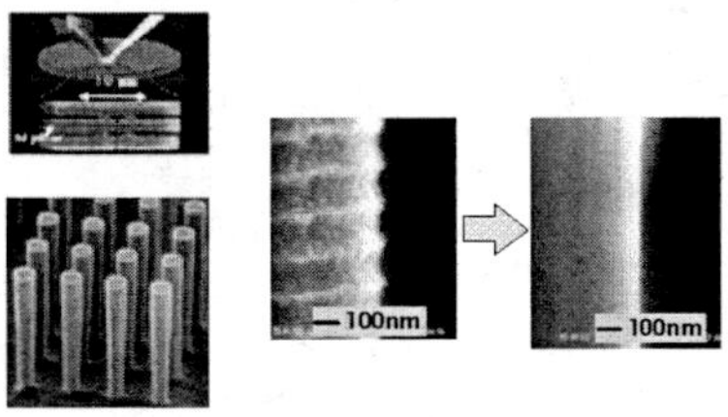

(a) (b)

FIGURE 3. GCIB application for photonic devices,
(a) Si pillar type photonic device,
(b) Surface smoothing of the pillar.

Kawasaki Heavy Industries, Ltd. has concentrated on the equipment development especially for industrial applications. According to the study of the effect of cluster size in the GCIB processing, it has been suggested that cluster size selection is an important factor in realizing non-damaging nano-processing. However, GCIB typical consists of a wide range of cluster sizes, between hundreds to thousands of atoms. It is not easy to achieve heavy ion mass beam selection by mean of conventional systems such as permanent magnets or quadrupole mass filter systems. A new type of cluster size selection system using essentially transient time effect in a high frequency deflection system has been developed. Figure 4 shows the concept of cluster size selection system and cluster size selection as a parameter of applied frequency. The mass resolution target is ±1 eV/atom. The device consists of two pairs of deflection electrodes that produce the same intensity of high frequency-electric field, but these phases are different by 180 degrees. This device utilizes the fact that the cluster ions for different sizes drift at the different velocities even though the cluster ions are accelerated at the same potential. When the cluster velocity is synchronized by the period of sinusoidal electric field, the force given to the selected size cluster ions can be canceled in both deflection-angle and deflected distance, whereas this does not occur for clusters of other sizes. This new type of cluster selection system has been operated to evaluate damage effects on a Si target. The results have agreed well with the data obtained using GCIB with mass selection by a permanent magnet system.

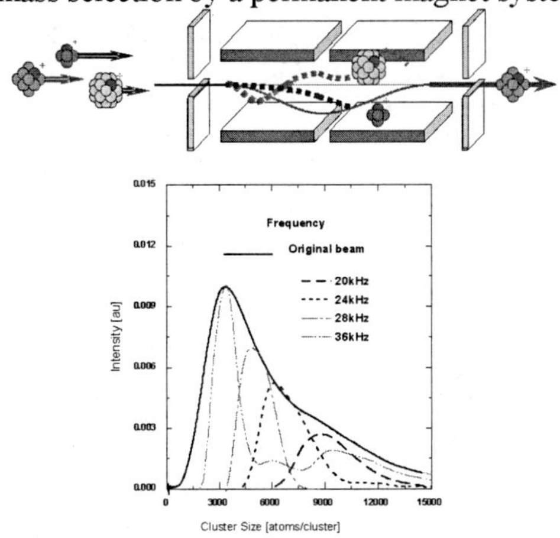

FIGURE 4. Concept of cluster size selection system and Cluster size selection as a parameter of applied frequency

Epion Corporation has developed production GCIB equipment. Cluster ion beams are very different than monomer ions, not only in terms of the surface interactions which have been described before, but also in the details of ionization, beam focusing and transport, beam deflection, beam metal and particle contaminations, etc. During the project, it has been shown that important subjects to be studied are parameters of the cluster beam energy (E), charge (Q), and mass (M). By developing a new diagnostic measurement tool called a QEM analyzer, the charge, energy and mass of the cluster ions can be examined independently. The QEM results showed that for Ar clusters of the same size, energy loss due to residual gas collisions is greater for higher energy clusters than for clusters at lower energies. Also at the same total acceleration energy, the energy loss of small size Ar-GCIB clusters is greater than it is from the larger size clusters.

ACKNOWLEDGEMENTS

The authors wish to express their thanks to the many individuals who have been associated with NEDO projects. The projects have been generously supported by New Energy and Industrial Technology Development Organization (NEDO).

REFERENCE

* Emeritus Professor, Kyoto University.

NEDO contract number P2048, Project leader I.Yamada, sub-leaders J.Matsuo and N.Toyoda. This paper is based on the information from the NEDO final report (in Japanese), Public version, October 31, 2007.

URL:www.nedo.go.jp/iinkai/kenkyuu/bunkakai/19h/jigo/8/1/5-1-1.pdf

Production of Liquid Cluster Ions by Nozzle Beam Source with and without He Gas

G.H. Takaoka, H. Ryuto, T. Okada, and K. Sugiyama

Photonics and Electronics Science and Engineering Center, Kyoto University, Nishikyo, Kyoto 615-8510, Japan

Abstract. We developed a new type of cluster ion source which could produce various kinds of liquid clusters such as water, methanol, ethanol and octane clusters. When the vapor pressure was larger than one atm, the water and ethanol clusters could be produced by an adiabatic expansion phenomenon without adding He gas. The peak size of the cluster ions increased with the increase of the vapor pressures. When the source temperature was at room temperature, the water and ethanol clusters were also produced by adding He gas. In another case of producing liquid clusters such as methanol and octane clusters, He gas was added to mix up with vapors of liquid materials. When the He gas pressure was larger than a few atms, the methanol and octane clusters were produced at a vapor pressure of two atm. The peak size increased with increase of the vapor pressure as well as the He gas pressure.

Keywords: cluster ion, ion beam, water cluster, ethanol cluster, methanol cluster, octane cluster, TOF analysis
PACS: 36.40.Wa, 61.46.Bc, 82.80.Rt

INTRODUCTION

Development of new materials including liquid materials has recently received much attention from various kinds of viewpoints. Control of material properties themselves as well as their surface and interface properties is crucial to the development of advanced materials from a technical viewpoint. As another crucial viewpoint, demand for the fabrication of advanced semiconductor devices has been increasing [1-3], and it is required to break through limitations embedded in device fabrication processes. For example, for the surface cleaning process by a wet process, it is required to develop advanced functional solvents, which have not only cleaning ability but also other functions such as etching and smoothing abilities [4-6]. Among various kinds of liquid materials, water and alcohol are well known solvents, and their chemical reactivity has received much attention from the fundamental viewpoint. For example, water is widely distributed in nature and human body, and both experimental and theoretical studies have recently been carried out to investigate the microscopic structure of water in terms of clusters of water molecules [7-9].

A cluster is a tiny particle, which is in a position to link between the atomic state and bulk state. The physical and chemical properties of the cluster are different from those of bulk state [10-13]. For example, the meting and freezing points of materials are not the same for the cluster state, although they are the same for the bulk state. As well as the properties of the cluster itself, the impact process of an energetic cluster ion on target surface represents specific features [14,15]. The cluster ion beam process has attracted much interest from both fundamental and applied viewpoints.

While focusing on the development of new liquid materials as well as the advantages of the cluster ion beam technique, we have developed the liquid cluster ion source [16-18]. The characteristics of liquid clusters are expected to be much different from those of bulk state materials used in a conventional wet process. In this article, the production of liquid clusters such as water, ethanol, methanol, and octane clusters is investigated, and the adding effect of helium (He) gas on the cluster formation is discussed.

EXPERIMENTAL PROCEDURE

Liquid materials such as water, ethanol, methanol and octane were introduced into the cluster source, and they were heated up to 180°C by a wire heater attached around the source. The vapors of liquid materials with or without He gas were ejected through a nozzle into a vacuum region. The nozzle was made of glass, and it was a converging-diverging supersonic nozzle. The diameter of the nozzle at the throat was 0.1 mm. The ejected vapors passed through a skimmer and entered an ionizer. In the ionizer, they were ionized by electron bombardment. The cluster ions produced were accelerated by applying an extraction voltage to the extraction electrode. The extraction voltage was adjusted between 0 kV and 2 kV. When the cluster size was measured by a time-of-flight (TOF) method, the extracted cluster ions entered a flight tube, in which

CP1066, *Ion Implantation Technology,* edited by E. G. Seebauer, S. B. Felch, A. Jain, and Y. V. Kondratenko

they were deflected by a pulsed negative-voltage. As shown in Figure 1 the pulse voltage was applied, and the cluster ions drifted toward a Faraday cup mounted on the end of the flight tube. The flight ions were detected by the Faraday cup. The acceleration voltage for the cluster ions was 6 kV, and the pulse voltage was 2 kV. The duration time and the repetition rate of the pulse voltage were 10 μsec and 100 Hz, respectively. The drift distance was 0.51 m. The cluster size was calculated based on the drift time ranging from μsec to msec, which was different depending on the mass.

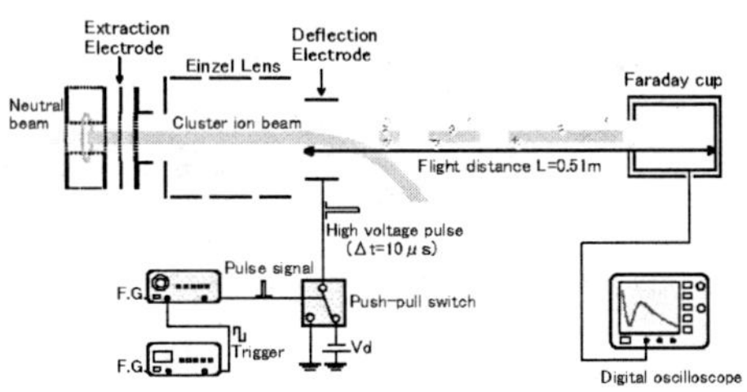

FIGURE 1. Schematic of the TOF system for the measurement of cluster size distribution.

RESULTS AND DISCUSSION

Figure 2 shows the size distribution for (a) ethanol and (b) water cluster ions as a parameter of vapor pressure. The acceleration voltage (Va) was 6 kV. The electron voltage for ionization (Ve) was 200 V, and the electron current for ionization (Ie) was 200 mA. As shown in the figure, ethanol and water clusters are produced, when the vapor pressure is larger than one atm. The beam intensity of the cluster ions produced increases with increase of the vapor pressure. The cluster size is distributed between a few hundreds and a few tens of thousands, and the average size (peak size) is about 750 to 1050 molecules for the ethanol and about 2500 molecules for the water, respectively. According to the thermodynamics and the gas dynamics, the cluster beam intensity is proportional to the intermolecular collision frequency during supersonic expansion, which depends on source pressure P_0, and source temperature T_0. It is proportional to $P_0(T_b/T_0)^{\frac{\gamma}{\gamma-1}}$, where γ and T_b are the specific heat ratio and the boiling temperature of the source materials, respectively. With regards to the value of $(T_b/T_0)^{\frac{\gamma}{\gamma-1}}$, T_b and γ are 351 K and 1.13 for ethanol, and 373 K and 1.33 for water, respectively. When the pressure P_0 is kept at the same value by adjusting the source temperature such as three atm, the value of $(T_b/T_o)^{\gamma/(\gamma-1)}$ for water is larger than that for ethanol.

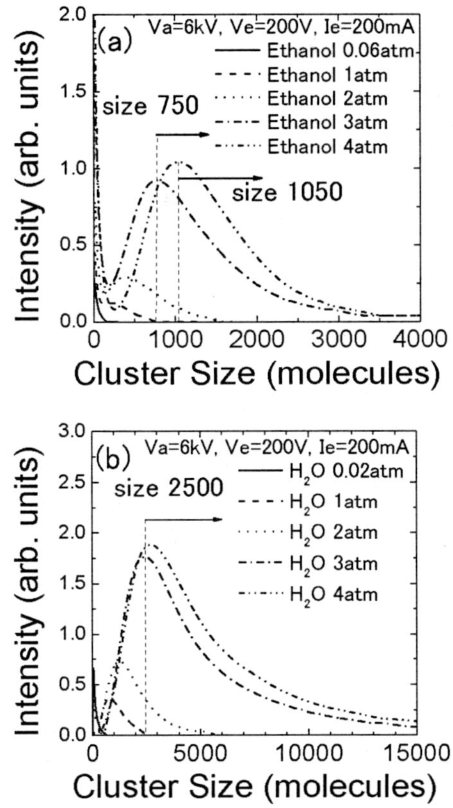

FIGURE 2. Size distribution for (a) ethanol and (b) water cluster ions as a parameter of vapor pressure.

In general, clusters with a smaller radius than the critical radius are not stable, and they are dissociated due to the evaporation. In contrast, larger aggregates (clusters) grow to become stable due to the condensation. The critical radius r^* of the cluster is given by [19]

$$r^* = \frac{2\sigma V_c}{kT \ln S} \tag{1}$$

where σ is the surface energy, V_c is the molecular volume in a cluster, T is the vapor temperature, k is the Boltzmann constant, and S is the saturation ratio (P/P_∞) of the vapor pressure (P) to the equilibrium vapor pressure (P_∞). In addition, with regard to the cluster growth, the nucleation rate is proportional to the concentration N^* of the critical nuclei, which is described as follows [20];

$$N^* = N_0 \exp\left(-\Delta G^* / kT\right) \tag{2}$$

where N_0 is the concentration of vaporized molecules, and ΔG^* is the maximum value of the Gibbs free energy ΔG at the critical radius r^*. The value ΔG^* is described as follows;

$$\Delta G^* = \frac{16\pi\sigma^3}{3\left(\frac{kT}{V_c}\ln S\right)^2} \quad (3)$$

Based on these equations (1) to (3), the critical radius r^* and the barrier height ΔG^* of water clusters are larger than those of ethanol clusters, because the surface energy of the water such as 72.8 dyne/cm is larger than that of the ethanol such as 22.8 dyne/cm [21]. Therefore, the nucleation growth of water clusters is rapidly enhanced, if the critical nuclei are formed. This indicates that the average size of water cluster ions is larger than that of ethanol cluster ions.

In another case of producing liquid clusters, He gas was added to mix up with vapors of the liquid materials. Figure 3 shows a peak size of ethanol and water cluster ions as a function of He gas pressure. The vapor pressures were kept at two atm as well as 0.06 atm for ethanol and 0.02 atm for water, respectively, which corresponds to the source pressure at room temperature. As shown in the figure, the peak size of ethanol and water cluster ions increases with increase of the He gas pressure. In particular, when the source temperature is room temperature, the clusters are not produced without adding He gas. However, when the He gas pressure is larger than 0.5 atm, the ethanol and water clusters are produced. The peak size of cluster ions is almost constant at 250 molecules for the ethanol cluster, and it increases with increase of the He gas pressure for the water clusters. The peak size of the water clusters formed without heating is saturated at 1250 molecules at higher He gas pressures. When ethanol and water vapors are ejected through the nozzle into the vacuum region together with the He gas, they are cooled down by collision with the He gas, which results in the effective formation of ethanol and water clusters even at the vapor pressure less than 0.1 atm.

Figure 4 shows the size distribution for (a) methanol and (b) octane cluster ions as a parameter of He gas pressure. The vapor pressure of methanol and octane was kept at 2 atm. When the He gas is not added, the clusters are not produced by an adiabatic expansion of the pure vapors. Even if clusters with a larger radius than the critical radius are

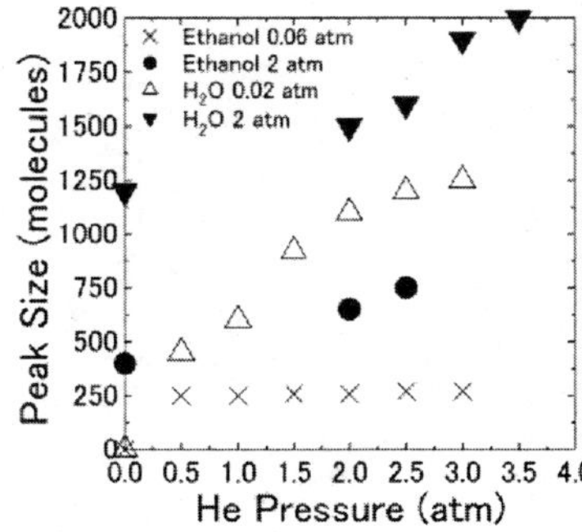

FIGURE 3. Dependence of peak size of ethanol and water cluster ions on He gas pressure.

instantaneously formed by the adiabatic expansion of pure vapors, they have possible fragmentation resulting in the unstable clusters. Therefore, higher vapor pressures than a few atms are required in order to enhance the intermolecular collision frequency during the supersonic expansion. On the other hand, for the case of He gas addition, stable clusters are produced, when the He gas pressure is larger than 2.0 atm for the methanol and 3.5 atm for the octane, respectively. The peak size increases with increase of the He gas pressure. Compared with water and ethanol clusters, methanol and octane clusters are hard to produce. Since the specific heat ratio is small as 1.06 for octane, the intermolecular collision frequency becomes very low. As a result, the octane vapors are not cooled down sufficiently during the supersonic expansion. However, the introduction of He gas assists cooling down the octane vapors, and the effective production of octane clusters is performed. On the other hand, with regard to the methanol, T_b and γ are 338 K and 1.26, respectively, and the value of $(T_b/T_0)^{\frac{\gamma}{\gamma-1}}$ is larger than that of the ethanol. Since the saturation ratio S is smaller than that of the ethanol, the critical radius and the barrier height of the methanol cluster are larger than those of the ethanol cluster. Therefore, compared with the ethanol clusters, the methanol clusters have a larger peak size, providing that the critical nucleus is formed.

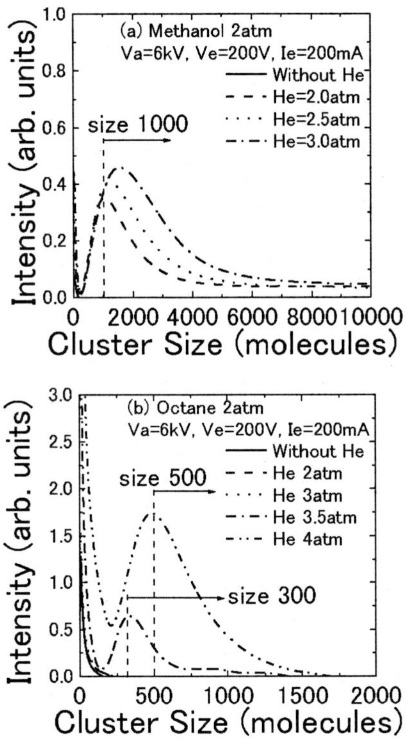

FIGURE 4. Size distribution for (a) methanol and (b) octane cluster ions as a parameter of He gas pressure.

CONCLUSION

In order to clarify impact processes of liquid cluster ions on solid surfaces, we developed a new type of cluster ion source which could produce various kinds of liquid clusters such as water, methanol, ethanol and octane clusters. The liquid materials were heated up to 180 degrees, and the vapors were ejected through a nozzle into a vacuum region. When the vapor pressure was larger than one atm, water and ethanol clusters could be produced by an adiabatic expansion phenomenon without adding He gas. The peak size of the cluster ions increased with the increase of the vapor pressures. In another case of producing liquid clusters such as methanol and octane clusters, He gas was added to mix up with vapors of liquid materials. When the He gas pressure was larger than 2.0 atm for methanol and 3.5 atm for octane, respectively, the liquid clusters were produced at a vapor pressure of 2 atm. The peak size increased with increase of the vapor pressure as well as the He gas pressure. The methanol and octane vapors were cooled down by the collision with the He gas, which resulted in the effective formation of the liquid clusters.

In addition, it should be noted that the water and ethanol cluster ions could be produced at the He gas pressures larger than 0.5 atm, even if the source temperature was at room temperature.

REFERENCES

1. J.E. Greene, *Mater. Res. Bull.* **26** (2001) 777.
2. Y. Xia and N.J. Halas, *Mater. Res. Bull.* **30** (2005) 338.
3. J.H. Zhao, *Mater. Res. Bull.* **30** (2005) 293.
4. A. O'Neil and J.J. Watkins, *Mater. Res. Bull.* **30** 967 (2005).
5. S.C. Lim, S.H. Kim, J.H. Lee, M.K. Kim, D.J. Kim and T. Zyung, *Synthesis Metals* **48** 75 (2005).
6. W. Puyrenier, V. Rouessac, L. Broussous, D. Rebiscoul and A. Ayral, *Micrelectronics Engineering* **83** 2314 (2006).
7. K. Gregory, D.C. Clary, K. Liu, M.G. Brown, C. J Carter and R.J. Saykaliy, *Science*, **275** (1997) 814.
8. D.M. Upadhyay, M.K. Shukla and P.C. Mishra, Int. *J. Quant. Chem.*, **81** 90 (2001).
9. I.B. Muller and L.S. Cederbaum, *J. Chem. Phys.* **125** 204305 (2006).
10. Ph.Buffat and J-P.Borel, *Phys. Rev.*, **A13** 2287 (1976).
11. M.D.Morse, *Chem. Rev.*, **86** 1049 (1986).
12. W.D.Knight, W.A.de Heer and W.A.Saunders, Z. *Phys.*, **D3** 109 (1986).
13. R.S. Berry, *Sci. Am.*, **263**(2) (Aug. 1990) 68.
14. W. Eckstein, "Computer Simulation of Ion-Solid Interactions" (Springer, New York, 1991)
15. D. Beyson, X. Campi, E. Pfefferkorn (Eds) "Fragmentation Phenomena" (World Scientific, Singapore, 1995)
16. G.H. Takaoka, H. Noguchi, T. Yamamoto and T. Seki, *Jpn. J. Appl. Phys.* **42** (2003) L1032.
17. G.H. Takaoka, H. Noguchi, K. Makayama, Y. Hironaka and M. Kawashita, *Nucl. Instrum. Methods,* **B237** (2005) 402.
18. G.H. Takaoka, K. Nakayama, T. Okada and M. Kawashita, Proc. 16th Int. Conf on Ion Implant. Technol. (AIP, 2006) p.321.
19. P. Wegener, "Nonequilibrium Flows" (Marcel Dekker, New York, 1969)
20. J. Frenkel, "Kinetic Theory of Liquids" (Dover Publications, New York, 1955)
21. R.C. Weast and M.J. Astle (Eds.), "CRC Handbook of Physics and Chemistry, 63rd Edition" (CRC Press, Boca Raton, Florida, 1982).

Investigation of Damage with Cluster Ion Beam Irradiation Using HR-RBS

Toshio Seki[1], Takaaki Aoki[2], Jiro Matsuo[3]

[1]*Department of Nuclear Engineering, Kyoto Univ., Gokasyo, Uji, Kyoto, 611-0011. Japan*
[2]*Department of Electronic Science and Engineering, Kyoto Univ., Nishikyo-ku, Kyoto, 615-8510, Japan*
[3]*Quantum Science and Engineering Center, Kyoto Univ., Gokasyo, Uji, Kyoto, 611-0011, Japan*

Abstract. Cluster ion beam can process targets with shallow damage because of the very low irradiation energy per atom. However, it is needed to investigate the damage with cluster ion beam irradiation, because recent applications demand process targets with ultra low damage. The shallow damage can be investigated from depth profiles of specific species before and after ion irradiation. They can be measured with secondary ion mass spectrometry (SIMS) and Rutherford backscattering spectroscopy (RBS). High resolution Rutherford backscattering spectroscopy (HR-RBS) is a non destructive measurement method and depth profiles can be measured with nano-resolution. The cluster ion beam mixing of thin Ni layer in carbon targets can be investigated with HR-RBS. The mixing depth with cluster ion irradiation at 10 keV was about 10 nm. The mixing depth with cluster ion irradiation at 1 keV and 5 keV were less than 1 nm and 5 nm, respectively. The number of displaced Ni atoms with cluster ion irradiation was very larger than that with monomer ion irradiation of same energy. This result shows that violent mixing occurs with single cluster impact.

Keywords: Ar cluster, HR-RBS, mixing depth
PACS: 82.80.Yc, 36.40.-c, 34.00.00, 79.20.Rf

INTRODUCTION

A cluster is an aggregate of a few to several thousands of atoms. When many atoms constituting a cluster ion bombard a local area, high-density energy deposition and multiple-collision processes are realized, which does not occur with other beam process. Because of the unique interactions, cluster ion beam processes can produce new surface modification effects, such as surface smoothing [1–4], high rate sputtering [5] and very shallow implantation [6,7]. To date, this technique has been applied to photonic [8], magnetic [9], electronic [10], and biological materials processing [11]. Cluster ion beam can process targets with shallow damage because of the very low irradiation energy per atom. For example, when a cluster with the size of 2000 atoms is accelerated at 10 keV, energy per atom is only 5 eV. However, it is needed to investigate the damage with cluster ion beam irradiation, because recent applications demand to process targets with ultra low damage. The shallow damage can be investigated with scanning tunneling microscope (STM), transmission electron microscope (TEM), secondary ion mass spectrometry (SIMS), and Rutherford backscattering spectroscopy (RBS) measurements. Cross-sectional TEM images can give the damage structures directly. However, sample preparation for TEM observation needs a long time and the sample is destroyed. Depth profiles of specific species before and after ion irradiation can be investigated with SIMS and RBS. In case of SIMS measurement, the ion beam mixing during etching samples with probe ion irradiation causes degradation of depth resolution. High resolution Rutherford backscattering spectroscopy (HR-RBS) is a non destructive measurement method and depth profiles can be measured with nano-resolution [12]. In this study, the ion beam mixing of thin layers in targets were investigated with HR-RBS. The cluster ion beam mixing damage will be discussed.

EXPERIMENTAL

In order to investigate the mixing depth with cluster irradiation, three kinds of triple layer structure (C/Ni/C) were fabricated on Si surfaces with RF magnetron sputtering. The thickness of each layer is shown in Table 1. In order to investigate the shallow mixing depth (<10 nm), the thickness of top carbon layer was controlled in the range of 1to10 nm. The shallow damage can be investigated from depth

CP1066, *Ion Implantation Technology,* edited by E. G. Seebauer, S. B. Felch, A. Jain, and Y. V. Kondratenko

profiles of Ni layer before and after cluster ion irradiation (Figure 1).

TABLE 1. Thickness of each layer of C/Ni/C structure.

	Sample A	Sample B	Sample C
top C layer (nm)	1	5	9.5
2nd Ni layer (nm)	1	1	1
3rd C layer (nm)	40	40	40

The depth profiles of Ni atoms in the carbon film can be obtained by simulation fitting of RBS spectra to the observed HR-RBS ones. HR-RBS spectra were measured with 155 keV ^{3}He ions. Incident angle was 20° and scattering angle was 90°. In the simulation, the concentration of Ni atoms in the carbon layer was set for every 0.1 nm interval of depth and principal energy spectra were calculated. The energy straggling function of the measurement system was estimated from HR-RBS spectrum of ultra thin Ni film (thickness 0.5 nm) on a Si substrate and RBS spectra were obtained by convolution of principal energy spectra and energy straggling function. If the mixing depth with cluster ion irradiation is deeper than the Ni layer, Ni atoms in the layer are displaced by ion mixing and the broad depth profile of Ni atoms is obtained. On the other hand, if the mixing depth is shallower than the Ni layer, Ni atoms in the layer are not displaced and the depth profile of Ni atoms is similar to initial depth profile. The three samples (A, B, C) were irradiated with Ar cluster ions at energy range of 1to10 keV. The mean size of clusters was about 2000 atoms and the ion dose was 1x10^{14} ions/cm^2. In order to compare mixing strength between cluster and monomer ion irradiation, sample A was irradiated with cluster and monomer ions at 3 keV.

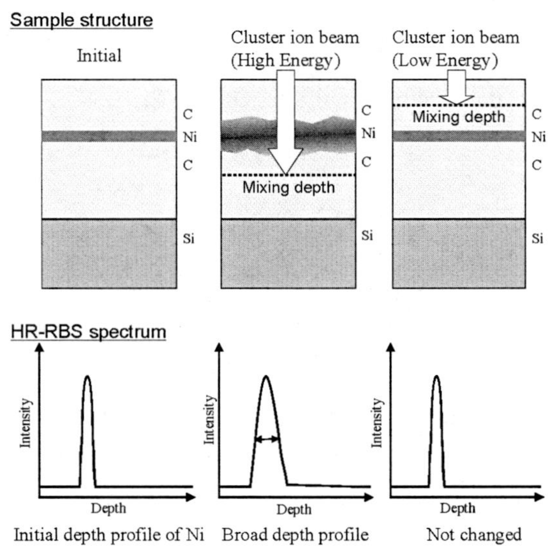

FIGURE 1. Mixing depth investigation method with HR-RBS.

RESULTS AND DISCUSSION

Figure 2 shows the initial depth profile of Ni atoms and the cross-sectional TEM image of sample A. The half width at half maximum (HWHM) of depth profile was about 1.2 nm. The dark field in the cross-sectional TEM image showed the Ni layer and the width of depth direction was about 2.5 nm. This value is similar to the width of depth profile at the concentration of

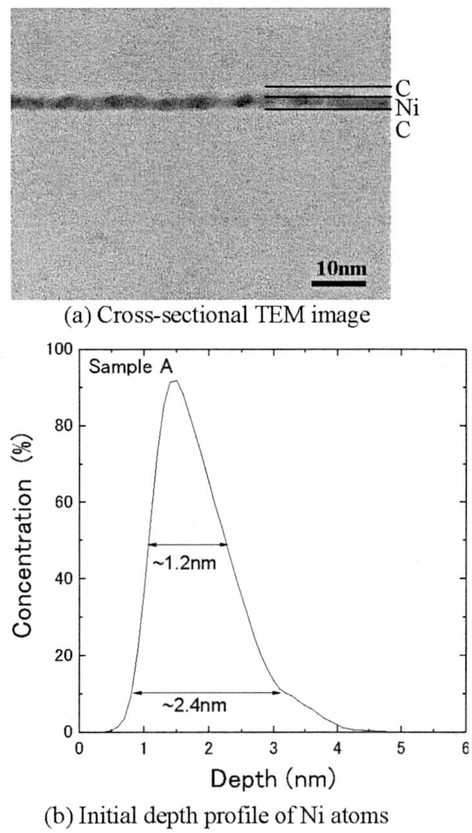

FIGURE 2. Initial depth profile of Ni atoms and the cross-sectional TEM image of sample A.

10 %. It is considered that the area of more than 10 % Ni concentration was observed as dark field in the TEM measurement. On the other hand, the thickness of top carbon layer was estimated about 1 nm.

Figure 3 shows the depth profile of three samples (A, B, C) before and after Ar cluster ion irradiation at 10 keV. In case of sample A and sample B, a large number of Ni atoms were displaced with cluster ion irradiation. On the other hand, a smaller number of Ni atoms were displaced in case of sample C. As the position of Ni layer of sample C didn't change after cluster ion irradiation, the sputtered depth of top carbon layer was very small. These results show that the mixing depth with cluster ion irradiation at 10 keV

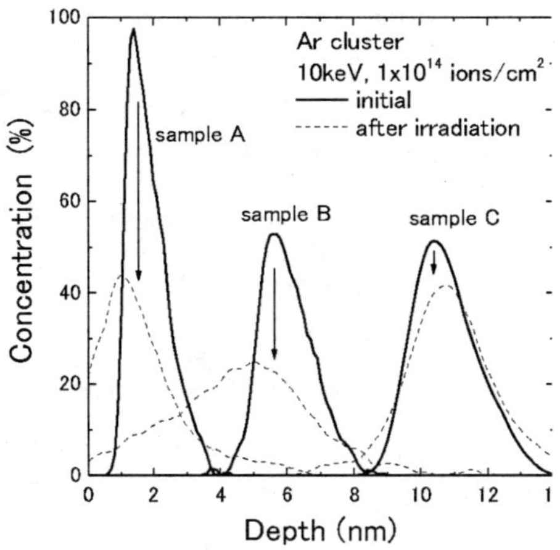

FIGURE 3. Depth profile of three samples (A, B, C) before and after Ar cluster ion irradiation at 10 keV.

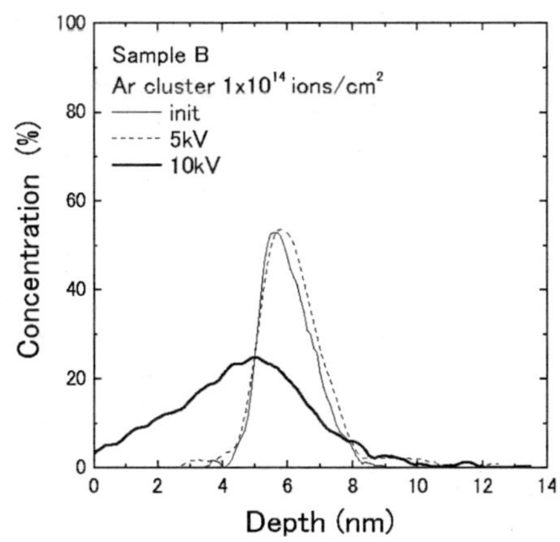

(a) Sample B at 5 and 10 keV

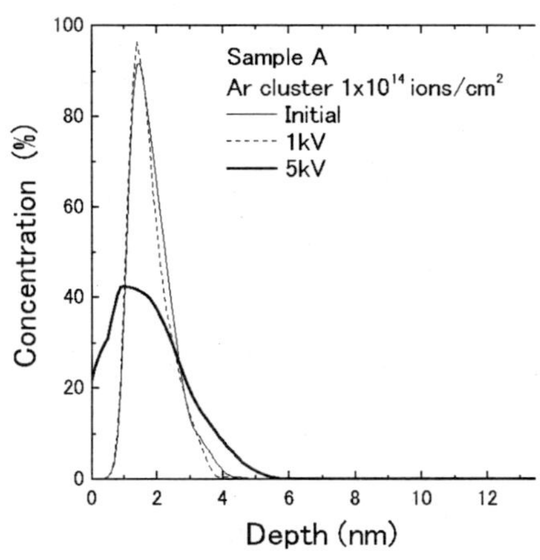

(b) Sample A at 1 and 5 keV

FIGURE 4. Depth profile of sample A, B before and after Ar cluster ion irradiation at energy range of 1-10 keV.

is about 10 nm. Because the mean cluster size was about 2000, the mean energy per atom was only 5 eV. The projection range of Ar monomer at 5 eV is very shallow (< 1 nm). When a cluster has enough large incident energy and penetrates into a solid target, a large number of collisions occur between cluster and surface atoms. After such multiple collision process, the most of incident energy of cluster is transferred to the target, and it can be quantified as the volume of damaged region [13].

Figure 4(a) shows the depth profile of sample B before and after Ar cluster ion irradiation at 5 and 10 keV. A large number of Ni atoms were displaced with cluster ion irradiation at 10 keV. However, a few number of Ni atoms were displaced at 5 keV. Figure 4(b) shows the depth profile of sample A before and after Ar cluster ion irradiation at energy range of 1 and 5 keV. A large number of Ni atoms were displaced with cluster ion irradiation at 5 keV. However, a few number of Ni atoms were displaced at 1 keV. These results show that the mixing depth with cluster ion irradiation at 1 keV and 5 keV are less than 1 nm and 5 nm, respectively.

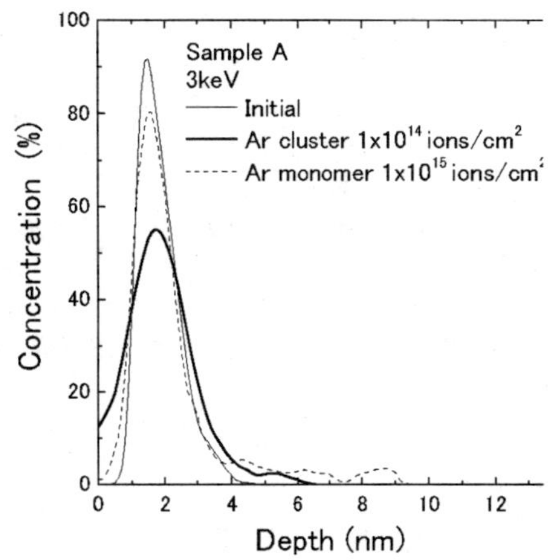

FIGURE 5. Depth profile of sample A before and after cluster and monomer ion irradiation at 3 keV.

In order to compare mixing strength between Ar cluster and Ar monomer ion irradiation, sample A was irradiated with cluster and monomer ions at 3 keV (Figure 5). The ion doses of cluster and monomer ion irradiation were 1x10^{14} ions/cm^2 and 1x10^{15} ions/cm^2, respectively. In case of cluster irradiation, a large number of Ni atoms were displaced. The ratio of moved Ni atoms calculated from differential area of depth profile before and after cluster ion irradiation was more than 23 % with ion dose of 1x10^{14} ions/cm^2. However, the number of displaced Ni atoms with monomer irradiation was smaller than that with cluster irradiation. The ratio of moved Ni atoms calculated from differential area of depth profile before and after monomer ion irradiation was less than 14 % with ion dose of 1x10^{15} ions/cm^2. Some Ni atoms only moved into deeper region in spite of ten times larger ion dose. These results indicate that a violent mixing occurs by single cluster impacts.

SUMMARY

The cluster ion beam mixing of thin Ni layer in carbon targets can be investigated with HR-RBS. The mixing depth with cluster ion irradiation at 10 keV was about 10 nm. The mixing depth with cluster ion irradiation at 1 keV and 5 keV were less than 1 nm and 5 nm, respectively. The number of displaced Ni atoms with cluster ion irradiation was very larger than that with monomer ion irradiation of same energy. This result shows that violent mixing occurs with single cluster impact.

REFERENCES

1. H.Kitani, N.Toyoda, J.Matsuo and I.Yamada, *Nucl. Instr. and Meth. in Phys. Res.* **B121**, 489 (1997).
2. N.Toyoda, N.Hagiwara, J.Matsuo and I.Yamada, *Nucl. Instr. and Meth. in Phys. Res.* **B148**, 639 (1999).
3. A.Nishiyama, M.Adachi, N.Toyoda, N.Hagiwara, J.Matsuo and I.Yamada, *AIP conference proceedings (15th International Conference on Application of Accelerators in Research and Industry)* **475**, 421 (1998).
4. T.Seki and J.Matsuo, *Nucl. Instr. and Meth. in Phys. Res.* **B216,** 191 (2004).
5. I.Yamada, J.Matsuo, N.Toyoda, T.Aoki, E.Jones and Z.Insepov, *Mat. Sci. and Eng.* **A253**, 249 (1998).
6. D.Takeuchi, J.Matsuo, A.Kitai and I.Yamada, *Mat. Sci. and Eng.* **A217/218**, 74 (1996).
7. N.Shimada, T.Aoki, J.Matsuo, I.Yamada, K.Goto and T.Sugui, *J. Mat. Chem. and Phys.* **54**, 80 (1998).
8. E.Bourelle, A.Suzuki, A.Sato, T.Seki and J.Matsuo, Jpn. *J. Appl. Phys.*, **43**(10A), pp. L 1253 (2004).
9. S.Kakuta, S.Sasaki, T.Hirano, K.Ueda, T.Seki, S.Ninomiya, M.Hada and J.Matsuo, *Nucl. Instr. and Meth. in Phys. Res.* **B257**, 677 (2007).
10. H.Tokioka, H.Yamarin, T.Fujino, M.Inoue, T.Seki and J.Matsuo., *Nucl. Instr. and Meth. in Phys. Res.* **B 257**, 658 (2007).
11. S.Ninomiya, Y.Nakata, K.Ichiki, T.Seki, T.Aoki and J.Matsuo, *Nucl. Instr. and Meth. in Phys. Res.* **B 256**, 493 (2007).
12. K.Kimura and M.Mannami, *Nucl. Instr. and Meth. in Phys. Res.* **B 113**, 270 (1996).
13. T.Aoki and J.Matsuo, *Nucl. Instr. and Meth. in Phys. Res.* **B 242**, 517 (2006).

Cluster Size Dependence of Etching by Reactive Gas Cluster Ion Beams

Noriaki Toyoda and Iaso Yamada

Graduate school of engineering, University of Hyogo
2167 Shosha, Himeji, Hyogo, 671-2280, Japan
ntoyoda@incub.u-hyogo.ac.jp

Abstract. Mass-selected reactive gas cluster ion beams (GCIB) were formed using a permanent magnetic filter. Irradiations of CO_2 GCIB on amorphous carbon films and irradiations of SF_6 and SF_6/Ar mixed GCIB on Si surfaces were performed to study the cluster size dependence on etching yields by reactive GCIB. The reactive sputtering yield of carbon by CO_2 GCIB was almost ten times higher than that by Ar GCIB. In the case of $(CO_2)_{20000}$ GCIB with energy of 20 keV (1 eV/atom), it showed the high sputtering yield of 200 atoms/ion, however, there was little crater formation on the carbon surface. It is thought that very soft etching without crater formation would take place in this condition. In the case of SF_6 GCIB on Si, the etching depth of Si showed maximum value when the fraction of SF_6 to Ar was around 50 %. As the etching yield was higher than pure SF_6 GCIB, there was a strong ion assisted etching effects in the case of Ar/SF_6 mixed cluster ion irradiations.

Keywords: Gas cluster ion, sputtering, size dependence
PACS: 79.20.Rf, 81.65.Cf, 36.40.-c

INTRODUCTION

Gas cluster ion beams (GCIB) bombardments have received considerable attention due to their unique impact processes on solid surfaces. By utilizing characteristics of cluster ions, industrial applications have been started such as shallow ion implantation [1], surface smoothing [2,3,4], low-damage etching[5,6] and high quality thin film assisted depositions [7,8,9]. However, it is very difficult to understand the impact processes with conventional linear collision theory because of complex collisions between the target and cluster atoms near surface. From molecular dynamics (MD) simulations, it was demonstrated that cluster size (number of atoms or molecules in a cluster) plays an important role in damage formations or surface modifications induced by energetic cluster impacts [10].

We have reported the experimental results of the cluster size dependence on the damage formation and the sputtering yield using size-selected Ar-GCIB [11]. With increasing cluster sizes, the thickness of the amorphous layer formed on the Si substrates decreased due to low energy per atom of the cluster ions. For example, there was no damage formation in the case of the Ar cluster ions with cluster size of 10000 and the acceleration voltage of 5 keV (0.5 eV/atom). Also, sputtering yields of Au by one Ar atom in a cluster showed a linear relationship with the energy per atom of cluster ions. Threshold energy for sputtering was far below that with Ar monomer ions due to the near surface energy deposition by Ar cluster ion bombardments. However, there is no study about the cluster size effects using reactive cluster ions.

In this study, reactive cluster ions were formed from SF_6, Ar, and CO_2 gases and irradiated on Si substrates and amorphous carbons (diamond like carbon : DLC film). By using the size-selected GCIB system, cluster size or energy per atom dependence on the etching yield was studied.

EXPERIMENTS

The cluster size selection was performed using a strong permanent magnet. The details of the size-selected GCIB system have been reported in the previous paper [11]. Using this system, the cluster size distribution after size selection can be reduced to 1/20 of the original one. The source gases used in this study were He, Ar, CO_2, and SF_6. As the vacuum pumps of this system had been cryo-pumps, the species of the source gas for GCIB had been limited. In this study, a large turbo molecular pump (2400 L/sec) was installed in the source chamber to use various gases. Figure 1 shows the schematic diagram of the configuration of the size-selected GCIB system used in this study. CO_2 and Ar GCIB can be generated from pure gases; however, it is difficult to form SF_6 GCIB without carrier gas. Pure SF_6 GCIB can be formed from the

CP1066, *Ion Implantation Technology,* edited by E. G. Seebauer, S. B. Felch, A. Jain, and Y. V. Kondratenko

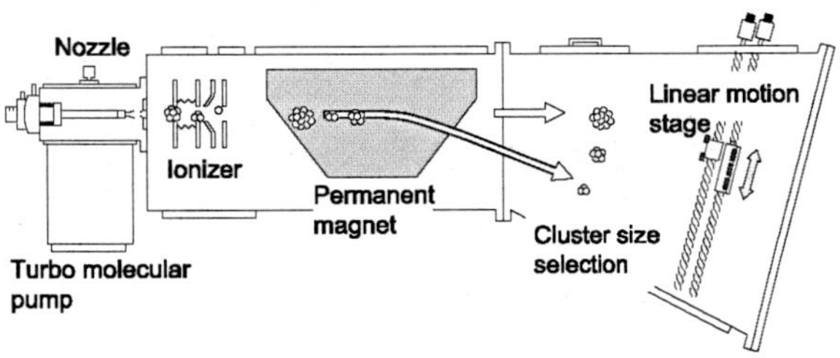

FIGURE 1. Size-selected GCIB system for reactive GCIB

mixing of SF_6 and He gases. He gas is used as a carrier gas and it is not included in the gas cluster ions due to its low boiling point. Also, SF_6 and Ar gases were mixed in the nozzle and Ar/SF_6 mixed GCIB was formed. The ratio of the flow rate of SF_6 to Ar was controlled by two mass flow controllers. As there is no linear relationship between the ratio of the flow rate of SF_6 to Ar and the fraction of each gases in the cluster produced, a residual gas analyzer (RGA) was installed at the end of the beam line in the target chamber. The ratio of the gas species in the cluster was determined from the peak intensities of Ar and SF_5.

The irradiated samples were DLC films for CO_2 GCIB and Si (100) substrates for SF_6/Ar GCIB. After irradiations, the etching depth was measured using a contact surface profiler. The acceleration voltage was 20 kV. For carbon and CO_2 system, irradiations of size-selected CO_2 GCIB on highly oriented pyrolitic graphite were performed. Subsequently, surface morphology was observed using a scanning tunneling microscope (STM).

RESULTS AND DISCUSSIONS

To observe the reactive sputtering of CO_2 GCIB on amorphous carbon films, the ion dose dependence on the etching depth was measured. Figure 2 shows the relationship between the etching depth of DLC films and the ion dose of Ar and CO_2 GCIB. The acceleration voltage was 20 keV for both GCIB. The sputtering yields in this condition were 42 atoms/ion for Ar GCIB and 493 atoms/ion for CO_2 GCIB. It is almost ten times higher sputtering yields than Ar GCIB for amorphous carbon films. This result clearly indicates reactive sputtering of carbon with CO_2 GCIB.

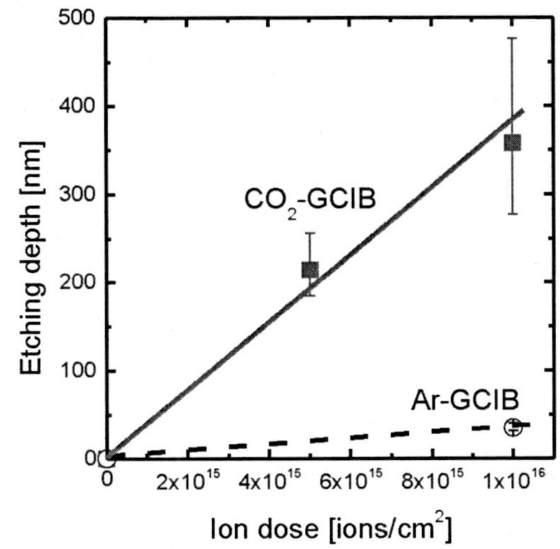

FIGURE 2. Relationship between the etching depth of DLC film and the ion dose of Ar and CO_2 GCIB with the acceleration voltage of 20 kV.

Figure 3 shows the cluster size dependence of sputtering yield of carbon by CO_2 GCIB using the size-selected GCIB system shown in Figure 1. The acceleration voltages were 10 and 20 kV. The ion dose of CO_2 GCIB was 3 x 10^{15} ions/cm^2. The cluster size was changed from 500 to 20000, which corresponded to 0.5 to 40 eV/molecule. By increasing the acceleration voltage from 10 to 20 kV, the sputtering yield of carbon increased. In the case of the acceleration voltage of 20 kV, the sputtering yield was almost constant with cluster size between 500 and 5000. Then it gradually decreased by increasing the cluster size. However, even though the cluster size was 20000 (1eV /molecule), the sputtering yield showed still high (200 atoms/ion).

To observe the surface morphology or crater formations on a carbon surface, CO_2 GCIB irradiation on highly oriented pyloritic graphite (HOPG) was performed. Figure 4 shows STM images of an HOPG surface irradiated with CO_2 GCIB. The cluster sizes were 500, 5000, and 10000. The total acceleration voltage was 20 kV. Therefore, the energy per atoms were 40, 4, and 2 eV/molecule, respectively. The scan area was 200 nm x 200 nm. The ion dose was 2 x 10^{11}

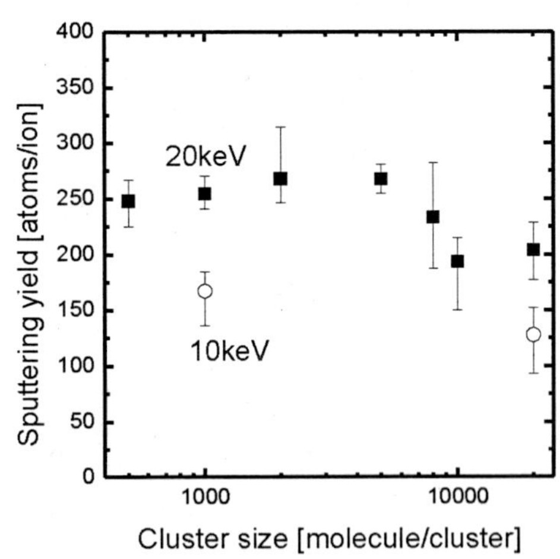

FIGURE 3. Cluster size dependence on the sputtering yield of carbon by CO_2 GCIB with the acceleration voltage of 10 and 20 kV.

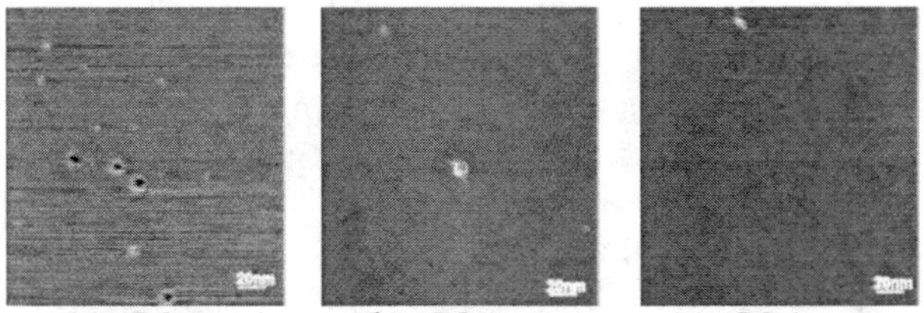

(a) $(CO_2)_{500}$ 40 eV/molecule (b) $(CO_2)_{5000}$ 4 eV/molecule (c) $(CO_2)_{10000}$ 2 eV/molecule

FIGURE 4. STM images of highly oriented graphite irradiated with size-selected CO_2 GCIB. The total acceleration voltage was 20 kV.

ions/cm^2. In the case of $(CO_2)_{500}$ (40eV/molecule), there were several craters formed by individual cluster impacts. However, there were very few craters in the case of $(CO_2)_{5000}$ (4 eV/molecule) and there was no craters in the case of $(CO_2)_{10000}$ (2 eV/molecule). As shown in Figure 3, $(CO_2)_{10000}$ with an acceleration voltage of 20 keV showed high sputtering yield (190 atoms/ion). In this condition, the cluster collapsed on the surface; however it did not create a crater. Dissociation of a molecule in a cluster ion occurred at the surface without crater formation and these atoms reacted with target atoms.

Similarly, when SF_6 GCIB bombards a Si surface, the breaking of the SF_6 cluster ion and dissociation of SF_6 molecules occur at the same time, as a result, reactive etching becomes dominant process. We have reported that SF_6 GCIB shows two or three orders of magnitude higher reactive sputtering yield for Si [12]. However, it has not been studied about the cluster size effects on reactive sputtering for Si by SF_6 cluster ions.

Figure 5 shows the etching depth of Si after Ar/SF_6 mixed GCIB irradiation (acceleration energy 20 keV, ion dose 1.0x10^{15}ions/cm^2). The horizontal axis represents the fraction of the number of SF_6 molecule to that of Ar in a cluster ion. The cluster sizes were 500, 2000 and 5000. The SF_6 molecule fraction of 0 % represents the results using pure Ar-GCIB. Similarly, that of 100 % does the Si etching depth using pure SF_6 GCIB. In Figure 5, etching depth by SF_6 GCIB without cluster selection was also plotted. In the case without cluster sizes selection, the etching depth of Si increased with increasing the SF_6/Ar ratio from 0 %. Then it exhibited a maximum around the SF_6 fraction of 50 % and decreased gradually towards SF_6 fraction of 100 %. In Figure 5, the SF_6 cluster size was changed at 500, 2000 and 5000. At the same SF_6 fraction to Ar, larger cluster size showed higher etching depth than that of smaller one.

Figure 6 shows the cluster size dependence on the etching depth of Si. The acceleration voltage and the ion dose were 20 kV and 1x10^{15} ions/cm^2, respectively. The SF_6 fractions in a cluster ion were 1.5, 10, 20, 50, and 80 %. Under the condition that the SF_6 fraction

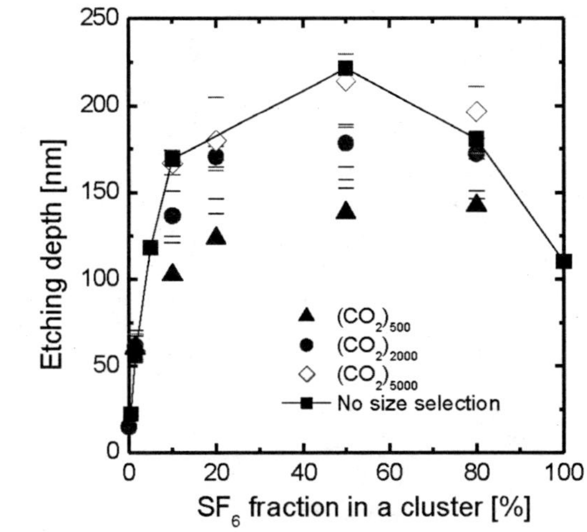

FIGURE 5. Relationship between the fraction of SF_6 in a Ar/SF_6 mixed cluster and the etching depth of Si. The acceleration voltage and the ion dose were 20 kV and 1 x 10^{15}ions/cm^2, respectively.

was over 10 %, the etching depth increased with cluster size. In this study, Ar/SF_6 mixed cluster ions were used. As shown in Figure 5, the etching depth of Ar/SF_6 mixed cluster ions were higher than pure those with SF_6 GCIB or Ar GCIB.

Figure 7 shows the relationship between the sputtering yield of Si per one SF_6 molecule and the energy per atom. When the SF_6 fraction was 50 or 80% and the energy per molecule was 40 eV/molecule, the sputtering yield by one SF_6 molecule was around 2 to 3. By decreasing the SF_6 fraction in a cluster, the Si sputtering yield by one SF_6 molecule increased. It can be estimated that large amount of Ar atoms assist the dissociation and reaction of SF_6 molecules in the

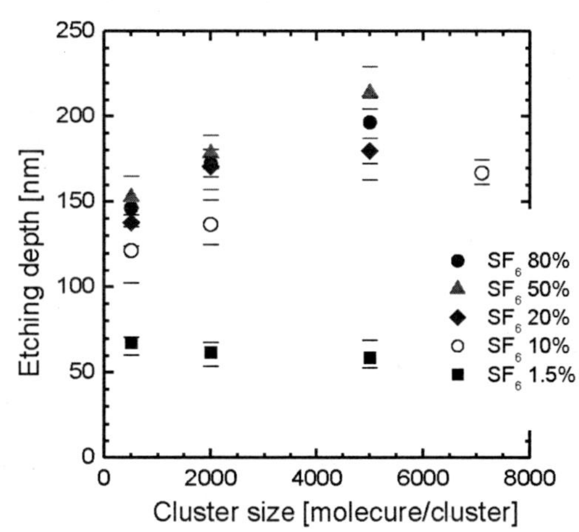

FIGURE 6. Cluster size dependence on the etching depth of Si. The SF_6 fraction in Ar/SF_6 cluster ion varied from 1.5 to 80 %.

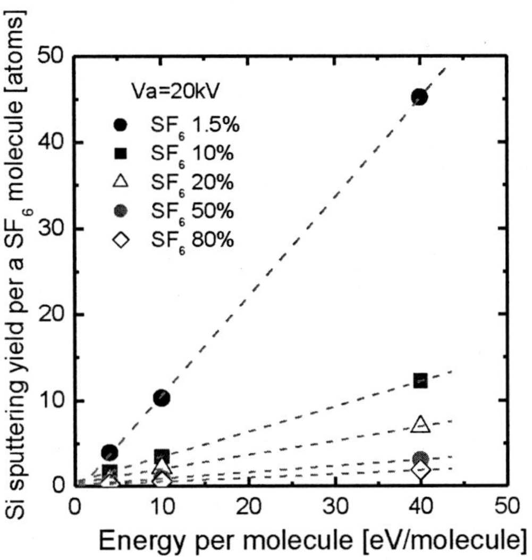

FIGURE 7. Energy per molecule dependence on the Si sputtering yield by one SF_6 molecule. The acceleration voltage was 20 kV.

cluster ions. In the case of condition such as 1.5 % SF_6 fraction, one SF_6 molecule will sputter 45 Si atoms at the energy per atom of 40 eV/atom. There is a contribution of Ar ion bombardment on etching of Si. Since very high chemical reactivity can be obtained by mixing the small fraction of SF_6 to Ar, very effective etching process can be realized using mixed GCIB.

SUMMARY

Cluster size dependence on the reactive sputtering of CO_2 and SF_6 GCIB was studied. Amorphous carbon showed reactive sputtering by CO_2 GCIB and very high sputtering yield was obtained without forming craters by using large size of cluster ions. Dissociation of reactive cluster and molecules occurred on the target surface and the target atoms etched very softly. When cluster was formed from Ar and SF_6 gases, the reactive sputtering yield became higher than the case with pure SF_6 GCIB. Irradiation of energetic Ar in a mixed Ar/SF_6 cluster would enhance the etching of Si by SF_6. The study of the Fetching mechanism by these mixed cluster ions would be a next challenge to understand these phenomena.

REFERENCES

1. K.Goto, J.Matsuo, T.Sugii, H.Minakata, I.Yamada and T.Hisatugu, IEDM Tech. Dig. IEEE (1996) 435.
2. N.Toyoda, H.Kitani, N.Hagiwara, J.Matsuo and I.Yamada, *Mat. Chem. and Phys.*, **54** (1998) 106-110.
3. N.Toyoda, N.Hagiwara, J.Matsuo and I.Yamada, *Nucl. Instr. and Meth. B*, **148** (1999) 639-644.
4. J.A.Greer, D.B.Fenner, J.Hautala, L.P.Allen, V.DiFilippo, N.Toyoda, I.Yamada, J.Matsuo, E.Minami and H.Katsumata, *Surface and Coating Technol.*, **133-134** (2000) 273-282.
5. N.Toyoda, H.Kitani, J.Matsuo and I.Yamada, *Nucl. Instr. and Meth. B*, **121** (1997) 484-488.
6. M.Nagano, S.Yamada, S.Akita, S.Houzumi, N.Toyoda and I.Yamda, *Jpn. J. Appl. Phys.*, **44**(4) (2005) L164.
7. K.Shirai, Y.Fujiwara, R.Takahashi, N.Toyoda, S.Matsui, T.Mitamura, M.Terasawa and I.Yamda, *Jpn. J. Appl. Phys.*, **41**, (2002) 4291-4294.
8. N.Toyoda and I.Yamda, *Appl. Surf. Sci.*, **226** (2004) 231-236.
9. T.Kitagawa, I.Yamada, N.Toyoda, H.Tsubakino, J.Matsuo, G.H.Takaoka and A.Kirkpatrick, *Nucl. Instr. and Meth. B*, **201** 405-412 (2003).
10. T.Aoki, T.Seki, J.Matsuo, Z.Insepov and I.Yamada, *Nucl. Instr. and Meth. B*, **153** 264-269 (1999).
11. N. Toyoda, S. Houzumi and I. Yamada, *Nucl. Instr. and Meth. B*, **242** 466-468 (2006).
12. N.Toyoda, H.Kitani, J.Matsuo and I.Yamada, *Nucl. Instr. and Meth. B*, **121** 484-488 (1997).

Non-Contact Wafer Fabrication Process Using Gas Cluster Ion Beams

Noriaki Toyoda[1], Hiromichi Isogai[2], and Iaso Yamada[1]

[1]*Covalent Materials Corporation*
6-861-5, Higashikou Seirou-machi, Kitakanbara, Niigata, 957-0197, JAPAN
ntoyoda@incub.u-hyogo.ac.jp
[2]*Graduate school of engineering, University of Hyogo*
2167 Shosha, Himeji, Hyogo, 671-2280, Japan

Abstract. Gas cluster ion beam (GCIB) was used for precise wafer fabrication process. GCIB realizes a quite low-energy ion beam and shows very precise and good repeatability. To obtain thickness uniformity of Si over the whole wafer, small beam diameter (~ 4mm) of GCIB was used. Thickness variations on the wafer can be reduced by location specific irradiation of collimated GCIB. By controlling the scan speed of GCIB irradiation based on the removal thickness at each irradiation position, thickness and height uniformity of Si can be improved to several tens of nm. In addition, etching enhancement by using Ar/SF_6 mixed cluster was studied.

Keywords: Gas cluster ion, reactive sputtering, location specific processing, wafer process
PACS: 61.72.Ff, 81.40.-Z, 81.10.Jt, 68.37.Lp

INTRODUCTION

With shrinking the gate length, the flatness of the semiconductor wafer has to be improved because the focusing point of the lithography process will be different if the wafer height is varied. From international technology roadmap for semiconductor (ITRS) 2007, the site flatness and nanotopography will be below 40 and 10 nm in 2011, respectively [1]. Also, it is predicted that the wafer diameter will become large from 300 to 450 mm in future. In the present, mechanical processing technology such as lapping or polishing is used for semiconductor wafer fabrications. However, these mechanical processes are difficult to control the thickness variations in nanometer level and it will be a big challenge to employ it in large wafers.

To solve these problems, dry etching processes are expected to be alternative wafer fabrication processes because of their good controllability. Ion beam is one of the candidates for these, however, the etching rate is relatively slow and the damage inside the wafer can not be avoided.

Therefore, we have studied the use of gas cluster ion beams (GCIB) for wafer fabrication process. As the GCIB is an aggregate of gaseous materials, it exhibits unique impact processes which are completely different from that of atomic or molecular ions (monomer ion beams). Very dense energy was deposited near the surface and multiple collisions with target atoms will take place with very low energy per atom. By utilizing characteristics of cluster ions, industrial applications has been started such as shallow ion implantation [2], surface smoothing [3], low-damage etching [4] and high quality thin film assisted depositions [5]. Since the GCIB is an ion beam process, any desired poison on a wafer can be processed by controlling the location of the irradiation point. In this study, thickness variation corrections for semiconductor wafers were studied using location specific processing of GCIB to provide the fabrication technology for next generation semiconductor wafers. Surface morphology and reactive etching with Ar/SF_6 GCIB were also studied.

EXPERIMENTS

Figure 1 shows the schematic diagram of the GCIB system with location specific processing (LSP). Neutral gas cluster beam can be formed from various gases such as Ar, O_2, N_2 and SF_6. There are two gas lines controlled by mass flow controllers, and mixed gas cluster beam can be easily formed. In this study,

CP1066, *Ion Implantation Technology,* edited by E. G. Seebauer, S. B. Felch, A. Jain, and Y. V. Kondratenko

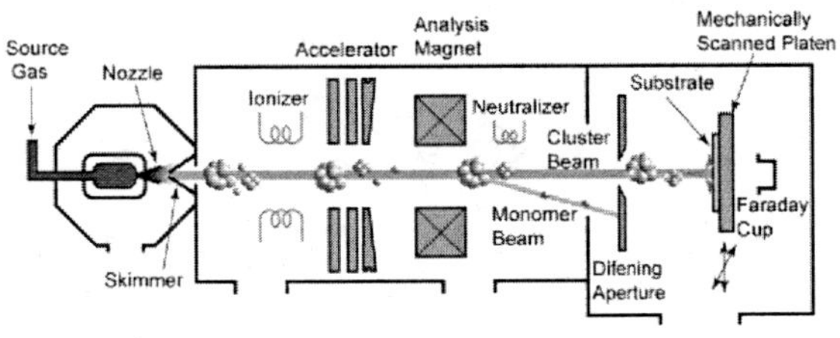

FIGURE 1. Schematic diagram of the GCIB system with location specific processing.

Ar and SF_6 gases were used for Si wafers. For irradiation position control, two linear motion stages in horizontal and vertical directions were used. The typical GCIB current was 200 μA at the acceleration voltage of 30 kV.

Figure 2 shows a schematic diagram of the location specific processing for semiconductor wafers. At first, the initial thickness map was measured in advance. Subsequently, GCIB current through a small aperture was measured to obtain the beam diameter at the irradiation condition. By positioning the aperture in horizontal and vertical axis, beam diameter or profile can be measured. Figure 3 shows the typical beam profile of SF_6 GCIB. The acceleration voltage was 30 kV. In this case the full width of the half maximum was 4 mm. As the beam profile usually shows Gaussian distribution, the overlaps at each irradiation position have to be considered. The ion dose required at the each position can be obtained from the thickness distribution over the wafer, the target thickness, beam profile and the ion dose dependence of the etching depth for the particular material of the wafer. The ion dose required is converted to the speed of the scanner at each position.

In this study, SF_6 GCIB was mainly used. As it is difficult to form SF_6 GCIB from pure SF_6, He gas was mixed as a carrier gas. Not only the pure SF_6 GCIB, but also Ar and SF_6 mixed gas was used in some experiments. The target wafers were silicon on insulator (SOI), grounded or polished eight inch Si wafers. The thickness variation of SOI wafers were measured with spectroscopic ellipsometer. Thickness variations of bulk Si wafers were measured with capacitance sensor probe (ADE corp.) In addition, the etching rate enhancement by Ar/SF_6 mixed cluster beam was studied from the etching depth measurements.

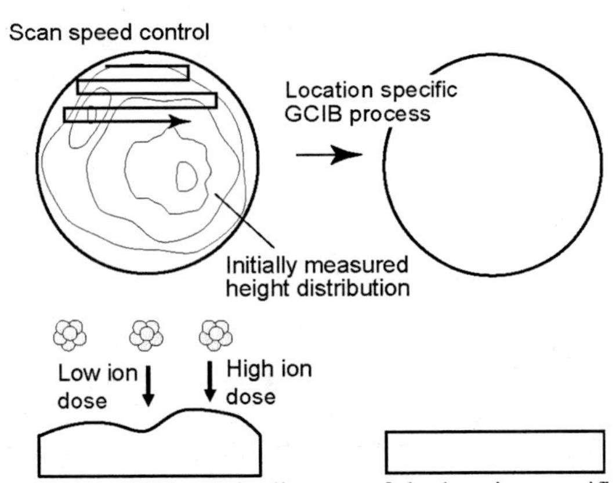

FIGURE 2. Schematic diagram of the location specific GCIB process for semiconductor wafers.

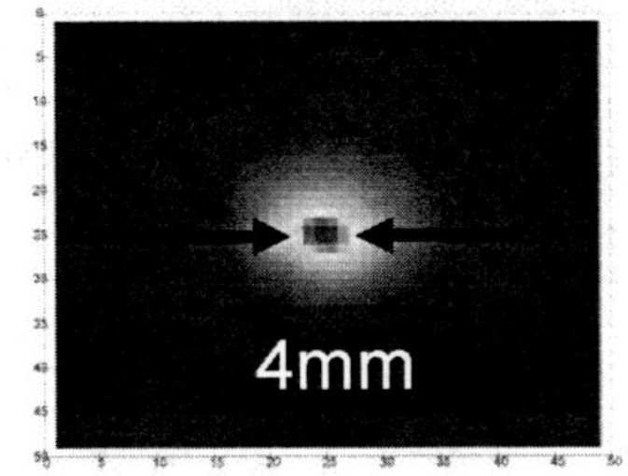

FIGURE 3. Beam profile of the SF_6 GCIB at acceleration voltage of 30 kV.

RESULTS AND DISCUSSIONS

Figure 4 shows the thickness variation of SOI wafer before and after location specific SF_6 GCIB processing. The acceleration voltage was 30 kV. The wafer diameter in this case was 100 mm. The maximum, minimum, mean and peak-to-valley of Si layer thickness before and after GCIB process is shown in Table 1. This SOI wafer was chosen because it had a wide thickness variation initially. Before irradiations, the peak-to-valley of the initial wafer was 540 nm. After location specific GCIB process, it was reduced to 77 nm. As the mean thickness became 866 nm to 483 nm, the average etching depth was 383 nm. Uniform thickness on wafer can be realized without deep etching.

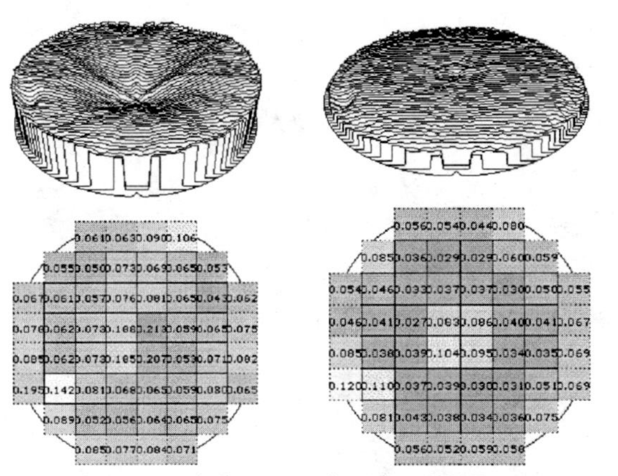

(a) As grounded (b) After GCIB process

FIGURE 4. Location specific GCIB process for bulk Si wafers 8 inch in diameter. (Upper: Height distribution, Bottom; site flatness distribution)

Figure 5 shows an example of the location specific GCIB process for bulk Si wafer (eight inches in diameter). This wafer was mechanically grounded before GCIB process. Due to the grounded process, the center of the wafer was hollowed. The bottom figure in the Figure 5 shows the site flatness. After the location specific GCIB process, the height of the Si wafer became uniform and the site flatness distribution became narrow. As these results were obtained at one time process, the thickness variation would be improved if multiple times location specific GCIB process was performed.

To study the effect of gas mixing in the source gas, SF_6 and Ar mixed cluster ions were formed. SF_6/Ar mixed cluster ion irradiation on Si was performed and the etching depth of Si at various mixing conditions was measured. Figure 6 shows the etching depth of Si with Ar, pure SF_6 and Ar/SF_6 mixed GCIB. The

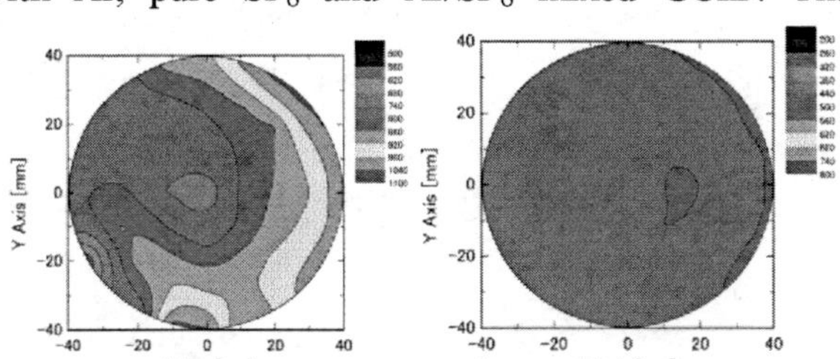

(a) Before irradiation (b) After GCIB process

FIGURE 5. Thickness variation of Si layer of SOI wafer before and after GCIB location specific process.

TABLE 1 : Thickness variation before and after GCIB.

	Maximum thickness	Minimum thickness	Mean thickness	Peak to valley
Before irradiation	1075 nm	536 nm	866 nm	540 nm
After GCIB process	519 nm	442 nm	483 nm	77 nm

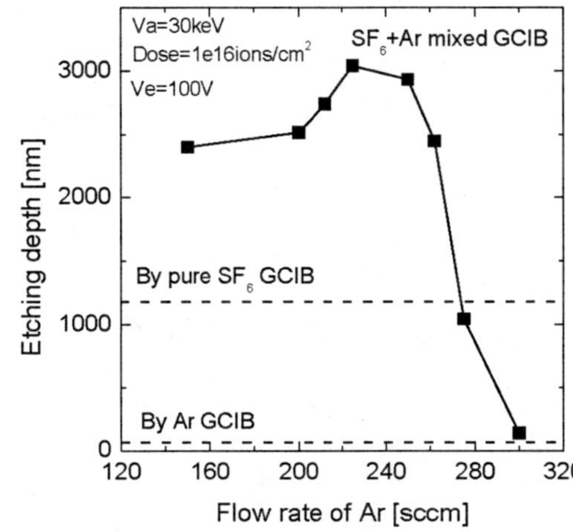

FIGURE 6. Etching depth of Si irradiated with Ar, pure SF_6 and Ar/SF_6 mixed cluster ions. The acceleration voltage and the ion dose were 30 kV and $1x10^{16}$ ions/cm^2, respectively.

acceleration voltage was 30 kV and the ion dose was 1 x 10^{16} ions/cm^2. The flow rate of SF_6 was fixed atn that of Ar was changed from 150 to 300 sccm. Thus the fraction of SF_6 in GCIB was changed. From the Figure 6, the etching depth of Si by SF_6 GCIB was 1181 nm although that by Ar GCIB was 62 nm at the same ion dose and acceleration voltage. On the other hand, the etching depth reached over 3000 nm by using Ar/SF_6 mixed cluster beam. This result indicates that there is an enhancement of etching process by Ar ion bombardments.

SUMMARY

A GCIB process was employed to improve the wafer height and thickness variations. Thickness variation of SOI and bulk Si wafers can be reduced by using the location specific irradiation of SF_6 GCIB. As GCIB is a beam process, basically it can be utilized for any size of wafers. Ar/SF_6 mixed cluster ion showed enhanced etching of Si.

REFERENCES

1. http://www.itrs.net/Links/2007ITRS/2007_Chapters/2007_FEP.pdf
2. K.Goto, J.Matsuo, T.Sugii, H.Minakata, I.Yamada and T.Hisatugu, *IEDM Tech. Dig. IEEE* 435 (1996).
3. N.Toyoda, H.Kitani, N.Hagiwara, J.Matsuo and I.Yamada, *Mat. Chem. and Phys.*, **54** 106-110 (1998).
4. M.Nagano, S.Yamada, S.Akita, S.Houzumi, N.Toyoda and I.Yamda, Jpn. *J. Appl. Phys.*, **44**(4) (2005) L164.
5. N.Toyoda and I.Yamda, *Appl. Surf. Sci.*, **226** 231-236 (2004).

Characteristics of Cluster Implantation and Low Diffusion Annealing

Tsutomu Nagayama[1], Nariaki Hamamoto[1], Sei Umisedo[1], Masayasu Tanjyo[1], Fumio Ootsuka[2,3] and Takayuki Aoyama[2]

[1]Nissin Ion Equipment Co., Ltd.,
575 Kuze-Tonoshiro-cho, Minami-ku, Kyoto, 601-8205 , Japan
Tel: +81-75-922-4611, Fax: +81-75-922-4615, e-mail: Nagayama_Tsutomu@nissin.co.jp
[2]Semiconductor Leading Edge Technologies, Inc, (Selete), 16-1 Onogawa, Tsukuba-shi, 305-8569, Japan
[3] Present affiliation: Renesas Technology Corp.

Abstract. Beyond 45 nm device fabrication, it is recognized that the ultra-shallow junction (USJ) formation is required, which is achieved by the ultra-low energy ion implantation and ultra-low diffusion annealing. For this purpose, Cluster ion implantation (CII) and the Flash lamp annealing (FLA) activation processes have been developed. In this paper, we introduce characteristics of activation using $B_{18}Hx^+$ implantation with FLA, which are applied to a source drain extension (SDE) for pMOSFET in Hgh-k/metal process. We will prove that the combined process of $B_{18}Hx^+$ and FLA can make the low sheet resistance and the fine crystal recovery for USJ formation such a SDE. Moreover, the combined process of C_7Hx^+ implantation and FLA can make the high stress region for nMOSFET by the high substitutional Carbon. If both of $B_{18}Hx^+$ implantation and C_7Hx^+ implantation with FLA are applied to High-k/metal process, it will give the highest performance. The combination of FLA and cluster ion implanter "CLARIS" which is developed for IC fabrication, will promise the most powerful performance beyond 45 nm node device.

Keywords: Implanter, cluster, USJ, millisecond annealing
PACS: 85.40.Ry, 61.46.Bc, 61.82.Fk

INTRODUCTION

According to ITRS2007, it is predicted that IC fabrication beyond 45 nm node semiconductor devices will begin in 2010 and a formation of the ultra-shallow junction (USJ) such a source drain extension (SDE) will have lengths below 10 nm. In parallel with IC integration, the semiconductor material and structure will be modified to Metal/High-k,device, FD-SOI device, and Fin-FET device. These new devices will be also using Stress Engineering processes which are required to optimize their processes for the USJ.

Cluster ion implantation (CII) for p-MOS is one of the key technologies, which has some advantages of device characteristic like low USJ depth Xj and low Sheet resistance Rs.[1]. To commercialize Cluster implantation technology, we are developing a CII system named "CLARIS" ion implanter, which can perform a precisely controlled implantation dose, uniformity, and incident beam angle. Some process advantages of CII are also reported, which are enhanced SDE, SD, Halo and poly-Gate implantation processes [2]. In addition, Carbon Cluster implantation ($C_7H_x^+$) is reported as the Stress Engineering process for n-MOS which makes a SiC layer having a high substitutional ratio [3].

In this paper, we introduce the characteristics of the device and the process of combining CII and FLA for IC production.

a) The feature of the high activation of FLA using CII ($B_{18}H_x^+$) is reported comparing with BF_2^+ implantation.
b) The characteristics of the Si:C for the stress engineering of n-MOS, combined with the carbon Cluster implantation and FLA.
c) The pMOSFET device characteristics with CII ($B_{18}H_x^+$) applied to SDE and FLA, which is used in a Metal/High-k device. Because of super steep implantation profile and non diffusion annealing, the improvement result of the device and process characteristics are compared with the Ge pre-amorphizaion (PAI) and monomer B implantation.

EXPERIMENTAL

Molecular ions ($B_{18}H_x^+$, $C_7H_x^+$) were implanted by a cluster implanter, which is controlled with parallel beam and mechanical wafer scan system. The beam profile monitor system is equipped with the front and the back faraday cups which are utilized in the measurement of the beam XY size and the divergence on a wafer [4]. Implanting dopants were activated using the rapid thermal annealing (RTA) and the FLA

CP1066, *Ion Implantation Technology,* edited by E. G. Seebauer, S. B. Felch, A. Jain, and Y. V. Kondratenko

method.

BF_2^+ and B^+ were implanted in a medium current implanter or high current implanter. Ge^+ was implanted for pre-amorphizaion implantation (PAI).

The doping profile and activation profile were measured by SIMS and the junction depth (Xj) was also determined. Rs was measured with four-point probe (4 pp). Thermal wave (TW) measurement was used for evaluation of the residual damages after implantation and annealing.

The surface amorphous layers and crystal defects were observed with XTEM. High resolution X-ray diffraction (HRXRD) was used to measure substitutional ratio of Si:C layer and the thickness of stress generation area. Metal/High-k pMOSFET was fabricated in Selete CMOS line. We used $B_{18}H_x^+$ implantation for source/drain extension (SDE) formation.

RESULTS AND DISCUSSION

Cluster Ion Implanter Characteristics

SemEquip's ClusterIon™ source built in Cluster ion implanter, which has a large source aperture to extract high current ion beam, is installed and typically more than 30 mA @ 1 keV equivalent in case of $B_{18}H_x^+$ are extracted and 4 mA @0.3 keV [4], respectively. The source material vapor delivery system and in-situ deposition cleaning system are located below the ion source.

The beam divergence are taken from 2 keV to 0.2 keV boron cluster beam divergence and the parallelism on a wafer, which shows less than 0.5 deg values are achieved. This exceptional beam quality is due to the medium current ion implanter-base hybrid scan beam control system. The details are introduced in this conference Oral presentation (Nissin New Cluster Ion Implanter : CLARIS).

Characteristics of Cluster Ion Implantation with FLA

For the USJ beyond 45 nm node devices, CII and FLA processes without diffusion are some advantages about diffusion length and activation, however FLA has some problems of re-crystallization in case of having thick amorphous layer and end of range damage (EOR). We researched the characteristics of activation and re-crystallization about typical SDE and SD conditions using $B_{18}H_x^+$ ion implantation with FLA.

Figure 1 shows the relationship of FLA power and Rs at $B_{18}H_x^+$ typical SDE ion implantation condition (B–equibalent @ 300eV, 1E15/cm^2, 0/0). The Rs of $B_{18}H_x^+$ is almost half value of that of BF_2^+ at same FLA condition (bias temp. 450°C + 30J)

Figure 2 shows the relationship of FLA power and TW units after FLA at typical SDE ion implantation condition. TW value of $B_{18}H_x^+$ is lower than that of BF_2^+ at same FLA condition (bias temp. 450°C + 30J). We estimated with this result and the SIMS data that FLA at 30J could make finer re-crystallization for $B_{18}H_x^+$ implantation comparing with BF_2^+ implantation. After FLA, $B_{18}H_x^+$ implantation has advantages of high activation and fine re-crystallization for USJ formation, comparing with BF_2^+ implantation.

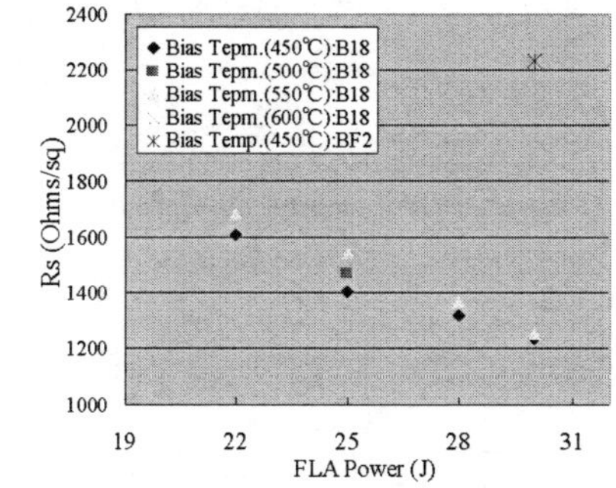

FIGURE 1. FLA Power vs Sheet resistance at SDE condition. (B-Equivalent@300 eV, 1E15/cm^2, 0/0).

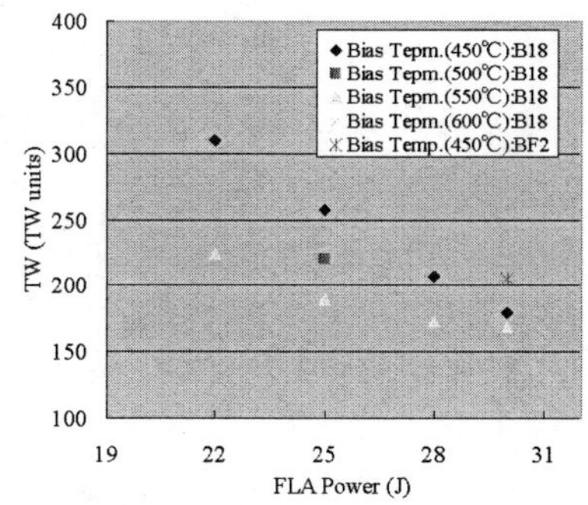

FIGURE 2. FLA Power vs TW units at SDE condition. (B-Equivalent@300 eV, 1E15/cm2, 0/0).

Figure 3 shows the relationship of FLA power and Rs at the $B_{18}H_x^+$ typical SD condition (B-Equibalent@2 keV, 5E15/cm^2, 0/0). Activation of $B_{18}H_x^+$ is two times larger than that of BF_2^+ at same FLA condition (bias temp. 550°C + 30J), which is not same to SDE condition.

Figure 4 shows relationship of FLA power and TW units after FLA at the $B_{18}H_x^+$ typical SD ion implantation condition. While the TW value of $B_{18}H_x^+$ is lower than that of BF_2^+, it keeps a high TW value even though FLA is done at a 30J condition.

We estimated that FLA at 30J was difficult to make fine re-crystallization when using $B_{18}H_x^+$ implantation at SD condition, because a TW value after FLA shows the amount of the crystal defects.

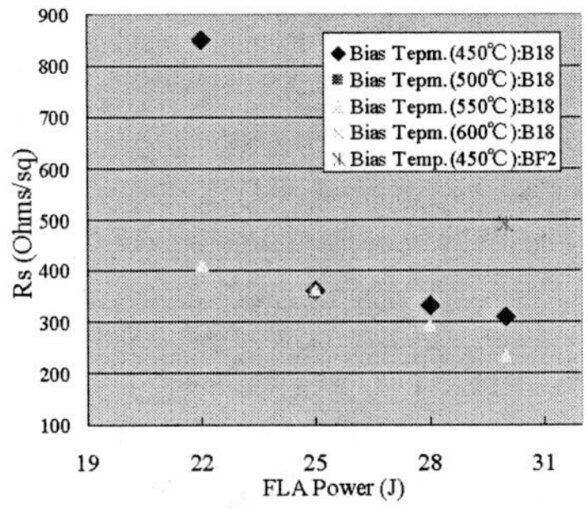

FIGURE 3. FLA Power vs Sheet resistance at SD condition. (B-Equivalent@2keV, 5E15/cm^2, 0/0).

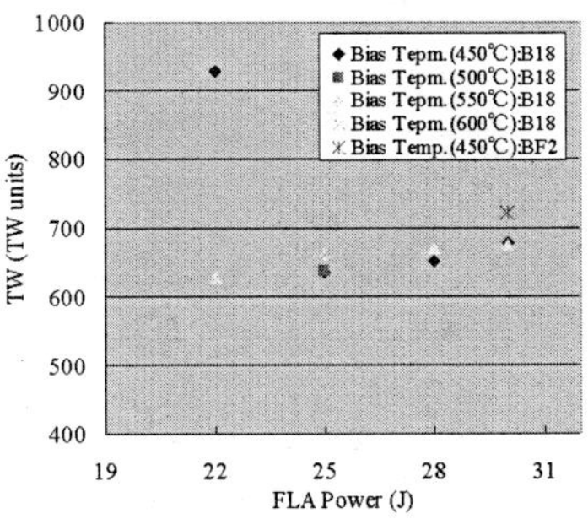

FIGURE 4. FLA Power vs TW units at SDE condition. (B-Equivalent@2keV, 5E15/cm2, 0/0).

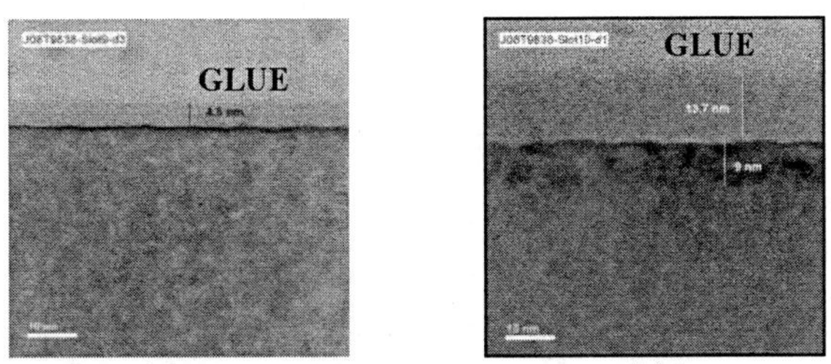

(a) **(b)**

FIGURE 5. Cross-sectional XTEM images of B18Hx+ . (a) Equivalent 300eV, 1E15/cm2 (b) Equivalent 2keV, 5E15cm2

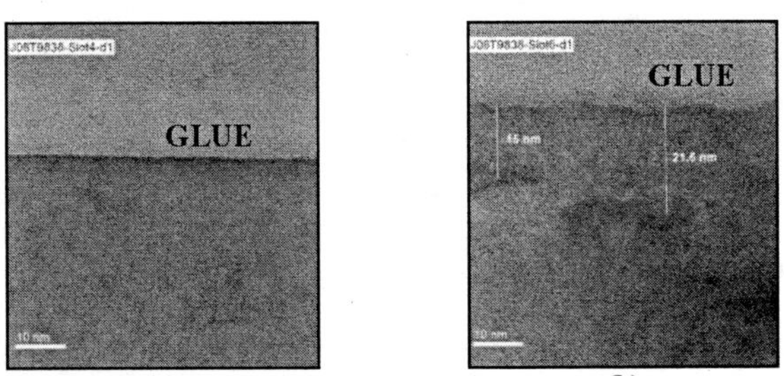

(a) **(b)**

FIGURE 6. Cross-sectional XTEM images of B18Hx+ after FLA. (a) Equivalent 300eV, 1E15/cm2 FLA: 450°C + 30J (b) Equivalent 2keV, 5E15cm2 FLA: 550°C + 30J.

Thus, to research in detail, the XTEM images of $B_{18}H_x^+$ ion implantation surfaces after FLA were observed. The thickness of the amorphous layers was measured and inspected for the crystal defects in the layers. As shown in Figure5, the amorphous/crystalline (a/c) interface is very smooth and any defect such as EOR-defect is not observed in case of $B_{18}H_x^+$ at B-equivalent 300 eV. On the contrary, in case of $B_{18}H_x^+$ at B-equivalent 2 keV, the a/c interface is a little bit rough and some defects are observed in the crystal region near the amorphous layer. We speculated that the high damage was created at 2 keV, resulted in High TW value. As shown in Figure6, full re-crystallization and no substrate crystal defects from XTEM image after FLA at B-equivalent 300 eV, is observed. On the other hand, the full re-crystallization and defects at 15-21 nm depth after FLA B-equivalent 2 keV, was observed. Therefore, it was possible to apply low thermal budget annealing at 450°C + 30J and to achieve fully re-crystallization by FLA when a thin amorphous layer (about 4.5 nm) was formed by implanting $B_{18}H_x^+$ at B-equivalent 300 eV.

We consider that it is required to develop a process with long pulse FLA and second annealing such SPE without diffusion for no residual damage and fine re-crystallization when a thick amorphous layer (about 13.7 nm) and defect region (9 nm) are formed by implanting $B_{18}H_x^+$ even at B-equivalent 2 keV.

Stress Engineering for n-MOS

Beyond 45 nm node devices, increasing of mobility using stress engineering makes enhancement of the ion in the channel region. For p-MOS, SiGe epi-layer will be put to practical use, however, for n-MOS, SiC epi-layer is of no practical use. We propose carbon cluster implantation and msec annealing for SiC formation, which makes higher carbon substitution. This results in a higher stress on channel region [3]. Figure 7 shows the Cluster Carbon™ mass spectrum ($C_7H_7^+$) from $C_{14}H_{14}$ source material. The HRXRD measurement data of the stress induced lattice layer are shown in Figure 8. High substiutional ratio (2.30%) and thick layer (46 nm) are achieved by Multiple Cluster carbon implantation with FLA. The conditions of implantation and annealing also affect the carbon substitutional ratio and the stress value. This method has a possibility to reduce the process steps and cost compared to the SiC epi-process.

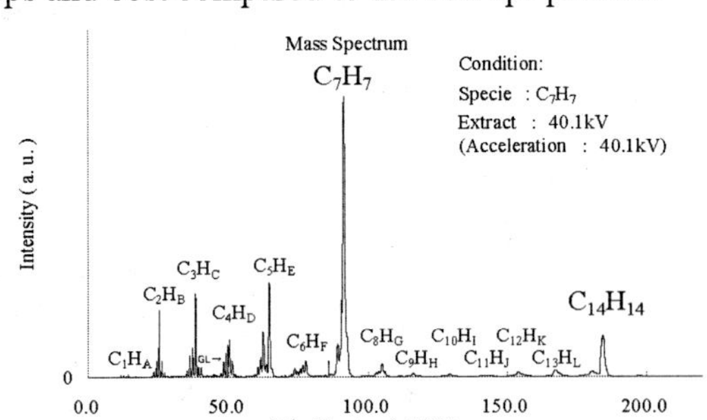

FIGURE 7. Mass spectrum of $C_7H_7^+$ produced by $C_{14}H_{14}$.

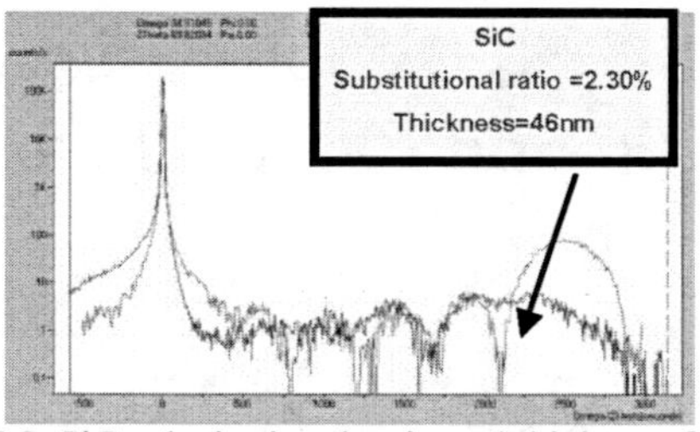

FIGURE 8. SiC substitutional ratio and thickness from HRXRD: Chained or Multiple implantation (1.5 keV :1E15/cm^2 + 10keV:6E15/cm^2) + Flash Annealing.

Bulk Metal/High-k MOSFET

Metal/High-k MOSFET are near the mass production stage, but W full metal Gate FET less than 45 nm are still under development [5]. It is difficult to achieve 10 nm USJ and get practical Ion-Ioff characteristics simultaneously. To solve the problem, we adopted tilted $B_{18}H_x^+$ implantation for p-MOS SDE formation with flash lamp annealing (FLA) for activation. In Figure 9, SIMS depth profiles of p-MOS and n-MOS are shown. By optimizing the tilt angle for suitable gate overlap length, Full-metal-gate HfSiON transistors whose gate length is less than 50 nm are fabricated with superior electrical characteristics.

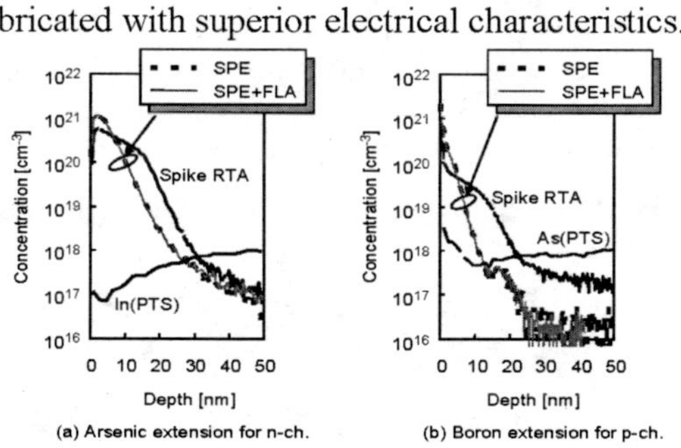

FIGURE 9. Dopant profile at S/D extension region. FLA does not affect the junction profiles fabricated by SPE growth alone. Junction depth is 26nm and 13nm for n-channel (a), and p-channel (b), respectively, in the case of SPE + FLA process.

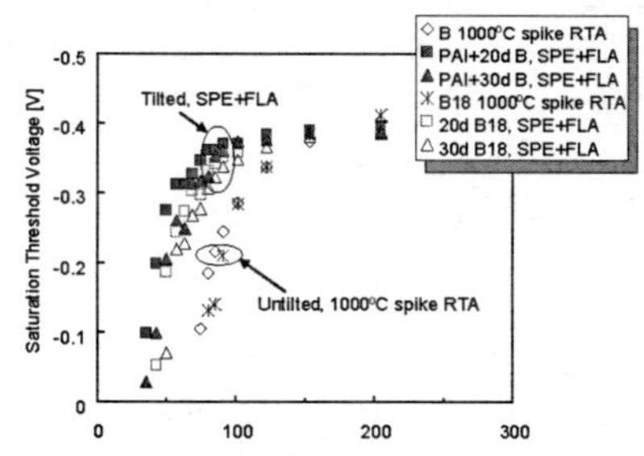

FIGURE 10. Saturation threshold-voltage roll-off of pFETs for the case of W/WFM/HfSiON gate when extension tilt-angles and activation conditions are varied. Threshold voltage roll-off is improved by tilted extension combined with SPE+FLA.

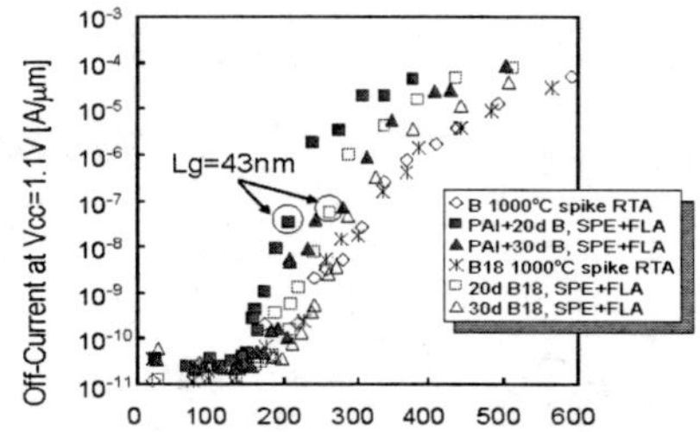

FIGURE 11. Ion-Ioff of pFETs for the case of W/WFM/HfSiON gate when extension tilt-angles and activation conditions are varied, using B18 and B implantation to fabricate drain extensions.

Figure 10 shows the roll-off characteristics of p-MOS for implantation, anneal conditions, and tilt angle, with both Cluster and monomer implantation having good roll-off characteristics. Figure 11 shows the Ion-Ioff's characteristics, for tilted implants cluster implantation shows a 20% improvement in I_{ON} current compared to monomer boron implantation with PAI implant.

CONCLUSION

Cluster ion implantation ($B_{18}H_x^+$) has some advantages of USJ formation beyond 45 nm node devises, which is capable of high activation and fine recrystallization with FLA, compared with $BF2^+$ implantation at SDE condition.

Cluster carbon ion implantation can make high substitutional ratio (2.30%) as SiC formation with FLA, which will be useful as stress engineering for n-MOS instead of epi-process.

We believe that Cluster ion implanter will be possible to facilitate fabrication of Full metal/High-k MOSFET for mass production by tilt Cluster ion implantation ($B_{18}H_x^+$) and FLA.

ACKNOWLEDGEMENTS

This work is supported by Japan Science and Technology Agency (JST) for the equipment development and The New Energy and Industrial Technology Development Organization (NEDO) of Japan for the process development. The authors are grateful to Hiroki Kiyama of Screen for FLA.

REFERENCES

1. T. Aoyama et.al, IWJT(2005) p. 28.
2. Uejima *et al.*, IEDM(2007) p. 151.
3. W. Krull , Proc. of International Conf. on INSIGHT in Semiconductor Device Fabrication, Metrology and Modeling, May 6-9, Napa, USA (2007) p. 189.
4. N. Hamamoto *et.al,* IIT p. 125. (2004)
5. F. Ootsuka *et.al,* IEEE Trans. Ed,**155**(4) p. 1042. (2008)

Investigation of pMOS Device Matching and Characteristics Using $B_{18}H_{22}$ Implantation

Joonho Lee[1], Jaewoong Choi[1], Jungsoo An[1], Seonho Ryu[1], KyungWon Lee[3], Jonghoon Kim[3], GeumJoo Ra[2], Steve Kim[3], and H.T.Cho[4]

[1]*Samsung Electronics Co., Ltd., San#24 Nongseo-dong, Kiheung-ku, Yongin-city, Kyungki-do, Korea*
[2]*Axcelis Technologies, Inc. 108 Cherry Hill Drive, Beverly, MA, United States*
[3]*Axcelis Korea, Ltd., 1024-16, Youngtong-dong, Suwon-city, Kyungki-do, Korea*
[4]*SemEquip, Inc 34 Sullivan Road North Billerica, MA, United States*

Abstract. Implantation of the molecular ion $B_{18}H_{22}^+$ has significant productivity benefits since very high particle current densities are possible with a low electrical beam current. We evaluated several applications of $B_{18}H_{22}$ in sub-70 nm DRAM manufacturing, including polysilicon doping, source/drain doping, and p+ contact resistance reduction. For p+ poly doping, the process control available from molecular implant is up to six times superior than that from plasma doping. We also characterized $B_{18}H_{22}$ in high dose implantation processes (p+ source / drain and p+ contact plug implantation) to improve the productivity of these steps.

Keywords: Molecular boron ($B_{18}H_{22}$), poly doping, BF_2, P+ source and drain, P+ contact plug

INTRODUCTION

Molecular ion implantation has been developed as a new method of forming high dose, low energy implants with high productivity and no energy contamination. Molecular implants of $B_{18}H_{22}$ have the potential to improve productivity on all low energy p-type doping applications. These include polysilicon doping, source/drain doping, and contact plug doping.

EXPERIMENTAL AND RESULTS

We investigated $B_{18}H_{22}$ implants in sub 70 nm devices with a gate poly thickness of 700 Å. Molecular ion implant and plasma doping were performed using the Axcelis Optima Imax and a plasma doping system respectively. The Imax can extract ions up to 80 keV using a conventional beam line equipped with a SemEquip 350 source. All devices received the same PR strip & anneal after implantation. The PR strip was performed by a wet cleaning system. The RTA anneal was 1000°C for 10 seconds. Electrical measurements were performed at the end of line.

Poly Gate Doping

Figure 1 shows the Vt variations at each implant dose and energy conditions, which were measured at the thirteen different channel lengths on a device wafer. The pMOS Vt is inversely proportional to the implant dose and energy of $B_{18}H_{22}$, because the poly depletion is reduced by a deeper junction and higher dose at the same gate voltage. This shows that molecular implantation has wide process windows in energy and dose.

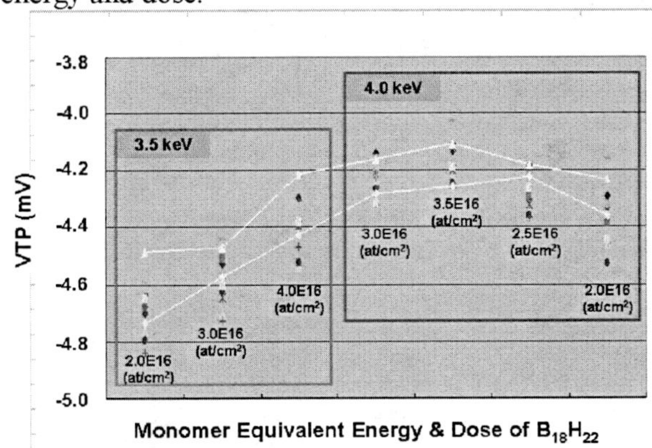

FIGURE 1. Vt variation for each dose and energy.

Threshold voltages were well-behaved with no dopant penetration into the channel in the long channel transistor under all doses and energies investigated (Figure 2). A boron molecular implant at 4 keV could be matched with the process of record (POR) using a much lower dose as compared to plasma doping.

CP1066, *Ion Implantation Technology,* edited by E. G. Seebauer, S. B. Felch, A. Jain, and Y. V. Kondratenko

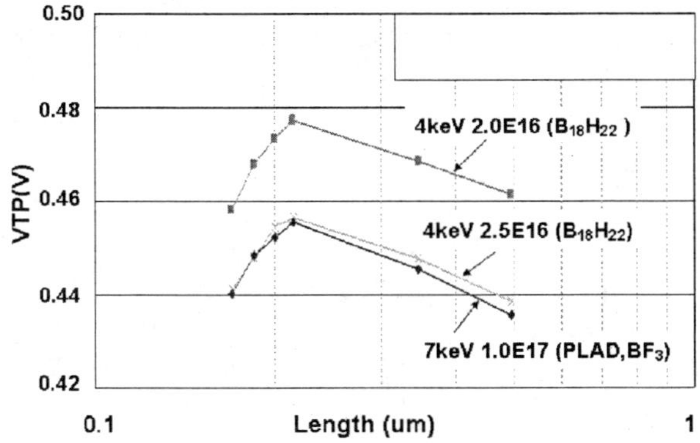

FIGURE 2. PMOS Threshold voltage roll-off characteristics shows that 4keV 2.5E16 at/cm^2 condition is well matched with the POR condition.

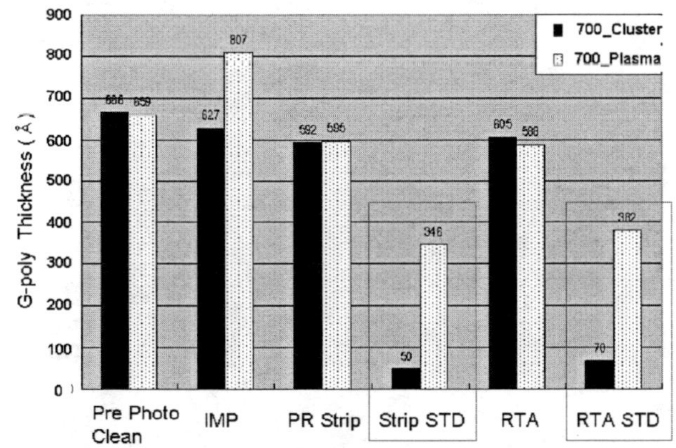

FIGURE 3. The gate poly thickness and within wafer thickness variation (boxes) from molecular implant and plasma doping.

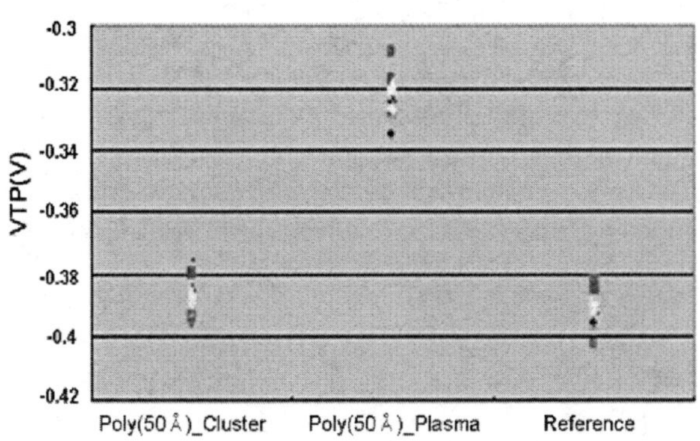

FIGURE 4. Vt comparison and variation for -50 Angstrom gate poly thickness.

Figure 3 shows the poly thickness and within wafer standard deviation of the molecular implant and plasma doping tools. The implanter has better within wafer poly thickness variation than the plasma doping tool after both cleaning and RTA. The within wafer variation of the plasma doping is six times higher than that of the molecular implanter. This is attributed to the boron oxide film deposition caused by the higher surface dopant concentration of plasma doping.

Figure 4 shows the pMOS Vth with a 700 Å gate poly thickness for molecular boron and plasma doping. At the same poly thickness, the threshold voltage of the plasma doping is much lower than the POR (monomer boron), while the threshold voltage of molecular boron is the same as the POR. This shows the molecular boron doping is less dependent on gate poly thickness than plasma doping. Most important, the threshold voltage variation is lower for the molecular implant, because of its superior process control as compared to plasma doping.

Source / Drain Formation

$^{49}BF_2$ has been traditionally used as the dopant for the source and drain. We investigated the possibility of molecular boron as a BF_2 replacement for this implant. At the equivalent dose and energy, BF_2 diffuses less than $B_{18}H_{22}$. Figure 5 shows that molecular boron makes a deeper junction depth at the equivalent energy and dose. This is because of the boron trapping by fluorine in the BF_2 case.

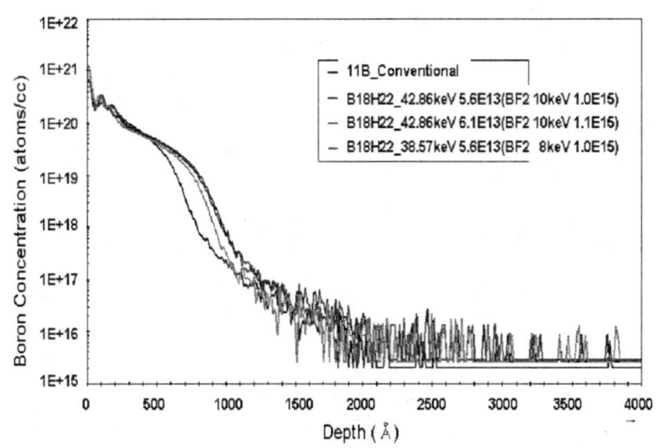

FIGURE 5. Boron SIMS profile of BF_2 vs. $B_{18}H_{22}$.

Two correlation tests were performed. The first was optimizing Vt after targeting the junction breakdown voltage (BV) and the second was optimizing Vt by only increasing the p+ S/D implant dose. This is expected to simultaneously increase BV.

The split conditions for the correlation analysis of Vt and BV are shown in Figure 6 (SIMS profile: BF_2 vs. $B_{18}H_{22}$). Figure 7 shows the electric characteristics of $B_{18}H_{22}$ applied in the P+ source and drain formation implantation process.

As the dose and energy increase, BV increases and the pMOS Vt is decreases due to the deeper junction depth.

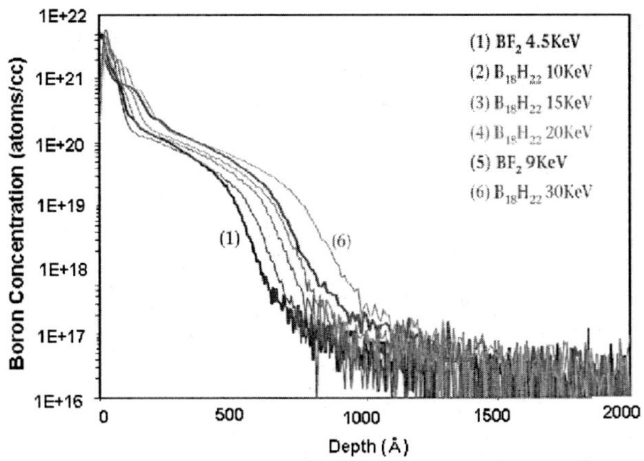

FIGURE 6. Boron ($B_{18}H_{22}$) SIMS at each energy and $^{49}BF_2$ POR condition for S/D implant.

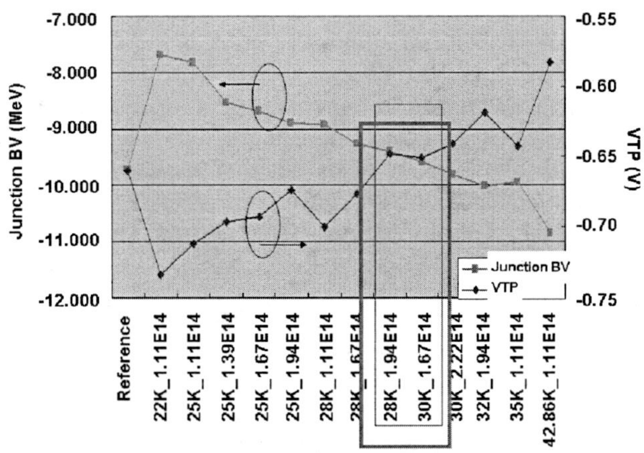

FIGURE 7. Vt and junction BV characteristics

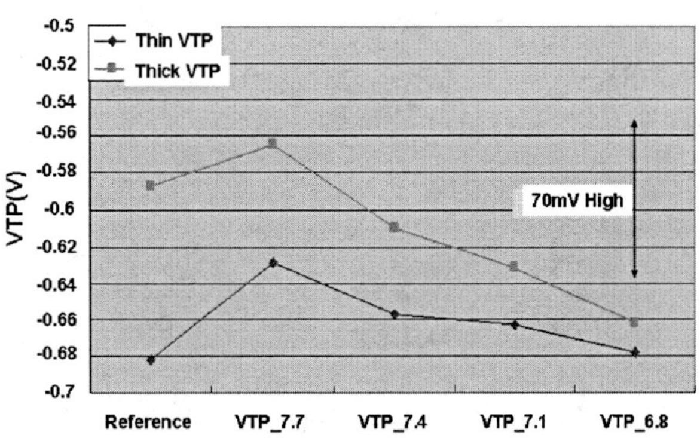

FIGURE 8. Vt at each pMOS adjust implantation dose

Figure 7 shows the proper dose and energy region which meet the targets of BV and thin oxide device Vt There were some problems in matching the thin and thick oxide device threshold voltages simultaneously.

Figure 8 shows the thin and thick pMOS Vt for each pMOS Vt adjust implant dose. By decreasing the Vt dose, we could achieve the thin Vth target but not the thick Vt simultaneously. The thick Vth was 70 mV higher than the reference. Figure 9 shows Vt roll-off characteristics. At the same level of the POR's BV, the L_{EFF} reduction is worse due to $B_{18}H_{22}$'s higher lateral diffusion as compared to BF_2.

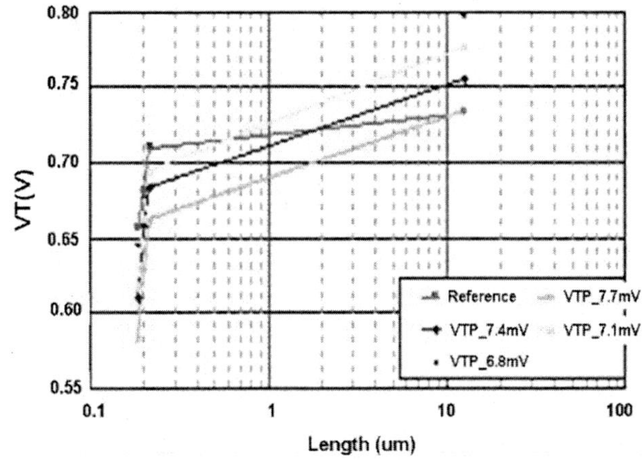

FIGURE 9. Vt roll-off characteristics.

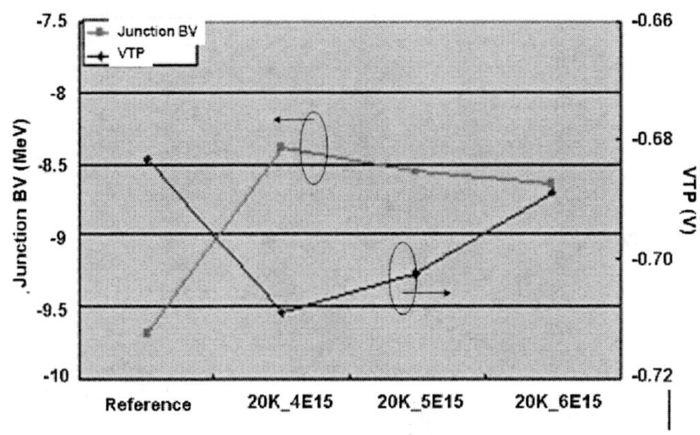

FIGURE 10. Vt and BV in each p+S/D dose.

Figure 10 shows the pMOS Vt & BV at each P+S/D dose implanted with molecular boron. Even though there is no problem matching the Vt, the BV is ~1.0 volt lower than that of the reference condition. In summary, differences in boron diffusivity make differences in the Vt roll off and BV.

Contact Plug Doping

The POR condition of the contact plug implant is BF_2, 35 keV, 4.0E15 atoms/cm^2. The equivalent implant condition of $B_{18}H_{22}$ is 150 keV, 2.22E14 atoms/cm^2. But the maximum extraction energy in the machine was limited to 80 keV, so we compensated by increasing the $B_{18}H_{22}$ dose.

Figure 11 shows that as the implantation energy and dose increase, BV is increases and contact resistance decreases. The energy has a smaller effect than dose on the BV and Rc.

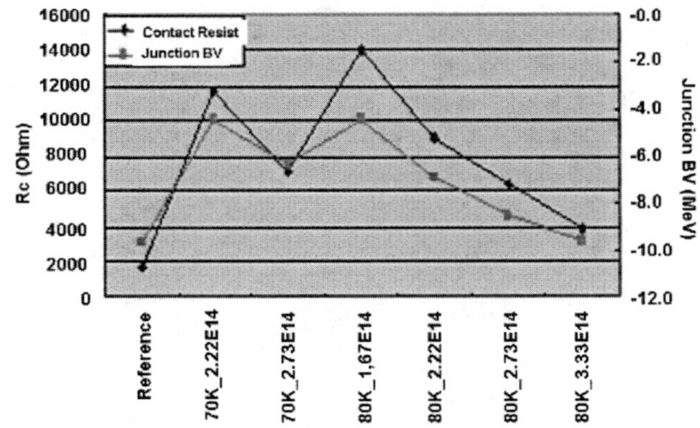

FIGURE 11. The Junction Bv and contact resistance at each $B_{18}H_{22}$ contact plug implantation dose and energy.

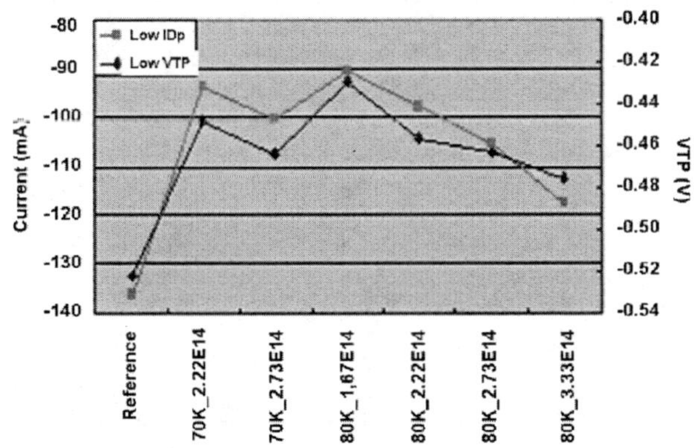

FIGURE 12. The Vt & Idsat in each $B_{18}H_{22}$ contact plug implantation dose and energy.

Figure 12 shows the Vt, and the drive current (I_{ON}) are both lower than the reference. This is caused by high contact resistance. The dose needs to be increased from 3.33E14 to 6.11E14 atoms/cm^2 (BF_2 1.1E16 atoms/cm^2,) to achieve the same contact resistance (Rc) as BF_2.

CONCLUSIONS

We compared molecular boron implantation and plasma doping for poly gate doping, p+ S/D implant, and p+ contact plug implant.

Many semiconductor manufacturers have adopted dual doped poly gates to enhance the performance of pMOS transistor and need precise control and high throughput for the poly doping implant. We have developed a p+ poly gate process with a molecular boron tool (Optima - Imax) for sub 70 nm devices. The molecular boron implant process has advantages which make it superior to plasma doping tools with respect to process control. Significantly, molecular boron implantation is free from the gate poly thickness variation which plagued the plasma doping. We did identify some differences in the electronic characteristics between $B_{18}H_{22}$ and BF_2 with respect to poly thickness variation.

For p+ S/D implant, the source and drain formation of a MOS transistor needs a high dose (over 1.0E15 atoms/cm^2) doping process. Molecular ion implantation is an attractive alternative to conventional ion implantation due to the fact that it can reduce the process time. Molecular boron does diffuse more than BF_2 after anneal. To overcome this difference, carbon or halo implantation should be added or a lower thermal budget used for the post annealing process.

Molecular boron implantation also substitutes well for BF_2 in the high dose contact plug implantation process. The use of molecular boron is expected to shorten the doping process time. Contact resistance has been a critical issue for deep sub micron devices. We found that molecular $B_{18}H_{22}$ implant for contact plug implantation reduces implantation time while improving the contact resistance.

REFERENCES

1. T. Aoyama *et al.*, "Advantageous decaborane ion implantation for ultra-shallow junction of PMOS FETs compared with boron monomer implantation into germanium preamorphized layer," in *Proc. Int. Workshop Junction Technology*, 2006.
2. D. Henke, *et al.*, "P-Type Gate Electrode Formation Using $B_{18}H_{22}$ Ion Implantation," *in Proceedings of the 16th International Conference on Ion Implantation Technology*, 2006.

Emissions Monitoring and Exhaust Analysis for the Axcelis Optima HD IMAX Ion Implanter

Shaun M. Wilson[1], Paul Marganski[1], Joseph Sweeney[1], and Steve Roberge[2]

1. ATMI, 7 Commerce Drive, Danbury, CT 06810.
2. Axcelis Technologies, 108 Cherry Hill Drive, Beverly, MA 01915.

Abstract. As integrated circuit features continue to become smaller, it has become difficult to meet the challenges of shallow junction implant using conventional ion implant dopant species. To meet this challenge, Axcelis has developed the Optima HD IMAX, a new ion implant tool that uses the molecular dopant species octadecaborane (ClusterBoron®, $B_{18}H_{22}$) and also has the capability to perform carbon implants using at least two molecular carbon-based species (ClusterCarbon™, $C_{14}H_{14}$ and $C_{16}H_{10}$). An additional feature of the tool is an integrated chemical clean based on NF_3 remote plasma technology. The use of new materials in this tool has generated the need for understanding the potential emissions and release vectors of the process materials and their byproducts. Presented here are data on exhaust emissions obtained using Fourier Transform Infrared (FTIR) spectroscopy. Additionally, FTIR spectroscopy has been used to measure concentration levels of residual species that might be found in the source and process chambers during maintenance operations.

Keywords: implanter, FTIR, exhaust, emissions, octadecaborane, ClusterBoron®, ClusterCarbon™.
PACS: 52.77.Dq and 78.30.-j

INTRODUCTION

The most common atomic dopant species used in implantation of silicon are arsenic, phosphorous, and boron, and to a lesser extent carbon, germanium, antimony, and indium. Many of these species are delivered to the implanter source as gases in their hydrogenated or halogenated forms. The hazardous nature of most of the dopants used in ion implanters leads to various safety concerns regarding their toxicity, corrosivity, and ability for emission control systems to efficiently abate any unused dopant or dopant precursor. Past studies describing the type and concentration of gas-phase effluents from the implantation process have focused on the aforementioned dopants. Arnó *et al.* [1] first reported the characterization of the exhaust from an Applied Materials xR-80 implanter during implantation using AsH_3, PH_3, BF_3, and SiF_4. The study measured the make up and concentration of gas-phase species from roughing and cryo pumps using Fourier Transform Infrared (FTIR) spectroscopy. Goolsby and coworkers [2] used the same methods to analyze the exhaust of the roughing pump of a high energy implanter during implantation using arsine and phosphine. Later, Arnó and [3] coworkers performed further work on an Axcelis GSD-VHE ion implanter running AsH_3, PH_3, and BF_3.

As integrated circuit features become smaller and smaller with each new process node, the need has arisen for new materials to be utilized in the implantation process. For smaller and more advanced transistors, conventional implanters using BF_3 as the source material suffer from poor productivity at the low implant energies required for advanced devices. To address this Axcelis Technologies has developed the Optima HD IMAX, a new ion implant tool that utilizes the molecular dopant species octadecaborane (ClusterBoron®) and also has the capability to perform carbon implants using two different molecular based carbon species (ClusterCarbon™). This tool also has a built in chemical cleaning process using remote NF_3 plasma technology.

With the advent of this new technology and these new dopant species, there comes a need to understand the potential emissions and release vectors of the process materials and their byproducts. To that end this paper focuses on the characterization of any gas phase species present in the implanter's exhaust system. The testing done at the OEM site in Beverly, MA also included monitoring of the effluent after being abated by the house scrubbing systems as well as measurement of any residual species that were

CP1066, *Ion Implantation Technology,* edited by E. G. Seebauer, S. B. Felch, A. Jain, and Y. V. Kondratenko

found in the source and process chambers during routine preventative maintenance activities.

EXPERIMENTAL

The tool used in these studies was an Axcelis Optima HD IMAX running the molecular dopant species ClusterBoron® and ClusterCarbon™. A schematic of the tool is shown in Figure 1. Various sampling ports before and after the roughing pumps (RP1 and RP100) and after the house scrubbing system were used to sample the exhaust. RP1 was utilized during the NF_3 plasma cleans, while RP100 was used during all of the implant recipes for ClusterBoron® and ClusterCarbon™.

A sampling pump was used to continuously draw sample effluent into the gas cell of the spectrometer while the temperature and pressure of the gas cell were kept constant during experimentation to allow for quantification of all species present in the exhaust stream. Mass flow controllers were also used to measure the flow rate of process gas through the pumping system. These flow rate values were then used to calculate the total mass of different species released during the implant process and the NF_3 cleaning procedure.

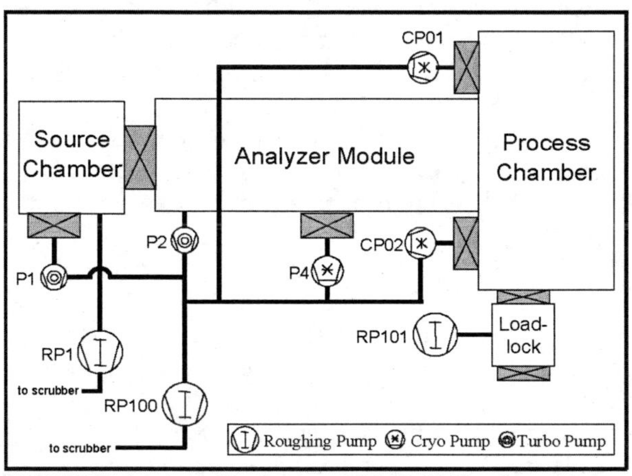

FIGURE 1. Axcelis Optima HD IMAX schematic showing the pumping system and vacuum chambers.

The analytical instrument used was a Midac I2000 FTIR spectrometer outfitted with a 10m path-length multi-pass gas cell. This instrument has the ability to measure in a continuous fashion and was configured to take a new sample approximately every 30 seconds. Spectra were taken at a resolution of 0.5cm^{-1} over the region from 650 to 4000cm^{-1}. Calibration spectra of known concentrations and path-lengths were used to quantify the species present in the exhaust system before sending the sampled effluent to the house scrubbing system.

RESULTS AND DISCUSSION

Emissions During ClusterBoron® Processes

Each experiment conducted began with the FTIR cell evacuated so that a background spectrum could be obtained. Process gas was then allowed to flow through the gas cell from one of several monitoring points while different dopant recipes were run. As seen in Table 1, during ClusterBoron® implantation both BF_3 and CF_4 were observed in the exhaust system despite the lack of any fluoride based dopants being used in the process. The concentration of both species seems to be dependent on the beam current and beam energy of $B_{18}H_{22}$ being implanted; the gentler the beam conditions, the less likely any BF_3 or CF_4 will be present in the exhaust. The very low levels of both compounds are most likely due to residual fluoride species remaining in the system from an NF_3 plasma cleaning procedure that was performed prior to running the octadecaborane recipes. Post scrubber monitoring showed no presence of BF_3 indicating the scrubber was properly abating the corrosive fluoride.

Of great importance from an environmental, health, and safety standpoint is the lack of any borohydride species in the FTIR spectra. Shown in Figure 2 are the results from monitoring pre RP100 for the maximum beam current recipe of $B_{18}H_{22}$. For comparison purposes a spectrum of pure pentaborane is also shown. Borohydride species including diborane, pentaborane, decaborane, and octadecaborane have strong B-H stretching bands in the 2500 to 2700cm^{-1} region of the infrared spectrum. The lack of any features in this region of the spectrum indicates that none of these compounds are present in the exhaust system.

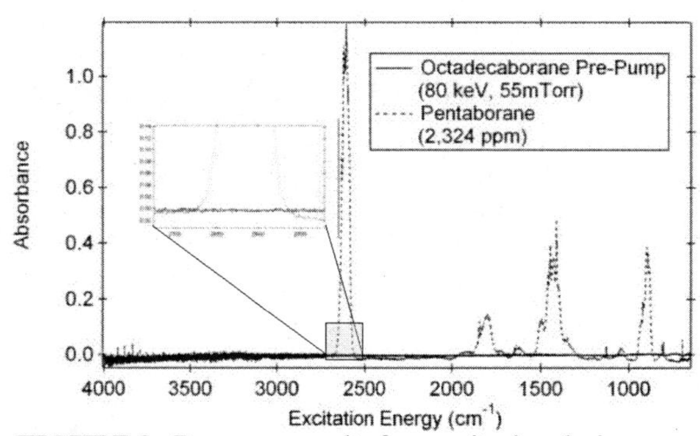

FIGURE 2. Pre pump results for monitoring during ClusterBoron® implants.

Emissions During ClusterCarbon™ Processes

The results for the monitoring performed while ClusterCarbon™ was being implanted are given below in Table 2. Two different molecular based carbon

Table 1: Characterization of Effluent During $B_{18}H_{22}$ Implants																
Delivery Pressure (mTorr)	Extraction Energy (keV)	$B_{18}H_x$ Beam Current (mA)	Average Pre-Scrubber BF_3 Concentration (ppm)			Average Post-Scrubber BF_3 Concentration (ppm)			Average Pre-Scrubber CF_4 Concentration (ppm)			Average Post-Scrubber CF_4 Concentration (ppm)		
55	80.0	High	1.39	±	0.24	ND	±	N/A	0.316	±	0.012	0.169	±	0.007
37	38.2	|	1.26	±	0.23	ND	±	N/A	0.230	±	0.009	0.134	±	0.005
44	5.73	↓	0.113	±	0.120	ND	±	N/A	0.013	±	0.002	0.010	±	0.001
45	2.86	Low	N/A	±	N/A	ND	±	N/A	ND	±	N/A	ND	±	N/A

Table 2: Characterization of Effluent During Cluster Carbon Implants																
Starting Material	Species Implanted	Delivery Pressure (mTorr)	Extraction Energy (keV)	Average Pre-Scrubber C_2H_2 Concentration (ppm)			Average Post-Scrubber C_2H_2 Concentration (ppm)			Average Pre-Scrubber CF_4 Concentration (ppm)			Average Post-Scrubber CF_4 Concentration (ppm)		
$C_{14}H_{14}$	C_7H_x	35	7.58	7.91	±	0.43	2.27	±	0.13	0.17	±	0.01	0.18	±	0.01
$C_{16}H_{10}$	$C_{16}H_x$	31	60.0	2.33	±	0.15	ND	±	N/A	ND	±	N/A	ND	±	N/A

species were used as precursors ($C_{14}H_{14}$ and $C_{16}H_{10}$). As was the case with ClusterBoron®, there were minimal species present in the exhaust system during ClusterCarbon™ implants. Acetylene was the main effluent observed when both $C_{14}H_{14}$ and $C_{16}H_{10}$ were used as the starting material. C-C stretching vibrations were observed in the post pump spectra during the $C_{14}H_{14}$ runs but were absent during the $C_{16}H_{10}$ runs. These vibrations have been attributed to unidentified linear carbon chain species. Finally, the extremely small presence of CF_4 is likely due to residual fluoride species remaining in the system from an earlier NF_3 cleaning procedure.

Emissions During NF_3 Plasma Cleaning

Extensive monitoring was also performed during an NF_3 plasma cleaning cycle that lasted for approximately 30 minutes. During this cleaning procedure NF_3 was flowed at a rate of 1 slpm into a plasma generator and was then injected into the source chamber of the implanter. The fluoride species present in the plasma then react with any metal (i.e. B) deposits present in the source. Periodic removal of these metal deposits is necessary in order to reduce glitch rates and reduce the amount of time between preventative maintenance routines.

Table 3 summarizes the species found in the exhaust stream during the cleaning process. Not surprisingly BF_3 is the main component present. Since the total flow through the exhaust system is known (via measurement with an MFC) and the BF_3 concentration of that exhaust stream is also known, a total volume and/or mass of BF_3 exiting the implanter can be calculated. Over the first 10 minutes of the cleaning cycle BF_3 was entering the scrubber at a rate of 70.4 mg/min. As demonstrated by Figure 3, as the cleaning cycle continued the concentration of BF_3 in the exhaust system dropped by approximately 33%. Note that post pump data was not collected during the middle portion of the NF_3 clean since the post scrubber data was being taken at that time.

By examining the last column in Table 3 the efficiency of the scrubbing system can be evaluated. Greater than 99% abatement of the three toxic gases present in the exhaust stream was achieved. Non-toxic CF_4 and NF_3 were most likely temporarily physi-adsorbed to the scrubber resin and eventually released. By integrating the mass exiting the scrubber over the entire cleaning cycle it was found that during the 27 minute process ~1160mg of BF_3 and hence ~185mg of B were removed from the source chamber. Considering the ~1.5g/hour usage rate of octadecaborane rate during normal tool operation, this accounts for less than 2% of the boron input into the source. This calculation was done assuming the tool

Table 3: Characterization of Effluent During NF_3 Cleaning													
Effluent	Average Pre-Scrubber Concentration (ppm)			Volume Entering Scrubber (cm^3/min)	Mass Entering Scrubber (mg/min)	Average Post-Scrubber Concentration (ppm)			Volume Exiting Scrubber (cm^3/min)	Mass Exiting Scrubber (mg/min)	Abatement Efficiency (%)		
BF_3	4305.7	±	54.3	25.4	70.4	1.9	±	0.05	0.011	0.030	99.96	±	0.001
SiF_4	139.1	±	6.4	0.8	3.5	1.2	±	0.3	0.007	0.030	99.14	±	0.219
COF_2	115.6	±	0.6	0.7	1.8	ND	±	N/A	ND	N/A	100.0	±	N/A
CF_4	114.9	±	10.3	0.7	2.4	14.9	±	0.7	0.088	0.317	87.03	±	1.312
NF_3	512.2	±	13.7	3.0	8.8	60.6	±	0.6	0.358	1.038	88.17	±	0.337

Note: Average concentrations given above were calculated based upon the first ~10 minutes of the cleaning procedure.

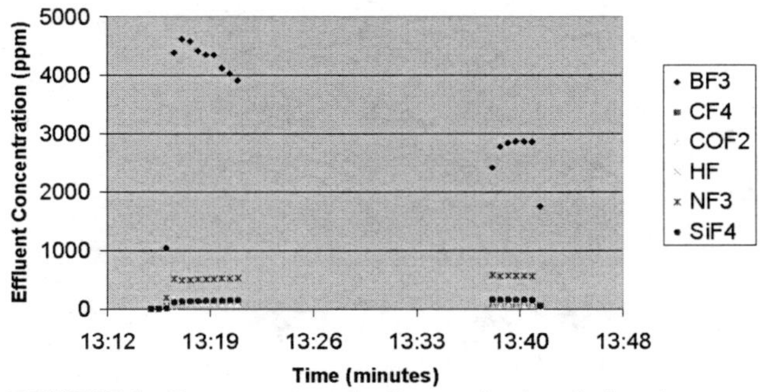

FIGURE 3. Post pump results for monitoring during the NF_3 plasma cleaning procedure.

was run for 8 hours before the cleaning cycle was performed. Given the small portion of source material actually implanted into a wafer (which is true for all dopant species) and the lack of borohydride species in the exhaust stream during processing, this data illustrates the deposition of octadecaborane and byproducts throughout the vacuum system of the implanter. Deposition of octadecaborane and byproducts make it necessary for the Optima HD IMAX to incorporate additional features to control the location of deposition in vacuum forelines and to protect personnel from deposit vapors during maintenance operations.

The final experiments performed were to monitor the exhaust system during cryo pump regeneration and to monitor the source and process chambers of the implanter when opened for preventative maintenance routines. Sampling of the cryo pump exhausts was done in the same manner as the experiments previously discussed. The only species found during these tests were CO, CO_2, and H_2O, and to a lesser extent C_2H_2 and CF_4. A sampling line was then placed at the entrance of the source and process chambers and any process gases present were pulled through the FTIR cell using a sampling pump. The only species found in the FTIR spectra during these tests were atmospheric H_2O and CO_2. However, a Honeywell MDA toxic gas monitor (TLD-1) calibrated for diborane did detect 114ppb in the process chamber. This concentration of diborane is greater than the ACGIH Threshold Limit Value (TLV) of 100ppb, though below the detection limit of the FTIR. Maintenance personnel must ensure that local exhaust ventilation or supplied air respirators are utilized to prevent exposure to borohydride vapors when accessing the source, beam line, process chamber, vacuum forelines, and exhaust ductwork.

CONCLUSIONS

Process effluents from the Axcelis Optima HD IMAX were fully analyzed using FTIR spectroscopy. This characterization was achieved for ClusterBoron® and ClusterCarbon™ implant recipes as well as an NF_3 plasma cleaning procedure. At no time during the monitoring was any diborane, pentaborane, decaborane, or octadecaborane observed in the pumping system. All toxic effluents present were also found to be abated by the dry scrubbing systems currently employed on the tool. Because maintenance procedures at or near TLV concentrations could not be characterized in this case using FTIR spectroscopy, users should continue to proceed with caution during maintenance activities.

REFERENCES

1. J. Arnó, M. Hayes, W.J. Mecouch, W. Boyd, M.J. Rendon, T. Romig, *Solid State Technology*, **41**, 113 (1998).
2. B. Goolsby, V. Vartanian, M. Rendon, M. Wagner, and L. Mendecino. *Electrochemical Society Proceedings*. Vol. 2001-6, pp 59-67 (2001).
3. J. Arnó, J. Sweeney, P. Marganski, R. Faller, S. Roberge, and M.C. Dolan, *IIT2002 Proceedings*, pp 475-478 (2002).

PLASMA IMMERSION

Conformal Doping of FINFETs: a Fabrication and Metrology Challenge

W. Vandervorst[1,2], J. L. Everaert[1], E. Rosseel[1], M. Jurczak[1], T. Hoffman[1], P. Eyben[1], J. Mody[1,2], G. Zschätzsch [1,2], S. Koelling[1,2], M. Gilbert[1], T. Poon[3], J. del Agua Borniquel[3], M. Foad[3], R. Duffy[4], B. J. Pawlak[4]

[1] *Imec, Kapeldreef 75, B-3001, Leuven Belgium*
[2] *Instituut voor Kern- and Stralings Fysika, K.U.Leuven, Leuven Belgium*
[3] *Applied Materials, 974 E. Arques Avenue - Sunnyvale – CA 94087, USA*
[4] *NXP-TSMC Research Center, Kapeldreef 75,B-3001 Leuven, Belgium*

Abstract. Whereas the introduction of 3D-dimensional devices such as FINFETs may be a solution for next generation technologies, they do represent significant challenges with respect to the doping strategies and the junction characterization. Aiming at a conformal doping, classical beam implants fail due to the differences in impact angle, ion incorporation efficiency and the effect of wafer rotation. Moreover shadowing represents an additional limitation for larger tilt angles when considering closely spaced FINs. Plasma immersion doping is an alternative approach which holds the promise of conformality but is also quite challenging and relies on secondary processes such as resputtering, deposition and in diffusion etc. Its implementation is compromised by concurrent artifacts, sputter erosion being the most important one. In support of these developments the measurement of the 3D-dopant distribution and the identification of conformality is essential requiring adequate metrology such as Scanning Spreading Resistance Microscopy, SIMS through FINs, resistors and the Tomographic Atomprobe.

Keywords: FINFET, conformal doping, plasma doping, SSRM, ion implantation
PACS: 52.77 Dq, 85.40.Ry, 81.70.Jb, 79.20.Rf

INTRODUCTION

For meeting the ITRS goals for the devices in the 32nm technology node and beyond, FinFET-based structures have shown immense potential due to their scalability by maintaining high drive current at scaled voltages and smaller gate dimensions, and in particular by their beneficial short channel effects and leakage control [1]. As in planar devices, the performances of these FinFET-based structures are set by the spatial distribution of the electrically active dopants (carriers) thereby controlling the gate – source/drain overlap and the series resistance of the source/drain regions. Due to the 3D-geometry of FINFETs and the need to obtain identical lateral dopant profiles on the top and the sidewall of the FINs, the classical doping strategies might not be applicable and alternative doping strategies such as plasma immersion doping may represent an interesting alternative. Although shown to be conformal for macroscopic features, the latter is less obvious on nanoscale objects with dimensions smaller than the plasma sheath thickness [2]. Hence their implementation still requires an extensive exploration of the parameter space in order to arrive at a high degree of conformality. The 3D-nature of FINFETs also implies that the development of these processes requires the availability of methods to probe doping conformality in a very efficient way.

The present paper describes the efforts towards conformal FINFET-doping using beam line implantation and plasma immersion doping as well as the development of metrology concepts used for its diagnostics. The results are restricted to the case of SOI-based FINFETs.

JUNCTION PROPERTIES AND METROLOGY: PLANAR VERSUS 3D-JUNCTIONS

The development of concepts for junction formation and characterization for planar transistors has been dominated for a very long time by a simple metric i.e. sheet resistance (R_S) versus junction depth

CP1066, *Ion Implantation Technology*, edited by E. G. Seebauer, S. B. Felch, A. Jain, and Y. V. Kondratenko

(X_j) [3]. Sheet resistance was typically obtained from a four point probe measurement and the junction depth extracted from a SIMS-profile. With the advent of shallower junctions, the Rs-measurement encountered problems related to probe penetration and alternative methods based on optical probes (RsL) or micromachined probes were introduced as well [3,4]. At the same time one started to recognize that the 1D-metric (Rs – Xj) did not catch the complex two-dimensional interactions (such as the interplay between extension and halo implants, lateral diffusion etc..) and additional 2D-profiling became necessary. In view of the required spatial resolution (1-2 nm) combined with the necessary sensitivity, 2D-profiling has been explored intensively and a few metrology concepts have emerged as being really useful [3-7]. At present Electron holography provides the requested spatial resolution for junction delineation [5] but suffers from a limitation in concentration sensitivity whereas Scanning Spreading Resistance Microscopy (SSRM) has emerged as the sole scanning probe based method combining sub-nm spatial resolution with adequate concentration sensitivity and quantification accuracy [6]. The complexity of junction engineering for next generation technologies has introduced additional metrics such as the lateral junction depth (source - gate overlap) and finally the lateral profile steepness [8]. Both can be extracted from lateral profiles as measured with SSRM. Cross correlation with TCAD and device characteristics provides essential feedback for further optimization of the process parameters [9].

In a (SOI)-FINFET the situation is different as the vertical junction depth is no longer a relevant parameter (but rather set by the Si-thickness) and only the lateral profile characteristics (gate overlap, lateral steepness) will determine the transistor operation. Rs is still important in the sense that the (external) FIN resistance in the source/drain region may limit the maximum drive current. Obviously fine tuning junctions based on Rs-Xj plots, is no longer relevant.

An additional parameter to consider is the requirement for doping conformality. The latter addresses the need to have identical doping characteristics on the top and the side wall of the FIN. Indeed differences in lateral extent (overlap) and steepness (top vs sidewall) will translate in differences in turn-on and current flow on the top and the sidewall of the FIN and thus less ideal operation [10]. From a metrology point of view, this is a very difficult measurement since it requires a full 3D-analysis (with nm-resolution) to probe for instance the source/gate overlap on the sidewalls and the top of the FIN.

An intermediate solution reducing the metrology requirements from 3D- towards 2D-resolution, can be found by looking at the degree of conformality on the cross section of a source/drain region (Figure 1). The conformal doping of the Fin in this region, is indeed the first step towards a fully conformal 3D-doping

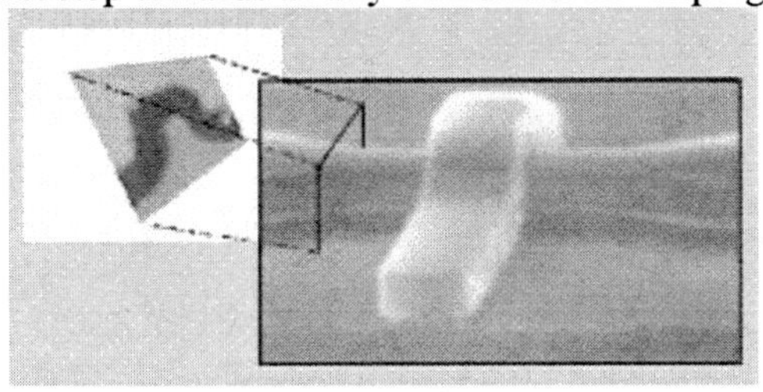

FIGURE 1. Conformal doping profile (as measured with SSRM) in the source/drain region of a FINFET.

profile, as one could expect that full 3D-conformal diffusion under the gate can only be achieved when the region outside the gate is conformally doped.

The summary presented in Table I clearly indicates that there are substantial differences between the junction properties for Planar versus FINFETs and as such, also in the required metrology.

Table I. Junction and metrology characteristics

PLANAR	*FINFET*
Junction properties	
Doping level	
Rs	Rs
Junction depth	
Vertical : Xj	Vertical : NON relevant
Lateral : X_l	Lateral : X_l
Profile steepness	
Vertical	Vertical : NON relevant
Lateral	Lateral
Conformality	
NON relevant	3D requirement
Metrology	
1D (Rs, SIMS)	NON relevant
2D (EH, SSRM)	2D (SSRM)
	3D (Atomprobe)

METROLOGY FOR FINFET JUNCTIONS

As already indicated in section 1, metrology for FINFETs deviates from the standard approaches for planar junctions in the sense that the 3D-geometry needs to be considered as well. One of the most important parameters to be measured is the degree of conformality which can be done (in theory) using the Tomographic Atomprobe [11, 12] or 3D-SSRM [13] (see further). Both methods are still in their early development phase and far too slow to provide the rapid feedback necessary for process development and optimization. Hence several concepts have emerged looking at less complicated approaches to probe the degree of conformality. Rather than tackling the 3D-problem they focus on probing the conformality of the source/drain region as a first step in the process development. The full range of metrology concepts for this study are : SIMS through FINs [14], Resistors [15], X- and 3D-SSRM [6, 13] and 3D-Atomprobe [11,12].

SIMS Through FINs

In view of the limited spatial resolution of SIMS relative to the dimensions of the FINs and the extent of the doping profile on the sidewalls, a regular SIMS profile on this 3D-structure is not possible. Moreover under ion beam erosion the 3D-structure will rapidly deform smearing out all relevant spatial resolution [16]. Therefore an alternative SIMS concept was developed (SIMS through FINs) which provides, at least, information on the sidewall dose. The concept uses an approach explored earlier when attempting 2D-profiling with SIMS [17]. In the present application the space in between an array of FINs which have been exposed to the doping/anneal process, is filled with undoped polysilicon and subsequently planarized by polishing (Figure 2a). When

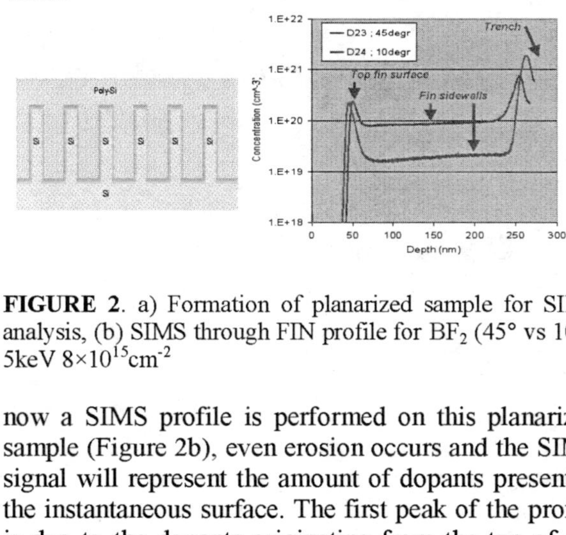

FIGURE 2. a) Formation of planarized sample for SIMS analysis, (b) SIMS through FIN profile for BF_2 (45° vs 10°), 5keV $8\times10^{15}cm^{-2}$

now a SIMS profile is performed on this planarized sample (Figure 2b), even erosion occurs and the SIMS signal will represent the amount of dopants present at the instantaneous surface. The first peak of the profile is due to the dopants originating from the top of the FIN whereas in the flat part of the SIMS profile, dopant atoms originate from the sidewall regions. The third peak relates to the region doped in between the FINs. When converting the SIMS signal to concentrations using the regular relative sensitivity factor approach, the following equation applies :

*Side wall dose (at/cm^2) = Sims concentration (at/cm^3) * FIN pitch*

The example shown in Figure 2b relates to the incorporation of B following a BF_2-implant (5 keV, 8 10^{15} at/cm^{-2}) and illustrates quite clearly the angular dependent incorporation efficiency and reduced dose (cfr section on junction fabrication). The advantage of this concept is that it provides information on the chemical dopant distribution whereas the other concepts discussed further, are more related to the active carrier fraction. Thus a comparison between both will lead to information on the activation behavior as well. It is also clear that no information on the under diffusion under the gate can be extracted from this concept.

Resistors

A limitation of most methods is that they are based on manual measurements which makes it very difficult to do full wafer analysis, sample within wafer non-uniformity and explore more extensive designs of experiments. Again for probing the degree of sidewall doping or conformality, a concept based on resistors has been proposed recently [15]. Basically a FIN is made out of polysilicon (on SOI) and partially covered with a hardmask. During the doping step the hard

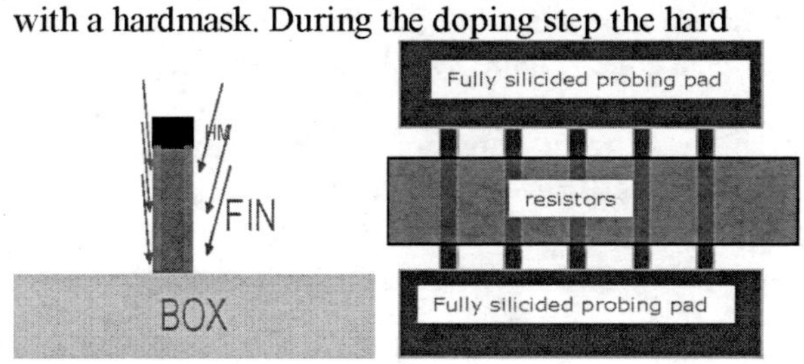

FIGURE 3. a) FIN covered with hardmask limiting the dopant incorporation to the sidewall process.

b) fully processed structure with silicided probing pads and non-silicided regions forming the resistors

mask will shield the top from being doped and only the sidewall gets exposed to the doping step. After annealing the entire region outside the hard mask region is fully silicided thereby creating a resistor whose value is proportional to the sidewall dose incorporated below the hard mask. When compared to the resistor value when no hard mask is used (i.e. when top and sidewall are exposed simultaneously), information on the sidewall doping efficiency can be extracted. Since the side wall represents a much larger area than the top, one should ideally reach a situation where the resistor value with or without a hard mask is almost the same. Being a full wafer process with dedicated test pads for automated measurements, statistical information (such as within wafer uniformity, wafer to wafer uniformity,..) on the sidewall doping process can be obtained.

Scanning Spreading Resistance Microscopy

Whereas the resistor approach provides information on the active dopants on the sidewall, no detailed information is present on their spatial distribution. SSRM is a scanning probe based method sampling the local carrier concentration with sub-nm spatial resolution [6]. It has already been applied to probe 2D-carrier distributions in planar transistors [6,8,18], Si-FINs [19] and thus can equally well be used to study the junction formation and conformality in the source/drain regions of the FINFET (Figure 1). The value of SSRM is that it not only provides information on the junction location but equally well on the distribution of the carriers and their respective concentrations. Recently a concept for 3D-SSRM has been proposed aimed at sampling not only the carrier profile in the source/drain region but equally well the under diffusion below the gate. Details are discussed elsewhere [13] but suffices it to say that the method relies on a (dedicated) array of staggered transistors such that each cross sectional image corresponds to a section taken at a (slightly) different position relative to the gate edge (Figure 4). The spatial resolution (i.e. the distance between two of those sections) is set by the ability to form transistors whereby the gates are staggered over a very small distance (1, 2 nm?) relative to each other. An initial experiment with a 50nm spacing has demonstrated the feasibility of the concept (Figure 4b). The results in Figure 4b show the carrier

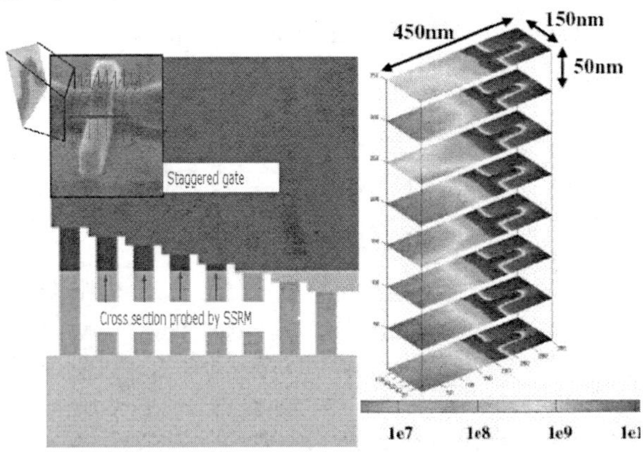

FIGURE 4. (a) Basic concept of 3D-SSRM. SSRM –images are taken on the cross sections of staggered transistors approximating the sectioning of one single transistor (cfr insert), (b) SSRM images taken on identical FINs with 50 nm staggering.

distribution in a series of identical FINs whereby the sampling planes are spaced with 50 nm [13]. Repeatability across the entire dopant profile is in the order of 15%. A refined structure with 1 nm and 3 nm spacing is presently under preparation.

Tomographic Atomprobe

Although the previous concepts provide a wealth of information, true 3D-information on one (individual) FINFET can only be obtained by methods like the Tomographic Atomprobe (AP) which have the intrinsic 3D-spatial resolution [11, 12]. The basic 3D-resolution originates from the atom by atom evaporation from a needle shaped specimen under the combined action of a very high electrical field and a short laser pulse. Atoms are sorted to their mass and spatial position with a proclaimed spatial resolution of 0.5 nm and a depth resolution of 0.2 nm. The resolution can be inferred from Figure 5a which shows the the location of the individual Si-atoms from a <111>-Si-substrate and their position onto the lattice planes. The main limitation of the AP lies in the very limited counting statistics linked to the small analyzed volume [3]. This becomes clearly visible in Figure 5b where we superimpose a 20nm gate on a Si-substrate doped with B. Each B-atoms shown in Figure 5b, corresponds roughly to an equivalent concentration of $\sim10^{19}$ at/cm^3. Hence accurate analysis of dopant profiles in FINs will face

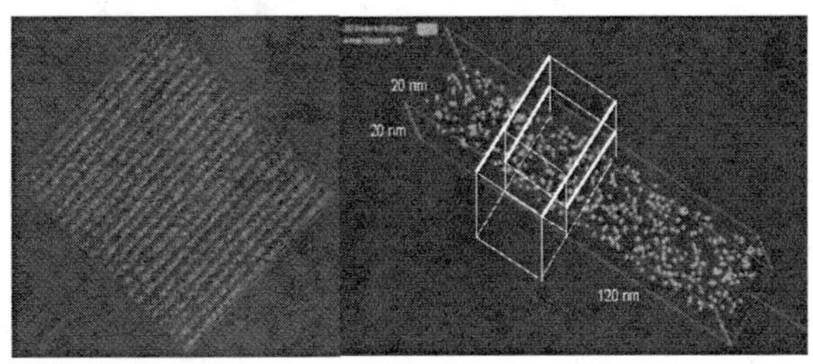

FIGURE 5. (a) Si-atoms and their location on the lattice planes of Si (Imec), (b) schematic view of 20 nm gate on Si-substrate doped with B ($\sim10^{19}$at/cm^3) (only B-atoms are shown for clarity) ([20]).

counting limitations and may require suitable averaging over multiple samples in order to obtain statistically meaning full results on junction depths and profile abruptness. Although the techniques is very appealing in view of its 3D-analysis capability, and presently is experiencing a lot of interest from the semiconductor community [21,22], still many practical problems related to the details of the sample preparation, the analysis conditions and 3D-data reconstruction (local magnification effects) need to be resolved before it will become a routine method. One important advantage is that it does not suffer from the matrix effects like SIMS, when performing multilayer analysis such as metal gate-high k stacks [22].

JUNCTION FORMATION

Ion Implantation

Classical beam line implantation is a very well established method providing unique control of implanted species, dose, wafer uniformity and repeatability. However translating ion implantation technology solutions from planar to FINFET junction formation is not so straightforward due to the (confined) 3D-structure. For instance as FINs form inherently a 3D-structure which will be closely spaced, a serious limitation on the tilt angle (angles <10-20°) is imposed due to shadowing induced by the proximity of nearby FINs. It is also important to note that diffusion in a small volume is not necessarily the same as diffusion in a thin film. The proximity of a large surface actually influences the interstitial concentration and leads to a reduced diffusion for B in the FIN-edges [19].

The isolation from a larger crystalline volume also leads to problems not seen in planar devices. The most notorious one is the difficulty in recrystallizing the FIN after amorphization (such as after a BF_2 or As implant) [23]. If completely amorphized, only a very small seed for recrystallization is present under the gate and defective (lateral, polycrystalline) growth can occur. Moreover even if a small seed is left at the bottom of the fin, recrystallization will still be hindered by differences in regrowth velocities in the bulk of the fin and its (sidewall) surface which induces a high density of twin defects. The latter impacts on the FIN resistivity and limits the drive current for narrow FINs [24].

In terms of conformality ion implantation is a poor method as the ion beam hits the sidewall and the top of the FIN under very different angles. Typical values will be around 10-20° for the top versus 80-70° for the sidewall. Under those conditions the ion incorporation efficiency becomes very different due to the effects of particle reflection, selfsputtering etc which are more pronounced for light elements at low energy bombardment conditions [15]. This is shown in Figure 6 using simple SRIM-calculations. An additional reduction follows from the geometrical effect related to the different angles relative to the beam. Indeed when the impact angle for the top is close to normal, the one for the sidewall will be close to 90° and thus the sidewall will receive a much smaller dose because of the redistribution of the (normal) beam density over the inclined surface area.

Finally the need to dope the FIN on both sides necessitates a wafer rotation halving the dose on each sidewall relative to the top dose. Due to these

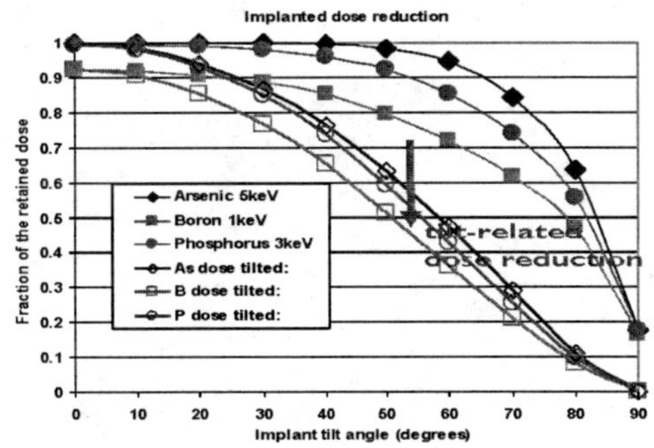

FIGURE 6. Calculated retention efficiency for various implant conditions (using SRIM) (solid symbols). The effect is magnified by the dose reduction due to the geometrical effect of tilting. Wafer rotation is not yet included and will add a factor 2 top versus sidewall.

phenomena the sidewall dose can be as small as 10% of the dose at the top of the FIN, leading to a strong non-conformality and thus poor device performance. Although it has been shown that these effects (and their impact on device performance) can be reduced by implanting at larger tilt angles (~ 45°), this is only practical for isolated devices and not in a dense array where shadowing limits the tilt angle to 10-20°.

Plasma immersion doping

Whereas ion implantation fails for conformal doping due to the directionality of the ion beam, one may wonder whether plasma immersion doping (PID) can provide an adequate solution. At least for macroscopic objects (relative to the sheath dimensions) conformal implantation of irregular shaped objects has been achieved [2]. Unfortunately due to the small dimensions of the FINFETs (40-60 nm height on top of a large planar surface) conformal doping can not be expected when considering only the electrostatic configuration [24]. This is confirmed by numerous SSRM-images showing a dramatic difference between top and sidewall doping (Figure 7c). At the same time some images (different processing conditions) reveal conformal doping (Figure 7a) [26]. Hence alternative mechanisms leading to conformal doping beyond simple electrostatic considerations need to be considered. The SIMS profiles (Figure 7) related to these SSRM-images are identical as well as their sheet resistance values (~450 Ohm/sq).

The obvious case to achieve conformal doping is through the transition from the implantation mode towards deposition mode by lowering the acceleration bias drastically (to zero?) and/or primarily using neutral atoms to form a doped layer on the top and the sidewall [27]. (In essence this transforms the PID process into a CVD-process). As this solely leads to a surface layer of dopant atoms, it needs to be combined

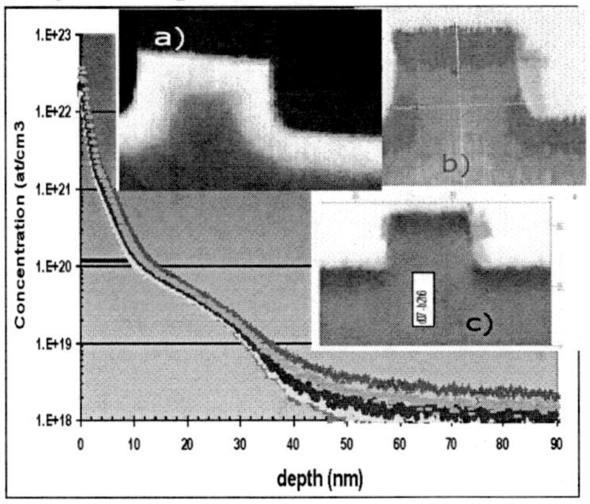

FIGURE 7. SSRM images of PID-doped (BF_3, 1.7keV bias) FIN structures illustrating the different degrees in conformality and corresponding (identical) SIMS profiles.

neutral atoms to form a doped layer on the top and the sidewall [27]. (In essence this transforms the PID process into a CVD-process). As this solely leads to a surface layer of dopant atoms, it needs to be combined with a proper thermal treatment in order to drive the dopants into the FIN. Whereas this works well for p-type dopants, it becomes less trivial for n-type species as they have the tendency to desorb (already at 350 °C) prior to their indiffusion (Figure 8), leading to very poor doping and high resistance values [28].

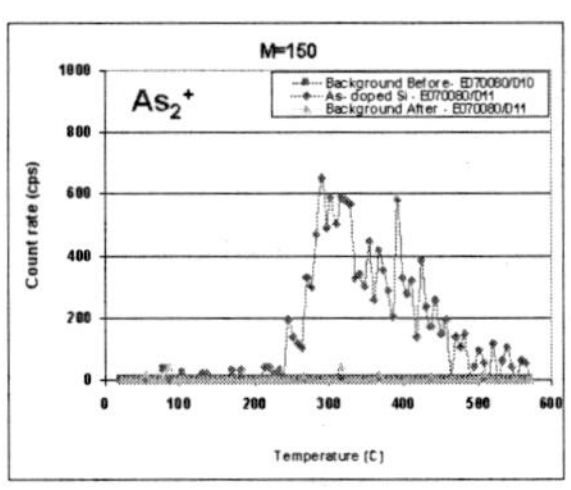

FIGURE 8. Thermal desorption mass spectra of As after As deposition in a PID-tool.

The absence of an energetic impact also implies that there will be no lateral scattering inside the solid and thus the source/drain – gate overlap needs to be created by the same thermal diffusion step as well. As diffusion proceeds from the outside inwards, the lateral diffusion in the center of the FIN will be smaller than at the outer surfaces and thus a non-conformality under the gate might still be present. The latter can not be captured by the 2D-profiles reported so far and a full 3D-analysis is still required to assess the importance of this effect.

At intermediate bias voltages (1-3 keV) deposition is no longer effective but nevertheless conformality can be achieved (cfr Figure 7a). In this case its origin lies in a secondary process i.e. selfsputtering of the implanted dopants. Indeed an ion implant process will always arrive at a stationary state (for high enough fluences) whereby the additional accumulation of dopant species at the sample surface by an extra fluence is balanced by dopant loss due to self sputtering (Figure 9) [29].

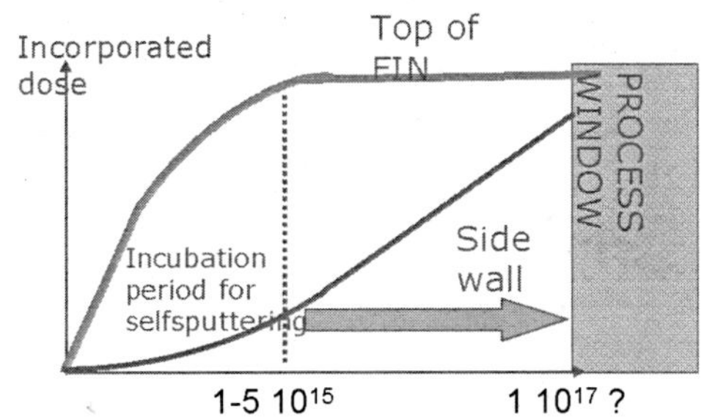

FIGURE 9. Evolution of incorporated dose on top and sidewall of FIN as a function of nominal dose.

Hence in case of FINs with the large difference in ion incorporation efficiency between top and sidewall, this means that the top of the FIN (and the region in between FINs) will reach this saturation level much sooner than the sidewall. The re-emission of the dopant atoms from the region in between the FINs now creates a secondary beam of dopant atoms which may impact on the sidewall as well and raise the sidewall doping in addition to the impact of the primary beam. This secondary implantation can clearly be seen in the SSRM–image shown fig 7.b with an accumulation starting to appear at the bottom corner of the FIN.

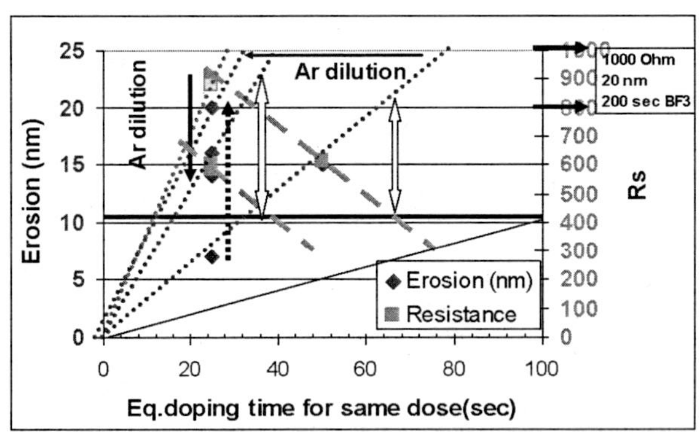

FIGURE 10. FIN resistance (right axis) and erosion (left) versus BF_3-PID doping time. Parameter is the Ar-dilution. Pure BF_3-PID requires 200 sec to reach 1000 Ω with 20 nm erosion.

The efficiency of the sidewall doping depends on the details of the plasma processing and can be (drastically) improved by adding a dilution gas like Ar, Xe. These atoms most likely lead to an enhanced knock-on of the deposited dopants thus promoting their incorporation and activation in the subsequent thermal indiffusion steps. Using the resistor value as metric for conformality (Figure 10, whereby 400 Ohm can be viewed as the target value) it is clear that this can be achieved for all gas- dilution gas ratio's. For a fixed doping time (corrected for the gas dilution ratio) a reduced resistance can be obtained (squares) with higher gas dilution. For a fixed dilution ratio, the observed resistance always scales with implantation time (long dashed lines in Figure 10) and the target of 400Ω can be reached.

Unfortunately the high dose required for this process also impacts on the FIN and its surroundings and FIN erosion can be observed (diamonds, short dashed lines). The fact that the erosion scales similarly with dilution ratio and exposure time suggests that physical erosion (sputtering) is the dominant process and not a chemical attack by F-containing species. Lowering the bias energy may reduce the sputter yield but at the same time leads to a smaller incorporation (Figure 11). Comparing the resistor value with/without hard mask, no substantial energy dependence on the conformality can be seen with only a small reduction in erosion for the lower bias voltages. The fact that the resistor value increases with reduced energy most likely results from the lower saturation dose, thereby limiting the amount of indiffused dopant atoms. As indicated by the double sided arrows, it is possible to achieve close to 100% conformality (by increasing the plasma exposure time) unfortunately at the expense of increased erosion.

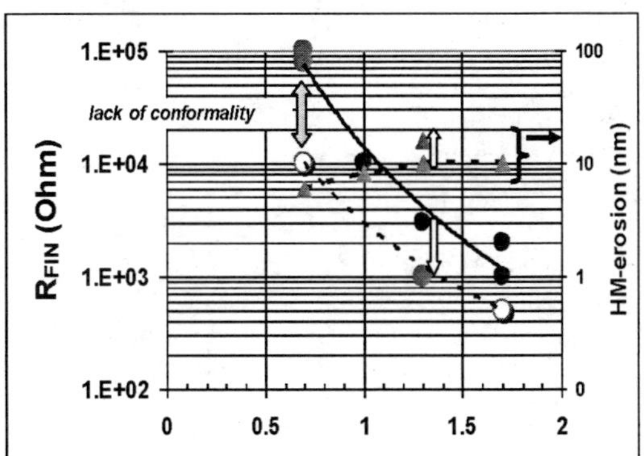

FIGURE 11. FIN resistance (circles) and erosion (triangles) versus bias voltage in BF_3-PID. Open symbols represent the (target) values achieved without hard mask.

When implementing these conditions into a full FIN pmos-transistor (FIN width 45 nm), I_{on}- I_{off} performance can be obtained equivalent with a device fabricated using a tilted implant (45°) (Figure 12). The impact of the erosion becomes visible when a smaller FIN is used (25 nm instead of 45 nm) where the 45° titled implant slightly outperforms the plasma doped device. However since the 45° tilt outperforms the normal 10° tilt considerably [29], one can conclude that even for these FIN dimensions superior performance is obtained when using the plasma doping process. The work of Lenoble [29] has also indicated that isolated and dense devices have identical PID-doping profiles and thus no effects of local loading can be observed.

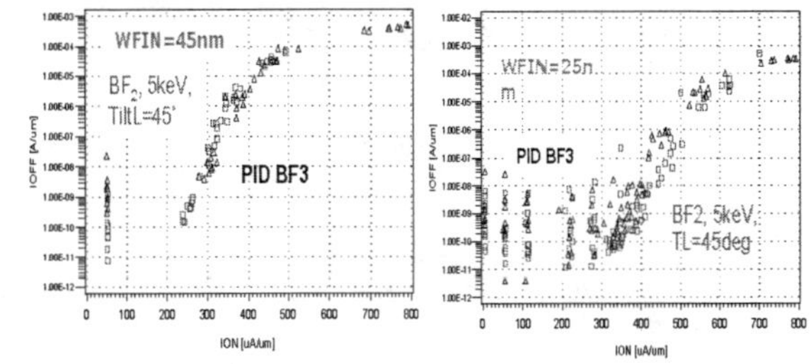

FIGURE 12. I_{onn}-I_{off} characteristics for PMOS transistors processed with PID-BF_3 and tilted implant (45°). For a wide FINs (45 nm) identical performance is obtained whereas for a narrower FIN (25 nm) FIN erosion reduces the performance slightly.

CONCLUSIONS

The formation of junctions in a FINFET structure requires a re-engineering of the doping approaches due its 3D-structure and the requirement for conformality. Considering that most applications will use densely spaced FINs, beam line implantation is restricted to small tilt angles leading to large differences in ion incorporation efficiencies (top vs sidewall) and thus a high degree of non-conformality. Moreover the recrystallization of amorphized FINs leads to less than ideal material quality which results in increased resistances. Due to the small FIN dimensions plasma doping is also not conformal when considering the classical electrostatic interactions. Conformality can only be achieved by going to a deposition mode with a concurrent thermal indiffusion step and/or a high dose implant whereby secondary processes aid the sidewall doping. Unfortunately one is then faced with a concurrent FIN erosion process. Despite this erosion devices based on plasma doping still outperform the one fabricated using the standard implant approach (10° tilt) and match the one fabricated with large tilt angles (45°). A further reduction of the erosion will however be necessary to apply PID to even narrower FINs.

Optimization of conformal junctions needs to be done on 3D-structures and no information can be extracted from 1D-experiments. Suitable metrology has been developed to probe the sidewall dopant dose

(SIMS through FINs), the electrically active fraction on the sidewall (resistors) and detailed carrier profile shape in the source/drain regions (SSRM). Details on the source/drain – gate overlap require methods with 3D-resolution. Methods such as the Tomographic Atomprobe or 3D-SSRM are presently emerging as candidates for this metrology problem

ACKNOWLEDGEMENTS

A part of this work was conducted in Kyoto-Advanced Nanotechnology Network, supported by "Nanotechnology Network" of the Ministry of Education, Culture, Sports, Science and Technology (MEXT), Japan.

REFERENCES

1. H. K. Choi, N. Lindert, P. Xuan, S. Tang, D. Ha, E. Anderson, T. J. King, J. Bokor, and C. Hu, Technical Digest of the IEEE International Electron Devices Meeting, (2001), p.421
2. A.Anders, *Surface and Coatings Technology* Volume **156**, Issues 1-3, pp. 3-12 (2001)
3. W.Vandervorst, AIP Conference Proceedings 931, American Institute of Physics, Melville, NY, (2007), pp. 233-245
4. T. Clarysse,.; Dortu, F.; Vanhaeren, D.; Hoflijk, I.; Geenen, L.; Janssens, T.; Loo, R.; Vandervorst, W.; Pawlak, B.J.; Ouzeaud, V.; Defranoux, C.; Faifer, V.N.; Current, M.I. *Materials Science & Engineering B*, vol.B114-115 : 166-73, (2004)
5. M.A.Gribelyuk, A.G.Domenicucci, P.A.Ronsheim, J.McMurray and O.Gluschenkov : *Journal of Vacuum Science & Technology B*26 (1) (2008) 408-14
6. P.Eyben, W.Vandervorst, D.Alvarez,, M.Xu. and M.Fouchier, in "Scanning Probe Microscopy, Electrical and Electromechanical Phenomena at the Nanoscale", edited by S. Kalinin and A. Gruverman (Springer, New York, 2007), Vol. 1, p. 31
7. H.Fukutome, Y.Momiyama, Y.; Kubo, T.; Tagawa, Y.; Aoyama, T.; Arimoto, H IEEE Transactions on Electron Devices **53**(11) : 2755-63, (2006)
8. S. Severi, E. Augendre,.; Pawlak, B.; Eyben, P.; Noda, T.; Felch, S.; Falepin, A.; Parihar, V.; Schreutelkamp, R.; Vandervorst, W.; De Meyer, K. Materials Research Society Symposium Proceedings Vol.912 : pp 39-44, (2006)
9. T. Noda, W.Vandervorst, Felch, S.; Parihar, V.; Vrancken, C.; Severi, S.; Falepin, A.; Janssens, T.; Bender, H.; Van Daele, H.; Eyben, P.; Niwa, M.; Schreutelkamp, R.; Nouri, F.; Absil, P.P.; Jurczak, M.; De Meyer, K.; Biesemans, S. , IEDM- Digest (2007) 955-8
10. N.Collaert, N.; De Keersgieter, A.; Dixit, A.; Ferain, I.; Lai, L.-S.; Lenoble, D.; Mercha, A.; Nackaerts, A.; Pawlak, B.J.; Rooyackers, R.; Schulz, T.; Sar, K.T.; Son, N.J.; Van Dal M.J.H., Verheyen, P.; von Arnim, K.; Witters, L.; De Meyer, Biesemans, S.; Jurczak, M., Proc. 37th European Solid-State Device Research Conference (2007) pp 143-6
11. K.Thompson, D.J.Larson, Ulfig, R.M.; Bunton, J.H.; Kelly, T.F *Solid State Technology* **49**(6) : 65-6, 68, 70, 72, 2006
12. D.Blavette ; C.Pareige, E.Cadel, Auger, P.; Deconihout, B. *Chinese Journal of Physics* **43**(1) pp. 132-44, (2005)
13. J.Mody, P.Eyben, W.Polspoel, M.Jurzack and W.Vandervorst, *Proc MRS* Spring 2008, Symp E (accepted for publication)
14. R.Duffy, E. Augendre, C. Jonville, M. J. H. van Dal, D. Baute, B. van Daele, W. Vandervorst, E. Rosseel, B. J. Pawlak, M. Kaiser, R. G. R. Weemaes, (to be published)
15. W.Vandervorst; Jurczak, M.; Everaert, J.-L.; Pawlak, B.J.; Duffy, R.; Del-Agua-Bomiquel, J.-I.; Poon, T., *Journal of Vacuum Science & Technology B*26 (1) (2008) 396-401
16. Subhendu Sarkar,_ Benny Van Daele, and Wilfried Vandervorst, *New Journal of Physics* (2008) (accepted)
17. G.A. Cooke, R.Gibbons and M.G.Dowsett, *AIP Conference Proceedings* 449 (1998) 766-70
18. S.Severi, L.Pantisano, E.Augendre, E. San Andres,.; P.Eyben and K.De Meyer, *IEEE Transactions on Electron Devices* **54**(10) (2007) 2690
19. S.M. Kluth, D.Alvarez, St.Trellenkamp, St., J.Moers, S.Mantl, J.Kretz and Vandervorst, W, *Journ. Vac. Sci. Techn. B*23(1). 76, (2005)
20. M.Gilbert, Ph.D thesis (2008), Rouen (France)
21. K.Jones, presented at IIT-2006 (Marseille)
22. P.Ronsheim, J.McMurray, P.Flaitz, C.Parks, K.Thompson, D.Larson and T.F.Kelly, AIP Conference Proceedings 931, American Institute of Physics, Melville, NY, 2007, pp. 129-136
23. R.Duffy, M.J.H. Van Dal,, B.J. Pawlak, M. Kaiser, R.G.R. Weemaes, B. Degroote, E. Kunnen, E. Altamirano, *Applied Physics Letters*, **90**(24) 241912/1-3 (2007)
24. M.J.H. van Dal,, Collaert, N.; Doombos, G.; Vellianitis, G.; Curatola, G.; Pawlak, B.J.; Duffy, R.; Jonville, C.; Degroote, B.; Altamirano, E.; Kunnen, E.; Demand, M.; Beckx, S.; Vandeweyer, T.; Delvaux, C.; Leys, F.; Hikavyy, A.; Rooyackers, R.; Kaiser, M.; Weemaes, R.G.R.; Biesemans, S.; Jurczak, M.; Anil, K.; Witters, L.; Lander, R.J.P, IEEE Symposium on VLSI Technology (2007) 110
25. B.Mizuno, K.Okashita.; K. Nakamoto,.; Jin, C.G.; Y.Sasaki, 8th International workshop on Junction Technology (2008) 22-6
26. W.Vandervorst Jurczak, Malgorzata, M.Foad,, J. del Agua Borniquel, P. Porshnev, J.L. Everaert, T.Poon, Nanochip Technology Journal, 5(1), 2-5 (2007)
27. K. Tsutsui, K.; R.Higaki, R.; Y.Sasaki; Sato, T.; Tamura, H.; Okashita, K.; Mizuno, B.; Twai, H., Japanese Journal of Applied Physics, Part 1 44 (6A) 3903, (2005)
28. J.L. Everaert, G. Zschaetzsch, E Vecchio, W.Vandervorst, IEEE RTP (2008) (accepted)
29. W. Vandervorst, T. Janssens, B. Brijs, R. Lindsay, E. J. H. Collart, David A. Kirkwood, G. Mathot, G. Terwagne, Mat. Res. Soc. Symp. Proc. Vol. 717 (2002) C7.2.1
30. D.Lenoble, Anil, K.G., De Keersgieter, A., et al. 2006 Symposium on VLSI Technology, p. 2-5.

Plasma Doping – Enabling Technology for High Dose Logic and Memory Applications

T. Miller, L. Godet, G. D. Papasouliotis, and V. Singh

Varian Semiconductor Equipment Associates, 35 Dory Rd., Gloucester, MA 01930 USA

Abstract. As logic and memory device dimensions shrink with each generation, there are more high dose implants at lower energies. Examples include dual poly gate (also referred to as counter-doped poly), elevated source drain and contact plug implants. Plasma Doping technology throughput and dopant profile benefits at these ultra high dose and lower energy conditions have been well established [1,2,3]. For the first time a production-worthy plasma doping implanter, the VIISta PLAD tool, has been developed with unique architecture suited for precise and repeatable dopant placement. Critical elements of the architecture include pulsed DC wafer bias, closed-loop dosimetry and a uniform low energy, high density plasma source. In this paper key performance metrics such as dose uniformity, dose repeatability and dopant profile control will be presented that demonstrate the production-worthiness of the VIISta PLAD tool for several high dose applications.

Keywords: Plasma Doping, High dose and Low Energy Implant, Precise Dose Control, Equipment Design.
PACS: 52.77.Dq, 85.40.Ry

INTRODUCTION

The fabrication of advanced CMOS devices calls for production-worthy solutions to meet the low energy, high dose doping challenge. The challenge is particularly acute in advanced DRAM for implants including dual poly gates, elevated source drains and contact plugs. With the elimination of traditional beamline implanter transport limitations, PLAsma Doping (PLAD) is quickly being adopted as a means of obtaining exceptional throughputs at ultra high dose and low voltage. A similar challenge is pending in logic processes for source drain extension implants (SDE) where doping requirements are becoming exceedingly shallow and abrupt with increasing doses in order to meet series resistance requirements. Additionally, semiconductor manufacturers are looking to PLAD as a potential solution for 3D doping requirements such as FINFETs and CMOS imagers [4]. In these advanced processes, precise and repeatable dopant placement is critical in order to maintain control over device parameters. In this paper we describe the key architectural features in the PLAD system that enable this control, as well as demonstrate relevant results.

KEY FEATURES

To achieve precise and repeatable dopant placement, several key features have been incorporated onto the PLAD chamber. These features enable uniform plasma generation and subsequent implantation, precise depth control with normal ion incidence, and repeatable dose control wafer to wafer.

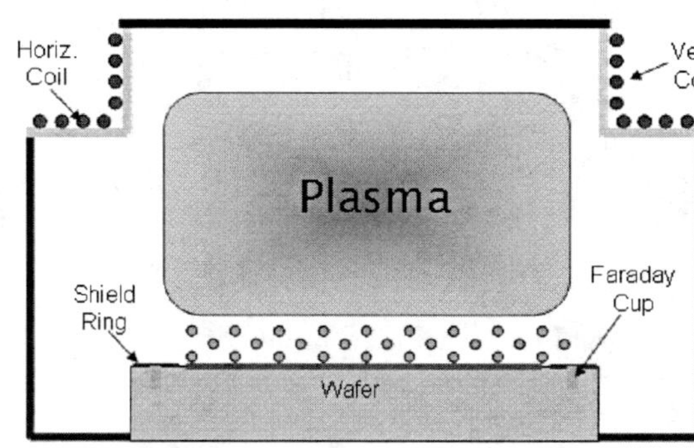

FIGURE 1. Schematic of the PLAD tool.

A uniform plasma generation and ion implantation has been achieved by reactor optimization through a combined modeling and diagnostic approach. The well established Hybrid Plasma Equipment Model (HPEM) developed at the University of Illinois at Urbana– Champaign was utilized for this optimization.

CP1066, *Ion Implantation Technology,* edited by E. G. Seebauer, S. B. Felch, A. Jain, and Y. V. Kondratenko

The use of the model along with chamber diagnostics allowed for off-line chamber optimization of reactor design.

Precise ion depth control at normal incidence is obtained through the reactor pedestal design. Key features that enable this are the pulsed DC power supply and shield ring.

Finally, repeatable dose control wafer to wafer is obtained with a closed loop Faraday system. In addition to providing repeatable dose control this system also provides a method for monitoring chamber health.

Uniformity - Combined Model and Diagnostic Approach

Using the chamber geometry shown in Figure 1 the HPEM was used to model PLAD's inductively coupled plasma. Two sets of coils, one horizontal and the other vertical, are rf-biased to both create the bulk plasma as well as control its uniformity. The distance between the wafer plane and the bottom surface of the dielectric beneath the horizontal coils, h_{w-d}, is used as a variable to investigate its consequence on plasma uniformity.

To verify calibration of the model electrostatic probe measurements are compared to HPEM results. A feed gas of Ar:NF3=80:20 is used in this investigation due to its well-known cross sections and similarities to doping process gasses. The conditions for the base case are as follows: gas flow of 100 sccm; pressure of 5 mTorr; 1 kW of rf power delivered through the horizontal coils and no DC bias. Total ion densities as a function of radius as calculated from probe measurements and as obtained from the HPEM are given in Figure 2. The predictions of the model match reasonably well with experimental data, especially in the central region ($r \leq 15$ cm).

To optimize plasma uniformity three values of h_{w-d} are modeled – a, b and c (with c being the largest). In this study, the vertical coils are used and the input power is 500 W. Other conditions are as for the base case. The total ion densities as a function of radius for different h_{w-d} are plotted in Figure 3. With increasing h_{w-d}, the ion density profile changes from being peaked at the edge to being center-high.

Additionally, the shape of the ion density profile can also be affected by using the horizontal coils instead of the vertical coils or a combination of the two. The resulting ion density profile when h_{w-d} = b and when powering the horizontal coils is given in Figure 3. With the horizontal coils the resulting ion density at the center of the chamber is higher than that at a radius of 15 cm. In comparison, the vertical coils have a higher ion density at a radius of 15 cm compared to the center of the chamber.

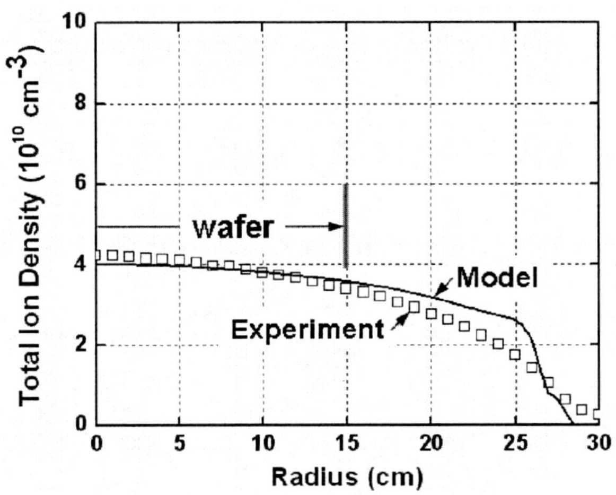

FIGURE 2. Total ion density – model vs. experiment. Conditions: Ar:NF3=80:20; 5 mTorr; 100 sccm; 1 kW rf power through the horizontal coils; no DC bias

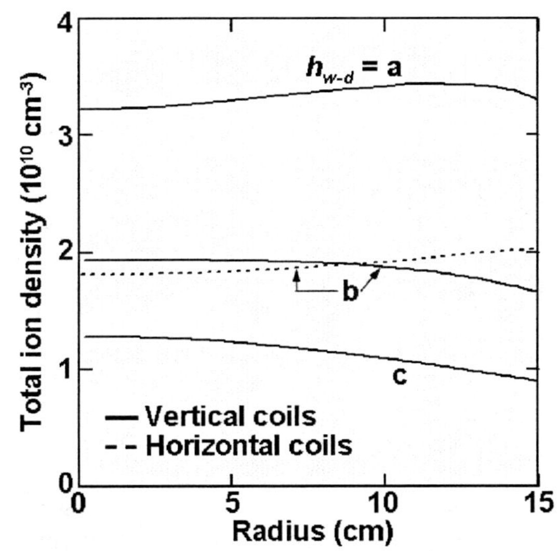

FIGURE 3. Total ion density vs. radius for varying distances between wafer plane and the bottom surface of dielectric beneath the horizontal coils (h_{w-d}) and for different coil configurations.

With chamber geometry optimized and plasma density modulation established through two sources of plasma generation, a uniform and controllable plasma density has been established. This has enabled uniform implantation and subsequent sheet resistance (Rs) values for sensitive low voltage implants. Figure 4 shows example implants in the energy range of typical poly gate doping and advanced source drain extension implants. Figure 4a shows Rs for a 7 kV 2E15 B implant with a 1000°C 20 second anneal in N_2.

Figure 4b shows Rs for a 325V 1E15 B implant with a 1050°C spike anneal in 100 ppm O_2.

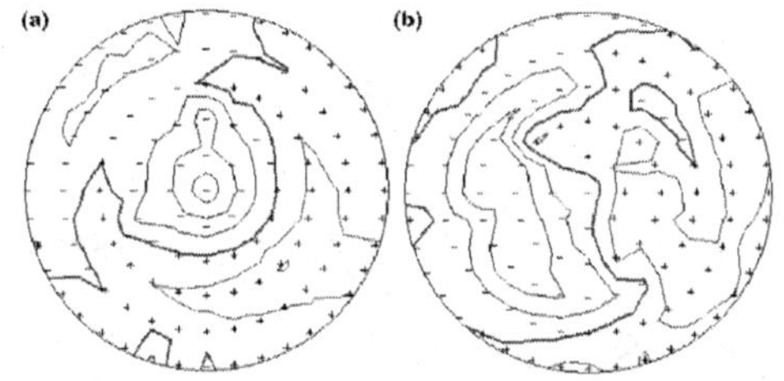

FIGURE 4. Rs maps using 121 pts and 3 mm edge exclusion showing (a) 1.0% uniformity at 7kV 2E15 B with 1000 °C 20 second anneal in N_2 (b) 1.2% uniformity at 325V 1E15 B implant with a 1050 °C spike anneal in 100 ppm O_2

Depth Control

Depth control in PLAD is achieved through the biasing of the wafer and surrounding shield ring with a negative pulsed voltage (Figure 5). A sheath is formed above the wafer accelerating ions to the wafer potential. The surrounding shield ring ensures that the sheath extends beyond the wafer plane allowing uniform zero degree implant over the entire wafer.

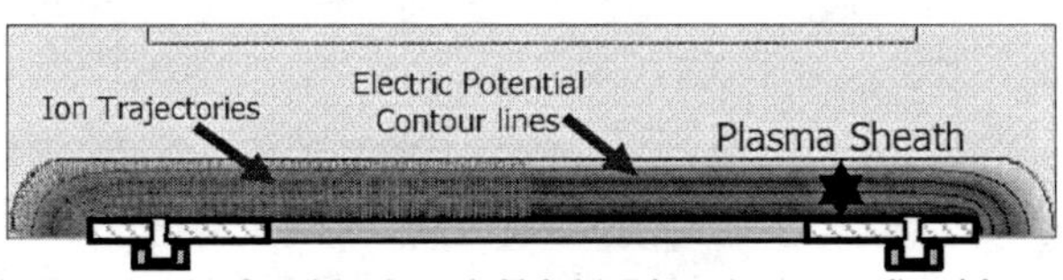

FIGURE 5. Wafer (blue) and Shield Ring (patterned) with modeled plasma sheath and ion trajectories

While this ion acceleration can be accomplished by means of a pulsed DC bias or an RF bias, pulsed DC offers a unimodal ion energy distribution. In an RF biased system with ion transit times on the order of the bias frequency a bimodal ion energy distribution is created. Pulsed DC implantation offers a wide window of operation by allowing a high degree of flexibility in the duty cycle. This flexibility allows the throughput to be maximized by allowing the pulse to be off only long enough for that required for charge dissipation. The fast, < 1 μS, rise and fall times and very square profiles of the pulse shown in Figure 6 enable this control without degrading the unimodal ion energy distribution

The energy controlled offered by this pulsed DC bias has enabled the formation of ultra-shallow abrupt dopant profiles suitable for the most advanced source drain implants. Figure 7a shows and examples of 1E15 B as implanted in the voltage range of 250 V to 750 V. Figure 7b shows comparison of msec annealed 1E15 B implants with and without Varian's advanced energy control feature. Profile abruptness, measured between 1E19 and 1E18 atm/cm^3, improves to 1.1 nm/dec from 1.6 nm/dec when the energy control feature is employed.

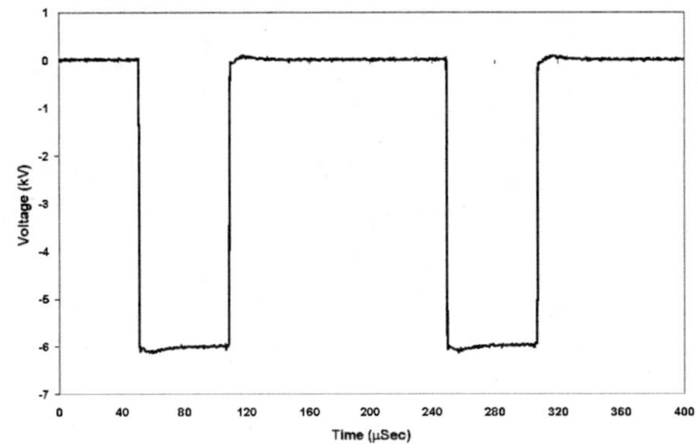

FIGURE 6. Two -6 kV Pulses, 50 μs pulse width

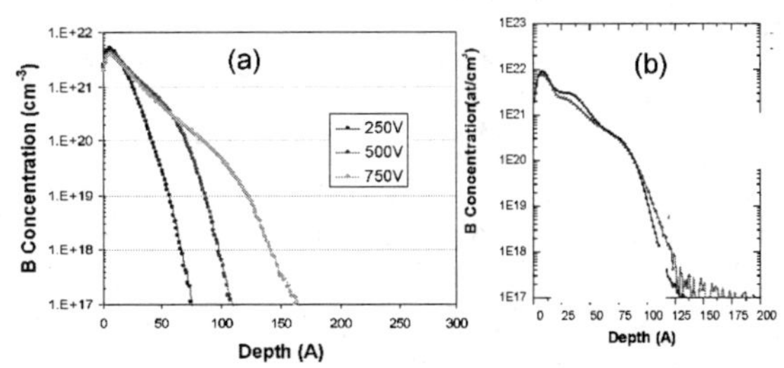

FIGURE 7. (a) As-implanted 1E15 B SIMS profile at 250 V (blue), 500 V (green), and 750 V (orange). (b) msec annealed samples with (blue) and without (green) Varian's advanced energy control feature having 1.1 nm/dec and 1.6nm/dec abruptness, respectively.

Dose Repeatability

The shield ring in Figure 5 also has a series of apertures in close proximity to the wafer, providing a known area through which ion flux can be measured. Ions accelerated through these apertures will impinge on a magnetically suppressed Faraday cup ensuring an accurate measurement of ion current. Synchronizing the current integration of the Faraday cup with the pulsed DC on time ensure that only ions at bias potential are counted. Analyzing the accumulated dose on a pulse by pulse basis allows for precise real time closed loop dose control.

The implementation of a closed loop dose control system on PLAD has enabled repeatable dose control and process stability. This is demonstrated in Figure 8 showing the results of a 2000 wafer manufacturing simulation at a major DRAM manufacturer prior to placing a tool into production. . Monitor wafer implant conditions were 8 kV 3E15 ion/cm^2 B with a anneal conditions of 1000°C for 20 sec. Repeatability of 0.64% and uniformity of 1.2% demonstrates the tool's readiness for mass production.

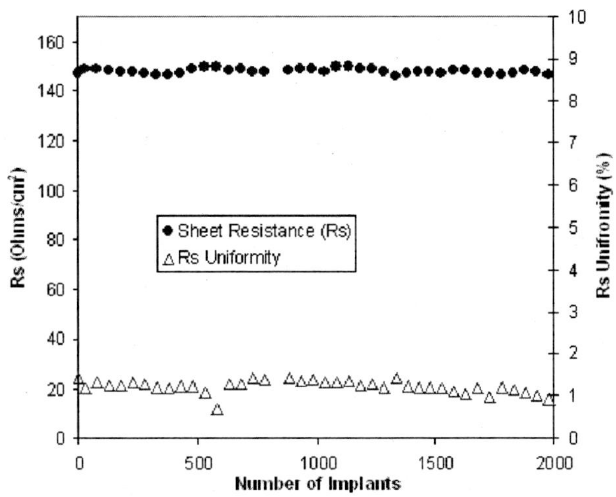

FIGURE 8. Sheet resistance measurements over 2000 wafers cycled. Repeatability 0.64%, average uniformity 1.2%.

The closed-loop dose control not only adjusts for ion flux variations to give a more repeatable implant but it is also gives a simple and accurate sensor for determining chamber readiness and health. It is well known that to create a stable plasma the walls of the chamber must come to steady state. This chamber conditioning process is demonstrated before and after optimization in Figure 9.

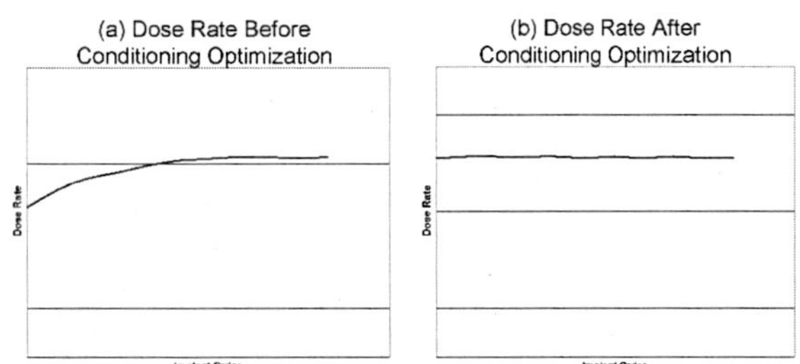

FIGURE 9. Dose rate measurements wafer to wafer before (a) and after (b) conditioning optimization.

CONCLUSION

In this paper we have identified the key features that have enabled the production worthiness of PLAD. A combined modeling and diagnostic approach for plasma generation was reviewed. This approach has allowed PLAD to quickly achieve a uniform and controllable source of ions needed for implantation. Pulsed DC bias was reviewed along with its advantages over rf biasing. Additionally, the ability of this biasing technique to create abrupt and retrograde profiles was demonstrated. Finally, the closed loop dosimetry system was reviewed and its ability to not only achieve highly repeatable tool performance but also determine chamber readiness and tool health.

ACKNOWLEDGEMENTS

The authors would like to thank Rajesh Dorai for his simulation work and many technical discussions.

REFERENCES

1. E. C. Jones, N. W. Cheung, "Plasma doping optimization for ultra-shallow junctions", *Nuclear Instruments and Methods in Physics Research B* **121**, 216-220 (1997).
2. W. Skorupa, R. A. Yankov, W. Anwand, M. Voelskow, T. Gebel, D. F. Downey, E. A. Arevalo, "Ultra-shallow junctions produced by plasma doping and flash lamp annealing", *Materials Science and Engineering B* **114-115**, 358-361 (2004)
3. S. R. Walther, S. Mehta, U. Jeong, D. Lenoble, "Formation of extremely shallow junctions for sub-90 nm devices", *Surface and Coatings Technology* **186**, 68-72 (2004).
4. Chang-Rok Moon, Jongwan Jung, Doo-Won Kwon, Jongryeol Yoo, Duck-Hyung Lee, Kinam Kim, "Application of plasma-doping PLAD technique to reduce dark current of CMOS image sensor", *IEEE EDL*, Vol. 28, No 2 (2007).
5. R. Liebert, S. Walther, S. Felch, Z. Fang, B. Pedersen, O. Pedersen, and D. Hacker. in: *Proceedings of the International Conference on Ion Implantation Technology*, Marseille, France, 2000 (American Institute of Physics,2000), pp. 472.
6. M. J. Kushner, *J. Appl. Phys.* **58**, 4024 (1985).
7. R. A. Stewart and M. A. Lieberman, *J. Appl. Phys.* **70**, 3481 (1991).
8. S. Qin, C. Chan, and N. E. McGruer, *Plasma Sources Sci. Technol.* 1 (1992).
9. M. A. Lieberman, *J. Appl. Phys.* **66**, 2926 (1989)
10. B. Briehl and H. M. Urbassek, *J. Appl. Phys.* **93**, 4420 (2003).
11. V. Vahedi, M. A. Lieberman, M. V. Alves, J. P. Verboncoeur, and C. K. Birdsall, *J. Appl. Phys.* **69**, 2008 (1991).
12. P. L. Kellerman, S. Qin, M. P. Bradley, and K. Saadatmand, *Rev. Sci. Instrum.* **73**, 837 (2002).
13. J. Brutscher, R. Gunzel, and W. Moller, *Plasma Sources Sci. Technol.* 54 (1996).
14. W. G. En and N. W. Cheung, *Trans. Electron Dev.* **44**, 379 (1997).
15. R. L. Kinder and M. J. Kushner, *J. Vac. Sci. Technol. A* **19**, 76 (2001).
16. A. Agarwal and M. J. Kushner, *J. Appl. Phys.* **101**, 063305 (2007).

Basic Aspects of the Formation and Activation of Boron Junctions Using Plasma Immersion Ion Implantation.

G. Zschätzsch[1,2], W. Vandervorst[2], T. Hoffmann[1], J. Goossens[1], J.-L. Everaert[1], J. I. del Agua Borniquel[3], T. Poon[3]

[1] *IMEC, Kapeldreef 75, B-3001 Leuven, Belgium*
[2] *also Instituut for Kern- en Stralingsfysica, K.U. Leuven, B-3001 Leuven, Belgium*
[3] *Applied Materials, 974 E. Arques Avenue - Sunnyvale – CA 94087, USA*

Abstract. This study investigates the basic aspects of junction formation using Plasma Immersion Ion Implantation using BF_3 and addresses the role of (pre)amorphization, C(F)-co-implantation, plasma parameters (bias, dose) and the thermal anneal cycle (spike versus msec laser anneal). The basic physics are studied using Secondary Ion Mass Spectrometry, sheet resistance and using four point probe and RsL. Profiles with junction depths ranging from 10-12 nm and sheet resistance values below 800 Ohm/sq are readily achievable.

Keywords: PIII, plasma immersion ion implantation, BF3, SIMS, Boron activation, Laser annealing, sheet resistance
PACS: 52.77 Dq

INTRODUCTION

As a result of the ever progressing downsizing of the dimensions of MOS device [1] it becomes more and more important to investigate new strategies to fabricate ultra-shallow, highly doped, highly activated and very abrupt box shaped doping profiles. These profiles are required to optimize the extension part of the source/drain contacts of MOSFET devices relative to the total series resistance and the control of the short channel effect.

Although traditional beam-line Ion Implantation (I/I) can reach the required target values for sub-32 nm technology [2] by combining low energy implantation with cocktail implants and msec anneals [3], alternatives such as cluster implantation and plasma immersion doping are explored in view of their potentially higher throughput and productivity. Plasma doping in particular differs significantly from monomer I/I in terms of angular spread, energy spread and beam species composition (ranging from B to BF_2 when a BF_3 plasma is used). A detailed study on its implementation and its response to the "known" recipes (pre-amorphization, cocktail implants, spike vs msec anneal,...) is required.

This study wants to provide the basic information related to that question and address the understanding of the underlying physical mechanisms which eventually will enable/limit the applicability of plasma ion immersion implantation for ultra-shallow junction formation.

EXPERIMENTAL

The wafers used for these experiments were 300mm, N-type with a resistivity ranging from 0.5 - 100 Ohm*cm. The Boron doping was carried out with an Applied Materials P3i plasma doping chamber using BF_3 as doping gas. The plasma is sustained with source power generators and an AC bias voltage connected between the wafer, and the plasma is used for accelerating the dopants towards the sample. In the present setup no absolute dose measurement is present and all implants are controlled by their plasma exposure time. Where used, pre-amorphization and co-implantation were done with regular beam line implantation using Ge (12 keV, 5 10^{14} at/cm^2) and C (1.5 – 3 keV, 10^{15} at/cm^2). Activation anneals were performed using spike anneals (1010-1050°C, 1.2 sec in He) or millisec laser anneal (peak temperature 1225 or 1300 C).

In order to characterize the doping profiles SIMS-analysis was performed using an Atomika 4500 (with a non-tilted O_2^+ beam at 500 eV) whereas electrical activation was inferred from the sheet resistance as measured with the non-contact RsL tool. In order to quantitatively asses the activation performance of the

CP1066, *Ion Implantation Technology*, edited by E. G. Seebauer, S. B. Felch, A. Jain, and Y. V. Kondratenko

various approaches, an electrically active carrier concentration level (at/cm^3) was derived from the combination of the sheet resistance and the SIMS profile. In the present estimate we use mobility values which are scaled relative to the ones for mono-crystalline silicon [4] as recent studies indicate their reduction for high doping concentrations.

RESULTS

As-implanted profiles

In Figure 1 we show the as-implanted B-profiles resulting from P3I-doping with three different bias voltages. Unless states otherwise, all profiles received a Ge pre-amorphization implant (12 keV, 5 10^{14} at/cm^2 leading to an amorphous layer thickness ~ 20 nm). The profiles differ significantly from the classical Gaussian-like profiles from I/I (B 500 eV, 10^{15} at/cm^3 +PAI, shown for comparison) and have a much more peak behavior at the surface followed by an exponential like decay (Figure 1b). This can be understood by realizing that within the plasma doping concept with BF_3 as doping gas, the implant species are a mixture of B, BF, BF_2, BF_3, ... ions with an energy distribution ranging from 0 up to the bias voltage [5] (disregarding potential gas phase collisions which might induce also an angular distribution). When comparing the as-implanted profile with or without PAI, one can clearly observe an important channeling tail up to a very high concentration level (~ 10^{20} at/cm^3). Obviously self-amorphization by the BF_x species does not appear to be very efficient at these low energies.

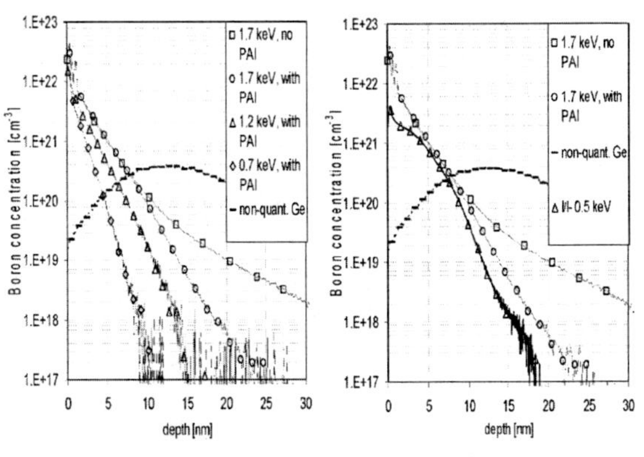

FIGURE.1 (a) As-implanted B-profiles from plasma doping with different bias voltages (with/without PAI). (b) Comparison between regular B I/I (500 eV) profile (with PAI) and plasma doping (1.7 kV, PAI).

The significance of this tail is consistent with the larger critical angle for channeling at low implant energies. Obviously an additional PAI will remain essential in order to limit the long (unpredictable) tails on these profiles.

Annealing

Although msec anneal will be the prime anneal concept for next generations, we will make a few comparisons to spike anneal in this paper in order to gain some further understanding. The differences in terms of indiffusion between the spike and msec anneal are clearly evidenced in Figure 2. The higher thermal budget for the spike obviously drives the B-atoms much deeper. Despite this increased depth distribution, the difference between the observed sheet resistances for msec (902 Ω) and spike anneal (799 Ω) remains relatively small due to the enhanced carrier concentration level in the msec case.

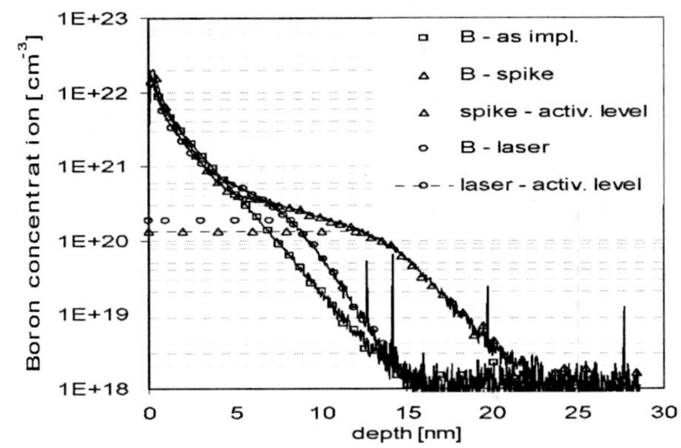

FIGURE 2. SIMS Boron profiles of samples doped by P3i (with PAI (Ge) and C) as-implanted, after spike or msec anneal. Also indicated are the electrically active carrier concentration levels after the anneals.

The correlation between anneal temperature, sheet resistance, junction depth and carrier concentration level is more exemplified in Figure 3.

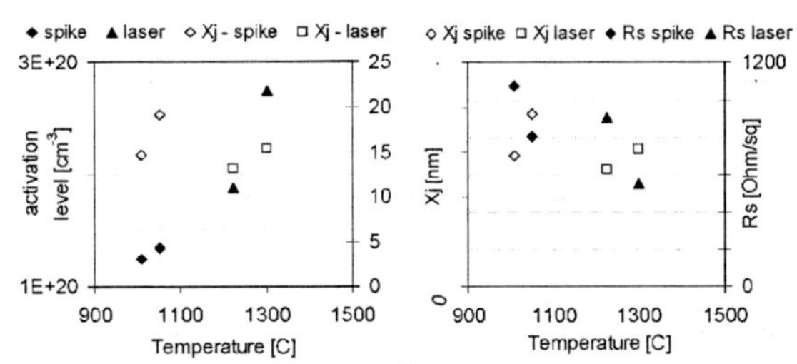

FIGURE 3. SIMS B-profiles for plasma doping (1.2 kV) with Ge PAI and C co-implantation. (a) Carrier concentration and junction depth versus max. anneal temperature. (b) Sheet resistance and junction depth versus max. anneal temperature

From Figure 3 it is very clear that the reduction in sheet resistance (Figure 3b) with increased anneal temperature during the msec anneal results primarily from an enhanced carrier concentration (Figure 3a), whereas in the spike anneal case the enhanced indiffusion (larger Xj) is the dominant factor. The

550 Ω, 15 nm Rs, Xj combination (without any detailed optimization) is already a promising value comparable with results based on cocktail implants in combination with msec anneal [2]. The route towards further improvement (lower sheet resistance, lower Xj) lies in a further increase of the maximum carrier concentration by a higher laser temperature with a concurrent control of the indiffusion.

Role of the implant parameters

In view of the interest to lower the sheet resistance and the junction depth, we also investigated the impact of plasma parameters (bias voltage, plasma dose) on the final result. A reduction of the bias voltage clearly reduces the junction depth (Figure 4, arrow A). In the case of spike anneal without PAI, the effect is substantial (arrow B). The sheet resistance increases concurrently from 500 to 916 Ω because of this reduction and a small concurrent decrease in the carrier concentration (from ~1.8 10^{20} at/cm^3 to 1 10^{20} at/cm^3). Adding a Ge PAI has no impact on the junction depth after annealing (not shown) and only the Ge/C combination (arrow C) leads to a clear improvement, similar to the classical I/I cases [6].

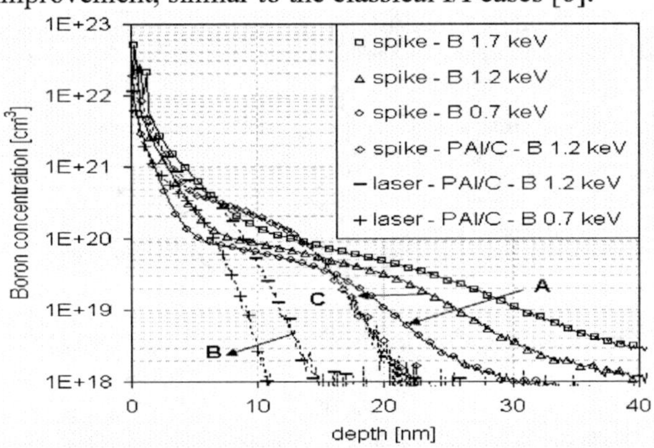

FIGURE 4. Impact of plasma bias voltage on B-profiles and carrier concentration level after spike and msec anneal. The addition of a Ge/C I/I strongly reduces the B-outdiffusion.

It is important to note here that the maximum carrier level is slightly reduced (~1.35 vs 1.8 10^{20} at/cm^3) and is now only weakly dependent on the anneal temperature. When the spike temperature is reduced to 1010°C (vs 1050°C), only a ~10% reduction in carrier concentration level (1.2 vs 1.3 10^{20} at/cm^3) is observed. The reduced indiffusion/junction depth (from 19 to 14 nm) of course leads then to an increased sheet resistance (1074 vs 799 Ω). Since the msec anneal exhibits only a small diffusion (cfr Fig 1), lowering the bias voltage directly leads to a reduced junction depth (Figure 4). At one bias energy one can observe a strong dependence of the sheet resistance on the total plasma exposure time (i.e. implant dose) (Figure 6a). The observed incorporation is still linearly related with an increase in dose (no self-sputtering yet) [7] but obviously predominantly occurs in the form of immobile clusters in the near surface without any (strong) impact on the junction depth (Figure 5).

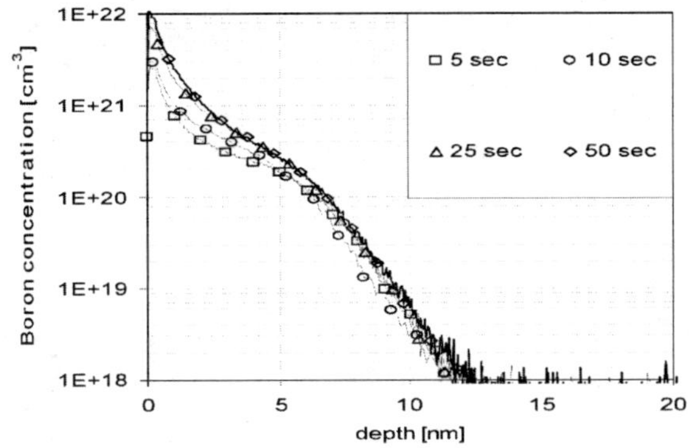

FIGURE 5. SIMS profiles of Boron profiles (0.7 kV, PAI/C) with different plasma exposure times after msec laser anneal.

Notwithstanding its immobility, the increased B-incorporation leads to a strong reduction in sheet resistance due to a significant increase in carrier concentration level (Figure 6b).

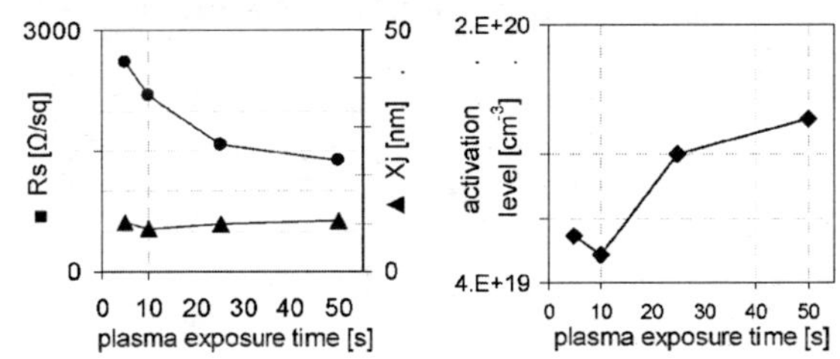

FIGURE 6. a) Sheet resistance and junction depth as a function of plasma dose for msec annealed B-implants (0.7 kV) with PAI/C
b) Corresponding carrier concentration level (0.7 kV)

Based on this trend one must assume that the active carrier concentration during msec anneal is determined by the probability to dissolve B-clusters and not by an actual solubility level [4]. Due to enhanced scattering at a highly increased amount of non-activated Boron, the actual carrier concentration level might be even higher than our present estimate.

Co-Implantation with Carbon

As already indicated in Figure 5, a Ge PAI together with a C co-implantation reduces the diffusion during annealing significantly. At the same time some indications for reduced activation (or mobility), concurrent with C-implantation, are observed. The impact of C was studied in more detail by analyzing a

B-plasma doped sample (PAI, 0.7 kV, msec anneal) which was co-implanted with C at different energies (1.5, 2 and 3 keV). The SIMS profiles (Figure 7) indicate that the C suppresses in all cases the outdiffusion leading to <10 nm junction depths (@5 10^{18} at/cm^3) with an abruptness close to 2-2.5 nm/dec.

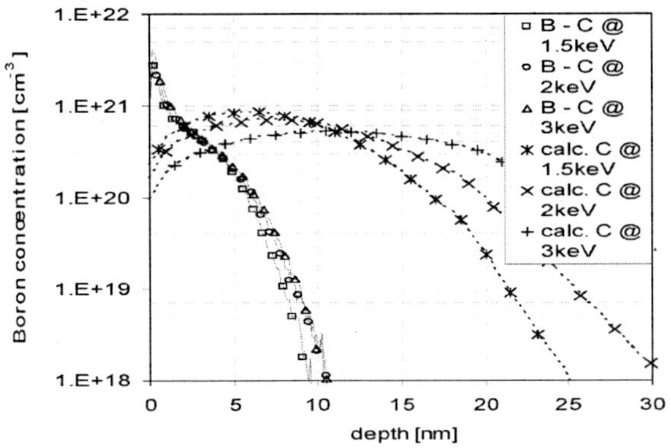

FIGURE 7. Plasma doped Boron profiles (PAI, 0.7 kV) with C co-implants at different energies after msec anneal. Calculated (SRIM) C-profiles are included for comparison.

The decreased C-energy reduces the B-diffusion (table I) slightly but has a dramatic impact on the sheet resistance and the carrier concentration level. The latter is reduced from 1.3 10^{20} at/cm^3 (3 keV) to 7 10^{19} at/cm^3 (2 keV) and even 3.6 10^{19} at/cm^3 (1.5 keV). An alternative explanation could be that the mobility is strongly reduced due to the increased amount of C-clusters. A similar impact was already reported previously [8] with indications that increased C-concentrations were leading to poor electrical conductivity. The present rapid decrease when going from 2 to 1.5 keV remains somehow surprising as the C-concentration profile varies only with 10-20% marginally in the B-doped region. Clearly a certain threshold C-cluster density is exceeded in the low energy case leading to enhanced deactivation (B-C formation?) and/or mobility degradation.

TABLE 1. Rs, Xj and profile abruptness as function of C co-implant energy.

C energy keV	Junction depth (@ 5e18) nm	Abruptness (nm/dec)	Sheet resistance Ω/sq
1.5	8.9	2.2	4426
2	9.5	2.3	2191
3	9.6	2.4	1540

It will be worthwhile to investigate C-dose dependence at these low C-energies in order to establish whether such a critical concentration exits and if a lowering of the dose can restore the activation levels while maintaining the reduced diffusion.

Comparison to beam line I/I

Two profiles with a compatible dose where implanted using I/I (at 1.2 keV) and P3i (at 0.5 keV). A reduced implantation energy was used for I/I to compensate the energy distribution of the P3i.

TABLE 1. Rs, Xj and activation conc. for I/I and P3i

Impl. Tech.	Activation concentration level (cm^{-3})	Junction depth (@ 5e18) nm	Sheet resistan ce Ω/sq
P3i	1.92 10^{20}	13.1	900
I/I	1.89 10^{20}	15.1	750

Both techniques show an identical activation level and the seen difference in Rs is a result of the slightly deeper profile of I/I.

CONCLUSIONS

The present investigations on the parameters determining junction formation with plasma immersion doping and msec anneal indicate that plasma doping behaves very similar to a regular beam line implant. The plasma process leads to a less Gaussian like profile and a channeling tail similar to regular ion implant. Hence a pre-amorphization implant remains inevitable. Co-implantation with C also leads to a suppression of B-diffusion. Optimum positioning of the C-profile is necessary as it leads to B deactivation. Whereas the final junction depth is set by the bias voltage, high (saturation) doses are required to achieve the minimum sheet resistance.

ACKNOWLEDGEMENTS

With many thanks to Eric Rosseel for the Laser anneals and Alain Moussa for the RsL measurements.

REFERENCES

1. ITRS roadmap (http://www.itrs.net/).
2. T. Noda, W. Vandervorst, S. Felch, *et al.*; Proceedings IEDM 2006
3. W. Lerch, S. Paul, J. Niess, *et al.*; IJWT 2007 p. 129 (2007)
2. T. Clarysse, J. Bogdanowicza, J. Goossens; *et al.*; Proceedings e-MRS 2008
5. L. Godet, Z. Zang, S. Radovanov, et at.; J.Vac.Sci.Techn. B24 (5) p2391 (2006).
6. B. Pawlak, T. Janssens, E.J.H. Collart; *et al. Appl. Phys. Lett.* **89**, p. 62110 (2006);
7. W. Vandervorst, T. Janssens, B. Brijs, *et al.*; Proceedings MRS 2002
8. A. Cacciato, J.G.E Klappe, N.E.B. Cowern, *et al.*; *J. Appl. Phys.* **79**(5), p 2314 (1996)

Performance Enhancement of PFET Planar Devices by Plasma Immersion Ion Implantation (P3I)

Claude Ortolland[1], Naoto Horiguchi[1], Christoph Kerner[1], Thomas Chiarella[1], Pierre Eyben[1], Jean-Luc Everaert[1], Jose Ignacio del Agua Borniquel[2], Tze Poon[2], Kartik Santhanam[2], Peter Porshnev[2], Majeed Foad[2], Robert Schreutelkamp[2], Philippe Absil, Wilfried Vandervorst[†], Susan Felch[2‡] and Thomas Hoffmann[1]

[1] IMEC, Kapeldreef 75, 3001 Leuven, Belgium
[†]Also IKS – Dept. Physics, KULeuven, Belgium
[2] Applied Materials, 974 E. Arques Avenue – Sunnyvale – CA USA
[‡] Now at Spansion – Sunnyvale – CA USA
Ph: +32 16 28 72 20 – Fax: +32 16 28 17 06 – Email: Claude.Ortolland@imec.be

Abstract. A study of doping the pMOS Lightly Doped Drain (LDD) by Plasma Immersion Ion Implantation (P3i) with BF3 is presented which demonstrates a better transistor performance compared to standard beam line Ion Implantation (I/I). The benefit of P3i comes from the broad angular distribution of the impinging ions thereby doping the poly-silicon gate sidewall as well. Gate capacitance of short channel devices has been measured and clearly shows this improvement. This model is clearly supported by high resolution 2D-carrier profiles using Scanning Spreading Resistance Microscopy (SSRM) which shows this gate sidewall doping. The broad angular distribution also implies a smaller directional sensitivity (to for instance the detailed gate edge shape) and leads to devices which are perfectly balanced, when Source and Drain electrode are switched.

Keywords: PMOS, BF3, P3i, LDD.
PACS: 52.77.Dq

INTRODUCTION

Plasma doping has been already reported as a good alternative to beam line implant to achieve aggressive Ultra-Shallow Junction (USJ) [1]. Some reports [2] already proposed a comparison of devices implanted by standard beam line I/I and plasma doping, like Plasma Immersion Ion Implantation (P3i) [3]. For the fist time we report a full analysis, from device performance to process control, of plasma doping for Lightly Doped Drain (LDD).

DEVICE INTEGRATION

PMOS devices have been integrated on a 300mm wafers in a 65nm process flow described in FIGURE 1.

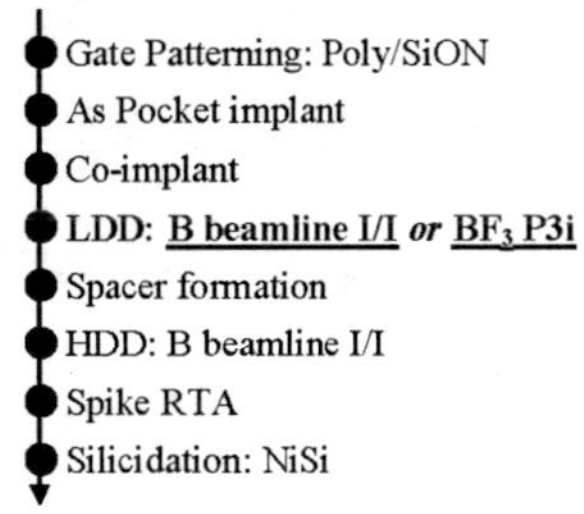

FIGURE 1. PMOS process flow used in this experiment.

Devices were fabricated on a (001) substrate with a <110> channel orientation. We have used a standard process flow (STI and Well formation) followed by the gate stack which is composed of a nitrided oxide and poly-silicon. After gate patterning, Arsenic was implanted as pocket followed by co-implantation (different cocktails of atoms were tried) in order to have good Short Channel Effect (SCE) control. Standard Boron beam line Ion Implantation (I/I) was used at LDD step for the reference and BF_3 in case of plasma doping. Recipe of BF_3 implanted by P3i, used in this experiment, has been matched, in term of junction depth, with standard B I/I (FIGURE 2).

CP1066, *Ion Implantation Technology,* edited by E. G. Seebauer, S. B. Felch, A. Jain, and Y. V. Kondratenko

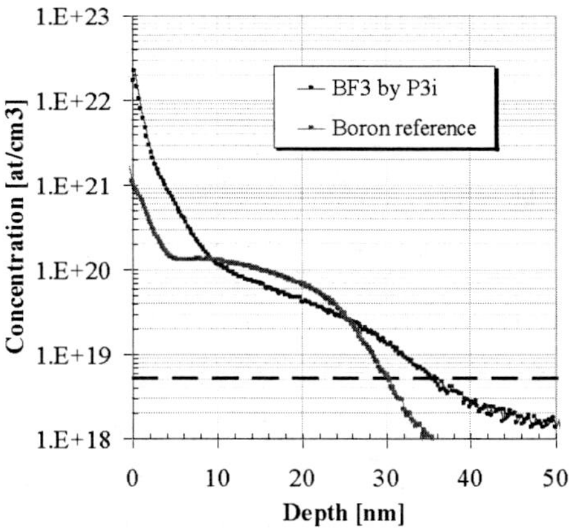

FIGURE 2. SIMS of BF3 implanted by P3i and B implanted by standard beam line I/I after anneal

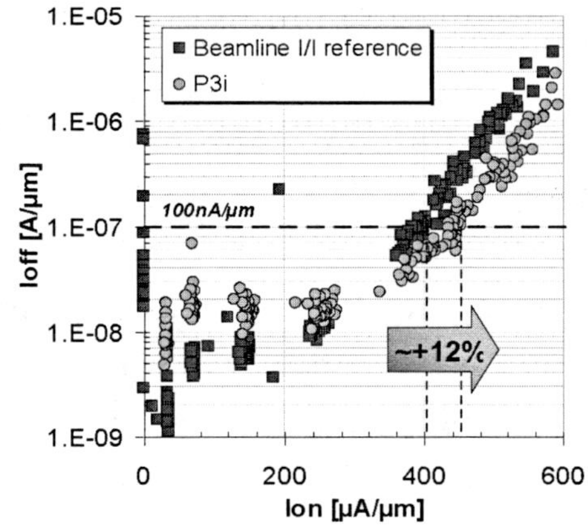

FIGURE 4. PMOS Ioff-Ion for different gate length and for a width = 1µm

DEVICE ELECTRICAL OVERVIEW

FIGURE 3 represents the threshold voltage (Vth) and the Drain Induced Barrier Lowering (DIBL) as a function of Gate length for PMOS devices processed by these two different implant techniques.

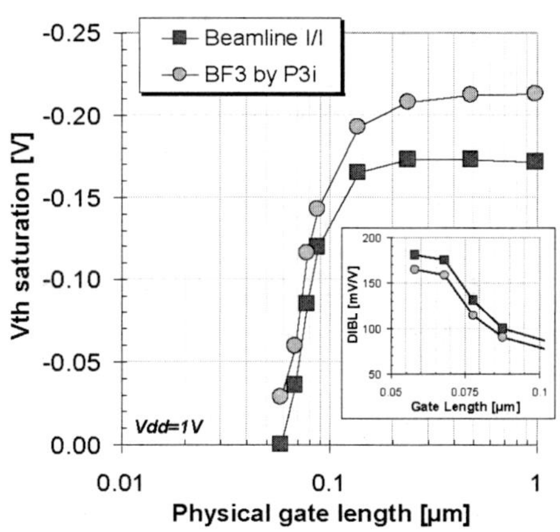

FIGURE 3. PMOS Vth sat Vs. Physical gate length for LDD implanted by beam line I/I or P3i. Inset: DIBL Vs. gate length. W=1µm, Vdd=1V

No difference in short channel devices is observed between LDD implanted by beam line I/I or by P3i, indicating that similar USJ were realized. FIGURE 4 shows a performance comparison between reference devices and those processed with plasma doping. For same Off-state current of 100nA/µm an On-state current gain of 12% have been obtained with BF3 implanted devices.

We first focus on the origin of this improvement. The Ion gain obtained is as a function of the gate length: the shorter the gate is, the higher the gain (FIGURE 5). This suggests that it comes from the edge of the channel.

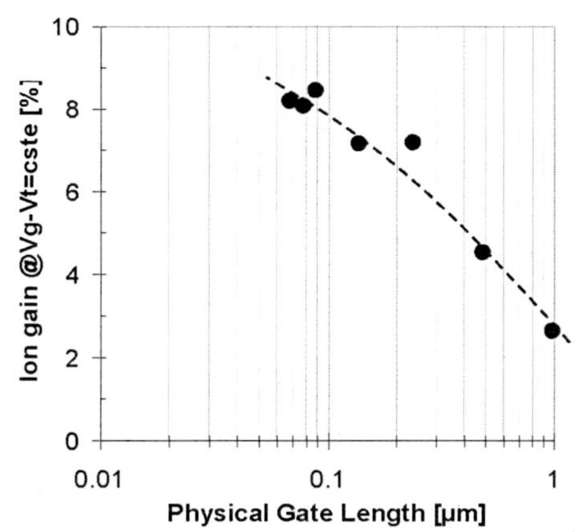

FIGURE 5. PMOS Ion gain (at Vg-Vt=cste) due to plasma doping use for LDD, as a function of gate length. Width =1µm

ORIGIN OF THE IMPROVEMENT

Model description

In order to explain the increase of drive current as a function of the layout, we need to go through a schematic process flow (FIGURE 6). After the gates are etched, Arsenic halo implantation is performed. Then we need to form the LDD regions. In case of beam line no tilt was used to keep a good control of SCE so we will dope only the top of the poly-silicon and the Source and Drain region (S/D). In the case of plasma doping, gate sidewall is doped as well due to broad angular distribution of the impinging ions. Spacer formation follows and Heavily Doped Drain (HDD) is implanted by beam line: S/D and top of poly-silicon are doped again, but deeper with a higher dose. Finally 1050°C spike anneal is done to activate and diffuse the dopants.

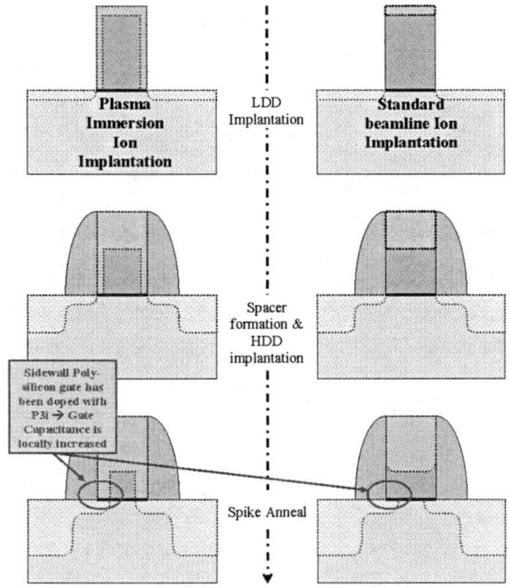

FIGURE 6. Schematic process flow: LDD with beam line Ion Implantation versus Plasma Immersion Ion Implantation

In silicon, similar junction position is obtained after anneal (same SCE on FIGURE 3): Junction depth of beam line I/I and P3i is matched, which were also confirmed by SIMS. However, the dopant diffusion in the poly-silicon gate is not the same due to the fact that sidewall has been doped by plasma doping. Better gate capacitance will be achieved locally and this edge effect will become preponderant for the short channel device. This will permit a performance improvement on the small gate length.

RFCV measurement

In order to confirm this model RFCV electrical characterization has been done on transistors with different layout. This measurement technique permits extraction of gate capacitance with a good accuracy on short channel devices. Results are summarized in FIGURE 7.

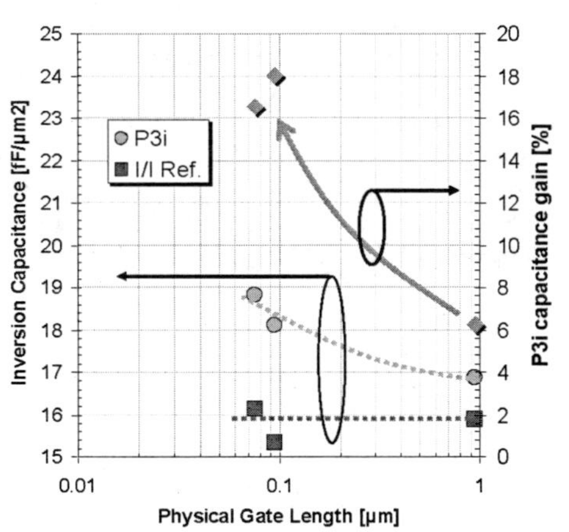

FIGURE 7. On the left: Inversion capacitance extracted at Vg=-1V for devices implanted by beam line I/I or P3i, on the right: Capacitance improvement obtained with the help of P3i, as a function of physical gate length.

Inversion capacitance was extracted at gate voltage equals 1V. We see that in the beam line case we keep a similar value for different gate lengths. For devices which have been implanted with BF_3 by P3i, inversion capacitance is increased for the smallest gate length. This result confirms the model of poly-silicon sidewall doped and the consequence is an improvement of gate capacitance as a function of gate length compare to reference which will increase the performance (FIGURE 7 & 5).

SSRM characterization

Physical characterization has also been done to confirm the model. Scanning spreading resistance microscopy (SSRM) is a powerful method for the characterization of Silicon semiconductor devices which permits visualization of activated dopants [4]. We have analyzed two different samples to check if the poly-silicon sidewall is effectively doped. FIGURE 8 shows the picture of SSRM. We could clearly see that the sidewall of the gate has a higher doping with plasma doping (on top) compared to with standard implantation (on bottom).

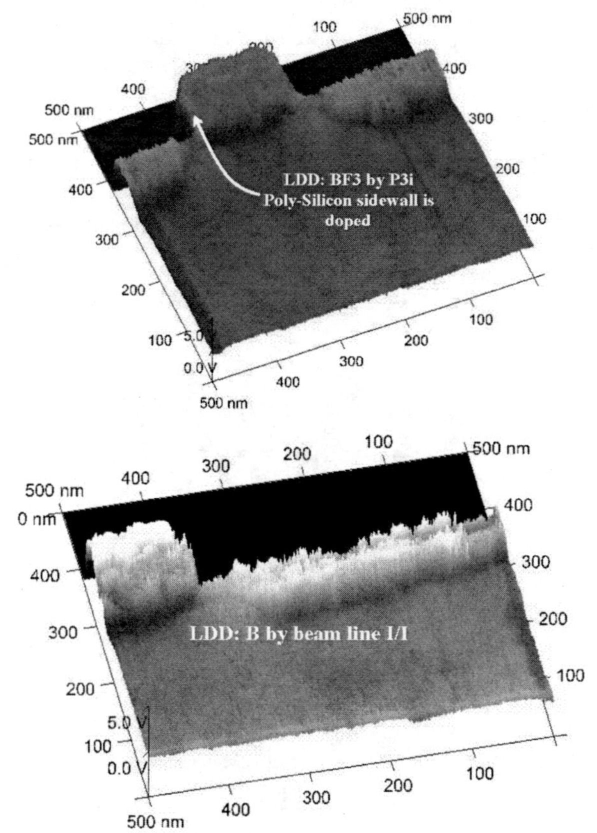

FIGURE 8. SSRM characterization: On the top on devices implanted by P3i, on the bottom on devices implanted by standard beam line I/I

Conductivity has been also measured (FIGURE 9) and the lateral doping from P3i is around 15nm. No peak of conductivity has been observed in standard beam line I/I case.

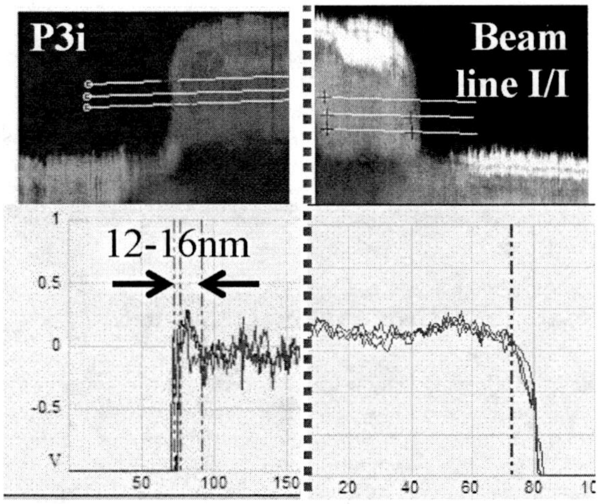

FIGURE 9. Conductivity measurement from SSRM: On the left on P3i implanted device and on the right on beam line implanted device

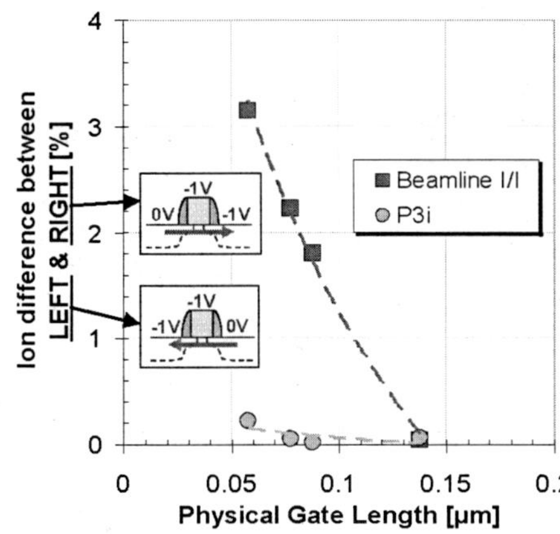

FIGURE 11. Ion symmetry of the devices vs. physical gate length for different LDD implant technique

PROCESS CONTROL

A study of the process control on 80 dies across the wafer has also been done to check the manufacturability of plasma doping. One of the most important parameters for the recent technology node is the threshold voltage variability: to have a good circuit control, e.g. SRAM, low device Vth spread need to be achieved. FIGURE 10 proposes the Vth standard deviation comparison of device implanted by beam line or P3i. No degradation is observed with plasma doping, which means that a good process control is workable.

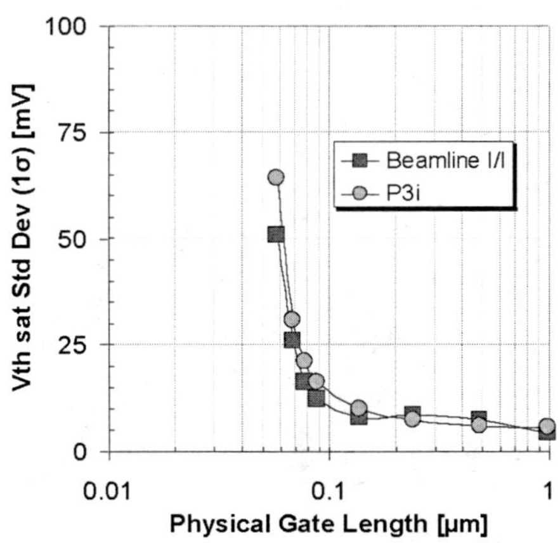

FIGURE 10. Standard deviation of Treshold voltage versus physical gate length with different LDD implant technique

One of the issues with beam line I/I could come from a slight error on the implantation tilt which will create an asymmetric device: On state current could be different on a single device if we switch the Drain and the source side where, ideally, it will be identical. This effect is shown in FIGURE 11 where Ion difference has been plotted versus the gate length. So as this difference is increased, for beam line, as a function of gate length, a slight error on LDD implant was occurring. Concerning LDD implanted by P3i, perfect balanced devices have been achieved due to, again, the broad angular distribution.

CONCLUSIONS

A better transistor performance with no SCE control loss compared to standard beam line I/I has been obtained on planar PMOS devices doped at LDD implantation step by P3i. This improvement, which comes from the sidewall poly-silicon doped due to broad angular distribution of plasma doping, has been confirmed by electrical and physical characterizations. No uniformity degradation has been observed and the advantage compared to the I/I reference has been also demonstrated showing the degradation of Ion symmetry of the devices as a function of physical gate length.

REFERENCES

1. G. Zschätzsch *et al.*., "Formation and Activation of Boron Ultra-Shallow Junction Using Plasma Immersion Ion Implantation", IIT Symposium 2008
2. F. Lallement *et al.*, "Ultra-Low cost and high performance 65nm CMOS device fabricated with Plasma doping", VLSI technology Symposium 2004 pp. 178-179
3. "Handbook of Plasma Immersion Ion Implantation and Deposition", Editor: A. Andres, Wiley Interscience (2000)
4. S.Kalinin and A.Gruvermann, "Probing semiconductor technology and devices with scanning spreading resistance microscopy" in *Scanning Probe Microscopy: Electrical and Electromechanical Phenomena at the Nanoscale* pp 31-87 (2007), Springer

Operation and Applications of the Boron Cathodic Arc Ion Source

J. M. Williams[1], C.C. Klepper[1,2], D. J. Chivers[3], R. C. Hazelton[2]
J. H. Freeman[1]

1. Brontek Delta Corporation, 6580 Valley Center Drive, Radford, VA 24141, USA
2 .HY-Tech Research Corporation, 105 Centre Court, Radford, VA 24141, USA
3. Ion Links Int. Ltd., 32 St. Mary's Place, Bathgate, Scotland.

Abstract. The boron cathodic arc ion source has been developed with a view to several applications, particularly the problem of shallow junction doping in semiconductors. Research has included not only development and operation of the boron cathode, but other cathode materials as well. Applications have included a large deposition directed toward development of a neutron detector and another deposition for an orthopedic coating, as well as the shallow ion implantation function. Operational experience is described and information pertinent to commercial operation, extracted from these experiments, is presented.

Keywords: boron, cathodic arc, vacuum arc, ion implantation, shallow junction doping, coatings
PACS: 61.72Tt

INTRODUCTION

General

The present development of the cathodic arc technique for semiconductor and other high value applications has been a stepwise process based almost entirely on funding from Small-Business-Innovation-Research (SBIR) funding from several federal agencies. These programs have enabled large and small increments of improvement in the technology, to the extent compatible with goals of the respective contracts. As a result, individual parts of the research have been determined by strict contractual obligations; large funding for semiconductor applications *per se* has not formed part of the program. Some low-energy implantations have been done and will be discussed. This paper will review operating experience and certain data, while drawing relevance to the doping function. In addition to shallow doping, other tentative successes have been for wear preventative boron coatings for the orthopedics industry and a "boron in-fill" process for neutron detectors. Since the present conference series has, in the past, taken note of commercial applications other than semiconductor processing, these results will be presented.

The Technology

We refer to Figure 1 for the characteristics of the cathodic arc source. Another name for such a source is the *vacuum arc* ion source, since it is the only ion source that works, in principle, in the presence of absolutely *no gas*. The electron arc current may be struck in vacuum by various strategies [1]. Physical size and power requirements of the source are not significantly different from those of a Freeman or Bernas source. The arc source is potentially more compact. Potential system strengths include compactness, no corrosion, "greenness," and relative simplicity.

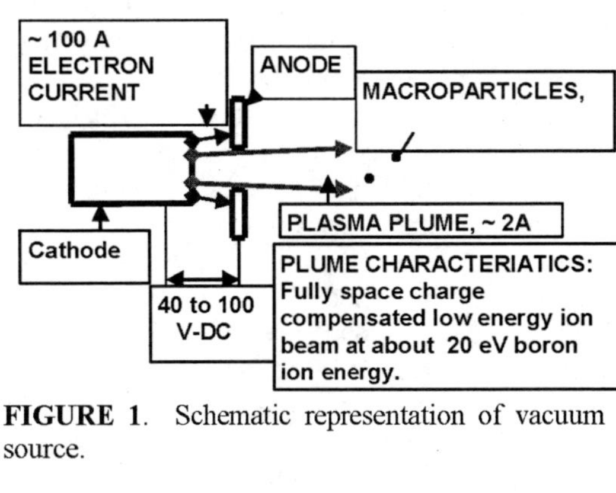

FIGURE 1. Schematic representation of vacuum arc ion source.

CP1066, *Ion Implantation Technology,* edited by E. G. Seebauer, S. B. Felch, A. Jain, and Y. V. Kondratenko

The signal feature of the source, in comparison with gas sources, is the very large ion current. Such currents consist entirely of ions from the solid cathode. They are derived by virtual explosion of highly localized regions of solid cathode at the electron arc anchor sites. For the present research, the emergent ion plume (Fig 1) typically has an effective B ion current of about 2A. After filtering of macroparticles [2] the current is about 400 mA. The plume is also already highly directional, as though it were an extracted beam, and is fully space charge compensated as it propagates. The plume has been depicted pictorially previously [2]. The ion energy is in the tens of eV range (see below).

The most important negative aspect of arcs is the presence of copious "macroparticles," so called. These are pieces of the cathode of some 10 μm in dimension. Development of an excellent filter for these particles was an important aspect of the present program [2,4], and results will be referred to again in the present paper. Depictions of the entire system have been presented before [3,4].

Motivation

The decision to develop the boron arc was based on a menu of possible applications. In addition to shallow junction doping, other applications based on favorable properties of B as a coating were anticipated. Good methods of synthesis of boron products did not exist, in part because of these favorable properties themselves.

1. Having a hardness value of 33 GPa, B is second only to diamond in hardness among elements. The hardness is the same as that of TiB_2 and is higher than that of most other industrial hard compounds, including boron carbide and most DLC values that are published.

2. The Pourbaix compilation on corrosion [5] reports excellent resistance to boiling HCl and to corrosion by HF. Thus, good performance against halide corrosion in general was expected.

3. Boron is a 3-valent elemental hard material rather than a covalently bonded hard compound. The high reactivity and tendency to form high-melting and hard interfacial compounds with transition metals was expected to contribute to adhesion to substrates.

Ideas for application included:

1. Shallow junction doping.

2. Treatment of semiconductor tool surfaces of stainless steel for prevention of corrosion and sputtering in the presence of halide gases and derived ions. Boron does not bind to aluminum.

3. Wear resistant coatings for orthopedics prosthetics. Because of the saline environment of the body, resistance to halide corrosion is a necessary but not a sufficient condition for biocompatibility. Biocompatibility was nevertheless expected.

A major prior research effort influencing the present effort was the work of Richter *et al.*, who first operated a nominally pure B arc [6,7]. Potential problems identified were the presence of 11 % of doubly charged boron in the plume and great difficulty obtaining cathodes. Apparently the 11 % doubly charged problem is not appearing in the present research [2,4] (see Fig 4 below).

RESULTS

Cathodes and Operation

Development of cathodes, particularly the nominally pure B cathode, has been an important aspect of the program and improvement is an ongoing part. Cathodes used thus far, and operated at length, are:

1. nominally pure B
2. boron carbide
3. silicon-semiconductor grade
4. carbon

As the Richter group reported, obtaining a satisfactory pure B cathode was difficult indeed and was one of the major and crucial achievements of the present effort. The patented process [8] involves sintering of the green ware by use of ultra-high frequency (24 GHz) microwaves. The only sintering aid added is C, most of which is removed in the synthesis process. The boron carbide cathode can be synthesized by use of microwaves of the common frequency of 2.45 GHz [9].

There are several other ideas for cathodes, either for n-doping, improvement on the above types of cathodes, or for other functions. For example, germanium, if needed for preamorphization instead of Si, is probably easier to use than Si because of a lower bandgap and still adequate thermal properties. Nevertheless, Si works well and could easily be used for preamorphization at the large current values of the cathodic arc. A wafer processor might have several cathodes on a carousel for quick rotation into position with the anode. Such cathode systems have been operated commercially. Such a system could provide for rapid changing among functions such as preamorphization, doping with C for diffusion inhibition [10, 11], and doping.

The boron in-fill process, described below, was first piloted by use of a C cathode and then executed with a cathode of B with natural isotopic abundance, and then again with a cathode of isotopic ^{10}B.

Boron In-fill

The boron in-fill program was part of development of a neutron detector. The substrate for the device consisted of a matrix of "pillars" and surrounding trenches formed on a substrate of Si. The pillars were of 6 μm in height and of square cross section with 2 μm sides. They extended upward from the surrounding flat valley, with 2 μm-wide trenches separating the rows of pillars in both directions. Thus, the area consisted of 25 % pillars and 75 % floor area. The need was to fill the valleys to the top of the pillars, while leaving minimal residual deposition on top of the pillars. Depending on deposition parameters, it might have been possible for the pillars to grow in height by the same amount as the deposition in the valleys or to fan out in to "mushrooms" of deposition, which would have cut off deposition in the valleys.

The strategy for the in-fill was devised by Othon Monteiro, formerly of Lawrence Berkeley Laboratory. Figure 2 shows the result. The depositions were accomplished by means of 600 each of cathode pulses of 0.5 s duration. The target bias cycle consisted of alternate periods of a few μs each between each of two conditions: no polarization and bias to -1 kV, where the PIII concept of target bias applies.

The reason for success may not be precisely known and may in fact be partly related to the role of the transients. It seems clear that little or no deposition occurs during the -1 kV phase at the outset, when deposition is needed at the bottoms of the trenches. That deposition occurs during the 0-bias part of the cycle. The high negative polarization part of the cycle is needed to sputter the deposit off the tops of the pillars.

The pyramidal structure on top of the pillars is a saturation structure and is acceptable. Without the sputtering phase, the deposition on top of the pillars would be much larger. TRIM [12] calculations help rationalize the idea that the pyramids saturate at a

FIGURE 2. Cross sectional micrograph showing boron in-fill. The black columns are the Si. The top is covered with Pt, which reaches down between the sputtered B pyramids on top of the Si pillars. Focused ion beam analysis by John McIntosh, Nanoscale Characterization and Fabrication Laboratory, Virginia Tech University, Blacksburg, VA.

height and angle such that the sputtering yield is equal to the retained fraction of incident ions. There are reasonable speculations as to the reasons for other details, such as the layering near the bottom of the trenches and the change in character near the tops of the fill. These will not be discussed at present.

The area of the target was 3 cm^2. The depths of B in the trenches and on the pillars represented 32 C of charge in B ions remaining as deposited in the 300 s. The estimated total in-flux, derived from the current values, was minimally 60 C and possibly as high as 100 C. There were only four particles for the deposit, and it is not necessarily clear that these came from the ion source, since the sample had been cleaved.

Orthopedic Coatings

The development of the orthopedics program has been published in detail [13]. The hypothesis of the program was that the overwhelmingly favorite orthopedics alloy, CoCrMo, has a topographical issue upon polishing that cannot be overcome. As a result, wear of the mating ultra-high-molecular-weight polyethylene (UHMWPE) was a serious problem in a number of failures. It was proposed that a suitable coating of B would remedy the problem.

Figure 3 shows results of a 24-h pin-on-disk test, for which a film of UHMWPE of 150 nm in thickness has wiped off on to the alloy for uncoated alloy, whereas for the boron coated, there is no transfer. These results were also confirmed by Rutherford backscattering with nuclear reaction analysis. In a first screening, the expected biocompatibility was confirmed by a subcontractor [14], and B was tentatively added to the slim menu of materials that can be safely implanted in to the body.

The application of this relatively thick B coating allowed for a better impurity analysis than would have otherwise been possible. The principal impurity was C, as expected. The concentration was at a maximum of 3 % of the B deposited [2]. A patent application has been submitted [15].

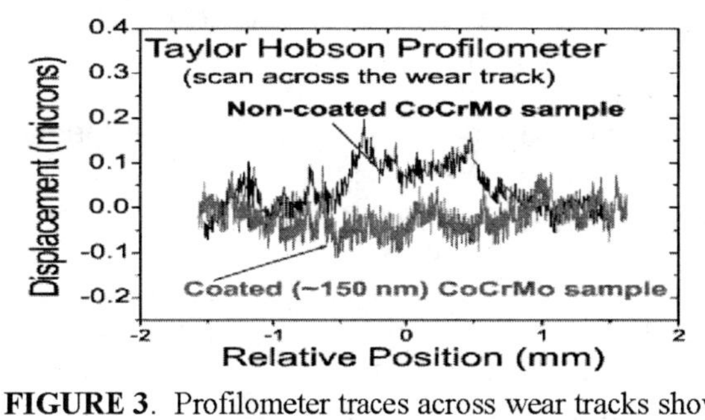

FIGURE 3. Profilometer traces across wear tracks showing adhesion of UHMWPE to uncoated alloy, while coated alloy remains unaffected.

DOPING

The SIMS implantation profiles are presented in Figure 4 for implantation in to crystalline silicon. Junction depths are 59, 67, and 155 angstroms, respectively, for the bias voltages of 60, 100, and 500 V. These profiles have been discussed before [2,4] and are fully competitive with all demonstrated other techniques of doping in terms of containment within depth for the given ion energy. They are internally self consistent.

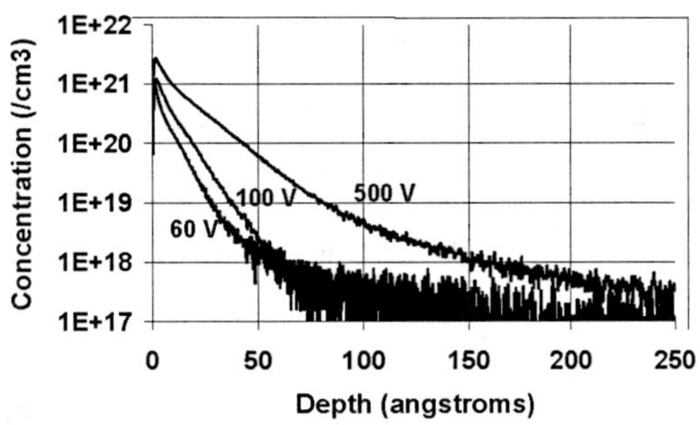

FIGURE 4. Concentration versus depth by SIMS for implanted crystalline Si at the bias voltages indicated. 250 eV O sputtering for the 60 V, 400 eV O for the other two. Incoming ion plume energies may need to be added to the bias values.

CONCLUSION

The vacuum arc ion source with heated elemental B cathode has demonstrated strong promise and unique suitability for possible commercial applications, including shallow junction doping. Further developments and maturation of the technology are expected.

Important developments implicit and explicit in the above research and prior research [2,3,4], include, but are not limited to:

- Development of cathodes and feasibility for ancillary functions such as preamorphization and C implants, with the isotopically pure B cathode being another example.
- Identification of commercial applications complementary to doping.
- Apparent elimination of macroparticles as a processing issue.
- No apparent effect of doubly charged B ions in doping profiles, possibly due to recombination in transport.
- Putative ability to control impurities.

Large improvements in firing reliability and current stabilization during pulses due to the in-fill depositions.

ACKNOWLEDGEMENTS

Sponsors: US Department of Energy, US Department of Defense, National Science Foundation, National Institutes of Health, Domestic Neutron Detection Office of DHS, Virginia's Center for Innovative Technology.

REFERENCES

1. I. G. Brown, "The Metal Vapor Vacuum Arc Ion Source," in *The Physics and Technology of Ion Sources,* New York, John Wiley and Sons, 1989, pp 331-354.
2. J. M. Williams, C. C. Klepper, D. J. Chivers, R. C. Hazelton, J. J. Moschella, M. D. Keitz, "Boron Ion Implantation in to Silicon by Use of the Boron Vacuum-Arc Plasma Generator," in *Ion Implantation Technology,* edited by Karen J. Kirkby, Russell Gwilliam, Andy Smith and David Chivers, AIP Conference Proceedings 866, American Institute of Physics, Melville, NY, 2006, pp 261-265.
3. J.M. Williams, C. C. Klepper, and R. C. Hazelton, *Nucl. Inst. Methods in Phys.Res,* **B 237,** 278-283 (2005).
4. J. M. Williams, C.C. Klepper, D. J. Chivers, R. C. Hazelton, J. J. Moschella, *J. Vac. Sci. Technol.* **B,** to be published, (INSIGHT 2007 Proceedings).
5. M. Pourbaix, *Atlas of Electrochemical Equilibria in Agueous Solutions*, James A.Franklin Translation, London, Pergamon, 1966.
6. F. Richter, G. Krannich, J. Hahn, R. Pintaske, M. Friedrich, S. Schmidbauer, D. R. T. Zahn, *Surf. and Coatings Technology,* **90,** 178-183 (1997).
7. Frank Richter, Siegfried Peter, Volodymyr, Gert Flemming and Michael Kuhn, *IEEE Transactions on Plasma Science,* **27,** 1079-1083 (1999).
8. B Marvin S. Morrow, Donald E. Schechter, Chin-Chi Tsai, C. Christopher Klepper, John Niemel, U.S. Patent No. 6,562,418 (13 May, 2003).
9. Cressie E. Holcombe and Marvin S. Morrow, U. S. Patent No. 5,252,267 (12 October, 1993).
10. Wade Krull, Brian Haslam, Tom Horsky, Kurt Verheyden, Klaus Funk, "Simplifying the 45 nm SDE Process with ClusterBoron® and ClusterCarbon™ Implantations," in *Ion Implantation Technology,* edited by Karen J. Kirkby, Russell Gwilliam, Andy Smith and David Chivers, AIP Conference Proceedings 866, American Institute of Physics, Melville, NY, 2006, pp 182-185.
11. Aaron Vanderpool, Mitch Taylor, *Nucl. Inst. Methods. Phys. Res.* **B 237,** 142 (2005).
12. J. F. Ziegler, M. D. Ziegler, J. P. Biersack, *SRIM, The Stopping and ranges of Ions in Matter*, Chester, MD, SRIM.com., 2007.
13. C. C. Klepper, J. M. Williams, J. J. Truhan, J. Qu, L. Riester, R. C. Hazelton, J. J. Moschella, P.J. Blau, J. J. Anderson, O. O. Popoola, M. D. Keitz, *This Solid Films,* **516,** 3070-3080 (2008).
14. M. Anabtawi, P. Beck, J. Lemons, *J. Biomed. Mater. Res. B, Applied Biomaterials,* **84B,** 478-485 (2008)
15. James M. Williams and C. Christopher Klepper, U. S. Patent Application 20080027431 (31 January 2008).

Ultra-Shallow Junctions Fabrication by Plasma Immersion Implantation on PULSION® Followed by Laser Thermal Processing

Frank Torregrosa[1], Hasnaa Etienne[1], Guillaume Sempere [1], Gilles Mathieu[1], Laurent Roux[1], Vanessa Vervisch[2], Philippe Delaporte[2], Thierry Sarnet[2], Aron Pap[3], Krisztián Kis-Szabo[3], Tibor Pavelka[3], Catherine Grosjean[4]

[1] *ION BEAM SERVICES, ZI Peynier-Rousset, rue Gaston Imbert Prolongée, 13790 Peynier, France*

[2] *UMR CNRS LP3, Université de la méditerranée, Campus de Luminy, Case 917,13288 Marseille cedex 9-France*

[3] *SEMILAB Semilab, Inc., 2 Prielle K. str., H-1117 Budapest, Hungary*

[4] *CIMPACA Caracterisation Plateform, ZI Peynier Rousset, F13790 Rousset*

Abstract. In order to achieve the requirements for P+/N junctions for < 45nm ITRS nodes, ultra low energy and high dose implantations are needed. Classical beamline implantation is now limited in low energies, compared to Plasma Immersion Ion Implantation (PIII) which efficiency is no more to prove for the realization of Ultra-Shallow Junctions (USJ) in semiconductor applications : this technique allows to get ultimate shallow profiles (as implanted) due to no lower limitation of energy and high dose rate.Electrical activation is also a big issue since it has to afford high electrical activation rate with very low diffusion. Laser annealing is one of the candidates for the 45 nm node. This paper presents electrical and physico-chemical characterizations of junctions realized with BF3 PIII followed by laser thermal processing with aim to obtain ultra-shallow junctions. Different implantation conditions (acceleration voltage/ dose) and laser conditions (laser types, fluence/ number of shots) are used for this study. Pre-amorphization is also used to confine the junction depth, and is shown to have a positive effect on junction depth but leads in higher junction leakage due to the remaining of EOR defects.The characterization is done using Optical characterization tool (SEMILAB) for sheet resistance and junction leakage measurements.SIMS is used for Boron profile and junction depth.

Keywords: Plasma Immersion Ion Implantation, Ultra-Shallow junctions, Laser thermal process, Laser anneal
PACS: 52.77.Dq ; 85.40.-e

INTRODUCTION

The formation of ultra-shallow junctions with high electrical activation and low junction leakage is one of the most difficult challenge to take up in order to match the continuing scaling down of submicron CMOS in silicon. Indeed, due to the miniaturization of integrated circuits, the semiconductor industry needs to find new doping techniques for the realization of USJ as low as 20 nm [3]. As classical beam line ion implantation is now reaching its limits in terms of low energy, Plasma Immersion Ion Implantation (PIII) or Plasma Doping (PLAD) seem to be promising solutions to obtain the junction specifications given by the International Technology Roadmap for Semiconductor (ITRS) for 45 nm node and beyond [4-8]. These requirements are: a sheet resistance for PMOS source/drain extension lower than 518 ohms/sq and a junction depth smaller than 11.2 nm [3].

Different annealing processes have been developed in order to activate the junctions with respect to the given critical specifications. Among them, the Laser Thermal Processing (LTP) appears to be a key technique. Many studies have shown the potentiality of this technique as a solution for the requirements of the ITRS nodes <45 nm [9-12].

This paper presents electrical and physico-chemical characterizations of junctions realized with BF3 PIII followed by laser thermal processing with aim to obtain ultra-shallow junctions. Different implantation conditions (acceleration voltage/ dose) and laser conditions (laser types, fluence/ number of shots) are used for this study. Pre-amorphization is also used to confine the junction depth, and is shown to have a positive effect on junction depth but leads in higher junction leakage due to the remaining of

CP1066, *Ion Implantation Technology,* edited by E. G. Seebauer, S. B. Felch, A. Jain, and Y. V. Kondratenko

EOR defects.The characterization is done using Optical characterization tool (SEMILAB) for sheet resistance and junction leakage measurements.SIMS (CAMECA 7F with low energy primary Oxygen beam) is used for Boron profile and junction depth.

PLASMA IMMERSION ION IMPLANTATION TOOL

The PIII tool used in this study is named PULSION®. It uses a pulsed plasma and works at low pressure to minimize Plasma / surface interactions. Figure 1 shows a photographie of the 200 mm/ 300 mm compatible machine developped for advanced R&D applications.

FIGURE 1. PULSION® 200mm / 300mm Bridge tool : Advanced R&D tool with a single process chamber.

In this study we have used BF3 plasma on N type substrates to realize P+ N Junctions. Depending on acceleration voltage put on the substrate holder implantation depth can go down to 2 nm up to several hundreds of nm. [2]. In this study, acceleration voltage was fixed to 1 kV (implantation depth of about 10 nm at 5E18/cm^3) and saturation dose was used (dose above which no more boron can be introduced in silicon due to sputtering effects).

On some samples a beam line Germanium pre-implantation was performed (Ge 10 keV 1E15/cm^2) to build an amorphized layer of 20 nm.

LASER THERMAL PROCESS

The experimental set up of the laser thermal process is presented on Fig 2. Three kinds of UV laser sources have been tested: ArF (193 nm, 15 ns), KrF (248 nm, 35 ns) and XeCl (308 nm, 50 ns).

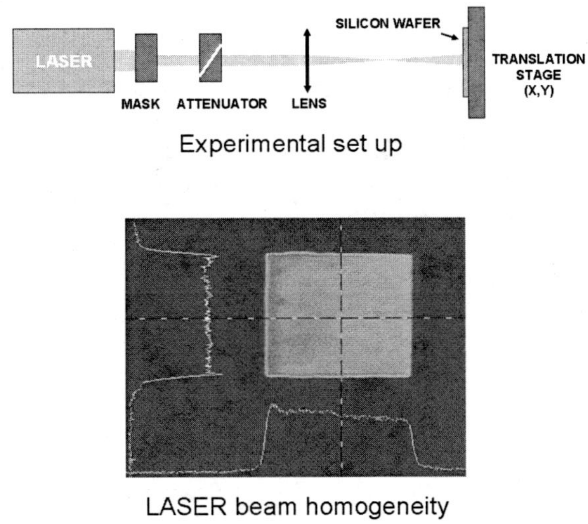

FIGURE 2. Laser Processing experimental set up and image of the beam homogeneity (using CCD).

For each laser source, the effect of fluence on sheet resistance, junction leakage, and junction depth is studied. Sheet resistance was measured both by 4 point probe and using JPV (junction photo voltage) tool by SEMILAB. This tool was also used to map junction leakage current. D-SIMS (CAMACA 7F) was used to measure junction depth.

RESULTS

Effect of Laser Source and Fluence

The effect of laser source and laser fluence on sheet resistance is presented in Figure 3. For each laser, a threshold is observed. Upon this threshold, activation is partial and sheet resistance highly decreases when increasing laser fluence (sub-melt mode). Above the threshold, activation is total and sheet resistance slowly decreases when increasing laser fluence (melt-mode). ArF laser source shows the lowest threshold with a full activation at 470 mJ/cm^2, whereas XeCl and then KrF need respectively 600 mJ/cm^2 and 750 mJ/cm^2.

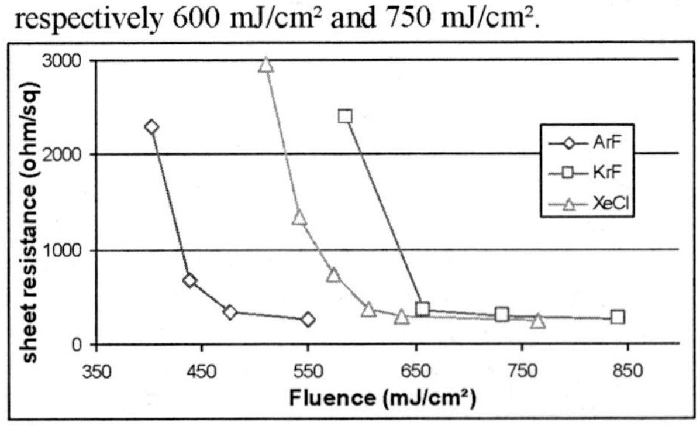

FIGURE 3. Effect of laser source and laser fluence on sheet resistance.

The effect of laser fluence on boron profile is shown on Figure 4: the higher the fluence is, the deeper the junction: this is due to the propagation in depth of the liquid phase when fluence is higher.

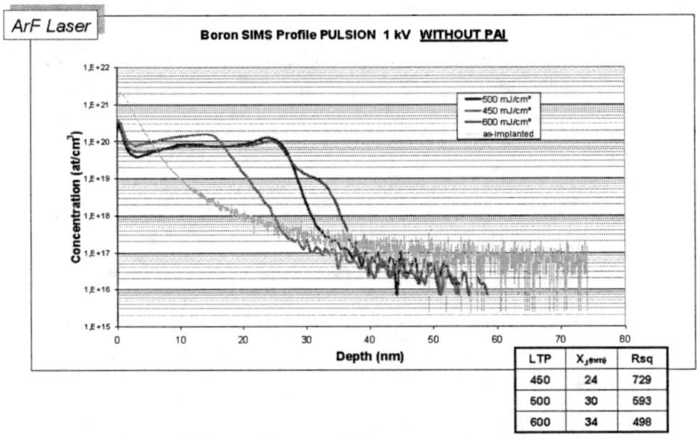

LTP	X_j	Rsq
450	24	729
500	30	593
600	34	498

FIGURE 4. Effect of Laser Fluence on boron profile.

Figure 5 shows the boron profiles on ArF and KrF annealed samples having the same sheet resistance of 500 Ohm/sq. Even if implantation depth on the as implanted sample is 10 nm (@5E18/cm^3), Junction depth after laser annealing is higher: respectively 40 nm for KrF and 110 nm for ArF. Due to its higher absorption in silicon ArF laser source allows a sooner activation and thus leads in a deeper fused depth when increasing fluence.

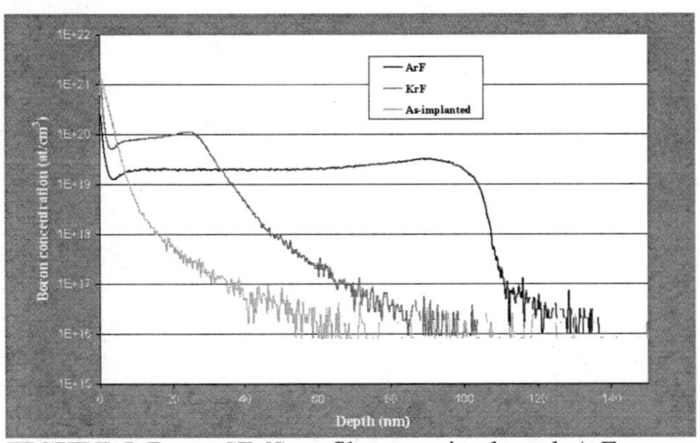

FIGURE 5. Boron SIMS profiles on as implanted, ArF and KrF annealed samples having the same sheet resistance of 500 Ohm/sq.

Effect of Ge Preamorphization

The effect of Ge preamorphization implantation (PAI) on sheet resistance and junction leakage current is shown on figure 6 (ArF laser source).

Ge PAI allows to reach the activation threshold sooner, but if on non PAI samples the leakage current is very low as soon as the junction is activated, on PAI samples it remains high and higher fluences are needed to reduce it (> 500 mJ/cm²). This may be due to the presence of EOR defects due to Ge PAI that needs a higher laser fluence to be solved.

Finally the best obtained junction were :

- With PAI : Xj = 23 nm , Rsq = 561 ohms/sq but high leakage current
- Without PAI : Xj = 24 nm , Rsq = 729 ohms/sq with a leakage current of 1E-6 A/cm²

For the same low leakage current, junction with PAI needs more fluence and is thus deeper than on the non PAI sample.

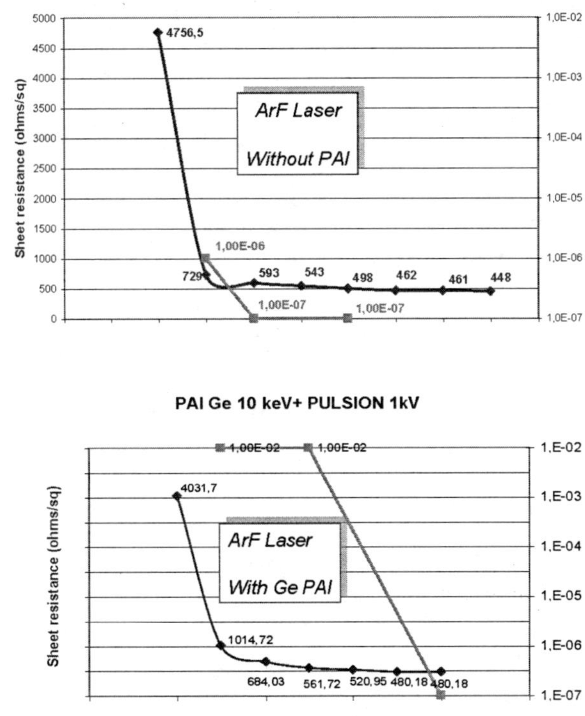

FIGURE 6. Effect of Ge Pre amorphization implantation (PAI) on sheet resistance and leakage current as function of Laser fluence: for samples without PAI, leakage current is low as soon as the junction is activated. For PAI, even if a lower fluence is enough to activate the junction a high fluence is needed to reduce the leakage current.

Then an additional RTP annealing process has been tried in order to dissolve the EOR defects and reduce junction leakage. Different temperatures have been tried: 600°C, 700°C, 800°C, 60 sec. It appears that junction leakage begins to decrease at 800°C. Junction depth variation induced has still to be investigated by SIMS. The study is on going.

CONCLUSION

We have shown that PULSION® implantation using BF3 plasma, followed by Laser Thermal process allows to build shallow, highly activated junction. It was demonstrated that the use of Ge Pre-amorphization implantation had a negative effect on leakage current. The comparison with ArF, KrF and XeCl laser source shows that ArF needs lower fluence to activate.

ACKNOWLEDGEMENTS

We want to acknowledge the French Ministry of Industry for its support through the FCE ALDIP Project.

REFERENCES

1. F.Torregrosa C. Laviron, F. Milesi, M. Hernadez, H. Faik, J. Venturini, *Proc. 14th International Conference on Ion Implant Technology*, (2004).
2. F.Torregrosa, H.Etienne, G.Mathieu, L.Roux, *Proc. 16th International Conference on Ion Implant Technology*, p. 6 (2006).
3. ITRS International Technology Roadmap for Semiconductors 2005. http://www.itrs.net
4. P.K Chu *et al.*. *Materials Science and Engineering*, **R17** (1996) 207-280.
5. P.K Chu *et al. Solid State Technology*. Oct 1999.
6. F.Le Coeur *et al.*. Surf. Coat. Technol. **125** (2000) 71-78.
7. D.Lenoble *et al. Surf. Coat. Technol.***156** (2002) 262
8. D.Lenoble, Etude, réalisation et intégration de jonctions P+N ultra fines pour les technologies CMOS inférieures à 0.18 µm. Thèse de 3ème cycle soutenue le 13/12/2000.
9. G.Fortunato, L. Mariucci, M. Stanizzi, V. Privitera, S. Whelan, C. Spinella, G. Mannino, M. Italia, C. Bongiorno, A. Mittiga. *Nucl. Instrum. Methods. Phys. Res.*, Sect.B **186** (2002) 401. 408
10. K.S.Jones, E. Kuryliw, R. Murto, M.Rendon, S.Talwar, IEEE CNF, Ion Implantation Technology (2000) p 111-114.
11. X.Zhang *et al.*, Int. *J.Heat Mass Transfer* **39** (1996) 3835.
12. C.Laviron *et al.*, *Jpn. Soc. Appl. Phys* (2001) 91.

Fabrication of Ultra-Shallow Junctions on 300 mm Wafers Using the Plasma Immersion Implanter PULSION® Followed by Spike Annealing Using LEVITOR Furnace

Frank Torregrosa[1], Hasnaa Etienne[1], Guillaume Sempere[1], Gilles Mathieu[1], Laurent Roux[1], Frederic Milesi[2], Frederic Gonzatti[2], and Xavier Pages[3]

[1] ION BEAM SERVICES, ZI Peynier-Rousset, rue Gaston Imbert Prolongée, 13790 Peynier, France
[2] CEA LETI – MINATEC; DRT / DPTS / SDOT;17 rue des Martyrs 38054 Grenoble cedex 9, France
[3] ASM Europe B.V., Versterkerstraat 8, 1322 AP Almere, The Netherlands

Abstract. Thanks to the European Project SEA-NET, an industrial version of the IBS Plasma Ion Implantation tool has been installed in LETI for the fabrication of Ultra-Shallow Junction for 45 & 32 nm CMOS on 200 and 300 mm wafers. In this study, we present first doping results on 300 mm wafers using BF_3 plasma. Annealing after PULSION implantation is perfomed by the ASM furnace LEVITOR. This tool, which works on conduction heating is emissivity independent for temperature control and allows reaching high ramp up and ramp down speeds (up to 900°C/s). Main characteristics of as implanted wafers are presented (metal contamination < 2E10/cm², SIMS depth profiles down to 5 nm). Then the effect of convection gas in the annealing process on sheet resistance, junction depth and uniformity is presented.

Keywords: Plasma Immersion Ion Implantation, Ultra-Shallow junctions, Spike Anneal
PACS: 52.77. Dq, 85.40.-e

INTRODUCTION

For more than 15 years now, Plasma Immersion Ion implantation or plasma doping technologies are forecasted to be a solution on applications where beam line implanters are limited [1 to 16]. The technology is now mature and the first PIII industrial systems are availables on the market.

Whithin the frame of European Integrated Projects "SEA-NET" and "PULLNANO", the French company ION BEAM SERVICES (IBS) has built an Industrial tool of Plasma Immersion Ion Implantation. This tool named PULSION®, which uses a pulsed plasma, is fully 200 mm / 300 mm compatible. It has been installed in CEA-LETI clean room to be tested for the achievement of doping processes for advanced microelectronics. In this paper, we present first results on shallow junctions obtained on 300 mm wafers using BF_3 PULSION® implantation followed by spike annealing using the convection heating tool LEVITOR from ASM. The effect of convection gas during activation is studied. We also compare beam line implantation of BF_2^+ performed on 300 mm industrial implanter with BF_3 Plasma implantation peformed on PULSION®.

ION IMPLANTATION

Plasma immersion ion implantation was performed on the PULSION® equipment that can be seen in figure 1. This equipment has already been described in ref [17,18]. It uses a pulsed plasma and is 200mm/300mm fully compatible.

FIGURE 1. PULSION® 200mm / 300mm Bridge tool: Advanced R&D tool installed in LETI with a single process chamber

CP1066, *Ion Implantation Technology,* edited by E. G. Seebauer, S. B. Felch, A. Jain, and Y. V. Kondratenko

For this study N type 300 mm wafers were implanted using BF_3 plasma using 1 kV acceleration votage and a dose of 1E15/cm². For this study, no specific surface preparation (native oxide removing) was performed before implantation.

As implanted boron profiles were measured by SIMS (CAMECA 7F using an O_2 primary beam at 430 eV) and can be seen of figure 2. With the used recipe of PULSION® the implantation depth was about 16 nm at 5E18/cm².

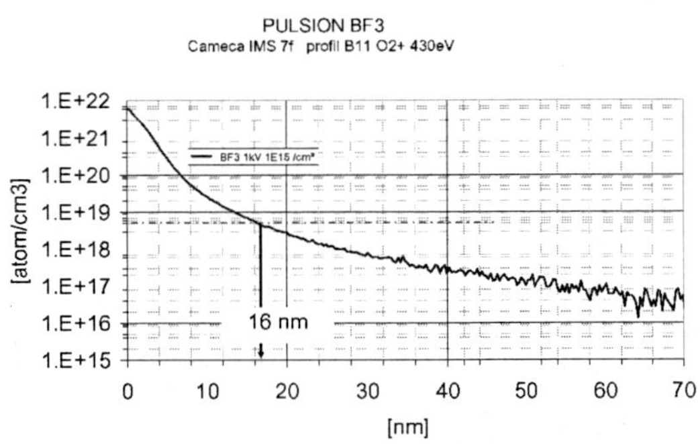

FIGURE 2. As Implanted Boron SIMS Profiles showing an implantation depth of 16 nm @ 5E18/cm³

It is interesting to note that the retained dose of boron measured by SIMS is 1.1E15 /cm², which is equal to the recipe dose if we consider that for USJ profiles, dose estimation accuracy from SIMS is about 10%. To compare results with beam line implantation some 300 mm wafers were implanted on an industrial 300 mm implanter at ST Crolles using the following implantation conditions: BF_2^+ ions, 1keV, dose of 1E15/cm², tilt 0°, without rotation. Unfortunately, implanted boron profiles were not measured on these samples.

SPIKE ANNEAL

The spike anneal process was performed on the LEVITOR furnace from ASM; the Levitor is an RTP system for single wafer rapid thermal processing. The system is based on conductive heat transfer. Therefore the processing in the Levitor system is independent of wafer emissivity, which is a known problem with lamp-based RTP equipment. In addition, the Levitor system is equipped with a very simple and reliable temperature control system based on traditional heater elements and thermocouples. This enables straightforward annealing at all temperatures between 200 and 1100°C. Within the Levitor system, the wafer is processed between two preheated blocks. The Levitor system uses a gas bearing to hold the wafer at a distance of 150 μm from the surfaces of the blocks. The narrow distance ensures an effective conductive heat transfer from the reactor blocks to the wafer. A schematic drawing of the reactor part of the Levitor system is shown in figure 3

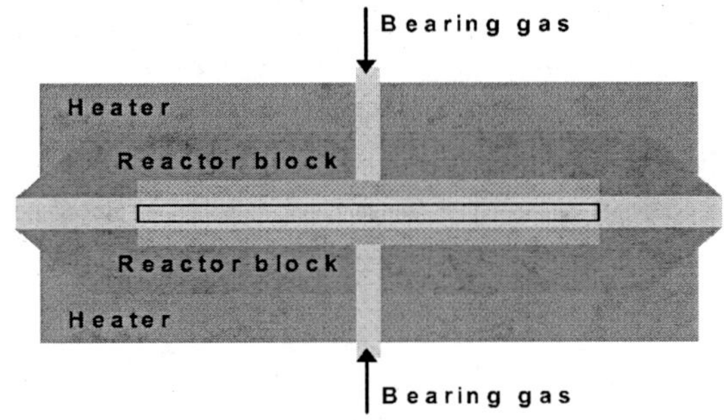

FIGURE 3. Schematic drawing of the reactor part of the Levitor system

An anneal process in the Levitor system is defined by three parameters:

- The block temperature,
- The bearing (carrier) gas,
- The reactor-closed time.

TABLE 1. Experimental table and main obtained results

	Equipment	Process step	1	2	3	4	5	6
Process			300 mm BEAM LINE IMPLANTER			PULSION		
N type 300 mm wafer	NA	Silicium CZ / N / 3-6 ohms*cm	x	x	x	x	x	x
Beam Line Implant	300 mm industrial implanter	BF2+ 1keV 1E15 Tilt0° Twist0° No rotation	x	x	x			
PULSION Implant	PULSION	BF3 1kV 1E15				x	x	x
Spike anneal	LEVITOR	Spike 1050°C under N2	x			x		
Spike anneal	LEVITOR	Spike 1050°C under He		x			x	
Spike anneal	LEVITOR	Spike 1050°C under He with 0,1% O2			x			x

			1	2	3	4	5	6
SHEET RESISTANCE	4 points probe	ohm/sq	985	1316	1399	686	898	938
		Standard déviation (%)	3,13	4,75	4.54	2,50	4,00	3,80
SIMS	Cameca 7F	Integrated Dose (at/cm²)	4.45E+14	4.65E+14	4.50E+14	4.51E+14	9.05E+14	7.76E+14
		Junction depth @ 5E18/cm3 (nm)	23	17	17	28	22	22

The combination of these three parameters determines the wafer temperature and the thermal budget of the process. The reactor-closed time is roughly the sum of the heat up time and the desired anneal/soak time. The heat up time is determined by the thermal conductivity of the bearing gas. For silicon wafers the heat up time to 900°C is approximately 1 second in helium and approximately 4 seconds in nitrogen (ramp up rates around 900°C/s and 200-300°C/s respectively). After annealing in the high temperature reactor unit, the wafer is transferred to a similar cooldown unit at room temperature.

Effect of convection gas was studied. Three gases were used:

- Pure helium which is known to be very efficient for thermal convection.
- He with 0.1% O_2 which is generally used to avoid pitting effects in high temp processes.
- Pure nitrogen.

Temperature during spike was 1050°C during approximatly 1 second.

After annealing wafers were characterised using 4 point probe mapping for sheet resistance measurements and Dynamic SIMS for boron profiles. Experimental matrices and main obtained results are summarized in table 1.

RESULTS

Sheet resistance

For all convection gases the sheet resistances obtained using PULSION® are about 45% lower than the ones obtained using beam line implanter. This is mainly due to the higher retained dose of boron. It is interesting to notice here that the boron retained dose on beam line implanter is only about 4.5E15 /cm² for a 1E15 /cm² recipe dose. This is due to the sputtering saturation effect. He and He + 0.1% of O_2 lead in quite the same sheet resistance with a smaller value when O_2 is used. Nitrogen allows a reduction in sheet resistance by 30%.

Sheet resistance non uniformity is between 3.1% and 4.75% on the beam line implanted wafers and 2.5% and 4% on PUSLION® implanted ones. These relatively high values may be explained by the fact that native oxide was not removed before implantation. Nevertheless, it is important to notice that PULSION® implanted wafers exhibit a better uniformity than the one implanted on the production 300 mm beam line implanter. Again nitrogen seems to have a benefit effect on sheet resistance uniformity with about 1.3% improvement.

Junction depth

Boron SIMS profiles after spike anneal for beam line implantation and PULSION® implantation are shown respectively in figures 4 and 5.

It's surprising to notice that whatever the convection gas, boron profiles from BF_2^+ beam line implantation and from BF_3 PULSION® are very similar. However as the retained dose of boron in PULSION® wafers is higher, the junction depth at 5E18/cm^3 is 5 nm deeper.

When we focus on the convection gas effect, it appears that He and He + 0.1% O_2 lead in exactly the same profile, whereas nitrogen is responsible for a 6 nm deeper diffusion.

This enhanced diffusion associated with the lower sheet resistance can be explained by the fact that nitrogen is a less efficient convection gas than helium. Thus, in the cooling phase of the Levitor process, temperature of the wafer might have stayed longer at high temperature. In other words, ramp up and ramp down slope rates are certainly less abrupt using nitrogen than using helium.

CONCLUSION

We have demonstrated with the use of PULSION® (BF_3 1kV 1E15/cm²) followed by LEVITOR® spike anneal that it is possible to make shallow junction on 300 mm wafers with the same quality as with the use of an industrial 300 mm beam line implanter (BF_2^+, 1keV, 1E15/cm²). More, PULSION® allows the ability to reach lower sheet resistance and better uniformity.

The use of nitrogen in LEVITOR® allows the ability to reduce sheet resistance and improve homogeneity by leads in a higher boron diffusion compared to He or He + 0.1% O_2.

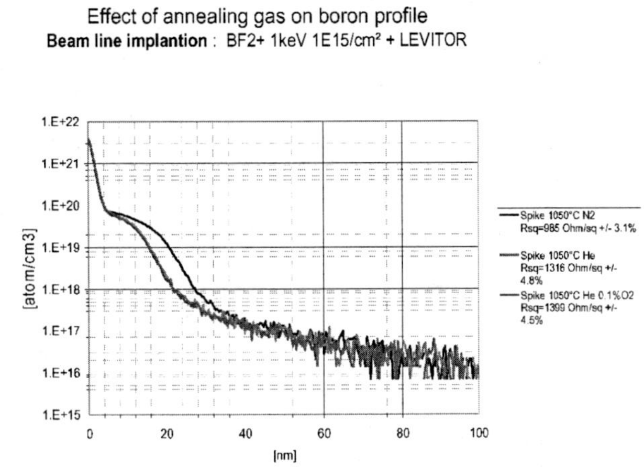

FIGURE 5. Boron SIMS Profiles of beam line implanted samples after Spike anneal

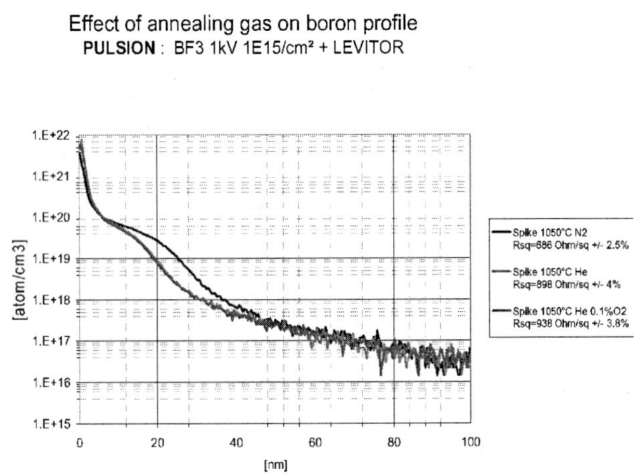

FIGURE 6. Boron SIMS Profiles of PULSION® implanted samples after Spike anneal

ACKNOWLEDGEMENTS

We want to acknowledge the European Commission for its support through PULLNANO and SEA-NET Integrated projects. Many thanks also to the CIMPACA Characterization platform for their work on SIMS.

REFERENCES

1. P.K. Chu *et al.*. *Materials Science and Engineering*, **R17** 1996 207-280
2. P.K. Chu *et al.*. *Solid State Technology*. Oct 1999
3. F. Le Cœur *et al. Surface and Coating Technology* **125** (2000) 71-78
4. M.J. Goeckner *et al. J. Vac. Sci. Technol. B* **17**(5), Sep/Oct 1999
5. Susan Felch *et al.*: Optimised BF_3 P^2LAD Implantation With Si-PAI For Shallow, Abrupt And High Quality P+/N Junctions Formed Using Low Temperature SPE Annealing.
6. P.K.Chu : Recent Applications of Plasma Immersion Ion Implantation. *Semiconductor International* / 165, June 1996.
7. S.B.Felch *et al.* : Plasma doping for the fabrication of ultra-shallow junctions. *Surface and Coatings Technology* **156** (2002) 229-236.
8. Woo Sik Yoo *et al.* : Results from SRTF show promise for shallow junction implant anneal. *Solid State Technology*. Dec 2002
9. A. Anders, *Handbook of Plasma Immersion Ion Implantation and Deposition*, Wiley-interscience publication, 2000
10. M.J Goeckner, S.B. Felch *et al. J. Vac. Sci. Technol. B* **13**(5) Sept/Oct 1999,p. 2290-2293
11. M. Takase, B. Mizuno : New doping technology-Plasma Doping for Next Generation CMOS process with Ultra-shallow Junction- LSI yield and surface contamination issues. IEEE 0-7803-3752-2 /97.
12. D. Lenoble *et al.*. *Surface and Coatings Technology* **156** (2002). 262-266.
13. D. Lenoble : Etude, réalisation et intégration de jonctions P+/N ultra-fines pour les technologies CMOS inférieures à 0.18 μm. Thèse de 3ème cycle soutenue le 13/12/2000.
14. D. Lenoble : Direct comparison of electrical performance of 0.1μm pMOSFET's doped by plasma doping or low energy ion implantation. International conference on ion implantation technology 2000.
15. M. Takase *et al.* : Shallow Source / Drain extensions for pMOSFET's with High Activation and low Process Damaged Fabricated by Plasma Doping. IEEE 0-7803-4100-7 /97.
16. K.S Jones *et al. Appl. Phys.Lett* **75** (1999) 3659.
17. F. Torregrosa *et al.* PULSION®: a versatile 200-300 mm bridge tool plasma immersion ion implanter for Ultra-Shallow Doping and nanotechology applications. IIT 08 Proceedings (2008).
18. F.Torregrosa, H.Etienne, G.Mathieu, L.Roux, Proc. 16th International Conference on Ion Implant Technology, p6 (2006)

Boron Profile Sharpening in Ultra-Shallow p^+-n Junction Produced by Plasma Immersion Ion Implantation from BF_3 Plasma

V. Lukichev[1], K. Rudenko[1], and A. Orlikovsky[1]
A. Pustovit[2] and A. Vyatkin[2]

[1]*Institute of Physics and Technology (FTIAN), Russian Academy of Sciences, Nakhimovsky prosp. 34, Moscow, Russia*

[2]*Institute of Microelectronics Technology (IMT), Chernogolovka, Russia*

Abstract. We have investigated plasma immersion ion implantation (PI^3) of boron with energies of 500 eV (doses up to $2x10^{15}$ cm^{-2}) from BF_3 plasma with He pre-amorphizing implantation (PAI) (energy 3 keV, dose $5x10^{16}$ cm^{-2}). Implanted samples were subjected to RTA (T = 900 to 1050 °C, t = 2 to 24 sec and spike anneal). SIMS analysis of boron profiles revealed its anomalous behavior. For short RTA times the profile tail (below $5x10^{19}$ cm^{-3}) moves toward the surface and then, as in the usual diffusion, toward the bulk at longer annealing times.

Keywords: plasma immersion ion implantation, pre-amorphizing implantation
PACS: 52.77.Dq

INTRODUCTION

The demand of ITRS (2007 Ed.) for ultra-shallow p-n junction depth is about half of the channel length. In the sub-100 nm technology, node plasma immersion ion implantation (PI^3) is a promising technique to meet the demand and is considered as potential solution of doping for Front End Processes.

EXPERIMENTAL

The processes of ultra-shallow plasma doping were studied in an experimental implanter designed at IPT RAS. The processing chamber (Figure 1) uses an ICP-plasma source with a planar inductor designed in this laboratory. It possesses a scalability property and is equipped with a permanent magnetic system to tune plasma uniformity in the plane near the chilled wafer chuck. The feeding of small helium flow under the wafer significantly improves the thermal contact and prevents the heating of the wafer over 100 °C.

The samples in the experiments were blank 150 mm wafers with n-doping up to the level of $2x10^{18}$ cm^{-3}. It is well known that to avoid boron channeling the step of pre-amorphizing implantation (PAI) is needed. In our previous work [1] we used Ar^+ or Xe^+ for this purpose. In these experiments we have investigated the PI^3 process with He^+ as PAI because of specific properties of amorphized silicon under He ion implantation [2, 3]. The process parameters for PAI step are shown in Table 1. A required amorphization dose of $5x10^{16}$ cm^{-2} was reached within 100 seconds.

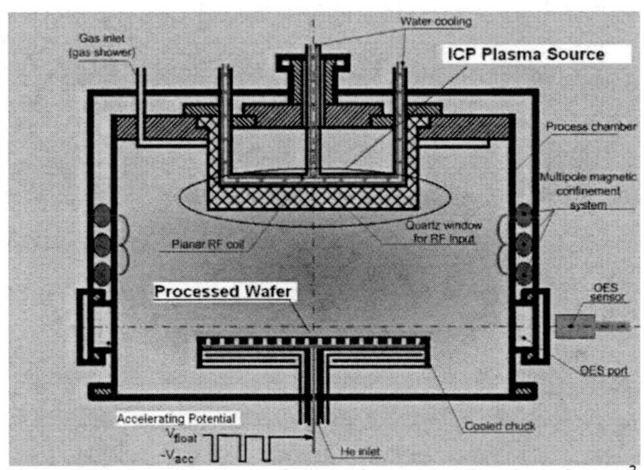

FIGURE 1. Scheme of Experimental Chamber for PI^3 Processing.

TABLE 1. Pre-Amorphizing Implantation Step

Process parameter	Value
PAI Source	He Plasma (Ions He^+)
Accelerating Pulse Voltage	3 kV
Pulse Duration	10 μsec
Pulse Repetition Rate	1 kHz
Dose for Si Amorphization	$5x10^{16}$ cm^{-2}

Boron implantation was carried out immediately after the PAI step by changing the feeding gas by BF_3. The process conditions are represented in Table 2. The time for a $2x10^{15}$ cm^{-2} boron dose was 20 seconds.

The plasma composition was monitored by optical emission spectroscopy (OES).

CP1066, *Ion Implantation Technology,* edited by E. G. Seebauer, S. B. Felch, A. Jain, and Y. V. Kondratenko

TABLE 2. Boron Implantation Step

Process parameter	Value
Boron-Containing Source	BF_3 Plasma (Ions BF_2^+, BF^+, B^+)
Accelerating Pulse Voltage	0.5 kV
Pulse Duration	10 μsec
Pulse Repetition Rate	1 kHz
Dose for Boron Implantation	$2x10^{15}$ cm^{-2}

Implanted samples were subjected to rapid thermal annealing (RTA) at different temperatures in the range of 900 to 1050 °C. We have applied the different annealing duration of 2 to 24 sec at flat temperature section, and have tested spike annealing with ramp-up rate of 250 °C/s and down rate of 100 °C/s). The details of RTA process are listed in Table 3.

TABLE 3. RTA Process Step

Process parameter	Value
Steady Temp. of Annealing	900 - 1050 °C
Time of Annealing	2 - 24 sec
Ramp-up Rate	250 °C/sec
Ramp-down Rate	100 °C/sec
Protective Ambient	High Purity Ar

RESULTS AND DISCUSSION

Both as implanted and post annealed boron profiles were analyzed by the SIMS technique with an IMS 4f Cameca instrument. The primary beam used was O_2^+ ions with energy of 2 keV to increase the depth analysis resolution and to minimize the redistribution of implanted boron by beam sputtering.

The initial implanted profile (Figure 2, 3 – as implanted curve) contains a pronounced tail in boron distribution.

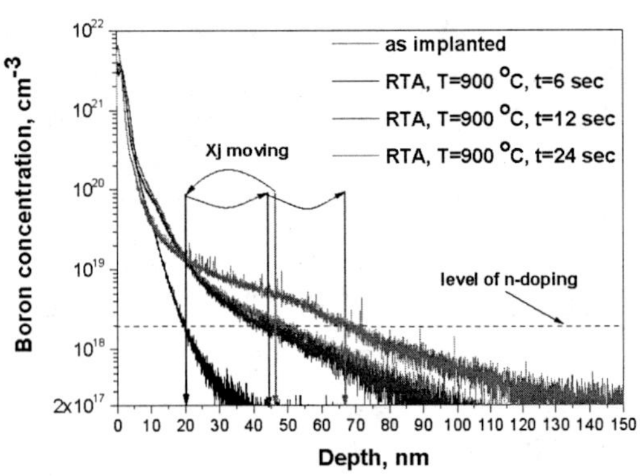

FIGURE 2. SIMS Boron profiles for series of RTA process time illustrating retrograde and forward movement of profile tails for T = 900 °C.

Our previous analysis [1] carried out by fitting of boron PI[3] experimental profiles from BF_3 plasma with the Pearson distribution shows two distinguished regions with different ranges and straggling. The ratio of these two ranges ratio is due to simultaneous implantation of molecular BF_2^+ ions and atomic B^+ from plasma without separation. This leads to the deterioration of doping profile steepness in junction location (**Xj**) formed by this technique.

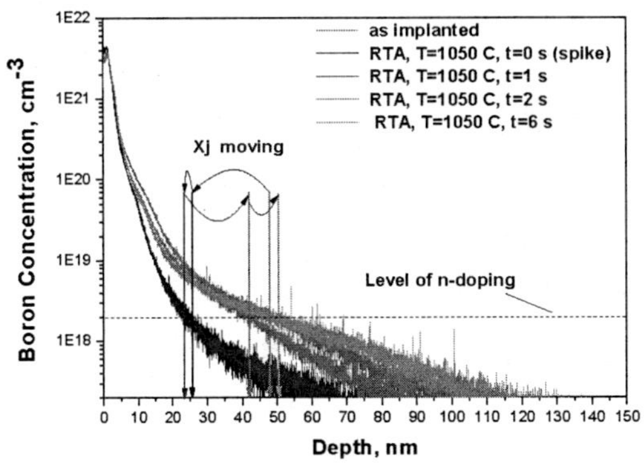

FIGURE 3. SIMS Boron profiles for series of RTA process time illustrating retrograde and forward movement of profile tails for T = 1050 °C.

A detailed SIMS investigation of annealed samples has shown an abnormal behavior of boron profile evolution for implanted structures using PAI with He^+ in comparison with samples when Ar^+ PAI was applied for silicon [1]. The main unusual effect is that at short RTA times the profile tail (below $5x10^{19}$ cm^{-3}) moves toward the surface, and then the tail diffusion changes its direction toward the bulk because it becomes controlled by ordinary diffusion laws (Figure 2, 3).

The alternation of B-diffusion direction was observed in the whole annealing temperature range, but the diffusion rate significantly increased at high temperatures both for the retrograde and forward diffusion. At T = 1050 °C the shift of the boron tail toward the surface takes place instantaneously even in spike RTA for the ramp-up and ramp-down time. It can be seen that estimated **Xj** are shifted to the surface due the retrograde diffusion approximately by a constant value of 20 nm and do not depend on annealing temperature.

The second consequence from the SIMS data is the dependence of the retrograde diffusion rate on the Boron concentration level. The lower the concentration, the faster the diffusion is. This results in the increase of estimated p+ - n junctions steepness from 20 nm/dec to 8.5 nm/dec even under non-optimal implantation conditions (long impurity distribution tails).

This behavior can be interpreted due to specific properties of amorphized layer produced by He^+ PAI. We carried out the structural comparison of a "classical" amorphous layer of silicon and the surface layer after He^+ PAI by fast electron diffraction (FED). It was revealed that the latter lost the first-range crystal order after a high dose Helium implantation by 3 keV ions while the ordinary amorphous silicon retained it. The previous investigation [2] of the properties of amorphous layer produced by 100 keV He^+ ion irradiation of silicon at room temperature and a dose of $5x10^{16}$ cm^{-2} showed that the treatment leads to the formation of a buried layer at the depth of maximum damage distribution. This layer is very unstable, contains a great excess of high mobility point defects (vacancies) and is transformed even at 150^0C. Under certain conditions, the formation of vacancy clusters and nano-pores is possible [3].

One of the possible explanations of the effect observed can be the surface transit effect accompanying SIMS analysis using O_2^+ ions as a primary ion beam. It is known [4] that at the normal incidence angle and relatively high ion energy of the primary ion beam (8 keV) the surface damaged layer arising due to sputtering reaches high thickness and the error in determinations of shallow junction depths can be as large as 30 nm. However, at the conditions of the present SIMS analysis the 2 keV O_2^+ and 64^0 incidence angle have been used. That minimized the probable surface transit effect of the SIMS analysis.

We propose the following phenomenological model for the retrograde boron diffusion in our experiments. There is evidently an excess concentration of vacancies in the surface layer, and neighboring surface serves as a natural sink for them. This gradient forms a flow of vacancies outside of silicon bulk. The high mobility of vacancies and instability of the helium amorphized layer lead to that the process takes place only for a short time at the initial period of annealing. It is known that there exists boron-vacancy complexes which are important in heavily doped materials [5] and could be enforced to the surface while vacancies go to sink. After this process is finalized, boron atoms diffuse in an ordinary manner according to the concentration gradient. The presence of additional impurities (e.g. fluorine co-implanted atoms) complicates the picture, which apparently needs more detailed investigations.

The revealed retrograde diffusion effect essentially differs from the uphill diffusion of boron [6] which takes place at high impurity concentration (order of solid solubility) for reason of boron precipitation to the near-surface region.

SUMMARY

The effect of retrograde motion of the tail of boron profile during RTA after PIII from BF3 plasma was observed. The effect is more pronounced at the very beginning of annealing. The retrograde motion effect becomes apparent after using He+ as PAI only. Ar+ and Xe+ do not produce this effect. Fast electron diffraction shows a specific structure of a silicon layer amorphized by He+ ions at the PAI step: "amorphous state with loss of the short-range of the crystal order".

The revealed retrograde motion of the boron tail at a short time of RTA is useful for sharpening of implanted profiles in the region of p+ - n junction. Experimental results show the boron profile steepness to increase from 20 nm/dec to 8.5 nm/dec in the location of p^+-n junction.

The investigated effect can be applied for impurity profile engineering.

ACKNOWLEDGEMENTS

The authors thank A. Lomov and M. Chuev for FED data and helpful discussion on the crystal structure of amorphized layers.

The investigation was supported by RFBR grant #05-02-08053.

REFERENCES

1. K. Rudenko, S. Averkin, V. Lukichev, A. Orlikovsky, A. Pustovit, A. Vyatkin. Proceedings of SPIE, Vol. 6260, p.626003-1 – 626003-9 (2005).
2. Buravlyov A.V., Vyatkin A.F., Egorov V.K., KireikoV.V., Zuev A.P. *Nucl. Instrum. and Meth. in Phys. Research* **B55**, p. 642 (1991). .
3. D. Alquier, F. Roqueta, L. Ventura *et al. Jpn. J. Appl. Phys. Part 1*, **41**(6A), p.3625 (2002).
4. Huyung-Ik Lee, Dae Won Moon, Suhk Kun Oh, Hee Jae Kanf, Hyun Kyong Kim and Jeong Yun Won, e-*J Surf. Sci. Nanotech.* **2**, 22-27, (2004)
5. J. Adey, R. Jones, D.W. Palmer, P.R. Briddon, S. Oberg. *Phys. Rev. B.*,**71**, 165211 (2005)
6. R. Duffy, V.C. Venezia, A. Heringa, T.W.T. Husken, M.J.P. Hopstaken, N.E.B. Cowern, P.B. Griffin, C.C. Wang. *Appl. Phys. Lett.*,**82**(21), p.3647 (2003).

PULSION®: A Versatile 200 to 300 mm Bridge Tool Plasma Immersion Ion Implanter for Ultra-Shallow Doping and Nanotechology Applications.

Frank Torregrosa[1], Hasnaa Etienne[1], Guillaume Sempere[1], Gilles Mathieu[1], Laurent Roux[1], Frédéric Milesi[2], Frédéric Gonzatti[2]

[1]*Ion Beam Services, ZI Peynier-Rousset, rue Gaston Imbert Prolongée, 13790 Peynier, France*
[2]*CEA LETI – MINATEC; DRT / DPTS / SDOT; 17 rue des Martyrs 38054 Grenoble cedex 9, France*

Abstract. Thanks to the European Projects SEA-NET and PULLNANO, an industrial version of the IBS Plasma Ion Implantation tool has been installed in LETI for the fabrication of Ultra-Shallow Junctions for 45 and 32 nm CMOS on 200 and 300 mm wafers. In this paper, we present the main machine characteristics (layout, contamination, homogeneity, reproducibility, uptime) and we show that implanted wafers are compatible with the 32 nm nodes requirements. Then doping results (sheet resistance vs junction depth and leakage current) using several annealing technics are presented (laser, flash and spike anneals). In the second part of the paper, versatility of the tool is demonstrated with its use for other applications (3D doping, hydrogenation, solar cells…)

Keywords: Plasma Immersion Ion Implantation, Ultra-Shallow junctions, Trench doping, Flash.anneal, Spike Anneal, Laser anneal
PACS: 52.77. Dq, 85.40.-e

INTRODUCTION

For more than fifteen years now, Plasma Immersion Ion implantation or Plasma doping technologies are forecasted to be a solution on applications where beam line implanters are limited [1 to 5]. The technology is now mature and the first PIII industrial systems are available on the market.

Whithin the frame of European Integrated Projetcs "SEA-NET" and "PULLNANO", The French Company ION BEAM SERVICES (IBS) has built an Industrial tool of Plasma Immersion Ion Implantation. This Tool, named PULSION®, uses pulsed Plasma and is fully 200 mm / 300 mm compatible. It has been installed in CEA-LETI clean room to be tested for the realization of doping processes for advanced microelectronic. In this paper, we present the principal machine characteristics (layout, contamination level, homogeneity…). Then we show the main obtained results for ultra-shallow junctions using BF3 plasma followed by several annealing technics (SPER, Spike, Flash, and Laser). Even if the main application of this USJ development is focused on Source/Drain extensions of advanced CMOS devices, we show in the last part of this paper that it can also be applied to photodetector or solar cells. In the second part of the paper, as it has already been proved on other Plasma doping tools [6], we show that PULSION® can also be used for polysilicon doping or 3D doping (conformal or trench doping).

MACHINE PRESENTATION

The advanced R&D version of PULSION® installed in LETI clean room can be seen on Figure 1. It's a cluster compatible bridge tool (200 mm/ 300 mm fully compatible) with a small layout loading robot and which can support 2 PIII process chambers plus an additionnal process station (for resist stripping or annealing). The tool has also the possibility to have three process chambers, which allows reaching high throughput for high volume production.

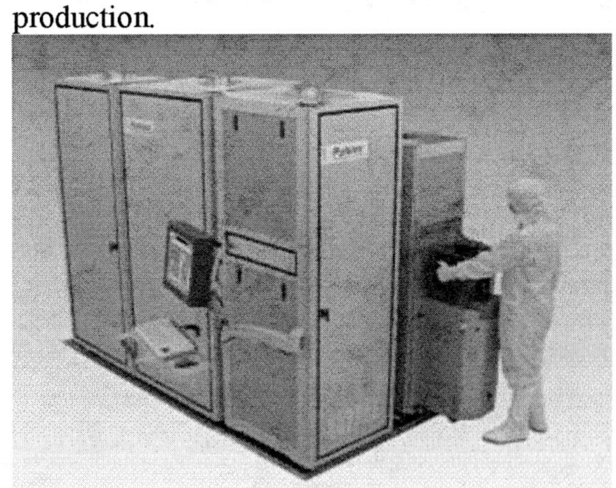

FIGURE 1. PULSION® 200mm / 300mm Bridge tool: Advanced R&D tool installed in LETI with a single process chamber.

Thanks to its pulsed and confined plasma and its ability to work at low pressure (down to 1E-4 mbar) with small gas flows, the machine minimizes plasma

CP1066, *Ion Implantation Technology*, edited by E. G. Seebauer, S. B. Felch, A. Jain, and Y. V. Kondratenko

/ surface interactions and thus reduces risks of etching or deposition on wafer or on process chamber. This reduces maintenance and consumable costs.

The main machine characteristics are the following:

- Cluster Compatible with 1, 2 or 3 process chambers
- For R&D or High Volume production
- 200mm / 300 mm compatible
- Low cost of ownership, low consumable costs
- Reliability: Up time = 95%
- Small footprint: 4.3 m^2 for a single chamber R&D machine to 8.9 m^2 for a 3 chambers production tool.
- Low Metal Contamination: Compatible with ITRS Specification for 32 nm node (Al : < 2E11/cm^2, Other < 2E10 /cm^2) : This level of metal contamination was checked on BF3 1kV 1E15/cm^2 and Ar 1kV 1E15/cm^2 processes using the last generation of TXRF from RIGAKU : all the values were under the detection limit of this tool.
- Non Uniformity: better than 1% (BF3 Doping process, 10 nm implantation depth
- Reproducibility: 1% (BF3 Doping process, 10 nm implantation depth)
- Available plasmas from: BF_3, B_2H_6, PH_3, AsH_3, H_2 and other gases...

TABLE 1. Effect of BF_3 pressure on sheet resistance: PULSION® BF_3 1kV 1E15/cm^2 followed by RTA (1050°C 20s)

BF_3 Pressure	2E-4 mbar	8E-4 mbar	2E-3 mbar
Sheet resistance	630 (Ohm/sq)	700 (Ohm/sq)	725 (Ohm/sq)

ULTRA-SHALLOW JUNCTION APPLICATION

Typical doping profiles of boron using BF3 plasma have already been presented [7] and can be seen on fig 2. Implantation depth is linear with acceleration voltage applied on the subtrate holder. One can notice that even if the "machine dose" was the same, the retained dose of boron reduces when the voltage goes down. As for beam line implantation this saturation effect is due to surface sputtering which increases when energy is reduced. Nevertheless in PIII chemical etching can also appear: we show here that reducing the BF3 pressure allows to reduce surface etching. Increasing the retained dose of boron lowers sheet resistance after activation (see table 1).

For BF_3 plasma with acceleration voltage from 0.5kV to 5 kV typical etching rates of Si (mesured by AFM) are between 0.03nm (at 500V) and 0.2 nm (at 5kV) per 1E15/cm^2. In any case surface roughness is about 0.2 nm (RMS).

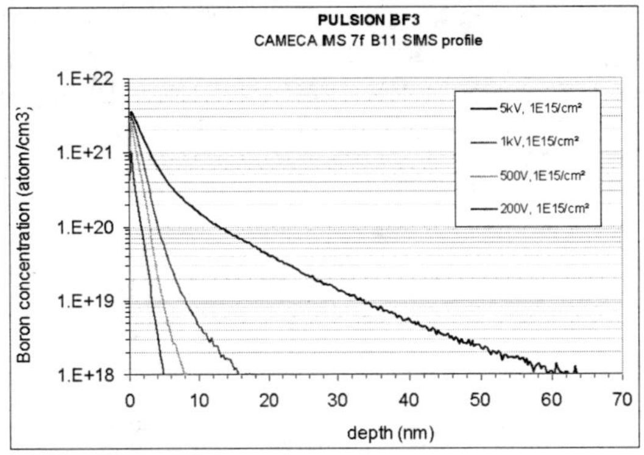

FIGURE 2. Boron SIMS Profiles as a function of the polarization voltage.

Annealing Process

For the fabrication of ultra-shallow junction, the post implant annealing process must ensure a high activation level while minimizing diffusion. The most used technologies are SPER (about 750°C during 10s, after a Pre-amorphisation step), SPIKE (around 1050°C during 1s), Flash (1300°C, some ms) and laser (>1300C, 10 to 200 ns). We have compared these annealing technologies on PULSION® implanted wafers (200 & 300 mm wafers) [8, 12]. The doping plasma was BF3, with acceleration voltage from 500V to 1 kV. On some samples a Ge preamorphisation implantation and/or a carbone pre-implantation was done (on beam line implanter). 4 point probe and junction photovoltage was used to measure sheet resistance and leakage current. Dynamic SIMS was used to measure junction depth.

If PULSION® + Laser allow to obtain the smallest sheet resistance (593 Ohm/sq / 30 nm) the junction depth is high due to the melt mode that was used. Ge preamorphisation allowed to reduce the junction depth to 11nm but with an increase of the sheet resistance to 1300 Ohm/sq. Due to its lower temperature (1050°C), spike was no efficient to highly activate dopant, and SPER was often a cause of high leakage current due to remaining defects. Finally Flash was the best annealing technique to achieve very shallow junction. Table 2 and figure 3 show the effect of Ge preamorphisation and Carbon pre-implantation on junction characteristics for Flash Anneal.

It appears that, as for conventional implanted junctions, the PAI leads to a decrease in sheet resistance while it slightly increases junction depth.

But on the other hand, it has a negative effect on the junction leakage. This is due to the fact that even with a high temperature annealing, the EOR defects are not completely dissolved. Carbon implantation added to a Ge PAI allows decreasing junction depth as well as Rsq. This may be due to the fact that Carbon atoms have a positive effect on the interstitial capture efficiency which leads to a decrease of Rsq and also blocks the Boron diffusion. Unfortunatly, leakage current due to the Ge PAI still remains high.

TABLE 2. Effect of PULSION® polarisation voltage, with and without Ge Pre-amrphisation implantation.

Implantation condition	**PULSION BF3 1kV saturation dose**	**PULSION BF3 0.5kV saturation dose**	**PAI Ge 10keV 5E14/cm²+ PULSION BF3 1kV saturation dose**	**PAI Ge 10keV 5E14/cm²+ PULSION BF3 0.5kV saturation dose**
Annealing process	Flash RTA 1300°C	Flash RTA 1300°C	Flash RTA 1300°C	Flash RTA 1300°C
Rsq (ohm/sq)	1527	4678	885	1309
Xj @ 5E18/cm3	11	6	14	8
Junction leakage (µA/cm²)	0.021	0.015	0.264	0.037

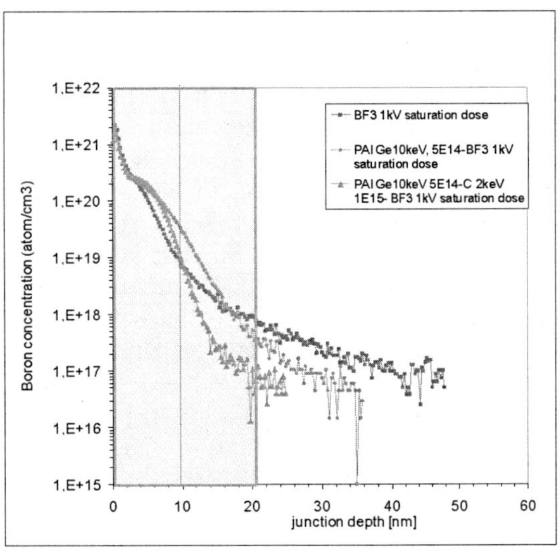

FIGURE 3. Boron SIMS profiles for flash annealing with 3 process conditions: without PAI, with PAI Ge 10keV and with PAI Ge 10keV + C 2keV.

Decreasing the acceleration voltage on PULSION® from 1kV to 0.5 kV has allowed to highly reduce the junction depth (see Table 3). But this has also the drawback to reduce retained dose of boron and thus increases Sheet resistance. In this case, Ge pre-amorphisation implantation had a deep positive impact and surprisingly did not lead in a higher leakage current (maybe defects are too far from the junction). Thus the best Shallow junction obtained had a junction depth of 8 nm for a sheet resistance of 1309 Ohm/sq and a leakage current of 0.037 µA/cm² which is compatible with relaxed target for 22 nm node.

TRENCH DOPING, CONFORMAL DOPING

Another great advantage of PIII, already published [6], is that under certain process conditions (rather high pressure and low acceleration voltage) it is possible to make conformal doping (thanks to a collision mecanism). In IIT 06 we have shown that PULSION was able to dope conformaly deep trenches [9] (200µm deep) (see fig 4).

But we have shown that conformal doping can also be performed on highly texturized surfaces. A PULSION® BF3 2 kV annealed using RTA on a laser texturized surface leads to a 60% increase in photocurrent, studied by LBIC, whereas it leads only to a 40% increase with a classical beamline implantation. This improvement is due to the good conformal doping of this surface obtained with PULSION.

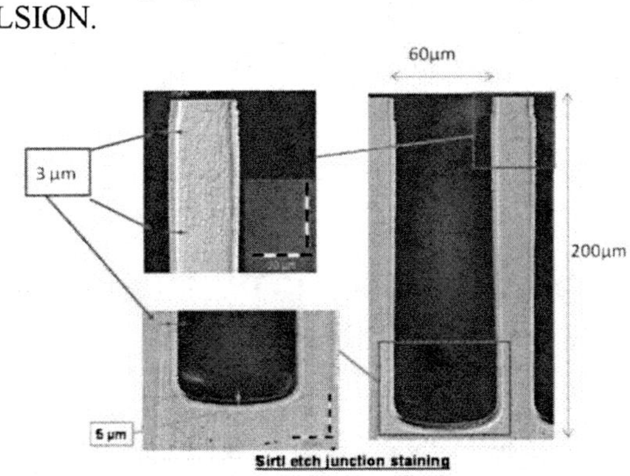

Conformal doping of 60µm wide/200µm deep silicon trench

FIGURE 4. Conformal doping of deep trenches using BF3 Plasma followed by a long annealing.

POLYSILICON DOPING

Several articles have already been published showing the economical interest to use Plasma Doping for polysilicon implantation, mainly for Flash memory applications [13]. B2H6 is often the chosen gas because it allows to put more boron into the surface (less sputtering or surface etching), but B2H6 plasmas are often responsible of undesirable polymer deposition on the walls of the process chambre. Thanks to the ability to work at lower pressure, we have shown that it was possible to reach low sheet resistance on a 100 nm polysilicon (deposited on top of a 100nm oxyde) using BF3. Wafers were implanted by PULSION® at 5kV and followed by RTA (1055°C 60s). On 200 mm wafers the obtained sheet-resistance was 287 Ohm/sq with a non uniformity of 1.7% (for an initial polythickess homogeneity of 0.6%).

OTHER APPLICATIONS

Thank to its ability to implant high dose of ions in a small depth of material, and with an acceptable cost, PULSION® is likely to fullfill a lot of requirements for niche markets. Among them , for exemple, the fabrication of shallow junction for emitters in photodetectors or solar cells application (with high sensitity in UV and blue spectrum) or hydrogenation have already been proved [10, 11].

CONCLUSION

The plasma immersion ion implantation tool PULSION® is now commercially available. It can be used either for volume production or for advanced R&D applications on 200 and 300 mm wafers where high doses on thin layers are needed.

We have shown that associated with different annealing technics it can fullfill the requirements for Source/Drain extension doping on 32 nm and even 22 nm node devices.

Thanks to the high thoughput of PIII technology, and its ability to work at low pressure, PULSION® is an economically efficient solution for Polysilicon doping even using BF3 plasma.

Trench and conformal doping were also demonstrated, this can have several applications from advanced devices to texturized surface doping.

Finally its unique capabilities for niche markets applications as photodetectors, power devices, solar cells, hydrogenation… were also demonstrated.

ACKNOWLEDGEMENTS

We want to acknoledge the Eurorepean Commision for its support through PULLNANO and SEA-NET Integrated projects. Many thanks also to the CIMPACA Characterisation plateform for their work on SIMS. Acknowledgements to LP3 and IM2NP Laboratories from Marseille University for their studies in Laser activation and solar cells applications.

REFERENCES

1. F. Le Cœur *et al. Surface and Coatings Technology* **125** p. 71-78 (2000)
2. S.B.Felch *et al. Surface and Coatings Technology* **156** 229-236. (2002)
3. Woo Sik Yoo *et al. Solid State Technology*. Dec 2002
4. D. Lenoble *et al. Surface and Coatings Technology* **156** 262-266. (2002).
5. D. Lenoble. "Direct comparison of electrical performance of 0.1µm pMOSFET's doped by plasma doping or low energy ion implantation." in *Proceedings of the International Conference on Ion Implantation Technology-2000*.
6. D. Lenoble *et al.* VLSI Technology Symposium Proceedings (2006)
7. F. Torregrosa *et al.* "Down to 2 nm Ultra-shallow Junctions : Fabrication by IBS Plasma Immersion Ion Implantation Prototype PULSION®." in *Proceedings of the International Conference on Ion Implantation Technology-2006*
8. H. ETIENNE at all. Ultra-shallow Junctions fabrication by Plasma Immersion Implantation on PULSION®® followed by different annealing processes. IJWT 08, Shangaï
9. S. Nizou *et al.* "Deep Trench Doping by Plasma Immersion Ion Implantation in Silicon." in *Proceedings of the International Conference on Ion Implantation Technology-2006.*
10. V. Vervisch. PhD thesis, 2007
11. F. Torregrosa, C. Laviron, H. Faik, D. Barakel, F. Milesi, S. Beccaccia. Realization of ultra-shallow junctions by PIII, application to solar cells. (PBII-2003), San Antonio, USA, 16-19 Sept. 2003. *Surface & Coating Technology* **186** pp. 93-98. (2004)
12. F. Torregrosa, C. Laviron, F. Milesi, Miguel Hernadez H. Faik, Julien Venturini. "Ultra-shallow P+/N junctions using Plasma immersion ion implantation and laser annealing for sub 0.1 µm CMOS devices." in *Proceedings of the International Conference on Ion Implantation Technology-2005.*
13. Z.Fang, T.Miller, E. Winder, H. Persing, E. Arevalo, A. Gupta, T. Parrill and V. Singh. "B2H6 PLAD Doped PMOS Device Performance". in *Proceedings of the International Conference on Ion Implantation Technology-2006.*

SIMS/ARXPS – A New Technique of Retained Dopant Dose and Profile Measurement of Low Energy Doping Processes

Shu Qin, Kent Zhuang, Shifeng Lu, Allen McTeer, Wendy Morinville, and Kari Noehring

Micron Technology, Inc., 8000 S. Federal Way, Boise, Idaho 83707, USA

Abstract. A newly-developed surface analysis technique, which combines Secondary Ion Mass Spectrometry (SIMS) with Angle-Resolved X-ray Photoelectron Spectroscopy (ARXPS), was used to achieve more accurate results of the retained impurity doses and profiles for ultra-low energy implants, including conventional beam-line implant and plasma immersion ion implantation (PIII). Using this method, it has been found that the B_2H_6 (Diborane) PIII demonstrates thicker native oxide and much more B dose loss during RTP and surface clean treatments than conventional beam-line ion implantation, due to the higher surface B concentration. In order to match the electrical parameters of the device, PIII must consider higher nominal dose to compensate the B loss.

Keywords: Secondary Ion Mass Spectrometry (SIMS), Angle-Resolved X-ray Photoelectron Spectroscopy (ARXPS), plasma immersion ion implantation (PIII), plasma doping.
PACS: 85.40.Ry, 52.77.Dq, 61.72.U-, 68.49.Sf, 82.80.Ms

INTRODUCTION

The device size must continue to shrink in order to increase the performance and decrease the cost of ULSI (Ultra-Large Scale Integrated circuit). In order to minimize the so-called short-channel effect, the ultra-shallow junction depth (x_j) and low enough sheet resistance (R_S) are required for the source/drain (S/D) fabrication of the MOS transistors. For example, for 45nm technology node, based on the International Technology Roadmap on Semiconductor 2005 (ITRS2005) [1], the junction depth of S/D extension must be shallower than 6.5nm and the activated R_S of S/D extension must be lower than 650Ω/sq. To meet these requirements, ultra-low energy ion implantation, or an alternative doping method, plasma immersion ion implantation (PIII) or plasma doping has to be used to overcome the physical limits of the conventional beam-line ion implants. PIII is a doping technique offering many applications in materials and semiconductor processing [2-4] that has been extensively investigated due to its unique advantages including simpler system, lower cost, higher throughput, and improved device performance [5]. However, PIII is a more complicated process than conventional ion implant because of its plasma environment and non-mass-analyzed feature. PIII process is a combination of ion implantation, deposition, and etching (including chemical etching and physical sputtering) depending upon the plasma species and processing conditions. It is very critical to accurately determine the doping results including the retained dopant profiles, which determine x_j and abruptness, and the retained dopant doses, which determine R_S, for the ultra-low energy implants. These critical requirements are particularly relevant for PIII because PIII have some unique properties compared with the conventional beam-line implants. These unique properties include a different impurity profile and more surface treatments, with potential deposition and etching effects besides the ion implant. PIII is a plasma-based process and without mass-separation mechanism, so that the impurity profiles are an exponential-like function (with an impurity peak at the surface) rather than the Gaussian-like function (with an impurity peak beneath the surface) for the beam-line ion implant [5,6]. The PIII involved surface treatments, such as deposition and etching effects, cleaning, thermal annealing, *etc.*, will much more significantly impact the impurity profile and dose.

SIMS is a widely used technique to measure implants, both for the implant impurity dose and profile shape. SIMS is able to provide a relatively accurate dose and profile for the conventional beam-line implant in a high energy regime. However, SIMS has shown its limitations to measure ultra-shallow implant profiles, mainly due to its inherent sputtering process. SIMS sputter process does not reach equilibrium state near the surface region due to its initial beam penetration and interaction with the matrix composition. This is known as the SIMS transient effect, which causes ion yield variations near the surface. In addition, the SIMS sputter rate will not be constant due to the presence of native oxide at the surface. When an average sputtering rate is used to calibrate the depth scale, the measured profile near the surface will be distorted. This is particularly true for PIII processes, where the impurity profile near the surface will dominate, as described previously. Hence, SIMS cannot accurately determine the surface concentration for the PIII process. Moreover, SIMS is not able to distinguish the dose retained within the surface oxide layer.

As a non-destructive, non-interfered, quantitative surface analysis technique, ARXPS [7] is a well-developed technique for the surface chemical

CP1066, *Ion Implantation Technology,* edited by E. G. Seebauer, S. B. Felch, A. Jain, and Y. V. Kondratenko

composition analysis. It is particularly well-suited for studying material deposition processes, although it has some physical limitations for ion implantation, in which the impurity can distribute much deeper than the maximum ARXPS sampling depth (~100Å with Al Ka X-ray source). While ARXPS provides important information about the chemical composition of a film, it provides measurements of film thickness and the distribution of components or depth profiles within the film. The generation of depth profiles from ARXPS is not a direct process but involves calculating the ARXPS response from trial depth profiles and comparing this with the ARXPS experimental data. The trial profile is then adjusted to improve the fit. This process is continued until the optimum fit is achieved.

In this paper, we optimize the surface measurement by combining SIMS technique with ARXPS technique, to overcome the transient effect of the SIMS measurement and achieve more accurate results of the impurity profiles and doses for low energy implant including conventional beam-line implant and PIII. A maximum entropy algorithm is used to extract the ARXPS compositional depth profiles within the oxide layer for boron implanted poly-Si. The conventional SIMS technique is used to determine the doping profile and dose. Because the SIMS results near surface may not be correct due to the sputtering effects mentioned above, we use ARXPS to determine the surface chemical composition, and in turn, the surface material, for example, a possible native oxide, and the doping profile and dose near the surface. The required information can be obtained by including or extracting the surface results dependent upon the applications.

EXPERIMENTS AND DISCUSSIONS

SIMS/ARXPS technique is used to determine the retained impurity profile and dose for as-implanted and after anneal of the conventional beam-line ion implant and PIII processes. The starting materials are 300mm blanket wafers with 700Å in-situ Phosphorus-doped n-type poly-Si film/1kÅ SiO_2/Si substrate. The process conditions of the conventional beam-line ion implant are as follows: ion species is $^{11}B^+$; energy is 4keV; dose is $1.5 \times 10^{16}/cm^2$; implant angle is 0°. The PIII system is a pulsed, RF-excited continuous plasma-doping system, described in detail elsewhere [6]. The PIII process conditions are as follows: the doping gas is B_2H_6 (Diborane), the nominal implant voltage and dose are -7kV and $3\times10^{16}/cm^2$, respectively. After implant, the wafers were cleaned by standard strip/clean processes to mimic the real production photoresist removing and cleaning processes. The wafers then were processed by rapid thermal-annealing process (RTP) to activate the impurities. There are three RTP cycles and the conditions are 965°C/20s + 995°C/20s + 1050°C/0s, and all RTPs are in N_2 ambient. The process conditions are listed in **Table 1**.

TABLE 1. Process conditions and ARXPS analysis of beam-line implants and PIII implants.

Wafer ID	Implant Type	Energy	Dose (/cm²)	RTPs	Surface concentrations (atomic %) B	C	O	Si	SiO_2 Thick (nm)	B in SiO_2 (/cm²)
1	Beam-line	4keV	1.5E+16	No	3.7	6.7	39.9	49.8	0.44	1.00E+13
2	Beam-line	4keV	1.5E+16	Yes	2.3	2.4	47.6	47.7	0.99	1.00E+14
3	PIII	7kV	3.0E+16	No	17.7	10.3	34.2	37.8	1.6	2.44E+15
4	PIII	7kV	3.0E+16	Yes	19.3	2.5	45.4	32.9	3.13	6.10E+15

All processed wafers, including the as-implanted and annealed wafers, were sent on for SIMS/ARXPS measurements to determine the B retained profiles/doses in poly-Si layers. The annealed wafers were also sent on for four-point probe measurement to determine the sheet resistance (R_S), and for Hall-effect measurement to determine the carrier (hole) doses and mobility (μ).

Figure 1 shows SIMS profiles of the conventional beam-line ion implant and B_2H_6 PIII on **(A)** as-implanted and **(B)** annealed wafers. SIMS measurement used an O_2 primary beam with energy of 1keV and normal incidence. The values beside the curves in **Figure 1** are SIMS/ARXPS measured B retained doses in Si, which have extracted the B loss in the surface oxides.

PIII differs from beam-line ion implant due to plasma environment and non-mass separation. The different B profiles between conventional beam-line implant and PIII are demonstrated by the SIMS measurements shown in **Figure 1**. The conventional beam-line ion implantation shows a Gaussian-like function with a B concentration peak beneath the surface due to mono-atomic, mono-energetic ion implantation. However, the PIII shows an exponential-like function with a B concentration peak at the surface due to the combination of multi-species, multi-energetic, collisional, and deposition phenomena [6]. As shown in **Figure 1**, the B concentrations at the surface
of PIII processed material are ~five times higher than that of the beam-line implantation counterpart.

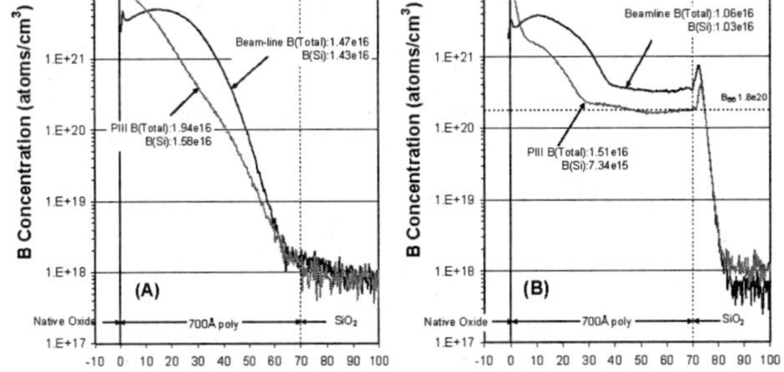

FIGURE 1. SIMS profiles of beam-line B ion implant and B_2H_6 PIII on **(A)** as-implanted and **(B)** annealed wafer.

Figure 2 shows the results of the ARXPS measurement of a beam-line ion implant on **(A)** as-implanted and **(B)** annealed wafers. **Figure 3** shows the results of the ARXPS measurement of a B_2H_6 PIII on **(A)** as-implanted and **(B)** annealed wafers. As shown in **Figure 2**, the beam-line implant wafer has 0.44nm thick native oxide for the as-implanted wafer, and 0.99nm for the annealed wafer. As shown in **Figure 3**, the PIII wafer has 1.6nm thick native oxide for the as-implanted wafer, and 3.13nm (also doubled the as-implanted) thick native oxide for the annealed wafer. The oxides on as-implanted wafers were grown natively by the environmental temperature and moisture after clean process. The oxides on the annealed wafers were roughly doubled the as-implanted wafers because they were formed from both native oxidation and the oxidation by the oxygen residue in the annealing ambient during RTPs. The surface oxides on PIII wafers are much thicker than beam-line implant wafers due to doping dependence of oxidation, because the surface B concentration of PIII is much higher than that of the beam-line implant [8]. The B doses within the native oxides were measured by ARXPS directly. The ARXPS results including surface oxide thickness and chemical composition in the oxide are listed in **Table 1.**

PIII shows much more significant surface impact than beam-line implant, including thicker surface native oxides and much higher B content within the oxides. The B content in the surface oxide also can be measured by SIMS technique, but it is not accurate because of the transient effect. We used the oxide thickness determined by ARXPS to separate the B doses in the oxide and in the bulk Si for the SIMS measurements. The results are listed and compared in **Table 2.**

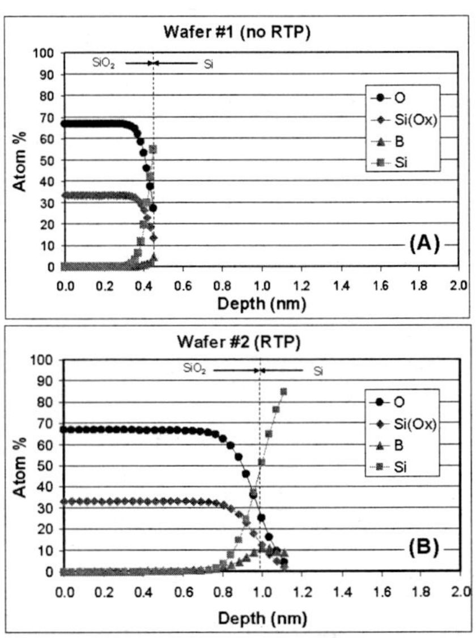

FIGURE 2. ARXPS data of a beam-line B ion implant on **(A)** as-implanted and **(B)** annealed wafers.

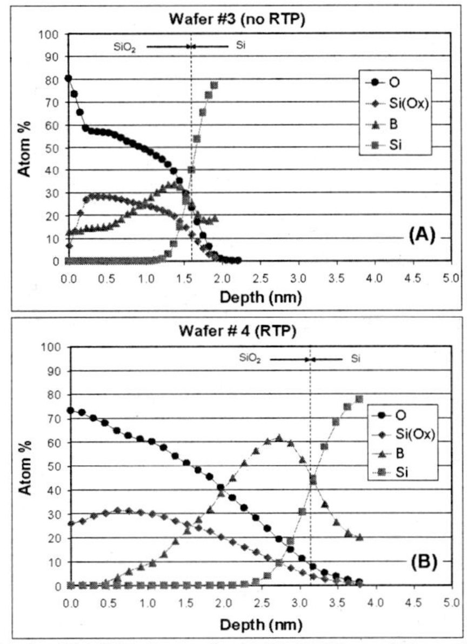

FIGURE 3. ARXPS data of a B_2H_6 PIII on **(A)** as-implanted and **(B)** annealed wafers.

TABLE 2. Comparison of beam-line and PIII implants.

Wafer ID	Implant Type	RTPs	R_S (Ω/Sq.)	SIMS Dose (/cm^2) B(Total)	SIMS Dose (/cm^2) B(Si)	SIMS Dose (/cm^2) B(SiO_2)	ARXPS Dose (/cm^2) B(SiO_2)	Hall Effect Dose (/cm^2) carrier	Hall Effect Dose μ (cm^2/Vs)
1	Beam-line	No		1.47E+16	1.43E+16	3.72E+14	1.00E+13		
2	Beam-line	Yes	525.37	1.06E+16	1.03E+16	3.15E+14	1.00E+14	1.44E+15	8.5
3	PIII	No		1.94E+16	1.58E+16	3.59E+15	2.44E+15		
4	PIII	Yes	592.59	1.51E+16	7.34E+15	7.76E+15	6.10E+15	1.11E+15	9.5

Table 2 demonstrates that the surface native oxides of beam-line implant are thinner and the B content in the oxide of beam-line implant can be negligible under the current implant conditions. However, for the PIII wafers, the B loss in the surface native oxide and the impact to the total retained B dose are very significant. Compared with the beam-line ion implant, PIII wafers have much thicker native oxide and much higher, at least one order of magnitude, B dose in the native oxides. The B doses in the oxide result from the segregation phenomenon [8,9] for both as-implanted and annealed wafers. It is well-known that B impurity has a segregation coefficient m < 1 during oxidation process [8,9]. The definition of the segregation coefficient is: m = (equilibrium impurity concentration in Si)/(equilibrium impurity concentration in SiO_2) near SiO_2/Si interface. The effect of the segregation phenomenon of B impurity near the SiO_2/Si interface is that during oxidation B are absorbed from Si and piled-up on the SiO_2 side, and B on the Si side are depleted. Therefore, the B content in SiO_2 is higher than that in Si near the SiO_2/Si interface.

Figures 4 and **5** show SIMS/ARXPS B profiles near SiO_2/Si interface of a beam-line B ion implant and a B_2H_6 PIII, respectively. ARXPS data shows more accurate B profiles and quantitatively demonstrates the B segregation phenomenon in SiO_2/Si system. The B components in the oxide must be extracted from the total B dose because the B in the oxide cannot be

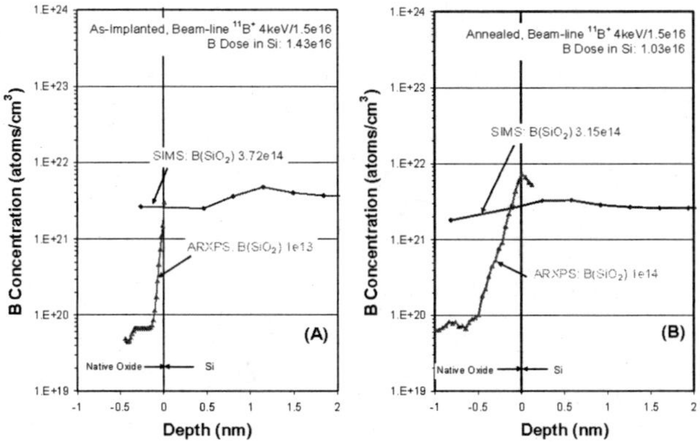

FIGURE 4. SIMS/ARXPS profiles of a beam-line B ion implant on **(A)** as-implanted and **(B)** annealed wafers.

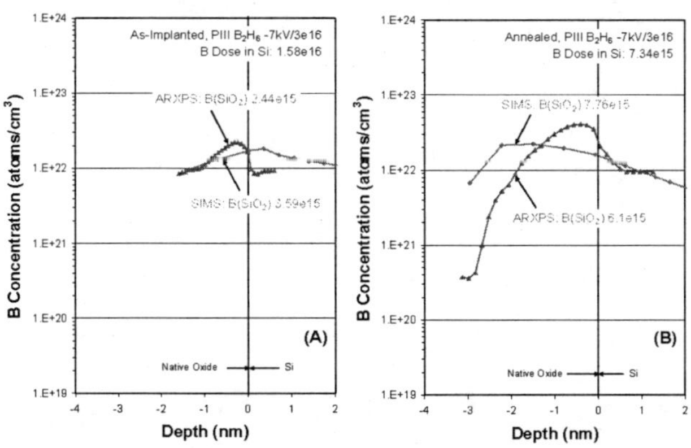

FIGURE 5. SIMS/ARXPS profiles of a B_2H_6 PIII on **(A)** as-implanted and **(B)** annealed wafers.

activated and will be lost by the follow-up clean processes as well.

The B content in the native oxide is one B loss mechanism for both as-implanted and annealed wafers. For the annealed wafers, additional B is lost due to out-diffusion. Therefore, there are two root causes for B loss mechanism for the annealed wafers: out-diffusion during RTPs and B segregation in surface native oxides.

For out-diffusion B loss, as shown in **Table 2**, conventional beam-line implant lost 28% B dose after RTPs, from 1.47e16 to 1.06e16/cm^2. PIII lost 22% B dose after RTPs, from 1.94e16 to 1.51e16/cm^2. For the B loss to the native oxides, conventional beam-line implant only lost ~3% B dose to the native oxide on the as-implanted wafers, from 1.47e16 to 1.43e16/cm^2, and only lost ~3% B dose to the oxide on the annealed wafer, from 1.06e16 to 1.03e16/cm^2. However, PIII processes show much more B loss to the native oxides due to unique impurity profiles of PIII as mentioned previously. PIII process, lost 19% B to the native oxide on the as-implanted wafers, from 1.94e16 to 1.58e16/cm^2, and lost 51% B to the oxide on the annealed wafers, from 1.51e16 to 7.34e15/cm^2.

The results of SIMS/ARXPS technique confirm our previous hypothesis in which PIII has much more significant surface impact on doping results than the conventional beam-line implant. In order to confirm this new method on the impurity profile/dose measurements, R_S values measured by four-point probe method, and carrier (hole) doses and mobility measured by Hall-effect mobility measurement for both doping methods are listed and compared in **Table 2**. The results in **Table 2** show correct correlations among R_S, retained B doses in Si, and carrier doses for both beam-line ion implant and PIII. By the conventional SIMS technique, PIII demonstrates higher total implant B doses than those of the beam-line implant, with 32% and 42% higher for as-implanted and annealed wafers, respectively. However, by SIMS/ARXPS technique, PIII demonstrates lower B retained doses in Si layer than those of the beam-line implant, with 29% lower for annealed wafers. This is because PIII loses more B from the native oxides by the B segregation phenomenon. These results can be confirmed by a lower carrier dose and higher R_S value of PIII, as shown in **Table 2**, than those of the conventional beam-line implant.

CONCLUSION

SIMS/ARXPS - a newly-developed surface analysis technique, which combines SIMS with ARXPS, was used to achieve more accurate results of the impurity doses and profiles for ultra-low energy implants. Using this method, it has been found that the B_2H_6 PIII demonstrates thicker native oxide and much more B dose loss during RTP and surface clean treatments than conventional beam-line ion implant, due to the higher surface B concentration. In order to match the electrical parameters of the device, PIII must consider higher nominal dose to compensate the B loss.

REFERENCES

1. ITRS2005 – Front End Processes, http://www.itrs.net/Links/2005ITRS/Home2005.htm, Semiconductor Industry Association, pp. 23, 2005.
2. C. Chan and S. Qin, US Patent, No.5,508,227, April 16, 1996.
3. T. T. Sheng, US Patent, No.6,632,482, October 14, 2003.
4. P. K. Chu, S. Qin, C. Chan, N. W. Cheung, and L. A. Larson, *Mat. Sci. & Eng. Reports*, **R17**, 207 (1996).
5. S. Qin and A. McTeer, *IEEE Electron Dev.*, **54**,2497 (2007).
6. S. Qin and A. McTeer, *Surf. Coat. Tech.*, **201**, 6759 (2007).
7. R. Champaneria, P. Mack, R. White, and J. Wolstenholme, *Surface and Interface Analysis*, **35**, 1028 (2003).
8. S. Wolf and R. N. Tauber, "Silicon Processing for the VLSI Era", in *Vol. 1: Process Technology*, Lattice Press, Sunset, CA, 1986, pp. 213-228.
9. A. S. Grove, O. Leistiko, and C. T. Sah, *J. Appl. Phys.*, **35**, 2695 (1964).

NON-DOPING AND NOVEL APPLICATIONS

Automotive SOI-BCD Technology Using Bonded Wafers

H. Himi and S. Fujino

DENSO CORPORATION, Ashinoya, Kota-cho, Nukata-gun, Aichi Pref., 444-0193 Japan

Abstract. The SOI-BCD device is excelling in high temperature operation and noise immunity because the integrated elements can be electrically separated by dielectric isolation. We have promptly paid attention to this feature and have concentrated to develop SOI-BCD devices seeking to match the automotive requirement. In this paper, the feature technologies specialized for automotive SOI-BCD devices, such as buried N^+ layer for impurity gettering and noise shielding, LDMOS with improved ESD robustness, crystal defect-less process, and wafer direct bonding through the amorphous layer for intelligent power IC are introduced.

Keywords: SOI, BCD, Automotive, buried layer, ESD, Well design, Bonding, Amorphous layer.
PACS: 85.30.De, 81.65.Tx, 68.35.Ct, 85.40.Ry.

INTRODUCTION

For automotive semiconductor devices, higher toughness against environment (temperature, humidity, etc.), higher reliability and noise immunity are demanded compared with consumer application. For example, life-times for the gate oxide and for the metallization are guaranteed more than about 20 years and Tjmax is recently needed to be 175 ℃ [1].

The SOI wafer is excelling in high temperature operation and noise immunity because the integrated elements can be electrically separated by dielectric isolation; hence, it is suitable for the automotive semiconductor. We promptly paid attention to this feature and put the SOI-BCD (BiCDMOS) process to practical use for power IC in 1994.

After that, we have concentrated to develop SOI-BCD devices seeking to match the automotive requirement.

In this paper, some examples in which ion implantation technology plays an important role, will be introduced. Concretely, improvement of gate dielectric breakdown characteristics and shielding of voltage fluctuation by buried N^+ layer, ESD immunity improved LDMOS by optimum well design, crystal defect-less process, and wafer bonding through the amorphous layer for intelligent power IC are discussed.

SOI-BCD AND AUTOMOTIVE REQUIREMENT

Figure 1 shows the block diagram of the electronic control unit (ECU), which is the brain of the various kinds of control systems used in an automobile. It is composed of a microcomputer, peripheral ICs and elements [2].

SOI-BCD devices are used as an IC, which integrates peripheral ICs and discrete elements in ECU, an IC which integrates all elements in the small scale ECU, and an IC added directly on the actuator.

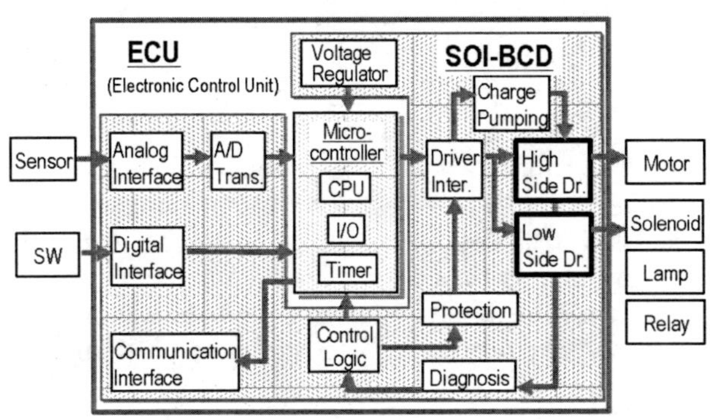

FIGURE 1. Application field of SOI-BCD in automobile

As the method of making SOI-BCD devices, using the thick film bonded SOI wafer and separating each element on the chip with the trenches is best. The reason is that the SOI-BCD devices made by the trench dielectric isolation method have no parasitic

CP1066, *Ion Implantation Technology*, edited by E. G. Seebauer, S. B. Felch, A. Jain, and Y. V. Kondratenko

operation shown in Figure 2 [2], have toughness against a negative input, and are excellent in automotive environmental conditions (high temperature operation and high breakdown voltage, etc.). By using this method, it is possible to integrate the multi-voltage devices, which operate under the different common-voltage conditions, and it is possible to have both high voltage and high integration.

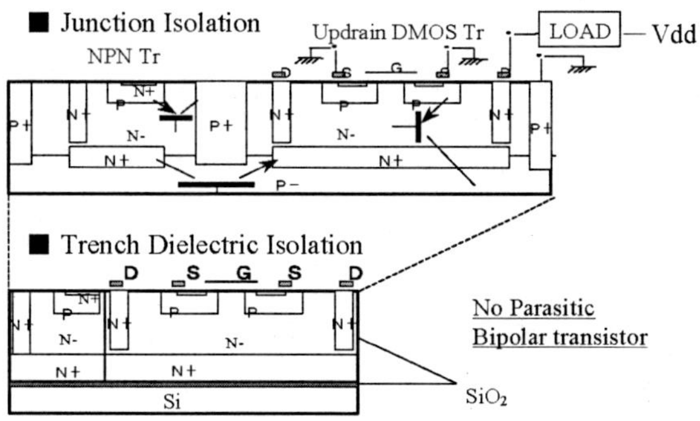

FIGURE 2. Merit of SOI structure: no parasitic bipolar transistor exists by the trench dielectric isolation.

FEATURES OF AUTOMOTIVE SOI-BCD DEVICES

Figure 3 shows the schematic cross section of our 4th generation SOI-BCD [1]. It has bipolar, CMOS, LDMOS, non-volatile memories, etc. The thickness of the SOI layer is about 15 μ m and the thickness of the buried oxide layer is 1.3 μ m. Each element is dielectrically separated using trench isolation. We have advanced the refinement of this TD(Trench Dielectric-isolation)-BCD process, and the devices specialized for automotive such as the CrSiN thin film resistor (which has a low TCR and is adjustable by laser irradiation) and ESD immunity improved LDMOS have been integrated [2]. In addition, the buried N^+ layer is prepared in the bottom of the SOI layer, which has important roles such as high current drivability, high surge immunity, improvement of gate dielectric breakdown characteristics and shielding of voltage fluctuation.

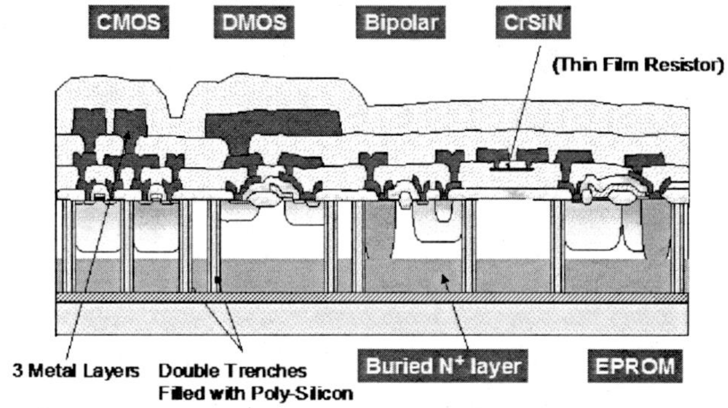

FIGURE 3. Cross sectional view of TD-BCD

Device Structure

Buried N+ Layer

A bonded SOI wafer with buried N^+ layer has been made as follows. First, N-type impurity ions such as arsenic or antimony, which have small diffusion constants, are implanted into the bonding surface of the device forming the wafer to a predetermined concentration level. Then it is bonded to the supporting wafer. Ion implantation technology is used for keeping the surface roughness small to make bonding.

Figure 4 shows the issue of impurity gettering in SOI wafers schematically.

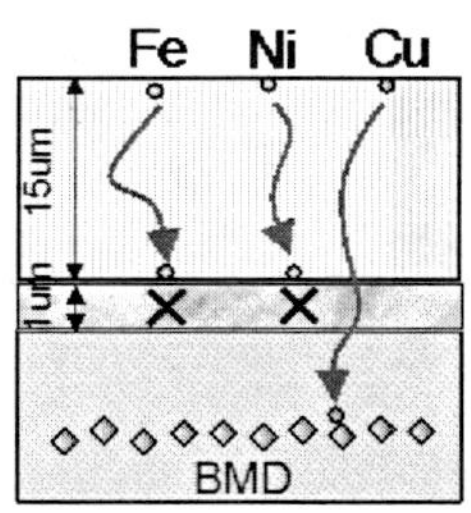

FIGURE 4. Singularity of SOI Structure: bulk micro defects are not valid due to existence of buried oxide layer

Differing from the bulk wafer, because of the existence of about 1um thick buried oxide layer, bulk micro defects or back side strain in the supporting wafer are not valid. On the contrary, when the buried N^+ layer exists, it is known that the buried N^+ layer has the function of the site for gettering. We have compared the 33nm gate dielectric breakdown characteristics in the cases of with and without a BN^+ layer. As a result, the gate dielectric breakdown voltage of the SOI wafer with a BN^+ layer is the same of the bulk wafer; however as for the SOI wafer without a BN^+ layer, ~1% of measured samples have the gate breakdown voltage less than 6MV/cm.

Next, we discuss the shielding of voltage fluctuation by a buried N^+ layer. Figure 5 shows the schematic model of noise interference. For example, the voltage fluctuation of power-GND, substrate voltage and supply voltage causes the voltage interference through the parasitic capacitance of the buried oxide. Although, the parasitic capacitance of the buried oxide is smaller than the junction capacitance of the PN-junction isolation, the AC component of steep voltage change occurring in automotive circumstances such as load damp can easily pass through the buried oxide.

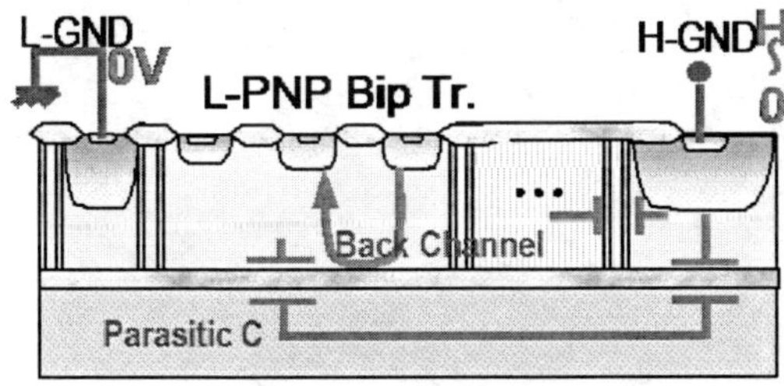

FIGURE 5. Schematic model of noise interference

As shown in figure 6, to evaluate the interference effect on the devices in SOI layer, the latch-up immunity of CMOS structure is measured when the substrate voltage falls down steeply from Vp to 0 in τ seconds [3,9]. As a result, shown in figure 6(c), the limitation of normal operation is in the line of 8 V/μ sec in the case of SOI wafer without buried N^+ layer. On the contrary, when the buried N^+ layer is inserted, the operation area expands with the decrease in resistance and has reached to 250V/ μ sec at Rs=10Ω/cm2. [9]

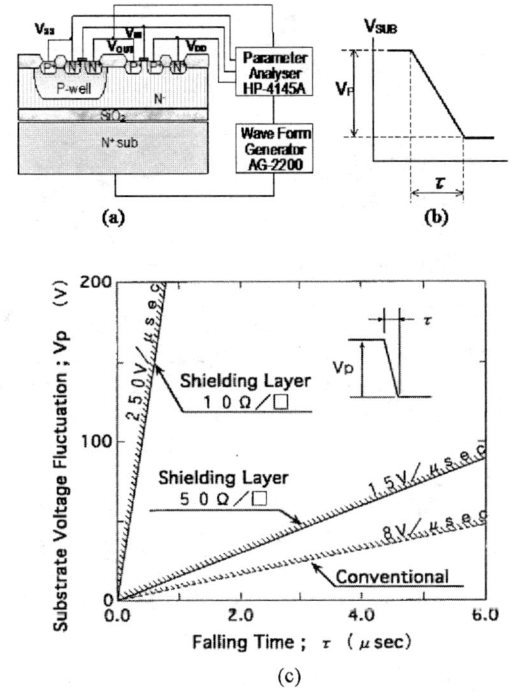

FIGURE 6. Evaluation of voltage fluctuation effect on device operation (a) evaluation system by measuring SOI-CMOS Latch-up, (b) supplied substrate voltage and (c) relationship between Vp and τ on the latch-up occurrence. Faulty operation occurs in the upper side of each line. Each line corresponds to the resistance of BN^+ layer.

ESD Immunity Improved LDMOS

LDMOS has the desirable feature of low on-state resistance. However, it has low immunity against EDS surge. When the LDMOS is integrated into SOI-BCD devices as an output terminal, it is exposed to ESD circumstances. Hence, ESD immunity improved LDMOS has been developed [4,5,6]. From the simulation of breakdown phenomena caused by the ESD serge, it is found that in the conventional LDMOS structures, due to the existence of local non-uniformity such as metallization resistance, once the current increases in one particular cell of the LDMOS (which is made up of a large amount of transistor cells), localized high current crowding occurs and ultimately thermal destruction.

Based on this mechanism, we have proposed in figure 7 the novel device structure of adding an adaptive N layer to the drain region to relax the electric field and adding an adaptive P layer under the source region to suppress the current generation.

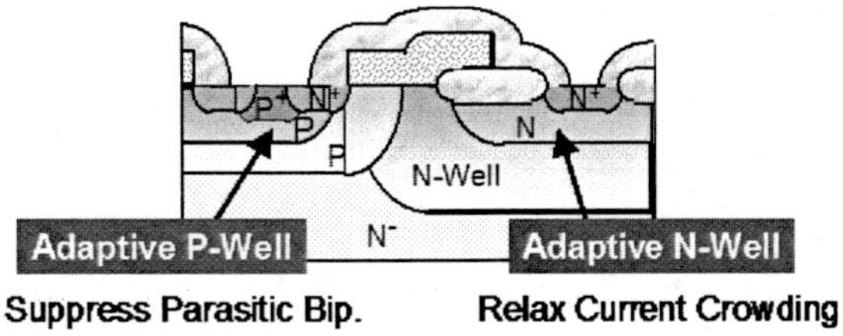

FIGURE 7. Structure of ESD immunity improved LDMOS

The effect of Impurity concentration of adaptive N and P layers on the sustain characteristics has been studied. It has been found that the higher concentration of adaptive N (to about 5×1017 cm-3 in our experiments) has resulted in the higher current density, at which the snapback occurs. However, the curvatures of their sustain characteristic are not enough positive over the entire range of ESD serge current. When the adaptive P layer is added with the adaptive N layer, the non-snapback characteristic is obtained. To optimize the impurity concentration of adaptive layers, the high amount of ESD robustness grater than 15kV/mm2 has been achieved as shown in figure 8.

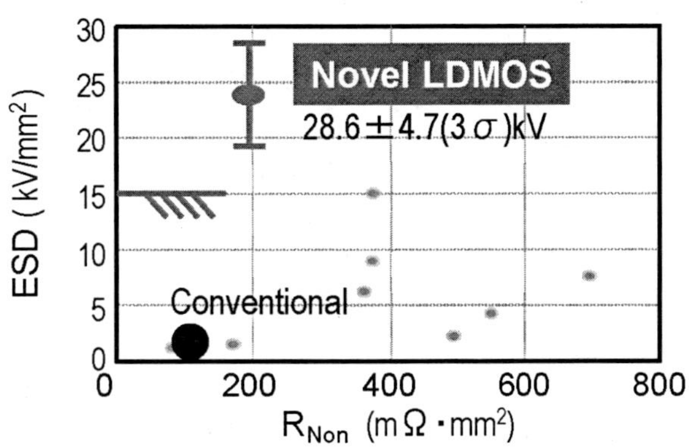

FIGURE 8. ESD robustness of novel LDMOS: optimum well-design resulted in the robustness greater than 15kV.

Processing Technology

Crystal Defect-less Process

In the case of SOI-BCD devices using thick SOI wafer and trench dielectric isolation, dislocation defects are easily induced by the stresses of trench and LOCOS oxidation. For example, in figure 9 dislocation is observed on a (111) sliding plane from LOCOS edge in the boron implanted base region of bipolar transistors [7]. For developing the no defect thermal process, careful thermal design must be necessary.

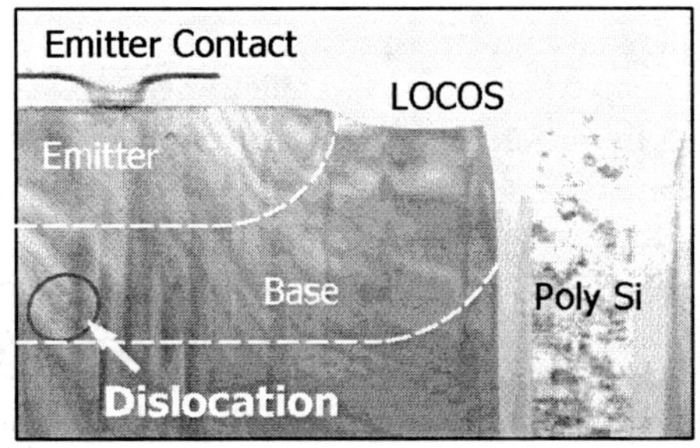

FIGURE 9. Example of crystal defect occurred in SOI structure. Dislocation is observed in the boron implanted base region of bipolar transistors.

To the above mentioned way, we have evolved the SOI-BCD technology seeking to match the automotive requirement. As an example of SOI-BCD devices, control IC for the airbag is shown in figure 10 [1].

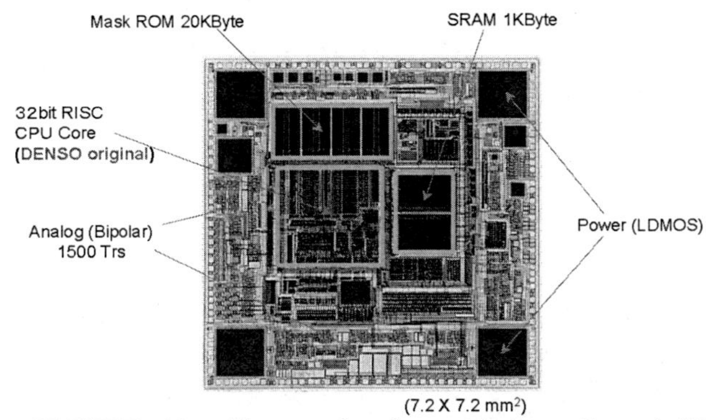

FIGURE 10. Photograph of an Air Bag Control IC fabricated by SOI-BCD technology

The 32bit RISC original microcomputer specialized in the automotive, and 4ch LDMOS transistors that improve the ESD tolerance for the power outputs, etc., are installed. By adopting this, it is possible to reduce the number of parts in the ECU by half, and the miniaturization of 40% has been achieved as an ECU.

Substrate

Wafer Bonding Through Amorphous Layer For Intelligent Power IC

The ideal substrate structure for the intelligent power IC is the partial SOI structure [8,9] shown in figure 11 because it can integrate the control circuits in the dielectrically isolated SOI region and the low on-state resistance vertical power element for more current capability in the non-SOI region [10].

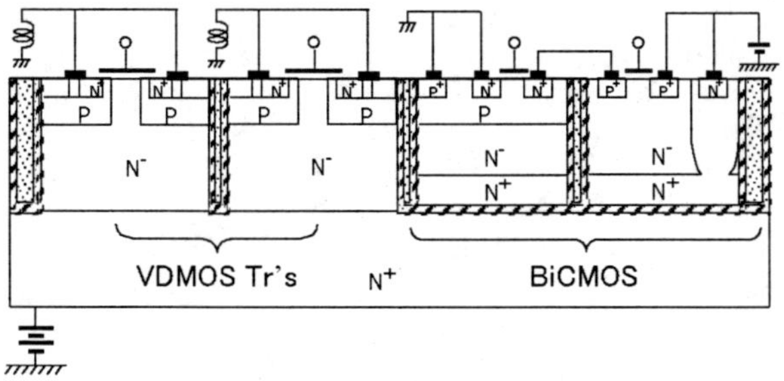

FIGURE 11. Partial SOI structure suitable for intelligent power IC

As a fabrication technology, wafer bonding is suitable because the thermal budget of wafer bonding is lower than epitaxy. In this work, the wafer bonding using the intermediate poly crystalline layer is proposed. In this method, the surface of the bonding wafer on which the oxide layer is formed in the predetermined region is covered with a poly-crystalline silicon layer, then it is planarized, and finally it is bonded to the single crystalline silicon wafer. In the case of these surface combinations, however, there was a problem of generating micro-voids in the bonding interface. We achieved the void-less interface by developing the bonding technology through the amorphous layer made by implantation of silicon ions into the poly-crystalline silicon surface.

It has been shown previously [11] that the amorphous layer at the bonding interface of single crystalline wafers is effective on the suppression of void generation. That is the remarkable void suppression has been observed in the case of implantation energy of 120keV and more.

In this work, the same implantation condition of 120keV has been applied, and the effect of the amorphous layer on the bonding of the poly-crystalline silicon and the single crystalline silicon has been examined. Figure 12 shows the cross sectional TEM images of the bonding interface (a) without implantation, and (b) with implantation. It is clear that in the case of (b), micro-void is not observed as expected.

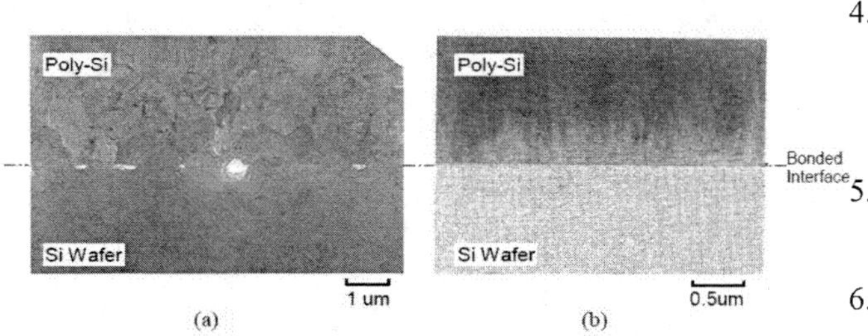

FIGURE 12. The cross sectional TEM images of bonded interfaces, (a) without implantation and (b) with implantation

Using this wafer bonding technology through amorphous layer, vertical power device integration has been realized.

CONCLUSION

The dielectric isolation structure using SOI wafer is the most suitable for various, severe demands to automotive semiconductors.

The feature technologies installed in SOI-BCD, such as

1) BN^+ Layer for current capability, gettering and noise-immunity,
   LDMOS with improved ESD robustness,
2) crystal defect-less process,
3) wafer direct bonding through amorphous layer for intelligent power IC

have markedly improved the integration of various functions.

The system integration in SOI-BCD is accelerated and further development of device and process technologies matching to the automotive is required.

ACKNOWLEDGEMENTS

Authors would like to sincerely thank members of Electronics Device R&D Dept. for their various collaborations in this work. Authors would also like to thank Senior Managing Director S. Shirasaki and Senior General Manager of Electronics Device Product Division K. Ookura for their useful advice and kindly support.

REFERENCES

1. S. Fujino, SEMI Tech. Symp., Yokohama, 2006, 5-27-39
2. S. Fujino, H. Himi, T. Fukada and H. Yamaguchi, "Device Applications Using Bonded SOI Wafers", in 7th International Symp. Semiconductor Wafer Bonding, Electrochemical Society Proceeding Vol.2005-2, Quebec, 2005, pp. 20-33.
3. H. Yamaguchi, H. Himi, S. Fujino and T. Hattori, IEICE Trans. Electron., **E75-C**, No.12, 1447-1452 (1992).
4. S. Fujino, S. Takahashi, T. Fukada, H. Himi and K. Kawamoto, "Device Application Using Bonded Thick SOI Wafers", in 5th International Symp. Semiconductor Wafer Bonding, Electrochemical Society Proceeding Vol.2001-27, San Francisco, 2001, pp.214-219.
5. K. Kawamoto, S. Takahashi, S. Fujino and I. Shirakawa, IEEE Trans. Electron Devices, **49**, No.11, 2047-2053 (2002).
6. H. Himi, Y. Nakayama, S. Fukatsu and S. Fujino, "Intellgent Power IC with ESD Robustness for Automobile", FISITA 2006 World Automotive Congress, SAE, Yokohama, 2006
7. Y. Nakayama, K. Eguchi, S. Mizuno, K. Kawamoto, K. Nakashima and Y. Watanabe, "No Defect Thermal

Process for Thick SOI Bipolar Transistors with Trench Dielectric Isolations", in the 2002 international conference on solid state devices and materials, Japan Society of Appl. Phys., Yokohama, 2002, poster.
8. S. Fujino, H. Yamaguchi, H. Himi and T. Hattori, Automitive Technology Papers, **26**, No.4, 152-156 (1995).
9. H. Yamaguchi, H. Himi, S. Fujino and T. Hattori, "Intelligent Power IC with Partial SOI Structure", in the 1994 international conference on solid state devices and materials, Extended Abstract, Japan Society of Appl. Phys., Yokohama, 1994, pp. 307-309. Also in *Jpn. J. Appl. Phys*., **34**, 864-868(1995).
10. H. Himi, M. Matsui, S. Fujino and T. Hattori, *Jpn. J. Appl. Phys*., **33**, No. 1A, 6-10(1994).
11. S. Fujino, M. Matsui, T. Hattori and Y. Hamakawa, Jpn. *J. Appl. Phys*., **34**, L1322(1995)

Threshold Voltage Engineering by Lanthanide Doping of the MOS Gate Stack

A. Fet[1], V. Häublein[1], and H. Ryssel[1,2]

[1] *Fraunhofer Institut für Integrierte Systeme und Bauelementetechnologie Schottkystrasse 10, 91058 Erlangen, Germany*

[2] *Lehrstuhl für Elektronische Bauelemente, Cauerstrasse 6, 91058 Erlangen, Germany*

Abstract. With the replacement of traditional polysilicon and silicon dioxide by metal gates and high-k dielectrics, respectively, in the MIS gate stack for the 45 nm technology node, higher than expected device threshold voltages have been observed due to the effect of Fermi-level pinning. While the debate as to the exact cause of Fermi-level pinning is ongoing, several attempts (capping layers, new gate metal compositions) have been made to curb this effect. In this paper, the tuning of metal gate work function by ion implantation is investigated as a tool for controlling the threshold voltage. Lanthanide incorporation is used to achieve a flat-band voltage shift of more than -1 V for n-MOS capacitors. It is shown that by adjusting dose and energy, the flat-band voltage shift can be tuned to a desired value, without substantial damage to the insulating quality of the gate. This translates to an effective shift in the threshold voltage.

Keywords: lanthanide, threshold voltage engineering, ion implantation.
PACS: 85.40.Ry, 61.72.U

INTRODUCTION

In order to circumvent the effect of Fermi-level pinning (FLP) [1, 2], capping layers of La_2O_3 and Al_2O_3 for n- and p-MOS transistors [3, 4], respectively, novel gate metal compositions [5], novel channel materials [9], oxygen enrichment [6] or fully-silicided (FUSI) gate doping [7] have been proposed to engineer the threshold voltage. These approaches often require extra process-intensive steps. Controlling the threshold voltage (V_t) by means of ion implantation permits a conventional process flow in device fabrication. The lanthanides lanthanum (La) and dysprosium (Dy) are implanted to engineer the work function of the n-type gate. By proper selection of the dose and implant energy, an appreciable shift in the flat-band voltage (V_{fb}) is achieved while maintaining reasonable oxide insulation.

EXPERIMENTAL

Figure 1 shows the samples used in the experiment. Terraced oxide MOS capacitor structures with SiO_2 thicknesses of 8 nm, 11 nm, and 16 nm were thermally grown on a Silicon wafer with field oxide isolation. Subsequently, a 40 nm n-polysilicon gate is deposited and implanted with dysprosium or lanthanum at energies of 30 keV and 40 keV. Two doses, 10^{14} cm^{-2} and $2{\cdot}10^{15}$ cm^{-2}, were implanted followed by rapid thermal annealing at 900°C for 20 s then 1070°C for 5 s. 500 nm aluminum was deposited for gate contact and capacitor structures were structured by lithography and dry etching. The samples were then characterized by capacitance-voltage (C-V) and gate current-voltage (I_g-V_g) measurements.

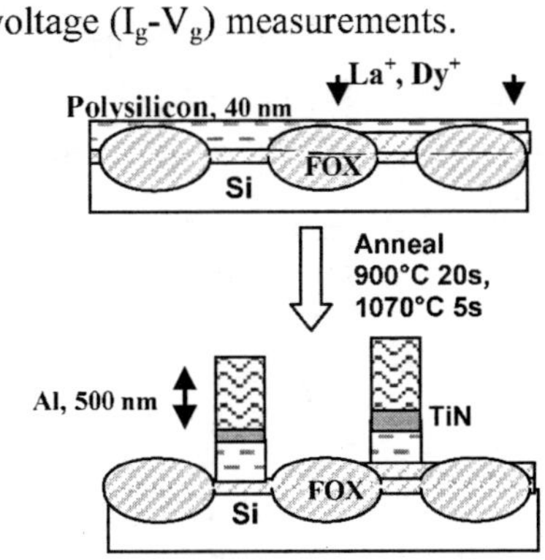

FIGURE 1. Experimental sample preparation

CP1066, *Ion Implantation Technology*, edited by E. G. Seebauer, S. B. Felch, A. Jain, and Y. V. Kondratenko

RESULTS

Dysprosium implantation shifts the C-V curve to more negative values, as shown in Figure 2 for capacitor structures with a 17 nm SiO_2 gate oxide. A shift of the order of approximately -1 V is observable for a dose of $1{\cdot}10^{14}$ cm^{-2}. A 20-fold increase in the dose approximately doubles the strength of the C-V characteristic shift. A similar shift is observed for lanthanum implantation. Figure 3 shows the shift of the C-V curve observed for lanthanum implantation in a device of oxide thickness 11 nm. With an implantation dose of 10^{14} cm^{-2} a shift of about -3 V is obtained. A 20-fold increase in the implanted dose for lanthanum, results in a slight increase in the size of the shift of about – 0.5 V. The shift obtained with dysprosium tends to have a stronger dependence on the implanted dose than for lanthanum.

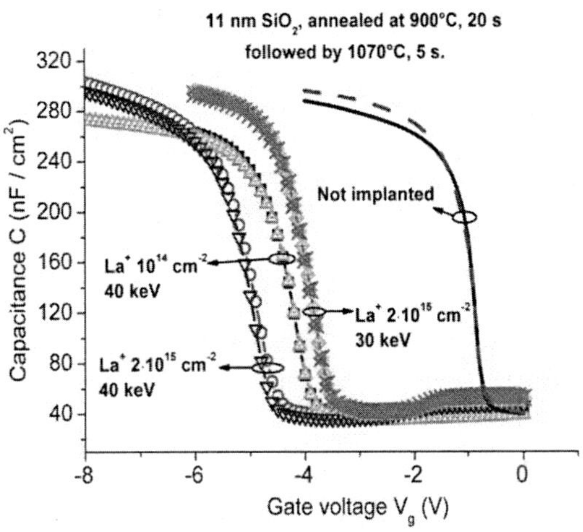

FIGURE 2. C-V curves for Dy-implanted MOS capacitors

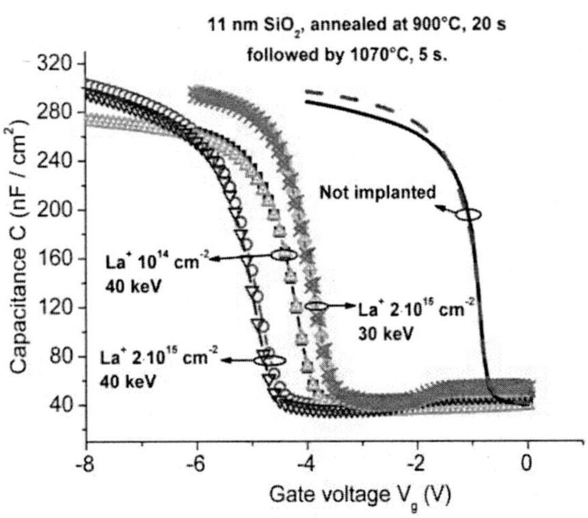

FIGURE 3. C-V curves for La-implanted MOS capacitors

Comparing results of Figure 2 and Figure 3, for the same dose of $1{\cdot}10^{14}$ cm^{-2}, a shift in the C-V curve of about -3 V for lanthanum implantation and -1 V for dysprosium is observed. Similarly, for other doses and implant energies, the shift obtained with lanthanum is stronger than for dysprosium.

The shift of the C-V curve is also dependent on the implantation energy. As shown in Figure 3, a lower dose ($1{\cdot}10^{14}$ cm^{-2}) at higher energy (40 keV) and a higher dose ($2{\cdot}10^{15}$ cm^{-2}) at lower energy (30 keV) do produce similar shifts of the C-V characteristic. The energy of implantation as well as the dose determine the strength of shift and will be discussed in the next section.

DISCUSSION

Extrapolating the V_{fb} shift for different oxide thicknesses, as shown in Figure 4 and Figure 5, shows that the V_{fb} shift does not only result from oxide charging due to implantation but from the chemical bonding of the implanted species at the interface. The contribution of oxide charging to the V_{fb} shift is seen to be relatively small.

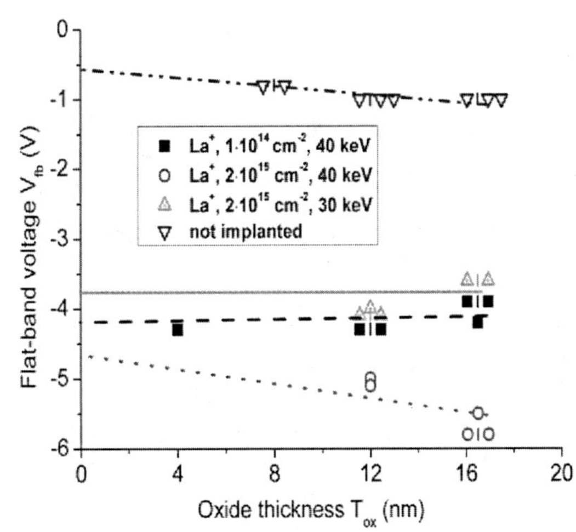

FIGURE 4. V_{fb} vs T_{ox} for La implanted MOS stacks

A dose of $2{\cdot}10^{15}$ cm^{-2} at 30 keV shifts V_{fb} by about -1 V for dysprosium and about -3 V for lanthanum. By increasing the implant energy to 40 keV while reducing the dose to 10^{14} cm^{-2}, a comparable V_{fb} shift is still obtained with a discrepancy of about +0.5 V and -0.5 V for dysprosium and lanthanum, respectively. Higher implant energy increases the concentration at the polysilicon-SiO_2 interface as shown in Figure 6. Thus, an increase in the strength of the shift at higher implantation energy is a result of the change in the dopant concentration at the polysilicon / SiO_2 interface. This underscores the importance of the interface in the V_{fb} shift.

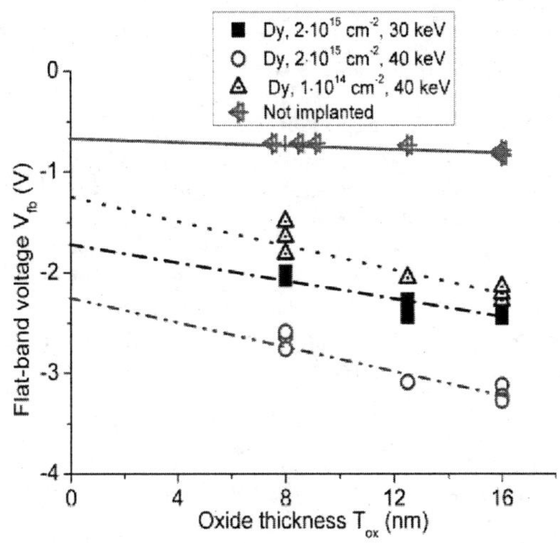

FIGURE 5. V_{fb} vs T_{ox} for Dy-implanted MOS stack

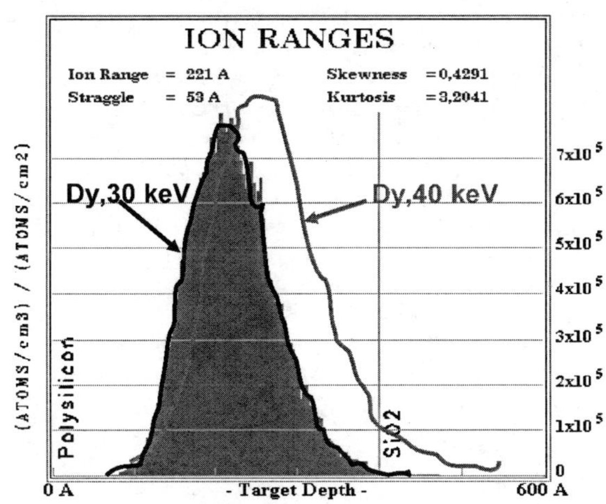

FIGURE 6. Simulated ion distribution for Dy implantation.

A stronger shift for lanthanum compared to dysprosium for similar doses and implant energies can be explained firstly by the fact that the lanthanum dose at the polysilicon-SiO_2 interface is higher because lanthanum has a lower mass. As a result, the projected range R_p of ions for the implantation in the 40 nm thick polysilicon gate is larger for lanthanum than for dysprosium at the same implantation energy. Values for R_p simulated with SRIM [10] are, 26.3 nm and 28.4 nm for dysprosium and lanthanum, respectively, implanted with an energy of 40 keV. Thus, for the same implantation energy, more lanthanum ions reach the polysilicon-SiO_2 interface. Secondly, as the shift has been attributed to dipole formation at the interface [8], lanthanum builds a stronger dipole due to its lower electronegativity (La: 1.1, Dy: 1.2) and the larger atomic radius (La: 187 pm, Dy: 178 pm).

The gate leakage current for lanthanum and dysprosium-implanted devices are shown in Figure 7 and Figure 8, respectively. Ion implantation results in a slight, almost negligible increase in the oxide leakage current. This is tolerable within operating voltage regimes. The increase in the gate current is stronger for dysprosium-implanted samples.

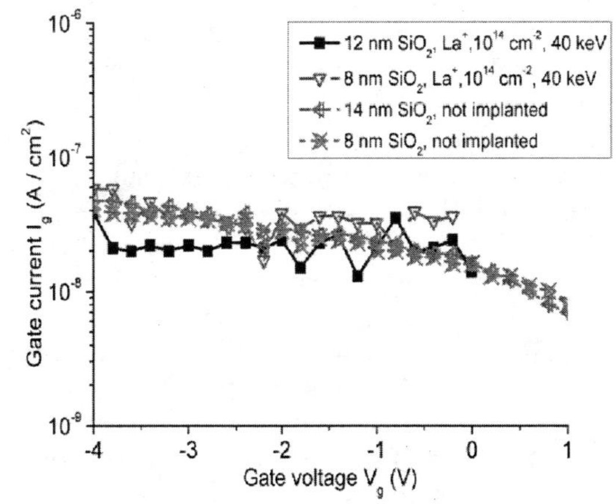

FIGURE 7. Gate leakage for La-implanted MOS gate stacks

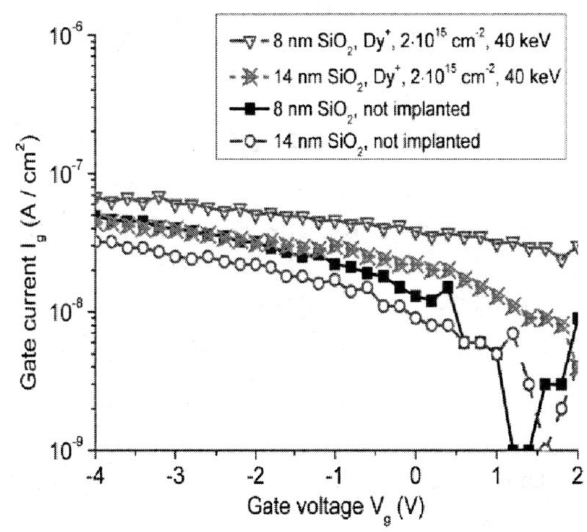

FIGURE 8. Gate leakage for Dy-implanted MOS gate stacks

CONCLUSION

Lanthanide implantation promises to be a less process-intensive alternative to capping layers for engineering n-MOS device threshold voltages. It has been shown that ion implantation of the gate stack can cause considerable flat-band voltage shifts without significantly compromising the insulating quality of the oxide. Important also in the selection of the dopant atoms is that the achieved shift in flat-band voltage is not entirely due to oxide charging. Lanthanide doping of the gate stack, therefore, presents an interesting tool to be investigated for threshold voltage tuning in high-k dielectric gate stacks, where the effect of Fermi-level pinning is strong. Important is also the strong V_{fb} shift, even after a high temperature anneal. This implies the obtainable flat-band voltage shift would persist after a typical source-drain anneal during device fabrication.

The trend of the V_{fb} shift can be partially explained by attributing the mechanism to the formation of interface dipoles at the implanted interface. The introduction of a pinning level by the dopant atom which leads to the V_{fb} shift is another possible explanation, but would require intensive investigation to validate. The concept of interface dipoles provides an intuitive approach, which can be easily applied in selection of dopant elements.

Preliminary investigations of the flat-band voltage shift due to implantations using the high-k dielectric ZrO_2 have show promising results. Implanting a polysilicon-ZrO_2 gate stack with lanthanum produced a flat-band voltage shift of about -0.5 V. Further investigations on the effectiveness of implantation with high-k dielectrics are being carried out.

ACKNOWLEDGEMENTS

We would like to thank E. Birnbaum of the ion implantation research team at Fraunhofer IISB for the cooperation on this project.

REFERENCES

1. C.C. Hobbs, L.R.C. Fonseca, A. Knizhnik, V. Dhandapani, S.B. Samavedam, W.J. Taylor, J.M. Grant, L.G. Dip, D.H. Triyoso, R.I. Hegde, D.C. Gilmer, R. Garcia, D. Roan, M.L. Lovejoy, R.S. Rai, E.A. Hebert, Hsing-Huang Tseng, S.G.H. Anderson, B.E. White and P.J. Tobin, IEEE Transactions on Electron Devices ,**51**(6), June 2004, pp. 971-977.
2. C.C. Hobbs, L.R.C. Fonseca, A. Knizhnik, V. Dhandapani, S.B. Samavedam, W.J. Taylor, J.M. Grant, L.G. Dip, D.H. Triyoso, R.I. Hegde, D.C. Gilmer, R. Garcia, D. Roan, M.L. Lovejoy, R.S. Rai, E.A. Hebert, Hsing-Huang Tseng, S.G.H. Anderson, B.E. White and P.J. Tobin, IEEE Transactions on Electron Devices, **51**(6), June, (2004), pp. 978-984.
3. Hyung-Suk Jung, Jong-Ho Lee, Sung Kee Han, Yun-Seok Kim, Ha Jin Lim, Min Joo Kim, Seok Joo Doh, Mi Young Yu, Nae-In Lee, Hye-Lan Lee, Taek-Soo Jeon, Hag-Ju Cho, Sang Bom Kang, Sang Yong Kim, Im Soo Park, Dongchan Kim, Hion Suck Baik and Young Su Chung, Symposium on VLSI Technol., (2005), pp. 232-233.
4. H. N. Alshareef, M. Quevedo-Lopez, H. C. Wen, R. Harris, P. Kirsch, P. Majhi, B. H. Lee, R. Jammy, D. J. Lichtenwalner, J. S. Jur, and A. I. Kingon, *Appl. Phys. Lett.* **89**, 232103 (2006).
5. H.-C. Wen, S.C. Song, C.S. Park, C. Burham, G. Bersuker, K. Choi, M.A. Quevedo-Lopez, B.S. Ju, H.N. Alshareef, H. Niimi, H.B. Park, P.S. Lysaght, P. Majhi, B.H. Lee and R. Jammy, 2007 IEEE Symposium on VLSI Technology, 12-14 June, (2007), pp. 160-161.
6. S.C. Song, C.S. Park, J. Price, C. Burham, R. Choi, H.C. Wen, K. Choi, H.H. Tseng, B.H. Lee and R. Jammy, IEDM 2007, 10-12, Dec. 2007, pp.337-340.
7. J.A. Kittl, A. Lauwers, M.A. Pawlak, A. Veloso, H.Y. Yu, S.Z. Chang, T. Hoffmann, G. Pourtois, S. Brus, C. Demeurisse, C. Vrancken, P.P. Absil and S. Biesemans, *Microelectronic Engineering*, **84**(9-10), (2007), pp. 1857-1860.
8. P. Sivasubramani, T.S. Boscke, J. Huang, C.D. Young, P.D. Kirsch, S.A. Krishnan, M.A. Quevedo-Lopez, S. Govindarajan, B.S. Ju, H.R. Harris, D.J. Lichtenwalner, J.S. Jur, A.I. Kingon, J. Kim, B.E. Gnade, R.M. Wallace, G. Bersuker, B.H. Lee and R. Jammy, 2007 IEEE Symposium on VLSI Technology, 12-14 June, (2007), pp. 68-69.
9. H.R. Harris, P. Kalra, P. Majhi, M. Hussain, D. Kelly, Jungwoo Oh, D. He, C. Smith, J. Barnett, P.D. Kirsch, G. Gebara, J. Jur, D. Lichtenwalner, A. Lubow, T.P. Ma, Guangyu Sung, S. Thompson, Byoung Hun Lee, Hsing-Huang Tseng and R. Jammy, IEEE Symposium on VLSI Technology, 12-14 June, (2007), pp. 154-155.
10. SRIM, http://www.srim.org/

Improvement of Contact Resistance with Molecular Ion Implantation

KyungWon Lee[1,3], JinKu Lee[2], JaeGeun Oh[2], TaeHoon Huh[1], MinAe Ju[2], SeungJoon Jeon[2], JaChun Ku[2], SungKi Park[2], Steve Kim[1], DaeHo Yoon[3], GeumJoo Ra[1], Mark A. Harris[1], Ronald N. Reece[1]

1. Axcelis Technologies Inc., 108 Cherry Hill Drive, Beverly, MA 01915, USA
2. Hynix Semiconductor Inc., San 136-1 Ami, Bubal, Ichon, Kyoungki-do, 467-701, Korea
3. Sungkyunkwan University., 300 Cheoncheon, Jangan, Suwon, Kyoungki-do, 440-746, Korea
E-mail: kyungwon.lee@axcelis.com

Abstract. Basic characteristics of ClusterBoron™ ($B_{18}H_{22}$) implantation were investigated for improving contact resistance in DRAM devices. Generally, $^{49}BF_2$ has been widely used for contact implant application in DRAM manufacturing because of its higher productivity compared to monomer boron (^{11}B). However, because of limited activation in a low thermal budget (~800°C) anneal, the sheet resistance was saturated for doses over $5x10^{15}$ ions/cm^2. Although many investigations have been reported, such as ^{30}BF implant mixed implant with monomer boron etc., no practical solution has been found for dramatic improvement of contact resistance in a productive manner. $B_{18}H_{22}$ was developed to overcome the productivity limitations encountered in low energy, high dose boron implantation and the limited activation of $^{49}BF_2$ due to co-implanted fluorine. In this study, basic characterization of the $B_{18}H_{22}$ contact implant was performed through sheet resistance, SIMS (Secondary Ion Mass Spectrometry) and XTEM (cross-sectional transmission electron microscopy). The $B_{18}H_{22}$ implants showed lower sheet resistance than conventional $^{49}BF_2$ for $5x10^{15}$ ions/cm^2 on bare wafer tests. Through XTEM study, we found the activation behavior of both $B_{18}H_{22}$ and $^{49}BF_2$ were directly related with the amorphous layer thickness and residual defects from low thermal budget anneal. PMOS contact resistance in the sub-70nm device by $B_{18}H_{22}$ implantation showed considerable improvement (about 30%), showing $B_{18}H_{22}$ could replace the BF_2 for contact implant in contact resistance implant.

Keywords: Octadecaborane, Ion Implantation, Contact Resistance.
PACS: 52.77.Dq, 82.45.Vp, 85.40.Ry

INTRODUCTION

As semiconductor devices continue to shrink, sensitivity to material and process environments increases [1-3]. For DRAM devices, the contact resistance becomes one of key parameters for control of the device speed. Generally, to minimize the PMOS contact resistance, a p-type dopant has been implanted on the bottom of a contact hole. However, low energy boron implant has a limitation in terms of beam current leading to insufficient productivity for manufacturing demands. $^{49}BF_2$ has been widely used for the contact implant application in DRAM manufacturing because of its higher productivity compared to monomer boron (^{11}B). However, because of the limited activation in low thermal budget (~800°C) anneal process, the sheet resistance was saturated at doses of $5x10^{15}$ ions/cm^2. Although many investigations into alternatives have been reported, such as ^{30}BF implant, mixed implant with monomer boron and etc. [4], no productive solution has been found for dramatic improvement of contact resistance. Recently, cluster boron ($B_{18}H_{22}$), developed to overcome the productivity limitations of conventional boron implantation at low energies, was investigated for the contact application. In this study, a basic characterization of implanted cluster boron was performed through sheet resistance, SIMS analysis and XTEM study. Following was an investigation in sub 70nm to prove if $B_{18}H_{22}$ could replace the conventional BF_2 contact plug process.

EXPERIMENTAL

At first, monomer boron equivalent energy had to be calculated for both $^{49}BF_2$ and $B_{18}H_{22}$ because of the difference in mass for each molecule. Based on the

CP1066, *Ion Implantation Technology*, edited by E. G. Seebauer, S. B. Felch, A. Jain, and Y. V. Kondratenko

calculation results plotted in Figure 1, the equivalent energy of 10 keV BF_2 was 42.8 keV $B_{18}H_{22}$.

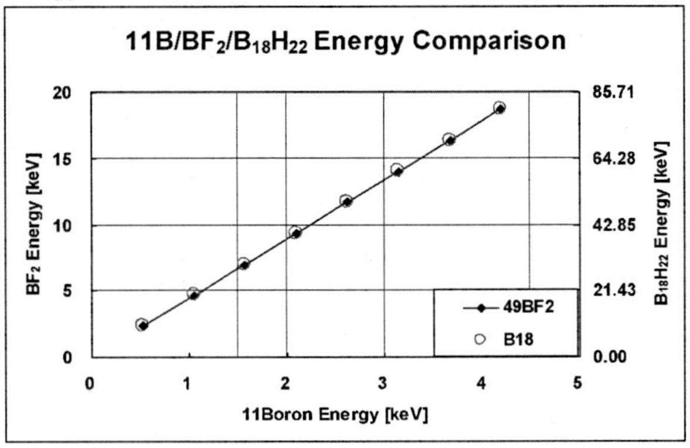

FIGURE 1. $^{11}B/^{49}BF_2/B_{18}H_{22}$ Energy comparison

To find a matching condition for the dopant profile of both as-implanted and annealed state, $B_{18}H_{22}$ implant energy was varied from 25.7 keV to 42.8 keV.

TABLE 1 Test matrix on bare wafers.

Dopant	Energy [keV]	Anneal	SIMS	TEM	Rs
BF_2	10	O	O	O	O
		X	O	O	X
$B_{18}H_{22}$	42.8 (BF_2 10)	O	O	O	X
		X	O	O	X
	34.3 (BF_2 8)	O	O	O	O
		X	O	O	X
	25.7 (BF_2 6)	O	O	O	X
		X	O	O	X

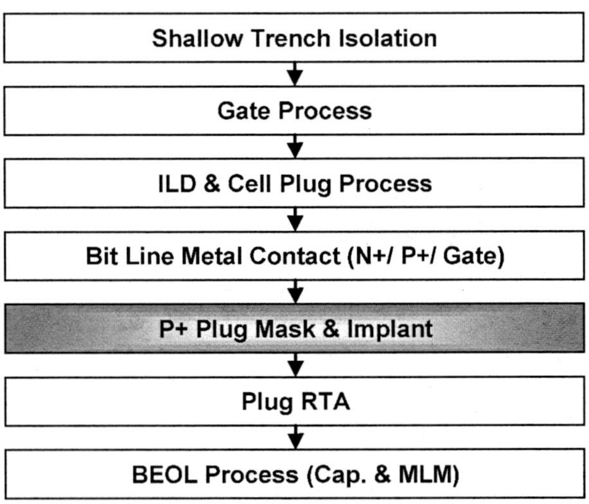

FIGURE 2. Standard process flow for p+ contact resistance measurement.

Anneal condition was matched with standard post anneal process for device flow. Details of the test matrix for the bare wafer test are included as Table 1.

The dopant profiles were analyzed by SIMS, and through XTEM, the amorphous layer thickness and the density of the residual defects were analyzed. A conventional four point probe was used for sheet resistance measurement.

Based on bare wafer test results, both $^{49}BF_2$ and $B_{18}H_{22}$ were applied to sub-70nm device as a PMOS contact implant. Figure 2 showed the standard process flow for p+ contact resistance measurement.

RESULTS AND DISCUSSION

Figure 3 shows the boron SIMS profiles for as-implanted condition comparing the BF_2 and $B_{18}H_{22}$. The $B_{18}H_{22}$ energy was varied from 6 keV to 10 keV $^{49}BF_2$ equivalents to match the 10 keV BF_2 implant.

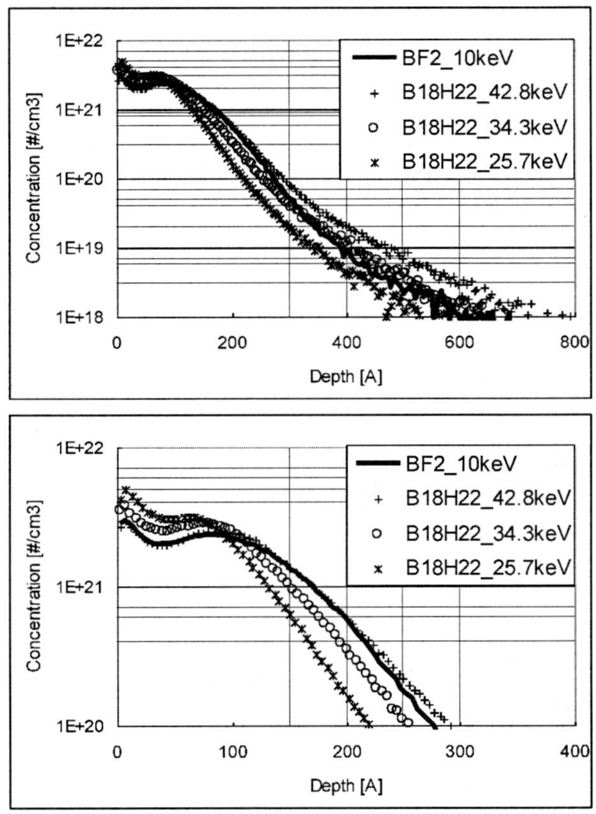

FIGURE 3. As-implant SIMS profiles for $B_{18}H_{22}$, $4x10^{15}cm^{-2}$ implants with various energy conditions.

The as-implanted profile of $B_{18}H_{22}$, 42.8 keV (10 keV BF_2 equivalent) is well matched at projected range with BF_2, 10 keV. But it has slightly deeper tail than BF_2, 10 keV. The $B_{18}H_{22}$, 34.3 keV (8 keV BF_2 equivalent) is close to the BF_2 profile at the tail region.

Figure 4 shows annealed profiles. The $B_{18}H_{22}$, 34.3 keV was well matched with the BF_2, 10 keV implant for the chosen anneal condition.

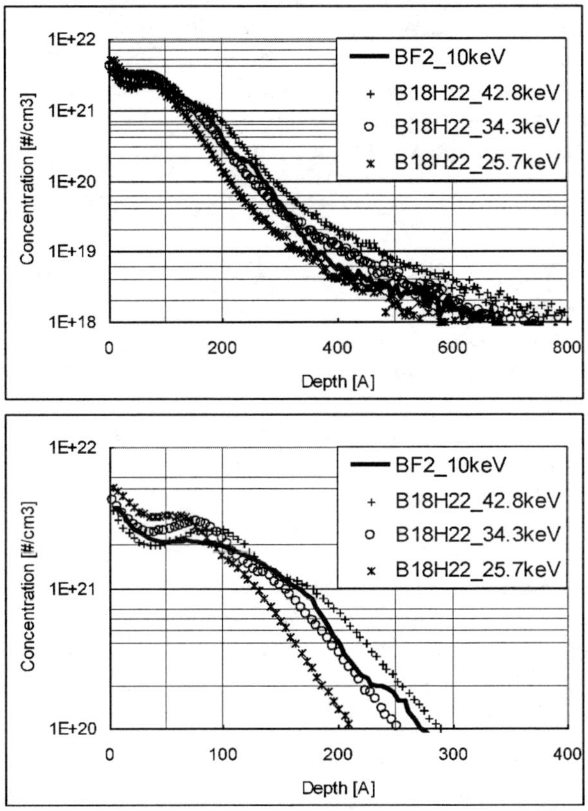

FIGURE 4. Annealed SIMS profile for $B_{18}H_{22}$, $4x10^{15}cm^{-2}$ implants with various energy conditions.

The retained dose was calculated by SIMS analysis and is included in Table 2. The $B_{18}H_{22}$ was implanted at 42.8 keV, 34.4 keV and 25.7 keV. The retained dose of $B_{18}H_{22}$ showed 102 % to 104 % for as-implanted samples, and dopant loss after annealing in nitrogen ambient was around 1~2 %. In case of BF_2, the retained dose of as-implanted sample was less than $B_{18}H_{22}$ [5]. The dopant loss after annealing was around 4 %.

TABLE 2 Proportion of the retained dose comparison.

Dopant	Energy [keV]	BF2 Equivalent	Proportion of Retained dose	
			As-implant	Anneal
$B_{18}H_{22}$	42.8 (BF_2 10)	18.7	102%	103%
	34.3 (BF_2 8)	10.0	104%	102%
	25.7 (BF_2 6)	7.0	103%	101%
BF_2		10.0	99%	95%

Figure 5 shows amorphous layer thickness as measured by XTEM analysis. The amorphous layer thickness of $B_{18}H_{22}$ was about 60% thinner than that of BF_2 at equivalent energy. That is, the BF_2, 10 keV formed ~250 Å amorphous layer and $B_{18}H_{22}$, 42.8 keV formed ~155 Å.

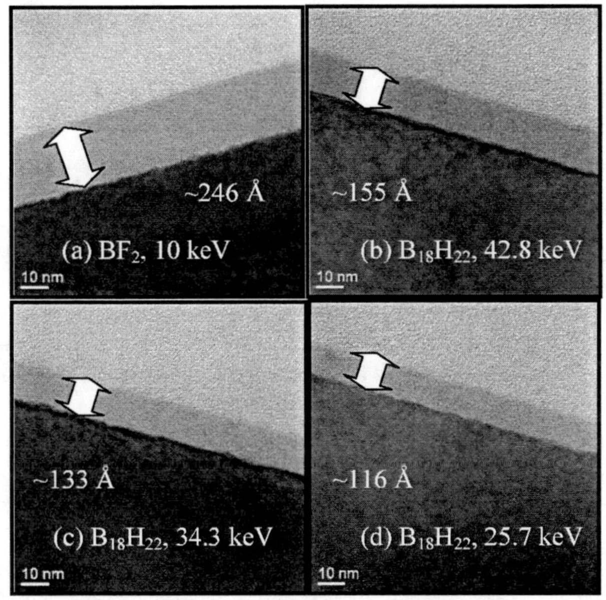

FIGURE 5. As-implanted TEM image (a) BF_2 10 keV, (b) B18H22, 42.8 keV, (c) B18H22, 34.3 keV, (d) B18H22, 25.7 keV

FIGURE 6. Annealed TEM image (a) BF_2 10 keV, (b) B18H22, 42.8 keV, (c) B18H22, 34.3 keV, (d) B18H22, 25.7 keV

Figure 6 shows the XTEM image of the residual defects, or so called EOR (End of Range) defects after the anneal process. The XTEM image shows that the defect density of BF_2 was much higher than that of $B_{18}H_{22}$. In addition, by decreasing the energy of $B_{18}H_{22}$implant, the defect density also decreased. A defect free surface was obtained by $B_{18}H_{22}$ 25.7 keV. The location of residual defect was almost same as amorphous/crystal interface in all cases.

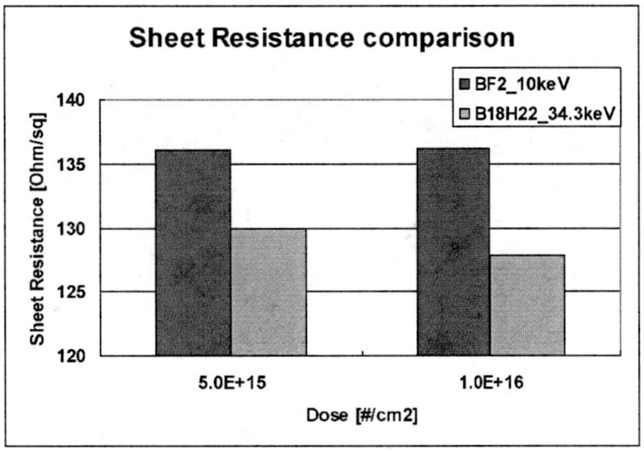

FIGURE 7. Sheet resistance comparisons after standard post-implant anneal process.

Before applying $B_{18}H_{22}$ to actual device wafer, the sheet resistance was measured for BF_2, 10 keV and $B_{18}H_{22}$, 34.3 keV, and is included as Figure 7. Doses of $5x10^{15}$ cm^{-2} and $1.0x10^{16}$ cm^{-2} were implanted and processed with the standard post-implant anneal process. The sheet resistance of BF_2 showed similar values for both $5x10^{15}$ cm^{-2} and $1.0x10^{16}$ cm^{-2} indicating saturation, while $B_{18}H_{22}$ showed lower sheet resistance at the higher dose. The results showed that the dopant activation ratio of $B_{18}H_{22}$ was higher than that of BF_2 at high doses for a low thermal budget anneal process. The hypothesis is the activation barrier which was enhanced by fluorine was eliminated at $B_{18}H_{22}$ implant case.

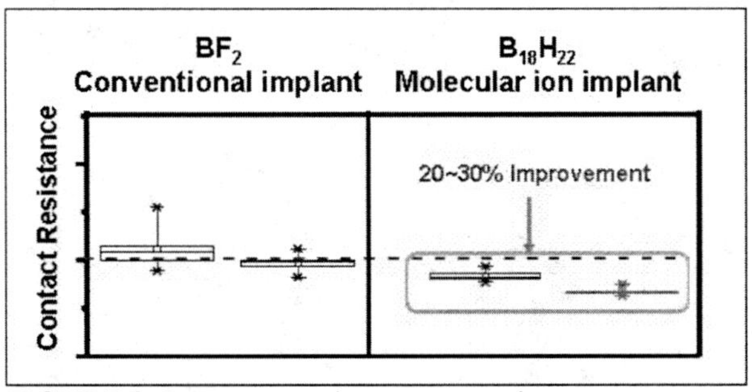

FIGURE 8. P+ contact resistance comparison.

Based on the bare wafer test results, $B_{18}H_{22}$ was applied to a sub-70nm DRAM device for p+ contact implantation. The average value of the contact resistance was dramatically reduced, by about 20 ~ 30% by implanting $B_{18}H_{22}$ as replacement for BF_2. Furthermore, the within wafer uniformity of the contact resistance was also improved. Based on this analysis $B_{18}H_{22}$ could replace the BF_2 for contact implant in contact resistance point of view with a significant improvement in device parameters.

CONCLUSIONS

The characteristics of cluster boron ($B_{18}H_{22}$) implantation for contact resistance application were compared with BF_2 at 10keV. The SIMS analysis results showed that boron profile of $B_{18}H_{22}$ was matched at 34.3 keV which was 20 % lower than equivalent BF_2 energy. The retained boron dose of $B_{18}H_{22}$ was higher than BF_2 for both as-implanted and annealed condition. The amorphous layer thickness of $B_{18}H_{22}$ was 60% thinner than that of BF_2 case. The amount of residual defect could be significantly reduced by $B_{18}H_{22}$ implantation. The sheet resistance could be also reduced by $B_{18}H_{22}$ implantation at higher doses due to its higher doping efficiency rather than BF_2. The device results suggested that $B_{18}H_{22}$ could replace the BF_2 with improvement in contact resistance for contact implant application.

ACKNOWLEDGEMENTS

The authors would like to thank Axcelis Korea Field Service Team, and Axcelis Product Support Team, especially Sebastian Seo, YK Jung, YC Seo and SK Kim for support of this study.

REFERENCES

1. M. Taylor, *et al.*, "Material Challenges for CMOS Junctions," in *MRS Symposium. Proc.* Vol C.1.1. pp. 810
2. A. Agarwal, in *Proceedings of the 13th International Conference on Ion Implantation Technology* 2000, p.293
3. A. J. Murrell *et al.*, *J. Vac. Sci. and Technol.* **B 18**, 462 (2000)
4. Y. S. Sohn, *et al.*, "The effects of low fluorine p+plug ion implantation on metal/p+ contact resistance of sub 80nm memory device", in *Proceedings of the 15th International Conference on Ion Implantation Technology*, Part II, 2004, pp. 6-10.
5. M. A. Harris, *et al.*,."Dose Retention Effects in Atomic Boron and ClusterBoron™ B18H22. Implant Processes" in *Proceedings of the 16th International Conference on Ion Implantation Technology*, 2006, pp. 155-158.

Development of Compact Electron Cyclotron Resonance Ion Source with Permanent Magnets for High-Energy Carbon-Ion Therapy

M. Muramatsu[1,2], A. Kitagawa[2], Y. Iwata[2], S. Hojo[2], Y. Sakamoto[2], S. Sato[2], Hirotsugu Ogawa[2], S. Yamada[3], Hiroyuki Ogawa[4], Y. Yoshida[5], T. Ueda[6], H. Miyazaki[6], and A. G. Drentje[7]

[1] *Graduate school of Mechanical Engineering, Toyo University, 2100 Kujirai, Kawagoe, Saitama 350-0815, Japan*
[2] *National Institute of Radiological Sciences, 4-9-1 Anagawa, Inage, Chiba 263-8555, Japan*
[3] *Gunma University, 3-39-22 Showa-machi, Maebashi, Gunma 371-8511,Japan*
[4] *Accelerator Engineering Corporation, Ltd., 2-13-1 Konakadai, Inage, Chiba 263-0043, Japan*
[5] *Department of Mechanical Engineering, Toyo University, 2100 Kujirai, Kawagoe, Saitama 350-0815, Japan*
[6] *Sumitomo Heavy Industries, Ltd., 5-2 Soubiraki, Niihama, Ehime, Japan*
[7] *K.V.I, University of Groningen, 9747 AA Groningen, The Netherlands*

Abstract. Heavy-ion cancer treatment is being carried out at the Heavy Ion Medical Accelerator in Chiba (HIMAC) with 140 to 400 MeV/n carbon ions at National Institute of Radiological Sciences (NIRS) since 1994. At NIRS, more than 4,000 patients have been treated, and the clinical efficiency of carbon ion radiotherapy has been demonstrated for many diseases. A more compact accelerator facility for cancer therapy is now being constricted at the Gunma University. In order to reduce the size of the injector (consists of ion source, low-energy beam transport and post-accelerator Linac include these power supply and cooling system), an ion source requires production of highly charged carbon ions, lower electric power for easy installation of the source on a high-voltage platform, long lifetime and easy operation. A compact Electron Cyclotron Resonance Ion Source (ECRIS) with all permanent magnets is one of the best types for this purpose. An ECRIS has advantage for production of highly charged ions. A permanent magnet is suitable for reduce the electric power and cooling system. For this, a 10 GHz compact ECRIS with all permanent magnets (Kei2-source) was developed. The maximum mirror magnetic fields on the beam axis are 0.59 T at the extraction side and 0.87 T at the gas-injection side, while the minimum B strength is 0.25 T. These parameters have been optimized for the production of C^{4+} based on experience at the 10 GHz NIRS-ECR ion source. The Kei2-source has a diameter of 320 mm and a length of 295 mm. The beam intensity of C^{4+} was obtained to be 618 eμA under an extraction voltage of 30 kV. Outline of the heavy ion therapy and development of the compact ion source for new facility are described in this paper.

Keywords: carbon ion, radiotherapy, ECR, ion source
PACS:29.25.Ni, 29.27.Ac, 87.56.-v

HEAVY ION CANCER THERAPY

The particle beams being currently used for treatment in humans are proton and carbon beams. Other particle beams that have been used for humans include negative pi meson beams, helium beams, neon beams, silicon beams, and argon beams. Each of these beams has specific characteristics, but only proton and carbon beams are now used for treatment [1]. In general, particles heavier than the electron that are accelerated to a high speed are called heavy charged particle beams [2], but the names of particle beams are not always internationally unified and require special attention. In some EU countries, charged particle radiotherapy is collectively called hadron therapy or, with particle masses around or less than the carbon nucleus mass, light ion therapy. In Japan, the term heavy ion beam often refers to carbon ion beam unless otherwise specified.

The depth-dose distribution of charged particles is completely different from photon and neutron as shown in Figure 1. The energy loss curves have a rather low flat plateau region after particles enter the body and then a sharp peak appears near the end of

CP1066, *Ion Implantation Technology,* edited by E. G. Seebauer, S. B. Felch, A. Jain, and Y. V. Kondratenko

their range (Bragg Peak). If the energy of the proton and carbon ion are adjusted to the depth of the tumor, far more excellent dose localization can be achieved. Actually, the tumor has its own thickness along the beam path. In order to irradiate the tumor volume with uniform dose distribution, various techniques for the enlargement of the Bragg peak and also scanning the transversal plane has been developed in case of charged particle.

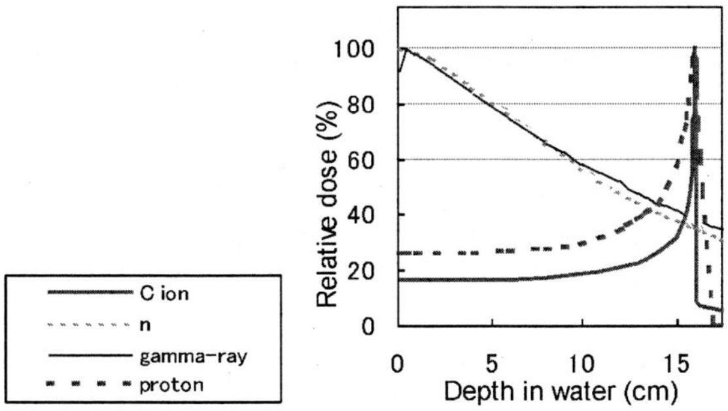

FIGURE 1. Depth-dose distribution in tissue of various radiation.

The biological effect of ionizing radiation is caused by damage to DNA depending on the ionization density producing along the radiation track. Average energy transferred per unit length of the track is called a liner energy transfer (LET) and becomes larger as the mass of the particle increases or the particle goes deeper. Since LET in the peak of a carbon beam is larger than that of a photon beam and proton beam, it has biological features beneficial to cancer radiotherapy: relative biological effectiveness (RBE) being 2 to3 times larger, less recovery from radiation damage in the target volume, less dependence of the radiation effect on oxygen concentration in the tissue, and a small difference in radiological sensitivity through the cell cycle. The RBE of particle beams becomes larger as the mass of the particle increases. When the ratios of the RBE in the plateau area to that in the peak area are compared, the carbon beam have the largest value among various ion beams [3], suggesting it is the best balanced particle beam [4]. This is the main reason we have selected the carbon beam for cancer radiotherapy at NIRS.

CLINICAL TRIAL AT HIMAC

Heavy-ion cancer treatment is being carried out at the Heavy Ion Medical Accelerator in Chiba (HIMAC) with 140 to 400 MeV/n carbon ions at NIRS. HIMAC was constructed as part of the Japanese government's "Comprehensive 10-years Strategy for Cancer Control" that started in 1984. HIMAC was the first heavy ion medical dedicated accelerator in the world and its aim was to verify effectiveness and safety of heavy ion therapy. HIMAC has been operated as a multipurpose facility open to international and domestic researchers for biological, physical, and accelerator engineering studies as well as cancer treatment. Figure 2 shows schematic view of the HIMAC facility. The HIMAC facility is composed of the following parts; three ion sources (one PIG type and two ECR type), RFQ liner accelerator, Alvarez liner accelerator, two synchrotron rings, beam transport system and three treatment rooms. The three treatment rooms have different features. Treatment room A with vertical beam, C with horizontal beam and both horizontal and vertical beam ports are provided to treatment room B. ECR ion sources has been proven to be reliable and stable for production of various ions at HIMAC.

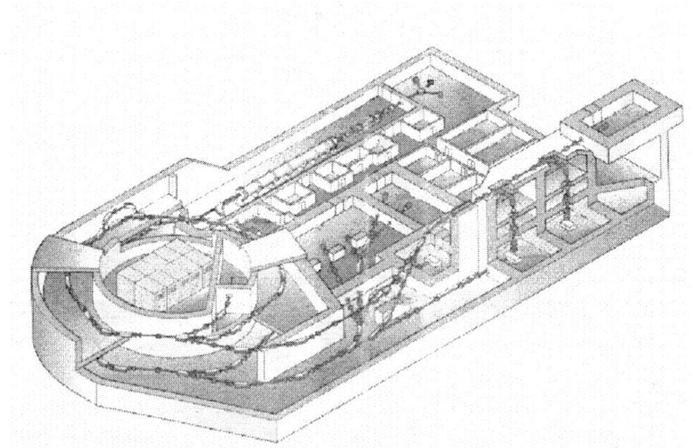

FIGURE 2. Schematic view of the HIMAC facility

Carbon ion radiotherapy started on June 21, 1994 at NIRS and has mainly focused on the group of diseases that are difficult to cure using conventional radiotherapy, and among them are head and neck tumor, brain tumor, lung cancer, hepatocellular carcinoma, prostate cancer, uterine cervix cancer, bone and soft tissue tumor, and esophageal cancer. The total number of patients enrolled by February 2008 was 3,819 where various types of tumors were treated (Figure 3). In the phase I/II studies to confirm the safety of carbon ion radiotherapy and to obtain a clue to an anti-tumor effect, the number of fractions and treatment period were fixed for each disease and the total dose was gradually increased by 5 to 10%. When the recommended dose was determined in the phase I/II studies, they were incorporated into the phase II studies. The number of the patients enrolled has increase year after year. This increase occurred as the treatment regimen became established and could be smoothly executed, thereby the number of fractions and treatment period per patient was reduced.

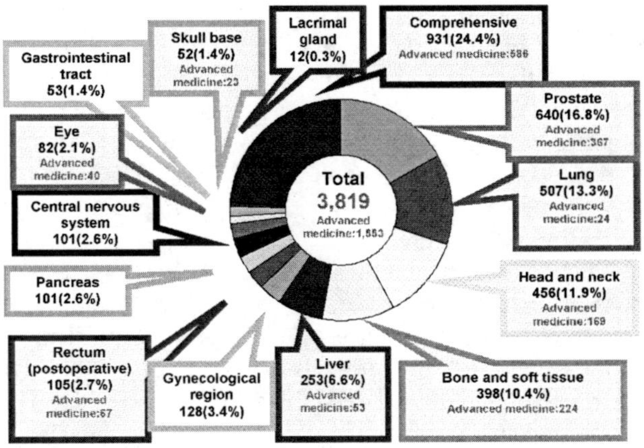

FIGURE 3. Number of patients enrolled for carbon ion radiotherapy (June 1994 to February 15, 2008).

Since carbon ion beam has the therapeutically advantageous feature of higher RBE at deeper sites from the surface of the body, treatment can be completed in a shorter period. For stage I lung and liver cancers, for example, treatment can be completed in one or two irradiation fractions. For prostate cancer and bone/soft tissue tumors, it is completed in 16 to 20 fractions, which is about half the number of fractions employed in X-ray and proton beam radiotherapy. In the case of lung cancer, clinical trial with one fraction is now in progress. It means the irradiation is finished within one day.

COMPACT ECR ION SOURCE FOR NEW FACILITY

HIMAC facility is useful for clinical trial and still powerful for various research progresses. However it is too big for hospitals because accelerators were designed for acceleration a silicon ion to energy of 800 MeV/n. In order to establish in many hospitals, we had to optimize and reduce the size of the facility to carbon ion therapy. A more compact accelerator facility for cancer therapy had been designed at the NIRS [5]. Important points of reducing the size of accelerators are as follows; 1) maximum acceleration energy of main accelerator is optimized for carbon ion for the medical use. From the results of the treatment, the residual range of 250 mm covers almost of all patients. The residual range depends not only on the beam energy, but also on the irradiation method. Using the spiral-wobbler method [6], the carbon energy is estimated to be 400 MeV/n in order to obtain the residual range of 250 mm. 2) output energy of carbon ion from injector Linac is reduced compare than HIMAC injector for optimizing for charge exchange to C^{6+}. HIMAC injector Linacs can be acceleration more than charge to mass ratio = 1/7 ions from energy of 8 keV/n to 6 MeV/n. Usually, HIMAC injector accelerates C^{2+}. In order to reduce the size of injector, we need charge to mass ratio 1/3. It means ion source has to produce C^{4+}. 3) Ion source is optimized for C^{4+} ion production for therapy. In order to reduce the size of the injector (consists of ion source, low-energy beam transport and post-accelerator Linac include these power supply and cooling system), an ion source requires production of highly charged ions and lower electric power for easy installation of the source on a high-voltage platform. A compact Electron Cyclotron Resonance Ion Source (ECRIS) with all permanent magnets is one of the best types for this purpose. An ECR ion source has advantage for production of highly charged ions. A permanent magnet is suitable for reduce the electric power and cooling system. The required beam intensity and emittance for the source are on the order of 500 eμA and 1.0 π mm mrad (normalized) for C^{4+}.

DESIGN OF THE KEI2-SOURCE

At first, a prototype compact ECRIS with all permanent magnets (Kei-source) was designed to study its performance in producing C^{4+} [7]. The fixed magnetic field profile of the Kei-source was copied from that of the 10 GHz NIRS-ECR [8] source. This particular field profile seems to be suitable for the production of C^{4+}, but not for C^{5+} or C^{6+}. The Kei-source has satisfied the basic requirements for the medical applications. The performance of the new source is expected to be close to that of the present NIRS-ECR source (C^{4+}:430 eμA). Our previous articles reported on the detail of the Kei-source [9,10]. We obtained an output current of 280 eμA for C^{4+}. The performance of the Kei-source could not reach that of the NIRS-ECR source, because the mirror magnetic field was less than the design value, due to a difference between the calculation and the real magnets. Therefore, it is necessary to increase the magnetic field in order to reach the performance of NIRS-ECR. Based on the results of the Kei-source, we have designed a new all permanent-magnet ECRIS for practical use (Kei2-source), in which the magnetic field and the extraction region have been modified to improve the performance.

The size of the magnets and their arrangement were determined in such a way that both the peak and minimum values in the mirror field would become close to those of the NIRS-ECR source. The magnetic field was calculated using the POISSON/SUPERFISH code [11] and measured by using a Hall probe. As shown in Figure 4, the differences between the calculated and measured values are 5% at most in the maximum mirror field and in the hexapole field.

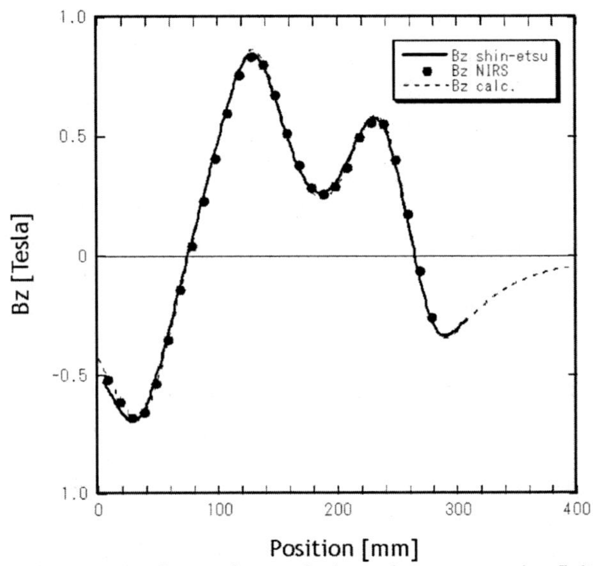

FIGURE 4. Comparison of the mirror magnetic field between the calculated and measured values. The solid line was measured by the manufacturing company (Shin-etsu). The circle points were measured by NIRS. The dotted line is the calculated value.

An ion source requires a high extraction voltage because the transmission efficiency in low-energy transport depends strongly on the injection energy due to the space-charge effect. Based on experience 18 GHz NIRS-HEC [12], designed with an extraction voltage of 60 kV, the Kei2-source has basically a similar structure from this point of view, as shown in Figure 5. In order to realize such a high voltage, two improvements are necessary. First, the extraction electrode would be cooled directly by water. This is very effective to reduce any outgasing from the electrode, and to keep a good vacuum at the extraction region. Second, all components of the source, including the plasma chamber, permanent magnets, all power supplies, and the vacuum system, would be put into high-voltage box at the extraction voltage. This is necessary to maintain good insulation. These two improvements should significantly improve the reliability of the source.

The general structure of the Kei2-source (size of magnets, vacuum chamber, gas injection, waveguide, plasma chamber, and extraction system) is almost the same as that of the previous Kei-source. Table 1 shows a comparison between the specifications of Kei-source and the Kei2-source. The source has an outer diameter of 320 mm and a length of 295 mm. The axial field is set to 0.87 T, 0.59 T, and 0.25 T at the injection side, the extraction side, and the minimum B region, respectively. The hexapole magnetic field on the plasma chamber surface is 0.66 T. A commercial model, N45M and N32Z (made in Shin-Etsu Co.), are employed as a ring magnets for mirror field and hexapole magnet for radial confinement field, respectively. A Traveling-Wave-Tube (TWT) amplifier is used in the new source, in order to find the optimum frequency under a fixed and uncontrollable magnetic field. Microwave power is fed into the plasma chamber through a rectangular wave guide from the axial direction. The TWT amplifier operated at a frequency of 8 to 10 GHz, and can be driven both in cw and pulse modes with a maximum output power of 300 W. A biased disk is also used for optimizing. Many groups have reported that the biased disk method is useful for increasing the beam intensity of highly charged ions (HCI) [13-17]. The diameter of the disk was 8 mm, and was made of molybdenum.

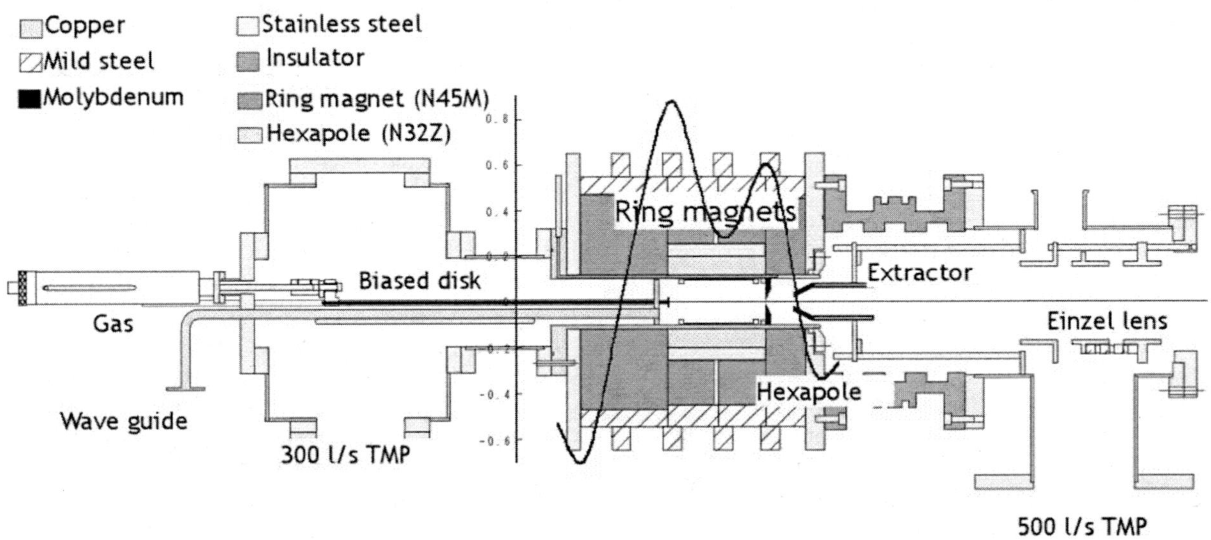

FIGURE 5. Schematic view of the Kei2-source.

The disk position is movable between 7 mm upstream and 25 mm downstream in the direction from the peak of the mirror field (injection side). From our experience, this method seems to be suitable for such a compact ECRIS, since the disk does not need a large space for installation. The plasma chamber is made of copper for a good cooling efficiency, in order to avoid a decrease in the magnetic field due to high temperature. It has an inner diameter of 50 mm and 120 mm in length. The vacuum pressures of the gas injection side and beam extraction side are 7.0E-6 Pa and 8.0E-5 Pa, respectively. The estimated vacuum in the plasma chamber is around 1.0E-4 Pa.

TABLE 1. Specifications of the Kei-source and the Kei2-source.

	Kei	Kei2
Mirror magnets		
Material	NdFeB	NdFeB
Maximum field strength		
Extraction sid	0.51 T	0.59 T
Gas injection side	0.75 T	0.87 T
Minimum B strength	0.25 T	0.25 T
Hexapole magnet		
Material	NdFeB	NdFeB
Maximum field strength on the surface of chamber	0.76 T	0.66 T
Length	105 mm	105 mm
Inner diameter	60 mm	58 mm
Microwave amp.		
Frequency	9-18 GHz	8-10 GHz
Maximum power	270 W	300 W

PERFORMANCE OF THE KEI2-SOURCE

The ion source was operated to test an injection line for the prototype Linac. The injection line consists of an analyzing magnet, electro static triplet quadruple lens and two monitor boxes. The analyzing magnet has a bending angle of 90 degree, edge angles of 28.64 degrees for both ends, and radius of curvature of 0.25 m. The maximum magnetic rigidity is 0.277 Tm. Monitor boxes consists of Faraday cups, horizontal and vertical slits, two couples of emittance monitor, and two vacuum pumps (both 450 liter/sec turbo molecular pump). A couple of horizontal and vertical slits and a Faraday cup (FC1) were installed at the focusing point of the analyzing magnet. The emittance monitor was also installed downstream the FC1. The beam performance of the Kei2-source was summarized in Table 2. We obtained an maximum output current of 618 eμA for C^{4+} at an extraction voltage of 30 kV. The typical charge state distribution (CSD) of the carbon shows in Figure 6. Parameters of the source (microwave power and frequency, gas flow, biased disk voltage and position) were optimized for C^{4+} production. The methane was used for ionization gas.

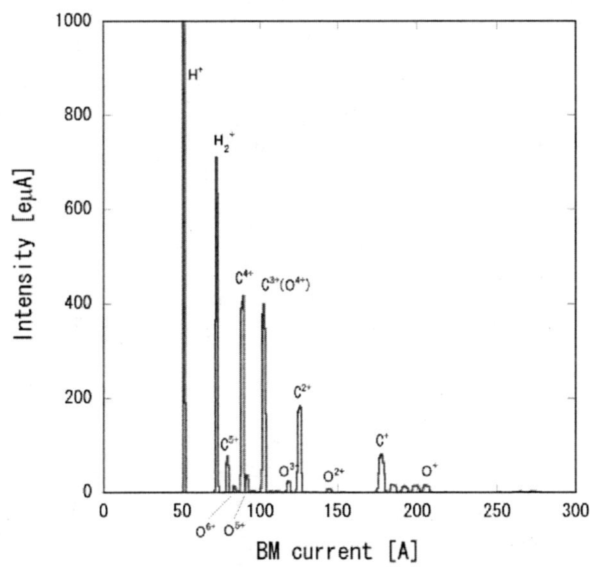

FIGURE 6. Typical CSD of carbon with CH_4 gas.

TABLE 2. Summary of maximum beam intensity at extraction voltage of 30 kV. [eμA]

	1+	2+	3+	4+	5+	6+	7+	8+	9+	10+	11+
He	>1000	540									
C	308	550		618	90						
N	405	505	330	167	75	5					
O	760	740	455		125	42.4					
Ar		445	295	225	135	80	75	88	30		2.5

The emittance is one of the important parameters to judge beam qualities. The horizontal and vertical emittance for C^{4+} ions produced with CH_4 gas was measured using the emittance monitor. The emittance monitor consists of double slits and multi wire monitor. Distance of the two slits is 20 mm and the width of each slit are 0.3 mm. Multi wires monitor has 64 wires and 0.615 distances. The vertical multi wire monitor was located 306 mm downstream from the slit. Distance of horizontal multi wires monitor and slits is 224 mm. The parameters of the source were optimized for production of C^{4+} and the beam intensity was 400 eμA in pulsed mode (2 Hz, 50 msec), extraction voltage was 30 kV. Figure 7 and 8 show horizontal and vertical emittance of C^{4+}, respectively. The 90% emittance for horizontal and vertical were 1.43 and 1.0 π mm mrad, respectively.

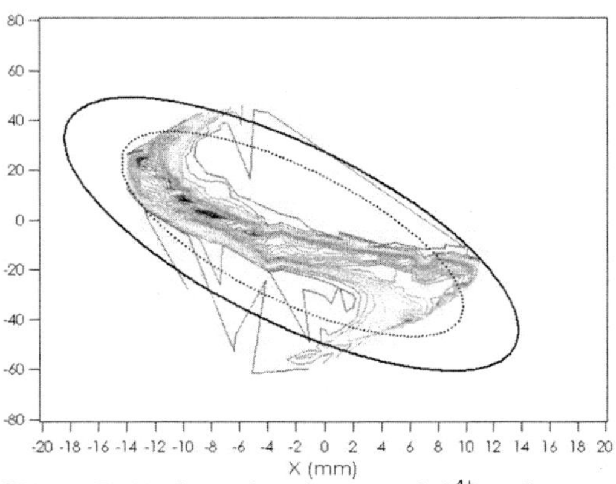

FIGURE 7. Horizontal emittance of C^{4+} under extraction voltage of 30 kV.

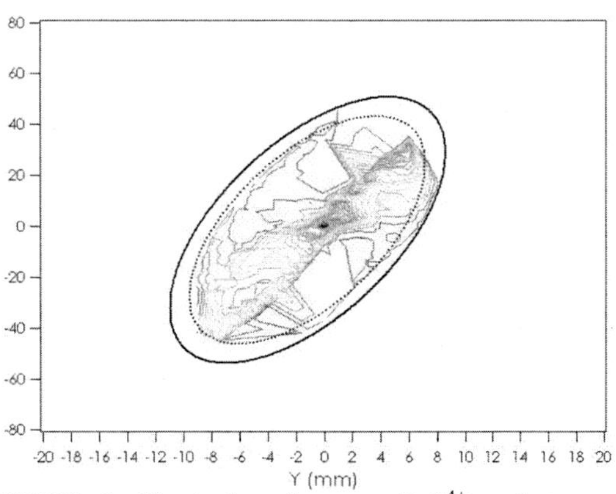

FIGURE 8. Vertical emittance of C^{4+} under extraction voltage of 30 kV.

IMPROVEMENT OF BEAM INTENSITY

In order to obtain highly charged carbon ion (5+, 6+), we investigate optimization of compound carbon gases. Usually, carbon molecular gases (e.g. CH_4 gas) are applied for production of the carbon ions, in particular C^{4+}. If CO_2 gas is used the yield is lower than with CH_4 gas [18]. This phenomenon is similar than what one observes in well known gas mixing experiments [17]: oxygen is not a good mixing gas (too heavy) for production of highly charged carbon. In previous experiments, the CH_4, CD_4, C_4H_{10}, C_4D_{10}, C_2H_2 and C_2D_2 gases were already tested at Kei2-source [19,20]. CD_4, C_4D_{10} and C_2D_2 gases are isotopic gas of methane, butane and acetylene. In these measurements it was found that in C_2H_2 the ratio of carbon to hydrogen is the best for production of C^{4+} ; isotopic gases is better than normal gases for production of C^{5+}.

In order to know the best ratio of carbon and hydrogen, we tested Naphthalene ($C_{10}H_8$) powder. A simple container for making naphthalene vapor was coupled to the plasma chamber. The source could be operated at the standard operation setting (300 W of RF power at around 10 GHz, extraction voltage 30 kV). Parameters of the source were optimized for the production of C^{4+}. Figure 9 shows the spectrum of carbon with $C_{10}H_8$, together with those using CH_4 and C_2H_2 gases obtained under the same source conditions. CSD of the carbon using $C_{10}H_8$ was shifted a bit to lower charges as compared to the one using C_2H_2. From this quick test one can conclude that - in terms of the gas-mixing explanation - the carbon to hydrogen ratio is too high for optimal carbon 4+ production with $C_{10}H_8$. Additionally, a maximum beam current of 768 eμA for C^{4+} at an extraction voltage of 40 kV.

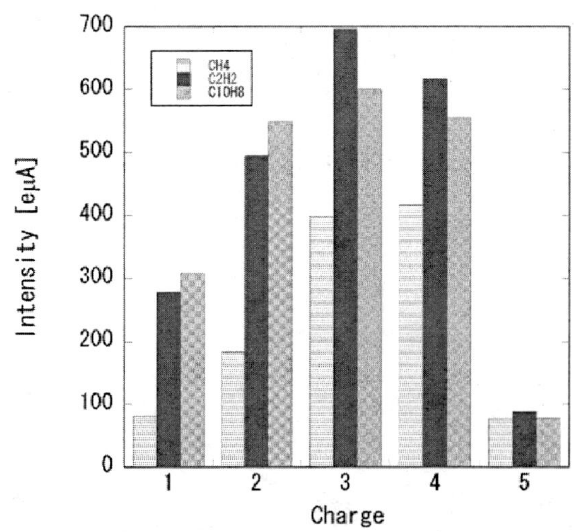

FIGURE 9. Charge state distributions of carbon.

BEAM TEST OF PROTOTYPE INJECTOR

We constructed a prototype injector consisting of compact ECR ion source, RFQ Linac, and IH-DTL (Interdigital H-mode Drift Tube Linac) [21,22] and test its performance. Injection energy of RFQ Linac is 10 keV/n and output energy is 600 keV/n. Output energy of IH-DTL is 4 MeV/n. Total length of prototype injector is less than 10 m. For test in March 2006, the Kei2-source was operated with C_2H_2 gas giving about 500 eμA C^{4+}, which was accelerated in the system. The intensity downstream of IH-DTL was 390 eμA. Transmission efficiency between source to output of Linac was 79%. Figure 10 shows result of the beam stability test. Beam intensity of 400 eμA at stripped C^{6+} was obtained. Beam stability is better than 10% during eight hour without any adjustment of parameters. From these results, new injector has acceptable performance for main accelerator synchrotron.

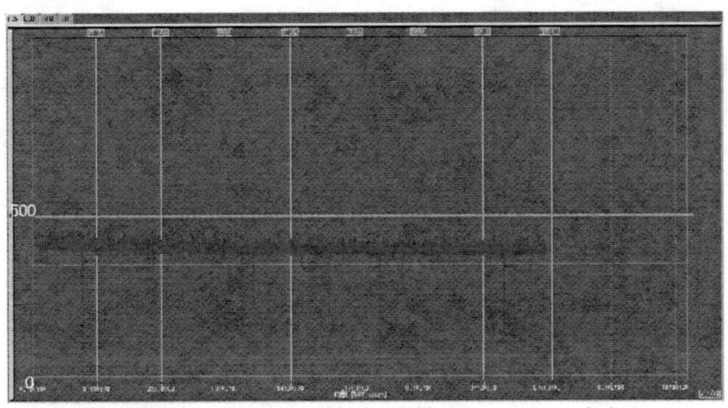

FIGURE 10. Result of beam stability test at new injector.

CONSTRUCTION OF NEW FACILITY AT GUNMA UNIVERSITY

A compact ECR ion source for Gunma University has been manufactured by Sumitomo Heavy Industries. General structure include magnetic field was copied from Kei2-source. Figure 11 shows photo of compact source with low energy beam transport. Preliminary beam test was already finished in March 2008. Beam intensity of 520 eμA at extraction voltage of 30 kV with CH_4 gas was obtained.

The construction of the facility will be finished until spring of 2009. Gunma University plans to start clinical trial in 2009 financial year.

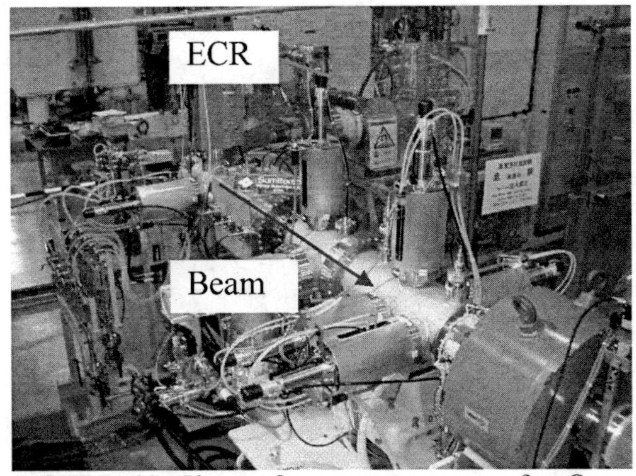

FIGURE 11. Photo of a compact source for Gunma University with beam transport.

ACKNOWLEDGEMENTS

The authors are thankful to Drs. E. Takada, K. Noda, and other members of the Department of Accelerator Physics and Engineering of NIRS for their continuous encouragement and useful advice. We would like to give our thanks to the members of Accelerator Engineering Corporation for their assistance in the data-taking process.

REFERENCES

1. Tujii H, Mizoe J, Kamada T, *et al.*, Clinical Results of Carbon Ion Radiotherapy at NIRS. *J. Radiat. Res.* 48, **A1-A13**, 2007.
2. Raju MR, Heavy Particle Radiotherapy, 1980.
3. Shoulte R, "Early and Late Response to Ion Irradiation." In *Ion Beam in Tumor Therapy* (ed. by Linz U), 53-62, 1995.
4. Kawashima K, Radiobiology On Cancer Therapy From Heavy Particle Physics. 20th Series of symposium at NIRS. (ed. by Tsunemoto H, Ohara H), 214-224, (1988).
5. K. Noda, *et al.*, Proceedings of EPAC2004, Lucerne, Switzerland, 2634 (2004)
6. M. Komori, *et al.*, *Jpn. J. Appl. Phys.*, **47**(9A) , pp. 6463-6467, (2004).
7. M. Muramatsu *et al.*, *Review of Scientific Instruments*, **71**(2), 984-986 (2000).
8. A. Kitagawa *et al.*, *Review of Scientific Instruments*, Vol. **65**, 1087 (1994)
9. M. Muramatsu *et al.*, *Review of Scientific Instruments*, **73**(2), 573-575 (2002).
10. M. Muramatsu *et al.*, *Proceedings of the 15th International Workshop on ECRIS*, 2002, p, 59.
11. Reference Manual for the POISSON/SUPERFISH Group of Codes, LA-UR-87
12. A. Kitagawa *et al.*, *Rev. Sci. Instrum.* **69**, 674 (1998)
13. G. Melin *et al.*, *Proceedings of the 10th International Workshop on ECRIS*, 1990, p, 1.
14. S. Gammino *et al.*, *Review of Scientific Instruments*, **63**(4), 2872-2874 (1992).

15. S. Biri *et al.*, *Proceedings of the 14th International Workshop on ECRIS*, 1999, p. 81.
16. S. Biri *et al.*, *Nucl. Instrum. Methods Phys. Res. B* **152**, 386 (1999).
17. A. G. Drentje, *Review of Scientific Instruments*, **74**, 2631 (2003)
18. A. Kitagawa *et al.*, *Rev. Sci. Instrum.* **71**, 1061 (2000).
19. A. G. Drentje *et al.*, *Rev. Sci. Instrum.* 77, 03B701 (2006). and A. G. Drentje *et al.*, Proceedings the 17th International Workshop on ECRIS, High Energy Physics and Nuclear Physics (Journal of the Chinese Physical Society C) 31, 94 (2007).
20. A.G.Drentje, A. Kitagawa and M. Muramatsu, to be published in IEEE Transactions on Plasma Science, August 2008
21. Y. Iwata, *et al.*, *Proceedings of PAC2005*, Knoxville, Tennessee, 1084-1086 (2005).
22. Y. Iwata, *et al.*, *Nucl. Instr. and Meth. B* **572** (2007) 1007-1021

The Effect of the Dose and Energy of a Pre-Silicide Implant on Nickel Silicide Formation

Jeffrey H. Rice

Fairchild Semiconductor, West Jordan, UT, U.S.A.
jeff.rice@fairchildsemi.com

Abstract. Pre-silicide implants have been used to increase the thermal stability of nickel silicide (NiSi) and to improve device performance. This study evaluates the effect of the dose, energy and species of a pre-silicide ion implant on NiSi phase formation. The resulting silicide was evaluated using sheet resistance, scanning electron Microscope (SEM) cross-sections, and Rutherford Backscattering Spectroscopy (RBS) analysis. It was found that a high dose argon implant will completely inhibit the silicide formation.

Keywords: Nickel, silicide, fluorine, argon, implantation
PACS: 61.72.U-, 68.55.Nq, 81.05Je

INTRODUCTION

Metal silicides of titanium and cobalt have been used in the integrated circuit industry to form low resistance source/drain contacts and gates. As device sizes have shrunk, materials that are not subject to line-width effects due to silicon consumption (e.g., titanium silicide) are required. For this reason, nickel silicide (NiSi) has gained acceptance due to its low resistance, low formation temperature, low silicon consumption and improved Schottky diode leakage performance due to its barrier height [1].

Nickel silicide is known to form in three phases. The first reaction occurs near 250 °C. This higher resistance phase is metal rich (typically Ni_2Si). After the temperature reaches 350 °C, a low-resistance monosilicide phase (NiSi) forms. At temperatures near 750 °C, a high resistance silicon rich phase forms ($NiSi_2$) [2].

Unfortunately, NiSi suffers from thermal instability in that post silicide processing at high temperatures can be problematic. Nickel silicide can agglomerate at temperatures as low as 600 °C, and it forms a high resistance $NiSi_2$ phase at a relatively low temperature [3]. To prevent thermal instability, several types of pre-silicide implants have been evaluated. Nitrogen has been used to improve the film's thermal stability, shifting the transformation from NiSi to $NiSi_2$ to a higher temperature. It has also been reported that the nitrogen implant will suppress oxidation at the nickel to silicon interface, creating a smoother surface which will result in lower junction leakage current [4]. Fluorine and hydrogen implants have also been found to reduce the thermal instability of NiSi and improve junction leakage [5, 6].

This study evaluates the effects of ion implantation species, dose and energy on the formation of nickel silicide. Argon, fluorine and nitrogen implants were used as their differing masses were expected to have different effects at the silicon to nickel interface.

EXPERIMENTAL

In order to gain a basic understanding of the relationship between the species, dose and energy of the pre-silicide implant, samples were prepared on <100> 3-20 ohm, p-type silicon wafers and implanted with the combinations of energy, dose and species shown in Table 1. Next, 1300 Å of nickel was evaporated on the wafers, followed by a 450 °C anneal.

TABLE 1. Initial test wafer parameters.

Implant Species	Low Dose ($ions/cm^2$)	High Dose ($ions/cm^2$)	Low Energy (keV)	High Energy (keV)
Fluorine	5E15	2E16	10	20
N_2	2E14	2E15	10	20
Argon	5E15	2E16	35	70

CP1066, *Ion Implantation Technology,* edited by E. G. Seebauer, S. B. Felch, A. Jain, and Y. V. Kondratenko

Sheet resistance measurements were taken after nickel deposition and after thermal annealing. In addition, scanning electron microscope (SEM) cross-sections and Rutherford Backscattering Spectroscopy (RBS) analysis was performed on selected wafers.

RESULTS AND DISCUSSION

Sheet resistance was initially used to detect the phase formation of nickel silicide. The low resistivity NiSi phase was expected to form at the 450 °C annealing temperature (in the absence of any ion implant).

Figure 1 shows the final sheet resistance reading, normalized to a starting value of 1.0 ohm/sq. Using only sheet resistance values as an indicator of the nickel silicide phase, the results for fluorine and nitrogen implants indicate that a high dose implant created a low resistance NiSi phase and a low dose implant created a high resistance Ni_2Si phase. The low dose argon implant sample behaved differently in that the sheet resistance decreased. The high dose argon implant result was similar to that of the fluorine and nitrogen implants.

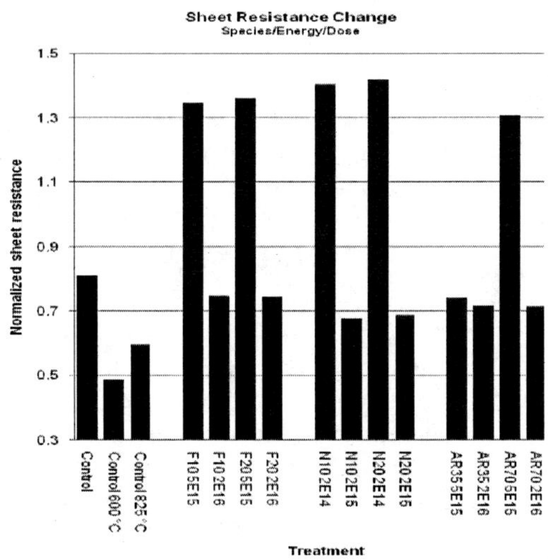

FIGURE 1. Sheet resistance of initial test wafers after 450 °C anneal.

RBS Analysis

Rutherford Backscattering Spectroscopy (RBS) was used to evaluate the phase of the resulting silicide formed on the 70 keV argon implanted samples. A non-implanted control sample was used for comparison.

Comparing the ratio of nickel to silicon in Figure 2 shows that the NiSi phase formed in the majority of the sample. This corresponds with the reduction in sheet resistance. Control samples were also annealed at 600 °C and then at 825 °C. The sheet resistance first decreased as the Ni_2Si converted to NiSi and then increased as the NiSi began fully converting to $NiSi_2$ (see figure 1).

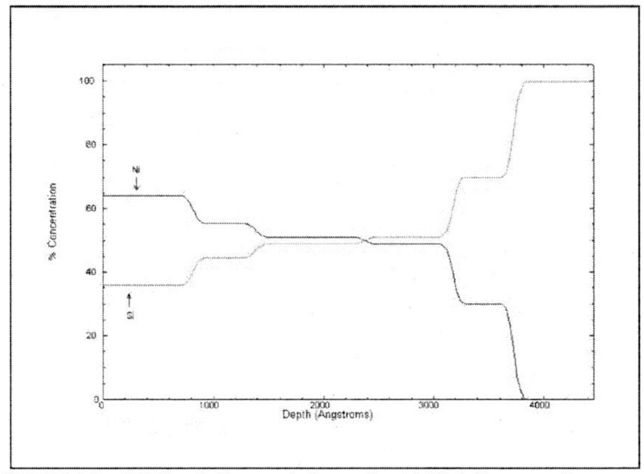

FIGURE 2. RBS of Control Sample (450 °C anneal).

Figure 3 shows the RBS plot of the wafer implanted with argon at 70 keV and 5E15 ions/cm^2 and then annealed. The sheet resistance increased on this sample, implying that a high resistance phase of nickel silicide formed. The RBS ratio of nickel to silicon confirmed that the Ni_2Si phase was present. It also showed the presence of argon throughout the sample. The results indicate that the implant partially inhibited the silicide formation. This is presumably a result of the implant impeding the diffusion of the nickel into the silicon.

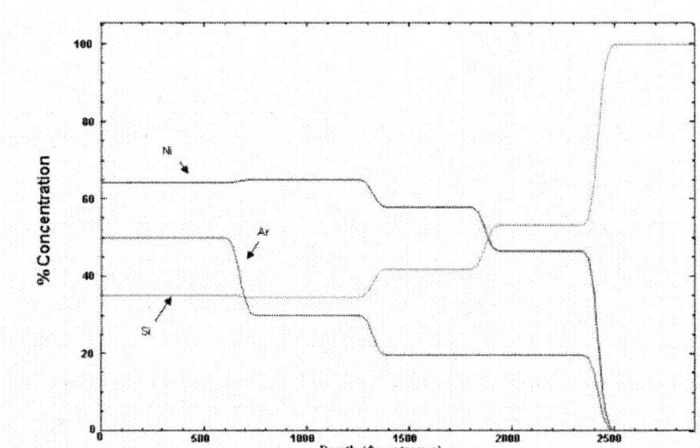

FIGURE 3. RBS of 70 keV, 5E15, argon implant sample (450 °C anneal).

Next, the 70 keV, 2E16 ions/cm^2 argon implanted sample was analyzed using RBS. The sheet resistance of this sample decreased after annealing, which would normally indicate formation of NiSi. However, as seen in figure 4, the RBS revealed that the silicide reaction was completely inhibited.

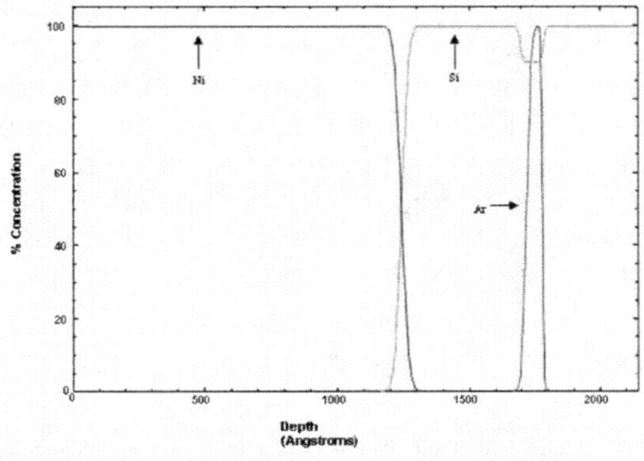

FIGURE 4. RBS of 70 keV, 2E16, argon implant sample (450 °C anneal).

The decrease in sheet resistance without a subsequent formation of NiSi is attributed to a change in the nickel grain size as a result of the annealing process [8].

Additional RBS data was collected on the high dose, low energy fluorine and nitrogen samples, also showing that no silicide was formed. Figure 5 shows the fluorine implant RBS results.

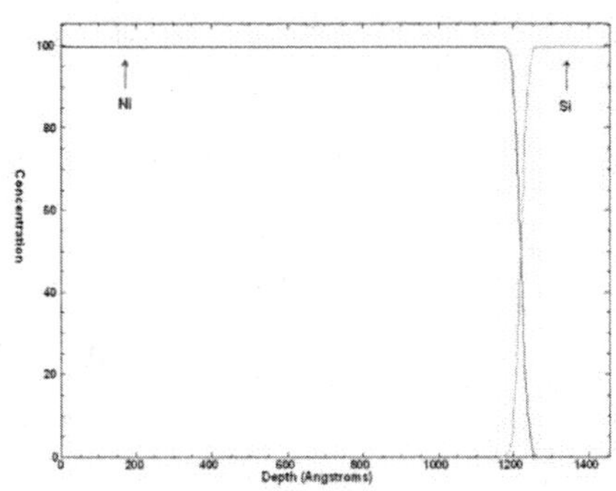

FIGURE 5. RBS of 20 keV, 2E16, fluorine implant sample (450 °C anneal).

SEM Analysis

Scanning Electron Microscope (SEM) cross-sections of the samples were taken to evaluate and confirm the RBS results. The samples were given a "Wright" etch stain (silicon defect etch) to highlight the film structure. Figure 6 shows the control sample, where the silicon to silicide interface can be seen with minimal etch of the film.

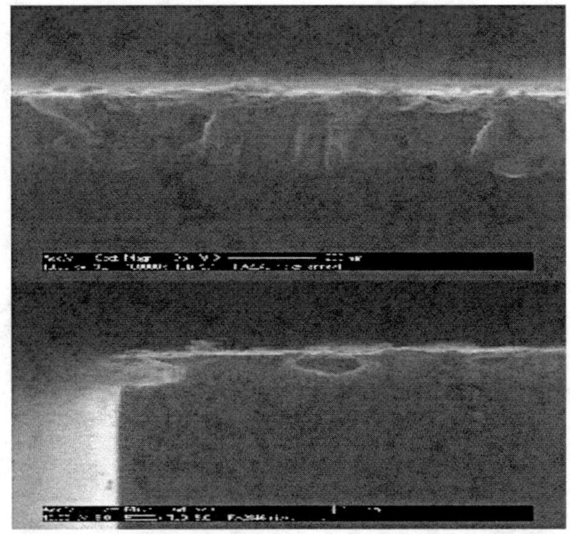

FIGURE 6. SEM cross-sections of control sample.

The cross-section of the 70 keV, 5E15 ions/cm^2 sample is shown in figure 7. The sample has smaller grains and more silicon etched than the control wafer.

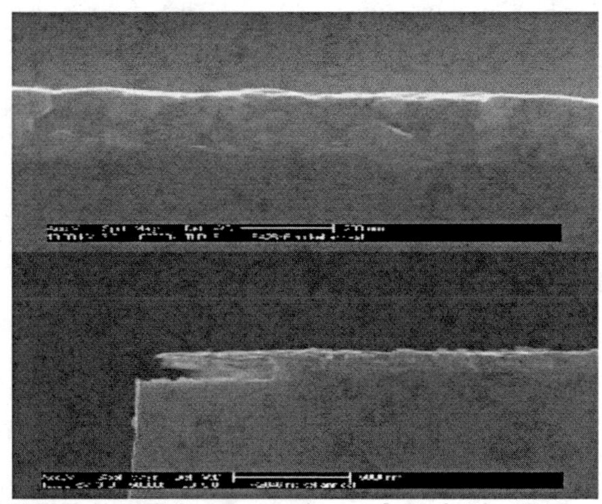

FIGURE 7. SEM cross-sections of 70 keV, 5E15, argon implant sample (450 °C anneal).

Figure 8 shows the 70 keV, 2E16 ions/cm^2 fluorine and the 35 keV, 5E15 ions/cm^2 sample. These SEM cross-sections showed that the silicon underneath the nickel was removed, confirming the RBS data shown in figure 5. The similar sheet resistance change for the argon sample suggested that silicide also did not form on it.

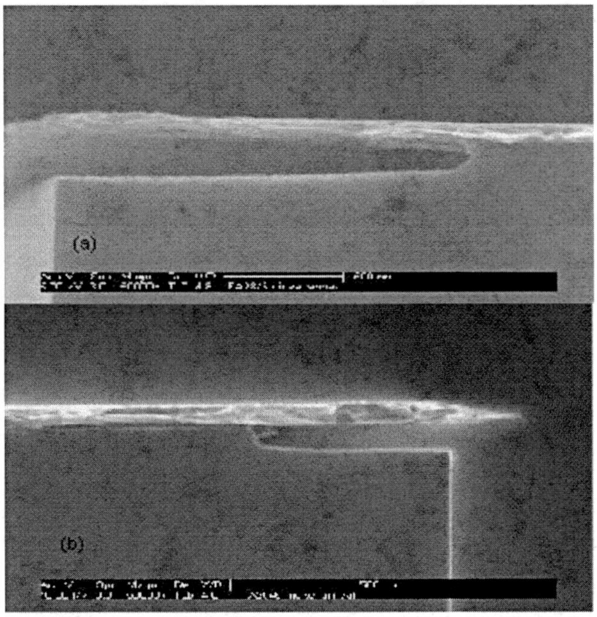

FIGURE 8. SEM cross-sections of 70 keV, 2E16, fluorine (a) and 70 keV, 5E15 argon (b) samples (450 °C anneal).

The SEM cross-sections correlated well with the RBS results. The etch profile was related to the silicide formation, as samples with more silicide formation (according to RBS results) were less aggressively etched. Since the Wright etch highlights silicon defects, the implant damaged silicon was more aggressively etched.

The combined RBS and SEM results suggest that as the surface amorphization from the implant increases the opportunity for silicide formation decreases. In the case of the argon implant, it can be hypothesized that the larger argon ion caused more silicon damage, fully amorphizing the surface and inhibiting the diffusion of the nickel into the silicon. The fluorine and nitrogen caused less damage and therefore allowed limited diffusion to occur. This explains why the low energy, low dose argon inhibited silicide formation but the low dose, low energy fluorine and nitrogen did not.

CONCLUSIONS

Nickel silicide formation using a pre-silicide implant is dependent on the species, dose and energy of the implant. A large ion with a low energy and high dose will create a fully amorphized surface that will impede the diffusion of the nickel into the silicon, completely inhibiting the formation of nickel silicide. Conversely, a small ion implanted with a low dose and high energy will partially impede the nickel diffusion, allowing some silicidation to occur. The nickel silicide phase formation can be controlled by selecting the appropriate implant parameters.

ACKNOWLEDGEMENTS

The author would like to thank Fairchild Semiconductor for their support of this evaluation.

REFERENCES

1. M. Tsuchiaki, K. Ohuchi and A. Nishiyama, *Japanese Journal of Applied Physics* **44**, 1673-1681 (2005).
2. L.W. Cheng, S.L. Cheng, J.Y. Chen, L.J. Chen and B.Y. Tsui, *Fourth International Workshop on Junction Technology IWJT '04*, 113-118 (2004).
3. D.Z. Chi, R.T.P. Lee and S.J. Chua, *Thin Solid Films* **355-356**, 412-416 (1999).
4. P.S. Lee, K.L. Pey, D. Mangelinck, J. Ding, A.T.S. Wee and L. Chan, *IEEE Electron Devices Letters* **21**, 566-568 (2000).
5. Jang-Gn Yuh, Soon-Young Oh, Hee-Hwan Ji, Bin-Feng Huang, Young-Ho Park, Jin-Suk Wang and Hi-Deok Lee, *Fourth International Workshop on Junction Technology IWJT '04*, 131-134 (2004).
6. C.-J Choi, S.-A. Song, Y.-W Ok and T.-Y. Seong, *Electronics Letters* **40**, 391-393 (2004).
7. S.P. Murarka, "*Silicides for VLSI Applications*, Orlando, FL: Academic Press, 1983, p. 130.
8. A.K. Kulkarni, K.H. Schulz, T.A. Lim and M. Khan, *Thin Solid Films* **345**, 273-277 (1999).

Development of an Ion Beam Irradiation System for Liquid Crystal Alignment Layer Production

Takeshi Matsumoto, Yuko Kinoshita, Masahiro Tanii, Junichi Tatemichi, Masashi Konishi, and Masao Naito

Nissin Ion Equipment Co., Ltd., 29, Hinokigaoka, Minakuchi-cho, Koka, Shiga, 528-0068 Japan

Abstract. Ion beam irradiation was employed to produce alignment layers for liquid crystal (LC) displays. The alignment characteristics were compared with those by the conventional rubbing method. Ion incident angle to the films played an important role in LC sample optical qualities. A new ion irradiation method to realize a multi-domain structure for a wide viewing angle was demonstrated.

Keywords: Liquid crystal, Alignment, Ion beam, Light leakage, multi-domain
PACS: 61.30.Gd, 41.75.Ak, 42.88.+h

INTRODUCTION

Alignment layers, which are LC display components contacting LC molecules and causing their alignment at the interface, have been made by the conventional rubbing method. Following polyimide alignment film coating on a substrate, a metal roller covered with a fluffy cloth rubs the surface. This method has several drawbacks such as frictionally generated electrostatic charges, dust fibers coming off from the cloth and difficulty in keeping the cloth stable. Alignment quality depends on the condition so much that LC display manufacturers pay careful attention to the cloth condition management. However, the management has been of great concern to the manufacturers because the condition is not easily quantified. What is more serious is that fibers make scratches in some places of the surface, posing disturbance of LC alignment. This results in light leakage through the vicinity of the scratches and a drop of a contrast ratio of LC displays [1].

As an alternative, oblique-directed ion beam irradiation to form an alignment layer attracts much attention, since no scratch is made and keeping the ion beam in a good condition which ensures alignment quality is easy. Ion beam for this application reportedly requires good parallelism and uniform distribution to realize a high contrast ratio [2].

We have developed an ion beam irradiation system for 400 x 500 mm^2-sized alignment layer productions [3]. The system has been designed for parallel ion irradiation with uniform distribution and small divergence, equipped with a beam parallelism monitor.

We reported that alignment films irradiated with argon ions showed a uniform LC aligned texture [3]. In this study, we examined ion-irradiated LC samples and compared them with those by the rubbing method.

Another advantage of the ion irradiation is a possibility of multi-domain alignment because it is a non-contact alignment method. Multi-domain alignment techniques have been studied intensively to pave the way for a wide viewing angle of LC displays

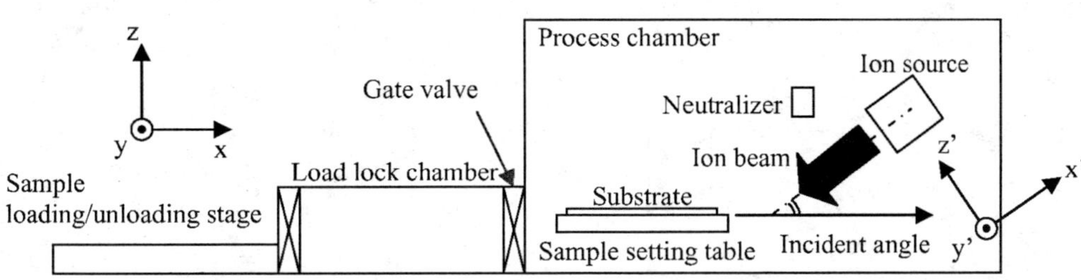

FIGURE 1. Schematic of the ion beam irradiation system.

CP1066, *Ion Implantation Technology,* edited by E. G. Seebauer, S. B. Felch, A. Jain, and Y. V. Kondratenko

[4-6]. We describe LC samples which had a striped multi-domain structure.

THE ION BEAM IRRADIATION SYSTEM

Figure 1 shows a schematic diagram of the ion beam irradiation system. A substrate is placed on the sample setting table at the loading stage and is transferred into the load lock chamber. The chamber is evacuated by a vacuum pump. After the load lock chamber is evacuated, the gate valve opens and the ion source runs with a low-energy beam. A hot filament neutralizer is included to neutralize ion beam charges accumulated on the substrate surface. The ion source is placed with its longitudinal dimension parallel to the y-axis. The substrate is moved into the process chamber and transferred along the x-axis at a constant speed during the ion irradiation. Ion incident angle to the substrate surface in the x-z plane is controlled from 15 to 90 deg by tilting the ion source inside the chamber.

In the process chamber, a multiple Faraday cup array is placed to measure the distribution of ion beam current density. In addition, the system has a beam monitor which enables us to measure the distribution of ion beam incident angles in the x'-y' plane along the y'-axis. The performance of the monitor is discussed elsewhere [3].

EXPERIMENTAL

In order to investigate the performance of LC alignment layer produced by the ion irradiation, LC samples were assembled in the following procedure. A pair of polyimide films on indium-tin-oxide (ITO) coated glass substrates was prepared. The films were irradiated with argon or nitrogen ions by the ion beam irradiation system. The ion energy, dose and incident angle to the substrate in the x-z plane were typically 250 eV, 1E+14 ions/cm^2 and 20 deg, respectively. The pair of the substrates was bonded with an adhesive to make a rigid structure with a certain gap, followed by LC injection into the space.

In order to make LC samples with a striped multi-domain structure, we took the following procedure. First, a pair of polyimide films was irradiated with ions similarly. One of the films was then 90 deg rotated around the z-axis and irradiated again. The second ion treatment was conducted with the use of a 3 mm thick metal mask which had a 50 μm wide single slit at the center. The film was placed beneath the slit with a gap of 4 mm and kept there during the irradiation. A single striped domain was created by the irradiation. To make another stripe, the ion source was shut down and the substrate was shifted to the next position where the second domain should be made. The ion source then started running again to make the second stripe. The intermittent ion beam irradiation and the substrate transfer were repeated. The remaining part of the procedure is the same mentioned above.

RESULTS AND DISCUSSION

First of all, we compared LC sample alignment texture between the ion irradiation and the conventional rubbing methods. Figure 2 shows the texture observed by a polarizing microscope. These photographs were taken under conditions of a crossed polarizer configuration and no voltage applied to the electrodes of the LC samples. The samples had an anti-parallel configuration. No light leakage should be ideally expected. The ion-irradiated sample exhibited a uniform dark texture while the rubbing sample had scratch-shaped bright areas which mean existence of light leakage. In the rubbing method, polyimide films are rubbed with a cloth. The fibers make scratches on the surface and cause disturbance of LC molecule alignment around them. The disturbance poses an

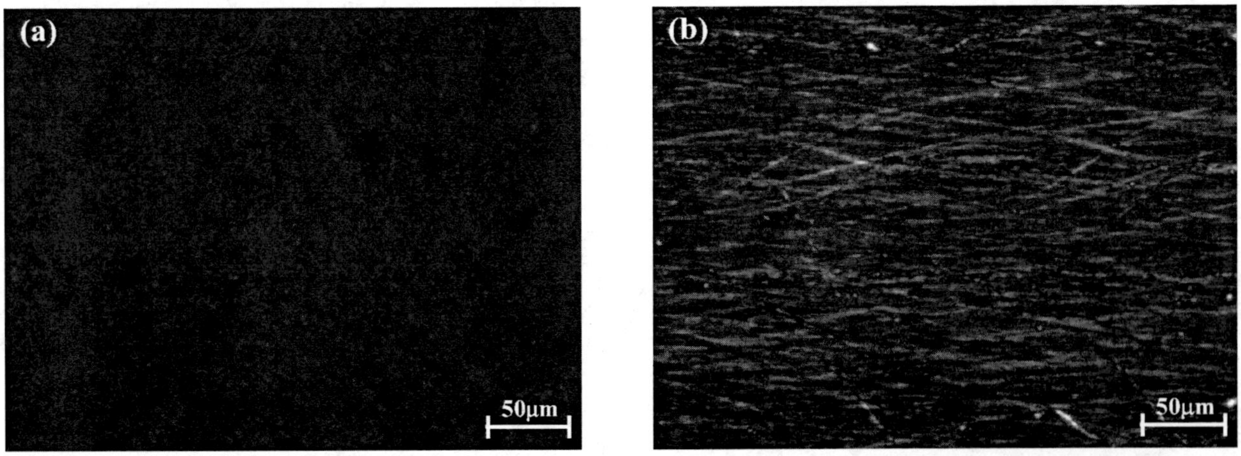

FIGURE 2. Polarizing microscope photographs of LC alignment texture of (a) ion-irradiated and (b) rubbed samples.

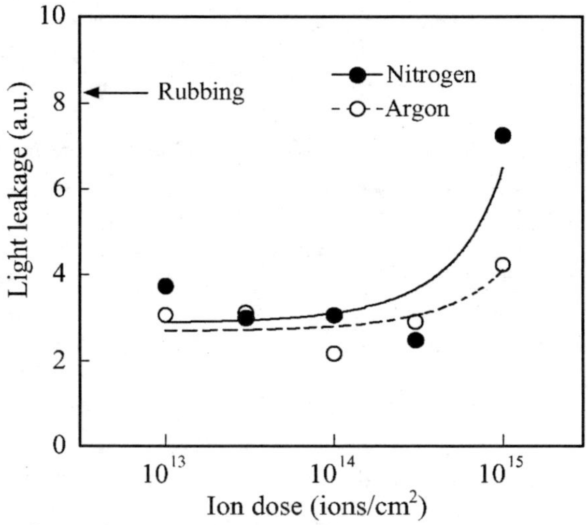

FIGURE 3. Light leakage dependence on ion dose and species.

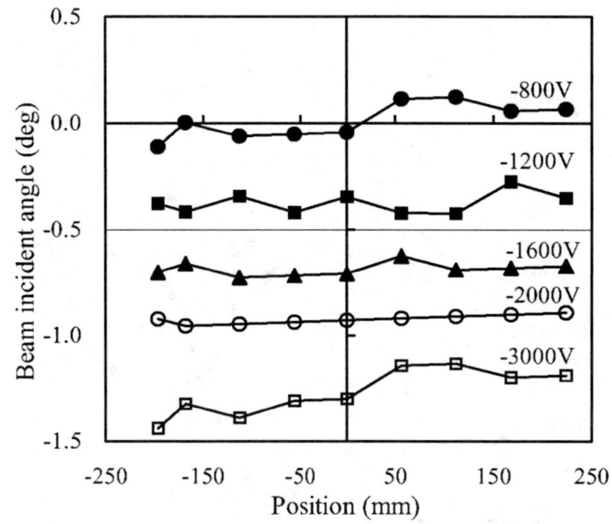

FIGURE 4. Distribution of the ion incident angle at various suppressor electrode voltages.

undesirable change in a polarizing axis, leading to the light leakage. On the other hand, the ion irradiation is a non-contacting LC alignment method and makes no scratch. The comparison of the photographs clearly indicates that the ion irradiation method has an advantage of no light leakage over the rubbing. This new LC alignment technique is promising to realize high contrast ratio LC displays.

We examined light leakage dependence on ion dose and species to optimize the irradiation condition, shown in Figure 3. With respect to argon ion beam, light leakage was almost constant at an ion dose ranging from 1E+13 to 3E+14 ions/cm^2 while it jumped at 1E+15 ions/cm^2 ion dose. That is also true for samples irradiated with nitrogen ions. This increase is probably due to film damage caused by too much ion irradiation. The figure also shows comparison between the ion irradiation and the rubbing methods. Light leakage of the rubbed sample was more than twice as many as that of samples irradiated at 1E+13 through 3E+14 ions/cm^2. This is attributed to scratches and LC alignment disturbance, as mentioned above.

We found that ion beam incident angles in the x'-y' plane change depending on suppressor electrode voltage. In order to emit ions perpendicularly to ion source electrodes, holes in each electrode should be arranged coaxially. However, the holes have a displacement in reality, leading to deflection of the beam [7]. The angle is critical to contrast ratio of LC displays because LC alignment direction is directly linked to the angle.

We measured the angle using the beam incident angle monitor to optimize the electrode voltage for a better contrast ratio. Figure 4 shows distributions of the angle along the y'-axis at various voltages. The angle at -800 V was about 0 deg. In this figure, 0 deg means that the ion trajectory is parallel to the x'-axis. The angle gradually decreased with a decrease in the voltage and dropped to approximately -1.3 deg at -3000 V.

We also measured the contrast ratio of LC samples produced at various suppressor voltages. The ratio showed the same trend as the ion beam incident angle. It exhibited highest value at -800 V and was on the decrease with a decrease in the voltage. These experiments indicate that the angle is one of important factors which must be held under control for the ion irradiation method.

We found that the optimized voltage was not persistent, and the value changed after overhaul of the ion source and reassembling the electrodes. The ion irradiation system has the incident angle monitor so that we can measure the angle and optimize the ion source operation parameters easily if necessary.

Figure 5 (a) shows a polarizing microscope photograph of a LC sample with a single stripe domain. The stripe domain had an anti-parallel LC alignment while the outside had a twisted one. The width was about 100 μm, which was nearly comparable to the size of pixels. Multi-domain structures are generally realized using a shield material like photoresist or a stencil mask on the surface of alignment films [4-6]. The problem is that the photoresist process and the mask are expensive. This method, however, does not need the shield, having an advantage over the other methods.

Figure 5 (b) shows a photograph of a LC sample with a multi-striped domain structure which was made by the intermittent ion beam irradiation and the substrate transfer described in the experimental section.

We estimated that the total time required to produce 1280 stripes was about 250 sec under conditions of

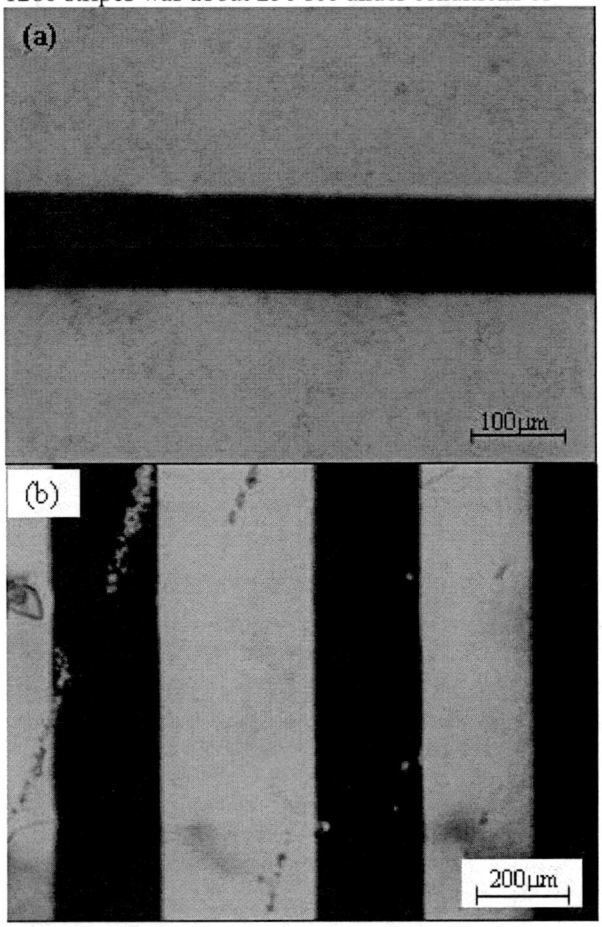

FIGURE 5. Polarizing microscope photographs of LC samples with (a) single stripe domain and (b) multi-striped domain structures.

3E+13 ions/cm^2 dose and 0.1 sec substrate transfer time from one stripe to another. This method enables not only a striped domain structure but four domains inside a single pixel with a help of the procedure reported by Yoshida *et al.* [4]. In addition, Son *et al.* reported a pretilt angle control for vertical alignment (VA) mode LC displays by the ion beam irradiation [8]. Our method has a potential to produce a multi-domain VA mode structure, which is generally realized by a bump formation process [9] with the costly photography technique.

CONCLUSIONS

Ion beam irradiation was employed to produce alignment layers for LC displays. The alignment characteristics were compared with those by the conventional rubbing method. The rubbed LC samples had scratches on the surface of the alignment film while the ion-irradiated samples had no one. Since the scratches induced disturbance of LC molecule alignment, the latter samples showed low light leakage, resulting in a high contrast ratio of LC displays. In the ion beam irradiation method, an accurate control of the incident angle to the films was necessary to obtain high contrast ratio. A new ion irradiation technique to realize a striped multi-domain alignment structure was demonstrated. The width of the stripe was approximately 100 μm which was nearly comparable to the size of pixels. The technique has a potential to achieve wide viewing angle LC displays with a four-domain structure.

ACKNOWLEDGEMENTS

We would like to express our gratitude to Mr. Fukuba of Tokyo University of A & T for his experimental support.

REFERENCES

1. P. Chaudhari *et al.*, *Nature* **411** 56-59 (2001).
2. A. Asahara, A. Horibe, H. Kimura, J. Nakagaki, T. Nishiwaki, H. Tokushige, T. Yamada, H. Kitahara and Y. Yoshida, "Evaluation of the performance of an ion beam using the planar distribution of a twist angle,"in *Proceedings of the 12th International Display Workshops in conjunction with Asia Display 2005*, The Institute of Image Information and Television Engineers, Minato-ku, Tokyo, 2005, pp. 309-312.
3. T. Matsumoto, T. Matsuda, M. Tanii, M. Konishi and Y. Andoh, "Development of an Ion Beam Aligner for Liquid Crystal Displays," in *Proceedings of the 16th International Conference on Ion Implantation Technology 2006*, pp. 429-432.
4. H. Yoshida, T. Seino and Y. Koike, *Jpn. J. Appl. Phys.* **36**, L1449-L1452 (1997).
5. K. H. Yang, *Jpn. J. Appl. Phys.* **31**, L1603-L1605 (1992)
6. R. Yamaguchi, Y. Goto and S. Sato, *Jpn. J. Appl. Phys.* **41**, L889-L891 (2002).
7. A. J. T. Holmes and E. Thompson, *Rev. Sci. Instrum.* **52**, 172-179 (1981).
8. P. K. Son, S. S. Cha, J. H. Seo, S. P. Lee, J. C. Kim, T. H. Yoon, S. J. Rho, B. K. Jeon and K. H. Kim, "Vertical Alignment of Liquid Crystal on a-SiO$_x$ Thin Film Surface by the Ion Beam Exposure," in *Proceedings of the 12th International Display Workshops in conjunction with Asia Display 2005*, The Institute of Image Information and Television Engineers, Minato-ku, Tokyo, 2005, pp. 157-158.
9. "High-Image-Quality MVA-Premium Liquid Crystal Display," in *FIJUTSU MICROELECTRONICS NEWS FIND* vol. 20, Japan, Fujitsu, 2002, pp. 12-17.

Development of Electron Cyclotron Resonance Ion Source for Synthesis of Endohedral Metallofullerenes

K. Tanaka[1,2], M. Muramatsu[2,3], T. Uchida[2], S. Biri[4], T. Asaji[1], K. Shima[1], T. Hanajiri[2], A. Kitagawa[3], Y. Kato[5] and Y. Yoshida[2]

[1]*Tateyama Machine Co., Ltd., 30 Shimonoban, Toyama, Toyama, 930-1305, Japan*
[2]*Bio-Nano Electronics Research centre, Toyo University, 2100 Kujirai, Kawagoe, Saitama 350-0815, Japan*
[3]*National Institute of Radiological Sciences (NIRS), 4-9-1 Anagawa, Inage, Chiba 263-8555, Japan*
[4]*Institute of Nuclear Research (ATOMKI), H-4026 Debrecen, Ben ter 18/c, Hungary*
[5]*Division of Electrical Engineering, Osaka University, 2-1 Yamada-oka, Suita, Osaka 565-0871, Japan*

Abstract. A new electron cyclotron resonance ion source (ECRIS) has been constructed for synthesis of endohedral metallofullerenes. The main purpose of the ion source is to produce new biological and medical materials. The design is based on ECRIS for production of multicharged ion beams with a traditional minimum-B magnetic field. An 8–10 GHz traveling wave tube (TWT) amplifier and a 2.45 GHz magnetron have been applied as microwave sources. Fullerene and metal vapor are introduced with a filament heating micro-oven and an induction heating oven, respectively. In preliminary ion-extraction test, Ar^+ is 54 μA. Many broken fullerenes such as C_{58} and C_{56} are observed in fullerene ion beams.

Keywords: Ion source, ECR, Microwave frequency, Magnetic field, Endohedral Metallofullerene.
PACS: 29. 25. Ni, 29. 27. Ac

INTRODUCTION

Endohedral fullerenes are fascinating new materials which may have considerable potential as new biological and medical materials or as building blocks for nanotechnology. One possible application is to shield radioactive "tracer" atoms inside fullerene cages, then inject them into people's blood in order to observe blood flow. Endohedral fullerenes are fullerenes that have additional atoms, ions, or molecules enclosed within their inner spheres [1].

The first lanthanum-fullerene complex was synthesized in 1985 represented La@C_{60} by J. R. Heath *et al.* [2]. Since then a variety of endohedral metallofullerenes have been reported, but their investigations have been severely limited because they are typically formed in extremely low yields (<0.5%). Moreover, only a handful of heavy metals can form stable endohedral compounds (a few of these are Sc [3], Y [4] and Ti [5]).

The production of the endohedral metallofullerenes uses usually laser evaporation or arc discharge. The Institute of Nuclear Research of the Hungarian Academy of Sciences (ATOMKI) has been studied the injection technique of gas atoms into fullerene in an electron cyclotron resonance (ECR) plasma [6]. Especially, there were carbons-loss fullerenes, e.g. C_{58}, C_{56}, …, in the ECR plasma. We aim at the efficient preparation of carbons-loss fullerenes, for mass production of the metallofullerenes. In comparing with other methods such as laser evaporation or arc discharge the ECR ion source has some special features:

- To generate dense fullerene plasmas and extract high-strength fullerene-ion beams.
- To produce and maintain the metal-fullerene plasma without support or background gas.
- To produce not only M@C_{60}-type metallofullerenes but also M@C_{60-x}-type carbons-loss metallofullerenes.
- To extract pure metallofullerenes from ion beams by the mass separator, such as a sector magnet type.

Therefore, the ECR plasma is suitable to form and study new endohedral metallofullerenes.

In 2006, a prototype magnetic structure (the ATOMKI-ECR) was tested to produce fullerene and iron ions, and it demonstrated that the required magnet fields were feasible [7]. However, the ECRIS is not optimized for metallofullerene synthesis because that

CP1066, *Ion Implantation Technology,* edited by E. G. Seebauer, S. B. Felch, A. Jain, and Y. V. Kondratenko

is suitable for highly charged ion production. In 2007, an evaporator using induction heating was tested to produce iron ions, and it demonstrated that this heating method was suitable for operation in the good vacuum [8]. Consequently, a mechanical design for optimum operation to produce endohedral metallofullerenes in ECR ion source was developed and this ion source was named Bio-nano ECRIS.

We aim to synthesize mainly Fe@C_{60} using the ECRIS. In this paper, we describe the concept and design of the ECRIS, and present preliminary results of ion beam production.

DESIGN OF BIO-NANO ECRIS

Device Configuration

Figure 1 shows the main components of Bio-nano ECR ion source. The specification of the ion source is shown in Table 1. The dimensions of the plasma chamber are 140 mm in diameter and 350 mm in length. We designed the chamber diameter which is considerably larger than that of the ECR ion sources for multicharged ion production, because the loss of fullerene ions to the chamber wall will be reduced. An extractor consists of a plasma electrode and an extraction electrode. Plasma chamber and electrode

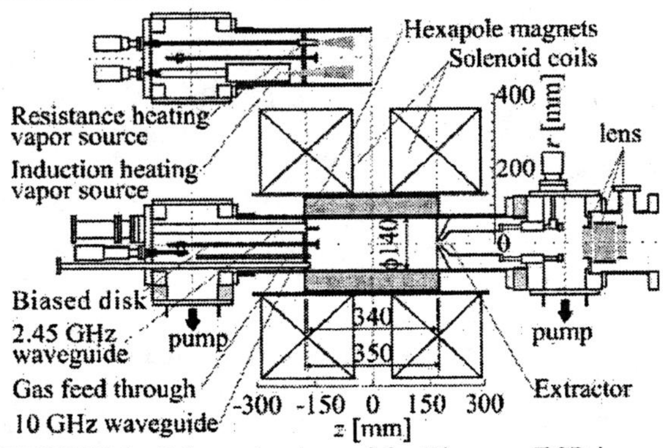

FIGURE 1. Schematic view of the Bio-nano ECR ion source.

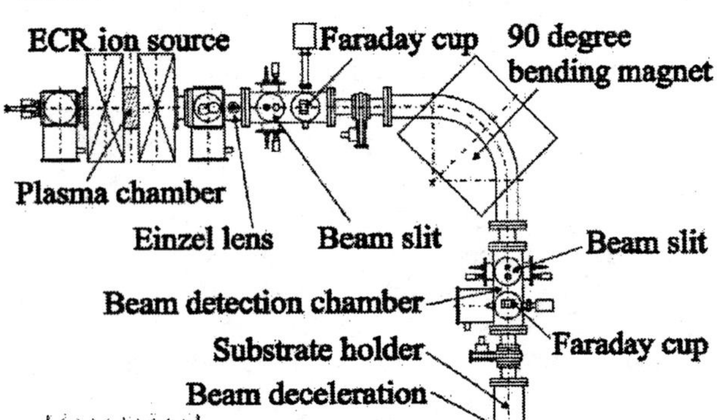

FIGURE 2. Schematic view of the Bio-nano ECR ion source and beam line.

apertures are 10 mm in diameter. They can impress high voltage. Maximum extraction voltage is ±10 kV. The extraction electrode is ground potential. The plasma electrode is fixed the position at z=170 mm and is same electrical potential as the plasma chamber. The position of the extraction electrode can be varied from z=180 mm to 230 mm. A biased disk is set at the opposite side of the extractor. The electric potential of the disk controls the diffusion of ions and electrons toward the axial direction.

A high-frequency wave is injected into the plasma chamber fed with gas, metallic, or fullerene vapors. The plasma chamber is immersed in a "minimum-B" magnetic structure achieved with two solenoid coils and hexapole magnets. To obtain the optimum impedance matching with the high-frequency generator, two types of generators are used. At the higher frequencies, the generator is a traveling wave tube (TWT) amplifier at 8–10 GHz. The magnetic field strength for ECR condition is 0.286–0.357 T. The microwave power applied to the ion source is up to 350 W. The microwaves are directly introduced through a rectangle waveguide (WR-90). At the lower frequency, the generator is a magnetron at 2.45 GHz. The magnetic field strength for ECR condition is 0.0875 T. The microwave power applied to the ion source is up to 300 W. The microwaves are launched from a rod antenna through a coaxial waveguide. It is possible to run the source under a two frequency mixing.

Two evaporators are installed in the plasma chamber. One is a resistance heating vapor source based on the RIKEN micro-oven [9] which produces a continuous beam of fullerenes. The fullerene sample is heated up to 500°C by the radiation heat of the wires and sublimated. The other is a high temperature evaporator which produces a continuous beam of metal. Metal rods are heated by a multilayer induction coil. A prototype of this evaporator has been developed and successfully tested at Osaka University for temperature up to about 1550°C. Therefore it should be well suited for the production of iron beams, which need to operate at about 1400°C in order to evaporate iron without melting [8].

TABLE 1. Specification of the Bio-nano ECRIS

Mirror magnets	Solenoid coils
Maximum field strength	0.642 T
Minimum B strength	0.265 T
Mirror ratio	2.42
Hexapole magnets	Nd-Fe-B
Maximum field strength	0.72 T
Inner diameter, Length	ϕ150 mm, 350 mm
8-10 GHz microwave	TWT amplifier
Maximum power	350 W
2.45 GHz microwave	Magnetron
Maximum power	300 W

The ion beam transport system shown in Figure 2 consists of the extraction system (operation at voltages up to 5 kV), an einezel lens, a large gap, 90° double focusing analyzing magnet, and a substrate holder. The beam transport system was designed for high current, heavy molecular extraction. Consequently, a large gap (80 mm) magnet was chosen to accommodate the ion beam size at high intensities and low energy. The maximum magnetic field in the center of beam line is as high as 0.8 T. The curvature radius is 500 mm. Assuming that ion energy is 5 kV, the ions which have mass number less than about 1500 can be bended by the magnetic field. After the mass analyzing section, a beam slit, a Faraday cup, the substrate holder, and a beam deceleration system, in this order. Three turbo-molecular pumps (450 L/sec) are used for the evacuation of the chamber. The background gas pressures of the plasma chamber and a beam detection chamber are 6.9×10^{-5} and 6.3×10^{-6} Pa, respectively.

In the early commissioning phase, when total helium gas currents 15 μA have been extracted, the gas pressures of the plasma chamber and the detection chamber are 4.3×10^{-3} and 5.5×10^{-5} Pa, respectively.

Magnetic Field

A "minimum-B" Magnetic field was designed and constructed with two solenoid coils and hexapole magnets. The traditional magnetic fields have been used for ECR ion sources. The magnetic field was designed aiming at strength equal with that of ATOMKI-ECRIS [10] in order to efficiently confine plasma although the Bio-nano ECRIS has the large-bore chamber. Figure 3 shows the distributions of the magnetic field strength on the *z*-axis formed with the two solenoid coils. The coils form axial mirror fields in order to confine charged particles. The maximum magnetic field is as high as 0.64 T at the coil current of 750 A. The ratio of the maximum to minimum magnetic field, that is mirror ratio, is about 2.4. At the current of 200A, the maximum and minimum magnetic fields are about 0.176 and 0.070 T, respectively. Figure 3 shows that ECR zone for 8–10 GHz does not exist on the z-axis.

The hexapole magnets consist of 12 pieces trapezoidal Nd-Fe-B magnets (NMX-S45SH, Hitachi Metals, Ltd.) which are mounted into a nonmagnetic stainless steel cylinder as shown in Figure 4. The two magnets which are combined form a magnetic pole toward the center of the plasma chamber. These combined magnets are made so that the magnetic field lines concentrate at a center line of a combined block surface by canting each magnetic pole. The three south and three north poles are alternately arranged on the outer side of the plasma chamber. Figures 4 and 5 show magnetic field lines and distributions of magnetic field strength by the hexapole magnets on a cross-section of the plasma chamber, respectively. Six arched mirror fields are produced between adjacent magnets. The mirror fields reduce the diffusion of charged particle toward the chamber wall. The solid lines and the dotted lines in Figure 5 show the radial distribution of the calculation and measurement values of magnetic fields, respectively. The heavy and the narrow lines are the distributions on the line OA and the line OB in Figure 4, respectively. The hexapole field on the plasma chamber wall can reach 0.72 T on the line OA. ECR zones for 2.45 GHz and 8–10 GHz microwaves are formed at around r=25 mm and 50 mm. Therefore "minimum-B" field (Figure 6), which is minimum magnetic field exists in the center region

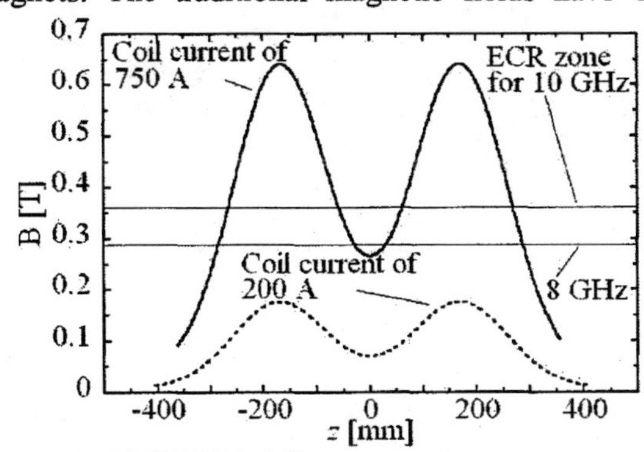

FIGURE 3. Mirror magnetic field on *z*-axis.

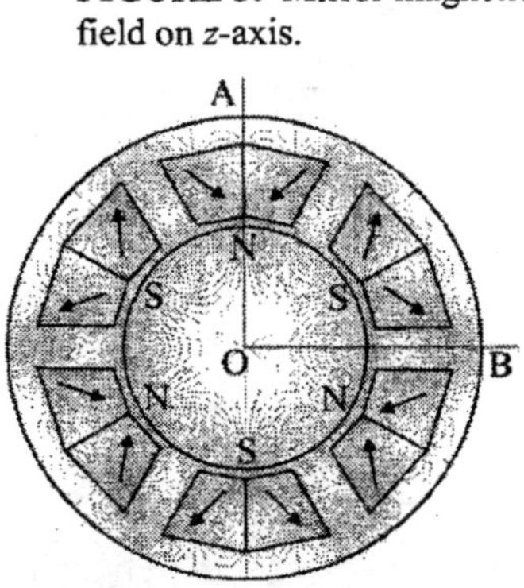

FIGURE 4. Hexapole magnetic field strengh.

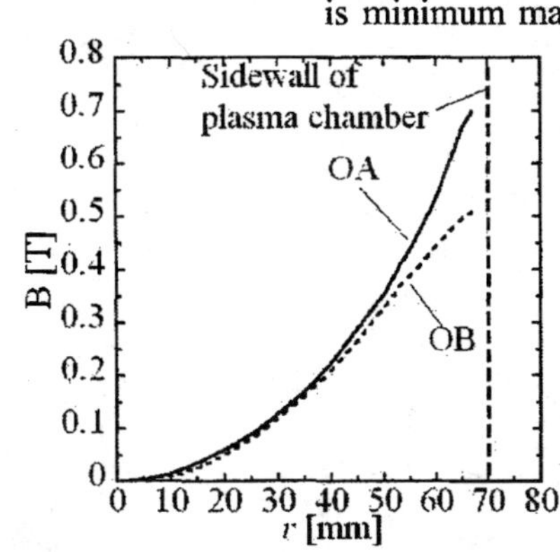

FIGURE 5. The distributions of hexapole magnetic field in radial detections.

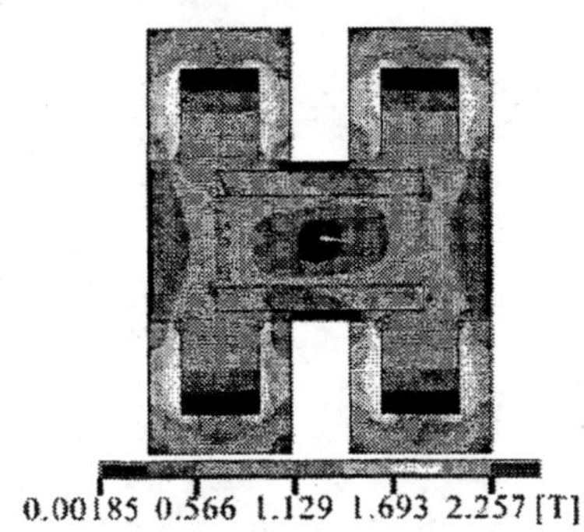

FIGURE 6. Mirror magnetic field strength.

(around z=0 mm and r=0 mm) of the chamber, can strongly confine charged particles by the gradient of the magnetic field.

RESULTS AND DISCUSSION

Argon Ion Beams

Multicharged Ar ions have been produced as the preliminary experiment with the ECR ion source. The source was operated at the magnetic field strength under ECR condition because of first run. We extracted the Ar ion beams at the extraction voltage of 5 kV, and measured the ion beam currents after the bending magnet. The typical charge state distribution of the extracted Ar ions is shown in Figure 7. The microwave power and frequency were 70 W and 8.5 GHz, respectively. The Ar pressure was 1.0×10^{-2} Pa. When each coil current was 250 A, the two peaks of the magnetic mirror field are estimated at approximately 0.21 T. We could identify the Ar^{+}–Ar^{4+}, at this point, Ar^{+} is 54 μA. The peaks of O_2, H_2O, etc. were observed as impurities.

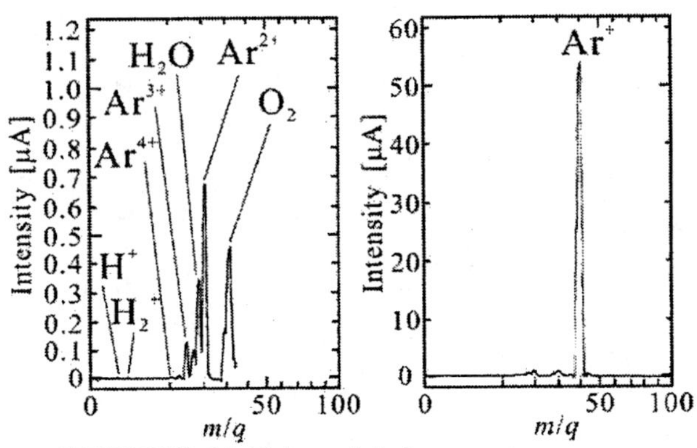

FIGURE 7. Multicharged Ar ions spectrum.

Fullerene Ion Beams

Fullerene vapor was introduced into the plasma chamber using the micro-oven, and the ECR plasma could be generated at the pressure of 1.5×10^{-3} Pa using support gases of Ar. In this experiment, fullerene is C_{60} powder (99.9 %) manufactured by Frontier Carbon Corporation. The other experimental conditions were the same as those of the Ar ions production. Figure 8 shows a fullerene beam spectrum. We identified C_{60-x} (x=2, 4, 6, ...) peaks as well as main C_{60} peak. This result is similar in spectrum to the previous works. It seems that high energy electrons can break stable fullerene molecules.

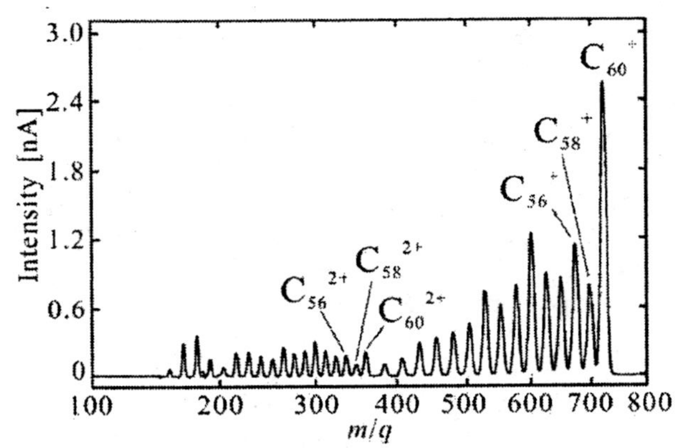

FIGURE 8. Typical fullerene spectrum.

CONCLUSIONS

A new ECR ion source (Bio-nano ECRIS) has been designed and constructed for synthesis of endohedral metallofullerenes. The ion source has a minimum-B magnetic field consisting of the two solenoid coils and the hexapole permanent magnets. The 8–10 GHz TWT amplifier and the 2.45 GHz magnetron are adopted as microwave sources. Fullerene and metal vapor are introduced with the micro-oven and the induction oven, respectively. We have observed multicharged Ar ions and broken fullerene ions such as C_{58} and C_{56}.

ACKNOWLEDGEMENTS

Part of this study has been supported by a grant for the 21st Century's Centre of Excellence (COE) Programme since 2003 and a Bilateral Programs; Joint Research Project, since 2006. Authors would like to thank Professor T. Maekawa of Toyo University, and Mr. J. Saito and Mr. M. Kanazawa of Tateyama Machine Co., Ltd. for technical assistance.

REFERENCES

1. R. Kitaura and H. Shinohara, *Jan. J. Appl. Phy*, **46**, 2007, pp.881-891
2. H. W. Kroto, J. R. Heath *et al.*, *Nature*, **318**, 162-163 (1985).
3. H. Shinohara, H. Yamaguchi *et al.*, *J. Phys. Chem.*, **97**, 4259 (1993).
4. J. H. Weaver, Y. Chai *et al.*, *Chem. Phys. Lett.*, **190**, 460 (1992)
5. B. Cao, M. Hasegawa *et al.*, *J. Am. Chem. Soc.*, 123,9679 (2001).
6. S. Biri, A. Valek *et al.*, *Rev. Sci. Instrum.* **73**, 2 (2002).
7. S. Biri, É. Fekete *et al.*, *Rev. Sci. Instrum.* **77**, 03A314 (2006).
8. Y. Kato, T. Kubo *et al.*, *Rev. Sci. Instrum.* **79**, 2 (2008).
9. M. Kidera, T. Nakagawa *et al.*, *Proc. of the 15th International Workshop on ECRIS*, 74-76 (2002).
10. S. Biri, J. Vámosi, *Rev. Sci. Instrum.* **67**, 3 (1996).

Pure Material Vapor Source by Induction Heating Evaporator for an Electron Cyclotron Resonance Ion Source

Y. Matsui[1], T. Watanabe[1], T. Satani[1], M. Muramatsu[2], K. Tanaka[3], A. Kitagawa[2], Y. Yoshida[4], F. Sato[1], Y. Kato[1], T. Iida[1]

[1] *Devision of Electrical, Electronic and Information Engineering, Graduate School of Engineering, Osaka Univ. 2-1 Yamada-oka, Suita-shi, Osaka 565-0871, Japan.*
[2] *National Institute of Radiological Sciences,4-9-1 Anagawa, Inage-ku, Chiba-shi, Chiba, 263-8555, Japan*
[3] *Tateyama Machine Co., Ltd., 30 Shimonoban, Toyama, 930-1305, Japan*
[4] *Department of Mechanical Engineering, Toyo Univ., 2100 Kuzirai, Kawagoe-shi, Saitama, 350-0185, Japan*

Abstract. Multiply charged iron ions are produced from solid pure material in an electron cyclotron resonance (ECR) ion source. We develop an evaporator by using induction heating with the induction coil which is made from bare molybdenum wire and surrounding the pure iron rod. We optimize the shape of induction heating coil and operation of rf power supply. We conduct experiment to investigate reproducibility and stability in the operation and heating efficiency. Induction heating evaporator produces pure material vapor, because materials directly heated by eddy currents have non-contact with insulated materials which are impurity gas sources. The power and the frequency of the induction currents range from 100 to 900 W and from 48 to 23 kHz, respectively. The working pressure is about 10^{-4} to 10^{-3} Pa. We measure temperature of iron rod and film deposition rate by depositing iron vapor to crystal oscillator. We confirm stability and reproducibility of evaporator enough to conduct experiment in ECR ion source. We can obtain required temperature of iron under maximum power of power supply. We are aiming the evaporator higher melting point material than iron.

Keywords: Ion source, ECR, Evaporator, iron, Induction heating
PACS: 29.25.Ni, 37.20+j,

INTRODUCTION

Multiply charged ions to be used prospectively are produced from solid material in an electron cyclotron resonance (ECR) ion source. Recently a pure iron source is also required for the production of caged iron ions in the fullerene in order to control cells in vivo in bio-nano science and technology [1].

We have already produced the multicharged iron ions by directly sputtering and by directly ohmic heating, *i.e.*, evaporating with an iron filament and a BN crucible. Because the working pressure is about 10^{-4} to 10^{-3} Pa, it is a very serious problem in experiments in ECR ion source. The BN crucible releases outgas at high temperature. While the iron filament doesn't have stability and reproducibility, the vapor purity is superior to the BN crucible because of direct heating. We adopt directly heating iron rod by induction heating because it has non-contact with insulated materials which are impurity gas sources [2]. The iron is heated directly by eddy current. Copper pipe is usually used for induction heating coil but we choose molybdenum because it doesn't need water cooling. We have already succeeded in producing the multicharged iron ions by induction heating evaporator in ECR plasma (TAIKO device), extracting them, and then producing β-$FeSi_2$ by implantation [3]. We have also already conducted experiment in Kei2-source; that is the ECR ion source of the advanced high-energy carbon-ion therapy in National Institute of Radiological Sciences (NIRS) and succeeded in producing multicharged iron ions up to Fe^{13+} [4].

We optimize the induction heating evaporator and made new two type ones. We investigate the reproducibility and the stability in their operation. We measure the temperature of iron rod and film deposition rate by depositing iron vapor to crystal oscillator. We confirm that induction heating evaporators have stability and reproducibility enough to conduct experiment in ECR ion source. We investigate the temperature characteristics of two type induction heating evaporators. In order to confirm iron vapor flux,

CP1066, *Ion Implantation Technology*, edited by E. G. Seebauer, S. B. Felch, A. Jain, and Y. V. Kondratenko

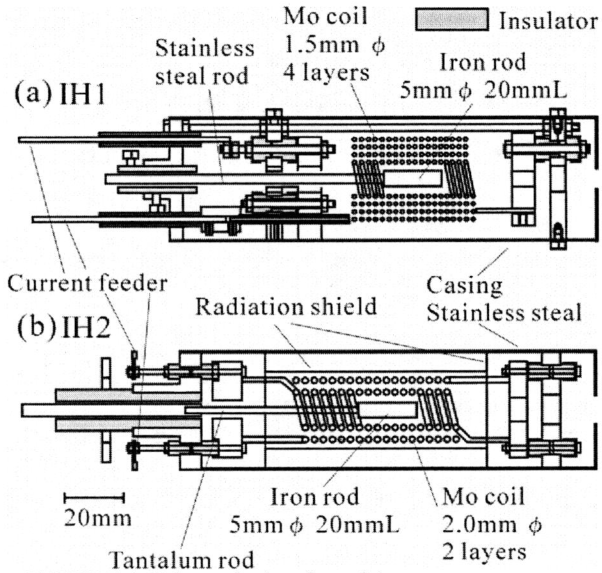

FIGURE 1. Schematic drawing of the evaporator

we observe film deposition rate at the temperature of iron rod about 1450°C.

EXPERIMENTAL APPARTUS

The schematic drawing of the evaporator is shown in Figure 1.

The evaporators consist of induction coils, pure iron rods and casing. The induction coils are made from molybdenum wire. The insulator is far from heated materials. The pure iron rod is 5 mm in diameter and 20 mm in length. We make two evaporators which are called by the IH1(a) and the IH2(b), respectively. The IH1 is composed by four layers induction coils and the IH2 is composed by two layers induction coils. The outer diameter of the IH1 and the IH2 are 48 mm and 45 mm, respectively. The diameters of molybdenum wire of the IH1 and the IH2 are 1.5 mm and 2.0 mm, respectively. We can change number of layers of induction coils. The casing is demountable. By selecting coil layers, the inductance and the resistance of the IH1 range from 2.49 μH to 43.7 μH, and from 0.16 Ω to 1.81 Ω, respectively. Those of the IH2 range from 1.86 μH to 5.29 μH, and from 0.20 Ω to 0.43 Ω, respectively. The IH2 has radiation shield which is made from stainless steel inside casing.

We use the rf power units of induction heating for home electric alliances [5]. The rf power supply is controlled by current resonance, and the power increase as the frequency decreases. Typical rf power and frequency range from 100 to 900 W and from 48 to 23 kHz, respectively. Maximum power of rf power supply is about 6 kW. Operating range of the rf power supply depends on the inductance and the resistance of whole circuit including induction coils. Therefore, we need to optimize the inductance and the resistance in the circuit. In order to protect rf power supply, we connect serial dummy and parallel dummy coils to rf power supply taking account of occurring accidents of short and open circuit, respectively. The iron rod inserted to the induction coil enhances the inductance of the induction coil due to ferromagnetic substance by itself at room temperature. This is effective at the initial heating, but it changed from ferromagnetic to paramagnetic property at the temperature above Cúrie point about 770°C. Then the inductance of the induction coil also changes with heating temperature of iron rod. Furthermore the resistance of induction coil is also changed by joule and radiation heating. However the resistance of the induction coils change, we make the resistance and the inductance in the circuit range within available operation condition for rf power supply.

Our previous experiments indicate the vapor pressure required above 0.1-0.5 Pa enough to obtain iron vapor from solid material [1,2]. Required temperature to attain this vapor pressure for iron is about 1350-1450°C.

We measure input power, input current and induction current. We measure the temperature of the iron rod in induction heating by using non-contact thermometer and the film thickness by depositing iron vapor to crystal oscillator to confirm iron film deposition rate.

EXPERIMENTAL RESULTS AND DISCUSSIONS

A. Temperature Characteristics of Induction Heating Evaporators

We compare iron temperature in each number of layer and each evaporator. The dependences of the iron rod temperature on the induction current in various cases are shown in Figure 2. Broken line (Open circles, squires and triangles) indicate experimental results of the IH1. Solid line (Closed circles, squires and triangles) indicate experimental results of IH2. IH1(2) in Figure 2 indicates IH1 with two layers coils. The notation of c and nc in Figure 2 indicate evaporator with and without casing, respectively. The $I_{rf\,max}$ is half to peak to peak value of the temporal revolution of induction current. The frequency of induction current range from 48 kHz to 23 kHz depending on output power.

We use the IH1 which number of layers change from one layer (IH1(1)) to four layers (IH1(4)). We can obtain the required temperature by using the IH1

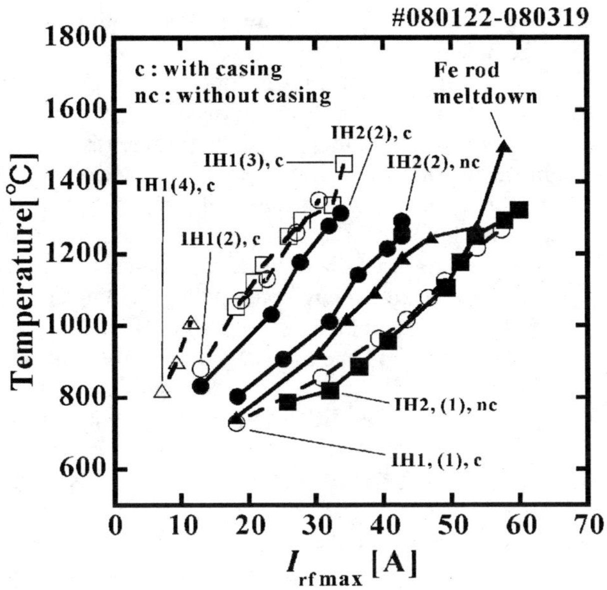

FIGURE 2. Temperature characteristics of evaporators

evaporator with even one layer. The iron rod temperature increases in proportion to induction current. Induction current needed for the required temperature of the iron rod increases as the layer decreases. The experiment by using the IH1 evaporator shows that two layers have good heating efficiency. But the coil strain in the operation because the wire rod is too thin. Insufficient insulation occurs due to deposition iron vapor on insulator. We make the IH2 evaporator which is composed by two layers coils with 2.0 mm in diameter. The diameter of coils of IH2 is thicker than IH1 and the number of insulators of IH2 is fewer than those of IH1.

We use the IH2 which number of layers change from one layer to two layers and the IH2 with or without casing. It is found that IH2(2) with casing have heating efficiency equivalent of IH1(3) (see Figure 2). When we compare two layers with one layer, two layers is more efficient than one layer. When we compare having casing or not, the evaporator with casing is more efficient than one without casing.

B. Stability and Reproducibility of the Induction Heating Evaporator

We use IH2 with two layers in experiment of confirming stability of the evaporator. Stability of the operation is typically shown in Figure 3. Figure 3 shows temporal evolution of iron rod temperature. We increase induction current and measure corresponding temperature of the iron rod. We increase input power to rf power supply step by step. We measure input power

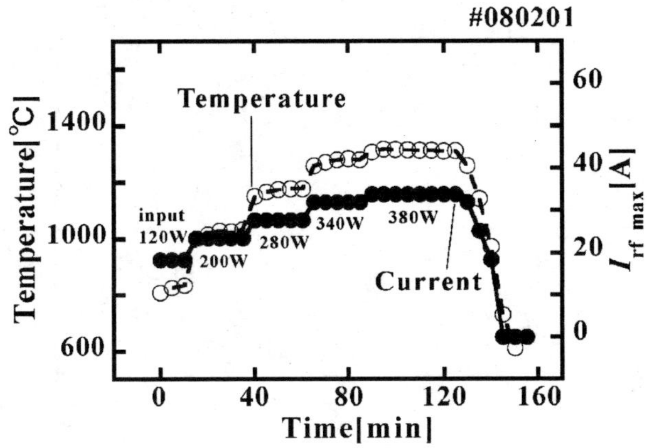

FIGURE 3. Stability of the induction heating evaporator

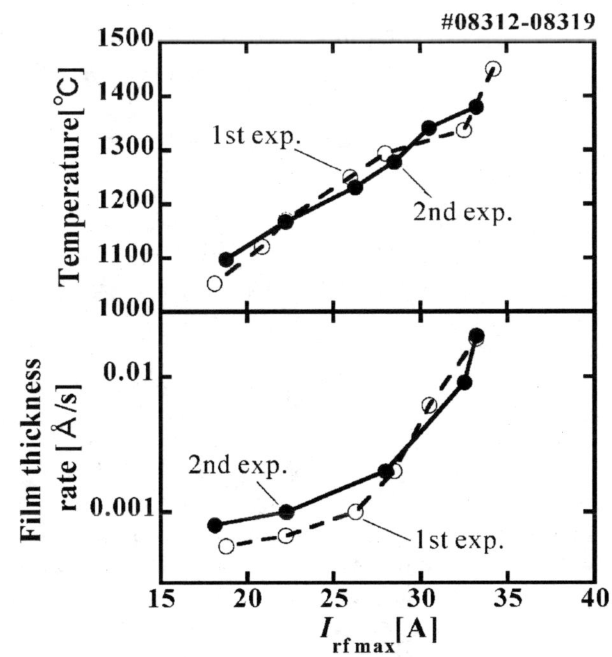

FIGURE 4. Reproducibility of the induction heating evaporator

to rf power supply by using power meter. Input power values are embedded in each step in Figure 3.

We can attain the required temperature at 33.7 A and 380 W in this experiment. It is found that the temperature can be sustained at 1350°C during 30 minutes. Changes in temperature in this duration are within 0.5 %.

We use the IH1 with three layers in experiment of conforming reproducibility in the performance of the evaporator. Reproducibility of the operation is typically shown in Figure 4. Upper figures of in Figure 4 show the dependence of the iron temperature against the induction current. Lower figures show the dependences

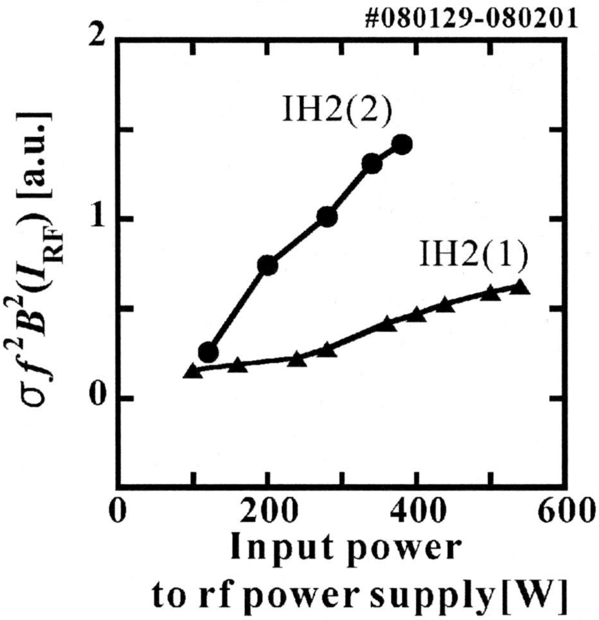

FIGURE 5. The dependence of the heating efficiency on the input power to rf power supply

of film deposition rate of iron film against the induction current. We conduct experiment twice. Open circle and broken line indicate experimental results of the first experiment. Closed circle and solid line indicate the second experiment.

We can attain the required temperature in each experiment. We confirm that the vapor pressure depends on the temperature. The iron rod temperature rises in the same way in proportion to the induction current. We confirm observed film deposition rate of 0.02 Å/s at about 1450°C in each experiment. As the results of this experiment, it is found that the evaporator has good reproducible performance.

We confirm that induction heating evaporators have stability and reproducibility enough to conduct experiment in ECR ion source.

C. Comparison of Heating Efficiency and Future Planning

We estimate input power to iron rod. Input power is described as following;

$$P_{\text{eddy}} \propto \sigma f^2 B^2, \qquad (1)$$

where σ is electrical conductivity, f is the frequency of rf current, B is magnetic flux density inside induction coil. The change of electrical conductivity depends on the temperature. The relative dependences of the P_{eddy} on the input current to rf power supply are shown in Figure 5.

Input power to the iron rod increases in proportion to input power to rf power supply. As the number of layers increases, heating efficiency increases. As the number of turns decreases, input power to rf supply increases. We can obtain required temperature of iron under maximum power of rf power supply. We aim at higher temperature with another material.

At the present, we are constructing the tandem type ECR ion source which is consist of first stage and second stage [6,7]. The first stage and the second stage can independently make plasma. We will insert the IH2 evaporator into the ECR ion source, first stage of tandem type ion source. Then we will extract and analyze iron ion currents. We will apply this tandem type ECR ion source to several applications. For example, we will make fullerene plasma in ECR ion source, second stage of tandem type ion source to synthesis of encapsulated iron atom into the fullerene sand extract it. We will apply the evaporator to synthesis of encapsulated iron atom into the fullerene in new ECR ion source in Toyo university [8].

We are also planning insert the IH2 evaporator in 18 GHz ECR ion source in HIMAC in NIRS and apply it to extract various kinds of ions.

We are aiming the evaporator higher melting point material than iron, *i.e.*, semiconductor and rare materials. We will choose tungsten for material of induction coil to obtain higher temperature.

ACKNOWLEDGEMENTS

The authors wish to thank Mr. Sugawara and Mr. Shinohara, Hitachi Home and Life Solutions, Inc., for making the rf power supply available for this study.

REFERENCES

1. F. Éva, S. Biri I. Ivan, Fullerenes, Nanotubes, Carbon Nanostruct. **15**, 249 (2007); S. Biri, A. Valek, L. Kenez, A. Janossy, and A. Kitagawa, *Rev. Sci. Instrum.* **73**, 881 (2002)
2. S. Sugiyama, Y. Kato and S. Ishii, *Rev. of Sci. Instrum.*, **73**, 542(2002).
3. M. Tomida, Y. Kato, and T. Asaji, *Nucl. Instrum. Methods Phys. Res. B* **237**, 83 (2005).
4. Y. Kato, *et al.*, *Rev. Sci. Instrum.* **79**(1) (2008)
5. Y. Kato, M. Tomida, et.al, *Rev. Sci. Instrum.* (2006), 77, 03A335.
6. Y. Kato, *et al.*, in these Proceedings.
7. T. Watanabe, *et al.*, in these Proceedings.
8. K. Tanaka, *et al.*, in these Proceedings.

Ion Beam Extraction from Large Bore 11 to 13 GHz ECR Ion Source with a Pair of Cylindrically Comb-Shaped Magnetic Fields

T. Watanabe, T. Satani, Y. Matsui, F. Sato, Y. Kato, T. Iida

Division of Electrical, Electronic and Information Engineering, Graduate School of Engineering, Osaka Univ. 2-1 Yamada-oka, Suita-shi, Osaka 565-0871, Japan.

Abstract. We have developed a large bore electron cyclotron resonance (ECR) ion source with cylindrically comb-shaped magnetic field configuration to realize high current density and comparatively large bore ion beam which can be used for beam processes under low pressure and low microwave power. It is found that the plasma confinement of cylindrically comb-shaped magnetic fields is more effective than that of simple multipole magnetic fields. We have constructed a large bore ion beam extractor and set to the ECR ion source. The electron density attained to $10^{17\sim18}$m^{-3} order, ion saturation current density is about 80～100 A/m^2 under low pressure 0.03 Pa and low microwave power 200 W. We conduct the first trial of ion beam extraction from the ECR ion source. We investigate the ion beam characteristics for extraction voltages and Ar gas pressures. The extracted ion beam current increases in proportion to extraction voltages, however at present the beam current is lower than those estimated from ion saturation current. We observed the self-similar distribution of the extracted ion beam which is comparatively large bore. We are now investigating the ion beam extraction and going to optimize the electrode gap, the electrode position and shape of multi-holes of the extractor.

Keywords: ion beam extraction, large bore, ECR, comb-shaped magnetic fields.
PACS: 52.50.Sw

INTRODUCTION

A novel magnetic field configuration for plasma production and confinement has been proposed in order to enhance efficiency of the electron cyclotron resonance (ECR) plasma as for broad and dense ion beam source. We have developed the original type ECR ion source with a pair of cylindrically comb-shaped magnetic fields. The ECR ion source has been studied by using 11 to 13GHz traveling wave tube (TWT) amplifier and investigated the characteristics of the plasma parameters in both of simple multipole and comb-shaped magnetic fields configurations [1-5]. It is found that the plasma confinement of cylindrically comb-shaped magnetic fields is more effective than that of multipole magnetic fields. The ECR plasma reached the plasma density of about 10^{18} m^{-3} at the microwave power of only 350 W in continuous operating [5].

We have investigated the plasma parameters *i.e.* the electron density n_e, the electron temperature T_e, the ion saturation current density J_{is} in the ECR ion source. The n_e and the T_e rise around second harmonics area [6,7]. The n_e attained to $3 \sim 4 \times 10^{17}$m^{-3} at the experimental conditions of the microwave frequency 11 GHz, the microwave power 200 W, and the various Ar gas pressures within 0.02 to 0.1Pa [7]. We are constructing a tandem type ion source. We made the comb-shaped ECR ion source with large bore ion beam extractor for the first stage of tandem type ion source. The diameter of the large bore extractor is about 120 mm. We conduct preliminary extraction and forming of large bore ion beam from the ECR ion source in various experimental conditions of extraction voltages and Ar gas pressures. We investigate the correspondence between the profiles of the extracted ion beam, and the n_e, the J_{is} in the ECR ion source. We observed self-similar distribution of the extracted ion beam which is comparatively large bore. The extracted ion beam current increases in proportion to extraction voltages, however the beam current is lower than the estimated values from measured J_{is} in the ECR ion source. We are now investigating the ion beam extraction and plasma parameters in the ECR ion source. We are also optimizing the electrode gap lengths and the electrode positions of the extractor in

CP1066, *Ion Implantation Technology,* edited by E. G. Seebauer, S. B. Felch, A. Jain, and Y. V. Kondratenko

order to enhance high current density and to form comparatively large bore ion beam for the source beams to second stage.

chamber wall (y=100mm) to the center of the chamber (y=0mm). The operating conditions are microwave

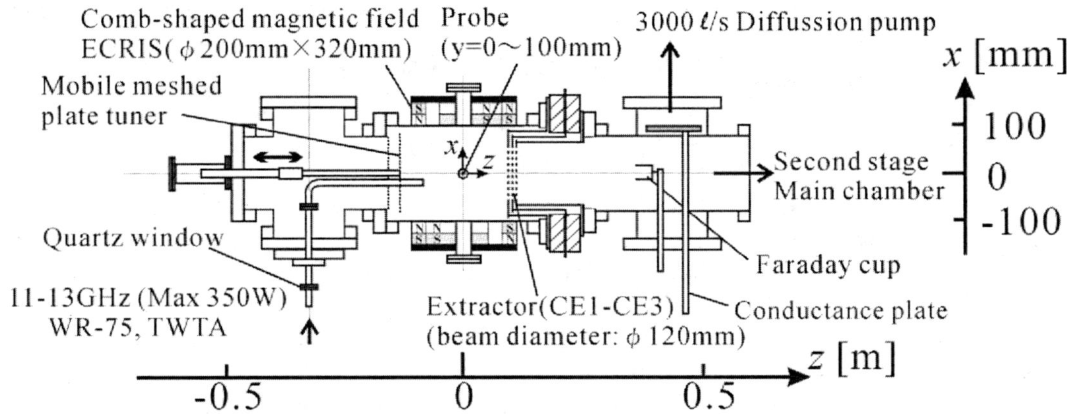

FIGURE 1. The schematic diagram of the comb-shaped magnetic field ECR plasma/ion source using 11 to 13GHz TWTA as the first stage of tandem type ion source and the set up the large bore extractor.

EXPERIMENTAL APPARATUS

The 11 to 13 GHz ECR ion source with the cylindrically comb-shaped magnetic field has been described in Ref. 5. A schematic diagram of the ECR ion source as the first stage of tandem type ECR ion source and the large bore extractor is shown in Figure 1. We use the coordinates (x, y, z) with origin at the center of the chamber. The z axis is the central axis of the chamber. The large bore ion beam extractor consisted of three plates with multi-holes (10 mm in diameter×97 holes) and placed at the end plate of the chamber opposite side of microwave feeder (Figure 1). The diameter of the multi-holes is about 120 mm. The cylindrical chamber is 197.2 mm in diameter and 320 mm in length. The microwaves are launched from WR75 waveguide through quartz window for vacuum sealing. The microwave frequencies and the maximum power of the TWT amplifier are 11 to 13 GHz and 350 W, respectively. The microwave is tuned by using the stainless steel mobile meshed plate. The mobile meshed plate moves along to z axis within 30 mm. The magnetic field strength for ECR is formed by using Nd-Fe-B permanent magnets.

We produce ECR plasma and investigate profiles of the n_e and the J_{is} in various experimental conditions of microwave powers, and Ar gas pressures. The n_e and the J_{is} are measured by using a Langmuir probe. The Langmuir probe is set at x=0 mm, z=0 mm. The distribution of the n_e and the J_{is} are measured along the y-axis. The position ranges vertically from the frequency 11 GHz, microwave power 200 W, and Ar gas pressures 0.02 to 0.1 Pa.

The extractor is consisted of three plates, *i.e.* the plasma electrode (CE1), the intermediate electrode (CE2) and the ground electrode (CE3). The electrodes are supported by stainless steel screws and electrically independent with each other by the insulated flange made from Teflon®. The ion beam is extracted by applying the extraction voltage for the chamber of the ECR ion source and the CE1. The CE2 is used as reflector of the electron by applying the negative voltage.

At the first trial of beam extraction, we conduct the ion beam extraction in case of the electrode gap length d=6.5 mm and the CE1 electrode position z_e=140 mm. Figure 2 shows the relation ship between position of the electrode and the field lines by cylindrically comb-shaped magnets.

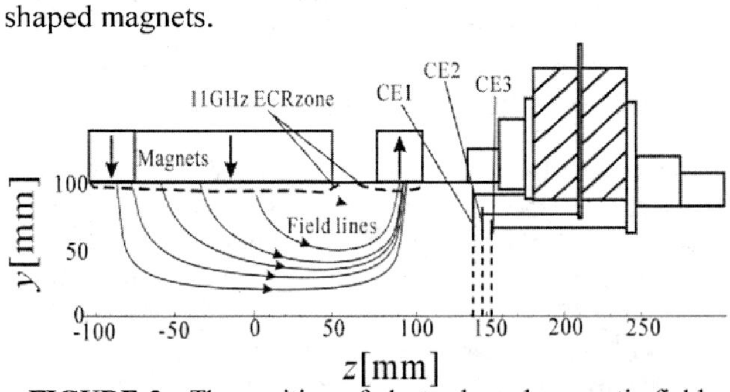

FIGURE 2. The position of electrode and magnetic field lines in comb-shaped magnets.
(a) Electrode gap d = 6.5 mm, position z_e = 140 mm

The extracted ion beam is measured by a faraday cup (FC). The FC is 20 mm in diameter and 35 mm in length is placed at the position of z=360 mm, *i.e.* apart from CE3 about 220 mm. The beam current is measured along to the x-axis (x=-65to 65mm). The ion beam is measured sequentially after the plasma measurements. The extraction voltage is 0.2 to 3 kV.

EXPERIMENTAL RESULTS

ECR Ion Source Plasma Characteristics

Figure 3 shows the profiles of the n_e and the J_{is}. The experimental conditions are microwave frequency 11 GHz, microwave power 200 W, Ar gas pressure 0.03 Pa. The horizontal axis denotes the position of Langmuir probe inserted from the chamber wall (y=100 mm) to the center of the chamber (y=0 mm). The closed and opened circles indicate the n_e and the J_{is}, respectively. The broken lines at y=100 mm along to vertical axis means the position of the chamber wall. The dashed line with arrows at both sides along to horizontal axis indicates the multi-holes radius of the extractor.

The profiles of the n_e and the J_{is} gradually increase from the center of the chamber y=0 mm to y=70 mm and decrease nearby the chamber wall. The peak positions of the n_e and the J_{is} are observed around the second harmonics resonance zone at y=80 mm.

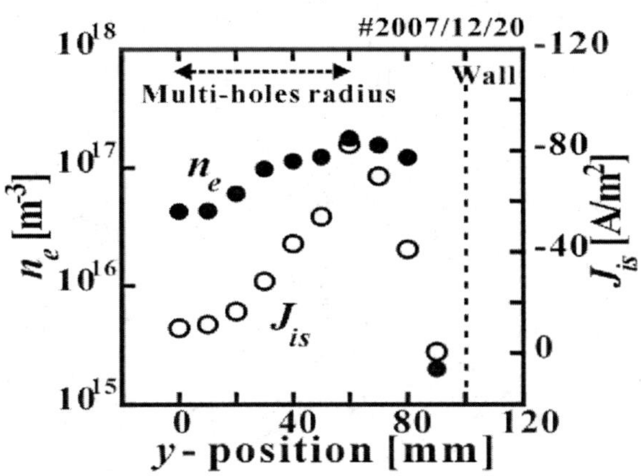

FIGURE 3. The distribution of the electron density (n_e), ion saturation current density (J_{is}) The experimental condition is microwave frequency 11 GHz, microwave power 200 W, Ar gas pressure 0.03 Pa.

The n_e reaches 10^{17} m^{-3} and the J_{is} is about 80 A/m^2 at the maximum, respectively. The J_{is} gives the maximum value of the ion beam to be able to be extracted from the ECR ion source. It is considered that the value of J_{is} in the ECR ion source can be index of the value of the ion beam extracted from the ECR ion source.

Distribution of the Extracted Ion Beam

Figure 4 shows the distribution of the extracted ion beam current density J_{FC} from the ECR ion source as the first stage of tandem type ion source. The experimental conditions are the same in the Figure 3. The J_{FC} distribution is measured in three extraction voltages 0.2, 0.4, and 0.8 kV. The horizontal axis means the position of the FC inserted from the chamber wall (x=-80 mm) to another side wall (x=80 mm).

We observed comparatively large bore and self-similar distribution of the extracted ion beam current within the three extraction voltages. The beam current increases in response to the extraction voltages. The each J_{FC} profile of Figure 4 is flat distribution around the center from x=-40 mm to x=40 mm and drop to a lower value nearby the both sides. It is considered that the radius of the extractor is about only 60 mm. The extracted ion beam current is lower to the estimated values from measured J_{is} in the ECR ion source (see Figure 3). It is considered that the position of the extractor is to far from the cylindrically comb-shaped magnets in the Figure 2.

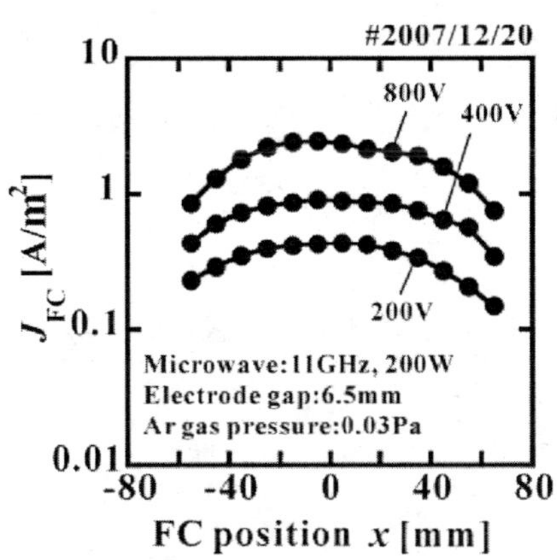

FIGURE 4. The distribution of the extracted ion beam (J_{FC}) in the extraction voltages 0.2, 0.4, and 0.8kV. The experimental condition is microwave frequency 11GHz, microwave power 200W, Ar gas pressure 0.03Pa.

Dependence of Extracted Ion Beam on Extraction Voltage

Figure 5 shows the dependence of the J_{FC} on the various extraction voltages V at the FC position is x=0 m. The experimental conditions are microwave frequency 11 GHz, microwave power 200 W, and Ar gas pressure 0.016 to 0.075 Pa. The horizontal axis denotes the extraction voltages. The closed and opened circles indicate the case of Ar gas pressure 0.03 Pa and 0.016 Pa, respectively. The opened squares indicate the case of 0.75 Pa. The broken line shows the slope in proportion to $V^{3/2}$ given from Child-Langmuir low. The J_{FC} increases in proportion to the V within 0.2 to 1.6 kV in various Ar pressures. The relativity of the beam current for V is similar in each Ar gas pressure. The J_{FC} reaches 7.7 A/m^2 at the maximum in the experimental condition of low pressure 0.016 Pa and low voltage V=1.5 kV at present.

We observe self-similar distribution of extracted ion beam which is comparatively large bore in V=0.2, 0.4, 0.8 kV. It is found that the extracted ion beam increases in proportion to the V within 0.2 to 1.6 kV. However the value of the J_{FC} is lower than the estimated value s from measured J_{is} in the comb-shaped ECR ion source. It is considered that we need to set the extractor inside the comb-shaped magnets and to enlarge the diameter of the multi-holes.

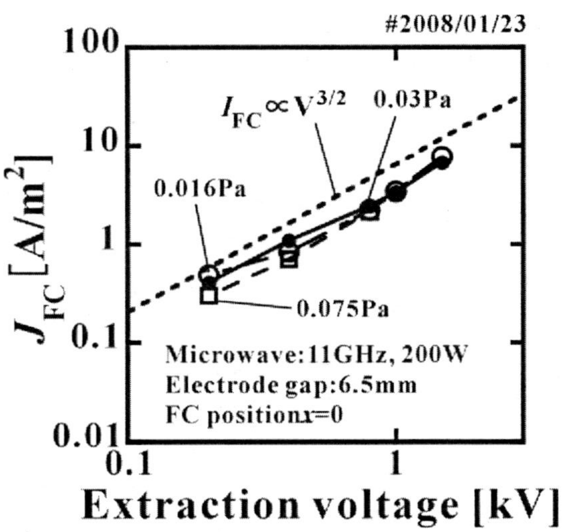

FIGURE 5. Dependence of the J_{FC} on the extraction voltages. The experimental condition is microwave frequency 11 GHz, microwave power 200 W, Ar gas pressure 0.03 Pa. The FC position is x=0 m.

Future Planning

We are now investigating the distribution of the extracted ion beam in detail and correspondence between the J_{FC} and the J_{is} in the comb-shaped ECR ion source as the first stage of tandem type ion source. We will optimize the electrode gap length, the electrode position, and the shape of multi-holes of the extractor to extract the higher current and larger bore ion beam from the comb-shaped ECR ion source. We prepare to transport the broad ion beam from this first stage plasma and transfer like a shower to plasma generated by the second stage ECR ion source which can produce another plasma independently for synthesized various materials. We will produce various ion beams, *e.g.,* from multi charged ions to cluster ions, fullerenes, etc and conduct mass analysis for various ion beams at downstream of the second stage of tandem type ion source.

ACKNOWLEDGEMENTS

The authors wish to thank T. Asaji, Tateyama Machine Co. Ltd., for making extractor and helpful discussions.

REFERENCES

1. Y. Kato, H. Furuki, T. Asaji, H. Sasaki, and S. Ishii, *Nucl. Instrum. Methods Phys. Res. B* **237**, 256 (2005).
2. Y. Kato, H. Furuki, T. Asaji, F. Sato, and T. Iida, *Rev. Sci. Instrum.* **77**, 03A336 (2006).
3. T. Asaji, H. Sasaki, H. Furuki, Y. Kato, S. Ishii, M. Kanazawa, and J. Saito, *Nucl. Instrum. Methods Phys. Res. B* **237**, 262 (2005).
4. T. Asaji, H. Sasaki, Y. Kato, F. Sato, T. Iida, and J. Saito, *Rev. Sci. Instrum.* **77**, 03C104 (2006).
5. T. Asaji, Y. Kato, F. Sato, T. Iida, and J. Saito, *Rev. Sci. Instrum.* **77**, 113503 (2006).
6. T. Asaji, PhD thesis 2006.
7. Y. Kato, T. Satani, T. Asaji, F. Sato, and T. Iida, *Rev. Sci. Instrum.* **79**, 02A323 (2008).

Ion Implantation into Nanoparticulate Functional Layers

S. Walther[1], M. P. M. Jank[2], A. Ebbers[3], and H. Ryssel[1,2]

[1] Chair of Electron Devices, University of Erlangen-Nuremberg, Cauerstrasse 6, 91058 Erlangen, Germany
[2] Fraunhofer Institute for Integrated Systems and Device Technology, Schottkystrasse 10, 91058 Erlangen, Germany
[3] Evonik Degussa GmbH, Paul-Baumann-Strasse 1, 45772 Marl, Germany

Abstract Layers of silicon nanoparticles with resistivity down to 10 mΩcm are demonstrated using chemical and thermal treatment as well as ion implantation. Morphology and crystallinity of the layers are characterized using AFM and XRD. After chemical and thermal treatment AFM measurement revealed a fine grain structure, where the grains are of good crystallinity as measured by XRD. Implantation of $5 \cdot 10^{15}$ cm^{-2} arsenic at 100 keV yields partially amorphous layers that can be recrystallized by post implantation anneal. By this process, surface roughness of layers increases as grain growth takes place.

Keywords: printable electronics, silicon nanoparticles, ion implantation
PACS: 81.40.Rs

INTRODUCTION

Metallic and semiconducting nanoparticles have received much attention in recent years not only for their unique physical properties, such as quantum confinement, but also for possible utilization in printable low-cost electronics. Manufacturing of electron devices with printing techniques offers a route for reduction of processing time, cost, and waste. Fabrication of printed electronics has already been demonstrated with organic materials. For example TFTs fabricated with pentacene as active layer have reached mobilities of almost $1 cm^2/Vs$ [1]. However, sensitivity to moisture and UV light and the lack of readily forming gate insulator materials are serious restrictions for organic electronics materials leading to low achievable mobility even in optimized systems. To overcome these problems, great effort has been devoted in recent years to the fabrication of printable electronics utilizing inorganic materials [2,3,4]. Silicon and germanium nanoparticles are promising candidates for both n- and p-type TFTs because of long established knowledge of their bulk properties and appropriate dopant species. However, printed layers of nanoparticle suspensions form porous layers with weak inter-particle contacts, thus current paths are limited. Moreover, controlling defect states in printed and, therefore, highly disordered layers is a challenging issue. Defect states result from various factors including silicon dangling bonds on the surface of the particles as well as metal and organic contaminations introduced by suspension processing and solvents. As was already indicated by Lechner *et al.* [5], in silicon nanoparticle layers there exists a threshold doping level at which conduction rises abruptly. It is assumed that this threshold level is in the order of the defect state density of the layer and dopant concentration has to exceed this level to affect conductivity. Moreover, silicon rapidly builds an oxide layer when exposed to air or other sources of oxygen, leading to a degradation of the electrical contacts between particles.

To form n and p-type silicon nanoparticle layers, it is necessary to achieve densely packed films, remove oxide shells, and to realize doping levels higher than the defect state density. In this paper, we present a way to achieve both doping and densification by ion implantation.

EXPERIMENTAL

We have used silicon nanoparticles produced by gas phase synthesis in a microwave reactor using SiH_4. These were dispersed in ethanol and ball-milled with zirconia beads to break weak agglomerates. Layers of approx. 500 nm in thickness were realized by spin-coating the suspension onto thermally oxidized silicon substrates and solvent was driven out on a hotplate at 120°C for 10 min. An optimized thermal and chemical treatment (OTCT) of the samples was conducted afterwards. The OTCT was established to effectively remove the oxide shells of the particles in the layer avoiding depletion of acid in the pores of the nanoparticle layer. It consists of a sequence of an oxide etch in diluted HF (1%) for 2.5 min followed by a rinse in DI

CP1066, *Ion Implantation Technology,* edited by E. G. Seebauer, S. B. Felch, A. Jain, and Y. V. Kondratenko

water and drying. This sequence is repeated to achieve a total of 5 min of effective etching. Directly after the second etch step, samples are heated in a RTA equipment at 1100°C for 20 s to enable sintering between the now oxide free silicon nanoparticles. The porosity of these layers can be estimated by secondary electron microscopy (SEM) analysis to be over 50%.

Ion implantation of arsenic was performed into the layers after the OTCT. The dose was varied from $5 \cdot 10^{15}$ cm^{-2} to $1.2 \cdot 10^{16}$ cm^{-2}. We used the SRIM simulation tool [6] to estimate the energy at which the nuclear interaction of the implanted species with the silicon and, therefore, the maximum energy transfer concentrates at half depth of the layer. Assuming a target of porous silicon with only half the density of the single crystal leads to an energy of 100 keV. Annealing was varied using the RTA equipment and a furnace, respectively, in an inert ambient.

Metal contacts were deposited by evaporation of aluminum through a shadow mask allowing for two and four point measurements of resistivity with a HP4156 parameter analyzer. All resistivity values are given at room temperature. Morphology of the layers was characterized measuring particle size and surface roughness with atomic force microscopy (AFM) and crystallinity with grazing incident x-ray diffraction (GIXRD).

RESULTS AND DISCUSSION

An AFM measurement of a silicon nanoparticle layer after OTCT is depicted in figure 1. A granular structure can be seen and primary particles with size 30 nm to 50 nm can be identified. The superstructure consists of agglomerated and coagulated primary particles. The root mean square of surface roughness (RMS) is 23 nm.

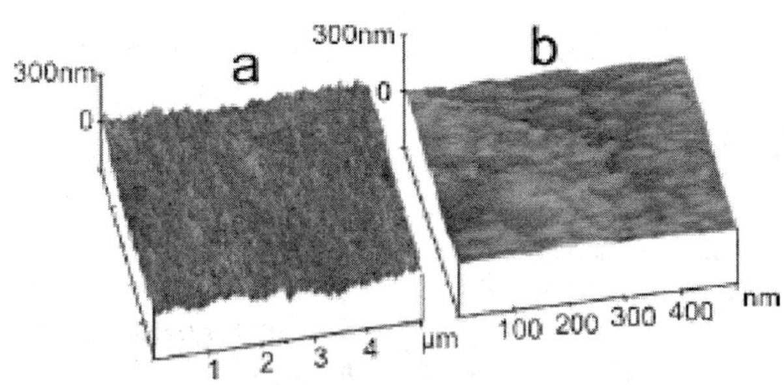

FIGURE 1. (a) AFM measurements of an as deposited nanoparticle layer; (b) enlarged detail. Film has fine grained structure with RMS 23nm.

We implanted layers of silicon nanoparticles with $1 \cdot 10^{16}$ cm^{-2} arsenic and varied the annealing time and temperature to investigate the activation of dopants and the effect on layer morphology. Figure 2 compares AFM measurements of layers with and without implantation, respectively, after OTCT and additional heat step at 800°C for 20s. In the layer without implantation in figure 2(a), primary particles can still be identified, whereas the grains of the layer with implantation in figure 2(b) are slightly enlarged and the root mean square of the surface roughness increased from 31 nm to 98 nm.

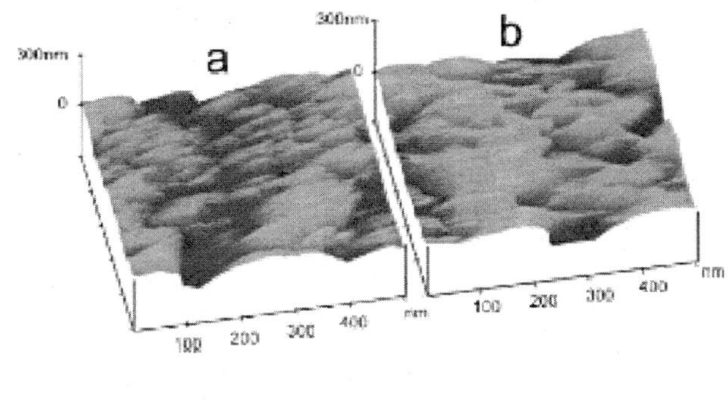

FIGURE 2. AFM measurements of nanoparticle layers a) optimized chemical and thermal treatment and post implantation anneal 20 s 800°C; b) optimized chemical and thermal treatment, implantation of $1 \cdot 10^{16}$cm^{-2} arsenic, and post implantation anneal 20 s 800°C. Increase of grain size and RMS is obtained after implantation

Figure 3(a) shows a layer with OTCT and implantation but without PIA. In comparison to layer 2(a) larger grains of size exceeding the primary particle size seem to appear. The surface roughness has increased to 52 nm. After a heat step at 800°C for 30 min, the connected volumes remained about the same size (figure 3(b)) and the roughness increased further to 73 nm, whereas the layer annealed for 20 s at 1100°C depicted in figure 3(c) shows grain size of more than 500 nm with surface roughness of 53 nm.

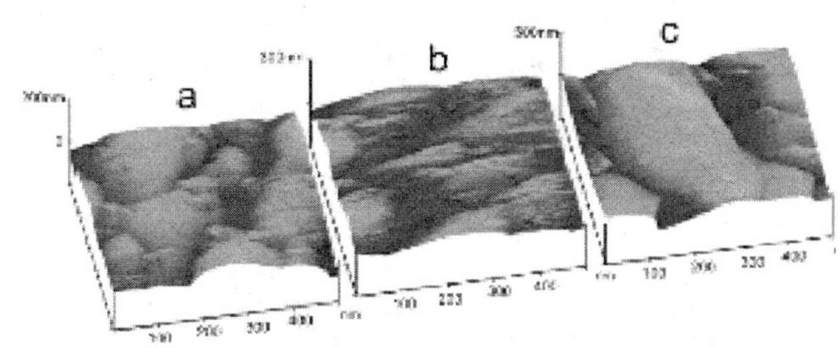

FIGURE 3. AFM measurements of nanoparticle layers after optimized chemical and thermal treatment, implantation of $1 \cdot 10^{16}$ cm^{-2} arsenic and (a) no post implantation anneal; (b) post implantation anneal 30 min at 800°C; (c) post implantation anneal 20 s at 1100°C.

XRD measurements on these samples depicted in figure 4 were conducted in grazing manner to avoid substrate influence on the measurement. The pattern of the layer with OTCT shows all expected peaks for crystalline silicon (upper spectrum). After ion implantation, the (111) peak has weakened and all other peaks have nearly vanished (spectrum in the middle). The layer is almost completely amorphous. After a heat step at 1100°C for 20s, however, all silicon crystal peaks can be seen in the XRD pattern again (lowest spectrum). In all measurements the signal at 34° is not unambiguously identified but originates most likely from zirconia abrasive from the milling process.

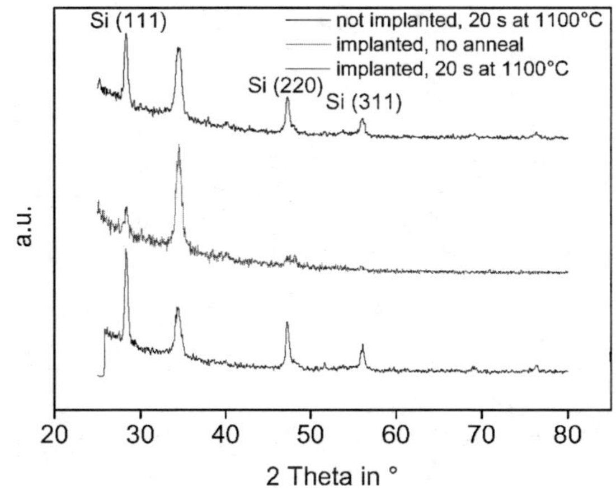

FIGURE 4. XRD pattern of samples before implantation (top), after implantation without anneal (middle), and implanted and annealed sample (bottom).

We suggest that during ion implantation amorphization takes place enabling the silicon atoms/grains to reorganize at room temperature and form large amorphous grains. Thus, porosity is reduced and all heat steps performed during this work are sufficient for the amorphous grains to crystallize.

TABLE 1. Crystal phase (c crystalline, a amorphous) and resistivity of silicon nanoparticle layers.

As Dose @100 keV	PIA RTA/furnace	phase	Resistivity (Ωcm)
none	none	C	10^{09}
none	30 min 800°C	C	$5 \cdot 10^5$
none	20 s 800°	C	$3 \cdot 10^5$
none	20 s 1100°C	C	$1 \cdot 10^6$
$1 \cdot 10^{16}$ cm^{-2}	none	A	$1 \cdot 10^3$
$1 \cdot 10^{16}$ cm^{-2}	30 min 800°C	C	0.1
$1 \cdot 10^{16}$ cm^{-2}	20 s 800°C	C	0.08
$1 \cdot 10^{16}$ cm^{-2}	20 s 1100°C	C	0.01

Resistivity was measured with four-point probe and with twoprobe method for very high resistivity values, respectively. Table 1 summarizes crystal phase and measurement results for all samples. It can be seen that after implantation, resistivity decreases even without PIA. This is likely due to the densification of the layer by ion implantation. After furnace anneal at 800°C for 30 min, resisitivity decreases further to 100 mΩcm and to 80 mΩcm after RTA at 800°C for 20 s. The lowest resistivity value of 10 mΩcm was achieved by RTA at 1100°C for 20 s. So, both crystallization and activation of dopants are believed to take place during PIA. Unexpectedly, the sample annealed at 800°C for 20 s by RTA has lower resistivity then the sample annealed at the same temperature but for 30 min in the furnace. This observation is probably due to a reaction of the silicon nanoparticles with residual oxygen in the furnace. The decrease of resistivity with increasing PIA temperature is attributed to the increasing grain size but also to the temperature dependent activation of arsenic as a dopant.

We have also varied the implantation dose using the PIA at 1100°C for 20 s. Figure 5 shows AFM measurements of the surface of the samples and table 2 summarizes RMS values and resistivities of the samples.

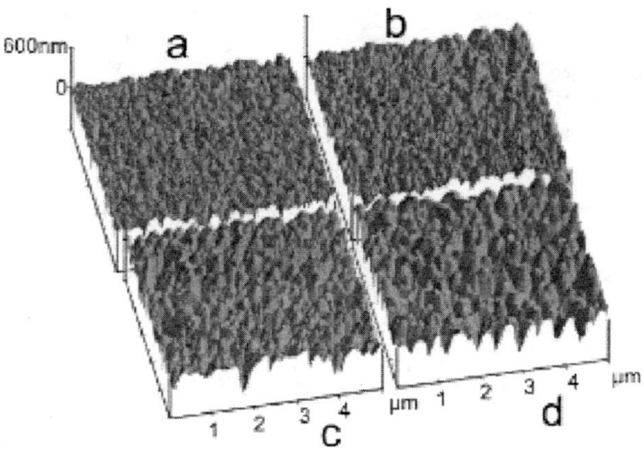

FIGURE 5. AFM measurements of nanoparticle layers implanted and annealed, doses (a) $5 \cdot 10^{15}$ cm^{-2}; (b) $7 \cdot 10^{15}$ cm^{-2}; (c) $9 \cdot 10^{15}$ cm^{-2}; (d) $1.2 \cdot 10^{16}$ cm^{-2}. Grain size and RMS increase with increasing dose. Vertical scale bar applies to all graphics.

TABLE 2. Resistivity of silicon nanoparticle layers after ion implantation and PIA.

As Dose @100 keV	RMS (nm)	Resistivity (mΩcm)
$5 \cdot 10^{15}$ cm^{-2}	60	10
$7 \cdot 10^{15}$ cm^{-2}	52	8
$9 \cdot 10^{15}$ cm^{-2}	88	6
$1.2 \cdot 10^{16}$ cm^{-2}	98	10

A dependence of the resistivity on implanted doses could not be observed. This could be due to the very high implantation dose regime we have chosen. Assuming 100% activation of dopant atoms and the implantation profile calculated with SRIM, the limit of solubility could have been reached with the lowest dose already. The surface topology seems to have minor influence on electrical conduction.

CONCLUSION

We have investigated the impact of ion implantation on nanoparticulate layers regarding surface topology, crystallinity, and electric resistivity. With $5 \cdot 10^{15}$ cm^{-2} arsenic at 100 keV amorphization of the layers took place as can be seen in the XRD pattern. AFM measurement shows enlarged grains after implantation. Apparently reorganization of silicon atoms takes place at room temperature during implantation. Recrystallization of the layers with large crystalline silicon grains could be observed after post implantation anneal. From this fact, we conclude that layer porosity is reduced.

A dependence of the resistivity on the implanted dose could not be observed in this work. In the very high implantation dose regime we have chosen, the solubility limit of arsenic could have been reached with the lowest dose of $5 \cdot 10^{15}$ cm^{-2} assuming similar behavior of the nanoparticle layer as known for single crystalline silicon. Segregation of the dopant atoms on grain boundaries would reduce the effective doping level. However, this effect could not be seen in this work. We conclude strong activation of the dopant atoms from the dramatic decrease of resistivity with implantation of arsenic and the insensitivity to the dose variation above $5 \cdot 10^{15}$ cm^{-2}.

ACKNOWLEDGEMENTS

The authors would like to thank Dr. Elke Meißner for XRD measurement and Thomas Scharrer.

Financial support by the Deutsche Forschungsgemeinschaft (DFG) under Grant No. GRK 1161/1 "Disperse Systems in Electronics" is gratefully acknowledged.

REFERENCES

1. S. Molesa, M. Chew, D. Redinger, S.K Volkman, and V. Subramanian, Mat. Res. Soc. Spring Meeting (2004)
2 S.K. Volkman, B.A. Mattis, S.E. Molesa, J.B. Lee, A. de la Fuente, T. Bakhishev, and V. Subramanian, International Electron Devices Meeting IEDM (2004)
3 I. Gur, N.A. Fromer, M.L. Geier, A.P. Alivisatos, *Science* **310**, 462 (2005)
4 D.V. Talapin, C.B. Murray, *Science* **310**, 86 (2005)
5 R. Lechner *et al.*, *Phys. Stat. Sol.* (RRL) **1**(6), 262– 264 (2007)
6 J.F. Ziegler, J.P. Biersack and U. Littmark, The Stopping and Range of Ions in Solids, Pergamon, New York (1985)

Depth Profiling of N and C in Ion Implanted ZnO and Si Using Deuterium Induced Nuclear Reaction Analysis

John Kennedy, Peter Murmu, and Andreas Markwitz

National Isotope Centre, GNS Science, P.O. Box 31312, Lower Hutt, New Zealand

Abstract. Nuclear Reaction Analysis (NRA) with deuteron ion beams has been used to probe for ion implanted nitrogen and carbon with high sensitivity in zinc oxide and silicon single crystals. The ion implanted N was measured using 1.4 MeV deuteron ion beams and was found to be in agreement with calculated values. The limit of detection for N in ZnO is 8 x 10^{14} ions cm^{-2}. Raman measurements of the ion implanted samples showed three additional modes at 275, 504, and 644 cm^{-1} compared to the un-implanted ZnO crystals. The NRA and Raman results provided information on the N concentration, depth distribution, and structural changes that occur in dependence on the nitrogen ion fluences. The deuterium induced $^{12}C(d,p)^{13}C$ reaction was used to measure the carbon impurity/dose in ion implanted silicon. It was found that the use of a large cold shield (liquid nitrogen trap) in the ion implanter chamber greatly reduces the amount of carbon impurity on the surface of ion implanted silicon. Various implantations with N_2, O_2, NO, NO_2 and Pb ions were performed with and without cooling of the liquid nitrogen trap. Simultaneous detection of ppm-level concentrations of ^{12}C, ^{16}O and ^{14}N enables highly sensitive measurement of impurities that may be incorporated during the fabrication process, transport of the samples and/or storage of the samples in air.

Keywords: Ion beam analysis, Ion implantation, nuclear reaction analysis, ZnO, Si, liquid nitrogen trap
PACS: 61.72.U-; 82.80.Yc; 81.05.Hd; 61.46.-w; 68.55.ag; 61.80.Jh

INTRODUCTION

Rutherford Backscattering Spectrometry (RBS) is a commonly used analysis technique for quantitative determination of bulk composition and of the concentration of implanted ions [1]. Nuclear reactions induced by bombardment with energetic protons, deuterons or alpha particles that cause the emission of other energetic particles are used to measure low-Z elements in high-Z matrices with high sensitivity [1,2]. Nuclear Reaction Analysis (NRA) using deuteron beam is used routinely to probe for light elements, such as C, N and O with high sensitivity (in the order of 0.01 atomic percent), in semiconductor materials due to high cross section values.

In recent years, we have used deuteron beam NRA to probe for light elemental concentrations in several series of group III-V nitride thin films [3], high temperature superconducting thin films [4] and helium ion implanted Ti and Ti alloys [5]. In this contribution, we present new results of (1) surface-near p-type doped Zinc oxide (ZnO) via low-energy nitrogen ion implantation and (2) N_2, O_2, NO, NO_2 and Pb ion implantation into wafer Si to study the impact of the liquid nitrogen cold trap in the implanter chamber on the C contamination of Si implanted under 10^{-7} mbar. This is important information to better understand the formation of SiC nanocrystals on the surface after electron beam annealing.

NITROGEN IMPLANTATION INTO ZINC OXIDE

ZnO is a natural n-type semiconductor because of the presence of intrinsic defects such as Zn interstitials and O vacancies which cause deviation from stoichiometry [6]. However, the key challenges in developing ZnO technology for optoelectronic applications is the need for p-type conductivity and discovering what kind of doping process results in p-type material and the subsequent fabrication of a p-n junction. The fundamental difficulty in p-type doping of ZnO arises from a number of reasons. For example, the donor level may be especially deep so that thermal excitation into the valence band is low. Other reasons could be a low solubility of the dopant or induced self compensating processes [7]. Theoretical studies have suggested that zinc vacancies in the ZnO matrix act as possible deep acceptors. However, nitrogen (among group V elements of P, As, N) atoms substituting for oxygen in the ZnO film have shown to be the most promising shallow acceptor in ZnO [7]. The aim of the present study is to establish an implantation protocol for nitrogen as an acceptor ion, into n-type ZnO crystals to produce p-type ZnO for various optoelectronic applications.

Twenty-three keV nitrogen ions were implanted as an acceptor ions into n-type ZnO crystals (obtained from Tokyo Denpa, Japan) with various doses between $1x10^{14}$ and $5x10^{16}$ ions cm^{-2} using the newly

CP1066, *Ion Implantation Technology*, edited by E. G. Seebauer, S. B. Felch, A. Jain, and Y. V. Kondratenko

established ion implantation facility at GNS [8]. Prior to implantation, DYNAMIC-TRIM [9] simulations were performed to calculate N depth profiles for the various doses which are shown in Figure 1. The calculations predict a mean projected range of 45 nm coupled with a maximum implantation depth of around 80 nm. For these fluences, the N peak concentration varies between 0.05 and 17 at.%. At fluences larger than 1 x 10^{16} N cm^{-2}, the N profile intersects with the surface.

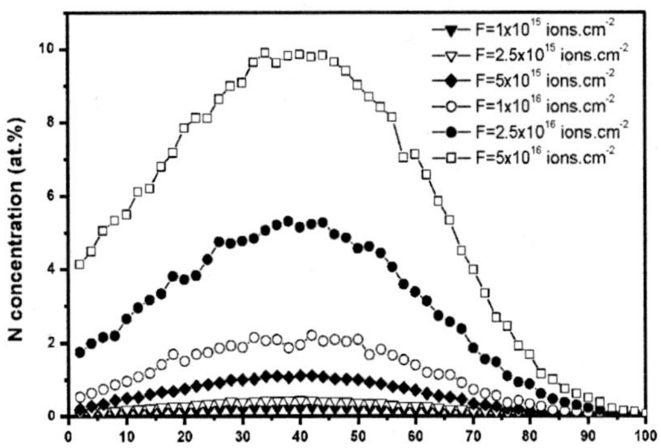

FIGURE 1. N depth profiles calculated with DYNAMIC-TRIM for N implanted at 23 keV into ZnO with fluences from $1x10^{15}$ to $5x10^{16}$ cm^{-2}.

A 1.4 MeV deuteron beam (beam current ~ 20 nA, beam spot diameter = 0.5 mm) was used for the nuclear reaction analysis with the reactions $^{14}N(d,\alpha_1)^{12}C$ ($d\sigma/d\Omega$ = 1.5 mb sr^{-1}) and $^{14}N(d,\alpha_0)^{12}C$ ($d\sigma/d\Omega$ = 0.7mb sr^{-1}) to determine the N concentration in the ZnO crystals. These reactions were chosen because their peaks occur in the background free region of the NRA spectrum. Standard material of 500 nm thick silicon nitride on Si was used for calibration purposes. The particles generated by the above mentioned nuclear reactions were detected with a surface barrier detector (SBD) detector (active area = 300 mm^2, mylar absorber thickness = 14.6 μm) mounted at a backscattering angle of 150°.

Typical NRA spectra obtained from N implanted ZnO crystals are shown in Figure 2. Most of the N related nuclear reaction peaks along with oxygen and carbon are seen in the spectra. The measured N concentrations using the higher cross section reaction of $^{14}N(d,\alpha_1)^{12}C$ are tabulated in Table 1. Most of the measured N concentrations are close to the theoretical values. It was found that the detection limit for N is around 8 ± 1 x 10^{14} ions cm^{-2}. The measured depth profiles shown in Figure 3 agree with the theoretical depth profiles calculated with DYNAMIC-TRIM as shown in Figure 1.

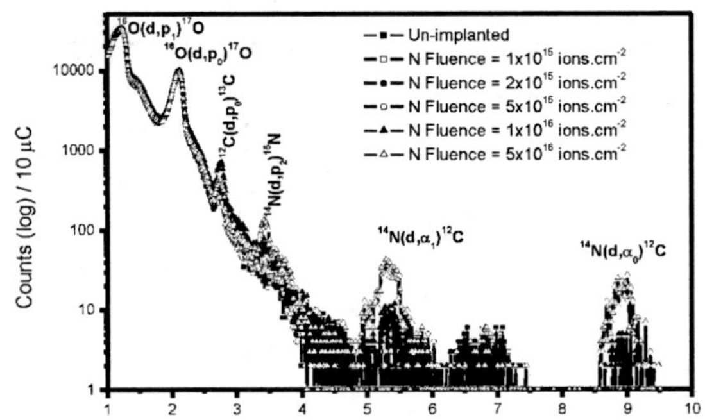

FIGURE 2. NRA spectra of nitrogen implanted ZnO

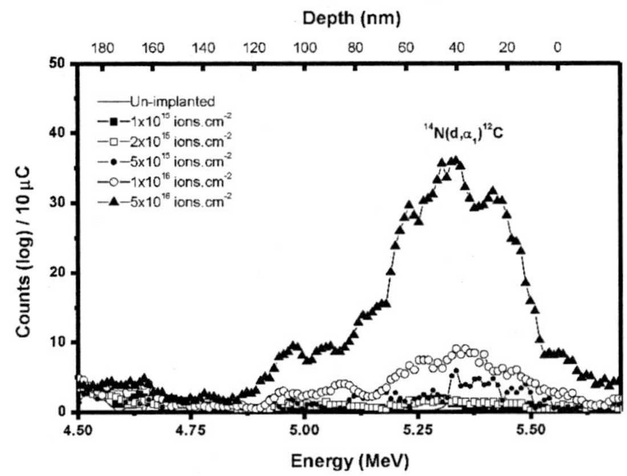

FIGURE 3. Nitrogen profiles derived from $^{14}N(d,\alpha_1)^{12}C$ peak for the implanted samples

TABLE 1. N concentration measured by NRA

Sample code	Fluence (ions cm^{-2})	Measured dose (ions cm^{-2})
A	$1.0x10^{-15}$	$0.91x10^{-15}$
B	$2.0x10^{-15}$	$1.9x10^{-15}$
C	$5.0x10^{-15}$	$4.7x10^{-15}$
D	$1.0x10^{-16}$	$9.4x10^{-16}$
E	$5.0x10^{-16}$	$4.6x10^{-16}$

RAMAN CHARACTERISATION OF NITROGEN IMPLANTED ZnO

Raman spectroscopy is a versatile technique for fast and non-destructive study of dopant incorporation, particularly when impurity-induced modes can be traced back to individual constituents and their concentration, as demonstrated, for example with Mg and As in hexagonal and cubic GaN [10].

With a wurtzite structure ZnO belongs to the C6v symmetry group. The group theory predicts two A1, two E1, two E2 and two B1 modes, i.e. the potential optical modes for ZnO are as follows: $1A_1(TO)+1A_1(LO)+2B_1+1E_1(TO)+1E_1(LO)+2E_2$. Among them, two B1 modes are not Raman active. The observed phonon frequencies in the Raman spectra of ZnO are E2(high)=437 cm^{-1}, E2(low)=101 cm^{-1}, A1(TO)=380 cm^{-1}, A1(LO)=574 cm^{-1}, E1(TO)=407 cm^{-1} and E1(LO)=583 cm^{-1}, respectively. On condition that the incident light is exactly normal to the surface of the ZnO samples, only A1(LO) and E2 modes are observed, and the other modes are forbidden according to the Raman selection rules [11, 12].

We have performed Raman scattering measurements at room temperatures at the Raman lab, School of Chemical and Physical Sciences, Victoria University of Wellington. The 514 nm line of an Ar^+ laser was used as an excitation source. The scattered light was detected in backscattering geometry and recorded with a liquid nitrogen-cooled charge coupled device detector. Figure 4 shows Raman spectra of both unimplanted and N implanted ZnO. The spectra have been shifted vertically for comparison reasons. The Raman spectra were recorded in backscattering configuration with incident light in the direction of the hexagonal *c*-axis and without polarization. In the Raman spectrum of the unimplanted sample, two Raman lines due to the *E*2 high frequency at 437 cm^{-1} and *A*1 longitudinal optical LO mode at 575 cm^{-1} are observed as expected from the Raman selection rules in wurtzite crystal structure. The feature at 332 cm^{-1} is the second-order structure of ZnO, which was interpreted as $2E2_M$ by Calleja and Cardona [13].

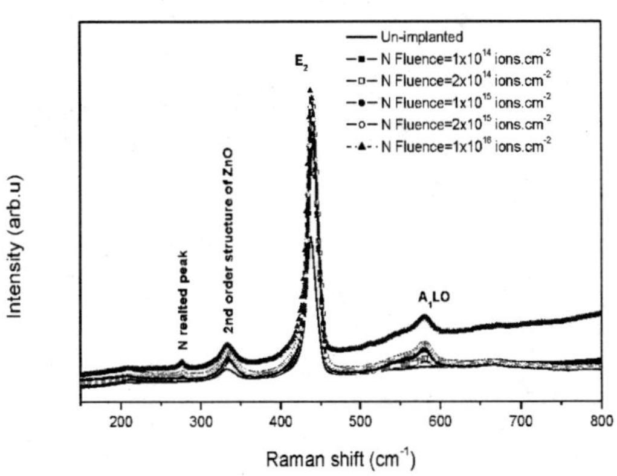

FIGURE 4. Raman spectra of un-implanted and nitrogen implanted ZnO

With increasing implantation concentration, an additional peak arises at 275 cm^{-1} as well as two broad and weak structures at 504 and 644 cm^{-1}. The appearance of the new features is accompanied by an increase of asymmetrical broadening of the E2 (high) line widths. It is worthwhile to note that, with increasing nitrogen concentrations, the peak at 575 cm^{-1} becomes stronger, and it also becomes much broader. This is contrary to the change of the LO phonon because it should decrease due to the increment of lattice damage. Therefore, this suggests that the broad peak at about 575 cm^{-1} in the ion-implanted ZnO is due to the density of phonon states (DOPS) of the high-energy phonon branch [14]. In our experiment, the intensity of this mode increases as the implanted nitrogen dose increases, which are in good agreement with the results in Ref. 14. Such correlation indicates that the occurrence of the additional mode is related to the incorporation of N atoms into the ZnO crystal.

The structural, electrical and optical properties were determined by XRD, Hall effect, and PL. It is found that the N implanted and electron beam annealed targets show p–type carrier concentration. PL revealed exciton peaks in the region between 300 and 450 nm. These exciton peaks correspond to p-type ZnO formed as a result of the acceptor doping process of low-energy acceptor ion implantation and EB annealing. Detailed results presented in Ref. 15

CARBON, N_2, O_2, NO, NO_2 AND Pb ION IMPLANTED SILICON

A series of low-energy ion implantations were performed on p-type (100) Si wafer under high vacuum conditions. The residual gas pressure in the implantation chamber was 4 x 10^{-7} mbar without using the liquid nitrogen trap that surrounds the target at a distance of 70 mm. The ion implanted sample area was set to 13 x 13 mm. The ion implantations were performed at the ion energy of 20 kV, except for lead where the ion energy was set to 7.5 kV to introduce Pb into the subsurface for producing metal contacts [16, 17]. Subsequent carbon ion implantation and electron beam annealing enables the control of the Pb diffusion into the depth of the material specified by the carbon ion energy [18]. The residual gas pressure improves to 6 x 10^{-8} mbar as soon as the liquid nitrogen trap is measured to have cooled to -190 °C. This residual gas pressure is maintained during the ion implantation process. Under these conditions, clean ion implantations are performed as shown in Table 2 where a summary is shown of the carbon concentration measured for various ion implantation series.

As can be seen from the data shown in Table 2, the cold trap (LN_2 trap) has a major effect on the C concentration in ion implanted Si. Using the LN_2 trap avoids contamination of the sample during ion implantation. The surface of the samples becomes slightly contaminated with C if the LN_2 trap is not used. In this case, the sputtering of the surface during ion implantation using low-energy ion beams, where a significant sputtering of the surface occurs during the implantation process, produces a C contamination in the order of 3 x 10^{15} cm^{-2}. This number represents the equilibrium C contamination that is independent on the ion species and ion fluence.

TABLE 2. Average C concentration in ion implanted (100) Si

Ion species	Fluence (ions cm^{-2})	LN_2 cold trap	C concentration (ions cm^{-2})
-			$6 \pm 1 \times 10^{14}$
Pb	3×10^{15}	+	$6 \pm 1 \times 10^{14}$
N_2	2×10^{15}	+	$6 \pm 3 \times 10^{14}$
N_2	2×10^{15}	-	$3.2 \pm 1 \times 10^{15}$
N_2	3×10^{15}	-	$3.4 \pm 1 \times 10^{15}$
O_2	3×10^{16}	-	$3.1 \pm 1 \times 10^{15}$
NO	6×10^{16}	-	$2.4 \pm 1 \times 10^{15}$
NO_2	3×10^{16}	-	$2.8 \pm 2 \times 10^{15}$

SUMMARY

Deuterium induced NRA was used to probe for carbon impurity in ion implanted Si. To investigate the incorporation of C in ion implanted samples, a series of Si wafer samples were implanted with various single and molecular ion beams under different vacuum conditions. It was found that an additional large liquid nitrogen cooled shield surrounding the samples during implantation prevents the build-up of carbon during the implantation. This information was used to ion implant ZnO materials. The nuclear reaction analysis study determined the dose of the nitrogen implanted ZnO single crystals. The N peak concentration was varied between 0.05 and 17 at.%. The measured N concentrations was found to be very close to the calculated values indicating that the sputtering yield calculations used in dynamic TRIM are correct for the N/ZnO system. Raman spectra of N implanted ZnO crystals with different implantation concentrations resulted in three additional Raman modes at 275, 504, and 644 cm^{-1} compared to un-implanted ZnO. The origin of the mode at 275 cm^{-1} has been attributed to vibration of Zn atoms, where parts of its first nearest neighbor O atoms are replaced by N atoms in the crystal lattice. With increasing implantation concentration, two broad and weak structures at 504 and 644 cm^{-1} also appeared. It is shown that NRA and Raman spectroscopy can be successfully used to obtain information of the implanted N concentrations and its structural modifications. This enabled us to establish the implantation protocol for p-type doping of Zinc oxide materials.

ACKNOWLEDGEMENTS

The work was carried out with the research contract from FRST, NZ.

REFERENCES

1. J.R. Tesmer and M. Nastasi, *Handbook for Modern Ion Beam Analysis,* Pittsburgh: Materials Research Society, (1995).
2. V. J. Kennedy, A. Markwitz and G.V. White, I.W.M. Brown. *Mod.Phy.Lett.B***15**, 1332 (2001).
3. J. Kennedy, A. Markwitz, B.J. Ruck, S.M. Durbin and W. Gao, *J. Elect.Mat* **36**, 472-482 (2007).
4. V.J. Kennedy, A. Markwitz, A. Bubendorfer, N. Long and N. Dytlewski. *Curr.Appl.Phy.* **4,** 292 (2004).
5. P.B. Johnson, V.J. Kennedy, A. Markwitz, C. Varoy, *Nucl. Instr. and Meth. in Phys. Res. B* **206**, 1056 (2003).
6. T. Yamamoto, H. Yoshida, *Physica B* **302-303**, 155 (2001).
7. H.C. Park, S.B. Zhang, S-H.Wei, *Phys.Rev B* **66,** 073202 (2002).
8. A. Markwitz, B. Barry, J. Kennedy, C. Purcell, M. Rudolphi and B. Trompetter, (private communication).
9. J. P. Biersack, *Nucl. Instr. and Meth. in Phys. Res. B* **27**, 21 (1987).
10. G. Kaczmarczyk, A. Kaschner, A. Hoffmann, and C. Thomsen, *Phys. Rev.B* **61**, 5353(2000).
11. A. Kaschner, U. Haboeck, M. Strassburg, M. Strassburg, G. Kaczmarczyk, A. Hoffmann, C. Thomsen, *Appl.Phys.Letters* **80**, 1909 (2002).
12. C. Bundesmann, N. Ashkenov, M. Schubert, D. Spemann, T. Butz, E. M. Kaidashev, M. Lorenz, and M. Grundmann, *Appl. Phys. Letters* **83,** 1974 (2003).
13. J. M. Calleja and M. Cardona, *Phys. Rev. B* 16, 3753(1977).
14. J.B. Wang, H.M.Zhong, Z.F. Li and W. Li, *Appl. Phys. Letters*, **88**, 101913 (2006).
15. John Kennedy and Andreas Markwitz, PCT International patent WO2007/117158 A1.
16. A. Markwitz, H. Baumann, P. Davy, C. Beckert, J. Kennedy, *Nucl. Instr. and Meth. in Phys. B***266/8**, 1553 (2008).
17. A. Markwitz, P. Davy, J. Kennedy, H. Baumann, *Surf. and Int. Anal.* **40**, 931 (2008).
18. A. Markwitz, H. Baumann, P. Davy, P. B. Johnson *Vacuum*, doi: 10.1016/j.vacuum.2008.02.006

Ion Implantation Technology 2008   June 5-13, 2008 Monterey, CA USA

Kevin Jones teaching, Tom Seidel & T.C. Smith at dinner during the IIT school.

Richard Lindsay accepts his McKenna Young Investigator Award beer mug, Sachiko Nakagawa checks the Internet and Della Miller (AVS coordinator) & Susan Felch (IIT08 lead-chair) confer at the conference start.

Dave Armour gives the conference keynote address and Takahiro Shinada talks about single-ion implantation.

Richard Lindsay gives the McKenna Young Investigator Award talk and Larry Larson clarifies a point with Yoji Kawasaki.

Jiro Matsuo (IIT10 chair) speaking on cluster ions and Dale Jacobson coordinates questions for John Hautala.

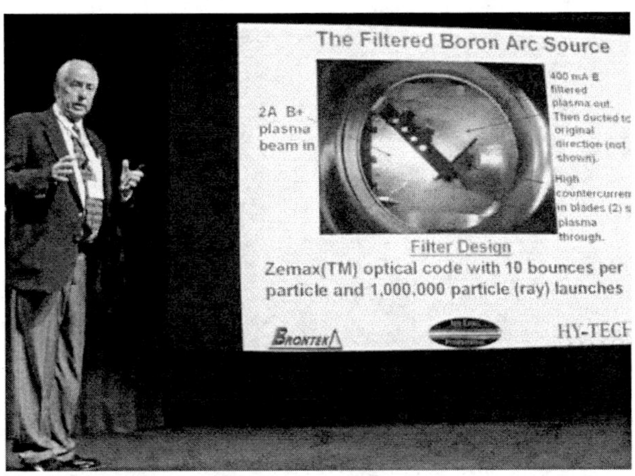

Jim Williams speaking on intense Boron beams and John Borland discussing shallow junction issues.

Ron Eddy & Tom Horsky and Takashi Kuroi, Tony Peaker & Leah Edelman discuss poster papers.

Andy Smith explains his poster paper for Richard Lindsay & Konstantine Bourdelle and Si Prussin & Wade Krull discuss new data.

Dave Chivers & Heiner Ryssel in conversation. Dale Jacobson points out key items in his poster paper for Robert Kaim & John Schuur.

Michael Current & Isao Yamada and Bob Brown & Ron Eddy enjoy the posters.

Ken Steeples and Noriaki Toyoda & Takaaki Aioki debate points at the posters.

Chuck Yarling & Jim Ziegler and Hiro Ito & Bunji Mizuno enjoy quiet conversations during the poster sessions.

John Borland & Vladimir Faifer review a paper and Masao Naito, Jiro Matsuo & Isao Yamada prepare the successful bid to host IIT10 in Kyoto.

Anatoly Viatkin & Paul Chu and Tony Renau & Russ Gwilliam prepare for the international committee meeting.

Mikio Takai & Masataka Kase relax and Susan Felch (IIT08 lead-chair) receives a "Buzz Lightyear" ion gun at the international committee meeting.

Mike Ameen & Dan Tieger enjoy the evening at the Monterey Aquarium and Si Prussin lectures at the concluding session.

Walt Johnson coordinates the scuba diving outing and a scene from a wine tour.

Bobbie McKenna & Heiner Ryssel at the reception honoring Chuck McKenna and IIT08 co-chairs Ed Seebauer, Susan Felch & Amitabh Jain at the aquarium.

(Photos by M.I. Current, K.S. Jones, M. Ameen & W. Stockwell).

Author Index

A

B

C

D

E

F

G

H

I

J

K

L

M

N

O

P

Q

R

S

T

U

V

W

X

Y

Z